Instrumentation for Environmental Monitoring

INSTRUMENTATION FOR ENVIRONMENTAL MONITORING

VOLUME 1, RADIATION

Second Edition

LAWRENCE BERKELEY LABORATORY
ENVIRONMENTAL INSTRUMENTATION SURVEY
Berkeley, California

Principal Authors: Robert J. Budnitz
Anthony V. Nero
Donnie J. Murphy
Robert Graven

Principal Investigator: Mary S.Q. Hunt

Editorial Committee: George A. Morton
Ralph D. McLaughlin
Charles W. Case
Alexandre T. Quintanilha

A Wiley-Interscience Publication

JOHN WILEY & SONS

New York Chichester Brisbane Toronto Singapore

Library of Congress Cataloging in Publication Data:

Main entry under Title:

Instrumentation for environmental monitoring.

Includes index.
Contents: v. 1. Radiation.
1. Environmental monitoring—Instruments.
I. Budnitz, Robert J. II. Hunt, Mary S. Q. (Mary S. Quinby) III. Morton, George A. IV. Lawrence Berkeley Laboratory. Environmental Instrumentation Survey.
TD193.I56 1983 363.7'0028'7 82-21832
ISBN 0-471-86880-9 (v. 1)

Printed in the United States of America

10 9 8 7 6 5 4 3 2 1

PROLOGUE

I often say that when you can measure what you are speaking about and express it in numbers you know something about it, but when you cannot measure it, when you cannot express it in numbers, your knowledge is of a meager and unsatisfactory kind. ...it may be the beginning of knowledge, but you have scarcely, in your thoughts, advanced to the stage of science whatever the matter may be.

Lord Kelvin, 1883

The wilderness and the parched land shall be glad, and the desert shall rejoice, and blossom as the rose. It shall blossom abundantly, and rejoice even with joy and singing... And the parched land shall become a pool, and the thirsty ground springs of water.

Isaiah 35:1,2,7

ACKNOWLEDGMENTS

This volume is the result of the participation of scores of interested persons at laboratories and companies across the country. We are indebted, especially to the following:

Dick A. Mack, the original principal investigator;

Nathaniel A. Greenhouse, who updated Chapter 9 (Dosimetry;

Lloyd D. Stephens, who updated Chapter 8 (Measurement of Neutrons).

We would especially like to thank posthumously Craig D. Hollowell, one of the original members of this project, who provided strong leadership and encouragement for many years. We also gratefully acknowledge the generous advice and review of the following:

Edmond J. Baratta
Harold R. Beck
Robert T. Beckman
Hanson Blatz
Seymour Block
Donald C. Bogen
Alfred J. Breslin
Thomas F. Budinger
E. Alfred Burrill
John R. Cameron
D. C. Camp
Evan Campbell
George C. Campbell
Kenneth W. Crase
Richard L. Douglas
Edward H. Essington
Eric B. Fowler
Peter G. Groer
Paul H. Gudiksen
William S. Johnson
Bernd Kahn
William P. Kirk
Thomas R. Lee
Edward T. Lessard
Leon Leventhal
Carl A. Lindeken
John M. Matuszek
Richard C. McCall
Robert Miltenberger
Alan A. Moghissi
David S. Myers
Ernest B. Nieschmidt
Robert V. Osborne
Vincent Panesko
H. Wade Patterson
Richard Perkins
Suzanne Pitter
Thomas H. Pigford
William S. Properzio
Peter D. Randolph
Ben G. Samardzich
Kenith J. Schiager
Jacob Sedlet
Douglas P. Serpa
Andrew M. Sessler
Claude W. Sill
Mike R. Singh
Joseph Steimers
William Steffan
Lloyd D. Stephens
David L. Stevenson
James Stoffels
Rex Trammell
Eric Vold
Sanford Wagner
Charles W. Williams

We appreciate the encouragement of Elton J. Cairns, Head of the Lawrence Berkeley Laboratory's Energy & Environment Division, and David A. Shirley, Director of the Lawrence Berkeley Laboratory.

We expecially acknowledge the support of our colleagues at the Lawrence Berkeley Laboratory. Our heartfelt thanks go to the many scientists, engineers, technicians, editors, secretaries and others who gave so generously of their time and knowledge. This project could not have been completed without their help, and their continued support is deeply appreciated.

PREFACE

Understanding the delicate balance between man and his environment requires the accurate and efficient measurement of the substances affecting that environment. In many important areas the present state of the art of instrumentation for environmental monitoring is inadequate. There are many reasons for this inadequacy; more relevant, however, is the question: where are we now and where do we go from here? The Lawrence Berkeley Laboratory, under grants from the National Science Foundation and the Department of Energy, is conducting a survey of instrumentation suitable for the measurement of the quality of our environment. The results of the survey appear in three volumes which discuss monitoring of radiation, water, and air. This volume describes ionizing radiation and the instruments used for measurement. Part I introduces units, sources, and effects of radiation, discusses basic detection methods, and describes measurement of alpha, beta, gamma and x-radiation. Part II discusses, in more detail, sources of anthropogenic radiation: nuclear power plants, nuclear spent-fuel reprocessing, and mining and milling. Part III describes measurement of specific radionuclides.

A number of environmental contaminants are well identified, others are, at present, only suspect, still others are potentially insulting to the ecological balance, and the effects of some substances are unknown. Upon the advice of recognized authorities, the effort expended in compiling this survey has been distributed in proportion to the severity of the problem. If problems are well identified and existing instrumentation is considered satisfactory, presently available instruments are emphasized. If current methods of detection are inadequate—because, for example, of insensitivity, the time consumed in making the measurement, or the ambiguity resulting from interfering species—present methodology is surveyed and new methods of detection and analysis are explored. Recommendations are made for development and exploitation of techniques that are employed in other disciplines and appear to be applicable in the environmental field.

Each volume contains (a) an overview of the basic problems, (b) comparisons among the basic methods of sensing and detection, and (c) instrument notes that summarize the characteristics and evaluate the presently available instruments and techniques. Steps that should be taken to alleviate the problems are recommended if appropriate.

Contact has been made with the staffs of federal, state, university, regional, and local laboratories as well as with manufacturers interested in instrumentation for environmental monitoring. Since there is such great nationwide concern in this field, it is obvious that not all the experienced professionals in any given specialty could be contacted; however, as the survey continues, an endeavor will be made to use all the experience and expertise that comes to the attention of the group.

This survey deals predominantly with instruments commercially available from U.S. sources. The information is derived from manufacturers' catalogs and specifications, the periodic literature, and scientific and engineering reports. Instruments from outside the United States have been included whenever they have been found to be widely used or to merit attention by our readers because of some particular characteristics.

The survey has been compiled from the aforementioned information sources. The limited scope of this program, however, has not allowed the in-house evaluation of operating instruments at this laboratory.

It is realized that no survey can be exhaustive and complete. If any existing or potentially useful instrument or technique has been omitted, it is urged that information about it be communicated to the Environmental Instrumentation Survey.

It is important to state that the explicit function of this survey is to aid in, but not be the sole information relied upon, the choice of instrumentation. Manufacturers' information and the periodic literature in particular must be considered by any potential user.

The technical information in this survey has been compiled from the best available sources and is believed to be correct. The mention of commercial products, their source, or their use in connection with material reported herein is not to be construed as either an actual or implied endorsement of such products. The opinions expressed herein represent those of the authors and do not necessarily express the opinion of the Lawrence Berkeley Laboratory, the University of California, the Department of Energy, or the National Science Foundation.

Mary S. Quinby Hunt
April 1983

CONTENTS

PART III

APPENDICES

PART I

1 IONIZING RADIATION

Ionizing radiation is any radiation capable of removing a charge (electron) from an atom or molecule. The radiation may be in the form of a stream of particles or electromagnetic waves.

To remove a charge, that is to ionize an atom, the radiation must be capable of delivering sufficient energy to the atom to overcome the binding energy. This is usually on the order of 10 electron volts (eV).

Particulate radiation consists of:

1. Electrons or β-rays.
2. Helium nuclei or α-rays.
3. Hydrogen nuclei or protons.
4. Neutrons.
5. Heavy ions, that is, heavier than helium (rare).

Neutrons are an interesting exception to the rule that the kinetic energy of each beam particle must be sufficient to cause ionization. Low-energy neutrons can interact with atomic nuclei and cause them to become radioactive and thus self-ionizing. In other words, a beam of thermal (low-energy) neutrons is a beam of ionizing radiation.

Electromagnetic radiation can be made up of wavelengths ranging from many kilometers to the tiniest fraction of a nanometer. Quantum theory has shown that, although electromagnetic radiation is determined by its wave characteristics, when it interacts with matter to give up energy, it behaves like particles. The energy of each particle, or photon, of radiation is inversely proportional to its wavelength and is given by:

$$E = \frac{hc}{\lambda} = \frac{12{,}350}{\lambda} \quad ,$$

where E = the energy in electron volts,
h = Planck's constant,
c = the velocity of light, and
λ = the wavelength in angstroms.

Electromagnetic radiation of wavelengths shorter than 1000 Å thus falls within the definition of ionizing radiation. The electromagnetic radiation spectrum has been loosely subdivided into regions according to wavelength as indicated in Fig. 1.1.

Some of these regions overlap (e.g., wavelengths of ultraviolet and x-rays). In this case the radiation source determines the designation. Similarly x-rays and γ-rays overlap and are distinguished by their source. Because ionizing electromagnetic radiation has wavelengths shorter than about 1000 Å, it thus includes some ultraviolet, together with x-rays and γ-rays. Ultraviolet radiation is not discussed in any detail in this book.

Man has been exposed continuously to natural radiation since his appearance on earth but has been aware of this penetrating ionizing radiation only for the last 80 years. During that time, the use of radiation and radioactive materials, often man-made, has steadily increased, and a substantial body of evidence has accumulated to indicate that radiation poses risks to human health. For this reason, a major effort has been made to understand these risks and to formulate standards for limiting radiation exposures. This requires the development of means to monitor radiation and radionuclides in the environment.

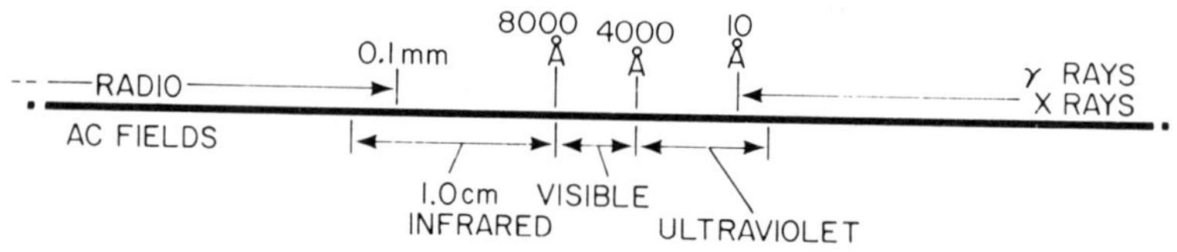

Figure 1.1. Electromagnetic radiation.

1.1. SOURCES OF RADIATION

Man's environment has two sources of radiation, natural and man-made. Natural sources, for example, cosmic rays and natural radioactive materials, are the greater contributor to human radiation exposure [1–5], averaging about 100 mrem/yr (1 mSv/yr) per capita. The important man-made causes of radiation exposures (doses in Table 1.1) are the following:

1. Increased exposure to natural background due to man's activities (high-altitude flights, use of naturally radioactive substances in construction, etc.), commonly referred to as technologically enhanced exposure.
2. Nuclear weapons (fallout of radionuclides from weapons, testing, and direct exposure to a nuclear event).

Table 1.1. Summary of Average Annual Exposures from Various Man-Made Sources (data presented in Nero and Wong, Ref. 2)

Man-Made Sources	Individual Annual Dose	Average Annual Dose in mrem (mSv)
Weapons test fallout[a]	Variable	4 (4×10^{-2})
Medical exposures[b]	Highly variable	60 (6×10^{-1})
Nuclear power[a]		0.003 (3×10^{-5})
Miscellaneous (consumer products, jet travel)[a]	Highly variable	2 (2×10^{-2})
Occupational (mostly medical) exposures	Highly variable	0.8 (8×10^{-3})
Total		~70 (7×10^{-1})

[a]From BEIR (Ref. 4), whole-body exposures. Not including exposures to mining and milling.
[b]From NCRP 39 (Ref. 3) and BEIR (Ref. 4); number given is for "abdominal dose," roughly corresponding to genetically significant dose.

3. Nuclear fuel mining, processing, and reprocessing operations (exposure of workers to radionuclides; release of gaseous, liquid, and solid radiation during processing, transportation, and storage).
4. Nuclear reactors used for either power or research (release of radiation during normal operation as well as accidental releases).
5. Diagnostic and therapeutic procedures (x-rays, radionuclides).
6. Industrial use of radionuclides and x-rays for processing or quality control.
7. Research instruments, such as x-ray diffractometers and photoelectric spectrometers.
8. Particle accelerators (whole-body external exposure).

Only medical procedures contribute average doses comparable to that from natural background radiation. In the past few decades, substantial effort has gone into studying these sources to understand, isolate, control, and reduce exposure to as many of them as possible. The problem of monitoring radiation levels is different in each case, requiring the development of many techniques. This book examines these techniques, categorized both by the radiation source and by the type of instrument.

1.2. OBJECTIVES OF ENVIRONMENTAL MONITORING PROGRAMS

The radioactive materials man uses must be adequately contained for the protection of humans and the broader biosphere. Environmental monitors (including those functioning at the source) assess the effectiveness of controls on radioactive releases and, where an abnormality occurs, the character of the failure and of the release.

The International Atomic Energy Agency (IAEA), Safety Series No. 41, lists the following as the primary objectives of environmental monitoring near installations using radioactive materials in *normal* situations [6]:

1. Assessment of the adequacy of controls on the release of radioactive materials to the environment.
2. Assessment of the actual or potential exposure of man to radiation or to radioactive materials present in his environment or, at the very least, the estimation of the probable upper limits of such exposure.
3. Demonstration of compliance with the applicable regulations, environmental standards, and other operational limits.
4. The possible detection of any long-term changes or trends in the environment resulting from the operation of the installations.

According to the IAEA, benefits received from such environmental surveys may include:

1. An increase in knowledge concerning the correlation of levels of discharge and their environmental effect, thus improving the basis for future predictions and for the estimation of maximum or other levels that might arise in an emergency.
2. The provision of data that may be used to provide the public with adequate information on environmental surveillance.
3. The maintenance of a capability that will go at least some way to provide a means of dealing with a foreseeable emergency situation.
4. The provision of data related to the behavior of elements in the environment, including the ecological, hydrological, geochemical, and meteorological aspects of such behavior.

The objectives of a survey during emergency situations differ from those for normal monitoring. The primary objectives of *emergency* surveys, as listed by the IAEA, are:

1. The rapid compilation of information, on a timely basis, on the magnitude and location of the possible hazards to the public, for the purpose of defining the type and extent of any necessary countermeasures or other emergency procedures.
2. The assessment of any hazard from inhalation or external radiation to serve as a basis for the planning of immediate countermeasures.
3. The rapid determination of the possible contamination of foodstuffs, including milk and drinking water, as a basis for decisions on rejection or continued use.
4. The provision of data necessary to assess the radiation doses actually received by members of the public, taking into account any countermeasures that have been applied.
5. The provision of timely information for the public regarding the emergency situation.
6. The collection of scientific information on the results of the emergency and on the behavior of the released radioactive material in the environment. Such information may be useful in checking the adequacy of the monitoring and accident mitigation systems and to determine whether any changes are required in the routine monitoring programs.

The most prominent example of a broad environmental monitoring program is that established initially to measure fallout from nuclear weapons testing [7]. As a direct outgrowth of a program initiated by the U.S. Naval Research Laboratory in 1954, the Department of Energy (DOE) Environmental Measurements Laboratory (EML, formerly the Health and Safety Laboratory), has been conducting the Surface Air Sampling Program since January 1963. The objective of this DOE program is to study the spatial and temporal distribution of specific natural and man-made radionuclides and of trace elements in the surface air. The raw data are summarized in *Environmental Quarterly*, published by EML. DOE also publishes the environmental monitoring reports of its contractors [8].

The U.S. Environmental Protection Agency (EPA) operates the Environmental Radiation Ambient Monitoring System (ERAMS) through its Office of Radiation Programs [9]. This is a surveillance program for measuring levels of radioactivity in air, air particulates, surface deposits, surface water, drinking water, and milk in the United States and its territories. A tabulation of all raw data from each sampling network is reported quarterly in *Environmental Radiation Data* by the Eastern Environmental Radiation Facility. Other federal agencies conduct smaller and more specific programs in areas of concern to them (e.g., the monitoring of mines by the Mine Health and Safety Administration). Moreover, many state agencies conduct broad surveys of radiation and radionuclide levels, usually concentrating their efforts on identifiable radioactivity.

1.3. RADIATION MEASUREMENT AND DOSE ASSESSMENT

Monitoring at a source or in the environment can have two different, but closely related, goals: either to characterize the radionuclide or radiation distribution, or to characterize the resulting burdens or doses to which tissue would be exposed. The purpose of a radiation protection program is to limit radiation doses. However, the monitoring employed to indicate the effectiveness of this program may focus on the use, production, or emission of radioactivity, on the radiation fields resulting from these materials, or on the doses from this radiation.

As a practical matter, a distinction is made based on the manner of exposure. Body tissue may be exposed to radiation from either *external* radioactivity, outside the body, or from *internal* sources, primarily ingested radioisotopes. The instrumentation used to determine radiation fields and absorbed doses depends on the mode of exposure.

1.3.1. External Exposures

For external radiation there are two generally accepted approaches:

1. An area may be monitored to determine radionuclide activity concentrations or alternatively, to characterize the radiation field in terms of spatial distribution, particle flux, energy and angular properties, and intensities.
2. The dose of radiation received by an individual can be determined by personnel dosimetry. The aim is either to determine the integrated dose equivalent (see Section 1.5) received over a period of time (e.g., the use of film badges), or to determine the instantaneous rate, type, and quality of radiation the individual is receiving. In the first case, the measurement is of exposure already received, whereas in the second, the monitoring information is used as a measure of exposure in the present or immediate future.

1.3.2. Internal Exposures

For internal radiation exposure, each or both of the following approaches can be used to determine the dose equivalent:

1. Radioisotope concentrations in water, foodstuffs, or respirable air can be analyzed and compared with maximum permissible concentration (MPC) levels.
2. Whole-body counting, specific-organ activity determinations (e.g., thyroid activity after ingestion of radioiodine), or excrement analyzers can be used to measure postexposure levels.

This book will discuss methods of measuring radiation levels from analyses of radionuclide distributions in water and rock to counts of whole-body or specific-organ activities. An introduction to basic radiation detection methods is given. The ultimate purpose, in any event, is dose assessment.

1.4. ORGANIZATION AND USE OF INSTRUMENT NOTES

This book contains instrument notes, located after the textual material, giving specifications of commercially available instruments, discussions of the operating principles of monitoring instrumentation, and giving background on monitoring problems. The Instrument Notes in the Appendix have been grouped according to radiation type and radionuclide so that the best instrument for the task can be selected. If an instrument performs more than one function, that is noted, and the reader can easily make comparisons of power, size, weight, cost, and other relevant information for each particular application.

At the beginning of some groups of instrument notes, summary sheets are included. These tabulate all the available instrumentation listed in the notes, compiled according to function or type. These summary sheets also list some of the more important characteristics of each device, and can be a useful starting point for persons interested in performing a specific task. After finding the instrument of choice, the note on that instrument may be located by using the key mnemonic which appears both on the summary sheet and on the upper right-hand corner of the note. Each instrument will be identified by its primary application or "class." Light-weight, portable, hand-held, rugged, field-survey monitors, designed for use by semiskilled personnel, are distinguished from heavy, sophisticated, ultrasensitive instruments used by highly trained scientists.

Each instrument note also contains a brief description of the "Principle of Operation." A general discussion of the various detection methods has been provided in Chapter 4. In this chapter a general familiarity with the physical, chemical, and engineering concepts which form the basis of instrument operation is assumed. For more detailed discussions of these concepts as applied to instrument operation, the interested reader is referred to the extensive bibliographies which accompany the various sections.

1.5. UNITS OF RADIATION AND RADIOACTIVITY

Radiation protection requires the development of quantitative concepts and their application to a complex biophysical process. In the early days of radiation research, much confusion existed about the best ways of quantifying radiological doses. Eventually, a set of units was developed and accepted by workers in the field [10]. Today, these units are being replaced by units of the *Système Internationale d'Unités* (SI) [11]. Both sets of units will be used in this book, although it is expected that one day only the SI units will be used. In radiation monitoring it is important to determine both the level (intensity) of radiation and the degree of hazard. Levels are measured in terms of amount of energy dissipated in air (e.g., from accelerators or x-ray machines). Decay rates (e.g., disintegrations per second from radioactive material) are also important in the determination of radiation levels. The unit of energy dissipation in air is the roentgen and the unit of decay rate is the curie (becquerel).

Degree of hazard is described in terms of the amount and type of tissue destroyed which is related to the energy deposited. The unit of energy dissipation in tissue is the rad (gray). Measurement of tissue destruction is complicated by differing effects caused by high levels for short periods compared to low levels for long periods and by differences in radiation type (e.g., alpha particles compared to gamma rays; neutrons are a special case). In addition, biological damage in the human body differs depending on the particular site of tissue destruction. The concept of relative biological effectiveness and quality factor are used to determine the degree of hazard from various types of radiation. The rem is used as a measure of hazard that takes into account the biological effectiveness.

This section introduces the units and concepts used to quantify the hazards of various types of exposure. Chapter 2 elaborates on them and uses them to describe biological effects and to present radiation guides. Because radiation and its interaction with matter is a random process at the microscopic level, certain radiation quantities can only be adequately described in statistical terms. This mode of description is analyzed in some detail in various reports [10,12,13].

1.5.1. Exposure—The Roentgen

Historically, there was a need to quantify the effect of x- and γ-radiation on tissue. Because various tissues absorb such radiation at different rates, x-ray fields were defined by the amount of ionization produced in a unit mass of standard air, the roentgen. Exposure is defined by the International Commission on Radiation Units and Measurements (ICRU) as [10]:

> the quotient [of Q/m] where Q is the absolute value of the total charge of the ions of either sign produced in the air when all the electrons (negatrons and positrons) liberated by photons in a volume of air having the mass m are completely stopped in air. The unit of exposure is the roentgen (R), defined as:
>
> $$1\ \text{R} = 2.58 \times 10^{-4} \text{ coulombs per kilogram of air.}$$

Since the average energy required to produce one ion pair in air is ~34 eV (5.5×10^{-11} ergs), and since each charge in the pair is 1.6×10^{-19} coulomb, one finds that

$$1 \text{ R} = 87.6 \text{ ergs per gram of air.}$$

1.5.2. Absorbed Dose—The Rad

The roentgen does not address the problem of energy absorbed by living tissue. To respond to this need, a basic unit of absorbed dose called the rad (radiation-absorbed-dose) was defined. The ICRU currently defines the rad as:

1 rad = 100 ergs of absorbed energy per gram of tissue
(0.01 joules per kilogram of tissue).

Under the SI system the rad is replaced by the gray (Gy), which is defined as:

1 gray = 1 joule per kilogram of tissue.

Therefore 100 rads = 1 gray.

1.5.3. The Rad–Roentgen Relationship

The roentgen is defined only for photons (x- and γ-rays), and in fact is not used widely except below about 3 MeV. On the other hand, the rad or gray, which measures energy absorbed in tissue, is applicable to all kinds of radiation. The question arises: when tissue is exposed to one roentgen, how much energy is absorbed? Measurements in various tissues (see Fig. 1.2) indicate that in the 1-MeV region one roentgen deposits 85 to 95 ergs/g in fatty or muscular tissue and about 75 ergs/g in bone. This is so close to 100 ergs/g, corresponding to 1 rad

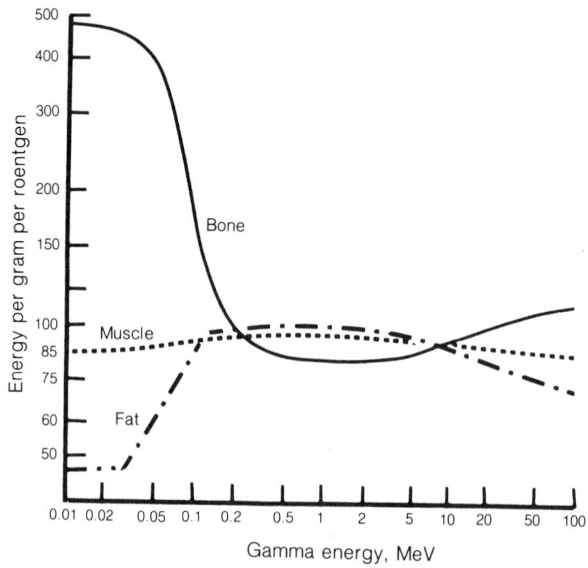

Figure 1.2. Energy absorption per roentgen for various tissues (reproduced from Cember, Ref. 14, with permission of Pergamon Press © 1969).

(0.01 gray), that an exposure to 1 roentgen has often incorrectly been considered equivalent to an absorbed dose of 1 rad. Figure 1.2 [14], indicates how significant the error becomes when bone, for example, is exposed to x-rays in the range of 0.01 to 0.05 MeV.

Almost all x-ray and γ-ray survey instruments measure levels of radiation in units of roentgens. This is because it is the charge generated in a gaseous chamber that they are designed to detect. The conversion of this unit to dose is described in what follows. For field survey work the approximate assumption is made that a rad (0.01 gray) is equal to a roentgen.

1.5.4. Activity—The Curie

A useful quantity in many contexts, including the specifications of radionuclide concentrations, is "activity." The activity of a quantity of material is defined as the number of spontaneous nuclear transformations occurring in this quantity per unit time. The unit of activity is the curie (Ci), where

$$1 \text{ Ci} = 3.7 \times 10^{10} \text{ transformations per second.}$$

In the SI system the curie is replaced by the becquerel (Bq):

$$1 \text{ Bq} = 1 \text{ transformation per second.}$$

Thus 1 Ci $= 3.7 \times 10^{10}$ Bq.

Survey meters for monitoring alpha activity display levels in terms of count transformations per minute. In some contexts, the specific activity (i.e., activity per unit mass) is also useful. Thus, one gram of cesium-134 (half-life = 2.06 yr) has 15 times the decay rate of one gram of cesium-137 (half-life = 30.2 yr). Cesium-134's specific activity is 15 times greater.

1.5.5. Neutral Particles—The Kerma

The interaction of neutrons with matter differs from that of charged particles and photons. Energy loss (in the form of ionizations) does not occur *directly* from the passage of neutral particles through matter. However, neutral particles produce local ionization through indirect means, such as nuclear recoil or decay, resulting from neutron collisions. Neutron collisions causing electron-positron (e^+/e^-) pair production activate some nuclei which then release high-energy γ-rays. The concept used to quantify neutral-particle transfer of energy to matter is the kerma. The following definition comes from ICRU Report 19 [10]:

> The kerma, K, is the quotient dE_{+v}/dm where dE_{+v} is the sum of the initial kinetic energies of all the charged particles liberated by indirectly ionizing particles in a volume element of the specified material (erg/cm^3) and dm is the mass of the matter in that volume element (g/cm^3)
>
> $$K = \frac{dE_{+v}}{dm} .$$
>
> The unit of kerma is the rad or gray.

As defined, the incident neutral particles can impinge upon the mass of interest from all directions. The definition also includes within dE_{+v} the energy of all charged particles (such as Auger electrons) produced in secondary processes within the volume element. The kerma requires a specification of the type of matter being irradiated, and for example one may speak of "the kerma rate of 100 keV x-rays in fatty tissue."

1.5.6. Relative Biological Effectiveness and Quality Factor

The concepts introduced as the rad (gray) still cannot completely describe the impact of energy on tissues, because the absorption of a given amount of energy in a given mass of tissue (rad or gray, ergs/g, or J/kg) does not by itself determine the effectiveness of the dose (dose-effectiveness) of a given radiation field. Tissue damage due to radiation generally increases as the rate of linear energy transfer (LET), that is, the instantaneous ionization per unit length of matter traversed by the radiation increases. For purposes of estimating the effect of such a dose or relative biological effectiveness (RBE) of various radiations, a reference dose was to be chosen: the effectiveness of 200-keV x-rays for creating the observed effect. The RBE allows the assignment of a quality factor Q to each type of radiation. In principle, a high Q value specifies radiation that is very effective in producing damage, and therefore probably has a high LET value. The relation of LET to Q is shown in Table 1.2. Table 1.3 shows a summary of Q numbers of various types of radiation. It should also be noted that Q numbers will depend not only on the radiation type but also on its energy. The Q numbers as a function of energy for neutrons is shown in Table 1.4.

Table 1.2. Relationship Between Quality Factor and Linear Energy Transfer (data presented in ICRU, Ref. 10, with permission of International Commission on Radiation Units and Measurements © 1971)

LET (keV/μm in water)	Q
3.5 (and less)	1
7.0	2
23	5
53	10
175 (and above)	20

Table 1.3. Quality Factor Values for Various Radiations (data presented in ICRP, Ref. 15)

Radiation	Q
Gamma rays	1
X-rays	1
Beta rays and electrons	1
Neutrons	10
Protons	10
Alpha particles	20
Heavy ions	20

Table 1.4. Quality Factor for Various Neutron Energies (data from NCRP, Ref. 16, with permission by National Council on Radiation Protection and Measurements © 1971)

Neutron Energy (MeV)	Q	Neutron Energy (MeV)	Q
2.5×10^{-8} (thermal)	2	5	8
1×10^{-7}	2	7	7
1×10^{-6}	2	10	6.5
1×10^{-5}	2	14	7.5
1×10^{-4}	2	20	8
1×10^{-3}	2	40	7
1×10^{-2}	2.5	60	5.5
1×10^{-1}	7.5	1×10^{2}	4
5×10^{-1}	11	2×10^{2}	3.5
1	11	3×10^{2}	3.5
2.5	9	4×10^{2}	3.5

1.5.7. Dose Equivalent—The Rem

Since the biological effectiveness of a given absorbed dose depends on the factors discussed previously, the absorbed dose is insufficient by itself to predict either the severity or the probability of the deleterious effects on health resulting from irradiation under unspecified conditions. Using the quality factor obtained from determining RBE, an indication of the effect on a given organ is inferred by defining a quantity that is weighted by certain modifying factors. This quantity, the *dose equivalent*, correlates better with important delayed stochastic effects of exposure to radiation (those for which the *probability* of an effect occurring, rather than its *severity*, is regarded as a function of dose, without threshold). Because of this, permissible levels for radiation protection are stated in terms of *dose equivalent*.

This quantity, the dose equivalent (H) is the product $D \times Q \times N$ at the point of interest in tissue, where D is the absorbed dose, Q is the quality factor, and N is the product of any other modifying factors (now assigned a value of 1 by the ICRP).

The most frequently used unit of dose equivalent is the rem, equal to 100 erg/g; thus, H is in rem if D is expressed in rads. In the SI system the rem is replaced by the sievert (Sv); 1 Sv = 1 J/kg (= 100 rem).

It should be noted that: (a) dose equivalent is limited to radiation protection applications and may be used when H is fairly low, that is, in the region of, or below, the applicable permissible dose equivalent (it should not be used for high-level accidental exposures); (b) the dependence of Q on LET is specified by the International Commission on Radiological Protection [16] and shown in Table 1.2; (c) in the usual case where D is delivered by particles having a range of LET values, the dose equivalent is the total dose weighted by an average quality factor, $H = D\bar{Q}N$. The rem (Sv) is applicable to all types of radiation and for both external and internal exposures. It is crucial to recognize, however, that to determine the rem (Sv) *properly* requires a knowledge of the absorbed dose (ergs/g or J/kg), the composition of the radiation, and its energy distribution.

1.5.8. Unit Conversion

Because the two systems of units are currently in common use, we include a table of conversion factors. Table 1.5 summarizes the SI units and their counterparts in the conventional usage of previous literature. In this text the traditional unit is used along with the corresponding SI unit in parentheses. The tabular material, however, is presented in the same units as were used in the original work.

Table 1.5. Relationship Between SI and Traditional Units

Physical Quantity	SI Unit	Value	Older Unit	Value of Older Unit	Conversion
Absorbed dose	gray (Gy)	1 joule/kilogram The gray is the absorbed dose when energy per unit mass imparted by ionizing radiation equals 1 joule/kg.	rad	100 ergs/g	1 Gy = 100 rad
Dose equivalent	sievert (Sv)	1 joule/kilogram The sievert is the dose equivalent, obtained by multiplying the numerical value of the absorbed dose (in grays) times the dimensionless factors Q (quality factor) and N (product of other multiplying factors). Q and N are specified by the ICRP.	rem	100 ergs/g $\times Q \times N$	1 Sv = 100 rem
Activity	becquerel (Bq)	1 decay per second The becquerel is the activity of a radionuclide decaying at the rate of one spontaneous nuclear transition per second.	curie (Ci)	3.7×10^{10} per second	1 Ci = 3.7×10^{10} Bq 1 Bq = 27.027 picocuries (pCi)

REFERENCES

1. National Council on Radiation Protection and Measurements, "Natural Background Radiation in the United States," NCRP Report No. 45, 1975.
2. A. V. Nero and Y. C. Wong, "Radiological Health and Related Standards for Nuclear Power Plants," Lawrence Berkeley Laboratory Report LBL-5285, 1977.
3. National Council on Radiation Protection and Measurements, "Basic Radiation Protection Criteria," NCRP Report No. 39, 1971.
4. Advisory Committee on the Biological Effects of Ionizing Radiations, "The Effects on Populations of Exposure to Low Levels of Ionizing Radiation," National Academy of Sciences–National Research Council, 1972.
5. Donald T. Oakley, "Natural Radiation Exposure in the United States," Office of Radiation Protection, ORP/SID 72-1, 1972.
6. International Atomic Energy Agency and World Health Organization, "Objectives and Design of Environmental Monitoring Programs for Radioactive Contaminants," Safety Series No. 41, Vienna, 1975.
7. E. P. Hardy, Jr., *Environmental Quarterly,* Environmental Measurements Laboratory Report, EML-349, 1979, Environmental Studies Division, U.S. Department of Energy, New York, NY 10014.
8. Energy Research and Development Administration, "Environmental Monitoring at Major U.S. Energy Research and Administration Contractor Sites Calendar Year 1976," ERDA 77-104, 1977; Two volumes by Division of Operational and Environmental Safety, ERDA, Washington, DC
9. Environmental Protection Agency, "Radiological Quality of the Environment in the United States," EPA-520/1-77-009, 1977, Office of Radiation Programs, Washington, DC 20460.
10. International Commission on Radiation Units and Measurements, "Radiation Quantities and Units," ICRU Report No. 19, 1971, 7910 Woodmont Avenue, Washington, DC 20014.
11. International Commission on Radiation Units and Measurements, "Radiation Quantities and Units," ICRU Report No. 33, 1980, 7910 Woodmont Avenue, Washington, DC 20014.
12. International Commission on Radiation Units and Measurements, "Dose Equivalent," ICRU Report No. 19S (Supplement to ICRU Report No. 19), 1973, 7910 Woodmont Avenue, Washington, DC 20014.
13. International Commission on Radiation Units and Measurements, "Conceptual Basis for the Determination of Dose Equivalent," ICRU Report No. 25, 1976.
14. Herman Cember, *Introduction to Health Physics,* Pergamon Press, New York, 1969.
15. International Commission on Radiological Protection, "Recommendations of the International Commission on Radiological Protection," *Ann. ICRP,* **1**(3), 1977, ICRP Publication 26.
16. National Council on Radiation Protection and Measurements, "Basic Radiation Protection Criteria," NCRP Report No. 39, 1971.

2 BIOLOGICAL EFFECTS OF RADIATION STANDARDS

Quantitative concepts have been introduced in Chapter 1 to express radiation doses and their effects on living organisms so that limits on radiation exposures can be set. This chapter discusses some of the biological factors that have influenced the process of setting guidelines for radiation exposure and protection. Two standing committees review the biological effects of ionizing radiation—the United Nations Scientific Committee on Effects of Atomic Radiation (UNSCEAR) and the National Academy of Sciences/National Research Council Advisory Committee on Biological Effects of Ionizing Radiation (BEIR). This discussion is based on the reports of these committees [1–5], as well as reports of the International Commission on Radiological Protection (ICRP) and National Council on Radiation Protection and Measurements (NCRP). Other sources are the International Atomic Energy Agency (IAEA) and the World Health Organization (WHO) [6,7].

After elaborating further on the units necessary to quantify the effects of radiation exposure, this chapter discusses biological radiation effects. Only effects in humans, and related animal and cellular studies are described. Table 2.1 summarizes the types of radiation that must be considered and some important characteristics of each type. Detailed information on the biological effects and permissible radiation levels is contained in Refs. 8–10. Regulatory agencies, including the ICRP [11] and NCRP, have stated that effects on forests [12], birdlife [13], and other parts of the biosphere [14–16] are adequately controlled when humans are given proper protection.

This chapter then describes the development of radiation protection philosophy which has lead to the promulgation of standards or guides. The NCRP and ICRP and other organizations charged with radiation protection are discussed in detail at the end of the chapter. After reading this text, the reader may wish to refer to several standard texts; see Taylor [17,18], Glasstone [19], Morgan and Turner [20], and Cember [21] and publications of the NCRP and ICRP. This discussion

Table 2.1. Basic Types of Radiation and Instrumentation

Radiation	Symbol	Sources and Characteristics	Instrumentation
Alpha particles	He^{++}, α	Highly ionizing, positively charged particles emitted by many naturally occurring radionuclides; discrete energies; short range; atomic mass 4	Ionizing chambers, proportional counters, ZnS scintillators; semiconductors
Beta particles	e^-, e^+, $\beta^{\pm}$	Electrons/positrons emitted by all fission products and most thermal neutron activation products; continuous distribution of energies from zero to characteristic maximum values; more penetrating than α-particles (for energies $>$ 0.1 MeV). Mass $\sim$ (1/1850)1	Gas counters in Geiger-Müller and proportional voltage regions; solid and liquid scintillation counters; photographic film; semiconductor
Gamma and x-rays	γ X	Relatively penetrating electromagnetic radiation; interacts with matter via the photoelectric effect, pair production or Compton scattering, exhibiting characteristic peak energies in the first two instances; absorption follows Beer's Law: $$I = I_o e^{-\mu x}$$ where I_o = initial intensity I = intensity after traversing absorber of thickness x (cm) μ = attenuation coefficient (absorption coefficient) of the absorber	Photon spectrometers; large crystal scintillators (e.g. NaI(Tl); $>$ 0.1 MeV); thin crystal scintillators; Ge(Li) detectors; proportional counters (xenon for $\sim$ 0.01 MeV x-rays; thermoluminescent detector (TLD)
Neutron	n	Indirectly ionizing, neutral particle emitted in fission	BF_3 gas-filled detector; ^6Li loaded TLD; pn junction diode; activation techniques

stresses standards aimed at the general population, and, where appropriate, occupational standards.

2.1. RADIATION EFFECTS

The preceding chapter introduced units of measurement for radiation and its effects. However, before this section can describe the effects of ionizing radiation on biological organisms it is necessary to examine and clarify the units used to quantify radiation effects and the factors such as dose rate and LET which can affect the amount of damage. The section also describes health effects and assessment.

2.1.1. Units for Measurement of Radiation Effects and Assessment

Each type of radiation induces biological effects in living tissue. These effects differ considerably, depending on many factors including the amount of absorbed energy and its actual effectiveness. The absorbed dose provides a measure of the amount of energy absorbed by tissue (introduced in Chapter 1). Absorbed dose is measured in rad (gray); one rad (1 rad = 10^{-2} Gy) corresponds to the absorption in any substance of energy to the extent of 100 ergs/g. (One gray corresponds to the absorption in any substance of energy to the extent of 1 joule/kg.)

While the rad (gray) is a general concept, experiments show that one rad (10^{-2} Gy) of absorbed dose does not always produce the same biological effect in all tissues. There is considerable evidence that the amount of damage produced by a charged particle depends on the instantaneous ionization per unit length of matter traversed, or linear energy transfer (LET); see Section 2.1.2. Biological effects also depend on the *rate* at which the absorbed dose is delivered.

The radiobiologist must address the problem that the same absorbed dose can have different relative biological effectiveness (RBE). Usually x-radiation near 200 keV is used as the reference for specifying RBE. The RBE of two radiations, 1 and 2, is defined in terms of absorbed doses, D_1 and D_2, that produce the same biological effect under identical experimental conditions:

$$\frac{(\mathrm{RBE})_1}{(\mathrm{RBE})_2} = \frac{D_2}{D_1} .$$

This definition does not imply a quantitative relationship between absorbed dose and biological damage. It merely defines the relative amounts of radiation required in an experiment with two radiations to make them biologically equivalent so that they produce the same specified amount of biological damage.

To permit assessment of the effect of differing RBE for various types of radiation a quality factor (Q) is established for each. Actual standards for radiation protection are set down in terms of the rem (sievert), which was defined in Chapter 1. The discussions that follow give overviews of the various types of biological effects and of the parameters that influence RBE.

2.1.2. Linear Energy Transfer

Linear energy transfer (LET) describes the local energy loss of a charged particle as it traverses through matter, producing ionization as well as undergoing certain other interactions. LET is usually expressed in units of kiloelectron volts per micron of path traveled in water. In general, the LET of a charged particle depends on its velocity, and for the particles of concern here the LET increases as the velocity decreases, reaching very large values just before the particle stops. This is because in passing through matter the probability of collision is greater for a slower moving particle because of the movement of atoms (of molecules) in the media. More collisions produce higher LETs.

A relationship exists between biological effectiveness and the LET of an incident charged particle. When the incident radiation is neutral (a neutron or a photon), the LET of the *secondary charged particles* correlates with their biological effectiveness. A few illustrative experiments will show the effects of different LETs on different types of biological organisms.

Figure 2.1 compares the RBE of relatively hard (high-energy) gamma rays, soft (low-energy) x-rays, and densely ionizing alpha particles on the inhibition of cell division. The horizontal axis gives the LET of the secondary electrons produced by photons and the LET of the alpha particles. Compared with gamma rays, x-rays produce lower-energy secondary electrons having somewhat higher values of LET. The data show an increase in RBE by a factor of about 3 between gamma rays and alpha particles. Alpha particles which have a higher LET inhibit cell division at much lower absorbed doses and hence a higher RBE.

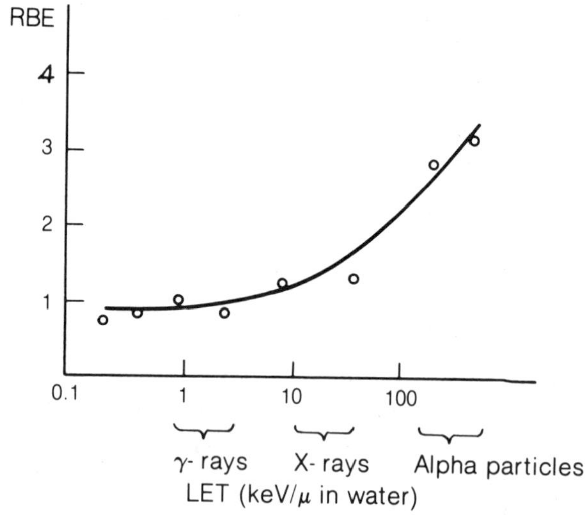

Figure 2.1. A comparison of RBE (relative biological effectiveness) among relatively hard gamma rays, soft x-rays, and alpha particles. The biological effect shown is inhibition of cell division, but the qualitative features are common to most systems (reproduced from Zirkle, Ref. 22, with permission of McGraw-Hill © 1954).

The results of another series of RBE experiments are shown in Fig. 2.2. Mice were exposed to acute doses of monoenergetic neutrons at several energies, and a number of biological endpoints were studied. Most of the neutron energy (e.g., at 10 MeV, 85%) is deposited in soft tissue through elastic collisions with hydrogen nuclei (protons). The average proton recoil energy $\bar{E}_p$ in this energy range is one-half the incident neutron energy E_n or: $\bar{E}_p = E_n/2$. Lower-energy neutrons, therefore, on the average produce lower-energy protons, which have a higher LET. As seen from Fig. 2.2, lower-energy neutrons are more efficient in producing the effects. The absorbed dose required from the lowest-energy neutrons is only about one-fourth of that from the fastest neutrons. Usually, RBE values of ionizing radiation fall in the range from 1 to 10.

The variation of RBE with LET does not follow a universal rule, although several types of variation can be described. Curve A in Fig. 2.3 shows in a stylized fashion the type of behavior often exhibited by mammalian cells. As in the previous two examples, RBE increases with increasing LET, but only up to a point. Evidently, some biological damage is produced most efficiently (i.e., with the lowest absorbed dose) when the average rate of energy loss along a particle track has a particular value. Additional energy deposited at higher LET is seen to be less effective. Curve B, in which the RBE is never greater than unity, is typical of some effects produced in microorganisms (for example, the inactivation of virus and enzyme molecules). For these small organisms, the response of the organism per unit absorbed dose is not increased by increasing the number of energy-loss events occurring along a particle track.

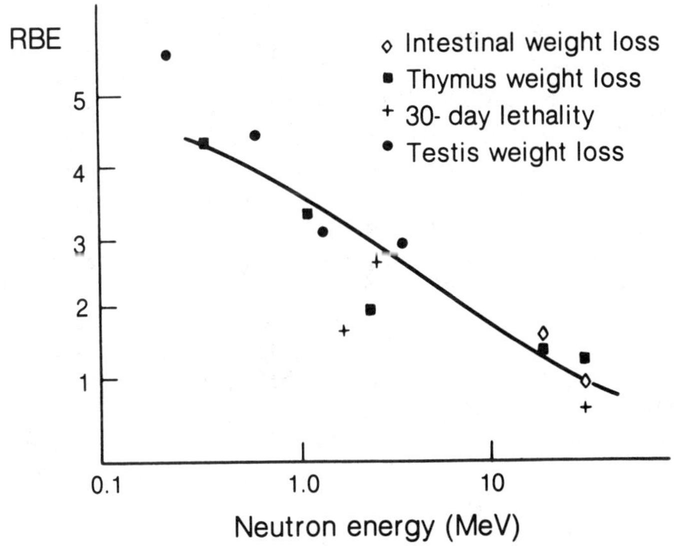

Figure 2.2. The RBE (relative biological effectiveness) of monoenergetic neutrons at several energies; mice were exposed to acute doses and a number of indicated biological endpoints were studied (reproduced from Bateman and Bond, Ref. 23, with permission of NY Academy of Sciences © 1964).

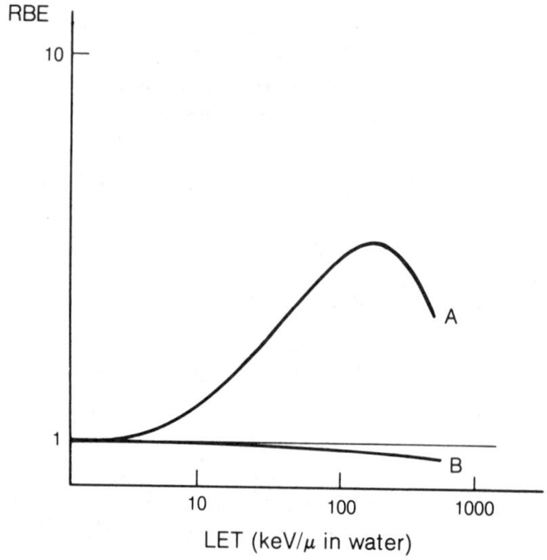

Figure 2.3. The variation of RBE (relative biological effectiveness) with LET (linear energy transfer), for typical systems. Curve A shows the qualitative behavior often exhibited by mammalian cells, while Curve B is typical of some effects produced in microorganisms (reproduced from Zirkle, Ref. 22, with permission of McGraw-Hill © 1954).

2.1.3. Dose Rate

The dependence of RBE on the time over which a dose is received (the dose rate) has been demonstrated in a number of experiments. Two examples are given here.

The data presented in Fig. 2.4 show the variation with dose rate of the 30-day LD-50 (the total absorbed dose required to produce 50% lethality in 30 days) in a population of mice. At rates above about 250 rad/hr (2.5 Gy/hr), the LD-50 is about 800 rad (8 Gy). At lower dose rates the LD-50 is larger [about 1600 rad (16 Gy) when the rate is 6 rad/hr (0.06 Gy/hr)].

Genetic studies with mice and fruit flies have shown that fewer mutations per rad (gray) occur when the dose rate is low than when it is high. Figure 2.5 shows experimental data for the mouse.

Sometimes the opposite dose-rate effect occurs: that is, a given dose can be less effective when delivered at once than when administered at intervals. The response of an organism appears to depend in a complicated way on many factors. A number of theories on damage and recovery have been proposed to account for observed dose-rate effects.

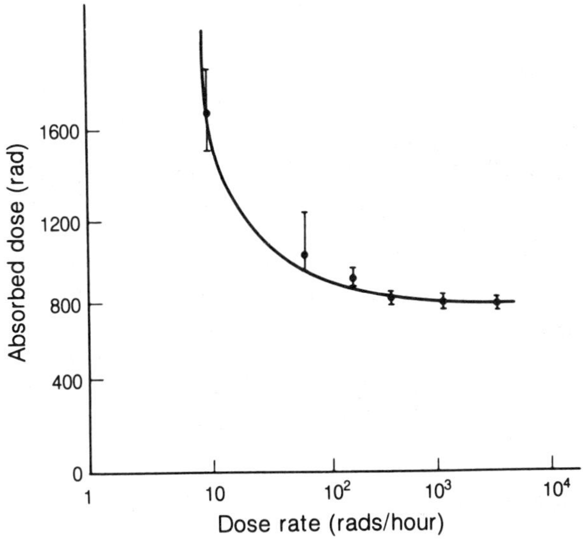

Figure 2.4. The effect of dose rate. The variation with dose rate of the LD-50 (the total absorbed dose required to produce 50% lethality) in 30 days in a population of mice (reproduced from Thompson and Tortellotte, Ref. 24, with permission of Williams and Wilkins Co. © 1953).

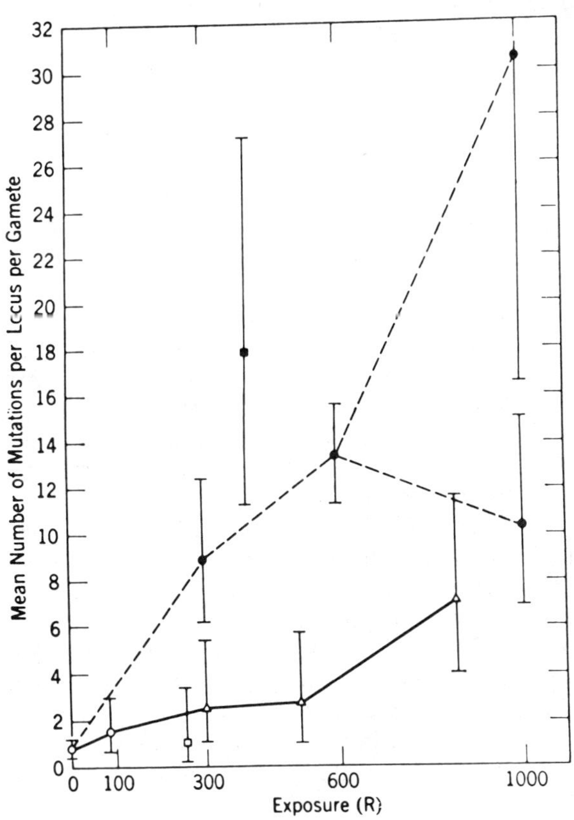

Figure 2.5. Dose curves for specific locus mutations in the mouse. Solid points (90% confidence) represent results with acute x-rays (80-90 R/min); open points are chronic gamma doses (10-90 R/week). Top 1000-R point represents a single exposure; lower one, successive exposures of 600 and 400 R, 15 weeks apart. Reproduced from Russell, Russell and Kelley, Ref.25, with permission of Taylor and Francis, Ltd. © 1960).

2.1.4. Cellular Effects

Radiation causes immediate physical changes in a cell and its environment and may alter biological behavior immediately or at some later time. Specific cellular changes following irradiation include the breakup of molecules, the production of chemically active free radicals (atoms or molecules with unpaired electrons), enzyme inactivation, alteration of deoxyribonucleic acid (DNA) synthesis, and chromosome breakage. Gross physical changes can occur, such as increased viscosity of cellular fluids, increased permeability of cellular membranes, swelling, or even death. When enough cells are damaged, the normal functioning of an organ or of an entire organism can be impaired. Although many aspects of radiation injury have been studied in experiments and clinical observations, the detailed mechanisms and steps that lead from the absorption of radiation to the appearance of symptoms are not fully understood. The production of radiation damage in biological systems is under intensive research and study.

Different kinds of cells and organs show different sensitivities to radiation. As a general rule, rapidly dividing cells, such as those often occurring in a malignant growth, are most sensitive. Cells of a fetus are also especially sensitive to radiation. The most radiosensitive cells in the normal body are those of the bone marrow, lymphoid, and epithelial (skin and other surface) tissues. Blood changes, resulting from bone marrow damage, and the destruction of lymphocytes, are among the earliest observed effects following the absorption of radiation. At low doses, changes in the blood may be the only evidence of irradiation. Damage to epithelial cells produces injury in the gastrointestinal tract and in the reproductive organs. Epithelial cell damage may also result in loss of hair. Bone, muscle, and blood vessel cells are less radiosensitive. Nerve cells show the most resistance to radiation damage.

2.1.5. Acute Radiation Syndrome

The characteristic group of symptoms shown by a person whose whole body, or most of it, is exposed to a large dose of radiation in a short time is called the acute radiation syndrome. These symptoms have been studied in animal experiments and, more important, in cases of radiation administered to persons in hospitals, in the atomic bomb survivors at Hiroshima and Nagasaki, and in persons exposed accidentally to large doses of radiation. This section discusses the acute radiation syndrome for gamma radiation, which, because of its penetrating power, gives a more or less uniform dose throughout the body. Because biological response to radiation varies from one individual to another, the numbers given here represent only an average expected among a number of exposed persons [26].

With doses below about 25 rad (0.25 Gy), little or no detectable effects occur. Doses from 25 to 100 rad (0.25 to 1 Gy), cause detectable changes in the blood. Near 100 rad (1 Gy), the lymphocyte count decreases within a matter of hours after exposure to the radiation. After one or more days the white blood count falls and may subsequently fluctuate. The count of blood platelets, which are necessary for clotting, also begins to decline shortly after exposure. The change in the red blood cell picture is not as rapid and may not show a decrease for several days.

A dose of 100 to 300 rad (1 to 3 Gy) can result in some radiation sickness but is rarely fatal. Nausea and vomiting usually occur within a few hours following a dose of 200 to 300 rad (2 to 3 Gy). Symptoms such as malaise, fatigue, and loss of appetite can be expected to appear. The blood will change dramatically in the manner already described, and many months may elapse before it returns to normal. Antibiotics may be useful to offset the danger of infection that arises from the decline in the number of white blood cells. If no complications develop, recovery of the exposed individual can be expected.

Doses in excess of 300 rad (3 Gy) may be fatal. In addition to changes already noted for lower doses, diarrhea and hemorrhaging can occur, and the dangers of infection are compounded by lowered resistance of the body. The gastrointestinal tract is severely damaged: epithelial cell linings ulcerate, and uncontrolled bacterial infections may develop. Loss of hair may occur in about two weeks, and temporary sterility can be expected. Therapeutic measures include the use of antibiotics, blood transfusions, and, at high dose levels, bone marrow transplantations.

Only a small percentage of individuals would survive an acute whole-body dose of 1000 rad (10 Gy). Doses up to several thousand rad (several tens of grays) cause death within one or two weeks. Still larger doses result in death within days. Besides affecting the gastrointestinal tract, high doses damage the central nervous system, and an exposed individual can be almost immediately incapacitated. Death can also be caused by circulatory or respiratory failure.

2.1.6. Delayed Somatic Effects

Somatic effects of radiation are those seen or detected in exposed individuals (in contrast to genetic effects, which show up in progeny). In addition to the acute somatic changes already discussed, radiation can cause delayed effects, with the latent period between exposure and the appearance of symptoms ranging from months to many years. Because somatic and genetic effects produced by radiation also have a normal spontaneous incidence, it is often impossible to ascribe to radiation the occurrence of a symptom in a particular exposed individual. Delayed somatic effects from radiation in a population occur statistically; an individual in an exposed population is presumably placed at a higher risk of injury.

Figure 2.6 shows the correlation between high radiation doses and the probability per year of the occurrence of leukemia. The peak incidence of the disease among

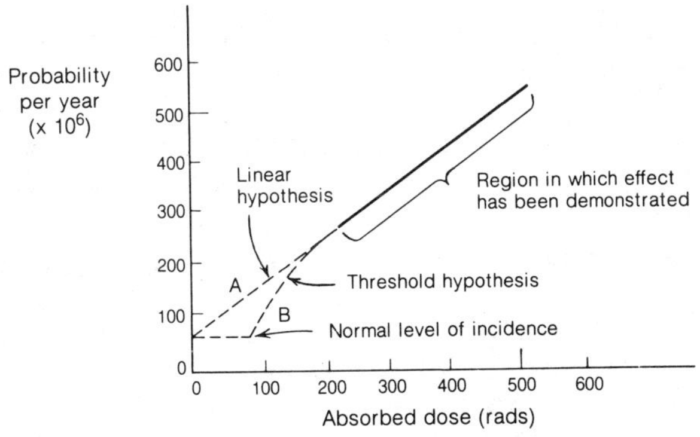

Figure 2.6. Incidence of radiation-induced leukemia in humans. Curve A illustrates a linear hypothesis for the effect of low doses; Curve B represents a threshold-type hypothesis (reproduced from Hurst and Turner, Ref. 27, with permission of Wiley © 1970).

the survivors at Hiroshima and Nagasaki after the bombings in 1945 occurred during the period 1950 to 1952, with an abnormally high incidence continuing beyond this time. Further studies of these exposed populations show evidence of an increased incidence of other types of neoplasia [26].

An example of disease produced in humans as a result of radiation exposure over a long period of time is bone cancer in radium dial painters [26]. Like leukemia, such effects have also been demonstrated in laboratory experiments with animals [26].

Radiation has induced lens opacities and cataracts. Acute exposure to radiation appears to be more serious than protracted exposure in producing cataracts, which may appear several years after exposure. The x-ray dose required to produce clinically significant cataracts is probably at least 400 rad (4 Gy); fast neutrons are much more effective. RBE values for cataract production in excess of 10 have been reported.

Pregnant women exposed to radiation show adverse effects, some of which appear many years later in the child [26]. In addition to abnormally high still-birth and infant mortality rates among Japanese survivors pregnant at the time of the atomic bombings, their offspring appears to exhibit an above-normal amount of mental retardation. Small children who were exposed may lag somewhat behind their contemporaries in overall physical development. These effects are noticeable only among survivors who received doses of at least several hundred rad.

Beta-ray doses of a few thousand rad to the skin of the body do not seem to effect deeper-lying organs, although in many cases a chronic abnormal condition of radiodermatitis developed over a period of years after exposure. The skin became thin and susceptible to ulceration and the formation of cancer.

2.1.7. Genetic Effects

Gene mutations occur spontaneously in nature. The rate of mutation in a population can be increased by elevated temperatures, ultraviolet light, and certain chemical agents. Müller, in 1927, discovered that ionizing radiation also produces mutations.

Although mutations can be produced in any cell of an individual, only those produced in a germ cell can be transmitted to the progeny. A mutant gene introduces changes that may or may not be apparent in the next generation. Genetic changes in an offspring may be minor or inconsequential, or they may be so serious as to cause a severe handicap.

The frequency of radiation-induced mutations depends on the absorbed dose and the dose rate. An increase in mutation rate in a large experimental population becomes noticeable with absorbed doses of a few tens of rads and thereafter it appears to increase linearly with increasing dose. The effect of dose rate for the mouse is illustrated in Fig. 2.5.

Both the frequency and the type of mutation depend on the time of mating after exposure. When mating takes place soon thereafter, the germ cell involved has a high probability of having been mature at the time of irradiation. Irradiated mature germ cells carry a relatively high incidence of dominant lethal mutations. When the mating occurs later, there is a relatively high probability that it will involve a cell that was immature when irradiated. Mutations arising in late matings following exposure are predominantly recessive.

2.1.8. Is There a Minimum Dose for Damage?

At present, it is not known whether an absorbed dose greater than a certain minimum, or threshold, amount is necessary for the production of biological damage from radiation. Although the effects just described are demonstrable with large amounts of radiation, doses can be made so low that it is virtually impossible to measure any effects, even with very large study populations. As an example, Fig. 2.6 shows how the probability per year (incidence) of leukemia in man varies with radiation dose. Studies of Japanese survivors from the atomic bombs at Hiroshima and Nagasaki and of patients receiving therapeutic x-ray treatments indicate that the solid curve describes the relationship between dose and incidence when the dose is several hundred rad (Gy) or greater. Below about 100 rad (1 Gy) too few cases of leukemia occur in the populations studied to demonstrate clearly an increased incidence due to radiation.

At doses of 1 to 10 rad (0.01 to 0.1 Gy), based on the observed incidence at high doses, the size of an exposed population needed to show a significant statistical increase would be enormous, and therefore experiments with animal populations are not feasible. At present, researchers can do no more than hypothesize what effect low doses might have on leukemia incidence.

One assumption frequently made is that the approximately linear solid curve at high doses should be extrapolated to zero dose, as shown by the dashed curve A in Fig. 2.6. According to this hypothesis, any amount of radiation, no matter how small, increases the probability of leukemia above its normal value by an amount proportional to the dose equivalent. Another assumption is that the body can repair damage done by small amounts of radiation. According to this hypothesis, the body tolerates low doses, and a certain threshold dose is required before the probability of leukemia is increased above its normal value. This threshold hypothesis is illustrated by the dashed Curve B in Fig. 2.6.

Curves with shapes other than those of A and B have also been postulated for connecting the high-dose region, in which data are statistically significant, to the region of low dose. Differentiating between a linear or a threshold response for any biological effect is complicated by the fact that apparent radiation-induced effects occur, from non-radiation-induced sources as well as from low levels of natural or anthropogenic radiation

2.1.9. Assessment of Radiation Effects

After World War II, routine operations in the atomic energy industry exposed more and more workers to neutrons and to radiation from radioactive materials as well as to x-radiation. Recognizing the need for a uniform set of guidelines to assess effects of radiation, the International Commission on Radiation Units and Measurements (ICRU) and ICRP recommended that quality factors (Q) be used for different kinds of radiation. The values of Q given in Table 2.2 based on the linear energy transfer (LET) of the radiation, are for long-term, occupational exposure to radiation. To assess radiation exposure for control and protection purposes, the absorbed dose from radiation in each different LET range is multiplied by the appropriate quality factor. As mentioned earlier, the resultant number is called the dose equivalent (DE), expressed in rem (sievert). When all of the absorbed dose D_a is from a single range of LET, the dose equivalent is given by

Table 2.2. LET-Q Relationship (recommended by ICRP and ICRU, Ref. 28)

LET (keV/μm in water)	Q
3.5 or less	1
3.5–7.0	1–2
7.0–23	2–5
23–53	5–10
53–175	10–20
γ-rays, x-rays, electrons and positrons of any LET	1

$$DE = Q \times D_a$$

$$(\text{rem} = Q \times \text{rad})$$

$$(\text{Sv} = Q \times \text{Gy})$$

The Q value for x-, γ-, and β-radiation is unity. For other radiations the Q's are generally two or three times larger. The dose equivalents in different LET ranges are added when monitoring different kinds and energies of radiation exposure.

2.2. RADIATION PROTECTION

If ionizing radiation were not useful, or were not a consequence of useful operations, it would not be produced by man and would not be an environmental problem. Whether natural radiation causes environmental problems is often hard to determine and almost always impossible to avoid. Some of the best known benefits of ionizing radiation are in the medical field where much of modern medicine would be at a virtual standstill without diagnostic x-rays. In addition, x-rays and γ-rays have considerable therapeutic value in the treatment of cancer and other diseases. Other types of radiation, including neutrons and protons, have shown promise in the treatment of a number of ailments.

X-rays and γ-rays are both used quite extensively in industry for nondestructive testing. To a much lesser extent both types of radiation are used in chemical processing and for food and equipment sterilization. Nuclear reactors, both for power plants and other industrial operations, are a major source of radioactive materials, and monitoring them and their products is a major environmental problem. Nuclear reactors are treated in Chapter 10. That ionizing radiation results from so many useful activities has made it necessary for both federal and state governments to set up strict standards for allowable amounts of radiation in areas where people may be exposed and for methods of measuring this radiation.

2.2.1. Development of Radiation Protection Criteria

After radiation was discovered by Becquerel and began to be used, it was quickly realized that the new phenomenon had definite biological effects. In one of the earliest reported incidents, Edison and his assistants complained of a smarting of the eyes after doing an experiment with a cathode-ray tube. For a long time, however, only the effects of high doses were recognized. Low-dose effects were not recognized until the middle of this century. Early medical use resulted in some unfortunate overdoses, leading sometimes to immediate results and thus the discovery of definite drastic biological effects. Up until 1928, however, there was no large-scale effort in radiation protection activities, and most of the responsibility for such protection fell to the societies and organizations of active radiologists.

Initial concerns were focused primarily on protecting radiologists from gross biological effects of external radiation. As a result, shielding was the primary method recommended to reduce exposures. Lead was found to stop x-rays more effectively than any other common metal and was quickly adopted for radiation protection. Because of the availability of effective shielding material, and the absence of any knowledge of radiation dosimetry, all early protection recommendations were based on shielding factors rather than limiting dose.

During the 1920s, researchers became aware of and were able to examine the effects of a bone-seeking α-emitter, ^{226}Ra. Radium dial painters used their lips and tongues to keep a sharp tip on their brushes, thus ingesting radium. From long-term studies of these workers, it has been learned that such an α-emitter can cause bone cancer, with a latent period of about 15 years. By estimating the body burdens of ^{226}Ra, a rough dose-effect curve began to take shape.

It was largely occupational experiences of radiologists and radium dial painters at high dose levels that caused an awareness of the danger of high doses of radiation. However, these experiences were not put on a quantitative basis, such as is done today, because no measure of radiation exposure had been agreed upon. See for examples Refs. 25 and 29. It was not until the 1930s that the roentgen was adopted as a measure of exposure. (An early attempt to quantify radiation dose was based on reddening of the skin, which clearly exhibited a threshold.) In 1934, the ICRP adopted a recommendation specifying a tolerance dose expressed in roentgens, thereby replacing shielding as the sole protection concept. The first standards put forth by the ICRP after its establishment in 1928 were intended only for use with x-rays, since it wasn't certain the effects for γ-rays and α-particles were related to those of x-rays. Radiation protection had developed haphazardly before 1930, but with the creation of the ICRU, ICRP, and NCRP, national and international supervision helped the field to develop more smoothly.

The experience of the radium dial painters showed that radiation posed an internal hazard. As the scope of the radiation hazard was realized, attention shifted from simply protecting radiation workers to considering the safety of people receiving medical or diagnostic doses. Except for scientific research (of which there was not as much as there is today), medical workers were the primary users of radiation up to about World War II. This situation changed rapidly in the 1940s and 1950s. Even so, until the late 1950s, radiation protection philosophy was aimed at preventing the immediate acute effects of radiation exposure. The latent effects of radiation were only slowly recognized, largely because of follow-up studies of the Hiroshima-Nagasaki survivors.

The development of fission and its uses in weapons and electricity production meant that many members of the general public might be exposed to fallout and effluents from nuclear fuel cycles. With the advent of the atomic age, it became obvious that, with such widespread use, the public needed protection from ionizing radiation. Because long-lived radioactivity was being widely dispersed in waterways and the environment, a major mode of exposure for the general public was internal. In the late 1950s, regulatory bodies began limiting internal exposures. Concentration standards for water and air were derived from the maximum permissible body burden calculated from the particular internal dosimetry of various radionuclides. Also during the 1950s, qualitative cost-benefit considerations were introduced, and the assumption that the biological effects of radiation have a definite threshold was formally dropped, largely because no threshold for genetic damage could be identified.

At the present time, the primary standards for radiation protection are given in terms of limiting radiation dose to the whole body, or particular body organs, from all sources of radiation other than natural background. Medical exposures are among the sources of radiation limited [3,11]. Limits of annual intake are being developed that will keep organ and body burdens below certain levels, even after a lifetime of intake. The maximum permissible concentrations (MPC) in air and drinking water are generally derived using only single modes of exposure. Whenever there is more than one activity present, the MPCs must be altered appropriately.

In determining the contribution of a particular source to human exposure both the present and the future contribution must be considered because some radionuclides remain active far into the future. A concept gaining increasing favor is to base the restriction on this "dose commitment" from the release rather than on the dose rate.

2.2.2. Current Radiation Protection Philosophy

A significant feature of existing protection philosophy is the recognition that radiation protection criteria depend not only on purely biomedical and physical considerations, but also on value judgments [18]. These criteria presently include the concept that some degree of risk may exist at any level of exposure. As the understanding of the radiation hazard grew, the acceptable levels grew smaller, as can be seen by looking at how radiation protection standards evolved numerically over the years (Table 2.3).

Table 2.3. Evolving Radiation Standards Recommended by the NCRP

1934	0.1 R/day	(or 0.5 R/wk)	(=25 R/yr)
1949		0.3 rem/wk	(=15 rem/yr)
1957		0.1 rem/wk	(= 5 rem/yr)
Present			5 rem/yr

Until 1971, the NCRP explicitly reviewed only the basic principles underlying the recommendations for *external* sources of ionizing radiation. The release of NCRP Report No. 39 [30] rectified that apparent anomaly. This report and others are quoted to review the general criteria under which the advisory and regulatory agencies operate. (Reference numbers in the quote have been changed to correspond to the reference numbers at the end of this chapter.)

> Radiation protection recommendations must set a pattern as scientific as possible, but in practice partially empirical; ideally, the rational use of radiation should conform to levels of safety to users and the public which are at least as stringent as those achieved for the other powerful agents. It is important to avoid unnecessary rigidity in any protection recommendations so that the advantages of new knowledge can be exploited quickly and easily. This basic posture must be supplemented by somewhat detailed guidelines for control in numerous specific applications; these guidelines may change frequently in the light of practical experience, availability of improved measuring methods, and the complexity of competing applications involving radiation. Bodies promulgating recommendations must maintain cognizance of advances in the understanding of radiation effects so that the basic criteria may become better founded in scientific principles and less dependent on value judgements, whenever possible [from Ref. 30].

From the NCRP review [31]:

> The NCRP position is centered on the principle that the "lowest practicable" radiation level is the fundamental basis for establishing radiation standards, and on the assumption that the most important radiation health hazards do not have a dose threshold. On this basis, the setting of radiation protection standards requires consideration of compensatory trade-offs between currently assumed hazards and benefits. A basic assumption of conventional benefit-cost analysis is that human welfare (which this procedure seeks to maximize) may be measured in monetary or materialistic terms. This assumption has fundamental problems in that it cannot handle the situations in which the risks and benefits do not accrue to the same people. The BEIR Committee has examined the problem of evaluation of the total benefits derived from exposure to ionizing radiations for comparison with the total risks. The reader is referred to their recently completed report [3] for a more detailed consideration of the myriad factors involved in health benefit-cost analyses.
>
> The NCRP continues to hold the view that risk estimates for radiogenic cancers at low doses and low dose rates derived on the basis of linear (proportional) extrapolation from the rising portions of the dose-incidence curves at high doses and high dose rates...cannot be expected to provide realistic estimates of the actual risks from low-level, low-LET radiations, and have such a high probability of overestimating the actual risk as to be of only marginal value if any, for the purposes of realistic risk-benefit evaluation.

And from the ICRU [32]:

> Numerical radiation protection guides or dose limits for the exposure of radiation workers or the general public are provided only as upper limits; all exposure should be kept to a practicable minimum [14].

The NCRP and the ICRP both base their recommendations on a system of dose limitations, the basic features of which are as follows [11]:

1. No practice shall be adopted unless the introduction produces a positive net benefit.
2. All exposure shall be kept as low as reasonably achievable, economic and social factors taken into account.
3. The dose equivalent shall not exceed the limits recommended for the appropriate circumstances by the Commission.

Two conservative positions are the basis of the ICRP and NCRP recommendations [18]:

1. All low doses delivered to an organ are completely additive, no matter at what rate they are delivered or what intervals there may be in delivery.
2. There is no biological recovery of radiation effect from low doses.

If indeed these conservative assumptions are true, then the root of the problem in setting radiation protection standards is what level of damage and deaths is acceptable and to whom.

2.2.3. Radiation Protection Organizations

This section introduces radiation protection organizations and their recommendations when appropriate, explaining where and how they are applicable. International bodies are discussed first as they are influential in setting guidance for national bodies.

International Agencies. The *International Commission on Radiological Protection* (ICRP) was established in 1928, as the International X-ray and Radium Protection Committee, by the Second International Congress of Radiology. It assumed its present name and organizational form in 1950 in order to cover more effectively the rapidly expanding field of radiation protection. The ICRP has continued its close relationship with succeeding Congresses of Radiology, and it has also been looked to as the appropriate body to give general guidance on the more widespread use of radiation sources caused by the rapid developments in the field of nuclear energy. The ICRP maintains its traditional contact with medical radiology and the medical profession generally, and it also recognizes a responsibility to other professional groups and an obligation to provide guidance within the field of radiation protection as a whole. The policy adopted by the ICRP in preparing its recommendations is to consider the fundamental principles on which appropriate radiation protection measures can be based while leaving to the various national protection bodies the responsibility of formulating the specific advice, codes of practice, or regulations that are best suited to the needs of their individual countries.

The ICRP has an official relationship with the World Health Organization (WHO) and the International Atomic Energy Agency (IAEA). Close working relationships are maintained with the United Nations Scientific Committee on the Effects of Atomic Radiation (UNSCEAR), The International Labour Office, the United Nations Environment Program, the Organization for Economic Cooperation and Development's Nuclear Energy Agency, and the European Economic Community by sending observers to meetings organized by these bodies. These organizations are in turn invited to send representatives to the ICRP's meetings with its committees.

As the purpose of the ICRP is not to promulgate dose limits per se, but only to suggest recommendations which can be used by various bodies who then promulgate dose limits or standards, their recommendations are not best represented in a set tabular format. The current ICRP recommendations can be found in ICRP Publication 26 [11]. Basically, their recommendations are to the effect that individual members of the public should not be exposed to a dose that entails a risk greater than that arising from 0.5 rem (5 mSv) uniform irradiation of the whole body, even when individual organs are irradiated.

The *International Commission on Radiological Units and Measurements* (ICRU) is the international advisory body which, since its inception in 1925, has been primarily concerned with the development of acceptable recommendations regarding:

1. Quantities and units of radiation and radioactivity.
2. Procedures suitable for the measurement and application of these quantities in clinical radiology and radiobiology.
3. Physical data needed in the application of these procedures, the use of which tends to assure uniformity in reporting [32].

The ICRU also makes recommendations for radiation protection. It does this in close cooperation with the ICRP. The ICRU collects and evaluates the latest data and information pertinent to the problems of radiation dosimetry and measurement and recommends the most acceptable values and techniques for current use. In its operating policy [32],

> The ICRU feels it is the responsibility of national organizations to introduce their own detailed technical procedures for the development and maintenance of standards. However, it urges all countries to adhere as closely as possible to the internationally recommended basic concepts of radiation quantities and units.

The ICRU maintains close relationships with other organizations in much the same fashion as the ICRP. It has formal working arrangements with WHO, IAEA, and UNSCEAR, and informal connections exist with many other organizations, such as the International Organization for Standardization.

The *United Nations Scientific Committee on the Effects of Atomic Radiation* (UNSCEAR) has, since the 1950s, regularly reviewed the scientific literature concerning biological effects of ionizing radiation and trends in the radiation exposure pathways (e.g., natural radiation background, nuclear fuel cycle, medical radiation, occupational and patient, etc.). These reviews have in the past caused some restructuring of the ideas underlying radiation protection guidelines. An example is the recognition that carcinogenesis is the main danger of low doses [5]. This reversed the previous view that the genetic component of damage should be the overriding concern in setting dose limits for the general population. These reviews contribute to ICRP development of radiation protection recommendations by providing estimates of the frequency of health effects.

The *International Atomic Energy Agency* (IAEA) was organized in 1956 to "accelerate and enlarge the contribution of atomic energy to peace, health, and prosperity throughout the world." Its membership is composed of approximately 105 member states of the United Nations. It was empowered to provide for the application of standards of safety for protection against radiation to its own operations and to operations making use of assistance provided by it or with which it is otherwise directly associated. To this end, authorities receiving such assistance are required to observe relevant health and safety measures prescribed by the Agency [6].

The recommendations of the IAEA rely heavily on the basic recommendations of the ICRP and ICRU, and are set forth in manuals and codes on health and safety published in the Safety Series or Technical Report Series. These manuals serve as basic reference documents for the preparation of national health and safety standards covering the use of radionuclides.

Some work of the IAEA is done in collaboration with other U.N. agencies. For instance, with the International Labor Organization the IAEA has issued general safety guidance for the mining and milling of radioactive ores (see Chapter 12). In addition, in collaboration with the WHO, the IAEA has issued guidance on handling radionuclides, environmental monitoring, and other topics. With the Food and Agricultural Organization, it monitors isotopes in precipitation.

Other international organizations also address nuclear issues. Some of these are Euratom, Inter-American Nuclear Energy Agency, and the Organization for Economic Cooperation and Development's Nuclear Energy Agency (an international professional organization that sponsors pertinent symposia is the International Radiation Protection Association).

United States Agencies. The *National Council on Radiation Protection and Measurements* (NCRP) provides for the United States the same sort of general guidance and recommendations that the ICRP and ICRU provide at an international level. The NCRP started in 1928 as an informal unit of the National Bureau of Standards (NBS) to provide a focus for U.S. input to the ICRU, and later the ICRP. It was chartered in 1964 by Congress as a nonprofit corporation to:

1. Collect, analyze, develop, and disseminate in the public interest information and recommendations about (a) protection against radiation and (b) radiation measurements, quantities, and units, particularly those concerned with radiation protection.
2. Provide a means by which organizations concerned with scientific and related aspects of radiation protection and of radiation quantities, units, and measurements may cooperate for effective utilization of their combined resources, and to stimulate the work of such organizations.
3. Develop basic concepts about radiation quantities, units, and measurements, about the application of the concepts, and about radiation protection.
4. Cooperate with the International Commission on Radiological Protection, the Environmental Protection Agency, Nuclear Regulatory Commission, International Commission on Radiation Units and Measurements, and other national and international organizations, governmental and private, concerned with radiation quantities, units, and measurements and radiation protection [30].

Like ICRP and ICRU, the NCRP maintains working relationships with many national and international organizations.

The current recommendations of the NCRP, presented in Table 2.4., are designed only to serve as reasonable guidance, not to be adopted verbatim by state and local agencies.

Table 2.4. NCRP Dose-Limit Recommendations (NCRP, Ref. 30, with permission of National Council on Radiation Protection and Measurements, © 1971)

Type of Exposure	Maximum Permissible Dose Equivalent
Combined whole-body occupational exposure	
Prospective annual limit	5 rems in any one year
Retrospective annual limit	10-15 rems in any one year
Long-term accumulation to age N years	$(N - 18) \times 5$ rems
Skin	15 rems in any one year
Hands	75 rems in any one year (25/qtr)
Forearms	30 rems in any one year (10/qtr)
Other organs, tissues, and organ systems	15 rems in any one year (5/qtr)
Fertile women (with respect to fetus)	0.5 rem in gestation period
Dose limits for the public, or occasionally exposed individuals	
Individual or occasional	0.5 rem in any one year
Students	0.1 rem in any one year
Population dose limits	
Genetic	0.17 rem average per year
Somatic	0.17 rem average per year
Emergency dose limits — life saving	
Individual (older than 45 years, if possible)	100 rems
Hands and Forearms	200 rems, additional (300 rems, total)
Emergency dose limits — less urgent	
Individual	25 rems
Hands and Forearms	100 rems, total
Family of radioactive patients	
Individual (under age 45)	0.5 rem in any one year
Individual (over age 45)	5 rems in any one year

The *Environmental Protection Agency* (EPA) is one of the primary regulatory bodies in the United States concerned with radiation protection. The authority of the Federal Radiation Council (FRC) to provide guidance for federal agencies in developing radiation standards was transferred to the EPA with its formation on December 2, 1970, by Executive Reorganization Plan No. 3.

Under this authority, the EPA provides generally applicable guidance in the form of recommendations similar to those developed by the NCRP (Table 2.4), as well as guidance on maximum permissible concentrations. The EPA, in addition, formulates "protective action guides" applicable to accidental releases of radioactivity. The EPA is also required to establish radiological standards for the protection of the general environment through a transfer of authority from the Atomic Energy Commission. "Standards mean limits on radiation exposure of levels, or concentrations or quantities of radioactive materials in the general environment outside the boundaries of locations under the control of persons possessing or using radioactive material" [33]. The EPA recently promulgated standards [34] to limit radiation doses to the general public and quantities of long-lived radioactive materials in the environment attributable to planned releases from operations contributing to the nuclear generation of electrical power from the uranium fuel cycle. These standards apply to nuclear power plants and to most other operations within the fuel cycle, but exclude mining, transportation, and waste management. The standards limit: (a) the annual dose equivalent to the whole body or any internal organ, except the thyroid, to 25 mrems (250 μSv), and the annual dose equivalent to the thyroid to 75 mrems (750 μSv); and (b) the quantities of ^{85}Kr, ^{129}I, and plutonium and other alpha-emitting transuranic elements with half-lives greater than one year released to the environment per gigawatt-year of electrical power produced by the entire fuel cycle to 50,000 Ci (1.85 PBq), 5 mCi (185 MBq), and 0.5 mCi (18.5 MBq), respectively (1 curie = 3.7×10^{10} becquerel).

The full statement of the rule is contained in the Code of Federal Regulations [34]. These standards were proposed to achieve two main objectives: (a) to protect members of the public against radiation doses from fuel cycle operations; and (b) to limit the environmental burden of long-lived radioactivity that may accumulate from producing electricity in order to limit their long-term impact on both current and future generations. See the final environmental statement [34] for detailed discussions of the data base and judgments upon which these standards are based.

The EPA has issued proposed guidance [29] on dose limits for individuals exposed to transuranium elements from presently existing cases of environmental contamination or from future unplanned release. The recommendations state that: the annual alpha-radiation dose rate to members of the critical segment of the exposed population as the result of exposure to transuranium elements in the general environment should not exceed either:

1. 1 mrad (10 μGy) per year to the pulmonary lung, or
2. 3 mrad (30 μGy) per year to the bone.

In addition, the EPA's jurisdiction includes the environmental radiation protection activities of the Department of Health and Human Services covered under the Public Health Service Act, except those functions kept by the Bureau of Radiological Health (discussed later). In carrying out these functions, the EPA on July 9, 1976, promulgated Interim Primary Regulations for Radioactivity in Drinking Water [35]. These regulations, which replace the Public Health Service Drinking Water Standards of 1962, set forth maximum contaminant levels for ^{226}Ra, ^{228}Ra, and gross alpha-particle radioactivity as follows [36,37]: (a) Combined ^{226}Ra and ^{228}Ra limited to 5 pCi/liter (0.19 Bq) in community water systems; (b) gross alpha-particle activity (including ^{226}Ra, but excluding radon and uranium) limited to 15 pCi/liter (0.55 Bq).

In addition, maximum contaminant levels for beta-particle and photon radioactivity from man-made radionuclides in community water systems were set as follows [37]:

(a) The average annual concentration of β particle and photon radioactivity from man-made radionuclides in drinking water shall not produce an annual dose equivalent to the total body or any internal organ greater than 4 millirem/year [40 μSv/yr].

(b) Except for the radionuclides [^{3}H and ^{90}Sr], the concentration of man-made radionuclides causing 4 millirem [40 μSv] total body or organ dose equivalents shall be calculated on the basis of a 2 liter per day drinking water intake using the 168 hour [7 day] data listed in "*Maximum Permissible Body Burdens and Maximum Permissible Concentrations of Radionuclides in Air or Water for Occupational Exposure,*" NBS Handbook 69 as amended August 1963, U.S. Department of Commerce. If two or more radionuclides are present, the sum of their annual dose equivalent to the total body or to any organ shall not exceed 4 millirem/year [40 μSv] [35].

For ^{3}H and ^{90}Sr, the limits are 20,000 pCi/liter (740 Bq/l) and 8 pCi/liter (0.30 Bq/l), respectively. These standards are of environmental importance, and are presented in detail in the Code of Federal Regulations, 40 CFR 141.

The *Nuclear Regulatory Commission* (NRC), formed by the Energy Reorganization Act of 1974, is, broadly speaking, the regulatory part of the old Atomic Energy Commission (AEC). The NRC is responsible for the licensing and regulatory functions relating to commercial nuclear facilities. It implements radiation protection standards by defining specific requirements in the licenses of individual plants and also by enforcing them. As its other major responsibility the NRC regulates the following:

1. "Source material," meaning uranium or thorium, or ores with above a certain concentration of either or both.
2. "By-product material," meaning any radioactive material (except as noted in "special nuclear material") that is yielded in, or made radioactive by, producing or using special nuclear material.
3. "Special nuclear material," meaning plutonium, uranium-233, uranium enriched in the isotope 233 or 235, any material enriched by the foregoing, and other material designated by the NRC.

Much of the regulation of these materials is actually carried out by the states rather than by the NRC. About half the states are "agreement" states that meet certain federal stipulations on this regulatory function. The role of the NRC with respect to nuclear power reactors is presented in Chapter 10.

The NRC is responsible for occupational standards for workers involved in an NRC-licensed activity. These standards are given in Title 10, Parts 19 and 20 of the Code of Federal Regulations and limit each worker to a quarterly dose of 3 rem (30 mSv) and an annual dose of 5 rem (50 mSv) to the whole body. The NRC is currently considering amendments [38] to remove a provision which allows certain workers in licensed commercial nuclear activities to receive a radiation dose of as much as 12 rem (120 mSv) in one year.

The *Department of Energy* (DOE) superseded the Energy Research and Development Administration in October 1977. As the agency designated to take over the AEC's research and development functions, DOE is responsible for the great bulk of research on the biomedical, environmental, physical, and safety aspects of nuclear and other kinds of energy. It is also responsible for radiation health and safety, and for environmental protection at DOE-owned facilities, such as the national laboratories. It has statutory power over radiation guidelines in its own areas of concern, namely, the national laboratories, DOE licensees and contractors, and the weapons testing program. The publications in DOE's Technical Information Center provide valuable reference materials for radiation protection as a whole, and they are referenced where appropriate. The DOE Environmental Measurement Laboratory (EML) [formerly Health and Safety Laboratory (HASL)] performs routine measurements of radiation contamination in food, air, water, and fallout. These data are published in environmental quarterlies [39]. EML also published a procedures manual [40] for radioactivity measurements.

The *Food and Drug Administration's* (FDA) *Bureau of Radiological Health* (BRH) in the Department of Health and Human Services has many general public health responsibilities associated with radiation protection [41]. It conducts an electronic product radiation control program, including the development and administration of performance standards. As the agency primarily responsible for radiation used in the healing arts, the Bureau develops criteria, recommendations, and standards relative to radiation use and exposure, as well as techniques, procedures, and users' qualifications in order to reduce unnecessary exposure. BRH also provides advice to the Bureau of Foods and the Bureau of Drugs on the control of radioactive materials and radiation in food and drugs. Other functions include research, technical assistance and training in occupational radiation exposure, research on health effects of radiation exposure, and participation in the development of model codes and recommendations.

The *Mining Safety and Health Administration* (MSHA) of the Department of Labor regulates occupational exposures in the mining industry. It is also responsible for developing and enforcing safety and health standards for mineral extraction operations. Its role is discussed in Chapter 12.

The *Department of Transportation* (DOT) regulates the transportation of all hazardous materials, including radioactive materials. DOT regulates both shippers and carriers who are engaged in interstate commerce, and prescribes packaging, marketing, labeling, loading, and storage regulations.

Several other federal government agencies serve specialized functions in providing services or data useful to the aforementioned agencies. For instance, the Soil Conservation Service, part of the Department of Agriculture, monitors soil for radionuclides. The Occupational Safety and Health Administration, part of the Department of Labor, develops and enforces health and safety standards for those occupationally radiation-exposed groups not covered by other federal agencies. The National Bureau of Standards, part of the Department of Commerce, serves a very important role in providing a standardization of national measurement systems by providing standards for radioactivity measurement and standard reference materials for analysis.

State Agencies. Under the U.S. Constitution the ultimate responsibility for most aspects of health protection, including radiological health, is vested in the states. Many states accept this responsibility by protecting their citizens through basic public health law and by specific laws concerning specialized fields, for example, radiological health and occupational health. To standardize the radiation control regulations of the states, the Council of State Governments, assisted by the AEC and BRH, initially published in 1962 the first *Suggested State Regulations for*

Control of Radiation. In 1964, 1966, 1970, and 1974, the suggested regulations were revised and updated. Naturally occurring isotopes (radium, thorium, etc.) or machine-produced (e.g., accelerator) isotopes are under the direct control of states.

To facilitate material assistance and cooperation between the states and various agencies of the federal government, a National Conference of Radiation Control Program Directors was established in 1968. The Conference collects and makes accessible to all radiation control program directors information and data that might be of assistance to them.

In 1959, Section 274 of the Atomic Energy Act was enacted to spell out a state role in radiation activities and to produce a statutory basis under which the federal government could relinquish to the states portions of its regulatory authority. The mechanism to transfer some of NRC's authority to a state is an agreement between the governor of the state and the NRC. The programs of "agreement" states are routinely reviewed by the NRC to assure that they are effective. States may use NRC assistance through NRC's Office of State Programs.

Working cooperatively with various professional societies, many states are also involved with educational programs on radiation protection, providing activities such as special seminars, pamphlets, books, and lectures in schools of radiologic technology. Several states have minimum standards for the training and education of radiologic technologists.

Local Agencies. Local agencies have a definite contribution to make in radiation protection, particularly through emergency preparedness activities for any radiation facilities within their jurisdiction.

Nongovernmental Organizations. There are also nongovernmental organizations that provide useful criteria for measuring ionizing radiation and/or exposure. Of course, the NCRP, ICRU, and the IAEA also address this area in some of their reports. The American National Standards Institute (ANSI), alone or in collaboration with the Institute for Electrical and Electronic Engineers (IEEE), provides voluntary standards. The American Society for Testing Materials (ASTM) published a set of reference methods for measuring radiation in various media. Other organizations that provide a variety of information on radiation protection matters are professional organizations, such as the American Chemical Society, Health Physics Society, IEEE Nuclear Science, American Physical Society, and the American Nuclear Society.

REFERENCES

1. Advisory Committee on the Biological Effects of Ionizing Radiation, "The Effects on Populations of Exposure to Low Levels of Ionizing Radiation," Division of Medical Sciences, National Academy of Sciences—National Research Council, Washington, DC, 1972. (Revision in progress, May 1979.)
2. Ad Hoc Committee on "Hot Particles" of the National Academy of Sciences Advisory Committee on the Biological Effects of Ionizing Radiation, "Health Effects of Alpha-Emitting Particles in the Respiratory Tract," National Academy of Sciences—National Research Council, Washington, DC, 1976.

3. Advisory Committee on the Biological Effects of Ionizing Radiation, "Considerations of Health Benefit-Cost Analysis for Activities Involving Ionizing Radiation Exposure and Alternative," BEIR II Report, EPA-520/4-77-003, 1977.
4. United Nations Scientific Committee on the Effects of Atomic Radiation, *Ionizing Radiation: Levels and Effects,* 2 vols., United Nations, 1972.
5. United Nations Scientific Committee on the Effects of Atomic Radiation, *Sources and Effects of Ionizing Radiation,* United Nations, 1977.
6. International Atomic Energy Agency and World Health Organization, "Safe Handling of Radionuclides," IAEA Safety Series No. 1, 1973.
7. International Atomic Energy Agency and World Health Organization, in *Proceedings of the Biological and Environmental Effects of Low-Level Radiation,* Symposium on Biological Effects of Low-Level Radiation Pertinent to the Protection of Man and His Environment, Chicago, November 3–7 1975, 2 vols., IAEA, Vienna, 1976.
8. E. Browning, *Harmful Effects of Ionizing Radiation,* Monograph on Industrial Toxic Agents, Elsevier Publishing Company, New York, 1959.
9. A. Martin and S. A. Harbison, *An Introduction to Radiation Protection,* Chapman and Hall, Ltd., London, 1972.
10. U.S. National Bureau of Standards, "Maximum Permissible Body Burdens and Maximum Permissible Concentrations of Radionuclides in Air and Water for Occupational Exposure," NBS Handbook No. 69, 1959 with Addendum I, 1963.
11. International Commission on Radiological Protection, "Recommendations of the International Commission on Radiological Protection," *Ann. ICRP*, **1**(3), 1977, ICRP Publication 26.
12. Howard T. Odum, Ed., *A Tropical Rain Forest: A Study of Irradiation and Ecology at El Verde, Puerto Rico,* TID-24270 (PRNC-138), 1970.
13. D. J. Mellinger and V. Schultz, "Ionizing Radiation and Wild Birds," *CRC Critic. Rev. Environ. Control*, **5**(3), 397, 1975.
14. Colbert E. Cushing, Jr., Ed., in *Proceedings of the Radioecology and Energy Resources,* Fourth National Symposium on Radioecology, Oregon State University, May 12–14, 1975, Dowden, Hutchinson, and Ross, Inc., Stroudsburg, 1975, The Ecological Society of America Special Publication No. 1.
15. V. Schultz and F. W. Whicker, "Ecological Aspects of the Nuclear Age: Selected Readings in Radiation Ecology," TID-25978, Technical Information Center, 1972.
16. V. Schultz and F. W. Whicker, "Radiation Ecology," *CRC Critic. Rev. Environ. Control*, **4**(4), 423, 1974.
17. L. S. Taylor, *Radiation Protection Standards,* CRC Press, Cleveland, Ohio, 1971.
18. L. S. Taylor, "The Origin and Significance of Radiation Dose Limits for the Population," WASH-1336, 1973.
19. S. Glasstone, *Sourcebook on Atomic Energy,* 3rd ed., D. Van Nostrand Company, Inc., Princeton, 1967.
20. K. Z. Morgan and J. E. Turner, *Principles of Radiation Protection,* Robert Kreiger Publishing Co., 1973.

21. H. Cember, *Introduction to Health Physics,* Pergamon Press, Maxwell House, New York, 1969, Volume 105 of International Series of Monographs in Nuclear Energy.
22. R. E. Zirkle, "Radiobiological Importance of Linear Energy Transfer," A. Hollander, Ed., *Radiat. Biol.*, **1**, 315, 1954.
23. J. L. Bateman and V. Bond, "The Effects of Radiations on Different LET on Early Responses in the Mammal," *Ann. NY Acad. Sci.*, **114**, 32, 1964.
24. J. F. Thompson and W. W. Tortellotte, "The Effect of Dose Rate in the LD-50 of Mice Exposed to Gamma Radiation for Co60 Sources," *Am. J. Roentg., Radium Therapy Nucl. Medic.*, **69**, 826, 1953.
25. W. L. Russell, L. B. Russell, and E. M. Kelley, "Dependence of Mutation Rate on Radiation Intensity," in *Symposium on the Immediate and Low-Level Effects of Ionizing Radiation,* Taylor and Francis, Ltd., London, 1960.
26. Advisory Committee on the Biological Effects of Ionizing Radiation, "The Effects on Populations of Exposure to Low Levels of Ionizing Radiation" Natl. Acad. of Sciences, Washington, DC 20006, 1972.
27. G. S. Hurst and J. E. Turner, *Elementary Radiation Physics,* Wiley, New York, 1970.
28. International Commission on Radiation Units and Measurements, "Radiation Quantities and Units," ICRU Report No. 11, 1968.
29. Environmental Protection Agency, "Proposed Guidance on Dose Limits for Persons Exposed to Transuranium Elements in the General Environment," Summary Report EPA-520/4-77-016, 1977; "Response to Comments on Proposed Guidance on Dose Limits for Persons Exposed to Transuranium Elements in the General Environment," Summary Report EPA-520/4-78-010, 1977.
30. National Council on Radiation Protection and Measurements, "Basic Radiation Protection Criteria," NCRP Report No. 39, 1971.
31. National Council on Radiation Protection and Measurements, "Review of the Current State of Radiation Protection Philosophy," 1975.
32. International Commission on Radiation Units and Measurements, "Conceptual Basis for the Determination of Dose Equivalent," ICRU Report No. 25, 1976.
33. Environmental Protection Agency, "Radiation Protection Activities 1976," EPA-520/4-77-005, 1977.
34. Environmental Protection Agency, 40 CFR 190, "Environmental Radiation Protection Requirements for Normal Operations of Activities in the Uranium Fuel Cycle," Final Environmental Statements, 2 vols., EPA-520/4-76-016, 1976.
35. Environmental Protection Agency, "National Interim Primary Drinking Water Regulations," EPA-570/9-76-003, 1976.
36. Environmental Protection Agency, 40 CFR 141.15, "Maximum Contaminant Levels for Radium-226, Radium-228, and Gross α Particle Radioactivity in Community Water Systems," 1981.
37. Environmental Protection Agency, 40 CFR 141.16, "Maximum Contaminant Levels for β Particle and Photon Radioactivity from Man-Made Radionuclides in Community Water Systems," 1981.
38. Nuclear Regulatory Commission, "NRC Proposes Changes to Occupational Radiation Dose Standards," Office of Public Affairs, Washington, DC, News Release No. 79-35, **5**(8), 1979.

39. For recent examples see Department of Energy, Environmental Measurements Laboratory Environmental Quarterly, EML-334, prepared by E. P. Hardy, 1978.
40. J. H. Harley, Ed., "HASL Procedures Manual," HASL-300, 1972.
41. Bureau of Radiological Health, "Progress in Radiation Protection," HEW Publication (FDA) 77-8015, 1975.

3 RADIATION SOURCES

This chapter divides radiation sources into four categories: natural background, nuclear reactors, x-ray equipment, and particle accelerators. X-ray equipment and particle accelerator monitoring are discussed here; the monitoring of nuclear reactors is described in Chapter 10.

3.1. NATURAL BACKGROUND

This section briefly summarizes background doses contributed from natural sources, including those whose levels are altered by human activities, and from fallout from nuclear explosions. The section first characterizes average natural background, that is, dose levels from radiation that is neither man-made nor greatly affected by technology. The section next indicates how the dose from natural sources may be technologically enhanced by human endeavors.

In recent years, several publications have undertaken to collect information on natural background, fallout, and—in some cases—technological enhancement. For worldwide data, the 1977 report of the United Nations Scientific Committee on the Effects of Atomic Radiation, *Sources and Effects of Ionizing Radiation* [1], is most useful. The National Council on Radiation Protection and Measurements has published a document, *Natural Background Radiation in the United States* [2]. The EPA has published annual reports on the *Radiological Quality of the Environment* [3], treating all three of the areas summarized in this section. For many years, DOE's Environmental Measurements Laboratory (EML) [previously the Health and Safety Laboratory (HASL)] has been actively involved in environmental radiation measurements, including those of the natural radiation environment [4] and of fallout (issued quarterly). Finally, a very extensive source of information has been a series of symposia on *The Natural Radiation Environment,* held in Houston, in 1963, 1972 [5,6], and 1978.

Cosmic radiation and terrestrial radioactivity provide a background of radioactivity and resulting exposures that exist largely independently of human activities and that must be considered if measurements of human-caused radiation are to be performed or their significance understood. Natural human radiation doses can be divided into three types: that caused by direct exposure to cosmic rays (cosmic radiation); that arising from terrestrial radionuclides external to the body (primarily gamma radiation); and that caused by materials inside the body (internal exposure).

3.1.1. Cosmic Radiation

Cosmic radiation arises mostly from galactic sources (but may also refer to high-energy solar radiation) and is taken to include both the energetic primary particles impinging on the earth's atmosphere and the secondary particles generated by resulting interactions. Galactic radiation at the top of the atmosphere is composed largely of protons (87%) and alpha particles (11%), with about 1% each of heavier nuclei and electrons. Energies range at least to 10^{20} eV, but the bulk are at lower energies (10^8 to 10^{11} eV). The spectrum falls rapidly with increasing energy, roughly as $1/E^2$ for the higher energies. At the lower end of this energy range (especially 10^6 to 10^{11} eV) galactic radiation may be augmented substantially by particles ejected from the sun during periods of flare activity [2,5,6].

The cosmic radiation field at ground level is almost entirely secondary particles from highly energetic ($\geq 10^{10}$ eV) primary particles from galactic sources. The flux near the surface is predominantly muons from the decay of pions produced in the upper atmosphere; these muons are accompanied by an equilibrium mixture of electrons and photons. In addition, there is a small flux of neutrons and protons [2,5,6].

As shown in Fig. 3.1 [2], the dose equivalent at ground level is dominated by muon exposure. However, the total dose rate increases rapidly with altitude, as do the relative contributions of various other components of the flux. In addition, at higher altitudes, dose rate significantly depends on latitude, due to the earth's magnetic field. At ground level, this dependence is much reduced, because the dose there arises largely from highly energetic particles that are little affected by the earth's field. Such data indicate a dose equivalent from cosmic radiation at sea level of about 28 mrem/yr (280 μSv/yr) [1,2]. Considering the altitudinal population distribution, [7] as well as shielding by structures (assigned an average decrease of 10%), the average U.S. population dose is approximately 28 mrem/yr (280 μSv/yr [2]. This is somewhat smaller than doses cited by the EPA [3] which estimated doses varying from 35 to 120 mrem/yr (350 to 1200 μSv/yr); Oakley and Goldin [8] estimate that doses vary from about 40 to 80 mrem/yr (400 to 800 μSv/yr) from one state to another. (The average elevation of U.S. residents is approximately 200 m [8]. Near sea level, a kilometer increase in elevation corresponds to an increase in dose equivalent of approximately 10 mrem/yr (100 μSv/yr) [2]. However, at high altitudes, the neutron-proton flux increases more rapidly which means the component with higher quality factor increases, so that at altitudes above 10 km, the dose equivalent rates are orders of magnitude higher than at sea level. At the altitude of jet flights (about 11 km), the dose rate is approximately 0.5 mrem/hr (5 μSv/hr), so that doses of a few mrem (a few tens of microsieverts) could be sustained from long-distance flights. As noted before, the

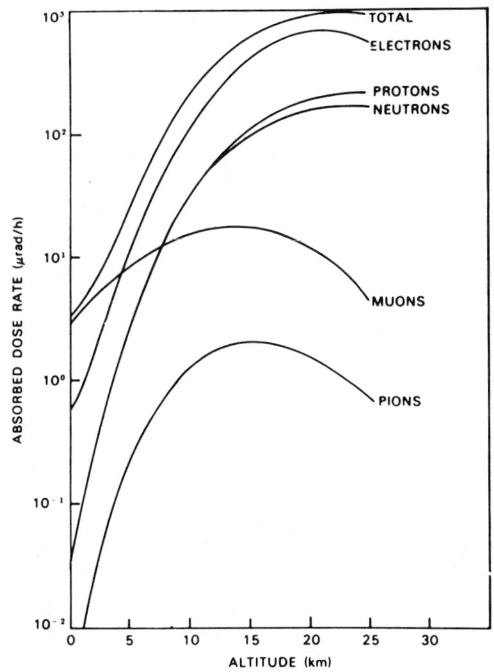

Figure 3.1. Absorbed rates at 5-cm depth in a 30-cm-thick slab of tissue, from various components of the cosmic radiation at solar minimum and at 55°N (reproduced from NCRP, Ref. 2, with permission by National Council on Radiation Protection and Measurements © 1975).

dose at high altitudes is also more dependent on latitude and the level of solar activity; see Ref. 2 for further discussion.

3.1.2. Natural Radionuclides

Naturally occurring radionuclides produce external γ-ray exposures as well as much smaller β-ray exposures, and negligible external α-particle exposures. The dominant contributors to the γ-ray field are potassium-40, thorium-232 and its daughters, and uranium-238 and its daughters, all of which are present in soil, rocks, and other earth materials. The distribution of radionuclides in the United States is discussed in Ref. 2. Methods for calculating γ-ray fields and dose rates are set forth by Beck [9]. Much of the dose from the uranium and thorium series arises from Rn and its daughters.

External dose estimates for the United States are summarized by the Nuclear Regulatory Commission (NRC) and EPA in Refs. 2,3. Oakley's data [7] indicate a population-averaged, outdoor adsorbed air dose of 40 mrad/yr (400 μGy/yr). To obtain average dose equivalent requires consideration of screening by structures and tissue, yielding a dose of about 26 mrem/yr (260 μSv/yr). However, most cited doses do not apply this correction. Beck, estimated [10] a range of 30 to 95 mrem/yr (300 to 800 μSv/yr) for 90% of the U.S. population, with an average of 55 mrem/yr (550 μSv/yr), divided into 17, 25, and 13 mrem/yr (170, 250, and 130

μSv/yr) for ^{40}K, the ^{232}Th series, and the ^{238}U series, respectively. A similar range was given by Oakley [7] varying—for the sites examined—from about 23 mrad/yr (230 μGy/yr) for the coastal plain to about 90 mrad/yr (900 μGy/yr) for the Colorado Plateau area (Denver). Worldwide external dose estimates are summarized in the UNSCEAR Report [1], which states an estimated average absorbed dose of 32 mrad/yr (320 μGy/yr).

As has been mentioned, the radiation field is affected by structural shielding. The degree of shielding (or increase, in some cases) depends on structure type. In addition, soil and meteorological conditions affect the outdoor levels. In particular, moisture in the soil shields gamma rays emitted from radionuclides in the soil and affects the migration of radon from the soil. For greater detail see the summary discussion in Ref. 2.

Radionuclides present in the environment may be ingested or inhaled, and subsequently deposited at various sites in the body. Radiation from internally deposited emitters constitutes a significant portion of natural background exposure. Radionuclides of particular importance are ^{40}K, ^{210}Po, ^{222}Rn, and ^{226,228}Ra. Carbon-14 (produced by cosmic rays) and ^{87}Rb are also contributors. The heavy isotopes, all from the ^{238}U or ^{232}Th series, are α-particle emitters that, because of their deposition sites, expose bone cells. The lighter isotopes, γ-ray or β-ray emitters, distribute their dose more uniformly. Tables 3.1 and 3.2 [2] summarize the dose equivalent from the important internally deposited radionuclides. Reference 2 discusses the pathways for deposition of these radionuclides. See also Refs. 1 and 3. Table 3.3 shows doses (in mrad/yr) of internally deposited radionuclides.

Although the dose (in mrad/yr) is dominated by γ-rays, β-rays, and ionizing cosmic rays, the α-particle dose can make an important contribution to the dose equivalent (in mrem/yr) for lung and bone, especially for exposure due to radon daughters. Concentrations of airborne radioactivity, including radon and its daughters, exhibit substantial variations with time, place, and meteorological conditions. The average dose to lung is about 100 mrem/yr (1000 μSv/yr) [2], (largely arising from the radon daughter ^{218}Po) for typical concentrations in the United

Table 3.1. Annual Internal Beta and Gamma Dose Equivalent Rates (mrem/yr) in Tissue from Internally Deposited Naturally Occurring Radionuclides (data presented in NCRP, Ref. 2, with permission of National Council on Radiation Protection and Measurements © 1975)

		Cortical Bone		Trabecular Bone	
Radionuclide	Soft Tissues (Gonads)	Osteocytes	Haversian Canals	Surfaces[a]	Marrow
^{3}H	~0.001	~0.001	~0.001	~0.001	~0.001
^{14}C	0.7	0.8	0.8	0.8	0.7
^{40}K	19	6	6	15	15
^{87}Rb	0.3	0.4	0.4	0.6	0.6
Total	20.0	7.2	7.2	16.4	16.3

[a]Cells close to surfaces of bone trabeculae.

Table 3.2. Annual Alpha Dose Equivalent Rates (mrem/yr) from Naturally Occurring Radionuclides (data presented in NCRP, Ref. 2, with permission of National Council on Radiation Protection and Measurements © 1975)

			Dose Equivalent Rates				
	Concentration		Cortical Bone			Trabecular Bone	
Radionuclide	In Air (pCi/m^3)	In Bone[a] (pCi/kg)	Gonads	Osteocytes	Haversian Canals[b]	Surfaces[c]	Marrow
$^{238\text{-}234}U$[d]	—	6.9	0.8	12.4	7.7	4.8	0.9
^{236}Ra[d]	—	7.8	0.2	16.4	10.2	6.6	1.2
^{228}Ra[d]	—	3.8	0.3	19.0	11.0	8.0	1.0
^{223}Rn[e]	150	—	0.4	0.2	0.2	0.4	0.4
^{220}Rn[e]	1	—	0.01	0.1	0.1	0.2	0.2
^{210}Po[d]	—	60	6	60	36	24	4.8
Total			8	110	65	44	8.5

[a]The alpha-emitting nuclides are assumed to be uniformly distributed in mineral bone, although this may not be the case.
[b]Cells lining the Haversian canals.
[c]Cells close to surfaces of bone trabeculae (dose) averaged over the first 10 μm.
[d]Calculated as described in [2].
[e]Derived from UNSCEAR [1].

Table 3.3. Estimated Annual Tissue Absorbed Dose from Natural Sources in "Normal" Areas (data presented in UNSCEAR, Ref. 1).

	Annual Tissue Absorbed Dose (mrad)			
Source of Irradiation	Gonads	Lung	Bone Lining Cells	Red Bone Marrow
External irradiation				
Cosmic rays				
Ionizing component	28	28	28	28
Neutron component	0.35	0.35	0.35	0.35
Terrestrial radiation (γ)	32	32	32	32
Internal irradiation				
Cosmogenic radionuclides				
^{3}H (β)	0.001	0.001	0.001	0.001
^{7}Be (γ)	—	0.002	—	—
^{14}C (β)	0.5	0.6	2.0	2.2
^{22}Na ($\beta^{+}\gamma$)	0.02	0.02	0.02	0.02
Primordial radionuclides				
^{40}K ($\beta^{+}\gamma$)	15	17	15	27
^{87}Rb (β)	0.8	0.4	0.9	0.4
^{238}U-^{234}U (α)	0.04	0.04	0.3	0.07
^{230}Th (α)	0.004	0.04	0.8	0.05
^{226}Ra-^{214}Po (α)	0.03	0.03	0.7	0.1
^{219}Pb-^{210}Po ($\alpha^{+}\beta$)	0.6	0.3	3.4	0.9
^{222}Rn-^{214}Po (α) inhalation	0.2	30	0.3	0.3
^{232}Th (α)	0.004	0.04	0.3	0.04
^{228}Ra-^{208}Tl (α)	0.06	0.06	1.1	0.2
^{220}Rn-^{208}Tl (α) inhalation	0.008	4	0.1	0.1
Total (rounded)	78	110	86	92
Fraction of absorbed doses delivered by alpha particles or neutrons (%)	1.2	31	8.5	2.1

States. For more detailed discussion of Rn and its daughters, see Chapter 14. It is also important to note that increased exposure to Rn and its daughters can be the most significant radiological result (aside from medical procedures) of routine human activities.

Based on exposures from cosmic radiation and natural radionuclides, estimates may be made of total natural background exposures. Table 3.4 summarizes average U.S. doses [2]. The whole-body dose of 80 mrem/yr (800 μSv/yr) arises from almost equal contributions from cosmic radiation, external terrestrial radiation, and internally deposited radionuclides. Bone surfaces and lungs receive substantial

Table 3.4. Summary of Average Dose Equivalent Rates (mrem/yr) from Various Sources of Natural Background Radiation in the United States (data presented in NCRP, Ref. 2, with permission of National Council on Radiation Protection and Measurements © 1975)

			Bone		
Source	Gonads	Lung	Surfaces	Marrow	G.I. Tract
Cosmic radiation[a]	28	28	28	28	28
Cosmogenic radionuclides	0.7	0.7	0.8	0.7	0.7
External terrestrial[b]	26	26	26	26	26
Inhaled radionuclides[c]	—	100[d]	—	—	—
Radionuclides in the body[e]	27	24	60	24	24[f]
Rounded totals	80	180	120	80	80

[a]"Cosmic radiation" includes 10% reduction to account for structure shielding.
[b]"External terrestrial" includes 20% reduction for shielding by housing and 20% reduction for shielding by the body.
[c]Doses to organs other than lung included in "Radionuclides in the Body."
[d]Local dose equivalent rate to segmental bronchioles is 450 mrem/yr.
[e]Excluding the cosmogenic contribution shown separately.
[f]This does not include any contribution from radionuclides in the gut contents.

additional doses from internal α-particle emitters. However, as indicated previously, estimates vary substantially and also have large geographical and temporal dependences (see Table 3.3 for estimates).

3.1.3. Technologically Enhanced Natural Radiation

Man's activities can change the amount of naturally occurring radiation to which people are exposed. These include not only the construction of structures (which, as noted previously, can decrease exposure to external cosmic or terrestrial radiation), but also more specific industrial activities, particularly those for energy production.

This section briefly discusses several of the principal possibilities for increased exposure, many of which are discussed in Ref. 3, as they pertain to the United States. The reader is also referred to Ref. 1 for a survey of worldwide information. The potential for increasing exposure to natural radiation arises largely from radon (see Chapter 14) generated in earth materials with which humans work or live. As noted in Section 3.1.1, exposure to cosmic rays can also be increased technologically, that is, by jet flights which, over long distances, contribute doses of a few mrem (a few tens of microsieverts).

Occupational exposure of uranium miners has given one of the most unambiguous examples of the effects of elevated exposure to radioactivity, in this case radon daughters. Exposures at mines and mills, as well as control methods and monitoring instrumentation, are discussed in Chapter 12.

These activities also subject members of the general public to increased radiation exposures, largely via the radon emanating from tailings piles associated with milling operations. The most notorious increased exposures have arisen when tailings were used in landfill operations. However, even without such activities, some portion of the short-lived Rn may reach human-occupied areas, even at a distance. Because radon ultimately arises from a long-lived parent (^{230}Th, half-life 80,000 years), inactive tailings piles constitute a low-level, but very long-lived source of general population exposures. The integrated population dose can exceed other exposures resulting from the nuclear industry (with the possible exception of ^{14}C, which also results in a long-term dose commitment). See, for example, estimated doses in Ref. 11. Reference 1 also gives a brief summary of the major considerations as well as doses and dose estimates from active and inactive mining and milling operations.

Uranium, thorium, and their daughters occur in phosphate ores. Substantial attention has recently been drawn to the phosphate industry due to the possibility of occupational and general public exposures, particularly to radon and daughters. Exposures may occur as a direct result of mining, from the processing plant, or from disposition of the product or tailings. Thus workers are subject to some increased exposure, possibly comparable to dose limitations for radiation workers [12]. In addition, radioactivity may be released to the air (see, for example, Ref. 13), subjecting members of the general public to slightly increased radiation exposures. Exposures may also result from use of by-products [14] or from construction of residences on reclaimed land [15]. Construction on reclaimed land could be particularly hazardous if the structure retains Rn (discussed later). See Ref. 3 for further discussion of radiological questions associated with the phosphate industry.

Non-nuclear energy industries can also release naturally occurring radionuclides into the human environment. For example, fossil-fuel fired power plants can release trace amounts of Ra and Th present in these energy resources. Amounts released from coal-fired plants without particulate control are in the range of curies (tens of gigabecquerels) per gigawatt-year; amounts from oil-fired plants are somewhat less. Geothermal plants release Rn that has diffused into geothermal resources within the earth's crust and the amounts vary substantially with the resource that is developed. Case et al. [16] briefly discuss radionuclides from fossil-fuel and geothermal power generation. See Ref. 3 for preliminary dose estimates from coal-fired plants. In a simple model for coal-fired power plants without good particulate control, the United Nations Scientific Committee on the Effects of Atomic Radiation (UNSCEAR) [1] estimates dose commitments of 10^4 man-rad (100 man-gray) per gigawatt-year from inhalation directly from the passing cloud in a region with a uniform population density of 100 per km^2.

Residential, commercial, or industrial structures can provide significant shielding from external radiation sources. However, particular types of material may themselves constitute significant sources of external radiation and of Rn and daughters. Moreover, depending on ventilation rates, buildings can effectively concentrate levels of pollutants that arise from sources inside the buildings, including Rn and daughters. These radionuclides may arise from building materials such as concrete, from the soil on which the building is constructed, and, in some cases, from water or gas utility systems. The NRC briefly surveyed studies of indoor Rn concentrations [2], and it is clear that these concentrations can be substantially greater than outdoor levels. The question of whether increased Rn exposures to indoor air is hazardous may have wider significance as energy conservation measures that would reduce air infiltration or ventilation rates are implemented on a large scale [17]. The question of radon levels in buildings has already received wide attention in connection with the uranium and phosphate industries.

3.1.4. Fallout

This section contains a discussion of the various types of radioactive fallout, and their relative intensities and concentrations as a function of time. It is meant primarily as background information and the techniques and problems of fallout measurement will not be discussed in detail.

The term *fallout* describes the radioactive component of the debris resulting from atmospheric nuclear (fission) and thermonuclear (fusion) explosions. Since the 1940s, the United States has maintained a nationwide (and with other nations, a worldwide) fallout measurement program. The AEC maintained both a Fallout Studies Branch and a Fallout Program at the HASL in New York [18], now DOE's EML. From these sources, from Fowler's compilation [19], and from Ref. 1, a comprehensive picture of the fallout situation can be obtained.

Distinctions have generally been made between three different kinds of fallout. Local fallout, confined to the vicinity of a detonation, usually occurs within about the first 48 hours. Tropospheric fallout, occurring in a band around the world near the general latitude of the detonation, occurs in the first 2 to 4 weeks. Finally, stratospheric fallout occurs over a longer period. Small weapons contribute all of their fission products to local and tropospheric fallout; the truly worldwide distributions come from stratospheric fallout after the largest ($>$ 1 megaton) detonations. Only these large weapons are capable of depositing significant debris in the stratosphere. Once products are carried into the stratosphere, they become widely dispersed and return to earth over periods of 3 months to several years. Because approximately 90% of all weapons-test fission products has come from multimegaton weapons, stratospheric fallout is the dominant vehicle for worldwide contamination.

Environmentally, the most important fallout constituents are ^{90}Sr and ^{137}Cs due to their long half-lives (28 and 30 years, respectively) and relatively high uptake by and hazard to biosystems. Strontium is chemically similar to calcium, which is used for bone formation, and cesium simulates potassium, crucial for the function of cells. Total weapons tests have produced about 9 and 14 million curies (MCi, 0.33 and 0.52 EBq) of ^{90}Sr and ^{137}Cs, respectively, to date. This activity is not distributed uniformly in space or time after a detonation. Figure 3.2 shows the latitudinal variation in ^{90}Sr surface deposition levels. Both the large peak in the northern temperate zone and the secondary peak in the southern temperate zone are due to the dynamics of stratospheric-tropospheric mixing: the poorly defined or discontinuous tropopause in the temperate latitudes provides a vehicle for preferential stratospheric deposition.

The time distribution for ^{90}Sr depositions in New York City is shown in Fig. 3.3. The time distribution shows the very obvious effects of the large-scale weapons tests of the 1960s. The ^{137}Cs/^{90}Sr ratio is found to be between 1.5 and 2.0 for almost all measurements [22].

Human exposure to fallout is from three sources: ingestion, inhalation, and whole-body external exposure. Of the nuclides ingested, the main contributors to dose are ^{90}Sr and ^{89}Sr (to whole body and to bone); ^{131}I (to thyroid); ^{137}Cs (to

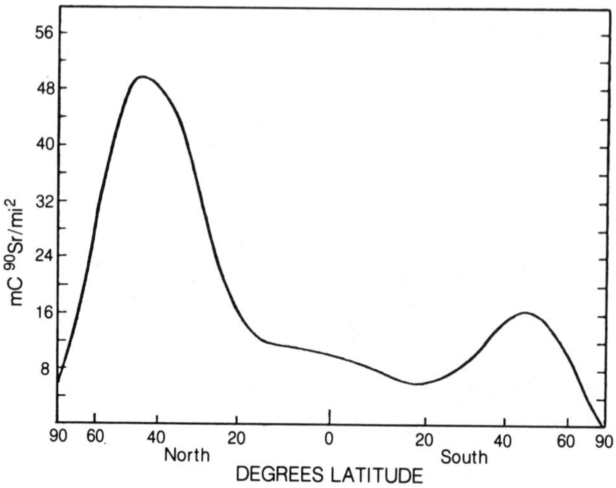

Figure 3.2. Ground level inventory of ^{90}Sr in late 1959 (reproduced from Kulp, Ref. 20, with permission by Elsevier-North Holland Biomedical Press, Amsterdam © 1965).

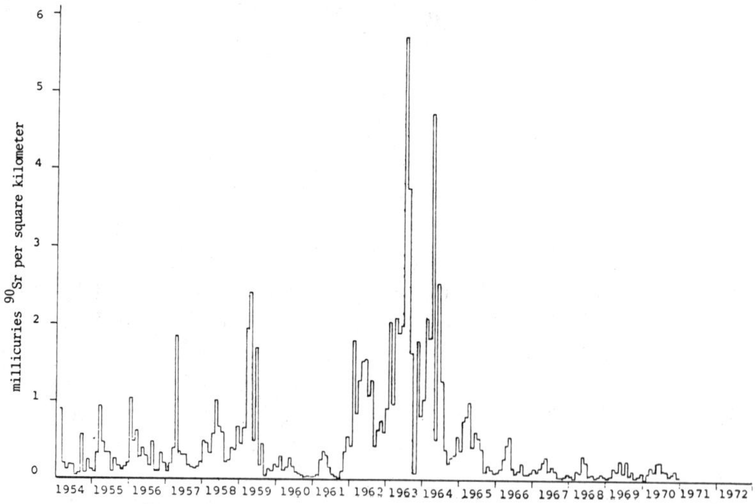

Figure 3.3. Monthly deposition of ^{90}Sr in New York City (reproduced from HASL, Ref. 21).

whole body); and ^{14}C (to whole body and to bone). Because Sr is deposited permanently in bone, ^{90}Sr doses have been estimated based upon a 50-year (or lifetime) delivery of dose. All other doses are assumed to be delivered in the year of ingestion; Table 3.5 [23] shows the 1963, 1965, and 1969 estimated annual fallout dose per capita from ingestion. Figure 3.4 [21] shows semiannual averages of the ^{90}Sr intake in New York City from 1961 to 1970.

The estimated dose from fallout via inhalation is also shown in Table 3.5. While the whole-body dose from inhalation is quite small, the 50-year respiratory lymph-node dose from Pu is large indeed.

Table 3.5. Estimated Average U.S. Dose from Fallout (mrem per capita per year) [23][a]

Source	Organ	1963	1965	1969
Ingestion				
^{14}C	whole body	0.3	0.7	0.6
^{89}Sr	bone	1.3	0.2	0.2
^{90}Sr	whole body	73.0*	72.0*	23.0*
^{90}Sr	bone	7.3	7.2	3.2
^{131}I	thyroid	25.0	4.0	3.0
^{137}Cs	whole body	4.3	2.3	0.4
Inhalation				
^{144}Ce	lung	8.2	0.4	0.3
^{144}Ce	bone	10.6	0.5	0.4
^{238}Pu	respiratory lymph nodes	79.0*	14.0*	19.0
^{239}Pu	respiratory lymph nodes	1000.0*	175.0*	36.0*
^{239}Pu	lung	4.5*	0.8*	0.2*
Several others	lung	4.3	0.4	0.1
External Gamma Dose				
^{137}Cs	whole body	5.9	1.8	0.9
Others	whole body	7.0	5.0	3.0

[a]Values followed by asterisks denote a 50-year dose from nuclide inhaled or ingested in that year.

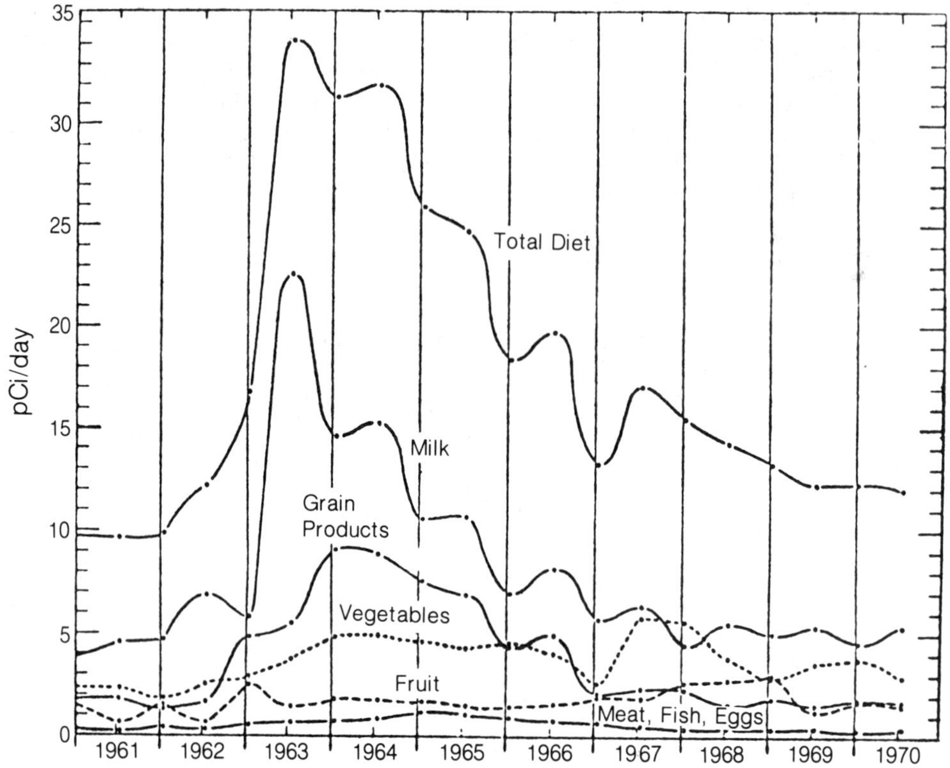

Figure 3.4. Average ^{90}Sr intake in New York City (pCi/day) during the 1960s (reproduced from HASL, Ref. 21).

For external γ-ray dose rates, the dominant contributor was ^{137}Cs. Including shielding factors (0.4 due to buildings, 0.8 due to body screening), the 1963, 1965, and 1969 estimated annual average external γ-ray doses [23] were 5.9, 1.8, and 0.9 mrem/capita (59, 18, and 9.0 μSv/capita).

In recent years, the average radiation dose from fallout has been about 4 mrem/yr (40 μSv/yr), considerably smaller than during the mid-1960s a period of heavy weapons testing (Table 3.5).

Table 3.6 shows the dose commitment arising from radionuclides released in testing before 1976. As for dose rate, the long range dose commitment is dominated by exposures from ^{90}Sr and ^{137}Cs. However, short-lived nuclides also made a substantial contribution. Moreover, were the ^{14}C contribution to be integrated over its long half-life, the contribution from ^{14}C alone would be comparable to the summed contribution from the other radionuclides, for example, of the order of 200 mrad (2.0 mSv).

Table 3.6. **Summary of Dose Commitment in mrad from Radionuclides Produced in All Nuclear Tests Carried out Before 1976 (data presented in UNSCEAR, Ref. 1, with permission of United Nations © 1977)[a,b]**

	Northern Temperate Zone				Southern Temperate Zone				World Population			
Source of Radiation	Gonads	Bone Marrow	Bone Lining Cells	Lung	Gonads	Bone Marrow	Bone Lining Cells	Lung	Gonads	Bone Marrow	Bone Lining Cells	Lung
External												
Short-lived nuclides	48	48	48	48	11	11	11	11	11	30	30	30
^{137}Cs	62	62	62	62	18	18	18	38	38	38	38	38
Internal												
^{3}H	2	2	2	2	0.2	0.2	0.2	0.2	2	2	2	2
$^{14}C^{c}$	7	32	29	9	7	32	29	9	7	32	29	9
^{54}Mn				1								1
^{55}Fe	1	0.6	1	1	0.3	0.2	0.3	0.3	0.7	0.4	0.7	0.7
^{90}Sr		84	120			24	33			52	71	
^{89}Sr		0.4								0.3		
^{106}Ru				41				10				24

Table 3.6. Continued

Source of Radiation	Northern Temperate Zone				Southern Temperate Zone				World Population			
	Gonads	Bone Marrow	Bone Lining Cells	Lung	Gonads	Bone Marrow	Bone Lining Cells	Lung	Gonads	Bone Marrow	Bone Lining Cells	Lung
^{137}Cs	27	27	27	27	8	8	8	8	17	17	17	17
^{144}Ce				65				15				38
^{239}P[d]			1	1			0.3	0.3			0.9	0.9
Total[e]	150	260	290	260	45	93	100	72	94	170	190	160

[a]The dose commitments for ^{85}Kr, ^{136}Cs, and ^{140}Ba, are not shown in this table because they are negligible compared with the values included.
[b]For internal irradiation, where body activities have been measured, uncertainties are probably within a factor of 2. For external irradiation and for lung dose estimates, where the models used were not checked by direct measurement, the uncertainties are probably within a factor of 5.
[c]Dose accumulated up to the year 2000. The total dose commitment to gonads and lung is about 120 mrad, to bone lining cells 414 mrad and to bone marrow 455 mrad, delivered over some 8300 yr.
[d]These dose commitments appear to be the same in different organs because of rounding. Only inhalation contributions are shown.
[e]Rounded to two significant figures.

3.1.5. Summary

The total average radiation dose from natural sources is 100 mrem/yr (1.0 mSv), with considerable variation, depending on place, time, and conditions. This constitutes an irreducible exposure background which, along with fallout radiation in some cases, can often hinder measurements of doses from other sources, particularly in view of the variation in the natural levels [24]. An important part of formulating an environmental monitoring program to measure increases in concentration or dose due to human activities is to characterize the natural radiation environment in the vicinity of interest. For the most part, instruments are available for measuring the individual contributions to exposure. Both are discussed in the chapters on detecting radionuclides (Chapters 13–20) and in chapters discussing instrumentation by measuring γ- and x-radiation (Chapters 5,6) and α- and β-radiation (Chapter 7). A further reference for environmental radiation monitoring is a recent NCRP report, No. 50 [25].

3.2. NUCLEAR REACTORS

Nuclear reactors are potentially the source of the greatest amount of radiation and radioactive materials. Under normal operation a reactor emits only a very small amount of radiation. This will be discussed in some detail in Chapter 10. In the unlikely event of certain types of accidents large amounts of radioactive materials can be released. Under these circumstances normal environmental monitoring instrumentation will not be adequate, and properly equipped experts must be called in.

Reactors also are used to produce radioactive isotopes for medical and industrial purposes. The maintaining of an uncontaminated, radiation-free environment will be the duty of a monitoring manager. A further discussion of this will be given in various chapters in Part II.

Frequently research reactors have windows in their shielding for the express purpose of allowing beams of neutrons and/or other radiation to escape to be used for various purposes. All of the area around such windows must be carefully monitored to reduce the chance of accident or injury.

3.3. X-RAY EQUIPMENT

The number of x-radiation sources in the world today defies accurate quantification. Since the discovery of x-rays by W. K. Roentgen in 1895, their use in industrial and medical applications has become widespread, and new applications are continually being developed or exploited.

X-ray equipment comes second as a potential source of dangerous radiation. For the most part medical x-ray equipment is very well shielded and there is little danger to either the patient or the operator. Even so, monitoring should be thorough and effective to avoid possible accident and the workers should carry personal dosimeters (e.g., film badges). Industrial x-rays which usually involve much more powerful equipment must be even more carefully monitored. This is particularly true where temporary setups are involved. Section 3.3.4 treats the problem of industrial x-rays in detail. For discussion of measurement techniques and instrument comparisons the reader is referred to Chapter 5.

3.3.1. Medical and Dental Exposures

The U.S. Public Health Service surveyed medical and and dental x-ray exposures of the U.S. population in 1964 [26] and 1970 [27]. Based on these surveys, researchers estimated doses to members of the population [28,29]. They are of particular importance because medical and dental x-ray exposures are the only peacetime anthropogenic sources that contribute an average radiation dose of the same order of magnitude as natural background doses [30,31,32]. UNSCEAR [31] summarized results from the United States as well as from other countries.

The methods of quantifying medical x-ray doses differ among various researchers. One concept is the "whole-body dose," which is the easiest to understand, but difficult to interpret because exposures are seldom to the whole body. Another is the "genetically significant dose," in which exposures that are genetically significant are weighted heavily, while exposures to body extremities or to individuals of nonreproductive age receive little weight. A third concept is the dose to individual organs (e.g., thyroid doses from dental diagnostic procedures) or to a more general organ (e.g., bone). No attempt will be made to give detailed data here on all of these categories or their significance but some of the findings which demonstrate the extent and nature of medical and dental x-ray exposures will be highlighted. Medical and dental x-ray exposures will be compared with doses from use of radiopharmaceuticals (but not with doses from accelerator therapy).

Genetically significant dose (GSD) is an index of radiation received by the genetic pool. This index permits comparisons between diverse national surveys and serves as an important measure of x-ray exposure. Mathematically the GSD is expressed [33] as:

$$\mathrm{GSD} = \frac{\Sigma D_i \hat{N}_i \hat{P}_i}{\Sigma N_i P_i},$$

where

$D_i =$ the average gonad dose to persons of age i who receive x-ray examinations,

$\hat{N}_i =$ the number of persons in the population of age i who receive x-ray examinations,

$\hat{P}_i =$ the expected future number of children for persons of age i who receive x-ray examinations,

$P_i =$ the expected future number of children for persons of age i in the population, and

$N_i =$ the number of persons in the population of age i.

Thus GSD weights the dose to individuals by their expected future number of children.

Because GSD has not generally been known for radiation sources other than medical radiology (e.g., not for environmental and occupational exposure), and because it gives no indication of somatic dose, the exclusive use of this index in reporting population exposures is not adequate for many purposes. (Somatic dose is that which affects the individual directly.) For this reason, although several other indices are roughly proportional to the GSD over wide ranges of variables, it is useful to report such indices separately. It is also possible to formulate a "somatic dose index," as is done by Oakley and Goldin [8].

In 1970, 130 million individuals made 179 million visits, and received 212 million x-ray examinations in which 661 million films were exposed [2], an increase of some 20 to 30% over 1964. The U.S. population increased 7% during this period. Table 3.7 gives the estimated annual rate of persons x-rayed during 1970. Sixty-five percent of the population had an x-ray examination of some kind. Figure 3.5 displays the age distribution for 1964 and 1970. Figures 3.6 through 3.9 give, respectively for both 1964 and 1970: the annual rate of medical x-ray procedures by body area; the distribution of examinations by number of films per examination; ratios of beam area to film area; the distribution of films by beam diameter for dental x-rays. Many of these parameters are significant when evaluating possibilities for reducing x-ray exposure. Table 3.8 [31] shows some changes in the data between the two U.S. surveys.

Table 3.7. Estimated Annual Rate of Persons X-rayed by Type of Procedure, Sex, and Age, United States, 1970 (data presented in DHEW, Ref. 27)

	Rate per 100 Persons					
			Medical			
Sex and Age	All Types	Total	Radiographic[a]	Fluoroscopic	Therapeutic	Dental
Both sexes	65.0	38.6	38.3	4.6	0.2	29.6
Under 15	42.4	17.8	17.7	0.7	(*b*)	25.9
15-29	77.3	41.1	40.8	2.0	(*b*)	40.8
30-44	76.9	47.1	46.8	6.4	(*b*)	34.3
45 and over	70.9	52.1	51.6	9.3	(*b*)	22.0
45-64	76.4	53.6	52.9	8.8	(*b*)	26.7
65 and over	58.8	49.1	48.5	10.3	(*b*)	11.7
Male	62.6	38.6	38.2	4.6	(*b*)	27.2
Under 15	44.1	19.8	19.8	1.0	(*b*)	25.9
15-29	73.4	43.6	43.3	2.0	(*b*)	34.7
30-44	71.5	44.1	43.8	5.8	(*b*)	31.7
45 and over	68.8	51.6	50.7	9.8	(*b*)	19.8
45-64	71.4	51.0	50.0	9.0	(*b*)	23.6
65 and over	62.4	53.2	52.4	11.7	(*b*)	10.5
Female	67.2	38.6	38.3	4.6	(*b*)	31.9
Under 15	40.6	15.8	15.6	(*b*)	(*b*)	25.9
15-29	80.8	38.9	38.5	2.1	(*b*)	46.3
30-44	81.8	49.8	49.6	6.9	(*b*)	36.7
45 and over	72.7	52.6	52.3	8.9	(*b*)	23.8
45-64	81.0	55.9	55.6	8.7	(*b*)	29.5
65 and over	56.1	46.0	45.7	9.3	(*b*)	12.5

[a]Includes persons with photofluorographic examinations.
[b]Figure does not meet standards of reliability or precision (more than 25% relative standard error).

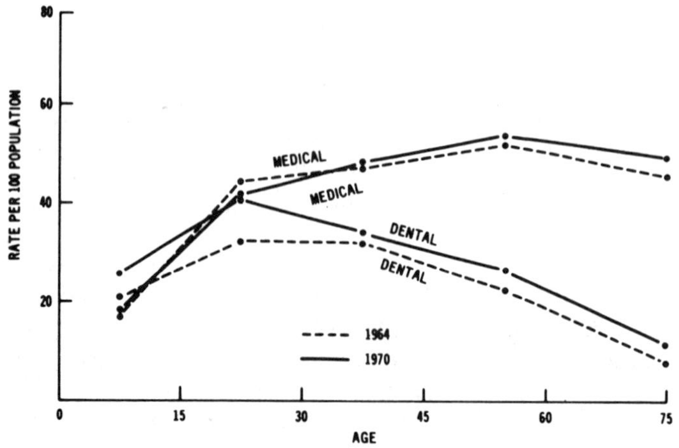

Figure 3.5. Estimated annual rate of persons x-rayed during medical and dental visits by age, United States, 1964 and 1970 (reproduced from DHEW, Ref. 27).

Based on the 1964 and 1970 surveys, various doses from medical and dental x-rays were estimated. Recent estimates of GSD have changed significantly from those calculated earlier. In 1972 UNSCEAR [1] estimated the total GSD for medical radiography to the U.S. population in 1964 to be 53 mrem/person (530 μSv/person). More recently the U.S. Department of Health, Education and Welfare estimated [28] that in 1964 and 1970 the population received about 17 and 20 mrad (170 and 200 μGy) annual GSD per capita, respectively. The apparent increase from 1964 to 1970 is not statistically significant [28].

However, other estimates of doses due to medical and dental x-rays show a statistically significant increase. In particular, the mean annual active bone marrow dose to the adult U.S. population from diagnostic radiology increased from 83 ± 4 mrad (830 ± 40 μGy) per capita in 1964 to 103 ± 5 mrad (1030 ± 50 μGy) in 1970. Table 3.9 gives the principal contributors to this dose, and the average dose

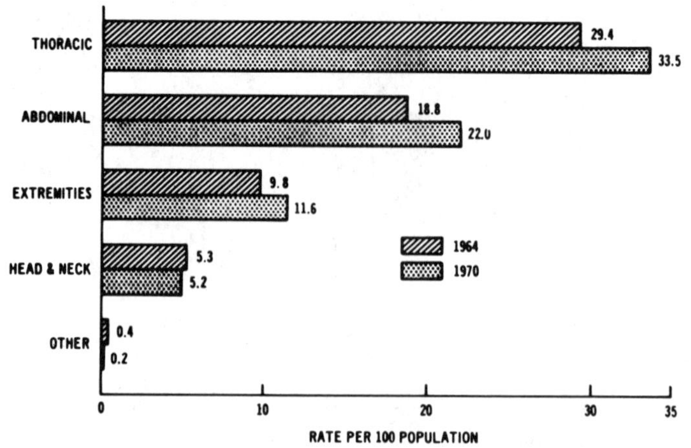

Figure 3.6. Estimated annual rate of medical x-ray procedures by body area, United States, 1964 and 1970 (reproduced from DHEW, Ref. 27).

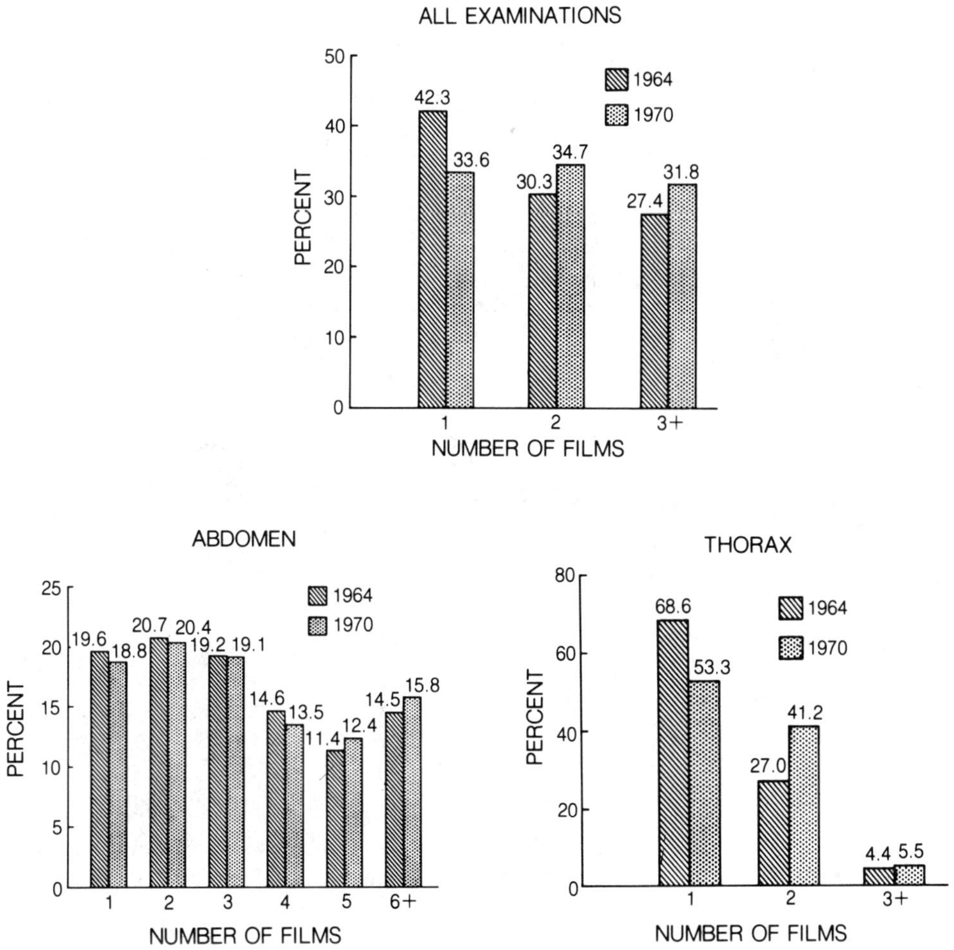

Figure 3.7. Estimated percent distribution of radiographic examinations of number of films per examination for selected body areas, United States, 1964 and 1970 (reproduced from DHEW, Ref. 27).

per examination. The variation in x-ray examination rate by age group implies a significant increase in per capita dose with age. Assuming that for each particular type of examination the dose does not vary with age group, Shleien et al. [29] estimate a per capita mean active bone marrow dose (1970) ranging from 52 mrad (520 μGy) for the 15–24-year age group to 151 mrad (1510 μGy) for those 65 or older. On the other hand, about 90% of the GSD is contributed by exposures of individuals less than 30 years old [28].

As shown in Table 3.9, dental x-rays contributed about 3% of the mean active bone marrow dose. They contribute an even smaller portion of the genetically significant dose.

Although therapeutic radiography was included in the 1964 and 1970 surveys, the corresponding contributions to GSD and bone marrow dose were not estimated by Shleien et al. [29] or the Department of Health, Education and Welfare [28]. The portion of x-ray procedures employed for such purposes is relatively small, less than

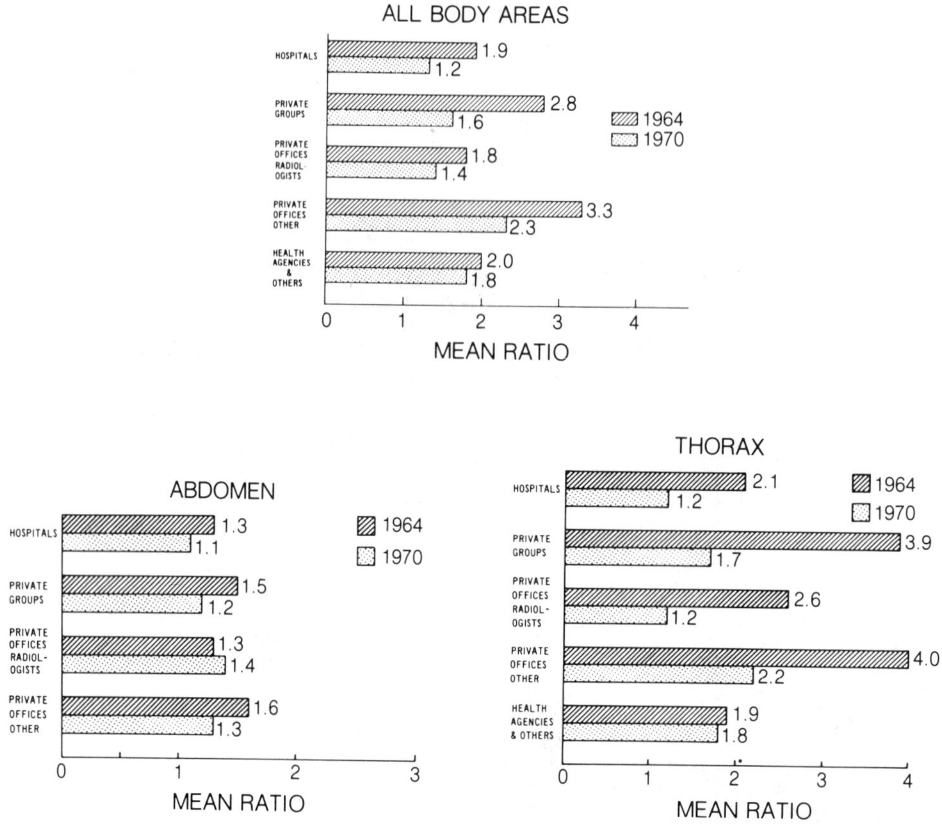

Figure 3.8. Estimated mean ratio of beam area to film area for radiographic films for selected body areas by type of facility, United States, 1964 and 1970 (reproduced from DHEW, Ref. 27).

1% of the total. These procedures, however, contribute a somewhat larger proportion of dose, since therapeutic radiological procedures typically entail larger average doses than diagnostic procedures [31].

Radiopharmaceuticals are also used for diagnosis and therapy. However, resulting annual per capita doses were estimated on a different basis from those from diagnostic radiology. As noted by the EPA [30], radiopharmaceuticals and radiation therapy contribute far less per capita dose than diagnostic x-rays. Data on U.S. use of diagnostic radiopharmaceuticals are given in the UNSCEAR report [31, Table 3, p. 339] based on research by Villafana et al. [34]. The dose to specific organs depends on the examination type; the total contribution to per capita gonad dose is approximately 0.5 mrad/yr (5 μGy/yr).

These estimated doses may be compared, to some extent, with natural background doses. Assuming an RBE of 1, the GSD of 20 mrad/yr (200 μGy/yr) from diagnostic radiology in 1970 is considerably less than the 80 mrem/yr (800 μSv/yr) gonad

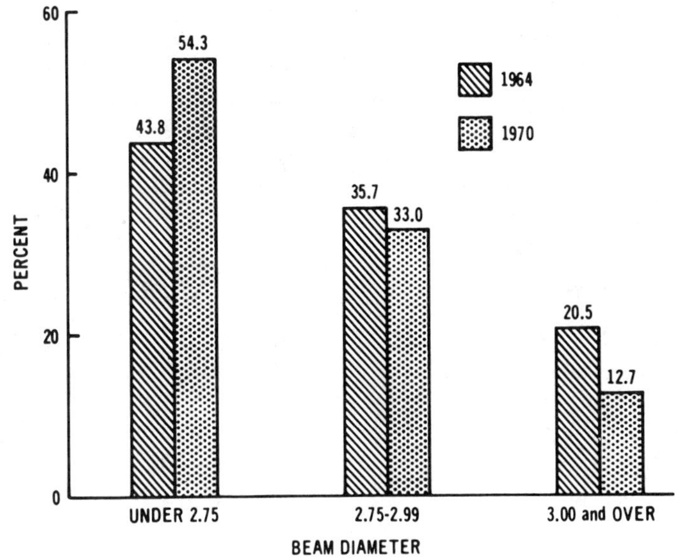

Figure 3.9. Estimated percent distribution of dental films by beam diameter, United States, 1964 and 1970 (reproduced from DHEW Ref. 28).

dose estimated [32] from natural sources. On the other hand, the 103 ± 5 mrad (1030 ± 50 μGy) 1970 mean bone marrow dose exceeds the estimated whole-body dose of 80 mrem/yr (800 μSv/yr) from natural sources. Even though these doses are not strictly comparable, they suggest that, in Fig. 3.10, the dominant contributions to U.S. average whole-body doses, that is, natural and medical, should be reversed in relative importance. Extrapolations to the year 2000 in Fig. 3.10 are, of course, only estimates; the indicated total could be strongly affected by changes in the medical contribution (due to changes in the frequency of or average dose from diagnostic procedures) or in the natural background contribution (which may, for example, be effectively altered by changes in housing construction practices).

UNSCEAR [31] reviewed medical exposures from a number of countries, and found substantial variation. For example, a 1969 Japanese survey gave a per capita annual bone marrow dose of 189 mrad (1.89 mGy) from diagnostic radiology; the value had decreased to 132 mrad (1.32 mGy) by 1974.

Until a detailed survey is completed it will not be clear how average U.S. population doses have changed since 1970. The increase that had taken place since 1964 may have continued, due to the introduction of new techniques such as computerized axial tomography (CAT). By 1979, 700 CAT scanners were in use, examining approximately 2 million patients per year, with typical doses to the head of 1 to 5 R [35]; a study of one scanner concluded that an average head dose of 2.1 or 2.4 rad (21.0 or 24.0 mGy) (depending on the kVp—kilovolt peak) resulted from a typical procedure [34]. On the other hand, attention to x-ray equipment and procedures has limited average doses, as discussed later.

Table 3.8. Changes in Data Pertaining to Diagnostic X-ray Procedures in a Six-Year Period, United States, 1964 and 1970 (data presented in U.N., Ref. 31)

	1964	1970	Change (%)
Number of persons having x-ray examinations	108×10^6	130×10^6	+ 20[a]
Number of x-ray procedures	173×10^6	212×10^6	+ 22[a]
Number of films exposed	506×10^6	661×10^6	+ 30[a]
Average number of films per examination	2.2	2.4	+ 9
Fraction of thoracic examinations with two or more films	31%	47%	+ 52
Mean ratio of beam area to film area (in hospitals)	1.9	1.2	− 37
Estimated mean skin exposures for posterior-anterior and anterior-posterior views of the abdomen	480 mR	620 mR	+ 29[b]
Mean skin exposure per dental film	1140 mR	910 mR	− 20[c]

[a]The population of the United States increased by 7% over the 6-year period.
[b]This increase may be due to the increased frequency of high-exposure examinations, the increase in the use of grids, and the use of higher tube potentials and currents without a corresponding increase in filtration.
[c]This decrease is due to the increased use of faster film.

Table 3.9. Mean Bone Marrow Dose to the Adult U.S. Population (1964 and 1970)[a]

	Annual per Capita Dose[b] (mrad)		Dose per Examination[b] (mrad)	
Type of Examination	1964	1970	1964	1970
Hip and upper femur	0.7	0.7	97	72
Pelvis	1.4	1.1	116	93
Lumbosacral joint	4.0	5.7	418	450
Lumbar spine	6.7	8.1	336	420
Urography	9.9	10.1	453	420
Stomach, upper GI tract	17.9	24.3	408	535
Small intestine	0.7	1.0	271	422
Colon	13.7	21.2	624	875
Abdomen	3.6	2.9	183	147
Cholecystography	3.2	3.7	183	168
Dorsal spine	2.0	2.5	232	247
Lung	2.0	3.2	10	10
Lung (photofluorography)	7.8	3.2	65	44
Head	1.0	1.6	65	78
Dental	3.5	2.9	13.2	9.4
Other	4.9	10.8	—	—
Total	83 ± 4[b]	103 ± 5[b]		

[a]Based on Refs. 29 and 31.
[b]Of the standard errors given in Ref. 29, we cite only those for the total annual per capita dose.

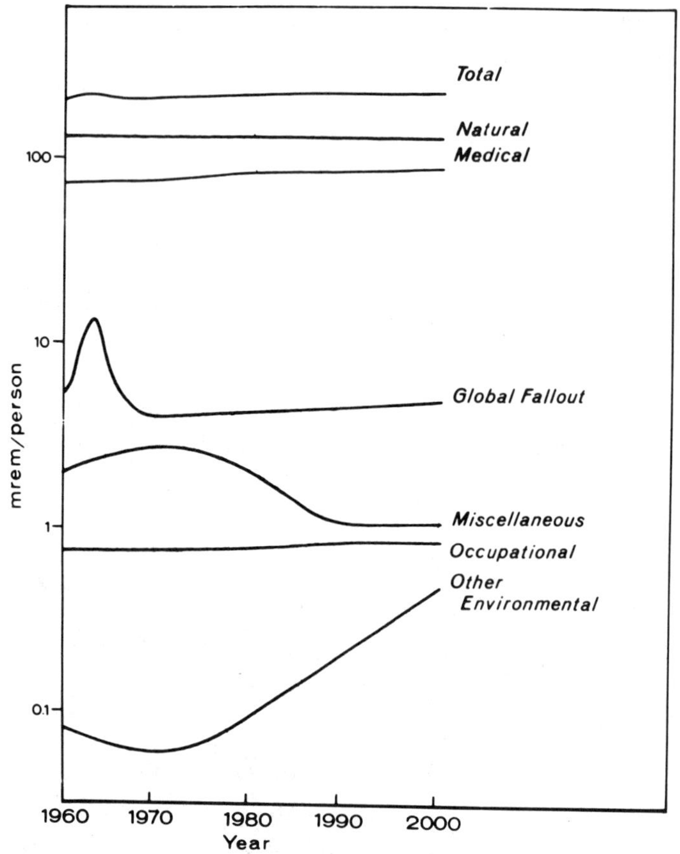

Figure 3.10. Summary of estimated average whole-body radiation doses in the United States (mrem/person) (reproduced from EPA, Ref. 23).

3.3.2. Occupational Exposures

The widespread and diverse use of x-radiation in medicine and industry makes it difficult to estimate occupational exposures. Both the EPA and UNSCEAR [30,31] surveyed available information for the United States and the world. In both cases, much of the U.S. data is drawn from Beck [9] who based his estimates of U.S. exposures on very limited data.

Based on available data on medical exposures around 1970, the EPA [23] projected average annual per capita whole-body doses of 320 mrad (3.2 mGy) for about 195,000 nonfederal medical x-ray workers and 125 mrad (1.25 mGy) for 171,000 nonfederal dental x-ray workers. These projections are roughly consistent with the data cited by UNSCEAR [31]. For other radiation-related occupations, the value of such averages is diminished by ambiguities in defining who is an x-ray worker, as opposed to an auxiliary worker. Based on data from a small survey, Barrall, Lanzl, and Hilbert [36] projected an annual dose of 540 mrad/yr (5.4 mGy/yr) for 38,000 radium workers. The value projected for 50,000 radioisotope workers was 262 mrad/yr (2.62 mGy/yr). A survey of 47 nuclear medicine departments found average annual per capita doses to employees of about 470 mrad (4.7 mGy) for the period 1970–1973, with somewhat lower values in 1968–1969 [36]. Many of the surveys showed that about 10% of the individuals included receive annual doses in excess of 1 rad (10 mGy) (see Ref. 31).

Similar difficulties arise in characterizing industrial dose rates. NRC licensees, including industrial workers not connected with nuclear power, reported average doses on about 9000 industrial radiographers of 0.41 rem/yr (4.1 mSv/yr) in 1973 and 0.30 (3.0 mSv/yr) in 1975. During this period the average annual dose to the 50 to 60% of these radiographers with measurable doses was consistently 0.6 rem (6 mSv); about 10% of the 9000 received doses in excess of 1 rem (10 mSv) during 1975. The data were similar for a smaller group of workers employed by NRC "manufacturing and distribution" licensees [30,31]. Most industrial workers exposed to x-rays are not NRC licensees, but they are often "agreement state" licensees (see discussion in Section 3.4.). The EPA [23] estimated that about 24,500 workers for such licensees (from about half the states) were exposed to a mean dose of 274 mrem/yr (2.74 mSv/yr), which is consistent with the NRC data.

Some types of accidental exposures have been examined. For example, Lindell [37] and Lubenau et al. [38] discuss exposures from units used for x-ray analytical work (diffraction studies, x-ray spectrography, x-ray fluorescence) in Pennsylvania (as a sample of the United States) and Sweden. The data suggest a probability of severe injury to be of the order of 1/100 per unit per year (for 1968), and 3 or 4 times higher for less serious injuries with clinical symptoms. These exposures are insignificant to the average population, but are very significant to the x-ray worker.

The total contribution of occupational exposures to the average population dose is relatively small. Barrall, Lanzl, and Hilbert [36] estimated to be slightly less than 1 mrem/yr (10^{-2} mSv/yr) (see Fig. 3.10). This is roughly consistent with the fact that about 0.4% of the U.S. population works in occupations exposed to radiation [31] at typical dose rates (as noted previously) of 200 to 300 mrem/yr (2.0 to 3.0 mSv/yr).

3.3.3. Other Sources of X-Ray Exposure

X-ray exposures that are neither medical nor occupational are not well quantified. Some of the specific sources are: color television receivers in private homes; cold cathode gas discharge tubes used for demonstrations in school science classes; accelerators and medical facilities that expose passers-by to small doses; high-voltage electronic devices (e.g., klystrons) that function improperly so as to leak x-radiation; and baggage x-ray systems.

Some limited surveys of these potential sources have been completed, often by the Bureau of Radiological Health (BRH), which, as discussed in the next section, has the responsibility for formulating electronic product radiation safety performance standards. NCRP Report 56 [39] summarizes information on doses from many products. Color televisions received specific attention when a significant percentage were found to have exposure rates exceeding the NCRP recommended limit of 0.5 mR/hr at 5 cm distance, which was adopted by the FDA in 1970 (discussed later). Average exposures were much lower. It has been estimated that a rate one-tenth of the limit would cause annual exposures averaging about 1 mrad/year (10 μGy/yr); newer televisions would yield even less then 1/10 the limit [39]. A BRH in-plant survey of television sets is described in Ref. 40. The BRH surveyed cold cathode gas discharge tubes in 1969 [41], finding that some tube types emit excessive radiation (see also Ref. 40). Few data are available on exposures of passers-by near medical and accelerator x-ray sources, although these would be controlled through proper engineering practice; Ref. 30 cites values of 0.04 to 4 mrem/yr (0.4 to 40 μSv/yr) at accelerator site boundaries (not necessarily from x-rays). See also Section 3.4, on particle accelerators. Typical exposures from x-ray baggage inspection devices are about 0.001 mR or less per inspection [39].

3.3.4. Licensing and Design for X-Ray Facilities

Most medical and industrial *sources* of x-radiation in the United States are subject to license by the several states. Practices have varied widely in the past, ranging from regular licensing and surveillance of essentially all sources to almost complete lack of regulation. Similarly, there have been wide differences in the licensing or certification of *operators* of x-ray equipment. Some states have certification programs for all medical x-ray workers; in California, even physicians are subject to this certification [42]. The licensing of individual industrial x-ray workers has been practically nonexistent, which is particularly unfortunate because most large accidental exposures arise from human error rather than from mechanical failure.

Progress is being made. Federal performance standards have been set for x-ray equipment. State programs for inspecting x-ray machines are assuring an increasing degree of compliance with applicable codes. Increasing attention is being given to proper use of x-ray equipment, partly as a result of State–Federal cooperative programs. To an increasing degree, recommendations on equipment and operation are available from national (and international) bodies concerned with radiation protection.

Under the U.S. Radiation Control for Health and Safety Act of 1968 (PL 90-602), the Bureau of Radiological Health of the Food and Drug Administration is empowered to set performance standards for electronic products that emit radiation as discussed in Ref. 43. Table 3.10 indicates the applicable FDA standards as of

Table 3.10. FDA Electronic Product Radiation Safety Performance Standards and Date of Applicability (see Ref. 43)

Performance Standard	Date of Applicability
Television receivers	Jan. 15, 1970
Cold cathode gas discharge tubes	May 19, 1970
Microwave ovens	Oct. 6, 1971
Diagnostic x-ray systems and their major components	Aug. 1, 1974
Cabinet x-ray systems baggage x-ray	April 10, 1975 April 25, 1975
Laser products	Aug. 2, 1976
Ultrasonic therapy products	Feb. 17, 1979
Mercury vapor discharge lamps	March 9, 1980

1979; many of these pertain to x-ray emitting equipment, of which that for diagnostic x-rays is the most important contributor to average population doses.

Proper design and maintenance of equipment, together with correct use, can have substantial effect on doses from diagnostic x-rays. Key questions include the voltage and current at which the x-ray beam is generated, the collimation of the beam, use of image intensifiers, and choice of film. Often these questions and the exposure used are determined, not by equipment design or condition, but by the procedure chosen by radiology personnel. Efforts to survey and control such factors are indicated in the next section.

Aside from the recent FDA performance standards, a substantial basis for the design and operation of x-ray equipment and facilities is provided by recommendations of radiation protection organizations. Relevant NCRP publications include: *Medical X-Ray and Gamma Ray Protection for Energies up to 10 MeV, Equipment Design and Use*, Report No. 33 [44]; *Structural Shielding Design and Evaluation for Medical Use of X-Rays and Gamma Rays of Energies up to 10 MeV*, Report No. 49 [45]; *Dental X-Ray Protection*, Report No. 35 [46]; *Radiation Protection for Medical and Allied Health Personnel*, Report No. 48, [47]. (See also ICRP Publication No. 16, Ref. 48.) Recommendations that apply specifically to industrial use of x-rays include: NCRP Report No. 61, *Radiation Safety Training Criteria for Industrial Radiography* [49]; NBS Handbook 114, *General Safety Standard for Installations Using Non-Medical X-Ray and Sealed Gamma-Ray Sources, Energies up to 10 MeV* [50]; NBS Handbook 111, *Radiation Safety for X-ray Diffraction and Fluorescence Analysis Equipment* [51]; and BRH's *Radiation Safety Recommendations for X-Ray Diffraction and Spectrographic Equipment* [52].

Adequately protecting patients and operators from excessive x-ray exposures or from accident requires a systematic program of monitoring and training. Many elements of such programs were mentioned in the last section, particularly for protection of operators of x-ray equipment. On the other hand, most of the population dose from x-rays is to patients in medical procedures. Monitoring patient exposure and its dependence on various factors requires a well-designed program, and such a program—together with operator training—has a great potential for leading to significantly reduced patient exposure.

One of the most important contributors to unnecessary patient exposure can be inadequate collimation of the x-ray beam. For example, the ratio of beam area to film area has been larger than necessary in some cases. This ratio decreased significantly between the 1964 and 1970 surveys and is still a parameter of great interest in current monitoring and training programs. An increasing percentage of films are taken with beams no larger than the 2.75-in. maximum diameter recommended by the American Academy of Dental Radiology.

Other factors, too, are important. These include the kilovolt peak (kVp) of the x-ray generator, the film speed, and so on. In some cases, the impact of changing these factors is correlated. For example, with the relatively fast film now used in most dental radiography, use of a higher kVp reduces exposure [53]. Assurance of photographic quality is an important related consideration [54].

Because patient exposures can be significantly affected by x-ray procedures, state and national agencies are making substantial efforts to monitor patient exposures and to effect reductions through programs of education and training. A useful avenue for effectively communicating information among interested parties has been the series of Annual Conferences on Radiation Control. Since the 1970 exposure study, new data on the effects of training have become available through the

Nationwide Evaluation of X-Ray Trends (NEXT), a cooperative x-ray survey program between the Bureau of Radiological Health and state radiation control programs. In this survey, the x-ray equipment operator is asked by an inspector to set up the machine for performing an examination of a patient of standard size. The measurements taken permit determination of the dependence of exposure and other parameters on training and equipment. For example, an examination of 1972-1975 data indicates that training improves collimation and decreases exposures [53].

Such surveys, as well as radiation control efforts on the part of individual institutions, depend on effectively calibrated monitoring instruments. Some surveys, such as the 1964 and 1970 exposure studies, depend directly on use of standard films. On the other hand, direct measurements can be taken with instruments, including ion chambers, that are sensitive to the x-ray spectrum being employed.

With a few exceptions, most medical diagnostic procedures are in the region above 50 kVp. Only in soft-tissue radiography (such as in mammography) is lower-energy radiation common: in such cases, 20 to 30 kVp is typical. Another low-energy application is superficial therapy (such as acne treatment, where energies as low as 5 kVp are used.) At the high end, energies above 150 or 175 kVp are rarely found, although there is some use of several MeV linear accelerators for diagnostic chest studies. Dental diagnostic equipment is also almost exclusively in the 50–100-kVp range.

From these comments, it is clear that a large portion of the surveys require instruments whose energy response is flat down to at least 50 keV and that lower-energy monitors are needed for special applications. Actual x-ray monitoring instrumentation is discussed in Chapter 5.

3.4. PARTICLE ACCELERATORS

This section provides background information about the extent of radiation hazards due to increased use of accelerators in modern technology. Sources of information on accelerator design and principles of operation are listed in Ref. 55. Included in this discussion are the types of radiation present and their measurement. For direct instrument comparisons the reader is referred to the appropriate chapter elsewhere in this book. In this spirit, the following section must be viewed as complementary to those chapters.

International and national conferences are regularly held on accelerator technology and applications. Several of the most important series are described in the following proceedings: the 1979 Particle Accelerator Conference at San Francisco [56]; the 1978 conference on Applications of Small Accelerators [57]; and 1974 International Conference on High Energy Accelerators, held at Stanford University, Palo Alto, California [58]; and an international conference on accelerator dosimetry at Stanford in 1969 [59].

A useful general reference on radiation protection at accelerators is the book by Patterson and Thomas [60]. Standards for radiation protection and monitoring are: ICRU Report No. 20, *Radiation Protection Instrumentation and Its Application* [61]; ANSI Standard N43.4—1969, *Radiological Safety in the Design and Operation of Particle Accelerators* [62]; and NCRP Report No. 51, *Radiation Protection Design Guidelines for 0.1–100 MeV Particle Accelerator Facilities* [63].

Particle accelerators in the world today number in the thousands, with the great majority having maximum energies below 100 MeV. Figure 3.11 indicates numeri-

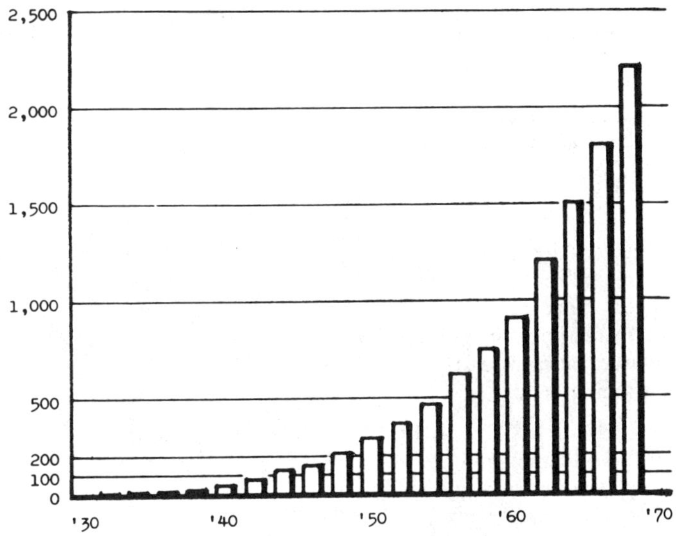

Figure 3.11. Worldwide particle accelerator population, 1930–1968 (reproduced from Burrill, Ref. 64).

cal growth until 1968. Since the early 1940s, the number has grown by about 10% annually. Approximately half the accelerators are located in the United States. The 1968 breakdown of U.S. accelerators by type and application is given in Table 3.11. In 1972, the U.S. number was found to be about 2600, as indicated in Fig. 3.12 [65]. Although no recent survey comparable to that of Morgan [65] is available, it is clear that the total number of accelerators has continued to increase, perhaps doubling during the 1970s.

Accelerators are applied in various ways. Table 3.11 gives a more detailed look at the expanding use of accelerators as of 1968. About 28% of the accelerators in the United States were used for nuclear science and engineering, about 35% in medical therapy and radiography, and 29% in the study of radiation effects. While the growth in nuclear science and engineering has been small in recent years, the medical usage is growing very rapidly, as is use in industry. Accelerator technology has just begun to be exploited in such diverse areas as hot-atom chemistry research, ion-beam polishing and surface-coating, dispersive x-ray analysis, ion implantation, and chemical processing. The increasing portion of low-energy accelerators used in industry has made systematic survey of numbers and application more difficult.

The accelerated particles are of two basic types: electrons and ions (including protons). Although there are a few machines (perhaps two or three dozen in this country) with energies in excess of 100 MeV, these are primarily for research, used in large, well-equipped laboratories. These machines range in energy all the way up to those of the Stanford linear accelerator (typically 22-GeV electrons, with a 40-GeV maximum), the Brookhaven Alternating-Gradient Synchrotron (33-GeV protons), and the Fermi National Accelerator Laboratory (500-GeV protons, soon to be increased to about 800-GeV).

The problems of radiation measurements around these few high-energy machines, which are both difficult and special, are being handled by teams of skilled experts; thus this discussion concentrates on the lower-energy machines. There are three reasons for this emphasis: (a) the numbers of low-energy machines are expanding

Table 3.11. U.S. Particle Accelerators, 1968 (adapted from Ref. 64)

Categories of Accelerator Application	1964	1968	Increase (%)	1968 Application (%)
By Application				
Nuclear science and engineering				
Nuclear structure and reaction research	108	108		
Neutron and photo-nuclear research	87	80		
Reactor physics, shielding, and engineering	23	41		
Injection into larger accelerators	21	22		
Teaching and training	43	46		
Total	282	297	5.0	27.5
X-rays (neutrons) for therapy and radiography				
Radiation therapy (with x-rays, electrons, etc.)	101	198		
Industrial radiography (with x-rays)	133	176		
Neutrons for therapy or radiography	—	—		
Total	234	376	60.0	34.7
Radiation effects				
Research in chemistry, biology, foods technology	41	52		
Activation or scattered-particle analysis	58	104		
Radioisotope production	—	7		
Space-radiation simulation and effects	31	46		
Other research in radiation effects	95	106		
Total	225	315	40.0	29.1
Atomic and solid-state physics				
Beam-foil optical spectroscopy	—	3		
Ion implantation and channeling	—	17		
Solid-state research and development (miscellaneous)	5	10		
Atom-ion interactions; hot-atom chemistry/physics	—	5		
Total	5	35	600.0	3.2
Radiation processing, industrial				
Processing of chemicals	21	46		
Processing for biological effect	1	—		
Processing of solid-state devices and materials	4	2		
Radiation service facility	10	12		
Total	36	60	66.0	5.5
Totals	782	1083	39% Av.	100.0%
By Accelerator Type				
Electron Accelerator		598		55.0
Ion Accelerator		485		45.0
Total		1083		100.0%

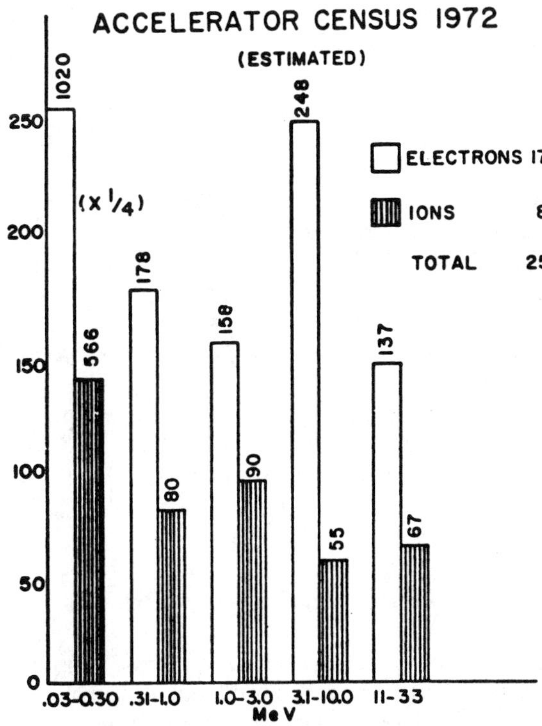

Figure 3.12. U.S. particle accelerators, estimate for 1972 (reproduced from Morgan, Ref. 65, with permission of IEEE Editorial Department, © 1973).

rapidly; (b) in many cases they are being installed in medical or industrial sites where personnel have had little previous experience with high-energy ionizing radiation; and (c) with their special problems, the largest accelerators tend to have the best monitoring equipment in the hands of the most experienced personnel.

3.4.1. Sources of Radiation Exposure

The sources of exposure from accelerators can conveniently be divided as follows:

1. Exposure from the direct beam or its halo or scatter.
2. Exposure from leakage of radiation through shielding walls.
3. Exposure to induced radioactivity from the activation of items such as the beam pipes, targets, and shielding walls.
4. Exposure from respiration of air containing radioactive particulates, or unintended ingestion of radioactivity from various sources.

In this section the first three of the sources are emphasized. Except for a brief treatment in connection with the environmental impact of accelerators, the ingestion and respiration of radioactive nuclides is discussed primarily in Part III. Thus, the text here deals primarily with external exposure. Although such exposure is primarily occupational in nature, some installations do produce small external levels in the general environment, as discussed at the end of this section. Another principal nonoccupational exposure, not discussed here, is that received intentionally by a medical or dental patient from accelerator-beam diagnostics or therapy. Section 3.3 discusses sources of x-radiation, which contribute the major dose from medical diagnostics. In many cases x-radiation is also the major dose from accelerator therapy. However, therapy can take various forms, and only generic considerations that pertain to occupational exposure are stressed here. Patient exposures, on the other hand, are a medical consideration. Reference 66 discusses medical electron linear accelerators and associated radiation safety considerations and guidelines.

At the large research accelerators, at which both interlock systems and shielding are well designed and maintained, the most important source of exposure is usually the induced radioactivity around beam pipes and targeting areas, rather than the beam or its secondaries [60].

In the case of occupational exposure, the relevant Maximum Permissible Exposure (MPE) guidelines (discussed in more detail in Chapter 2) are basically a limit of 5 rem/yr (50 μSv/yr), with no more than 3 rem (30 mSv) to be received in any 13-week quarter. As always, exposures are to be kept as low as is reasonably achievable.

3.4.2. Safety Procedures—Licensing and Compliance Practices

Accelerators fall into three broad categories: first, those operated by DOE and its prime contractors and licensees; second, those operated in "agreement states" under state supervision; and third, those in "non-agreement states." An agreement state is one where the state has agreed to certain NRC stipulations [67] on the licensing and supervision of nongovernmental radiation usage, and has undertaken this supervisory task. There were 25 such states in 1979. In non-agreement states, the state has either not requested the authority or has been unable to obtain such NRC approval (the latter is rare); even in these states, associated use of radioactivity is typically overseen to some degree by the state. Bailey [68] surveyed state programs for particle accelerators as of 1976.

Agreement states typically have programs of registration, licensing and compliance, but, as discussed by Bailey [68], procedures vary from state to state, as may the state agencies involved. For example, a state may require licensing only for those accelerators from which significant activation (radioisotope generation) occurs, thereby excluding many of the smaller installations, such as electron accelerators below about 8 MeV. The responsibility for supervising compliance may be divided according to application, for example, medical, instructional, industrial. And finally, registration of new accelerators may be required of the owner upon delivery, independent of a license requirement. After registration (or upon licensing), an appropriate state agency may conduct an inspection to ensure that proposed shielding is adequate, that appropriate monitoring will exist, that records will be kept properly, and that the responsibility for operation rests with well-informed personnel. Subsequently, routine visits at fixed intervals may be made to ascertain that compliance with licensing requirements is continuing.

The responsibility for safe installation of an accelerator rests ultimately with the employer or user, not with the manufacturer. However, the manufacturer may be liable for product defects and for design inadequacies and is responsible for meeting applicable local, state, and federal rules, regulations, and design standards.

The adequacy of the shielding and the resulting radiation levels in various places on and near the installation are often determined with the aid of a consulting health physicist. The consultant will usually make quite complete survey measurements with sophisticated instrumentation and offer advice as well on the choice of monitoring instrumentation for later (routine) survey work. The day-to-day monitoring is then performed with less complex instrumentation by the appropriate on-site personnel, relying heavily upon the basic measurements of the original consultant.

The arrangement just described is a judgement based on a number of factors. First, a complete survey by a qualified health physicist is essential; it is required for initial licensing and it is not done by the state itself due to the lack of financial resources. Second, the typical safety officer at a hospital or industrial plant is often an industrial hygienist by training, whose experience with ionizing radiation may not be adequate or up-to-date for the initial licensing requirements, but may suffice for routine monitoring. Third, the budget for monitoring instrumentation for such radiation is sometimes much less than would be desirable. Finally, the types of radiation fields which exist in these situations usually change very little with time, so that frequent extensive reevaluations of the levels are often unnecessary.

The exceptions are in applications where the operating conditions change frequently. Thus, research accelerators use a variety of beam configurations and a variety of targeting and detection equipment, in contrast to a typical medical therapy installation where there may be few important changes in operation over a period of several years.

Perhaps the most compelling reason for engaging a consulting health physicist is the complexity of the considerations in making meaningful measurements of the radiation levels and associated hazards. It is therefore important for every installation to retain the services of a consulting health physicist. Measurement problems will be discussed in the next section.

3.4.3. Special Measurement Aspects

Duty Cycle. Duty cycle, defined as the fraction of the time in which the beam is on, is the term introduced because many accelerators only deliver fully accelerated beams during a fraction of the available time. One of the worst duty cycles is that typical of the high energy electron linear accelerators (linacs); thus at the 22-GeV Stanford Linear Accelerator the beam occurs in bursts of 1 to 2 μsec or at a repetition rate of 360 bursts/sec. The beam is present for less than one-thousandth of the time. Another typical accelerator type is the high-energy proton synchrotron, with useful beams for perhaps 5 to 10% of the time. At the other extreme are the low-energy Van de Graaff and Cockcroft-Walton accelerators, in which the beam is usually on continuously. Superimposed on the beam-pulse structure just described is the rf-structure present in essentially all of the machines whose operation relies upon the synchronous principle. The rf accelerating voltage imposes its own characteristic periodic structure upon the accelerated beam, which can range in frequency from a few to a few hundred megahertz. The current in each rf packet may occupy only a small fraction of the available time in an rf cycle; 5% to 30% is typical.

The time structure of the beam is a crucial consideration in the choice and operation of monitoring instrumentation. Typical radiation fields during beam operation, even outside of shielding walls, are very closely correlated in time with the pulse structure of the primary beam, that is, within a few nanoseconds. The main exception to this rule is thermal neutrons; depending upon geometrical and shielding variations, as much as several milliseconds can elapse before the thermal neutron flux has died away. Counters (for example gas proportional, scintillation, G-M counters) are particularly susceptible to incorrect measurements of the dose with low duty cycle, since in some cases they can have dead-times of many microseconds. Ionization chambers are somewhat less sensitive if operated so that the voltage is high enough to overcome much of the columnar recombination (recombination of positive ions and electrons within a single track before they are separated sufficiently by the applied voltage). In using ionization chambers, it is important to determine the extent of this latter problem and of volume recombination by calibrating with an actual pulsed source and varying the collector voltage to insure saturation. Instruments with no sensitivity to duty cycle are activation/counting systems, such as those counting the radioactivity resulting from the activation of foils containing gold, indium, carbon, and other elements.

There are two methods for circumventing the problem of poor duty cycles. The first is to use a measurement technique that is intrinsically an integrating method (such as a photographic film, a thermoluminescent dosimetry chip, or perhaps an ion chamber or other electronic instrument operated in current-integration mode rather than pulsed mode). The second is to alter the source appropriately: a sufficient reduction in intensity or an artificial lengthening of the beam pulse structure might be feasible, although these factors are usually governed by considerations other than radiation protection monitoring.

Mixed Fields. Another key concept in the discussion of measurement problems around accelerators is the idea of mixed fields. External radiation may consist of a number of particle species over a wide range of energies, with fluxes of variable spatial distribution, and sometimes with unusual duty cycle. All of these considerations immensely complicate the job of measuring dose equivalent levels.

Fortunately, most shielding is sufficiently thick that exposure to the direct beam or its secondaries does not present a problem. A proper interlock system should prevent access to the beam areas while the beam is on, and shielding is usually sufficient so that charged particle leakage is extremely small. Thus, outside the shielding walls of well shielded machines, the main radiation hazard during beam-on operations is usually from neutral radiation; the two dominant species are neutrons and photons (x- or γ-rays). It is also generally true that photons dominate the problem at electron accelerators, while at proton and ion machines the neutron flux is the dominant feature of the radiation field [14]. The difficulty in monitoring neutral particles is due to their behavior when they interact with matter; indeed the differences between neutrons and photons as classes may often be less troublesome than the differences as a function of energy of either one of the species by itself.

This leads to the following statement: the most important task in the monitoring of external radiation levels around an operating accelerator is the evaluation of the dose equivalent rate due to neutrons and photons by the quantitative measurement of their flux densities as a function of energy. To this must be added the task, usually both easier and less important, of measuring the flux density of charged particles.

With regard to accuracy, a useful viewpoint has been expressed by ICRU Report No. 20 [61]:

Because the dose limits recommended by ICRP have been conservatively derived, great accuracy in radiation protection measurements appears unwarranted. It is suggested that when the MDE (maximum dose equivalent) is comparable to the Maximum Permissible Dose (MPD), an accuracy of ± 30% be achieved.

The ideal detector (of photons and neutrons in a mixed field) would:

1. Separate the two species and determine their flux densities independently.
2. Weigh the fluxes by the appropriate quality factors (Q), which have an energy dependence in the case of neutrons.
3. Provide a direct readout of the flux densities (particles/cm^3), the absorbed dose (rad), and the dose equivalent (rem or sievert) for each species separately.

Unfortunately, no ideal instrument exists. The closest approximation is the use of several instruments, which are sensitive to one or more of the particles or energy regions but are reliably insensitive to others. In combination, these can be used to give a reasonable picture of the radiation field.

There are two basic pieces of information that one wants to know: (a) the particle constituents of the radiation field as a function of energy (for example, neutrons/$cm^2 \cdot$ sec $\cdot$ MeV); (b) the total dose equivalent (rem). While it is true that sufficiently accurate knowledge of (a) permits quantitative determination of (b), the reverse is not true [61]. However, considerable effort has been devoted in recent years toward the development of instrumentation that reads directly in rem (Sv), especially for the fast neutron fields which are so troublesome around most proton accelerators. It should be emphasized that a rem-meter is useful in only a limited class of circumstances and is not a substitute for more detailed knowledge of the constituents contributing to the total rem (Sv) being measured. There are several reasons for this; the most obvious is that if radiation levels need to be reduced, one has no indication of the predominant constituents. Is increased shielding necessary? If so, what kind and where? Is it primarily thermal neutrons? Is low-energy photon radiation properly accounted for? One real danger is that the composition of the radiation might change without a correspondingly significant change in the rem (Sv) value (for example, the thermalization of fast neutrons by some change in operating conditions). For these reasons, this text deals separately with neutron and photon monitoring instrumentation, and then follows with discussions of other particles and of induced radioactivity.

Neutrons. One key design problem for neutron-monitoring in the vicinity of accelerators is to make an instrument which is insensitive to photons, but with great neutron sensitivity. Discussion of such instruments is deferred to Chapter 8, Measurements of Neutrons. However, a few comments are pertinent here. For example, there are several types of accelerators whose radiation fields outside of the shielding walls normally contain very few neutrons of any energy. This is almost always the case around electron machines of low energy (~10 MeV) and is even the case at the few very-high-energy (GeV-range) machines [69]. The reason for this is that in electron machines the dominant process of primary-beam energy loss is cascade bremsstrahlung, in which electromagnetic showers build up in the shielding. Shielding is usually designed primarily to cope with this problem, but even so, the dominant remaining backgrounds are photons. At low energies, neutrons result primarily from electro- or photo-disintegration of nuclei, but the thresholds are high (several MeV to several tens of MeV) for the nuclei that comprise most shielding walls, and the disintegration cross sections are relatively small. However, this must be applied with some caution. For example, the use of a deuterium or tritium target (binding energies 2.3 and 8.5 MeV, respectively) is a guaranteed way to increase the neutron component of the radiation field, and if measurements with other targets show little or no neutron component, this will not necessarily remain true when a target of deuterium or tritium is placed in the beam. Likewise, change of collimation material can be a problem; thus while lead is commonly used in collimators, it is important to realize that substitution of U for Pb can increase neutron backgrounds manyfold, due to beam-induced fission.

When changes are made in operating procedures and one requires a rapid check on the possible increase in thermal neutron levels, a well-known method is the indium-foil technique, because In has an extremely large thermal-neutron capture cross section. Also, intermediate energy and fast components can be rapidly assessed by surrounding the In with a moderator, such as polyethylene. Both these techniques are described in Chapter 8.

Finally, a quick method for ascertaining an upper limit estimate of dose equivalent is the use of a tissue-equivalent ion chamber (with various absorbers to study the problem of particle equilibrium) to measure total absorbed dose (rad or gray). This can then be assigned a Q of 10 (essentially always an overestimate) to determine dose equivalent (rem or sievert) in a rough way.

Photons. For more detailed information on instrumentation, the reader should refer to the chapter discussing measurement of photons, Chapters 5 and 6. Only some aspects of the problem which are unique or relevant to the case of accelerators are discussed here.

For photon measurements one does not have the problem of quality factor dependence, since $Q = 1$ for all photon energies. However, the possible energy spectrum of γ-rays and x-rays has to be considered in the choice of instruments. Thus, while the ion chamber is the most common x- and γ-ray measuring instrument in the energy region from about 50 keV to about 3 MeV, it is not usually appropriate outside of this range. At higher energies, it is limited when it becomes impractical to make chamber walls of a thickness sufficient to establish particle equilibrium. At low energies, the opposite is true: the limitation occurs when the wall cannot be made thin enough to establish electronic equilibrium. These limitations will be discussed in more detail in the chapter on γ- and x-ray measurements. Ion chambers are often used improperly when photon energies extend from well above 3 MeV down even into the 10 to 15-keV range, although there is usually little need around MeV-sized accelerators to measure the low-energy x-radiation separately.

Although the response of most ion chambers is not proportional to absorbed dose for energies above a few MeV, ion chambers are still usually used when higher energy photons are present or suspected. When very-high-energy (even GeV-level) electromagnetic showers develop in shielding, the primary energy is successively degraded into more and more particles of lower and lower energy and eventually the process reaches a peak with regard to the number of particles in the shower. The absorption processes of photons in shielding are such that most low-energy photons are absorbed in rather short distances after their production. However, for high-energy photons, that is, above about 5 or 10 MeV, the absorption coefficients are both small and relatively constant. When one of these higher-energy photons does interact, energy is lost quite quickly due to production of lower-energy secondaries of shorter range.

Once equilibrium is reached, the fall-off of intensity with increasing penetration is slow, and governed by the high-energy attenuation probabilities, despite the fact that at any point the dominant particle flux (and absorbed dose) are from low-energy particles. The low-energy flux, the one of most interest from the radiological point of view, remains proportional to the high-energy photon progenitors, and it is often not necessary to measure the latter directly. This is why ionization chambers, whose sensitivity is poor above a few MeV, are nevertheless adequate instruments for measuring absorbed dose in equilibrium photon radiation fields.

If it is suspected that nonequilibrium high-energy components are present, the best check is to add absorber around the ion chamber in stages, up to about 30 g/cm^2. In equilibrium fields, there should be a small, monotonic decrease in instrument response, while in nonequilibrium situations an actual buildup of response may occur as mass is added.

Charged Particles. For all well-shielded accelerators, the charged particle fluxes are almost insignificant compared to the neutral fluxes. This section discusses only those few situations in which charged particle backgrounds can be important. It does not consider low-energy charged particles that are in equilibrium with the dominant flux of neutral particles.

In a very-high-energy regime, there is a significant yield of muons. These come mainly from decay of pions (at proton accelerators) or from pair production (at electron accelerators). Muons are a problem because they interact with shielding wall materials almost exclusively through ionization loss. Below a few GeV, muons present little problem, since the shielding necessary for protection against nuclear cascade leakage is sufficient to stop muons by range. The higher the machine intensity below a few GeV, the more nearly true this statement becomes because the shielding requirements become more stringent. At energies above a few GeV (this includes only about five machines in this country, as well as several abroad) muons have such long ranges that their effective removal by shielding is difficult and expensive. At the Brookhaven AGS (33-GeV proton accelerator) there were some initial operational difficulties due to muon backgrounds [70] and at Fermilab, a penetrating muon beam is one of the more significant sources of radiation off site [71].

Another source of charged particles is pin-holes or cracks in the shielding. These are especially difficult to monitor, and in common practice the only way to determine charged particle fluxes (above, say, a few MeV) is through use of a scintillation-counter coincidence telescope. This is very seldom done, even in large, sophisticated laboratories, because it is seldom a significant problem. Even in these cases, telescopes are essentially always assembled from instruments at hand. Integral counter telescopes are not commercially available.

To summarize, there is very little instrumentation presently available for the measurement of charged particle fluxes; this is due to the rather low level of typical backgrounds (except in a few rare cases). It would appear that despite this lack in instrumentation, there is probably no need to assign high priority to its development.

3.4.4. Universal Dose Equivalent Instruments

Instruments for determining radiation levels near accelerators, x-ray machines, or radioactive sources (e.g., accelerator targets or reactor fuel rods) are almost always survey instruments. They respond by generating an electrical pulse every time an ionizing particle (x-ray, α-ray, β-ray, γ-ray, etc.) is detected. The number of pulses can easily be converted to absorbed dose (J/kg) by the correct multiplicative factor. However, the major concern is the amount of biological damage caused by this radiation.

As discussed earlier, the linear energy transfer or quality factor must be measured to assess biological damage. Instruments designed to make this measurement (called universal dose equivalent instruments) are particularly valuable in mixed fields. Three are described in the ICRU Report No. 20 [63]. In each case a separate measurement of absorbed dose must be made. The following has been taken directly from Report No. 20. The references cited in the quotation are listed as Refs. 71 through 83 of this text. Even as of 1978, instruments of this type were not available commercially.

> For accelerator applications, an instrument giving dose equivalent for all types of radiations would be very useful. Such instruments are not commercially available but have been developed and used at a number of laboratories. Three types have been developed. The first is a spherical linear energy transfer (LET) spectrometer, the second is based on columnar recombination in a gas, and the third on the same phenomenon in a liquid. Although these methods are presently difficult to apply, improvements resulting in practical field instruments are to be anticipated.
>
> The distribution of dose with LET may be determined by means of the spherical LET spectrometer (Rossi and Rosenzweig, 1955 and 1955a; Rossi, et al., 1962). The detector employed is a spherical proportional counter of tissue equivalent plastic containing tissue equivalent gas at low pressure. The energy deposited by individual charged particles traversing the interior of the counter depends on the product of LET and the length of the track intercepted. The resulting spectrum of pulse heights can be converted to a spectrum of LET by appropriate mathematical treatment. A limitation is imposed on the operation of the instrument by the requirement that particles traversing the cavity incur only small changes in LET and that the particle trajectories be essentially straight lines. This limitation and the fact that the proportional counter has a finite recovery time, which limits the particle fluence rate (density of particles) which can be measured, restrict the applicability of the device. Since the instrument is functional for essentially all charged particles with a quality factor of 1 or more, and since the total tissue absorbed dose may be determined with a tissue equivalent ionization chamber, a combination of these two instruments may be used to determine dose equivalent for any radiation. This method falls short of being ideal in two respects: first, it is quite complicated in application and, second, it does not give values of dose equivalent directly. Considerable progress, however, has been made towards overcoming these difficulties (Baum, 1967; Baum et al., 1970).
>
> The phenomenon of columnar recombination has been utilized to obtain an indication of effective quality factor for an unknown mixture of radiations (Zielczynski, 1963; Sullivan and Baarli, 1963). The method uses a simple parallel plate ionization chamber which is operated at 6 atmospheres of pressure of tissue equivalent gas and field strengths of up to 2,000 V/cm. The response of such a chamber can be deduced from existing theories of columnar ionization. The corrected ionization current (I) is given by the equation $I = KV^n$, where K is a function of absorbed dose rate and type of radiation, V is the electric field strength and n is an index depending on the LET of

the radiation used. The index n, within limits acceptable for practical application, is independent of dose rate and the direction of the radiation, and directly proportional to quality factor. In mixed radiation fields the measured value of n would be an average value for the various components, if the collected ionization current were directly proportional to the rate of energy absorbed. This, however, is not the case because the collection efficiency varies with the LET of the ionizing particles. However, the error due to this effect is thought to be no more than 20% for unknown fields of mixed radiations. More recently, values of quality factor and dose equivalent have been obtained by comparison of the current in two such ionization chambers operated with widely different collecting potentials (Sullivan, 1964; Distenfeld and Markoe, 1965); the use of multiplate chambers has also been tested (Zielczynski, et al., 1964). While these developments are still experimental and not commercially available, they are promising and, like the LET spectrometer approach discussed above, may lead to a practical dose equivalent meter applicable to any type of radiation.

Work on the basic properties of liquid filled ionization chambers (Blanc, et al., 1963; Ladu, et al., 1967), indicated that, by use of a property of liquid dielectrics, it may be possible to determine the effective quality factor of a mixed radiation field. For this type of detector, the ionization current (I) is said to be a linearly increasing function of the collecting voltage V for sufficiently high values of the applied electric field i.e., $I = R(hV + k)$, where R is the intensity of the radiation and h and k are two constants whose values depend upon the nature of the radiation, the chamber geometry and the dielectric. Plots of I vs V, when extrapolated to zero current, will intersect the V axis at $-(k/h)$ and this point of interception has been shown to depend on quality factor. Dielectrics that have been used are *n*-pentane, and *iso*-octane, the latter being some three times more sensitive than other dielectrics. The system has potentialities but suffers from the inherent instability experienced with current measurements in liquids. Also, further development is required to study the consequences of different energy spectra and dose ranges.

3.4.5. Induced Radioactivity

A major source of personnel exposure around accelerators is from radioactivity induced in such materials as beam pipes, targets, and shielding walls. This exposure problem usually requires portable monitoring instrumentation: the task is to determine the level of activity and the dosage received by personnel at work on and around these "hot" areas. The particles principally involved in this case are the familiar products of nuclide decay, that is, β-, γ-, and x-rays of low energy and well-defined character. Heavily ionizing α-particles have such small range that, if not ingested, they are only a possible hazard to the skin upon contact; moreover, except for nuclear chemistry experiments, there is typically little α-particle activity at accelerators. The typical nuclear beta particles also have short range. The γ-rays, of course, have penetrating power but are not difficult to measure. There are normally no neutron emitters among these sources. Many types of instruments can be used to solve this measurement problem, and a discussion of these can be found elsewhere in this volume (Chapters 5–8).

A few comments are in order, however. First, there are no duty cycle considerations when measuring induced activity; this simplifies the problem considerably. Second, care must be taken to measure activity exactly where the personnel might be exposed to it. Thus, if "hot" beam pipes or targets are being handled, one needs to measure activity at the surface. Third, radiation levels are highest immediately after beam turn-off, so that it may be wise to delay access to the area in question to allow some of the high-activity but short-lived components to decay. In this regard, one ought to carry a portable meter whenever access is made to an area hot enough to be even close to maximum permissible levels, inasmuch as previous measurements may have missed a particular area. Alternatively, an accident (e.g., a small piece of material may have slipped into a beam line) can produce dangerously high levels not previously present. Finally, as mentioned earlier, care must be taken whenever a new or different material is exposed because of the likelihood of new and different activation processes.

3.4.6. Environmental Impact

To the extent that radiation or radionuclides can escape from an accelerator facility, these machines can have a radiological impact on the environment. For very-low-energy machines, shielding can control the amount of prompt radiation reaching the environment. Moreover, radionuclides produced in the course of operation have little potential for reaching the environment if ordinary caution is exercised.

High-energy accelerators, on the other hand, produce very intense radiation fields, so that even with massive, carefully designed shielding, sufficient radiation can escape to require monitoring. Moreover, these fields can produce radioactivity either in the environment or in a position to escape thereto [84]:

> In summary, the four possible radiological environmental impacts due to high energy accelerators, in order of apparent importance are:
>
> 1. The production of "prompt" radiation fields during accelerator operation;
> 2. The production of radionuclides and noxious chemicals (such as ozone and nitric oxide) in the air in the accelerator vault and their subsequent release;
> 3. The production of radionuclides in the soil and ground water near the accelerator, with the possibility of infiltration in the underlying ground water systems and subsequent migration from the accelerator site into wells or other sources of potable water;
> 4. The induction of long-lived radionuclides in accelerator components which may subsequently be "recycled" with the consequent transmission of radionuclides into the general environment; this would be particularly true in the case of valuable metals such as copper, brass, iron, or aluminum.
>
> Thus, we see that the environmental impact of high-energy accelerators is different in character from most types of nuclear installations. The predominant source of population exposure is due to the radiation field produced during operation rather than due to the leakage of radionuclides into the environment.
>
> It is only in the past 5 or 6 years that the relative importance of radiological impact has been placed in this order. For many years, it was believed that the production of radionuclides in the soil and ground water would be the second most important source of impact. The fact that we now understand that the potential for radioactive contamination of ground water is very small is an important step forward in our understanding of accelerator-produced radiation.

High-energy accelerators are intense neutron sources, so that associated environmental monitoring programs are usually devoted largely to neutron monitoring (discussed in Chapter 8). The dose to surrounding populations is dominated by prompt radiation, usually consisting largely of neutrons, although in some cases photons or muons may contribute significantly. Shleien, Tucker, and Johnson [29] note that, for large accelerators a population dose of 50 to 1000 person-rem (0.5 to 10 personsieverts) is incurred for each rem (sievert) of dose equivalent at the fence post. This typically implies on the order of 10 person-rem/yr (0.1 person-sievert/yr) due to a large accelerator, about the same population dose as is typically ascribed to routine operation of a new large nuclear power plant (see Chapter 10). Like nuclear power plants, environmental impact statements are required for new high-energy accelerators in the United States. In addition, existing U.S. high-energy accelerators, which are operated for DOE, file annual environmental monitoring reports (see, for example, Ref. 85).

3.5. RADIOACTIVE MATERIALS

Natural radioactive materials have been discussed in the first section of this chapter, however a very important area, and one which is likely to become increasingly important, is that of man-made radioactive materials. At present these are largely fission fragments although neutron activated atoms are also important. If nuclear power becomes important in our society, the important radioactive materials will be either fission products (if fission reactors are the power source) or activated atoms (if the power sources are fusion reactors). Part III of this volume discusses in detail radioactive materials describing their source, their effects, and methods of detecting and measuring them.

REFERENCES

1. United Nations Scientific Committee on the Effects of Atomic Radiation, "Sources and Effects of Ionizing Radiation," New York, 1977. An earlier UNSCEAR report, "Ionizing Radiation: Levels and Effects," 2 vols., appeared in 1972.
2. National Council on Radiation Protection and Measurements, "Natural Background Radiation in the United States," Report No. 45, Washington, 1975.
3. U.S. Environmental Protection Agency, "Radiological Quality of the Environment in the United States," EPA 520/1-77-009, 1977, published annually.
4. Health and Safety Laboratory, "Workshop on Natural Radiation Environment (March 1972)," Technical memorandum, HASL 269, 1972; "Second Workshop on the Natural Radiation Environment (February 1974)," HASL-287, 1974.
5. J. A. S. Adams and W. M. Lowder, Eds., *Proceedings of the 1963 International Symposium on The Natural Radiation Environment,* Univ. of Chicago Press, Chicago, 1964.
6. J. A. S. Adams, W. M. Lowder, and T. F. Gesell, Eds., *Proceedings of the 1972 International Symposium on The Natural Radiation Environment,* Energy Research and Development Administration Conference Proceedings CONF-720805, 1975.
7. D. T. Oakley, "Natural Radiation Exposure in the United States," U.S. Environmental Protection Agency, ORP/SID 72-1, 1972.
8. D. T. Oakley and A. S. Goldin, "Cosmic Ray Population Exposure," p. 91 of Ref. 6.
9. H. L. Beck, "The Physics of Environmental Gamma Radiation Fields," p. 101 of Ref. 6.
10. H. L. Beck, "Environmental Gamma Radiation from Deposited Fission Product, 1960-1964," *Health Phys.*, **12**, 313, 1966.

11. U.S. Nuclear Regulatory Commission, "Final Generic Environmental Statement on the Use of Recycle Plutonium in Mixed Oxide Fuel in Light Water Cooled Reactors (GESMO)," NUREG-0002, Executive Summary and five volumes, 1976.
12. U.S. Environmental Protection Agency, "Radiation Dose Estimates to Phosphate Industry Personnel," EPA-520/5-76-014, 1976.
13. U.S. Environmental Protection Agency, "Radiation Dose Estimates Due to Air Particulate Emissions from Selected Phosphate Industry Operations," Technical Note ORP/EERF-78-1, 1978.
14. U.S. Environmental Protection Agency, "Radioactivity Distribution in Phosphate Products, by Products, Effluents, and Wastes," Technical Note ORP/CSD-75-3, 1975.
15. U.S. Environmental Protection Agency, "Preliminary Findings: Radon Daughter Levels in Structures Constructed on Reclaimed Florida Phosphate Land," Technical Note ORP/CSD-75-4, 1975.
16. G. D. Case, T. A. Bertolli, J. C. Bodington, T. A. Choy, and A. V. Nero, "Health Effects and Related Standards for Fossil-Fuel and Geothermal Power Plants," Lawrence Berkeley Laboratory Report LBL-5287, 1977.
17. R. J. Budnitz, J. V. Berk, C. D. Hollowell, W. W. Nazaroff, A. V. Nero, and A. H. Rosenfeld, "Human Disease from Radon Exposures: The Impact of Energy Conservation in Buildings," Lawrence Berkeley Laboratory Report LBL-7809, 1978.
18. Health and Safety Laboratory, Fallout Program Quarterly Summary Report Series, published quarterly since 1960.
19. E. B. Fowler, Ed., *Radioactive Fallout, Soils, Plants, Food, and Man,* Elsevier, Amsterdam, 1965.
20. J. L. Kulp, "Radionuclides in Man from Nuclear Tests," in *Radioactive Fallout, Soils, Plants, Food, and Man,* E. B. Fowler, Ed., Elsevier, Amsterdam, 1965, p. 247.
21. Health and Safety Laboratory, "Fallout Program Quarterly Summary Report, April 1, 1971," HASL-242 and Appendix.
22. Health and Safety Laboratory, "Fallout Program Quarterly Summary Report, July 1, 1967," HASL-182.
23. U.S. Environmental Protection Agency, "Estimates of Ionizing Radiation Doses in the United States, 1960-2000," ORP/CSD 72-1, 1972.
24. Advisory Committee on the Biological Effects of Ionizing Radiation, National Academy of Sciences—National Research Council, "The Effects on Populations of Exposure to Low Levels of Ionizing Radiation," 1972.
25. National Council on Radiation Protection and Measurements, "Environmental Radiation Measurements," Report No. 50, 1976.
26. J. N. Gitlin and P. S. Lawrence, "Population Exposure to X-Rays, U.S. 1964, Report on the Public Health Service X-Ray Exposure Study," Public Health Service Publication No. 1519, 1966; "Population Dose from X-Rays, U.S. 1964," Public Health Service Publication No. 2001, 1969.
27. Department of Health, Education and Welfare, "Population Exposure to X-Rays, U.S. 1970, Report on the Public Health Service X-Ray Exposure Study," DHEW(FDA) 73-8047, 1973.
28. Department of Health, Education and Welfare, "Gonad Doses and Genetically Significant Dose from Diagnostic Radiology, U.S. 1964 and 1970," DHEW(FDA) 76-8034, 1976.

29. B. Shleien, T. T. Tucker, and D. W. Johnson, "The Mean Active Bone Marrow Dose to the Adult Population of the United States from Diagnostic Radiology," *Health Phys.*, **34**, 587, 1977.
30. U.S. Environmental Protection Agency, "Radiological Quality of the Environment in the United States," PB-274/229/4BE, (especially Chapter 8), 1977.
31. United Nations Scientific Committee on the Effects of Atomic Radiation, "Sources and Effects of Ionizing Radiation," 1977.
32. National Council on Radiation Protection and Measurements, "Natural Background Radiation in the United States," Report No. 45, 1975.
33. Stanford Research Institute, "Survey of the Use of Radionuclides and Medicine," U.S. Department of Health, Education and Welfare Report BRH/DMRE 70-1, 1970.
34. T. Villafana, J. Scouras, L. Kirland, N. McElroy, and P. Paras, "Aspects of the EMI Computerized Tomography Brain Scanner," *Health Phys.*, **34**, 71, 1978.
35. Interagency Task Force on the Health Effects of Ionizing Radiation, "Report of the Work Groups on Exposure Reduction," U.S. Department of Health, Education and Welfare, 1979; for exposure range, see P. Shrivastava, S. Lynn, and J. Ting, "Exposure to Patient and Personnel in Computed Axial Tomography," *Radiology*, **125**, 411, 1977.
36. R. C. Barrall, L. H. Lanzl, and J. W. Hilbert, "A Survey of Personnel Exposure in Nuclear Medicine," American Association of Physicists in Medicine, 1975.
37. B. Lindell, "Occupational Hazards in X-Ray Analytical Work," *Health Phys.*, **15**, 48, 1968.
38. J. O. Lubenau, J. S. Davis, D. J. McDonald, and T. M. Gerusky, "Analytical X-Ray Hazards: A Continuing Problem," *Health Phys.*, **16**, 739, 1969.
39. National Council on Radiation Protection and Measurements, "Radiation Exposure from Consumer Products and Miscellaneous Sources," NCRP Report No. 56, 1977.
40. U.S. Department of Health, Education and Welfare, "In-Plant Survey of TV Receivers for X-Radiation Emission Capability," Report No. BRH/DEP 70-24, 1970.
41. U.S. Department of Health, Education and Welfare, "X-Ray Emission from Cold-Cathode Gas Discharge Tubes Used for Educational Purposes," Report No. BRH/DEP 70-1, 1970.
42. S. Kinsman, "Second Annual Conference on Radiation Control, Palm Beach, 26–29 April 1970," U.S. Department of Health, Education and Welfare Report No. BRH/ORO 70-5, 1970.
43. U.S. Department of Health, Education and Welfare, "Regulations for the Administration and Enforcement of the Radiation Control for Health and Safety Act of 1968," DHEW (FDA) 79-8035, 1978.
44. National Council on Radiation Protection and Measurements, "Medical X-Ray and Gamma Ray Protection for Energies up to 10 MeV, Equipment Design and Use," NCRP Report No. 33, 1968.
45. National Council on Radiation, "Structural Shielding Design and Evaluation for Medical Use of X-Rays and Gamma Rays of Energies up to 10 MeV," Report No. 49, 1976.
46. National Council on Radiation Protection and Measurements, "Dental X-Ray Protection," NCRP Report No. 35, 1970.

47. National Council on Radiation Protection and Measurements, "Radiation Protection for Medical and Allied Health Personnel," NCRP Report No. 48, 1976.
48. International Commission on Radiological Protection, "Protection of the Patient in X-Ray Diagnosis," ICRP Publication No. 16, Pergammon Press, New York, 1970.
49. National Council on Radiation Protection and Measurements, "Radiation Safety Training Criteria for Industrial Radiography," NCRP Report No. 61, 1978.
50. American National Standards Institute, "General Safety Standards for Installations Using Non-Medical X-Rays and Sealed Gamma-Ray Sources, Energies up to 10 MeV," ANSI N543-1974, NBS Handbook 114, 1975.
51. American National Standards Institute, "Radiation Safety for X-Ray Diffraction and Fluorescence Analysis Equipment," ANSI N43.2-1977, NBS Handbook 111, 1978.
52. T. M. Moore and D. J. McDonald, "Radiation Safety Recommendations for X-Ray Diffraction and Spectrographic Equipment," U.S. Department of Health, Education and Welfare Report MORP 68-4, 1968.
53. J. F. Wochos, N. A. Detorie, and J. R. Cameron, "Patient Exposure from Diagnostic X-Rays: An Analysis of the 1972–1975 NEXT Data," *Health Phys.*, **36**, 127, 1979.
54. J. E. Gray, "Photographic Quality Assurance in Diagnostic Radiology, Nuclear Medicine, and Radiation Therapy," U.S. Department of Health Education and Welfare Publication (FDA) 76-8043, 1976.
55. References on accelerator design and principles of operation include: J. J. Livingood, *Principles of Cyclic Particle Accelerators,* D. Van Nostrand, 1961; M.S. Livingston and J. P. Blewett, *Particle Accelerators,* McGraw-Hill, New York, 1952; E. M. McMillan, "Particle Accelerators," Part XII in *Experimental Nuclear Physics,* Vol. III, E. Segrè, Ed., Wiley, New York, 1959; E. A. Burrill, Sections 2.2.1.1 and 2.2.1.2, and R. Wideroe, Section 2.2.1.3, in *Engineering Compendium on Radiation Shielding,* Vol. 1, Springer-Verlag, Berlin, 1968.
56. "1979 Particle Accelerator Conference-Accelerator Engineering and Technology," San Francisco, March 12–14, 1979, *IEEE Trans. Nucl. Sci.*, **NS-26**(3), Part 1, 1979.
57. "Proceedings of the Fifth Conference on the Applications of Small Accelerators in Research and Industry," North Texas State University, Nov. 6–8, 1978, *IEEE Trans. Nucl. Sci.*, **NS-26**(1), Part 2, 1979.
58. "Proceedings of the Ninth International Conference on High Energy Accelerators," May 2–7, 1974, CONF 740522.
59. "Proceedings of the Second International Conference on Accelerator Dosimetry and Experience," Stanford, California, 1969, CONF-691101.
60. H. W. Patterson and R. H. Thomas, *Accelerator Health Physics,* Academic Press, New York, 1973.
61. International Commission on Radiation Units and Measurements, "Radiation Protection Instrumentation and Its Application," Report No. 20, 1971.
62. American National Standards Institute, "Radiological Safety in the Design and Operation of Particle Accelerators," ANSI N43.1-1969, NBS Handbook 107, 1970.

63. National Council on Radiation Protection and Measurements, "Radiation Protection Design Guidelines for 0.1–100 MeV Particle Accelerator Facilities," Report No. 51, 1977.
64. E. A. Burrill, "The Expanding Use of Particle Accelerators in Research, Medicine, and Industry," p. 1 in Ref. 59.
65. I. L. Morgan, "Impact of Accelerators on Technology," in "1973 Particle Accelerator Conference, San Francisco, March 5–7 1973," *IEEE Trans. Nucl. Sci.*, **NS-20**(3), 36, 1973.
66. U.S. Department of Health, Education and Welfare, "The Use of Electron Linear Accelerators in Medical Radiation Therapy: Physical Characteristics," DHEW (FDA) 76-8027, 1976; see also A. P. Turner, "Regulations, Standards, and Guidelines for the Use of Medical Electron Linear Accelerators," p. 1608 of Ref. 2.
67. U.S. Atomic Energy Act of 1954.
68. E. D. Bailey, "State Registration and Inspection Programs for Particle Accelerators," p. 363 of Fourth Conference on the Scientific and Industrial Applications of Small Accelerators, North Texas State University, Oct. 27–29, 1976, Proceedings 76CH 1175-9 NPS, J. L. Duggan and J. L. Morgan Eds., IEEE, New York, 1976, p. 363.
69. H. W. Patterson and R. H. Thomas, "Radiation Fields," Chapter 2 of Ref. 5.
70. J. Kastner, J. Rose, and F. Shonka, *Science*, **140**, 1100, 1963.
71. S. I. Baker, "Environmental Monitoring Report—Calendar Year 1975," Fermi National Accelerator Laboratory, TID-27020, 1976.
72. J. W. Baum, "Nonlinear Amplifier for use in Mixed Radiation Rem-Responding Radiation Meters," *Health Phys.*, **13**, 775, 1967.
73. J. W. Baum, A. V. Keuhner, and R. L. Chase, "Dose Equivalent Meter Designs Based on Tissue-Equivalent Proportional Counters," *Health Phys.*, **19**, 813, 1970.
74. D. Blanc, J. Mathieu, and L. Torres, "Relation Entre les Courbes Caracteristiques des Chambres D'ionisation Remplies D'hexane Liquide et L'ionisation Specifique du Rayonnement Incident. Application Possible a la Measure du Facteur de Qualité d'un Rayonnement Complexe," *Nucl. Instrum. Meth.*, **27**, 353, 1963.
75. C. H. Distenfeld and A. M. Markoe, "Determination of Quality Factor through the Utilization of a Balanced Tissue Equivalent Ionization Chamber," in *Proceedings of the First Symosium on Accelerator Radiation and Experience,* CONF 651109, p. 181, 1965.
76. M. Ladu, M. Pelliccioni, and M. Roccella, "Determination of Quality Factor in a Mixed Field of Gamma Rays and Neutrons by an Ionization Chamber Filled with a Dielectric liquid, II," *Nucl. Instrum. Meth.*, **53**, 71, 1967.
77. H. H. Rossi and W. Rosenzweig, "Measurements of Neutron Doses as a Function of Linear Energy Transfer," *Rad. Res.*, **2**, 117, 1955.
78. H. H. Rossi and W. Rosenzweig, "A Device for the Measurement of Dose as a Function of Specific Ionization," *Radiology*, **64**, 404, 1955.
79. H. H. Rossi, W. Rosenzweig, M. H. Biavati, L. Goodman, and L. Phillips, "Radiation Protection Surveys at Heavy Particle Accelerators Operating at Energies Beyond Several Hundred Million Electron Volts," *Health Phys.*, **8**, 331, 1962.

80. A. H. Sullivan, "An Approach to a Rem Dosimeter Using Ionization Chambers," CERN Report DI/HP/29, 1964.
81. A. H. Sullivan and J. Baarli, "An Ionization Chamber for the Estimation of the Biological Effectiveness of Radiation," CERN Report 63-17, 1963.
82. M. Zielczynski, "Recombination Method for Determination of Linear Energy Transfer (LET) of Mixed Radiation," in *Neutron Dosimetry,* Vol. 2, International Atomic Energy Agency, Vienna, 1963, p. 397.
83. M. Zielczynski, V. N. Lebedev, and M. I. Salatskaya, "A Device for Determining Relative Biological Radiation Efficiency," *Pribory: Teknika Eksperimenta*, **6**(6), 73, 1964; English translation in *Instrumentation and Experimental Techniques*, **6**, 1217, 1964.
84. R. H. Thomas and A. Rindi, "Radiological Environmental Impact of High-Energy Accelerators," *CRC Critical Reviews in Environmental Control,* 1979, p. 51.
85. A. Hull and J. A. Ash, Eds., "1974 Environmental Monitoring Report," Brookhaven National Laboratory Report ERDA-54, 1975.

4 RADIATION DETECTION METHODS

Late in 1895, Professor Roentgen, working with high voltage electrical discharges in rarified gasses, discovered that a screen coated with barium platinocyanide would light up even when protected from the direct light from the discharge tube by black paper. He named this penetrating radiation x-rays. Thus one can say that a phosphor was the first means of detection for this ionizing radiation. The following spring, Henri Becquerel working with a uranium salt, discovered that it emitted rays that would blacken a light-shielded photographic plate, much in the same way x-rays would. This property of uranium was called radioactivity. The discovery of radioactivity in uranium led to the discovery of radium by Marie Curie. The photographic plate, then, was the second means of detecting ionizing radiation. Soon after, it was discovered that radiation had the power to discharge an electroscope. This then was a third means of detecting ionizing radiation. Two others were discovered before 1910, the spinthariscope, in which flashes were produced by a phosphor as it was bombarded with an alpha particle (the forerunner of the scintillation counter, to be described later) and the Geiger counter.

These early devices employed most of the basic principles used in modern instruments as will become apparent later in this chapter. Before discussing the instruments themselves, the text examines in greater detail the various types of ionizing radiation, and their interactions with matter.

4.1. PROPERTIES OF IONIZING RADIATION

Ionizing radiation can be grouped as particulate rays and electromagnetic rays. Because of the quantum nature of all radiation it becomes difficult to distinguish between the two at high energies [1–4].

4.1.1. Particulate Radiation

Alpha Particles. Alpha (α) particles are helium nuclei ($^4_2He^{++}$) principally emitted by heavy radionuclides such as naturally occurring uranium and thorium and their daughters. The kinetic energy of the alpha particles generally lies between 2 and 7 MeV, and the particles are typically emitted with discrete energies characteristic of the radionuclide.

Beta Particles. Beta (β) particles are high-energy electrons or positrons emitted by certain radionuclides. Beta particles are emitted with a continuous spectrum (see Fig. 4.1), and tables of β-energies commonly refer to the maximum energy of emission characteristic of the particular radionuclide. For many purposes, the approximation of a mean energy equal to one-third of the maximum energy is satisfactory.

Neutrons. Neutrons (n) are uncharged particles. They and protons are the major constituents of nuclei. Neutrons have no Coulomb interaction with atomic electrons or nuclear protons.

1. Thermal neutrons have the same average kinetic energy (and therefore the same temperature) as atoms in the surrounding matter. At room temperature, this corresponds to a Maxwellian distribution with an average kinetic energy of about 0.025 eV.
2. Intermediate neutrons are those with energies from 0.5 eV to 200 keV. Neutrons having energies less than 100 eV are referred to as *slow* neutrons.
3. Fast neutrons are those having energies between 20 keV and 10 MeV.
4. Relativistic neutrons have energies greater than about 20 MeV, so that the mass equivalent of the kinetic energy becomes a significant fraction of the total mass-energy of the neutrons. Relativistic corrections should be applied in analyses of interactions involving this type of neutron.

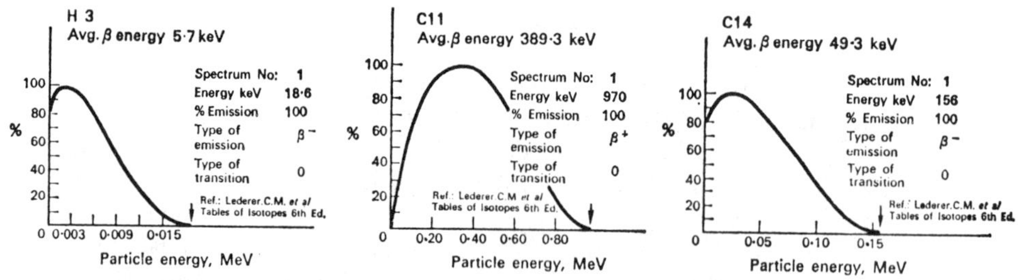

Figure 4.1. The Bragg spectrum and average β-energy of selected isotopes (reproduced from Mantel, Ref. 5, with permission of Pergamon Press © 1972).

4.1.2. Electromagnetic Radiation

Both x- and gamma (γ) rays are electromagnetic radiation of very short wavelength, making them no different than visible light except they are constituted of photons of much higher energy. Gamma rays are emitted with discrete energies characteristic of the radionuclide. x-ray emission is of two types: characteristic radiation (discrete energies) and continuous spectra (such as bremsstrahlung). Characteristic x-rays result from transitions between energy levels of the orbital electron surrounding an atom, whereas γ-rays are produced by transitions from higher to lower energy states within the nucleus of the atom. (Alpha and beta decay also result from transitions in the nucleus but they yield a different radionuclide upon decay.)

4.1.3. Interactions of Ionizing Radiation with Matter

Directly Ionizing Radiations. Included in this category are all charged particles of sufficient energy to produce ionization in matter (e.g., α-particles, heavier ions, and β-particles). Such particles lose energy by interaction with the orbital electrons or nuclei of atoms in the materials they traverse. These collisions may be either elastic (Rutherford scattering) or inelastic. Any or all of the possible collision processes may take place, reducing the energy and altering the direction of the particles. (Nuclear reactions and strong interactions require the particle to overcome the nuclear Coulomb barrier, usually tens of MeV and will be ignored because the energies of the principal radioactivities with which we are concerned are much lower.) Because the electron clouds surrounding atoms are much bigger than atomic nuclei, collisions with electrons are far more probable. Orbital electrons in atoms or molecules may be excited by these collisions resulting in the emission of electromagnetic radiation when the excited electrons subsequently return to their ground states. Alternatively, the atom may be ionized, resulting in the creation of an ion pair. In air an average of 34 eV is expended in the creation of each ion pair. If the charged particle is much heavier than the electron, its direction will not be appreciably altered by any interaction, and it will travel through matter in almost a straight line. The electrons that are liberated (sometimes called delta rays) may have sufficient kinetic energy to induce further ionization or excitation in atoms along their paths.

Because heavy charged particles move in almost straight lines and lose energy in a large number of small steps (allowing fluctuations in energy loss to "average out"), they have definite ranges. The range of such a particle is a function of its initial energy, its mass and charge, and of the properties of the stopping medium. The particle's energy loss per unit path length is given in classical terms as [1–4]:

$$-\frac{dE}{dx} \approx \frac{8\pi z^2 e^4 NZ}{mv^2} \ell n \frac{2mv^2}{E_i},$$

where E = energy of the moving particle,
x = path length,
z = charge of particle,
v = velocity of particle,
m = mass of electron,
e = charge of electron,
N = number atoms/cm^3 in stopping medium,
Z = atomic number of stopping medium,
E_i = average energy required for ionization.

For a given stopping medium this equation can be represented in the following approximate form:

$$-\frac{dE}{dx} \approx \frac{Mz^2}{E^{0.73}},$$

where E = energy of moving particles,
M = mass of moving particles.

Thus, the rate of energy loss increases as the particle slows down, so that the density of ionization along the path increases. However, when its velocity is very low, the particle can capture an electron from the stopping medium. It may lose the electron later, but the average value of z will decrease causing $-dE/dx$ to decrease, and the ionization becomes *less* dense. Figure 4.2, which is called a Bragg curve, shows the density of ionization along the path of an α-particle moving in air.

Although electrons induce ionization and excitation in the stopping media in the same way as heavy charged particles, an electron can lose up to half its energy in a single encounter and may also be deflected considerably. Therefore, because electrons lose energy in a random process and their paths are not straight lines, their

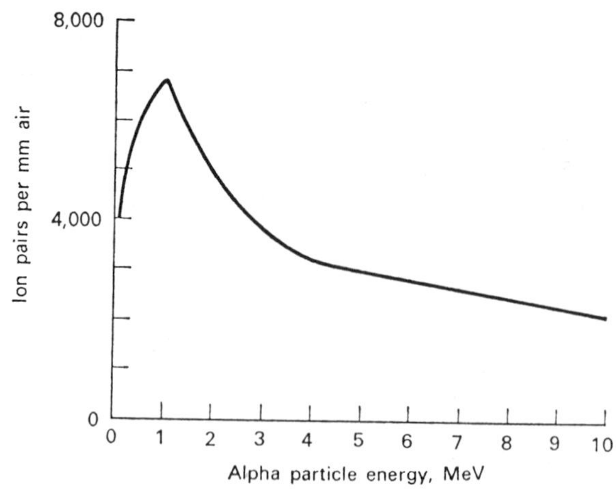

Figure 4.2. Bragg curve of specific ionization by alpha particles in air at standard temperature and pressure (reproduced from Cember, Ref. 6, with permission of Pergamon Press © 1969).

range is not well defined. (Another source of energy loss in a stopping medium is emitted bremsstrahlung, but because it is only important at high energies, it will not be discussed here.)

Because β-decay yields a continuous energy spectrum, even when the maximum energy is high, the spectrum will contain low-energy electrons that a thin absorber can stop. Thus, a thin absorber can cause a reduction in the measurement of the electron flux. For monoenergetic heavy particles, a reduction does not occur until the absorber thickness approaches the range of the particles.

As a positron slows, it may interact with an electron in an annihilation process whereby their total mass energy is converted to electromagnetic radiation.

Indirectly Ionizing Radiation. Some types of electromagnetic radiation and neutrons are indirectly ionizing. These radiations interact with matter by giving rise to secondary radiation that is ionizing. Indirect ionizing radiations lose energy by interaction with electrons or atomic nuclei, and the resulting excited particles collide in turn with the orbital electrons or nuclei.

Electromagnetic Radiation. There are three processes of importance in the interaction of photons with matter: the photoelectric effect, the Compton effect, and pair (positron-electron) production.

The most important energy loss mechanism for low-energy photons (in the range of hundreds of electron volts) is the *photoelectric effect.* In this process, a low-energy photon interacts with a bound electron in an orbital shell of the atom and transfers its entire energy to the electron, which is then ejected from the atom as a photoelectron. This photoelectron then produces most of the positive and negative ions formed. The energy of the photon must be greater than the binding energy of the electron in order for the photoelectric effect to occur. The cross section for this process decreases with increasing photon energy until, at high energy, it no longer plays a significant role. However, the cross section increases with increasing atomic number (Z) of the absorber. Very roughly, the photoelectric absorption probability is proportional to

$$\frac{Z^5}{E^{7/2}} ,$$

where Z = atomic number of stopping medium,
E = energy of photons.

Therefore, it is most important for the absorption of low-energy γ-rays in heavy elements. Although photon energy must be higher than binding energy, the photoelectric effect is most likely to occur when the photon energy is only slightly higher than the binding energy. The most likely electron to be removed is the most tightly bound one in the atom (i.e., the K electron) provided the photon energy is sufficient to remove it from its orbit.

In the *Compton effect,* the incident photon undergoes an elastic collision with a free or loosely bound electron, giving up energy. The probability of this process depends on the number of electrons present in the material. It is the dominant absorption process for γ-rays in the energy region from 0.6 to 2.5 MeV. The Compton cross section decreases monotonically with increasing photon energy.

Pair production occurs for photon energies exceeding 1.02 MeV. Above this threshold the photon may interact with the Coulomb field of the nucleus or, less frequently, with that of an electron to produce a positron-electron pair (the inverse of positron-electron annihilation). The energy of the photon over 1.02 MeV appears as the kinetic energy of the two particles created. The cross section varies approximately as the square of the atomic number (Z^2) and increases with photon energy to become the predominant mode of interaction at about 10 MeV for elements of high Z (e.g., lead) and at much higher energies for low-Z elements (e.g., aluminum). Figure 4.3 illustrates how these three processes interact throughout a large range of photon energies.

Neutrons. Neutrons interact with matter in quite different processes from those of other ionizing radiation, in each case reacting directly with nuclei:

1. Elastic scattering: The neutron shares its initial kinetic energy with the nucleus, which recoils without being excited. The smaller the mass of the nucleus, the larger the fraction of kinetic energy given to it. This process is, for example, dominant for fast neutrons in hydrogenous material.
2. Inelastic scattering: The neutron collides with a nucleus which is left in an excited state, and the neutron emerges with a reduced energy.
3. Capture: The neutron is captured by a nucleus, forming a compound nucleus that is usually excited and emits γ-rays with typically a total energy of about 6 to 18 MeV.
4. Particle emission: The neutron interacts with a nucleus, causing the emission of particles such as protons, neutrons, or alpha particles. Because outgoing

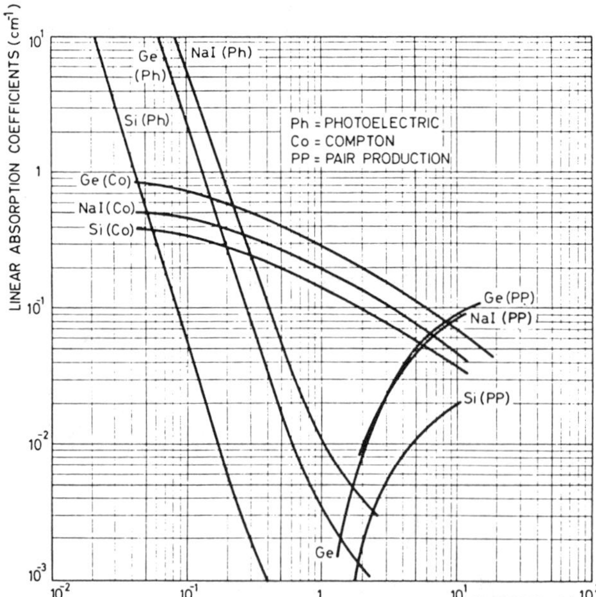

Figure 4.3. Linear absorption coefficients (in cm^{-1}) versus γ-ray energy for silicon, germanium, and sodium iodide (reproduced from Cappellani and Restelli, Ref. 7, with permission of North-Holland Publishing Co. © 1968).

charged particles must overcome the Coulomb barrier, this type of reaction is most probable for light nuclei (with a small Coulomb barrier) and fast neutrons (with a large amount of energy).

5. Fission: The compound nucleus that is formed with the incoming neutron splits into two lighter nuclei (fission fragments) with the emission of, on the average, about two neutrons. Fission in ^{235}U, ^{239}Pu, and ^{233}U may even be induced with thermal neutrons. It may be induced in a larger group of heavy nuclides with fast neutrons.

4.2. BASIC DETECTION METHODS

The detection of ionizing radiation may depend on: (a) the collection of the electrons and ions freed in the ionization process; (b) amplification of this ionization to give a more detectable pulse; (c) the fluorescence of a phosphor that has absorbed radiation; (d) a chemical reaction that is radiation induced; or, more rarely, (e) the measurement of the heat produced by absorption of radiation. This discussion will be restricted primarily to the first three methods because most commercial instruments for environmental and source monitoring are based on them. The ionizing radiation field to be detected and measured may be characterized by four parameters: radiation type, the energy of its photons or particles, their flux or rate of arrival, and the time/spatial distribution. Usually the type of radiation to be detected is known and determines the type of instrument required. The following discussion is based on generic instrument types. For more detailed information, the reader is referred to Chapters 6–9, Part III, and to the extensive literature on radiation detectors [8–19].

4.2.1. Gas-Filled Detectors

Gas-filled detectors are based on one or both of the first two processes just mentioned. A voltage is placed across two physically separated electrodes, and the region between them is filled with an easily ionized gas. Positive ions and electrons formed therein are collected by the electrodes and read as a current or a series of electronic pulses, depending upon the associated electronics. The number of ion electron pairs initially formed (by the interaction processes mentioned previously) is generally proportional to the energy of the radiation.

The three common gas-filled detectors are referred to as ionization chambers, proportional counters, and Geiger-Müller (G-M) tubes. Their difference can be explained with the aid of Fig. 4.4. At voltages below Region I there is competition between the loss of ion pairs by recombination and their collection on the electrodes. With an increasing electric field the drift velocity of the ions increases, decreasing the time available for recombination, thereby causing the fraction of ion pairs lost to become less. In Region 1 (saturation region or ionization chamber region) the recombination loss is negligible and the charge collected is directly related to the number of ion pairs formed. In Region 2, the collected charge is increased by a factor through the phenomenon of gas multiplication: electrons released in the primary ionization are accelerated sufficiently to produce additional ionization adding to the collected charge. At the onset of Region 2, the multiplication factor for a given voltage is independent of the initial ionization, allowing

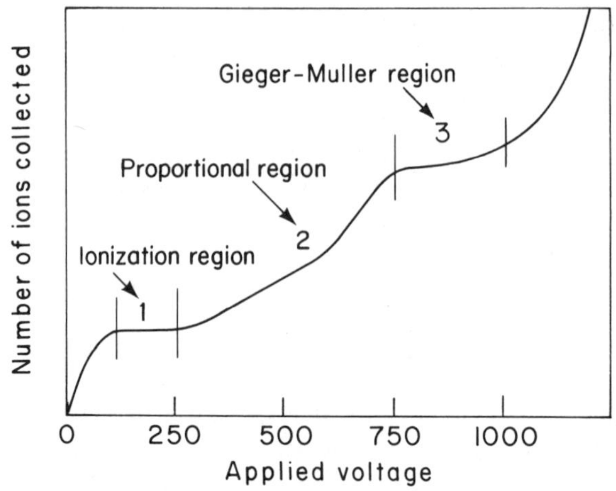

Figure 4.4. Ideal response versus voltage.

preservation of the proportionality of charge collected. This is known as the proportional region. In Region 3, the charge collected is independent of the ionization initiating it. This is the G-M region.

Ionization Chamber. Because many of the principles involved in the ionization chamber are applicable to other detectors, these chambers are discussed in some detail. The principal advantage of this instrument is that the exposure or absorbed dose can be calculated directly from the instrument reading provided the design is optimized and there is rudimentary knowledge of the spectrum of the radiation to which the chamber is exposed. Equally, since the roentgen is the unit of exposure and is defined in terms of the release of total electric charge resulting from the irradiation of a unit volume of air at STP (standard temperature and pressure), the ion chamber method is the most direct method of measuring exposure in these units. A simple ionization chamber is presented in Fig. 4.5. The electrons would be collected on the upper electrode. For ion chambers of a practical size the collection time is on the order of microseconds. For filling the chamber, almost any gas is suitable, so long as it can be ionized.

As a particle passes through the chamber, ion pairs, consisting of positive and negative charges of size e, are formed. Each pair corresponds to a loss in particle energy of approximately 34 eV. If particles pass through the chamber at a rate N, each losing a total energy E, then (positive or negative) charge is formed at the rate

$$Q = \frac{NEe \times 10^6}{34} .$$

For N in particles per second, E in MeV, and e in coulombs, Q will have the dimensions coulombs/sec. If the strength of the electric field is sufficient, it is possible to collect all of the electrons and positive ions so that a current, $I = Q/z$, flows in the circuit. Usually a few hundred volts is sufficient to collect all of the charge.

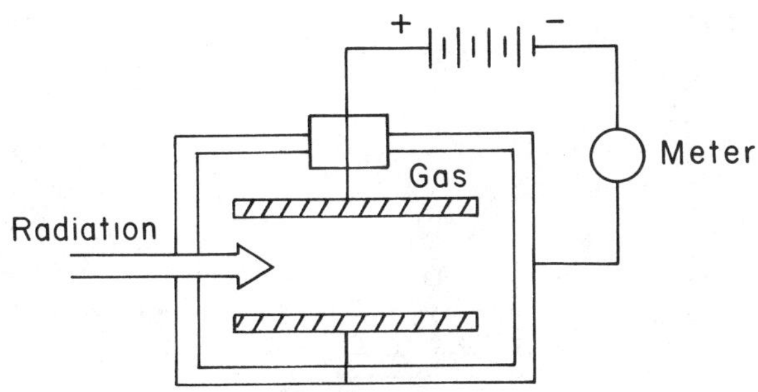

Figure 4.5. Simple ionizing chamber.

An ammeter used to measure the current, as shown in Fig. 4.5, will not count the arrival of each group of electrons corresponding to the instantaneous arrival of particles, because its response is usually too slow. Instead it is calibrated to respond to the effect of many particles by giving a reading that is proportional to the number of particles arriving per second, and to the average energy loss in the gas.

The roentgen is defined in terms of the total ionization produced by secondary electrons formed in a specific volume of air. Because some of these secondary electrons may have ranges of several meters, large and cumbersome apparatus would be required to measure their total ionization and hence the exposure in roentgens. To avoid this problem, air-equivalent, or thimble chambers, have been developed. Their operation is based on the principle that when a tiny cavity, such as a small ionization chamber, is placed in a large homogeneous absorbing medium that is uniformly irradiated, the flux of secondary electrons in the cavity is identical to the flux that existed in the medium before the cavity was introduced. If the chamber gas is air and if the walls are composed of materials having an atomic number near that of air, then the energy loss per gram of air in the chamber will be substantially the same as the energy loss per gram of air at the point where the chamber is located. An ideal chamber can cover only a limited range of energies, because of the factors that enter into the construction and use of such an instrument. For example, the use of "approximately" air-equivalent walls, the use of metallic electrodes, the absorption of low-energy radiation in the chamber walls, and the lack of sufficient wall thickness for electron equilibrium at high energies are some of the factors.

Sometimes, instead of a flux measurement, individual particles need to be measured; this is particularly true for α-particles. If the ionization chamber of Fig. 4.5 is connected to the circuit shown in Fig. 4.6, it may be used to detect the arrival of individual particles.

When operated in this way the electrical action of the system is quite complicated. As soon as the free electrons start to move, a displacement current starts to flow in the circuit. However, because the capacity of the ionization chamber electrodes is high and the resistance R is large (to maximize the voltage input to the amplifier) the time constant of this input circuit is long compared to that of the electrons' transit time, and usually even to that of the ions which travel much slower. Consequently the displacement current builds to a charge and the voltage at the amplifier input will be approximately:

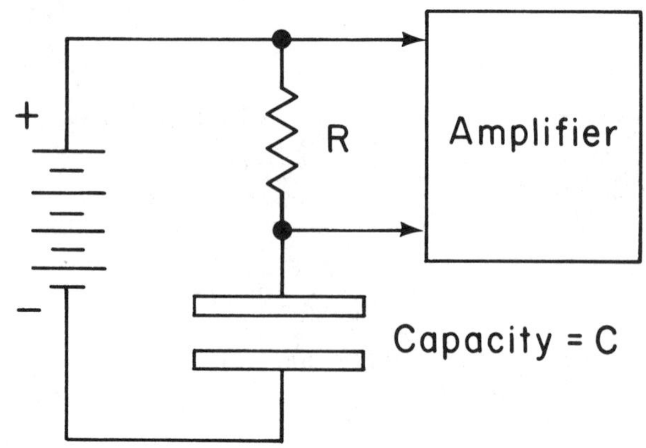

Figure 4.6. Ionization chamber and circuit for detection of individual particles.

$$V = Q/C = e \times N/C \ ,$$

where C = the electrode and wiring capacity,
e = the charge on an electron,
N = the number of ions formed.

If the circuit time constant ($R \times C$) is short enough the capacitor will be essentially discharged before the next ionizing radiation particle enters the ion chamber and the output of the amplifier will be a series of pulses. The rate of occurrence of the pulses is a measure of the intensity of the radiation, while the height of the pulse measures the energy of the particle. If the interval between arriving particles is less than the time constant of the input circuit the output will be a series of steps or saw teeth. The output circuit can be designed to count properly even under these circumstances. The circuit problems for this and many of the other detectors are discussed in a later chapter.

Although the ionization chamber and circuit of Fig. 4.6 are reasonably satisfactory for the detection of charged particles, they are not satisfactory for measurement of the energy of these particles. Therefore a more elaborate apparatus is needed, for example to measure the energies of α-particles from a source placed on the negative electrode. The particles are emitted into the gas in all directions, as shown in Fig. 4.7 (lines A and B). The electrons are quickly removed from the tracks and travel rapidly towards the collecting electrode C. The positive ions remain behind and they induce a small charge on C whose magnitude depends on the distance between C and the "center of gravity" of the positive ion track. The distance is clearly more for A than B. The pulse produced will be greater for track A (because of the smaller induced charge), even if each particle initially produced the same number of ion pairs. To minimize this effect, the grid G is placed as an electrostatic shield between the positive ions and the collector. It is operated at a potential just sufficiently negative to prevent electrons from being attracted to it, but because it consists of very fine wires (0.005-in. diameter) spaced well apart, the electrons can readily pass through it. Thus a grid ionization chamber gives an output pulse that is proportional to the number of ion pairs produced by an alpha particle, and thus to its energy, regardless of the orientation of the track.

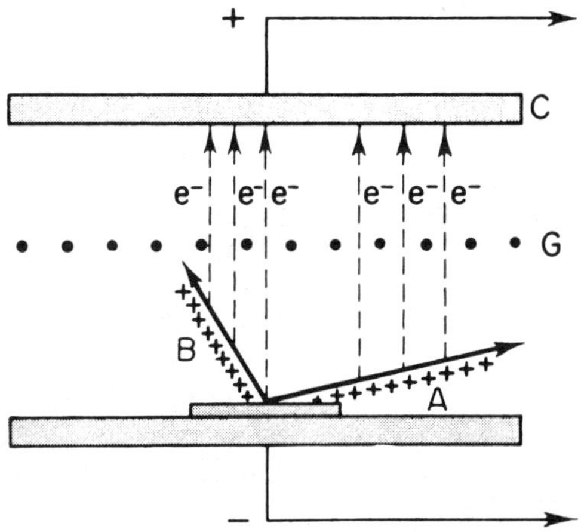

Figure 4.7. Operation of ionization chamber with grid (reproduced from Harvey, Ref. 2, with permission of Prentice-Hall © 1969).

If the counter and amplifier are designed to detect only a fast-rising electronic pulse, electrons must be free to move rapidly through the gas of the counter. In some gases, such as those containing oxygen and halogen vapors, electrons rapidly become attached to neutral molecules to form slow-moving negative ions. These gases must be carefully excluded from the chamber.

The gas most commonly used in an ionization chamber with grid is argon, usually mixed with a few percent of methane or carbon dioxide. The time required for collection of the electrons depends on their drift velocity through the gas toward the collector. Initially fast electrons are slowed down by collisions with the gas atoms, but argon is a poor stopping gas for low-energy electrons. Argon is a tightly bound atom with a large ionization potential, so that there are no low-energy states that can be formed by inelastic collisions with the electrons (the lowest excited state of argon is at 11.5 eV). A quantum mechanical phenomenon called the Ramsauer effect produces a very large increase in the average distance between collisions with argon atoms for electrons with kinetic energies below 10 eV, but in argon it is not easy for electrons to slow down to energies this low. The velocity v with which electrons drift through a gas of pressure P under the influence of an applied electric field of E volts/cm is

$$v = \frac{e E \lambda}{P\sqrt{2m \epsilon}} ,$$

where m = electron mass,
ϵ = average kinetic energy of electrons,
λ = mean free path of electrons.

If a few percent of methane or carbon dioxide is added to argon, inelastic collisions become more probable because these molecules have many low-energy (vibrational and rotational) states that can take up energy in collisions with electrons. The average kinetic energy, ϵ, of the electrons is only one-tenth as great in argon containing the proper concentration of these molecules as it is in pure argon. Thus, because of the Ramsauer effect, λ becomes much larger, and both the reduction in ϵ and the increase in λ cause v to increase. The addition of about 10% of CH_4 or CO_2 is commonly used to decrease the time required for collection of electrons in an argon-filled ionization chamber, which gives a corresponding decrease in the resolving time.

More information on the design of ionization chambers for specific applications and on α- and β-particle detection can be found in the IAEA Technical Reports Series [12].

The electrometers associated with ionization chambers are discussed in Section 4.3.1.

Proportional Counters. The proportional counter has the same basic design features as an ionization chamber, except that the electric field strength near the central electrode is much higher. This high field strength causes the electrons released in ionization to be accelerated toward the central electrode, and to gain sufficient energy to ionize further gas atoms by collisions. Thus, the charge collected, and hence the initial voltage developed, is amplified considerably. This type of detector is particularly valuable because the magnitude of the voltage pulse is proportional to the amount of the primary ionization. Therefore, such a detector can be used to measure the ionizing power of radiations.

In order to get a high electric field the electron-collecting electrode in a gas-filled counter is made of thin wire (typically 0.001 in. in diameter). If the counter has the design shown in Fig. 4.8, the electric field strength E at a distance r from the wire is

$$E = \frac{V}{r \,\ell n(d_1/d_2)} ,$$

where V is the positive voltage (in volts) of the central wire relative to the outer cylinder, and d_1 and d_2 are the counter and wire diameter (in cm). At small distances (r), large values of E are obtained, but the smallest value of r is limited by the wire radius. Hence the field strength is greatest adjacent to a thin wire, even though thick wires give greater values of E for a fixed value of r.

By this arrangement an avalanche of ionization can thus be initiated by even a single electron. Most of the avalanche occurs very close to the wire where the electric potential gradient is high. The size of the output pulse is proportional to the gas multiplication factor, and to the number of primary electrons (and hence to the energy lost by the particle). The gas multiplication factor (maybe 10^2 or more) depends on V, the voltage on the wire, as shown in Fig. 4.9. However it cannot be increased without limit as the gas becomes unstable at very high voltages. If the energy of a particle is to be measured in a proportional counter, the gas multiplica-

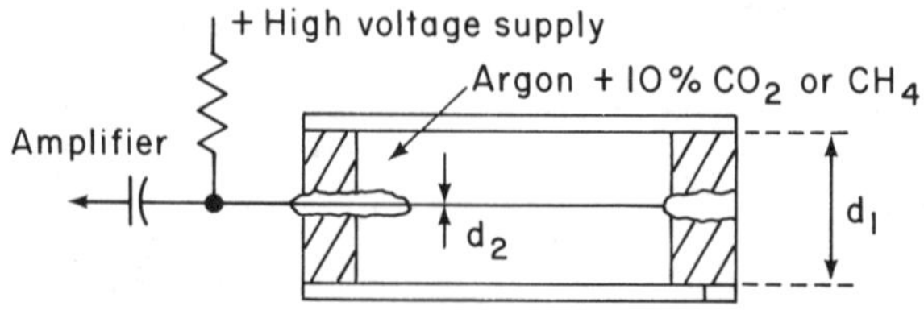

Figure 4.8. Proportional counter (schematic).

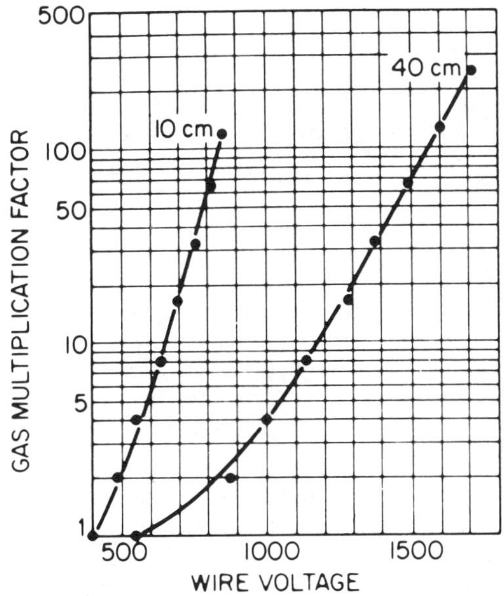

Figure 4.9. Gas multiplication factor as function of applied voltage for argon pressures 10 and 40 cm Hg in counter 0.87 in. in diameter and wire 0.01 in. in diameter (reproduced from Rossi and Staub, Ref. 17).

tion factor must be kept constant by using a very stable high voltage supply. Energy resolution of about 1% can be achieved. This counter is occasionally used for dosimetry but is more commonly used for x-ray spectrometry up to about 20 keV, above which energy the low absorption of the gas causes the efficiency to become very low.

Proportional counters are often used for the detection of α-, β^--, β^+-particles, and electron capture (EC) decays even when measurement of the energy is not required. The sample may be placed inside the counter and the air removed by means of a continuous flow of gas, usually methane. Figure 4.10 shows a typical arrangement with sample external to the counter. Figure 4.11 shows the additional circuits used with such a counter. The discriminator rejects pulses unless they exceed a certain voltage, which can be adjusted. The number of pulses that arrive are recorded by the scaler and register. The counter can detect as many as 10^5 particles per second and since a mechanical system may be far too slow, an electronic register must be used.

The proportional counter can be designed for internal counting (sample in the chamber) of both radioactive gases and solid sources. Special designs have also been made for surface monitoring [12].

Proportional counters are used for the detection of beta or gamma radiation, and can be used for alpha radiation as well, but for the latter they have poorer energy resolution than ion chambers. Proportional counters, while less sensitive than the Geiger counter, will give a good indication of the particle or photon energy. Furthermore, their response to photon energy can be made to resemble that of human tissue. This is advantageous when precise physiological measurements are to be made. However, the proportional counter is as complicated and expensive to use as the scintillation counter (discussed later) and very much less sensitive.

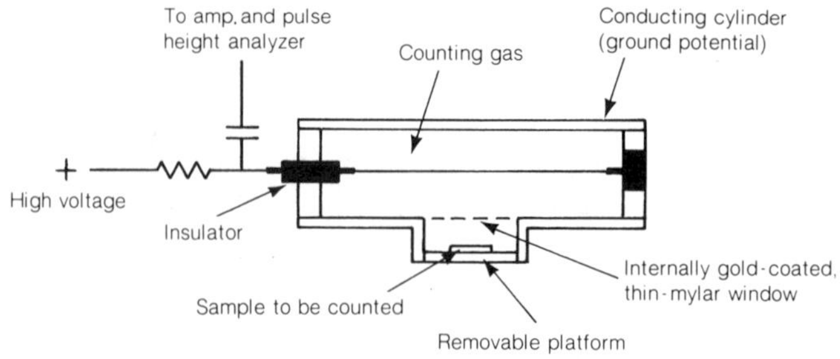

Figure 4.10. Cut-away view of a thin-window proportional counter.

Amplifiers, pulse shaping circuits, pulse-height analyzers and counting systems for these detectors will be discussed in some detail in Section 4.3.

Geiger Counters. The Geiger-Müller tube is the simplest and most widely used detector. It is moderately sensitive, but gives no indication of the energy of the radiation being detected.

The Geiger counter is one of the oldest counting devices, and it has many shortcomings. In this type of detector, the amplification of the voltage pulse described in the case of proportional counters is carried one step further by increasing the potential of the central electrode. The amplification is sufficient to cause a breakdown discharge along the entire length of the central electrode (usually in the form of a wire) being used in the chamber, thus making the size of the ultimate pulse completely independent of the extent of the primary ionization.

As the initial avalanche of electrons approaches the central wire, the positive ions produced in the counter gas move toward the outer wall. When these ions hit the wall, electrons are ejected, resulting in a second avalanche and similarly in successive avalanches. Special precautions called "quenching" are required to stop the avalanche from turning into a continuous discharge.

One way to quench a tube in order to restore it to its original quiescent state is to remove the high voltage momentarily. This can be done electronically and is called external quenching. Another method more commonly used today is to employ internal quenching by mixing a small quantity of a polyatomic gas with the counter gas.

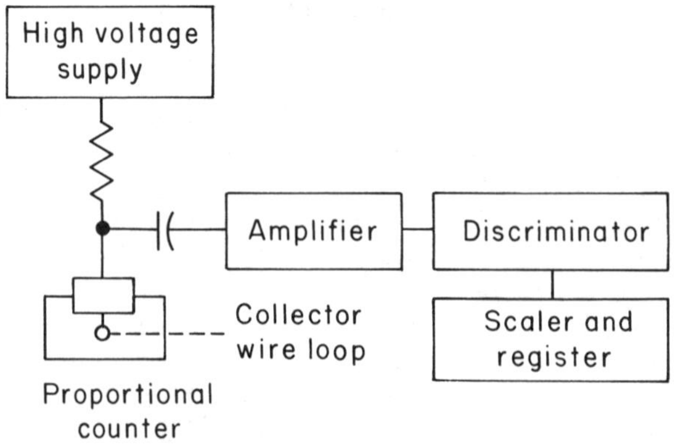

Figure 4.11. Block diagram of typical circuit used with proportional counter.

The positive argon ions become neutralized by removing electrons from the polyatomic molecules. The polyatomic ions then drift relatively slowly away from the wire toward the counter wall, but their energy of recombination with electrons from the counter wall goes into decomposing the polyatomic atom rather than into the emission of new electrons. In this way, the avalanche is prevented from becoming self-sustaining. If a substance like ethanol or butane is used, the tube is said to be organic-quenched. Such a tube has a useful life of about 10^8 or 10^9 counts because the molecules are decomposed irreversibly. Halogen-quenched tubes utilize chlorine, bromine, and their compounds as quenching agents. These tubes have a much longer life because the atoms normally recombine. These counters give larger pulses and operate at considerably lower voltages than organic-quenched G-M tubes.

Once ionization of a G-M tube is initiated, the tube becomes insensitive for a short interval of time. This is due to the time it takes for the positive ions to drift away from the central electrode ($\sim 10^{-3}$ sec). This interval is called the dead time and sets a lower limit to the resolving time. It represents the time during which two or more ionizing particles striking the sensitive portion of the tube will be counted as a single particle. As a result of this phenomenon, the number of counts will be less than the actual number of particles passing into the tube. This could be a potential problem in a very-high radiation field. The difference between the true and observed count is known as the coincidence loss.

After a G-M tube has been exposed to a high-intensity gamma source, the background of the tube may be abnormally high for some time after the source has been removed. The reason for this "memory" is not entirely understood but it is probably due to the formation of activated molecules of gas within the tube.

G-M tubes are not equally sensitive to α-, β-, and γ- radiation. To initiate discharge of the tube, the radiation must first reach the sensitive volume. Alpha particles, being the least penetrating, may be absorbed by the window unless it is very thin. Beta particles are more penetrating, and gamma rays are very penetrating. Thus thicker windows can be tolerated with the latter two types of radiation.

Considering only the radiation reaching the sensitive volume of the tube and assuming zero coincidence loss, it is found that the efficiency of the G-M tube is essentially 100% for alpha particles, nearly 100% for beta particles but only a few percent for gamma radiation. Ironically the property of γ rays which enables them to penetrate the window of the tube so efficiently to reach the sensitive volume, that is, their ability to pass through relatively thick layers of matter without interacting with the matter, now works against their detection. Most of them pass through the sensitive volume of the gas without interacting.

G-M tubes are available in a variety of forms, a common form being the end-window variety. End windows consist of a thin film ($\sim$2 mg/cm^2) of mica, Mylar, or other suitable substance especially selected to retain the gas in the tube while offering minimum interference to the passage of radiation. Obviously this window is extremely delicate and must not be touched with the fingers. Broken windows frequently account for G-M tube casualties in the laboratory. Side windows are normally about ten times thicker than end windows, because the window itself serves to support the end of the tube. Side-window tubes are useless for alpha particle detection, but are quite useful for the detection of energetic betas. On the other hand side-window tubes are more sensitive to gamma radiation than are end-window tubes presenting the same cross section to the radiation. The thicker window increases the probability of interaction of the gamma rays to eject an ionizing electron. Certain G-M tubes, especially designed to be gamma sensitive, have bismuth-coated windows which also serve as cathodes. Steel windows may also be used. For these tubes the gamma efficiency is often as high as 6 or 8%.

Geiger-Müller tubes can be compensated in several ways to make them exhibit a reasonably flat energy response from 60 keV to 2 MeV; Uncompensated G-M tubes do not have anything like a uniform energy response. Examples are shown in Fig. 4.12. If any accurate exposure rate is to be determined with the G-M tube, the energy response of the particular G-M tube as well as the energy distribution of the source must be known. Moreover, it should not be assumed that the radiation detected by the G-M tube is monoenergetic just because the source is monoenergetic. In most applications, measurements will be taken of scattered radiation rather than a direct beam measurement from a monoenergetic source. Under most conditions, the G-M tube will detect radiation that has bounced off everything from concrete to steel beams. Under typical scattered radiation conditions, the actual dose rate, when using an uncompensated G-M tube, can be off by a factor of 3 or 4. However, tube compensation normally eliminates the detector's sensitivity to beta particles.

The advantage of the G-M tube is that it produces a large pulse requiring little further amplification, but the pulse size is independent of the nature of the incident radiation. Geiger counters are cheap and simple. They find their main use as portable instruments in surveying laboratories for possible radioactive contamination, and in mineral prospecting. Their other advantages are high sensitivity, small size, and simple reliable electronic circuitry.

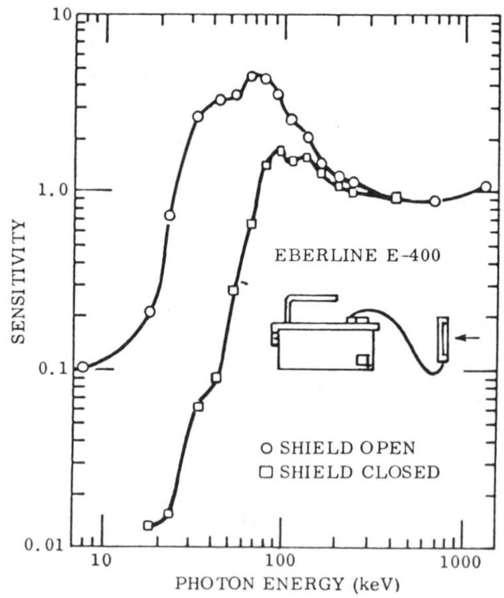

Figure 4.12. Energy response of a Geiger counter system (reproduced from Storm, et al., Ref. 20).

4.2.2. Scintillation Counters

In 1947, researchers in Germany [16] and in the United States [18] described a new type of radiation detector, the scintillation counter. This detector was at least an order of magnitude more sensitive than the Geiger counter, many orders of magnitude faster, and indicated the energy of the detected particle or photon. In spite of its sensitivity the instrument was rugged, flexible, and reliable. For many years it was the dominant detector in the field [19].

The scintillation counter consists of a clear phosphor scintillator, a multiplier phototube, and the required presentation circuitry as illustrated in Fig. 4.13. Incident nuclear radiation excites light from the phosphor. This light releases photoelectrons in the multiplier which, after secondary emission amplification by the cascaded dynodes, appear as a current pulse at the anode. The presentation circuitry analyzes and displays or records the results.

The instrument's high sensitivity depends on the availability of high-absorbing, efficient phosphors and on photomultipliers that can detect very small light flashes. These requirements have been responsible for the development of scintillators of a wide range of sizes and characteristics, and of more sensitive multipliers ranging in cathode diameter from 3/4 to 16 in.

The very high speed of the scintillation counter is a consequence of the extremely short fluorescent lifetimes of certain organic phosphors and the very small time dispersion in multiplier phototubes. With presently available phosphors and multipliers, durations can be measured in the 100-psec range.

The two principal components of the scintillation counter, namely, the multiplier phototube and the scintillator, must be examined in some detail.

The Multiplier Phototube. The photomultiplier provides a means for amplifying the minute photocurrent involved in scintillation counting up to the point where it can be handled by existing amplifiers and circuitry. Any attempt to collect the original photocurrent pulse produced by a scintillation and to amplify this current by means of a thermionic amplifier would be unsuccessful because the size of the pulse is so small that it would be completely lost in amplifier noise. In a photomultiplier, the photoelectrons released from the cathode are directed onto a surface so sensitized that each electron which strikes it causes the release of several secondary electrons. The electrons emitted by this first surface or first dynode are focused onto a

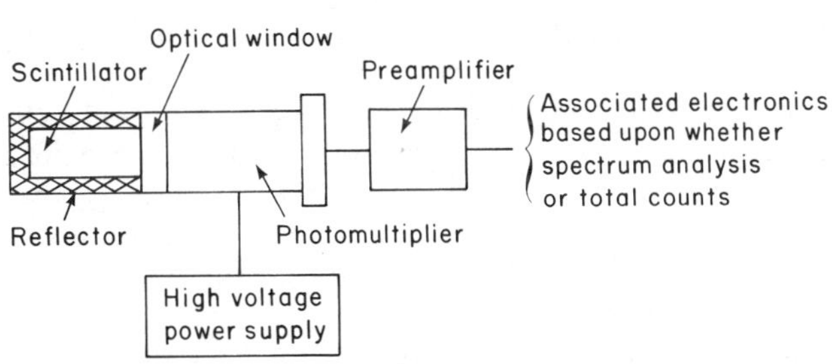

Figure 4.13. Schematic diagram of scintillator detector.

second similar surface where again each incident electron produces a number of secondary electrons. This process is repeated on successive dynodes until the current is built up to the desired level. It is evident that if there are k dynodes, each having a secondary emission ratio σ, a primary photocurrent i_p will produce an output current $i_o = i_p \sigma^k$. As a result of the stimulus provided by the scintillation counter, a number of different photomultipliers have been developed.

The photocathode is of great importance to the photomultiplier. The reasons for this are elaborated in the discussion of energy and time measurements with scintillation counters. The photocathode's importance has prompted a great deal of work aimed toward a better basic understanding of the phenomenon of photoemission and toward the development of improved cathodes.

It is important that the spectral response of the cathode match the light output from the phosphor or scintillator. Any mismatch will result in a loss of photons, which is an irretrievable loss of information or will lead to excess thermionic emission and noise if the response is broader than required.

The problem of directing photoelectrons from the cathode of a photomultiplier into the dynode structure is not negligible. The general philosophy is to design an electron lens system in such a way that the first dynode is at the exit pupil of the system. The solution is complicated by the large spherical aberration found in most simple electrostatic electron lens systems. However, adequate systems have been nearly worked out.

The dynode system itself is an electron optical system for drawing secondary electrons away from one dynode and causing them to impinge upon the next succeeding dynode with a minimum loss of electrons [19]. A number of different electron optical systems have been devised for doing this.

A number of different types of secondary emitting surfaces are used in present commercial multipliers. These include silver-oxygen-cesium, magnesium-oxide-cesium, and cesium antimonide. Special very-high-gain surfaces have been developed using gallium and phosphorous, processed to have a negative electron affinity [21]. These are used for the first dynode of high pulse-height resolution multipliers because it is very important to have the highest possible gain in the first stage to improve resolution [22]. The high voltage required to obtain this high gain makes it impractical to use these stages throughout.

Recently a new type of structure has been developed making use of the high gain that can be obtained with plates made up of a vast array of very fine resistive tubes whose inner surface is responsible for secondary emission. These microchannel plate multipliers are very fast and have excellent pulse-height resolution under certain conditions but at present have other serious limitations.

A discussion of the processes involved in the production of scintillations when high-energy particles are incident on a phosphor can be conveniently divided into two parts: the manner in which the material is excited by the energy of the incident particle, and the mechanism of fluorescence in the excited material.

It is probable that almost all, the conversion of energy of the incident nuclear particles into excitation energy is the result of the production of high-energy electrons in the scintillator which, in turn, produce the excitation. For β-rays, passage of the fast electron through the material raises electrons in the scintillator into excited states from which they decay later to produce fluorescence. For gamma rays, the conversion is less direct. The γ-ray can give up its energy to the material by a Compton collision, photoelectric capture, or pair production. The Compton process is the predominant one for light materials (e.g., organic phosphors) and low-energy γ-rays, that is, up to 2 or 3 MeV. Here the γ-ray photon interacts with a relatively free electron imparting energy to it. Photoelectric absorption occurs when the γ-ray photon interacts with a tightly bound electron. The chance of photoelectric absorption increases rapidly with the atomic number of the atoms of the material, the probability being proportional to the fourth power of the atomic number. Pair production is impossible for γ-ray energies below 1 MeV and is relatively unimportant for γ-ray energies under 5 MeV. For very-high-energy γ-rays, however, it may be one of the major processes involved in absorption of energy.

The excitation of phosphors by heavier particles such as protons, mesons, and α-particles is a good deal more complicated. In all these cases, probably high-energy electrons serve as an intermediate between the excitation of the crystal and the giving up of energy to the material on the part of the incident particle.

Excitation of scintillators by neutrons is a rather special problem since the neutron has no electric field associated with it. For low-energy neutrons, the process usually involves a nuclear reaction between the neutron and an atom of the phosphor with the subsequent production of a β-particle, γ-ray, or fission fragment [23]. For high-energy neutrons, the transfer may be effected by using a scintillator having hydrogenous material, for example, as a constituent and using the knocked-on energy of the protons with which the neutrons collide as a means of the transfer of energy from the neutron to the lattice [24,25]. Lithium iodide activated with europium (particularly ^{6}Li enriched) is a very effective neutron-sensitive scintillator. ^{6}Li has a neutron capture cross section of 950 barns (950×10^{-24} cm^2/nucleus) as compared with values of less than 10 barns for Na and I.

Scintillators can be conveniently divided into two broad classifications, namely, inorganic and organic scintillators. This separation is made because of the very different mechanisms involved in the two types of material. Organic scintillators, in turn, may be subdivided into single crystal phosphors and solution scintillators. Solution scintillators include a solvent in which the primary excitation occurs and an activator or fluorescent solute which emits the light actuating the photomultiplier. The solvent may be a crystal, plastic or liquid, all three having been used very successfully in scintillation counting.

Inorganic Scintillators. The excitation-fluorescence process in inorganic phosphors involves the crystal lattice as a whole rather than individual molecules. For photo conductive phosphors the excitation energy lifts electrons from the valence band to the conduction band. The electrons or holes then move through the crystal until they encounter fluorescent recombination centers. Where a nonphotoconductive phosphor is involved, the excitation probably produces excitons which again move through the crystal until they encounter an appropriate activator center. This process can be quite efficient but, in general, leads to fluorescence having a rather long time constant.

Thallium-activated sodium iodide is by far the most widely used of the inorganic scintillators. This material, which was introduced by R. Hofstadter, can be quite easily grown into large, clear, single crystals [26]. These crystals are now in commercial production and can be obtained in a wide range of sizes and shapes.

Sodium iodide is one of the most efficient phosphors known for scintillation counter work. A good crystal will yield one useful photon per each 30 to 50 electron volts of energy from the incident particle; in other words, a 1-MeV beta particle will generate 20,000 to 30,000 photons. The spectral response output of this phosphor is a band some 800 Å wide centering around 4100 Å. This corresponds very favorably to the spectral region of highest sensitivity of a cesium antimony photocathode. The decay time of this material, though small compared to a good many inorganic phosphors such as zinc sulfide, zinc oxide, and others, is very much longer than that of organic materials. The measured decay time is of the order of 200 or 300 nsec.

One of the important attributes of NaI for γ-ray detection is its very high density. A relatively small amount of the sodium iodide will completely absorb γ-rays in the 1/2 to 1 MeV range. Furthermore, the high atomic number of iodine means that a fairly large percentage of the energy given up to the crystal by a γ-ray will be in the form of photoelectric excitation. This, in turn, leads to light flashes which are strictly proportional to the energy of the incident photon. As discussed later, sodium iodide is for this reason very frequently used in γ-ray spectrometers.

Sodium iodide, however, does have two disadvantages. The first is that its optical index of refraction is very high, being of the order of 1.8. The problem of making efficient optical coupling between the photocathode and the scintillator is therefore quite difficult. It is a common practice to cement the crystal to a glass or lucite "lightpipe" which, in turn, is cemented to the multiplier cathode. Even here the reflection loss at the crystal lightpiper interface is severe. In order to conserve some of the light lost by the first reflection, the sides and end of the crystal are coated with a highly reflecting matte finish. The second difficulty is that the material is chemically unstable in the presence of water vapor. Even when the humidity is quite low, the material cannot be exposed to air for more than a very short time without risk of damage. The moisture causes the dissociation of iodine which colors the crystal slightly yellow. Since the emission is entirely short-wavelength blue light, even a small amount of yellow discoloration makes the material quite opaque to its fluorescent radiation. Techniques, however, have been developed for mounting these crystals in air-tight containers sealed with a glass or plastic disk. A properly mounted commercial crystal will retain its efficiency satisfactorily for a long time.

A large number of other inorganic crystals have been studied as scintillators. Cadmium and calcium tungstates may offer interesting possibilities. Both of these have a high density and contain a high atomic number constituent [27]. Measurements to date indicate that although their scintillation efficiency is high, the time constant of the materials is very long, which make them rather unsatisfactory for scintillation counter use. The difficulty of growing these crystals is rather formidable because they are very refractory. More experimental work on the preparation and purification of materials must be carried out before the question can be answered as to whether the long time constant is fundamental to the materials. Cesium iodide has some interesting properties as a scintillator [28]. It is a good deal faster than sodium iodide. In fact, its speed is such as to indicate that a different mechanism may be involved. The high atomic number of its two components enhances its photoelectric excitation and gives it a high absorption coefficient for γ-rays.

The lithium halides have been investigated as possible scintillators for thermal neutrons. Here the excitation depends upon a nuclear interaction between lithium and slow neutrons.

Inorganic phosphors have been used in scintillators for the detection of alpha particles [29]. Here the penetration is very small and the specific ionization is large. Where high speed is not necessary, screens formed of a rather thin layer of silver-activated zinc sulfide have given very good performance.

Organic Scintillators. Fluorescent processes in organic scintillators are quite different from those in the inorganic materials discussed. Most of the organic phosphors contain in their molecules carbon rings having conjugated double bonds involving π electrons. The molecules in organic crystals are rather loosely held together so that bonding between molecules has little effect on the emission spectra of the material.

As with inorganic materials, the excitation may occur at one point in the crystal and be propagated to another molecule at some distance away where the fluorescence is actually emitted. The excitation transfer in organic materials appears to be due to photons of considerably shorter wavelength than the fluorescent photons or by some form of intermolecular coupling. In either case, the velocity of propagation is essentially that of light and, as a consequence, the fluorescent decay times are very short. The efficiency of some of the organic phosphors is fairly high but is lower than that of sodium iodide. Their reasonably high efficiency coupled with the extremely short decay time makes them very valuable scintillators wherever time measurements are involved. Because of their importance, these phosphors have been investigated extensively by such workers as Kallmann [30], Birks [31], Hayes, Williams, and Rigers [32], Liebson, Bishop, and Elliot [33], Reynolds [34], Swank [35], and others. This work continues to provide a fairly good fundamental and practical understanding of this type of scintillator.

Organic scintillators may be in the form of single crystals or in the form of an activator solute in a liquid or plastic solvent. Both play important roles in scintillation counting, the former being the more efficient but with a longer decay time. Crystals are used for small and moderate sized, fairly fast, efficient counters, whereas solutions are used for the ultimate in speed and for very large counters. Before proceeding with an account of practical organic scintillators, it would be well to review some of the hypotheses concerning the physical mechanisms in solution scintillators.

One of the characteristics of these scintillators is that they show a relatively high efficiency when only a minute amount of the fluorescent solute is present. An estimate shows immediately that, if the fluorescence is the result of interaction of the nuclear particle with the solute molecules, their quantum efficiency must be far in excess of 100% [35]. This is, of course, contrary to all expectation and is not consistent with their behavior otherwise. Consequently, it is necessary to assume that the excitation is the result of the interaction between the nuclear particle and the solvent molecules and that this excitation is transferred to the solute molecule where it is converted into visible light. The mechanism of the transfer of energy from the solvent molecules to the solute has been studied extensively by Kallmann, Swank, Birks, and others. There seems to be little doubt that photon exchange is in some measure involved. There is some question as to whether the entire energy transfer can be accounted for on the basis of photon transport, and some workers feel that at least part of the energy is transferred by some form of intermolecular coupling. However, evidence is accumulating, indicating that the whole effect can be accounted for on the basis of photon exchange.

Birks has discussed this matter in some detail and shows that all of the effects which have been observed in solution phosphors can be accounted for on the basis of photon exchange if it is assumed that at least a fraction of the photons are of an extremely short wavelength (i.e., in the neighborhood of 100 Å) where the penetration through the solvent is extremely small [36]. The mechanism postulated is very similar to that used by Umberger and LaMer [37] to explain the fluorescent behavior of fluoresceine solutions. In this model, short-wavelength ultraviolet photons are emitted by the solvent as a result of the excitation from the nuclear particle. Each photon is absorbed by the solvent after traveling a short distance through it and then, after a short time delay, is reemitted. This process is repeated until eventually the photon excites an activator solute molecule. The return of the activator molecule to ground state gives rise to the longer-wavelength photon, which is one of the photons making up the scintillation.

The quantum efficiency of excitation and reemission by the solvent is less than unity, and also each reemission has a finite decay time. Consequently, when the cascade process is repeated many times, the efficiency of the scintillator decreases and its time constant increases. The relationship between scintillator efficiency and decay time is given by Birks [38]:

$$\log S = \log A + Bt_{\mathrm{I}} \ ,$$

$$B = \log q_0/t_0 \ ,$$

where S = pulse height,
q_0 = reemission quantum efficiency,
t_0 = reemission time,
t_{I} = decay time.

Birks has used this relationship to explain the variation of scintillation decay time with temperature for certain organic scintillators as determined by Liebson [39]. (The decay and reemission times are a function of temperature).

The efficiency of a scintillator increases as the concentration of activator solute is increased up to a certain point. During this time, the time constant of the material decreases. Beyond this point, the efficiency of the scintillator decreases as the concentration of activator solute is increased. In this range of concentration, the time constant remains essentially constant. Figure 4.14 illustrates this behavior for tetraphenylbutadiene in polystyrene [40]. The falling off of the efficiency with high concentration of the solute is probably the result of absorption of the visible photons by the solute with little or no reemission.

Birks' assumption of photons of two different energies involved in the radiative transfer is adequate to explain the results obtained by Swank and Buck [40] in their experiment comparing the variation of scintillation efficiency with concentration for tetraphenylbutadiene in polystyrene. It showed an abrupt saturation when the solution phosphor was covered with a layer of polystyrene impervious to the exciting alpha particles, whereas when the alpha particles were allowed to strike the solution scintillator directly, the saturation was much more gradual and at a higher concentration of the solute. In this instance, even the thin layer of pure polystyrene was sufficient to absorb the short wave component of the transfer photons.

It is very probable that this same photon transfer mechanism occurs in single crystal organic phosphors. Evidence for this is the variation of fluorescent decay time for ultraviolet excitation that has been shown by Liebson and co-workers [33] to be invariably shorter than the decay time for gamma-stimulated fluorescence.

Single Crystal Organic Scintillators. Single crystal anthracene is a widely used organic crystal scintillator. The material forms a monoclinic crystal having a specific gravity of 1.25 and an index of refraction of 1.5948. The melting point of

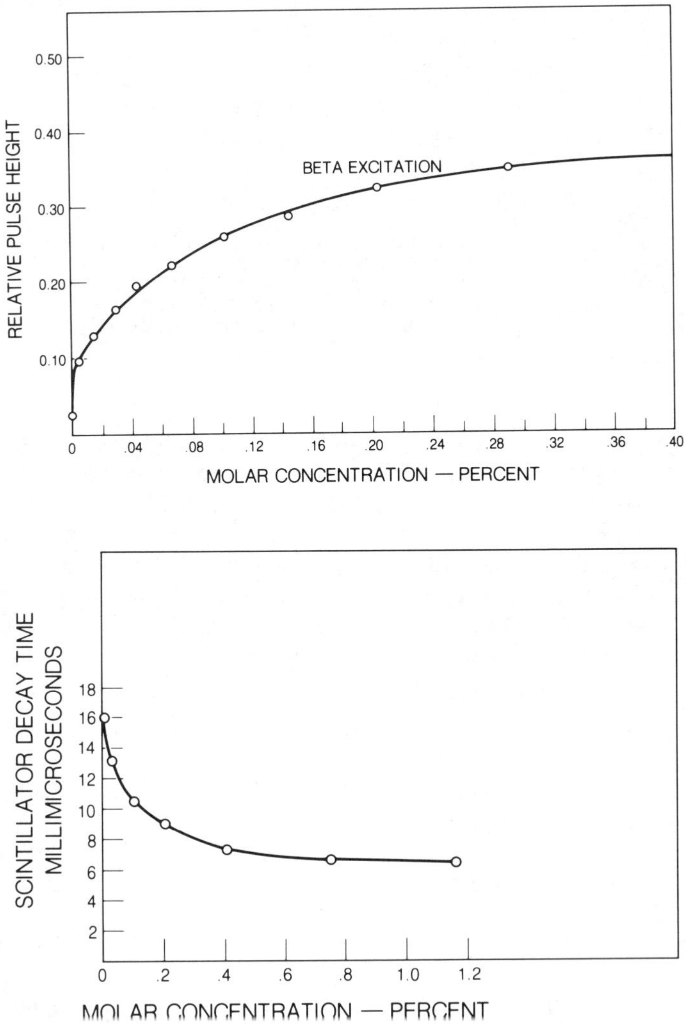

Figure 4.14. Response and time constant of tetraphenylbutadiene in polystyrene (reproduced from Swank and Buck, Ref. 40, with permission of Physical Reviews © 1953.

the material is high enough to be practical for most scintillation counter applications but low enough so that the material can be easily melted for crystal growing. The growing procedure most commonly used is the Stockbarger method, the material being sealed in a vacuum crucible. If care is taken with the purification and cleanliness of the materials, large water-clear crystals can be readily grown. The efficiency of this material for β-ray detection is about 40 to 50% that of sodium iodide. Its decay time is about 36 nsec. The decay is temperature dependent, and Fig. 4.15 shows a variation of decay time with temperature as measured by Liebson. The emission spectrum is a series of bands centering around 4400 Å. It is, therefore, well suited to the spectral response of a cesium antimony activated multiplier phototube.

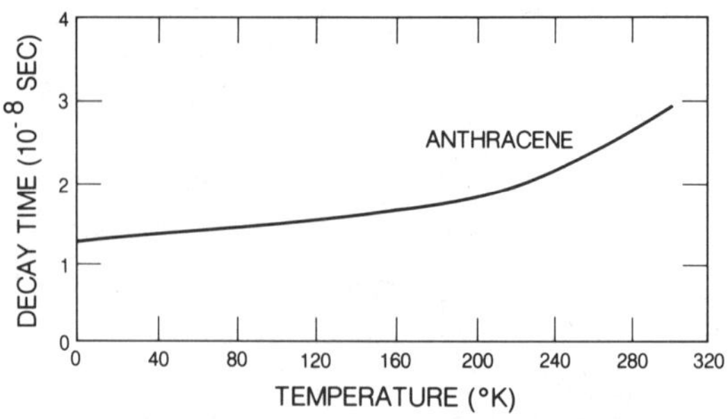

Figure 4.15. Temperature variation of decay time (reproduced from Liebson, Ref. 39).

A second organic crystal that finds wide application in scintillation counting is *trans*-stilbene. Again, it is a material which can be easily crystallized into large water-clear masses. The scintillation efficiency of stilbene is considerably lower than that of anthracene, but for many applications this is compensated for by its very short decay time. The efficiency of *trans*-stilbene relative to anthracene is about 30%, while its decay time is of the order of 6 nsec at room temperature. As with anthracene, the decay time decreases as its temperature is lowered. Because of its high speed, it has been widely used for coincidence work.

Solution Scintillators. High-energy particle physics research, both on cosmic rays and on particles from high-energy machines, frequently requires scintillators occupying many cubic feet. Similarly, there are a number of problems in biophysics where it is desirable to have very large scintillators. It is both expensive and impractical to grow crystals large enough for such applications. However, solution scintillators appear to be a very practical alternative and are coming into wide use. Kallmann and Furst at New York University [41] were among the first to describe a number of types of this interesting scintillator. He and other workers have found many organic liquid solvents which will serve as the scintillator base and an almost equal number of phosphor activators. Toluene, xylene, and phenylcyclohexane have all been used very effectively as the solvent, and terphenyl, diphenyloxazole, and tetraphenylbutadiene are all effective as solutes. The efficiency of a solution scintillator can frequently be improved by the addition of a so-called wavelength shifter. For example, a liquid scintillator with phenylcyclohexane as the solvent and 3 g/liter terphenyl as the solute plus 0.01 g/liter diphenylhexatriene as wavelength shifter yields about 30% greater pulse height than does the material without the wavelength shifter. However, the use of the wavelength shifter decreases the speed of the scintillator in almost every case.

As has already been mentioned, rather than being liquid the solvent may be a plastic with a solute dissolved in it. Polyvinyltoluene and polystyrene are very effective plastic solvents. Figure 4.16 shows the pulse height as a function of concentration for a number of solutes in polystyrene [35]. In polyvinyltoluene, the response is in general similar but slightly higher. These plastic phosphors have a very great practical value because they can be made into rigid masses of very clear material of any size or shape desired.

A summary of the more important scintillators together with the wavelength of their peak emission, relative efficiency to β-ray excitation and fluorescent decay time is given in Table 4.1.

Further information on the amplifiers and analyzers used with scintillation counter spectrometers is to be found in Sections 4.3.2 and 4.3.3.

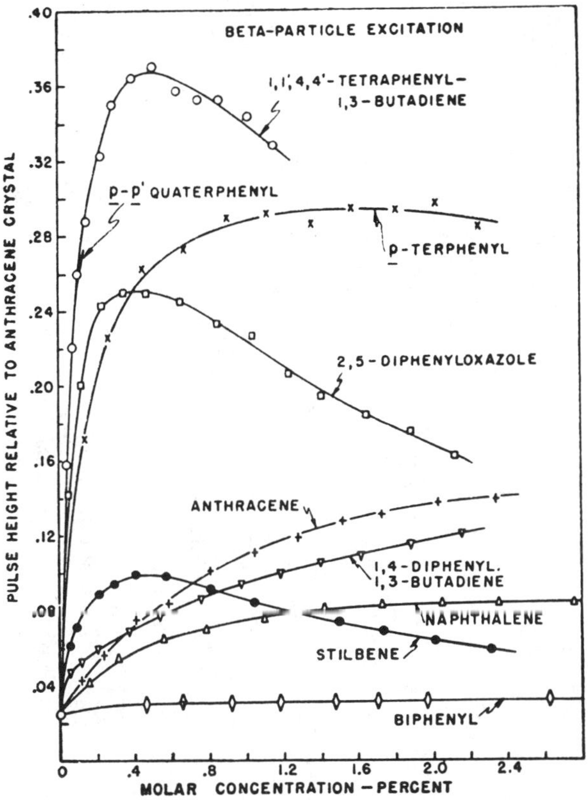

Figure 4.16. Response of various solutes in polystyrene (reproduced from Swank, Ref. 35).

Table 4.1. Scintillator Characteristics

Scintillator	Density	Index of Refraction	Spectral Maximum (Å)	Time Constant (nsec)	Efficiency
Sodium iodide (NaI:Tl)	3.67	1.7745	4100	250	100
Cadmium tungstate	7.90	2.2–2.3	5200	10^3	100
Anthracene	1.25	1.59	4400	36	48
trans-Stilbene	1.16	1.622	4100	6	28
Xylene + terphenyl + diphenylhexatriene	0.86	1.500	4500	5	23
Xylene + terphenyl	0.86	1.500	4000	3	16
Polyvinyltoluene + terphenyl + diphenylstilbene			4400	5	23
Polystyrene + tetraphenyl-butadiene	1.06	1.595	4000	3	17

4.2.3. Semiconductor Detectors

In terms of solid-state theory when atoms group together to form a solid the energy levels of the atoms coalesce to form bands. The ground-state levels form a filled band, generally referred to as the valence band. Because it is completely filled with electrons, electrons cannot move in it. In semiconductors and insulators, the next level, corresponding to the first excited state, forms a conduction band above the valence band through which electrons can travel. If the valence band and the conduction band touch, electrons easily go into the latter where they can move under the influence of an electric field. Such a material is a metal, a good conductor of electricity. If on the other hand the two bands are separated by an energy gap of several electron volts, electrons will not enter the conduction band and the material is an insulator. There is a third class of material where the conduction band is separated from the valence band by only a small energy gap, small enough so that some electrons can be excited from the valence band into the conduction band by room temperature thermal energy. Such a material is known as a semiconductor. At present two semiconductors, germanium and silicon, have been found most useful as radiation detectors. Silicon has a band gap of 1.1 eV and has a very high resistance at room temperature. Germanium, with a band gap of 0.7 eV, has a low resistivity (18 ohms/cm) at room temperature; however, when cooled its resistance becomes very high.

With appropriate electrical contacts and a voltage applied between them, a block of pure silicon or of cooled germanium will conduct electricity when exposed to ionizing radiation [42]. The ionizing radiation gives up its energy to electrons in the valence band, exciting them into the conduction band where they can move under the influence of the applied voltage. In principle, a semiconductor detector can be referred to as a solid-state ionization chamber. The charge carriers produced by the ionizing radiation instead of being ion pairs are electron-hole pairs. The electrical signal is obtained by separating and collecting the electrons and holes, whose numbers are proportional to the energy deposited by the incident radiation in the sensitive region of the detector.

The problem in connection with the use of these materials (or indeed any semiconductor) as radiation detectors is that of minimizing the unwanted charge carriers. Impurity atoms with electrons near or in the conduction band release such carriers into the conduction band. Similarly those with vacant levels near the valence band can accept electrons from this band leaving mobile holes that serve as charge carriers. Such foreign atoms are known as *n*- and *p*-type impurities, respectively. A second consideration, and one which greatly limits the choice of suitable materials is that the mobility and lifetime of charge carriers must be high or else the carriers freed by radiation cannot conduct much current.

Thin detectors can be made by forming a sharp *p-n* junction near the surface of a disk of the material by using a semiconductor doped with an *n*-type impurity and then doping the surface layer with a *p*-type impurity. If a voltage is applied with such a polarity as to try to draw more electrons from the *n*-type material and holes from the *p*-type, there will be a thin region at the junction of the two that is depleted of unwanted charges. In other words the junction is back-biased and nonconducting. Ionizing radiation absorbed in their region can excite hole-electron pairs which the applied voltage will separate, producing an electric current pulse. (Similarly a *p*-type semiconductor may have an *n*-type surface layer applied.)

Also, thin detectors can be made by applying a thin gold layer to *n*-type material, the gold layer acting like the *p*-type doping just described.

For thicker detectors a process has been worked out where acceptor (*p*-type) impurities in lightly doped germanium or silicon can be compensated by drifting (diffusing) Li ions through the material under appropriate conditions of electric field and temperature. This procedure automatically compensates the acceptor producing a drifted region having essentially zero net carriers. Silicon detectors up to about 5 mm thick can be formed in this way. Because of the low density and low atomic number of silicon, detectors of this thickness are not very satisfactory for γ-ray detection, but they are very useful for other types of radiation.

Germanium detectors up to 15 mm thick can be made by lithium drifting which are excellent γ-ray detectors. However they have the disadvantage not only of having to be cooled to liquid nitrogen temperature when used, but also they must be cooled in storage to prevent the lithium from back-diffusing.

Recently it has been possible to grow germanium crystals of sufficient purity that lithium drifting is not required. This is a major step forward. The material can be handled and stored at room temperature but still must be cooled when operated. However, even germanium, which is the most satisfactory semiconductor material available at present for γ-ray spectrometry, has too low an atomic number and density to be optimal for γ-ray detection.

Active work is being done on such semiconductor materials as mercury iodide and cadmium telluride, and promising results have been obtained.

Regardless of the particular semiconductor detector, general comparisons can be made with a gas ionization chamber:

1. The energy required to form an electron-hole pair in silicon is about 3.7 eV and for germanium 2.96 eV compared with 34 eV per ion pair in air, making possible greater energy resolution in spectrometry.
2. The higher density of the semiconductor detector material makes possible higher detection efficiencies than in a gas counter; the larger atomic number also helps.

To recap, detectors as discussed above can be formed with depletion depths up to 2 mm and can be made with very thin windows, making them suitable for the spectrometry of low-energy β- and α-radiation. Typically the detection efficiencies of surface barrier detectors are 100% for α-particles, without rejection. Lithium-drifted detectors can be made with a large volume ($\sim$100 cm^3) making them suitable for γ-ray and x-ray spectroscopy. In addition, hyperpure Ge detectors are available for photon spectrometry from about 1 to several MeV. For spectrometric purposes, the energy resolution of these detectors is an important advantage.

More information on the electronics associated with semiconductor detectors is given in Sections 4.3.2 and 4.3.3.

For a more in-depth discussion of nuclear radiation detection with semiconductors see the book by Bertollini and Coche [14] and a more recent article by Goulding and Pehl [42].

4.2.4. Thermoluminescence Dosimeter

In contrast to gas counters, scintillation, and semiconductor detectors, which are devices that give an immediate indication of the presence of radiation, the devices that use thermoluminescence integrate the effects of all the doses received and only produce a signal when heat is applied. Basically, in thermoluminescent (TL) materials, the electrons and holes produced by radiation are trapped and later released by heating to a temperature of a few hundred degrees centigrade. During the heating process light is emitted and detected with a photomultiplier tube. Similarly, in thermally stimulated electron emission (TSEE) the heating process releases electrons from the surface of the material which can be measured by a proportional counter. Only TL dosimeters are discussed here.

Exposing a thermoluminescent phosphor to radiation causes electrons to be raised from the valence band into the conduction band, and holes are left in the valence band. These electrons (and holes) can move through the crystal lattice until they either return to the valence band or are captured by lattice imperfections. Those electrons that are caught in shallow traps (imperfections) are returned to the conduction band fairly rapidly by thermal effects at room temperature, whereas those in the deeper traps require considerably more energy to escape and only do so when the material is heated. These released electrons return to the ground state with the emission of visible light. It is the deep traps that are least affected by a change in ambient conditions. Because a single rapid rise in temperature does not generally empty all the traps, it is normally necessary to anneal the TL material before reusing.

The TL material may be in the form of a powder, crystal, or polycrystalline disc. The readout instrument for TL material consists of a heater with a thermocouple attached to measure the temperature, a photomultiplier tube with perhaps a filter to shift or remove certain wavelengths of the radiation given off, an amplifier, and generally an *x-y* plotter. A curve plotted with the *x*-axis as the temperature measured by the thermocouple and the output current from the photomultiplier displayed on the *y*-axis is known as a glow curve and is illustrated by Fig. 4.17. For dosimetry purposes, a fixed heating cycle can be used and the output of the photomultiplier tube converted to a digital signal which is integrated between suitable points on the glow curve.

In dosimetry, the user should be aware of the response of the TL material to different energies of radiation. For instance, below 50 keV LiF exhibits an enhancement of 50% in response. The dynamic range of response is also important; for LiF it is from 10 mrad (100μGy) to ~1 krad (10 Gy). The response should be independent of dose rate, but in any case should be known. For LiF, the response is independent of the dose rate up to at least 10^{11} rads/sec (1 GGy/sec). Above an integrated total dose of about 100 rads (1 Gy), the response is supralinear, and it starts to saturate at a limit of 10^5 rads (1 kGy), beyond which radiation damage results in a loss of reading. It is also important to know how the TL material responds to different radiations (gamma, neutron, etc.).

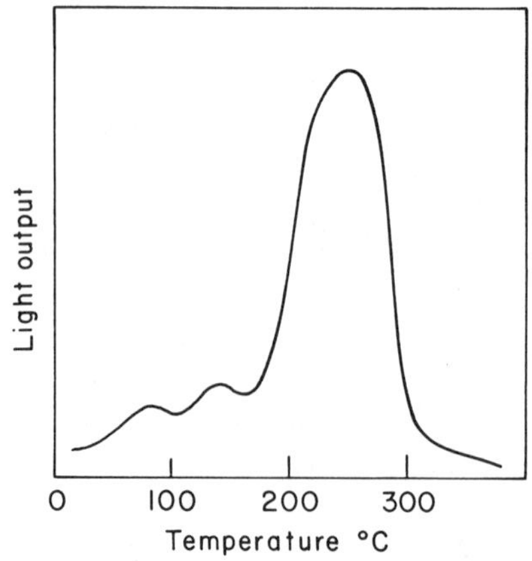

Figure 4.17. Typical glow curve for TLD.

4.2.5. Photographic Detection

The latent image of the track from ionizing radiation traversing a silver halide emulsion is relatively permanent. Thus when the film is developed a record is obtained. Depending on the technique used, a photographic emulsion can be used as a single event detector by recording the track of a single ionizing particle or can be used to locate sites of radioactivity on a surface. Its most prominent use in dosimetry is as a film badge where it is used to integrate many events over a period of time, after which the change in optical density is measured.

An important property of the emulsion is its chemistry and grain size, since this determines how much and what type of radiation is measured. The user should, as always, be aware of the energy response, dose-rate dependence, the dynamic range, and any radiation type dependence to response.

4.2.6. Solid-State Track Detectors

Charged particles impinging on an object such as a film of a high polymer or glass plate produce trails of damage which can be chemically etched into tracks that can be examined individually or counted with an optical microscope or by spark counting. Since these detectors find primary use for measuring heavy charged particles, fission products, and other ions from accelerators or in cosmic radiation, they will not be discussed here. Solid-state track detectors are discussed in some detail by Granzer, Paretzke, and Schopper [15].

4.3. ASSOCIATED ELECTRONICS

As discussed earlier, a wide variety of detectors may be used to convert radiation into electrical signals which may then be processed to provide an appropriate measure of the radiation. Much of the associated electronics used to process the signals is the same, independent of the type of detector, although the details may differ. For example, the large signals associated with Geiger counters mean that little or no electronic amplification is needed, whereas electronic amplification of signals is crucial in using ionization chambers, or semiconductor detectors. Also, for example, Geiger counters produce fixed-amplitude pulses independent of the energy of the events producing the Geiger discharges. Therefore, the associated electronics is designed only to measure the number of counts in a predetermined time or to measure the counting rate. This contrasts with many applications of scintillation detectors, ionization chambers, and semiconductor detectors where individual pulses are both analyzed and counted to provide an energy spectrum of the measured radiation.

Two quite distinct approaches are used for processing signals from radiation detectors. The first method, used mainly in radiation monitors and nuclear reactor flux measuring systems, involves direct measurement of the current flowing in a detector as a result of radiation absorbed in it. The current may be measured directly to provide a prompt measure of the radiation flux or it may be integrated to provide a measure of total fluence. Although the currents in detectors vary widely, they are generally quite low (often in the range below 10^{-12} A) and special electronic amplifiers, called *electrometers,* which are capable of measuring such low currents, are employed. Where integration of the current is required, it may be achieved either by applying charge integration in a capacitor incorporated in the electrometer or by integrating the output of an electrometer which measures the instantaneous detector current. Section 4.3.1 will discuss both such electrometers.

The second basic processing method treats each individual signal from a radiation event (i.e., produced by absorption of a photon or particle) as an entity and processes it in the appropriate manner to produce the required output information. At its simplest, this involves *amplifying* the pulse signal, *discriminating* against noise, and *counting* the signal pulses that exceed the noise threshold. The pulses may be counted for a predetermined time in a scaler or their "instantaneous" rate may be determined by a *ratemeter*. The scale of the ratemeter may be linear or logarithmic. The situation becomes much more complex when the type of spectrometry discussed in the next chapter is to be performed. Here each pulse must be *amplified* and its height must be determined. A *pulse-height analyzer* (PHA) is used to compile a histogram of the pulse amplitudes which will then represent the energy spectrum of incident radiation. In some cases, *single-channel analyzers* are used to accumulate counts in preselected amplitude intervals—these may feed ratemeters or scalers. In more sophisticated applications only those pulses are processed which are coincident with signals from other detectors or, for example, which satisfy conditions that identify a particle as being of a specific type (e.g., specific reaction products produced in targets bombarded at an accelerator may be selected for study). Although pulse processing systems are complex from an electronic point of view, the desired results can be accomplished by combining a few elementary operations and by assembling systems from building blocks (modules) designed to accomplish these operations.

4.3.1. Electrometers

Depending on the application, electrometers are required to amplify dc input currents ranging down to values well below a picoampere (10^{-12} A) and sometimes as low as 10^{-15} A. In the great majority of electrometers this current passes through a high-valued resistor (up to 10^{12} ohms) and the voltage developed across the resistor is measured.

A more sophisticated alternative is to use a capacitor to integrate the charge, then to use a circuit that resets the capacitor voltage to its reference value when it reaches a predetermined level and to measure the reset rate. This rate is directly proportional to the input current. Whichever of these basic arrangements is used, the key to the electrometer must be an amplifying element exhibiting very high input impedance ($\gg 10^{12}$ ohms) and very low parasitic input current.

Many older instruments used vacuum tubes as the amplifying element and special electrometer tubes were developed especially for this purpose. Common radiation monitors used ion chambers feeding such electrometer tubes with a high-valued resistor to ground as the chamber load. In some cases the resistor was omitted and a logarithmic response resulted from the logarithmic nature of the grid current versus grid voltage relationship.

The operation of vacuum tubes to reduce the grid current to an adequately low value ($\sim 10^{-14}$ A) involved careful study of control of the effect of various sources of current in the grid. The many factors involved are beyond the scope of this discussion, and the reader requiring more information detail should refer to Ref. 43. Suffice it to say that the best tubes, operated under appropriate conditions, can achieve a grid current smaller than 10^{-14} A.

The circuit techniques developed using vacuum tubes are still employed in electrometers employing modern semiconductor amplifying devices. Therefore, before describing the more modern electrometers it is appropriate to discuss some aspects of the overall electronic design of electrometers.

Input Circuit Precautions. Apart from the problems of achieving the required performance in the electrometer, the high impedance requirements involve very special considerations throughout the input circuit. For example, excellent insulating materials, such as Teflon®, must be employed for component mounting in the input circuit and the use of grounded guard rings is recommended around the input circuit points to prevent leakage across insulator surfaces from high voltage points to the input circuit. Coating of insulator surfaces with silicone sprays (such as DRI-FILM) to prevent formation of surface leakage paths under humid conditions is also advisable. Finally, the switching of input resistors required to change the current range of the instrument must be carried out with great care because mechanical strain on insulators often results in charge generation and transients which settle down only very slowly in such high impedance circuits.

Use of Negative Feedback (Operational Amplifiers). The conventional electrometer, illustrated in Fig. 4.18, uses an input load resistor R_L to ground and the electrometer amplifies the voltage V_{SIGNAL} produced across the resistor. The input circuit also contains inevitable capacity C_{in} to ground where the value of C_{in} is generally in the range of 10 pF or more depending on the size and geometry of the signal (detector) source and the capacity of the input circuit amplifying element. A consequence of this is that the response time (time constant $= R_L C_{in}$) in the input circuit can be quite substantial (e.g., 100 sec if $R_L = 10^{12}$ Ohm and $C_{in} =$ 100 pF). The long response time is inconvenient and becomes a serious problem when input circuit switching produces long-lasting transients. To overcome some of these problems the feedback electrometer technique shown in block form in Fig. 4.19 was introduced and is now used almost universally.

In the feedback electrometer (a configuration now known as an *operational* amplifier that is employed widely in conventional electronic systems) the load resis-

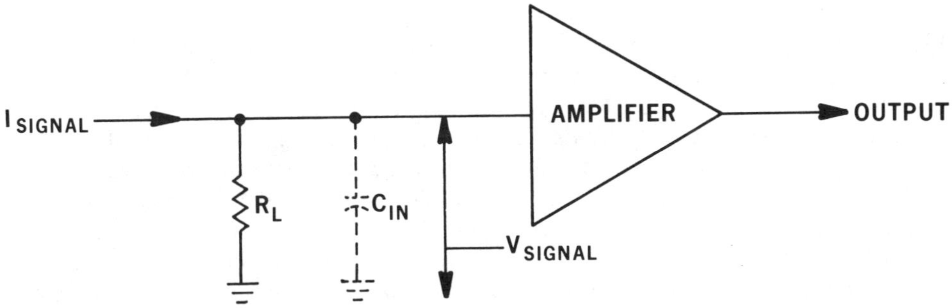

Figure 4.18. Simple electrometer amplifier.

tor is connected between the output point of a high-gain inverting amplifier (gain = A) and its input point. Neglecting any input current in the electrometer amplifier input, the input current I_{SIGNAL}, flows into the feedback resistor R_L. Since the amplifier forces its output to move by an amount $A\ V_n$ (where V_{in} is the input voltage movement) we can write:

$$V_{OUT} = -I_{SIGNAL}\ R_L\ A/(1 + A) \ .$$

If the gain A is made very large the entire voltage $R_L \times I_{SIGNAL}$ appears essentially as a negative output signal and the input point does not move in voltage. Consequently no charging current is taken by the input capacitance C_{in} and the response time of the system to changes in the input current is equal to $R_L \times C_S$ where C_S consists only of capacitive strays across R_L. Since C_S is generally very small (< 1 pF) the resulting response time is many times smaller than that of the standard electrometer of Fig. 4.18.

Another major advantage of the feedback system is that the overall gain is, to first order, independent of the value of the internal gain A (if A is large). This makes the output signal virtually independent of gain drifts and nonlinearities in the amplifying elements.

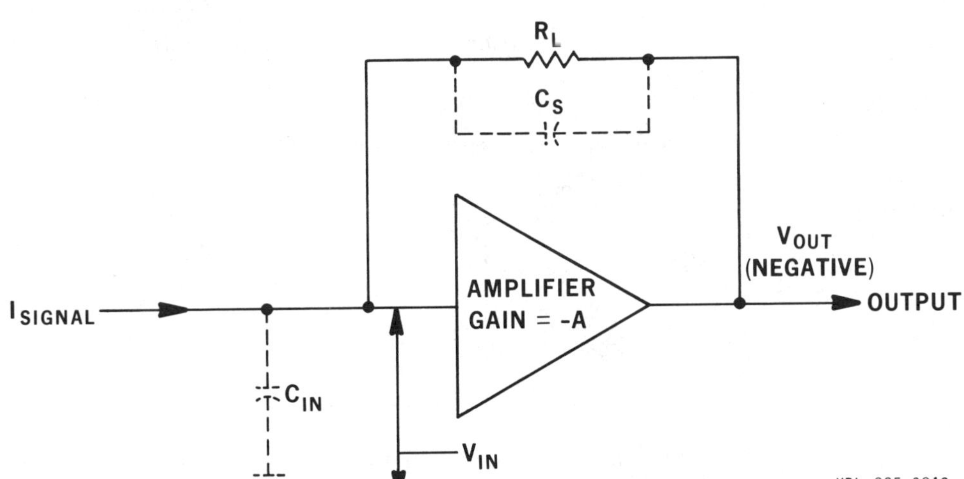

Figure 4.19. Feedback electrometer.

Time-Varying Capacitor Electrometer. In simple circuits such as those shown in Figs. 4.18 and 4.19 long-term changes and temperature effects cause shifts in the operating point of the input electrometer element with consequent zero drift in the output level. Although this effect can be made very small in modern balanced semiconductor devices, it is very troublesome in vacuum tube circuits where short-term drifts in the range of a few percent of full scale are common (corresponding to shifts of millivolts per day in the grid voltage of the tube). Consequently the vibrating-reed electrometer was developed to eliminate the dc drift in electrometers. This consists of a time-varying capacitor shunted across the input of an amplifier circuit as shown in Fig. 4.20. The capacitor C_A is a vibrating-reed capacitor driven by a drive coil excited by a sinusoidal voltage supply. The ac signal developed across the capacitor is coupled to the input of the high-gain amplifier A whose output is synchronously rectified in phase with the drive to the vibrating capacitor. The circuit, as shown here, applies feedback to the vibrating capacitor via the feedback resistor R_F.

The operation of the vibrating-reed system depends on the fact that a sinusoidal variation $\Delta C \sin \omega t$ in a capacitor whose average value is C_0, with an average voltage V_0 across it, produces an ac signal equal to $V_0/[1 - (\Delta C/C_0) \sin \omega t]$ across the capacitor. As with any negative feedback system the effect of feeding the dc output of the synchronous rectifier back to the input is to make the dc voltage across C_A very small. Therefore the output is simply given, to a close approximation by $-I_n \times R_F$.

The voltage drift of a vibrating-reed electrometer is at least an order of magnitude smaller than that of vacuum tube electrometers and is determined mostly by the stability of the surfaces of the capacitor. Therefore, care is taken in cleaning surfaces and coating with an inert metal (usually gold). Furthermore, the basic circuit eliminates the effect of any input current in the input amplifier element, so the vibrating-reed electrometer can be used to measure currents as small as 10^{-16} A. This compares with approximately 10^{-14} A or more in standard electrometers.

Semiconductor Device Electrometers. The advent of semiconductor devices has resulted in analogs of the types of electrometer just discussed. The most common version of these involves the replacement of the vacuum tube electrometer by a

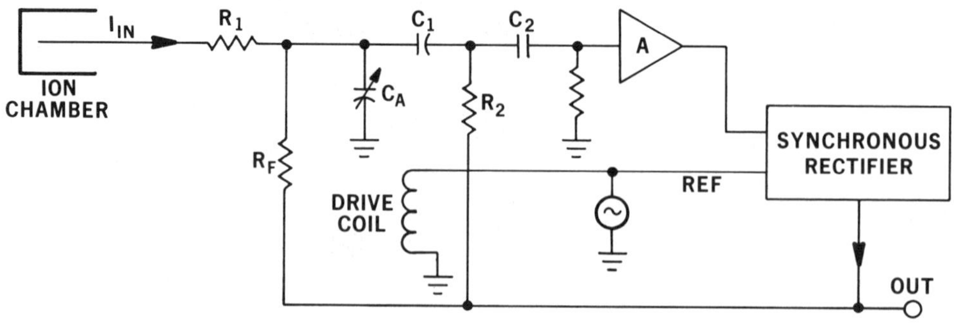

Figure 4.20. Vibrating-reed electrometer.

Metal-Oxide-Silicon (MOS) Field Effect Transistor (FET). In this device a metal electrode deposited on an insulating silicon dioxide/nitride layer is used to modulate the surface conductivity of the silicon beneath the insulating layer. The very high impedance of the silicon dioxide/nitride layer makes these devices ideal in electrometer applications. Furthermore, unlike thermionic devices, drifts in the operating point are low, so substantial improvements in zero drift can be realized.

Many commercial instruments and survey meters (such as the Victoreen Model 470A) use MOSFET input stages in feedback configuration similar to that shown in Fig. 4.18.

Semiconductor analogs also exist for the vibrating-reed capacitor electrometers. These employ the variation in capacitance that occurs in semiconductor diodes as the reverse voltage across them changes. Varactor diodes with very low leakage are employed in such electrometers. The block diagram of a typical commercial varactor-diode electrometer package is shown in Fig. 4.21; the entire circuit is contained in a small potted module approximately 3 in. × 2 in. × 0.7 in. Such units can exhibit zero drifts in the range of 100 μV and input leakage currents of 10^{-14} A which means currents of this order can be measured with them. As with all electrometers, great care is needed in the external connections in the input circuit if this performance is not to be degraded.

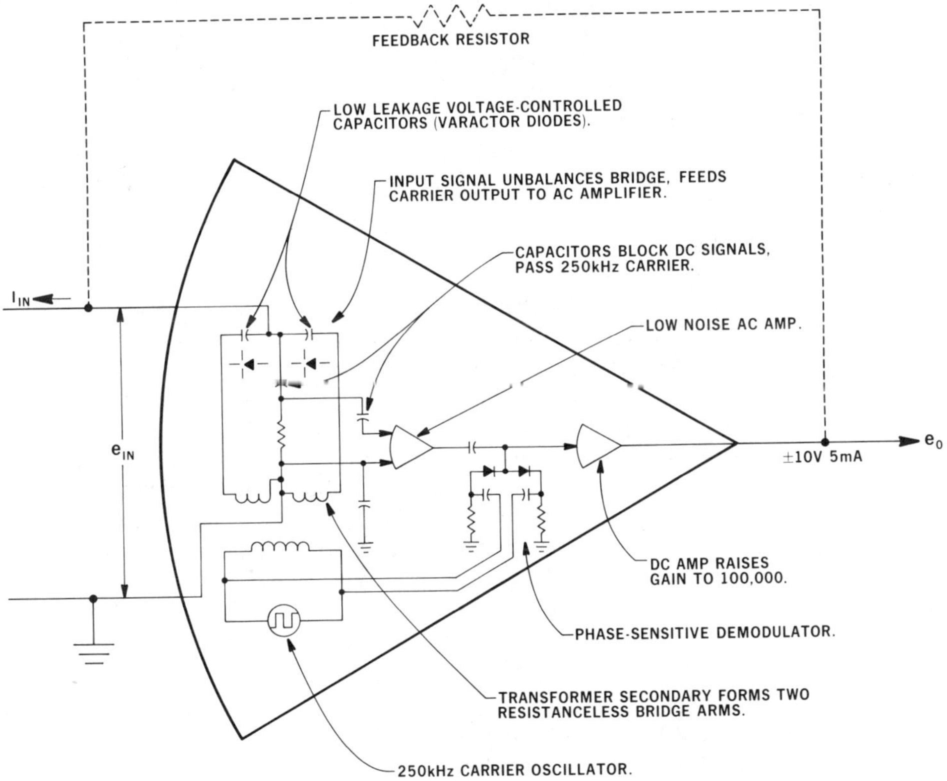

Figure 4.21. Modern varactor-diode electrometer.

4.3.2. Pulse Amplifiers

In many applications, particularly those where the quality (e.g., type of radiation or its energy) of the radiation is to be determined, individual detected events must be processed separately. The transient electrical signal produced in the detector must, in general, be amplified to a size sufficient to allow its amplitude to be measured. The amplification is provided by a pulse amplifier which usually consists of two parts, a preamplifier followed by a shaping amplifier which is designed to optimize the signal/noise ratio in the system and thereby to optimize the accuracy of amplitude measurement. On occasion, the optimization may be aimed at improving the measurement of the event timing.

The key role played by pulse amplifiers in nuclear spectrometry justifies this section. It should also be noted that nuclear spectrometry is generally performed only with types of detectors that inherently provide good energy resolution. This includes, to a small degree, gas ionization chambers for short-range particles such as natural alpha particles and gas proportional counters mainly for x-ray spectrometry. However, scintillation detectors used for gamma-ray spectrometry and semiconductor detectors used for particle, x-ray and gamma-ray spectrometers are the main modern tools. In the case of scintillation detectors, the high gain of the photomultiplier normally makes the amplifier noise essentially negligible and the energy resolution (i.e., the width of a peak produced by monoenergetic gamma rays) is determined entirely by basic physical mechanisms in the scintillator and phototube. The situation is much more complex in semiconductor detectors (or ionization chambers) because such devices do not provide the almost unlimited internal gain of a photomultiplier and because the inherent resolution capabilities of the detector (determined by physical processes) are so outstanding that electronic noise in the amplifier chain contributes significantly to resolution. This difference is reflected in the fact that gamma-ray scintillation counters exhibit resolutions in the 10% range while semiconductor detector spectrometers are in the range well below 1% and often below 0.1% (full width at half-maximum). Therefore, this discussion on pulse amplifiers will focus on semiconductor detector spectrometers, although much of the material is equally applicable to other detector systems.

The limitation in resolution of semiconductor detector spectrometers imposed by the physical processes in the detector are illustrated in Table 4.2, which shows the resolution limit for silicon and germanium detectors at different energies, assuming zero contribution from electronic noise. The basic physical mechanism causing this limitation in resolution is the statistical sharing of the absorbed radiation energy between excitation of lattice vibrations in the semiconductor crystal and the production of hole-electron pairs which can be collected to produce an external signal. Clearly, since a competition exists between these two physical processes, a resulting spread in the external signal occurs even for monoenergetic absorption processes in the detector. In any practical spectrometer a contribution to resolution also occurs due to electronic noise which must be added in quadrature to the resolution caused by the charge production statistics in the detector [$\Delta E_{\text{total}} = (\Delta E^2_{\text{Noise}} + \Delta E^2_{\text{Detector}})^{1/2}$]. Depending on the particular system and its constraints, the electronic noise contribution is generally in the range of a few hundred eV to a few keV. Comparing this with the results of Table 4.2, it can be seen that electronic noise is generally dominant at the lower energies, whereas the detector charge production statistics dominate at high energies. It is in this framework that electronic noise will be considered.

Table 4.2. Energy Resolution versus Energy (Si and Ge) (Fano Factor = 0.12)

	ΔE_{FWHM} (eV)	
Energy	Si	Ge
1 keV	50	44
2	70	62
5	111	99
10	157	140
20	222	198
50	351	313
100	496	442
200	701	626
500	1,109	990
1 MeV	1,568	1,401
2	2,217	1,980
5	3,506	3,132
10	4,960	4,420
20	7,010	6,260
50	11,090	9,900
100	15,680	14,010
200	22,170	19,800
500	35,060	31,320

Figure 4.22 shows, in block form, a typical spectrometric system. In addition to the detector, preamplifier, and pulse amplifier, there is also shown the pulse pile-up rejection system used to prevent distortion of pulses (and therefore spectra) that would otherwise occur due to pulse pile-up at high random counting rates. It should also be noted that the preamplifier output consists of a step function for each detected event and that the pulse amplifier forms an output pulse shape (roughly Gaussian) designed to provide a pulse of limited duration so that individual events can be separately processed at high rates. The shaping circuits in the main pulse amplifier always consist of a differentiator (usually an RC high-pass filter) and some type of integrator (multiple RC low-pass filters in the simplest case). The fast channel, where sharply differentiated pulses are formed, is used to sense events spaced too close together and, when this occurs, the slow (amplitude-measuring) channel pulses are gated off by an inhibit pulse generated by this pile-up rejector. Noise sources present in a detector/preamplifier system are illustrated in Fig. 4.23. Two representations of the sources are given here, the lower one shows their time behavior while the upper one shows their representation in the frequency domain. The step noise generator (or current noise) results mainly from fluctuations in the detector leakage current, while the delta noise generator (or voltage noise) arises mainly from fluctuations in the current in the input amplifying element—which is an FET in modern spectrometers. It is clear from this figure that the step noise pulses are reduced by increasing the input capacity C in the same proportion as are

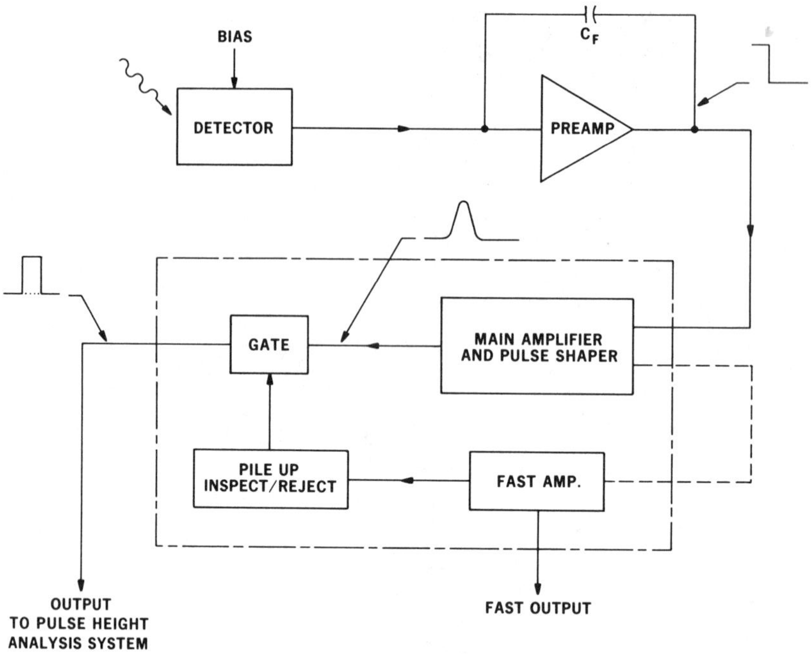

Figure 4.22. Elements of a nuclear spectroscopy system.

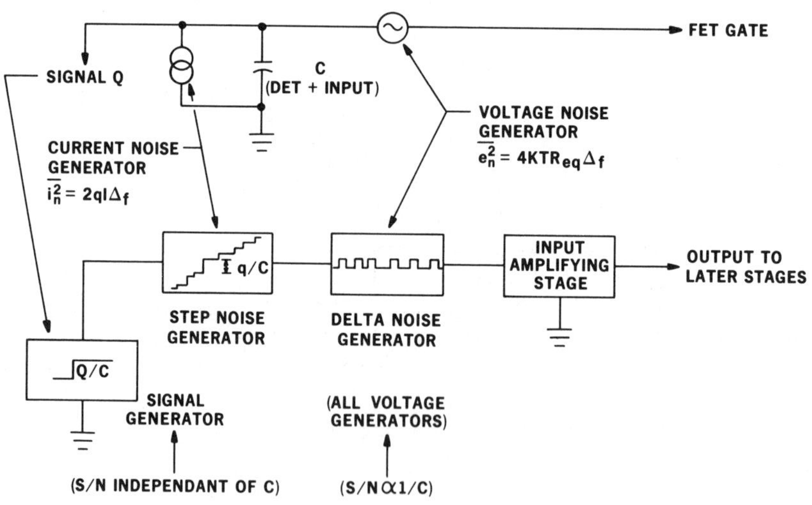

Figure 4.23. Noise sources in a nuclear detector/amplifier system.

signals, but the delta noise is unaffected by input capacity. Therefore, increasing detector capacity (or other capacity) degrades the signal/noise ratio as far as delta noise is concerned, but does not affect the signal/step noise ratio.

It is also important to recognize the effect on the two basic noise mechanisms of varying the measurement time. For present purposes, measurement time can be regarded as the width of the shaped pulse at the pulse amplifier output (it is assumed that the integration and differentiation times used in the amplifier are approximately equal). As the measurement time is increased, the delta noise impulses are better arrayed and the signal/delta noise ratio improves, but the step noise pulses accumulate and the signal/step noise ratio degrades. Figure 4.24 shows the overall behavior of the noise performance (energy resolution) as the measurement time T_m is changed. It may be noted that this figure also indicates the presence of a third type of noise ($1/f$) which may be present and which remains constant with varying measurement time. As indicated in Fig. 4.24, an optimum measurement time exists as far as noise is concerned—in practical systems this is usually in the range of 1 to 100 μsec. Performance at high rates, where pile-up is a consideration, is obviously degraded if long measurement times are employed. So, in general, a compromise must be made between noise performance and rate requirements.

The detailed design of amplifiers is beyond the scope of this book, and the reader is referred to the published literature for more details. Commercial amplifiers exist and are used in most spectrometers. The foregoing comments should allow the reader to select intelligently the appropriate operating conditions for such equipment in a given application.

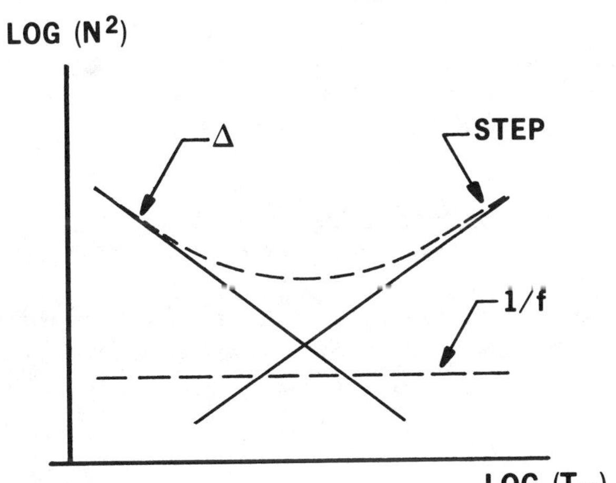

Figure 4.24. Variation in resolution as a function of measurement time.

4.3.3. Pulse Amplitude Analysis

The output of a nuclear pulse amplifier is used to feed some type of amplitude analysis equipment. In its simplest form this might be an amplitude *discriminator* which selects only those events whose amplitude exceeds a preset threshold. In a slightly more complicated analog circuit, the *single-channel analyzer,* signals are selected if their amplitude falls within a selected range lying between two discriminator levels. These simple analog devices are rarely used at the present time and multichannel analyzers are much more common.

Multichannel analysis systems are all based on digitizing signals on an individual basis to produce numbers representative (usually linearly-related) to the signal amplitudes. An *analog to digital converter* (ADC) performs this function, producing digital outputs whose maximum range is usually in the range 2^{13} to 2^8 (i.e., 8192 to 256). A unit charge of the ADC output is referred to as a channel in the analyzer—therefore, typical systems contain 256 to 8192 channels.

The output value of the ADC for each signal is generally used to provide an address in a memory and each time a certain address is selected, a count is added to that address (channel). The result is to generate a histogram in the memory of the pulse amplitudes represented to the input of the ADC. The complete system of ADC, memory, display, and controls, is usually called a *pulse-height analyzer* (PHA). As described here the analyzer provides a spectrum of pulses from a single detector. In other cases a pair of ADCs associated with two detectors might be used to generate a combined address in a memory and a two-dimensional spectrum can be obtained.

In many research applications multidimensional data is produced (i.e., an event is represented by simultaneous signals in several detectors) and more complex spectral accumulation and analysis methods must be employed. In some cases this type of problem is handled by *on-line* computer data acquisition; in others the multiple ADC outputs are recorded as digital data on magnetic tapes (or disks) and the tapes are then processed and analyzed in an *off-line* computer. On occasion, timing information (e.g., the delay between signals in two detectors) must also be analyzed; this can be accomplished using a *time-to-amplitude converter* to generate a pulse whose amplitude is proportional to the delay and by feeding this signal to an amplitude analysis system as described earlier.

Only the main components of nuclear spectrometry systems have been outlined in this discussion. Such systems may be very complex and some special units (such as devices for multiplexing signals into ADCs) may be used. However, as indicated earlier, the large complex systems can generally be broken down into the few functional units discussed here.

4.4. FURTHER INFORMATION

Ongoing developments in radiation detection are reported in the journals of the Institute for Electrical and Electronics Engineers.

For further information the reader is referred to the IEEE Transactions of the Scintillation Counter Symposia and to IEEE Transactions of the Symposia on Nuclear Science.

REFERENCES

1. R. D. Evans, *The Atomic Nucleus*, McGraw-Hill, New York, 1955.
2. B. G. Harvey, *Introduction to Nuclear Physics and Chemistry*, 2nd ed., Prentice-Hall, New Jersey, 1969.
3. E. Segrè, *Nuclei and Particles: An Introduction to Nuclear and Subnuclear Physics*, 2nd ed., Addison-Wesley, New York, 1977.
4. P. Marmier and E. Sheldon, *Physics of Nuclei and Particles*, 3 vols., Academic, New York, 1969.
5. J. Mantel, "The Beta Ray Spectrum and the Average Beta Energy of Several Isotopes of Interest in Medicine and Biology," *Int. J. Appl. Radiat. Isot.*, **23**, 407, 1972.
6. H. Cember, *Introduction to Health Physics*, Pergamon Press, New York, 1969.
7. F. Cappellani and G. Restelli, "X-Ray and Gamma-Ray Spectrometry," *in* G. Bertollini and A. Coche, *Semiconductor Detectors*, Wiley-Interscience, New York, 1968, p. 365.
8. W. J. Price, *Nuclear Radiation Detection*, 2nd ed., McGraw-Hill, San Francisco, 1964.
9. E. Fenyves and O. Haiman, *The Physical Principles of Nuclear Radiation Measurements* (revised and enlarged version of *Die Physikalishen Grundlager Der Kernstrahlung Messungen)* Academic, New York, 1969.
10. J. S. Handloser, *Health Physics Instrumentation*, Pergamon, New York, 1959.
11. P. J. Ouseph, *Introduction to Nuclear Radiation Detectors*, Vol. 2 of Laboratory Instrumentation and Techniques, G. W. Ewing, Series Editor, Plenum Press, 227 West 17th Street, New York, N.Y. 10011, 1975. (For the nonspecialist, with little reliance on mathematics or formulas)
12. International Atomic Energy Agency, "Measurement of Short-Range Radiations," Technical Reports Series No. 150, United Nations, Vienna, 1973.
13. National Council on Radiation Protection and Measurements, "A Handbook of Radioactivity Measurements Procedures," NCRP Report No. 58, 1978. (Available from NCRP, 7910 Woodmont Avenue, Washington, DC 20014).
14. G. Bertollini and A. Coche, *Semiconductor Detectors*, Wiley-Interscience, New York, 1968.
15. F. Granzer, H. Paretzke, and E. Schopper, Eds., *Solid State Nuclear Track Detectors,* 2 Vols., Proceedings of the 9th International Conference, Neuherberg/Munchen, 30 September–6 October 1976, Pergamon, New York, 1978.
16. H. Kallmann, "Quantitative Measurements with Scintillation Counters," *Phys. Rev.*, **75**, 623-626, 1949. Initially, in *Natur und Technik*, 1947.
17. B. B. Rossi and H. H. Staub, *Ionization Chambers and Counters*, McGraw-Hill, New York, 1949.
18. J. W. Coltman and F. H. Marshall, "Some Characteristics of the Photomultiplier Radiating Detector," *Phys. Rev.*, **72**, 528, 1947.
19. G. A. Morton, "Photomultipliers for Scintillation Counting," *RCA Rev.*, **10**, 525, 1949.
20. E. Storm, B. J. Krohn, and W. B. Chambers, "The Photon Energy Response of Several Commercial Ionization Chambers, Geiger Counters, Scintillators, and Thermoluminescent Detectors," Los Alamos Scientific Laboratory Report LA 4052, P.O. Box 1663, Los Alamos, NM 87545, 1969.

21. R. E. Simon and B. F. Williams, "Secondary-Electron Emission," *IEEE Trans. Nucl. Sci.*, **15**(3), 167, 1968.
22. G. A. Morton, H. M. Smith, and H. R. Krall, "Pulse Height Resolution of High Gain First Dynode Photomultipliers," *Appl. Phys. Lett.*, **13**(10), 356, 1968.
23. J. W. Draper, "A Scintillation Counter for Thermal Neutrons," *Rev. Sci. Instrum.*, **22**, 543, 1951.
24. W. S. Emmerich, "A Fast Neutron Scintillator," *Rev. Sci. Instrum.*, **25**, 69, 1954.
25. J. W. Draper, "A Fast Neutron Scintillator Spectrometer," *Rev. Sci. Instrum.*, **25**, 558, 1954.
26. R. Hofstadter, "The Detection of Gamma-Rays with Thallium Activated Sodium Iodide," *Phys. Rev.*, **75**, 796, 1949.
27. R. H. Gillette, "Calcium and Cadmium Tungstate as Scintillation Counter Crystals for Gamma Ray Detection," *Rev. Sci. Instrum.*, **21**, 294, 1950.
28. J. Bonanomi and J. Rossel, "Luminescence of Alkaline Iodides," *Physica*, **18**, 486, 1952.
29. J. D. Graves and J. P. Dyson, "Scintillation Counter for Laboratory Counting of Particles," *Rev. Sci. Instrum.*, **20**, 560, 1949.
30. H. Kallmann, "Scintillation Counting with Solutions," *Phys. Rev.*, **78**, 621, 1950.
31. J. B. Birks, *Scintillation Counters*, McGraw-Hill, New York, 1953.
32. F. N. Hayes, D. L. Williams, and E. Rigers, "Liquid Scintillation Counting of Natural C^{14}," *Phys. Rev.*, **92**, 512, 1953.
33. S. H. Liebson, M. E. Bishop, and J. W. Elliot, "Fluorescent Decay of Scintillation Crystals," *Phys. Rev.*, **80**, 907, 1950.
34. G. T. Reynolds, "Solids and Liquid Scintillation Counters," *Nucleonics*, **10**, 46, 1952.
35. R. K. Swank, "Recent Advances in Theory of Scintillation Phosphors," *Nucleonics*, **12**, 14, 1954.
36. J. B. Birks, "Energy Transfer in Organic Phosphors," *Phys. Rev.*, **94**, 1567, 1954.
37. J. Q. Umberger and V. K. LaMer, "The Kinetics of Diffusion Controlled Molecular and Ionic Reactions in Solutions as Determined by Measurements of the Quenching of Fluorescence," *J. Am. Chem. Soc.*, **67**, 1099, 1945.
38. J. B. Birks, "Temperature Dependence of Organic Scintillation Materials," *Phys. Rev.*, **95**, 277, 1954.
39. S. H. Liebson, "Temperature Effects in Organic Fluors," *Nucleonics*, **10**, 41, 1952.
40. R. K. Swank and W. L. Buck, "The Scintillation Process in Plastic Solid Solutions," *Phys. Rev.*, **91**, 927, 1953.
41. H. Kallmann and M. Furst, "Fluorescence of Solutions Bombarded with High Energy Radiation," *Phys. Rev.*, **76**, 857, 1950.
42. F. S. Goulding and R. H. Pehl, "Semiconductor Radiation Detectors," in *Nuclear Spectroscopy and Reactions*, Part A, J. Cerny, Ed., Academic, New York, 1974.
43. E. Fairstein, *Nuclear Instruments and Their Uses*, A. H. Snell, Ed., Wiley, New York, 1962, p. 194.

5 MEASUREMENT OF GAMMA AND X-RAYS

Chapter 2 shows that exposure to ionizing radiation can be harmful, even lethal. On the other hand we are continuously exposed to a certain amount of this radiation from cosmic rays and natural radioactivity. This natural background may vary by factors up to 3 or 4 from place to place and no difference in health or longevity has been observed in people living in these different areas. In other words humans may tolerate a certain amount of radiation in their environment without detectable harm. Guidelines have been established for safe levels of both continuous and intermittent irradiation. The best means for measuring the amount of radiation present in an environment remains to be determined.

The general techniques for detecting and measuring ionizing radiation were presented in Chapter 4. This chapter discusses the specific instrumentation for detection and monitoring of γ- and x-rays. Chapter 6 will consider radiation spectrometry of γ- and x-rays which is valuable for identifying which radioactive sources may be present.

5.1. THE FILM BADGE

The most widely used means for monitoring radiation is the film badge. Almost everyone working with or around x-rays, nuclear devices, accelerators, or radioactive materials uses these badges. In many situations they are required by law.

The film badge consists of a small piece of photographic film wrapped in an opaque cover and supported by a metallic backing, which can be pinned to the wearer's clothing or worn as a ring. The film emulsion is thicker than that used in normal photography and is not dye-sensitized. The film wrapping may be made radiation absorbing so that the exposure of the film more nearly matches the radiation dosage received by the wearer. The badge is removed at regular intervals and the film developed. This film serves as a permanent record of radiation dose received.

Further information on film badges, including literature on their construction and their legal aspects, is contained in Chapter 9, Dosimetry, and Refs. 1–4.

5.2. ELECTRONIC INSTRUMENTATION, MEASUREMENT CONSIDERATIONS

From the point of view of radiation protection instrumentation (as opposed to instruments for radiation detection), it is useful to divide the photon energy spectrum into three broad regions, delineated approximately in the following way:

1. Below 150 keV.
2. 150 keV to 3 MeV.
3. Above 3 MeV.

Below 150 keV the number of different types of useful instruments is somewhat limited due to difficulties in window construction, whereas at higher energies a much wider variety is available. Extremely thin windows (< 0.05 g/cm^2) are required to measure the maximum ionization loss for photons of < 150 keV. As discussed earlier, it is this ionization loss that must be determined to obtain the dose equivalent, which is important for radiation protection measurements.

Between 150 keV and 3 MeV, the measurement of exposure with an air-equivalent ionization chamber gives nearly the same value as measurement of absorbed dose with a tissue-equivalent ion chamber or almost any other device designed to measure absorbed dose. In Chapters 1 and 2 it was pointed out that the roentgen (an air dose unit) is equivalent to the rad (0.01 Gy) (tissued dose units) below 3 MeV. This is the energy region in which the roentgen has been defined, in which ionization in air has a meaningful relationship to ionization in tissue, and in which the ranges of the charged secondaries from photon interactions are short.

Above about 3 MeV, the pair-production cross section becomes increasingly important, and its dependence on the square of the atomic number requires that any ionization detector match closely the composition of tissue. This leads to the need for tissue-equivalent instruments for measurements at higher energies.

The difference between dose and dose rate is another important consideration. For many applications (such as at pulsed accelerators or medical x-ray machines) the relevant radiation occurs in extremely short bursts, from a few milliseconds down even to a fraction of a microsecond in duration. Many instruments cannot respond in such short times, and will seriously underestimate the intensity of the radiation field if used in these situations. There are two ways to circumvent this problem. The first is to alter the source appropriately: thus one can often decrease the intensity and lengthen the time of a medical x-ray pulse until the measuring instrument can respond properly. The second method is to use a measuring device which uses an integrating technique. A photographic film, a thermoluminescent dosimeter chip, or an ion chamber in the charge mode (assuming that it is operated so that recombination does not occur) satisfies this requirement. Other electronic instruments can be designed to operate in a current-integration rather than instantaneous mode. An extremely valuable feature of some electronic instruments is their flexibility which allows optional use in either pulse-counting or current-integration mode.

Also important is the *directional sensitivity* of response and the sensitive *aperture* of any instrument. Any commercial instrument should specify these two parameters; unfortunately, this is not always the case. The importance of the directionality of response is obvious. Fortunately, in most γ- and x-ray measurement situations the direction from which the radiation is coming is known. The aperture-size consideration is more complex. For example, an ion chamber with 10 cm^2 effective aperture may be calibrated so that a given output indication corresponds to 1 mR/hr, in a general diffuse photon radiation field of a size much larger than 10 cm^2. Now suppose instead of a large radiation field, a focused beam of only 1 cm^2 in diameter is being measured. If the instrument response is 1 mR/hr, then it is easy to see that within the beam the radiation field actually corresponds to an absorbed dose rate of 10 mR/hr, because the amount of ionization produced in the instrument is identical to that that would have been produced by a 1 mR/hr general field. For this reason, both the effective sensitive aperture of the instrument and the effective size of the radiation field must be known if a meaningful measurement is to be made.

Another concern is the *stability of calibration.* Calibration can be affected by changes in temperature, pressure, or humidity. Scintillation counters are especially sensitive to changes in temperature because of increase of noise from the photomultiplier with increase in temperature. Geiger tubes and proportional counters show little temperature effect when operating over normal temperature ranges. Pressure corrections are especially important for ion chambers that operate open to the atmosphere. High humidity can have adverse effects on the calibration of many types of instruments.

Most ionization chambers are read through an electrometer. Initially gold-leaf and vibrating-reed electrometers were used. The vibrating-reed electrometer, while expensive, has low noise and excellent resistance to overloading. Then the electrometer tube was a standard device for many years, but it suffers from several drawbacks, perhaps the most important of which are poor shock sensitivity and poor stability, both long- and short-term. However, the output of ionization chambers may now be measured by using MOSFET (metal oxide semiconductor field effect transistor) electrometer. The MOSFET's advantages are low cost, shock insensitivity, and high reliability; its disadvantages are that it has high noise and can sometimes be burned out by overloading.

5.3. X- AND GAMMA RADIATION BELOW 150 keV

After reading this chapter, the reader may wish to refer to three general references for discussion of instrumentation in this energy region: ICRU Report No. 17 [5], NCRP Report No. 57 [6], and IAEA Technical Report Series No. 150 [7]. In addition, the proceedings of two conferences (on color television x-radiation measurement [8] and radiation safety in x-ray diffraction and spectrometry [9]) contain good summaries of work in these areas.

There are two quite different measurement situations frequently encountered in this low-energy region: (a) the measurement of exposure levels from short, high-intensity bursts of radiation from accelerator or pulsed x-ray equipment; or (b) the measurement of exposure from continuous emitters such as from various radioactive materials (e.g., the 17-keV x-radiation from decay of plutonium) or from unintentional x-ray leakage of color TV receivers, klystrons, and other high-voltage sources. In the first case, the total absorbed dose can be significant (a few mrad to a few rad) (10 μGy to 10 mGy), but the pulse is usually short (5 to perhaps 500 msec); the "signal" is therefore large enough to be detected easily, and measurement problems are due to time- or energy-dependent response of the instruments. In the second case, the signal is extremely small (accurate readings of 0.1 mrad/hr may be required) (1.0 μGy/hr), but there is no pulse-time problem. These two areas shall be dealt with separately. If no other radiation types are present (as is the case in many medical or industrial applications) the instrument reading of exposure (roentgens) is sufficiently close to the reading of absorbed dose (rads or grays), and therefore to dose equivalent (rem or sievert), that they may be used interchangeably (see Chapter 1).

5.3.1. Detection Methods Below 150 keV

When properly applied there are many different detection techniques that can be used to detect low-energy photons. Some of these are:

1. Ionization chambers.
2. Geiger-Müller counters.
3. Scintillators.
4. Films and emulsions.
5. Solid-state detectors (thermoluminescence, fluoroluminescence, and induced conductivity in semiconductors).

In deciding on the most appropriate technique three critical parameters must be examined: the energy-dependence of the response, the time characteristics, and the sensitivity.

To gain insight into problems of measuring γ- and x-rays with *air ionization chambers*, consider that the exposure of 1 roentgen brings about the release of about 3×10^{-10} coulombs of charge from 1 cm^3 of air. A radiation rate of 1 mR/hr thus yields an ion current of about 10^{-16} A/cm^3 [10] which is very difficult to detect. The only way to increase the current is to increase the sensitive volume. This introduces other problems, such as variations of sensitivity within the volume. Basically the ultimate sensitivity of an ion chamber depends on electronic circuitry capable of detecting these extremely small currents or charges.

ICRU Report No. 17 [5] discusses in detail the use of free-air ionization chambers below 150 keV, along with the corrections needed due to attenuation and scattering of photons in air, recombination of ions, and various other effects. A standard instrument for absolute measurement of exposure (roentgens) within about 0.5% is described, as well as the relationship of exposure to absorbed dose.

In a *cavity ionization* chamber, the low-energy sensitivity may be limited because the majority of initial interactions occur in the chamber wall. Chambers with walls as thin as 5 mg/cm^2 have been constructed [11] in order to overcome this problem. Figure 5.1 shows a typical energy response for a well-designed ion chamber, whose walls are approximately air-equivalent; the fall-off at low energies is due to wall thickness and the rise above unity response is due to an energy region in which the effective atomic number of the walls exceeds that of air.

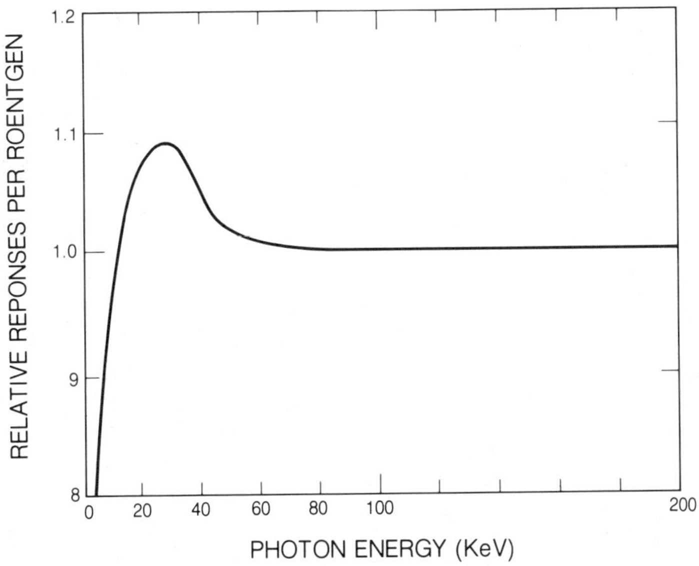

Figure 5.1. Variation of response with photon energy of a typical cavity ionization chamber whose wall material is approximately air-equivalent (reproduced from Bates, Ref. 10).

Of the other possible detection means, the most common is *Geiger-Müller* (G-M) tubes. G-M counters can detect a single ionization event, because of the avalanche resulting from the very great multiplication of the initial ionization, which is then detected as a pulse. The sensitivity is about two orders of magnitude greater than that of a cavity ion chamber with comparable dimensions. However, there is a very significant energy dependence at low energies, because a single x-ray at low energies will give one "count" even though its energy deposition is vastly less than a single photon of high energy. Figure 5.2 shows the variation in response of typical G-M tubes with various wall materials.

Scintillation Counters have been found to be very useful They are the most sensitive of all detectors; a NaI(Tl) crystal of even modest size will detect individually every low-energy x-ray or gamma-ray photon incident upon it. Because of this, the detector may produce a significant perturbation of the radiation field, and in some situations large detectors are thus ruled out. The energy dependence of the inorganic scintillators [such as NaI(Tl)] for radiation is nonuniform. Because their effective atomic number differs considerably from that of air or tissue, inorganic scintillators cannot be made to read directly in roentgens or rad (gray). On the other hand, the organic plastic scintillators are composed essentially of $(CH)_n$, which is reasonably tissue-equivalent. An inorganic scintillator/photomultiplier combination used as an energy spectrum analyzer will give a measurement of absorbed dose.

Films and *emulsions* are also sensitive to low-energy x-radiation. (Indeed, W. K. Roentgen discovered x-radiation in 1895 by the darkening of film; and film is the most commonly used technique in medical diagnostic applications.) Up to a least a few roentgens of exposure, the linearity of response of film (density versus exposure) is quite good As an example, Fig. 5.3 shows this linearity for two kinds of

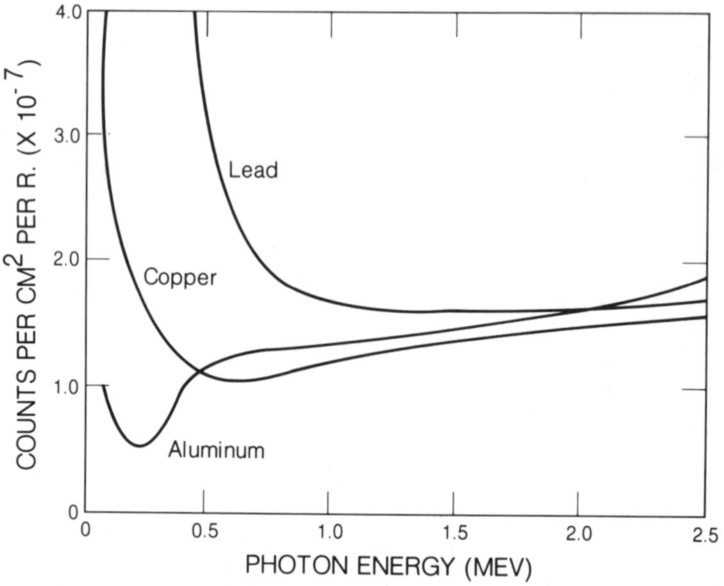

Figure 5.2. Variation of response of a G-M tube with photon energy for various wall materials (reproduced from International Commission on Radiation Units and Measurements, Ref. 12).

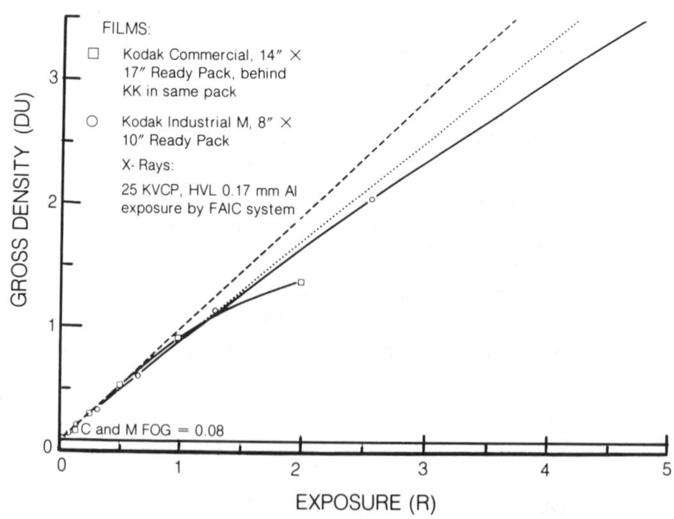

Figure 5.3. Optical density against roentgen exposure plotted on linear grid for Commercial Film and for Type M Films in Ready Packs (reproduced from Stewart, Ref. 13).

Kodak film. The spatial resolution of film is excellent, and far exceeds the requirement of any typical radiation-protection application. Its drawbacks are twofold: first, the processing of film must be done with great care if absolute exposure values are desired from absolute density measurements. Second, due to fading, the time delay between film exposure and processing plus analysis sometimes makes its use inappropriate. Thus, film finds its most common use in attempts to quantify intense, spatially complex x-ray sources (for example, slit leakage from the sides of tubes or voids in shielding) or in long-term, integrating dosimetry (where, for example, the film is placed on a wall and developed only periodically). It is seldom used for rapid surveys if only because of inconvenience and cost.

Although the use of *solid-state detectors* in the low-energy region is hampered by the fact that the atomic numbers of the detectors are very different from those in tissue (producing too rapid a variation of response with energy) they are used for surveys of exposure due to diagnostic radiology. To quote from ICRU Report No. 17 [5]:

> Two thermoluminescent materials, lithium fluoride and lithium borate may prove to be of value for dosimetry at low energies. The effective atomic number of the former is higher than that of soft tissue whereas that of lithium borate is lower. Attempts have been made to produce a dosimeter with tissue-like x-ray absorption characteristics by mixing them in a suitable proportion. (see Kirk et al. [14])

In Fig. 5.4, the energy-dependence of the commercially available TLD-700 the thermoluminescent material is shown. It can be seen that it is quite satisfactory down to perhaps 25 keV, below which its response varies significantly. With appropriate calibration, however, it can be used below this energy.

As far as the application of TLDs in the x-ray domain is concerned, many of the same comments made about film apply: it is most useful as an integrating dosimeter (for example, on a wall for weekly information) and is not often used for real-time survey work.

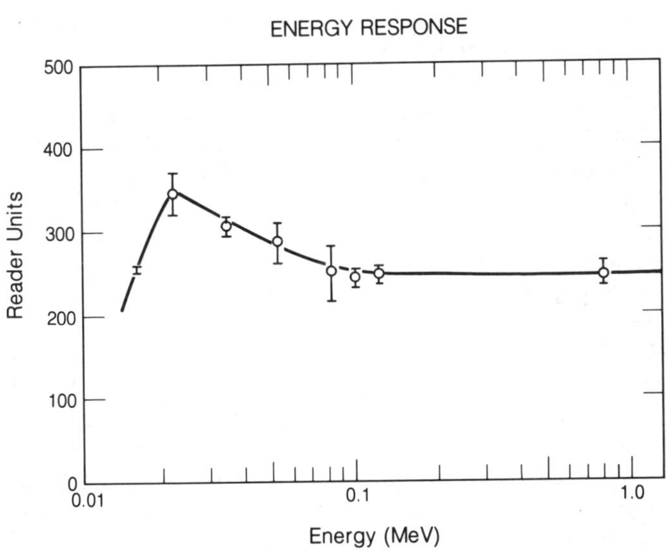

Figure 5.4. Energy response of TLD-700 thermoluminescent material (reproduced from Endres, Ref. 15).

5.3.2. Medical and Industrial X-Rays

This section discusses measurements of photons with energies below about 150 keV that are mainly from pulsed sources such as *medical (and dental) x-ray machines* (for therapeutic and diagnostic applications) and *industrial x-ray machines*. It is not intended that this section apply only to these categories; low-duty-cycle, high-voltage klystrons may emit radiation similar in quality to that considered here. Indeed, any device with high voltage is capable of being a source of x-rays. While this discussion deals explicitly with the medical x-ray problem other situations are implied throughout. Of course, as noted in Section 3.3., the medical use of x-rays accounts for more than 90% of all human exposure to man-made sources of ionizing radiation.

The important properties of the radiation under consideration are the following: (a) energy between about 5 and 150 keV; (b) pulsed time structure; (c) large intensities in each pulse (at least a few mR); (d) radiation directed only at a relatively small target. Although the x-ray diffraction and spectroscopy applications do not commonly involve pulsed radiation, they are also treated here.

When one examines the radiation protection aspects of x-radiation from industrial or medical usage, there are several aspects of the problem to be considered:

1. Proper functioning of the x-ray machine itself.
2. Proper design of collimation systems, beam stops, and shielding.
3. Proper operation of the machine by the operator (proper alignment, correct exposure setting, removal of impediments from beam).
4. Proper radiation protection procedures (ensuring that accidental exposures do not occur; routine surveying of direct and stray levels; appropriate personnel monitoring and record-keeping; properly installed and operating interlock systems; etc.).

This discussion deals with the task of monitoring direct and stray radiation and is limited by excluding both personnel monitoring of occupational exposures and patient dosimetry in medical therapy (these subjects are covered in Chapter 9). Thus, this discussion deals only with measurements of direct and stray radiation for the purposes of a radiation protection survey. The interested reader is referred to several basic sources for guidance on shielding and interlock design and other general considerations on x-radiation [5,16,17,18].

There are two aspects of the measurement problem: first, the measurement of possible stray radiation levels far outside of the main beam; and second, the problem of an unintended halo around the intended beam.

Of the many types of instruments sensitive in this region, only two are in common use for this type of work: ionization chambers and solid state integrating devices (film, TLDs). The latter are used mainly in two applications: film is used for determining the exact pattern of a beam profile and halo, and both film and TLDs are used for long-term integration of levels where expected exposure rates are exceedingly small.

For measurement of *absorbed dose* around x-ray machines, ionization chambers of various kinds predominate. There are five parameters of interest in the choice of an instrument: (a) the energy dependence of the chamber; (b) the limiting sensitivity, together with the time required to obtain a reading at that limit; (c) the stability of the instrument against drift; (d) the sensitive area of the instrument; (e) resistance to electrical interference [19].

On the first point (energy dependence), most of the Instrument Notes on ionization chambers contain energy response curves, in many cases taken from the work of Storm at Los Alamos [20]. Medical x-radiation occurs predominantly between about 10 and 100 keV and for many of the instruments there is a very large variation in instrument response over this region. Of those tested by Storm, only the Victoreen 440, 440RF, 444, and 666 have reasonably flat responses ($\pm$ 30%) down to 50 keV. Since the work of Storm, many chambers with flatter responses have been developed. Others have a serious fall-off in sensitivity in the 50–100 keV region; still others have too much sensitivity. However, if the approximate energy spectrum is known, many of them can be used without gross error in this region. It is important to remember that no beam is really monochromatic; there is always a low-energy tail, making low-energy response a desirable feature.

The low limit of response is an important parameter mainly for measurements at a distance from the main beam, because the intensity in and near the beam is usually high enough to provide plenty of signal. The low duty cycle of many x-ray machines can produce stray radiation or radiation behind shielding. Many sources cannot produce a radiation pulse more often than one every several seconds, so that ionization chambers that attempt to read in dose *rate* (mR/hr) will have difficulties in this situation. The solution is an instrument such as some type of ionization chamber that can *integrate* several short bursts and read total exposure (mR).

The importance of knowing the area of sensitivity of any instrument has been discussed. For ionization chambers this can be a problem. Large volumes are used to achieve great overall sensitivity, but if it is obtained for large windows such volumes are of no help if a narrow-cross-section beam is being monitored. Indeed, a large-size window is a handicap if the beam is smaller than the instrument volume, because corrections are then required. For small enough beams, even rather large intensities may produce low readings, inducing a dangerous false sense of security. Large size in general may prevent access to some inconvenient locations where survey information might be needed.

For rapid searches for hotspots, it is not necessary to have an instrument with response proportional to absorbed dose. Devices such as G-M tubes, proportional counters, and scintillation counters are useful. These instruments are preferable to the ion chamber when searching for small cracks because they can be intrinsically more sensitive by as much as two orders of magnitude and because their response times are much faster.

If one then needs absorbed dose information, an ion chamber can be called upon. This fact was brought out in a striking way during the Conference on Radiation Safety [9]. In the conference proceedings, Dr. J. W. Thomas discussed the use of G-M tubes for survey work around x-ray diffraction units:

> In most cases of protection surveys about diffraction equipment we actually are doing a go/no-go type of survey. We are not trying to measure the radiation level in the direct beam because we really don't care what the direct beam is, only where it is. So, if you are approaching apparatus from a position where there is essentially no radiation, all of this group of instruments would start to repond at about the same point in a radiation field. This point is low enough as to read mR/hr to serve our yes/no survey needs. If the meter responds but doesn't go banging off the lower scale, the region in question is safe enough for any reasonable and necessary operations. If the

meter goes banging off the most sensitive scale and perhaps the next most sensitive scale too, you really want to turn off the x-ray unit and figure out what's wrong before going on with the work. This means to me that the instruments can serve a valuable function, and they don't require highly skilled radiation measurements men to do the job. You have to remember that when the meter dial says mR/hr, that may be wrong; you may need a correction factor of 2, of 10 or 30 or more if you really want to know mR/hr. But for an awful lot of protection work you don't really care about mR/hr; you want an indication of where there may be a problem and help in locating and correcting the source of a problem. This is the very great value of these meters in operational radiation safety such as we are discussing.

One of the main goals of a radiation protection survey around a medical or industrial setup is to insure the radiation safety of the system as routinely operated by a technician. The problem of the integrity of shielding is particularly difficult. In many installations, workers in adjacent rooms may be exposed to radiation far above permissible levels because of poor design or improper maintenance of shielding. It is essential, as standard operating procedure, to perform an adequate test of shielding and maintain its integrity against the inroads of convenience or expediency. One of the most common failings of shielding design [21] is that the shielding is only overdesigned in the region where the main beam is intended to be stopped; other areas are shielded by the minimum thicknesses required for small stray radiation levels. If beam setup changes, or a leak develops in the housing, the shielding may no longer be sufficient. The New York City Department of Health now recommends shielding tests by directing the main x-ray beam at areas which ordinarily cannot be expected to receive it; shielding improvements are then suggested until the levels are small even under those circumstances [21].

The detection of pinholes and cracks in shielding may be quite difficult. Because many industrial and medical x-ray machines cannot put out bursts at high repetition rates, the task of scanning for cracks is time-consuming beyond the patience of many surveyers. The recommended practice here [21] is the use of a strong radioactive x-ray source (such as Technetium 99m with 140-keV x-rays, 6-hr half-life), placed in the correct location. A survey with any sensitive rate meter (e.g., a G-M or proportional counter) will detect the presence of leaks without trouble. In this case, one wants to fix the leak, not quantify it.

As already mentioned, another problem of radiation protection is a possible halo around the main beam. This halo can serve as a secondary source itself if it scatters from objects along the path. The way to measure this type of effect is to quantify the actual main beam size with exposure of film, and then to search near the beam edges with a sensitive electronic instrument of small sensitive area. An important problem occurs when beam halo from a medical application strikes a patient outside of the intended area of exposure. The seriousness of this problem cannot be overemphasized: since more than 90% of all exposure to man-made radiation comes from medical x-radiation, it is estimated that dramatic decreases in total exposure could easily be made by proper collimation of the beams of those machines that still do not possess adequate collimation systems [22].

In this regard, the U.S. Bureau of Radiological Health (BRH) has published requirements for x-ray equipment manufacturers governing machine operation at time of sale [23]. One of the BRH's functions is to conduct an electronic product control program to protect the public health and safety. They have as a part of this role issued performance standards for ionizing radiation emitting products, specifically for cold-cathode gas discharge tubes, diagnostic x-ray systems and their major components, radiographic equipment, fluoroscopic equipment, and cabinet x-ray systems [24].

Another function the BRH has performed is to insure that adequate instrumentation exists for determining compliance with the Federal Performance Standard for Diagnostic X-Ray Source Assembly and Their Major Components. The BRH performed a preliminary design of the needed instrument and contracted for it to undergo final design and packaging commercially [25]. The instrument represents the culmination of work on the development of a charge digitizer, a form of electrometer which is inherently digital. By exploiting the speed of response of such devices, timing information, as well as radiation information, is derived from ion chamber currents. This permits a large amount of testing with a single instrument and eliminates the need for a separate instrument to evaluate time performances.

5.3.3. Continuous Low-Level Leakage

In this section, emissions from γ-ray sources, color TV receivers, klystrons, and other low-level, low-energy ($<$ 150 keV) x-radiation sources will be considered. Industrial application of x-ray diffraction is also treated here, although the intensity levels may in this case be very high.

Continuous, low-level emissions generally have a known energy spectrum: for isotope sources this is certainly true, and for leakage radiation at least the maximum energy is known, since x-radiation significantly above the accelerating voltage does not occur. This simplifies the measurement problem enormously. As an example of the problems of measurements of radiation from *leakage* sources, the particular case of color TV receivers will be treated since this problem has received much attention recently. More stringent regulation and increasing use of solid-state circuitry has now reduced the x-ray emission from recently built color TVs to negligible levels under conditions of normal operation and proper servicing [24,26,27].

The requirements for monitoring leakage from color TVs is similar to that for monitoring leakage from other sources. Most x-radiation emitted from the active elements within a typical color TV set is limited to the energy region up to 45–50 keV, and in the past it emerged through narrow cracks and slits [28,29]. This is no longer so; now it only emerges through the face of the cathode ray tube (CRT) [30]. The levels which must be detected are very low: the recommended NCRP maximum emission level and BRH performance standard is 0.5 mR/hr over a 10 cm^2 area, at a distance of 5 cm from the receiver surface [24,31,32].

Figure 5.5 shows the energy spectra of the three important x-ray sources: picture tube, rectifier, and from a typical 25-in. rectangular color TV receiver (manufacturer undisclosed), operating at 25 kV. The x-rays peak at about the energy of the tube, with a secondary peak at about 12 keV from the shunt regulator (the use of these regulators was discontinued about 10 years ago). Figure 5.6 shows the spatial distribution from a typical receiver when de-tuned and adjusted so that the radiation level is exactly at the 0.5 mR/hr maximum recommended by the National Council on Radiation Protection and Measurements (NCRP).

In order to measure radiation from these sources and other low-level x-ray sources, several specific instruments have been designed. One of the earliest described by Braestrup [33], built in 1949, is a flat, cylindrical ion chamber run either in condenser mode or attached to an electrometer. It has a 0.1-mm window of nylon, coated on the inside with Aquedag to make it conducting. Its principal disadvantage is that the measurement is time-consuming (several minutes for

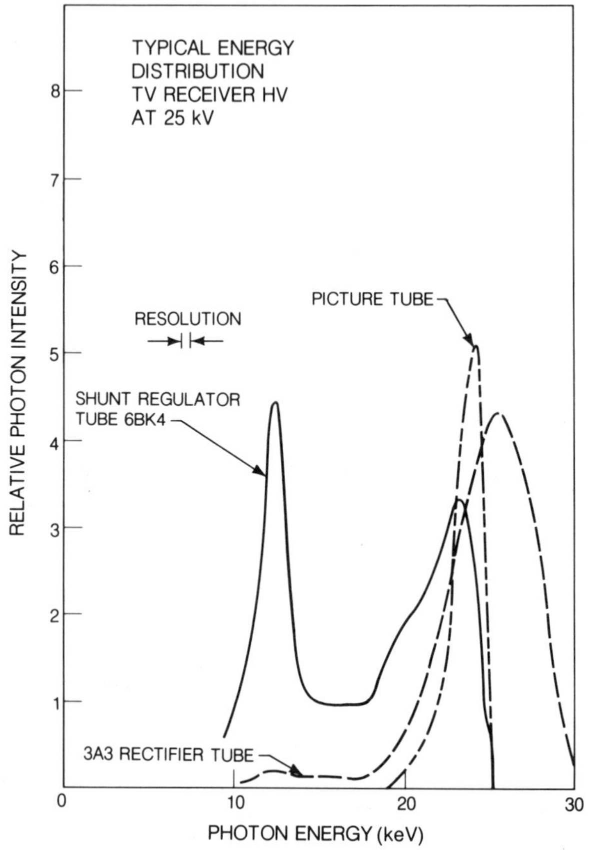

Figure 5.5. Typical energy distribution for TV receiver HV at 25 kV (reproduced from Wang, Ref. 29).

0.1 mR/hr levels) and represents an average over a large (50 cm^2) area. The latter problem is very serious when searching for narrow-beam leaks.

In 1968, the U.S.P.H.S. National Center for Radiological Health (now the Bureau of Radiological Health) designed a TV set x-ray survey meter [30] using six aluminum 1B85 G-M tubes connected so that the meter circuit responds only to the signal from that G-M tube with the highest momentary output. The most sensitive scale responds to a product of exposure rate of 0.5 mR/hr over a detector area of 13 cm^2; more intense exposure levels over smaller areas are integrated over the 13-cm^2 effective area. Its energy dependence is explicitly compared to several other instruments by Lee [30].

Another BRH study [23] examined the relative merits of film (Kodak Royal Blue in Ready Pack), an end-window G-M tube (Eberline RM-3A), and an ion chamber (Victoreen 440RF), when used to test leakage from an ensemble of about 100 misaligned Type GEF4 shunt voltage regulator tubes. The size of the leakage spot varied from a few mm^2 to a few cm^2, and the effective energy was 10 to 16 keV. It was found that, using the film as the standard, the two electronic instruments were unable to distinguish the intense spotlike characteristic of the leakage beam from a diffuse source of lower intensity over a bigger area. The average "area correction

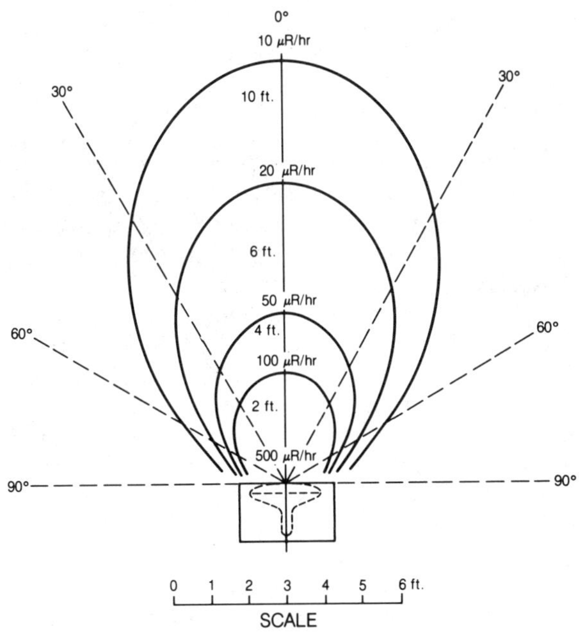

Figure 5.6. Spatial distribution from a typical 25-in. color receiver (full raster) when adjusted to 500 R/hr level at 5 cm from tube front (reproduced from Wang, Ref. 29).

factor" for the two cases was a factor of 18.5 for the G-M tube and 30 for the ionization chamber. The conclusion of the study was that the data

> suggest a need for additional investigation and development of instrumentation for surveying small-area x-ray beams which are also very intense and soft. [13]

Stewart [13] concludes:

> Based on the work conducted at the X-ray Exposure Control Laboratory, several statements can be made and verified by experimental data concerning the calibration of radiation detection devices used in measuring x-ray emissions from color television receiver components.
>
> Film has been found to be the best detector for permanently recording and measuring in detail the area distribution of x-ray exposures, especially for small x-ray beams such as found from the components investigated. Calibration of film for each quality tested is necessary for accurate results, but the quality-dependence over the range 20 to 30 kev, 0.1 to 0.5 mm Al HVL, is at most 30%, so only the extremes need be calibrated and an average conversion curve used for the purpose of surveying high voltage shunt regulator tubes and the like.
>
> End-window G-M tubes are useful in probing for narrow x-ray beams in that they can locate rapidly with relative ease the x-rays emitted from television components and from sets. A drawback, however, to the ordinary large-aperture end-window probe is that only a small fraction (5% on the average for our data) of the actual intensity will be read when surveying narrow x-ray beams.

> Ion chamber survey meters are considered the best for accurate exposure measurements of broad beams that uniformly irradiate the entire volume of the ion chamber and are of high enough exposure rate to give statistically valid readings. In narrow beam use the resulting readings were too low by an average factor of 30.

For narrow-beam, high-intensity studies, an electronic instrument with as small as possible a sensitive volume should be used. These instruments have recently become available. Several other techniques have been proposed and the use of a small scintillator shows promise for measuring slitlike sources.

The use of a NaI(Tl) scintillation counter, in the Eberline Model PRM-5 with PG-1 probe, has been described by Rechen and Williams [34]. The PRM-5 unit houses a battery-operated high voltage supply for the phototube, a double pulse-height discriminator, and a count rate meter. The discriminator must be adjusted to a pulse height "window" to suit the wavelength of interest. The detector, a thin (1 mm) chip of NaI of area 8 cm^2, has a $> 80\%$ efficiency from 10 to 70 keV. The sensitivity is adequate: 0.5 mR/hr produces 110,000 cpm. It has an extremely large dynamic range, making it "very useful for plotting approximate iso-exposure rates around television receivers and other electronic products." [34] The instrument is easily calibrated and convenient to use. However, the small detector area makes a survey for minute cracks difficult, and it should therefore be used in conjunction with a larger-area detector. Rechen and Williams [34] present a complete analysis of this instrument's performance and should be consulted for details.

For the problem of sources distributed broadly in space, a very useful study has been made at Zenith Radio Corporation by Savic et al. [35], who tested the following instruments:

1. Victoreen 440RF shielded ionization chamber.
2. Victoreen Thyac III G-M Counter with a 489-35 α-, β-, γ-particle probe.
3. Victoreen Scintillator "γ probe" which was used with the Thyac III count rate meter.
4. Eberline Model PRM-5 pulse rate meter having a single-channel analyzer capability, in conjunction with a sodium iodide [NaI(Tl)] crystal Plutonium Gamma Probe (Model PG 1).

Although these may not be the best choice of instruments today, a discussion of their characteristics illustrates basic concepts in such instrumentation. In this study [35] the Victoreen 440 RF ionization chamber was used as a standard, and Fig. 5.7 shows the energy dependence of the other four instruments compared to it. Here the rate of emission is 1 mR/hr of approximately monochromatic radiation with the energy on the abscissa. Data were taken 18 in. from the front of a color TV picture tube. All except the Thyac III (with γ-ray probe) can be used in this application with only minor corrections required for energy dependence in the 25 to 35 keV energy region. The Victoreen 1B85 G-M tube is the detector used in the NCRH (BRH) instrument described earlier [36].

Throughout the literature on this subject, the Victoreen 440RF with the 440RFC ionization chamber is considered a reference instrument. Victoreen's 208A/239A Minometer chamber has also been a common reference laboratory standard. The responses of these two instruments are shown in Figs. 5.8 and 5.9.

To summarize, it seems clear that although instrumentation exists for the task, there are difficulties which must be understood by the user. For measurements of

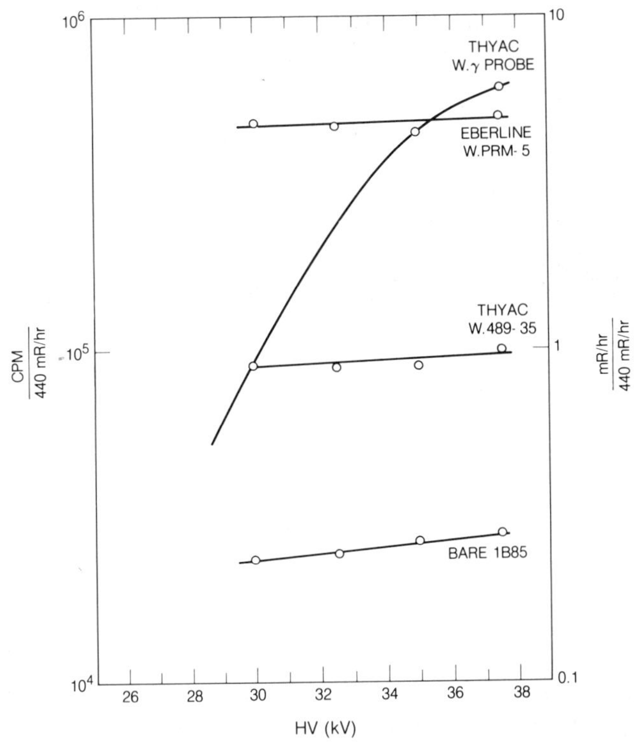

Figure 5.7. Response versus voltage of tube for four detectors, normalized to 1 mR/hr as read on a Victoreen 440RF ionization chamber (reproduced from Savic, et al., Ref. 35).

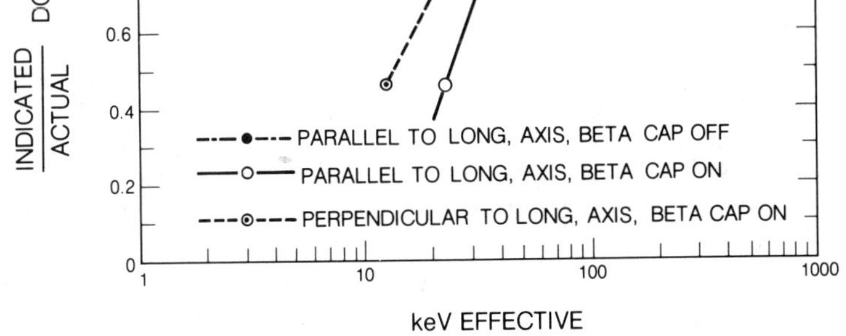

Figure 5.8. Energy response of Victoreen 239A and 208A ionization chambers (reproduced from Bellian, Ref. 37).

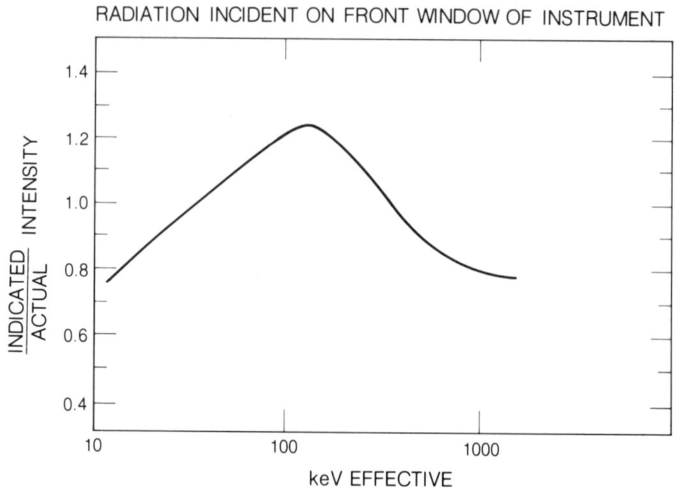

Figure 5.9. Energy response of Victoreen 440RF ionization chamber (reproduced from Bellian, Ref. 37).

broad-area, low-level radiation the ionization chamber technique is the most reliable; of course, chambers with slowly varying energy dependence at low energies must be used. The use of G-M counters is recommended only when the energy spectrum is understood. For studies of slitlike, high-intensity leakage, the scintillation counter technique is appropriate, because its sensitive area can be made small. Recently, ionization chambers with small volume (down to 0.2 cm^3) have become available and can be used also. Alternatively, an instrument like that described by Rechen [36] (with six G-M tubes, connected so that only the one with the greatest instantaneous count rate is read out) can be useful. The use of TLDs and films or emulsions is very satisfactory, but only in those situations where an instantaneous answer on exposure levels is not required.

5.4. X- AND GAMMA RADIATION FROM ABOUT 150 keV TO SEVERAL MeV

Photons in this energy region arise principally from two sources: particle accelerators and radionuclides. Although only a very few radionuclides produce γ-rays of energies above 2 MeV, particle accelerators can yield photons of much higher energies; the latter will be discussed separately (Section 5.5). An important distinction between these sources is that whereas accelerator photons often have an unknown energy spectrum and can be accompanied by other species of particles (principally neutrons), radionuclide sources usually present neither of these problems.

The basic references on radiation protection in this energy region are NCRP Reports 33, 35, 36, 40, and 49 [16,17,38,39,40], ICRU Report 14 [41], and especially NCRP Report 25 [42]. These reports discuss fundamental aspects of the problem (e.g., shielding, handling, and protection instrumentation) which should be incorporated into the basic operating philosophy of any worker in this field.

For the detection of x-rays and γ-rays in the 0.15 to 3 MeV range, the Geiger counter was by far the most widely used instrument until the early 1950s. Since then the scintillation counter, because of its greater sensitivity, has replaced it for much of this work. There is a variety of other types of instruments that can be used for radiation detection in the energy range in question.

There is also a wide range of instrumentation available for health hazard measurements in this energy region. However, the instrument most suited for handheld survey measurements (and most commonly used) is probably the *cavity ionization* chamber, in one of several designs. If absorbed dose must be determined, the other kinds of common instruments (G-M counters, gas proportional counters, scintillation counters) have serious, energy-dependent, sensitivity problems.

In this energy region most of the comments made in the previous section about lower energies still apply. One basic difference in photons at these higher energies is that they penetrate materials (especially low-Z) to significant depths, and Compton scattering (gamma-photon electron scattering) overtakes the photoelectric effect as the dominant interaction process [43,44]. An additional complication for radiation protection is that these γ-rays can backscatter, because the Compton process has such a wide angular distribution [17,18]. Thus, the energy spectrum reaching a detector from a sealed γ-ray source can contain components from many directions besides the source itself, and with the wide energy spectrum characteristic of the Compton process. The instrumentation required for the measurement of such a spectrum needs a wide, flat energy response, and the same comments apply as for the lower energies.

Fortunately, at these higher energies the wall thickness of the cavity detectors becomes less critical. Also, the uniformity of energy response is better for G-M and proportional counters, especially above about 1 MeV, so that these instruments become useful for rough absorbed-dose measurements.

Neutron interference can be a problem in this energy region. In this energy region, air and tissue are nearly equivalent in terms of ionization (ergs/g) (joules/kg) deposited by a given incident photon. The use of air-equivalent and tissue-equivalent materials has been common; these all work fairly well. The use of TLDs, film, and emulsion for personnel health protection purposes is discussed elsewhere in this volume (see Chapter 9, Dosimetry). Ionization chambers with rather small sensitivity to neutrons can be constructed. However, sensitivity to neutrons varies from one material to another. In particular, the response of hydrogenous materials to neutrons is often more than an order of magnitude greater than that of nonhydrogenous materials. This latter fact makes possible the independent measurement of γ-ray and neutron absorbed doses by comparing the response of two differently constructed chambers.

To illustrate this, consider the comparison [42] between a tissue-equivalent chamber and a chamber with conducting Teflon lining and CO_2 gas. Exposure of each of these chambers to 1 R of gamma radiation of about 1 MeV produced a reading of 0.97 rad (9.7 mGy). However, 8 to 10 rem (80 to 100 mSv) were required to produce a reading of 1 R in the tissue-equivalent chamber; response was nearly rad-proportional, since 8 to 10 rem (80 to 100 mSv) is equal to 1 rad (10 mGy) for fast neutrons. For the Teflon-CO_2 chamber, 30 to 60 rem (300 to 600 mSv) (depending on the energy) were required for a response of 1 R, and below about 100 keV, the neutron response was completely negligible.

5.5. HIGH-ENERGY GAMMA-RAYS

Photons with energies above a few (say 10) MeV are almost never found in quantities of significance to health except near high-energy accelerators (see Chapter 3, Section 3.4, Particle Accelerators). In these installations, the radiation fields outside of the shielding walls tend to be dominated by neutrons [45,46]. Even when high-energy components are present, ionization chambers are most commonly used for health hazard assays, despite the fact that their response is not proportional to absorbed dose beyond a few MeV. The reason is that when very-high-energy (even GeV-level) electromagnetic showers develop in the shielding, the primary energy is successively degraded into more and more particles of lower and lower energy; eventually the primary photon produces a peak number of particles in the shower. Most low-energy photons are absorbed in rather short distances after their production in the shielding. Above about 5 or 10 MeV, however, the absorption coefficients are both small and relatively constant. When one of these higher-energy photons does interact, energy is quickly lost due to production of lower-energy secondaries of shorter range. Thus, the fall-off of intensity with increasing penetration is slow but at any point the dominant particle flux (and absorbed dose) consists of and is due to low-energy particles, to which these instruments have good response.

This concept is crucial to an understanding of measurement of fluxes such as those from high-energy photons. This is why ionization chambers, whose sensitivities are poor above a few MeV, are nevertheless adequate instruments for absorbed doses in equilibrium photon radiation fields.

If there is a suspicion that nonequilibrium high-energy components are present, the best check is to add absorber around the ion chamber in stages, up to about 30 g/cm^2. In equilibrium fields, there should be a small, monotonic decrease in instrument response, while in nonequilibrium situations an actual buildup of response will occur as mass is added.

5.6. INSTRUMENT CALIBRATION

The two common sources of photons used for instrument calibrations are radioactive γ-ray emitters and x-ray tubes. Since the response of many instruments is a strong function of energy, particularly at the lower energies, the source should be as monoenergetic as possible.

Table 5.1 shows some common radionuclide γ-ray sources. Radionuclide sources are available even in the low-energy region where x-ray tubes are also used. The last column, taken from Nachtigall [47], shows the specific γ-ray constant in units of roentgen-meter^2/curie-hour, which specifies numerically the exposure rate (R/hr) at 1 m distance from 1 curie of activity. However, these values may not be accurate enough for calibration purposes in the presence of scattering and absorption problems.

Beams from x-ray tubes are useful in the lower energy region. Their main problem is that their energy spectrum usually is broad even as it emerges from the tube. The low-energy tail on an x-ray spectrum can be greatly reduced in magnitude by the use of absorbers. However, care is required. To quote from ICRU Report No. 20 [11]:

Table 5.1. Some Gamma-Ray Calibration Sources (Ref. 11, with permission of the International Commission on Radiation Units © 1971)

Radionuclide	Effective Photon Energy (keV)	Half-life (Physical)	Specific Gamma-Ray Constant ($R\text{-}m^2/Ci\text{-}hr$)
Am-241	60	458 years	0.016
In-114m	192	50 days	0.043
Hg-203	279	47 days	0.12
Au-198	412	2.7 days	0.23
Cs-137	662	29.9 years	0.32
Co-60	1170, 1330	5.23 years	1.30
Ra-226[a]	—	1608 years	0.825
Na-24	1370, 2750	15 hours	1.83

[a]0.5 mm Pt filtration, decay products included.

> Care should be taken that secondary peaks are not introduced from the radiation produced by fluorescence of the filter material. For example, with a lead filter it may prove necessary to follow the lead with a tin filter to absorb the K radiation from the lead. Likewise, a copper filter may be required to absorb the K radiation from tin; an aluminum filter may be used to absorb the K radiation from copper. Normally high purity materials should be used as absorbers.... For measurements of energy dependence below 100 keV, the K-fluorescence radiation of different elements can be used as an almost monoenergetic source of radiation [48,49]. For most radiation protection applications the better energy definition obtained by this method is not required.

Two geometrical effects are particularly important. First, the angle of incidence of the radiation can affect the sensitivity of the detector, and the effect can be dependent on energy. Second, backscatter from a human body can significantly affect the response of an instrument held in the hand or worn on the body. This can be checked by using a phantom positioned appropriately.

Radionuclide sources, with intrinsically monoenergetic γ-ray energy, can be degraded either by absorption or by scattering (in the source itself, its container, or the air). For calibration of instruments to be used below a few tens of keV, absorption is the dominant problem, and care should be taken to account for it in the calibration procedure. One also wants the detector to be far enough away that the finite instrument dimensions (depth and area) are not the major uncertainty in the measurement. Because it is difficult to know the effective "center" of a detector to better than one-third the relevant dimension, one usually needs to locate the detector several detector-diameters away from a source to reduce the $1/r^2$ uncertainties to 10%. Deviations from inverse-square fall-off of flux can be studied with a detector whose size is much smaller than the distances involved.

Compton scattering is very important in the MeV range because of its broad angular distribution [44]. This makes it desirable to locate the calibration setup as far as practically possible from large scattering surfaces such as walls.

The effect of humidity, temperature, and pressure during calibration should be included with the calibration information.

A detailed discussion of absolute national standards for instrument calibration is beyond the scope of this work. The National Bureau of Standards (NBS) has for some years performed calibrations of x- and γ-ray instruments against its own national standard [33]. However, the need for a larger number of calibration facilities has prompted the American Association of Physicists in Medicine (AAPM) to establish two certification centers. The primary motivation for these centers is for purposes of precise dosimetry in radiation therapy, but their use for radiation protection instrumentation is also important. The centers, with secondary standards calibrated against the NBS national standard, perform routine calibrations of instruments for users. Information about the effort is available from the American Association of Physicists in Medicine, 111 E. Nacker Dr., Rm. 620, Chicago, IL 60601. Two regional centers have been certified: M. D. Anderson Hospital in Houston and Memorial Hospital in New York City.

5.7. SUMMARY

Monitoring of x- and γ-radiation has four distinct classes: (a) the presence of ionizing radiation where it is generally not intended (e.g., klystrons and TV receivers); (b) industrial and medical x-radiation, usually at energies of about ~150 keV; and (c) higher-energy photons, usually from accelerators; or (d) sealed radioactive sources.

Two important guidelines should be kept in mind. First, it is essential that detailed radiation protection surveys be performed whenever an installation begins operation or makes substantial changes in operating conditions. This survey should identify and quantify the important sources of radiation, thus leading the way to reducing unnecessary exposures wherever possible. Second, routine surveillance is essential for maintaining the integrity of any operation.

The routine surveillance should have two goals: The quantification of absorbed dose (and dose equivalent) levels, and the determination of possible changes in operating conditions which might demand remedial action (e.g., the search for cracks in shielding). Instruments suitable for each of these tasks have been described in the text.

The ionization chamber is the basic survey instrument for x- and γ-ray absorbed dose measurements; the properties of and applicability of various ionization chambers have been discussed. For survey use where absorbed dose is not required, a variety of other instruments is available: G-M tubes, gas proportional counters, scintillation detectors, and in some cases film and thermoluminescent detectors are appropriate. Guidelines have been presented to aid in selecting the type of instrument which will best perform a particular function, and a discussion has been included concerning instruments and techniques which are not generally commercially available, but are worthy of widespread availability. The reader is referred to the original references for more details.

Finally, a comment about the electronics circuit design is necessary. Some monitoring instruments have not as yet incorporated the most recent advances in electronic technology. Properly exploited, these advances may lead to many improvements: an increase in stability and reliability; a reduction in frequency and maintenance; a decrease in size, weight and cost; and a better means of data readout.

REFERENCES

1. International Commission on Radiological Protection, "General Principles of Monitoring for Radiation Protection of Workers," ICRP Publication No. 12, Pergamon Press, Oxford, 1968.
2. M. Ehrlich, "The Use of Film Badges for Personnel Monitoring," International Atomic Energy Agency, IAEA Safety Series No. 8, Vienna, 1962.
3. International Atomic Energy Agency, "The Basic Requirements for Personnel Monitoring," IAEA Safety Series No. 14, Vienna, 1965.
4. D. E. Barber, "Standards of Performance for Film Badge Services," U.S. Department of Health, Education and Welfare, PHS No. 99-RH-20, Washington, DC, 1966.
5. International Commission on Radiation Units and Measurements, "Radiation Dosimetry: X-Rays Generated at Potentials of 5 to 150 kV," ICRU Report No. 17, 1970.
6. National Council on Radiation Protection and Measurements, "Instrumentation and Monitoring Methods for Radiation Protection," NCRP Report No. 57, 1978.
7. J. A. B. Gibson and J. U. Ahmed, "Measurement of Short-Range Radiations," International Atomic Energy Agency, Technical Reports Series No. 150, Vienna, 1973.
8. Proceedings of the Conference on Detection and Measurements of X-Radiation from Color Television Receivers, March 28–29, 1968, by U.S. Public Health Service, Bureau of Radiological Health, Rockville, Maryland, 1968.
9. Radiation Safety in X-Ray Diffraction and Spectroscopy, Proceedings of a Conference in Philadelphia, January 6–7, 1970, U.S. Department of Health, Education and Welfare, Publication No. (FDA) 72-8009; also available from U.S. Bureau of Radiological Health as Document BRH/DEP 72-3, 1971.
10. L. Bates, "Characteristics of Detectors," p. 105 of Ref. 8.
11. International Commission on Radiation Units and Measurements, "Radiation Protection Instrumentation and Its Applications," ICRU Report No. 20, 1971.
12. International Commission on Radiation Units and Measurements, "Dose Specification for Reporting External Beam Therapy with Photons and Electrons," ICRU Report No. 29, 1978.
13. H. Stewart, "Low Energy Calibration of X-Ray Film Used in the Measurement of X-Rays from Color TV Receivers," p. 185 of Ref. 7.
14. R. D. Kirk, J. H. Schulman, E. J. West, and A. E. Nash, "Studies on Thermoluminescent Lithium Borate in Dosimetry," in Proceedings of the Symposium on Solid State and Chemical Radiation Dosimetry in Medicine and Biology, (International Atomic Energy Agency, Vienna, 1966), p. 91.
15. G. W. R. Endres, "Characteristics of Teflon–LiF Thermoluminescent Dosimeter," Battelle Northwest Laboratories Report No. BNWL-741, 1968.

16. National Council on Radiation Protection and Measurements, "Medical X-Ray and Gamma Ray Protection for Energies up to 10 MeV, Equipment Design and Use," NCRP Report No. 33, 1968.
17. National Council on Radiation Protection and Measurements, "Structural Shielding Design and Evaluation for Medical use of X-Rays and Gamma Rays of Energies Up to 10 MeV," NCRP Report No. 49, 1976.
18. National Council on Radiation Protection and Measurements, "Radiation Protection for Medical and Allied Health Personnel," NCRP Report No. 48, 1976.
19. J. R. Frazier, T. R. Ohlhaber, and P. S. Ruggera, "Performance of X-Ray Measurement Instruments When Subjected to Environmental Level RF Fields," U.S. Department of Health, Education and Welfare, Publication No. (FDA) 78-8065, 1978.
20. E. Storm, B. J. Krohn, and W. B. Chambers, "The Photon Energy Response of Several Commercial Ionization Chambers, Geiger Counters, and Thermoluminescent Detectors," Los Alamos Scientific Laboratory Report LA-4052, 1969.
21. Hanson Blatz, Department of Health, City of New York, private communications.
22. J. C. Villforth, *Proceedings of Second Annual Conference on Radiation Control,* April 26–29, 1970, Palm Beach, Florida, U.S. Bureau of Radiological Health, BRH/ORO 70-5, 1970, p. 23.
23. U.S. Department of Health, Education and Welfare, "Diagnostic X-Ray Systems and Their Major Components: Proposed Performance Standard," *Fed. Register,* **36**, 196, 1971.
24. Department of Health, Education and Welfare, Public Health Service, "Regulations for the Administration and Enforcement of the Radiation Control for Health and Safety Act of 1968," HEW Publication No. (FDA) 76-8035, 1976.
25. T. R. Lee, "Evaluation of MDH Model 1015 X-Ray Monitor," HEW Publication No. (FDA) 77-8014, 1977.
26. United Nations Scientific Committee on the Effects of Atomic Radiation, *Sources and Effects of Ionizing Radiation,* 1977 report to the General Assembly, with Appendix, United Nations, New York, 1977.
27. U.S. Environmental Protection Agency, "Radiological Quality of the Environment," EPA-520/1-76-010, 1976 (May 1976).
28. H. D. Youmans, "Spectra and Percentage Depth Dose of X-Radiations Emitted by a Color Television Receiver," p. 31 of Ref. 8.
29. S. P. Wang, "Spectra and Spatial Distribution of X-Rays from Color Television Receivers," p. 53 of Ref. 8.
30. T. R. Lee, Bureau of Radiological Health, Rockville, Maryland, private communication, 1980.
31. National Committee on Radiation Protection and Measurements (NCRP), *Radiology*, **75**, 122, 1960. A further discussion by H. C. Wyckoff is on p. 19 of Ref. 8.
32. National Council on Radiation Protection and Measurements, "X-Ray Protection Standards for Home Television Receivers, Interim Statement of the National Council on Radiation Protection and Measurements," Washington, DC, 1968.
33. C. B. Braestrup, "Basic Instrumentation and Calibration for TV X-Ray Measurements," p. 143 of Ref. 8.

34. H. Rechen and C. Williams, "Energy Dependence and Linearity of Response to Soft X-Rays of Eberline PRM-7 Plutonium Gamma Survey Meter with PG-1 Probe," p. 213 of Ref. 8.
35. S. Savic et al., "Comparison of Commercially Available Instruments of Different Types," p. 223 of Ref. 8.
36. H. J. Rechen, "Calibration and Energy Dependence of the EPRL X-Ray Survey Meter," p. 175 of Ref. 8. Articles by R. K. Stroms and R. N. Schneider in Ref. 12 also discuss this instrument.
37. J. G. Bellian, "Calibration Techniques for Determining Color TV X-Rays Emissions," p. 147 of Ref. 8.
38. National Council on Radiation Protection and Measurements, "Dental Protection in Veterinarian Medicine," NCRP Report No. 35, 1970.
39. National Council on Radiation Protection and Measurements, "Radiation Protection in Veterinarian Medicine," NCRP Report No. 36, 1970.
40. National Council on Radiation Protection and Measurements, "Protection Against Radiation From Brachytherapy Sources," NCRP Report No. 40, 1972.
41. International Commission on Radiation Units and Measurements, "Radiation Dosimetry: X-Rays and Gamma Rays with Maximum Photon Energies Between 0.6 and 50 MeV," ICRU Report No. 14, 1969.
42. National Council on Radiation Protection and Measurements, "Measurement of Absorbed Dose of Neutrons, and of Mixtures of Neutrons and Gamma Rays," NCRP Report No. 25, 1961; also published as National Bureau of Standards Handbook 75.
43. N. W. Holm and R. J. Berry, *Manual on Radiation Dosimetry,* Dekker, New York, 1970.
44. Emilio Segrè, *Nuclei and Particles: An Introduction to Nuclear and Subnuclear Physics,* 2nd ed. Addison-Wesley, Reading, MA, 1977.
45. F. J. Coleman, D. C. Thomas, and G. Saxon, "An Experiment to Determine Shielding Requirements for a Multi-GeV Electron Synchrotron Ring," Daresbury Nuclear Physics Laboratory Report No. DNPL/P 72, Daresbury, England, 1971.
46. Ralph Thomas, Lawrence Berkeley Laboratory, private communication.
47. D. Nachtigall, *Table of Specific Gamma Ray Constants,* Thiemig–Taschenbucher Band 34, Karl Thiemig Verlag, Munich 1969.
48. E. Storm and S. Shaler, "Development of Energy-Independent Film Badges," *Health Phys.*, **11**, 1127, 1965.
49. A. Shambon, A. Zila, and R. Sanna, "Energy Dependence Determinations in the 60–180 keV Range with an X-Ray Machine of Varying Output," Health and Safety Laboratory Report HASL-133, New York, 1963.

6 GAMMA- AND X-RAY SPECTROMETRY

Spectrometry of ionizing radiation, like spectrometry of visible light, is the measurement of intensity as a function of wavelength, or as a function of photon energy. There are two techniques for making this measurement. One is by using a crystal as a diffracting means and measuring radiation intensity as a function of the angle of incidence and emergence, the radiation wavelength being given by Bragg's Law:

$$n\lambda = 2d \sin \theta \quad ,$$

where n = an integer (e.g., 1,2,3,etc.),
λ = radiation wavelength,
d = distance between crystal planes,
θ = angle between the direction of the incoming radiation and the crystal plane.

This method is not practical for radiation of wavelengths shorter than about 0.1 Å (i.e., 125 keV photon energy). Furthermore the radiation usually must be in the form of a well-collimated beam. For these reasons this method is not used to identify γ-ray sources, or x-ray leakages, and is not discussed further here. The second method is by measuring the energy given up by each photon and counting photons as a function of their energy. This second technique is widely used and is the subject of the rest of this chapter.

The need for γ-ray spectrometry has increased over the past decade in response to the needs of the nuclear power industry and of regulatory agencies that monitor the nuclear facilities and their environmental residues. This expanding use derives from the fact that many environmentally important radionuclides from the fission process are γ-ray emitters. Measurements of radionuclides that are primarily β-particle emitters (^{3}H, ^{14}C, ^{85}Kr) or α-particle emitters (transuranics) are discussed in other chapters. The most common reason for x-ray and γ-ray spectroscopy in the radiation-protection field is to identify radionuclides. A γ-ray spectrometer can be used as a laboratory device where samples are counted and also as a field device for performing *in situ* analysis. For example, the γ-ray spectrometer can be used at the source of radioactivity (such as a nuclear power reactor) in order to identify specific radionuclides in gaseous or liquid effluents.

There is a wide range of phenomena in which photons (γ-rays or x-rays) are emitted by atoms or nuclei.

Radioactive gamma-ray emission occurs when an excited nucleus emits a photon as it makes a transition to a lower energy level is generally monoenergetic. This gamma-ray emission is characterized by other features common to all radioactivity. Specific radionuclides have a definite half-life associated with them which, when coupled to the energy spectrum, helps to make an identification.

Although most γ-rays occur singly, there are many important cases in which two or more γ-rays emerge in sequence as the nucleus undergoes successive transitions. Usually, the times between these multiple transitions are much shorter (< picoseconds) than instrumental resolving times, and the multiple emissions can be thought of instrumentally as being in "time coincidence". Such coincidences, where appropriate, can be measured and used as unique signatures to identify particular decay schemes. Coincidences between a γ-ray and another particle such as a β-ray may also be useful in a similar way.

For many α-particle emitting, high-Z elements, α-particle decay is followed by both x-radiation and γ-radiation, as transitions occur in both the electronic and nuclear state of the daughter atom. The x-radiation, if sufficiently energetic, can provide a signature for the radioactive decay. The K x-rays for elements with $Z > 20$ are detected. For the elements heavier than $Z = 80$, such as uranium, the L x-rays are also useful. They are in the 10 to 20 keV range, which is near the lower end of the energy range for the type of instruments discussed here.

6.1. BASIC MEASUREMENT CONSIDERATIONS

In making accurate and sensitive measurements of photon energies in the keV to MeV energy range, a number of important considerations must be taken into account. Ideally, of course, one would like a system that: detects every photon with 100% efficiency; determines its energy with high resolution; has no spurious background indistinguishable from real events; covers the energy range of interest; accommodates all interesting samples; and costs very little to purchase, operate, and maintain. Field instruments should also be rugged and simple.

Actual instruments, of course, compromise between all of these features. Inevitably, as well, the better systems are frequently the more expensive ones. Yet precisely because the typical instrument user is trading one feature for another in his choice of instrumentation (e.g., efficiency versus resolution), it is important to determine the nature of the trade-offs and their interconnections. The key factors are as just indicated: efficiency, resolution, background, energy range, sample capability, and cost.

Usually, environmental radionuclide measurements must be made of a species in very low concentrations. This places severe operational requirements on the instrumentation. Other restraints are imposed by the type and form of sample. The media that require sampling are numerous. Some are directly ingestible (water, food); some are materials used in monitoring direct human exposure (air filters);

many are media making up the pathways through the environment to man (soil, animal tissues, plants); and some are biomedical specimens (bioassay, whole-body determinations). It is therefore difficult to generalize about the levels of concentrations that require measurement, and no attempt to do so is made in this discussion. However, many of these measurements are at levels requiring sensitivities that approach the limit available with the best systems. Hence there is nearly always a premium on sensitivity, and this is coupled with resolution requirements when a minor activity must be identified in the presence of larger activities.

First, the important characteristics of all detection and measurement techniques (efficiency, sensitivity, and background) are discussed. Subsequently, various specific techniques are examined in detail, followed by some aspects of data handling and analysis are described.

6.1.1. Efficiency

The efficiency of a gamma-ray spectrometer is the percentage of the desired photons analyzed by the instrument from the total photons incident on the detector.

The detector size and shape are important parameters in the overall efficiency: Clearly some photons will not interact in a small detector, so large detectors are definitely advantageous if only from considerations of volume or solid-angle. There is more to the size question than mere first-interaction volume, however. For larger detectors, there is an increasingly larger probability that Compton-scattered photons from a primary interaction will interact again in the crystal. Such multiple interactions effectively enhance the full-energy-peak-to-Compton ratio.

To illustrate some of the losses that may occur in an instrument Fig. 6.1 is presented. This figure shows a typical energy spectrum as detected by a high-resolution Ge(Li) detector. The radionuclide shown is aluminum-28, which, although not of consequence in many environmental situations, is good for illustrative purposes. This nuclide decays by emitting a β-particle with maximum energy of 2.8 MeV, followed by a 1.7789-MeV γ-ray. Sometimes the β-particle emits bremsstrahlung photons as it decelerates; sometimes the γ-ray produces e^+/e^- pairs, either in the source, in the detector, or elsewhere.

The peak at the highest energy (channel number 1700) is due to those events in which the full energy of the 1.7789-MeV γ-ray is converted within the detector to ionization (by the photoelectric effect or by multiple interactions). The broad continuum below the peak contains those events in which an incident photon scatters by Compton effect from an electron in the detector and then escapes from the detector. The signal is representative of the energy imparted to the electron. A head-on Compton collision yields the highest energy electrons, detected at the "Compton edge," clearly seen in Fig. 6.1 near channel 1600.

When γ-rays undergo conversion to an electron and positron pair, 1.02 MeV of the γ-ray energy is required to produce the pair and any additional energy is distri-

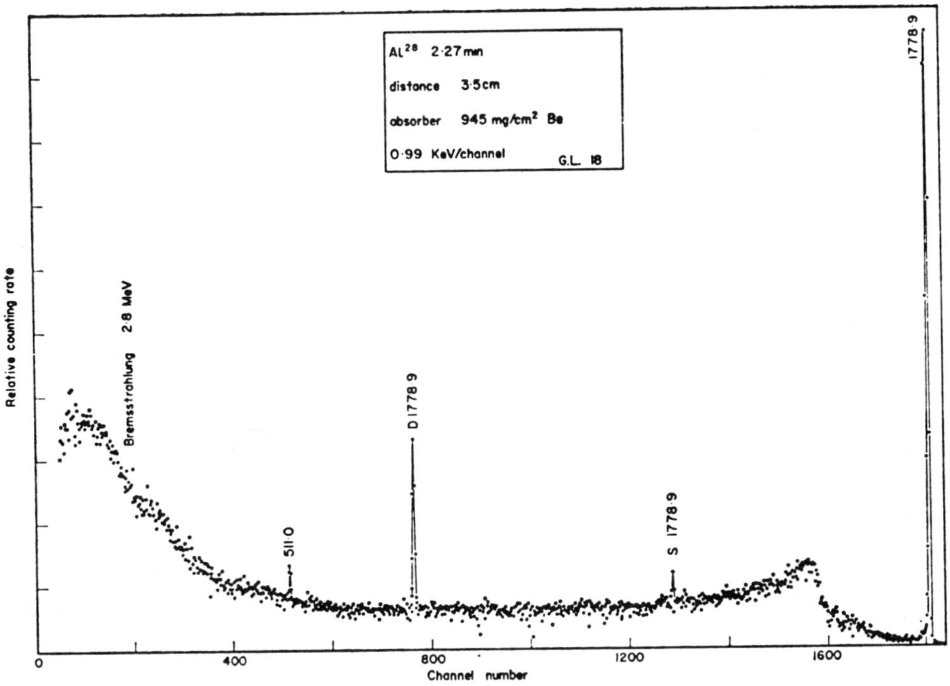

Figure 6.1. Spectrum at aluminum-28 as measured with a Ge(Li) detector (reproduced from Crouthamel, Ref. 1, with permission of Pergamon Press © 1970).

buted between the electron and the positron as kinetic energy. The positron will be annihilated in a short distance with the production of two γ-rays of 511-keV of energy. Peak S 1778.9 results when one of the 511-keV γ-rays has escaped from the detection region and represents the energy deposited by the electron from pair production plus 511-keV (single escape peak). The peak D 1778.9 represents the energy from the electron from pair production when neither annihilation γ-ray has been stopped in the detector (double escape peak).

The "511" peak represents total absorption of 511 keV γ-rays, presumably from annihilation of positrons mostly within the ^{28}Al source itself. Another feature is the prominent low-energy continuum labeled "Bremsstrahlung 2.8 MeV," from the β-particles decelerating mostly in the source. The ratio of the area under the photoelectric peak (channel 1700) to that under the Compton spectrum is primarily determined by physical parameters (the photon energy, the geometrical size and shape of the detector, and the Z of the material). As this ratio is fundamental to the ultimate system sensitivity, it is naturally desirable that it be maximized. Figure 6.2 shows this effect for two sizes of sodium iodide crystals.

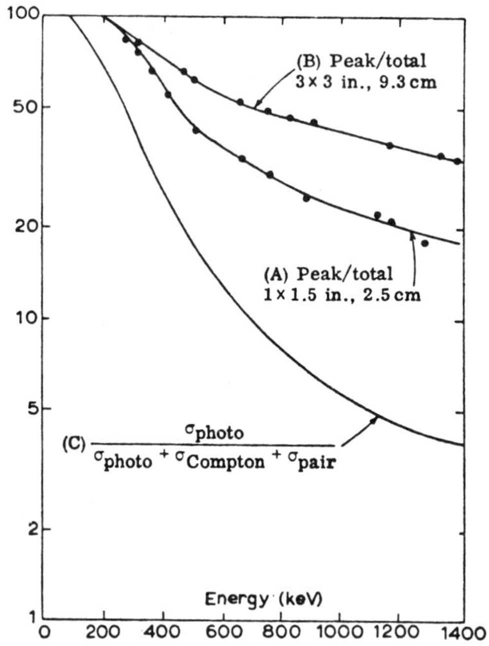

Figure 6.2. Peak-to-total ratios for two crystal sizes and the peak-to-total ratio that would be found if no secondary absorption occurs (reproduced from Neiler and Bell, Refs. 2,3, with permission of North Holland Publishing Co.© 1965).

6.1.2. Energy Resolution

The second key factor is energy resolution because for a given peak area the width of the peak determines its ability to stand out above a continuum background and its ability to be distinguished from other nearby peaks. Resolution is usually stated in terms of the full width of a peak at half-maximum (FWHM) and sometimes also by full width at one-tenth maximum (FWTM). The latter is an indicator of the extent of tail in the peak distribution. Because energy resolution is so extremely important, much of the discussion on the individual measurement techniques is devoted to it. Here, only a few general comments are made.

Resolution by itself is not a definitive parameter by which to compare performance of various systems. For example, resolution enters directly into the ability of a system to detect a small peak sitting on top of a large background continuum. Yet in many systems improving the resolution is not as important as suppressing continuum backgrounds. To combine these effects, various figures-of-merit have been proposed. One is the *full-energy-peak-to-Compton-edge ratio* defined as the ratio of full-energy-peak height to the height of the Compton edge, as measured in the final system energy distribution. This ratio is, in large part, inversely proportional to the peak energy resolution and also depends on many factors such as the overall detection efficiency and the ability of the system to suppress Compton events.

Such suppression is usually accomplished by surrounding the detector with an anticoincidence shield. The anticoincidence shield absorbs secondary lower-energy Compton photons produced by Compton scatter of photons in the detector and vetoes them in the detector circuit. If such Compton events are vetoed electronically, using such a shield, the full-energy-peak-to-Compton ratio can be enchanced by as much as two orders of magnitude. These veto systems are discussed in more detail later.

6.1.3. Background

The third important characteristic is the degree to which counts not arising from the radionuclides of interest constitute "noise," or background. There are several types of background: (a) cosmic radiation; (b) contaminating radioactivity in the counting area, the shielding, or the detector equipment itself; and (c) electronically induced background. Much effort has been expended devising techniques to minimize the radioactivity backgrounds by shielding and by choice of materials. This is particularly important when great sensitivity is desired. The instrumentation discussions will point out the performance of the various systems with respect to efficiency, resolution, and background.

6.2. DETECTION AND MEASUREMENT TECHNIQUES

There are two types of γ-ray spectrometers most used at present. One employs a scintillation counter and the other a semiconductor detector. The areas of applicability along with the basic properties of each detector type will be discussed, as well as areas of new developments. The gas proportional counter, having much lower resolution than the two first mentioned, is still used for special purposes and is discussed first.

6.2.1. Gas Proportional Counters

The use of gas proportional counters for γ- and x-ray spectrometry predates the development of scintillation spectrometry by a few years. For low-energy photons this was the most widely used technique until the development of semiconductor detectors. Today, gas proportional counters are used for low-level γ-spectrometry only when special problems arise.

The passage of photons through the gas produces pairs of electrons and positive ions. About 25 to 30 eV is required to make one such pair, which is about 10 times greater than the 2.9 eV required for an electron-hole pair in solid germanium, but a factor of about 10 smaller than required to produce one photoelectron from a thallium-activated sodium iodide [NaI(Tl)] scintillator-phototube instrument. Hence, in the region of low-energy photons where the absorption of the gas is high, the resolution of the gas counter is intermediate between that of the Ge and NaI(Tl) systems discussed later.

The total primary charge liberated in a gas by a low-energy photon is very small, so that the signal-to-noise ratio obtainable with a practical amplifier would be very poor. Gas counters operated in the ionization-chamber mode have therefore seldom been used for γ-ray spectrometry. It is usually necessary to operate the gas counters in the proportional mode, in which the primary signal is amplified by successive ionizing collisions of the electrons on the way to the collector anode. Such amplification can raise the output pulse size above the noise level of present amplifiers. Most commonly, a cylindrical geometry is used, with the positive anode wire on the axis of a cylindrical gas container.

Because of limitations on size and pressure, gas counters do not have the stopping power of denser detectors (solid or liquid) for high-energy γ-rays. Yet at sufficiently low energies, they are in many ways superior to the scintillation counters and good energy resolution is obtained. Lower energies can be detected and measured reliably. Another advantage is their inherent energy discrimination against high-energy background radiation. Proportional counters are most important for photon energies below 2 keV and they have been used down to 250 eV.

There are a number of factors which must be considered in the design and operation of a gas proportional counter as a photon spectrometer. The choice of window material is important: it must be very thin and of low atomic number (Z) to minimize interactions in the window, and it should be chosen to have low fluorescence (excitation of secondary x-radiation by the primary incident photons). For these reasons a window of beryllium, carbon, or a plastic is often chosen.

The gas fill is usually high-purity xenon, krypton, or argon. The choice is based on the energy region to be measured and primarily takes into account the gas absorption characteristics as a function of energy. Figure 6.3, for example, shows the absorption of photons for each of these three gases in a typical commercial counter. The x-ray "escape" peak from excitation in the gas may also be important: the K x-rays for the three gases occur at 29.5 keV (xenon), 12.5 keV (krypton), and 3.0 keV (argon), respectively. A quench gas, typically a few percent CO_2, is used to prevent continuous discharge, to improve pulse characteristics, and to lengthen counter lifetime.

The choice of wall material is also important. Interference from fluorescence in the walls should be avoided, and this often dictates using a low-Z liner such as aluminum if a higher-Z material such as stainless steel is used for the main wall structure.

The performance of the best gas proportional counters at low energies is very good. For example, resolutions in the range of 15% (FWHM) can be routinely achieved for the 5.9-keV line from iron-55 in the better counters, and sometimes even finer resolution is obtained. For example, resolution values of 11.6% and 12.2% for 5.9 keV have been obtained experimentally [5] for proportional counters with gas fillings Ne + 0.5% Ar and Ar + 0.5% C_2H_2. To achieve the best resolution, a very stable high voltage supply must be used. The resolution obtainable at various energies with instruments of this type is shown in Fig. 6.4.

Among the advantages of gas counters are that a large source area can be utilized, up to about 100 cm^2 for some specially designed counters. Backgrounds can be very low, with an especially important low sensitivity to high-energy ($\geq$ 100 keV) photons. The instruments are easy to operate and can be made rugged. A variety of configurations are commercially available with choices of side or end windows and various gases and pressures, and the costs are usually low. It might be noted also that often the same proportional counter can be used for obtaining spectra of β-particle energies. Two reviews of the use of gas proportional counters for photon spectrometry are those of Crouthamel [1] and Curran and Wilson [6].

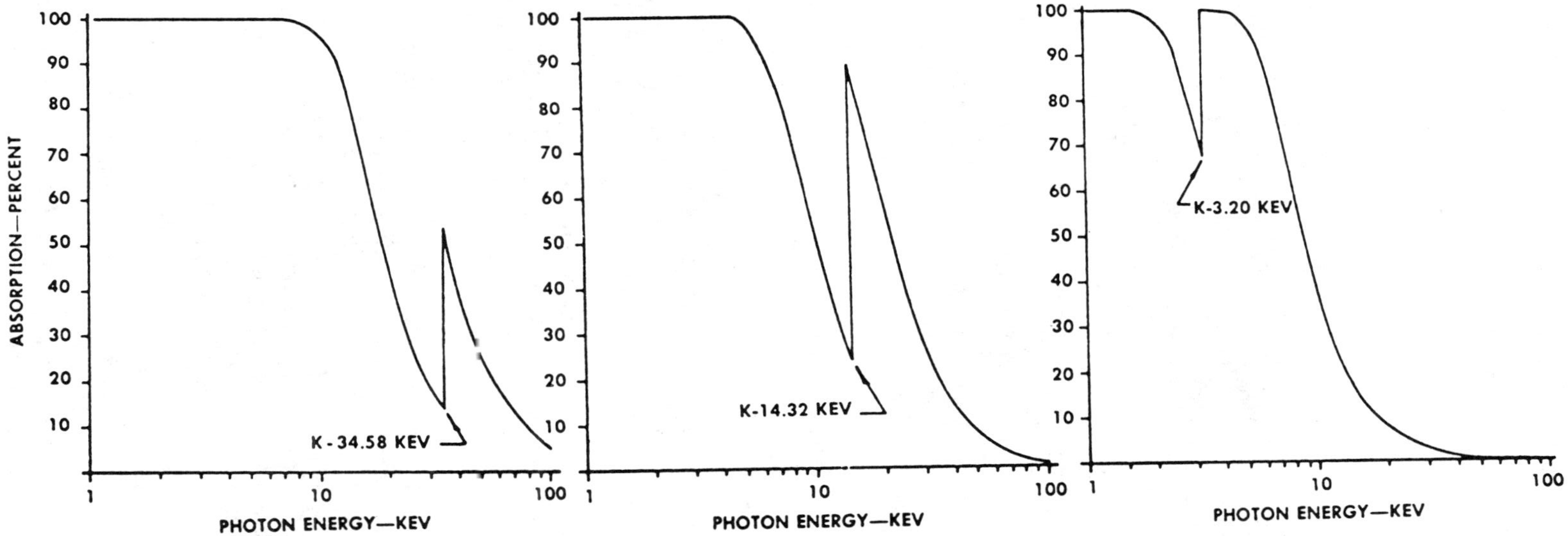

Figure 6.3. Gas absorption (%) at one atmosphere pressure in Reuter-Stokes RSG-61, a typical commercial gas proportional counter of 2-in. diameter, 4.7-in. sensitive length. The effect of window absorption is not included. Gases are (*a*) 97% xenon/3% CO_2; (*b*) 97% krypton/3% CO_2; (*c*) 97% argon/3% CO_2 (reproduced with permission of Reuter-Stokes Electronic Components, Inc., Ref. 4).

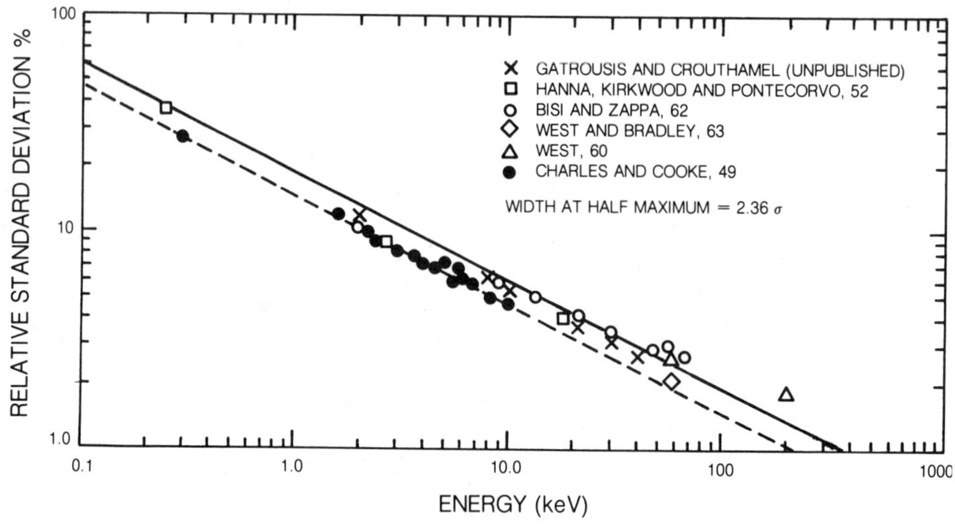

Figure 6.4. Resolution of x-ray proportional counter as a function of energy, from measurements of several investigators (reproduced from Crouthamel, Ref. 1, with permission of Pergamon Press © 1970).

6.2.2. Scintillation Counters: NaI and CsI

As described earlier, the scintillation counter consists of a clear phosphor scintillator, a photomultiplier tube, and electronic circuitry. An incident γ-ray photon produces ionization that excites the phosphor and causes emission of light. This light ejects photoelectrons from the photomultiplier cathode, which (after secondary-emission amplification by the cascaded dynodes) appears as a current pulse at the anode. The electronics analyze and display or record the results. Because the amount of light in the flash is proportional to the energy deposited in the phosphor, the size of the output pulse is a measure of that energy. If the electronics include a multichannel pulse-height analyzer, it will record the spectrum of the ionization energies.

Sodium Iodide Crystal Spectrometers. Scintillation counters employing NaI(Tl) have been a mainstay of gamma-ray spectrometry since the development of NaI(Tl) as a scintillator by Hofstadter in 1948 [7]. This discussion will concentrate on outlining the basic properties, uses, sensitivities, and limitations of sodium iodide in γ-spectrometry. For more detailed discussions, the reader is referred to the proceedings of the Scintillation Counter Symposia published as IEEE Transactions as well as several reviews [2,8,9,10,11,12].

Crystalline NaI is dense (3.67 g/cm^3) and has a high effective Z due to the iodine ($Z = 53$). The thallium activator, an impurity of about 0.1 to 0.3% introduced into the crystal, converts some of the absorbed γ-ray into visible light. The decay time of the light (time for light intensity from an instantaneous event to fall to $1/e$ of its peak value) is about 250 nsec. The pulse size is proportional to the ionization deposited. Although the light pulse from NaI(Tl) lasts several hundred nanoseconds, the time resolution achievable is much shorter. Using electronic analysis methods such as constant fraction discrimination or time-of-pulse-peak determination by differentiation and zero-crossing circuits, time resolution as low as a fraction of a nanosecond can be achieved.

Physically, it is desirable for every photon to interact by photoelectric effect, imparting all of its energy except the electron binding energy to the ejected electron. The binding energy subsequently goes into either characteristic x-rays or Auger electrons as the electron vacancy is filled. In NaI(Tl), the dominant photoprocesses occur in the iodine innermost shell, and the photoelectrons are nearly always accompanied by 28-keV K x-rays. When this radiation escapes from the crystal two peaks result, a full energy peak (when both the photoelectron and K x-ray are stopped in the crystal) and a lower-energy escape peak (when only the photoelectron is stopped in the crystal). This is most important for low-energy incident photons, both because they interact near the crystal surface where the chance of photoelectron escape is greater, and because the loss of 28 keV is a relatively large effect for low energies. This is shown in Fig. 6.5, in which the ratio of the iodine escape peak to the full energy peak is shown as a function of energy. The angle of incidence also makes a substantial difference.

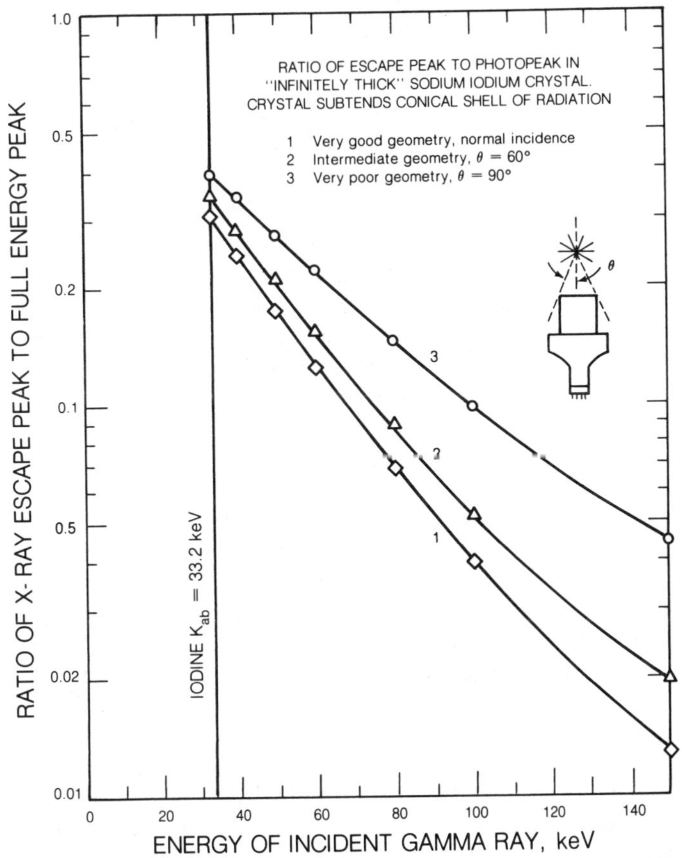

Figure 6.5. Ratio of escape peak to photopeak in "infinitely thick" sodium iodide crystal. Crystal subtends conical shell of radiation (reproduced from Crouthamel, Ref. 1, with permission of Pergamon Press © 1970).

There are a number of considerations in the choice of a NaI(Tl) system. Of great importance are the crystal size discussed previously, geometry, and the type of electronic instrumentation. The geometrical shape of the crystal, is also important; right cylindrical crystals are most commonly used, because both optical coupling to the phototube and edge-effect losses are minimized in this geometry. Another common geometrical arrangement is the well-type, in which the sample is placed in a well cavity in the center of a NaI(Tl) crystal. This provides increased solid-angle, but at a sacrifice in effective detector volume for the same overall size.

The energy resolution of a NaI(Tl) system is primarily dominated by statistical uncertainties. In a typical NaI(Tl) crystal, about 150 to 300 eV of γ-photon energy must be deposited on the average for every photoelectron created at the photocathode of the multiplier tube. Because the light emission, the photoelectron process, and the tube multiplication are all subject to statistical fluctuations, there is an inevitable "line-width" for monoenergetic energy deposition within the crystal.

If the statistical fluctuations were due only to the random deviation ΔN about its mean value N for a given photon energy the final electronic pulse would be proportional to the energy deposition E, and the fractional line width $\delta E/E$ would be proportional to $N^{1/2}/N = 1/N^{1/2}$. This is nearly the correct model at low energies but breaks down above about 200 keV because of the added contributions from Compton scattering and edge effects. This is shown in Fig. 6.6, where for a 2.5-in. diameter $\times$ 2-in. high cylindrical crystal, $(\delta E/E)^2$ is seen to be nearly linearly proportional to $1/E$.

The choice of electronic instrumentation is crucial to overall system performance. The fluorescent efficiency of the scintillator and the quantum efficiency of the multiplier tube's photocathode are most important in the system resolution, together with the uniformity of the photocathode area. Low tube noise and good multiplication statistics also must be considered, especially for detecting the lowest-energy photons. Sometimes two photomultipliers are used in coincidence, in which case the contribution of direct tube noise to the background coincidence counting rate is greatly reduced. Much has been written about the detailed considerations (optical, mechanical, and electronic) involved in phototube choice and use. For more information in this area the reader is referred to Refs. 1,2,9,10.

Cesium Iodide Crystal Spectrometers. While sodium iodide is by far the most widely used inorganic scintillator, thallium-activated cesium iodide, CsI(Tl), also has useful applications. CsI(Tl) differs from NaI(Tl) in three fundamental ways: (a) it produces only 40 to 50% as much light per eV of deposited energy; (b) its mean decay time for γ-rays is about 1600 nsec compared to about 250 nsec for NaI(Tl); and (c) cesium has a much higher Z (55) than sodium (11), so that the photoelectric interaction is somewhat enhanced.

CsI(Tl) is used widely for routine γ-spectrometry. This is probably because the cost of CsI(Tl) is much greater than that of NaI(Tl), even though its overall resolution and other performance characteristics are similar.

The use of CsI(Tl) most pertinent to this discussion is the phosphor sandwich or "phoswich." This arrangement, shown schematically in Fig. 6.7, places a NaI(Tl) crystal in front of a CsI(Tl) crystal, which in turn is in front of the photomultiplier tube. The entire system is optically coupled together. Low-energy γ-rays stopping within the NaI(Tl) crystal appear as a fast pulse, while higher-energy rays that penetrate to the CsI(Tl) yield a phototube pulse containing both short and long

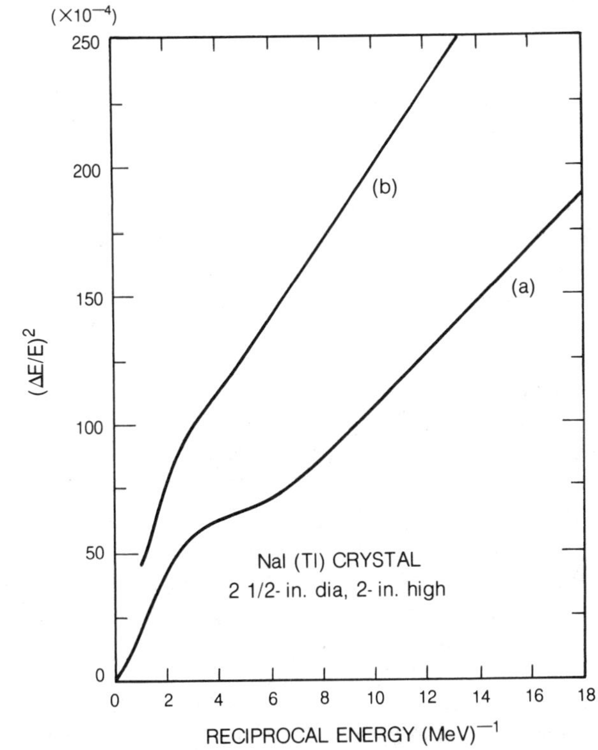

Figure 6.6. Calculated values of resolution for a 2.5 in. diam. × 2 in. high sodium iodide-thallium activated crystal. $(\Delta E/E)^2$ is plotted against reciprocal of energy $(1/E)$. Curve (a) includes only intrinsic broadening and photomultiplier statistics. Curve (b) includes also the contribution from the variance in photon transfer (reproduced from Crouthamel, Ref. 1, with permission of Pergamon Press © 1970).

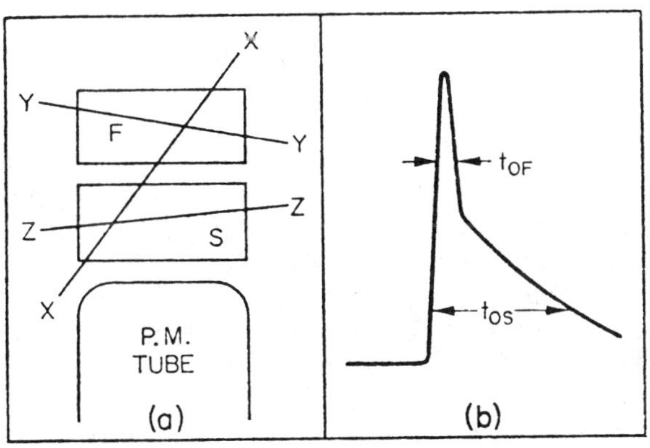

Figure 6.7. The "Phoswich". (*a*) Fast-slow phosphor sandwich; (*b*) Dual pulse from particle traversing XX in (*a*). T_{of} and t_{os} are the $1/e$ times for the fast and slow components of the resulting pulse (reproduced from Birks, Ref. 9, with permission of Pergamon Press © 1964).

components, as shown in the figure. This configuration allows the low-energy photons, which have very short penetration, to be separated electronically by pulse-shape analysis.

6.2.3. Semiconductor Detectors

The importance of semiconductor detectors in γ-spectrometry cannot be overemphasized. Since the use of these devices became widespread in the mid-1960s, many important fields of research and measurement have been revolutionized. This is mainly because of the good energy resolution achievable with these detectors, but also because of their stability and reliability. It is now possible to make routine measurements of many radionuclides at low concentrations which are beyond the capabilities of other systems.

The basic properties of semiconductor detectors were discussed in Chapter 5. Gamma-ray energy given up in a semiconductor will excite a number of electrons proportional to this energy from its filled band into its conduction band forming hole-electron pairs. If the semiconductor is sufficiently free from impurities so that (a) the number of free impurity carriers is small and (b) the charge trapping is small, the gamma-ray-produced carriers (i.e., the hole-electron pair) can, under the influence of an electric field, be made to move through the material and collect on appropriate electrodes. The charge collected will be proportional to the energy of the absorbed γ-photon.

To be suitable as detector material, in addition to freedom from impurities which will introduce unwanted charge carriers, it must meet three other requirements. First, the average ionization energy ϵ required to produce a hole-electron pair must be small, to minimize the fractional fluctuations from statistical uncertainties. (For germanium $\epsilon \simeq 2.9$ eV, while for silicon $\epsilon \simeq 3.7$ eV.) Second, the mobility must be sufficiently high that essentially all of the carriers are collected at the electrodes. Finally, the carrier lifetime must be great enough so that recombination is unimportant. Under these conditions and after suitable amplification, an electronic pulse may be produced that is proportional to the energy deposited in the active area of the detector.

If the electrical contacts on the material permit a free exchange of holes and electrons between the semiconductors and the electrodes, large currents will flow through the detector when voltage is applied across it. To overcome this "blocking," contacts must be used—that is, the positive contact must not inject holes, nor should the negative contact inject electrons into the detector material. This can be achieved either by careful choice and application of the metals used for the contacts, or by forming heavily doped n and p regions by high temperature diffusion or by ion implantation into the detector surfaces. The n^+ surface is used as the positive detector electrode, and the p^+ surface as the negative electrode. (The n^+ and p^+ symbols refer to heavily doped n and p material). The resulting junction is then operated in the reversed-bias mode, driving free holes in the p region and free electrons in the n region away from the junction. The depletion layer left is the actual detector. In the depletion layer, the supply of free carriers available to move in the applied field direction is small, and the bias causes only a very low leakage current. The applied field permits efficient collection of electron-hole pairs produced by the radiation.

The need for low leakage current prompted efforts to compensate for impurities that produce unwanted free carriers. The usual method for compensating is drifting with *n*-type lithium [1,13,14]. In recent years, it has been possible to produce ultra-high-purity semiconductor material, and large volume germanium detectors with 100 cm^3 crystals in common use.

Germanium. Because photoelectric photon conversion (Z^5 dependent) is a key detection mechanism, germanium ($Z = 32$) is superior to silicon ($Z = 14$) by a factor of about 40. This is particularly important at energies above about 30 keV. The actual observed peak results from both the Compton and photoelectric effects. Most germanium detectors in use today are lithium drifted, Ge(Li), but, as noted, intrinsic germanium detectors are increasingly available.

Various detector geometries have been developed for semiconductor γ-ray detectors. *Planar* or *parallel plate p-i-n* configurations were the earliest detector geometries to achieve wide use. These consist of an intrinsic (compensated) interior with a *p*-type substrate on one side and an *n*-type on the other surface through which the photons enter. Unfortunately, the practical limit of lithium drift (15 to 20 mm) limits the total volume which can be compensated in this way. To overcome this problem, the *coaxial p-i-n detector* was developed. High-quality *p*-type germanium is used. Lithium drifting compensates the major fraction of the volume, and the central core remains *p*-type; the lithium-diffused surface layer is *n*-type. Figure 6.8 schematically shows two kinds of coaxial detector, the true "open-ended" coax and the "closed-end" coax. A major advantage of the coaxial geometry is that large volumes can be compensated. Furthermore, leakage currents are low because the exposed junction area at the ends is constant even for very long cylinders and large sensitive volumes. The primary disadvantage is that very long cylinders have high capacitance which can degrade the resolution significantly.

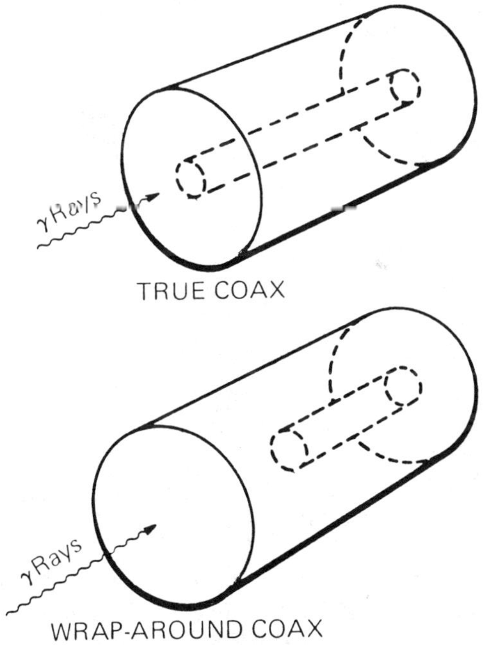

Figure 6.8. True coaxial and wrap-around coaxial geometries are shown schematically (reproduced with permission of EG&G ORTEC, Ref. 15).

The larger volumes obtainable with coaxial geometries are a clear advantage at high energies (> a few hundred keV), where γ-ray penetration beneath the thin surface layer is substantial and the entire volume is useful for detection. At low energies (< 100 keV), the exposed detector area is a key parameter. The choice of geometry thus depends on the energy range which is being measured. An example of the trade-off is illustrated in Fig. 6.9. At low energies, the counting efficiency is governed by the area, and at high energies by the volume. A closed-end coaxial configuration made with n-type germanium using an ion-implanted outer contact with about 0.3 μm thickness exhibits enhanced response to low-energy photons. Conventional closed-end coaxial detectors made from p-type germanium have entrance window thicknesses ranging from 300 to 1000 μm.

For some anticoincident measurements, the closed-end-coaxial geometry has a serious disadvantage compared to an open-ended configuration because it puts a dead region right at the "heart" of the measurement, near the source. The dead region can absorb many of the low-energy electrons generated near the Compton edge, preventing them from escape and subsequent detection in a NaI(Tl) anticoincidence detector.

Cooper has expressed a key geometrical consideration [17]:

> In selecting detectors for low-level counting applications, care should be taken to maximize the ratio of the detector's cross-sectional area facing the source to its volume. Since the natural background is generally isotropic, its detection efficiency depends mainly on the volume while the detection efficiency for source gamma rays depends strongly on the detector cross section. Thus, although long detectors may have greater high energy efficiencies than short detectors, their natural background interferences will be higher at all energies.

For many applications the short fat coaxial geometry (40 to 50 mm diameter, 25 to 30 mm length) is best, since it has fairly low capacitance (10 to 15 pF) but rea-

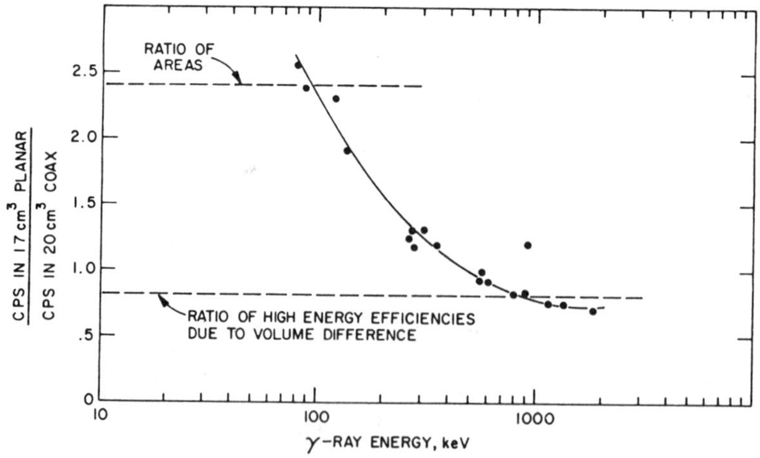

Figure 6.9. Ratio of counting efficiency as a function of energy for a 17-cm^3 planar and a 20-cm^3 coaxial germanium detector. At high energies the volume governs the efficiency, whereas at low energies the efficiency is governed by the area (reproduced from Armantrout, Bradley, and Phelps, Ref. 16, with permission of the Institute of Electrical and Electronic Engineers © 1972).

sonably high detection efficiency. However it must be long enough to absorb the γ-rays. Indeed, there are applications in which a detector either too short or too long would be equally undesirable.

Germanium semiconductor detectors must be operated at or near to liquid-nitrogen temperatures (77°K). This is because the Ge band-gap is not large enough to allow room-temperature operation. The low-temperature operation reduces the leakage current from thermal excitation of holes and electrons. In addition, Ge(Li) detectors must also be cooled to prevent movement of Li, which is highly mobile at room temperatures. For further information on germanium γ-ray detectors, the reader is referred to Ref. 18.

Silicon. Silicon is not widely used for γ-spectrometry except at low energies. The reason that Ge is preferred over Si in the 0.1 to 1.0 MeV region is Ge's much greater photopeak-to-Compton ratio. At energies below about 30 to 50 keV, the advantage of Ge becomes less important, and in fact at the lowest energies Si is now most widely used. Many of the considerations just set forth are generic to semiconductor detectors, applying to Si as well as Ge.

One important advantage of Si is that for some applications it can be operated at room temperatures. This eliminates the need for any windows and makes spectrometry of very-low-energy photons (a few keV) much simpler. For the best resolution, however, low temperatures are still necessary.

As pointed out earlier, a disadvantage is the lower density (2.33 g/cm^3) of silicon compared to germanium (5.33 g/cm^3). Another is that its intrinsic resolution is not quite as good, although when operated with only a small amount of cooling (near 0°C) the differences are slight. Figure 6.10 shows a typical very-low-energy spectrum taken with a Si(Li) detector, measuring the radioactivity on an air filter. Note that the 5.9-keV photon from ^{55}Fe is resolved. Today, commercially available Si(Li) detectors offer resolutions (FWHM) for ^{55}Fe in the vicinity of 160 eV.

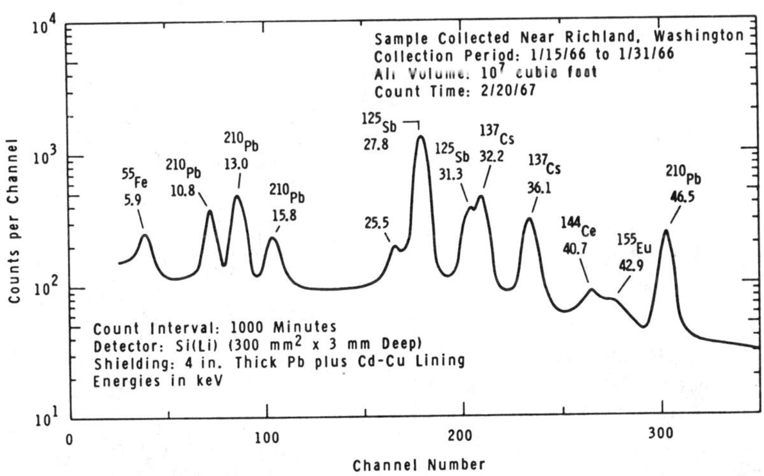

Figure 6.10. Si(Li) photon spectrum of the airborne radionuclides collected on an air filter at Richland, Washington (reproduced from Cooper et al., Ref. 19, with permission of Pergamon Press © 1968).

Silicon detectors of small size are often used when it is necessary to count low-energy photons in the presence of large backgrounds of high-energy photons, because the detection efficiency for the higher energies is small.

6.2.4. Comparison: Thallium-Activated Sodium Iodide versus Germanium

NaI(Tl) crystal spectrometry was the most important γ-ray measuring technique from the time of its discovery in 1948 until Ge systems began (in the 1960s) to play an increasingly larger role in many important types of measurements. Hence, it is appropriate to highlight the relative advantages and disadvantages of these two techniques. Such a comparison is particularly important because the spectrographic systems under discussion usually comprise the largest single initial expense in equipping a laboratory for environmental radionuclide measurements.

A first point to be noted in this comparison is that a germanium detector requires cooling to liquid nitrogen temperature whereas the scintillation counter operates at room temperature. The basic trade-off is between the high efficiency of NaI(Tl) and the high resolution of Ge. Semiconductor resolution is obviously crucial in some applications as can be seen, for example, in Fig. 6.11, in which the NaI(Tl)

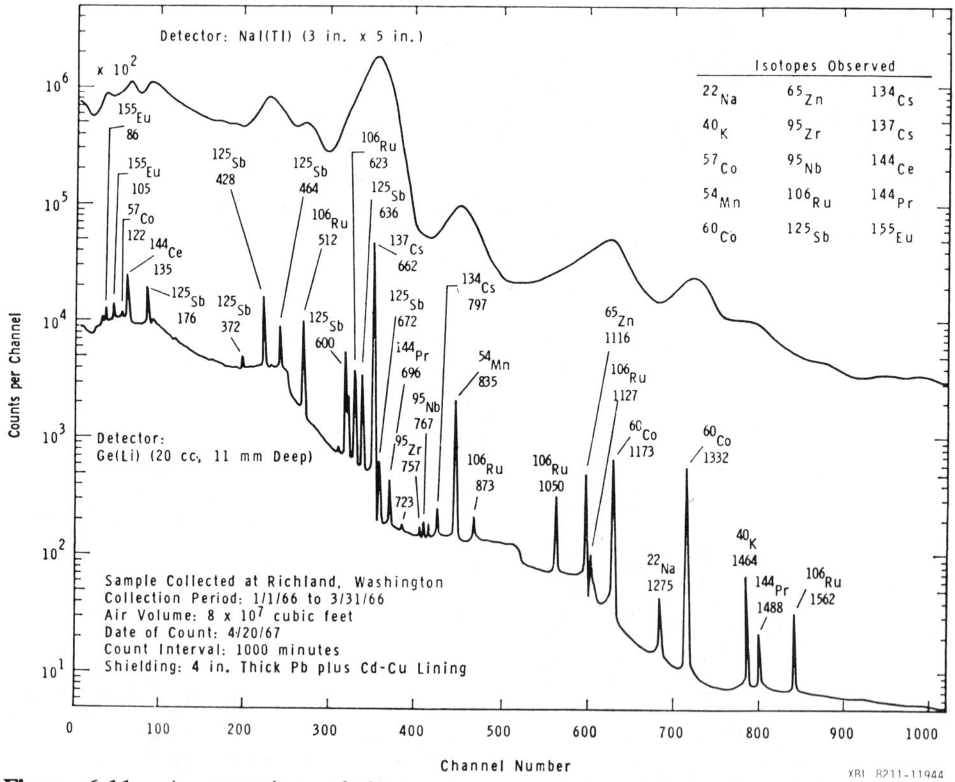

Figure 6.11. A comparison of the gamma-ray spectra of the radionuclides present one year after collection on an air filter at Richland, Washington, as measured with a 5-in. diam. by 3-in. thick NaI(Tl) scintillator and a 20-cc Ge(Li) diode (reproduced from Cooper et al., Ref. 19, with permission of the Health Physics Society © 1968).

system is totally incapable of resolving a number of closely spaced lines which appear clearly separated in the Ge(Li) measured spectrum.

The contrasting parameter is overall efficiency, determined both by intrinsic efficiency and by detector size. Even medium-sized NaI(Tl) crystals have an efficiency advantage over the largest Ge(Li) detectors. This is seen above in Fig. 6.12 [20] in which the overall efficiency of a 3-in. × 3-in. NaI(Tl) crystal is compared to that of several Ge(Li) detectors. Even the 54-cm^3 Ge(Li) detector has overall efficiency less by a factor of 3 or 4 than the 3-in. × 3-in. NaI(Tl) crystal.

Because even the largest Ge detector crystals are of limited size, they cannot achieve the performance of NaI(Tl) crystals in terms of overall detection efficiency. The best Ge detectors now produced commercially have a peak efficiency at 1.33 MeV which is 30% that of a 3-in. × 3-in. NaI detector [22]. Also Ge detectors cannot as easily accommodate large samples, or achieve the multiple interaction performance of NaI(Tl). Figure 6.13 illustrates the full-energy-peak-to-total-enhancement from use of Ge detectors. To utilize advantages of both systems, hybrid systems have been developed. For example, NaI(Tl) can be used to surround a Ge inner detector. NaI(Tl) detects in time coincidence many of the

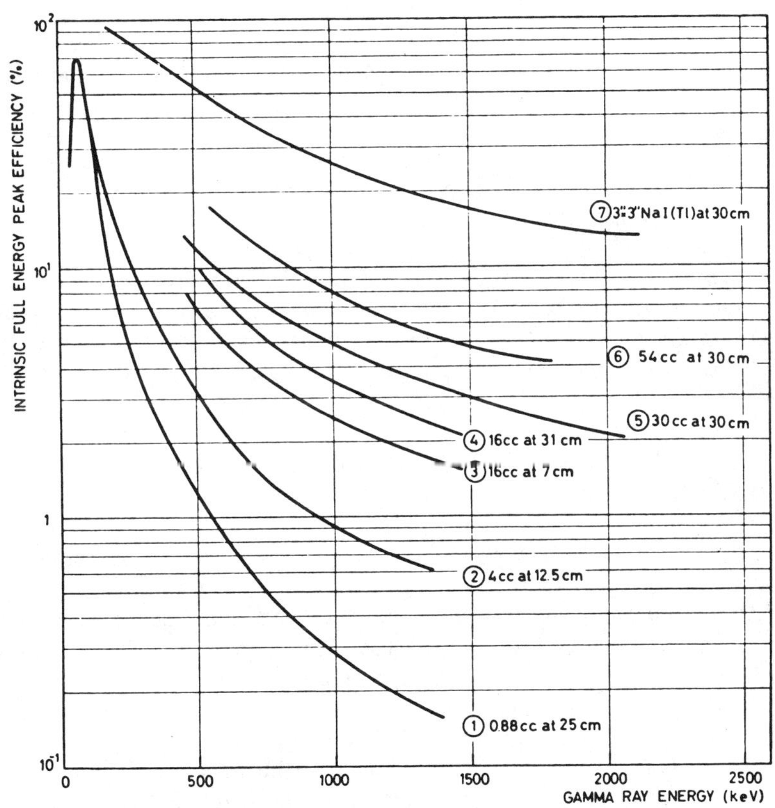

Figure 6.12. Experimental full energy peak efficiency for detectors of different volumes. Curves 1 and 2 refer to planar detectors, and curves 3, 4, 5, and 6 refer to coaxial detectors. Curve 7 is for a 3-in. × 3-in. NaI(Tl) detector. Distance of source from detector face is indicated (reproduced from Cappellani and Restelli, Ref. 20, with permission of North-Holland Publishing Co. © 1968).

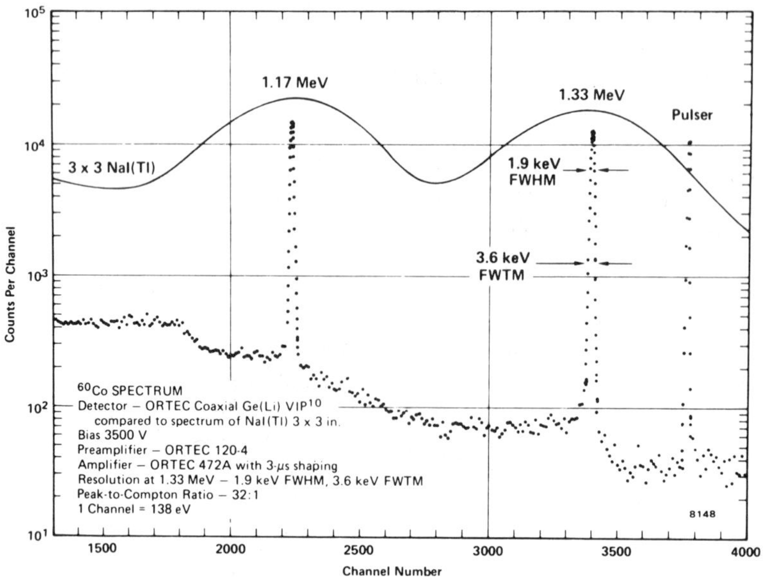

Figure 6.13. ^{60}Co spectrum showing resolutions and peak-to-Compton ratios for a Ge(Li) coaxial detector and a NaI(Tl) detector (reproduced with permission of EG&G ORTEC, Ref 23).

Compton-scattered photons escaping from the Ge crystal, enabling their rejection (to help low-energy photopeaks to stand out above a Compton continuum) or occasionally their inclusion (to improve the system sensitivity).

Another key consideration is the amount of effort required to obtain measured activity values from the raw energy spectrum. In this regard, NaI(Tl) systems usually require substantial computer processing for complex spectra, whereas Ge systems often can yield important information with only moderate data analysis.

6.3. INSTRUMENTAL LIMITATIONS DUE TO THE STATISTICS OF PULSE COUNTING

The problem of analyzing the pulse output from spectrometers is complex, and is described here briefly.

The pulse output from the detector is supplied to an amplifier system followed by a pulse-shaping unit if required. From there the signals go (usually) to a pulse-height analyzer. The amplitudes of the pulses measure the photon energy of the received ionizing radiation, and the number of pulses per unit time indicates the rate of photon arrival. With a perfect system all the pulses produced by a monoenergetic γ-ray would fall into a single channel of the pulse-height analyzer. However if the number of pulses is plotted against channel number a Gaussian distribution is obtained.

The reasons are many: (a) background radiation, including Compton radiation; (b) the random way in which radiation interacts with matter to give up its energy; (c) random emission of photoelectrons from the multiplier cathodes or of carrier in a semiconductor; and (d) noise in the electronic system amplifying and processing the pulses.

The Gaussian form of the pulse seen on the pulse-height analyzer is usually characterized by its full width at half-maximum (FWHM). The greater the FWHM the less certain is the value of the gamma-ray energy. This is because it is harder to determine the exact location of the peak. The FWHM is related to root-mean-square deviation or standard deviation of energy determination by a constant of proportionality, 2.355. The probable error is 1.59 × FWHM.

The largest contributor to the FWHM of a spectral line is the statistical variation of the carrier generation process in semiconductor detectors and gas proportional counters and the photoelectron in a scintillation counter. The average number N of electron-hole pairs created by deposition of energy E is E/ϵ (where ϵ = the energy needed to produce one pair), and the theoretical standard deviation σ_T due to the statistical fluctuations can be expressed as $\sigma_T = (E/\epsilon)^{1/2}$.

Experimentally with a semiconductor spectrometer the fluctuations turn out to be much smaller. This is determined by using the relationship FWHM = 2.355 σ_{exp}, where the experimentally determined standard deviation is denoted by σ_{exp}. Then the ratio σ_{exp}/σ_T serves to define the Fano factor, F:

$$F = (\sigma_{exp}/\sigma_T)^2$$

$$\sigma_{exp} = [(E/\epsilon)F]^{1/2}$$

The discrepancy between σ_{exp} and σ_T ($\sigma_T = N^{1/2}$) can be explained as follows: not all of the energy deposited in the semiconductor goes into creating electron-hole pairs; a fraction goes into exciting other modes such as lattice vibrations. If all of the energy loss were to go into creating electron-hole pairs, there would be no statistical fluctuations in the number N of pairs formed; FWHM, σ_{exp}, and F would all equal zero. On the other hand, if only a small fraction of the energy loss were to go into creating pairs, the fluctuations in this small fraction would be truly random and F would equal one. In semiconductors the true situation lies between these two extremes. Experimental values of F differ widely, depending probably on the techniques used in crystal fabrication. Fano factors as low as 0.08 have been measured [24] but values in the region of about 0.12 are typical of present detectors. The intrinsic resolution of such detectors can be read off of Fig. 6.14.

With scintillation counters the pulse formation is a multistep process. First, light photons are released from the scintillator, and these light photons in turn cause the release of photoelectrons from the multiplier cathode. Because the net efficiency of photoelectron release (taking into account light losses between the crystal and the multiplier) is only about 10%, the statistical distribution of the photoelectrons is the dominant factor.

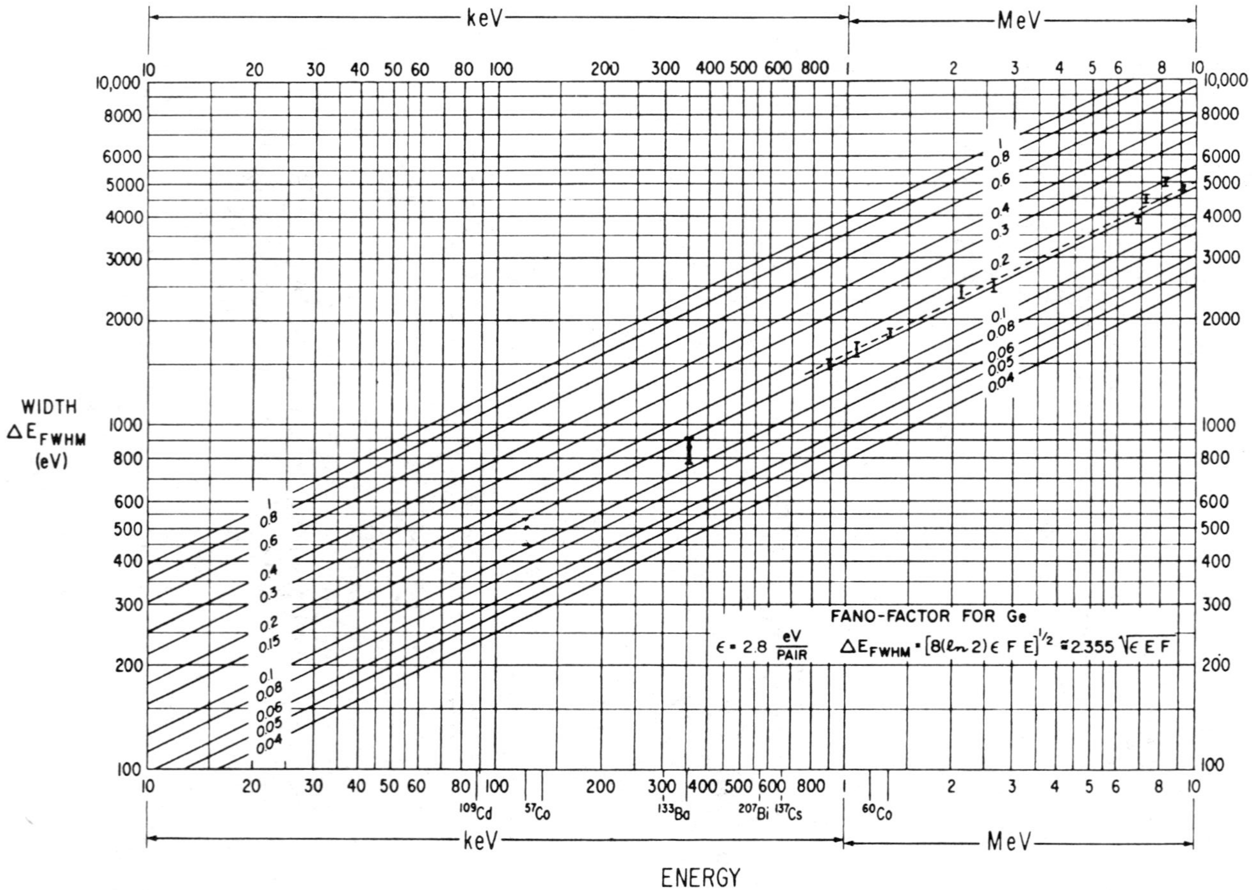

Figure 6.14. Resolution of x-ray proportional counter as a function of energy, from measurements of several investigators (reproduced from Crouthamel, Ref. 1, with permission of Pergamon Press © 1970).

6.4. DATA REDUCTION AND ANALYSIS

Data reduction can lead to difficulty and error if not carried out properly. Therefore, a number of sophisticated methodologies have been developed, some of which are almost fully computerized.

Two limitations of all detector-analyzer systems are responsible for most of the difficulties. First, finite resolution leads to peak overlap in complex spectra. This is particularly true with scintillation counters. Second, each spectrum invariably includes some background or "noise" counts (for example from the instrument itself, the Compton spectrum of a higher-energy γ-ray, or from external backgrounds) which degrades the resolution of the peaks.

The simplest procedure for dealing with a complex spectrum is to analyze successively the various photopeaks, beginning with that at the highest energy. Successive stripping subtractions are made. Unfortunately, error analysis is extremely difficult, and after even a few subtraction steps the uncertainties become large and very difficult to estimate for the lower energy peaks. For this reason, simple stripping is used only for relatively simple complicated spectra.

A somewhat more powerful method is to generate a set of simultaneous equations for a number of radionuclides in terms of their observed photopeak count totals. Again, uncertainties can become very large for the smaller photopeaks or those at lower energies above large backgrounds.

Both of the foregoing are unsophisticated methods, typically using only the full-energy-peak information and estimating backgrounds beneath the peaks by measurements in adjacent channels. More elaborate analysis techniques attempt to fit an entire spectrum including Compton-scattering contributions. One of the most widely used methods of computer analysis is the method of least-squares. Essentially, this method uses as input a set of standard spectra taken with the instrument in question and containing known activities of all radionuclides suspected of being present in the unknown sample. A least-squares fit is then performed on the data; the variances are minimized and the most likely concentrations of each sought-for radionuclide are determined. The principle as applied to γ-ray spectra has been thoroughly discussed [26–28] and a sample computer program is available [29]. Among the disadvantages of this method are the requirements that all sought-after radionuclides must have reference input spectra available, and that those thought to be present must be estimated as a starting point for the analyses.

Multiple regression is an analysis technique that combines the concepts of stripping and least squares. This procedure is described by Shumway [30], using methods developed by Efroymson [31]. Essentially, the method begins with a least-squares fit to the most prominent highest-energy photopeaks and their accompanying full spectra. The less prominent and lower-energy radionuclides are then introduced sequentially, with recomputation of least squares each time to improve the goodness of fit. Each successive recomputation can be programmed to eliminate a radionuclide previously included. Again, a library of standard spectra must be used.

The techniques just described are most useful for complicated, overlapping spectra such as those observed using NaI(Tl) containing several radionuclides. For spectra taken using Ge, the peaks are usually narrow enough that backgrounds beneath them can be determined reasonably well from measurements in nearby channels near the sides of each peak. This is usually what Ge analysis programs do, with less consideration being given to the full spectra. An additional refinement in

Ge analyses is the use of peak-shape information to determine more exactly the entire peak area. Some analysis programs, tied to particular measuring systems, are almost completely computerized and automated. They can find peaks, determine the peak sizes, and even correct for detector efficiency and energy calibration effects. These depend, of course, on access to libraries of standard spectra. Examples include analyses described by Routti and Prussin [32] and Gunnink and Niday [33]. Other sophisticated methods have been developed for analysis of Ge spectra in neutron activation analysis [34]. The reader is referred also to several other papers for more details [35–40].

6.5. PRACTICAL MEASUREMENT CONSIDERATIONS

In environmental studies most of spectroscopic work is done in the laboratory under conditions that can be controlled to some extent. Even in the field certain measures can be taken that will greatly improve and facilitate measurements. As will be explained, coincidence and anticoincidence techniques are some of the more important.

In laboratory work the most common complement to an ordinary system is availability of a low-background counting room. Two important sources of background can be minimized: cosmic radiation and natural terrestrial backgrounds. To minimize these, completely enclosed lead "caves" are used, and in some cases counting rooms have been built below ground. The gains can be substantial: Background reductions by factors up to a few hundred are common. For example, studies of backgrounds 110 m beneath Grand Coulee Dam [41] are shown in Fig. 6.15. Use of special low-activity lead also offers advantages particularly at the lower energies. Studies of natural backgrounds in cement [1] and concrete [42] have led to the discovery and use of special low-background materials in counting enclosures [43]. Background reductions by additional factors of 2 to 5 over backgrounds in normal concrete rooms can be gained by use of such materials, even if lead shielding is already in use around the detector system.

Another common improvement that can be made to a normal detector system is by the use of *anticoincidence counters*. In the simplest arrangement, thin plastic scintillation counters can be used to veto (cancel electronically) pulses from high-energy cosmic-ray charged particles traversing the detector system. A more costly but much more effective arrangement can be made by surrounding the detector with larger volumes of anticoincidence scintillators. In addition to eliminating cosmic ray events, these systems detect and veto events in which some of the energy escapes after a primary photon interaction in the inner detector. Large volumes are needed to achieve a high detection efficiency. The suppression of the Compton continuum can be particularly effective, along with the detection of escaping 511-keV photons from positron annihilation after pair production.

In one common arrangement, large NaI(Tl) detectors have been surrounded by additional NaI(Tl) scintillators. Figure 6.16 shows the effect of such a system on suppressing the Compton continuum below the 662-keV gamma peak from ^{137}Cs. The suppression of the cosmic-ray muon background by 3 to 4 orders of magnitude is shown in Fig. 6.17. It should be emphasized that when anticoincidence methods are used, the addition of lead shielding provides little extra benefit. This can be seen, for example, in Fig. 6.18.

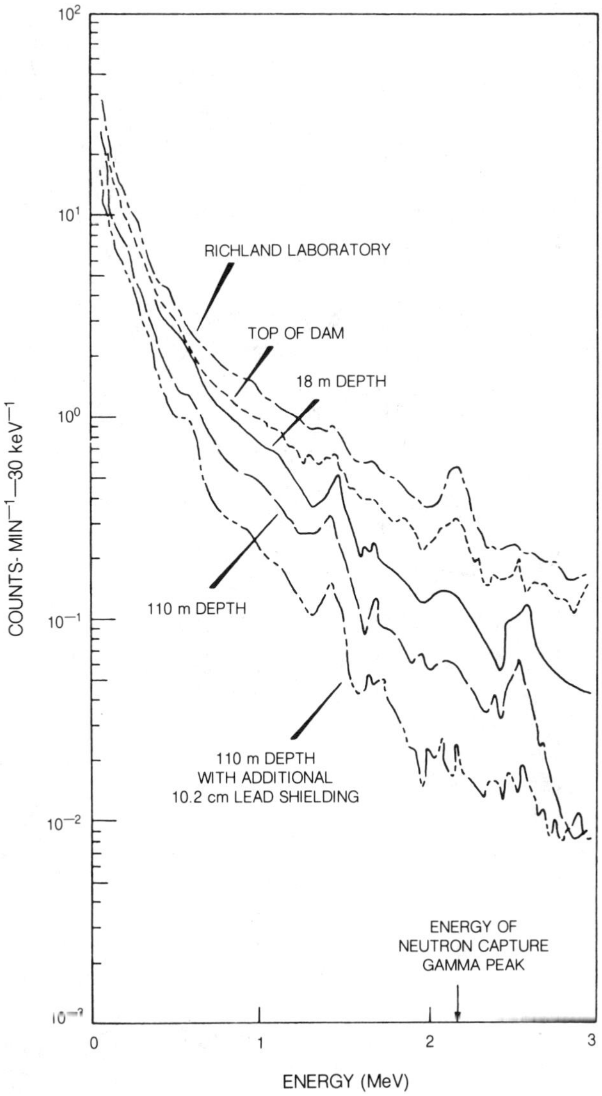

Figure 6.15. Effects of massive shielding on backgrounds of simple NaI(Tl) detector system. Comparison is between Richland, Washington, and various configurations at top of and inside Grand Coulee Dam, Washington (reproduced from Kaye et al., Ref. 41, with permission of Nucl. Instrum. Meth. © 1972). (reproduced from Wogman, Perkins, and Kaye, Ref. 44, with permission of Nucl. Instrum. Meth. © 1969).

Surrounding a Ge detector with anticoincidence counters provides even more impressive Compton suppression, mainly because the small size of even the largest Ge detectors results in very large Compton backgrounds. Figure 6.19 shows that a reduction by nearly an order of magnitude can be achieved by surrounding a 20-cm^3 Ge detector with plastic phosphor scintillation counter of 26-in. diameter and 24-in. thickness. The peak-to-Compton height ratio is 245:1, as indicated. The

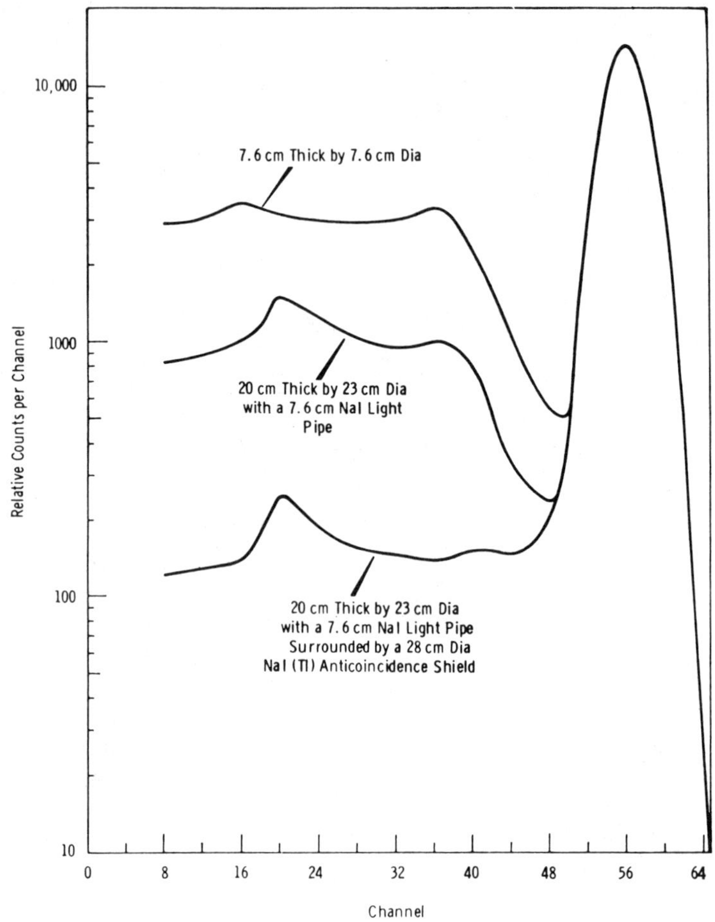

Figure 6.16. Comparison of the ^{137}Cs peak to Compton distribution for two sodium iodide detector systems; the bottom curve includes surrounding the larger NaI(Tl) detector with an anticoincidence shield. The peak is at 662 keV (reproduced from Wogman, Perkins, and Kaye, Ref. 44, with permission of Nucl. Instrum. Meth. © 1969).

detection sensitivity of this system under various conditions is shown in Fig. 6.20. Figure 6.21 shows a particular example in which the Compton continuum below the two ^{60}Co photopeaks normally would have completely obscured the lower-energy photopeaks of several other γ-emitters. The anticoincidence system, while reducing the sensitivity for the ^{60}Co lines by a factor of about 6, reduces the ^{60}Co Compton edges by about a factor of 50. By storing the spectra both with and without anticoincidence, no primary information is lost.

Another aspect of the problem is the ability of systems to measure large samples. With Ge systems, losses in efficiency with large samples can significantly offset other positive factors. As an example, Fig. 6.22 shows the efficiency of a 100-cm^3 Ge(Li) cylindrical detector for a point source and for a sample of 6 in. diameter and 1 in. thickness (sample volume of about 450 cm^3). With the larger sample, the efficiency is lower by a factor of 5 to 10.

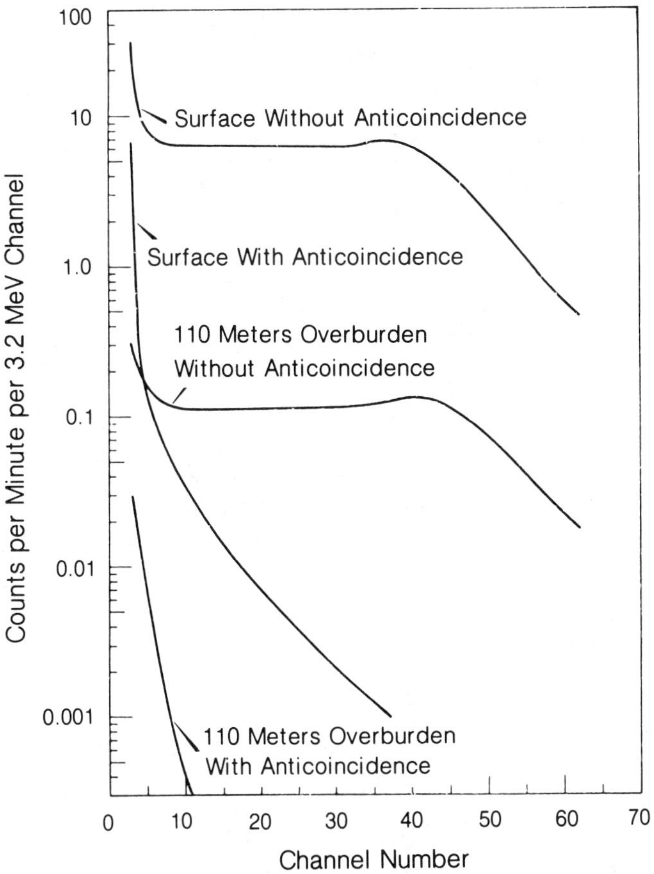

Figure 6.17. Comparison of the muon background of an all-sodium-iodide detector system to various mass and anticoincidence shielding conditions. The 110-m overburden is beneath Grand Coulee Dam (reproduced from Wogman, Perkins, and Kaye, Ref. 44, with permission of Nucl. Instrum. Meth. © 1969).

Recently the Marinelli beaker has been introduced to give better geometry for a large sample. Because the Marinelli beaker completely covers all but under the detector (see Fig. 6.23), greater sensitivity results. Environmental samples usually involve large-volume sources so it is important to determine the detectable activity in terms of γ-ray/min·ml of the sample material, as given in Table 6.1. Clearly the Marinelli geometry in this respect gives the higher sample sensitivity.

Well-type NaI(Tl) systems are now in common use. Insertion of a sample within an enclosed well increases the solid angle to nearly a complete sphere. However, losses in efficiency can still occur. Germanium well detectors have also been built and used to advantage [19]. With the availability of high-purity germanium, better well-type detectors can be made than is possible with Ge(Li) [48]. In an evaluation of well-type Ge(Li) detectors for low-level radiochemical analyses [49], well detectors were found to have better detection efficiencies and as low background as either large coaxial detectors or opposed coincident detector systems.

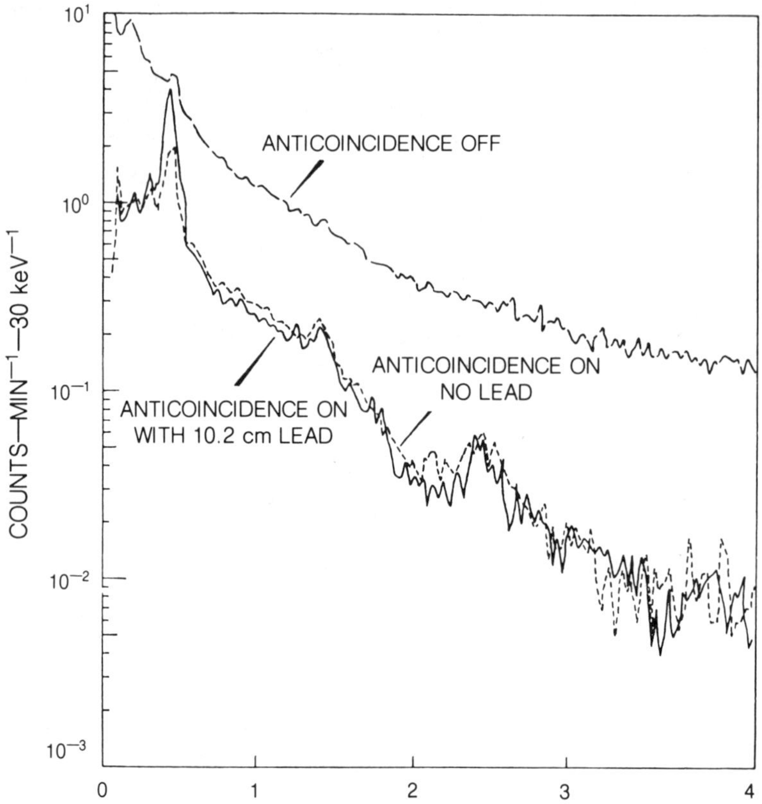

Figure 6.18. Background measurements at Richland, Washington, with an anticoincidence-shielded sodium iodide detector, showing effect of anticoincidence and of 10.2 cm of additional lead (reproduced from Kaye, Ref. 41, with permission of Nucl. Instrum. Meth., © 1972).

Coincidence counting is another variation on gamma spectrometry, using unique signatures to decrease backgrounds and identify certain radionuclides in the presence of others. The most common method is to use two NaI(Tl) crystals on opposite sides of a sample. Dual-Ge systems and hybrid Ge-NaI(Tl) coincidence arrangements have also found some application. Each has its advantages, the trade-offs being efficiency versus resolution. This trade-off is especially critical for coincidence work because the overall efficiency is the product of the two individual detector efficiencies.

The efficiencies of some dual-NaI(Tl) systems are shown in Fig. 6.24. Background suppression is shown in Fig. 6.25.

Another arrangement is to place a Ge detector inside a NaI(Tl) well. Here the NaI(Tl) can be used in both anticoincidence mode and coincidence mode. The effect of such an arrangement on a Ge spectrum from an air filter is shown in Fig. 6.26. A large number of radionuclides can be observed in the coincidence spectrum, in which a coincident NaI(Tl) pulse gates the Ge pulse height into a separate region of the pulse-height-analyzer memory.

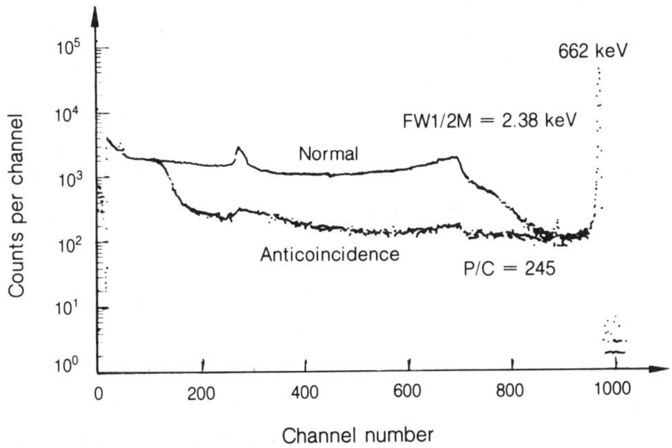

Figure 6.19. Gamma-ray spectra of ^{137}Cs recorded in the normal and anticoincidence modes, with a germanium detector. Full width half-maximum is noted as 2.38 keV, and peak-to-Compton height ratio as 245 (reproduced from Cooper, Rancitelli, and Perkins, Ref. 45, with permission of Journal of Radioanalytical Chemistry © 1970).

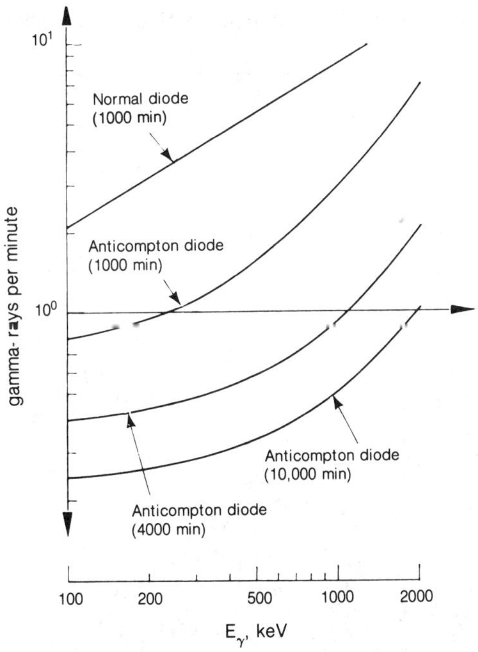

Figure 6.20. Gamma-ray detection sensitivity for a normal germanium diode (1000-minute counting time) compared to sensitivities of anticoincidence shielded diode for 1000, 4000, and 10,000 minute counting times (Reproduced from Cooper, Rancitelli, and Perkins, Ref. 45, with permission of Journal of Radioanalytical Chemistry © 1970).

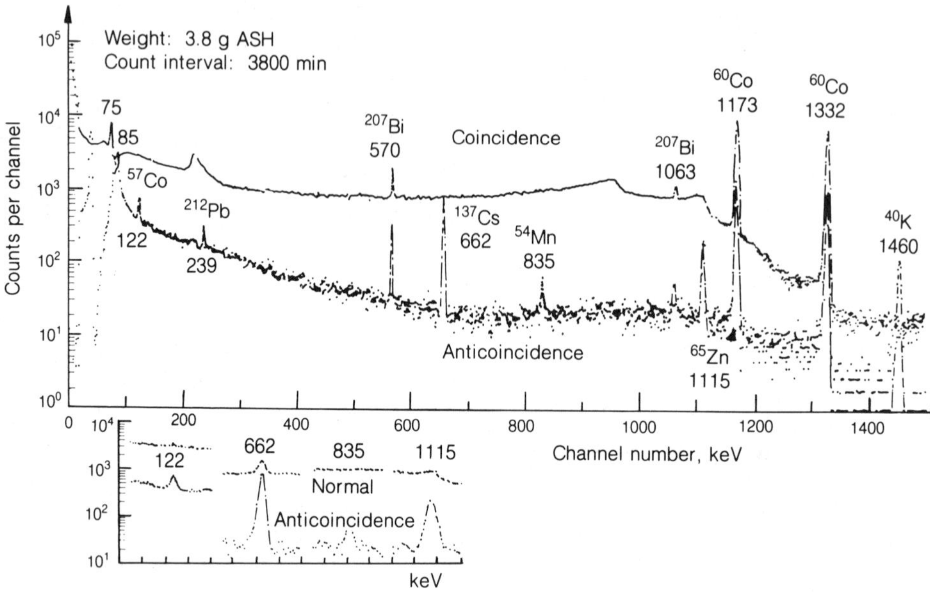

Figure 6.21. Coincidence and anticoincidence γ-ray spectra of muscle tissue from a tuna collected in the region of the Bikini atoll. Data taken with a 3800-minute counting time using a Ge(Li) diode (reproduced from Cooper, Rancitelli, and Perkins, Ref. 45, with permission of Journal of Radioanalytical Chemistry © 1970).

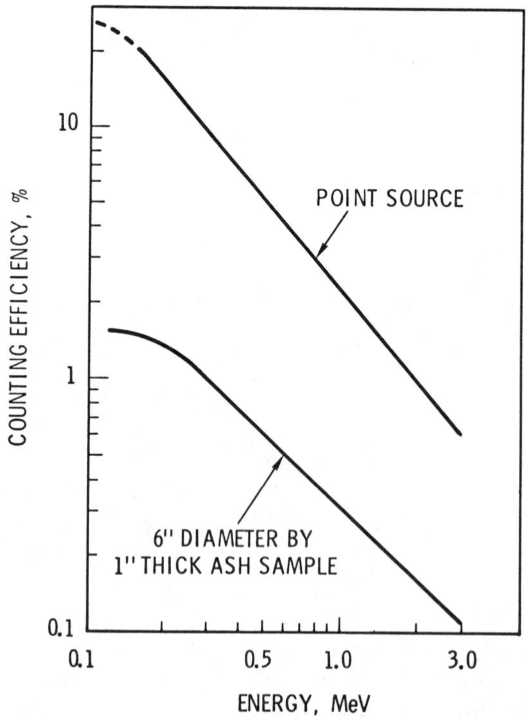

Figure 6.22. Counting efficiency of a large Ge(Li) detector for two source sizes (private communication from R. W. Perkins, Ref. 46).

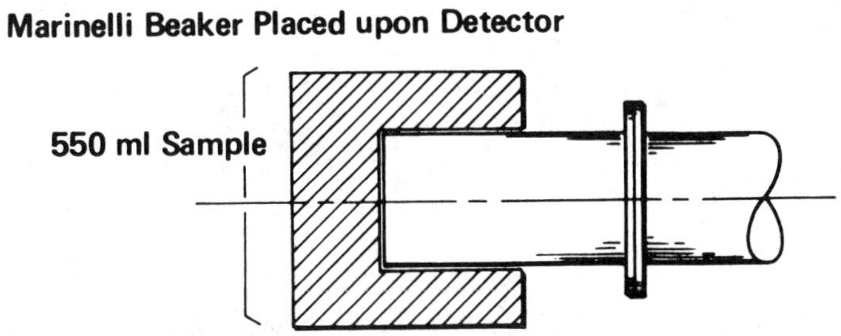

Figure 6.23. Marinelli sample-detector geometry used for measurement of weak activities (reproduced from Walford, Cooper, and Keyster, Ref. 47, permission of Institute of Electrical and Electronic Engineers © 1976).

Table 6.1. Minimum Detectable Activity in γ/min·ml of Sample Material for the Marinelli Beaker and for the Disk Sample (data presented in Walford, Cooper, and Keyster, Ref. 47, with permission of Institute of Electrical and Electronics Engineers © 1976)[a]

Energy (keV)	(For Disk Sample) γ/min·ml	(For Marinelli Beaker) γ/min·ml	Marinelli Vol. Reduced by Evaporation to 5-ml Disk[b]
		One VIP [10] *Detector (a 10% efficient Ge(Li) detector)*	
185	0.3	8.9×10^{-3}	2.7×10^{-3}
609	0.64	1.7×10^{-2}	5.6×10^{-3}
1238	1.1	2.7×10^{-2}	9.6×10^{-3}
1764	1.2	3.1×10^{-2}	1.1×10^{-2}
2204	1.5	3.8×10^{-2}	1.36×10^{-2}
		One WIN [15] *Detector (a 15% efficient Ge(Li) detector)*	
185	0.3	7.2×10^{-3}	2.7×10^{-3}
609	0.6	1.4×10^{-2}	5.4×10^{-3}
1238	0.9	2.0×10^{-2}	7.6×10^{-3}
1764	0.8	1.9×10^{-2}	7.6×10^{-3}
2204	1.2	2.7×10^{-2}	1.1×10^{-2}

[a]Count time = 40,000 sec. Statistical precision = 0.2 (standard deviation of peak area divided by peak area).
[b]Expressed as the activity in γ/min·ml of original sample volume.

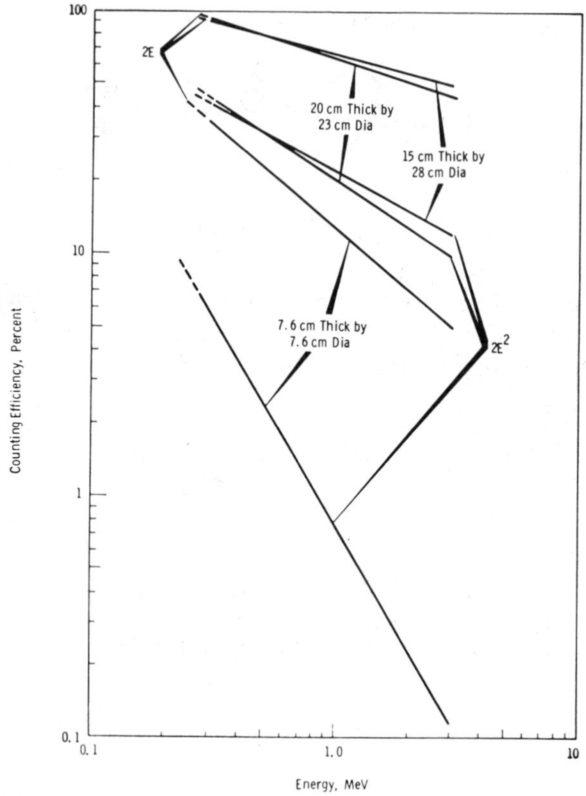

Figure 6.24. Total counting efficiencies and total coincidence counting efficiencies of three multidimensional gamma-ray detector systems. Curves labeled "2E" show summed efficiency of two identical detectors. Curves labeled "$2E^2$" show coincidence efficiency (reproduced from Wogman, Perkins, and Kaye, Ref. 44, with permission of Nucl. Instrum. Meth. © 1969).

Other types of coincidence arrangements have also been developed. For example, *beta-gamma coincidence counting* can be used to obtain substantial improvements in sensitivity for certain radionuclides. Using the coincidence configuration shown in Fig. 6.27, Brauer, Kaye, and Connally report measurements of iodine-131 at detection limits in the range of 0.01 pCi (0.37 mBq), with 1000-min counting times. A typical spectrum of a chemically separated ^{131}I sample is shown in Fig. 6.28. This spectrum was taken using a 0.75-in. × 0.020-in. plastic β-particle scintillator, inside a 5-in. × 5-in. NaI(Tl) well detector. The detection limit is a factor of 100 lower than that achievable with the 5-in. × 5-in. detector alone.

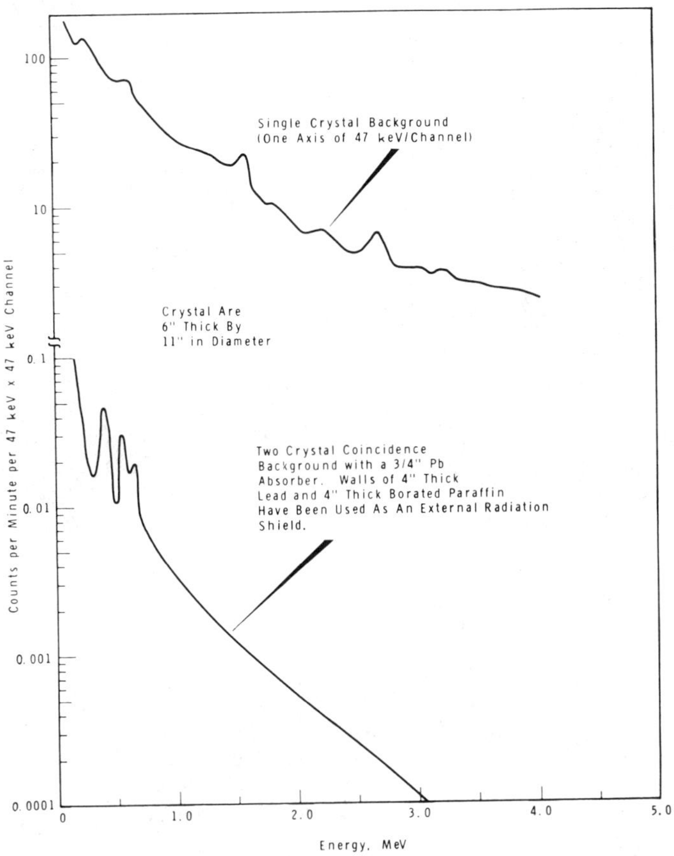

Figure 6.25. Comparison of the background of a single 6-in. thick × 11-in. diam. crystal with that of the two crystal multidimensional detector system. The detector system is housed inside a 4-in. thick lead and 4-in. thick borated-paraffin shield (reproduced from Wogman, Robertson, and Perkins, Ref. 50, with permission of Nucl. Instrum. Meth. © 1967).

Similar arrangements for γ-x-ray coincidence counting have also been reported by Brauer using a gas-flow proportional counter for x-ray detection inside a NaI(Tl) well.

Efficiency often is determined for the simpler systems by calibrating on a few isotopes with γ-rays of widely spaced energies; efficiencies at intermediate energies are interpolated. In using any of the anticoincidence or coincidence systems, however, precise quantitative analysis requires that efficiency calibrations be performed for each isotope to be measured.

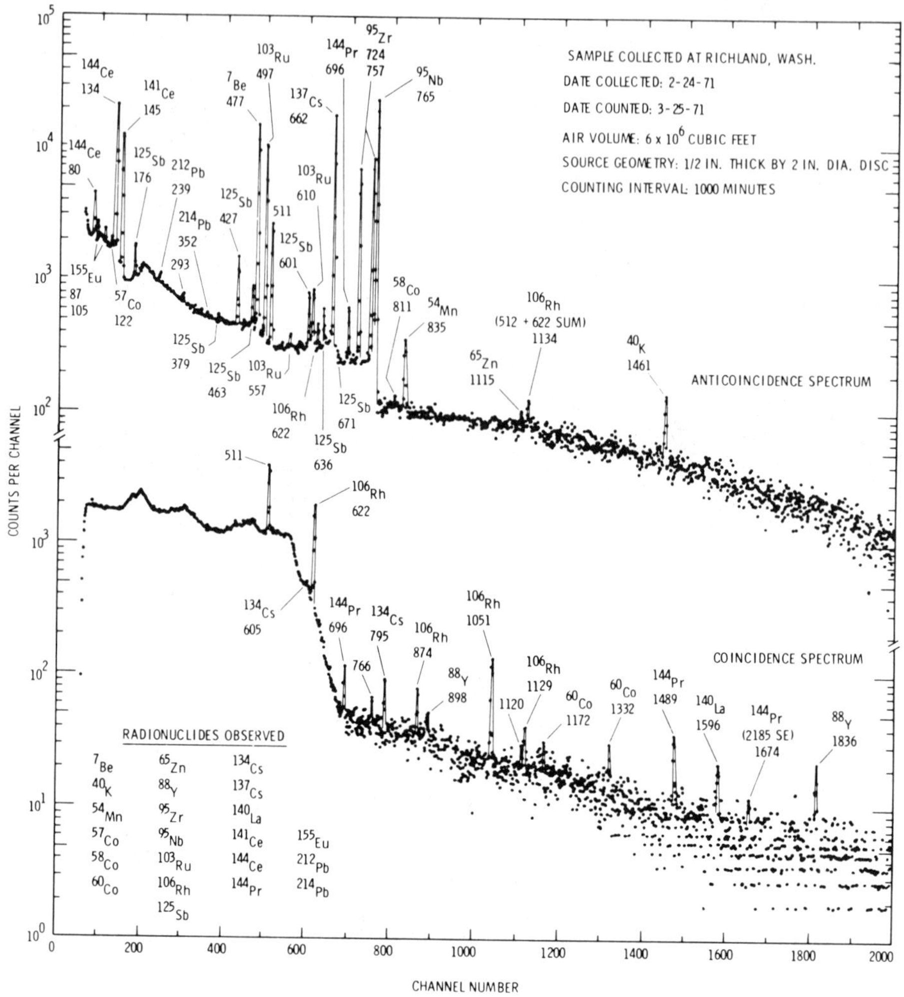

Figure 6.26. Coincidence and anticoincidence gamma-ray spectra of the radionuclides collected on an air filter from 6×10^6 ft^3 of air. Detector is Ge(Li) semiconductor, counting interval of 1000 minutes (reproduced from Cooper and Perkins, Ref. 51, with permission of Nucl. Instrum. Meth. © 1972).

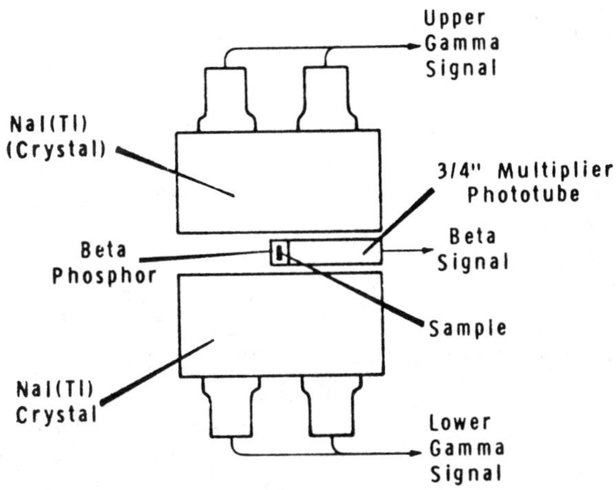

Figure 6.27. Low-level β-γ detector, showing beta phosphor inside well-type NaI(Tl) system (reproduced from Brauer, Kaye, and Connally, Ref. 52, with permission of American Chemical Society © 1970).

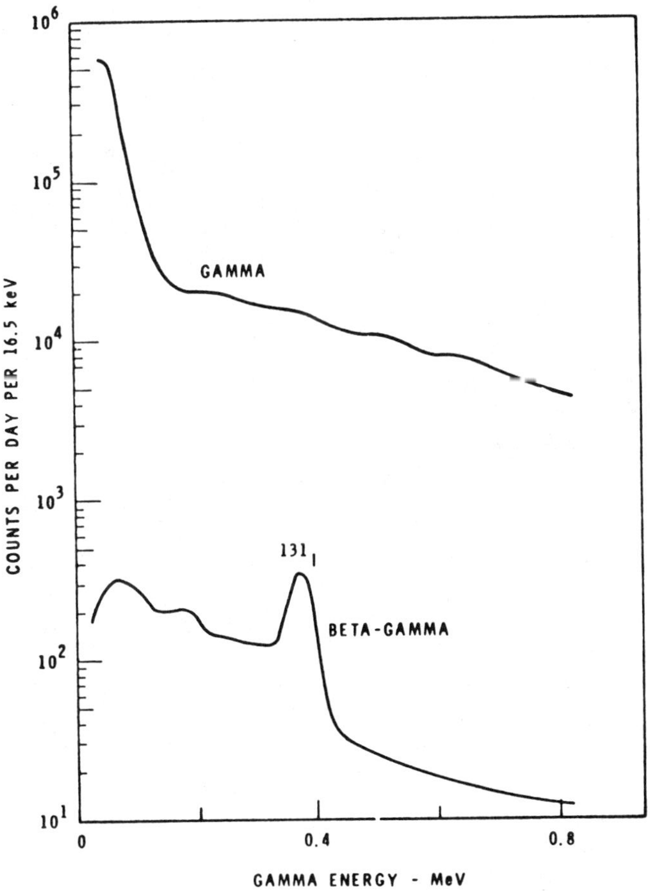

Figure 6.28. γ-Spectra of separated iodine sample taken in a 5-in. NaI(Tl) well system, with beta detection with a small plastic phosphor inside the well (see Fig. 6.27). Gamma and beta-gamma coincidence spectra shown separately (reproduced from Brauer, Kaye, and Connally, Ref. 52, with permission of the American Chemical Society © 1970).

6.6. LABORATORY MEASUREMENTS: SENSITIVITY

Although some types of γ-ray spectrometric measurements are made in the field, most are made in the laboratory on samples brought in from the field. This section discusses the factors that influence the ultimate sensitivity of laboratory systems. For information on quality control of environmental measurements using γ-ray spectrometry, the reader is referred to Reference 53.

There are two classes of measurement aims which are so different that one might imagine completely different measurement systems for them:

1. The sample to be measured may be known to contain only one radionuclide whose concentration must be determined.
2. The sample may be suspected of containing an unknown number of radionuclides which must be identified. Presumably the activities of the unknowns can vary widely in magnitude.

An example of the first category might be measurement on an air filter taken near the stack of a nuclear reactor, whereas the second might be to determine the radionuclides in waste from a nuclear biochemical laboratory.

Clearly, the choice of instrument may be different in these two cases. However, a typical radioanalytical laboratory is often confronted with both extremes as well as the entire range of measurement situations that lie in between.

6.6.1. Sensitivity for Measuring a Single Known Gamma-Ray Emitter

The most sensitive instrument system for measuring a single known γ-ray emitter with energies above about 100 keV uses a large NaI(Tl) crystal for its detector. The ultimate in sensitivity is then limited, essentially, by the background counting rate above which a signal must stand out. Many of the better low-background facilities are able to achieve background counting rates (in the 100 to 1000 keV range) on the order of 0.1 to 1 counts per minute per keV (cpm/keV) of energy acceptance. (Sensitivities as low as 0.01 cpm/keV are sometimes achieved.) As an example, the spectrum from NaI(Tl) scintillation counter for a typical monoenergetic line (say, ^{137}Cs at 662 keV) may be about 50 keV wide (FWHM), and the background counting rate over that energy acceptance may be 5 to 50 cpm (sometimes as low as 0.5 cpm). Clearly, activities yielding this count rate are at the minimum detectable level. Even with 1000-min counts, detection limits below 1 pCi (1 pCi = 2.2 disintegrations/min, 0.037 Bq) are achievable only in the best low-background facilities employing detectors for Compton-scattered events to aid in the detection efficiency. This estimated minimum detectable activity is increased by any number of inefficiencies (e.g., from the fact that not all photons convert their full energy into ion pairs in the detector, from solid-angle, and from other losses).

When more than one photon energy is present, all but the highest-energy photon may suffer additional sensitivity losses because the "background" beneath their full-energy peak(s) may be enhanced by the Compton continuum from the highest-energy line.

6.6.2. Sensitivity for Measuring a Large Number of Unknown Radionuclides

This measurement problem is often succinctly termed "peak hunting." NaI(Tl) scintillation counters are not suitable for resolving small closely spaced peaks in the presence of much larger ones (for example, in Fig. 6.13). Therefore for peak-hunting, the better resolution of a semiconductor spectrometer is preferred (i.e., Ge detectors).

The considerations are few in number but interdependent in a complicated way: better resolution makes a peak stand out more above background; higher detection efficiency yields more counts; lower background and small Compton-scattering efficiency both improve sensitivity; high peak-to-Compton and peak-to-total values are important. One trade-off is that although larger detector volumes yield higher efficiencies, they may make Compton tails worse, degrading sensitivity in the low-energy region. Also, large volumes usually imply some sacrifices in resolution.

Cooper [17,54] has combined many of these considerations into one overall "figure-of-merit" (FOM), which can be calculated for any Ge system as a function of energy being measured. The FOM was developed to aid in comparing the performance of systems measuring known lines, but has application for peak-hunting as well. Some of the results are surprising. For example, at lower energies ($\sim$122 keV) the FOM of even a very small 0.4-cm^3 detector is only 1.5 times poorer than that of detectors with 75-cm^3 volumes [17].

At low energies the smaller planar germanium detectors have definite advantages over the larger coaxial detectors for peak-hunting.

The detection sensitivity depends on so many factors that it cannot be fully covered in the brief discussion here. The situation is in many ways similar to that for NaI(Tl) as used to measure a single radionuclide. For Ge systems background counting rate from cosmic radiation tends to be smaller, but this gain is counter-balanced by lower volume and smaller efficiency. A reasonable rule-of-thumb is that Ge systems in peak-hunting mode (in the presence of many other peaks) are about one order of magnitude more sensitive than NaI(Tl) systems with comparable accoutrements (comparable low-background facilities, comparable electronic sophistication, etc.).

6.7. FIELD MEASUREMENTS

A significant effort has been devoted in recent years, to the development of measurement systems for use in the field. There are three types: portable, mobile, and stationary. The aim of the measurements can be to estimate either dose distributions or radionuclide distributions, and sometimes elemental distributions.

The reason for field measurements is usually to make a sensitive study of large areas. Because there are some measurements that cannot be made in any other way, these techniques have proven to be of value despite their drawbacks (chiefly lack of precision, logistical difficulties, and cost).

Beck, DeCampo, and Gogolak [55] have stated some of the advantages of field measurements:

In situ measurements of soil activity are more sensitive and provide more representative data than data obtained by sample collection and subsequent laboratory analysis. An unshielded detector placed about one meter above the ground detects gamma rays from an area within about a 10 meter radius, representing a large volume of soil compared to the typical soil sample, and comparable counting statistics can be obtained in only a small fraction of the time required for the laboratory analysis. For example, a field spectral analysis for the natural emitters, ^{40}K, ^{238}U, and ^{232}Th, can be carried out in approximately 15 minutes with a 4 in. by 4 in. NaI(Tl) detector. A comparable analysis in the laboratory, excluding transportation and sample preparation time, would require several hours. Furthermore, a single soil sample from a site may not be representative of the mean soil activity, so a number of samples or composite samples are required. A single field analysis averages out small local inhomogeneities in the sample.

The typical stationary system [55,56,57] is carried to the field, set up with the detector some distance ($\leq$ few meters) above the ground, and allowed to count whatever radiation may be present. An energy spectrum results, but the problem involves the unraveling of the true radionuclide concentrations in the media being studied. There are three different sources of events: the air, the ground, and the cosmic radiation. Because the penetration of γ-rays is greater at higher energies, the detector samples larger soil areas, deeper soil depths, and greater air volumes at high energies than low. Furthermore, because of the geometrical situation, any of the γ-rays being detected has a fair chance of coming from a Compton-scattering event rather than being a primary γ-ray from a monoenergetic source. These Compton γ-rays cannot be eliminated by alterations of the detector. This set of complications degrades the intrinsic accuracy, but perhaps the major difficulty is that the analysis depends on knowledge of the distribution of the radioactivity with soil depth, and to a lesser extent on soil composition [58]. Various analysis methods have been developed for this technique. That of Beck [55] is typical: given the soil density and chemical composition, and having calibrated the detector response as a function of energy using NBS standards, an integral is performed taking into account the absorption, angular-response corrections, and Compton-scattering corrections. The result may be expressed as activity per unit area ($\mu Ci/m^2$ or 3.7×10^4 Bq/m^2); activity per unit volume, assuming a uniform depth distribution (i/m^3); or a specialized activity distribution if detailed soil composition is known separately. A similar approach is described by Phelps et al. [59].

Field measurements are sometimes done on the gaseous and liquid effluents from nuclear facilities. This can be useful in correlating levels with normal operations and procedures [60–69]. Sakai, Terada, and Katagiri [63], for an example, discuss sensitivity for detecting certain radioactive gases. Kirby et al. [70] and Dickson et al. [71] describe the advantages and limitations of soil sampling versus *in situ* γ-ray counting.

With the development of high-purity Ge detector arrays, it has been possible to expand the use of photon spectrometry to measure low-energy x-ray photons such as those emitted by the transuranics [72,73].

Typical *in situ* measurements with Ge and NaI(Tl) systems are shown in Figs. 6.29 and 6.30; both spectra were taken at the same site [74]. Clearly, the detail observed in the Ge spectrum is vastly superior to that in the NaI(Tl) spectrum. Considerably more equipment is required for the Ge system: a large capa-

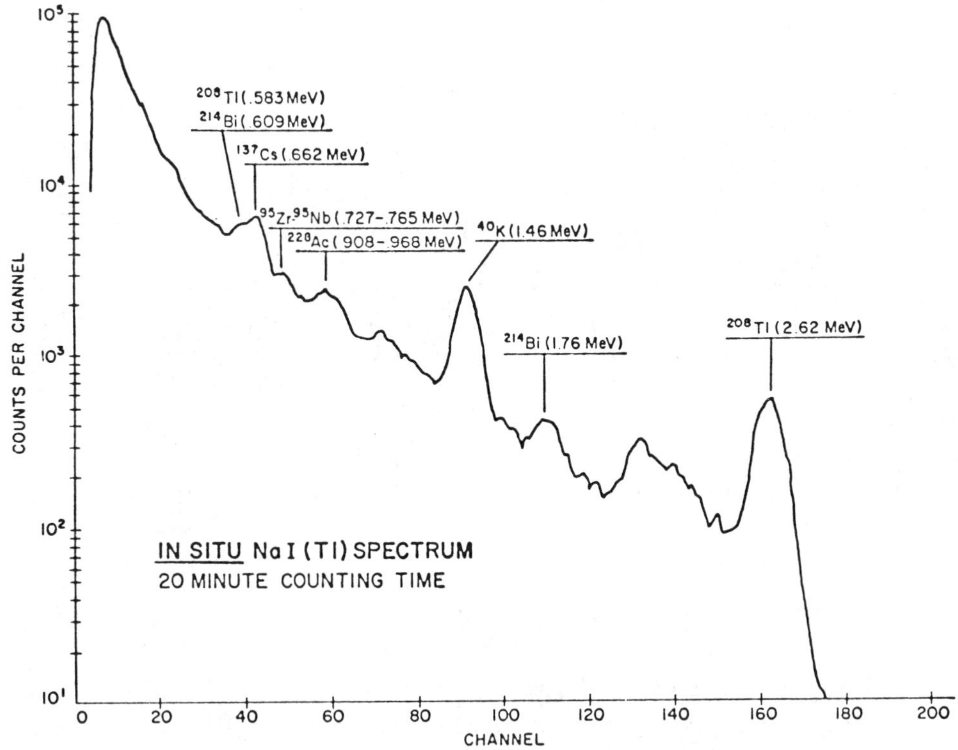

Figure 6.29. *In situ* spectrum, northeastern U.S. location, taken in 1971 with 10-cm × 10-cm NaI(Tl) crystal, 20-minute counting time (reproduced from Beck, Lowder, and McLaughlin, Ref. 74, with permission of International Atomic Energy Agency © 1971).

city multichannel analyzer, a liquid nitrogen system, special amplifier electronics, extra power, and a more bulky and fragile detector setup. This equipment restricts the convenience and mobility of the Ge system. Whenever a NaI(Tl) system can provide sufficient information (for example, to study backgrounds in remote areas), its use is preferable. However, with the availability of high-purity Ge detectors simpler cooling systems have been developed, so that more portable systems are becoming available [69,75]. For truly portable measurements (as contrasted to mobile measurements using a small van), a Ge system is ruled out.

There have been a number of portable NaI(Tl) systems developed for use in the field, some for general γ-spectrometry and some for special applications. An example of the latter is the FIDLER detector [76], developed to survey for the 60-keV γ-rays from americium-241 as a means of assessing plutonium deposition on soil. This device consists of a thin, disk-shaped NaI(Tl) crystal viewed by a phototube, the whole assembly is carried over the terrain about 50 cm above the soil to search for hot-spots of activity. Chapter 16, Plutonium, contains a more detailed discussion of this procedure. Several other NaI(Tl) instruments for measurements of natural environmental compilation are discussed by Adams and Lowder [77].

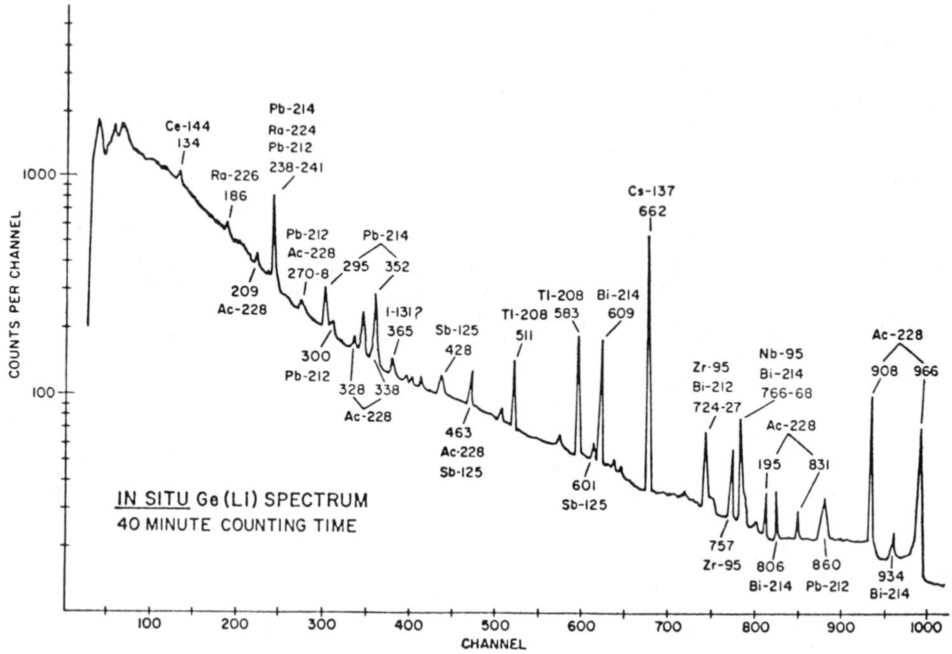

Figure 6.30. *In situ* spectrum, taken as in Fig. 6.29 with 60 cm^3 Ge(Li) detector, 40-minute counting time. Photon energies in keV (reproduced from Beck, Lowder, and McLaughlin, Ref. 74, with permission of International Atomic Energy Agency © 1971).

6.8. CALIBRATION STANDARDS

The most desirable method for calibrating any γ-ray spectrometer is to use a standard source containing known activities from each of several radionuclides. The source should reproduce exactly the geometrical and physical characteristics of the unknown sample to be measured with the same system. The purpose of using a known standard source is to reproduce as much as possible the same operating conditions used in the actual measurements.

A set of published γ-ray energies from 45 to 1300 keV that are useful for Ge spectrometer calibration and testing have been evaluated and found to be consistent with both a new reference value for the γ-ray energy scale and the 1973 least-squares adjustment of the fundamental constants. This new energy scale is based on recent measurements at the National Bureau of Standards (NBS) of the absolute γ-ray wavelength for the 411-keV line from ^{198}Au (actually 411.80441 ± 0.00108 keV) [78]. At approximately 1 MeV the proposed energies typically have measurement uncertainties of 4 eV and energy scale error of 3 eV compared to 14 eV and 17 eV in a previous publication [79].

For calibration, NBS furnishes some known standards [80]. Several of the NBS standards contain a known activity of a single radionuclide, and in addition there is a series of standards each containing several γ-ray emitters spanning the energy range from 88 to 1800 keV. The reader is referred to a method for calibrating Ge detectors suggested by the American National Standards Institute [81].

In calibration, care must be taken to avoid or minimize error from a wide variety of problems. Perhaps the most difficult to correct for are geometrical effects from differences in source location and source size. This is especially true for lower-energy photons, in which self-absorption can occur.

6.9. OTHER DETECTORS: PRESENT AND FUTURE

Research in detector technology is now proceeding mainly in the following areas: (a) to develop other semiconductor materials besides Ge and Si; (b) to improve and enlarge ultra-pure Ge crystals; (c) to improve the resolution of scintillation counter systems; (d) to develop improved proportional-counter techniques; and (e) to develop gas proportional scintillation counters. All of these efforts have a common goal: to make available at lower cost, significantly larger detector sizes, with resolution at least comparable to that obtainable with existing detectors. A second goal of semiconductor research is to find materials that do not require storage and operation at low temperatures Current progress includes:

1. The search for *new materials* has been going on since the beginning of semiconductor technology. Among the materials now being actively investigated are cadmium telluride (CdTe), gallium arsenide (GaAs), and mercuric iodide (HgI_2). A recent conference on CdTe [82] contains much important information. An article by Miller [83] gives an overview for these three materials.

Additional information is to be found in the proceedings of the International Workshop on CdTe and HgI_2 held in Jerusalem, Israel [84]. HgI_2 has an ϵ value (hole-electron pair formation energy) of 4.2 eV. Both CdTe and GaAs have ϵ values of about 4.3 to 4.6 eV/pair. For use at x-ray energies, these high-Z semiconductors offer efficient photon detectors at or near room temperature with energy resolution significantly better than NaI(Tl) scintillators [85]. However, due to size limitations their usefulness is not very great. The problems seem to be in the development of methods to make materials of sufficiently high purity, with good crystal perfection, and with accurate stoichiometry. A recent review [86] discusses the history, general properties, nuclear spectroscopic performance, and device application of CdTe and HgI_2:

> In spring 1978, CdTe and HgI_2 detectors [were] limited to areas of 100 mm^2 and to a thickness of 2 and 0.7 mm, respectively. This gives a gamma-ray interaction efficiency for CdTe and HgI_2, respectively, of about 99.9% and 99% for energies below 50 keV, 78% and 76% for 100 keV, and 6% and 4% for 662 keV. The optimal energy resolution is 18% and 8% at 6 keV and 1.3% and 0.7% at 662 keV for CdTe and HgI_2, respectively.

Other materials, such as AlSb, were discussed by Armantrout et al. [85].

2. Work on *high-purity germanium* has been actively continuing. A group at Lawrence Berkeley Laboratory [87,88] has defined the nature of the difficulty in obtaining high-purity Ge as being due to the poor segregation of the impurities boron and aluminum (because O_2 and Si form complexes with them that are electrically inactive). With proper attention to controlling Si and O_2 by appropriate choice of container and coating materials and ambient atmosphere, they have succeeded in reducing Al and B to levels below 10^{10} atoms/cm^3. Other studies [89,90] have shown the importance of freedom from hydrogen in keeping net impurity concentration of "fast defects" low.

Although high-purity Ge detectors are commercially available, they are still under development [91] and improvements can be expected in the near future. For example, it has been found [92] that having the p^+ contact on the coaxial periphery for high-purity Ge yields significant improvement in radiation damage resistance. Furthermore, it has been found [93] that large Ge detectors and an array of Ge detectors can operate at up to 160°K with good resolution. These results open doors to new applications and improvements. (See Pehl [18] for a discussion of some of these applications.)

Intrinsic Ge detectors with volumes as large as 180 cm^3 have now been fabricated, and resolution is comparable to that obtainable with conventional Li-drifted Ge detectors of similar size [22]. Research is aiming toward larger volumes, and even lower impurity concentrations. Among the advantages of high-purity Ge is ease of handling and the fact that it can be warmed up to room temperature without harm (except for possible surface contamination).

Active research is being undertaken to improve scintillation counter systems. There is the search for better scintillators: bismuth germanate ($Bi_4Ge_3O_{12}$) is being investigated because of the large atomic number of its major components, the wavelength of its light output (480 nm), and its chemical inertness and very primary results have been obtained. Much work is also being done on improving the efficiency and stability of plastic scintillators. Research is continuing near infrared phosphors which would decrease the amount of energy to produce a photon. This work is accompanied by a search for a suitable photocathode, possibly one employing a negative electron affinity surface such as those used to obtain very high secondary emission ratios. There is also work on photomultipliers, aimed at improving their efficiency and shape, increasing their size, and improving their speed. There is a great deal of interest regarding microchannel plate multipliers because they have position sensing and imaging properties as well as being excellent photomultipliers. Improvement in cathode and dynode efficiencies and stability are also being sought.

3. Research on *improved gas proportional counters* probably cannot lead to very significant improvements over current performance: the intrinsic limitation for low-energy x-rays is the value of about 25 eV/ion pair required in typical gases. For high-energy γ-rays, the detection efficiency cannot be very great because gases (even at high pressure) are not sufficiently dense. This has led to research on the use of liquids as the ionization medium. One research program in this area is that using liquid xenon [94], where feasibility of such detectors has been demonstrated. The goal is the potential availability of dense proportional counters of large size, high detection efficiency, and good resolution (for xenon, $\epsilon \simeq 2$ eV/ion pair).

4. Published work has demonstrated the feasibility of *gas proportional scintillation counters* [95–97]. The concept involves a gas counter operating in the ionization mode rather than the proportional mode. A photomultiplier detects the light emitted by the ionized electrons as they accelerate in the applied electric field. The counter relies on the existence of an accelerating electric field lower than that inducing ion multiplication but high enough to produce secondary light. The phototube then acts as the amplifying element. Resolutions of about 470 eV (FWHM) for the 5.9-keV ^{55}Fe x-rays have been obtained [97]. These counters, although still in the study/development stage, show considerable promise for even better resolution at the lowest energies. For example, resolutions of 100 eV have been obtained [97] for carbon x-rays (277 eV) and 230 eV for aluminum x-rays (1.49 keV).

6.10. SUMMARY AND CONCLUSIONS

This section has presented an overview of the main factors that affect the operating characteristics of γ-ray spectrometers. The emphasis has been on systems useful for measuring low-level γ-rays from radioactivity in various environmental media.

The two main considerations in choosing among the various system types are efficiency and energy resolution. These two factors are typified by the two most widely used kinds of systems: NaI(Tl) scintillation counters with high efficiency, and Ge semiconductor systems with excellent resolution. Other kinds of detectors are useful for special classes of measurements, in particular Si(Li) detectors and gas proportional counters are useful for measurements at low energies.

Because a γ-ray spectrometer can be one of the most expensive pieces of equipment in any radioactivity analysis laboratory, the user must be aware of the tradeoffs among the several parameters to be considered. This has been discussed in some detail in this chapter.

Perhaps the most revealing contrast is that between a NaI(Tl) scintillation counter system, on the one hand, used in the mode where one (or a few) γ-ray lines need to be measured at very low levels; and a Ge or Si spectrometric systems, on the other hand, used in the peak-hunting mode on a sample containing a large number of unknown radionuclides. Neither of these systems could adequately do the other's job!

It appears that the most important possible improvement foreseeable in the near future would be further development of large, ultra-pure Ge detectors. Larger detector sizes would give the semiconductor detectors sufficient overall efficiency to compete with NaI(Tl) for even the lowest low-level measurements; the superior resolution would enable even the largest detectors to resolve γ-ray lines far beyond the capability of any other existing detector type.

REFERENCES

1. C. E. Crouthamel, *Applied Gamma-Ray Spectroscopy,* 2nd ed. revised by F. Adams and R. Dams, Vol. 41 of the International Series of Monographs in Analytical Chemistry, Pergamon Press, Oxford, 1970.
2. J. H. Neiler and P. R. Bell, "The Scintillation Method," p. 245, Vol. I of Ref. 3.
3. International Atomic Energy Agency, "The Basic Requirements for Personnel Monitoring," IAEA Safety Series No. 14, Vienna, 1965.
4. Product Literature, Reuter-Stokes Electronic Components, Inc., 18530 S. Miles Parkway, Cleveland, OH 44128.
5. H. Sepida, "Energy Resolution of the Proportional Counter," *Nucl. Instrum. Meth.*, **133**, 252, 1976.
6. S. C. Curran and H. W. Wilson, "Proportional Counters and Pulse Ion Chambers," in *Alpha, Beta, and Gamma Spectroscopy*, K. Siegbahn, Ed., Vol. I, North-Holland Publishing Co., Amsterdam, 1965, p. 303.
7. R. Hofstadter, "Alkali Halide Scintillation Counters," *Phys. Rev.*, **74**, 100, 1948.
8. Beginning in 1962, the proceedings of the annual *Scintillation and Semiconductor Counter Symposia* have been published annually in *IEEE Transactions on Nuclear Science*.

9. J. B. Birks, *The Theory and Practice of Scintillation Counting,* Pergamon Press, Oxford, 1964.
10. G. A. Morton, "Recent Developments in the Scintillation Counter Field," *IEEE Trans. Nucl. Sci.*, **NS-3**(4), 122, 1956.
11. R. L. Heath, *Gamma-Ray Spectrum Catalogue,* Vol. I: Scintillation Spectrometry Gamma-Ray Spectrum Catalogue, 2nd ed., Phillips Petroleum Company, 1964; (available from NTIS as IDO-16880-1, Volume II: Gamma-Ray Spectrum Catalogue: Ge(Li) and Si(Li) Spectrometry, Aerojet Nuclear Company, 1964).
12. R. L. Heath, R. Hofstadter, and E. B. Hughes, "Inorganic Scintillators: A Review of Techniques and Applications," *Nucl. Instrum. Meth.*, **162**(1-3), Part II, 431, 1979.
13. E. Segrè, *Nuclei and Particles: An Introduction to Nuclear and Subnuclear Physics,* 2nd ed., W. A. Benjamin, New York (1977).
14. G. Dearnaley and D. C. Northrup, *Semiconductor Counters for Nuclear Radiation,* E. and F. Spon Ltd., London, 1966.
15. ORTEC, "Catalog 1002, Instruments for Research," 100 Midland Road, Oak Ridge, TN 37830, (1970).
16. G. A. Armantrout, H. E. Bradley, and P. L. Phelps, "Sensitivity Problems in Biological and Environmental Counting," *IEEE Trans. Nucl. Science*, **NS-199**(1), 107, 1972.
17. J. A. Cooper, "Figure-of-Merit Measurements for Ge(Li) Detectors," *Nucl. Instrum. Meth.*, **94**, 289, 1971.
18. R. H. Pehl, "Germanium Gamma-Ray Detectors," *Phys. Today*, **30**, 50, 1977.
19. J. A. Cooper, N. A. Wogman, H. E. Palmer, and R. W. Perkins, "The Application of Solid State Detectors to Environmental and Biological Problems," *Health Phys.*, **15**, 419, 1968.
20. F. Cappellani and G. Restelli, "X-Ray and Gamma-Ray Spectroscopy," in G. Bertolini and A. Coche, Eds., *Semiconductor Detectors*, North-Holland Publishing Co. © 1968.
21. National Council on Radiation Protection and Measurements, "Instrumentation and Monitoring Methods for Radiation Protection," NCRP Report No. 57, 1978.
22. G. T. Ewan, "The Solid Ionization Chamber," *Nucl. Instrum. Meth.*, **162**, 75, 1979.
23. ORTEC, *Experiments in Nuclear Science*, Application Note 34, 2nd ed., 100 Midland Road, Oak Ridge, TN 37830, 1976.
24. R. H. Pehl and F. S. Goulding, "Recent Observation on the Fano Factor in Germanium," *Nucl. Instrum. Meth.*, **81**, 329, 1970.
25. H. M. Mann, H. R. Bilger, and I. S. Sherman, "Observations on the Energy Resolution of Germanium Detectors for 0.1 to 10 MeV Gamma Rays," *IEEE Trans. Nucl. Sci.*, **NS-13**(3), 252, 1966.
26. R. G. Helmer, R. H. Heath, D. D. Metcalf, and G. A. Cazier, "A Linear Least-Squares Fitting Program for the Analysis of Gamma-Ray Spectra Including a Gain-Shift Routine," Phillips Petroleum Co., Report No. IDO-17015, 1964.
27. E. Schonfeld, A. H. Kibbey, and W. Davis, Jr., "Determination of Nuclide Concentrations in Solutions Containing Low Levels of Radioactivity by Least-Squares Resolution of the Gamma-Ray Spectra," Oak Ridge National Laboratory, Report No. ORNL-3744, 1965.

28. L. G. Kanipe, S. K. Scale, and W. S. Liggett, "Least-Squares Resolution of Gamma-Ray Spectra in Environmental Monitoring," U.S. Environmental Protection Agency Report No. EPA-6001 7-77-089, 1977.
29. C. R. Phillips, J. A. Stewart, and T. W. Athey, "A Computer Program for the Analysis of Gamma-Ray Spectra by the Method of Least Squares," USDHEW Publication No. 999-RH-21, Public Health Service Environmental Health Series, Washington, DC 20201, 1966.
30. R. H. Shumway, "Radionuclide Analysis of Gamma Ray Spectra by Stepwise Multiple Regression," USDHEW Publication No. 999-R-5, Public Health Service Environmental Health Series, Washington, DC 20201, 1963.
31. M. A. Efroymson, "Multiple Regression Analysis," in *Mathematical Methods for Digital Computers*, A. Talston and H. S. Wilf, Eds., Part V, (17), Wiley, New York, 1960, p. 191.
32. J. T. Routti and S. G. Prussin, "Photopeak Methods for the Computer Analysis of Gamma-Ray Spectra from Semiconductor Detectors," *Nucl. Instrum. Meth.*, **72**, 125, 1969.
33. R. Gunnink and J. B. Niday, "Computerized Quantitative Analysis by Gamma-Ray Spectrometry," LLNL Report UCRL-51061 (in 4 vols.), 1971, unpublished.
34. J. R. DeVoc, Ed., *Modern Trends in Activation Analysis*, Proceedings of the 1968 International Conference, NBS Special Publication 312 (in 2 vols.), 1969.
35. M. A. Mariscotti, "A Method for Automatic Identification of Peaks in the Presence of Background and Its Application to Spectrum Analysis," *Nucl. Instrum. Meth.*, **50**, 309, 1967.
36. R. G. Helmer, R. L. Heath, M. Putnam, and D. H. Gipson, "Photopeak Analysis Program for Photon Energy and Intensity Determinations with Ge(Li) and NaI(Tl) Spectrometers," *Nucl. Instrum. Meth.*, **57**, 46, 1967.
37. T. Inouye, T. Harper, and N. C. Rasmussen, "Application of Fourier Transforms to the Analysis of Spectra Data," *Nucl. Instrum. Meth.*, **67**, 125, 1969.
38. G. W. Phillips and K. W. Marlow, "Peak Search and Analysis of Gamma Ray Spectra with Very Low Statistics," *IEEE Trans. Nucl. Sci.*, **NS-24**(1), 154, 1977.
39. F. P. Brauer and J. E. Fager, "A Computor System for Environmental Sample Analysis and Data Storage and Analysis," *IEEE Trans. Nucl. Sci.*, **NS-21**(1), 743, 1976.
40. R. E. Wood and J. H. Palms, "A Gamma-Ray Spectrum Analysis Technique for Low-Level Environmental Radionuclides," *IEEE Trans. Nucl. Sci.*, **NS-21**(1), 536, 1974.
41. J. H. Kaye, F. P. Brauer, R. E. Connally, and H. G. Pieck, "Background Reductions Obtained with Gamma Detectors by Use of Massive Cosmic-Ray Shield," *Nucl. Instrum. Meth.*, **100**, 333, 1972.
42. H. A. Wollenberg and A. R. Smith, "Low-Background Concrete," Lawrence Berkeley Laboratory Report, UCRL-11674, unpublished, (1964).
43. H. A. Wollenberg and A. R. Smith, "A Concrete Low-Background Counting Enclosure," *Health Phys.*, **12**, 53, 1966.
44. N. A. Wogman, R. W. Perkins, and J. H. Kaye, "An All Sodium Iodide Anticoincidence Shielded Multidimensional Gamma-Ray Spectrometer for Low-Activity Samples," *Nucl. Instrum. Meth.*, **74**, 197, 1969.

45. J. A. Cooper, L. A. Rancitelli, and R. W. Perkins, "An Anticoincidence-Shielded Ge(Li) Gamma-Ray Spectrometer and Its Application to Radioanalytical Chemistry Problems," *J. Radioanal. Chem.*, **6**, 147, 1970.
46. R. W. Perkins, Battelle Pacific Northwest Laboratory, private communication.
47. G. V. Walford, J. A. Cooper, and R. M. Keyster, "Evaluation of Standardized Ge(Li) Gamma Ray Detectors for Low Level Environmental Measurements," *IEEE Trans. Nucl. Sci.*, **NS-23**, 734, 1976.
48. P. Ryge, "Improved Germanium Well Detectors," in *Radiation Instrumentation*, W. Wadman, Ed., Proceedings of the Health Physics Society Eleventh Mid-year topical Symposium on Radiation Instrumentation, January 16–19, 1978, Health Physics Society, 1978.
49. F. P. Brauer and W. A. Mitzlaff, "Evaluation of Well-Type Ge(Li) Detectors for Low-Level Radiochemical Analysis," *IEEE Trans. Nucl. Sci.*, **NS-25**(1), 398, 1978.
50. N. A. Wogman, D. E. Robertson, and R. W. Perkins, "A Large Detector, Anticoincidence Shielded Multidimensional Gamma-Ray Spectrometer," *Nucl. Instrum. Meth.*, **50**(1), 1967.
51. J. A. Cooper and R. W. Perkins, "A Versatile Ge(Li)–NaI(Tl) Coincidence-Anticoincidence Gamma-Ray Spectrometer for Environmental and Biological Problems," *Nucl. Instrum. Meth.*, **99**, 125, 1972.
52. F. P. Brauer, J. H. Kaye, and R. E. Connally, "X-Ray and Beta-Gamma Coincidence Spectrometry Applied to Radiochemical Analysis of Environmental Samples," *Adv. Chem. Ser.*, **93**, 231, 1970.
53. L. H. Zeigler and H. M. Hunt, "Quality Control for Environmental Measurements Using Gamma-Ray Spectrometry," Environmental Protection Agency Report EPA-660/7-77-14, (1977).
54. J. A. Cooper, "Factors Determining the Ultimate Detection Sensitivity of Ge(Li) Gamma-Ray Spectrometers," *Nucl. Instrum. Meth.*, **82**, 273, 1970.
55. H. L. Beck, J. DeCampo, and C. Gogolak, "*In Situ* Ge(Li) and NaI(Tl) Gamma-Ray Spectrometry," Health and Safety Laboratory, Report HASL-258, 1972.
56. S. P. Nielson, "In Situ Measurements of Environmental Gamma Radiation Using a Mobile Ge(Li) Spectrometer System," Riso National Laboratory, Report No. 367, 1977; available on exchange from Fiso Library, Riso National Laboratory, P.O. Box 49, DK-4000, Roskilde, Denmark.
57. D. C. Camp, "Mobile System for *In Situ* Gamma Ray Surveillance of Nuclear Facility Sites," *A Review of LLL-Developed Instruments, Techniques, and Methods Applicable to Alternative Fuel Cycle Technologies,* Lawrence Livermore Laboratory Report No. UCRL-5226, 1977. Livermore, CA 94550 (September, 1977).
58. H. L. Beck, "The Physics of Environmental Gamma Radiation Fields," in *Proceedings of the Second International Symposium on the Natural Radiation Environment, Houston, Texas, August 7–11, 1972,* U.S. Energy Research and Development Administration, 1972, p. 101.
59. P. L. Phelps, L. R. Anspaugh, J. J. Koranda, and G. W. Huckabay, "A Portable Ge(Li) Detector for Field Measurement of Radionuclides in the Environment," *IEEE Trans. Nucl. Sci.*, **NS-19**(1), 199, 1972.
60. P. L. Phelps, L. R. Anspaugh, S. J. Roth, G. W. Huckabay, and D. L. Sawyer, "Ge(Li) Level *In-Situ* Gamma-Ray Spectrometer Applications," *IEEE Trans. Nucl. Sci.*, **NS-21**(1), 543, 1974.

61. R. L. Ragaini, D. Jones, G. W. Huckabay, and T. Todachine, "Terrestrial Gamma-Ray Surveys at Preoperational Nuclear Power Plants Using An *In-Situ* Ge(Li) Spectrometer," *IEEE Trans. Nucl. Sci.*, **NS-22**(1), 636, 1975.
62. R. L. Heath and M. C. Dyer, "Applications of Gamma-Ray Spectrometry in Real Time Environmental Monitoring" in *Nuclear Methods in Environmental Research*, J. R. Vogt and W. Myer, Eds., Proceedings of the Second International Conference, CONF-740701, 1974, available from National Technical and Information Service.
63. E. Sakai, H. Terada, and M. Katagiri, "*In-Situ* Gamma-Ray Measurement Having Ge(Li) Detectors," *IEEE Trans. Nucl. Sci.*, **NS-23**(1), 726, 1976.
64. W. A. Jester, F. J. Hepbum, "A Ge(Li) System for Monitoring of How Level Radioactive Gases," *Trans. Am. Nucl. Soc.*, **24**, 489, 1977.
65. H. L. Nielson, N. A. Wogman, and R. L. Brodzenski, "*In-Situ* Subterranean Gamma-Ray Spectroscopy," *Nucl. Instrum. Meth.*, **143**, 385, 1977.
66. L. Lovborg, L. Botter-Jensen, E. M. Christiansen, and B. L. Nielsen, "Gamma-Ray Measurements in an Area of High Natural Radioactivity," in *Proceedings of the Third International Symposium on the Natural Radiation Environment, Houston, Texas, April 23–28, 1978.*
67. R. R. Finck and B. R. H. Persson, "*In-Situ* Ge(Li) Spectrometric Measurements of Gamma Radiation from Radon Daughters Under Different Weather Conditions," in *Proceedings of Third International Symposium on the Natural Radiation Environment, Houston, Texas, April 23–28, 1978.*
68. L. D. Stephens, A. R. Smith, and R. H. Thomas, "An Analytical Approach to Environmental Radiation Measurements," *IEEE Trans. Nucl. Sci.*, **NS-23**(1), 719, 1976.
69. R. F. Hill and W. Garber, "Portable X-Ray Analyzer for *In-Situ* Gold Qualifications, *IEEE Trans. Nucl. Sci.*, **NS-25**, 790, 1978.
70. J. A. Kirby, L. R. Anspaugh, P. L. Phelps, G. W. Huckabay, "A Comparison of *In-Situ* Gamma Soil Analysis and Soil Sampling Data for Mapping ^{241}Am and ^{239}Pu Soil Concentrations at the Nevada Test Site," *IEEE Trans. Nucl. Sci.*, **NS-24**(1), 587, 1977.
71. H. W. Dickson, G. D. Kerr, P. J. Purdue, and S. A. Abdullah, "Environmental Gamma Ray Measurements Using *In-Situ* and Core Sampling Techniques," *Health Phys.*, **30**, 221, 1976.
72. F. P. Brauer, J. M. Kelley, R. W. Goles, and J. E. Fager, "Measurement of Environmental ^{241}Am and the Pu ^{242}Am Ratio by Photon Spectrometry," *IEEE Trans. Nucl. Sci.*, **NS-24**(1), 591, 1967.
73. M. G. Strauss, I. S. Sherman, and E. J. Swanson, "Consideration in Measuring Trace Radionuclides in Soil Samples by L X-Ray Detection," *IEEE Trans. Nucl. Sci.*, **NS-25**, 740, 1978.
74. H. L. Beck, W. M. Lowder, and J. E. McLaughlin, "*In-Situ* External Environmental Gamma Ray Measurements Utilizing Ge(Li) and NaI(Tl) Spectrometry and Pressurized Ionization Chamber," in *Proceedings of the Rapid Methods for Measuring Radioactivity in the Environment,* International Atomic Energy Agency, Vienna, 1971.
75. W. Garber and R. F. Hill, "Hand-Held High Resolution Gamma Ray Spectrometer," in *Proceedings of the Health Physics Society Eleventh Midyear Topical Symposium on Radiation Instrumentation,* 1978.
76. J. F. Tinney, J. J. Koch, and C. T. Schmidt, "Plutonium Survey with an X-Ray Sensitive Detector," LLL Report UCRL-71362, 1969.

77. J. A. S. Adams and W. M. Lowder, Eds., *The Natural Radiation Environment*, Univ. of Chicago Press, 1964.
78. R. G. Helmer, R. C. Greenwood, and R. J. Gehrke, "Revaluation of Precise γ-Ray Energies for Calibration of Ge(Li) Spectrometers," *Nucl. Instrum. Meth.*, **155**, 189, 1978.
79. R. C. Greenwood, R. G. Helmer, and R. J. Gehrke, "Precise Comparison and Measurement of Gamma-Ray Energies with a Ge(Li) Detector I.50–420 keV," *Nucl. Instrum. Meth.*, **77**, 142, 1970.
80. U.S. National Bureau of Standard, "Catalog of Standard Reference Materials," NBS Special Publication 260, available from NBS, Washington, DC 20234.
81. American National Standards Institute, "Calibration and Usage of Germanium Detectors for Measurement of Gamma-Ray Emission Rates of Radionuclides," ANSI N42.14, 1978.
82. A. J. Strauss, "The Physical Properties of Cadmium Telluride," *Rev. Phys. Appl.*, **12**, 167, 1977.
83. G. L. Miller, "A Brief Review of Recent Advances in Compound Semi-Conductors for Radiation Detectors," *IEEE Trans. Nucl. Sci.*, **NS-19**(1), 251, 1972.
84. Proceedings of the International Workshop on CdTe and HgI_2, Jerusalem, Israel, June 7–8, 1977, *Nucl. Instrum. Meth.*, **150**(1), 1978.
85. G. A. Armantrout, S. P. Swieskowski, J. W. Sherohman, and J. H. Yee, "What Can be Expected from High-*Z* Semiconductor Detectors," *IEEE Trans. Nucl. Sci.*, **NS-14**(1), 121, 1977.
86. R. C. Whited and M. M. Schieber, "Cadmium Telluride and Mercuric Iodide Gamma Radiation Detectors," *Nucl. Instrum. Meth.*, **162**, 113, 1979.
87. E. E. Haller and W. L. Hansen, "Impurities in High-Purity Germanium" *IEEE Trans. Nucl. Sci.*, **NS-21**, 219, 1974.
88. S. Hubbard, E. E. Haller, and W. L. Hansen, "Zone Refining High Purity Germanium," *IEEE Trans. Nucl. Sci.*, **NS-25**(1), 362, 1978.
89. E. E. Haller, "A New Method to Determine the Chemical Composition and Structure of Non-Elemental Acceptor and Donor Centers in Ultra Pure Germanium," *IEEE Trans. Nucl. Sci.*, **NS-25**(1), 418, 1978.
90. R. N. Hall, "Interactions Between Quenched-In Defects in High-Purity Germanium," *Inst. Phys. Conf. Ser.*, **23**, 190, 1975.
91. E. E. Haller, "Recent Advances in Common Semiconductor Materials," *IEEE Trans. Nucl. Sci.*, **NS-25**(2), 1978.
92. R. H. Pehl, N. W. Maddem, J. E. Elliott, T. W. Raudorf, R. C. Trammell, and L. S. Darken, Jr., "Radiation Damage Resistance of Reverse Electrode Ge Coaxial Detectors", *IEEE Trans. Nucl. Sci.*, **NS-26**(1), 321, 1979.
93. G. H. Nakeno, D. A. Simpson, and W. L. Imhof, "Characteristics of Large Intrinsic Germanium Detectors Operated at Elevated Temperatures," *IEEE Trans. Nucl. Sci.*, **NS-24**(1), 68, 1977.
94. S. E. Derenzo, Lawrence Berkeley Laboratory, private communication.
95. A. J. P. L. Policarpo, M. A. F. Alves, M. C. M. DosSantos, and M. J. T. Carvalho, "Improved Resolution for Low Energies with Gas Proportional Scintillation Counters," *Nucl. Instrum. Meth.*, **102**, 337, 1972.
96. H. E. Palmer, "Recent Advances in Gas Scintillation Proportional Counters," *IEEE Trans. Nucl. Sci.*, **NS-22**(1), 100, 1975.

97. M. Alice, F. Alves, A. J. P. L. Policarpo, and M. S. S. C. P. Leite, "The Energy Resolution and Window Area Capabilities of the Gas Proportional Scintillation Counter," *IEEE Trans. Nucl. Sci.*, **NS-22**(1), 109, 1975.

7 MEASUREMENT OF ALPHA AND BETA PARTICLES

Charged particles include beta rays, alpha rays and protons. The reason for needing to detect and measure charged-particle radiation is the same as that for gamma rays: these particles can be a severe health hazard. The passage of charged particles through matter produces ions, but in a different way than do γ- or x-rays. Charged particles give up their energy in small increments to each atom in their immediate environment as they progress through matter, each increment being large enough to excite or ionize the atom. After this excitation the atom may fluoresce in a scintillation counter or conduct a current in a gas or semiconductor detector.

Another difference is that charged particles penetrate matter for much shorter distances than do γ- or x-rays of the same energy. Consequently, instruments for their measurement, though based on the same principles are rather different in construction. Also many radioactive nuclides emit charged particles but not γ-rays, making γ-ray detectors unusable as a means of assay.

This chapter discusses several classes of instruments and techniques for particle detection and identification, including:

1. Detectors for radiological protection, monitoring, and surveys. These instruments may measure absorbed doses or particle flux.
2. Spectrometers for identifying individual energy spectra to assist in the identification of the radiation source. The source may contain more than one emitter.
3. Techniques for measuring of specific activities in air or water.
4. Chemical separation techniques for preparing samples for measurement or spectrometry.

The chapter is divided into two parts discussing β- and α-ray instrumentation, respectively. Other charged-particle radiation is not specifically discussed because, except in high-energy laboratories, it does not represent a health hazard. It suffices to say that the problems of proton and heavy-ion measurement are very similar to that of α-ray measurement.

7.1. MEASUREMENT OF BETA RAYS

7.1.1. Physical Considerations

The term β-particle originally referred to those particulate emissions of radioactive nuclear decay which had small mass (to distinguish them from α-particles) but were charged (to distinguish them from γ-rays). Although naturally occurring β-ray emitters are sources of negative beta rays (β^-), artificially produced isotopes include a large number of positron (β^+) emitters as well.

A β-particle is an electron. This fact was not known when β-radiation was first discovered by Rutherford in the earliest days of nuclear physics, but as time went on, the properties (mass, magnitude of charge, interactions) of the β^+ particle were found to be identical. Nevertheless, some doubt as to the exact identity of the two particles existed for a long time. Two discoveries finally settled the issue: first, that β^+-particles annihilated with atomic electrons; and second, that β^--particles, when captured into atomic orbits, obey the Pauli exclusion principle just like electrons.

The term β-decay embraces all modes of nuclear decay in which the atomic number (Z, i.e., the nuclear charge) changes by one unit while the mass number, A, remains constant. Beta decay can be thought of as a mechanism that changes the relative numbers of neutrons and protons in a nucleus, bringing the neutron/proton ratio closer to the region of stability on the chart of the nuclides. In the β-decay process, the nucleus may emit a positive electron (changing Z to $Z-1$), emit a negative electron (changing Z to $Z+1$), or capture an orbital electron (Z to $Z-1$). For practical purposes, β-emission can be thought of as the conversion of a neutron to a proton within the nucleus. Indeed, the simplest case of nuclear β-decay is the decay of a free neutron into a proton, a beta particle, and an anti-neutrino (the anti-neutrino is defined later). Proton-to-neutron conversion produces β^+-particles.

The details of β-decay are not given here. For such a discussion, the reader is referred to Ref. 1. However, one key feature, essential to design of β-ray detection instruments, must be discussed. Beta particles emerge from nuclear decay with a spectrum of energies characteristic of each nuclide. The spectrum has a characteristic lopsided bell shape (see Fig. 7.1), and the average energy is approximately one-third of the maximum β-particle energy. It is now known that all β-emission is accompanied by the emission of a neutrino (ν) or anti-neutrino ($\bar{\nu}$). The neutrino is a massless particle with only very weak interactions as it traverses matter. The two particles (β^- and ν, or β^+ and $\bar{\nu}$) together with the recoiling nucleus share the available energy. It is this sharing that produces the wide β-particle energy spectrum. Thus β^-- and β^+-emission can be thought of as the following "virtual" processes within a nucleus:

$$n \rightarrow p + \beta^- + \nu$$

$$p \rightarrow n + \beta^+ + \bar{\nu} \ .$$

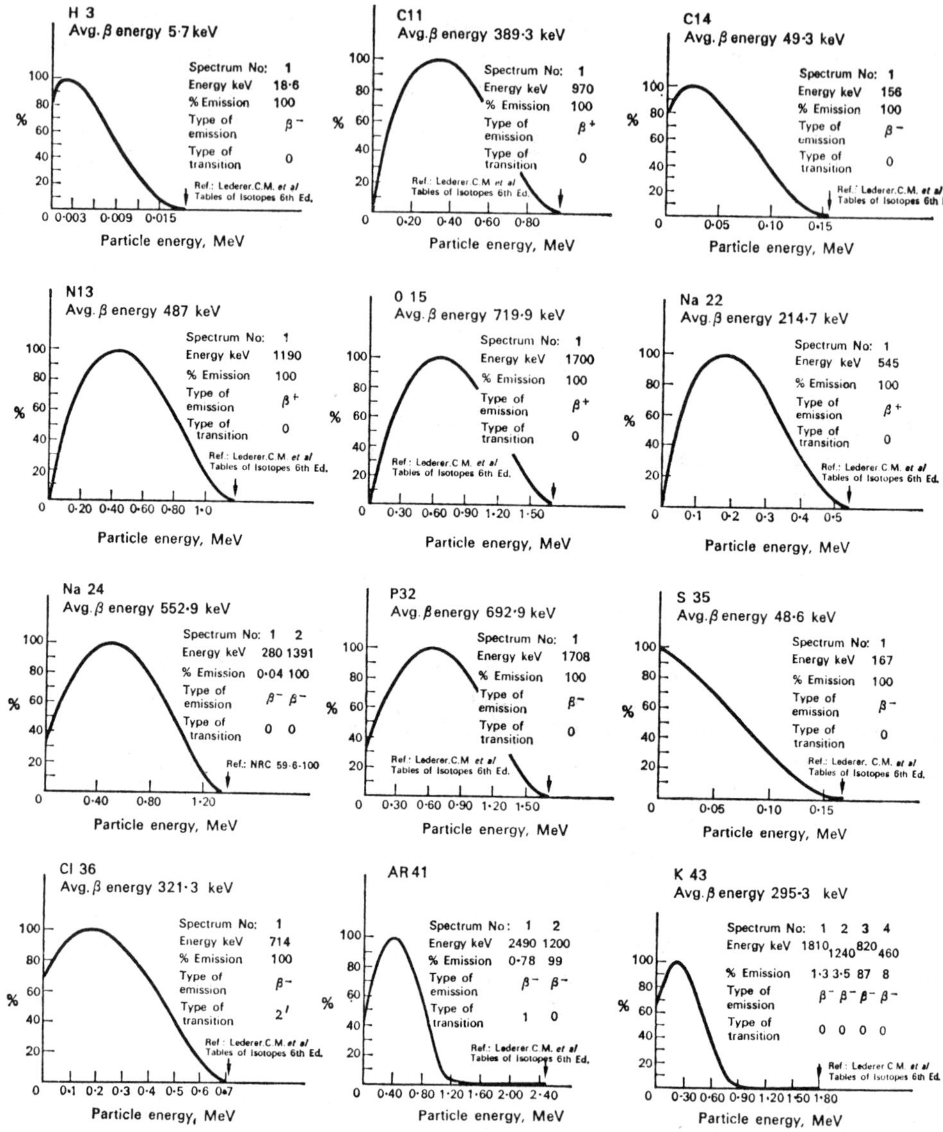

Figure 7.1. The beta-ray spectrum and average beta energy of selected isotopes (reproduced from Mantel, Ref. 2, with permission of Pergamon Press © 1972).

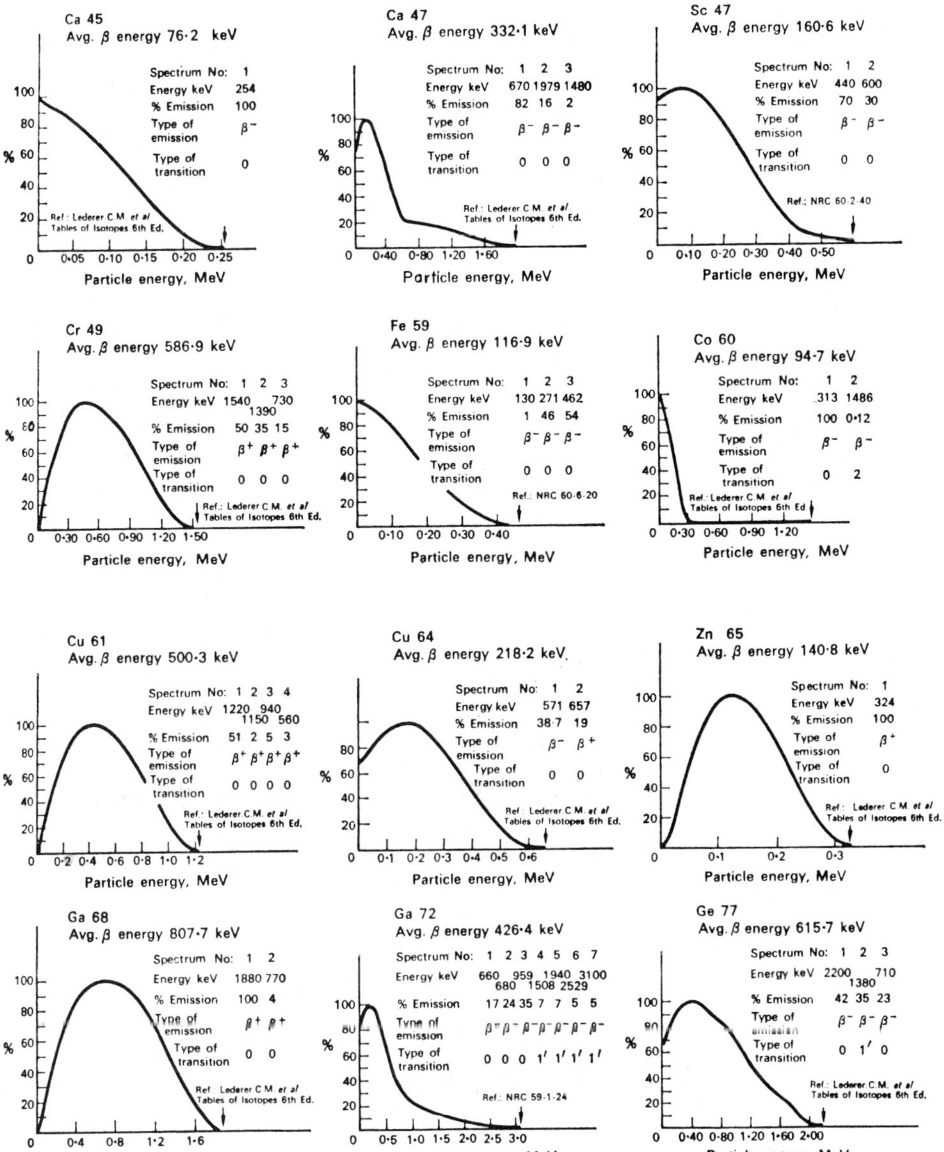

Figure 7.1. (Continued)

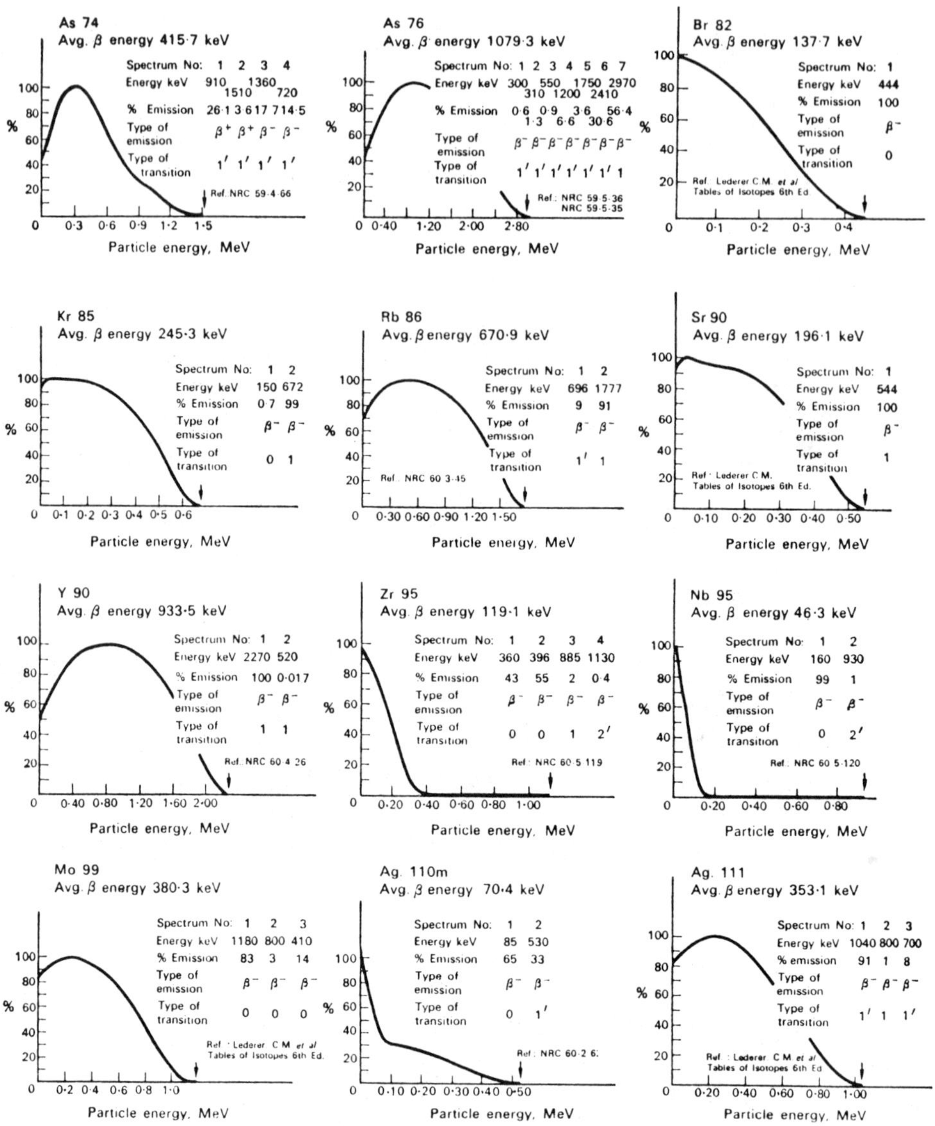

Figure 7.1. (Continued)

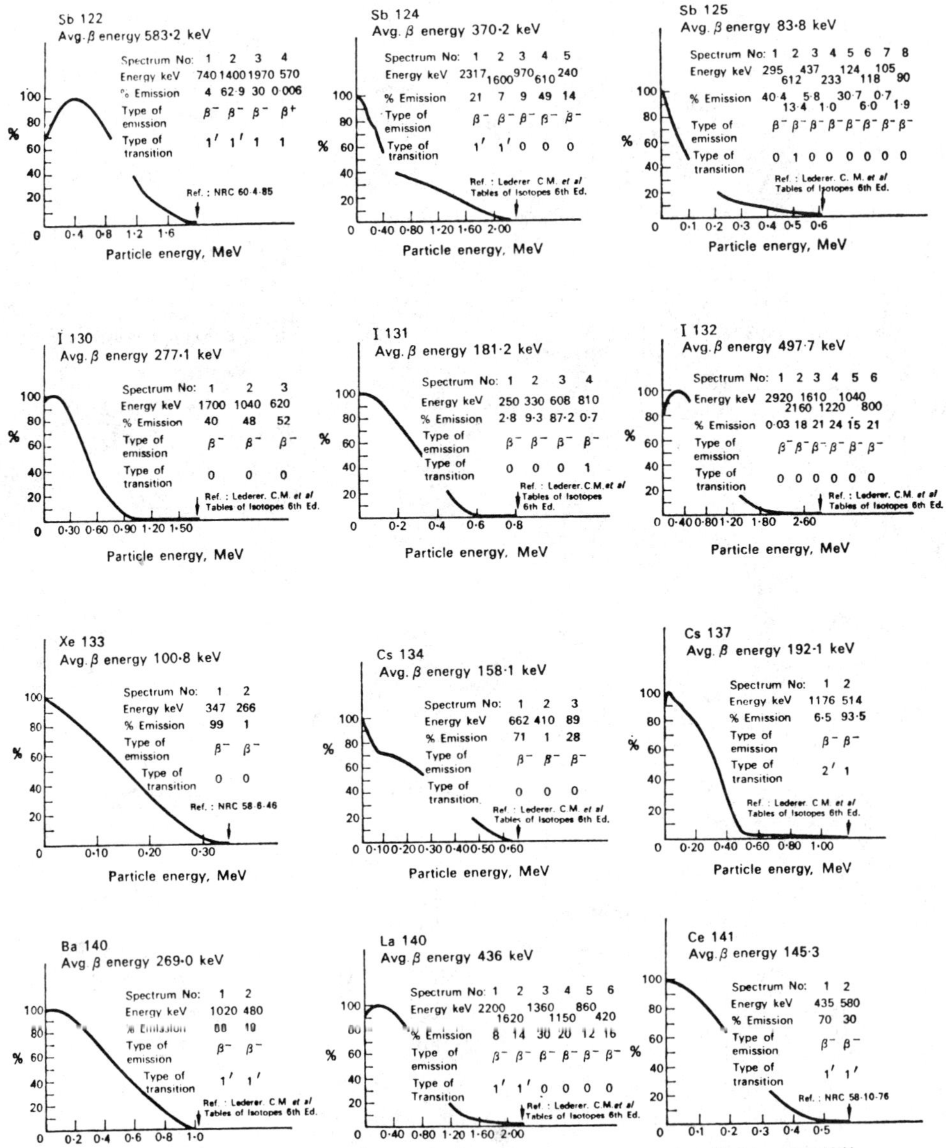

Figure 7.1. (Continued)

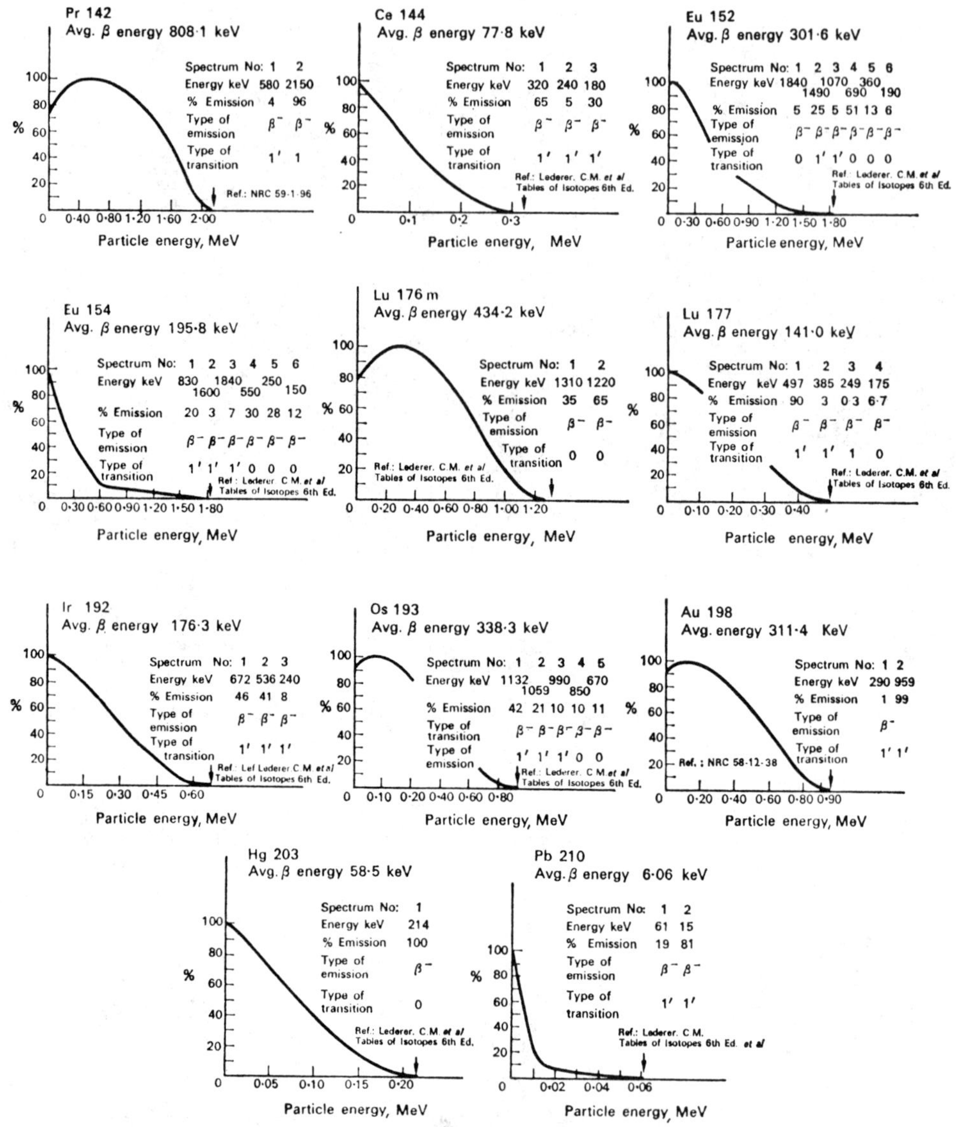

Figure 7.1. (Continued)

Beta emission is commonly tabulated and referred to by its maximum possible β-energy, E_{max}. The spectrum, however, ranges all the way from E_{max} to zero energy. The energy distribution reveals details about the dynamic processes (for example, nuclear spin-flip or non-flip) which occur within the nucleus during the decay. In Fig. 7.1 a large number of β-particle spectra are shown. These spectra demonstrate the wide variety in the types of spectra that characterize β-decay.

The maximum energy in β-decay can be very small (E_{max} = 18.5 keV for tritium) or large (E_{max} = 3.53 MeV for one of the β-rays from ^{106}Rh). However, most E_{max} values are in the energy region between 0.3 and 1.0 MeV. Because the stopping of electrons of this energy in different materials does not have the simple straight trajectory typical of heavier particles such as α-particles, it is not nearly as meaningful to speak of their range. Large deviations in the electron path are possible because the mass of the electron is equal to that of the orbital electrons with which it is interacting, and a large fraction of its energy can be lost in a single encounter. In addition, electron-nuclear interactions sometimes occur, which can lead to an abrupt change in the electron direction. However, range-energy curves for β-particles are used widely; they usually deal with what is called the maximum range. Normally, the electron range is taken from a plot such as that given in Fig. 7.2 by linear extrapolation of the portion of the curve at the inflexion near its end. The range curve for β-particles emitted by a radionuclide source differs significantly; this is due mainly to their continuous distribution in energy. An example is given in Fig. 7.3. The initial slope is much greater because the "soft" or low-energy β-particles are readily absorbed even in thin absorbers. As a practical matter, if the detector is sensitive to photons, the plot becomes more complicated for the β-particle emitters that also emit photons [3].

Cember [4] notes the useful rule-of-thumb that the absorber thickness that stops one-half of the β-particles is found to be about one-eighth of the range. Figure 7.4 shows the range of β-particles in various materials in centimeters. For all materials, especially with low atomic numbers, the range expressed in mg/cm^2 (see Fig. 7.5) is nearly independent of the absorber material.

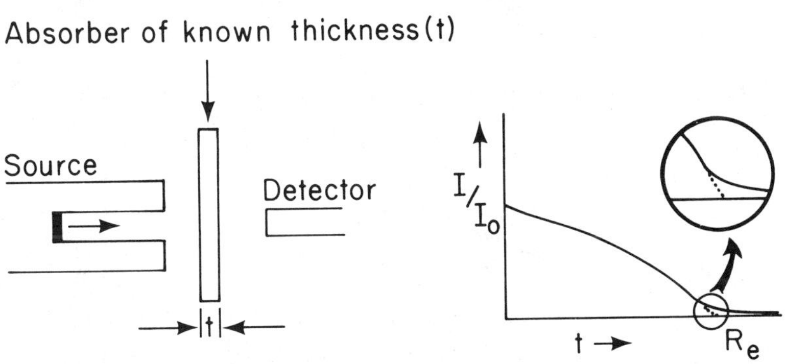

Figure 7.2. Usual method for the determination of electron range.

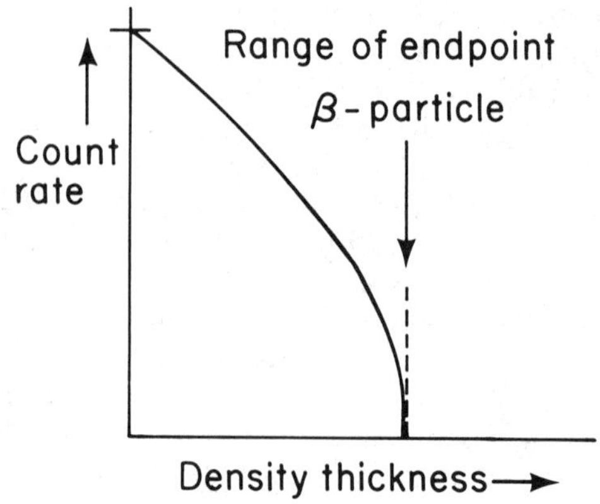

Figure 7.3. Typical range curve for electrons (beta particles) resulting from radioactive decay.

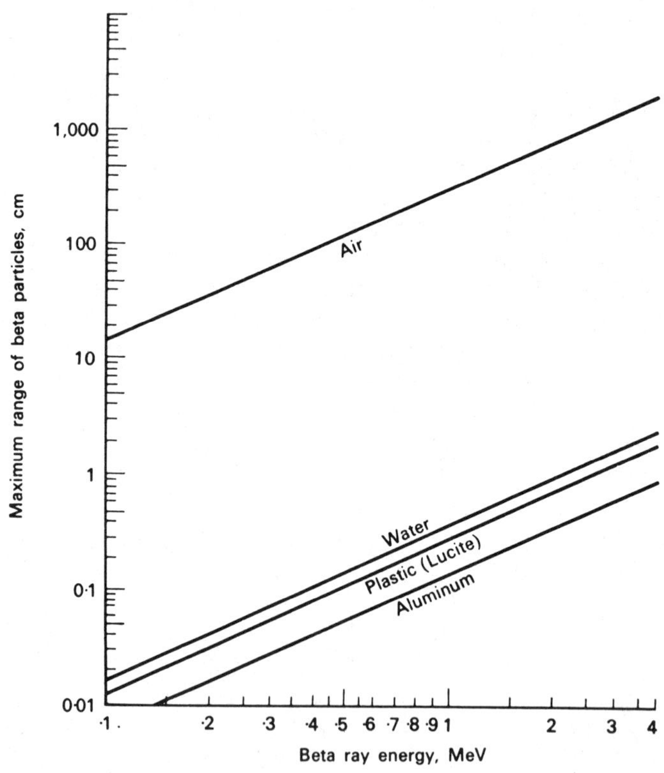

Figure 7.4. Range-energy curves for beta rays in various substances. Note that both scales are logarithmic (reproduced from Cember, Ref. 4, with permission of Pergamon Press © 1969).

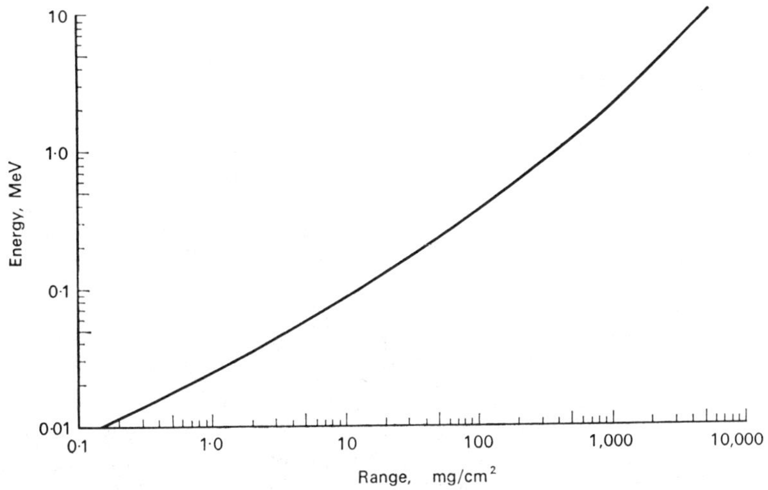

Figure 7.5. Range-energy curves for beta particles. The range is expressed in units of density thickness. Note that both scales are logarithmic (reproduced from Cember, Ref. 4, with permission of Pergamon Press, © 1969).

7.1.2. Measurement Considerations

The main detection mechanism for β-particles is ionization energy loss, that is, the interaction with the atomic electrons. Although ionization and nuclear scattering/deflection occur simultaneously for β-particles traversing through matter, the phenomena are sufficiently different that they can be considered separately.

Like all charged particles, β-particles lose more energy per unit material traversed as they come to the end of their range and stop. However, the effect does not become very significant until the kinetic energy is below about 100 keV. This can be seen in Fig. 7.6, which shows the specific ionization (number of ion pairs/cm) in air as a function of β-particle energy. The average energy lost per ion pair produced in air is about 34 eV.

Another feature of electron interactions is bremsstrahlung, which is the emission of photons by the electrons as they decelerate in a medium. The energy distribution of bremsstrahlung photons emitted by an electron of energy E is a $1/E$ spectrum, so that most bremsstrahlung photons from β-particles are soft. However, even behind an absorber thicker than the maximum range of the β-particles, penetrating bremsstrahlung photons will sometimes be detected. This is another reason why the range of β-particles is a concept without exact meaning. Bremsstrahlung also occurs for heavier charged particles, of course, but the probability is proportional to $1/M^2$ (where M = mass of particle). Thus it is less probable for α-particles than for β-particles by a factor of $m_\beta^2/m_\alpha^2 = 1.9 \times 10^{-8}$, a very small number indeed.

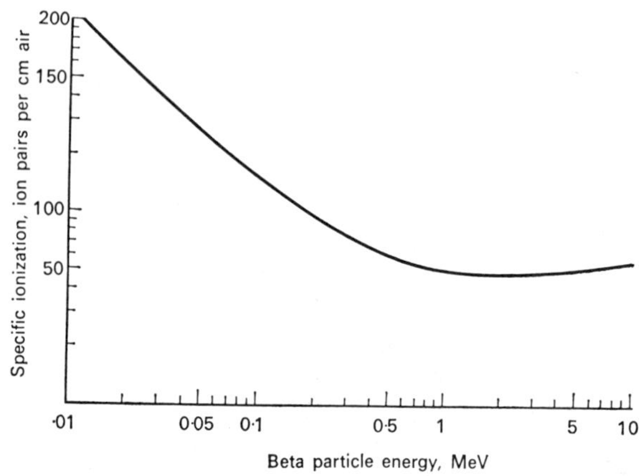

Figure 7.6. Relationship between beta-particle energy and specific ionization of air. Note that both scales are logarithmic (reproduced from Cember, Ref. 4, with permission of Pergamon Press © 1969).

Cember [4] finds:

> For purposes of estimating the bremsstrahlung hazard, the following approximate relationship may be used:
>
> $$f = 3.5 \times 10^{-4} ZE,$$
>
> where f = the fraction of the incident beta energy converted into photons,
> Z = atomic number of the absorber,
> E = maximum energy of the beta particle, in MeV.
>
> Because the likelihood of bremsstrahlung production increases with atomic number of the absorber, beta-ray shields are made with material of the minimum practicable atomic number. In practice, beta-ray shields of higher atomic number than 13 (i.e., aluminum) are seldom if ever used.

Another physical feature of β-decay that must be remembered when making measurements is backscatter. The fact that electrons often undergo large-angle deflections along their track leads to the possibility that an electron entering one surface of an absorber may undergo sufficient deflection that it may reemerge from the same surface. Such electrons do not deposit their full energy in the absorbing medium and can thus have a significant effect on the response of detectors designed to measure the energy of externally incident electrons. Electrons that backscatter in the detector entrance window, or dead layer, will totally escape detection. Backscatter is most pronounced for absorbers with high atomic number and electrons with low incident energy. Figure 7.7 shows the fraction of monoenergetic electrons that are backscattered when normally incident on the surface of various absorbers. Additional data for materials commonly used as electron detectors are given in Table 7.1.

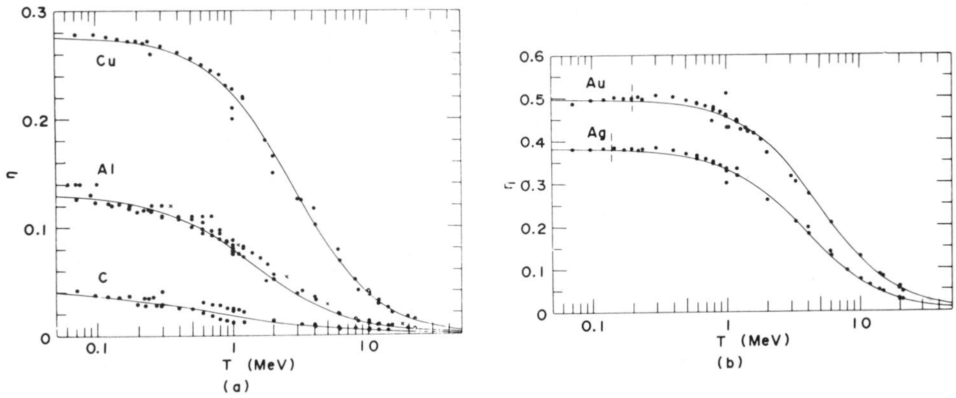

Figure 7.7. Fraction of normally incident electrons which are backscattered from thick slabs of various materials, as a function of incident energy T (reproduced from Tabata, Ito, and Okabe, Ref. 5, with permission of North-Holland Publishing Co, © 1971).

Backscatter can also influence the apparent yield from radioisotope sources of β-particles or conversion electrons [7]. If the source is mounted on a thick backing, electrons which are initially emitted into this backing may backscatter and reemerge from the surface of the source.

In all of the foregoing discussion, the behavior of β^+-particles (positrons) is essentially identical to that of β^--particles. Positrons exhibit nuclear scattering, ionization energy loss, backscattering, and bremsstrahlung. In these features, only minor deviations from the behavior of β^- occur. The one feature in which β^+ differs significantly from its negative counterparts is, of course, annihilation. Positrons essentially always annihilate at rest, and sometimes in flight, by interactions with the atomic electrons they encounter. The annihilation converts the rest energy of the e^-/β^+ pair into photon energy shared by the resulting two and sometimes three γ-rays. This total rest energy is 1.022 MeV, twice the 511-keV mass-energy of each particle. This γ-ray energy can be detected, of course, and can provide a unique signature for a stopping β^+-particle.

Some β-ray emitters are "pure," that is, there are no other accompanying radiations. Examples of pure β-ray emitters are tritium, ^{14}C, ^{32}P, ^{35}S, ^{45}Ca, and ^{90}Sr. More commonly, β-ray emission is preceded, accompanied, or followed by emission of photons (γ-rays, x-rays); ^{137}Cs, ^{85}Kr, ^{60}Co, ^{90}Y, and ^{147}Pm are examples.

Figure 7.8 shows the principal decay schemes of the isotopes tritium, ^{137}Cs, and ^{60}Co. Tritium exemplifies the case of a pure β-ray emitter. ^{137}Cs has two decay modes into ^{137}Ba: for 6.5% of the decays, a β^- (E_{max} = 1.176 MeV) leads directly to the ^{137}Ba ground state; for the remaining 93.5%, a β^- (E_{max} = 514 keV) leads to an isomeric state of ^{137}Ba with half-life 2.55 minutes, which goes to the ground state by emission of a 661-keV γ-ray. Cobalt-60 is a well-known γ-ray emitter because of the two prominent γ-rays (1.173 and 1.332 MeV) in the decay chain. However, it can be seen that a β^--ray always accompanies the decay: 99.88% of the time a β^--ray (E_{max} = 319 keV) precedes the sequential two γ-ray emission; 0.12% of the time a β^--ray (E_{max} = 1.48 MeV) leads directly to the intermediate state, followed by emission of the single 1.332-MeV γ-ray.

Table 7.1. Fraction of Normally Incident Electrons Backscattered from Various Detector Surfaces (data presented in Titus, Ref. 6 with permission of Nuclear Instruments and Methods © 1970)

Scintillator	Electron Energy (MeV)				
	0.25	0.50	0.75	1.0	1.25
Plastic	0.08 ± 0.02	0.053 ± 0.010	0.040 ± 0.007	0.032 ± 0.003	0.030 ± 0.005
Anthracene	0.09 ± 0.02	0.051 ± 0.010	0.038 ± 0.004	0.029 ± 0.003	0.026 ± 0.004
NaI(Tl)	0.450 ± 0.045	0.410 ± 0.010	0.391 ± 0.014	0.375 ± 0.008	0.364 ± 0.007
CsI(Tl)	0.49 ± 0.06	0.455 ± 0.023	0.430 ± 0.013	0.419 ± 0.018	0.404 ± 0.016

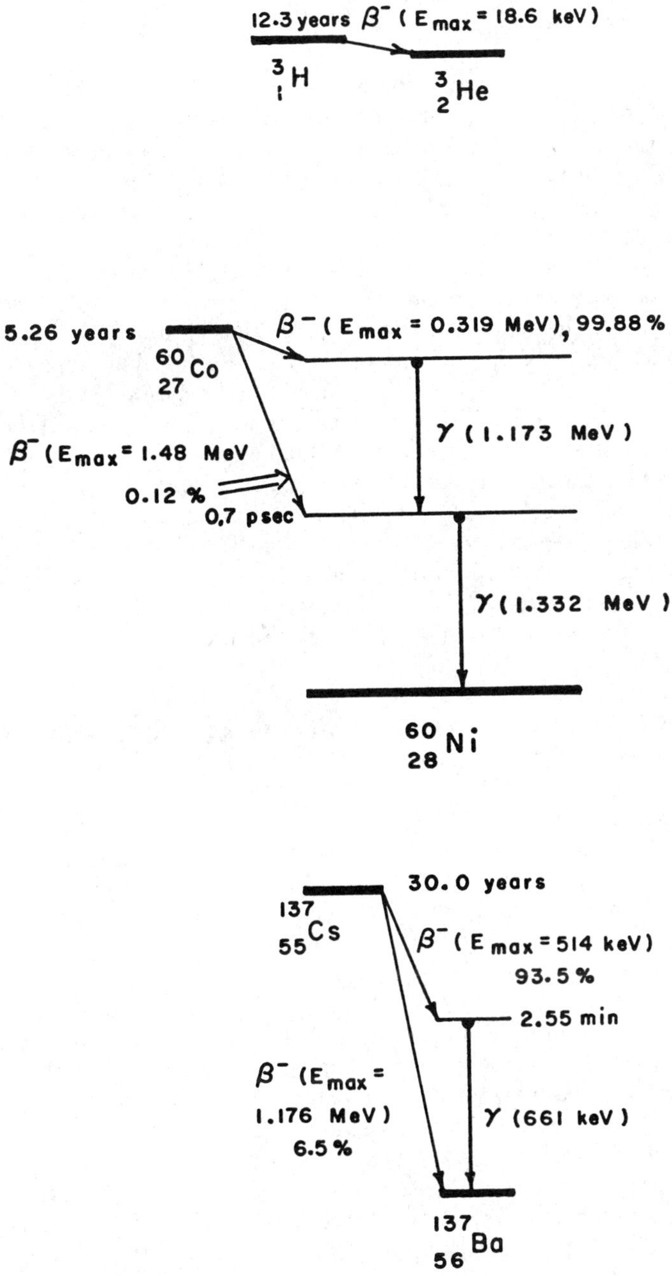

Figure 7.8. The dominant transitions in the decays of tritium, cobalt-60, and cesium-137 (reproduced from Lederer and Shirley, Ref. 8, with permission of Wiley-Interscience © 1978).

7.1.3. Measurement Techniques

For isotope identification, the total ionization deposited in stopping a β-ray is not as useful a signature as it is for γ-ray or α-particle detection, because of its distributed spectrum. Absolute isotope identification relies upon either precise spectral measurements, or upon chemical separation techniques. Since both of these pose special problems, each has been the subject of intense research and development efforts. They are discussed separately later.

In making measurements of either the β-radioactivity absorbed dose (rad or gray) or its concentration (picocurie/volume or mass, or becquerel/volume or mass), it is useful to remember the relevant radiation protection guides (dose equivalent limits are discussed in Section 2.2, Current Radiation Protection Philosophy). They set a limit of 5 rem/yr (50 mSv/yr) for occupational exposure and 0.5 rem/yr (5 mSv/yr) for a public individual exposure. Table 7.2 lists the MPCs (maximum permissible concentration) for several radionuclides of environmental importance.

Detection and measurement are quite different concepts, of course. For many purposes, detection is sufficient as is the case when a β-ray emitting material is spilled and needs to be located and cleaned up. Although the discussion of techniques that follows is divided into several parts, the techniques are interrelated. For example, detector types that are discussed as spectrometers may be used to measure the particle flux from a spill or measure β-activity of a chemically separated fraction.

Detectors which Measure Particle Flux or Dose Equivalent. Detectors may be designed for two different purposes:

1. Measurement of particle flux (particles/$cm^2 \cdot$ sec) (e.g., spills).
2. Measurement of dose equivalent. This is necessary when unavoidable exposure must be quantified for radiation protection purposes. Since the quality factor for β-rays is unity, dose equivalent (rem or Sv) is numerically equal to absorbed dose (rad or Gy).

In either of these cases, it is always preferable to ascertain the specific radionuclide(s) responsible for the activity. However, it is not always possible.

For detection of β-rays on or near a surface, an important requirement for a β-ray detector is a thin window or thin wrapping. Typical windows in β-ray counters have thicknesses of a few mg/cm^2. The thickness of a window on a survey instrument is often determined by the need to count α-particles as well; in this case windows of a fraction of a mg/cm^2 must be used. Alternatively, it might be desirable to screen out α-particles and detect only the β-rays, in which case thicker windows are used. Because the β-ray spectrum is a continuum, there will always be some low-energy particles on the which are not detected. For survey instruments, especially when the radionuclide of interest might be unknown, a thin window is desirable to minimize low-energy absorption loss.

Electrons with an energy of about 2 MeV or less are readily absorbed by the surface layers of the body. Because the outer part of the skin is an inert layer it is not affected by radiation; therefore, measurements are required only at depths in the body equal to (or greater than) the thickness of this inert layer. By convention this thickness is taken to be 7 mg/cm^2 [10]. The quantity measured at greater depths is the absorbed dose or absorbed dose rate. Naturally, the entrance window must be 7

Table 7.2. Maximum Permissible Concentration in Air and Water for Selected β-Emitters (data presented in Ref. 9)

Radionuclide		Occupational MPC μCi/ml		Unrestricted MPC μCi/ml	
		Air[a,b]	Water[a,b]	Air[a,b]	Water[a,b]
^{3}H	S	5×10^{-6}	1×10^{-1}	2×10^{-7}	3×10^{-3}
	I	5×10^{-6}	1×10^{-1}	2×10^{-7}	3×10^{-3}
	Sub	2×10^{-3}	—	4×10^{-5}	—
^{14}C	S	4×10^{-6}	2×10^{-2}	1×10^{-7}	8×10^{-4}
	(CO_2)Sub	5×10^{-5}	—	1×10^{-6}	—
^{32}P	S	7×10^{-8}	5×10^{-4}	2×10^{-9}	2×10^{-5}
	I	8×10^{-8}	7×10^{-4}	3×10^{-9}	2×10^{-5}
^{35}S	S	3×10^{-7}	2×10^{-3}	9×10^{-9}	6×10^{-5}
	I	3×10^{-7}	8×10^{-3}	9×10^{-9}	3×10^{-4}
^{85}Kr	Sub	1×10^{-5}	—	3×10^{-7}	—
^{99}Tc	S	2×10^{-6}	8×10^{-2}	7×10^{-8}	3×10^{-4}
	I	6×10^{-8}	1×10^{-2}	2×10^{-9}	2×10^{-4}
^{103}Ru	S	5×10^{-7}	2×10^{-3}	2×10^{-8}	8×10^{-5}
	I	8×10^{-8}	2×10^{-3}	3×10^{-9}	8×10^{-5}
^{106}Ru	S	8×10^{-8}	4×10^{-4}	3×10^{-9}	1×10^{-5}
	I	6×10^{-9}	3×10^{-3}	2×10^{-10}	1×10^{-5}
^{129}I	S	2×10^{-9}	1×10^{-5}	2×10^{-11}	6×10^{-8}
	I	7×10^{-8}	6×10^{-3}	2×10^{-9}	2×10^{-4}
^{133}Xe	Sub	1×10^{-5}	—	3×10^{-7}	—
^{135}Xe	Sub	4×10^{-6}	—	1×10^{-7}	—
^{135}Cs	S	5×10^{-7}	3×10^{-3}	2×10^{-8}	1×10^{-4}
	I	9×10^{-8}	7×10^{-3}	3×10^{-9}	2×10^{-4}
^{141}Ce	S	4×10^{-7}	3×10^{-3}	2×10^{-8}	9×10^{-5}
	I	2×10^{-7}	3×10^{-3}	5×10^{-9}	9×10^{-5}
^{144}Ce	S	1×10^{-8}	3×10^{-4}	3×10^{-10}	1×10^{-5}
	I	6×10^{-9}	3×10^{-4}	2×10^{-10}	1×10^{-5}
^{147}Pm	S	6×10^{-8}	6×10^{-3}	2×10^{-9}	2×10^{-4}
	I	1×10^{-7}	6×10^{-3}	3×10^{-9}	2×10^{-4}
^{151}Sm	S	6×10^{-8}	1×10^{-2}	2×10^{-9}	4×10^{-4}
	I	1×10^{-7}	1×10^{-2}	5×10^{-9}	4×10^{-4}
^{241}Pu	S	9×10^{-11}	7×10^{-3}	3×10^{-12}	2×10^{-4}
	I	4×10^{-8}	4×10^{-2}	1×10^{-9}	1×10^{-3}

[a]To convert to pCi/cm^3 divide numerical value by 10^{-6}.
[b]To convert to $\mu Ci/m^3$ divide numerical value by 10^{-6}.

mg/cm^2. The extension of the detector in the direction perpendicular to the field is limited to prevent averaging the dose over too great a cross section of the field. The base on which the window and detector are mounted is usually thick enough to simulate effectively a human body with respect to absorption and backscattering of the electrons. Hence, these instruments detect only the electrons that arrive from sources in front of the detector window. Electrons arriving nearly tangentially to the entrance window must penetrate a long, slanted path both through the window and the detector and many will be largely or completely absorbed. For these reasons the response of β-ray counters depends on the direction of the β-particle radiation as is shown in Fig. 7.9.

Once a β-particle penetrates the insensitive window or wrapping, its detection is possible using any of a number of detector types [ionization chambers, scintillation detectors, solid-state detectors, Geiger-Müller (G-M) tubes, gas proportional counters]. For a discussion of these detectors refer to Chapter 4, Radiation Detection Methods; also see [7].

For survey work the most common instrument is an ionization chamber with a thin window, through which β-particles penetrate. In a gas-filled detector the chamber is usually filled with gas at pressure slightly more than 1 atmosphere. The amount of gas is determined by the criterion that a β-particles has a high probability of producing at least one ion pair. This dictates a path length of 5 to 10 mm in

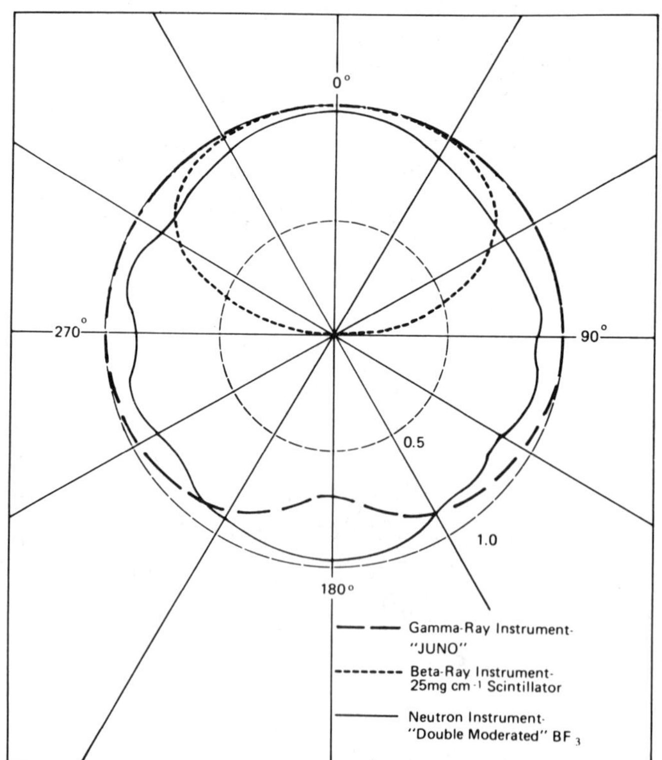

Figure 7.9. Directional dependence of response of gamma-ray, beta-ray, and neutron instruments (reproduced from NCRP, Ref. 10, with permission of National Council on Radiation Protection and Measurement © 1978).

the gas to assure ion-pair production probabilities in excess of 99% [11]. These instruments normally stop only the lowest energy β-particles, and are not able to provide spectral information.

Windows on gas-filled detectors today are commonly plastic film, such as Mylar, with a vacuum-evaporated conducting surface on the inside. For survey use, G-M tubes usually employ one of two special mixtures as the counting gas: 98.7% helium + 1.3% butane, or 99.05% helium + 0.95% isobutane. Proportional counters employ either pure methane or a mixture known as P-10 gas (90% argon + 10% methane) [11].

If a G-M counter is used, because it indicates the number of particles rather than their energy, its reading is only approximately proportional to dose. Its chief use should be to detect and establish the relative magnitude of electron fields.

Good electron dose meters can be made with very thin layers of scintillators, behind a thin entrance window, coupled to photomultipliers whose output current is measured. Because the scintillators are thin, the total light output, and hence the current, is small. However, they can measure dose rates above a few mrad/hr (μGy/hr). An opaque covering is necessary to exclude visible light from the scintillator, which may be sufficiently thick to exclude low-energy electrons also.

Ionization chambers made especially for electron measurements have a very small ratio of chamber thickness to diameter of the entrance window to maximize the number of electrons detected. However, measurements of electrons or β-particles are more often made with chambers designed primarily for measuring x- or γ-rays which have larger ratios. For small electron sources on the axis of the entrance window, the error is minimal, but for extended sources, such as one might find in a contaminated area, the error may be large. Although the walls of an ion chamber should be made of tissue-equivalent material, the exact composition is not critical.

For measuring absorbed dose of β-rays, the thin-window ionization chamber is the most frequently used instrument. It is also sensitive to γ-rays, and such β/γ ion chambers are widely available commercially. These instruments are usually calibrated using γ-rays sources (e.g., cobalt-60), and have readouts in absorbed dose (rad or gray) or less commonly in exposure (roentgen), although the roentgen is not a suitable unit for expressing β-ray measurements. Indeed, it is only defined for x- and γ-rays.

When a tissue-equivalent ion chamber is used, the measurement is fairly well representative of absorbed dose, provided the wall thickness allows the β-rays to penetrate. The problem with thin windows is that they may not be thick enough to achieve particle equilibrium by incident γ-rays in the MeV energy range. In situations where mixed-energy, mixed-species β/γ activity is present, the use of a series of window thicknesses is advised in order to estimate the true absorbed dose more accurately. This discussion points out a most important fundamental aspect of radiation protection measurements: more important than accurate measurements of some components of a mixed field is the assurance that all components are at least qualitatively identified. This requires attention to possible sources and may require use of more than one instrument.

Absorbed dose should be proportional to the total energy deposited in the detector sensitive volume. As such, the detector properly integrates over the β-ray energy distribution. Particle flux, on the other hand, merely measures all particles energetic enough to pass through the detector insensitive layer and then produce enough ionization to make a count. Even if the particular β-ray emitting nuclide is known, it is still not easy to calculate absorbed dose from flux because both the energy spectrum and various geometrical effects (thickness of the source, solid angles) enter into the calculation. If the isotope responsible is unknown, it is not possible to determine dose.

Calibration of ratemeters involves using a known radioactive source in a known geometrical configuration. This is a most difficult problem. Of major concern is overloading of ratemeters at high instantaneous rates. A well-designed instrument must indicate that it is being overloaded.

Beta Spectrometers. In environmental measurements, β-spectrometry may be used for differentiation between different β-particle emitters in a mixed or unknown sample. High-resolution β-spectrometry has been one of the most valuable research tools in understanding the detailed features of β-decay nuclear dynamics. The shape of the spectrum is sensitive to a number of parameters in the quantum-mechanical matrix elements which govern decay. Because of the broad energy spectra of β-particles, high resolution is not necessary to differentiate between different β-particle emitters. However, if detection of monoenergetic conversion electrons is required, then resolution becomes crucial.

Best energy resolution is achieved by spectrometers that use momentum selection in a magnetic field. They are not discussed here because they are seldom used in environmental monitoring measurements. Their role in research, however, is indispensable. The reader is referred to [12,13,14] for discussions, comparisons, and critiques of these spectrometers.

For environmental measurements, various ionization-detecting spectrometers are common. The following types will be discussed separately:

1. Gas proportional counters.
2. Solid scintillation counters.
3. Liquid scintillation counters.
4. Solid-state detectors.

Ionization-detecting spectrometric techniques such as cloud chambers, photographic-emulsion detectors, Cerenkov counters, and differential range/absorption measurements will not be discussed because they are not used commonly for environmental monitoring.

Gas proportional counters are the preferred gas-filled instruments for β-spectrometry.

Gas Proportional Counters. Gas proportional counters were not widely used for β-particle counting in the early days of nuclear physics, chiefly because of the difficulty in achieving the required high gains and good stabilities. However with the development of better electronic systems and improved gas purification, β-particle counting with proportional counters has become a common technique for routine measurement. Today, gains of 10^4 to 10^5 are typical, and even 10^6 is used in some counters for β-spectrometry at low energies.

The gas-filled proportional counter is distinguished from the ionization chamber because in a gas-filled proportional counter the primary ionization electrons are accelerated by the high fields to where they induce further ionization by collisions in their passage through the gas to the collector. For β-particle counting, the main requirement is high gain in order to overcome the low initial ionization of the primary β-particles. A discussion of pulse height variation with applied voltage can be found, for example, in the book by Curran and Craggs [15]. The operating characteristics of these counters has been well described by Curran and Wilson (see, e.g., Ref. 16).

The choice of filler gas is critical for β-particle counting (in contrast to α-particle counting, where gains of $\sim$100 are usually sufficient and a wide variety of gases will do). The amount of energy required to produce a single ion pair is an important factor in determining the resolution in a gas counter and consequently the choice of counting gas (and its purity). A common gas is P-10 (90% argon, 10% methane). As Curran and Wilson say [16]:

> Pure argon is unsuitable, owing to the existence of a long-lived excited state which gives rise to after-pulsing. On the other hand, higher concentrations of methane give much higher operating voltages. The methane may be replaced by nitrogen, carbon dioxide, or other gases without notable loss in performance. Many gases are unsatisfactory for use in proportional counters operating at high gas gain, principally because of their electro-negative character involving the capture of electrons to form negative ions.

The gas impurities that can be least tolerated are the halogens. Noble gases are chosen partly because of the low amounts of energy required to produce an ion pair (26.4 eV/pair for argon, 24.2 for krypton). Neon can be used, but it requires 36.6 eV/pair and it has a lower stopping power than argon at a given pressure. Methane is the most common mixing gas because it only needs 27.3 eV/pair, and because argon/methane can be easily purified [16].

The maximum range of a 1-MeV β-particle in air at STP is about 300 cm (Fig. 7.4); thus to stop β-particles of this energy in any gas, the vessel must be pressurized. Pressures of a few atmospheres have been employed, and pressures above 150 atmospheres have been examined in order to stop β-particles in a very small geometrical area.

The design of counters to minimize the problems with walls has been discussed by Curran and Wilson [16]. These authors also discuss detecting high-energy β-particles by using a magnetic field to confine the particles to the chamber volume and increase their path length.

Source preparation is a critical step. The problem is not as severe as for α-particles, where sample thicknesses as small as a few $\mu g/cm^2$ are required; however, for low-energy beta-spectrometry as little as a fraction of a mg/cm^2 is necessary (see Fig. 7.5). Techniques such as evaporation, sublimation, freeze drying, electrolytic deposition, and lacquering have been used successfully and are discussed by Curran and Wilson [16]. Another technique, described by Phillips and Kauffman [17], uses thin ($\sim\mu g/cm^2$) conductive films upon which the sample can be deposited. This enables the sample to be counted with almost total geometrical efficiency.

Solid Scintillation Counters. The advantages of using a scintillator coupled to a photomultiplier tube for β-spectrometry are: fast time response, high gain, and relatively low noise. Another advantage of solid scintillators is that they are available commercially in a variety of sizes and shapes. Because most β-particles will normally stop in thicknesses of the order of 1 cm of scintillator, detectors of modest size may be used.

Unfortunately, the widespread use of solid scintillators for spectroscopy is hampered by several factors. Perhaps the most important is intrinsic resolution; the ionization energy required to produce one photoelectron at the tube photocathode in these counters is about 300 eV, compared to about 30 eV/ion pair in a gas and only about 3 eV/electron-hole pair in semiconductors. This clearly limits scintillator resolution, making them uncompetitive for many applications. However, in β-spectrometry, high resolution is unnecessary because of the wide spectrum of the β-particle energies. Thus for many applications scintillator resolution is sufficient.

As mentioned earlier, backscatter is a problem in measuring β-rays. When low-energy electrons impinge on a solid (scintillator) surface, a significant fraction will backscatter before coming to rest (see Table 7.1), as the result of multiple Coulomb scattering interactions. The effect increases with atomic number. Quoting from Neiler and Bell [18]:

> About 8% of the β-rays entering a flat anthracene surface from a nearby source will scatter out before being stopped and hence will produce smaller pulses than they should. For NaI(Tl), on the other hand, 80% to 90% of the entering electrons will scatter out before stopping. It is consequently quite difficult to obtain a satisfactory β-spectrum using NaI(Tl) with an external source.

Although the backscatter may be reduced by collimating the β-particles in front of the phosphor, this is still not a sufficient remedy for higher-Z materials.

An advantage of solid scintillators is ease of calibration: using a γ-ray source, an absolute calibration can be relied upon to detect long-term drifts in pulse height due to phototube gain changes. Another advantage is the inherently low background of such detectors compared to gas-filled counters. This is due to their small size, and the ability to use electronic discrimination to bias against any minimum-ionizing high-energy cosmic rays that impinge on the scintillator.

Many β-particles are accompanied by coincident γ-rays or daughter x-rays, which can be measured by a two-phosphor coincidence arrangement and can be used as a unique signature for certain nuclide decays. To take advantage of this property, one places the unknown β-particle emitter between two crystals as described by Neiler and Bell [18]. Of course, one problem is that in these cases a single-phosphor detector will have a distorted pulse-height spectrum due to those events in which both the β-particle and the photon register in the same scintillator.

Solid scintillators are also used to measure of gaseous β-ray emitters (tritium, ^{85}Kr). Chapter 20 contains a detailed discussion. Backscatter and resolution are just as important problems in monitoring gaseous β-ray emitters, but are counterbalanced by the ease of sample preparation and handling.

Liquid Scintillation Counters. The use of liquid scintillation counters for β-rays is described in detail in Chapter 20. A liquid scintillator is applicable whenever the active nuclide is dissolved or immersed in the liquid. Detection efficiency is 100% except for the β-rays escaping near the edges.

This technique is very sensitive, and is commonly used for ^{14}C, ^{85}Kr, and ^{133}Xe, as well as for tritium. An important disadvantage is cost, since the liquid scintillator must be discarded after each sample and energy resolution is not great. The comments made about solid scintillators in the previous paragraphs apply equally well here.

Although resolution is not high, liquid scintillators can be used to distinguish more than one isotope in the same sample. In general, the activities of two isotopes can be measured separately if the β maximum energies differ by a factor of four or more [19]. To do this, two differential discriminators are used, each one set to select the optimum energy range for one of the isotopes. However, because they are continua, they will overlap, as shown in Fig. 7.10. The higher-energy spectrum can be measured with little interference, but the low-energy spectrum will contain an appreciable contribution from the high-energy one. In Fig. 7.10, the measured result in Region I must be corrected for the presence of the other isotope as well as for background. As with solid scintillators, liquid systems can yield distorted energy spectra for radionuclides where the β-emission is accompanied by or followed closely by x-ray or γ-ray emission.

Quenching is a possible problem that may occur with the introduction of varied or unknown samples into a liquid scintillator, either by coloration of the liquid, or by allowing chemical reactions to compete with light emission as a de-excitation mechanism for scintillator molecules [11].

Semiconductor Detectors. The entire field of nuclear spectrometry was revolutionized in the early 1960s by the introduction of solid-state semiconductor detectors for high-resolution particle detection. The most striking advances were in the field of γ-spectroscopy, where the resolution of previous techniques [such as NaI(Tl) scintillators] was vastly improved. For β-spectrometry, advances have been nearly as great, although precise energy resolution is not needed in much of β-spectroscopy research (because of the wide energy spectrum). However, for conversion electrons, this technique provides high selectivity for the monoenergetic line.

From Fig. 7.4, it can be seen that the maximum range in aluminum of a 1-MeV β-particle is 0.14 cm. This is sufficiently short that β-particles can be totally absorbed in solid-state detectors of quite modest dimensions. Silicon is nearly identical to Al in its stopping power for β-particles. Two problems are important, however. First, some backscattering will occur (as discussed previously in the section on solid scintillators), resulting in a degradation of energy resolution at low energies for a fraction (several percent) of the incident β-particles. This effect is not as serious for Si as it would be for a high-Z scintillator like NaI(Tl). Nevertheless, it contributes to the difficulty in measuring low-energy spectra in the presence of high-energy β-particles.

Second, it is essential to make the dead layer at the surface of the detector as thin as possible. For this reason, detectors with sensitivities almost directly at the surface, silicon surface-barrier detectors, have been developed. For a detailed dis-

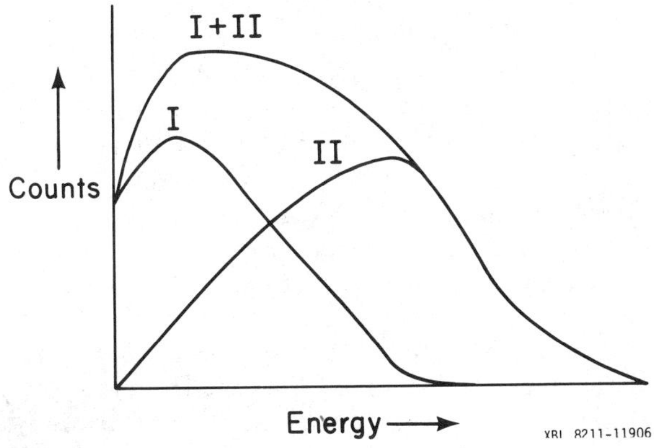

Figure 7.10. Spectrum that results from the decay of two beta-emitting isotopes.

cussion of the theory of operation of these devices, the reader is referred to Section 7.2.3 and Refs. 20–23.

A typical Si surface-barrier detector is a cylindrical diode with a very thin *p*-type Si layer on the face of an *n*-type Si wafer. Above the *p*-type layer, a thin surface of gold (Au, thickness $\leq 0.02\ \mu m$, $\leq 40\ \mu g/cm^2$) provides the negatively biased electrical contact; the positive bias is made through some metal contact (e.g., Al) on the back of the *n*-type Si. The Au layer also serves as a means of protecting the active elements from contamination. Detectors with depleted depths from about 50 μm to a few millimeters are now commercially available, with surface areas from a few to thousands of square millimeters.

A β-particle detector using Si or Ge has distinct advantages over a gas-filled detector. The two primary advantages are:

1. For a given amount of energy lost through ionization, the number of charge carriers (corresponding to ion pairs in the gas) is greater by a factor of as much as 10. This decreases the fractional statistical fluctuations in the ion collection process and is the main reason for the high resolution.
2. Collection time in the nanosecond range makes fast time resolution possible. This is due to the high carrier mobilities and short distances charge carriers have to travel.

Among the difficulties associated with surface-barrier detectors are:

1. They are fragile and require careful storage and handling.
2. Because the electrical signal is small, noise in the detector/amplifier system is often the most important characteristic around which a system is designed.
3. The detectors can become contaminated, and once this happens decontamination is essentially impossible.

Although output signals are small, the associated electronics necessary for amplification are readily available. The detection of β-particles that enter the sensitive area of the detector is essentially 100%; however, backscatter loss and window absorption will degrade the signal. Backgrounds from external sources (e.g., cosmic rays) or from detection of γ-rays are very small because of the small size of a typical detector.

Energy resolution is the main reason for the choice of a surface-barrier detector. Broadening due to random noise fluctuations in the detector and/or preamplifier is an intrinsic resolution limitation. In addition there is broadening from backscatter and from variations in energy lost in the dead initial layer. Line widths of about ± 5 eV for electrons in the 1 MeV region are typically obtained in the best systems.

Gross-Beta Measurements in Water. Measurements of gross β in water are now standardized. The standard technique is described in *Standard Methods for the Examination of Water and Wastewater, 15th ed.* [24]. The standard method deposits the sample on a counting pan by oven or hot-plate drying. Counting is carried out with an internal gas proportional counter. (Also see Ref. 25 for a tentative reference method suggested by the EPA.) This method is used as a screening technique throughout the United States and is one of the mainstays of public-health monitoring. If gross-β levels are shown to be sufficiently small, it is usually unnecessary to isolate and measure individual radionuclides with more complicated techniques.

If both α- and β-particle activity are present, they can be separated by measuring the counting rate versus voltage; a typical counter might respond as shown in Fig. 7.11. The first plateau is from α-particles, the second from the sum of (α- + β-particles).

In intercomparison studies, the EPA found that intra-laboratory precision for gross-β analysis was satisfactory. However, the accuracy of these measurements is unsatisfactory [26]. Only 44% of the labs were within the control limits. One of the key problems with measuring gross β is that with any instrument there is some effective β-ray energy below which the β-particles will be absorbed and not counted making gross-β measurements difficult to interpret.

Gross-Beta Measurements in Air. Measurements of gross-β levels in air have been performed for many years, but only recently the Intersociety Committee has attempted to standardize the method by adopting the "Tentative Method of Analysis for Gross-Beta Radioactivity Content of the Atmosphere." [27]

Air particulates collected on a filter paper are counted with an end-window G-M tube, thin-window proportional flow counter, scintillation counter, solid-state detector. The choice of detector depends partially on whether concurrent gross-α measurements are desired. In that case, an internal gas proportional counter would probably be chosen. External counting devices can be used to include or exclude α-particles by varying the window thickness.

The sensitivity of the method is proportional to the volume of air moved through the filter [27]:

> A sample of 10 m^3 of air with a beta particulate concentration of 3 pCi/m^3 (0.11 βq/m^3) will contain 66.6 dpm, sufficient to produce acceptable precision in ten minutes' counting time.

This method is sensitive to those airborne β-particles which are attached to particulate material in the air. Naturally occurring radon can interfere because its daughters are chemically active and attach themselves to particulate matter. The problem is easily resolved [27]. Since their half-lives are controlled by relatively

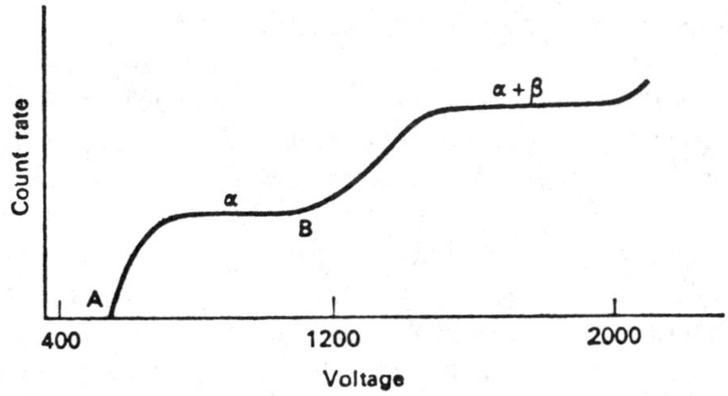

Figure 7.11. Alpha and alpha + beta counting rates as a function of voltage in a proportional counter (reproduced from Cember, Ref. 4, with permission of Pergamon Press © 1969).

short-lived radionuclides, namely ^{214}Pb ($t_{1/2}$ = 26.8 min) and ^{214}Bi ($t_{1/2}$ = 19.7 min) for ^{222}Rn and ^{212}Pb ($t_{1/2}$ = 10.6 hr) for ^{220}Rn, their interference can be nullified by counting after these nuclides have decayed, about 3 days [27]. Accuracy depends upon the uncertainty in the volume of air sampled, the possible error from self-absorption effects which may not be properly accounted for, and calibration difficulties.

Because β-spectra are so broad and because of the possible energy-dependent response of many counting instruments, it is desirable to calibrate with at least two or three β-ray sources of differing energies. If the dominant nuclide species being measured or sought is unknown,

> Cesium-137 is a desirable standard, since its average energy corresponds to that of aged mixed fission products in fallout. [27]

The main difficulty in this technique is assuring that a reliable sample is deposited on the filter. Large errors may result in sampling dust-laden air. Another problem is self-absorption. If the sample to be counted is thicker than a few mg/cm^2 a sizable fraction of the β-particles may not get out of the sample to be counted without significant energy degradation. This self-absorption is an important problem at the low end of the spectrum.

In a discussion of the results of an air filter cross-check program for 1973–1975, Easterly et al. [28] concluded that the participating laboratories had greater difficulties in analyzing gross β than gross α. Only 39% of the labs were within the control limits for accuracy.

7.1.4. Chemical Separation Techniques

In the foregoing discussion, one theme has been reiterated: because of the broad energy spectrum of β-rays from an individual radionuclide, the ability to identify individual emitters in a mixed sample is limited. Also, the relatively short range of β-rays, especially at low energies, may require that they be separated from the bulk material of the sample.

For these reasons, chemical separation techniques have been developed to concentrate one particular element (or one class of elements) prior to measurement. The chief problem is the uncertainty in yield as successive chemical procedures are performed. Of course, if a complete separation has been made, spectrometry is unnecessary; a gross β count may be attributed to the radionuclide in question.

Another problem is the ubiquitous nature of the naturally occurring β-ray emitter ^{40}K. It can cause difficulties in chemical separations because of its presence in the sample being analyzed or in the reagents being used. Before using any reagent the analyst should determine whether ^{40}K is present (this can be determined by flame photometry of γ-ray counting).

Decay chains and parent-daughter relationships can also be used in gaining greater specificity. Information can be obtained by following the ingrowth of a daughter or the relative quantities of nuclides following buildup and decay of a long chain (such as the chains of α- and β-particle emitters among the heavy elements). The ingrowth technique is particularly useful when the daughter's presence is more easily detected than the parent's (e.g., strontium-90 determinations by measurements of yttrium-90).

There is such a large number of chemical separation techniques in the literature that a comprehensive treatment of them here is prohibitive. This discussion summarizes only a few important techniques.

The most complete source on chemical separation is a series of monographs done by the Subcommittee on Radiochemistry of the Committee on Nuclear Science of the National Academy of Sciences [29]. For additional information see Ref. 30. Other compilations are the DOE Environmental Measurements Laboratory *Procedures Manual* [31]; the handbook of the EPA Environmental Monitoring and Support Laboratory in Las Vegas [32]; the U.S. Public Health Service compilation [33]; and the *Tentative Methods* published by the Intersociety Committee [27].

The β-particle emitting radionuclides treated here are:

1. Carbon-14.
2. Ruthenium-106 and 103.
3. Xenon-133 and 135.
4. Cerium-144 and 141.

Beta-emitting radionuclides treated in separate chapters of this book are:

1. Tritium (Hydrogen-3).
2. Krypton-85.
3. Strontium-89 and -90.
4. Iodine-129 and -131.

Some important contributors to gross-β measurements in environmental samples are daughters in the decay chain of ^{222}Rn. Among these daughter β-particle emitters are ^{214}Pb, ^{214}Bi, ^{210}Pb, and ^{210}Bi. These are treated, but only implicitly, in Chapter 14.

Many other contributors to gross-β levels are more easily determined by γ-spectroscopy, and these are not treated here. For many environmental samples, γ-spectroscopy is easier, more accurate, and enables the use of larger samples in the bulk. As an example, in natural rocks, where man's impact has been minimal, one important β-ray emitter is potassium-40. Potassium-40 has an 89% branching ratio to ^{40}Ca by β-particle emission (E_{max} = 1.314 MeV), but in rock samples it is almost always preferable to detect the 1.460-MeV γ-ray which follows the electron capture decay mode (11%) to ^{40}Ar.

A large number of β-emitting nuclides are not treated in this book because they are rarely found in environmental samples, their half-lives are very short and they are γ ray as well as β-ray emitters and are measured as such.

Carbon-14. Carbon-14 is one of the most well-known radionuclides. Its use in radiocarbon dating of archeological artifacts has made it a household word. Carbon-14 is a pure β-ray emitter, with a very weak β-ray (E_{max} = 155 keV) which is difficult to detect. The long half-life (5730 years), which makes it useful in dating, is a drawback for its detection because a long half-life means few disintegrations per minute (dpm) for detection. Carbon-14 is also widely employed in research, such as when organic compounds are tagged with ^{14}C to trace biochemical processes in living systems.

The primary mechanism for natural ^{14}C production is by cosmic-ray interactions with ^{14}N in the atmosphere. The reaction is ^{14}N (n,p)^{14}C. With its long half-life the radiocarbon ultimately reaches equilibrium with the stable carbon in the biosphere. For more information on ^{14}C production and distribution, refer to Libby [34] and Watt and Ramsden [35].

Carbon-14 is also produced in nuclear reactors via a number of reactions. The emissions expected from power plants and the rest of the fuel cycle constitute one of the most substantial dose contributions from nuclear power, if the ^{14}C half-life is considered.

The problems of measuring ^{14}C are in many respects similar to the problems with tritium, which also has a weak β. Preparation of biological or environmental samples for ^{14}C determination has been studied extensively, and many of the problems solved. For gas counting, organic samples are usually converted to CO_2 by combustion and inorganic samples by decomposition with acids. The CO_2 can then be used as a gas, or alternatively converted to $BaCO_3$ for counting as a solid. For liquid scintillation counting, CO_2 can be trapped in an organic base (for example, phenylethyl-amine or ethanolamine) and then introduced directly into the scintillator solution [11].

Liquid scintillation counting is the most common ^{14}C measurement technique today. This method is as sensitive for ^{14}C as it is for tritium: activities as low as ~0.2 pCi/cm^3 (7.4 mBq/cm^3) of organic solvent can be measured. Since the technique is also used for tritium refer to Chapter 20, which contains a full description of the various considerations pertaining to this technique.

The other common technique for ^{14}C measurements is the use of an internal gas counter (proportional, G-M) with radiocarbon CO_2 introduced into the counting gas or introduced as solid $BaCO_3$. Chapter 20 contains a more complete discussion of similar techniques used to measure tritium.

Ruthenium-106 and -103. Ruthenium (Ru, atomic number 44) is one of the important fission product elements which may be emitted by nuclear fuel-reprocessing plants. The Ru isotope of most interest in environmental media is ^{106}Ru. Its half-life is 367 days, and it is a pure β emitter (E_{max} = 39.2 keV) leading to a short-lived daughter (rhodium-106), which itself decays ($t_{1/2}$ = 30 seconds) to stable palladium-106 by a complicated β-ray spectrum dominated by a very energetic β-ray (E_{max} = 3.53 MeV) and several half-MeV-range γ-rays.

The most common β-ray detection technique for ^{106}Ru is measurement of the energetic β-ray from the rhodium (Rh) daughter. This can be done, after chemical separation, with very little interference. Because the 30-sec half-life of the daughter ^{106}Rh is so short, it is an ideal tag for the parent ^{106}Ru, whose own β-ray is so weak as to be almost undetectable in a mixed sample.

Ruthenium-103 ($t_{1/2}$ = 40 days) is also found in some samples. It has a multiple β-spectrum dominated by a low energy β-ray (E_{max} = 200 keV) followed by a 490-keV γ-ray (90% of the decays). It can be determined in a mixture with ^{106}Ru by separate counting with and without a heavy (~0.3 g/cm^2) Al filter: the ^{106}Rh daughter β-ray of ^{106}Ru β-ray will penetrate through the filter while the ^{103}Ru β-ray will not. Because as little as 5 mg/cm^2 of material will screen out the weak ^{106}Ru β-ray, it is not counted.

Ruthenium-103 and 106 can both be counted with γ-spectroscopy, and in many samples this is the preferred method; one may detect the 490-keV γ-ray from ^{103}Ru

or a 622-keV γ-ray (11%) from ^{106}Ru. Indeed, the β-ray detection technique in the EML Manual [31] recommends that the β-ray technique be employed only if γ-ray counting is not possible.

The EML radiochemical separation technique [30] employs a Ru carrier. The sample is

> transferred to a distillation flask with 1:1 H_2SO_4, cooled, and $KMn0_4$ crystals added. The ruthenium is distilled as $Ru0_4$ into a receiver containing 6 N NaOH. Ruthenium is further purified by successive precipitations as the hydrous oxide and metal.

Xenon-133 and -135. In nuclear reactors, radioxenons are among the important isotopes that can be discharged into the air. This is especially important in the boiling water reactors, where gas hold-up is short, or during an accident (see Chapter 10, Nuclear Power Plants).

Two of the four important radioxenons are β-ray emitters. These are:

1. Xenon-133: $t_{1/2}$ = 5.27 days, E_{max} = 346 keV. The β-ray is followed by an 80-keV γ-ray leading to stable cesium-133.
2. Xenon-135: $t_{1/2}$ = 9.2 hr, E_{max} = 920 keV. The β-ray is followed by a 250-keV γ-ray leading to ^{135}Cs, which itself decays ($t_{1/2}$ = 3×10^6 yr) to stable barium-135 by β-ray (E_{max} = 210 keV).

Two other important radioxenons are γ-ray emitters. Their properties are listed here for reference:

1. Xenon-131m: $t_{1/2}$ = 11.8 days, E_γ = 164 keV, leading to stable xenon-131.
2. Xenon-133m: $t_{1/2}$ = 2.26 days, E_γ = 233 keV, leading to ^{133}Xe, whose β-decay has just been discussed.

Xenon is, of course, a noble gas and as such is not amenable to normal chemical manipulation. Chemical separation can be accomplished in gaseous samples, if required, by purging the gas of its more reactive components (e.g., the radioiodines).

The β-ray emitting radioxenons are usually monitored by techniques very similar to those used for krypton-85. These techniques are either ionization chambers, gas proportional counters, or G-M tubes. The reader is referred to Chapter 19, Krypton, for details.

The γ-ray emissions are also easily detected, and where γ-spectroscopy is possible it is generally preferred over β-ray detection.

Cerium-144 and -141. Cerium-144 is one of the radionuclides most commonly associated with fresh fallout and general mixed fission products. Because of its 284-day half-life, it is sufficiently long-lived to be distributed in a wide band throughout the northern hemisphere whenever large nuclear bomb tests deposit fission products in the stratosphere. After reading this section, the reader may wish to refer to a recent NCRP Report [36] which summarizes the present knowledge of the relevant chemical, as well as physical and biological properties of radiocerium as a basis for establishing radiation protection guidelines. The radioactive isotopes of Ce of most concern are ^{141}Ce, ^{143}Ce, and ^{144}Ce. These three isotopes, all of which are β-ray emitters, are abundant products of fission and have relatively long half-lives.

Cerium-144 is a β-ray emitter with three principal β-ray lines leading to praseodymium-144, which itself decays to neodymium-144 by energetic β^--emission (E_{max} = 2.99 MeV) with a 17.3-min half-life. The three ^{144}Ce β-rays have energies and branching ratios as follows:

$$E_{max} = 309 \text{ keV}, 76\%$$
$$E_{max} = 240 \text{ keV}, 5\%$$
$$E_{max} = 175 \text{ keV}, 19\%$$

Cerium-141 frequently occurs in the same samples with ^{144}Ce. Because they are the same chemically, they cannot be separated chemically. Cerium-141 decays by β-ray emission with $t_{1/2}$ = 33 days; its two β-rays have E_{max} = 582 keV (30%) and E_{max} = 442 keV (70%). The daughter of ^{141}Ce is stable ^{141}Pr.

One common way of determining ^{144}Ce in the presence of nearly all other interferences is by counting the energetic decays of its daughter ^{144}Pr; after filtration with about 0.35 g/cm^2 of aluminum, only the 2.99-MeV maximum β-ray emission survives. Because of the short 17.3-min half-life of ^{144}Pr, equilibrium is reached very quickly, after which the disintegration rate of ^{144}Pr equals the rate for the parent ^{144}Ce.

If it is necessary to differentiate between ^{141}Ce and ^{144}Ce activities, the ^{144}Ce activity can be measured by determining the ^{144}Pr rate after filtration, as just described. In a mixture of ^{144}Ce and ^{141}Ce, the many overlapping β-spectra are characterized by several maximum energies: 442 and 583 keV from ^{141}Ce; 175 keV, 240 keV, and 309 keV from ^{144}Ce; and 2.99 MeV from ^{144}Pr.

The EML Manual [31] describes three methods for the radiochemical determination of ^{141}Ce and ^{144}Ce. In each method, the final sample is counted by the ^{144}Pr method, using a thin-window G-M counter. In the first method, ^{144}Ce and a Ce carrier are ignited to insoluble oxides and fused with Na_2CO_3 at 900°C; redissolved in HF, HNO_3, and H_2SO_4 acids; and precipitated as ceric iodate for counting. In the second method [37], Ce is leached from an ashed sample by digestion in a mixture of H_2O_2 and HCl; the rare earths and alkaline earths are collected as oxalates and ignited to carbonates for collection as hydroxides. Cerium is then precipitated as iodate for counting. In the third method, Ce(IV) is extracted into a 0.75 *M* di-2-ethylhexyl phosphoric acid in *n*-heptane solution from a 1 *M* potassium bromate/10 *M* nitric acid solution, then back-extracted with H_2O_2–HNO_3, converted to the oxalate, and determined gravimetrically before counting. Using one of these three methods, Ce can be determined in water, urine, soil, feces, vegetation, milk, food, and air filters [31].

The second method has been used by Sutton and Dwyer [38] to measure ^{144}Ce levels in wheat flour. Levels in the range of 10 to 20 pCi/kg (0.37 to 0.74 Bq/kg) were measured, simultaneously with ^{137}Cs, at levels of about 35 to 125 pCi/kg (1.3 to 4.6 Bq/Kg).

Sill and Willis [39] have described a chemical method for separating Ce and lanthanum for counting, by coprecipitating with $BaSO_4$ from strong H_2SO_4 solutions containing potassium salts and hydrogen peroxide. Recoveries in excess of 99% are reported.

7.1.5. Summary and Conclusions

In this section, an attempt was made to outline the various techniques for detection, measurement, and energy spectrometry of β-particles in various environmental media.

The current state of the art in β-particle instrumentation is generally satisfactory. Instruments for detecting β-particles are varied and well developed, and in the hands of competent personnel they are capable of defining β-radiation fields to sufficient accuracy. However in the area of low-energy β-emitters, more development work needs to be done in order to measure low levels in real time for situations where the identity of the β-emitter is known.

Beta spectrometry is also well developed. The requirement for high resolution is generally not severe, because most β-ray spectra are broad. Also, the best instruments (e.g., surface-barrier detectors) have far better resolution than is usually required. The principal exception is for detection at low levels of internal-conversion electrons, where resolution is important, being directly related to the ultimate sensitivity.

The principal problems needing further research and development are in the area of chemical separation techniques. At present, these techniques are often laborious, costly, and liable to error. Unfortunately, there does not seem to be any more promising approach to the identification of low-level pure β-emitters than the use of chemical separation followed by counting. The continuous β-spectra make it very difficult indeed to separate by spectrometry alone an individual β-ray emitter from a mixed sample of more than two at widely differing concentrations.

7.2. MEASUREMENT OF ALPHA PARTICLES

Many different types of instruments are capable of detecting α-particles. However, more than a mere α-particle detector is involved in measuring α-particles because the short range of the α-particles may prevent their reaching the detector. Also, many applications require measurements of specific activity (Ci/g or Bq/g) or identification of the emitter. In this section, the following instrumental considerations and techniques for measurements in specific media will be treated:

1. gross-α detection instruments (such as hand-held meters) for radiation survey work and radiological protection. These measure particle emission rate (particles/cm^2 sec) or, less frequently, dose equivalent (rem or Sv).
2. Alpha spectrometers for identifying individual α-particle energies in a sample that may contain more than one α-particle emitter.
3. Techniques for measurements of specific activity (pCi/m^3 or pCi/g, or Bq/m^3 or Bq/g), in air and water.
4. Chemical separation techniques, developed primarily for use with α-spectrometry. Because they are not treated elsewhere in this volume, chemical separation of polonium ($Z = 84$), protactinium (91), neptunium (93), and americium (95) are treated individually.

These categories overlap substantially for many applications, as will be seen in the discussions following.

Alpha-emitting elements treated more fully in separate chapters elsewhere in this volume are not discussed individually here. These are: radon (86) and its daughters, radium (88), uranium (92), and plutonium (94). The chapter on plutonium also discusses methods for monitoring other transuranics, particularly in air.

7.2.1. Physical Considerations

An α-particle is a helium-4 nucleus, with a mass of almost 4 times that of a proton and with charge +2. It can be thought of as 2 protons and 2 neutrons, bound together. Alpha particles were discovered, named, and studied by Ernest Rutherford in the earliest days of nuclear physics. Emission of an α-particle is only found among the heavy elements. With the exception of a few nuclides of mass number around 150 (e.g., ^{147}Sm, ^{152}Gd, ^{156}Dy, ^{148}Sm), naturally occurring α-particle emitters are all heavier than lead ($Z = 82$). This section will not describe in detail the characteristic features of α-particle emitters, except those features essential for their measurement.

The most characteristic feature of α-particles is their high specific ionization density as they traverse matter, and their consequent short range. This feature is relied upon in virtually all detection methods. It also limits the sensitivity and capability of α-particle detecting instruments.

Figure 7.12 shows the range of α-particles in air. Figure 7.13 [4] shows the distribution of ionization versus energy for α-particles slowing in air. It can be seen both that the range is short, and that the specific ionization is high and peaked at the end (the Bragg peak). Figure 7.14 shows the energies of all the known heavy α-particle emitters. From this figure it can be seen that nearly all α-particles have energies in the range from 4 to 9 MeV.

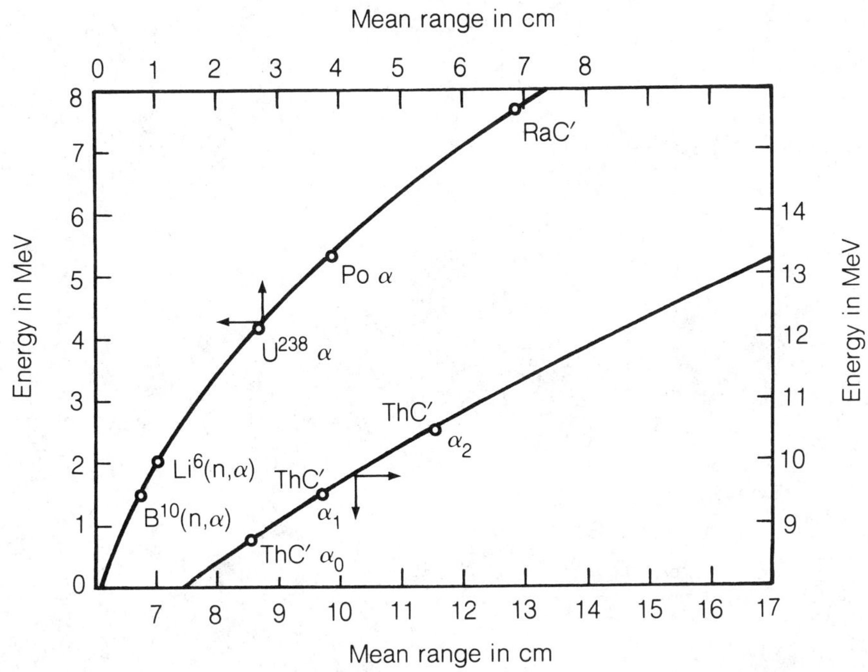

Figure 7.12. Range-energy curve for alpha particles in air at 15°C and 760 mm Hg (reproduced from Siegbahn, Ref. 12, with permission of North-Holland Publishing Co. © 1965).

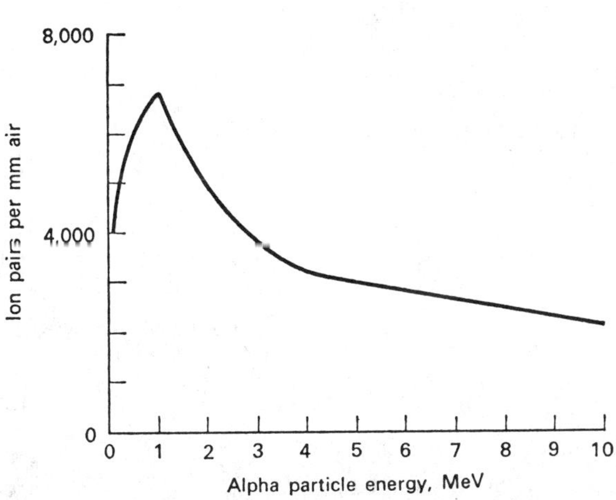

Figure 7.13. Bragg curve of specific ionization by alpha particles in air at standard temperature and pressure (reproduced from Cember, Ref. 4, with permission of Pergamon Press © 1969).

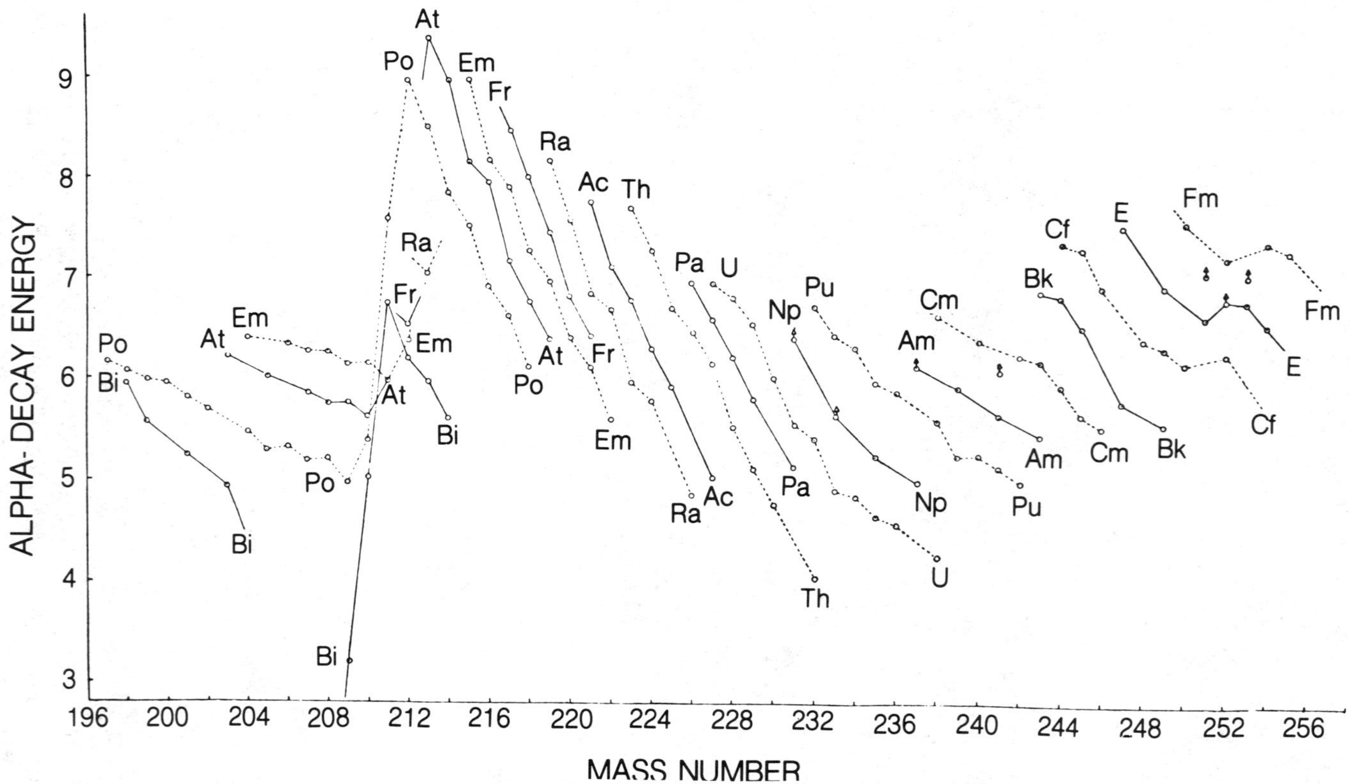

Figure 7.14. Alpha-decay energy versus radionuclide mass number (reproduced from Siegbahn, Ref. 12, with permission of North-Holland Publishing Co. © 1965).

7.2.2. Measurement Considerations

Because of their very short range, the hazard from α-particle emission is almost entirely the result of inhalation or ingestion; skin dose is usually a minor hazard. Inhalation is the most serious hazard for such important α-particle emitters as plutonium and radon daughters. Intake may be direct such as when food, water, or air containing α-particle emitting materials are taken into the body; or indirect such as when, for example, the active material from contaminated hands, clothing, or tools is transferred to food or water that is later consumed.

Alpha-particle survey techniques are often directed at indirect hazard assessment and hence are designed merely to indicate the presence of α-particle emitting material in an area, particularly on a surface. Survey techniques may also be intended to provide a rapid or convenient measure of the α-particle activity in some bulk material or even in air. Except for air monitoring, the short range of α-particles often makes direct α-particle monitoring difficult. Measurement capability may, however, be enhanced by chemical separation of the material (more often a prelude to quantitative measurements of α-particle activity than of surveys) or by reliance on γ- or x-ray emission associated with the α-decay sequence.

Alpha-particle activity (in pCi or Bq per unit volume or mass) may be determined in a number of ways for various media (e.g., air, water, soil, food) or for biological samples (e.g., blood, urine, tissue). Except for air, α-particle monitoring usually requires separation of the α-particle emitter from the bulk sample by chemical procedures, or by physical means such as filtering, and preparation of an effectively "thin" sample appropriate to the detector system being employed. To some extent, the inconvenience associated with chemical separation is mitigated by the resulting possibility of using α-spectrometry and by an inherent advantage of α-particle detection. Two salient physical properties of α-particles (relatively high energy and high specific ionization) allow design of detectors with extremely low backgrounds; indeed, backgrounds in the more sensitive detectors are typically dominated by α-particle contamination of one sort or another. Another background contribution can come from high-LET (high linear energy transfer) particles associated with cosmic radiation, but this usually is not important. Contamination is usually from the materials used in the construction of the instrument. Care in the choice of construction materials is essential for a low-background α-particle detector. An alternative form of interference occurs from radionuclides other than the one(s) to be detected or from recoiling α-particles during counting.

An alternative technique for identification of α-particle emitters is the use of decay-chain information to isolate one nuclide from a mixed sample. This can be done, for example, by the buildup of a daughter (if the daughter can be measured easily, perhaps by its own decay). Because γ-ray detection or spectroscopy is relatively easy, it is often used where appropriate. Alternatively, α-γ or α-β "coincidences" can be used when the daughter's lifetime is extremely short, and its γ- or β-ray emission can be counted in coincidence with the parent emission. This can be very specific, as well as providing a very low background.

Emission of x-rays by the daughter immediately following α-particle emission is a common occurrence for the heavier elements, particularly the transuranic elements. The α-particle emission leaves the daughter in an atomic excited state, which de-excites by x-radiation. The γ- and x-ray emissions are essentially simultaneous, and in some cases the x-rays are a better signature than the α-particle emission for elemental identification. This is especially true when samples cannot be made thin enough for good α-particle counting; the x-rays will penetrate further and can be detected instead. A good discussion of the use of x-radiation in detecting α-particle emitters has been given by Swinth [40], who has also tabulated ratios of x- to α-particle intensity for a number of important transuranic isotopes (^{237}Np; $^{238,239,240,241}Pu$; ^{241}Am; $^{242,244}Cm$). This technique is also commonly used for measuring Pu, which is discussed in a separate chapter.

Techniques for measuring α-particles are discussed in the following section, and those for measurings and x-rays are discussed in Chapters 5 and 6. Application of these techniques to monitoring of α-particle emitters is treated in the chapters on individual radionuclides in Part III.

7.2.3. Measurement Techniques

Techniques that Measure Particle Flux or Dose Equivalent. There are two common types of measurements for determining particle flux or dose equivalent:

1. A measurement of particles per second may suffice, such as when α-particle activity needs to be only roughly determined (for example, a spill of α-particle emitters).
2. A direct measurement of dose equivalent may be required, such as when a worker is unavoidably exposed to an α-particle environment that could result in a significant, measurable dose. In such situations the likelihood of inhalation should also be considered.

For spills, actual α-particle dose equivalent is usually not required because prudent practice demands immediate clean up rather than permitting exposure to it for any length of time. In either case, it may or may not be possible to ascertain the species of radionuclide responsible. For measurements of dose equivalent, knowledge of the species (more specifically, the α-particle energy) is important but not crucial. For measuring the counting rate, it is less often important.

Calibration of a ratemeter usually is not a problem. The major difficulty is assuring a known geometrical setup. For calibrating a dose equivalent instrument, a known source can be used, but care must be exercised because of the finite thickness of the source itself. Alpha particles of degraded energy, emerging from below the surface of the standard source, can influence the accuracy of the measurement. The source must be thin enough that none of the α-particle are too degraded to register in the counter.

Particle emission rates are commonly measured for two different kinds of media: surfaces with possible contamination, and media that can be specially prepared for measurement (usually by chemical preparations; see also α-spectrometers, following).

Instruments used for surveys, including detection of surface contamination, require a large-area thin window. Typical windows in α-particle detectors have thicknesses of a few mg/cm^2. Much thinner windows are occasionally used, but are fragile and usually not necessary. Thicker windows are too opaque. Once an α-particle penetrates the window, it can be easily detected because any of several ionization-sensitive instruments can be used (proportional counters, ionization chambers, scintillation counters, semiconductor detectors). Common survey instruments employ air proportional counters or zinc sulfide (ZnS) scintillators. Detector types, discussed later, are often α-spectrometers, although any of these could also be used for measuring particle flux or gross α.

The ZnS scintillator, used for surveys in air, as well as for measuring small, specially prepared samples, was designed especially for α-particle detection. Zinc sulfide crystals with silver impurities [ZnS(Ag)] can be prepared as screens of about 15 mg/cm^2 thickness and/or applied to a backing as a thin layer. The detection efficiency for α-particles should be at least 95% for a properly constructed screen [11]. The ZnS(Ag) phosphor is viewed by a photomultiplier tube, preferably with low noise. This combination is ideal for use as a ratemeter or even as an analysis device, as in a Lucas cell (see measurements in air). But it is not useful for spectrometry, because the energy resolution is so poor (poorer than for any of the systems discussed in the next section). However, as noted in the EML *Procedure Manual* [31], the activated ZnS phosphor gives a large α-particle pulse, and excellent discrimination against β- and γ-radiation can be obtained.

Details of sample analysis techniques are discussed by Hallden and Harley [41]. Techniques for measurement of airborne α-particle emitters are discussed later in this section; in their simpler form, these techniques may be used for general surveys. A general requirement for scintillator-phototube systems is, of course, that they be shielded from external light.

Integrating techniques sometimes employ materials that may be examined for total α-particle counts. Methods for measuring Rn and its daughters use thermoluminescent materials and plastics that show α-particle tracks after etching. The latter technique has been developed primarily for uranium surveys. See air measurements, later.

Alpha Spectrometers. Alpha spectrometry (i.e., the separation and measurement of an energy spectrum) has been used since the earliest days of nuclear physics; indeed, Rutherford and Geiger built a proportional counter for this purpose in 1880. There are two different approaches to α-spectrometry. The first measures ionization, and the second relies on momentum selection in a magnetic field. The latter, not discussed here, is predominantly used for research. An excellent discussion of this class of instruments is given by Siegbahn [12]. The following common alpha spectrometers that rely on ionization, are briefly treated here (see also Section 4.2 of Chapter 4).

1. Pulse ionization chambers.
2. Gas proportional counters.
3. Liquid scintillation counters.
4. Solid scintillation counters.
5. Semiconductor (solid-state) detectors.

Besides these, several other ionization techniques have found occasional use in various applications, but are not discussed here. Among them are nuclear emulsions, cloud chambers, and differential range/absorption techniques.

Pulse Ionization Chambers. The pulse ion chamber is widely used because of its excellent resolution. This is an ionization chamber operated so as to count individual α-particles as pulses of ionization. There are two different ways in which the α-particle source might be introduced into the counter, either as a gas or (more commonly) as a thin solid film located on or near the anode. A detailed discussion of the techniques for improving resolution of this instrument is given by Curran and Wilson [42].

Resolution in an ordinary plane-parallel chamber is limited because the total electronic signal depends on the distance of electrons from the collection plane. The signal then depends on the position, length, and orientation of the α-particle track. A grid inserted between the two planes substantially reduces this effect and makes possible much-improved resolution. This type of instrument is known as a Frisch Grid ion chamber, after its inventor, O. R. Frisch. Resolution of about 30 to 40 keV (FWHM) at about 5 MeV can now be obtained routinely, and even ~20 keV has been reported [43]. The most difficult part of this measurement method, and the part which limits its convenience and accuracy (and hence its widespread use), is source preparation (discussed later).

Gas Proportional Counters. The gas proportional counter uses an α-particle sample similar to that of the pulse ion chamber. The principal advantage over the pulse ionization chamber is its ability to discriminate more completely against background β- and γ-radiation. A drawback is that energy resolution is inherently poorer (no better than ~100 keV (FWHM) in a typical well-designed system), due to the resolution spread from the charge multiplication process. Moreover, if used for spectroscopy, choice of gas is important.

However, for α-particle counting, gas type and quality are not critical, although it is important that the density be high enough to stop the α-particles before they strike the chamber wall. This contrasts markedly with the case of β-particle proportional counting, where gas type and quality are critical. In a typical proportional counter for α-particles, a pulse-height gain of about 100 is sufficient, because the ionization is dense and the primary ion-pair count is high. Again, this is in contrast to β-particle counting, where gains of ~10,000 are common. Figure 7.15 shows the performance of a typical proportional counter when the activity is about one-half from α-particles and one-half from β-particles. At point A, the α-particle pulses are just big enough to trigger the discriminator. The "alpha plateau" occurs when all of the α-particle produce counts, and the β-particles are of insufficient pulse height to trigger the discriminator. At point B, the β-particles begin to count, and the higher plateau is the sum of (alpha + beta) counts. This relatively clean separation is characteristic of a typical proportional counter. Curran and Wilson [42] discuss the detailed problems and features of gas proportional counters.

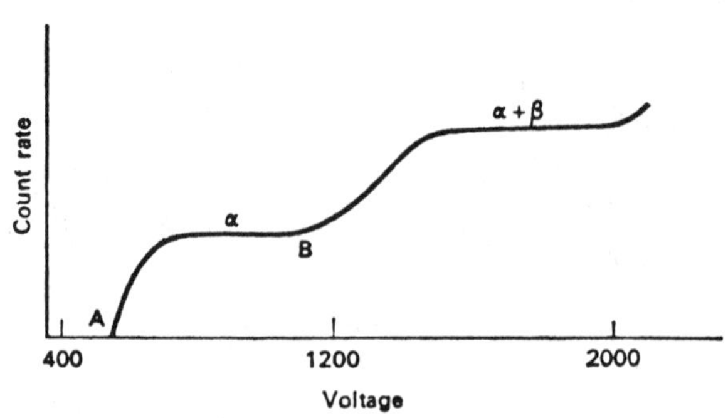

Figure 7.15. Alpha and alpha + beta counting rates as a function of voltage in a proportional counter (reproduced from Cember, Ref. 4, with permission of Pergamon Press © 1969).

Liquid Scintillation Counters. Liquid scintillation counting has been developed primarily for counting low-energy, low-activity β-particles, and its application to α-particles has not been widespread. Details on this technique are found in Chapter 20, Tritium. Aspects particular to α-particle monitoring are described here.

The basic technique requires incorporation of the material to be monitored into a small amount of liquid scintillator, which is then introduced into the detection system. Using a very small sample (250 μl of scintillator solution) viewed by a single phototube, Horrocks [44] has achieved energy resolutions of about 6% (FWHM) for 4- to 7-MeV α-particles. A typical spectrum is shown in Fig. 7.16, where 4.8-MeV α-particles of ^{233}U are relatively well resolved from the 5.5-MeV α-particles of ^{238}Pu. The advantages of a scintillator/photomultiplier technique include: ease of operation, particularly in sample preparation; high and well-understood efficiency (essentially 100%); and speed. This technique is particularly appropriate for water sampling or when chemical extraction is required.

A principal difficulty with α-particle counting in liquid scintillators is quenching, that is, decrease or variability in light output, due to interference in the scintillation mechanism, or absorption of the emitted light, by the sample itself. McDowell and co-workers [45,46] have developed sample preparation techniques that minimize quenching. In addition, pulse-shape discrimination techniques for liquid scintillators have been adapted [46,47] to α-particle counting in order to distinguish α-particle-induced pulses from electron-induced pulses originating from cosmic rays or from β-particle activity in the sample or counter. These techniques are based on dependence of the pulse shape on the ionization density. Using these techniques, the minimum background under an α-particle peak can be reduced from the 5 counts per minute (cpm), obtained with commercially available beta-particle liquid scintillator systems, to about 0.01 cpm [46].

A possible disadvantage of this technique is that the light output from α-particle scintillation is not proportional to particle energy because the scintillation process becomes saturated at the high specific ionizations involved. This is shown in Fig. 7.17 [44]: the pulse height is a linear, but not proportional, function of α-particle energy.

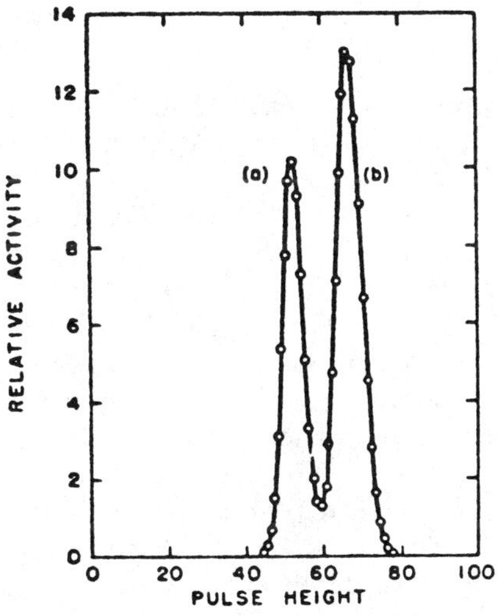

Figure 7.16. Pulse height spectrum for (a) 4.8-MeV alpha particles of ^{233}U and (b) 5.5-MeV alpha particles of ^{238}Pu in liquid scintillator (reproduced from Horrocks, Ref. 44, with permission of American Institute of Physics © 1964).

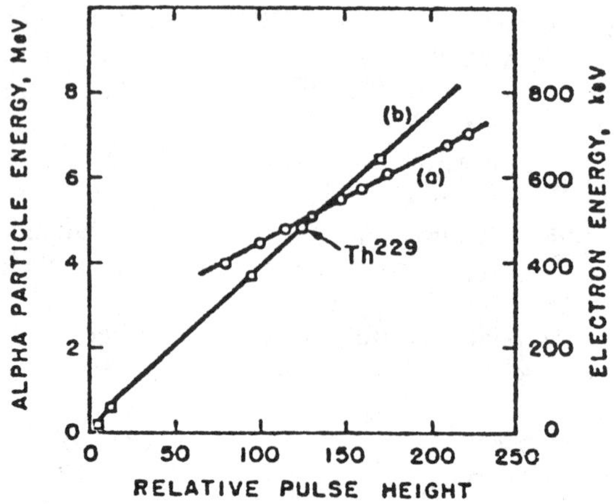

Figure 7.17. Pulse height-energy relations for alpha particles and electrons in a liquid scintillator, (*a*) alpha particles, (*b*) electrons (reproduced from Horrocks, Ref. 44, with permission of American Institute of Physics © 1964).

Solid Scintillation Counters. The main advantages of solid scintillators for α-spectroscopy are their availability in large sizes and their ability to identify particles by pulse-shape discrimination. However, solid scintillators have two disadvantages: (a) The resolution is poorer than for several other techniques by a significant factor. This is because the required energy of about 300 eV per photoelectron is so much higher than the ~30 eV per ion pair in gases or the 3 eV per electron-hole pair in semiconductors. Typical α-particle resolutions for scintillator detectors are in the few-percent range (several hundred keV out of several MeV). (b) Sources must be counted externally because scintillators ordinarily need to be wrapped or otherwise contained for light-tightness, making the introduction of the α-particle emitting active sample difficult. When these problems do not present insurmountable difficulties, the scintillator/photomultiplier system has much to recommend it, including many of the advantages of liquid scintillators. Neiler and Bell [48] discuss the utility and limitations of solid scintillator systems. Presently, they are not often used for α-spectroscopy. However, the zinc sulfide crystal (with silver impurities) scintillator is often used for measuring gross α, as discussed later.

Semiconductor (Solid-State) Detectors. Commercially available and laboratory spectroscopic systems depend principally on semiconductor (often called solid-state) detectors. Their introduction in the 1960s revolutionized high-resolution particle and photon spectroscopy for many research and practical applications. For detecting heavy charged particles relatively thin detectors are adequate, as opposed to detecting γ-rays, for which large volume detectors are often necessary. For α-particles, with their relatively short range and high specific ionization, the sensitive region of the detector must extend nearly to the surface where α-particles enter the detector, because any "dead" material absorbs energy from α-particles. Surface-barrier and diffused-junction silicon detectors have been found useful for detection of α-particles with energy up to tens of MeV.

Figure 7.18 shows schematic diagrams of semiconductor detectors based on a silicon wafers. The diffused-junction detector typically is formed from p-type material, that is, material rich in acceptor sites, by diffusing into the surfaces of the

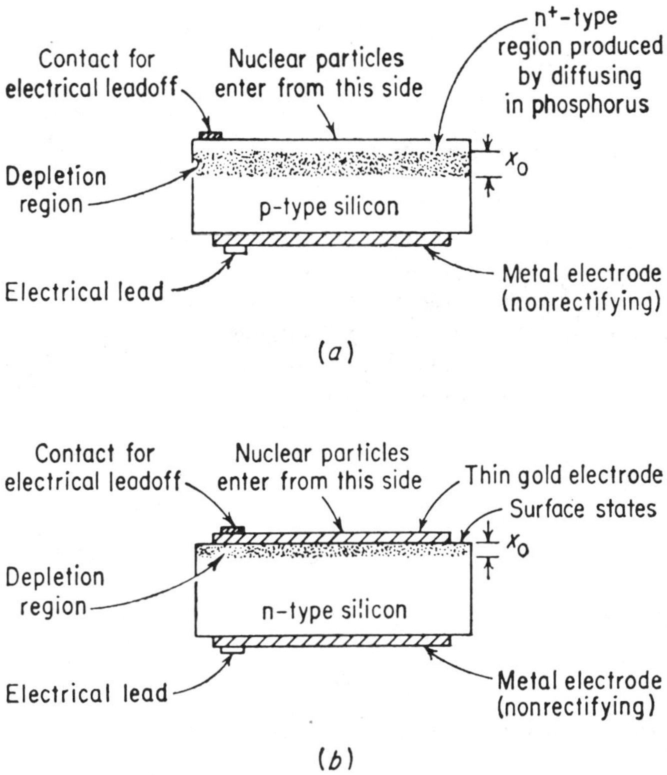

Figure 7.18. Schematic representations of diffused-junction (*a*) and surface-barrier (*b*) detectors (reproduced from Price, Ref. 49, with permission of McGraw-Hill © 1964).

wafer a donor such as phosphorous, thereby causing an *n*-type region, containing many free electrons. Application of a reverse bias to the *n-p* junction depletes a layer of material of free carriers, and this depletion region constitutes the sensitive volume of the detector. Holes and electrons produced in it by the ionizing action of incident radiation are collected in the electric field of the depletion layer, thereby producing signals. The surface-barrier detector similarly forms a depletion region, but the junction is normally formed by etching *n*-type thin silicon and depositing on this a thin layer of gold (i.e., about 50 $\mu g/cm^2$). For both detector types, ionizing particles traversing the depletion region lose energy by producing electron-hole pairs that are rapidly collected at the positive and negative detector electrodes. The theory, fabrication, and operation of semiconductor detectors is presented in more detail in Chapter 4 and in numerous publications [49–50]. A brief review of semiconductor detectors is given in Ref. 51. The electronic units required for operation of semiconductor detectors are widely available in modular form and their application is discussed in manufacturers' literature.

Detectors with depleted depths from less than 10 μm to several millimeters are commercially available. For naturally occurring α-particles, depths with a minimum of 20 μm (at 4 MeV) to 50 μm (at 9 MeV) are acceptable. Detectors with effective areas up to about 20 cm^2 are available, and their detection efficiency for α-particles is virtually 100% for particles entering the detector area. The small physical size of these detectors may be an advantage or a disadvantage, depending on the application. Their principle advantage is the combination of low background with high resolution. Background counting rates of only 0.003 cpm can be routinely obtained.

A semiconductor detector has several distinct advantages over a gas-filled detector:

1. For a given amount of energy lost through ionization, the number of charge carriers (corresponding to ion pairs in the gas) is greater by a factor of as much as 10. This decreases the fractional statistical fluctuations in the ion collection process and provides very good energy resolution.
2. With detectors now available, recombination effects are less important than for a gas; this is particularly important when ionization is locally dense, as is the case with α-particles (or fission fragments).
3. Charge collection times are in the nanosecond range, allowing fast time resolution where required.

Among the difficulties that may be associated with semiconductor detectors are:

1. The detectors can be fragile, requiring careful storage and handling.
2. The output signals are very small, so that low-noise electronic amplification is required.
3. They can become irreversibly contaminated, so that the background may slowly rise. (However, diffused-junction detectors are often marketed as decontaminable.)

Figure 7.19 shows a high-resolution α-particle spectrum. Resolutions of 15 keV (FWHM) are commonly obtained using junction or surface barrier detectors, but surface barrier detectors generally have slightly thinner entry windows, often an advantage for high-resolution α-particle measurements. Such high resolution is only possible by sacrificing counting efficiency.

Alpha-Particle Measurements in Air. Most α-particle emitters in air are in particulate form (an important exception being radon, which is discussed in Chapter 14). For this reason, measurements of α-particle activity in air often depend on collection of airborne particulates on a filter whose collected activity is measured. This measurement may be accomplished by several methods, ranging from simple gross-α techniques using scintillators to precise α-spectrometry that distinguishes individual radionuclides.

Integrating techniques may employ thermoluminescent material, or even film, sensitive to α-particles. The most straightforward techniques for measuring activity on particles involves placing the filter in front of a detector, as shown in Fig. 7.20. Ideally, all of the α-particles leaving the filter strike the detector. One difficulty in this technique is assuring that a reliable sample is deposited on the filter; large errors may result in dust-laden air (e.g., in mines). Another problem is self-absorption; if the sample to be counted exceeds a few mg/cm^2 in thickness, some α-particles will not leave the sample without significant energy degradation. In fact significant loss can sometimes occur in the individual particles containing the activity of interest, particularly after chemical separation (discussed later). Similarly, air between the filter and detector can degrade α-particle energy by amounts

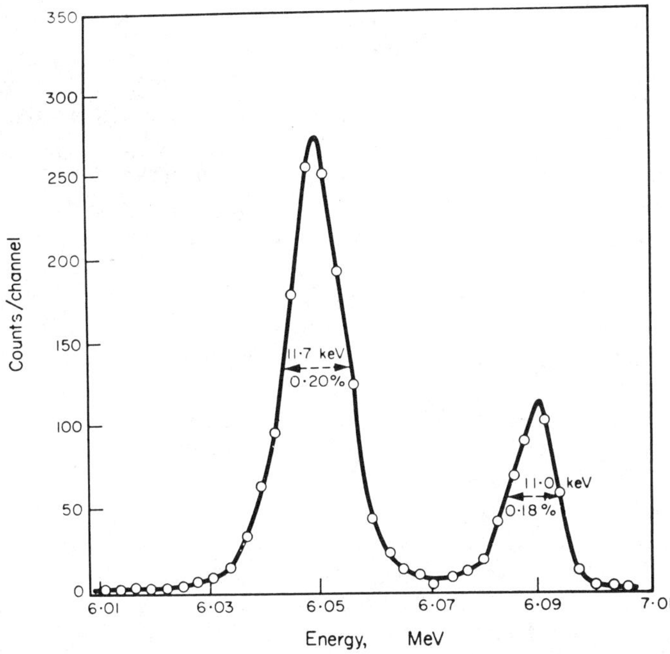

Figure 7.19. An early high-resolution spectrum of alpha particles from the decay of ^{212}Bi (reproduced from Watt and Ramsden, Ref. 52, with permission of the Pergamon Press, Ltd. © 1964).

that depend on the path length (see Fig. 7.20). However, for measurements where the α-particle energy is not determined, such degradation is unimportant unless the resulting signal is reduced below some system threshold.

Gross-α measurements, in the strict sense, do not require energy measurement. Any of the detection techniques discussed previously may therefore be applied in relatively simple form, including gas proportional counters, scintillators, and semiconductor detectors. The American Public Health Association's Intersociety Committee [27] describes a "Tentative Method for Analysis of Gross Alpha Radioactive Content of the Atmosphere." This method depends on filter collection followed by counting with any detector sensitive to α-particles.

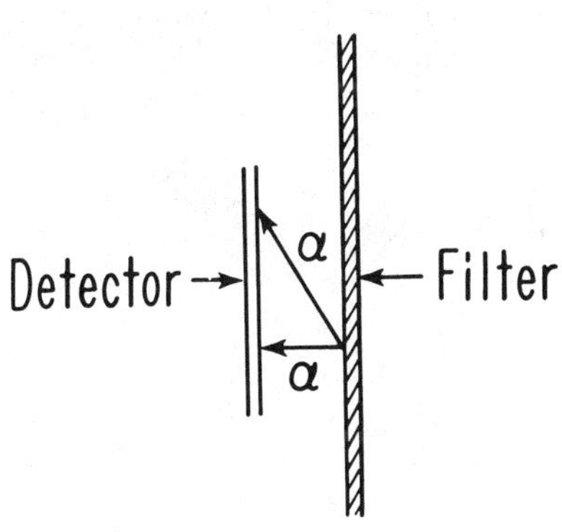

Figure 7.20. Schematic arrangement for alpha detection using a semiconductor detector and an air filter.

For most situations, a predominant contributor to total α-particle activity in air is radon and its daughters. Radon daughters typically attach themselves to particles, which are collected by air filter systems. To measure total activity contributed by radionuclides other than Rn and its daughters, discrimination between Rn and its daughters and other radionuclides may be effected on two bases. First, the controlling half-lives of Rn daughters are relatively short (less than 30 min for daughters of ^{222}Rn and up to 10 hr for daughters of the usually less important, ^{220}Rn), so that delaying counting for a period after collection reduces the degree to which they interfere. For long collection or holdup periods, the limiting source of background is ^{212}Pb which, because of its 10-hr half-life, remains after the shorter lived polonium isotope decays. Secondly, the α-particle energies from Rn daughters are greater than 6 MeV, which is greater than those of most other α-particle emitters of interest, such as radium, uranium, or plutonium. As a result, detectors capable of discriminating different energy levels, such as semiconductor detectors, may be used to discriminate against Rn daughters. Due to poor energy resolution in some detectors, Rn daughter counts can spill over into the low-energy region. This difficulty that may be reduced by subtraction techniques employing two energy "windows." More detailed discussion of subtraction techniques is given in Chapter 16, Plutonium. Plutonium and other transuranics have the lowest maximum permissible concentrations in air so accurate background corrections are especially important to permit accurate determination of the low levels involved.

Although instruments for measuring most gross α (or gross α except for Rn and daughters) monitor activity collected on a filter, a notable exception is the class of devices depending on a Lucas chamber, a small spherical or cylindrical cell lined with a ZnS scintillator viewed by a photomultiplier tube [53]. The sample, which may be air containing gaseous and/or particulate activity, is introduced into the volume and counted directly. Since the chamber size is small, nearly all of the emitted α-particles strike the scintillator on the wall. This method is discussed in Chapter 14.

Commercially available instruments that measure α-particle activity in air are of the more conventional type described previously, that is, based on a collection filter viewed by a proportional counter, scintillator (especially ZnS), or semiconductor detector. The more sophisticated of these, typically employing semiconductor detectors of the diffused-junction or surface-barrier type, discriminate to a large extent against Rn daughters.

Alpha-particle integrating techniques may also employ plastics in which etching shows α-particle tracks, as well as thermoluminescent materials. Fleischer, Price, and Walker [54] discuss nuclear tracks in solids. Track-etch materials and associated procedures have been developed based on detection of Rn and its daughters for U surveys and for detection of these short-lived radionuclides in air. TLDs have also been used in Rn and Rn daughter monitoring. See Chapter 15, Uranium and Thorium; Chapter 14, Radon; and Chapter 9, Dosimetry.

Semiconductor detectors (and others) may also be employed in a more complete spectroscopic system, one distinguishing between individual radionuclides. In systems that measure radiation from the collection filter directly, Rn daughters still interfere, although the interference can be substantially reduced by placing the filter sample in an evacuated counting chamber, thereby improving the energy resolution of the system. Such techniques are described in the chapters on Pu and Rn and its daughters (where the purpose is to distinguish between individual daughters).

Recently developed methods for monitoring long-lived α-particle emitters in air, particularly Pu and other transuranics when borne as particles, depend on their behavior in a moving air stream. Partial segregation of the activity of interest may be effected by impaction techniques such as those discussed in the chapter on Pu.

An alternative to examining the filter sample directly is to separate chemically the radionuclides of interest from the filter and then place them in an α-spectrometer system. Chemical separation techniques are discussed later in this chapter.

Alpha-Particle Measurements in Water. Measurements of α-particle activity in water can range in sophistication from simple measurements of gross-α activity to measurements of individual radionuclide concentrations using chemical separation techniques followed by α-spectroscopy. Chemical separation techniques are discussed separately later and in the chapter on individual radionuclides. The present discussion emphasizes measurements primarily of gross α in community water systems. Before describing techniques for measuring gross α, note that α-spectrometry of water samples may use any of the spectrometric techniques discussed previously, provided that a thin sample is prepared [55]. Direct evaporation may result in a residue that absorbs some alpha-particle energy and hence is not appropriate for high-resolution spectrometry. Moreover, spectrometry is not often necessary as the scintillation counter method described by Hallden and Harley [41] is well suited for measuring gross α, and is both reliable and simple.

Measurements of gross-α levels in water is a mainstay of public health monitoring systems. If these levels are shown to be small enough, which is usually the case, it is ordinarily unnecessary to identify and measure individual radionuclides using more sophisticated techniques. However, gross-α counts must be used with caution because efficiency depends on isotope and sample form.

Interim regulations for α-particle activity in drinking water from the Environmental Protection Agency (EPA) [56] specify maximum levels for ^{226}Ra, ^{228}Ra, and gross-α activity in community water systems as: 5 pCi/liter (0.18 Bq/liter) for combined ^{226}Ra and ^{228}Ra; and 15 pCi/liter (0.55 Bq/liter) for gross α, including ^{226}Ra but excluding radon and uranium. The Public Health Service had recommended a ^{226}Ra limit of 3 pCi/liter (0.11 Bq/liter) [57]. Thus a measured gross-alpha activity of less than 5 pCi/liter (0.18 Bq/liter) would preclude the need for measurement of Ra activity. (As will be discussed in the chapter on Ra, the latter measurements may be based either on chemical precipitation or on Rn emanation.)

A Standard Method for measuring gross α is specified by the American Public Health Association (APHA) [24]. This method is similar to tentative method proposed by the EPA [25]. The technique involves depositing a sample derived from the water on a counting planchet, which is then inserted into the detector volume or placed next to the detector, as appropriate. Sample preparation is the difficult part. Because of the low range of α-particles, only a very thin (≤ 3 mg/cm^2) sample can be used without needing a significant self-absorption correction. The water sample is dried by oven evaporation just below the boiling temperature, or on a hot plate, then counted. The main systematic error is from loss of particles which might become airborne during the drying process. The recovery of gross-α activity in this procedure is in the 85 to 90% range, with reproducibility (precision) of about $\pm 20\%$ at the 95% confidence level. If separate determinations of suspended and dissolved matter are desired, a separation using a Gooch crucible or membrane filter is possible.

In one study of the APHA method conducted by the EPA's Analytical Quality Control Service [58], unknown samples were distributed to about 15 participating laboratories. The gross-α activities in the sample ranged from about 25 to 100 pCi/liter (0.925 to 3.700 Bq/liter). Analysts used the Standard Method from the 12th Edition. The gross-α measurements were found to agree quite well (the average activity found was 0.92 $\pm$ 0.10 of the known activity) and showed that the Standard Method for gross alpha was acceptable method. In fact, Baratta and Knowles reported that "as a screening technique, the procedure was better than expected." [58]

For counting gross α, an internal gas proportional counter is recommended: it has high sensitivity, good geometrical efficiency, and reliability. In the presence of both α-particle and β-particle activity, the α-particle counts can be separated from β-particle counts by measuring the counting rate versus voltage; a typical counter might respond as shown earlier in Fig. 7.15. As noted previously, the first plateau is from α-particles, the second from the sum of (alpha + beta).

Alternative methods might employ either a thin-window proportional counter or a Geiger-Müller tube. However, both have a lower counting efficiency, and the G-M tube has the added limitation of not separating α-particle from β-particle activity. G-M tubes may be calibrated using standard α-particle emitters, such as uranium salts.

7.2.4. Chemical Separation Techniques

Alpha spectroscopy is capable of directly identifying individual α-particle emitters only under a restricted set of circumstances. In particular, the short range of α-particles often necessitates the use of chemical separation techniques to concentrate the α-particle emitter from the bulk material of the main sample. The aim of such a separation process is to reduce the sample mass to the microgram range without quantitative loss of α-particle emitters under investigation. The initial sample may be water, soil, biological tissue, or even an air filter. The final sample should be a thin layer deposited on a suitable substrate. In many cases, the fraction of interest is electroplated onto a metal planchet. In others (such as one of the Standard Methods for radium), a precipitate collected on a filter may be dried before measurement with an α-particle detector. (See, for example, Sill and Williams [59].) It is also possible, in principle, that the fraction of interest be incorporated into a liquid scintillator sample for spectroscopy.

When an energy discriminating spectrometer can be used, it is not generally necessary to separate the α-particle emitters from each other (unless there is overlap in the energy spectrum). However, chemical separation of individual radionuclides can make spectrometry unnecessary: a gross-α measurement can be ascribed to one and only one element after adequate chemical separation. In some cases, even if complete separation is not possible, it can simplify the spectrum sufficiently.

The fact that many of the α-particle emitters are members of natural or artificial radioactive series may also be used to advantage. Following buildup or decay, after chemical separations have been made, a check on the purity of a particular fraction may be carried out.

Unfortunately, the number and variety of chemical separation techniques prevent a comprehensive discussion of them here. There are 15 elements and almost 100 α-particle emitting isotopes from bismuth to the transuranic elements, and many possess a wide variety of chemical oxidation states which complicates separation.

There are several major concerns in chemical separations:

1. The fractional recovery of the individual α-particle emitter must be known to appropriate accuracy. Recoveries $> 95\%$ are desirable, because quantitative errors are less likely when recoveries are high. The use of radioactive tracers is common as a means for checking chemical yields, although differences in recovery may occur between tracers and unknowns if the exchange is not properly assured.
2. The α-particle emitting fraction must be separated from the bulk matrix in a way that allows accurate, convenient counting or spectroscopy at the final stage.
3. The presence of multiple oxidation states may make a complicated series of procedures necessary, especially in samples of unknown or mixed composition. For example, uranium may be found in both hexa- and quadrivalent states in soil samples, and these states have completely different chemical properties. Decomposition generally oxidizes the uranium to a single state. Much more difficult is the case of plutonium where oxidation potentials are such that states are always split.
4. Sometimes it is not necessary to completely separate all of the elements because α-spectroscopy may be used for the final discrimination amongst the unknowns. However, certain interferences can be important (e.g., the α-particle from ^{237}Np has an energy close to those of several of the uraniums).

Several handbooks and compilations detail chemical procedures used at some of the larger radiation laboratories. Among these are the DOE EML *Procedures Manual* [30]; the handbook of the EPA National Environmental Research Center [60]; an older standard USPHS Radiological Health compilation [61]; the Methods published by the APHA [24]; and the handbook of the Nevada Applied Ecology Group [62]. In addition, Sill and co-workers have described detailed procedures for separations of most of the significant α-particle emitting elements [63,64].

Most routine chemical separations, such as those described in this section, employ acid leaching to remove the radionuclides of interest directly from the bulk sample. Sill has described a total dissolution technique "by which virtually all alpha-emitting nuclides can be determined in soil" using a few grams of sample [64]:

> The sample is decomposed completely by a combination of potassium fluoride and pyrosulfate fusions with simultaneous volatilization of hydrogen fluoride and silicon tetrafluoride. The cake is dissolved in dilute hydrochloric acid and all alpha emitters are precipitated with barium sulfate. An option is provided by which uranium can be obtained in a separate fraction to eliminate mutual interference with ^{237}Np. The barium sulfate is dissolved in acidic aluminum nitrate and the elements from thorium through plutonium are extracted into Aliquat-336. Acid-deficient aluminum nitrate is then added to the aqueous phase to reduce the acidity, and americium, curium and californium are extracted into fresh Aliquat-336. Thorium is removed by scrubbing the first organic extract with 10 *M* hydrochloric acid, and protactinium, uranium, neptunium and plutonium are stripped with a mixture of perchloric and oxalic acids. Americium, curium, and californium are stripped from the second organic extract with 8 *M* nitric acid. Each fraction is electrodeposited and analyzed by alpha spectrometry with few interferences. Overall recovery excluding electrodeposition ranges from 88% for protactinium to 99% for plutonium [64].

Sill and Williams discuss detailed chemical methods of determining α-particle emitters in the environment [65]. This section will discuss methods for the following (atomic number in parentheses):

1. Polonium (84).
2. Protactinium (91).
3. Neptunium (93).
4. Americium (95).

Chemical separation techniques for radium (88), uranium (92) thorium (90), and plutonium (94) are given in the chapters on these radionuclides. Some α-emitting elements are not discussed here because: (a) they do not appear commonly in environmental samples; (b) they have very short half-lives; or (c) procedures for their individual separation have not come to the authors' attention. Among these are: astatine (85), francium (87), actinium (89), curium (96), and heavier elements.

Polonium. Polonium (Po) occurs in isotopic states with atomic weights from 193 to 218. Almost all of the Po isotopes are alpha emitters. In environmental media, Po isotopes occur most importantly as daughters in the ^{238}U decay chain, following ^{222}Rn. The radon-daughter isotopes of polonium are:

^{218}Po (E_α = 6.00 MeV, $t_{1/2}$ = 3.0 min)
^{214}Po (E_α = 7.69 MeV, $t_{1/2}$ = 164 sec)
^{210}Po (E_α = 5.30 MeV, $t_{1/2}$ = 138 days)

As progeny of ^{222}Ra, these isotopes may be gaseous or more commonly adsorbed on particles in the air. Measurements of Po species in air are discussed in the chapter on Rn.

The occurrence of Po in water, urine, or soils can be important when the parent (Ra, Rn, or a daughter such as ^{214}Pb) is found in these media.

The two similar techniques for chemical separation of polonium use a strong HCl solution and ascorbic acid to immobilize any iron that might be present [30,61], followed by plating (spontaneously induced by stirring) on a silver disk [61] or nickel disk [30]. The plating procedure takes 3 to 5 hr, after which the disk is removed and α-particles counted using an internal proportional counter or other α-spectrometer. This procedure is specific and can be done in the presence of many other radionuclides. A ^{208}Po tracer may be used to determine chemical yield. Both procedures are also applicable to those organic materials that can be converted to chloride solutions [30]. Johns [60] described a method for determination of ^{210}Po.

Protactinium. Protactinium (Pa) occurs naturally with mass number 231 as a daughter of ^{235}U. Protactinium-231 has a half-life of 3.2×10^4 yr, and its α-decay spectrum consists principally of three major groups within 50 keV of 5.0 MeV. Because this isotope is a likely constituent of uranium tailings, Sill [66] has developed a method to extract Pa. As with the Ra-to-Cf method described previously [64], the Pa method involves fusion of the sample with potassium fluoride and pyrosulfate. After dissolution of the resulting cake in HCl, Pa is extracted into diisobutylcarbinol either directly or following precipitation with barium sulfate. Extraction may be followed by electrodeposition and α-spectrometry, or by gross-α counting.

Neptunium. Neptunium (Np) is the first transuranic element. Its longest-lived isotope, ^{237}Np, has a half-life of 2.14 × 10^6 yr and a complex α-decay scheme with energies near 4.6 MeV. Although several Np isotopes are α-particle emitters, none except ^{237}Np is of much consequence for environmental monitoring purposes, because of either their short half-lives or their low production rates in processes of environmental concern.

A method for Np determination in urine, feces, tissue, and air filters is described in the EML *Procedures Manual* [30]. Reduced Np is coprecipitated with lanthanum fluoride from nitric acid solution. Neptunium is extracted into thenoyl-trifluoro-acetone from dilute hydrochloric acid, back-extracted into more concentrated acid, and electrodeposited for α-particle counting or α-spectroscopy. The procedure takes 6 to 8 hr. A similar method specifically for urine has been developed by Union Carbide at Paducah [67].

Americium. Americium (Am) is the third transuranic element. Its more commonly found isotopes are ^{241}Am (E_α = 5.3 MeV, $t_{1/2}$ = 440 yr) and ^{243}Am (E_γ = 5.3 MeV, $t_{1/2}$ = 7950 yr). None of the other isotopes of Am has an α-decay mode of significance. A description has already been given of the total dissolution chemical separation techniques for Am [63,64]. Another typical chemical technique is found in the EML *Procedures Manual* [30]:

> The actinide elements are concentrated by coprecipitation with lanthanum fluoride. Plutonium, if present, is removed by extracted with thenoyltrifluoractone (TTA) by the Hanford plutonium bioassay procedure. Excess sodium hydroxide is added to the aqueous phase to redissolve aluminum hydroxide, while coprecipitating americium on lanthanum hydroxide. Trivalent americium is oxidized to the americyl ion by silver catalyzed peroxydisulfate. Addition of ammonium fluoride stabilizes the americyl ion, while precipitating the lanthanum and any alpha emitting impurities as fluorides. Following reduction to the trivalent state with manganous ion, the purified americium is collected by coprecipitation with cerium fluoride. Measurement is by alpha counting for Am-241 and Am-243, the usual americium isotopes encountered.
>
> The method has been tested for urine, tissue, feces and air filters. The procedure gives excellent separation from neptunium, thorium, uranium, plutonium and cerium. It is completely compatible with the plutonium procedure....

Also discussed in the EML manual [30] are methods for Am in soil, water, air filters, and tissue that employ leaching with acids such as nitric and hydrochloric.

Americium-241 is an important isotope since it is the daughter of ^{241}Pu. The 60-keV γ-ray that accompanies 36% of the ^{241}Am α-particles is often used for detection. Detection using the 60 keV γ-ray is discussed fully in the chapter on Plutonium.

7.2.5. Calibration Sources

For calibration, the analyst needs known sources. To satisfy this need, the U.S. National Bureau of Standards (NBS) has for many years provided standard materials. The α-particle standard samples "consist of a practically weightless deposit of the nuclide on a thin platinum foil cemented to a monel disk." [68] Samples with activities from a few to thousands of nanocuries (from becquerels to kilobecquerels) can be purchased from NBS for the isotopes ^{210}Po (E_α = 5.305 MeV), ^{238}Pu (E_α = 5.50, 5.46 MeV), ^{241}Am (E_α = 5.49, 5.44 MeV), and ^{248}Gd (E_α= 3.18 MeV).

7.2.6. Summary and Conclusions

This section outlined the various techniques for the detection, measurement, and energy spectrometry of α-particles in various media. The most salient physical features of α-particles (their high energy, dense specific ionization, and short range) make their detection relatively easy as is discrimination against other radiations. The difficulties with α-particle detection are a result of these same physical properties. For example, it is often difficult to detect α-particle activity from an accidental spill on the floor, especially in cracks and near inaccessible corners. Also, α-particle emitters contained deep within a soil sample (below 50 μm or so) require chemical separation prior to counting, as do α-particles in water or urine. For this reason, x- or γ-radiation techniques are often used, if possible, for monitoring of α-particle emitters.

The problems of α-particle detection have led to a variety of instrumental and chemical techniques. For gross-α measurements, the standard methods in air and water are sufficiently sensitive and well understood to provide a satisfactory screening technique. Also, for measurement of emission rate (e.g., to detect spills), instruments now exist that can probe most surfaces and odd geometries.

The area most in need of further development is identification of individual nuclides. Alpha spectrometers with quite good resolution now exist, and their use makes possible the identification of many individual isotopic states whenever good samples can be prepared. Proper sample separation may be the crux of the problem.

Present wet-chemical techniques, required to isolate α-particle emitting nuclides from background matrices (soil, air filters, water, tissue, urine), are laborious, liable to systematic error, and expensive. The use of tracers to follow chemical-yield fluctuations through a long chain of analytical steps is crucial. However, tracers are not universally available in convenient form, nor is there any way to guarantee their effectiveness, especially when the α-particle emitters themselves are bound in multivalent states in a variety of organic compounds (e.g., plutonium or uranium in soil), unless the sample is totally dissolved and the tracer is totally exchanged.

In conclusion, α-particle detection is relatively well developed. However, except in cases where samples are easily prepared, quantitative measurements of specific activity, especially when isotopic identification is required, are handicapped by cost, time, and sample-size considerations.

REFERENCES

1. P. Marmier and E. Sheldon, *Physics of Nuclei and Particles,* Vol. I, Academic, New York, 1969.
2. J. Mantel, "The Beta Ray Spectrum and the Average Beta Energy of Several Isotopes of Interest in Medicine and Biology," *Int. J. Appl. Radiat. Isot.*, **23**, 407, 1972.
3. B. G. Harvey, *Introduction to Nuclear Physics and Chemistry*, 2nd ed., Prentice-Hall, New Jersey, 1969.
4. H. Cember, *Introduction to Health Physics*, Pergamon Press, Oxford, 1969.
5. T. Tabata, R. Ito, and S. Okabe, "An Empirical Equation for the Backscattering Coefficient of Electrons," *Nucl. Instrum. Meth.*, **94**, 509, 1971.

6. F. Titus, "Measurements of the Energy Response Functions of Scintillators for Monoenergetic Electrons," *Nucl. Instrum. Meth.*, **89**, 93, 1970.
7. Glenn F. Knoll, *Radiation Detection and Measurement*, Wiley, New York, 1979.
8. C. M. Lederer and V. S. Shirley, Eds., *Table of Isotopes*, 7th Ed., Wiley-Interscience, New York, 1978.
9. 10 CFR 20, Appendix B.
10. National Council on Radiation Protection and Measurements, "Instrumentation and Monitoring Methods for Radiation Protection," NCRP Report No. 57, 1978.
11. International Commission on Radiation Units and Measurements, "Measurement of Low-Level Radioactivity," ICRU Report No. 22, 1972.
12. K. Siegbahn, Ed., *Alpha-, Beta-, and Gamma-Ray Spectroscopy*, Vols. 1 and 2, North-Holland Publishing Company, Amsterdam, 1965.
13. Milorad Mladenovic, *Development of Magnetic β-Spectrometers*, Springer-Verlag, New York, 1976.
14. D. Baker and R. Brundle, *Electron Spectroscopy: Theory, Techniques and Applications*, Academic, New York, 1977.
15. S. C. Curran and J. D. Craggs, *Counting Tubes*, Butterworths, London, 1950.
16. S. C. Curran and H. W. Wilson, "Proportional Counters and Pulse Ion Chambers," p. 303 of Ref. 12.
17. C. R. Phillips and P. Kauffman, "A Simplified Method of Preparing Conductive Films for 4 Pi Counting," *Int. J. Appl. Radiat. Isot.*, **19**, 811, 1968.
18. J. H. Neiler and P. R. Bell, "The Scintillation Method," p. 245 of Ref. 12.
19. L. A. J. Venverdov, *Practical Measuring Techniques for Beta Radiation*, Philips Technical Library, MacMillan Press Ltd., London (distributed in the United States by Crane, Russak and Company, Inc., 52 Vanderbilt Avenue, New York, N.Y. 10017), 1971.
20. G. Dearnaley and D. C. Northrup, *Semiconductor Counters for Nuclear Radiations*, 2nd ed., Wiley, New York, 1966.
21. J. J. Taylor, *Semiconductor Particle Detectors*, Butterworths, Washington, DC, 1963.
22. F. S. Goulding, "A Survey of the Applications and Limitations of Various Types of Detectors in Radiation Energy Measurement," *IEEE Trans. Nucl. Sci.*, **NS-11**(3), 177, 1964.
23. G. Bertolini and A. Rota, "Beta and Electron Spectroscopy," in *Semiconductor Detectors*, G. Bertolini and A. Coche, Eds., Wiley, New York, 1958, p. 3410.
24. American Public Health Association, *Standard Methods for the Examination of Water and Wastewater*, 15th ed., 1015 Eighteenth Street, N.W., Washington, DC, 20036, 1976.
25. U.S. Environmental Protection Agency, "Tentative Reference Method for the Measurement of Gross Alpha and Gross Beta Radioactivities in Environmental Waters," EPA-680/4-75-005, Quality Assurance Branch, Environmental Monitoring and Support Laboratory, Las Vegas, Nevada 89114, 1975.
26. A. N. Jarvis, R. F. Smillinski, and D. G. Easterly, "The Status and Quality of Radiation Measurements of Water," EPA Report No. EPA-600/4-76-017, Environmental Protection Agency, Office of Research and Development, Environmental Monitoring and Support Laboratory, Las Vegas, Nevada 89114, 1976.

27. Intersociety Committee on Methods of Air Sampling and Analysis, *Methods of Air Sampling and Analysis*, 2nd ed. American Public Health Association, 1015 Eighteenth Street, N.W., Washington, DC 20036, 1977.
28. D. G. Easterly, R. R. Kinnison, A. N. Jarvis, and R. F. Smiecinski, "The Status and Quality of Radiation Measurements for Air," EPA-600/4-77-043, Environmental Protection Agency, Environmental Monitoring and Support Laboratory, Las Vegas, Nevada 89114, 1977.
29. NAS Monographs on Radiochemistry (e.g. Ref. 30).
30. National Academy of Sciences, "Source Material for Radiochemistry," Nuclear Science Series Report No. 42, 1970. Revision Subcommittee on Radiochemistry of Committee on Nuclear Science, available Printing and Publishing Office, NAS, 2101 Constitution Avenue, Washington, DC 20418, 1970.
31. J. H. Harley, Ed., *EML Procedures Manual,* Report No. HASL-300, 1972. USDOE Environmental Measurements Laboratory 376 Hudson Street, New York N.Y. 10014 (1972 edition, revisions supplied annually). Previous editions were issued as Report NYO-4700.
32. F. B. Johns, Ed., U.S. EPA National Environmental Research Center (now Environmental Monitoring and Support Laboratory, *Southwestern Radiological Health Laboratory Handbook of Radiochemical Analytical Techniques,* EPA-680/4-75-601, 1975.
33. U.S. Public Health Service, *Radioassay Procedures for Environmental Samples,* Pub. No. 999-RH-27, U.S. Public Health Service, Washington, DC 20201, 1967.
34. W. F. Libby, *Radiocarbon Dating,* University of Chicago Press, Chicago, 1955.
35. D. E. Watt and D. E. Ramsden, *High-Sensitivity Counting Techniques*, Pergamon Press, London, 1963.
36. National Council on Radiation Protection and Measurements, "Physical, Chemical, and Biological Properties of Radiocerium Relevant to Radiation Protection Guidelines," NCRP Report No. 60, 7910 Woodmont Avenue, Washington, DC 20014, 1978.
37. D. C. Sutton, "Determination of Cerium-133 in Vegetation and Soil," HASL Report No. HASL-143, USAEC Health and Safety Laboratory, 376 Hudson Street, New York, 1964.
38. D. C. Sutton and K. R. Dwyer, "Cerium-144 and Cesium-137 Measurements in the 1963 U.S. Wheat Crop and Milling Products," *Science*, **145**, 486, 1964.
39. C. W. Sill and C. P. Willis, "Determination of Radioisotopes of Cerium, Barium, Lanthanum, and Neptunium after Separation of Barium Sulfate," *Anal. Chem.*, **38**, 97, 1966.
40. K. L. Swinth, "Photon Intensities and Their Importance in Counting Transuranic Materials," BNWL Report No. BNWL-1648, 1972, Battelle Pacific Northwest Laboratories, Richland, WA 99352.
41. N. A. Hallden and J. H. Harley, "An Improved Alpha-Counting Technique," *Anal. Chem.*, **32**, 1861, 1960.
42. S. C. Curran and H. W. Wilson, "Proportional Counters and Pulse Ion Chambers," p. 303, Vol. 1 of Ref. 12.
43. L. Leventhal, LFE Environmental, private communication.
44. D. L. Horrocks, "Alpha Particle Energy Resolution in a Liquid Scintillator," *Rev. Sci. Instrum.*, **35**, 334, 1964.
45. W. J. McDowell, "Liquid Scintillation Counting Techniques for the Higher

Actinides," in *Organic Scintillators and Liquid Scintillators*, D. L. Horrocks and L. T. Peng, Eds., Academic, New York, 1971.
46. W. J. McDowell, "High-Resolution Liquid Scintillation Method for the Analytical Determination of Alpha-Emitters in Environmental Samples," *IEEE Trans. Nucl. Sci.*, **NS-22**, 649, 1975.
47. J. H. Thorngate, "A Simplified Pulse-Shape Discrimination Circuit for Use in Liquid-Scintillation Spectroscopy of Alpha Particles," *Health Phys.*, **34**, 103, 1978.
48. J. H. Neiler and P. R. Bell, "The Scintillation Method," p. 245, Vol. 1 of Ref. 12.
49. W. J. Price, *Nuclear Radiation Detection*, 2nd ed., McGraw-Hill, New York, 1964; see especially Chapter 8.
50. G. Bertolini and A. Coche, Ed., *Semiconductor Detectors*, Wiley, New York, 1968.
51. F. S. Goulding and R. H. Pehl, "Semiconductor Radiation Detectors," *Nuclear Spectroscopy and Reactions*, in J. Cerny, Ed., Part A, Academic, New York, 1974; see also the chapter on "Semiconductor Detector Spectrometer Electronics", by F. S. Goulding and D. A. Landis.
52. D. E. Watt and D. Ramsden, *High Sensitivity Counting Techniques*, Macmillan, New York, 1964.
53. H. F. Lucas, "Improved Low-Level Alpha Scintillation Counter for Radon," *Rev. Sci. Instrum.*, **28**, 680, 1957.
54. R. L. Fleischer, P. B. Price, and R. M. Walker, *Nuclear Tracks in Solids*, University of California Press, Berkeley, 1975.
55. American Society for Testing and Methods, "Standard Recommended Practice for Alpha Spectrometry of Wastes," ASTM D-3084-75, 1975.
56. U.S. Environmental Protection Agency, "National Interim Primary Drinking Water Regulations," EPA-570/9-76-003, 1976.
57. U.S. Public Health Service, "Public Health Service Drinking Water Standards," PHS No. 956, 1962.
58. E. J. Baratta and F. E. Knowles, "Statistical Limits of Accuracy and Precision of Gross Alpha and Beta Radioactivity Measurements in Water," *Radiol. Health Data Rep.*, **12**, 195, 1971.
59. C. W. Sill and R. L. Williams, "Preparation of Actinides for α Spectrometry Without Electrodeposition," *Anal. Chem.*, **53**, 412, 1981.
60. F. B. Johns, Ed., U.S. EPA National Environmental Research Center (now Environmental Monitoring and Support Laboratory), Las Vegas, "Laboratory Handbook of Radiochemical Analytical Methods," EPA-680/4-75-001, 1975 (avail. from NTIS and GPO).
61. U.S. Public Health Service, "Radioassay Procedures for Environmental Samples," PHS No. 999-RH-27, 1967. Bureau of Radiological Health, Rockville, MD 20852.
62. M. G. White and P. B. Dunaway, Eds., "Nevada Applied Ecology Group Procedures Handbook for Environmental Transuranics," ERDA Report NVO-166 (2 vols.), 1976. Nevada Operations Office, Energy Research and Development Administration, 1976 (NTIS).
63. C. W. Sill, "Separation and Radiochemical Determination of Uranium and the Transuranium Elements Using Barium Sulfate," *Health Phys.*, **17**, 89, 1969.

64. C. W. Sill, K. W. Puphal, and F. D. Hindman, "Simultaneous Determination of Alpha-Emitting Nuclides of Radium Through Californium in Soil," *Anal. Chem.*, **46**, 1725, 1974.
65. C. W. Sill and R. L. Williams, "Rapid Identification and Determination of Alpha Emitters in Environmental Samples," in *Rapid Methods for Measuring Radioactivity in the Environment*, International Atomic Energy Agency Symposium Proceedings, STI/PUB/289, 1971, p. 201.
66. C. W. Sill, "Radiochemical Determination of Protactinium-231 in Environmental and Biological Materials," *Anal. Chem.*, **50**, 1559, 1978.
67. J. A. Miller, G. E. Biggerstaff, R. W. Levin, and V. G. Katzel, "Determination of Neptunium-237 in Human Urine," Report No. KY-429, Union Carbide Corp. Paducah Plant, Paducah, KY, 1964.
68. U.S. National Bureau of Standards, "NBS Standard Reference Materials Catalog," NBS Special Publication 260, 1979.

8 MEASUREMENT OF NEUTRONS

There are a large number and variety of applications where a significant amount of radiation is due to the presence of neutrons. Among these are:

1. Charged-particle accelerators, particularly proton and ion accelerators, and high-energy accelerators.
2. Nuclear reactors, (both fission and fusion).
3. Sealed neutron sources.

These sources vary widely in the intensities, energy spectra, and temporal durations of the neutron fluences involved. From the highest-energy proton accelerators, primary beam interactions can produce neutrons with energies up to the full energy of the primary protons themselves, while in fission reactor situations there can be a heavy weighting toward thermal neutrons. Given this wide energy range, it is remarkable that one can give any unified treatment to the general problems of measurement of neutron fluences. However, a unification is indeed possible, for a large class of applications, primarily because of the characteristics of neutron interactions in shielding materials.

8.1. INTRODUCTION

Neutrons traversing through material can interact in one of five general ways:

1. Elastic scattering from nuclei.
2. Inelastic scattering through nuclear-level excitation.
3. Quasi-two-body interactions such as (n,α) or (n,np) interactions.
4. Nuclear excitation or catastrophic breakup.
5. The high-energy phenomena of many-particle production.

The relative importance of each of the five general types depends on both neutron energy and target material. Thus, elastic scattering from heavy nuclei is a very inefficient energy loss mechanism, as is easily seen from kinematic considerations. However, elastic scattering from hydrogen nuclei (protons) is one of the most useful processes in degrading neutron energies in shielding (hence, the common use of concrete). Also, at energies much higher than those of typical nuclear levels (a few tens of MeV) a major energy loss mechanism is nuclear breakup or (above a few hundred MeV) multiparticle interactions, such as pion production. These facts have long been well known, and have influenced shielding design around major sources of neutrons and other particles. The unified treatment mentioned above is possible because all of the energy-degradation processes mentioned quickly reduce the neutrons energy to less than a few tens of MeV in appropriately designed shielding.

Neutron interactions, then, generally tend to degrade energies continually, down to the region where the neutrons can be absorbed and eliminated by nuclear reactions. The ultimate energy degradation is *neutron thermalization;* thermal neutrons are those with an energy spectrum in thermal equilibrium with the ambient material. Thermal neutrons can scatter without energy loss in, for example, air or concrete for relatively indefinite periods; while thermalization of a several-MeV neutron can take from a few to a few hundred microseconds. Thermal-neutron lifetimes themselves can be as long as a few tens of milliseconds, until they are ultimately eliminated by nuclear absorption or reactions such as the (n,α) reactions mentioned.

Historically, it has been common to divide the neutron energy spectrum into the following approximate categories:

1. Thermal neutrons: energy from 0 to 0.5 eV.
2. Intermediate neutrons: energy from 0.5 eV to 200 keV.
3. Fast neutrons: energy from 200 keV to 20 MeV.
4. Relativistic neutrons: energy $>$ 20 MeV.

This division is convenient both because of the different types of processes which are typical of these energy regions, and because instrumentation is often sensitive in only one of the categories. This chapter discusses instrumentation partially according to these divisions, noting that in many applications one has *a priori* knowledge that the neutron energy extends up to only a particular energy region.

This chapter gives a fairly detailed discussion of the techniques involved in neutron dosimetry, but if further information is required the reader is referred to the several excellent papers and reference proceedings [1–7].

8.2. RADIATION PROTECTION CONSIDERATIONS

In the general case of *occupational* exposure to external radiation, the relevant Maximum Permissible Exposure guideline (discussed in more detail in Chapter 2 of this book) is a basic limit to dose equivalent of 5 rem/yr (50 mSv/yr), with no more than 3 rem (30 mSv) to be accumulated in any 13-week quarter.

The complication in measuring neutron rem (absorbed dose equivalent) is the variation of quality factor (Q) with energy. Figure 8.1 shows the International Commission on Radiological Protection (ICRP) recommended Q for neutrons as a

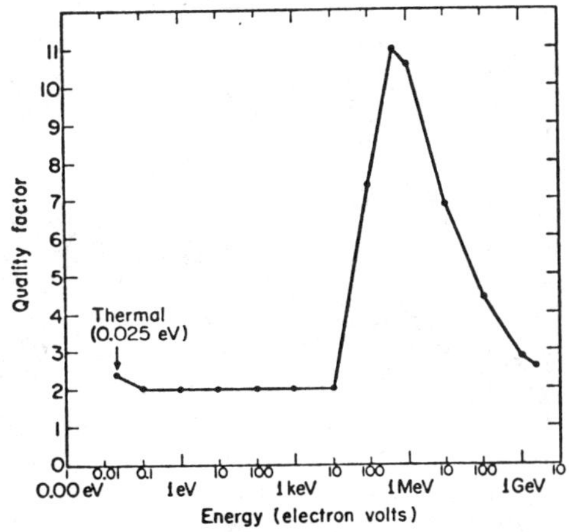

Figure 8.1. Quality factor versus energy for monoenergetic neutrons, as recommended in ICRU Report No. 20 (reproduced from The International Commission on Radiological Units, Ref. 1, with permission of the ICRU © 1971).

function of energy. Figure 8.2 gives the flux density of primary neutrons, in equilibrium with their secondaries, which is recommended by ICRP as being equivalent to 2.5 mrem/hr (25 μSv/hr) [8,9]. The inverse curve (Fig. 8.3) is the energy response required of a neutron detector such that different energies are weighted properly for a direct determination of dose equivalent (rem or sievert).

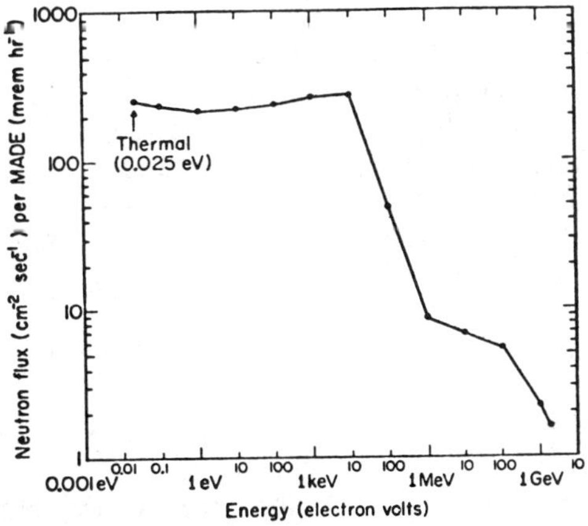

Figure 8.2. Monoenergetic neutron flux versus energy which yields MADE of 1 mrem/hr (reproduced from The International Commission on Radiological Units, Ref. 1, with permission of the ICRU © 1971).

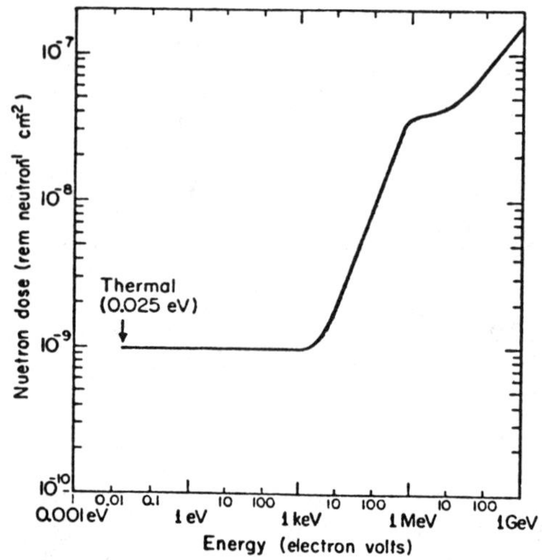

Figure 8.3. Neutron dose (rem) versus energy for 1 neutron/cm^2 for monoenergetic neutrons incident normally on a 30 cm-thick tissue-equivalent phantom (reproduced from The International Commission on Radiological Units, Ref. 1, with permission of the ICRU © 1971).

8.3. MEASUREMENT CONSIDERATIONS

There are three main tasks in neutron monitoring:

1. Measurement of the intensity spectrum as a function of energy (neutrons/cm^2 · sec · MeV).
2. Measurement of the total absorbed-dose equivalent (rem or rem/sec) (sievert or sievert/sec).
3. Determination of the level of other possible particle species, and the response of neutron instrumentation to them.

There does not exist a single instrument or instrument type that can perform completely all of these functions (or indeed, any one of them separately). However, there does exist instrumentation more or less adequate to the needs of the health physicist responsible for radiation protection around any of the sources just mentioned. The wide range of energies and fluences previously mentioned must be considered, along with several other concepts such as tissue-equivalence and particle equilibrium, which have been treated in Chapter 2 of this book.

8.3.1. Thermal-Neutron Measurements (Energies Below 0.5 eV)

Truly thermal neutrons, which have an energy spectrum in thermal equilibrium with the ambient environment, are centered at an energy of about 0.025 eV. However, in actual practice, very few situations exist in which the neutron fluence is solely thermal: one possible case is that in which neutrons are almost totally thermalized for some application such as in a thermal column. Many detection systems for higher-energy neutrons depend on the process of (partial or complete) thermalization, with subsequent detection of the thermal neutrons using one of the techniques to be discussed in this chapter.

There is usually both a truly thermal (Maxwellian) component and a component with a continuous energy spectrum above thermal, typically with a distribution inversely proportional to energy (epi-Maxwellian). Also, most of the reactions used to determine thermal neutron flux densities have cross sections that are almost exactly proportional to 1/velocity ($1/v$) in the range from thermal up to 0.5 eV, the cadmium cutoff energy. (Because a cadmium absorber prevents neutrons below about 0.5 eV from reaching a detector, it allows a convenient separation at about that energy.) Because flux density (neutrons/$cm^2 \cdot$ sec) is proportional to velocity (if they are going twice as fast, they cross any area in half the time), the reaction rate per atom for a pure $1/v$ detector is independent of velocity and proportional only to the total neutron density. Table 8.1 [5] shows most of the common thermal-neutron reactions; the coefficient g(20°C) in the right hand column indicates the departure from $1/v$ dependence for each reaction. For a more detailed

Table 8.1. Useful Thermal Neutron Detectors (reproduced from the International Commission on Radiation Units, Ref. 5, with permission of the ICRU © 1969)

Nuclide or Element	Reaction	Half-Life of Nuclide Produced	$\sigma_0 \times 10^{24}$ for Production (cm^2)	g(20°C)
^{3}He	$(n,p)^{3}He$	12.3 yr	5,327	1.0
^{6}Li	$(n,t)^{4}He$[b]	stable	945	1.0
^{10}B	$(n,\alpha)^{7}Li$	stable	3,837	1.0
^{23}Na(IS)[a]	$(n,\gamma)^{24}Na$	15.0 hr	0.534	1.0
^{45}Sc(IS)	$(n,\gamma)^{46}Sc$	85 days	22.3	
^{51}V	$(n,\gamma)^{52}V$	3.8 min	4.9	1.0
^{55}Mn(IS)	$(n,\gamma)^{56}Mn$	2.58 hr	13.3	1.0
^{59}Co(IS)	$(n,\gamma)^{60}Co$	5.24 yr	36.6	1.0
^{63}Cu	$(n,\gamma)^{64}Cu$	12.8 hr	4.5	1.0
^{115}In	$(n,\gamma)^{116}In^{m}$	54 min	157	1.019
^{157}Gd	$(n,\gamma)^{158}Gd$	stable	242,000	0.854
^{197}Au(IS)	$(n,\gamma)^{198}Au$	2.70 days	98.8	1.005
^{235}U	fission	many	577	0.976

[a](IS) denotes natural monoisotopic element.
[b]t denotes tritium.

discussion of thermal-neutron spectra, departures from Maxwellian distributions, and neutron temperature the reader is referred to ICRU Report No. 13 [5].

The most commonly used thermal neutron detector is the *Boron Fluoride (BF_3)-filled gas proportional counter,* which is commercially available in a wide variety of sizes and sensitivities. The reaction employed is $^{10}B(n,\alpha)^7Li$, in which the alpha particles are detected. The Q (kinetic energy released) is 2.8 MeV, but most of the capture goes through an intermediate excited state of ^7Li, which then decays to the ground state by emission of a 470-keV gamma (γ-ray). In most counters the BF_3 gas is enriched with ^{10}B to about 96%. (Natural abundance is 20% ^{10}B and 80% ^{11}B). The cross section of the foregoing reaction is about 3800 barns with a $1/v$ dependence in the thermal-energy range. Thus BF_3 counters are relatively insensitive to particles with energies significantly above that of thermal neutrons (0.025 eV). The sensitivity depends on counter size, gas pressure, and enrichment. The response to γ-fields is very low. In fields of equal γ- and neutron fluxes, the γ-response is typically two orders of magnitude smaller than the neutron response [1]. An exception is when the BF_3 detector is inside a moderator, and pulsed fields are being detected. Photons can pile up and count when these detectors are exposed to very high instantaneous exposure rates.

The *helium-3 gas proportional counter* is another commonly used detector, similar to the BF_3 counter. It relies upon the reaction ^3He (n,p) ^3H, with a Q value of 765 keV. Both the proton and tritium are available for ionization, but one of the problems is that near the walls efficiency can suffer, due to the rather low stopping power of the ^3He gas. This deficiency can be overcome either by increasing the volume or gas pressure, or (more commonly) by adding an impurity (usually argon or krypton) to improve the stopping power. The disadvantage of the impurity is that γ-ray sensitivity is increased, while pulse rise-time is decreased. Because of the small Q value, γ-ray sensitivity is greater than in the BF_3-filled counter, but neutron sensitivity is also greater [1], until the recoil particles become too long (at neutron energies above about 10 MeV).

For neutron detection in high proton fields, *boron-10-lined proportional counters* are used. The reaction is the same as for BF_3 counters, except that the ^{10}B is coated on the counter walls, in thicknesses that range from about 0.2 to 1.0 mg/cm^2. Sensitivity is proportional to coating thickness, but self-absorption of the α-ray prevents significant increases in sensitivity at greater thicknesses. These counters are extremely insensitive to γ-photons, and are capable of neutron-sensitive operation in photon fields as high as 10^4 or even 10^5 R/hr. However, their sensitivity to neutrons is perhaps an order of magnitude smaller than that of the BF_3-gas-filled counter [1]. A typical fill gas is CO_2 or Argon. The operating voltage is usually lower than that of the BF_3 or ^3He tubes, and typically larger output pulses with shorter rise times are produced.

Still another gas proportional neutron detector is the *fission counter*. A material such as ^{233}U, ^{235}U, or ^{239}Pu is coated on the walls. The kinetic energy of the fission fragments (approximately 200 MeV) produces large pulses and fast response time. The sensitivity of these counters is inherently lower by perhaps an order of magnitude than the ^{10}B-lined counters, and two orders of magnitude below that of the BF_3-filled tubes [1]. However, operation at temperatures as high as 700°C and in extremely high γ-ray fluxes (up to 10^6 R/hr) makes them attractive for applications such as inside reactor shielding. It should be noted here that special licensing is often required to obtain the fissionable material to use within these counters.

Ionization chambers are also used for thermal-neutron detection. Both ^{10}B-lined and BF_3-filled ion chambers are used. Many of the same characteristics apply which have already been discussed in the case of the gas proportional counters. The ion chambers are generally less sensitive than the gas proportional counters [1]. A more complete treatment of ionization chambers can be found later in this chapter.

Scintillation counters are also used for detection of thermal neutrons [1]. Among the most sensitive scintillators are lithium iodide crystal with europium activation [^{6}LiI(Eu)]; ^{6}Li-loaded, cerium-activated silicate glass; and ^{6}Li-loaded zinc sulfide crystals. In each case one counts the α-particles and tritium from the reaction $^6Li(n,\alpha)^3H$, with a Q of 4.8 MeV. If the detector is large there can be a significant perturbation of the field being measured. However, the γ-ray rejection is excellent, with γ-ray sensitivities typically one to two orders of magnitude below those for neutrons. The crystal has negligible anisotropy of sensitivity, except for self-shielding and the flux perturbations already mentioned. A description of some of the properties of Ce-activated glass has been given by Ginther [10], and a discussion of ^{6}Li-loaded glasses is given by Bormann et al. [11].

Another technique is the activation of a *thallium-activated sodium iodide [NaI(Tl)] crystal,* in which one counts the decay after activation of ^{24}Na (15-hr half-life, beta with E_{max} = 2.12 MeV). An advantage of this technique is the good detection efficiency for the decay beta rays. The sensitivity is very low, perhaps 10^4 below that of the others previously mentioned. Although common at one time, this technique is not often used today. Small crystals (say, 1 mm thick) must be used to minimize the problem of perturbation of the incident flux. For a good description of the NaI method, the reader is referred to Grimeland et al. [12].

Zinc sulfide (ZnS) glass with ^{10}B loading is another material used as a scintillator; however, it suffers from two problems [1] when compared to its counterpart, ^{6}Li-loaded ZnS: first, poorer γ-ray discrimination, because of the lower Q value of the (n,α) reaction; and second, typical sensitivities several orders of magnitude lower than for the ^{6}Li-loaded crystals. Thus, only in situations where neutron fluxes are high and photon levels are known to be small can the ^{10}B-loaded ZnS glass be useful.

Foil Activation is another common technique for detection of thermal neutrons. Table 8.1 showed the properties of some of the useful reactions. Perhaps the most widely used foil material when high sensitivity is required is indium, (In), due to its very large thermal-neutron cross section. Use of cobalt is common for the integration of neutron fluence over very long periods, because of the long (5.2-yr) half-life of ^{60}Co. Gold is used when high precision is required, because its capture cross section is well behaved for energies up to about 3 eV.

Thermoluminescence is still another detection technique for thermal neutrons [13]. The most commonly used thermoluminescence detector (TLD) for neutrons is ^{6}LiF, but this detector also has substantial γ-ray sensitivity [13]. Therefore, it is common practice when γ-ray backgrounds are present to use two or three TLDs with varying fractional concentrations of ^{6}Li and ^{7}Li, in order to determine the γ-ray and neutron contributions separately. TLDs are integrating devices which must be read out in a separate operation after exposure. The general discussion of TLDs is found in Chapter 9, Dosimetry. The sensitivity of ^{6}LiF is typically above 1000 neutrons/cm^2, and the sensitivity to fast neutrons is small. Another common TLD device is CaF_2, but it has a very small intrinsic neutron sensitivity and large

γ-ray response; addition of ^{10}B or ^{6}Li is used to enhance the neutron response, but with the exception of pure neutron fields it is not recommended.

To summarize the discussion on thermal neutrons, there are many methods used for their detection; however, complications can arise when higher-energy neutrons are present, because most thermal-neutron detectors have at least some sensitivity to higher energies. When one suspects the presence of higher energy fluences, a common technique to use is the *cadmium ratio* method. This involves shielding the detector with a layer of cadmium (an excellent absorber of neutrons below about 0.5 eV) and comparing detector response to the bare situation. To determine the flux in the region near 1.5 eV, one can surround an indium foil with Cd. The Cd will absorb neutrons below 0.5 eV, and the indium foil (with its huge capture resonance of 26,000 barns near 1.5 eV) can be β-counted by any low-background Geiger-Müller (G-M) counter to detect the β-ray from ^{116}In decay after the reaction $^{115}In(n,\gamma)^{116}In$. Accuracies within better than a factor of 2 in flux measurements can be obtained. A further description of this procedure can be found in the article by Hankins [14].

8.3.2. Intermediate-Energy Neutrons (Energies from 0.5 eV to 200 keV)

Intermediate-energy neutrons are most often detected by moderating them down to the thermal region and then employing one of the thermal methods just described.

Direct detection of recoils from neutron scattering does not work well in this energy region, because it is difficult to observe the recoils against even a small γ-ray background. Instead, the most common technique is to detect the energetic products of an exothermic reaction such as (n,α). However, the cross sections for many of these reactions decrease rapidly at energies above a few eV, and thus neutrons in the keV range are usually detected by rather specialized means.

One such technique is the so-called *black detector* which absorbs all neutrons within a certain energy region. The most commonly used black detectors are ^{10}B-doped liquid scintillators or detectors consisting of a boron plate which uses the associated photon from the (n,γ) reaction for detection [15]. The problem is that these instruments absorb all neutrons below a certain maximum energy. To derive absorbed dose equivalent from flux density involves a detailed knowledge of the energy spectrum (Fig. 8.3).

The moderator approach has been used with reasonable success. Moderators can be appropriately designed to yield either rem (sievert) equivalent response or flux density response weighted to yield values of absorbed dose. An instrument with a response as shown in Fig. 8.4 has been reported by Leake [16]; however, the figure indicates a substantial fall-off in sensitivity at high and low energies, necessitating a correction for neutrons outside of the sensitive range. This device uses a $^{6}LiI(Eu)$ crystal scintillator surrounded by a polyethylene moderator. Another instrument [17] uses a boron-loaded ZnS scintillator enclosed in a polyethylene sphere. Figure 8.5 shows the energy response of this instrument for various moderator thicknesses. Note that the response can be made quite flat. Both of these instruments have excellent γ-ray discrimination [1].

Instruments which cover the intermediate range in a rem (sievert) equivalent manner but are also sensitive over much broader energies are discussed in the section on dose equivalent instruments.

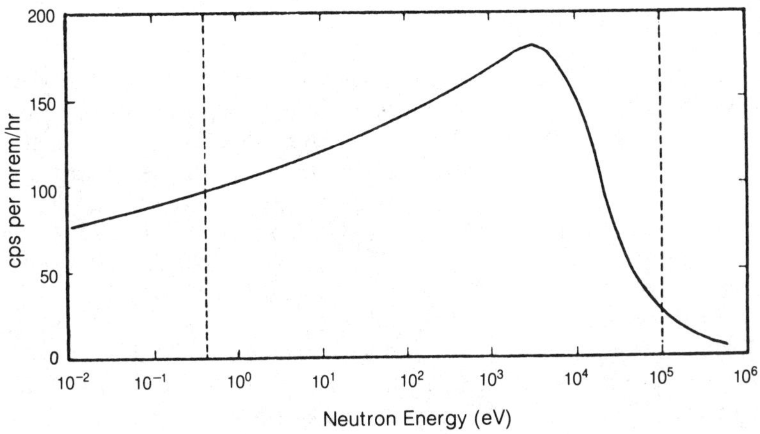

Figure 8.4. The response of a dosimeter designed to measure dose equivalent of neutrons in the intermediate-energy range (reproduced from Leake, Ref. 16).

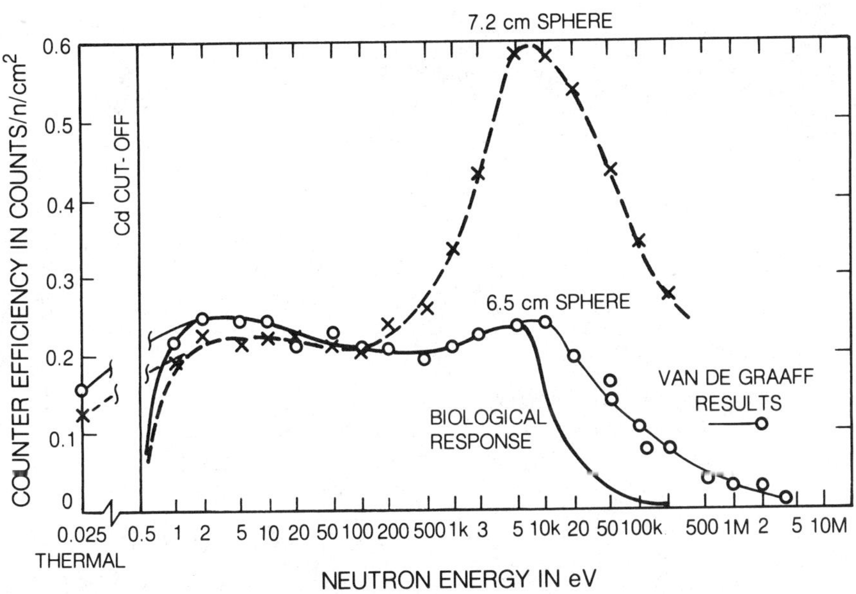

Figure 8.5. Variation of efficiency and dose dependence of the optimum arrangement with neutron energy (reproduced from Basson, Ref. 17, with permission of North-Holland © 1963).

8.3.3. Fast Neutrons (Energies from About 200 keV to 20 MeV)

In situations where fast neutrons require measurement, there is usually an accompanying flux of lower-energy neutrons, and there may be γ-ray backgrounds as well. For this reason, a frequently used technique for isolating the fast neutron component is foil activation, because one can choose processes with a variety of thresholds and sensitivities.

Activation of foils is also used for relativistic neutrons. Figure 8.6 shows the cross section as a function of energy for some of the commonly used reactions [18]. Table 8.2 gives some properties of the reactions, and the half-lives of the activation products. The last two columns in Table 8.2 show the sensitivity for 1 rad (10 mGy), and the background, for the detector used by Cross [19]. The sensitivity with In, Fe, and Zn is perhaps an order of magnitude better than for the fission detectors, because a decay γ-ray peak stands out above background to a greater extent than when counting a wide spectrum.

Gas proportional counters which rely upon the measurement of proton recoils have found widespread use for fast neutron detection because of their excellent γ-ray discrimination. Several such counters have been described in the literature. As an example, consider the counter built at Oak Ridge National Laboratory and described by Hurst and Wagner [20]. The counter, which reads out in rads, is completely lined with polyethylene and filled with ethylene to a pressure of 750 torr. Figure 8.7 shows its response to 1.3 MeV ^{60}Co γ-ray (at rates from 1 to 100 R/hr) and to the neutron spectrum from a polonium-beryllium source, at rates

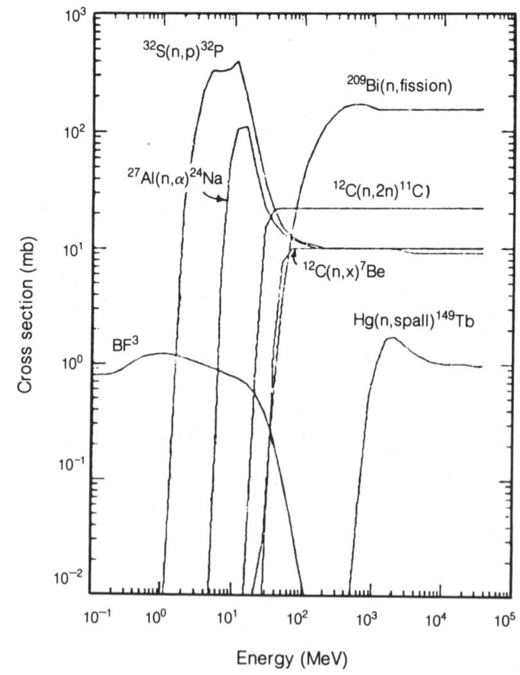

Figure 8.6. Response functions of high-energy neutron detectors (reproduced from Routti, Ref. 18).

Table 8.2. Properties of Some Threshold Reactions (reproduced from Cross, Ref. 19)

Reaction	E_{th} [a] (MeV)	$t_{1/2}$ [b]	$\bar{\sigma}$ [c] (b)	E_γ [d] (cpm/g)	Sensitivity [e] for 1 rad (cpm)	Background [f]
$In^{115}(n,n')In^{115m}$	1.2	4.5 hr	0.3	0.335	121	14
$Ni^{58}(n,p)Co^{58}$	2.6	72 days	0.6	0.810	0.63	8
$S^{32}(n,p)p^{32}$	2.9	14.2 days	0.35	β	17	1
$Zn^{64}(n,p)Cu^{64}$	3.3	12.8 hr	0.25	0.511	27	12
$Fe^{56}(n,p)Mn^{56}$	6.7	2.6 hr	0.08	0.845	56	8
$Mg^{24}(n,p)Na^{24}$	7.2	15 hr	0.15	2.76	13	1
$Al^{27}(n,\alpha)Na^{24}$	7.6	15 hr	0.1	2.76	9	1
$In^{115}(n,2n)In^{114}$	10.6	50 days	1.4	0.192	0.46	17
$Cu^{65}(n,2n)Cu^{64}$	11.4	12.8 hr	1.0	0.511	23	12
$Ni^{58}(n,np)Co^{57}$	12	267 days	0.7	0.122	0.58	20

[a] E_{th} = Effective threshold energy (in MeV) at which cross section reaches 25% of $\bar{\sigma}$.
[b] $t_{1/2}$ = Half-life of product.
[c] $\bar{\sigma}$ = Average cross section (in barns) over the most sensitive energy region.
[d] E_γ = Energy of decay γ (MeV).
[e] Sensitivity; counts per min/gram of detector for 1 rad of neutrons.
[f] Bkgd.; counter background (cpm).

of 0.001 rad/hr (0.01 mGy/hr). If one is willing to set a bias level to eliminate signals from neutrons below about 0.5 MeV, the γ-ray discrimination is excellent.

Another counter, filled with methane argon, has been described by Dennis and Loosemore [21]. Its energy response is shown in Fig. 8.8. Although this counter has sensitivity which is roughly dose equivalent (rem–sievert) in the fast neutron region, it suffers from inaccuracies due to a loss of counts below the bias level. Theoretical techniques have been developed to estimate corrections for these losses [22], but some independent measurements will probably be required when the neutron spectrum is unknown.

Other instruments of this type have been described by Andersson [23] and Murthy, Muthukrishnan, and Sunta [24]. Murthy's instrument can measure neutron fluxes up to 15,000 neutrons/cm^2 · sec. These instruments can tolerate [1] thermal/fast-neutron dose rate ratios of about 100/1 and γ-ray/fast-neutron dose rates of about 200/1.

Another fast-neutron detection technique is scintillation counting with plastic or liquid *scintillators*. Neutrons can be counted in a scintillator through their interaction and production of secondaries. The enhancement in accomplished by setting a rather high threshold. Figure 8.9 shows the neutron detection efficiency of a typical liquid scintillator for various threshold values.

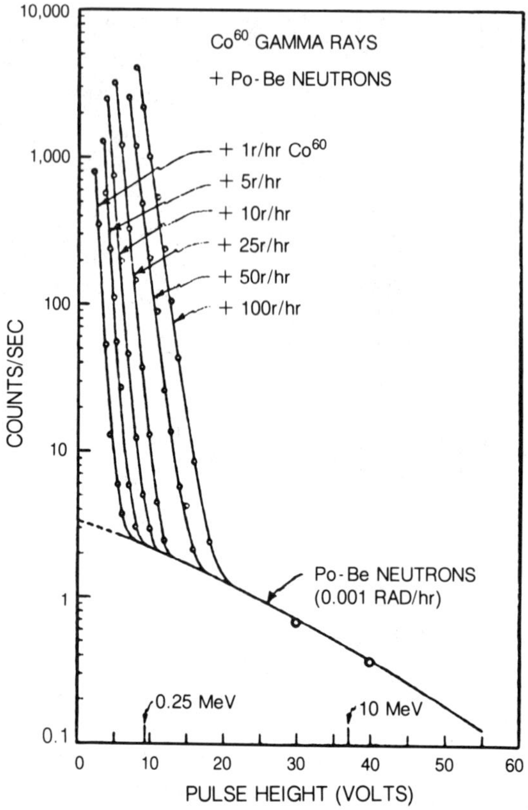

Figure 8.7. Integral pulse-height curves for mixtures of gamma rays and fast neutrons (reproduced from Hurst and Wagner, Ref. 20).

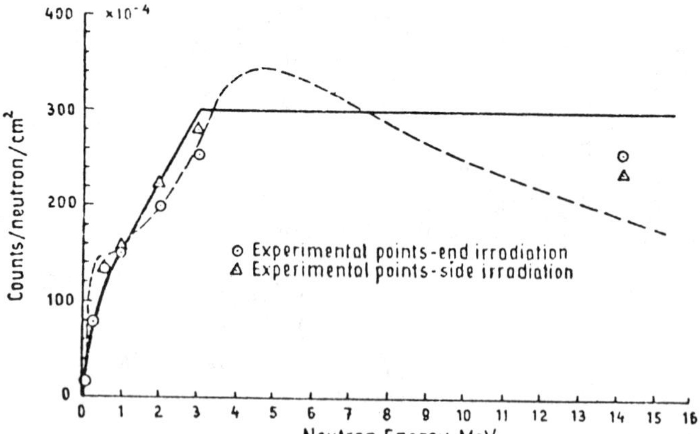

Figure 8.8. Average response of three type FN 2/3 counters as a function of neutron energy. The solid line is the response required according to the ICRP recommendations (ICRU Publ. 4) for a counting rate of 0.9c/sec/MPL. The dashed line is the theoretical response of the counter for end-on irradiation at a bias level equivalent to 0.09 MeV (Reproduced from Dennis and Loosemore, Ref. 21).

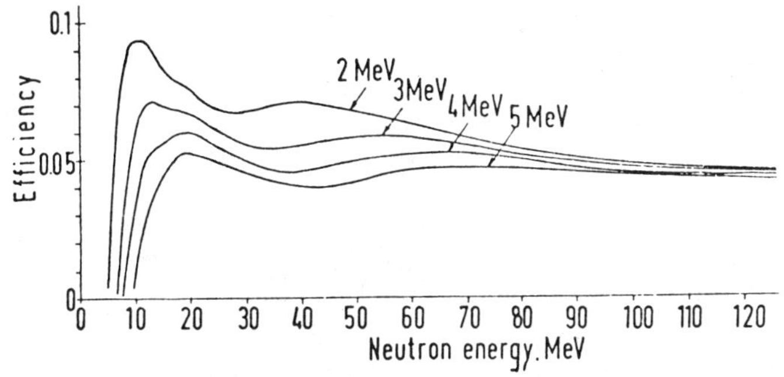

Figure 8.9. Neutron detection efficiency of a liquid scintillator, 4.7 cm × 4.7 cm, with thresholds of 2, 3, 4, and 5 MeV electron energy (reproduced from Tesch, Ref. 25).

A problem with scintillation counting is that if any relativistic charged particles are present, they will count with nearly 100% efficiency. Therefore, to isolate the neutron component one must either determine that charged particles are absent or measure them separately.

Unfortunately, the photon response of some plastics, especially the organic plastics which contain significant fractions by weight of hydrogen, is also very large, making them unsuitable where γ-ray backgrounds are important.

Efforts have been made to overcome the problem of γ-ray sensitivity. The most useful method for distinguishing neutrons from γ-rays is *pulse shape discrimination*. As an example, Verbinski et al. [26] have successfully differentiated the two species in the 1 to 10 MeV range using pulse shape discrimination on the response from NE-213 liquid scintillator manufactured by Nuclear Enterprises, Ltd., Edinburgh, Scotland. However, this technique suffers because it is delicate, and requires sophisticated electronic equipment.

Another fast-neutron detector is the so-called *Long Counter* originally developed in 1947 by Hanson and McKibben [27], and improved in 1966 by DePangher and Nichols [28]. The Long Counter is a BF_3 proportional counter surrounded by two concentric cylindrical hydrogenous moderators, with a thermal-neutron shield in between. It has high sensitivity, good γ-ray discrimination, and wide energy response (from about 25 keV to 14 MeV). The most significant drawback of the counter is its nonisotropy of response; it is insensitive except for neutrons incident along the axis.

As in the case of intermediate-energy neutrons, it is possible to moderate fast neutrons to thermal energies and apply any of the large number of thermal detection schemes previously described. One such detector system makes use of moderated indium foils and is characterized as follows: indium foils essentially detect only thermal neutrons; however, they may be used to measure fast neutrons by placing them in a cadmium-covered paraffin moderator of 6 in. diameter × 6 in. height. The effect of the Cd is to absorb the thermal neutrons present in the incident neutron flux. The paraffin thermalizes the high-energy neutrons, which then activate the indium foils. Stephens and Smith [29] have shown that such a detector has a fairly uniform energy sensitivity from a few keV to about 20 MeV.

In neutron measurements by induced activity, a stable element is exposed to the neutron flux for a measured period of time. Following the irradiation, the element is removed from the neutron field, and the activity induced in it is determined.

Neutron activation detectors have a number of advantages:

1. Many simultaneous measurements are possible.
2. The locations where foils may be placed are quite unrestricted.
3. The results are unambiguous, because the induced activity is caused solely by neutrons.
4. Foils are not disabled by high neutron flux rates, as conventional electronic counters may be.
5. They respond correctly in pulsed radiation fields.

Thermal-neutron flux densities are frequently measured by activation techniques, for which it is common practice to use a small piece of Au or In as the irradiated target.

When In is used, foils of mass 300 to 400 mg are convenient. The nuclear reactions for foil activation and decay are

$$^{115}_{49}\text{In} + ^{1}_{0}\text{n} \rightarrow ^{116}_{49}\text{In}$$

and

$$^{116}_{49}\text{In} \rightarrow \beta^- + ^{116}_{50}\text{Sn}.$$

The β^- decay of the ^{116}In follows a 54.0-min half-life, which may be detected in a gas-flow proportional counter with standard radiation laboratory counting equipment. The gas used in the proportional counter can be ordinary gas supplied by the local utility company.

The following formula can be readily applied to either In or Au foils to determine the counting rate, cpm/gm, at saturation and to or end of irradiation. This combined with the foil and counter calibration data will easily provide neutron flux density.

$$A_{\text{sat},\, t_0}\,(\text{cpm/g}) = \frac{N_\text{d} - Bt_\text{c}}{T\left(e^{-t_\text{d}/T} - e^{-(t_\text{d}+t_\text{c})/T}\right)\left(1 - e^{-t_\text{i}/T}\right)\cdot W}$$

where N_d = observed counts,
B = background counting rate (cpm),
t_c = counting time (min),
t_i = irradiation time (min),
t_d = decay time (min),
T = teamlife, $t_{1/2}/\ln 2$,
W = foil weight (gm),
$t_{1/2}$ = half-life.

8.3.4. Relativistic Neutrons (Energies Above 20 MeV)

These neutrons are almost never found at levels which can pose a potential hazard to personnel except at high-energy accelerators. *Carbon-11 activation* (20.4-min half-life) is one of the most widely used techniques for measuring the high-energy neutron components of a radiation field. A carbon-rich detector is exposed to the radiation field and the decay positrons (0.97 MeV E_{max}) are then counted by standard techniques. The threshold for ^{11}C excitation is about 20 MeV, and Fig. 8.10 shows cross sections for various particle types. Note that the (γ, n) cross section is very small. Although the neutron cross-sectional behavior overemphasizes the region just above threshold, and high-energy charged particles can also produce ^{11}C, this technique is often useful for a rough check on the possible contribution from high-energy neutron fluences. The neutron flux density above 20 MeV can be defined and estimated from the amount of ^{11}C activity, assuming an average production cross section of 22 mbarns [30]. The problem is to interpret this measured quantity in terms of dose equivalent, which depends on the assumed spectrum of neutrons. For 20-MeV neutrons, the ICRP recommended conversion factor [8] states that 4 neutrons/cm^2 · sec is equivalent to 1 mrem/hr (10 μSv/hr) (the number is 3.9 at 100 MeV). Various high-energy laboratories have recommended different conversion factors, which are summarized in the following table [30]:

Laboratory	Flux mrem/hr (10 μSv/hr)
Brookhaven	4
Daresbury	4
CERN	10
Rutherford	3.6
Berkeley	12.1/E(MeV)$^{1/4}$

This problem of exactly how to weight a measurement is inherent in all cases where activation is used.

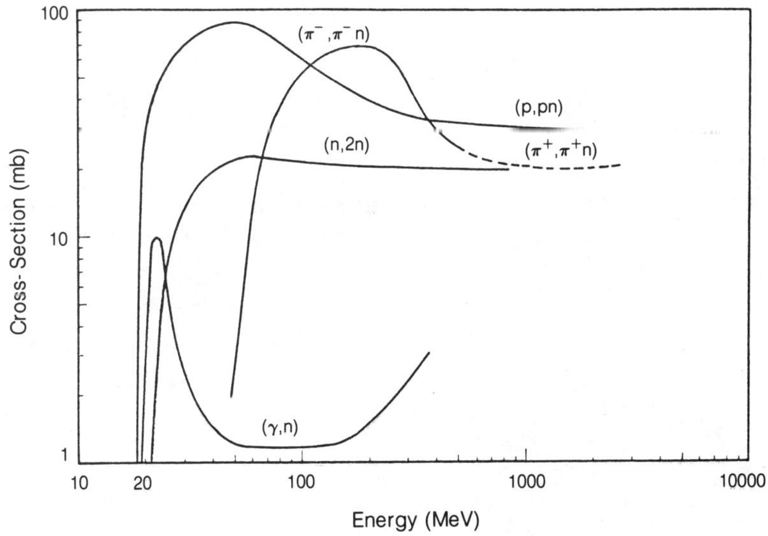

Figure 8.10. Cross sections for the production of carbon-11 from carbon-12 (reproduced from Sullivan, Ref. 30).

Another high-energy detection technique depends upon fission. Ionization chambers lined with bismuth are sometimes used for neutron measurements at very high energies [1]. Figure 8.11 shows fission cross sections for uranium, bismuth, gold, and tantalum as a function of energy. Another technique is to coat thin evaporated layers of these materials on mica, which is subsequently scanned for tracks from fission events. To determine the energy spectrum, ratios of fission probabilities are taken among the various materials (e.g., U/Au, Bi/Au, U/Ta), and the different cross sections of Fig. 8.11 are then used to determine the neutron energy spectrum.

Both foil activation and scintillation counting are commonly used for relativistic neutron detection; these have already been discussed under *fast neutrons.*

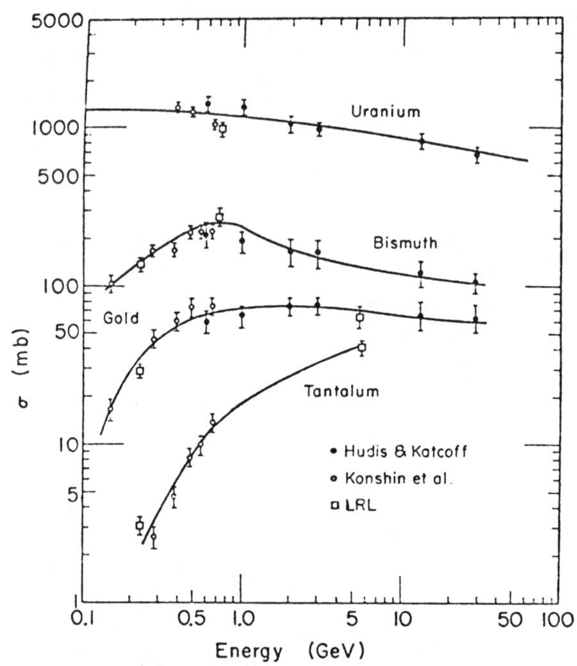

Figure 8.11. Fission cross sections versus energy of incident nucleon (neutron or proton) for targets of U, Bi, Au, and Ta (reproduced from Wollenberg and Smith, Ref. 31).

8.3.5. Dose Equivalent Instruments

In this section, instruments will be discussed which are intended to provide a direct measurement of dose equivalent (rem–sievert) over a wide range of neutron energies. Referring to Fig. 8.3, the neutron flux (particles/cm^2·sec) is shown as a function of energy which provides proper weighting for dose equivalent.

There are several different approaches to the direct measurement of rem (sievert). The most widely used method is the use of a suitably designed moderator surrounding a thermal-neutron detector. There are two geometries in common use: cylindrical and spherical. The cylindrical design was first described by Andersson and Braun [32], who calibrated it using both thermal neutrons and 25-keV to 15-MeV neutrons. A similar instrument was evaluated by Block and Petrock [33], who showed that in the intermediate-energy range the response was rem-proportional to within ± 10% and directionally isotropic to ± 15%. Another design, by Leake and Smith [34], was an improvement over the original Andersson–Braun instrument. The generic instrument design is as follows: a BF_3 proportional counter is surrounded by two cylindrical layers of polyethylene separated by a 200-mg/cm^2 layer of boron plastic. Similar disks cap the ends of the cylinder. The Andersson–Braun approach was to vary the two polyethylene thicknesses, and to drill a number of holes through the boron plastic layer, until the best empirical fit to the desired rem-response was achieved. Using a BF_3 tube, 2.5 cm in diameter and filled to a pressure of 600 torr, the counter was found [33] to be insensitive to ^{137}Cs photons (660 keV) up to above 500 R/hour, with neutron sensitivities down to 2 mrem/hr (20 μSv/hr) full scale. Also, the energy response was within ± 10% of rem-equivalent all the way from thermal to 15 MeV, except for a slight region of oversensitivity near 10 keV. This is shown in Fig. 8.12.

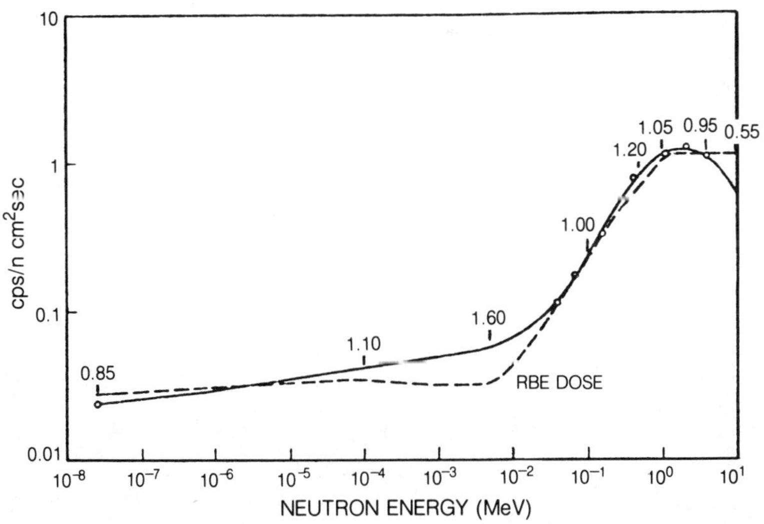

Figure 8.12. Neutron detection efficiency as a function of neutron energy. The figures give the counting rate relative to that from the Andersson–Braun counter with an ideal rem dose response of 19 cps/2.5 mrem hr (reproduced from Andersson and Braun, Ref. 32).

The cylindrical instrument is now a widely used survey instrument for mixed-field neutron dosimetry measurements. It is rugged, portable (about 11 kg), and sensitive. It has one important disadvantage: the electronics associated with the BF_3 tube cannot follow extremely brief, high-intensity radiation bursts such as those around low-duty-factor linear accelerators. Another Andersson–Braun type detector has been reported by Leake [35]. This instrument, specifically designed for use around pulsed-radiation sources, replaces the BF_3 counter with a BF_3-filled ion chamber run in current mode. The output is measured using a dc amplifier with a 30-sec integrating time constant. However, its greatest sensitivity is about an order of magnitude less sensitive than that of a BF_3 counter instrument.

The use of a number of different detectors in spherical geometry was first described by Hankins [36]. The basic instrument is commonly called the *Bonner sphere [37]: a ^{6}LiI(Eu) crystal,* 4 mm in diameter and 4 mm thick surrounded by polyethylene moderating spheres. "About 80% of incident thermal neutrons are absorbed in 1 mm of ^{6}LiI, so that thermal neutron detection is essentially a surface effect, whereas the efficiency for detection of gamma radiation and fast neutrons is roughly proportional to the volume of the crystal." [37] In Hankins' description [36] the detector is surrounded by a succession of polyethylene spheres; 2, 3, 5, 8, and 12 in. in diameter. Figure 8.13 shows the relative count rate versus energy for the various sphere sizes, and Fig. 8.14 shows the estimated response of a single 10-in. sphere. While the γ-ray discrimination is slightly less impressive than for the cylindrical instrument, it is still very good: 5 R/hour of photons produces negligible response. That is quite sufficient for most purposes, since one seldom needs to measure 1 mrem/hr (10 μSv/hr) of neutrons in the presence of photon exposure rates as large as 5 R/hr.

The directional sensitivity of the spheres is, of course, essentially isotropic except on the side where the electronic circuitry is located. From Fig. 8.14 it can be seen that if one can tolerate errors of about factors of 2 or 3 in the fast-energy range, a single 10-in. sphere will suffice; the multisphere technique must be used when greater precision is required. These spheres are now commercially available.

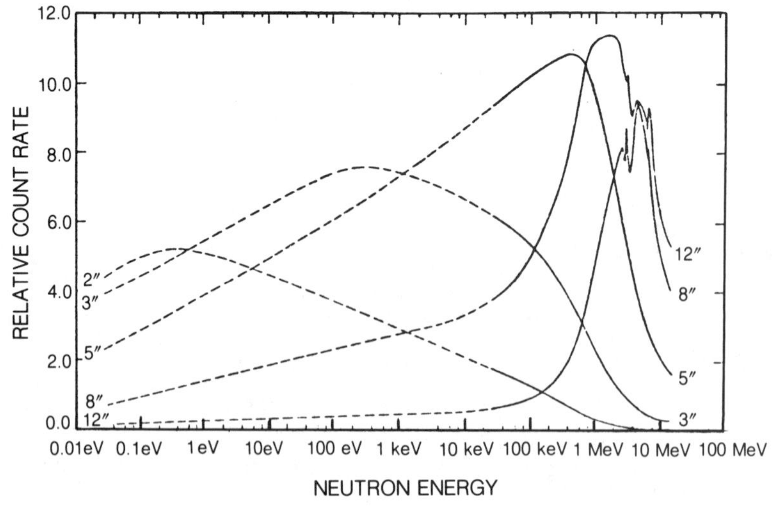

Figure 8.13. Relative counting rate plotted against energy for several moderating spheres (reproduced from Hankins, Ref. 36).

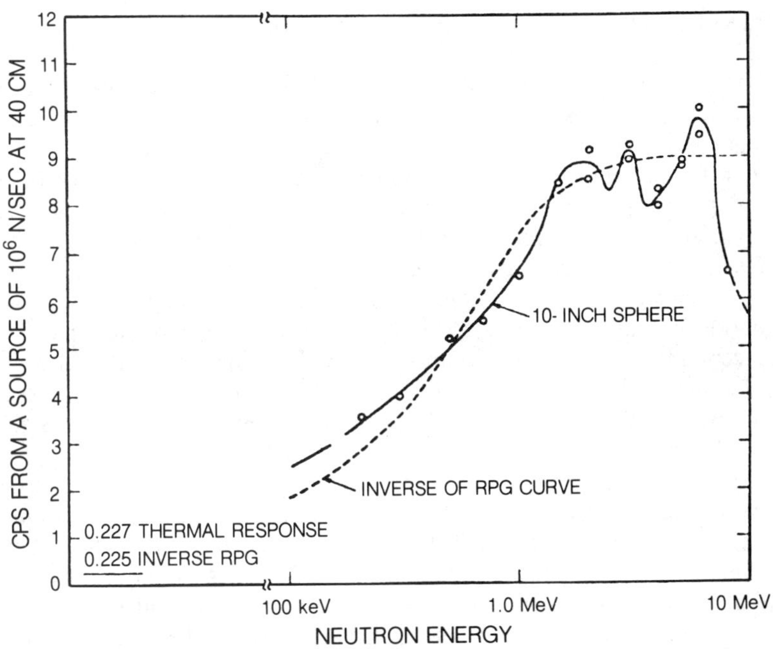

Figure 8.14. Estimated response of a 10-in. sphere plotted against energy. The points shown are experimental points (reproduced from Hankins, Ref. 36).

Because of the use of LiI with resolution times of about 1 μsec, these devices suffer from an inability to measure properly in low-duty-factor situations, a failing they share with the cylindrical (BF_3 counter) instrument.

Leake [35] has reported a spherical-geometry detector with a moderator similar to the Andersson–Braun design, except that the boron plastic is replaced by Cd between the polyethylene layers. The detector is LiI(Eu), similar to that of Hankins [36] and Bramblett, Ewing, and Bonner [37]. Commercial manufacture has been undertaken. Nachtigall and Rohloff [38] have succeeded in extending the multisphere technique up to an energy of 50 MeV, using four spheres of diameter 2, 5, 11, and 18 in.

Recent developments in the field of fusion energy research have made new demands on neutron rem (sievert) meters. Very high neutron dose rates result from reactions taking place in the Tandem Mirror and Magnetic Fusion machines. An instrument has recently been developed and tested [39]. Thorngate, Hunt, and Rueppel have reported on a rem meter capable of recording and displaying dose measurements from pulses 50 msec in length with dose per pulse as high as 100 mrem (1.0 mSv). The repetition rate is ~1.25/hr. Therefore they are able to accurately measure dose equivalent rates of the order 7200 rem/hr (72 Sv/hr).

This particular rem meter also has application for measuring low-level continuous fields due to its high sensitivity and low inherent background.

Dose-equivalent instruments which employ several other principles of operation have been reported in the literature. One of these is the use of gas proportional counters sensitive to proton recoils. These instruments have been discussed under Section 8.3.3. They follow the desired response reasonably well up to about 15 MeV, but are insensitive to neutrons below about 100 keV.

8.4. SOURCES FOR CALIBRATION OF NEUTRON INSTRUMENTS

Some of the general problems with calibration of α-, β- and γ-ray detectors have been discussed in various parts of this book. Here attention is focused on a few of the aspects specific to the use of *neutron sources* for calibration.

The central problem in any calibration is to understand the source of neutrons, its energy spectrum, directionality, presence or absence of equilibrium secondaries, and possible secondary sources from the scatter of the primary neutron flux. One common calibration technique in the fast-neutron region is the use of radioactive neutron sources from (α,n) reactions. Considerable research has been done on the determination of energy spectra from these sources, and a review paper by Hanson [40] contains an excellent summary of information up to 1960. More recent work is summarized in ICRU Report No. 13 [5]. The most common sources are Po-Be, Ra-Be, Pu-Be, and Am-Be. Figure 8.15, from ICRU 13, shows the energy spectrum of neutrons from ^{241}Am-Be as measured by several groups. The purpose of the figure is to show how much difference exists among various investigators for an Am-Be source; this situation is also true for each of the other sources mentioned. The spectral uncertainties imply that absolute accuracies better than about $\pm 10\%$ are probably not achievable with this method; but this is probably sufficient for most radiation protection work. One remaining problem is that the spectral studies have not generally extended below about 1 MeV, despite the fact that 20 to 30% of the neutrons from a typical source are in that low-energy region [5]. The Am-Be

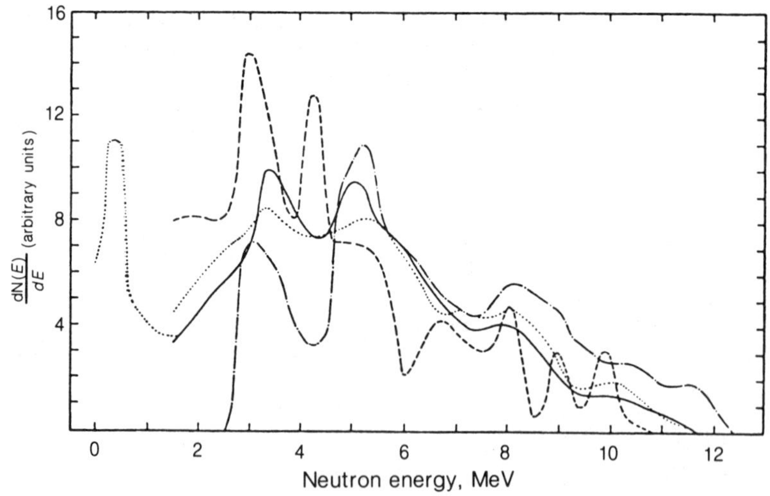

Figure 8.15. Energy distribution of ^{241}Am–Be neutrons according to different authors. The curves have been normalized to give the same total number of neutrons between energies 2.5 and 10.5 MeV: ——— Geiger and Hargrove (1964), proton-recoil counter telescope and time of flight; --------- Greiss (1968), nuclear emulsions; .... Thompson and Taylor (1965), stilbene crystal with pulse-shape discrimination;— · — · —Salgir and Walker (1967), proton-recoil spectrometer (silicon detector). The dotted low-energy peak has been drawn according to the suggestion of Geiger and Hargrove (reproduced from The International Commission on Radiological Units, Ref. 5, with permission of the ICRU © 1969).

source has one important advantage over the others in that there is insignificant contamination from decay of daughters subsequent to americium decay.

Fission is another common source of fast neutrons. Most data on energy spectra seem to fit the Maxwellian distribution function [5], where E is the fission neutron energy and E_n is an experimental parameter:

$$\frac{dN(E)}{dE} \sim E^{1/2} \exp \frac{-E}{E_n} .$$

Experimental numbers for E_n are in the range 1.2 to 1.6 MeV for several fission nuclei, such as ^{233}U, ^{235}U, ^{239}Pu, ^{240}Pu, ^{241}Pu, and ^{252}Cf. A compilation of studies on fission spectra was done by Barnard et al. [41], and summarized in ICRU No. 13 [5].

Use of *accelerators* for production of nearly monoenergetic neutrons has been common; a wide variety of reactions is available leading to a variety of neutron energies. Figure 8.16 and Table 8.3 show the properties of some of the reactions used.

Sources of *thermal neutrons* are usually made by moderating the fast neutron from the sources just mentioned [1]. Alternatively, a thermal column from a reactor can be used. The fluence rate of any of these thermal sources can be calibrated

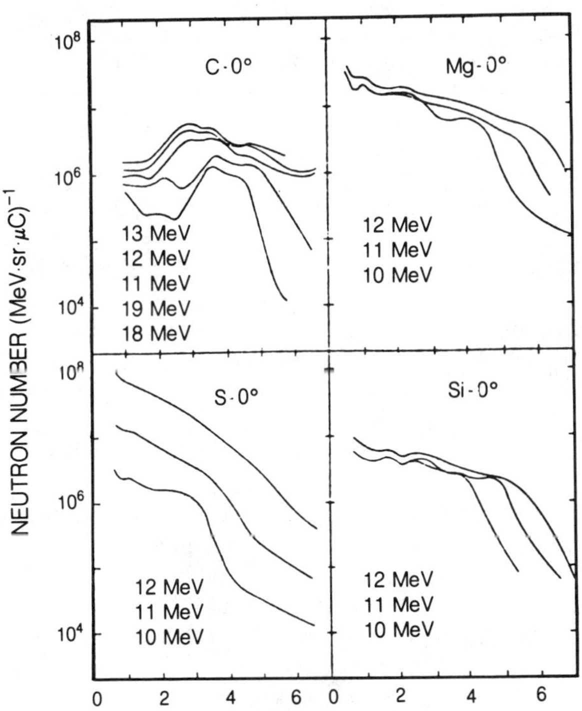

Figure 8.16. Neutron energy spectra resulting from the bombardment of carbon, magnesium, sulfur, and silicon with protons of several energies (reproduced from The International Commission on Radiological Units, Ref. 5, with permission of the ICRU © 1969).

Table 8.3. Monoenergetic Neutron Producing Reactions (all energies in MeV) (reproduced from The International Commission on Radiation Units, Ref. 5, with permission of the ICRU © 1969)

Reaction	Q Value	Threshold Energy	Minimum Neutron Energy at 0^3	Maximum Neutron Energy at Bombarding Energy 2 MeV	Maximum Neutron Energy at Bombarding Energy 3 MeV	Maximum Bombarding Energy for Monoenergetic Neutrons
$^2D(d,n)^3He$	3.265	—	1.8^a	5.24	6.26	—
$^3T(p,n)^3He$	−0.764	1.019	0.0639	1.20	2.21	—
$^3T(d,n)^4He$	17.6	—	12.4^a	18.25	19.57	3.7
$^7Li(p,n)^7Be$	−1.647	1.882	0.0294	0.228	1.3	2.38
$^{13}C(d,n)^{14}N$	−0.281	0.328	0.00195	1.6	2.6	3.0
$^{45}Sc(p,n)^{45}Ti$	−2.79	2.908	0.00140	—	0.112	3.67
$^{51}V(p,n)^{51}Cr$	−1.562	1.565	0.00059	0.215^b	—	2.36
$^{65}Cu(p,n)^{65}Zn$	−2.136	2.169	0.00051	0.214^c	—	2.29

^{a}Measured at 2.0 MeV, 150°.
^{b}At bombarding energy = 1.762 MeV.
^{c}At bombarding energy = 2.368 MeV.

by the foils-activation technique; gold foils are commonly used when high accuracies are required.

The directionality of the neutron source is sometimes difficult to determine, especially for thermal neutrons. The important problem here is the possible presence of scattered particles in addition to the primary flux. For accelerator-generated neutrons the beam direction is known well, but beam halo is difficult to understand or reduce. For isotopic sources of small size, one can use a small detector to study departures from inverse-square fall-off of the flux. To minimize scattering problems, the physical layout of the calibration setup should be carefully planned; for example, apparatus placed near a wall may be subject to large scattering corrections. A discussion of this, and of methods for measuring the possible scattering correction, can be found in ICRU Report No. 20 [1].

In all calibrations, cross-comparison with standard instruments is extremely important. This is necessary to complement the understanding of the properties of the neutron source, because of possible unknown changes in the source.

As of this time, there does not exist a standard mechanism of calibrations for neutron instruments through any of the national standards bodies. Further, there are large regions in the neutron energy spectrum for which no well-understood neutron sources exist [5]: this is true in the region around 1 eV, and also near 2 keV.

The available data are poor for many of the important cross sections in the fast-neutron region, such as

$$^{31}P(n,p)^{31}S \text{ and } ^{115}In(n,n')^{115}In \,.$$

ICRU Report No. 13 contains a summary of basic research (primarily cross-section measurements) needed to make possible accurate neutron instrument calibrations.

The proceedings of the International Atomic Energy Association (IAEA) Symposium on National and International Standardization of Radiation Dosimetry [7] provide the latest information on calibration sources, techniques, and facilities. The proceedings published in two volumes should prove helpful to those persons charged with this important responsibility.

8.5. SUMMARY AND RECOMMENDATIONS

Neutron monitoring comprises two distinct tasks: (a) making detailed area surveys to determine the properties of the radiation fields; and (b) conducting routine surveillance monitoring. The latter should be supplemented with detailed surveys, undertaken occasionally and whenever there have been important changes in operating conditions.

The purpose of this chapter has been to enumerate the various classes of instruments, the conditions under which each class is appropriate, and the changes in operation which require a detailed reevaluation. The reader is directed to the original references for relevant details about specific techniques.

It is important to remember that whenever a new installation begins operation, it is essential to perform a detailed survey. Thus, separate instruments dominantly sensitive to thermal, intermediate, fast and possibly to relativistic neutrons are needed, as well as an instrument sensitive to photons. The relative weighting of neutrons in the various energy ranges can be assessed using a combination of techniques discussed in the text. There should by all means be an overlap in the means of measuring the various neutron energies to provide some redundancy of information. Also, these spectral studies are an important aid in studies for reducing levels for radiation protection purposes.

For routine surveillance an instrument whose response is proportional to dose equivalent (rem–sievert) is recommended. Among those commercially available the cylindrical Andersson–Braun type is widely used. Also available are the Bonner-sphere instruments.

An instrument sensitive to thermal neutrons is frequently required: BF_3 and ^3He gas proportional counters, boron-lined chambers, LiI(Eu) scintillation detectors, as well as several other types are discussed. Activation of foils is another good technique for monitoring thermal neutrons because it is both simple and sensitive. A variety of foils applicable to a wide range of neutron energies is available commercially.

The indium/cadmium technique, together with any of the thermal neutron detectors, can be used to provide a rapid assessment of fluxes of thermal- and intermediate-energy neutrons.

In the text no attempt has been made to comment on the commercially available instruments individually: the advantages and disadvantages are discussed in the classes of detectors (e.g., BF_3 tubes). However, sufficient information is not available at this time to make valid comparisons concerning the regions of applicability of many instruments. Also, a brief discussion is included of some instruments and techniques which are not now commercially available, but are worthy of commercial development.

Finally, a comment should be made about the state of the art with regard to new instruments. Solid-state electronics development is not only changing but is accelerating in rate of change. New systems featuring improved stability, reduced maintenance, greater reliability, and reductions in size and weight appear with greater frequency with each passing year.

REFERENCES

1. International Commission on Radiation Units and Measurements, "Radiation Protection Instrumentation and Its Application," ICRU Report No. 20, 1971.
2. International Atomic Energy Agency, *Selected Topics in Radiation Dosimetry*, Proceedings of an IAEA Symposium in Vienna, June 7, 1960; Vienna, 1961.
3. International Atomic Energy Agency, *Neutron Dosimetry*, Proceedings of an IAEA Symposium in Harwell, England, December 10–14, 1962, 2 vols.; Vienna, 1963.
4. International Atomic Energy Agency, *Neutron Monitoring*, Proceedings of an IAEA Symposium in Vienna, August 29, 1966; Vienna, 1967.
5. International Commission on Radiation Units and Measurement, "Neutron Fluence, Neutron Spectra, and Kerma," ICRU Report No. 13, 1969.
6. U.S. Atomic Energy Commission, *Second International Conference on Accelerator Dosimetry and Experience*, CONF-691101, Washington, 1969.
7. International Atomic Energy Agency, *National and International Standardization of Radiation Dosimetry*, Proceedings of an IAEA Symposium in Atlanta, Georgia December 5–9, 1977, 2 vols.; Vienna, 1978.
8. International Commission on Radiological Protection, "Recommendations of the International Commission on Radiological Protection, Report of Committee IV," ICRU Report No. 4, Pergamon Press, Oxford, 1964.
9. Internatio al Commission on Radiological Protection, Recommendations of the International Commission on Radiological Protection, ICRP Publication 26, Pergamon Press, Oxford, 1977.
10. R. Ginther, "Cerium-Activated Scintillating Glasses," p. 521 of Ref. 2.
11. M. Bormann H. Neuert, H. Balzer, J. Geisler, and G. Magdanz, "Flux Measurements of Medium Energy Neutrons by Means of Stilbene Scintillators and ^{6}Li Glasses," p. 209 of Ref. 4.
12. B. Grimeland, E. Kjellsby, K. Bryhn-Ingebrigtsen, and C. Van Dijk "Measurement of Neutron Densities by Activation of NaI(Tl) Crystals," p. 377, Vol. I of Ref. 3.
13. J. R. Cameron, N. Suntharalingam, and G. N. Kenney, *Thermoluminescent Dosimetry*, University of Wisconsin Press, Madison, WI, 1968.
14. D. E. Hankins, "Monitoring Intermediate Energy Neutrons," *Health Phys.*, **9**, 31, 1963.

15. J. K. Basson, "Counting Intermediate Energy Neutrons," p. 241, Vol. II of Ref. 3.
16. J. W. Leake, "Calibration of the Neutron Survey Probe 95/0054," U.K. Atomic Energy Research Establishment, Report AERE-M-1564, Harwell, England, 1965. The figure is from p. 643 of Ref. 6.
17. J. K. Basson, "A Detector for Intermediate Energy Neutrons," *Nucl. Instrum. Meth.*, **22**, 339, 1963. Another description of the Basson detector is by J. W. Smith on p. 261, Vol. I of Ref. 3.
18. J. T. Routti, "Mathematical Considerations of Determining Neutron Spectra from Activation Measurements," p. 494 of Ref. 6.
19. W. G. Cross, "The Use of Mg, Ti, Fe, Ni, and Zn in Fast-Neutron Activation Dosimetry," p. 389, Vol. I of Ref. 3.
20. G. Hurst and E. Wagner, "Special Counting Techniques in Mixed-Radiation Dosimetry," p. 409 of Ref. 2.
21. J. Dennis and W. R. Loosemore, "A Fast-Neutron Counter for Dosimetry," p. 443 of Ref. 2.
22. R. H. Ritchie, "Calculations of Energy Loss Under the Bias in Fast Neutron Dosimetry," *Health Phys.*, **2**, 73, 1958.
23. I. O. Andersson, "A Proton-Recoil Proportional Counter for Fast Neutron Dosimetry", p. 453 of Ref. 2.
24. K. B. S. Murthy, G. Muthukrishnan, and C.M. Sunta, "A Counter for Fast Neutron Dosimetry," *Indian J. Pure Appl. Phys.*, **5**, 26, 1967. A brief description can be found on p. 462 of *Radiation Protection Monitoring*, Proceedings of Joint WHO-IAEA Regional Seminar in Bombay, 9–13 Dec. 1968, International Atomic Energy Agency, Vienna, 1969.
25. K. Tesch, "Radiation Problems Around the 7 GeV DESY Electron Accelerator," p. 595 of Ref. 6.
26. V. V. Verbinski, W. R. Burrus, R. M. Freestone, and R. Textor, "Proton Recoil Neutron Spectrometry with Organic Scintillators," p. 151 of Ref. 4.
27. A. Hanson and J. McKibben, "A Neutron Detector Having Uniform Sensitivity from 10 keV to 3 MeV," *Phys. Rev.*, **72**, 673, 1947.
28. J. DePangher and L. Nichols, "A Precision Long Counter for Measuring Neutron Flux Density," Battelle Pacific Northwest Laboratories, Report BNWL-260 Richland, Washington, 99352, 1966.
29. L. D. Stephens and A. R. Smith, "Fast Neutron Surveys Using Indium-Foil Activation," Lawrence Berkeley Laboratory UCRL-8418, 1958.
30. A. H. Sullivan, "The Present Status of Instrumentation for Accelerator Health Physics," p. 625 of Ref. 6.
31. H. A. Wollenberg and A. K. Smith, "Energy and Flux Determination of High Energy Nucleons," p. 586 of Ref. 6.
32. I. O. Andersson and J. Braun, "A Neutron Rem Counter," *Nukleonik*, **6**(5), 237, 1964. Also reported on p. 93, Vol. II of Ref. 3, from which Fig. 8.12 is taken.
33. S. Block and K. Petrock, "Evaluation of a Neutron Rem Dosimeter," Lawrence Livermore National Laboratory UCRL-12167, 1964.
34. J. W. Leake and J. W. Smith, "Calibration of a Neutron Rem Counter," Report AERE-R4524, Harwell, U. K. Her Majesty's Stationery Office, London, 1964.

35. J. W. Leake, "Portable Instruments for the Measurement of Neutron Dose-Equivalent Rate in Steady-State and Pulsed Neutron Fields," p. 313 of Ref. 4.
36. D. Hankins, AEC Idaho Operations Office Report IDO-16655, 1961; Los Alamos Scientific Laboratory Reports LA-2717, 1962; LAMS-2977, 1964, and LA-3700, 1968.
37. R. C. Bramblett, R. I. Ewing, and T. W. Bonner, "A New Type of Neutron Spectrometer," *Nucl. Instrum. Meth.*, **9**, 1, 1960.
38. D. Nachtigall and F. Rohloff, *Nucl. Instrum. Meth.*, **50**, 137, 1967; in German.
39. J. E. Thorngate, G. F. Hunt, and D. W. Rueppel, "REM Meter for Pulsed Sources of Neutrons," Lawrence Livermore National Laboratory, UCID-18792, 1980.
40. A. O. Hanson, "Radioactive Neutron Sources," in *Fast Neutron Physics*, J. B. Marion and J. L. Fowler, Eds., Wiley-Interscience, New York, 1960, Part 1.
41. E. Barnard, A. T. C. Ferguson, W. R. McMurray and I. J. Van Heerden, "Time-of-Flight Measurements of Neutron Spectra from Fission of ^{235}U, ^{238}U and ^{239}Pu," *Nucl. Phys.*, **71**, 228, 1965.

9 DOSIMETRY

This section discusses instrumentation for measuring integrated external radiation doses to personnel. The archetype of such instrumentation is the film badge. This discussion is broader though, dealing with the general measurement problem of integrating dose instruments designed to be pinned to clothing, carried in pockets, worn on fingers, or otherwise portable without the user's constant attention.

There are several reasons why such instruments are used. Their primary purpose is the routine assessment of external dose in order to maintain occupational doses below the prescribed limits, and as low as reasonably achievable (ALARA). Secondary purposes include compliance with legal requirements, the assessment of accidental exposures, provision of quality assurance data for the overall radiation protection program, and the psychological benefit afforded to those workers who somehow feel "safer" when wearing a dosimeter.

9.1. DESIGN CONSIDERATIONS

Different motivations place different sets of constraints on the dosimetry system. Thus, the whole question of accuracy is related both to legal considerations and the requirements imposed by decision making in postaccident situations. Further, the portability and unobtrusiveness required of the dosimeter impose size and weight restrictions, while economic concerns demand that large numbers of readings be possible at low unit cost.

Obviously, the more accurate the better. However, certain basic limitations prevent the attainment of extremely precise measurements. The most important is the problem of sampling. A dosimeter can only measure the radiation that strikes it, but the problem arises when one relates the dosimeter reading to an equivalent whole-body external dose. Several factors enter here; inhomogeneity of the radiation field due to collimation, partial shielding, or the inverse-square relationship, for example, will contribute to potentially large sampling errors. Also, the angle of incidence of the radiation is important: exposures from one side of the body may be misrepresented by some dosimeter geometries, especially if the dosimeter is at the front while radiation is incident at the back of the body.

All dosimeters have some degree of energy dependence in their response to both photon and corpuscular radiation. Significant errors will result in dose assessments without prior knowledge about the spectral characteristics of the radiation field, or the use of some device (e.g., filters), to "flatten" the dosimeter response as a function of energy. Similarly, special consideration must be given to cases of exposure to mixed radiation fields, where the dosimeter response functions differ greatly for γ-, β-, x-rays, and neutron radiation.

Dosimetry thus presents a nearly impossible problem if accuracies in the few-percent range were to be required. Fortunately, no such need exists, from the viewpoint of either postaccident treatment or legal responsibility.

To illustrate these points, here are quotations from the International Commission on Radiological Protection, Publication 12 [1]:

> A dosimeter carried on the surface of the worker's body is best regarded as a sampling device. It provides a sample of the dose received at the surface of the body during the movement of the worker through his environment. It does not provide a direct measurement of the radiation dose to organs or tissues, except for the dose to the skin in the immediate vicinity of the dosemeter. Neither does it necessarily provide an adequate assessment of the situation in the workplace because it measures the dose only at one point on the body of each worker.
>
> It is only in exceptional cases, usually associated with substantial accidental exposures, that attempts need be made to estimate the actual organ doses, rather than doses at the surface of the body. The greatest uncertainties in the assessment of organ doses from routine monitoring are caused by the fact that a small number of dosemeters, often only one, has to be taken as representative of the exposure of the whole surface of the body. Unless the dosemeter results are representative, there is little point in conducting detailed calculations of the depth dose in a composite, finite medium—the human body.
>
> The need to use many dosemeters and the complexity of the necessary interpretation to assess the actual organ doses prohibit the making of these assessments as part of routine individual monitoring. A simplified system of interpretation has to be adopted...and this will usually have the effect of assessing upper limits to organ doses rather than actual values....
>
> It has become conventional to note and record all radiation doses above the threshold of detection of the personal dosemeter. This detailed attention to low doses may, in some cases, focus unrealistic attention onto exposures which, if they cause any risk at all, cause risks which are trivial in comparison with other risks of life. Other exposures, often of the same order of magnitude, are not recorded, either because they are below the threshold of detection or because the individuals are not monitored. The Commission [ICRP] has now recommended that individual monitoring and associated records are not needed if the doses are most unlikely to exceed 3/10 of the annual Maximum Permissible Doses. It is possible to make use of this recommendation to make individual monitoring simpler and more logical....

> In practice, however, such a precise equivalence cannot be achieved because most individual monitoring for external radiation is carried out over periods substantially shorter than a year. It is therefore recommended that, for the whole body, the investigation level for penetrating radiation should be set somewhat lower—50 mrem for a dosimeter issued for 2 weeks, 100 mrem for 1 month, or 300 mrem for 3 months. In the rare cases when a substantial part of the dose to an individual results from neutrons or from internal contamination, the investigation level should be applied to the total recorded dose in a control period rather than to each of the separate contributions....
>
> The uncertainties acceptable in routine individual monitoring should be somewhat less than the investigation level and can best be expressed in relation to the annual dose. The uncertainty in assessing the upper limits to the annual dose equivalent to the whole body or to the organs of the body... should not exceed 50%. Where these doses are less than 2 rems an uncertainty of 1 rem is acceptable. This uncertainty includes errors due to variations in the dosemeter sensitivity with incident energy and direction of incidence, as well as intrinsic errors in the dosemeter and its calibration.

For the purposes of this discussion here, suffice it to say first that the obvious accuracy limitations on external personnel dosimetry must be recognized by the user. Second, when high accuracies are required other methods must be called upon to complement the dosimeters of concern. Finally, most workers in the field realize that the need for great intrinsic accuracy is not the most significant factor in the choice of dosimeters.

What *is* important is an assurance that no substantial component of the absorbed dose is missed entirely by the dosimetry system being used. Thus one needs to base a dosimetry system on detailed information about the types of radiation likely to be encountered. The discussion here concentrates on the *types* of dosimeters; their relative *sensitivities* to various particle species and energy ranges; and their *advantages and disadvantages*.

9.2. DOSIMETER TYPES

For routine daily monitoring, there are two basic types of dosimeters in common use today: *photographic film* and *thermoluminescent dosimeters* (TLDs). Historically film has been the mainstay of personnel dosimetry systems. TLDs are now achieving increasingly widespread use. A significant subject of debate today in this field seems to concern the relative merits of film and TLD.

Several other types of dosimeters are also in use, usually for special applications. Among these are pocket ionization chambers, electronic alarming dosimeters, activation detectors, and a broad category of devices known as track etch detectors. This discussion mentions the basic properties of each of these various types, their various regions of applicability, and new techniques which show promise but still require extensive investigations and development.

There are two basic types of external radiation fields that are commonly monitored: *beta-gamma* and *neutron* fields. These are discussed separately, because the latter involves techniques and considerations completely different from the former.

9.3. GAMMA-RAY AND BETA-RAY DOSIMETRY

9.3.1. Photographic Film

The basic property of film dosimeters is the ionization-induced darkening of emulsion. The intrinsic limitations of film are due to four problems: (a) some (unavoidable) background darkening (fog) occurs in the absence of any ionization and from background radiation; (b) film must be wrapped to avoid light leaks, and even with thin wrappings the dosimeter is typically insensitive to low-energy β-particles, and very-low-energy x-rays; (c) the high average atomic number of typical emulsions makes the response non-tissue-equivalent to low-energy photons; and (d) emulsion darkening ultimately saturates at high doses.

The advantages of film include the ability to re-scan emulsions after an exposure to determine details of dose depositions; use of filter shadowing and exposure-uniformity information to measure directionality of exposure and source distance; and the potential for a wide variety of film types and grades (although in actual fact the variety of films now commercially available is quite limited). Photon spectral information can often be derived from the use of several filters having different atomic numbers. The film provides a permanent record. Lastly, the image recording property of film often provides useful information on the interpretation of exposure from artifacts such as surface contamination on the badge, radium-dial wristwatches, and other similar objects.

Besides the physical or engineering considerations, two other properties of film are often cited: low unit cost, and long-term storage (for remeasurement and legal record-keeping).

The following section describes some typical films and their approximate ranges of operation. However, this section does not discuss or compare all of the various commercially available films and film services. Because of their large number, that would be an almost insurmountable task. Film badge services have come under the study of the International Atomic Energy Agency and the U.S. Public Health Service, and documents containing overviews of the important considerations have been in the open literature for several years [2–4]. Also, Becker [5] summarizes the basic properties and considerations in film dosimetry.

The three basic operational characteristics to consider are accuracy, dynamic range, and energy response. The *intrinsic* accuracy of film is quite good. In one study [6] under carefully controlled conditions x-ray exposures to Kodak Type 3 film of 10 to 200 mR/week (0.5 to 10.0 R/yr) have been determined to $\pm 20\%$ (standard deviation) for a single film and $\pm 10\%$ (standard deviation) on a quarterly basis. For a single 10-mR exposure standard deviations of $\pm 25\%$ are possible. Of course, such carefully controlled conditions are not typical of actual use. One important quality-control practice is to set aside and develop a few blanks and controlled-exposure films from each new batch of emulsions, to determine if any changes in sensitivity have occurred. The main point to be made here is that lack of intrinsic accuracy is not a significant contributor to error in the use of film dosimeters.

The dynamic range of the more sensitive films is typically from a few mrad (10 μGy) to a few thousand mrad (10 mGy), with films available which respond up to the Mrad region (10 kGy). The typical dosimeter system might use films of both the most and least sensitive types in order to cover as large a dynamic range as possible. Film's linearity of response (emulsion density versus dose) is excellent, until saturation sets in at the high dose end. This is illustrated in Fig. 9.1.

Humidity and *temperature* significantly affect instrument sensitivity. Studies of duPont 555 and Kodak Type 2 films have shown rapid losses of up to 40% in sensitivity at high humidity [8]. The effect is probably on latent image formation, possibly from "the action of the various radiolytic products of water upon the silver halide crystal, affecting either the migration of conduction band electrons, the fraction of silver ions free to migrate to the sensitivity centers, or both." [8] Figure 9.2 shows the effect of 75 to 95% humidity, 23 to 32°C on optical density for a typical film.

The nonuniformity of photon sensitivity of films as a function of energy is considered by some to be their greatest drawback. At low energies, the intrinsic sensitivity of emulsions is much larger than at medium energies, in some cases by big factors. Much effort has gone into the design of various types of filters to compensate for this energy dependence.

A typical example of the use of filters for decreasing the energy dependence is the work of Storm and Shlaer [10]. Figure 9.3 shows the way in which the unfiltered response is modified by using filters with one element (bismuth), two elements (bismuth and erbium), and six elements (bismuth, cadmium, erbium, gold, neodymium, and tantalum). The idea is that the incident x-rays are attenuated by the K-edge absorption of each successive material, decreasing the effective dosimeter sensitivity. Figure 9.4 shows the measured response of four commercial films using this technique. Use of a lead frame back was necessary to eliminate a scattering problem in the plastic encasing the filters.

The conclusion of Storm and Shlaer is that it is possible to eliminate much variation in sensitivity at the low energies in film dosimeters, at least down to energies of

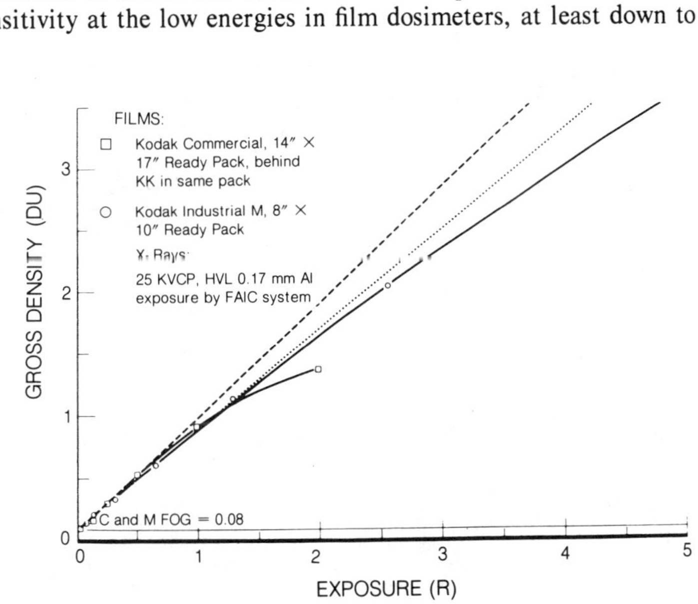

Figure 9.1. Optical density against roentgen exposure plotted on linear grid for Kodak Commercial Film and for Kodak Type M Films in Ready Packs (reproduced from Stewart, Modine, and Rolofson, Ref. 7, with permission of the Bureau of Radiological Health © 1968).

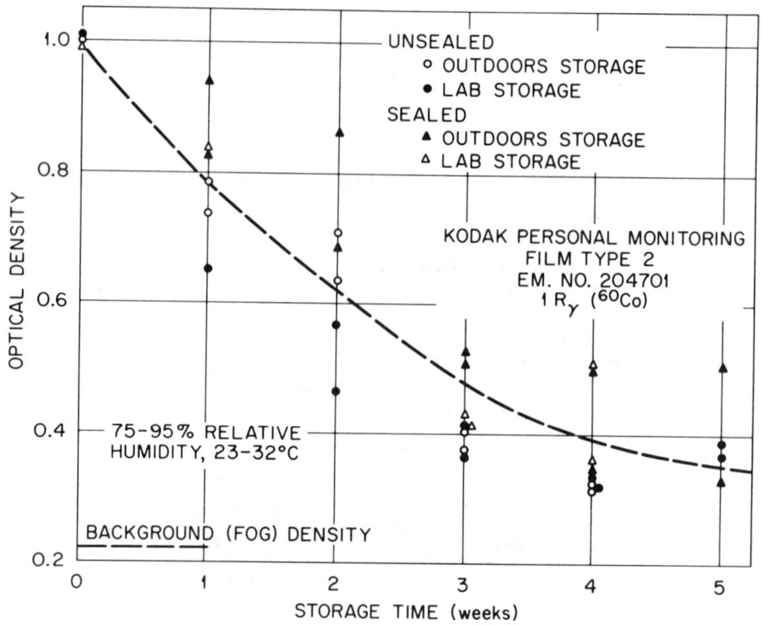

Figure 9.2. Effect of 75 to 95% humidity, 23 to 32°C on optical density of Kodak Type 2 Film, as a function of storage time (reproduced from the Danish Atomic Energy Commission, Ref. 9).

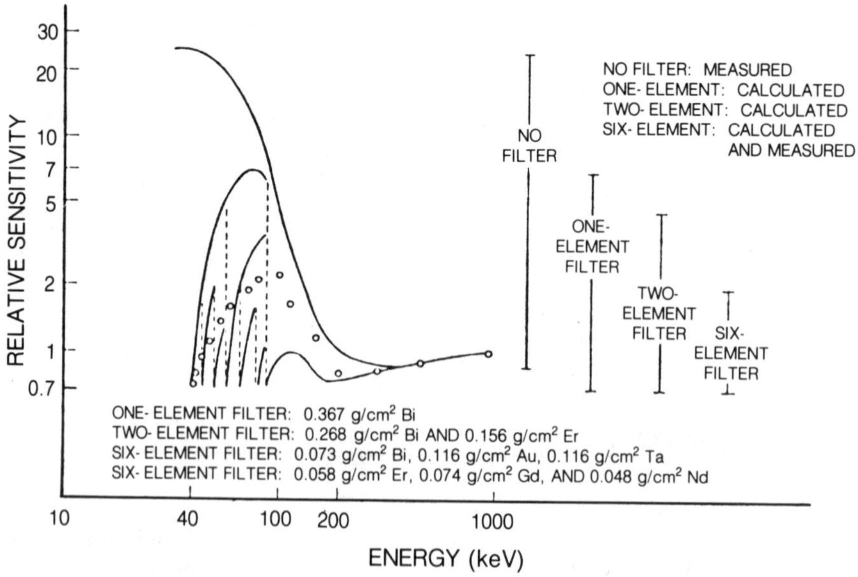

Figure 9.3. The photon response of DuPont 502 film unfiltered and filtered by one-, two- and six-element filters (reproduced from Storm and Shlaer, Ref. 10, with permission of Pergamon Press © 1965).

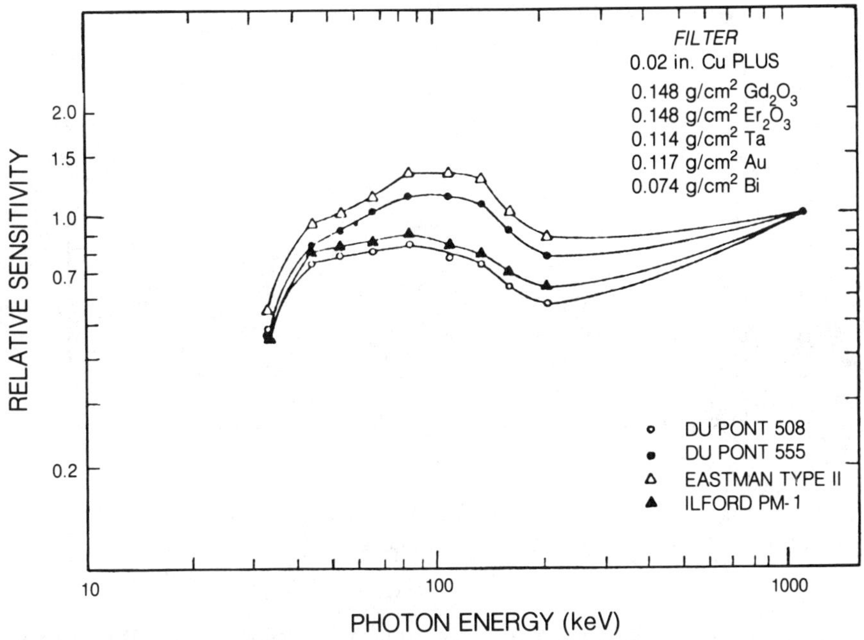

Figure 9.4. Response of DuPont 508 and 555, Eastman Type II, and Ilford PM-1 Films in the lead-frame plastic badge (reproduced from Storm and Shlaer, Ref. 10, with permission of Pergamon Press © 1965).

a few tens of keV. It should be noted that multifilter holders are relatively expensive, which has discouraged their wide use. Also, response becomes poor at energies high enough that charged-particle equilibrium is not established upstream of the dosimeter (this is true for all dosimeters under consideration in this section).

9.3.2. Thermoluminescent Dosimetry

Thermoluminescent dosimetry (TLD) is becoming more and more widely used for routine γ-ray and β-ray dosimetry. The basic physical principle is now reasonably well understood. Ionizing radiation incident on a crystal elevates an electron from the valence band to the conduction band, leaving a hole in the valence band. The electron and the hole move through the crystal until they are trapped in metastable states, usually associated with impurity sites. When the crystal is heated, sufficient energy may be given to the electron (or the hole) so that it wanders around, eventually recombining with a counterpart hole (or electron). The recombination energy is released as visible light (a thermoluminescent photon). There may be some recombination before the carriers are captured in a metastable state thus lowering the sensitivity of the detector.

A good description of basic TLD can be found in the book by Cameron, Suntharalingam, and Kenney [11]. Lin and Cameron have also compiled a bibliography on the subject [12].

The most common thermoluminescent crystal used today is lithium fluoride (LiF). There is extensive documentation of the properties of LiF; the following are some typical data. Figure 9.5 shows that the useful range of LiF extends beyond

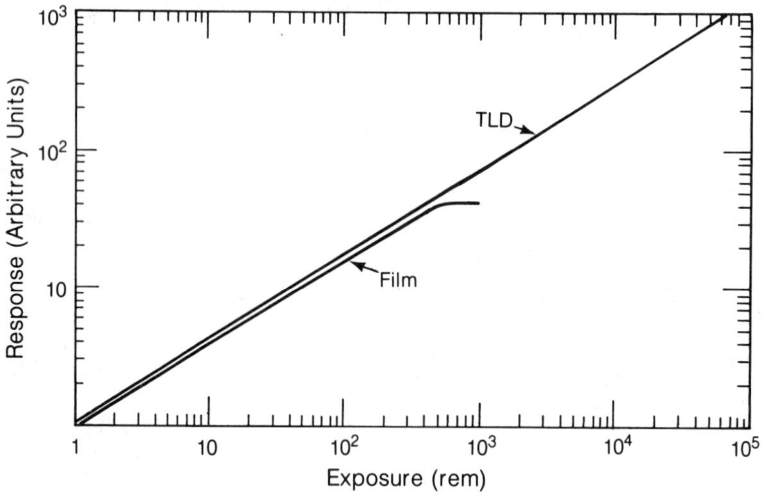

Figure 9.5. Useful range of TLD and film (reproduced from Cusimano and Cipperley, Ref. 13, with permission of Pergamon Press © 1968).

10^5 rem (kSv). A typical sensitive film is shown for comparison. The linearity of response is good, extending below 10 mrem (0.1 mSv) for low doses. The directional dependence is shown in Fig. 9.6; again, the uniformity is excellent. The energy response of a typical TLD is shown in Fig. 9.7. The enhanced sensitivity in the 20-keV region is much more easily handled by filtering techniques than is the (more extreme) problem with with film.

In addition to LiF, several other TLD materials have been studied and applied in personnel dosimetry. Among those in use today are CaF_2:Dy, CaF_2:Mn, $CaSO_4$:Tm, $CaSO_4$:Dy, Al_2O_3, and fluorite. Some of these are used in combination with or as backups for LiF. Some of them also have properties which are improvements over one or more of the less desirable characteristics of LiF.

For example, although LiF is subject to tribothermoluminescence (mechanically induced luminescence) and other possible spurious luminescence, $CaSO_4$:Dy is significantly less susceptible. A factor of 30 less tribothermoluminescent response

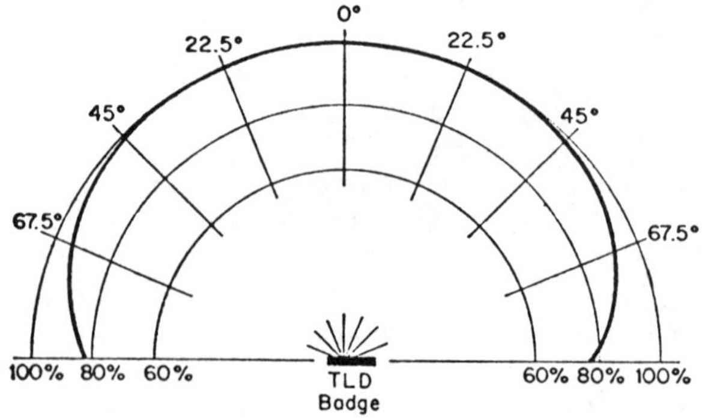

Figure 9.6. TLD dosimeter response versus angle of incidence (reproduced from Cusimano and Cipperley, Ref. 13, with permission of Pergamon Press © 1968).

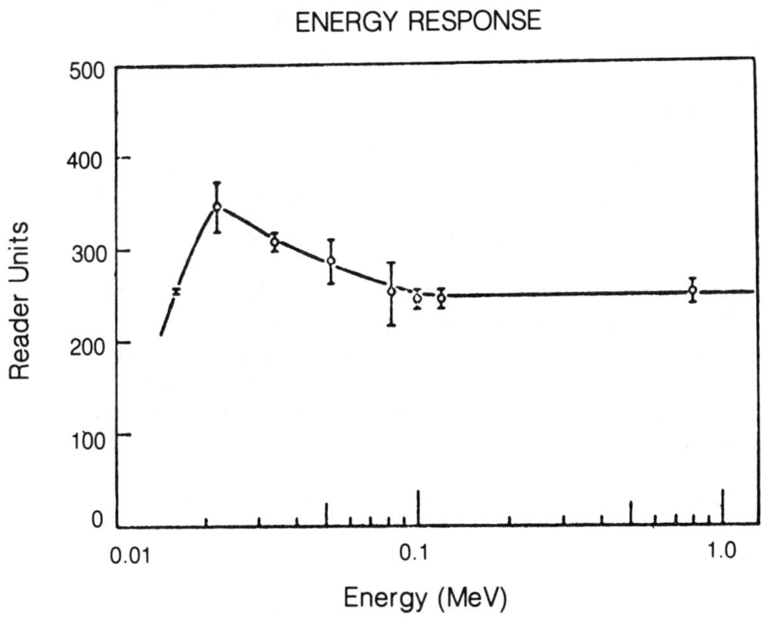

Figure 9.7. Energy response of TLD-700 thermoluminescent material (reproduced from Baumgartner, Ref. 14, with permission of the Bureau of Radiological Health © 1968).

was measured in one study [15]. Also, this study showed excellent low-dose sensitivity for $CaSO_4$:Dy (down to ~0.2 mR), which can be of occasional usefulness in dosimetry, especially in environmental monitoring situations.

In perusing the literature on TLD, one is impressed by the tremendous amount of enthusiasm for TLD among its proponents. This impression can be had from even a brief study of the proceedings of several international conferences on TLD [9,16,17]. However, both film and TLD seem to have some advantages in particular situations, and can be used to complement one another for mixed-field dosimetry.

The advantages of TLD over film are several: (a) Reproducible accuracies under laboratory conditions have been achieved at the ±5% level. (b) The dynamic range is large, with doses of from a few mrem (10 μSv) to many thousand rem (10 Sv) measurable in the same crystal. (c) Directional uniformity surpasses that of film; it is easier to design a dosimeter to reduce the directional dependence to ±20% even at low x-ray energies. (d) Fading is minimal; TLDs do not fade significantly with extreme climatic conditions. (e) Energy response is more uniform at low x-ray energies. (f) Thermoluminescent dosimeters are rugged and reusable. (g) Thermoluminescent dosimeters are sensitive to β-particles. Lithium fluoride can be fabricated in very thin layers and requires little wrapping, so that β-ray dosimetry is possible.

The disadvantages include the destructive readout (once read out, the information is lost); difficulties when neutrons are present (discussed later); the inability to determine directionality of dose deposition; and slightly higher costs than film.

The destructive readout is intrinsic to the method. One heats up the TLD, measures the integrated signal, and is then ready to anneal and reuse the chip. Systems are now commonly used to semiautomate the process; however, all information is lost, and in the event of some failure there is not second opportunity to fall back on. The most common way to avoid losing data by accident is by using one or more backup detectors. This is now found in many systems.

Another disadvantage of TLD compared to film pertains to the legal, record-keeping requirement. The argument is made that possession of the actual exposed film dosimeter is intrinsically preferable to the record from a TLD reading. However, it is now generally recognized that well-maintained records are sufficient evidence in any legal situation [18].

The problem of fading is an important consideration if dosimeters are not changed often. There are situations in which large groups of employees are monitored with no expectation that any detectable exposure will occur. For example, at Lawrence Livermore National Laboratory (LLNL) quarterly or semiannual monitoring cycles were desired for many employees whose exposure risk is minimal, but it was estimated that

> ...the maximum period that film should be used is three months. In practice, a one-month exposure period was never exceeded because of the significant aging and fading characteristics of the film dosimeter [19].

The fading of LiF, by contrast, is considered to be of only minor concern. However, in order to double-check for anomalous or incorrect readings, backup TLD chips are used at LLNL and read out in the event it is necessary to correct for such an anomaly [19].

Another drawback of LiF is that its thermoluminescence is strongly dependent on the thermal history of the material [20,21]. Therefore the annealing technique must be carefully controlled. This is much less true of CaF_2:Mn and $CaSO_4$ dosimeters.

9.3.3. Pocket Ionization Chambers

A frequently used backup or auxiliary dosimeter is a pocket ionization chamber, usually pen-sized. This chamber is typically used to assess *in situ* the dose being received by a worker in a high-radiation-level environment. These instruments are for x- and γ-radiation only (although many will respond as well to β-radiation) and capable of either direct read-out or immediate indirect read-out with a separate instrument.

Specifications for such dosimeters have been established by the American National Standards Institute, Subcommittee N13.5 [22]. The electrodes of the ion chamber form a capacitor which is charged to a predetermined voltage. Ionization occurs in the chamber surrounding the electrodes causing a decrease in the charging voltage which can be read by an electrometer; or alternatively causes motion of a movable needle, which can be read directly.

A typical instrument today can measure exposure (roentgen) to about $\pm 15\%$ of full scale for 200-mR full-scale settings and can undergo discharge (leakage) at rates of almost a few tenths of a milliroentgen per hour, for 200-mR full scale [22]. It is meant for use on at most a daily (8-hr or perhaps 24-hr) basis, and is a very useful instrument for this purpose or for hourly dose assessment. Pocket ion chambers can be obtained in a variety of rugged, sealed, environment-insensitive designs. Their disadvantage is a poor sensitivity to β-rays, and neutrons.

9.3.4. Other Dosimeters

Modern microelectronics have made possible the production of personal electronic alarm dosimeters which provide an audible tone ("chirp") in the presence of significant x- or γ-ray exposure rates. The more elaborate versions of these devices can be set to alarm after a predetermined integrated exposure has been received. These devices will also indicate total integrated exposure at any time on a digital readout. Most are about the size and shape of a pack of cigarettes, and are designed to be carried in a shirt pocket or clipped to the belt. The detector is usually a Geiger-Müller tube with some compensation to limit energy dependence to within $\pm 40\%$ from 50 to 1200 keV. The precision of dose-rate conversion (chirps/mR) tends to decline with increasing exposure rate; however, most of these devices will continue to provide a constant alarm in very high exposure rate situations.

9.4. NEUTRON DOSIMETRY

The problem of neutron personnel dosimetry is discussed separately from the γ- and β-ray problem for three reasons: (a) A large fraction of radiation exposure involves no neutron fluxes at all. (b) The present instrumentation for neutron dosimetry is in many important cases *not* adequate to the task, unlike that for γ- or β-ray dosimetry. (c) The dosimetry of high-linear energy transfer radiations (of which neutron dosimetry is but one part) has been of increasing concern in recent years.

For details on the various energy divisions, properties, and relative importances of neutron fluences, the reader is referred to Chapter 8, Measurement of Neutrons. In what follows here, a familiarity with that chapter is assumed.

Baarli and Dutrannois [23] have discussed the problem of personnel dosimetry in mixed radiation fields around high-energy accelerators. They concluded that present techniques are adequate only if the approximate composition and energy spectra of the component radiations are understood reasonably well.

At the present time, separate dosimeters must be used to record neutron doses in thermal-, intermediate- and fast-energy ranges. The most difficult of these ranges is the intermediate-energy region (between ~0.5 eV and ~100 keV) where neutrons must usually be detected by the use of a moderator and a thermal-neutron detector. In contrast, thermal neutrons are easily detected by capture in materials with high cross sections, and fast neutrons can be detected by charged secondaries produced, for example, in (n,p) reactions in hydrogenous materials. There are also activation reactions with sufficiently high cross sections at fast-neutron energies to be useful in personnel dosimetry.

As discussed in Chapter 8, relatively few situations exist in which a substantial fraction of the *dose* is from thermal neutrons; the main exception is near a thermal column at a reactor. However, thermal-neutron detectors are important because many indirect methods of assessing higher-energy neutron fluences actually rely on thermal-neutron detectors. Recent studies of neutron dose equivalent rates in commercial pressurized water reactor containments have indicated that the largest contribution is from neutrons in the intermediate-energy range, precisely the region in which good neutron dosimetry is most difficult [24].

This section will discuss fast-neutron and then thermal-neutron dosimeters, and finally body backscattering and albedo dosimeters.

9.4.1. Fast Neutrons

The basic technique most commonly used for fast neutrons (~200 keV to ~20 MeV) is the *nuclear track emulsion.* A typical emulsion of this type is Eastman Kodak Type A (NTA) emulsion, which has been described in detail by Lehman [25] and Becker [5]. NTA is a special fine-grained film, typically 25 to 35 μm thick. Neutrons are detected by the trail of ionization from charged particles released by one of three mechanisms: (a) elastic (n,p) interactions with hydrogen, which register in NTA above about 0.6 MeV; (b) exoergic (n,p) reactions with nuclei, which register in NTA below about 10 eV and then only on nitrogen nuclei; and (c) inelastic nuclear interactions resulting in a "star," significant above about 20 MeV.

The energy response of NTA is shown in Fig. 9.8. The sensitivity is seen to be significant only for fast and thermal neutrons. Standard practice is to report the response of neutron film in tracks per unit area, as seen through a microscope. There are difficulties in relating the measured quantities to neutron flux (and hence to dose equivalent), mainly because some of the response comes from interactions in the emulsion wrapping and some from interactions in the emulsion itself. Varying emulsion thicknesses (batch to batch) complicates calibration procedures. One problem with this class of detectors is that it is susceptible to hydration at high humidities, which increases the hydrogen content and hence the fast-neutron sensitivity. This and other problems (temperature effects, storage problems, etc.) have been discussed by Lehman [25] and others [26,27,28]. Figure 9.9 and 9.10 show the fast neutron response of NTA film as a function of incident angle.

Figure 9.11 shows the effect of 70% relative humidity on latent image stability of NTA film. Desiccation followed by the heat-sealing of Kodak NTA Type B film in moisture-proof bags has been successfully used for several years by a British group.

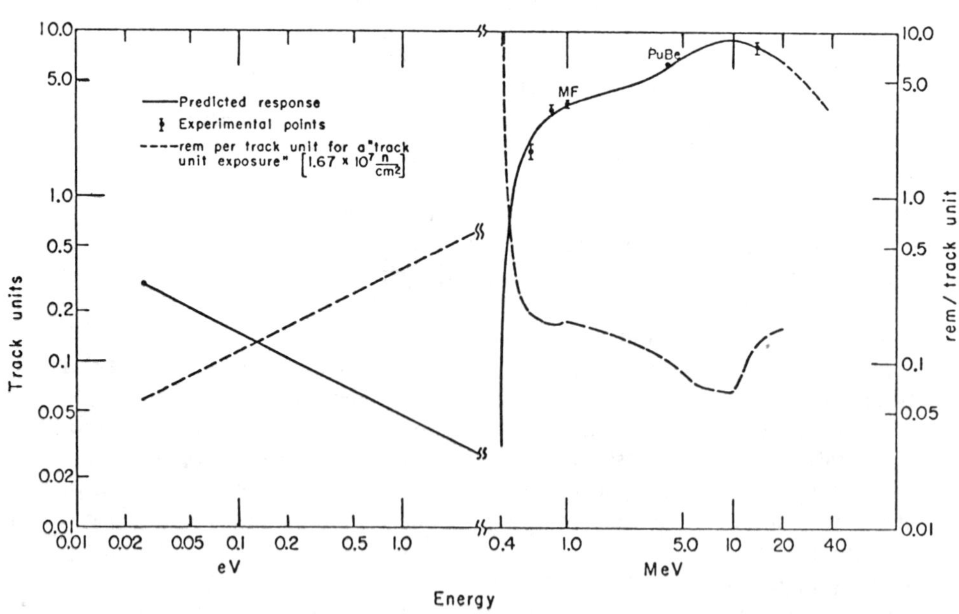

Figure 9.8. Response of NTA packets to neutrons of various energies (reproduced from Lehman, Ref. 25).

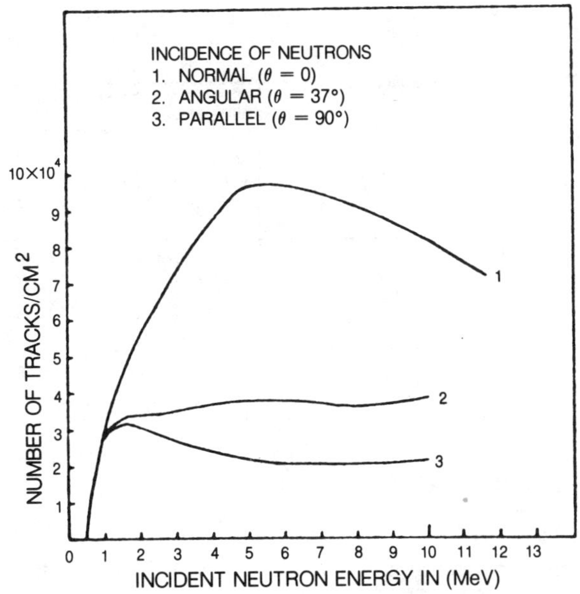

Figure 9.9. Estimated NTA film response (reproduced from Nagarajan, Ref. 27).

Their unpublished results [29] show that the 2- to 4-week latent-image half-life of NTA film can be extended to about 8 weeks using this technique. The bags are made of a laminate of paper/0.012-mm Al foil/polyethylene.

Marshall and Stevenson [30] have considered the edge-effects problem in scanning of emulsion, and Oshino [31] has studied the energy dependence of NTA worn on a phantom.

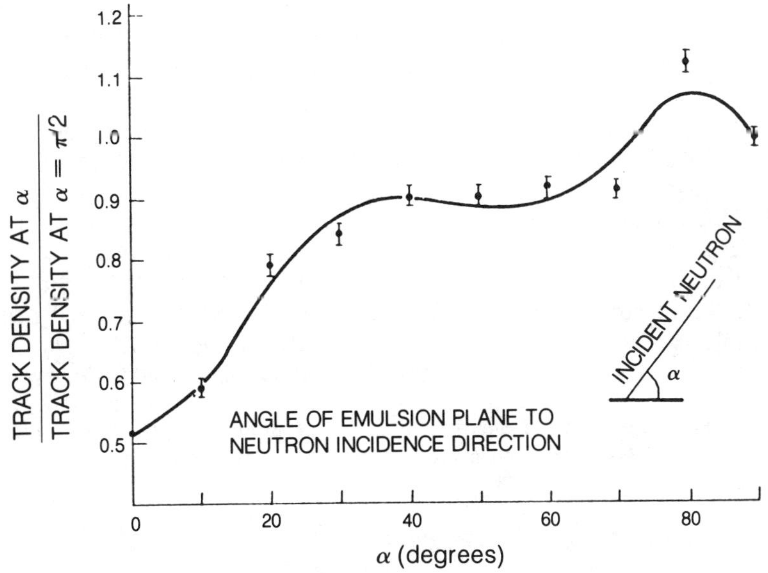

Figure 9.10. NTA film badge response as a function of neutron incidence angle (reproduced from Cavallini, Ref. 28).

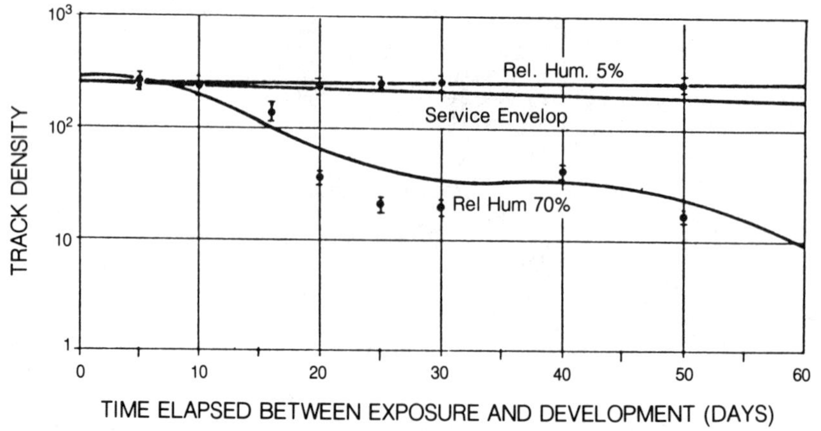

Figure 9.11. Effect of 70% relative humidity on latent image stability of NTA film (reproduced from Cavallini, Ref. 28).

Although fast-neutron dosimetry has been carried out for almost two decades using nuclear track emulsions, it is clear that some features are in need of improvement. However, in situations where neutron spectra are known to be dominated by fast neutrons, and where high humidity is not a problem, this technique can be used with reasonably good results.

Three other methods of fast-neutron dosimetry involve activation or fission foils combined with plastic films, or certain polymer films alone which when chemically etched produce visible tracks or pits at the sites of charged-particle interactions. These techniques are discussed next.

9.4.2. Thermal-Neutron Detectors

Thermal-neutron personnel dosimetry is commonly performed today with one of three means: Thermoluminescent dosimeters, fission foil dosimeters, or film surrounded by a foil that is activated and then decays.

The commonly used *TLD* method relies on two isotopic compositions of lithium fluoride, ^{6}LiF and ^{7}LiF. Lithium-6 has a thermal-neutron cross section of about 960 barns; the cross section of lithium-7 is smaller by a factor of about 25,000. The ^{7}LiF TLD responds to photons with the characteristics already described, while the ^{6}LiF responds to both photons and thermal neutrons. The two may be used in differential mode to separate γ-rays from neutrons. Harvey, Rudd, and Townsend report that the response of a ^{6}LiF TLD to 1 rem (10 mSv) of thermal neutrons is approximately 80 times the response to 1 R of γ-radiation [32], and that the minimum detectable thermal-neutron dose is 0.5 mrem (5.0 μSv). However,

> ...it is necessary to calibrate each batch of ^{6}LiF separately, as significant differences in sensitivity may occur between batches, ...the ratio of neutron to gamma sensitivity is not constant...[32].

Fast-neutron sensitivity of LiF is insignificant.

> The response in equivalent gamma rads per fast neutron tissue rad is roughly proportional to neutron energy, ranging from 1% at 1 MeV to 10% at 10 MeV...[32].

Many of the basic advantages of and problems with TLDs for thermal-neutron detection are identical to those of TLDs in general, which have already been discussed.

The *fission foil* technique exploits the property of certain crystalline materials, plastics, and glasses, whereby radiation damages the materials so that certain etchants will preferentially attack the damaged areas. In particular, some materials only produce tracks (etched pits or holes) when the linear energy transfer (LET) exceeds some critical rate. Neutrons themselves, and γ-rays, do not provide the minimum LET, whereas fission fragments do; hence, the fission fragment technique consists of placing a fissionable foil next to a plastic detector and the integrated thermal-neutron fluence is subsequently determined by etching and microscopic scanning or electronic counting. Typical track sizes of 0.1 μm are observed. Becker [33] has prepared a review of track etching.

These dosimeters are relatively unaffected by humidity, high temperature, and mechanical shock. Using ^{235}U, detection limits of as low as 10^{-6} rem (10 mSv) have been reported for thermal neutrons [34]; and using other nuclides, ^{238}U, ^{232}Th, and ^{237}Np, with thresholds in the MeV region, detection limits from 0.5 to 1 mrem (5.0 to 10.0 μSv) for fast neutrons are also reported. A discussion by Unruh et al., gives some of the important considerations in applying this technique [35]. Another discussion by Widell [36] indicates that mica is an excellent substance for image formation, but that residual trace uranium in some natural micas produces a high latent background, which must be erased before use by baking for several hours at 500°C.

Still another thermal-neutron technique is *activation*. One approach uses film or TLD surrounded by cadmium (Cd), with response from the γ-rays due to the reaction $^{113}Cd(n,\gamma)^{114}Cd$ [27]. Unruh et al. [35] describe a system in which a series of three foils is activated, with subsequent exposure of emulsion by foil decay. The three foils are tin-rhodium (sensitive to γ-rays, thermal, and body-moderated fast neutrons), cadmium-rhodium (γ-rays, body-moderated fast neutrons), and tin-iron (mainly γ-rays). The various radiations are determined independently using three film-darkening measurements, and by solving a set of simultaneous algebraic equations.

9.4.3. Albedo-Neutron Dosimeters

A fundamental consideration in neutron personnel dosimetry is that of *backscatter* or *albedo*. Quoting from Hankins [37]:

> When the human body is exposed to neutrons, some of the incident neutrons are backscattered to create a flux of neutrons of various energies leaving the body. These neutrons are called *albedo neutrons*, and a dosimeter placed on the body to measure this flux of backscattered neutrons is called an *albedo-neutron dosimeter*. Such dosimeters are usually designed to detect thermal neutrons, and when cadmium is used to eliminate incident thermal neutrons, the dosimeter response is largely from the thermal neutrons that return from the body.

The power of albedo-neutron dosimetry is that it has sensitivity in the energy range where very few other techniques work: for intermediate-energy neutrons (~0.5 MeV to ~200 keV). There are many situations in which a significant fraction of the dose equivalent is from intermediate-energy neutrons, and in which personnel dosimetry is desired.

Unfortunately, many of the albedo-neutron dosimeters in present use have a significant energy dependence, which decreases their usefulness in situations where the neutron energy spectrum is unknown. Where the spectrum is known, correction factors can be applied to compensate for energy dependence and arrive at a reasonably accurate ($\pm 30\%$) dose equivalent measurement.

A recent study by Hankins [37] examined the way in which various design variables affect dosimeter response. Hankins used small TLD chips (^{6}LiF) encased in various thicknesses of 2-in.-diameter polyethylene (CH), all surrounded by Cd. He studied the effects of placement of the TLD in front of, within, and behind the CH, and the effect of removing the Cd on the back of the badge. His most important conclusion was that almost none of the variables significantly affected the rather large energy-dependent sensitivity. On the positive side, he found the following: (a) Adding CH increased the dosimeter sensitivity. (b) With total encasement of Cd the sensitivity is not significantly affected by distance away from the body, at least up to 3 cm. (c) Sensitivity is greatly affected by shielding thickness between source and dosimeter. (d) If Cd is not present on the back (body side), most of the response is from albedo rather than incident neutrons, except for thermal-neutron exposures on a totally cadmium-enclosed dosimeter. (e) The placement of both ^{6}LiF and ^{7}LiF chips in the same or equivalent positions is critical.

The energy dependence as found by Hankins shows a monotonic decrease of a factor of about 10 between 200 keV and 3 MeV neutron incident energies, with comparatively high sensitivities for the intermediate and 200 keV cases.

Similar results have been reported by Hoy [38] at Savannah River, where one of the main monitoring problems is around a plutonium production plant, with a typical $^{238}PuF_4$ neutron spectrum mainly between 0.3 and 25 keV. The dosimeter (a 2-in.-diameter two-section concentric hemisphere, separated by Cd, using ^{6}LiF and ^{7}LiF TLD, and worn on a belt) gave adequate dose equivalent measurements.

Another albedo dosimeter has been developed in the United Kingdom [32]. TLDs (^{6}LiF and ^{7}LiF) are housed in a plastic cover loaded with boron carbide. The design criterion was as follows: for an unshielded 1/velocity ($1/v$) detector at the surface of the body, the response to the albedo from incident thermal neutrons exceeds that to the albedo from incident 10-keV neutrons by a factor of about 10. The dosimeter was designed to modify this response to decrease the thermal and epithermal sensitivity: hence the boron. The resulting sensitivity is shown in Fig. 9.12. Note that the sensitivity is reasonably flat from the thermal energies to 10 keV, which is a wide range indeed.

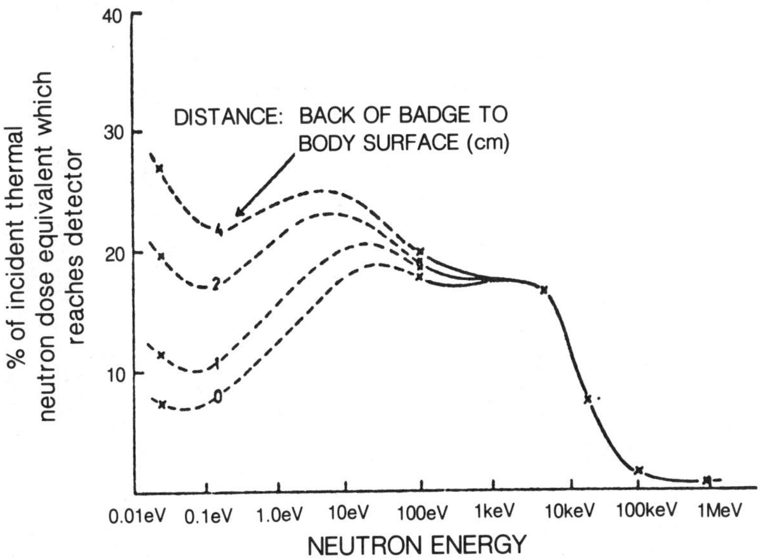

Figure 9.12. Response of albedo-neutron dosimeter worn on the body as a function of incident neutron energy (reproduced from Harvey, Rudd, and Townsend, Ref. 32, with permission of Berkeley Nuclear Laboratories © 1969.

9.4.4. Track Etch Plastic Dosimeters

Certain plastic materials will yield etchable damage tracks from recoil nuclei (typically protons from hydrogen) which are produced by neutron scattering interactions occurring within the material itself, or in an adjacent proton radiator. Polycarbonates and certain proprietary plastics such as Makrofol K, G, and E possess excellent dosimetric properties for neutrons above about 1 MeV. Several investigators have been seeking track etch materials which will permit substantial reductions in this energy threshold. A promising candidate is a carbonate-based homopolymer having the trade name, CR-39 [39]. When CR-39 is etched both chemically and electrochemically (by imposing a high-voltage high-frequency field across the material in the presence of the etchant) the threshold response to neutrons can be reduced to below 100 MeV [40].

Recent developments in neutron dosimetry technology make it possible now to provide adequate personnel monitoring for neutrons over all energies of practical interest. In situations where a broad range, 100 keV to 1 MeV, are encountered the dosimeter must utilize several different elements to cover the entire spectrum. Such a device might include the following, for example: an albedo element for thermal and low-intermediate neutron energies; a CR-39 track etch element for the approximate range 100 keV to 1 MeV; and a polycarbonate element for energies above 1 MeV. Neutron spectra measurements in the workplace must, perforce, accompany the personnel dosimetry results to correct for energy dependence, particularly in the range covered by albedo-neutron dosimeters.

A comprehensive review of recent neutron dosimetry developments may be found in a recent article by Griffith et al. [41]. The characteristics of several current neutron dosimetry approaches are summarized in Table 9.1 [41].

Table 9.1. Comparison of Characteristics of Currently Available Personnel Neutron Dosimeters (reproduced from **Griffith et al., Ref. 41, with permission of Health Physics © 1979**)

Dosimeter Type	Advantages	Disadvantages	Application
Neutron film, type A (NTA)	Fast-neutron response Low cost	Careful packaging necessary to minimize fading Large response to low-energy gammas Tedious optical counting Poor response to intermediate energy neutrons	Fast-neutron environments
Albedo	Sensitivity Ease of automation Low cost	Wide variation in response with energy Some gamma sensitivity	Environments with moderated spectra Environments with comparatively little variation in neutron spectra Large numbers of personnel
^{232}Th fission track	Fast-neutron response Lack of gamma sensitivity	Use of radioactive material Poor overall dose equivalent response Marginal sensitivity	Fast-neutron environments Use in combination with albedo detectors
^{237}Np fission track	Good dose equivalent approximation Good sensitivity with thick radiators Lack of gamma sensitivity	Use of radioactive material High gamma dose to wearer with thick ^{237}Np layer High unit cost	High-dose neutron environments Few monitored personnel Dosimeters retained at work site

9.5. SPECIAL DOSIMETRY PROBLEMS

Although the most common monitoring problem in personnel dosimetry is whole-body exposures, there are often requirements for dosimetry in other situations. Perhaps the most familiar of these is measurements of doses to the extremities (particularly to the hands). For this purpose, several specially designed classes of integrating dosimeters have been developed. Thus, ring dosimeters to be worn on the fingers are now in common use, and dosimeters for mounting in shoes have also been designed.

The main operational difficulties with such instruments are related to the validity of the measurement: The accuracy of the particular dosimeter used is by no means the most important consideration. For example, dose may be accrued through *contact* with radioactive material, a common problem in radiochemistry laboratories. Here even finger dosimeters can give incorrect results in many cases, because of local shielding and distance considerations.

The main point to be made here is that users must investigate very carefully the particular radiation field being monitored, to assure that the measured dose is meaningfully related to the dose received. A discussion of finger dosimetry can be found in Refs. 42 and 43.

9.6. SUMMARY

The aim of this section has been to discuss instrumentation for the measurement of integrated external radiation doses to personnel. The text outlines the types of measurements needed, the accuracies required, the intrinsic limitations imposed by the sampling problem, and the capabilities and drawbacks of presently existing instrumentation.

To reiterate several points:

1. The *accuracy* required of personnel dosimetry systems is not excessively great: the ICRP [1] has recommended maximum uncertainties of less than 50% [or less than 1 rem (10 μSv), whichever is greater] in assessment of annual dose equivalent.
2. One key requirement of any dosimetry system is to make sure that *no significant component* of the radiation field goes completely unmeasured: in cases of large accidental exposure, such an omission would be regrettable.
3. For routine monitoring of *external* beta and gamma radiations, adequate instrumentation exists today. While improvements are definitely possible (and in some cases imminent, or at least probable from the current research program in dosimetry), the basic task of β- and γ-ray dosimetry is now being performed reasonably well.
4. There is still need for a dosimeter useful in those situations dominated by x-rays below about 15 keV.
5. *Thermal-neutron* dosimetry can now be performed adequately with any of several dosimetry systems.
6. *Fast-neutron* dosimetry has largely relied for more than a decade on the nuclear-track-emulsion technique, which suffers from several important drawbacks. Research on a number of possible alternate systems ought to be encouraged.

7. *Intermediate-energy* neutrons (~0.5 eV to ~200 keV) are still a difficult measurement problem. Albedo-neutron dosimeters are not now sufficiently well understood to allow their use in neutron fields with unknown energy spectra, and no other systems now seem capable of performing this task adequately.
8. More work is required to develop a simpler way of assessing the *energy spectra* of external radiation fields to complement the measurement of *dose equivalent*. For cases of accidental exposure this is particularly important, and has not been given sufficient attention.
9. Multiple-element dosimeters with albedo and several track etch components show promise for personnel monitoring over a wide range of neutron energies. These devices require fairly detailed information on ambient neutron spectra in order to compensate for energy dependence in the dosimeter components.

REFERENCES

1. International Commission on Radiological Protection, "General Principles of Monitoring for Radiation Protection of Workers," ICRP Publication No. 12, Pergamon Press, Oxford, 1968.
2. M. Ehrlich, "The Use of Film Badges for Personnel Monitoring," IAEA Safety Series No. 14, International Atomic Energy Agency, Vienna, 1965.
3. "The Basic Requirements for Personnel Monitoring," IAEA Safety Series No. 8, International Atomic Energy Agency, Vienna, 1962.
4. D. E. Barber, "Standards of Performance for Film Badge Services," PHS Publication 99-RH-20, Environmental Health Series, USDHEW, Public Health Service, Washington, DC, 1966.
5. K. Becker, *Photographic Film Dosimetry*, The Focal Press, London and New York, 1961 (in German), 1966 (in English).
6. A. Brodsky, A. Spritzer, F. Feagin, F. Bradley, G. Karches, and H. Mandelberg, "Accuracy and Sensitivity of Film Measurements of Gamma Radiation—Part IV, Intrinsic and Extrinsic Errors," *Health Phys.*, **11**, 1071, 1965.
7. H. Stewart, N. Modine, and J. Rolofson, "Low Energy Calibration of Film Used in Measuring X-Rays from Color Television Receivers," in *Proceedings of the Conference on Detection and Measurement of X-Radiation from Color Television Receivers,* U.S. Public Health Service, 1968, p. 185, Bureau of Radiological Health, Rockville, MD 20852.
8. R. Kathren and A. Brodsky, "Accuracy and Sensitivity of Film Measurements of Gamma Radiation—Part III, Effects of Humidity and Temperature During Gamma Irradiation," *Health Phys.*, **9**, 769, 1963.
9. *Third International Conference on Luminescence Dosimetry,* Risö, Denmark, Risö Report No. 249, 1971 (available from Danish Atomic Energy Commission, Risö, DK-4000, Roskilde, Denmark).
10. E. Storm and S. Shlaer, "Development of Energy-Independent Film Badges with Multi-Element Filters," *Health Phys.*, **11**, 1127, 1965.
11. J. R. Cameron, N. Suntharalingam, and G. N. Kenney, *Thermoluminescent Dosimetry,* University of Wisconsin Press, Madison, 1968.
12. F. M. Lin and J. R. Cameron, "A Bibliography of Thermoluminescent Dosimetry," *Health Phys.*, **14**, 495, 1968.

13. J. Cusimano and F. Cipperley, *Health Phys.*, **14**, 339, 1968.
14. W. V. Baumgartner, "Capabilities and Methods to Measure X-Ray Radiation Originating in TV Receivers," in *Proceedings of the Conference on Detection and Measurement of X-Radiation from Color Television Receivers, 28 March 1968,* p. 265 (available from U.S. Public Health Service, Bureau of Radiological Health, Rockville, MD 20852).
15. T. Yamashita, N. Nada, H. Onishi, and S. Kitamura, "Calcium Sulfate Activated by Thulium or Dysprosium for Thermoluminescence Dosimetry," *Health Phys.*, **21**, 295, 1971.
16. F. H. Attix, Ed., *Luminescence Dosimetry,* Proceedings of the International Conference on Stanford, CA, June 21–23, 1965, (available as CONF-650637, 1967, U.S. Atomic Energy Commission, Division of Technical Information, Oak Ridge, TN 37830).
17. J. A. Auxier, K. Becker, and E. Robinson, Eds., *Proceedings of Second International Conference on Luminescence Dosimetry, 23 September 1968,* CONF-680920, 1967, U.S. Atomic Energy Commission, Division of Technical Information, Oak Ridge, TN 37830.
18. J. C. Hart, "Legal and Administrative Aspects of Personnel Dosimetry," Symposium on the Future of Personnel Dosimetry, Annual Meeting of the Health Physics Society, New York, July 1971; referred to by F. H. Attix in "A Current Look at TLD in Personnel Monitoring," *Health Phys.*, **22**, 287, 1972.
19. B. L. Rich and B. G. Samardzich, "Organizational Aspects of a Personnel Dosimetry Service," LLL Report UCRL-73254, 1972, Lawrence Livermore Laboratory, Livermore, CA 94550.
20. I. Bernstein, B. Bjarngard, and D. Jones, "On the Use of Phosphor-Teflon TLDs in Health Physics," *Health Phys.*, **14**, 33, 1968.
21. D. Zimmerman, C. Rhyner, and J. Cameron, "Thermal Annealing Effects on the Thermoluminescence of LiF," *Health Phys.*, **12**, 525, 1966.
22. American National Standards Institute, "Performance Specifications for Direct Reading and Indirect Reading Pocket Dosimeters for X- and Gamma Radiation," ANSI Standard N13.5 (1972) American National Standards Institute, 1430 Broadway, New York, NY 10018.
23. J. Baarli and J. Dutrannois, "Purpose and Interpretation of Personnel Monitoring Data for High Energy Accelerators," in *Proceedings of ENEA/OECD Conference on Radiation Dose Measurements, Stockholm, June 12, 1967,* European Nuclear Energy Agency/Organization for Economic Cooperation and Development, Paris, 1967, p. 275.
24. R. S. Sanna, F. Hajnal, J. E. McLaughlin, J. E. Gulbin, and R. M. Ryan, "Neutron Measurements Inside PWR Containments," USDOE Environmental Measurements Laboratory, EML-379, 1980.
25. R. L. Lehman, "Energy Response and Physical Properties of NTA Personnel Neutron Dosimeter Track Film," LBL Report UCRL-9513, Lawrence Berkeley Laboratory, Berkeley, CA 94720, 1961.
26. J. Cheka, "Fast Neutron Film Dosimeter," *Phys. Rev.*, **90**, 353, 1953; "Recent Developments in Film Monitoring of Fast Neutrons," *Nucleonics*, **12**, 40, 1954.
27. P. S. Nagarajan, "Neutron Personnel Monitors Currently in Use," in *Radiation Protection Monitoring,* Proceedings of Joint WHO-IAEA Regional Seminar in Bombay, 9 December 1968, International Atomic Energy Agency, Vienna, 1969, p. 103.

28. A. Cavallini, "Role, Interpretation, and Accuracy of Measurement of Fast Neutrons by Means of Photographic Emulsions," p. 293 of Ref. 23.
29. A. Knight, National Radiological Protection Board, private communication, 1972, Clifton Avenue, Belmont Sutton, Surrey, United Kingdom.
30. T. O. Marshall and G. R. Stevenson, "Edge Effects in the Scanning of Nuclear-Track Emulsions for Neutron Dose Assessment," in *Neutron Monitoring,* Proceedings of an IAEA Symposium in Vienna, 29 August 1966, International Atomic Energy Agency, Vienna, 1967, p. 549.
31. M. Oshino, "Response of NTA Personnel Neutron Monitoring Film Worn on Human Phantom," Lawrence Berkeley Laboratory Report LBL-342, 1972.
32. J. R. Harvey, W. H. R. Rudd, and S. Townsend, "A Personal Dosimeter Which Measures Dose from Thermal and Intermediate Energy Neutrons and from Gamma and Beta Radiation," Berkeley Nuclear Laboratories, Report RD/B/N1547, Central Electricity Generating Board, Berkeley Nuclear Laboratories, Berkeley, United Kingdom, 1969.
33. K. Becker, "Dosimeter Applications of Track Etching," in *Progress in Radiation Dosimetry,* F. Attix, Ed., in preparation.
34. K. Becker, J. S. Cheka, K. W. Crase, and R. B. Gammage, "New Exoelectron Dosimeters," in *Advances in Physical and Biological Radiation Detectors,* Proceedings of a Symposium in Vienna, 23 November 1970, International Atomic Energy Agency, Vienna, 1971, p. 25.
35. C. Unruh, W. Baumgartner, L. Kocher, L. Brackenbush, and G. Endres, "Personnel Neutron Dosimeter Developments," p. 433 of Ref. 30.
36. C. O. Widell, "Neutron Dosimetry by the Fission-Fragment Method," p. 417 of Ref. 30.
37. D. E. Hankins, "Factors Affecting the Design of Albedo-Neutron Dosimeters Containing Lithium Fluoride Thermoluminescent Dosimeters," Los Alamos Scientific Laboratory, Report LA-4832, Los Alamos, NM 87544, 1972.
38. J. E. Hoy, "An Albedo-Type Personnel Dosimeter," *Health Phys.*, **23**, 385, 1972; "Personnel Albedo Neutron Dosimeter with Thermoluminescent ^{6}LiF and ^{7}LiF," Savannah River Laboratory, Report DP-1277, Aiken, SC 29801, 1972.
39. B. G. Cartwright, E. K. Shirk, and P. B. Price, "A Nuclear-Track-Recording Polymer of Unique Sensitivity and Resolution," *Nucl. Instrum. Meth.*, **153**, 457, 1978.
40. S. A. R. Al-Najjar, R. K. Bull, and S. A. Durrani, "Electrochemical Etching of CR-39 Plastic: Applications to Radiation Dosimetry," *Nucl. Tracks*, **3**, 169, 1979.
41. R. V. Griffith, D. E. Hankins, R. B. Gammage, and Luigi Tommasino, "Recent Developments in Personnel Neutron Dosimeters—A Review," *Health Phys.*, **36**, 235, 1979.
42. B. E. Bjarngard and D. Jones, "A New Technique for Finger and Hand Dosimetry Using Solid TLD LiF-Teflon Dosimeters," Health Physics Society Annual Meeting, Houston, 29 June 1966, unpublished, (available from Controls for Radiation, Inc. Cambridge, MA 02139).
43. T. F. Johns, "Use of Energy-Independent Dosimeters for Measuring Partial and Whole-Body Doses," p. 241 of Ref. 23.

PART II

10 NUCLEAR POWER PLANTS

The purpose of this section is to provide background information on the types and levels of radioactivity and radiation fields associated with the use of nuclear power reactors and to describe the instrumentation available for measurement of these levels and fields. For detailed information on instruments, the reader is referred to the appropriate chapter or sections (such as Chapter 8, Measurement of Neutrons) in this volume.

The present discussion is also intended to acquaint the reader with the radiation levels associated with the operation of nuclear power plants, to put their hazards in perspective, and to serve as an introduction to instrumentation suitable for making measurements to determine the effects of reactor operation on radiation doses or radioactivity levels. A representative selection from the literature now available on the subject of radioactivity from nuclear power plants is cited.

10.1. INTRODUCTION

The introductory sections of this chapter (10.1.1 through 10.1.3) give a very brief review of the physics behind atomic fission and nuclear reactors. The material is not necessary for the understanding of the measuring and monitoring instrumentation. It is tutorial only and those not wishing this review can omit it without loss as far as the main purpose of this volume is concerned.

10.1.1. Historical Note

The discovery of neutrons by Chadwick in 1932 prompted active research on the effects of bombarding a wide variety of elements with these particles. Because of their lack of charge they could, at low energies, penetrate the positively charged nucleus of any atom. It was soon found that neutron bombardment caused profound changes in the bombarded material often leading to radioactive atoms that decayed by beta-ray emission, and thus transmuted into another element.

Enrico Fermi and his group, working in Italy, attempted to produce transuranic elements by neutron bombardment of uranium. He succeeded in producing a beta emitter but could not verify that the decay product was element 93. Shortly thereafter, Hann and Strassmann demonstrated that bombardment produced barium and lanthanum among others. Meitner, working closely with Hann and Strassmann, identified the process as the fission of uranium and demonstrated that the process was accompanied by the release of large amounts of energy. This chain of events triggered active work in this field and it was soon shown that it was the uranium isotope 235 that was being disrupted and that not only were the fragments carrying an enormous amount of energy but several neutrons were also freed. This was immediately recognized as the features required for a chain reaction.

By then the world was at war and all scientific work was greatly hampered by secrecy and lack of international exchange of information. However, intensive work continued both for its value as a power source and as a weapon. The first nuclear reactor put in operation was the Chicago uranium-graphite pile built by Fermi's group in 1942. Since then the development of nuclear rectors for power and other peaceful uses and also work on nuclear explosives for war has progressed rapidly in many parts of the world.

Reactors generate large amounts of radioactive material which must be carefully monitored because, as has already been made clear, radioactive material can be a serious health hazard. Its disposal is a major problem. A reactor in operation generates an enormous flux of neutrons and all areas where there might be humans must be shielded and monitored. Neutron activation can create radioactive materials in cooling water. Gases involved in reactor operation must be kept from escaping in reactor effluents. Effluents must therefore be constantly monitored.

10.1.2. Nuclear Forces

Atomic nuclei are bonded together primarily by the strong nuclear force. Other forces are also involved. As yet a complete theory of the structure on nuclei has not been developed but the theory as it stands at present explains much of the observed nuclear behavior. The reason for the inadequacy of the theory is the multiplicity of forces involved in the structure.

The strongest force is the "strong" nuclear force. This force, which binds nucleons, regardless of charge, has very short range dropping to a negligible value at distances greater than about 3×10^{-13} cm. This is a volume force and contributes an increasing amount of energy as the number of nucleons A increases for each nucleon added when A is small. However when A is greater than about 20, the range of the force is so small that the added energy per added nucleon does not increase very much. Coulomb repulsion, which affects protons but not neutrons, subtracts from the energy of the system. There remain the spin force and the exclusion principle. The latter says that only one nucleon of a kind can occupy a given quantum state. But each nucleon can have a spin of $+1/2$ or $-1/2$. This accounts for the shell nature of the nucleus and for the pairing of neutrons and protons.

Summing these terms, the binding energy B for nuclei above about $A = 20$ gives:

$$B = \underbrace{15.7A}_{\text{strong forces}} - \underbrace{17.8A^{2/3}}_{\text{allows for short range of strong forces}} - \underbrace{0.71Z(Z-1)A^{-1/3}}_{\text{Coulomb force}} - \underbrace{23.6A^{-1}(A-2Z)^2 \pm 132A^{-4}}_{\text{spin + exclusion giving shell form}} \text{ MeV}, \tag{10.1}$$

Last term: + for N, Z even; − for N, Z odd; 0 for A, odd

where A = number of nucleons,
Z = atomic number,
B = binding energy.

A plot of the binding energy per nucleon as measured and the values of B/A from Eq. 10.1 is shown in Fig. 10.1.

Two important facts can be concluded from this model: (a) the nucleus is a quasispherical entity with a diameter of about $2.4 \times 10^{-13} \times A^{1/3}$ cm which can be thought of as a liquid-drop; and (b) a nucleon in a nucleus $A = 92$ has a binding energy of 7.6 MeV, whereas one near the maximum of the curve in Fig. 10.1 has 8.7 MeV. This means that elements with nuclear weight of 60 to 120 are much more stable than uranium with $A = 235$. This difference in binding energy manifests itself in a considerable difference in mass for the same number of nucleons. For example the atomic weight of ^{235}U is 235.04 and that of ^{132}Xe and ^{103}Rh together is 234.21.

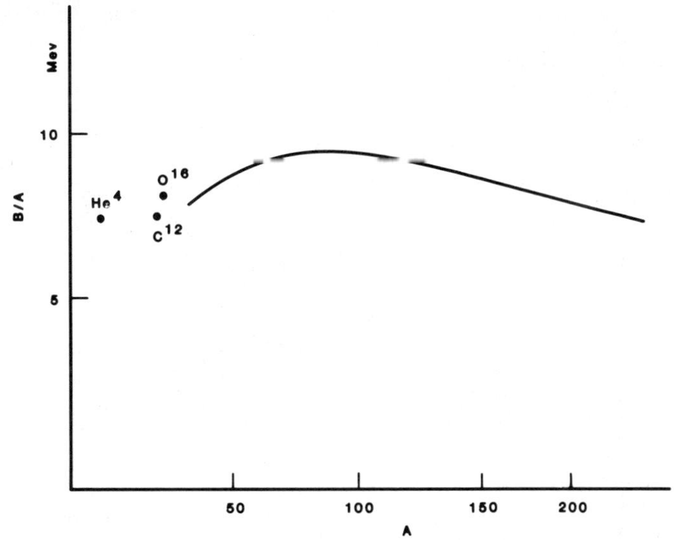

Figure 10.1. Binding energy per nucleon as a function of mass number.

10.1.3. Fission

A neutron absorbed by a heavy nucleus, on the basis of the liquid-drop model, sets the sphere into violent oscillation. For ^{235}U and ^{239}Pu, the oscillations are sufficiently violent that the nucleus fissions into two parts. The decrease in mass goes into kinetic energy and the two fragments have ~200 MeV kinetic energy. This energy is enormous compared with such nuclear processes as alpha or beta emission. Furthermore when the nucleus fragments it also emits several energetic neutrons.

This is, as has just been pointed out, exactly the condition required for a chain reaction. One uranium atom fissions and the neutrons from it cause a neighbor to fission, and so on. The process of fission can be pictured as shown in Fig. 10.2. The atom is held together in a condition of unstable equilibrium with a barrier perhaps as small as 6 MeV keeping it from spontaneously fissioning. A neutron added to the system sets the nucleus into violent oscillation and at some point the two portions of the asymmetrical configuration separate enough so that the electrical repulsion is sufficient to drive the two portions completely apart. The potential energy due to their close proximity is converted into kinetic energy which they carry away.

The process is (to an extent) random, so that the fragment pair is not the same for each fission, and the fragments have a fairly wide distribution of masses. For reasons not fully understood the distribution is not symmetric about either $A = 117$ or $Z = 46$. The actual peaks for the mass numbers are $A = 95$ and 140. Figure 10.3 shows the fraction of fission yield as a function of mass number. In any instant of fission, the sum of the mass numbers of the fragments less the number of free neutrons emitted must equal that of the fissioning atom. Similarly the sum of the charges must be the same. Since the ratio of neutrons to protons is very much higher in the parent atom than in the two lighter atoms formed, the fission fragments must emit either neutrons or beta particles. The emission of beta particles is very much more probable and all of the fission fragments are radioactive. Most are

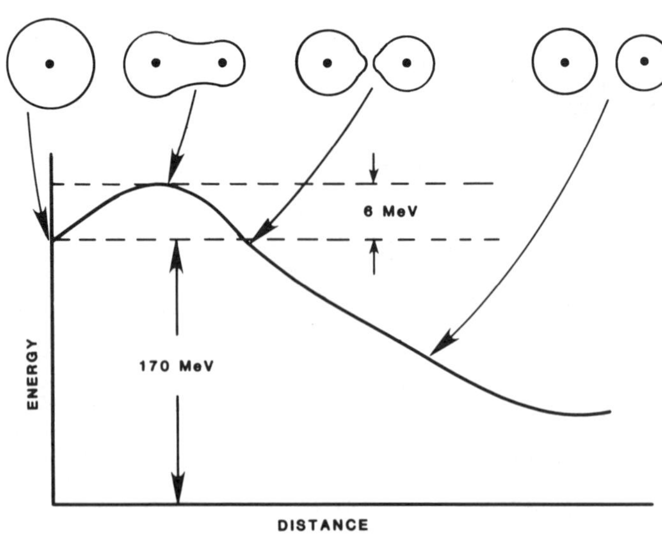

Figure 10.2. Fission of uranium.

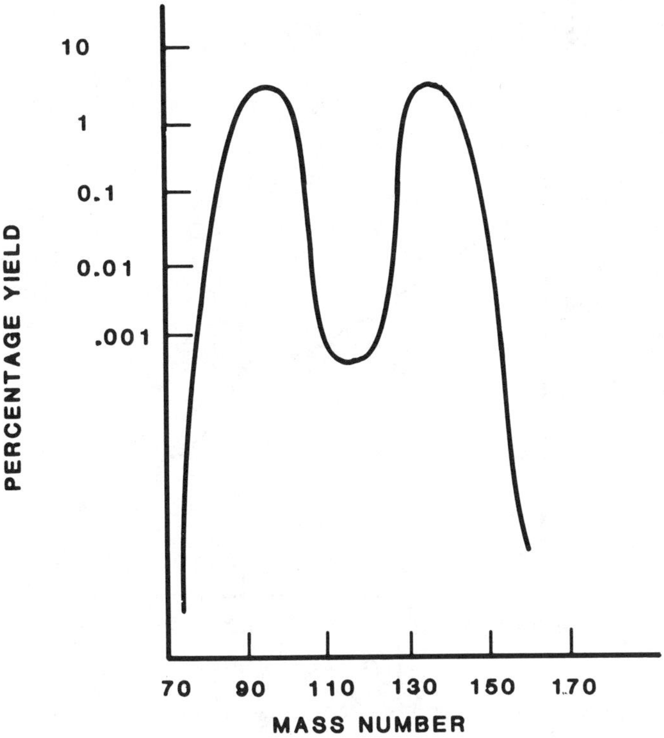

Figure 10.3. Distribution of fission fragments.

very short-lived and they will have decayed to half their initial radioactivity in about 10 sec. A typical example of the decay of a pair of fission fragments near two peaks of the distribution, ${}^{95}_{39}Y$ and ${}^{139}_{53}I$ is:

$$
{}^{235}_{92}U \nearrow {}^{95}_{39}Y \rightarrow {}^{95}_{40}Zr \rightarrow {}^{95}_{41}Nb \rightarrow {}^{95}_{42}Mo
$$

$$
{}^{235}_{92}U \searrow {}^{139}_{53}I \rightarrow {}^{139}_{54}Xe \rightarrow {}^{139}_{55}Cs \rightarrow {}^{139}_{56}Ba \rightarrow {}^{139}_{57}La
$$

Neutrons are emitted from the fissioning atoms with considerable energy, averaging about 2 McV, but having a broad distribution of energies. The time of emission is important in reactor design. Some 99% of the neutrons from the fissioning material are emitted in less than 0.01 sec. The remaining are distributed over periods of several seconds.

In describing nuclear reactions, it is necessary to give some designation of the probability that the bombarding particle will cause the reaction. This is usually done by assigning a target area to the bombarded particle. The unit selected for this is the barn, equal to $10^{-24}cm^2$, which was chosen because it is convenient in terms of the effective cross-section area of many atoms. It turns out that this area is approximately that for many reactions. The fission cross section of ${}^{235}U$ is $1.05 \times 10^{-27}m^2$, or 10.5 barns. Close to energy resonances the reaction cross section may be much larger. Figure 10.4 gives the fission cross section of ${}^{235}U$ as a

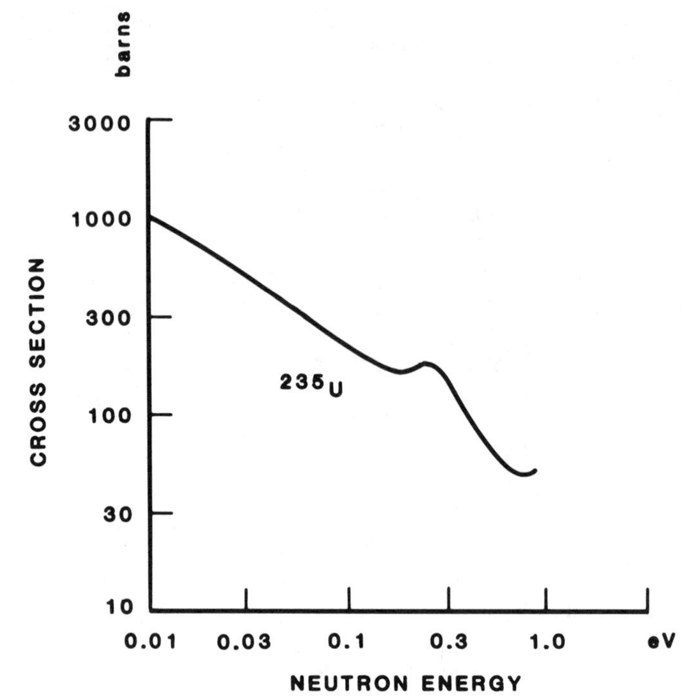

Figure 10.4. Fission cross section as a function of neutron energy.

function of neutron energy for neutron energies under 1 eV. Clearly one of the problems of an efficient reactor is that of slowing the neutrons resulting from fission.

10.1.4. Nuclear Reactors

It was shown in the preceding section that the conditions for a chain reaction are satisfied by ^{235}U. Because this was discovered in the midst of World War II, one of the first aims was to use fission in weapons, that is, to develop an uranium atomic bomb. This was first shown to be feasible by Frisch and Peierls when they calculated the critical mass for ^{235}U. Another critical effort was made to develop a reactor using natural uranium. The problem was that only 0.7% of natural uranium is ^{235}U. The major constituent of natural uranium is ^{238}U, a very strong absorber of neutrons which does not fission. Therefore, there is no critical mass for natural uranium, or its oxides. This is almost self-evident since if natural U could fission, any large assemblage of uranium ore would constitute a natural reactor. In passing it is interesting to note that natural reactors must have existed in the very distant past. In one region of Gabon, Africa, there is an area where an assemblage of stable elements just matches the fission fragment distribution in some uranium ore. It should be noted in connection with the disposal of nuclear waste that these fragments stayed within a few miles of this very large natural reactor in spite of the heavy rainfall in the area.

Calculations showed that if natural uranium were distributed in a three-dimensional array, and if the space between elements of the array were filled with a material which would slow the neutrons down to thermal energies without absorbing too large a fraction of the neutrons, a critical assemblage could be built. This task was undertaken by Enrico Fermi. The material used to slow the neutrons, known as the moderator, had to have a mass number matching as nearly as possible the neutron mass and at the same time have a small capture cross section for neutrons. The latter requirement rules out the use of ordinary water because hydrogen is a strong neutron absorber. Heavy water would be very effective, but at the time was too difficult to obtain. Fermi concluded that very pure graphite would meed the requirements. After overcoming many difficulties, Fermi succeeded in building, under a stadium in Chicago, and making critical, a reactor, known in those days as a "pile." Theorists had shown that such an arrangement could not possibly explode, but mathematicians are only human—if it had exploded it would have taken half of Chicago with it. It did not and the reactor was controlled with the control rods provided for the purpose.

The construction of a more practical reactor at Oak Ridge overlapped the building of the Fermi "pile" in Chicago. It was put into operation shortly after the feasibility of a natural uranium reactor was proven. The principle purpose of this installation was to test the feasibility of producing plutonium, which is also fissionable. Plutonium is formed by the neutron bombardment of ^{238}U. It is produced copiously in a natural uranium reactor. This, again, was a wartime measure and its objective was to produce weapon-grade fissionable material. For this purpose it was much more efficient than the very difficult task of separating the isotope ^{235}U from ^{238}U.

The Oak Ridge reactor was a low-power reactor suitable for making radioactive isotopes for medical use, experimenting with the sterilization of foods, and similar peacetime uses. Quite a number of similar low-power reactors were built both at universities and at other institutions. They performed well and did not lead to any serious accidents.

A major possible use for reactors is a source of power. In particular, for electrical power plants they might be economical sources of energy. Every fission occurring in a reactor releases about 200 MeV of energy, which mostly appears as heat given up when the high-energy fission fragment, neutrons, and other particles are stopped. Since 3×10^{10} fissions yield 1 W/sec, a 1000-MW power plant operating for one year would burn 9.8×10^{26} atoms of ^{235}U or about 385 kg, less than half a ton a year. The radioactive waste would have about the same weight. A certain amount of plutonium would be formed which should be salvaged for use in plutonium reactors.

It is possible to design reactors (breeder reactors) in such a way that they produce from the ^{238}U more plutonium than the ^{235}U burned. Neutrons are only slightly moderated in these breeder reactors and instead of using water as a coolant they use liquid sodium, which has little absorption and little moderating action. Except for the danger of the misuse of the plutonium produced they are a possible solution to the energy problem.

There is enough natural uranium to supply energy needs for many years; however, even setting aside the weapons hazard of such a program, there are many environmental hazards that must be overcome.

Certainly the most important sources of information on emissions from power plants are the utilities that operate them, and the federal agencies involved in their monitoring and licensing. The primary agency in this respect is the U.S. Nuclear Regulatory Commission (NRC), which is responsible for the licensing of commercial nuclear facilities and which issues an Environmental Statement for each proposed nuclear power plant. All such statements are available at NRC libraries and from the National Technical Information Service (NTIS). The NRC and the U.S. Environmental Protection Agency (EPA) have also issued Environmental Statements for important actions connected with the operation of nuclear power plants. Of particular interest are the EPA statement related to their standard on emissions from the uranium fuel cycle [1] and the NRC statement on the use of mixed oxide (i.e., mixtures of U and Pu) fuel in light water reactors [2].

Figure 10.5 shows the locations of commercial nuclear power plants in the United States as of December, 1977 [3]. The total operable nuclear electrical generating capacity given for that date was 46 gigawatts (GW), with another 163 GW under construction or on order. It is expected that the total U.S. electrical generating capacity will rise substantially in the near future and that nuclear power plants may constitute an important part of the increase. Precise predictions are not possible both because of alterations in the perceived need for new electrical capacity and because of the various possible responses to this need. However, planning for the future is important because of the long time between the decision to build a power plant and the completion of the plant. For nuclear power plants, this time is now about ten years.

Virtually all of the large power reactors in this country are cooled with ordinary (light) water. Light-water reactors (LWR) are made in two varieties: the boiling-water reactor (BWR) and the pressurized-water reactor (PWR). Other types of reactors include heavy-water reactors, gas-cooled reactors, liquid-metal fast breeder reactors, and a few others. Various types of power reactors are discussed briefly in the next section.

In addition to large power reactors, which now tend to be designed for about 1 GW output, there are hundreds of smaller reactors in the United States, plus a large number of operational military reactors (such as those used in submarines). The smaller nonmilitary reactors are typically used for experimental purposes, including research and teaching, not necessarily connected with the nuclear power program. Many of the early small power reactors served as pilot and demonstration plants for the large-scale reactors now being built. The considerations set forth in the following sections on reactor emissions and monitoring requirements pertain primarily to large commercial nuclear power plants. However, many of the same considerations apply to these smaller reactors.

Nuclear electric generation involves not only the operation of nuclear power plants, but also many associated auxiliary activities, including transportation, fuel reprocessing/waste management (see Chapter 11), and mining and milling (see Chapter 12). Each facility must comply with both radiation protection standards and more general standards on conventional emissions, heat, noise, occupational protection, and so on. A principal force behind careful analysis of the overall impacts of proposed facilities is the National Environmental Policy Act, which requires the preparation of environmental statements in support of major federal actions, including the licensing of nuclear facilities.

Various regulatory agencies have responsibilities where nuclear power is concerned. As has been pointed out, federal agencies regulating the operation of domestic nuclear facilities in the United States are the NRC and the EPA. The NRC has responsibility for licensing and regulating individual facilities, but the EPA has responsibility for limiting environmental impacts. As a result, the emission and public exposure limits imposed by the NRC are expected to be consistent

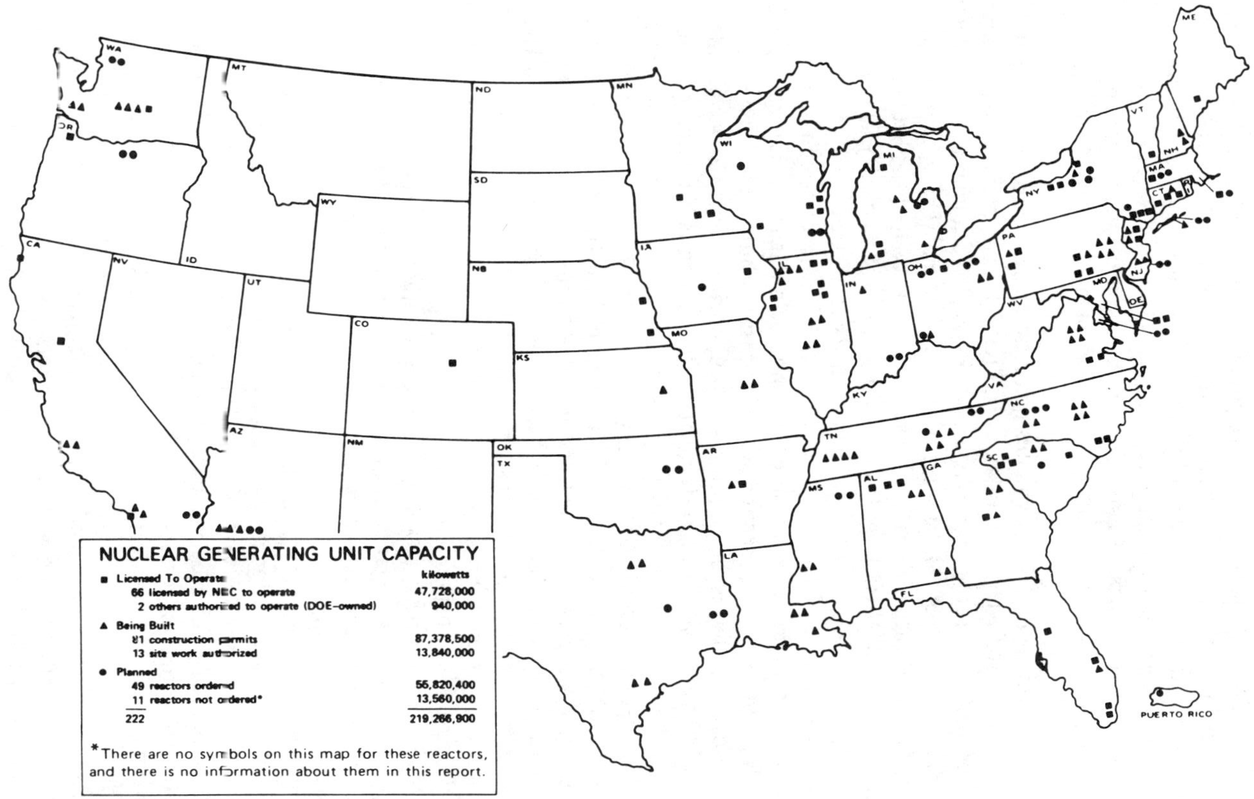

Figure 10.5. United States commercial nuclear power plants (reproduced from Department of Energy, Technical Information Center, Ref. 3).

with EPA regulations. The NRC also regulates exposure of workers at nuclear facilities. In many occupational matters, the Occupational Safety and Health Administration and state industrial safety departments also have jurisdiction. State and local agencies together with the foregoing organizations exercise authority over the siting of nuclear facilities. Many of the radiation protection standards applicable to nuclear power facilities, including power plants, are discussed in earlier chapters.

The manner in which the NRC regulates domestic nuclear facilities is specified in Title 10 of the Code of Federal Regulations (CFR) [4]. The parts of major interest in connection with nuclear power plants are Part 20 (Standards for Protection Against Radiation), Part 50 (Licensing of Production and Utilization Facilities), and Part 100 (Reactor Site Criteria). In addition to 10 CFR, the NRC review process depends heavily on NRC Regulatory Guides [5], which specify how the applicant for a license may comply with the requirements of 10 CFR.

The EPA is specifically charged with controlling the impact of emissions, including those which are radioactive, in order to limit their impact on the environment, and to protect humans. In connection with this responsibility, the EPA may monitor concentrations and levels of pollutants in the environment. However, the primary monitoring responsibilities reside with other agencies (including state and local) and with the operators of facilities. The EPA has published a guide for monitoring environmental radiation, the "Environmental Radioactivity Surveillance Guide." [6] In addition, they have promulgated, in Title 40 CFR Part 190, limitations on emissions and exposures from uranium fuel cycle activities [7]. This regulation places limitations on the overall fuel cycle, rather than on any particular facility (which would be the responsibility of the NRC). It requires a more stringent control of gaseous emissions from reprocessing plants than has been anticipated in the past (see Chapter 11).

The character of a nuclear power plant imposes rather specific requirements on monitoring equipment, as regards both radionuclides of interest and monitoring conditions. This discussion considers both monitoring at the source and environmental radiation monitoring.

Source monitors, in this context, are distributed within the reactor site and include: (a) devices to monitor the ambient radiation fields to which employees are exposed; and (b) fixed release monitors, designed for monitoring gaseous, liquid, or particulate radioactivity released via stack effluents or liquid discharges during normal operation, refueling, or accidents.

Environmental radiation monitors are designed to measure radioactive concentrations or radiation fields in the general environment, particularly along pathways to human exposure. These may include fencepost monitors surrounding a reactor site, instruments for sampling water or soil, and others. A third class of instruments, not the subject of this book, are process monitors, which keep track of the flow of materials within a facility. To the extent that these instruments may be used for site surveys and environmental monitoring, they are included in the Instrument Notes and discussion.

Although the radioactive content and emissions of a nuclear power plant may be characterized on a generic basis, the specific instrumentation required for monitoring a particular site is usually selected after careful consideration of the features of that site. The environmental monitoring program is designed to ascertain the amount of radioactivity released from the site. The release modes and, certainly, the ultimate hazards associated with any release depend strongly on the site characteristics. On-site monitoring programs, on the other hand, usually do not vary substantially from one plant to another. These programs may include stack and liquid effluent monitors, personnel dosimeters, area monitors, and special exercises (such as radiation emergency drills).

It is also important that natural background radiation levels be evaluated, particularly as a baseline for monitoring man-made environmental radiation and radioactivity. The total average natural dose to individuals in the United States is about 100 mrem/yr (1 mSv/yr) (see Chapter 3), most of which is contributed by cosmic ray and other external natural sources, but it includes a small contribution from radioactive fallout from weapons testing. This corresponds to an irreducible background dose of about 10 μrem/hr (1 μSv/hr), which can cause difficulties in making dose measurements below a few μrem/hr (a few μSv/hr). A significant problem, in fact, is that the natural levels may have diurnal and seasonal changes, masking the presence of man-made contributions to the total radiation environment. Similar difficulties may arise in determining radionuclide concentrations, rather than the overall radiation field. Measurements of man-made radionuclide concentration in environmental samples, such as air, water, or soil, can be significantly affected by natural background activities from such radionuclides as potassium-40 and the heavy natural α-emitting chains. As discussed in Chapter 1, the regulations applicable to nuclear power operations only limit the increases in radiation levels that are caused by human activities.

The discussion that follows characterizes reactors as sources of radiation, discusses routine and accidental exposure modes, and notes requirements on monitoring instrumentation. A brief discussion of the nuclear fuel cycle in this chapter serves as an introduction to fuel processing and uranium extraction that are discussed in following chapters.

10.2. NUCLEAR POWER PLANTS AS SOURCES OF RADIATION

Commercial nuclear power plants in the United States use a specific class of nuclear reactors, that is, those cooled and moderated with ordinary water (light water). Therefore the regulatory apparatus and associated monitoring requirements are directed primarily to this type of reactor (LWR) power plants and their associated fuel cycles. There are, however, numerous other reactor types, some of which may assume importance by the end of this century. Fortunately from the point of view of environmental monitoring, the radioactive emissions associated with nuclear power do not often depend drastically on the type of reactor used, except for specific differences in recycled materials (see Chapter 11).

Figure 10.6 provides a simplified diagram of the essential elements of nuclear-powered electrical generating plants which have been commercially available in the United States. The portion on the right shows the elements common to all of these plants. Steam produced in the reactor or in a heat exchanger associated with the reactor drives a turbine-generator to produce electricity. The turbine exhaust is condensed and the water pumped back to the reactor system, where it is vaporized once again. Typically about one-third of this energy generated by the heat source is converted to electricity, and two-thirds is rejected to the environment as waste heat. This heat is dissipated in the atmosphere (via cooling towers) or in a body of water, or both.

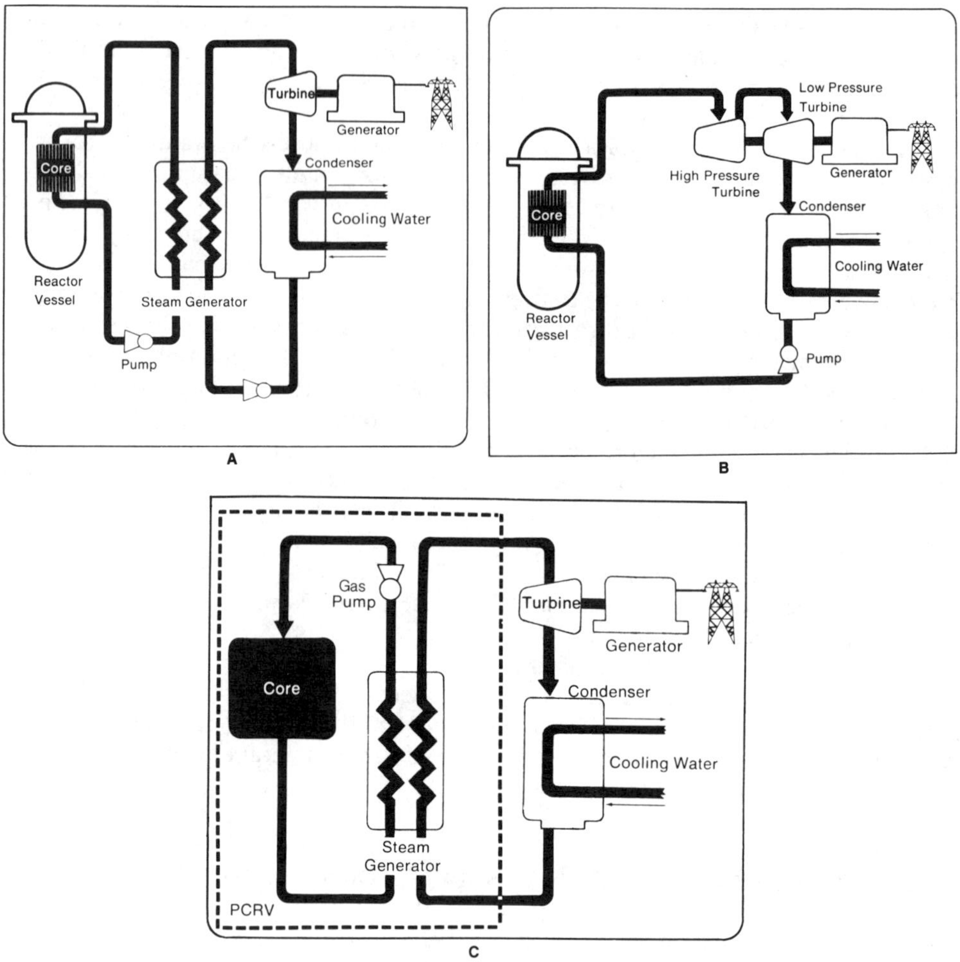

Figure 10.6. The three drawings represent three important types of power reactors built in the United States: (*a*) the PWR (Pressurized-Water Reactor); (*b*) the BWR (Boiling-Water Reactor); and (*c*) the HTGR (High-Temperature Gas-Cooled Reactor). They all boil water to generate steam. Following the generation of steam, the nuclear power plants are essentially the same, as indicated by the turbogenerator, condenser, and cooling water system. Holding ponds are used sometimes to cool the water before releasing it to the environment. An important point: the cooling water from the environment does not enter the reactor (reproduced from ERDA-76-107, Ref. 8).

It is a common feature of nuclear power plants that the primary coolant is recirculated to the reactor. Because of this, radioactive contaminants escaping from the reactor fuel into the coolant can be effectively confined. As a result, the routine emissions from a power plant are predominantly gaseous species; even these, however, can be rather effectively contained. Nevertheless, there are opportunities for other radioactive releases and, independently of this, effort is made to prevent workers at a nuclear power plant from exposures to radioactive materials which are within the plant.

10.2.1. Nuclear Power Plant Types

The pressurized-water reactor (PWR) and the boiling-water reactor (BWR), represented in Fig. 10.6, contribute the bulk of nuclear power generation in the United States and a substantial portion of that abroad. The high-temperature, gas-cooled reactor (HTGR), also shown, has been withdrawn from the commercial market after construction of only one demonstration plant, but interest in this reactor type continues. A heavy-water-moderated reactor (CANDU—Canadian deuterium-uranium reactor) has been successfully marketed in Canada and elsewhere. CANDU uses heavy water (wherein the hydrogen is the mass 2 isotope, deuterium) as moderator and coolant and a pressurized coolant system, as in a PWR.

Figure 10.6*a* shows a PWR. The energy generated in the core as a result of fission heats the water passing through the reactor vessel. The coolant is maintained at sufficiently high pressure that gross boiling does not occur in the primary loop. Instead, the coolant is pumped through a heat exchanger or steam generator in which heat is transferred to a secondary fluid, also water, which is permitted to boil. The resulting steam drives the turbogenerator.

In a BWR, represented in Fig. 10.6*b*, the coolant is recirculated within the reactor vessel and is permitted to boil. The steam that is formed within the vessel is used directly to drive the turbogenerator. Because of the differences in the cooling systems, the radioactive effluents associated with reactor operation differ between a PWR and a BWR.

Both PWRs and BWRs have cores loaded with uranium oxide pellets encased in a metal cladding made of zirconium alloy. This is characteristic of most commercial power plants, including the Canadian CANDU reactor. However, LWR fuel consists of enriched U, wherein the percentage of ^{235}U has been elevated to 2 to 4%, slightly above the natural 0.7%. The CANDU, on the other hand, uses natural uranium, a choice which is made possible by the use of heavy water, which absorbs fewer neutrons than ordinary water. It uses on-line, rather than batch, refueling.

Gas-cooled reactors, such as the HTGR, can use a different fuel type and moderator. The HTGR uses small fuel pellets of thorium and highly enriched uranium embedded in large blocks of carbon moderator. (Neutron capture by ^{232}Th leads to formation of ^{233}U, a fuel that is even superior to ^{235}U.) Functionally, though, the HTGR is similar to a PWR (see Fig. 10.6*c*).

All of the foregoing are thermal reactors, in which neutrons are moderated to low (thermal) energies in order to increase the probability of nuclear fission. Fast breeder reactors, in which the neutrons are not moderated, have undergone substantial development in the United States and abroad. As ordinarily envisioned, the cores of these reactors would contain ^{239}Pu and ^{238}U. They would use ^{239}Pu as the principal material to be fissioned, releasing enough neutrons so that not only is one available to split another nucleus (thereby continuing a chain reaction), but enough neutrons are captured by ^{238}U, leading to ^{239}Pu, that more ^{239}Pu is produced than is consumed. As a result, there would be little demand on U resources, even if the nuclear electric generating system were growing relatively rapidly. The breeder that is most highly developed is the liquid-metal fast breeder reactor (LMFBR).

Other types of reactor/fuel systems can be built. In many cases these rely on the Th–^{233}U fuel cycle, indicated above for the HTGR, in order to achieve effective U utilization in a thermal reactor. Depending on the design of the reactor and fuel cycle, as much ^{233}U may be produced as is destroyed, so that again little demand would be made on U resources, except to build up an inventory of ^{233}U. See Chapter 11 (Nuclear Spent Fuel Processing) for further discussion.

A more detailed discussion of nuclear reactor types and their principal features is to be found in Nero [9]. Hebel [10] treats associated questions of fuel reprocessing and waste disposal for conventional, as well as alternative, fuel cycles.

10.2.2. Sources and Types of Radiation

Nuclear power plants, as well as other parts of the fuel cycle, utilize or produce materials with a range of chemical, physical, and radioactive characteristics. Together these properties strongly affect both pathways to humans and the resulting radiation doses. In some circumstances, the materials from nuclear power plants may expose humans externally, particularly when their decay yields x-rays or γ-rays. The effects of less penetrating forms of radiation may be mitigated by small amounts of shielding; in many cases, even the air or the external layer of skin will suffice. However, such radionuclides may be taken up by the body and the decay radiation may expose the body internally. This constitutes the primary hazard of most alpha-particle and many beta-particle emitters, as well as some of the x- and γ-ray emitters [11].

Not all of the radiation doses associated with operations arise from radioactivity generated in reactors. Nor are reactors the only points in the nuclear fuel cycle where releases of reactor-produced material may occur. The next section summarizes other types of releases and indicates where some of these are discussed in this book.

The amounts of radioactivity released from nuclear power plants and/or the radiation doses to workers or members of the general population depend both on the reactor type and on the manner in which it is constructed and operated. In spite of this variation, some general statements can be made. This arises both because the fundamental nuclear processes occurring inside nuclear reactors do not vary greatly from one reactor to another and because regulatory practice has some degree of uniformity.

Table 10.1 gives estimates of the radioactive inventory (based on discharge fuel composition) of a LWR that has reached equilibrium; it also indicates some important physical and health considerations for the various elemental classes. In addition, the table specifies how the inventory changes for two time periods. (See Chapter 11.)

The total radioactive inventory of a large operating reactor amounts to billions of curies (EBq). However, only an extremely small portion of this can escape into the environment, even under very unusual circumstances. Reactors are ordinarily designed so that several barriers serve to confine this large amount of radioactivity. Moreover, the internal shielding of a power plant is designed to limit exposures of plant workers to acceptable levels.

More specifically the fuel material itself, wherein most of a reactor's radioactive inventory is generated, is the first intrinsic barrier. For conventional reactors, this material is a solid ceramic, uranium dioxide (UO_2) or, in reactors that use plutonium, plutonium dioxide (PuO_2), formed into pellets. Most of the products of the chain reaction taking place in this material remain in this solid matrix. This includes both the fission products and actinides (see Table 10.1). However, some portion of this radioactivity can migrate out of the fuel pellets. This is particularly true of the gaseous fission products, including tritium (^3H), krypton (Kr), xenon (Xe), and iodine (I). However, the fuel pellets are contained in long, thin metal tubes (cladding), which provide both structure to the nuclear core and serve as a second barrier. (Bundles or assemblies of these rods are the basic fuel handling unit.) Other possible sources of fission products—apparently minor—are fuel that contaminates the surface of new fuel elements and fuel that passes into reactor coolant water from failed fuel elements. Some radioactivity is also generated in the cladding itself, primarily from neutron activation (see Table 10.1), but this remains in the cladding except for minor corrosion.

For a very small percentage of cases the cladding develops leaks, permitting the more volatile fission products to escape from the fuel rods into the surrounding coolant. In addition, other radionuclides, including ^3H, ^{13}N and ^{16}N, and ^{14}C, are generated as activation products in the coolant itself. These activation products in reactor coolant water are formed by neutron irradiation of the water and its contents (including gases and dissolved or suspended solids) and materials in contact with the coolant (container and structural surfaces, fuel and control rod cladding) that subsequently corrode or erode. Additional radionuclides may be formed in other parts of the reactor than the core. However, the fact that the coolant is confined to a closed recirculating system serves largely to confine both these radionuclides and leaked fission products. Some of these are so short-lived that relatively brief retention in the primary coolant system permits enough decay to effectively reduce their concentration.

Not all of this remaining radioactivity is permitted to accumulate in the coolant. Every power plant has a system that controls the chemistry of the coolant by routinely bleeding off and treating a portion of it. Some of the radioactivity is collected in residue from this chemistry control system, later to be packaged for disposal. Some is discharged to the environment, often after storage in holdup tanks, permitting further decay. And a portion may be returned to the coolant system. As discussed in the next section, routine releases to the environment are strongly effected by the form of various radioactive control systems.

Table 10.1. Radioactivity in Reactor and Fuel Cycle for a 1000-MW_e Uranium-Fueled Light-Water Reactor[a,b] (adapted from Nero, Ref. 9)

Radionuclides		Half-Life	Reactor Inventory[c] (10^6 Ci)	In Discharge Fuel[d] (10^6 Ci/yr) At Discharge	150-Day Decay	10-Yr Decay	Health Consideration and Principal Radiation
			Fission Products				
Tritium[e]	^3H	12.26 yr	0.0723	0.0241	0.0239	0.0139	Internal hazard, β (0.019 MeV max)
Krypton	83m	1.86 hr	5.71	1.90	0	0	External irradiation, β (0.67 MeV max for ^{85}Kr)
	85m	4.4 hr	17.2	5.70	0	0	
	85	10.76 yr	1.16	0.383	0.373	0.201	
	87	76 min	34.0	11.3	0	0	
	88	2.8 hr	49.0	16.5	0	0	
	89	3.18 min	61.8	20.3	0	0	
	90	33 sec	58.5	22.7	0	0	
Total[f]			325	107.7	0.373	0.201	
Strontium	89	52.7 days	71.6	23.8	3.22	0	Internal hazard to bone and lung, β (0.55 MeV max for ^{90}Sr)
	90	27.7 yr	7.80	2.58	2.56	2.02	
Total[f]			526	174	5.78	2.02	
Iodine	129	1.7×10^7 yr	3.03×10^{-6}	1.0×10^{-6}	1.02×10^{-6}	1.03×10^{-6}	Internal hazard to thyroid, β (0.8 MeV max for ^{131}I) and γ
	131	8.05 days	71.9	23.9	6.01×10^{-5}	0	
	132	2.26 hr	103	34.2	0	0	
	133	20.3 hr	137	45.6	0	0	
	134	52.2 min	156	51.8	0	0	
	135	6.68 hr	123	40.7	0	0	
	136	83 sec	54.0	17.9	0	0	
Total			1,017	337	6.11×10^{-5}	1.03×10^{-6}	

Xenon	131m	11.8 days	0.582	0.193	9.05×10^{-5}	0	External irradiation, β (0.35 MeV max for ^{133}Xe) and γ
	133m	2.26 days	3.29	1.09	0	0	
	133	5.27 days	137	45.6	0	0	
	135m	15.6 min	36.8	12.2	0	0	
	135	9.14 hr	25.7	8.52	0	0	
	137	3.9 min	132	43.8	0	0	
	138	17.5 min	128	42.5	0	0	
	139	43 sec	107	35.6	0	0	
Total[f]			680	225	9.05×10^{-5}	0	
Cesium	134	2.046 yr	19.0	6.32	5.49	0.215	Internal hazard to muscle, β (1.176 MeV max for ^{137}Cs) and γ
	137	30.0 yr	9.92	3.29	3.26	2.61	
Total[f]			595	198	8.75	2.83	
Zr, Nb, Mo, Tc, Ru, Rh, Pd, Ag Cd, In, Sn, Sb			1,880	625	54.1	0.0450	External and internal hazard
Rare earths: La, Ce, Pr, Nd, Pm, Sm, Eu, Gd, Tb, Dy, Ho			4,140	1,374	48.8	0.416	External and internal hazard
Total fission products			11,970	3970 (1.067 Te/yr)	130	9.98	
			Actinides				
Uranium	237	6.75 day	108	35.9	7.42×10^{-6}	0	External and internal hazard, β (1.29 MeV max for ^{237}U) and γ
	239	23.5 min	1,708	566	0	0	
Total[f]			1,816 (99.7 Te)	601 (33.0 Te/yr)	2.72×10^{-5}	2.15×10^{-5}	

Table 10.1. (Continued)

Radionuclides		Half-Life	Reactor Inventory[c] (10^6 Ci)	In Discharge Fuel[d] (10^6 Ci/yr) At Discharge	150-Day Decay	10-Yr Decay	Health Consideration and Principal Radiation
			Actinides				
Plutonium	238	86.4 yr	0.138	0.0459	0.0483	0.0463	
	239	24,390 yr	0.0318	0.0105	0.0107	0.0107	Internal hazard to bone,
	240	6,580 yr	0.0500	0.0166	0.0166	0.0166	liver, lung, lymph,
	241	13.2 yr	12.4	4.11	4.04	2.42	α (about 5 MeV for
	242	3.79×10^5 yr	1.24×10^{-4}	4.14×10^{-5}	4.14×10^{-5}	4.14×10^{-5}	$^{238,239,240,242}Pu$)
	243	4.98 yr	22.2	7.35	0	0	and β (for ^{241}Pu)
Total[f]			34.9	11.57	4.11	2.49	
			(0.893 Te)	(0.296 Te/yr)			
Fissionable Pu ($^{239}Pu + ^{241}Pu$)			(0.627 Te)	(0.207 Te/yr)			
Americium (241, 242m, 242, 243) and Curium (242, 243, 244)			1.14	3.79	0.340	0.0611	Internal hazard, α (5.5 MeV for ^{241}Am)
Total actinides Th, Pa, U, Np, Pu, Am, Cm			3,614	1,198	4.45	2.55	

Activated Zircaloy cladding and Inconel spacers (Cr,Mn, Fe,Co,Ni,Zr Nb,Sb)	12.9 (28.2 Te)	4.28 (9.36 Te/yr)	0.967	0.109
TOTAL Fission products, Actinides, and Cladding[g]	15,600 (132 Te)	5,170 (43.9 Te/yr)	135	12.6

[a]Adapted from T. H. Pigford et al., "Fuel Cycles for Electric Power Generation," Teknetron report EEED 101 (January 1973, rev. March 1975).

[b]Overall thermal efficiency = 32%.

[c]Calculated on the basis that all fuel has the composition of discharge fuel. For nonsaturable species, whose half-lives are large compared with the typical fuel life of 3 years, the inventory within the reactor will be about one-half to two-thirds of the value listed, depending on the fuel management scheme. The inventories listed in the table are accurate for saturable (short-half-life) species. Material quantities in metric tonnes (Te) are listed in parentheses.

[d]Fuel cycle quantities are calculated on the basis of a fuel exposure of 33,000 MWd/Te 100% load factor. Material quantities are listed in parentheses.

[e]Quantities listed are for fission-product tritium only and do not include tritium formed by neutron-activation reactions in boron controls and contaminants.

[f]Totals include all isotopes of the element, including short-lived species not listed in the table.

[g]Does not include ^{14}C, which is produced in the fuel and cladding (the order of 50 Ci/yr), as well as the coolant (roughly 10 Ci/yr).

10.3. ROUTINE AND ACCIDENTAL RELEASES AND EXPOSURES FROM THE NUCLEAR FUEL CYCLE

The main focus of this section is the nuclear power plant itself. However, although such a power plant is the central object of a commercial nuclear power system, the facilities that support this plant by supplying fresh fuel and disposing of spent fuel are also of interest from an environmental point of view and are discussed briefly at the end of this section and more thoroughly in Chapter 11 and 12. Figure 10.7 presents a material and environment release flow sheet [12] for an LWR fuel cycle that reprocesses spent fuel and stores Pu. The NRC [2] estimates radioactive releases and exposures that may be associated with various portions of LWR fuel cycles, with and without reprocessing or Pu recycling.

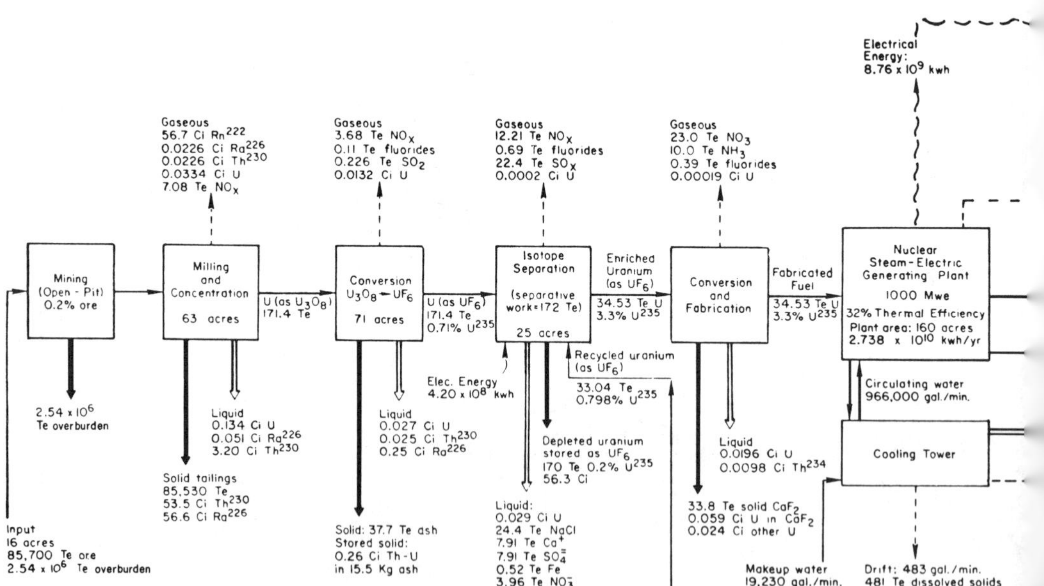

Figure 10.7. Material and environmental release flow sheet for a typical 1000 MW_e Light-Water Reactor nuclear power plant (reproduced from Pigford et al., Ref. 12).

10.3.1. Routine Power Plant Emissions

Turning first to routine emissions from reactor power plants, several radionuclides, alluded to in the last section, are of primary interest: ^{3}H, Kr, Xe, I, and ^{14}C. Except for ^{14}C and ^{3}H, these releases arise almost solely from radionuclides produced in the fuel. A substantial portion of the ^{14}C may be produced in the coolant. The major sources of released ^{3}H are (a) ternary fission in the fuel, (b) activation of deuterium (^{2}H) in the water, and (c) neutron-boron reactions in the boron (B) control curtain (see table on page 16 of Ref. 13). These nuclides are able to escape from the power plant largely because they are gaseous or volatile in elemental form or enter into compounds which are. However, the amounts which are released depend strongly on the form of the liquid and gaseous radioactivity control systems.

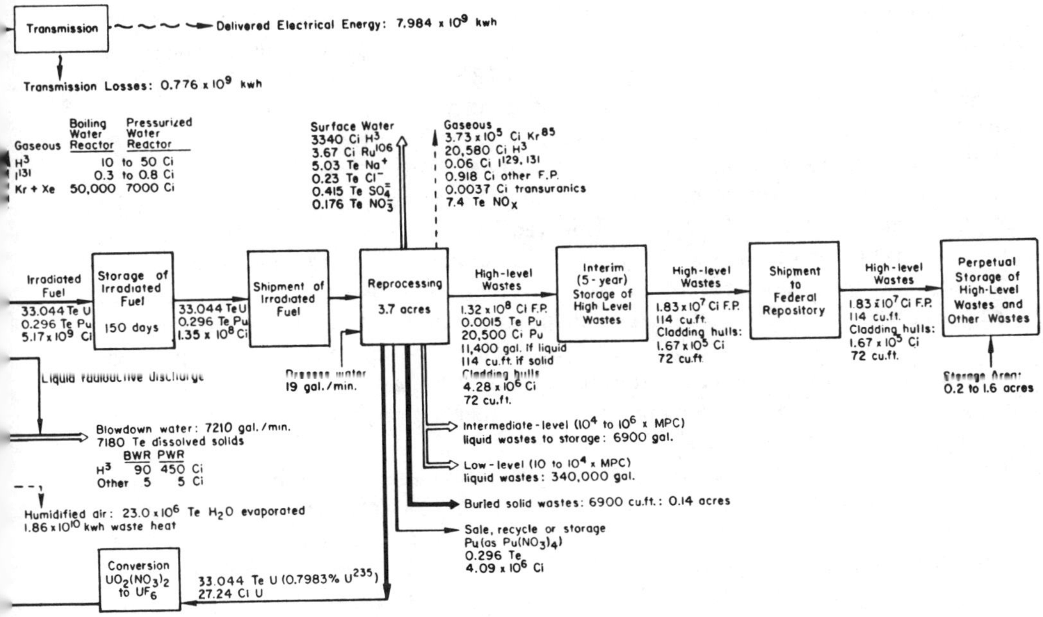

Questions concerning the fraction of a ^{14}C release in the form of CO_2 as opposed to other chemical species (e.g., CH_4) remain unanswered. (See Fig. 10.8 for examples of these systems in a recently completed nuclear power plant). Table 10.2 presents some estimates of annual releases from nuclear power plants.

For a 1-GW power plant, some tens to hundreds of curies (GBq) of ^3H as part of the water molecule, are released annually to the air in the form of water vapor or to the body of water being used for cooling. Most of the ^3H produced remains confined in the fuel and would only be released were the fuel to be reprocessed. (See Chapter 11, Nuclear Spent Fuel Processing, and Chapter 20, Tritium). Tritium, with a half-life of 12.3 years, decays by emitting a very soft β-ray (E_{max} = 18.6 keV), so that, for modest releases, its principle hazard arises from internal doses.

Tens of thousands of curies (100 TBq) of the noble gases Kr and Xe are routinely released per gigawatt-year of operation. The releases would be substantially higher, but for holdup of these gases to permit the short-lived radioactivity to decay. The most important releases are ^{85}Kr, which has an 11-year half-life and a 670-keV (max) β-ray, and ^{133}Xe, with a 5-day half-life and a 546-keV (max) β-ray. Much greater amounts of these materials are left in the fuel. Of the noble gas radionuclides, it is ^{85}Kr that, like ^3H, can cause significant population exposures if released at fuel reprocessing plants. (See Chapter 11.)

Releases of I are very small, less than 1 Ci/GW-yr (37 GBq/GW-yr), but are controlled very carefully because deposited iodine can be concentrated by an important food chain (grass–cow–milk–human) and, furthermore, it tends to collect in the thyroid. For short-term considerations, ^{131}I (8-day half-life and 364-keV γ-ray) has the same significance, but in terms of longer commitment, ^{129}I (17 million years half-life and 150-keV-max β-ray) can assume greater importance.

From the point of view of dose commitment, the several curies of ^{14}C released yearly from a large power plant can contribute the largest population dose over the medium term (hundreds of years). This radionuclide (5730-year half-life, 155-keV-max β-ray) escapes in the form of CO_2 or other compounds and, over centuries, can contribute hundreds of person-rem per gigawatt-year. It is difficult to decide over what period to estimate such dose commitments or to understand the significance, compared with other risks, of such long-term population exposures. The doses accumulated during the immediate operation of the plant are numerically much smaller, a few person-rem per gigawatt year. These population doses arise primarily from ^3H and noble gas releases.

The doses to individuals in the general population are ordinarily much smaller than the generally applicable limit of 500 mrem/yr (5.0 mSv/yr). The numerical guidelines used in examining the adequacy of proposed radioactivity control systems for nuclear power plants specify doses in the vicinity of 10 mrem/yr (100 μSv/yr) as design goals (see Table 10.3). These guidelines were chosen by comparing the availability and cost of control systems with the benefits of reduced radiation exposures [17]. Similarly, EPA regulations on overall fuel cycle emissions [1] were set on the basis of similar considerations. These annual doses are much less, by at least an order of magnitude, than the generally applicable standards of 500 mrem/yr (5.0 mSv/yr) for an individual member of the general public and substantially less than normal background dose levels which is roughly 100 mrem/yr (1.0 mSv/yr). Measuring reactor-caused increases against a background of this size can be a challenge for a monitoring system, particularly when the natural background varies as a function of both time and place. Several reports [18,19] are

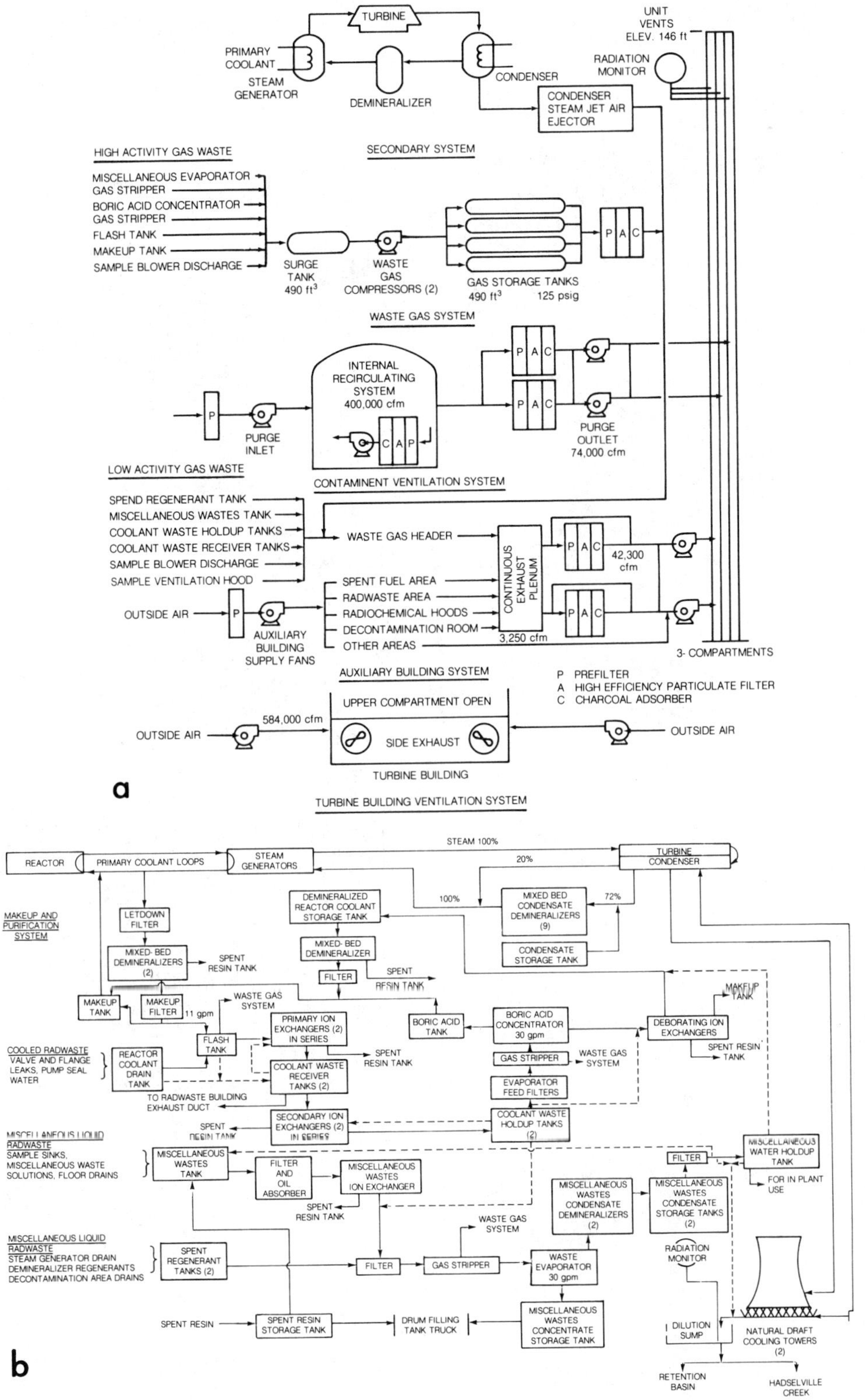

Figure 10.8. Radioactivity control systems: (*a*) gaseous; (*b*) liquid (reproduced from U.S. Atomic Energy Commission, Ref. 21).

Table 10.2. Projected Yearly Routine Gaseous Emissions from LWR Power Plants (Ci/reactor-year, uranium fuel cycle)

Radionuclides	PWR (BWR)[b]	Rancho[c] Seco (PWR)	Diablo[d] Canyon (PWR)	Kosh.[e] (PWR)	GESMO[f] (PWR)	Processing[g]
^{3}H[a]	10 to 50(same)	900	Not given	580	1100	2.1×10^{4}
^{131}I	0.016(0.016)	0.011	(0.28)	0.009	0.025	$0.06(^{129,131}I)$
Krypton, Xenon	7000(50,000)	12,000	(3700)	330	13.000	$1.8 \times 10^{5}(^{85}Kr)$
^{14}C	Not given	Not given	Not given	9	8	15
Transuranics					0.004	

[a]It should be noted that a trade-off can occur between tritium discharges into air versus water, so that these numbers can be highly variable, even aside from normal considerations of control technology.
[b]From Pigford et al., Ref. 12.
[c]From the 1973 Final Environmental Statement for the 900-MW_e PWR unit at Rancho Seco, Ref. 14.
[d]From the 1973 Final Environmental Statement for each 1060-MW_e PWR unit at Diablo Canyon, Ref. 15.
[e]From the 1976 draft environmental statement for each 994-MW_e PWR unit at Koshkonong, Ref. 16.
[f]PWR releases assumed by GESMO[2] for radwaste systems of the current type; releases calculated in accordance with Regulatory Guide 1.112.
[g]Approximate numbers for releases in curie per gigawatt-year from reprocessing plants; taken from Ref. 12, except for ^{14}C, which is taken from Ref. 2. For radionuclides other than ^{14}C, the two references broadly agree, except that the numbers are more difficult to extract from Ref. 2.

Table 10.3. Nuclear Power Plant Off-Site Dose-Limiting Objectives (10 CFR 50, Appendix I)

Type of Dose	Design Objective (per unit)
Liquid effluents	
Dose to total body from all pathways	3 mrem/yr
Dose to any organ from all pathways	10 mrem/yr
Gaseous effluents (only for noble gases)	
Gamma dose in air	10 mrad/yr
Beta dose in air	20 mrad/yr
Dose to total body	5 mrem/yr
Dose to skin	15 mrem/yr
Radioiodines and particulates released to the atmosphere	
Dose to any organ from all pathways	15 mrem/yr

available that summarize radioactive wastes released from, or shipped off site by, nuclear power plants in the United States; these summarize measurements of specific radionuclides both in gaseous and liquid effluents and in solid waste shipments.

10.3.2. Accidental Releases from Nuclear Power Plants

The ceramic fuel pellets, the sealed cladding, and the closed reactor coolant system typical of LWRs and most other reactor types serve as barriers to releases of radioactivity under both routine and accident conditions. However, accidents might occur that would breach one or more of these intrinsic barriers, possibly leading to increased, even massive, releases of radioactivity. To cope with the spectrum of failures that might occur, these intrinsic features are supplemented by several engineered safety systems. These systems do not prevent failures, but are designed to limit the course of the accident, either by preventing substantial release of radioactivity bound in the fuel or by containing radioactivity that succeeds in escaping from the primary coolant system.

Under ordinary conditions, almost all of the plant radioactivity is contained in the coolant system or in auxiliary systems, such as the radioactive effluent control systems. Releases from these auxiliary systems or even from the primary system may be regarded as relatively minor, provided the fuel does not overheat to such an extent that contained volatile fission products are released from the cladding or, worse yet, the fuel melts.

Reducing the possibility of such damage to the core is a major goal of engineered safety systems. Such damage, however, can occur—once a reactor has been operating for some time—by losing the coolant from the reactor vessel. Although for an LWR such loss of coolant terminates the chain reaction (due to lack of neutron moderation), the fission products continue to decay, releasing energy at a substantial rate. Immediately after shutdown, the decay power is about 7% of the full power level. This reduced power level is sufficient to melt the fuel cladding and/or fuel very rapidly if cooling is not available. To cope with the possibility that coolant may be lost, power reactors are equipped with emergency core cooling systems designed to replenish the coolant inventory under a wide range of conditions.

The primary and emergency cooling systems are designed with a high degree of redundancy and independence to prevent major damage to the core.

In the United States, the primary system is surrounded by a containment building designed to withstand overpressures that might be expected on breach of the primary system. This containment serves to confine radioactivity releases, whether large or small, from the reactor system. The containment includes cooling systems to reduce the internal pressure, filter systems to collect radioactivity, and spray systems.

Considering the variety of failures that may occur in the basic reactor systems and in the event of failure of the safety systems, a wide range of radioactive releases might result from possible accidents. These releases would vary as to size, radionuclide mixtures, and dispersion conditions (such as rate of release, amount of heat released, meteorological conditions). Finally, the population at risk varies markedly from one power plant site to another. Large uncertainties and substantial controversy exist over the probability and consequences of reactor accidents, arising directly from imprecise information on failure rates, the effectiveness of safety systems, and the details of accident sequences and subsequent exposures. Lewis [20] examines the state of LWR safety. In 1975, a major AEC/NRC study [21] of the risk from nuclear power was completed. Many of its results are uncertain and controversial, as indicated in a review [22] of the study. It is useful to note that the study estimates the average population dose from accidents from power plant operation is in the range of hundreds of person-rem per gigawatt-year. This is much greater than the yearly population doses in the vicinity of a nuclear power plant and comparable to the dose commitment estimated from ^{14}C. Nero and Farnaam [19] also review studies of reactor safety.

Extremely high doses could result from nuclear accidents. These doses would arise from substantial releases of a wider range of radionuclides than are given off under normal conditions. In particular, radionuclides such as cesium-137 and strontium-90 that ordinarily remain in the fuel may be released. Moreover, compared to normal conditions far greater amounts of I and other volatile radionuclides might escape. However, in addition to causing high doses to individuals, an accident can release radioactivity that persists, subjecting large numbers of people to chronic low levels of radiation over a large area. Either of these cases presents a serious monitoring problem; the first because ordinary environmental monitoring instruments are not designed to measure high radiation levels; at the other extreme, low levels of radiation may be difficult to measure, merely because of the interference from ordinary background radiation.

The present regulatory approach to nuclear accidents concentrates on assuring that doses to members of the general public do not exceed certain limits. The nominal accident dose of interest in examining a proposed site during licensing proceedings is 25 rem (250 mSv). (See Ref. 4, Part 100.11.) On the other hand, EPA protective action guides [23], specify 1 to 5 rem (1×10^{-2} to 5×10^{-2}Sv) as the dose at which emergency actions should begin.

10.3.3. Fuel Handling and Other Occupational Exposures

The operation of a nuclear power plant raises the ambient radiation levels nearby only slightly, but this occurs to a much larger degree inside the plant. As a result, plant workers receive radiation doses that are larger than those imposed on the general public. The bulk of these increased exposures arise, not when the reactor is operating, but during refueling and maintenance.

Radioactivity collects at several points in the plant systems. Most resides in the fuel, and, not surprisingly, the bulk of worker doses is associated with x-ray and γ-ray doses from handling fuel during refueling operations. This is true even though precautions are taken that control these doses, such as handling the fuel under water. In addition, radioactivity may collect in the primary system and in the radioactive effluent control systems. Maintenance associated with these systems can lead to substantial x-ray and γ-ray exposures unless proper precautions are taken. Plant operators take care that occupational exposure limits [such as the 5 rem/yr (5×10^{-2} Sv/yr) whole-body dose] are not exceeded, but often workers are permitted to receive doses close to this amount.

10.3.4. Transportation and the Balance of the Fuel Cycle

Of the few person-rem (person-sievert) per gigawatt-year general public exposures that are directly associated with power plant operations, about half is estimated to arise from transport of the spent fuel [14–16]. This fuel is transferred in large shipping casks, designed to provide physical protection, cooling, and radioactive shielding. Care must be taken that this shielding is adequate and particularly that it remain so after even serious accidents.

In the full nuclear fuel cycle, there are of course other possibilities for radiation exposures than the ones so far discussed. Release points are indicated in Fig. 10.7. Other parts of this book discuss some of the principal fuel cycle operations: uranium mining and milling; fuel reprocessing, and waste disposal. Although their average impact, too, is uncertain, each of these could cause population exposures comparable to those already discussed for nuclear power plants.

10.4. MEASUREMENT CONSIDERATIONS

To assure that nuclear power plants meet applicable regulations and specifications for emissions and exposures, the NRC has formulated regulatory guides that specify monitoring practice within the plant and in its environs [5]. Of particular interest are: RG 1.21 ("Measuring, Evaluating, and Reporting Radioactivity in Solid Wastes and Releases of Radioactivity in Solid Wastes and Releases of Radioactive Materials in Liquid and Gaseous Effluents from Light-Water-Cooled Nuclear Power Plants," Rev. 1, 6/74); RG 4.1 ("Programs for Monitoring Radioactivity in the Environs of Nuclear Power Plants," Rev. 1, 4/75); and various guides from Division 8 ("Occupational Health"). These guides, and the requirements that they are intended to fulfill, set forth programs for dose monitoring within the plant, effluent release monitoring, and environmental monitoring (with particular attention to critical pathways leading to human exposures). These programs must serve in a variety of situations, including normal operation, operational transients, and severe accidents. Both the level and type of radiation to be measured can depend on the situation at hand. For example, most of the radionuclides released during normal operation emit β- andγ-radiation; during accidents, significant amounts of α-particle emitters could also be released.

For the purposes of this discussion, measurements are regarded to be applied within the plant itself. Plant monitoring is taken to include programs for measuring radiation fields within the plant, as well as those for measuring releases at the stack or through other effluent streams. However, it does not include process radiation monitoring, even though process monitors may provide a reference for correlating possible sources of environmental releases. Monitoring operating conditions with sensors (especially those operating near the core) may require substantially different designs than those for use in making low-level measurements.

10.4.1. Plant Monitoring

One general requirement of plant monitoring is that sufficient capabilities be present to monitor radiation fields within the plant. Instrumentation of various types is available for such measurement. These have been discussed in early chapters. Measurements of gross β- and γ-activity provide an overall reference, although more specialized instruments may be employed to acquire more detailed information. In addition, (thermal) neutron detectors are placed to indicate large fluxes. Specific attention to the various components of radiation due to individual sources is not always necessary, although it is certainly useful information when attempting to diagnose a leak and document its history.

Within the plant, instruments may be fixed in position, used for special surveys, or designed for personnel dosimetry. Wall-mounted β-γ/neutron counters should have sufficient sensitivity (down to about 1 μR/hr) to indicate ordinary daily variations. A distinct set of instruments is required for measuring accidental releases; ranges up to about 10,000 R/hr are appropriate. For either type, fail-safe audible and visual alarms, as well as provision for recording measurements, are ordinarily available.

Portable instruments operating on similar principles as fixed units may also be used in surveys. On the other hand, surveys may be conducted by securing samples that are then analyzed in fixed instruments. As an example, surface radioactive contamination may be monitored by taking surface wipes for analysis by such instruments.

Dosimeters for measuring employee exposures may be both pocket ion chambers, worn outside of heavy clothing, and integrating dosimeters (of film or thermoluminescent type), changed periodically. These provide information, on short-term doses and on long-term (integral) doses received by employees.

A second requirement of plant monitoring systems is that they provide adequate information on radioactivity leaving the plant, whether as packaged waste material, or as liquid or gaseous effluents. Stack effluent monitors can provide a temporal record of emissions, in addition to integrating the total amount released. They can also serve as a warning of abnormalities. However, for operational occurrences and accidents, the instrument system should be designed for unusually high radiation levels [24].

A principal requirement for these instruments, as well as those indicated under environmental monitoring, is that periodic calibration and maintenance procedures be performed. For example, remotely controlled or portable radioactive sources can be used routinely to check the calibration and operation of permanently installed monitoring instrumentation. In addition, a calibrated electronic test pulse may be used for rapid identification of certain types of instrument failures. A part of this inspection and testing program is also normally devoted to assuring that sampling, such as in the stack, [25] is occurring properly.

As suggested by such considerations, monitoring instruments are not to be viewed in isolation, but rather as part of a monitoring system that is designed to meet certain objectives and must operate within certain specifications, all of which are reviewed during the licensing process. Furthermore, with recent advances in data-handling equipment, individual monitoring units may be incorporated into a system that not only assures proper functioning of the units, but reads data out of these instruments, and performs necessary conversions (e.g., from count rate to dose rate). The systems may also organize the resulting information for examination and, if necessary, response (see, for example, Ref. 26). Such systems would, of course, be designed to meet requirements specified in applicable regulations.

Information on satisfying NRC requirements for waste and effluent monitoring is given in Regulatory Guide 1.21 [5]. See also the American National Standards Institute recommendation on continuously monitoring effluents [27]. Requirements for in-plant radiation monitoring are also given in other regulatory guides, including Division 8 ("Occupational Health"). Other important considerations for plant monitoring, including program design, quality assurance, and inspection are discussed briefly in the section following.

10.4.2. Environmental Monitoring

The instruments for monitoring in the environs of a power plant are based on the same principles as those indicated previously. However, the requirements on sensitivity may be more stringent because of the relatively small increases in levels typically caused by nuclear power plant effluents. For the same reason, variations of background level with place and time, whether hourly, daily, or seasonally, are an important monitoring consideration. Local background radiation exposures are typically about 100 mrem (1 mSv), with $\pm 50\%$. It may therefore be difficult to measure the additional dose from a properly operating reactor, which is ordinarily expected to contribute less than 5 mrem/yr (50 μSv/yr) at the site fence. Another primary consideration for environmental monitors is that accuracy be maintained over a wide range of environmental conditions, including large variations in temperature, humidity, pressure, wind speed, precipitation, and dust.

Simple passive monitors, the thermoluminescent dosimeter and photographic film, serve as the backbone of environmental monitoring systems. They are easily used to provide a record of penetrating radiation at numerous points in the area to be monitored, and they are conveniently used for archival storage. However, more complex instruments or techniques may be required for specific monitoring purposes, including observation of particular radionuclides, monitoring of critical pathways, and emergency radiation monitoring. The choice of instruments depends on the purpose and the design of the monitoring program, and may include any of the detectors described in Chapters 4–8 both for survey and stationary devices.

For additional information the reader is referred to Ref. 5 for the NRC regulatory guide on environmental monitoring programs, RG 4.1. In addition, the EPA has published an Environmental Radioactivity Surveillance Guide [6]. References 28 and 29 are useful guides to surveillance around nuclear facilities. As illustrative of the types of measurements that may be made around nuclear facilities Refs. 13, 27, and 30 describe the monitoring in the vicinity of a PWR and a BWR, respectively. For a discussion of environmental impact monitoring, see Ref. 31. A guide to general environmental monitoring has recently been published by the

National Council on Radiation Protection and Measurements (NCRP) [32]. Questions of program design are the subject of the next section.

10.4.3. Monitoring Programs

The foregoing discussion has alluded to several generic questions that affect the design of a monitoring program. These include choice of instruments, sampling method and rate, data handling, and testing maintenance. These matters are discussed in moderate detail in the references cited previously, but it is useful to summarize them briefly.

Instrument type and *sampling method* or *frequency* depend strongly on the purpose of the monitoring program. Much of the previous discussion emphasized *in situ* measurements, but grab sampling techniques, where a sample is taken at one location, then analyzed elsewhere, are important, particularly for sensitive monitoring. *In situ* techniques may employ fixed instruments of various sorts, in the plant, at emission points, or in the environment. Also in use are various types of survey instruments and dosimeters. All of these are discussed in detail elsewhere in this volume.

In many cases, such as at the stack or liquid effluent stream, fixed instruments monitor continuously. This provides both an alarm capability and a record of emissions. Based on meteorological models and local meteorological data, emission data may be used to estimate doses resulting from plant releases.

On the other hand, a substantial part of the monitoring program, particularly in the environs of the plant, may depend on taking samples for subsequent analysis. Samples may be taken of environmental media and of plants and animals: air, water, fish, birds, corn, cows, milk, thyroid, and so on. Depending on the particular monitoring program, the sampling frequency may vary (daily to even yearly), and the radionuclides chosen may vary (^{3}H, ^{85}Kr, ^{90}Sr, ^{131}I). The purpose of such a program would be both to check the plant emission data and, more importantly, to evaluate radionuclide transport in the environment, particularly along critical pathways. Both for these pathways and for the more general environmental monitoring program, the site, type, and rate of sampling should be carefully chosen to meet the monitoring objectives. The techniques applicable in the vicinity of a nuclear facility will differ in some cases from those for monitoring far from such facilities not only because of the difference in intensity but also because the mix of radionuclides may vary with proximity and the radionuclides of interest depend on the monitoring objectives. Table 10.4 is a summary of the advantages and disadvantages of a variety of sampling methods for the principal types of environmental media [28].

Although samples may be examined only for gross activity or for overall α-, β-, or γ-activity, they may also be analyzed for radionuclide content. Techniques for specific radionuclide monitoring are discussed in Chapters 4–8 and also in Part III, Chapters 13–20. For gas and liquid phase measurements, as well as soil and biological samples, chemical separation techniques are employed, which may be followed by a variety of instrumental measurements. Airborne particulates are typically collected by air filters for subsequent analysis, perhaps using similar instruments. These instruments may include spectrum analyzers discussed in Chapter 4.

Designing a monitoring program also requires choosing an approach to data handling and use. As noted under plant monitoring, the instruments themselves may be incorporated into a monitoring system that collects the data, performs necessary conversions, and displays or otherwise transfers the results for designated purposes.

Table 10.4. Advantages and Disadvantages of Sampling Methods for Various Samples (Reproduced from Moeller, Ref. 28, Nuclear Safety © 1978)

Type of Sample and/or Method of Collection	Advantages	Disadvantages
	Atmospheric	
Particulates		
Constant air monitors	Real-time readout, chart record	High cost, special maintenance requirements, lack of sensitivity, interference (e.g., radon)
Pump and filter	Sensitivity, low cost	Time delay in obtaining results
Electrostatic precipitators	Sensitivity	High cost, variable efficiency
Halogens		
Charcoal	All forms collected	Not nuclide-specific
Molecular sieves		High cost, low capacity, not nuclide-specific
Bubblers	Sensitivity, low cost	Special maintenance requirements, not all forms collected
Noble gases		
Evacuated chamber	Portability	Lack of sensitivity, not continuous
Compressed air	Sensitivity	Not continuous
Cryogenic	Sensitivity	High cost, not portable
Direct radiation (thermoluminescent dosimeter)	Low cost, dose integration, direct dose measurement	Lack of sensitivity, high background, not nuclide-specific
Ion chamber	Real-time readout, chart record sensitivity direct dose rate measurement	High cost, special maintenance requirements
Tritium		
Cryogenic	No sample preparation	High cost, not portable, HT not collected, collection not ^3H-specific
Bubblers	Low cost, no sample preparation	Special maintenance requirements. Lack of sensitivity, HT not collected, collection not ^3H-specific
Molecular sieve	Portability, ^3H-specific (with distillation)	Sample preparation needed HT not collected, capacity varies with relative humidity

Table 10.4. (Continued)

Type of Sample and/or Method of Collection	Advantages	Disadvantages
	Atmospheric	
Drierite	Portability, ^{3}H-specific (with distillation)	Poor water capacity, capacity varies with relative humidity
Silica gel	Portability, ^{3}H-specific (with distillation)	Sample preparation needed HT not collected, capacity varies with relative humidity
	Terrestrial	
Milk	Direct dose vector, ease of data interpretation	Not always available
Foodstuffs	Direct dose vector, ease of data interpretation	Not always available, weathering effects
Wildlife	Direct dose vector	High mobility, not always available, data difficult to interpret
Vegetation	Availability, multiple accumulation modes	Not nuclide-specific, data difficult to interpret, weathering effects
Direct instrument reading	Low cost, speed	Lack of sensitivity, not nuclide-specific
Deposition collection	Defined time span for sampling period, area and collection factors known	Weathering effects
Soil sampling	Good time integrator	High cost, lack of representativeness, long time delay in obtaining results
	Aquatic	
Surface water	Availability	Not dose related
Groundwater	Indicator of waste management	Not always available, data difficult to interpret
Drinking water	Direct dose vector	Lack of sensitivity
Aquatic plants	Sensitivity	Data difficult to interpret
Sediment	Sensitivity, integration	Data difficult to interpret
Fish, shellfish	Direct dose vector, sensitivity	Frequent unavailability, high mobility
Waterfowl	Direct dose vector	Frequent unavailability, high mobility, data difficult to interpret

In addition, the system may partially check instruments to ensure proper functioning, an important requirement, as discussed later. To some extent, systems for environmental monitoring may also be integrated in a similar way. In reducing data, monitoring systems—whether for plant or environment—may even perform dispersion calculations (based on meteorological models and data supplied), as well as pathway analysis (based on appropriate pathway models).

Even if it is not appropriate to devise such a monitoring system, it is important that the overall program be organized to meet specified objectives. This must include assurance that the program operates properly. In Regulatory Guide 4.15 [5] the NRC specifies "Quality Assurance for Radiological Monitoring Programs (Normal Operation)—Effluent Streams and the Environment." A quality assurance program should include: acceptance testing of instrumentation, routine calibration and routine replicate sampling, laboratory cross-checking, procedural audits, and program documentation [29].

10.5. SUMMARY

The commercial nuclear power program in the United States is currently based on use of light-water reactors that use low-enriched U fuel, although other types of power plants and fuels are feasible and may later be used. The principal federal entities that regulate operation of these power plants for the protection of both workers and the general public are the NRC and the EPA. The design of a monitoring program to assure compliance with applicable requirements must consider both the limitations imposed on emissions or exposures and the possible sources of radiation or radioactivity at the particular facility. Substantial monitoring efforts are required both at the facility, internally and at emission points, and in the environment, which may be affected by plant emissions. Appropriate instrumentation has been designed for the important emissions from nuclear power plants, and the reader is referred to more detailed discussions of instruments designed for particular types of radiation, Chapters 6–9, or particular radionuclides, Chapters 13–20.

REFERENCES

1. U.S. Environmental Protection Agency, "Final Environmental Statement on 40 CFR 190, Environmental Radiation Protection Requirements for Normal Operations of Activities in the Uranium Fuel Cycle," EPA 520/4-76-016, 2 vols., 1976.
2. U.S. Nuclear Regulatory Commission, "Final Generic Environmental Statement on the Use of Recycle Plutonium in Mixed Oxide Fuel in Light Water Cooled Reactors," NUREG-0002, 5 vols., 1976, plus errata Nov. 26, 1976.
3. Technical Information Center, "Nuclear Reactors Built, Being Built, or Planned in the United States as of Dec. 31, 1977," Department of Energy TID-8200-R37 (NTIS), 1978.
4. Code of Federal Regulations, Title 10, 10 CFR 0.735-1, U.S. Government Printing Office, Washington, DC, 616 pp., 1978.
5. U.S. Nuclear Regulatory Guides, U.S. Nuclear Regulatory Commission, Washington, DC, 10 CFR 20, 50, and 100.

6. U.S. Environmental Protection Agency, "Environmental Radioactivity Surveillance Guide," Report ORP/SID 72-2, U.S. EPA, Office of Radiation Programs, Washington, DC 20460, 1972.
7. Code of Federal Regulations, "Subchapter F—Radiation Protection Programs," Title 40, Part 190, 40 CFR 104.1, p. 568, 1979.
8. Energy Research and Development Administration, "Advanced Nuclear Reactors," ERDA 76-107, May 1976.
9. A. V. Nero, *A Guidebook to Nuclear Reactors,* University of California Press, Berkeley, 1979.
10. L. C. Hebel, chairman, APS Study Group, "Report to the American Physical Society by the Study Group on Nuclear Fuel Cycles and Waste Management," *Rev. Mod. Phys.*, **50**(1), Part II, S1, 1978.
11. M. Eisenbud, *Environmental Radioactivity,* 2nd ed., Academic, New York, 1973.
12. T. H. Pigford, M. J. Keaton, B. J. Mann, and G. L. Sessler, "Comprehensive Standards: The Power Generation Case," Environmental Protection Agency Report PB-259-876, 1975.
13. U.S. Environmental Protection Agency, "Radiological Surveillance Studies at the Oyster Creek BWR Nuclear Generating Station", EPA-520/5-76-003, Washington, DC, 1976.
14. U.S. Atomic Energy Commission, "Final Environmental Statement Relating to Operation of Rancho Seco Nuclear Generating Station," DOCKET-50312-67, Draft Environmental Statement, Resources Agency of California, 1973.
15. U.S. Atomic Energy Commission, "Final Environmental Statement, Diablo Canyon Nuclear Generating Station," DOCKET-50275-42, Directorate of Licensing, 1973.
16. U.S. Nuclear Regulatory Commission, "Environmental Statement Related to Construction of Koshkonong Nuclear Plant," (NUREG-0079, 1976 Draft).
17. U.S. Atomic Energy Commission, "Final Environmental Statement Concerning Proposed Rule Making Action: Numerical Guides for Design Objectives and Limiting Conditions for Operation to Meet the Criterion 'As Low as Practicable' for Radioactive Material in Light-Water-Cooled Nuclear Power Plant Effluents," 3 vols., WASH-1258, 1973 (NTIS). Generally available in any of the public information libraries of the DOE Regional Offices.
18. U.S. Nuclear Regulatory Commission, "Radioactive Materials Released from Nuclear Power Plants, 1974," NUREG-0077, Office of Management Information and Program Control, Washington, DC, June 1976.
19. U.S. Environmental Protection Agency, "Summary of Radioactivity Released in Effluents from Nuclear Power Plants from 1973 through 1976", EPA-520/3-77-012, Washington, DC, 1977.
20. H. W. Lewis, chairman, APS Study Group, "Report to the American Physical Society by the Study Group on Light-Water Reactor Safety," *Rev. Mod. Phys.*, **47**, Supplement No. 1, S1, 1975.
21. U.S. Nuclear Regulatory Commission, "Reactor Safety Study: An Assessment of Accident Risks in U.S. Commercial Nuclear Power Plants," 9 vols., WASH-1400 (also called NUREG-75/014), 1975 (NTIS).
22. A. V. Nero and M. R. K. Farnaam, "A Review of Light-Water Reactor Safety Studies," Lawrence Berkeley Laboratory Report LBL-5286, 1977 (NTIS).
23. U.S. Environmental Protection Agency, "Manual of Protective Action Guides and Protective Actions for Nuclear Incidents," 1975.

24. J. W. Selby and C. M. Unruh, "Technological Considerations in Emergency Instrumentation Preparedness, Phase I: Current Capabilities Study," Battelle Pacific Northwest Laboratory Report No., BNWL-1552, P.O. Box 999, Richland, WA 99352, 1971. Available from NTIS.
25. B. P. Bergson and K. W. DeBaun, "Radiation Monitoring of Nuclear Power Station Stack Effluents," *IEEE Trans. Nucl. Sci.*, **NS-25**(1), 853, 1978.
26. W. C. McArthur and B. G. Kniazewycs, "Radiation Monitoring Systems: Current Status and Future Prospects," *IEEE Trans. Nucl. Sci.*, **NS-25**(1), 17, 1978.
27. American National Standards Institute, "Specification and Performance of On-Site Instrumentation for Continuously Monitoring Radioactivity in Effluents", ANSI N13.10-1974, IEEE, Inc., 345 East 47th Street, New York, NY 10017, 1974.
28. D. W. Moeller, J. M. Selby, D. A. Waite, and J. P. Corley, "Environmental Surveillance for Nuclear Facilities," *Nucl. Safety*, **19**(1), 66, 1978.
29. J. P. Corley, D. H. Denham, D. E. Michels, A. R. Olsen, and D. A. White, "A Guide for Environmental Radiological Surveillance at ERDA Installations," ERDA 77-24, ERDA Division of Safety, Standards, and Compliance, Washington, DC, 1977
30. U.S. Environmental Protection Agency, "Radiological Surveillance Study at the Haddam Neck PWR Nuclear Power Station," EPA-520/3-74-007, Washington, DC, 1974.
31. Battelle Pacific Northwest Laboratories and Columbus Laboratories, *Environmental Impact Monitoring of Nuclear Power Plants, Source, Book of Monitoring Methods,* 2 vols., AIF/NESP-004, Atomic Industrial Forum, 475 Park Avenue South, New York, NY 10016, 1975.
32. National Council on Radiation Protection and Measurements, "Environmental Radiation Measurements," NCRP Report No. 50, NCRP, 7910 Woodmont Avenue, Washington, DC 20014, 1976.

11 NUCLEAR SPENT FUEL REPROCESSING

Nuclear reactor fuel is routinely removed, after some period of use, from the reactor and replaced by fresh fuel. The "spent" fuel that is removed, however, still contains substantial amounts of useful fuel materials, including uranium (U), plutonium (Pu), and even thorium (Th) (should this come into use). Moreover, this fuel contains most of the radioactivity that was present at the nuclear power plant. Choosing what to do with this fuel directly addresses two related questions: how or whether to make use of the fissionable material remaining in the fuel; and what to do with the large amount of radioactivity contained therein.

As shown in Fig. 11.1, fuel removed from nuclear power plants is presently being left in storage, most often at the power plant sites. However, for light-water reactors (LWRs) it has been anticipated that this fuel would be reprocessed both to extract U and Pu and to concentrate the remaining large amounts of radioactive waste into a form suitable for disposal. Recycling the U and Pu into fresh fuel would reduce the demand on U resources by about 30%. Until a few years ago, it was thought that the cost of reprocessing was more than offset by the savings resulting from reduced purchases from U suppliers. In addition, it appeared that consolidating the radioactive wastes into a selected type of matrix would result in a superior disposal form, from the point of view of volume and, more importantly, stability. Finally, experience with reprocessing and recycling would provide a natural precedent for the liquid-metal fast breeder reactor (LMFBR) fuel cycle, which requires that Pu be recycled.

In recent years the perspective on what might be done with spent fuel has broadened considerably. In particular, substantial attention has been given both to the possibility of avoiding or deferring recycling (by storing or disposing of fuel directly; see Fig. 11.1) and even to alternative forms of reprocessing. This expansion of options has resulted from diverse factors, including: the recognition that at current U prices recycling is economically marginal; increased interest in the environmental aspects of spent fuel reprocessing and/or disposal; and substantially

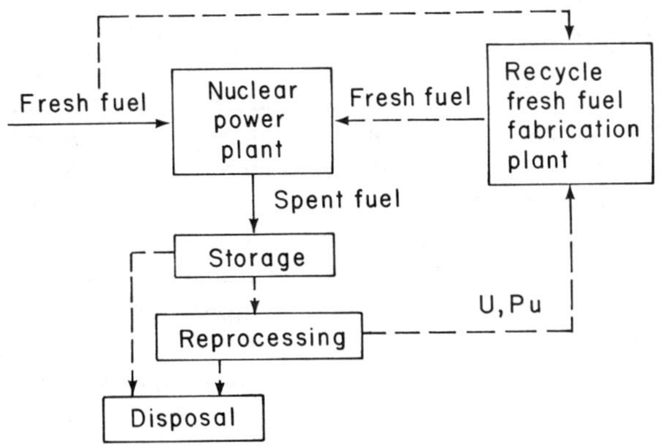

Figure 11.1. Possibilities for spent fuel handling. Spent LWR fuel is now being placed in storage; reprocessing, recycle, and disposal are not yet occurring on a large scale.

decreased projections of nuclear growth. This last factor could reduce the depletion rate of U resources and thereby the need for a breeder. It could also make alternative recycling schemes more attractive. The implications of basing nuclear power on a commerce in Pu, a material that may also be used in nuclear weapons, have also strongly affected the interest in alternatives to Pu recycling as conceived earlier.

United States commercial experience with spent fuel handling has included a large number of spent fuel storage sites, a single reprocessing plant, and no disposal facilities. That reprocessing plant, owned by Nuclear Fuel Services (NFS) and located at West Valley, New York, began operations in 1966 with a capacity of 1 TeHM (metric tonnes of heavy metal, i.e., U and Pu) per day. By 1972, this plant closed for improvements and expansion to 2 to 3 TeHM/day. However, a decision was subsequently made to close the plant permanently. This plant has provided the only commercial U.S. experience with reprocessing operations and the resulting effluents.

Two other plants have been constructed but not operated. The first was the General Electric Midwest Fuel Recovery Plant (MFRP) at Morris, Illinois. This plant was scheduled to begin operation during 1973. However, after difficulties discovered during preoperational testing, General Electric chose not to operate the plant. So far, it has only been used for spent fuel storage.

A 5-TeHM/day plant, the Barnwell Nuclear Fuel Plant (BNFP), has been built at Barnwell, South Carolina, by Allied-General Nuclear Services. Although the facilities for separating Pu, U, and wastes have been completed, construction of facilities for solidifying wastes, for converting Pu to the proper form, and for handling materials contaminated with transuranics has not begun, pending decisions on the recycling of Pu and on criteria for waste forms, as well as on the licensing of reprocessing per se.

Thus, further development of commercial reprocessing and related forms of fuel fabrication and waste disposal must await a decision in favor of this direction for commercial nuclear power. In the meantime, the need for spent fuel storage capacity increases, if only on an interim basis, while the need to develop and choose a satisfactory means for final disposal of nuclear wastes continues.

These proposed questions have been seriously considered during the last few years, both in the United States and abroad. This examination has led, in the case of the United States and Germany, to a substantial slowing of the development of fuel processing capabilities. On the other hand, several countries, such as France and the United Kingdom, have moved forward vigorously with such programs.

Information relevant to spent fuel processing has been set forth in great detail in a number of reports of government agencies. Plutonium recycling in LWRs was examined in Ref. 1, waste disposal in Ref. 2, and the LMFBR program in Ref. 3. However, these reports arose largely out of a milieu that presumed Pu recycling and the introduction of breeders. More recent studies [4–6] by other entities have examined a broader range of possibilities, with special emphasis on the implications of reduced growth rates on alternative forms of nuclear power and on the risks of nuclear weapons proliferation. Alternatives for spent fuel handling and storage are considered in Ref. 7. Nero [8] gives an elementary summary of various possibilities and attendant issues.

11.1. SPENT FUEL RADIOACTIVITY

As indicated in Chapter 10 on nuclear reactors, an LWR now discharges about 30 Te (metric tonnes) of fuel annually, containing billions of curies (10 EBq) of radioactivity. This radioactivity is a major concern, regardless of the ultimate form of the fuel cycle. The amounts discharged are extremely large and, as indicated in Fig. 11.2, substantial amounts persist over not just decades—the lifetime of a nuclear power plant, but millenia—the length of human history. However, the amount of radioactivity does decrease relatively rapidly after removal of the fuel from the reactor. To facilitate subsequent handling and processing (if any), it had long been presumed that LWR fuel would remain in at-reactor-storage for a cooling period of 150 days.

As seen in Table 10.1, most of the radioactivity present in the discharge fuel is fission products, and the bulk of the remainder is actinides, including U, Pu, Am, and Cm. The bulk of the inventory consists of radionuclides with very short half-lives. However, a sufficiently large portion have half-lives of years or more, so that large amounts of radioactivity persist. Of the predominant fission products, Xe and iodine decline very rapidly in amount (within months); the rare earths and the zirconium-antimony class (Fig. 11.2) then dominate until a few years pass, after which ^{90}Sr and ^{137}Cs contribute the largest amount for the first few centuries, with ^{85}Kr and certain of the rare earths making some contribution during this period. After the sharp decline in total radioactivity occurring in the first few centuries, a plateau is reached that lasts millenia. This arises from certain fission products, as well as from actinides such as ^{239}Pu (24,000-year half-life). Note, however, that Fig. 11.2 only displays 0.5% of the Pu, the amount that nominally would remain in the high-level wastes were reprocessing to occur.

Such considerations give a rough indication of the need for sequestering the radioactivity in the spent fuel. Note that some of the major fission product contributors to the total radioactivity are radionuclides of major concern for routine emissions (I, Kr, Xe) or accidental releases (Sr, Cs) from nuclear power plants. However, whether for emissions or for waste isolation considerations, amounts of radioactivity are not necessarily a direct measure of biological significance. Thus, for example, tritium (^{3}H) and ^{129}I are not major contributors to the total radioac-

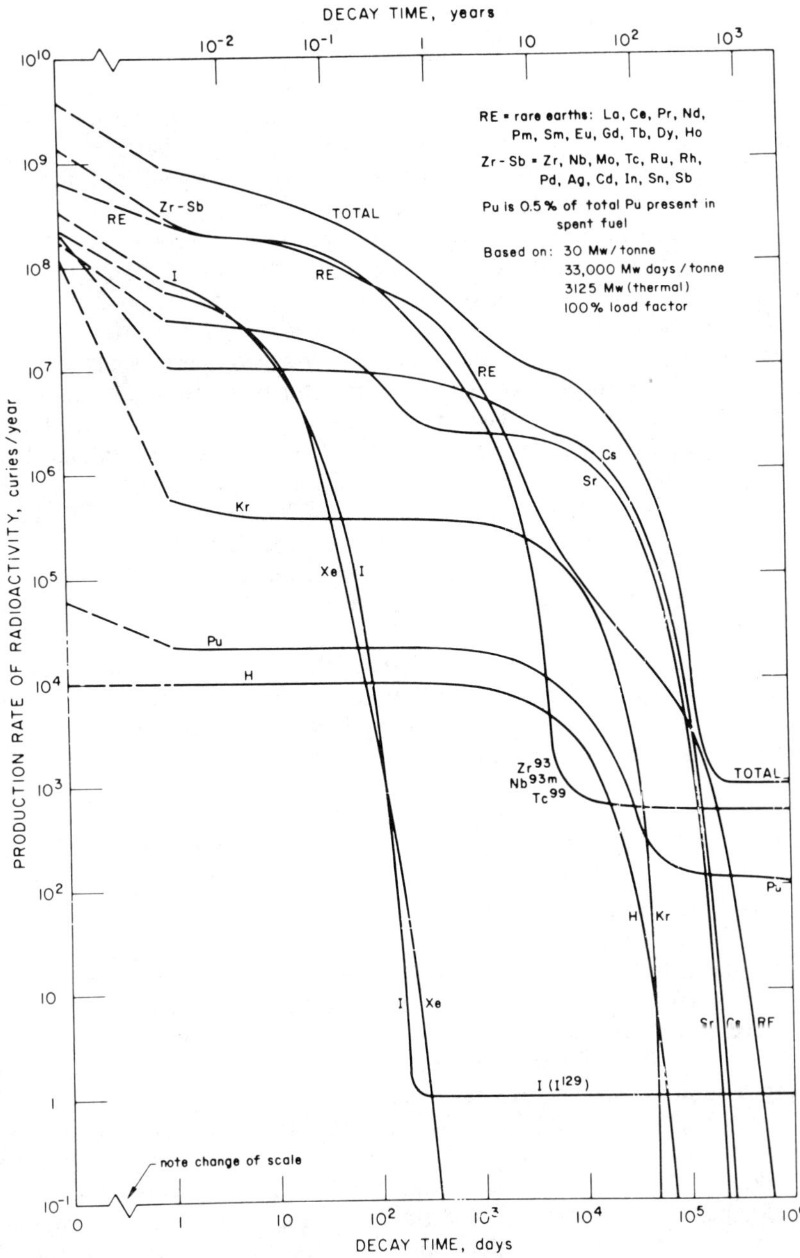

Figure 11.2. Radioactivity produced in year by a 1000 MW_e Light-Water Nuclear Power Plant (reproduced from Pigford et al., Ref. 9).

tivity, nor is carbon-14, which is not even shown in Fig. 11.2. At the very least, a hazard index must consider the relative biological significance of the various nuclides. The usual maximum permissible concentrations (MPCs) could serve as part of such an index. This simple approach would not, however, give any weight to the probability that the radionuclides considered might reach humans. As noted in Ref. 4, it is difficult to construct a completely satisfactory hazard index. It is interesting to note, though, that following a sharp decline that ends after about 1000 yr, MPC-based ingestion hazard indices change relatively slowly with time—only about an order of magnitude over the period 1000 to 1×10^6 yr (Ref. 4, particularly Figs. 7B1 and 7B2). This is consistent with the form of the curves in Fig. 11.2.

Consideration of the total problem indicates that for periods approaching the length of mankind's history, large amounts of reactor-produced radioactivity need to be isolated from the biosphere. The annual feed to a uranium-fueled LWR contains only about 100 Ci (3.7 TBq), in the form of ^{238}U and ^{235}U. The annual discharge contains billions of curies (millions of TBq) and this decreases to less than 1000 Ci (37 TBq) (exclusive of the Pu) only after about 1000 yr. The 4×10^6 Ci (15×10^4 TBq) annual Pu discharge, if not recycled, decreases to about 2×10^4 Ci (740 TBq) by 1000 yr. Thus from the environmental point of view several needs have to be considered: (*a*) short- to medium-term possibilities, whether spent fuel storage or reprocessing, that adequately handle the fuel for periods of years to decades; (*b*) medium-term waste disposal that effectively isolates the great amount of radioactivity, largely fission products, that remains for hundreds of years; and (*c*) very-long-term isolation of radioactivity, both fission products and actinides, that remains even after these periods.

11.2. SPENT FUEL, REPROCESSING, AND WASTE FACILITIES

11.2.1. Spent Fuel Storage

Most spent fuel is presently stored at reactor sites. However, this capacity is being pressed, and because of the much slower than anticipated development of reprocessing and recycling, a clear need has arisen for larger storage facilities at which spent fuel could remain for more than a year or so, perhaps decades. Such facilities, which may be away from reactor sites, need not differ greatly from reactor storage, which places spent fuel assemblies in water pools for cooling and shielding. However, for relatively long-term retrievable storage, methods for dry storage may be both feasible, because of lower cooling/shielding requirements, and attractive, because of simpler maintenance. A description of pool storage may be found in environmental statements and safety analysis reports for nuclear power plants. Reference 7 discusses an alternative for LWR spent fuel handling and storage.

11.2.2. Fuel Reprocessing

The prime function of a reprocessing plant is to dissolve the spent fuel and separate the resulting solution into waste and fuel streams. Such separation has been performed for more than three decades, but usually in military facilities whose aim is to extract Pu for use in weapons. The technique employed for light-water reactor fuel would be, after 5 or 6 months storage, to dissolve the fuel in nitric acid, then to strip out and purify U and Pu. The remaining nitric acid solution, containing both fission products and activation products, is often referred to as *high level wastes*. However, beyond this separation, a related function that would ordinarily be performed at a reprocessing complex is to convert both waste and product streams to forms that are suitable for their disposal or use. A functional diagram of a reprocessing plant that includes such conversions is given in Fig. 11.3.

For U the product would be uranium hexafluoride (UF_6), the chemical form needed for feed to uranium enrichment plants. The appropriate product for Pu would be plutonium dioxide (PuO_2), the chemical form used in fresh fuel. In fact the Environmental Protection Agency (EPA) 10 CFR 71.42 [10] requires that Pu be shipped in a solid form, so that such conversion is effectively required at the reprocessing plant. However, sufficient uncertainties still exist that the builder of a reprocessing plant (such as that at Barnwell) does not have design criteria for this conversion. For example, to reduce the weapons usability of transported material, reprocessors may be required to mix the Pu with a larger amount of U.

Similarly large uncertainties exist for the high level wastes. 10 CFR 50, Appendix F, specifies that high level liquid radioactive wastes be converted to solid form within 5 yr of generation and transferred to a federal repository within 10 yr.

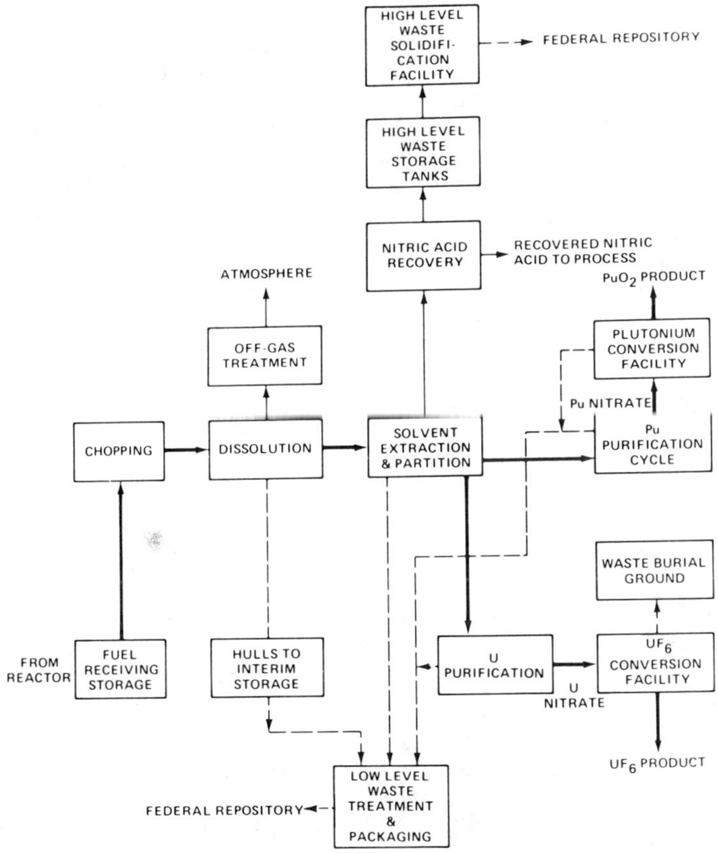

Figure 11.3. Simplified flow diagram of reprocessing plant complex (reproduced from Nuclear Regulatory Commission, Ref. 1).

However, a form for disposal has not been specified, so that—even if reprocessing itself were approved—it could not proceed because design criteria for waste conversion facilities are not available. The possibility given most attention is vitrification, that is, drying the wastes and incorporating them into a glassy material.

The processes for separation facilities have been summarized in Ref. 11 as follows:

> ...Typically, a fuel recovery process entails shearing the fuel (to rupture the corrosion-resistant sheath and expose the fuel), dissolution of the fuel in nitric acid, separation and purification of the uranium and plutonium by solvent extraction and ion exchange, and conversion of the product nitrates to oxides suitable for refabrication into fuel elements....
>
> The spent fuel is transported from the reactor to the reprocessing plant in heavy, shielded casks. The cask is unloaded in a water-filled pool, and the fuel is stored under water, which serves both as a transparent radiation shield and as a coolant. The fuel elements to be processed are transferred to a head-end cell and sheared into two-inch lengths to expose the inner core, which is then leached with nitric acid in batch-dissolving tanks. The leached hulls constitute a solid waste that is ultimately disposed of by land burial. The nitric acid solution of the fuel, containing the uranium, plutonium, and nearly all of the fission products, is the feed solution for the solvent extraction process.
>
> Solvent extraction processes exploit the wide difference in concentration distribution between two immiscible phases—the organic and the aqueous. Nearly all major fuel reprocessing facilities employ some form of the Purex process, which makes use of the organic complexing compound, tributyl phosphate (TBP), in an inert hydrocarbon diluent. When this organic mixture is brought into countercurrent contact with the aqueous feed solution, the TBP extracts both the uranium and the plutonium into the organic phase, leaving the fission and corrosion products behind in the aqueous phase.
>
> ...In addition to the primary process, there are many auxiliary operations: treatment of the solvent to provide for its reuse, recovery of nitric acid from the aqueous streams, management of the gaseous, liquid, and solid waste effluents, and the specialized techniques and equipment required for process control and personnel protection.

As suggested in this discussion, radioactive waste from a reprocessing plant would include, not only the high level waste that draws most attention, but the cladding hulls, a variety of low level waste streams, and various kinds of contaminated material. A final residue to be considered is the reprocessing equipment itself.

11.2.3. Waste Disposal

An approach to waste disposal has not been chosen. Although considerable effort has been expended on this question, both on methods for solidifying or encapsulating wastes and on selection of disposal sites, a sufficiently detailed understanding of the requirements or consequences of long-term disposal does not exist to specify criteria for waste form or site design. Moreover, most work has presumed that reprocessing would occur and that the material of concern is high level waste in the usual sense, i.e., radioactivity separated from the spent fuel. Reference 12 provides a useful summary of recent understanding of such high level waste management.

As typically conceived, such wastes would be incorporated into a glasslike matrix, encapsuled, then buried in salt domes, formations that testify to the lack of groundwater by their very existence and which may be appropriate sites, if undisturbed by human activity. On the other hand, serious attention is being given, both in the United States and abroad, to the possibility of storage in granitic formations (See, for example, Refs. 2 and 4). An artist's conception of a salt-based repository is given in Fig. 11.4.

Relatively little attention has been given, until recently, to the possibility of direct disposal of spent fuel. In principal, such disposal could occur, after encapsulation of the fuel, at sites such as mentioned previously. From the point of view of possible emissions, this approach would differ from that involving reprocessing in two major respects. First, the uranium dioxide (UO_2) pellets, encased in Zircaloy cladding, is a substantially different matrix for containing radioactivity than the glasslike substances that are the usual end product for solidifying liquid high level wastes. The effect of this difference on the risk from possible leaking or leaching of this radioactivity is not now understood. Secondly, the fuel would retain the Pu that, in reprocessing, is recovered to be recycled. Again, the significance of this for the biosphere is not understood. These uncertainties arise from insufficient study either of the stability of various waste forms and disposal methods for times and conditions of interest or of the modes of radionuclide transport from sites of the type being considered. See Ref. 2 for further discussion of disposal possibilities.

The choice of waste disposal mode, whether spent fuel or reprocessing wastes, depends on more than purely environmental considerations. It also depends on the perceived value of reprocessing from the point of view of nuclear demand and U resources and on external considerations, such as the effect that reprocessing and

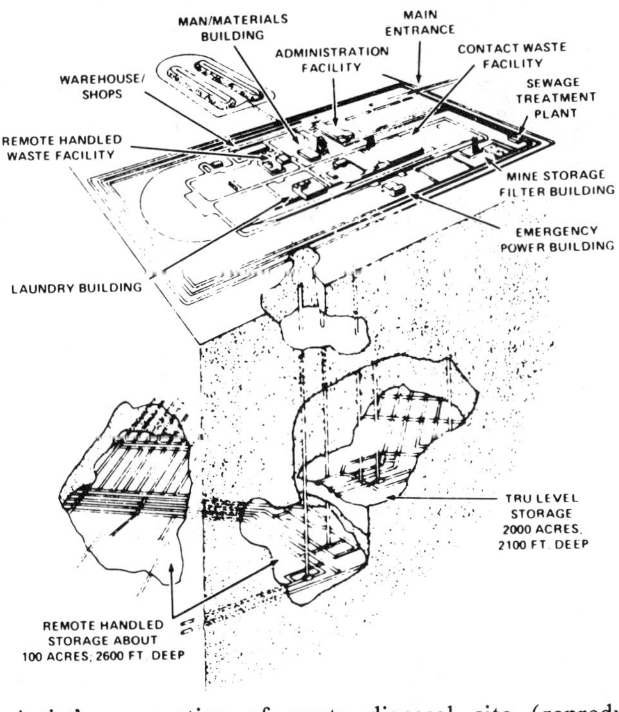

Figure 11.4. Artist's conception of waste disposal site (reproduced from the Energy Research and Development Administration, Ref. 12).

recycling may have on the proliferation of nuclear weapons. In any case, implementation of either reprocessing or spent fuel disposal would probably be preceded by interim storage of the sort just suggested: either pool storage or newer dry methods. This question is being given the urgent attention of concerned parties; see, for example, Ref. 7.

11.2.4. Recycle Fuel Fabrication

The products of a reprocessing plant would be used to produce fresh fuel. Basically the same processes are involved as for production of ordinary UO_2 fuel. The steps from oxides of U and Pu to fuel rods are [1]:

1. Blending of feed powders [PuO_2, UO_2, mixed-oxide (MOX), or recycled scrap].
2. Pretreating the mixture by comminution, compaction, and granulation.
3. Pelletizing.
4. Sintering of the pellets.
5. Grinding of the pellets to finished dimensions.
6. Cleaning and drying the pellets.
7. Loading the pellets into fuel rods, decontaminating the rod ends.
8. Welding the end caps.
9. Decontamination of the rod, if necessary.
10. Inspection of the rod for integrity.

These rods are then assembled into bundles for insertion into reactors.

Several MOX-fueled fabrication plants have operated in the United States (Table 11.1), although not all of these are now active. Some of these have used nitrate solutions as the feed material although, as noted previously, new regulations require shipment of Pu in a solid oxide form.

Table 11.1. Existing MOX Fuel Fabrication Plants (data presented in Nuclear Regulatory Commission, Ref. 1)

Licensee	Plant Location	Feed Material	Plant Product	Pu Possession Limit (kg)	Estimated production Capacity (Te/yr)
Babcock & Wilcox	Parks Township, Pa.	Nitrate solution or PuO_2, U nitrate or (U and Pu)O_2 $UO_2 + PuO_2$	(U,Pu)O_2 fuel rods		20
Exxon Nuclear	Richland, Wash.	$UO_2 + PuO_2$	(U,Pu)O_2 fuel assemblies	10 unencapsulated; 100 total	15
General Electric	Pleasanton, Calif.	Nitrate solution (U and Pu)	(U,Pu)O_2 rods	fuel	150
Kerr-McGee	Crescent, Okla.	Nitrate solution (U and Pu)	(U,Pu)O_2 fuel rods	360	5–10
Westinghouse	Cheswick, Pa.	Pu nitrate solution PuO_2, U nitrate, or UO_2	(U,Pu)O_2 fuel rods	120	10–15

11.3. RADIOACTIVITY CONTROL SYSTEMS

The primary emissions from stored spent fuel are the relatively volatile fission products discussed in the section on nuclear reactors. For extended storage, the emission of greatest significance is ^{85}Kr. At power plants, this emission is not controlled (although more short-lived noble gases would be reduced by holdup). Presumably ^{85}Kr would also be released from storage facilities. The release rate [1] is estimated to be rather low [about 1 Ci/yr · Te (3.7×10^{10} Bq/yr · Te) of stored fuel].

On the other hand, the reprocessing plant frees all the radioactivity of the fuel from its solid matrix. Strict measures must be taken to control this large amount of radioactivity, both to protect the operating personnel and to minimize releases to the environment. Because of the intense gamma radiation from the fuel material, chemical processing equipment is heavily shielded, and relies on remote control and maintenance. Moreover, process and waste streams are carefully confined and treated. Figure 11.5 shows the effluent control systems associated with the process flow shown in Fig. 11.3. The principal means of control are summarized by the following excerpts from Ref. 11:

> Radioactive gaseous wastes from these operations are treated chemically, as well as by filtration, sorption, and scrubbing in order to reduce their radioisotope content to levels that can be discharged to the atmosphere. The aqueous radioactive wastes that contain essentially all the fission products are generally concentrated by evaporation

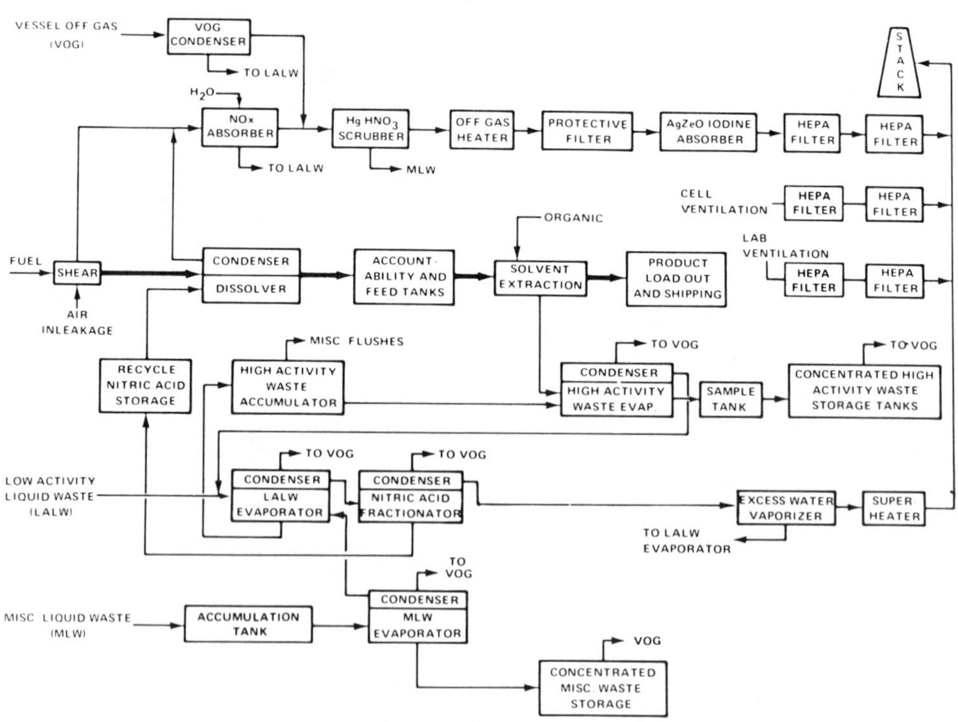

Figure 11.5. Generic reprocessing plant effluent systems (reproduced from Nuclear Regulatory Commission, Ref. 1).

and stored on an interim basis in underground tanks. The evaporator overheads are sufficiently decontaminated of radioisotopes to permit their discharge to the environment under existing regulations.

Processing plants are designed to ensure containment of airborne radioactivity by providing increasing levels of vacuum in three successive envelopes so that all air leakage flows from areas of low to those of high contamination potential.

Off-Gas Treatment. Because of its higher radioactivity and chemical fumes content, the dissolver off-gas in most radiochemical plants is treated, in turn, for nitric acid recovery, for iodine removal, and for removal of residual acid fumes before being blended with the off-gas from the vessels in the balance of the plant. The vessel off-gas is usually scrubbed with caustic, dried, and filtered through one roughing and two high-efficiency filters (HEPA, asbestos-glass fiber paper, 99.97% DOP efficiency).

Iodine in most of its chemical states is removed from gas streams by reaction with $AgNO_3$ impregnated on ceramic packing and by scrubbing with $Hg(NO_3)_2$–HNO_3 or caustic solutions. However, organic iodides, particularly methyl iodide, can be removed most efficiently by catalytic decomposition and sorption on silver, copper, or iodine-impregnated charcoal. The efficiency of iodine removal units is sharply dependent upon the concentration of the iodine, but 99.5% is a commonly quoted design efficiency in cases where organic iodides are not removed. The efficiency of charcoal impregnated with potassium iodide has been quoted at 99.99%.

All off-gas streams from the plant are blended with the cell ventilation streams and passed through a sand filter, a deep-bed fiberglass filter, or a bank of high-efficiency particulate air filters before being monitored and discharged up a stack.

Outgassing of Volatile Fission Products. One of the most difficult problems in reprocessing is that of containment of the volatile fission products, especially ^{131}I. The volatile fission products (iodine, ruthenium, tellurium, cesium, tritium, krypton, and xenon) have been found to volatilize from oxide fuel at moderate temperatures (450 to 750°C).

Although the noble gases are chemically inert, they can be concentrated by cryogenic distillation, absorption in fluorocarbons, or diffusion through permselective membranes. The concentrated noble gases could then be compressed and stored, or immobilized with foam (plastic, glass, or metal), assuming that this method can be shown to be practical.

Cell and Vessel Off-Gas. The cell-ventilation and vessel off-gas systems are primary sources of routine and accidental releases of radioactivity. The recycle of gas from both systems is feasible and will minimize the volume of off-gas needing routine treatment. Recycle will probably be economical, and the use of an inert cell atmosphere may become practical. This would practically eliminate the possibility of solvent fires in process cells.

Liquid and Solid Wastes. It can be anticipated that all radioactive liquid wastes from fuel reprocessing will be evaporated and blended to yield only two streams: a high-level waste consisting of a highly concentrated solution of fission products and actinides, and a low-level aqueous waste that has been sufficiently decontaminated of radioisotopes to permit it to be either discharged to the environment or recycled to the process. The conventional practice of accumulating aqueous solutions of fission products in underground storage tanks will probably be curtailed because of the difficulty in providing safe containment of the more concentrated solutions derived from reprocessing highly exposed power-reactor fuels. Adequately engineered storage systems may be so complex and expensive, as compared with the cost of early solidification of the wastes, that liquid storage systems will be limited in capacity merely to those surge volumes needed for plant flexibility. Under these circumstances, liquid waste inventories in storage would represent, at most, only a few weeks (or months) of plant operations.

Table 11.2 [1] summarizes the major waste systems for the three plants that have existed: the NFS, MFRP, and BNFP Facilities. Note that, although the NFS plant permitted both gaseous and liquid releases, the more recent plants are designed to release only volatile radionuclides, principally ^{3}H, noble gases, and ^{14}C. As discussed later, the amounts of ^{3}H and ^{85}Kr released are much greater than that emitted at nuclear power plants. Reference 1 has a brief summary of possible improvements in control technology, based on a more complete treatment given in Ref. 13. The EPA has considered such technology in detail [14] for use in setting standards (see below). An American Physical Society (APS) study group has recommended [4] that technology be developed for sequestering ^{3}H, ^{14}C, and ^{85}K in a stable form. Should a breeder be introduced, improved I removal would be required if, as has been anticipated, the fuel were reprocessed only a short time (less than 5 months) after discharge.

Disposal sites for spent fuel or solid high level waste from reprocessing have not, of course, been operated. The effluent control is expected to consist of HEPA filters

Table 11.2. Fuel Reprocessing Plants–Major Waste Systems (data presented in Nuclear Regulatory Commission, Ref. 1)[a]

		Liquid Wastes		
Plant	Gaseous Effluents	High Level	Other than High Level	Solid Wastes
NFS	To atmosphere via iodine scrubber, HEPA filters, 61-m stack	Stored on-site as neutralized liquid in tanks in vault[b]	Controlled release to creek	Stored on-site in drums in impervious clay deposits
MFRP	To atmosphere via iodine scrubber, sand filter, 91-m stack	Converted to solid. Stored on-site in containers under water in storage basin.[c]	Stored on-site in tanks in vault as immobile salt cake	Stored on-site in underground lined vault
BNFP	To atmosphere via iodine scrubber, silver zeolite adsorber, HEPA filters, 100-m stack	Stored on-site as acidic liquid in tanks in vault[d]	Stored on-site in tanks in vault	Stored on-site in concrete or steel drums

[a]Information in this table has been complied from the safety analysis reports for these facilities.
[b]Plans to add tanks for storage of high level radioactive liquids in acidic conditions for ultimate conversion to solids and transfer to a federal storage facility.
[c]Prior to shipment to federal storage facility.
[d]Plans for solidification of high level radioactive liquid wastes for transfer to federal storage facility.

to remove radioactive particulates. The principal routine effluent would be very small amounts of ^{3}H and ^{85}Kr.

The operations performed at the fuel fabrication plant are mechanical and leave the Pu in dioxide (powder or ceramic) form. Workers in the plant are protected by confining these operations to lightly shielded sealed (glove) boxes. Emissions are controlled by carefully designed HEPA filtration systems.

11.4. RADIOACTIVE EMISSIONS, EXPOSURES, AND REGULATIONS

Of the spent fuel handling facilities discussed above, the reprocessing plant emits by far the largest amount of radioactivity, an amount that, per gigawatt-year, is far greater than power plant emissions. Essentially all of the ^{3}H, ^{85}Kr, and ^{14}C contained in the fuel is released. As discussed in Ref. 14,

> The gases are released in the early stages of reprocessing, usually during chopping or dissolution. ^{85}Kr is released in the chopping and dissolving stages and goes into the off-gas system. Under current practice, some of the tritium is released from fuel or cladding during dissolution and goes into solution as HTO with only a small fraction (1 to 10%) going into the off-gas system. The remainder of ^{3}H is retained by the Zircaloy hulls. ^{129}I exists in a wide variety of chemical forms within the fuel and accordingly is released in many forms, but usually as elemental iodine or organic iodides. It has been found largely in the off-gas, but significant amounts can enter a liquid effluent stream if one exists. ^{14}C is produced as an activation product from the action of neutrons on the ^{17}O naturally occurring as a constituent of UO_2, and on ^{13}C and ^{14}N impurities. Its chemical form in the fuel or cladding is unknown, although it is believed that its release to off-gas would be as CO or CO_2.

In Table 11.3, the amounts of ^{3}H, ^{85}Kr, and ^{14}C that could be released without the incorporation of additional controls are estimated. Decontamination factors are presumed for the other radionuclides and these depend strongly on the control technology (see Refs. 13 and 14). The amounts cited are released in the gaseous effluent system. Current designs incorporate radwaste systems that, rather than making controlled releases to liquid effluent streams, retain and recycle liquid and solid radioactive materials, ultimately for shipment to waste sites. However, as noted in Table 11.2 and discussed later, older plants (such as the NFS facility) had both gaseous and liquid radioactive effluents. For the radionuclides that are not controlled, such as ^{3}H or ^{85}Kr, one would expect estimated releases such as those given in Table 11.3 to correspond closely to experience at operating reprocessing plants.

Because of their transport properties in the environment, the releases of ^{3}H, ^{14}C, and ^{85}Kr are relatively quickly dispersed generally throughout the earth's atmosphere. Thus they can result in worldwide exposure commitments. Oak Ridge National Laboratory (ORNL) [11] has made estimates of the worldwide buildup of ^{3}H and ^{85}Kr (Fig. 11.6), assuming a rapid growth of nuclear power and associated reprocessing activities. However, because of its much longer half-life, the relatively small curie (becquerel) amounts of ^{14}C released can eventually contribute a larger dose commitment [4].

Table 11.3 also gives estimates of (whole-body) population dose commitments associated with a year's operation of the model reprocessing plant of Ref. 1. The NRC [1] also estimated the annual dose to the individual living nearest the plant

Table 11.3. Estimates of Radioactive Releases from Reprocessing Ci/yr from 2000 TeHM/yr Model Plant (data presented in Nuclear Regulatory Commission, Ref. 1)

	UO_2 Fuel	MOX + UO_2 Fuel
Gases (including entrained matter)		
^{3}H	1,030,000	1,040,000
^{14}C	1,360	1,300
^{85}Kr	22,800,000	21,900,000
^{90}Sr	0.32	0.31
^{106}Ru	8.13	8.77
^{129}I	1.91	1.98
^{131}I	7.5	47.8
Other fission products	10	10
U	2.23×10^{-3}	2.69×10^{-3}
Puα	0.05	0.07
Puβ	1.28	1.76
^{241}Am	8.13×10^{-4}	1.33×10^{-3}
^{243}Am	6.44×10^{-5}	1.54×10^{-4}
^{242}Cm	8.49×10^{-2}	1.66×10^{-1}
^{244}Cm	6.11×10^{-3}	2.27×10^{-2}
Population Dose Commitments (person-rem)		
Occupational	1,250	1,350
U.S. Population	18,800	18,600
Foreign Population	12,200	11,700
Total	32,250	31,650

(assumed distance: 1300 m) to be about 8 mrem/yr (80 μSv/yr) (whole body) (See Tables IV E-9 to E-13 in Ref. 1 for details). Finally, Hebel et al. [4] estimated the average dose rate as a function of time of operation of a standard reprocessing plant, given in Fig. 11.7.

Although no reprocessing plants or large-scale MOX fabrication plants are now operating, some regulations applicable to their emissions or resulting exposures have already been formulated. The NRC review procedure for fuel reprocessing and fabrication plants use Section 3 of the Regulatory Guides, including 3.10, "Liquid Waste Treatment System Design Guide for Plutonium Processing and Fuel Fabrication Plants;" 3.12, "General Design Guide for Ventilation Systems of Plutonium Processing and Fuel Fabrication Plants;" 3.18, "Confinement Barriers and Systems for Fuel Reprocessing Plants;" and 3.32, "General Design Guide for Ventilation Systems for Fuel Reprocessing Plants."

The EPA standard for the uranium-fueled nuclear fuel cycle, 40 CRF 190, sets limits for emissions and public exposures, exclusive of mining, but including processing. It is evident that this standard, with its limit of 50,000 Ci/GW_e-yr (1.85 PBq/GW_e-yr) on ^{85}Kr emissions, presumes noble gas control equipment at the reprocessing plant. Moreover, the ^{129}I limit of 5 mCi/GW_e-yr (185 MBq/GW_e-yr) will require an order of magnitude improvement in decontamination (see

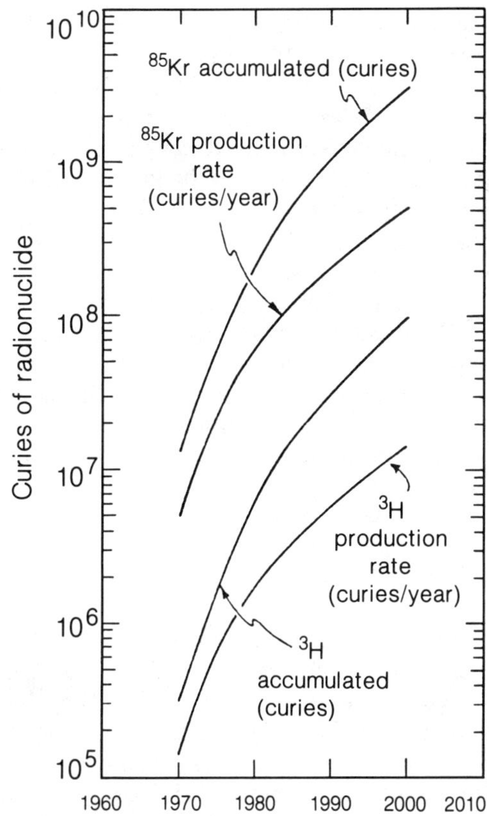

Figure 11.6. Estimated production of ^{85}Kr and ^{3}H (reproduced from Oak Ridge National Laboratory, Ref. 11).

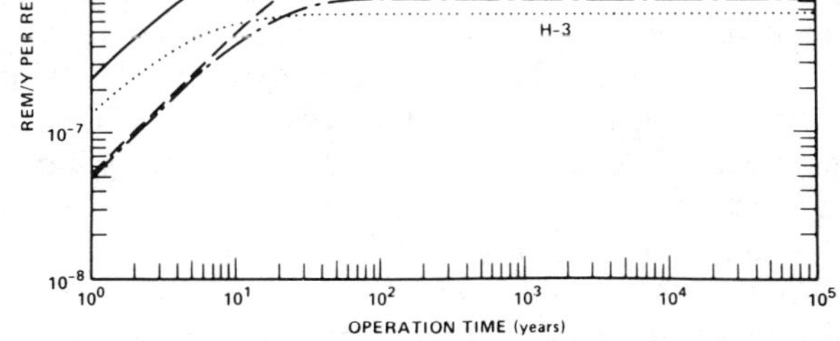

Figure 11.7. Global whole body dose rate increment due to T years of operation of a reference (1500 Mg/yr) LWR reprocessing plant with *no engineered retention* of effluents (reproduced from Hebel et al., Ref. 4, with permission of American Institute of Physics © 1978).

Table 11.3). The basis for this standard is discussed by the EPA in their Final Environmental Statement on 40 CFR 190 [15].

In addition to standards that apply to particular facility types, more generally applicable EPA standards, such as 10 CFR 20 (Standards for Protection Against Radiation) and Regulatory Guide Section 8 (Occupational Exposures), must be considered in the design and operation of fuel handling facilities. See Chapter 10, Nuclear Power Plants, for discussion.

In closing, radioactive emissions from the fuel cycle depend significantly on whether or not fuel is reprocessed. For example, in an LWR system without reprocessing, most of the ^{3}H, ^{14}C, and ^{85}Kr releases from the fuel cycle are avoided. On the other hand, the ^{222}Rn and other emissions associated with mining and milling (e.g., ^{238}U, ^{230}Th, ^{226}Re) are increased by about 30%, as are related occupational exposures. In a fast breeder system, the mining and milling type emissions are virtually eliminated, but the volatile fission products are released at reprocessing. A particular problem for fuel that is reprocessed soon after removal from the reactor is the high concentrations of ^{131}I, (8-day half-life) emitted.

11.5. MONITORING TECHNIQUES

The capability to monitor radiation fields and radionuclide emissions at fuel handling facilities is important, both for the protection of workers at these facilities, and for evaluating impacts on the environment and, ultimately, on members of the general public. The most immediate goal is to assure compliance of the facility with the applicable limitations on exposures or emissions, just discussed. Monitoring requirements at the facility are therefore oriented toward providing a long-term record, as well as a short-term warning of abnormal operation leading to unusually large exposures or releases. Monitoring the environment provides a direct record of the increases in radiation levels or radionuclide concentrations in the vicinity of those facilities.

The basic requirements for monitoring at fuel handling facilities are not dissimilar from those at nuclear power plants. This is particularly true of occupational exposure monitoring because dose limitations are the same at all nuclear facilities and, in the cases of concern, exposures arise from external radiation. In general the emissions will be found to be significantly greater at reprocessing plants where larger quantities of radioactivity are present and could be released (under different conditions) than at power plants. Similarly, a plant fabricating Pu-bearing fuel has a significantly greater potential for transuranic releases than a power plant. However, this is a relatively narrow problem, because of the relative uniformity of material at a fabrication plant, which can be controlled by relying on particulate filters and monitoring by α–particle detection systems (see Chapter 16, Plutonium). Because of the greater likelihood of emissions at fuel reprocessing plants plants this section emphasizes effluent and environmental monitoring.

The principal guidance for fuel reprocessing plant monitoring is the NRC Regulatory Guide RG 4.16, "Measuring, Evaluation, and Reporting Radioactivity in Releases of Radioactive Materials in Liquid and Airborne Effluents from Nuclear Fuel Reprocessing and Fabrication Plants." This general guide, in turn, depends on other documents (including publications of industry standards) for detailed information on monitoring program design and measurement requirements. Consideration also has to be given to the emission and dose limitations cited previously.

Some guidance on monitoring both emissions from, and the environmental concentrations around, reprocessing plants is available from the three commercial plants that have been built, as well as from many of the facilities that have been operated in the weapons program. However, it is important to note that future commercial reprocessing facilities would be operated in such a way that radionuclides are emitted only in the gaseous effluent stream (much as at the Barnwell facility). Furthermore, judging from present EPA regulations, control techniques would have to be applied to radionuclides (such as ^{85}Kr) that had not been controlled in earlier facilities. It is also useful to keep in mind that, although stack monitoring techniques for reprocessing plants may have general similarities to those for other facilities, including power plants, a significant difference is introduced by the fact that the gaseous effluent may be relatively corrosive. This corrosive property comes from the basic operation at a reprocessing plant which involves the dissolution of the spent fuel in nitric acid.

The discussion that follows emphasizes stack monitoring and environmental measurements. It also summarizes the effluent monitoring systems at the NFS, MFRP, and BNFP facilities, as well as the results of environmental monitoring at the first.

11.5.1. Airborne Emissions

Table 11.3 shows the expected emissions, in gaseous effluents from a large reprocessing plant. The only large plant now constructed in the United States is the BNFP. Its stack monitoring capabilities are known, although the plant has not been put into operation. All continuous stack measurements at the BNFP were to be on the stream from a single sampling nozzle, as shown in Fig. 11.8. Because the only commercial plant to have operated in the United States is the NFS facility, the results of environmental monitoring in its vicinity are important. The NFS facility released both gaseous and liquid radioactive effluents.

Krypton-85. If a reprocessing plant does not control noble gases, the ^{85}Kr monitor may be regarded to be a process monitor, because the amount released may be predicted solely on the basis of the fuel history. Hence a simple sodium-iodide-based monitor such as shown in Fig. 11.8 merely provides a check on the plant operation by measuring a weak γ ray of the large ^{85}Kr flux. On the other hand, for facilities that control ^{85}Kr, more sensitive monitors would be required. See Chapter 19 for ^{85}Kr instrumentation.

Studies [17] by the Northeastern Radiological Health Laboratory (NERHL), part of the U.S. Bureau of Radiological Health, confirm that virtually all of the ^{85}Kr for a particular reprocessing cycle was emitted directly from the stack at the NFS Facility. At the BNFP, conservative calculations yield 6 mrem/yr (60 μSv/yr) potential skin exposure at the site boundary [18]. Others have estimated similar doses on the basis of the NFS Facility emission data [19].

Tritium. For plants such as the BNFP, all the ^{3}H from the fuel is released via the stack, primarily as water vapor. At the BNFP, ^{3}H may be monitored off-line by drawing a sample of liquid condensate from the dehumidifier (Fig. 11.8), from which water can be separated for laboratory scintillation counting (see Chapter 20, Tritium). However, the ^{3}H emissions can also be estimated from fuel history because it is not controlled. At the NFS Facility, most of the ^{3}H was relegated to liquid wastes [17], discussed later.

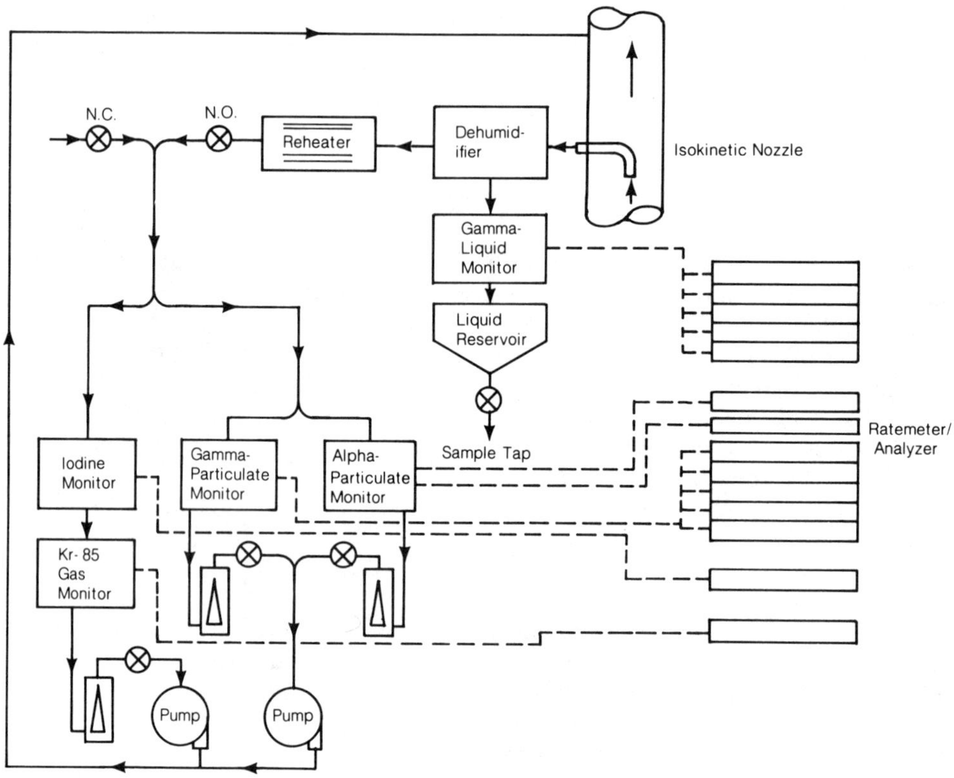

Figure 11.8. BNFP main stack monitor (reproduced from Hower, Hekkala, and Pence, Ref. 16).

Iodine-129 and -131. Because of the high efficiency of iodine scrubbing/absorption in the off-gas system, only a small fraction of these isotopes will normally be emitted from the plant. Because of the large amounts of iodine available, compared with the applicable standard, it is important that the iodine output be monitored to verify the continued effectiveness of the waste treatment system. The monitor at the BNFP consists of silver zeolite absorber, on which buildup of ^{131}I is observed by a sodium iodide detector. See Chapter 18, Iodine, for a discussion of stack and environmental iodine monitoring.

For spent fuel that has not been stored for a long period, the short-term iodine activity released at reprocessing is predominantly ^{131}I. This isotope could cause particular difficulties in a breeder system because the fuel may be reprocessed within a few months of discharge. In the long term, ^{129}I is a greater problem, because of its very long half-life, either in the environment or in the residue from the iodine control system. Particular attention has been given to environmental concentrations of ^{129}I in the vicinity of the NFS facility [20], where the iodine control system was neither highly efficient nor consistently operated.

Particulates. Airborne radioactive particles are usually removed from the gaseous effluents by filtration, the effectiveness of which is checked by the particulate monitoring system. A class of particular interest at reprocessing plants is the transuranics, especially Pu. Although it is expected that these transuranic emissions can effectively be controlled by filter systems, the strict limitations on their concentrations prompt effective monitoring. At the BNFP, only gross particulate monitoring would be performed: a γ-ray monitor using a sodium iodide crystal, and an α-particle monitor using zinc sulfide scintillators. A more sophisticated and sensitive system for transuranic monitoring using α-ray spectroscopy has been developed at the Lawrence Livermore Laboratory [21]. See Chapter 16, Plutonium, for a discussion of Pu-monitoring instrumentation.

Seefeldt, Mecham, and Steindler [22] report emission of approximately 4 mCi/GW-yr (150 MBq/GW-yr) of particulate alpha activity from NFS operations, substantially greater than the EPA limit of 0.5 mCi/GW-yr (19 MBq/GW-yr). In addition, Terpilak and Jorgenson [20] report measurements of α- and β-particulate activity at NFS site perimeter stations.

Other Radionuclides. Monitoring systems have ordinarily not been designed for continuous monitoring of radioactive species other than those just mentioned, although others may be of interest. Measurements of such radionuclides may be made by using laboratory analysis of samples taken at the stack. The emissions on a radionuclide such as ^{14}C may be estimated on the basis of fuel history if controls are not implemented. Overall estimates of other species (such as ^{89}Sr, ^{106}Ru) may be obtained from the liquid condensate monitors and the particulate monitors. These are often not very effective, so that laboratory analyses are necessary as part of a complete monitoring program. Hower, Hekkala, and Pence [16] summarize the status of airborne effluent monitoring at fuel reprocessing and waste management facilities.

11.5.2. Liquid Effluents

Large reprocessing plants such as the BNFP are designed to release no radioactive liquid effluent. The MFRP was similarly designed. The NFS Facility had no such control, this discussion indicates some results of environmental monitoring in the vicinity of NFS. However, even for plants as are being conceived now, effluent and environmental monitoring for possible liquid releases is necessary. These may depend on gross α-ray and β-ray monitors, as indicated previously for particulates, and on laboratory analyses of liquid and environmental samples.

At the NFS Facility, the major source of low level liquid waste was the "condensate from the series of evaporators which are used to concentrate high level liquid waste and to recover the acid." [23] The low level liquid wastes passed from the main plant into interceptor tanks, then into a series of three lagoons, and finally into the environment at Erdman's Creek, which flows eventually into Cattaraugus Creek, which flows downstream into Lake Erie some 39 miles away. Precipitation and sedimentation of insolubles, as well as holdup for the decay of short-lived species, are the important functions of the interceptor and lagoon systems.

The fate of radioactivity found in the liquid discharged from the system can be divided into several areas: some of the radionuclide activity flowed down Cattaraugus Creek to Lake Erie; some settled out onto the bottom of the creek bed; some made its way into the biosphere through plant and animal takeup (e.g. small animals which drink the water, fish, aquatic-plant takeup); and perhaps some became airborne through evaporation.

The NERHL, during 1969, took measurements at several points in the liquid system and also studied creek bed samples [24]. Table 11.4 shows radioactive discharges and concentrations as a percentage of maximum permissible concentrations for periods during this relatively active year at NFS.

Table 11.4. Radionuclides Contributed to the Environment in Liquid Effluent from NFS (reproduced from Magno, Reavey, and Apidianakis, Ref. 24)

Radionuclide	Quantity Discharged May to October 1969 (Curies)	Percent of 10 CFR 20 Concentration Limit in Cattaraugus Creek Water	
		June 17, 1969	November 4, 1969
Tritium	1700	1.1	0.10
Ruthenium-106	52	5.2	0.75
Cesium-137	8.0	0.008	0.15
Cesium-134	2.0	< 1	< 1
Strontium-90	8.3	17	23
Cerium-144	0.16	< 1	< 1
Promethium-147	0.092	< 1	< 1
Zirconium-95	0.0016	< 1	< 1
Cobalt-60	0.20	< 1	< 1
Antimony-125	0.59	< 1	< 1
Manganese-54	0.0027	< 1	< 1
Plutonium	0.0018	< 1	< 1
Uranium	0.032	< 1	< 1
Gross alpha	0.048	< 1	< 1

11.6. SUMMARY

The handling of spent fuel may involve a variety of facilities providing for spent fuel storage, fuel reprocessing, mixed oxide fabrication, and geologic disposal. Monitoring requirements for spent fuel are not dissimilar from those for power plants. However, because fuel reprocessing plants free the radioactivity contained in the fuel rods, the possibility of radioactive discharges and the corresponding monitoring requirements are more severe at such plants. This discussion summarized the anticipated reprocessing emissions, noting their dependence on control design, and also indicated the monitoring instrumentation that has been designed for these facilities. More detailed information on monitoring specific radionuclides is presented in Part III of this book.

REFERENCES

1. U.S. Nuclear Regulatory Commission, "Final Generic Environmental Statement on the Use of Recycled Plutonium in Mixed Oxide Fuel in Light Water Cooled Reactors: Health, Safety, and Environment," 5 vols., NRC Report No. NUREG-0002, 1976.
2. U.S. Energy Research and Development Administration, "Alternatives for Managing Wastes from Reactors and Post-Fission Operations in the LWR Fuel Cycle," 5 vols., ERDA Report No. ERDA-76-43, 1976.
3. U.S. Energy Research and Development Administration, "Final Environmental Statement, Liquid Metal Fast Breeder Reactor Program," 3 vols., ERDA Report No. ERDA-1535, 1975; "Proposed Final Environmental Statement," 7 vols., AEC Report No. WASH-1535, 1974.
4. L. C. Hebel, Chairman, American Study Group, "Report to the American Physical Society by the Study Group on Nuclear Fuel Cycles and Waste Management," *Mod. Phys.*, **50**, S1, 1978.
5. S. M. Keeney, Chairman, Nuclear Energy Policy Study Group, *Nuclear Power: Issues and Choices*, Report of the Ford Foundation/MITRE Corporation, Ballinger, MA, 1977.
6. Congress of the United States, Office of Technology Assessment, *Nuclear Proliferation and Safeguards*, Praeger, New York, 1977.
7. U.S. Nuclear Regulatory Commission, "Generic Environmental Impact Statement on Handling and Storage of Spent Light Water Power Reactor Fuel," (draft), 2 vols., NRC Report No. NUREG-0404, 1978.
8. A. V. Nero, *A Guidebook to Nuclear Reactors*, University of California Press, Berkeley, 1979.
9. T. H. Pigford, M. J. Keaton, B. J. Mann, and G. L. Sessler, "Fuel Cycles for Electric Power Generation"; "Fuel Cycle of 1000-MW Uranium-Plutonium Fueled Water Reactor"; and "Fuel Cycle for 1000-MW High Temperature Gas-Cooled Reactor"; in *Comprehensive Standards: The Power Generation Case*, U.S. Report No. PB-259-876, 1975.
10. Federal Register 20960, 1974.
11. Oak Ridge National Laboratory, "Siting of Fuel Reprocessing Plants and Waste Management Facilities," ORNL-4451, 1971.
12. U.S. Energy Research and Development Administration, *Proceedings of the International Symposium on the Management of Wastes from the LWR Fuel Cycle*, Denver, July 11–16, 1976, CONF-76-0701.
13. Oak Ridge National Laboratory, "Correlation of Radioactive Waste Treatment Costs and the Environmental Impact of Waste Effluents in the Nuclear Fuel Cycle for Use in Establishing 'As Low as Practicable' Guides—Nuclear Fuel Reprocessing," ORNL-TM-4901, 1975.
14. U.S. Environmental Protection Agency, "Environmental Analysis of the Uranium Fuel Cycle. Part III—Nuclear Fuel Reprocessing," EPA-520/9-73-003-D, 1973; "Environmental Analysis of the Uranium Fuel Cycle. Part IV—Supplemental Analysis—1976," EPA 520/4-76-017, 1976.
15. U.S. Environmental Protection Agency, "Final Environmental Statement, 40 CFR 190: Environmental Radiation Protection Requirements for Normal Operations of Activities in the Uranium Fuel Cycle," 2 vols., EPA 520/4-76-016, 1976.

16. R. B. Hower, B. Hekkala, and D. T. Pence, "Radioactive Airborne Effluent Measurement and Monitoring Survey of Reprocessing and Waste Treatment Facilities," COO-3049-9, 1977, prepared by Science Application, Inc., 4060 Sorrento Valley Boulevard, San Diego, CA 92121, for Harvard Air Cleaning Laboratory (ERDA).
17. J. A. Cochran, D. G. Smith, P. J. Magno, and B. Shleien, *An Investigation of Airborne Radioactive Effluent from an Operating Nuclear Fuel Reprocessing Plant*, Northeastern Radiological Health Laboratory Report No. BRH/NERHL 70-3, 1970.
18. Allied-Gulf Nuclear Services, *Barnwell Nuclear Fuel Plant, Environmental Report*, Barnwell, SC, 1971.
19. J. A. Martin, "Calculations of Environmental Radiation Exposures and Population Doses Due to Effluents from a Nuclear Fuel Reprocessing Plant," *Radiat. Data Rep.*, **14**, 59, 1973.
20. M. S. Terpilak and B. L. Jorgenson, "Environmental Radiation Effects of Nuclear Facilities in New York State," *Radiat. Data Rep.*, **15**, 375, 1974.
21. J. F. Kordas and P. L. Phelps, "A Review of Monitoring Instruments for Transuranics in Fuel Fabrication and Reprocessing Plants," Lawrence Livermore Laboratory Report No. UCRL-52123 Revision 1, 1977.
22. W. B. Seefeldt, W. J. Mecham, and M. J. Steindler, "Characterization of Particulate Plutonium Released in Fuel Cycle Operations," Argonne National Laboratory Report No. ANL-75-78, Chemical Engineering Division, 1976.
23. J. E. Logsdon and J. W. N. Hickey, "Radioactive Waste Discharges to the Environment from a Nuclear Fuel Reprocessing Plant," *Radiol. Health Data Rep.*, **12**, 305, 1971.
24. P. Magno, T. Reavey, and J. Apidianakis, *Liquid Waste Effluents from a Nuclear Fuel Reprocessing Plant*, Northeastern Radiological Health Laboratory Report No. BRH/NERHL 70-2, 1970.

12 MINING AND MILLING OF URANIUM ORES

This chapter gives an overview of the radiological problems associated with the mining and milling of uranium (U) ores, activities that are increasing with the rising worldwide demand for nuclear fuel. Requirements for radiation-measuring instrumentation are emphasized, noting how specific aspects of the mining and milling industry have created the need for unique instrumentation. Radiation protection guides and general radiological safety considerations are also summarized. Although this chapter is devoted to uranium mining and milling, many of the same considerations apply to the extraction of other materials, particularly thorium but also (because of the pervasive distribution of U and Th) non-nuclear materials, such as the Florida phosphates. Ore mining and processing enhances the contact and distribution of natural decay chain radionuclides in the biosphere.

This chapter emphasizes the important radiological impacts and the resulting instrumentation requirements for the uranium industry, giving little attention to measurement problems that are not unique to the industry or that are of very restricted interest. The text primarily discusses measurements inside underground mines of radon-222 (^{222}Rn) and its daughters (these measurements are sometimes described in terms of the "working level," a quantity defined for occupational exposures), and measurements *outside* mills (e.g., mill tailings) of radionuclides in air and water. Also external γ-ray measurements, bioassays, and airborne particulate monitoring are briefly discussed. In many respects, this chapter complements Chapter 14, Radon and its Daughters, which includes detail on the properties of radon, its decay chains, and other physical considerations. Chapter 13, Radium, is also of interest, largely because of the presence of Rn in mill tailings. Because these radionuclides tend to be α-emitters, the reader is referred to the parts in Chapter 7 on alpha-particle measurement.

12.1. ENVIRONMENTAL RADIATION HAZARDS

Although there are other environmental impacts of mining and milling [1], this text discusses only those that derive from radioactivity. The details of uranium mining and milling techniques may be found elsewhere [2,3]. Briefly, U is extracted principally from underground or surface (open pit) mines, although solution mining is becoming more widespread [4]. Milling ores from mines involves crushing, leaching, extraction, and conversion operations. Radiological hazards may arise from U, or its daughters, most importantly, ^{238}U, ^{235}U, ^{234}U, ^{230}Th, ^{226}Ra, and Rn and its daughters. (See Table 12.1 and Chapter 15, Uranium and Thorium.) In mining and milling, alpha, beta, and gamma radiation from these nuclides can be important with hazards arising primarily from internal exposure and to a lesser extent from external radiation.

Table 12.1. Some Important Radionuclides and Their Properties

Isotope	Type of Activity	Half-Life	α Energies, if any, (MeV) Branching Ratio
^{238}U	α,β^-,γ	4.51×10^9 yr	4.20
^{235}U	α,γ	7.1×10^8 yr	4.58
^{234}U	α,γ	2.45×10^5 yr	4.7
^{222}Rn	α	3.82 yr	5.48
^{218}Po	α	3.05 months	6.00
^{214}Pb	β^-,γ	26.8 months	
^{214}Bi	β^-,γ	19.7 months	
^{214}Po	α	1.6×10^{-5} sec	7.68
^{210}Pb	β^-,γ	22.3 yr	
^{210}Po	α,γ	138.4 days	5.3
^{230}Th	α	8.0×10^4 yr	4.68
^{232}Th	α	1.4×10^{10} yr	4.01 (76%)
			3.95 (24%)
^{228}Th	α	1.9 yr	5.42 (70%)
			5.34 (28%)
^{226}Ra	α,β^-,γ	16.02 yr	4.78
^{224}Ra	α	3.665 days	5.3
^{220}Rn	α	55.3 sec	6.29
^{216}Po	α	0.15 sec	6.78
^{212}Pb	β,γ	10.64 hr	
^{212}Bi	α,β,γ	60.60 months	6.09 (10%)
^{228}Ra	β,γ	6.7 yr	

The hazards from exposure to β- and γ-radiation, uranium dusts, and surface contamination in open pit mining for U are about the same as those encountered in underground mining [5], but underground mining has more significant problems associated with Rn and Rn progeny. Although mining operations will release radioactivity into the air, with proper ventilation it can be dispersed and will dilute rapidly in the atmosphere. Radon is a noble gas, unlike its precursors, and therefore it can easily become airborne. The airborne radionuclides consist of Rn and its daughters, which constitute the most serious hazard, and long-lived alpha-emitters, which are present in dusts of the mine atmosphere from the decay chain of natural U. Where underground mines are poorly ventilated and Rn can collect, Rn and its daughters have been found to pose substantial radiological hazards for uranium miners. Gamma-radiation levels greater than 2 mrem/hr (20 μSv/hr) sometimes pose an external hazard even in lower-grade mines due to buildup of Bismuth-214 (^{214}Bi) [6]. The other members of the uranium series, with less mobility and long half-lives, do not ordinarily pose a comparable hazard, either from external exposure or from internal exposure due to inhalation or ingestion of these nuclides in the mineral dusts.

Mine effluent wastes include liquids, solids (particulates and tailings), and gases. The liquids are mostly water from mine drainage and water used in drilling. Solid waste is primarily waste rock and low-grade ore. Leaching of mine tailings is an additional source of low-level liquid waste. Because Rn is soluble to a certain extent, sometimes water seeping into a mine may be a source of Rn. About 1% of Rn in the ore is soluble and can also be contained in water, resulting in contamination if it is discharged directly to the environment from mining operations. Contamination of surface or ground water can lead to an important drinking water exposure.

Although impacts on ground water supplies and disposal of liquid wastes are the primary environmental concerns associated with *in situ* solution mining, similar problems associated with conventional mining and milling operations are actually larger. When a 0.1% grade of ore is assumed, an estimated maximum of 16 tons of solid waste per ton of produced yellowcake (semirefined uranium compound which is the feed material for uranium hexafluoride) would be generated by solution mining compared with 1000 tons of radioactive wastes (tailings) associated with conventional milling methods. In addition, the tailings from conventional operations contain almost all the ^{226}Ra and other daughter products associated with the ore, whereas only $< 3\%$ of the Ra in the ore is brought to the surface during *in situ* solution mining [7]. To avoid negating advantages gained from bringing smaller quantities of radioactive materials to the surface by solution mining, extreme care should be taken in disposing of the solid wastes. The mining process exposes previously inaccessible U so that the daughter activities, particularly Ra and Rn, have a greatly increased probability of transport to the general environment. The primary population at risk is the uranium miner.

Uranium is extracted by a milling operation that uses sulfuric acid (or, less often, an alkali) to leach U from ore that has been crushed and ground. Uranium is subsequently recovered from solution by solvent extraction or ion exchange. Its product is the semirefined uranium compound, yellowcake. The uranium mill produces large volumes of both solid waste, basically tailing from the ore, and liquid waste, from the various mill processes. The bulk of the wastes initially discharged is a mixture of solids and liquids that form a slurry for easy movement to the disposal site. Most of the nonvolatile uranium decay daughters are relegated to these waste streams.

Volatile daughters, notably Rn, may escape throughout the milling operation. Airborne particulates may be released from yellowcake drying and packaging operations, crushing and grinding operations, and the ore storage pads. The principal source of Rn and its daughters is the decay of longer-lived uranium daughters, most importantly ^{230}Th and ^{226}Ra, present in the waste tailings [8]. These two radionuclides can also have direct radiological importance, either because of their presence at relatively high concentration in the liquid milling effluents [2] or because of their presence in tailings allowing them to become airborne in particulate form [8]. Airborne releases from the milling process thus include both particulate matter and gases. Low stacks are the means by which these dusts or gases are discharged to the atmosphere from milling operations.

The liquid effluent from a process mill consists of waste solutions from the leaching, grinding, extraction, and washing circuits of the mill. The radioactive daughter products of Rn may be present in small concentrations. Because the concentrations of ^{226}Ra and ^{230}Th are about an order of magnitude above the specified liquid limits in 10 CFR 20 [9], considerable effort must be exerted to prevent any release of the material from the site. The solid and liquid waste is stored as a slurry in a tailings retention pond which should be constructed to prevent discharge into the surface water system and to minimize percolation into the ground.

The tailings retention system or tailings pond will have a radiological impact on the environment through (a) the air pathway by continuous discharge of ^{222}Rn gas; (b) external exposure through γ-rays given off by decay of ^{226}Ra and ^{222}Rn daughters, primarily ^{214}Bi; and (c) through air, water, or food pathways if ^{226}Ra and ^{230}Th are blown off dried areas of the tailings ponds by wind or are leached from the pond into surface waters [8].

When a retention system involving a dam is initially put into use, because all the waste is under water, small losses of radioactive slurry waste as seepage are expected through and around the dam. Levels as high as 32 pCi/liter (1.184 Bq/liter) of ^{226}Ra have been found in seepage from operating mills [2]. Unless the impoundment area is lined with an impervious material, considerable quantities (10% to > 80%) of the liquid effluent from the mill will leak out through the bottom of the pond [10]. This problem requires continued monitoring programs to insure that there is no seepage of contaminated liquids off-site.

Toward the end of the operating lifetime of a tailing retention system, some of the tailings will no longer be under water and will dry out to form a beach. Wind can erode tailing off these beaches as airborne particulate matter. The amount of radioactive particulate material removed from the tailings beach by wind erosion depends on the area and topography of the beach, the wind velocity, and particle size distribution of the tailings [8]. Long-lived α-emitting radionuclides ^{238}U, ^{230}Th, and ^{226}Ra come from this source and are the most significant contributors to the lung dose. Because the airborne particulate effluent discharges of tailings dust (as well as calcinated yellowcake and ore dust) are considered insoluble aerosols when they are inhaled, these aerosols tend to remain in the lung. The lung, thus, is the critical organ for calculating radiation dose. A recent study [11] has indicated that windblown tailings from inactive unstabilized tailings piles have caused elevated γ-ray exposures of 25 mrem/yr (0.250 MSv/yr) at distances up to one mile from the pile. Inhalation is probably the critical method of exposure because levels of control sufficient to limit exposure through inhalation will also prevent exposures from external whole-body exposure.

Long-range atmospheric diffusion of the noble gas ^{222}Rn and resultant air concentrations of daughters is of concern as an external whole-body exposure. Radiation exposure to radon gas and radon daughters constitutes over 98% of the environmental impact of such piles [12]. A recent study [13] was undertaken that provided estimates of inhalation dose from ^{222}Rn, ^{210}Pb, and ^{210}Po to the population in the western United States resulting from unit releases of ^{222}Rn from tailings piles (see Table 12.2). Due to lack of proper treatment in the past, significant hazards (summarized in Table 12.3) have arisen from inactive uranium milling sites [14,15]. Compared with other types of nuclear waste, uranium mill tailings are generated in large volume, about 10 to 15 million tons annually [16] and are long half-life, low-level radioactive wastes [17].

In the absence of very large controlled areas, or unless individuals live more than one mile from the tailings pile, radiation from tailings must be kept very low by using preventive measures. Both active and inactive tailing piles have been studied to determine the source terms and ambient levels [18–23]. In addition, transport and distribution of radiation from tailing piles in the environment has been studied [24,25].

In summary, for the entire mining and milling process (including after shutdown) the critical pathways to be considered are: (a) inhalation of α-emitting radioactive airborne particles and radioactive gas; (b) deposition of airborne radioactive particles causing whole-body exposure from γ-rays; (c) radioactive contamination of surface and/or ground water by seepage from the tailings pond or by discharge of process water; and (d) ingestion of airblown material or irrigation with contaminated water. Mining and milling each impact different populations. Mining generally exposes workers whereas milling affects the general population.

Table 12.2. Population Dose per Unit Radon Release to Inhabitants of the United States from Inhalation of ^{222}Rn, ^{210}Pb, and ^{210}Pb in 1978. (data presented in Travis, Watson, and McDowell-Boyer, Ref. 13, with permission of American Nuclear Society © 1978)

	Population Dose per kCi ^{222}Rn Released in 1978			
	Person-rem	Organ-rem		
Release Site	Lung[a]	Bone[b]	Kidney[b]	Liver[b]
Casper, Wyoming	84.8	24.3	0.5	0.4
Falls City, Texas	109.5	21.3	0.5	0.4
Grants, New Mexico	71.5	19.9	0.4	0.4
Wellpinit, Washington	61.5	20.7	0.5	0.4

[a]Dose to bronchial epithelium of tracheobronchial tree.
[b]Dose from uniform distribution of radionuclides in organs.

Table 12.3. Principal Radionuclides Involved in the Exposure Pathways from Tailings Piles to Man. (data presented in Goldsmith, Ref. 14, with permission of Nuclear Safety © 1976)

Pathway	Radionuclide	Pathway Processes	Exposure Hazard
Radon diffusion	^{222}Rn	1. Diffusion to pile surface 2. Atmospheric transport 3. Inhalation	Inhalation of daughters of ^{222}Rn, ^{210}Pb, ^{214}Pb, ^{214}Bi, and ^{214}Po
Airborne activity	^{230}Th	1. Resuspension from pile surface 2. Atmospheric transport 3. Inhalation 4. Body surface contamination	Inhalation of airborne activity carried by respirable particles
Terrestrial	^{226}Ra, ^{210}Pb	1. Resuspension from pile surface 2. Atmospheric transport 3. Deposition on soil or foliage 4. Uptake by plants 5. Uptake by animals 6. Consumption of contaminated plants, meat, and milk 7. Consumption of contaminated soil	Ingestion of ^{226}Ra and ^{210}Pb
Aquatic	^{226}Ra, ^{210}Pb	1. Dissolution of radionuclide in pile 2. Migration through soil to water 3. Migration in water 4. a. Uptake by aquatic organisms b. Consumption of aquatic organisms 5. a. Use of water for irrigation b. Uptake by plants c. Consumption of plants 6. Consumption of contaminated drinking water 7. Immersion in contaminated water	Ingestion of ^{226}Ra and ^{210}Pb
External gamma	^{226}Ra daughters	1. Attenuation by pile material 2. Air-distance attenuation 3. External exposure	External exposure to gamma radiation produced by the decay of ^{226}Ra and its daughters

12.2. UNITS OF EXPOSURE

Radiation protection guides for most radiation exposures are expressed in familiar units. For external whole-body exposures or exposures of specific organs, the limits are in absorbed dose equivalent (rem or Sievert); and for radionuclide concentrations in air and water, the limits are as maximum permissible concentrations (in $\mu Ci/cm^3$ or Bq/cm^3, for example).

For the purposes of radiological protection of the lungs in the uranium mining industry, a specialized unit of exposure to ^{222}Rn and its daughters has been developed. This is the *working level* (WL), defined as: "any combination of radon daughters in one liter of air that will result in the ultimate emission of 1.3×10^5 MeV of alpha energy." [26] The reason for the specialized unit is operational: the WL is a concept having validity in any mixed concentration of Rn and its daughters, whether or not they are in equilibrium, and it lends itself to practical measurements in the mines.

The WL considers only radon-222 daughters and not radon gas itself, because radiation dose to the lung comes primarily from the retention of the daughters in the lung. Radon itself is chemically inert, is not preferentially retained, and hence gives substantially less dose to the lungs.

An extension of the WL concept is the *working level month* (WLM), which expresses a cumulative exposure. It is defined as: "inhalation of air containing a radon daughter concentration of one WL for 173 working hours results in an exposure of one WLM." [27]

12.3. EPIDEMIOLOGICAL DATA

Today, there seems to be little doubt about the epidemiological evidence connecting uranium mining and lung cancer. This is true despite some disagreements about the detailed conclusions that can be drawn from the available evidence. Holaday [28] has written an excellent historical account of the uranium miners' radon exposure problem in the May, 1969, *Health Physics,* an issue devoted entirely to various reports in this field. Epidemiological data were reviewed in 1967 by the Federal Radiation Council (FRC) [26], and again in 1971 by the National Institute for Occupational Safety and Health [29]. The FRC noted "a clear association between exposure to radon daughters in mine air expressed as cumulative WLM and the number of lung cancer deaths in the study group." This is confirmed in more recent reviews [30,31]. Figure 12.1 [26] shows some of this evidence, indicating increased mortality at cumulative exposures greater than 1000 WLM. The FRC found a possible synergistic association with cigarette smoking. No other possible lung carcinogens (silicon, hydrocarbons, longer-lived radioactive materials such as Th or U) were found to have a synergistic association with radon daughters [26].

One of the principal difficulties in these epidemiological studies was determining the average WL in uranium mines over the two decades studied. However, measurements in 108 U.S. mines in 1956 and 231 mines in 1966 revealed the following distribution by WL ranges [26]:

WL Range	Fraction of U.S. mines in WL Range 1956	1966
Less than 1.0	19%	44%
1.0 to 2.9	25%	42%
3.0 to 10	33%	13%
Greater than 10.0	23%	1%

In the mid-1950s a sizeable fraction of miners were exposed to atmospheres with cumulative levels in excess of 100 WLM per year. Substantial progress was made in reducing WL in mines (see Table 12.4). For 1977, about 89% of uranium mines had concentrations less than 1 WL, with most of the remainder less than 2 WL.

Although the association between radon daughters and lung cancer was first identified in underground uranium miners, Rn also occurs in other types of mines, and recent evidence (e.g., Ref. 32) links resulting exposures to increased incidence of lung cancer. Based on such evidence, radiation protection guides have been established.

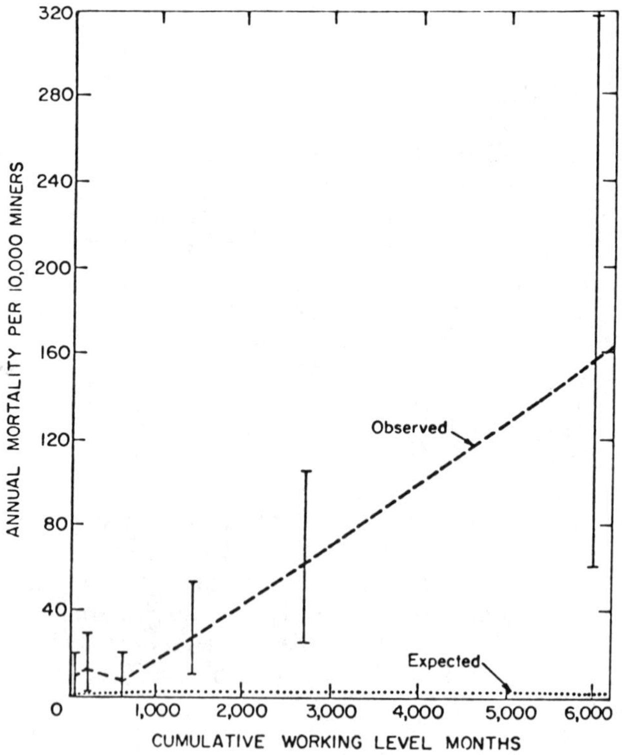

Figure 12.1. Observed and expected annual lung cancer mortality per 10,000 mines and 95% confidence limits in relation to exposure (data presented in Federal Radiation Council, Ref. 26).

Table 12.4. Radon Daughter Concentration and its Exposure in Uranium Mines (data presented in Environmental Protection Agency, Ref. 31)

Year	Average Concentrations (WL)	Average Exposures (WLM)
1975	0.71	5.68
1976	0.58	4.64
1977	0.51	4.08

12.4. RADIATION PROTECTION GUIDES

In the United States, the WL and WLM concepts are used for exposure to Rn and its daughters. Radon-222 gas itself is typically not limited separately, because the impact of the radon daughters is the more important consideration. Mining standards were promulgated under the authority granted to the Secretaries of Labor and of Health, Education, and Welfare (now the Department of Health and Human Services) by the Federal Mine Safety and Health Amendments Act of 1977 (Public Law 95-164). The operating standards for mines are given in 30 CFR 55, 56, 57, and 58. Standards regarding radiation in underground mines are given by 30 CFR 57.5-35a through 30 CFR 57.5-47a. Table 12.5 paraphrases the current radiation standards. Standards for exposure to radiation other than the inhalation of radon daughters must be considered separately. The usual occupational limits apply, that is, 5 rem/yr (50 mSv/yr) for whole-body external exposure, and so on (see Chapter 2 of this book).

Publication 24 of the International Commission on Radiological Protection (ICRP) [33] outlines the principles of monitoring and limitation of radiation exposure in uranium mines. The ICRP guide [34] recommends that the annual average concentration of ^{222}Rn and its daughters should not exceed 30 pCi (1.11 Bq) in occupational settings, or a little less than 4 WLM in a year.

Quoting from ICRP Publication 24 [33].

> For the purpose of radiation protection in mines, it is understood that this limit refers to ^{222}Rn in equilibrium with its short-lived daughters, where the daughters release, through their decay to ^{210}Pb, an alpha-energy equal to that released by the daughters of 30 pCi/l (1.11 Bq/l) of radon in equilibrium with its daughters (3.9×10^4 MeV/l).
>
> The annual limit of exposure to radon daughters can therefore be expressed as a time integral of alpha-energy concentrations: $2000\ \text{h} \times 3.9 \times 10^4\ \text{MeV/l} = 7.8 \times 10^7$ MeV h/l. An additional quarterly requirement is that the exposure in a quarter of a year should not exceed 3.9×10^7 MeV h/l.

In addition, the ICRP limits the annual exposure to pure ^{222}Rn, to 6×10^6 pCi hr/liter, (0.222 MBq hr/liter) [34], corresponding to an average concentration of 3×10^3 pCi/liter (0.111 kBq/liter). This limit is applicable in situations where the potential α-particle energy of the daughters is a small fraction of the value in equilibrium, as may result from the use of respirators, and from air filtration or electrostatic precipitation of daughters. Also, the ICRP limits uranium ore

Table 12.5. Current Radiation Standards, Underground Only (February 17, 1978) (data from Beckman, Ref. 6)

Section	Standard
57.5-37	*Mandatory.* Mine atmospheres shall be sampled to determine if hazardous concentrations of radon daughters are present. Where potentially hazardous concentrations are found, or known sources of radon exist, each active work area shall be sampled as often as necessary by a qualified person.
57.5-38	*Mandatory.* No person shall be permitted to receive an exposure in excess of 4 WLM in any calendar year.
57.5-30	*Mandatory.* Except as provided by standard 57.5-5, persons shall not be exposed to air containing concentrations of radon daughters exceeding 1.0 WL in active workings.
57.5-40	*Mandatory.* (a) Where uranium is mined, if measurements in areas indicate exposure to concentrations of radon daughters in excess of 0.3 working level, complete individual exposure records shall be kept for all employees entering these areas.
57.5-41	*Mandatory.* Smoking shall be prohibited in all areas of a mine where exposure records are required to be kept in compliance with standard.
57.2-42	*Mandatory.* If levels of permissible exposures to concentrations of radon daughters different from those prescribed in 57.5-38 are recommended by the Environmental Protection Agency and approved by the President, no employee shall be permitted to receive exposures in excess of those levels after the effective dates established by the Agency.
57.5-44	*Mandatory.* The wearing of respirators approved for protection against radon daughters shall be required in environments exceeding 1.0 WL and respirator use shall be in compliance with standard 57.5-5.
57.5-45	*Mandatory.* Inactive workings, in which radon daughter concentrations are above 1.0 WL, shall be posted against unauthorized entry and designated by signs indicating them as areas in which approved respirators shall be worn.
57.5-46	*Mandatory.* Where radon daughter concentrations exceed 10 WL, respirator protection against radon gas shall be provided in addition to protection against radon daughters. Protection against radon gas shall be provided by supplied air devices or by face masks containing absorbent material capable of removing both the radon and its daughters.
57.5-47	*Mandatory.* Gamma radiation surveys shall be conducted annually in all underground mines where radioactive ores are mined. (a) Surveys shall be in accordance with American National Standard ANSI N13.8-1973, entitled "Radiation Protection in Uranium Mines," Section 14.1 page 12, which is hereby incorporated by reference and made a part hereof. (b) Where average gamma radiation measurements are in excess of 2.0 mR/hr in the working place, gamma radiation dosimeters shall be provided for all persons affected, and records of cumulative individual gamma radiation exposure shall be kept. (c) Annual individual gamma radiation exposure shall not exceed 5 Rems.

dust; the limits are based on the total α-particle activity of the long-lived nuclides present in the dust, under the assumption that these nuclides are all members of the ^{238}U series (excluding Rn and its daughters) in secular equilibrium. It has taken into account the fact that only five-ninths of the relevant disintegrations are α-particles. The annual operational limit for α-activity in ore dust is 70 pCi hr/liter (2.59 Bq hr/liter); the quarterly limit is 35 pCi hr/liter, (1.295 Bq hr/liter), [32] corresponding to the average concentrations of 3.5×10^{-2} pCi/liter (1.295 mBq/liter) and 1.75×10^{-2} pCi/liter (0.648 mBq/liter), respectively.

There is no explicit guideline for exposure of the general public to Rn and its daughters. However, the ICRP has recommended that members of the general public should be limited to exposures that will not exceed the risk of 0.5 rem (5 μSv) whole-body exposure in a year [35].

The Environmental Protection Agency (EPA) regulations [36] require that mining and milling operations be conducted such that the annual dose equivalent to any member of the public does not exceed 25 mrem (250 μSv). The EPA further requires that all mining and milling operations which can possibly contaminate surface and/or ground water used for drinking water adhere to the standards set by the National Interim Primary Drinking Water Regulations for Radioactivity [37]. Uranium mill tailings are regulated under Uranium Mill Tailings Radiation Control Act of 1978 [38]. Under this act the EPA promulgates health and environmental standards for uranium mill tailings and the Nuclear Regulatory Commission (NRC) insures proper waste management.

12.5. SAFETY AND MONITORING CODES

The complexity of adequate radiation protection in uranium mines and mills has motivated several codes of practice. The U.S. Bureau of Mines has issued a two-volume handbook [39] which contains detailed recommendations on control of exposures in underground uranium mines. This compilation discusses details of ventilation practices, mine planning, sampling procedures for measurement, and the Kusnetz method for measuring WL (see Chapter 14, Radon). It is highly recommended to any reader interested in the fine points of exposure control.

The International Atomic Energy Agency (IAEA), jointly with the International Labour Office (ILO) has published a Code of Practice [40], which outlines the general duties of mine employees and employers for radiation safety. It recommends a maximum permissible dose equivalent of 15 rem/yr (150 μSv/yr) to the bronchial mucosa of the lung. It also describes procedures for medical surveillance, protective measures (ventilation, dust abatement, safety), and protection equipment. The IAEA/ILO Code is accompanied by a Technical Addendum giving detailed con siderations for control of Rn and daughters, for monitoring radiation in mines, and for radiation protection in uranium and thorium mines and mills.

In addition the IAEA and ILO have issued a manual [41] for the guidance of those responsible for the regulation, management, and operation of uranium and thorium mines and mills. The manual *(IAEA Safety Series No. 43)* follows and supplements IAEA *Safety Series No. 26,* and is intended as an operating manual with particular emphasis on implementation methods. The scope of the manual is limited to occupational hazards in mining and milling, but includes a section in which liquid, solid, and gaseous wastes produced by the various processes are identified although not treated in detail. For the monitoring and control of radioactive wastes, the reader should refer to the IAEA *Code of Practice on the Management of Wastes from the Mining and Milling of Uranium and Thorium Ores,* the first part of IAEA *Safety Series 44 [42].* The *Code of Practice* identifies wastes, presents environmental problems, and sets out minimum requirements for their competent management. The second part of Ref. 42 is a *Guide to the Code,* suggesting ways to meet the requirements of the Code.

Another code of practice, ANSI N7.1, was published in 1960 by the American National Standards Institute (ANSI) [43], with a revision, ANSI N13.1, in 1972 [44]. ANSI N7.1 contains details of air and breath sampling protective measures, and medical procedures. The 1972 revision contains more recent techniques and suggestions for ventilation, air sampling patterns, record-keeping, and control responsibilities. Most important, it contains suggested procedures for measuring γ-radiation, and Rn its daughters.

Another code of practice is the 1957 U.S. Public Health Service *Publication 494* [45]. Also, the 1963 IAEA *Symposium* [46] contains much information on control and monitoring procedures. The NRC has issued a guide [47] describing programs it finds acceptable for measuring, evaluating, and reporting releases of radioactive materials in liquid and airborne effluents from typical uranium mills. The radiological effluent monitoring program is summarized in Table 12.6. The NRC has also released a guide [49] useful in the preparation of environmental reports for uranium mills.

Table 12.6. Summary of Proposed NRC Radiological Effluent Monitoring Program for Uranium Mills (data presented in Hickey, Ref. 48, with permission of American Nuclear Society © 1977)

Sampling Location and Frequency	Type of Analysis
Airborne	
Site boundary, continuous	U-natural, ^{230}Th, ^{226}Ra, ^{222}Rn, ^{210}Pb
Drier stack, semiannual	U-natural, ^{230}Th, ^{226}Ra
Other stacks, semiannual	U-natural
Liquid	
Discharge point, continuous	U-natural, ^{230}Th, ^{226}Ra
Wells, quarterly	U-natural, ^{230}Th, ^{226}Ra
On-site water bodies, quarterly	U-natural, ^{230}Th, ^{226}Ra
Tailings seepage, quarterly	U-natural, ^{230}Th, ^{226}Ra

12.6. INSTRUMENTATION REQUIREMENTS

Five different classes of measurements are used in the uranium mining and milling industry which are discussed in the next section. These classes are:

1. Measurements of Working Levels, to ensure the radiological protection of miners.
2. Measurements of the individual radionuclides (^{222}Rn and each individual daughter separately) by ventilation engineers and safety officers to aid in the design of adequate safety systems. Also in this class is the determination of the uncombined fraction of ^{218}Po. These measurements have not become widely used in the uranium industry.
3. Measurements of other sources of occupational radiation exposure. These include external γ-radiation and particulate airborne radionuclides, Th, U, and Ra. These latter measurements are not widely performed.
4. Measurements of bioassay media (for example, urinalysis). These can be valuable indicators of radionuclide body burdens and are principally used at mills, and much less often at mines.
5. Measurements of the radiological impact of mines and mills on the environment. These include measurements in waste waters, in ambient air near mill tailings, in soil samples, and in plants and animals.

12.6.1. Measurements of Working Level

The most obvious requirement for occupational protection is the measurement of working level (WL). Indeed, as mentioned previously, the original choice of WL as a measure of exposure was largely operational.

Direct measurements of WL itself can now be made by the well-established but time-consuming and approximate Kusnetz method; by the Tsivoglou method (in Ref. 46); and by more rapid but less accurate methods using Instant Working Level Meters. In addition, when ^{220}Ra is present, the potential α-particle energy (working levels) from ^{220}Ra daughters should be determined by the method of the Mining Enforcement and Safety Administration (predecessor of MSHA) [50]. These are all discussed in Chapter 14, Radon.

Instrumentation for the measurement of WL should be sensitive down to at least 0.03 WL (10% of the present average guidelines of 0.33 WL), and be capable of measuring WL in the range of about 0.1 to 1.0 WL with accuracies of ± 10 to $\pm 20\%$. It should be reliable in the working conditions found in mines. This implies ruggedness, convenience, portability, and small size and weight. When determining exposure, a rapid readout is not always important; however, it can be important for ventilation engineers (the Kusnetz method requires 45 min, start to finish).

Several devices have been proposed [51] for monitoring individual exposure of persons to Rn and, more importantly, to radon progeny attached to aerosol particles. A working level dosimeter has been developed and tested [52]. Its response is linear over several decades of cumulative working level exposure and the limit of sensitivity is about 1.4 WL-hr as long as controlled conditions exist.

12.6.2. Measurements of Individual Radionuclides

Another class of measurements is the separate determination of individual radionuclide activities in the radon chain, principally, ^{222}Rn, ^{218}Po, ^{214}Pb, and ^{214}Bi. If simple instrumentation were available it would enable industrial hygienists and ventilation engineers to assess the extent to which a particular atmosphere is in equilibrium.

Two common methods for ^{222}Rn detection are the Lucas chamber and the two-filter method, although neither is entirely satisfactory. Sensitivities are adequate [better than 10 pCi/liter (0.37 Bq/liter) in each case] but each suffers from some drawbacks, and a truly satisfactory radon gas monitor is still unavailable.

Measurements of various other radon daughters are possible. Very good sensitivity for radon daughters can be obtained by the Thomas modification of the Tsivoglou method [53]. Sensitivities as low as 1 pCi/liter (0.037 Bq/liter) for each daughter can be obtained [54]. Unfortunately, neither method is simple or portable. Instruments using these methods should be capable of measuring individual daughter activities at levels in the range of 10 pCi/l (0.37 Bq/l). Use of air filters and α-spectroscopy is possible but difficult. The topic is discussed further in Chapter 14.

12.6.3. Measurements of Other Occupational Exposures

There are several other occupational measurement problems besides the measurement of Rn and its daughters. In most cases, instruments adequate for such occupational radiological surveillance are available. This includes:

1. *External beta and gamma radiation.* These can be adequately measured with personnel dosimeters or hand-held meters described elsewhere in this book (see Chapters 4 and 6). Gamma-ray measurements are required by 30 CFR, Part 57.5-37, for underground measurements only. Currently, there are no surface standards, but work is in progress for developing them [6].
2. *Beta radiation in water.* A variety of instruments can make this measurement. Chapters 7 and 12 describe them.
3. *Uranium and thorium.* Chapter 15 discusses the measurement of U and Th in air and water. Available instruments are adequate to monitor most occupational exposures, which occur mainly at uranium mills.

12.6.4. Measurements of Biomedical Media

Bioassays are commonly used to measure occupational uptake of radionuclides. Common procedures are analyses of urine, exhaled breath, and tissue (urinalysis is most common). Techniques are available to determine radioactive U, Ra, Po, and Pb in urine [46], Rn in exhaled breath and Ra and Th in tissue [55].

Measurement typically requires detection of low specific activities. For example, workers exposed to 20 to 30 WL for about 3 yr exhibited 5 to 10 pCi (0.185 to 0.370 Bq) of ^{210}Pb per 24-hr urine sample [56]. In Japanese workers chronically exposed to 1 to 10 WL [57], levels of urinary ^{210}Po were in the range of about 0.05 to 0.5 pCi/liter (1.85−18.5 to mBq/liter). Another study [58] found that exposure at 1 WL results in ^{210}Po urinary excretions of about 0.25 pCi/day (9.25 mBq/day).

Uranium concentrations in urine at levels near 100 μg/liter were reported in still another study [59]. Correlations of WL with ^{210}Pb and ^{210}Po in urine are difficult to interpret if the concentrations of ^{210}Pb and ^{210}Po in the ambient air are not known.

Other media, including blood, hair, and whiskers, have been studied for correlation to radiation exposure. None of these shows strong correlation: predictions of mean WL cannot be made from measurement of ^{210}Pb or ^{210}Po in these media. Estimates of mean WL from such measurements are accurate only in the $\pm 50\%$ range at best [60]. Fortunately, this is adequate. Errors almost as large are also found for urinalysis correlation with WL. Activity levels found in whiskers were a few pCi of ^{210}Pb per gram, for current exposure in the range of 0.2 to 0.7 WL [60].

Bioassays for low activity levels usually use wet-chemical methods. They are typically quite sensitive, but are expensive, susceptible to analytical errors, and very laborious. The NRC has approved a bioassay method for uranium mill workers [61].

12.6.5. Measurements of Environmental Media

Uranium mining and milling activities may produce measurable radiological impacts on the immediate vicinity. Among the measurements required are the following:

1. Measurements of radon gas emitted from mill tailings, exposed ore bodies, and occasionally from concentrated (processed) uranium products. Because radon gas is produced by decay from the parent chain, its activity level must be evaluated periodically. Radon measurements in the range of a few tenths to a few hundred pCi/liter of air are typically required.
2. Measurements of Ra, U, and Th in surface water, ground water, waste water, ores, and mill tailings. The processing of ore may augment the naturally occurring levels of these radionuclides in local water. Both the short-term and long-term effects must be quantified. In particular, leaching from long-standing tailings can sometimes produce significant contamination. Typical natural ^{226}Ra concentrations in surface waters near uranium-rich regions range from 0.05 to 0.3 pCi/liter (1.85 to 11.1 mBq/liter) (Tsivoglou, in Ref. 46). Natural uranium content is in the range of 3 to 10 μg/liter [approximately 2 to 6 pCi/liter (0.074 to 0.222 Bq/liter)]; its contribution to gross α-particle levels is typically ten times greater than that of ^{226}Ra. Sill has developed a method for determining Th and U isotopes in ores and mill tailings [62].
3. Long-lived α-particle emitters contained in airborne particulates can present a potential exposure. Some specific radionuclides that should be measured are ^{226}Ra, ^{230}Th, ^{238}U. Recently, Kordas and Phelps reviewed the measurement techniques for such radionuclides [63]. Yule has developed an on-line monitor capable of detecting less than one maximum permissible airborne concentration (MPC_a) of these hazardous aerosols in less than one-half hour [64]. Unfortunately these techniques involve bulky apparatus, not simple, portable instruments.
4. Measurements in biological media. Media such as aquatic biosystems, local grasses and other flora, animal species, and soils can be indicators of environmental reconcentration mechanisms. Measurements are more frequently performed around mills than around mines, and even then they are usually for

research rather than for routine surveillance. Typical Ra activities in soil of 0.1 to 5 pCi/g (0.0037 to 0.185 Bq/g) occur naturally in uranium areas, so measurement techniques must be sensitive to at least that level [37]. Reconcentrations of Ra have been studied in a number of aquatic biota, and reconcentration factors, (activity in medium)/(activity in water) of 500 to 1000 have been found in algae and aquatic insects, with a factor of 3 typical for fish flesh and 100 typical for fish bone [65].

12.7. SUMMARY AND CONCLUSIONS

Uranium mining can present an occupational hazard. Exposures as large as those commonly experienced decades ago will probably not be allowed to recur because present standards and radiation protection guides, if properly applied, should be adequate for the protection of the miners.

Because of ongoing research and development, a more sophisticated approach to measurement and monitoring around uranium mines has become available. This approach emphasizes: (a) automated measuring instruments with direct readout and recording capability; (b) improved portable survey instruments; and (c) sensitive personnel dosimeters.

At uranium mills, the principal measurement problems are determining occupational exposures and environmental impacts. Both bioassay procedures and techniques for determining radionuclides in the environment need further development to permit more widespread use of them without undue economic impact. Techniques for measuring radionuclides in the environment should be developed for routine surveillance. Today they are still largely used for research.

REFERENCES

1. A. K. Reed, H. C. Meeks, S. E. Pomeroy, and V. W. Hale, "Assessment of Environmental Aspects of Uranium Mining and Milling," U.S. Environmental Protection Agency, EPA-600/7-76-036, Industrial Environmental Research Laboratory, Office of Research and Development, Cincinnati, OH 45268, 1976.
2. U.S. Environmental Protection Agency, "Environmental Analysis of the Uranium Fuel Cycle Part IV—Supplementary Analysis: 1976," EPA-520/4-76-017, Office of Radiation Programs, Washington, DC, 1976.
3. D. A. Clark, "State-of-the-Art: Uranium Mining, Milling, and Refining Industry," EPA-660/2-74-038, Environmental Protection Agency, National Environmental Research Center, Office of Research and Development, Corvallis, OR, 1974.
4. A. J. Kuhaida, Jr., and M. J. Kelly, "Solution (In-Situ Leach) Mining of Uranium: An Overview," *Trans. Am. Nucl. Soc.*, **30**, 98, 1978.
5. H. T. Miller, "Radiation Exposures Associated with Surface Mining for Uranium," *Health Phys.*, **32**, 534, 1977.
6. R. T. Beckman, private communication, Supervisory Mining Engineer, Radiation Branch, Denver Technical Support Center, U.S. Department of Labor, Mine Safety and Health Administration, P.O. Box 25367, DFC, Denver, CO 80225 (October 1978, April 1979).

7. Nuclear Regulatory Commission, "Final Environmental Statement Related to the Operation of Irigaway Uranium, Solution Mining Project, Wyoming Mineral Corporation," NUREG-0481, Office of Nuclear Material Safety and Safeguards, Washington, DC 20555, 1978.
8. Ford, Davis and Bacon Utah Inc., "Radiation Pathways and Potential Health Impacts from Inactive Uranium Mill Tailings," FB&DU 130-41 prepared for DOE as GJT-22, 375 Chipeta Way, Salt Lake City, UT 94108, 1978.
9. Code of Federal Regulations, 10 CFR 20.
10. TRW for EPA, "Environmental Study on Uranium Mills, Part I," (February 1979); private communication, Dick Douglas (April 1979).
11. R. L. Douglas and J. M. Hans, Jr., "Gamma Radiation Survey at Inactive Uranium Mill Sites," ORP/LV-75-5, Environmental Protection Agency, Office of Radiation Programs, Las Vegas, NV 89114, 1975.
12. M. B. Sears, R. E. Blanco, R. C. Dahlman, G. S. Hill, A. D. Ryon, and J. P. Witherspoon, "Correlation of Radioactive Waste Treatment Costs and the Environmental Impact of Waste Effluents in the Nuclear Fuel Cycle for Use in Establishing as Low as Practicable Guide—Milling of Uranium Ores," ORNL-TM-4503, Vol. I, Oak Ridge National Laboratory, P.O. Box X, Oak Ridge, TN 37830, 1975. (Available NTIS).
13. C. C. Travis, A. P. Watson, and L. M. McDowell-Boyer, "Potential Population Dose from Radon-222 Released by Uranium Tailings Piles," *Trans. Am. Nucl. Soc.*, **30**, 90, 1978.
14. W. A. Goldsmith, "Radiological Aspects of Inactive Uranium-Milling Sites: An Overview," *Nucl. Safety*, **17**(6), 733, 1976.
15. D. R. Dreesen, "Uranium Mill Tailings: Environmental Implications," LASL-77-37, Los Alamos Scientific Laboratory, Los Alamos, NM 87545, 1978.
16. Interagency Review Group (IRG) on Nuclear Waste Management, "Report to the President by the IRG on Nuclear Waste Management," TID-28817 (Draft), Department of Energy, Washington, DC 20545, 1978.
17. L. C. Hebel, Chairman, APS Study Group "Report to the American Physical Society by the Study Group on Nuclear Fuel Cycles and Waste Management," *Rev. Mod. Phys.*, **50**, S1, 1978.
18. J. J. Swift, J. M. Hardin, and H. W. Calley, "Potential Radiological Impact of Airborne Releases and Direct Gamma Radiation to Individuals Living Near Inactive Uranium Mill Tailings Piles," EPA-520/1-76-001, Environmental Protection Agency, Office of Radiation Programs, Washington, DC 20560, 1976.
19. D. L. Duncan and G. G. Eadie, "Environmental Surveys of the Uranium Mill Tailings Pile and Surrounding Area, Salt Lake City, Utah," EPA-520/6-74-006, Environmental Protection Agency, Office of Radiation Programs, Las Vegas, NV 89115, 1974.
20. W. E. Clements, S. Barr, and M. L. Marple, "Uranium Mill Tailings Piles as Sources of Atmospheric ^{222}Rn," LA-UR 78-828, presented at NRE III, Los Alamos Scientific Laboratory, Los Alamos, NM 87545, 1978.
21. F. F. Haywood, W. A. Goldsmith, P. T. Perdue, W. F. Fox, and W. H. Shinpaugh, "Assessment of Radiological Impact of the Inactive Uranium-Mill Tailings Pile at Salt Lake City, Utah," ORNL/TM-5251, Oak Ridge National Laboratory, Oak Ridge, TN 37830, 1977.
22. D. L. Duncan, G. A. Boysen, L. Grossman, and G. A. Franz, III, "Outdoor Radon Study (1974-1975), An Evaluation of Ambient Radon-222 Concentra-

tions in Grand Junction, Colorado," ORP/LV-77-1 Environmental Protection Agency, Office of Radiation Programs Las Vegas, NV 89114, 1977.

23. G. G. Eadie, R. F. Kaufmann, D. J. Markley, and R. Williams, "Report of Ambient Outdoor Radon and Indoor Radon Progeny Concentrations During November 1975 at Selected Locations in the Grants Mineral Belt, New Mexico," ORP/L-76-4, Environmental Protection Agency, Office of Radiation Programs, Las Vegas, NV 89114, 1976.
24. W. A. Goldsmith, F. F. Haywood, and R. W. Leggett, "Transport of Radon Which Diffuses from Uranium Mill Tailings," in *Proceedings on Natural Radiation Environment III*, CONF-780422, Vol. 1, Technical Information Center, U.S. Department of Energy, 1980, p. 1584.
25. P. M. Markose, K. P. Eappen, S. Venkataraman, and P. R. Kamath, "Distribution of Radium and Chemical Toxin in the Environment of a Uranium Complex," in *Proceedings on Natural Radiation Environment III*, CONF-780422, Vol. 1, Technical Information Center, U.S. Department of Energy, 1980, p. 1655.
26. Federal Radiation Council, *Guidance for the Control of Radiation Hazards in Uranium Mining*, Report FRC No. 8 (Revised), Washington, DC, 1967.
27. Code of Federal Regulations, 30 CFR 57.2 Definitions, 57.2-35a through 57.5-47a.
28. K. Z. Morgan, Editor in Chief, entire issue of *Health Physics*, **16**(5), 545–660, 1969.
29. F. E. Lundin, J. K. Wagoner, and V. E. Archer, "Radon daughter exposure and respiratory cancer: quantitative and temporal aspects," Joint Monograph No. 1 (1971), National Institute of Occupational Safety and Health and Health National Institute of Environmental Health Sciences.
30. A. W. Hilberg, "A Review of the Uranium Miner Experience in the United States," in *Noble Gases*, Richard E. Stanley and A. Alan Moghissi, Eds., registered by DOE TIC as CONF-730915, 1975.
31. U.S. Environmental Protection Agency, "Radiation Protection Activities 1977," EPA-520/4-78-003, Office of Radiation Programs, Washington, DC 20460, 1978.
32. *Report of the Royal Commission on the Health and Safety of Workers in Mines*, Appendix C, "Radiogenic Cancers in Ontario Uranium Miners 1955–1974," Ministry of the Attorney General, Province of Ontario, Toronto, Canada, 1976.
33. International Commission on Radiological Protection, "Radiation Protection in Uranium and Other Mines," ICRP Publication 24, *Ann. ICRP*, **1**(1), 1977.
34. International Commission on Radiological Protection, "Recommendations of the ICRP Report of Committee II on Permissible Dose for Internal Radiation," ICRP Publication 2, Pergamon Press, Oxford, 1969.
35. International Commission on Radiological Protection, "Recommendations of the ICRP," ICRP Publication 26, *Ann. ICRP*, **1**(3), 1977.
36. Code of Federal Regulations, 40 CFR 190, Environmental Radiation Protection Requirements for Normal Operations of Activities in the Uranium Fuel Cycles.
37. Code of Federal Regulations, 40 CFR 141.15 "Maximum contaminant Levels for ^{226}Ra, ^{228}Ra, and Gross α Particle Activity in Community Water Systems."
38. Ninety-Fifth Congress of the United States of America, "Uranium Mill Tailings Radiation Control Act of 1978," H.R.13650, 1978.

39. R. L. Rock, D. K. Walker, R. W. Dalzell, and E. J. Harris, *Controlling Employee Exposure to Alpha Radiation in Underground Uranium Mines*, U.S. Bureau of Mines Handbook Vol. I and II, 1970 and 1971.
40. International Atomic Energy Agency and International Labour Office, *Radiation Protection in the Mining and Milling of Radioactive Ores, Code of Practice and Technical Addendum*, IAEA Safety Series No. 26, Geneva, 1968.
41. International Atomic Energy Agency and International Labour Office, *Manual on Radiological Safety in Uranium and Thorium Mines and Mills*, IAEA Safety Series No. 43, Vienna, 1976.
42. International Atomic Energy Agency, *Management of Wastes from the Mining and Milling of Uranium and Thorium Ores*, A Code of Practice and Guide to the Code, IAEA Safety Series No. 44, Vienna, 1976.
43. ANSI N7.1, "Radiation Protection in Uranium Mines and Mills (Concentrators)," American National Standards Institute, 1430 Broadway, New York, NY 10018, 1960.
44. ANSI N13.1, "Radiation Protection in Uranium Mines," American National Standards Institute, 1430 Broadway, New York, NY 10018, 1972.
45. U.S. Public Health Service, *Control of Radon Daughters in Uranium Mines and Calculations on Biologic Effects*, Publication No. 494, Washington, DC, 1957.
46. International Atomic Energy Agency, *Radiological Health and Safety in Mining and Milling of Nuclear Materials*, Vol. I and II, Vienna, 1963.
47. Nuclear Regulatory Commission, "Measuring, Evaluating, and Reporting Radioactivity in Releases of Radioactive Material in Liquid and Airborne Effluents from Uranium Mills," Regulatory Guide 4.14, Office of Standards Development, Washington, DC 20555, 1977.
48. J. W. N. Hickey, "Radiological Effluent and Environmental Monitoring for Uranium Mills: A Regulatory Viewpoint," *Trans. Am. Nucl. Soc.*, **27**, 149, 1977.
49. Nuclear Regulatory Commission, "Preparation of Environmental Reports for Uranium Mills," Regulatory Guide 3.8, Revision I, Office of Standards Development, Washington, DC 20555, 1978.
50. R. L. Rock, "Sampling Mine Atmospheres for Potential Alpha Energy Due to the Presence of Radon-220 (Thoron) Daughters," MESA Informational Report 1015, U.S. Department of Labor, Mine Safety and Health Administration, P.O. Box 25467, DFC, Denver, CO 80225, 1975.
51. K. Becker, "Personnel Dosimetry of Radon and Radon-Daughters," in *Noble Gases*, Richard E. Stanley and A. Alan Moghissi, Eds., registered by DOE TIC as CONF-730915, 1975, p. 630.
52. A. J. Breslin, S. F. Guggenheim, A. C. George, and R. T. Graveson, "A Working Level Dosimeter for Uranium Miners," EML-333, Environmental Measurements Laboratory (DOE), New York, NY 10014, 1977.
53. J. W. Thomas, "Modification of Tsivoglou Method for Rn Daughters in Air," *Health Phys.*, **10**, 691, 1970.
54. J. W. Thomas, "Measurement of Radon Daughters in Air," *Health Phys.*, **23**, 783, 1972.
55. W. Stahlhofen and A. Kaul, "Measurement of Ra^{226}, Th^{232} and Daughter Products in Biological Samples," p. 475, Vol. II of Ref. 46.

56. R. F. Bell and J. C. Gilliland, "Urinary Lead-210 as Index of Mine Radon Exposure," p. 411, Vol. II of Ref. 46.
57. T. Inouye, Y. Yamada and S. Akiyama, "On the Level of Polonium in Urine of Japanese Uranium Miners," p. 425, Vol. II of Ref. 46.
58. D. Djurić, D. Panov, M. Kilibarda, L. Noak, and M. Vukotić, "Polonium in the Urine of Miners as A Measure of Exposure to Radon," p. 431, Vol. II of Ref. 46.
59. D. Karajović, M. Kilibarda, D. Panov, D. Djurić, Medjedović, P. Raicević, and V. Delić, "Uranium in the Urine of Miners Exposed to Uranium Compounds," p. 443, Vol. II of Ref. 46.
60. R. L. Gotchy and K. J. Schiager, "Bioassay Methods for Estimating Current Exposures to Short-Lived Radon Progency," *Health Phys.*, **17**, 199, 1969.
61. Nuclear Regulatory Commission, "Bioassay at Uranium Mills," Regulatory Guide 8.22, Office of Standards Development, Washington, DC 20555, 1978.
62. C. W. Sill, "Determination of Thorium and Uranium Isotopes in Ores and Mill Tailings by Alpha Spectrometry," *Anal. Chem.*, **49**(4), 618, 1977.
63. J. F. Kordas and P. L. Phelps, "A Review of Measurement Techniques for Stack Monitoring of Long-Lived Alpha Emitters," UCRL-81780, Lawrence Livermore Laboratory, Livermore, CA 94550, 1978.
64. T. J. Yule, "An On-Line Monitor for Alpha-Emitting Aerosols," *IEEE Trans. Nucl. Sci.*, **RNS-25**(1), 762, 1978.
65. E. C. Tsivoglou, "Environmental Monitoring in the Vicinity of Uranium Mills," p. 231, Vol. II of Ref. 46.

PART III

13 RADIUM

Radium (Ra) is one of the most familiar radioactive elements. Its use in cancer radiotherapy was one of the earliest "miracles" among the applications of nuclear physics, and its toxicity when taken internally is equally well known, from the case of the radium watch-dial painters.

13.1. INTRODUCTION

This section discusses measuring Ra concentrations in environmental media, primarily for environmental radiological protection. The text concentrates on methods for determining low levels of Ra in water and certain food chains, but does not treat bioassay methods in detail. Although the main emphasis will be on radium-226, the other important isotopes (^{223}Ra, ^{224}Ra, and ^{228}Ra) are discussed where appropriate.

First a review of the radioactive properties of the important Ra isotopes is given; next, the radiation protection guides for Ra concentrations are discussed; and then the ranges of Ra concentrations in the natural environment are presented. Finally various Ra measurement techniques are discussed. Discussions that are important for a more complete picture of the measurement techniques dealt with in this chapter are found in Chapter 14, in which the ^{238}U decay chain and the instrumentation for ^{222}Rn measurements are discussed; and Chapter 7, in which instruments for detecting alpha radiation is treated in detail.

13.2. RADIOACTIVE PROPERTIES

Radium occurs naturally in four isotopic states, each of which is a member of a naturally occurring decay chain:

13.2.1. Radium-226

Radium-226 (half-life = 1602 yr) is a member of the ^{238}U chain. Its immediate parent is the α-emitter thorium-230 (^{230}Th) (half-life = 80,000 yr), and its daughter is the inert gas ^{222}Rn (half-life = 3.82 days). Radium-226 α decays to ^{222}Rn, 94.6% by a 4.78-MeV α-particle emission and the remaining 5.4% by a 4.60-MeV α-particle emission followed by a 187-keV γ-ray.

13.2.2. Radium-223

Radium-223 (half-life = 11.4 days) is a member of the ^{235}U chain. Its immediate daughter is ^{219}Rn (half-life = 4.0 sec.), an inert gas known historically as "actinon." Radium-223 α decays to ^{219}Rn. The most important α-particle energies and branching ratios are: 5.75 MeV α (9%); 5.71 MeV α (52%); 5.61 MeV α (25%); 5.54 MeV α (9%). Each α-particle is accompanied by a γ-ray emission before the ground state of ^{219}Rn is reached.

13.2.3. Radium-228

Radium-228 (half-life = 6.7 yr) is a β-ray emitting member of the ^{232}Th chain. Its low energy β (E_{max} = 55 keV) leads to actinium-228 (half life = 6.1 hr), which in turn decays to ^{228}Th by one of a large number of β-ray emissions, the most energetic of which is 2.15 MeV. Thorium-228 (half-life = 1.9 yr) then decays by α-particle emission (5.4 MeV) to ^{224}Ra.

13.2.4. Radium-224

Radium-224 (half-life = 3.64 days) decays by α-emission to the gas ^{220}Rn, historically "thoron" (half-life = 55 sec). The decay products are a 5.68-MeV α-particle (94%) or a 5.44-MeV α-particle followed by a 241-keV γ-ray (6%). This information is summarized in Table 13.1 [1].

13.2.5. Chemical Properties

Chemically, Ra is an alkaline earth element in the same family as beryllium, magnesium, calcium, strontium, and barium. Its chemical similarity to Ca leads to its principal toxic effect, the preferential lodging in bone upon ingestion.

An interesting historical note is that 1 Ci (3.7×10^{10} Bq) was originally defined as the activity of 1 g of ^{226}Ra. Small shifts in the definitions have occurred since then, so that now ^{226}Ra's specific activity is about 0.988 Ci/g. However, today ^{226}Ra is still sometimes measured in, for example, pg/liter instead of pCi/liter—these are numerically almost equal. The curie as a unit of radioactivity is now being replaced by the becquerel (= 1 disintegration/sec)

Table 13.1. Some Physical Properties of Radium Isotopes (data presented in Lederer, Hollander, and Perlman, Ref. 1)

Isotope	Half-Life	Member of Which Decay Chain	Immediate Parent	Immediate Daughter	Main Decay Products and Energies (for β, E_{max})	
					Branching Ratio (%)	Decay Products
Radium-223	11.4 days	U-235	Thorium-277 (18.2 days)	Radon-219 (4.0 sec)	9	5.75-MeVα + 338-keVγ
					52	5.71-MeVα + 270-keVγ
					25	5.61-MeVα + 154-keVγ
					9	5.54-MeVα + 122-keVγ
Radium-224	3.64 days	Th-232 chain	Thorium-228 (1.9 yr)	Radon-220 (55 sec)	94	5.68-MeVα
					6	5.44-MeVα + 241-keVγ
Radium-226	1602 yr	U-238 chain	Thorium-230 (80,000 yr)	Radon-222 (3.82 days)	94.6	4.78-MeVα
					5.4	4.60-MeVα + 187-keVγ
Radium-228	6.7 yr	Th-232 chain	Thorium-232 (14 billion yr)	Actinium-228 (6.1 hr)	100	55-keV (max β)

13.3. RADIATION PROTECTION GUIDES

Radium's toxicity depends on whether its chemical form is soluble, or insoluble. When ^{226}Ra is soluble, for example, it is a much more severe problem when ingested by mouth. These differences have been taken into account in the establishment of the radiation protection guides.

For nominal 168-hr/wk occupational exposures, the International Commission on Radiological Protection (ICRP) has established maximum permissible concentrations in air and water MPC_a and MPC_w. These concentrations are given in the ICRP Publication 2 [2]. The 40-hr/wk MPCs stated in the 10 CFR 20, and presented in Table 13.2, are, with a few exceptions, about 4 times these values. The maximum permissible body burdens (MPBB) for workers (ICRP Publication 2) are also given. In addition, Table 13.2 gives 10 CFR 20 MPCs for unrestricted area, those areas to which individuals in the general public have access. These values are, with the notable exception of ^{226}Ra, a factor of about 30 to 40 smaller than those for 40 hr/wk occupational exposures. For exposure of a suitably large sample of the general public, the MPC would be another factor of 3 smaller still.

Considering these factors, the public MPC for soluble ^{226}Ra in drinking water for a large population would be about 1% of 400 pCi/liter or 4 pCi/liter (0.148 Bq/liter). The 1962 U.S. Public Health Service Drinking Water Standard [3] adopted 3 pCi/liter (0.111 Bq/liter) as its recommended limit. But in its Interim Primary Drinking Water Regulations, promulgated in 1976 [4], the Environmental Protection Agency (EPA) adopted a concentration limit of 5 pCi/liter (0.185

Table 13.2. Maximum Permissible Concentrations in Air and Water and Maximum Permissible Body Burdens for Radium

	MPC[a] for: Workers/General Public		MPBB[b]
Isotope	In air (10^{-12}Ci/liter)	In water (10^{-9}Ci/liter)	Occupational (10^{-6}Ci)
Radium-223			
Soluble	2/0.06	20/0.7	50
Insoluble	0.2/0.008	100/4	
Radium-224			
Soluble	5/0.2	70/2	60
Insoluble	0.7/0.02	200/5	
Radium-226			
Soluble	0.03/0.003	0.4/0.03	100
Insoluble	0.05/0.002	900/30	
Radium-228			
Soluble	0.07/0.002	0.8/0.03	60
Insoluble	0.04/0.001	700/30	

[a]Code of Federal Regulations, 10 CFR 20, Appendix B. The MPC for workers assumes a 40-hr work week. The critical organ for soluble radium is bone; for insoluble radium in air or water, the critical organ is lung or GI tract (lower large intestine), respectively (Ref. 2).
[b]Reference 2.

Bq/liter) for the sum of ^{226}Ra and ^{228}Ra activity. The total α-particle limit (excepting Ra and U) was given as 15 pCi/liter (0.555 Bq/liter). This regulation also states that gross-α measurements may be substituted for ^{226}Ra and ^{228}Ra (not an alpha-particle emitter) analysis, provided that the activity does not exceed 5 pCi/liter (0.185 Bq/liter). Chapter 7 discusses gross-α determination techniques.

13.4. SOURCES AND LEVELS OF ENVIRONMENTAL RADIUM

Wherever U or Th occurs naturally in the environment, the various Ra isotopes are found as well. Generally, the decay of the parents in the chain proceeds down through several steps, during which time the decaying nucleus usually remains immobile in the original location. Each of the decay chains ultimately leads to Rn (222, 220, or 219), a gas that can and does migrate. Radon emanates from surface layers or liquids and is one of the more prominent external manifestations of the natural radioactivity present.

An important migration mechanism for Ra is leaching: water-soluble or reactive members of the chain, including Ra, can be leached out into ground water. Leaching can cause a significant radiological effect in some areas. For example, although ^{226}Ra concentrations in domestic tap water typically range from 0.01 to 0.2 pCi/liter (0.37 to 7.4 mBq/liter) [5], concentrations from 1 to 5 pCi/liter (0.037 to 1.85 Bq/liter) occur in several areas, and concentrations in the 10 to 50-pCi (0.37 to 1.85-Bq/liter) range are occasionally found [6]. Because of the shorter half-lives of the other Ra isotopes, the long-lived ^{226}Ra is the dominant isotope. However, ^{228}Ra and ^{224}Ra are sometimes found at levels comparable to those of ^{226}Ra [7].

In most cases, however, the principal dietary source of Ra is food. Radium enters the human food chain in many ways. Plant uptake of Ra directly from the soil, or from the water, produces small but measurable Ra concentrations in foods, commonly in the range 0.1 to 5 pCi/kg (0.0037 to 0.185 Bq/kg). Studies in Italy [8] of food from sources not known to be artificially contaminated showed the following approximate specific activities for ^{226}Ra: 0.4 pCi/kg (0.0148 Bq/kg) in fruits; 0.4 to 0.7 pCi/kg (0.0148 to 0.0259 Bq/kg) in potatoes; 0.7 to 1.2 pCi/kg (0.0259 to 0.0444 Bq/kg) in green vegetables and lettuce; 3.6 to 3.8 pCi/kg (0.133 to 0.141 Bq/kg) in flour; and 5.8 to 6.9 pCi/kg (0.215 to 0.255 Bq/kg) in eggs.

Holtzman [9] reviewed dietary levels of ^{226}Ra and ^{228}Ra. Intakes in normal areas of the United States are approximately 1 pCi/day (0.037 Bq/day) for each isotope, arising in roughly equal part from: cereals; meats and dairy products; and vegetables (from data of Ref. 10). Drinking water is normally a much less significant contributor, but this is not the case, of course, for areas with water Ra contents approaching 1 pCi/liter (0.037 Bq/liter) or greater.

In air, concentrations of Ra itself are usually minute. Mechanisms that might produce airborne (particulate) Ra concentrations include such local effects as dust storms. Normal airborne concentrations appear to be within a factor of 10 of 0.001 fCi/liter (0.037 μBq/liter) [11] which would imply daily respiratory intakes on the order of 10 fCi (0.37 mBq). Of course, the emanating gases (^{222}Rn, ^{220}Rn, and ^{219}Rn) and their daughters are widely distributed in much higher concentrations.

In soils, Ra concentrations correlate with the natural U/Th content. Pearson [12] reported ^{226}Ra concentrations from 0.1 to 0.8 pCi/g (0.0037 to 0.0295 Bq/g) (dry soil), in several samples scattered through the United States, but as high as 15 pCi/g (0.555 Bq/g) in U mining areas. Even in phosphate areas, U and Ra contents are significantly higher. Roessler et al. [13] found concentrations of 8 to 36 pCi/g (0.296 to 1332 Bq/g) in Florida phosphates. Report No. 45 of the National Council on Radiation Protection and Measurements [14] summarizes data on the concentration of the ^{238}U and ^{232}Th series in major rock types and soils. See also Chapter 15, Uranium and Thorium.

Building materials containing earth or soil have Ra concentrations comparable with soil, unless processing has altered the concentration. As reported in 1977 by the United Nations Scientific Committee on the Effects of Atomic Radiation [15], ^{226}Ra contents of 1 pCi/g (0.037 Bq/g) are typical, but variations by an order of magnitude are not unusual.

Of soils and rocks, the highest U series concentrations are found in U ores. The ^{238}U:^{235}U atomic ratio is constant at about 138.5 [1]. At equilibrium, the ^{226}Ra:^{223}Ra atomic ratio is about 20 [16], and one would thus expect a comparable ratio from uranium-mill effluents. This is seldom the case, however, because during the sulfuric acid leaching process used in most mills, Ra is barely soluble but ^{227}Ac and ^{227}Th dissolve more readily; these nuclides then give rise to their daughter ^{223}Ra in the aqueous phase [16]. The bulk of mill liquid effluent Ra activity turns out to be ^{223}Ra. Most of the ^{226}Ra remains with the solid tailings. Considering the tens of curies of U supplied annually for operation of a 1-GW nuclear power plant (see Chapter 3), a similar amount of Ra would be rejected to the tailings pond.

High-level nuclear waste from commercial nuclear electric generation initially do not contain Ra. However, long-lived transuranics often produce Ra (especially ^{228}Ra and ^{226}Ra), so that beginning about 10,000 years after removal from the reactor, Ra may dominate the biological hazard from these wastes [17].

This brief summary is intended only to point out the range of concentrations of Ra in typical sample media. Most of the levels mentioned are natural, in that they are not enhanced by man's use of radioactivity. However, for the most part man's activities have not significantly increased the natural levels of Ra in various media. Obvious exceptions are in the U mining and milling industry, and in the disposal or accidental release of Ra used in industrial or medical applications.

13.5. MEASUREMENT TECHNIQUES

Because Ra concentrations in most samples are small, and because the decay α-particles have such short ranges in matter ($\sim$5 to 10 mg/cm^2), direct α-particle counting of environmental samples is seldom practical. Instead, measurement techniques usually fall into two categories, which are treated separately in the next sections. In one, Rn gas (usually ^{222}Rn) is allowed to emanate from the Ra-containing medium, followed by measurement of the activity of the Rn or its daughters. In the other, chemical methods are used to separate Ra from the matrix in which it is found, followed by α-particle counting or in some cases α-spectroscopy. Sometimes the two approaches are combined: after chemical separation, the prepared sample may be stored for Rn ingrowth, followed by Rn measurements. A method distinct from these approaches employs γ-spectrometry of the Ra daughters *in situ*.

13.5.1. Radon Emanation Techniques

Radon emanation followed by measurement of ^{222}Rn is by far the most widely used method for measuring ^{226}Ra. For further discussions of ^{222}Rn measurements the reader is referred to Chapter 14.

There are three Ra-to-Rn decays of interest:

1. ^{224}Ra → ^{220}Rn.
2. ^{223}Ra → ^{219}Rn.
3. ^{226}Ra → ^{222}Rn.

Two of the three resulting Rn isotopes, ^{219}Rn and ^{220}Rn, are so short-lived (4-sec and 55-sec half-lives, respectively) that measurements of them after emanation are unsatisfactory. Hence, the emanation technique is used exclusively with ^{222}Rn (to determine its parent ^{226}Ra). Indeed, because the emanated gas is usually stored before counting, the interference from the other isotopes can be made unimportant.

Direct de-emanation is an old technique. A sample is placed in a vessel filled with helium (He) or "old air" (e.g., air stored long enough to allow decay of natural Rn) and put aside to allow ^{222}Rn ingrowth. For example, two half-lives (7.65 days) will give ~75% of the maximum buildup. The simplest procedure is transfer of the carrier gas plus radon, through a particulate filter to a Lucas chamber for α-particle counting. The Lucas chamber [18] is a small (~100 to 200 cm^3) cell coated on the inside with ZnS(Ag) scintillator. Through a flat window on one side, the scintillation light is viewed and counted by a photomultiplier. This instrument is described in more detail in Chapter 14.

There are several difficulties with direct de-emanation. Foremost is uncertainty in knowing the quantitative emanation rate. Emanation from a solid medium is a process intimately dependent upon geometrical idiosyncrasies. For example, Rn from Ra locked in crystal structures far from a surface simply does not get out at a rate that can be defined in advance. Other problems can occur because direct transfer to the Lucas chamber (even through a high-efficiency filter, which removes Rn daughters) can result in errors at high humidity, as well as in uncertainties in the electrostatic properties within the chamber, which can seriously degrade counting efficiency.

To solve this latter set of problems, one common procedure today is to pass the Rn through a purifying system before introduction to the Lucas chamber. The system [19,20], described in Chapter 14 and shown in Fig. 13.1, sends the Rn through a Drierite trap and a cooled water trap into a pretreated activated-charcoal Rn trap at −78°C (dry ice). The charcoal is subsequently de-emanated at 300°C and the Rn gas transferred to a Lucas chamber with a He carrier. This transfer/purification system is quantitative, nearly 100% efficient, and is almost error-free when properly operated.

To solve the problem of uncertainties in the emanation of Rn from the original (solid) medium, the commonest procedure today is dissolution. Even this process is subject to uncertainties that depend on the chemical form of the sample and on the

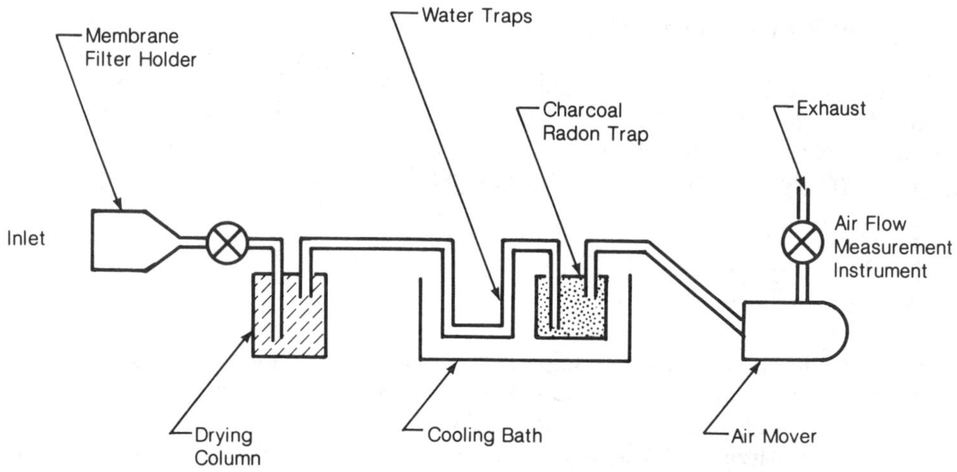

Figure 13.1. Sampling apparatus for collection of gaseous airborne radon (reproduced from Intersociety Committee on Methods or Air Sampling and Analysis, Ref. 20, with permission of the American Public Health Association © 1977).

dissolution technique. However, once Ra is in solution, Rn de-emanation occurs with essentially 100% efficiency. During Rn ingrowth, some of the Rn gas collects in the (He) gas above the liquid. The remainder is flushed from the liquid at the end by forcing He through the solution. The He-carried Rn is then run directly into a Lucas chamber, or alternatively through the Drierite/water-trap/activated-charcoal/Lucas chamber system.

Special flasks have been designed for de-emanation of the Ra-in-solution. One typical design is shown in Fig. 13.2 [21]: a 500-ml Pyrex flask is altered by insertion of a 10-mm Pyrex tube for the He. This flask can be easily decontaminated before introduction of the liquid sample.

Rushing, Garcia, and Clark [22] describe a system in which, after chemical dissolution and de-emanation, the He-Rn mixture is run directly into a Lucas chamber. After transfer, 3 to 4 hr are allowed to elapse before counting, to permit ^{222}Rn to come to transient equilibrium with its daughters. These investigators achieved minimum detectable Ra activities in the range of 0.03 pCi (in 300 ml) or about 0.10 pCi/liter (3.7 mBq/liter). An interference from ^{220}Rn may occur, but can be overcome by interim storage of the gas before introduction to the Lucas chamber. Even if storage is not used, a 1:1 ratio of ^{224}Ra:^{226}Ra in the original sample only produces a ^{226}Ra overestimate in the 0.5% range, while the effect of ^{223}Ra is an order of magnitude smaller still.

Blanchard [21] has described a complete system, including the Drierite/activated-charcoal purification procedure. Being careful to flush remnant Rn from various parts of the system, and using a 300-ml Ra containing sample with a 1-hr Rn emanation period, he measured an NBS standard [1.00 pCi (0.037 Bq) of Ra] to within ±0.04 pCi (1.5 mBq), with an average recovery of 100 ±4% for nine trials. The limit of detection (95% confidence) is calculated to be 0.016 pCi (0.6 mBq) which (in 300 ml) corresponds to about 0.050 pCi/liter (1.85 mBq/liter).

Another emanation technique, similar to the foregoing, is included as a Standard Method in the American Public Health Association (APHA) compilation [23]. The

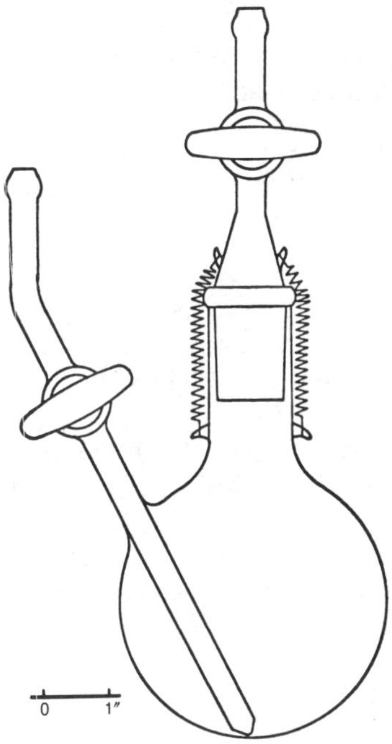

Figure 13.2. The de-emanation flask (reproduced from Blanchard, Ref. 21).

sensitivity is comparable, and this reference reports that "neither radium-224 at an activity equal to that of the radium-226 nor dissolved solids up to 610 mg/liter produced a detectable error in the results." This Standard Method satisfies nearly all requirements of sensitivity, accuracy, and versatility which might be desired for environmental monitoring. Its drawback is the expense and difficulty of operating and maintaining the equipment.

The importance of running reagent blanks through the entire system must be emphasized: minute Ra contamination in any of the reagents can sometimes occur, and a correction must be applied (or the contaminant eliminated) to achieve the ultimate in sensitivity.

13.5.2. Sample Preparation for Radon Emanation

The Rn emanation technique used today requires that the Ra be in solution. The following discussion outlines briefly a few of the various processes developed for dissolving Ra from the more important sample media.

In many media (e.g., water or urine) the Ra is already in solution. However, even for some liquid media, such as raw sewage, it may be desirable to process the Ra before emanation. Also, in some samples the dissolved Ra is at too low a concentration to permit measurement by emanation, so preconcentration is necessary.

A method applicable to water, urine, or fecal samples is described in the DOE Environmental Measurements Laboratory (EML) *Procedures Manual* [24]. Radium is coprecipitated with barium sulfate; silica is removed with hydrofluoric acid and reprecipitation of the sulfate. The precipitate is then dissolved in alkaline EDTA (ethylenediaminetetraacetic acid), and the solution used for emanation.

A method applicable to food, milk, vegetation, soil, and bone is also described in the EML *Procedures Manual* [24]. Some samples require a sodium carbonate fusion, after which the radium carbonate is dissolved in nitric acid.

> Successive fuming nitric acid separations remove the calcium and most other interfering ions. Traces of some fission products are scavenged with yttrium hydroxide. Finally, radium is coprecipitated with barium chromate, then dissolved in perchloric acid and water [24].

The *Handbook* of the EPA's Las Vegas laboratory [25] describes a slightly different method for ashed samples using ammonium chromate, and also a method for soil, sludge, and air filters in which fusion is performed with Nicholson's flux, followed by sulfuric acid dissolution and a barium-carrier addition for coprecipitation of barium-radium phosphate. The U.S. Public Health Service's (PHS) procedures compilation [26] describes a method of dissolving ashed samples with ammonium dichromate.

Standard Methods [23] describes separate methods for determining dissolved and undissolved ^{226}Ra in water, using a membrane filter to separate the undissolved (suspended) matter for separate dissolution using Nicholson's flux.

13.5.3. Radiochemistry with Radium Alpha Counting

The techniques discussed here employ chemical separation of Ra from the medium in which it is sampled, followed by counting or spectroscopy of the Ra α-particles.

Ordinary chemical techniques are intrinsically incapable of differentiating among Ra isotopes. Thus if simple α-particle counting is used, a single measurement can only yield total Ra activity (curies or becquerels). Alpha spectroscopy, on the other hand, enables the various α-particle energies to be separately identified quantitatively. Another technique for separating isotopes is to make activity measurements at several different times, so that the buildup and decay of activities from the various Ra/daughter series can be followed. The basis for this can be seen in Fig. 13.3 [27], which shows the theoretical growth and decay of total α-particle activity (parent plus daughters) from initially pure Ra isotopes. From this time variation it is relatively simple to determine ratios of $^{226}Ra:^{224}Ra:^{223}Ra$.

Chapter 7 contains detailed discussions of instruments for α-particle counting and spectroscopy. To summarize, the common instruments for α-particle measurement are internal gas proportional counters, semiconductor (diffused-junction or surface-barrier) detectors, pulse ion chambers, and scintillators. Because the range of α-particles in matter is so small ($\simeq$ 5 to 10 mg/cm^2), one problem is preparing a sample thin enough to be counted without important self-absorption corrections. As discussed in Chapter 7 this can be accomplished either by electrodeposition or precipitation and filtration.

Some techniques for radiochemical determination of specific Ra isotopes by measuring their daughters (Pb, Po, and Bi) are discussed later.

Standard Methods [23] describes a precipitation technique designed for screening drinking water. The method is also applicable to sewage or industrial wastes, provided that organic matter is destroyed.

> Lead and barium carriers is added to the sample containing alkaline citrate, then sulfuric acid is added to precipitate radium, barium and lead as sulfates. The precipitate is purified by washing with nitric acid, dissolving in alkaline EDTA, and re-

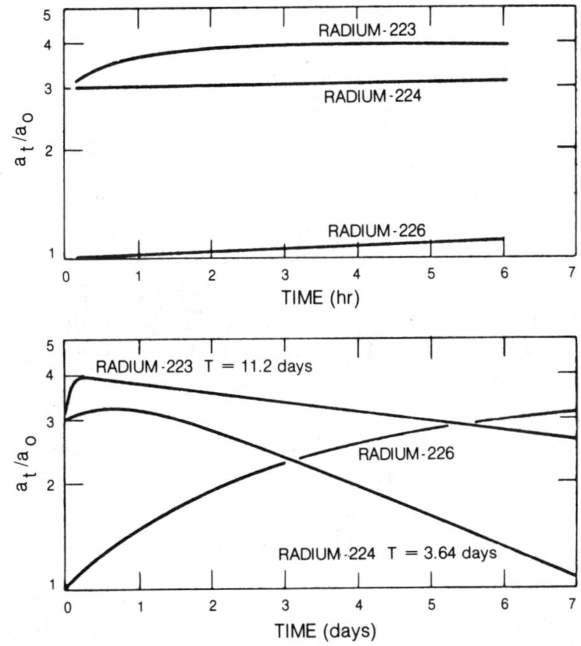

Figure 13.3. Growth of alpha activity into pure radium isotopes, (reproduced from the American Society for Testing and Materials, Ref. 27, with permission of the American Society of Testing and Materials © 1971).

> precipitating as radium-barium sulfate after adjusting the pH to 4.5. This slightly acidic EDTA keeps other naturally occurring alpha emitters and lead carrier in solution. [23]

The precipitate is counted with an internal gas proportional counter, after a 3 to 18-hr wait. To determine isotopic mixtures,

> additional counting at 7, 14 or 28 days is suggested, depending on the number of isotopes in mixture. The amounts of the various isotopes can be determined by solving a set of simultaneous equations. [23]

Average chemical recoveries are excellent ($\simeq$ 95%) in this method, and the technique is valuable as a screening method for routine Ra-in-water analysis.

An EPA tentative reference method [28] uses a similar method for Ra and ^{226}Ra in water. Total Ra is determined by α-particle counting of the precipitate. Radium-226 is determined by ^{222}Rn de-emanation. The American Society for Testing and Materials (ASTM) compilation [27] describes a similar method involving α-particle counting. The lower detection limit is about 1 pCi/liter (0.037 Bq/liter).

The APHA, EPA, and ASTM methods are based on a technique developed at Idaho Falls [29] described later by Goldin [30].

Another PHS chemical method [26] involves Ra coprecipitation with barium sulphate from a solution containing EDTA. The precipitate is converted to carbonate and dissolved in nitric acid. Radium daughters are extracted into thenoyltrifluoroacetone, and Ra-$BaSO_4$ is reprecipitated for counting.

A quite different method has been developed by Duquesne et al. [31]. The final sulfate precipitate is suspended in ethanol together with a powdered scintillator, ZnS(Ag), and this mixture is filtered and dried for counting with a photomultiplier. Sensitivity below 0.1 pCi/liter (3.7 mBq/liter) is claimed, and by studying the α-particle activity as it evolves in time, the ^{224}Ra:^{226}Ra ratio can be roughly determined.

Kelkar and Joshi [32] describe a system that is similar but even simpler. After a 1-liter sample is filtered, hydrochloric acid and barium chloride are added. Radium, Pb, and Po are coprecipitated with concentrated H_2SO_4; and ZnS(Ag) is added while stirring. The sulfates and the ZnS(Ag) are collected on a filter, which is dried under an infrared lamp, and counted. Chemical yields exceed 95%, and concentrations as low as 1 to 2 pCi/liter (0.037 to 0.074 Bq/liter) can be determined. This method, though less sensitive than the more sophisticated techniques, is quick and inexpensive.

Researchers at Idaho Falls have extended separation methods based on precipitation with Pb or $BaSO_4$ to the case of solid environmental samples. This is accomplished by preparing the sample with hydrochloric acid and pyrosulfate fusion, as described in Chapter 7. Percival and Martin [33] adapt this technique for ^{226}Ra, ^{228}Ra, ^{227}Ac, and Th in ores, mill tailings, soils, and air dusts. Sill [34] applies it to all the long-lived members of the ^{238}U series in the same media.

To summarize, there are a variety of chemical methods which take advantage of coprecipitating Ra with Pb or Ba as sulfate or carbonate, to separate it from other α-emitting elements (principally the various Ra daughters). After successive purifications in one of several chemical procedures, the methods either count the Ra salt in an internal gas proportional counter, or use ZnS(Ag) and a photomultiplier tube for scintillation counting.

In some of the methods that use α-particle counting, the use of α-spectroscopy enables a direct determination of Ra isotope ratios. However, resolution is a key consideration. Resolutions ($\pm$ half-width) that can in principle (i.e., with little regard for counting efficiency) be obtained in various kinds of α-spectrometers are about ± 15 keV with surface-barrier detectors; ± 20 keV with pulse ion chamber; and ± 50 keV with gas proportional counters. Liquid scintillators (± 150 keV) and solid scintillators (± 200 keV) are poorer. Any of these systems can resolve ^{226}Ra (4.60, 4.78-MeV α-particles) from ^{223}Ra (5.54, 5.61, 5.72, 5.75-MeV α-particles) and ^{224}Ra (5.44, 5.68-MeV α-particles) in principle (although the fact that practical systems have poorer resolution often reduces capabilities for separation). However, separating ^{223}Ra and ^{224}Ra from each other is more difficult because the energies are so intermingled.

13.5.4. Radiochemistry with Counting of Radium Daughters

Some methods involve counting of Ra daughters (other than Rn gas) after radiochemical separation.

One technique, described in the EML *Procedures Manual* [24], chemically separates ^{212}Pb in urine and measures the α-particles from ^{212}Bi and ^{212}Po after equilibration. This technique is specific for ^{224}Ra (the precursor of ^{212}Pb). In another EML technique applicable for Ra in urine or water [24], Ra-$BaSO_4$ precipitate is stored for buildup of ^{222}Rn, ^{218}Po, and ^{214}Po, which are measured in equilibrium using a ZnS(Ag) disk phosphor and a photomultiplier.

Petrow and Allen [16] described a specific determination of ^{223}Ra. After Ra is separated from a Pb carrier, it is stored to allow ingrowth in solution of ^{223}Ra's daughter, ^{211}Pb [36.1 min, 1.39-MeV (max) β]. The Pb is then extracted from solution with dithizone. Interference from ^{214}Pb (daughter of ^{226}Ra) is minimized by boiling the solution during the 3-hr period, in order to purge the 3.82-day half-life ^{222}Rn. The selection occurs because some of the ^{219}Rn (4-sec half-life) in the $^{223}Ra \rightarrow {}^{211}Pb$ chain decays before it too is purged in the boiling. The 6.62-MeV α from ^{211}Bi (^{211}Pb's daughter) is the activity that is counted.

A similar procedure can be used to determine ^{224}Ra. The separated Ra sample is aged for 16 to 24 hrs to enable ^{212}Pb ingrowth from ^{224}Ra. Lead-212 (10.6-hr half-life) is separated from ^{211}Pb (half-life = 36.1 min) by aging of the lead extract for 3 to 4 hrs before beta counting.

Petrow and Allen [16] also describe a method for ^{228}Ra by separating and counting its daughter, ^{228}Ac. To avoid an interference, the Ra sample must be purified of ^{227}Ac before ^{228}Ac ingrowth. This is achieved by using lanthanum hold-back carrier prior to the Ra-$BaSO_4$ precipitation. After ^{228}Ac has grown from ^{228}Ra, the sample is β-counted in an internal gas proportional counter.

To summarize, several chemical methods have been discussed which selectively determine one of the Ra isotopes by the chemical or physical properties of its daughter(s). These analytical methods are useful only in restricted circumstances, but they are a valuable tool when detailed knowledge of the various Ra isotopic ratios is required.

13.5.5. In Situ Gamma Counting of Radium Daughters

Gamma spectrometry of the ^{222}Rn daughters, ^{214}Pb and ^{214}Bi, is often used for determination of ^{226}Ra and, even more indirectly, of other members of the ^{238}U series. See Chapter 15 for gamma spectra of U and Th decay chains using sodium iodide and germanium-lithium spectrometers. Gamma spectrometry is treated in Chapter 6.

Sanderson [35] has described the determination of ^{226}Ra in food, soil, or biological samples by coincidence thallium-activated sodium iodide gamma spectroscopy of the 0.61-MeV and 1.12-MeV γ-rays from ^{214}Bi. Thorium-228 can also be determined by the coincident 0.58-MeV and 2.61-MeV γ-rays from its daughter, ^{208}Tl. Sanderson reports ^{226}Ra measurements in the range of 0.30 pCi/g (0.011 Bq/g) with $\pm 10\%$ errors, using 20 to 40-g samples. Minimum sensitivities from 0.03 to 0.06 pCi/g (1.11 to 2.22 mBq/g) are reported for both ^{226}Ra and ^{228}Th in food, soil, and fecal ash.

13.6. CALIBRATION SOURCES

There is an obvious need for sources of known activity to be used for calibration purposes. To fulfill this need, the NBS has developed a set of standards with ^{226}Ra content ranging from 0.1 to $\simeq$ 100 μg, in flame-sealed glass ampoules [36]. These emanation standards, in liquid form, are for use in calibrating an entire counting system.

13.7. SUMMARY AND CONCLUSIONS

This section discusses the typical levels at which Ra is found in environmental samples; the radiation protection guides; and the types of measurement techniques developed to measure Ra in the environment. The emphasis has been placed on ^{226}Ra, although the other main isotopes (^{223}Ra, ^{224}Ra, ^{228}Ra) are also discussed.

The activity levels requiring measurement are very small, and the measurement techniques are correspondingly sophisticated. Most routine measurements today use either an emanation technique or radiochemical separation followed by α-counting. Each of these types of measurement can be performed with sufficient sensitivity and accuracy to determine the Ra isotopes at their naturally occurring activity levels in the environment. Unfortunately, each suffers from drawbacks: the radiochemical methods require exacting analytical procedures; the emanation techniques are time consuming and the apparatus clumsy and relatively expensive.

However, in the overall picture of radionuclide measurement capabilities in environmental media, Ra measurements appear to be in a reasonably sound position. The quick screening (gross-α) measurement techniques for public health protection are adequate, and Ra levels can seldom be expected to increase suddenly to the danger point where both time and accuracy would be essential. Thus the present methods, though time-consuming, are probably adequate for the task of routinely monitoring environmental samples for Ra content.

REFERENCES

1. C. M. Lederer, J. M. Hollander, and I. Perlman, *Table of Isotopes,* 6th ed., Wiley, New York, 1967; a revised and enlarged 7th ed. appeared in 1979.
2. International Commission on Radiological Protection, "Permissible Dose for Internal Radiation," ICRP Publication No. 2, Pergamon Press, New York, 1959.
3. U.S. Public Health Service, "Drinking Water Standards, Revised 1962," PHS Publication No. 956, Washington, DC 20525, 1962.
4. U.S. Environmental Protection Agency, "National Interim Primary Drinking Water Regulation," 1976, (NTIS); the regulations occur in 40 CFR 141.15.
5. J. B. Hursh, "Radium Content of Public Water Supplies," *J. Am. Water Works Assn.*, **46**, 43, 1954.
6. L. D. Samuels, "A Study of Environmental Exposure to Radium in Drinking Water," in *The Natural Radiation Environment,* J. A. S. Adams and W. M. Lowder, Eds., Univ. of Chicago Press, 1963, p. 239.
7. R. C. Turner, J. M. Radley, and W. V. Mayneord, "Naturally Occurring Alpha Activity of Drinking Waters," *Nature,* **189**, 348, 1961.
8. M. DeBortoli and P. Gaglione, "Radium-226 in Environmental Materials and Foods," *Health Phys.*, **22**, 43, 1972.
9. R. B. Holtzman, "Normal Dietary Levels of ^{226}Ra, ^{228}Ra, ^{210}Pb, and ^{210}Po for Man," in *Symposium on Natural Radiation Environment III*, Houston, Texas, April 23–28, 1978.
10. I. M. Fisenne and H. W. Keller, "Radium-226 in the Diet of Two U.S. Cities," U.S. DOE Environmental Measurements Laboratory Report HASL-224, 1970.

11. H. E. Moore and S. E. Poet, "Background Levels of ^{226}Ra in the Lower Troposphere," *Atmos. Environ.*, **10**, 381, 1976.
12. J. E. Pearson, "Natural Environmental Radioactivity from Radon-222," U.S. Public Health Service Publication No. 999-RH-26, 1967.
13. C. E. Roessler, Z. A. Smith, W. E. Bloch, and R. J. Prince, "Uranium and Radium-226 in Florida Phosphate Materials," *Health Phys.*, **37**, 269, 1979.
14. National Council on Radiation Protection and Measurements, "Natural Background Radiation in the United States," NCRP Report No. 45, 1975.
15. United Nations Scientific Committee on the Effects of Atomic Radiation, "Sources and Effects of Ionizing Radiation," United Nations, New York, 1977.
16. H. G. Petrow and R. J. Allen, "Estimation of the Isotopic Composition of Separated Radium Samples," *Anal. Chem.*, **33**, 1303, 1961.
17. L. C. Hebel, Chairman, APS Study Group, "Report to the American Physical Society by the Study Group on Nuclear Fuel Cycles and Waste Management," *Rev. Mod. Phys.*, **50**, S1, 1978.
18. H. F. Lucas, "Improved Low-Level Alpha Scintillation Counters for Radon," *Rev. Sci. Instrum.*, **28**, 680, 1957.
19. G. E. Jones and L. M. Kleppe, "A Simple and Inexpensive System for Measuring Concentrations of Atmospheric Radon-222," Lawrence Berkeley Laboratory, Report UCRL-16952, 1966 (NTIS).
20. Intersociety Committee on Methods of Air Sampling and Analysis, *Methods of Air Sampling and Analysis,* 2nd ed., American Public Health Association, 1015 18th Street, N.W., Washington, DC 10036, 1977.
21. R. L. Blanchard, "An Emanation System for Determining Small Quantities of Radium-226," U.S. Public Health Service Publication No. 999-RH-9, 1964.
22. D. R. Rushing, W. J. Garcia, and D. A. Clark, "The Analysis of Effluents and Environmental Samples from Uranium Mills and of Biological Samples for Radium, Polonium, and Uranium," in *Radiological Health and Safety in Mining and Milling of Nuclear Materials,* International Atomic Energy Agency, Vols. 1 and 2, Vienna, 1963, p. 187.
23. *Standard Methods for the Examination of Water and Wastewater,* 15th ed., American Public Health Association, 1015 18th Street, N.W., Washington, DC 20036, 1980.
24. J. H. Harley, Ed., *EML Procedures Manual,* Report HASL-300, DOE Environmental Measurements Laboratory, 376 Hudson St., New York, NY 10014, 1972; revisions supplied annually. Earlier editions of this manual were published as Report NYO-4700.
25. F. B. Johns, Ed., "Handbook of Radiochemical Analytical Methods," U.S. EPA Las Vegas Environmental Monitoring and Support Laboratory, EPA-680/4-75-001, 1975 (avail. GPO and NTIS).
26. U.S. Public Health Service, "Radioassay Procedures for Environmental Samples," PHS Publication No. 999-RH-27, 1967, Bureau of Radiological Health, Rockville, MD 20852.
27. *Annual Book of ASTM Standards*, Part 23, 1971, American Society for Testing and Materials, 1916 Race Street, Philadelphia, PA 19103. The radium method is designated D2460-70.
28. U.S. EPA Las Vegas Environmental Monitoring and Support Laboratory, "Measurement of Total Radium and Radium-226 in Environmental Waters: A Tentative Reference Method," EPA-600/4-76-012, 1976 (NTIS).

29. E. R. Ebersole, A. Harbertson, J. K. Flygare, Jr., and C. W. Sill, "Determination of Radium-226 in Mill Effluents," Report IDO-12023, USAEC Idaho Operations Office, Idaho Falls, ID 83401, 1959.
30. A. S. Goldin, "Determination of Dissolved Radium," *Anal. Chem.*, **33**, 406, 1961.
31. H. Duquesne, E. Godfroi, J. Govaerts, R. Warin, and J. Michiels, "Low-Level Radium Content in Water Determined by Alpha Counting," *Health Phys.*, **9**, 927, 1963.
32. D. N. Kelkar and P. V. Joshi, "A Rapid Method for Estimating Radium and Radon in Water," *Health Phys.*, **17**, 253, 1969.
33. D. R. Percival and D. B. Martin, "Sequential Determination of Radium-226, Radium-228, Actinium-227, and Thorium Isotopes in Environmental and Process Waste Samples," *Anal. Chem.*, **46**, 1742, 1974.
34. C. W. Sill, "Simultaneous Determination of ^{238}U, ^{234}U, ^{230}Th, ^{226}Ra, and ^{210}Pb in Uranium Ores, Dusts, and Mill Tailings," *Health Phys.*, **33**, 393, 1977.
35. C. G. Sanderson, "Determination of Radium-226 and Thorium-228 in Food, Soil, and Biological Ash by Multidimensional Coincident Gamma-Ray Spectroscopy," *Health Phys.*, **16**, 747, 1969.
36. U.S. National Bureau of Standards, "NBS Standard Reference Materials Catalog," 1979–1980 ed., NBS Special Publication 260, 1970 (available from NBS, Washington, DC 20234).

14 RADON

Radon (Rn) gas arises from trace amounts of radium (Ra) present throughout the earth's crust. Radiological exposures to ^{222}Rn daughters are known to be an important occupational hazard in the uranium mining industry. They also contribute, largely through their presence in indoor air, a significant—if not dominant—portion of the radiation dose that members of the general public receive from natural radioactivity. Because of their importance, and because instrumentation for their measurement is unique, Rn and its daughters are discussed in this chapter.

First an outline of the physical characteristics of the radon decay chain is given, and followed with discussions of the radiation protection guides and of the special unit (working level) developed for measuring daughter activities. Finally, the measurement techniques and instrumentation discussed, covering both measurements in the natural environment (including indoors) and measurements for occupational radiation exposure monitoring and control, mainly in uranium mines.

14.1. PHYSICAL CONSIDERATIONS

14.1.1. Radon-222 and Its Decay Chain

The decay chain of which ^{222}Rn is a part is one of the most thoroughly studied phenomena in nuclear physics. It begins with the naturally occurring isotope ^{238}U and ends with stable ^{206}Pb. In between there are 8 α-decays and 6 β-decays. The first part of the series ^{238}U to ^{222}Ra is shown in Fig. 14.1, and Fig. 14.2 shows the second part (^{222}Ra to ^{206}Pb). More detailed data are given in Table 14.1. Note that in this table many historical names of the chain members are also given: Radium A–^{218}Po, Radium B–^{214}Pb, Radium C–^{214}Bi, and Radium C′–^{214}Po.

Radon-222 is an important radionuclide because it is a noble gas. As ^{238}U decays through five steps to ^{226}Ra, the heavy nucleus remains relatively fixed in place (in

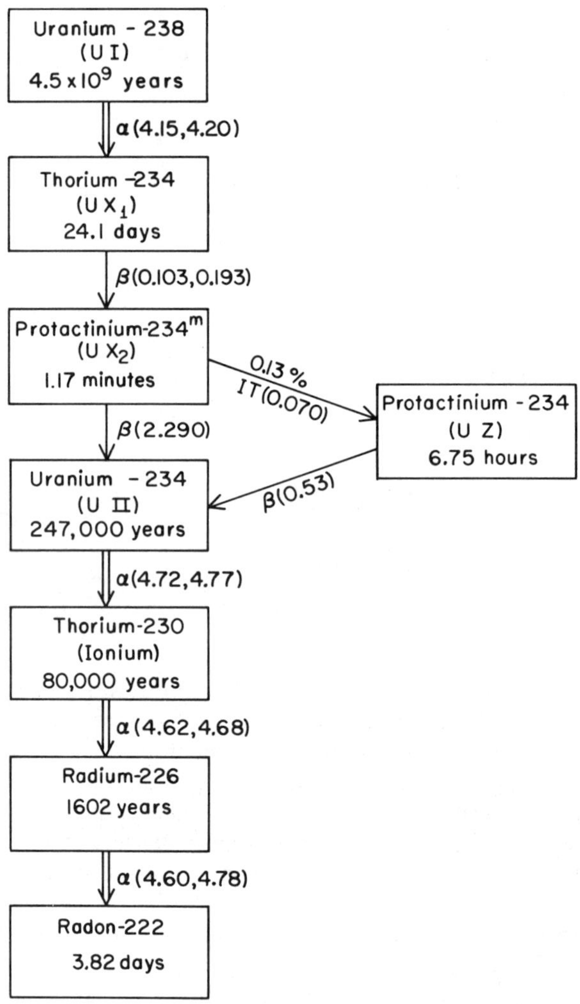

Figure 14.1. Decay chain, uranium-238 to radon-222 (α, β energies in MeV).

rock, for example). However, Rn gas can diffuse from the site of production, and in any particular (surface) rock some of the radon escapes into the surrounding atmosphere. This is the source of the historical name for ^{222}Rn, "emanation radon."

The properties of the chain Ra → Rn daughters are important. Radium-226 has a half-life of 1600 yr. Because this is much longer than the half-life of any subsequent daughter, ^{226}Ra can be viewed over a short time as a fixed-rate generator for ^{222}Rn (whose half-life is 3.8 days). Thus for many practical purposes one can think of radium-bearing materials as a source of radon gas, as well as a source of the many alpha, beta, and gamma radiations emitted by the constituent nuclei themselves. The equilibrium buildup of the succeeding daughters from ^{222}Rn (^{218}Po, ^{214}Pb, ^{214}Bi) is shown in Fig. 14.3. After the ^{214}Bi decay, ^{214}Po decays almost immediately (164 μsec) and that ^{210}Pb has a 22-yr half-life, which for short times

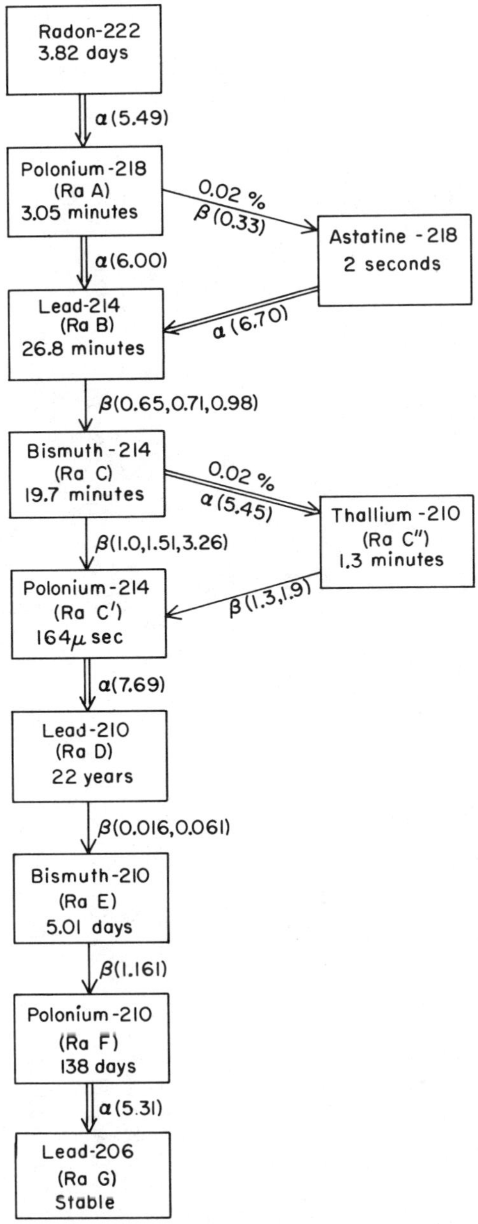

Figure 14.2. Decay chain, radon-222 to lead-206 (α, β energies in MeV).

effectively blocks the decay chain. Therefore this treatment almost exclusively discusses the first four daughters, ^{218}Po, ^{214}Pb, ^{214}Bi, and ^{214}Po.

Evans [2] has given a very useful summary of the physical and engineering considerations concerning the radon daughters. The chemical properties of the daughters are crucial. Although Rn is relatively unreactive, the three elements following it in the decay chain (Po, Bi, and Pb) are all chemically active. In particular, when Rn (gas) decays in air, most of the newly created ^{218}Po atoms, typically

Table 14.1. Main Sequence of Decays from Uranium-238 to Lead-206. Three Very Weak Collateral Branch Disintegrations, All With Branching Ratios Less Than 0.1%, Are Omitted (data from Ref. 1)

Historical Name or Symbol	Isotope	Half-Life	Principal Radiations	Alpha Energy (MeV)	Beta Maximum Energy (MeV)	Gamma-Ray Quanta per Disintegration	Average Gamma-Ray (MeV)
Uranium I	Uranium-238	4.49×10^9 yr	α	4.18	0.205 (80%)		
Uranium X_1	Thorium-234	24.1 days	β		0.111 (20%)		
Uranium X_2	Protactinium-234	1.17 min	β		2.32 (80%) 1.5 (13%) 0.6 (7%)		
Uranium II	Uranium-234	248,000 yr	α	4.76			
Ionium	Thorium-230	80,000 yr	α	4.68 (75%) 4.61 (25%)			
Radium	Radium-226	1,602 yr	α	4.78 (94.3%) 4.69 (05.7%)			
Radon	Radon-222	3.825 days	α	5.486			
Radium A	Polonium-218	3.05 min	α	5.998			
Radium B	Lead-214	26.8 min	β		0.65		
			γ			0.82	0.295
Radium C	Bismuth-214	19.7 min	β		3.13 (23%)		
				1.67 (77%)			
			γ			1.45	1.050
Radium C′	Polonium-214	164 μsec	α	7.68			
Radium D	Lead-210	22 yr	β		0.018		
			γ			1	0.047
Radium E	Bismuth-210	5.02 days	β		1.17		
Radium F	Polonium-210	138.3 days	α	5.298			
Radium G	Lead-206	Stable	Stable				

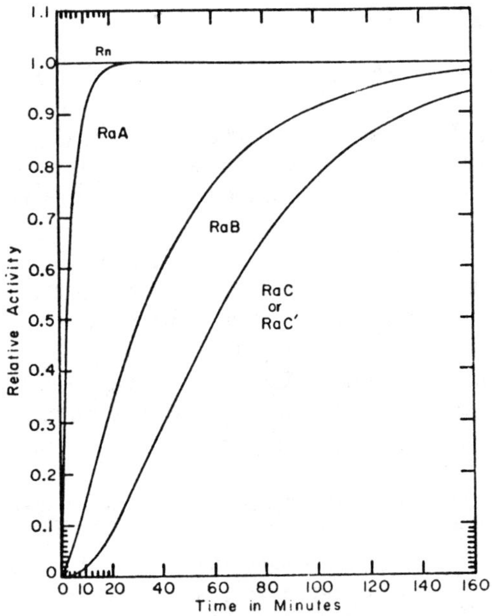

Figure 14.3. Growth of activity of the individual short-lived decay products from a constant source of radon having unit activity (reproduced from Evans, Ref. 2, with permission of the Health Physics Society © 1969).

ionized, attach almost immediately to any particulate matter present. Most ^{214}Pb, ^{214}Bi, and ^{214}Po atoms are also adsorbed at, or soon after, birth. In consequence, airborne particulate matter becomes radioactive. Whether or not they are attached to particulate matter, the daughters are hazardous when inhaled, because they tend to lodge in the lung mucosa.

The α-emitting daughters are ^{218}Po and ^{214}Po. Their decay energies are as follows:

$$^{222}Rn \rightarrow {}^{218}Po\text{: } 5.48 \text{ MeV}$$
$$^{218}Po \rightarrow {}^{214}Pb\text{: } 6.00 \text{ MeV}$$
$$^{214}Po \rightarrow {}^{210}Pb\text{: } 7.69 \text{ MeV}.$$

These energy differences can, of course, provide a basis for distinguishing the individual components by α-spectroscopy.

The two important β/γ-emitting daughters are ^{214}Pb and ^{214}Bi. Each has a rather complicated set of emissions, with three important β-rays and several γ-ray energies each (see Table 14.1). Because of this, β-ray detection alone as a means of precise measurement is complicated: the spectral efficiency of any β-ray detector is very hard to determine. Gamma-ray energies could be detected by γ-spectroscopic methods. This is not now common, except in research studies where the detector can be shielded adequately from other γ-rays or where high resolution systems may be used. Bismuth-214 γ-rays are, however, commonly detected as an indirect indication of U ore bodies, although some methods described below employ β-ray detection.

14.1.2. Radon-220 and Radon-219

When ^{222}Rn must be measured, two other isotopes of Rn can potentially interfere. The most important is ^{220}Rn, a member of the decay chain that originates with naturally occurring ^{232}Th. The other isotope is ^{219}Rn, from the chain originating with ^{235}U. Both of these gases behave as ^{222}Rn does: they emanate from the solid radioactive materials after their formation via α-decay of their immediate parents ($^{224}Ra \rightarrow {}^{220}Rn$, $^{223}Ra \rightarrow {}^{219}Rn$).

In uranium mines, the thorium content of the ores is low (typically $\leq 1\%$ of that of ^{238}U), and the ^{235}U:^{238}U ratio is uniform at about 0.72%. Thus the equilibrium production of both ^{220}Rn and ^{219}Rn gases is relatively small. The half-lives of both are quite short (55 sec for ^{220}Rn, 4.0 sec. for ^{219}Rn). In more typical rocks and soils, thorium and uranium contents are similar and the short half-life of ^{220}Rn and ^{219}Rn (as compared with ^{222}Rn) permits little of these isotopes to reach air to which humans are exposed. As a result, their direct radiological impacts are usually much less significant than that of ^{222}Rn.

Some of the ^{220}Rn daughters produce counting backgrounds when ^{222}Rn daughters are collected on air filters for measurement (^{212}Pb and ^{212}Bi have half-lives of 10.6 and 1.01 hr, respectively). Because of the ^{219}Rn short half-life, its ability to diffuse out of earth and rock is so limited that it is seldom, if ever, present at levels requiring measurement in the environment.

Throughout this section, the term "radon" (Rn) will be used to denote radon-222 (^{222}Rn), unless specifically stated otherwise.

14.2. RADIATION PROTECTION GUIDES

The risk from ^{222}Rn and its daughters arises primarily from the chemically active α-particle emitting members of the chain. Alpha particles have low penetrating power. However, if cells are directly in contact with the α-particle emitter, the energy released has a high relative biological effectiveness. Alpha particles from the ^{222}Rn decay chain have a significant chance to irradiate lung tissue because ^{218}Po and ^{214}Po, can—through their chemical activity—be collected directly or via dust particles, on the lining of the tracheobronchial tract. The corresponding quantity of interest for radiation protection and dosimetry is the airborne daughter activity *potential alpha energy concentration* (PAEC). This quantity, often (and confusingly) referred to as the "working level," is measured in units of Working Level (WL). WL is defined as: "any combination of [^{222}Rn] daughters in one liter of air that will result in the ultimate emission of 1.3×10^5 MeV of potential alpha-particle energy" [3] by decay as far as ^{210}Pb. This value is derived from α-particle energies released by the total decay of the short-lived daughters (^{218}Po through ^{214}Po) at radioactive equilibrium with 100 pCi (3.7 Bq) of ^{222}Rn per liter of air. The PAEC or "working level" considers only the α-particles from radon daughters and not from ^{222}Rn gas itself.

The reason for the specialized unit is mainly operational: the potential α-particle energy concentration, measured in WL, is a concept having validity in any mixed concentration of Rn and its daughters, whether or not they are in equilibrium. Just as important, it permits practical measurements in occupational and other situations.

The working level concept is extended to a unit of cumulative exposure, the Working Level Month (WLM). It is defined as follows: "Inhalation of air containing a radon daughter concentration of one WL for 170 working hours results in an exposure of one WLM." [3] The Working Level Hour is also used as a unit of exposure.

The notion of PAEC may also be employed in measuring the airborne concentration of ^{220}Rn daughters. For equilibrium situations, 7 pCi/liter (0.259 Bq/liter) of ^{220}Rn is associated with the same PAEC as 100 pCi/liter (3.70 Bq/liter) of ^{222}Rn, that is, 1 WL. Because of its relatively long half-life, the ^{220}Rn daughters PAEC is ordinarily dominated by ^{212}Pb (Thorium B). In most situations, the PAEC contributed by ^{220}Rn daughters is considerably smaller than that from ^{222}Rn daughters.

Radiation protection guides for ^{222}Rn and daughters are based primarily on the daughter concentration, measured in WL. In some cases, these guides permit either the radon concentration or an equilibrium daughter concentration to be adopted as the operational limit. The observed increased incidence of lung cancer among uranium (and other) miners has been correlated directly with exposure as measured in WLM [4].

The Secretary of Labor, acting under provisions of the Walsh-Healy Act, promulgated the following standard in 1968 [5]:

> Occupational exposure to radon daughters in mines shall be controlled so that no individual will receive an exposure of more than 2 WLM in any consecutive 3-month period and no more than 4 WLM in any consecutive 12-month period. Actual exposures shall be kept as far below these values as practicable.

The annual limit corresponds, for a 40-hr work week, to an average daughter concentration of 1/3 WL.

In 1969, the Department of the Interior issued the following standard calling for action on the basis of individual concentrations [6]:

> If samples show an atmospheric concentration of radon daughters of more than 1 WL but less than 2 WL, immediate corrective action shall be taken or the men shall be withdrawn. When concentrations higher than 2 WL are indicated, the men shall be withdrawn from the area until corrective action is taken and the radon-daughter atmospheric concentrations are reduced to 1 WL or less.... Smoking shall be prohibited where uranium is mined.

In 1971, the Environmental Protection Agency (EPA) reaffirmed the recommended individual annual exposure limit of 4 WLM for the uranium mining industry [7], and this is the limit now applied.

A similar limit has more recently been applied [7] to industries regulated by the Nuclear Regulatory Commission (NRC), principally the uranium milling industry. The NRC annual average concentration limits, stated in 10 CFR 20, for those exposed for a normal work week, are 30 pCi/liter (1.11 Bq/liter) of Rn and an equilibrium amount of daughters or 1/3 WL daughter concentration, consistent with the 4 WLM annual exposure limit.

These U.S. occupational limits represent a decrease from the previous levels permitted, when 1 WL was the maximum annual average daughter concentration for mines and other operations. However, the change to the 4 WLM annual exposure limit brought radiation protection for radon daughters roughly into accord with the occupational limits for other types of radiation exposure, in respect to the estimated risk of fatal cancer, and with the radon daughter limits recommended by the International Commission on Radiological Protection (ICRP) and the National Council on Radiation Protection and Measurements (NCRP).

The generally applicable occupational dose limit is 5 rem (50 mSv) whole-body dose equivalent per year or organ doses contributing the equivalent risk. However, the absorbed dose (rad or gray) and the dose equivalent (rem or sievert) resulting from 1 WL of exposure depend on the exact nature of the mine atmosphere, and, in any case, the size and distribution of dose to the lung is not well characterized. With modern ventilation procedures, the equilibrium condition is never approached for any but the first daughter, ^{218}Po. Discussed by Altshuler, Nelson, and Kuschner [8], an example is the "reference atmosphere" which contains in one liter of air the following activities totaling 200 pCi (7.4 Bq): 94 pCi (3.48 Bq) of ^{218}Po; 62 pCi (2.29 Bq) of ^{214}Pb; 44 pCi (1.63 Bq) of ^{214}Bi. This yields 7.4×10^4 MeV or 57% of a WL. According to the Federal Radiation Council [9]:

> Altshuler's reference atmosphere results in 69% of a "WL dose" to the bronchi, and has a value between 10 and 30 rads per year. In view of the ambiguities of conversion, Altshuler's reference atmosphere will be considered, with important reservations, to produce 20 rads in a normal working year.... Exposure for one year at...one WL would give 33 rads. The mean organ dose would be lower. In the absence of an appropriate factor for the RBE, the calculated dose cannot be converted to rem.

Ambiguities in the relationship between daughter concentrations and lung dose (equivalent) persist, and limits are still based on miner epidemiological evidence, rather than on calculated dose equivalent to the lungs. Exposure to radiation other than the inhalation of radon daughters must be considered separately, applying the usual occupational limits of 5 rem/yr (50 mSv/yr) for whole body external exposure, and so on, discussed in the introductory section of this book.

The ICRP attempted in its early recommendations to take explicit account of the dose from ^{218}Po. Based on the view that the free ions of this nuclide delivered all but a fraction of the dose to the bronchi, ICRP Report No. 6, written in 1959 and revised in 1964 [10], states:

> The maximum permissible concentration in air $(MPC)_a$ for exposure to radon and daughter products is found to be
>
> $$(MPC)_a = \frac{3000}{(1 + 1000\,f)} \text{ pCi}^{222}\text{Rn/l of air}$$
>
> where f is the fraction of the equilibrium amount of ^{218}Po ions which are unattached to nuclei.

If $f = 10\%$, as the ICRP indicates might be typical for mines, then the occupational MPC_a is 30 pCi/liter (1.11 Bq/liter), consistent with the limit now set by the ICRP and others.

The ICRP is evaluating currently available epidemiological evidence as a basis for radon daughter exposure limits. In the meantime, according to ICRP Publication 24 on miner exposures [11]:

> The recommendation included in ICRP Publication 2 is still applicable, namely that the annual average concentration of ^{222}Rn should not exceed 30 pCi/l. For the purpose of radiation protection in mines, it is understood that this limit refers to ^{222}Rn in

equilibrium with its short-lived daughters, and that it also applies to any mixture of radon and daughters, where the daughters release, through their decay to ^{210}Pb, an alpha energy equal to that released by the daughters of 30 pCi/l of radon in equilibrium with its daughters (3.9×10^4 MeV/l).

This daughter PAEC is 0.3 WL and equals the NCRP limit [12]. For exposures in buildings and other general environmental situations, it would be inappropriate in any case to base a limit solely on the unattached fraction, because in such situations the particles carrying the attached activity may be small enough to be collected efficiently by the lung.

There is no generally applicable guideline for exposure of the general public to ^{222}Rn and its daughters. Both the ICRP [13] and the NCRP [12] recommend in their general overviews that individuals in the general public be limited to exposures one-tenth of occupational exposure. Also, for a suitably large sample of the general population, the general guideline is another factor of 3 smaller still. The individual and population annual exposure limits are approximately 0.4 and 0.1 WLM, respectively. These exposures may not yield the same lung doses for the general public as for miners, because breathing rates, for example, may differ. For 168 hr/wk, they correspond, respectively, to daughter concentrations of 0.01 and 0.003 WL and, for equilibrium, ^{222}Rn concentrations limits of about 1 and 0.3 pCi/liter (0.037 and 0.011 Bq/liter). Considering that indoor ^{222}Rn levels often exceed 1 pCi/liter (0.037 Bq/liter), they are difficult limits to meet, were they applied. The NRC maximum permissible concentrations for uncontrolled areas adjacent to controlled industrial areas are 3 pCi/liter (0.111 Bq/liter) and one-thirtieth (1/30) of a working level. Practical limits and control strategies for more general application were actively considered during the late 1970s. In some cases, discussed next, remedial action guidelines have been associated with specific sources.

14.3. SOURCES OF RADON AND ITS DAUGHTERS

Radon and its daughters occur naturally as members of the ^{238}U decay chain. Two excellent reviews on the occurrence of these radionuclides and others are an NCRP report, *Natural Background Radiation in the United States* [14], and a United Nations review, *Sources and Effects of Ionizing Radiation* [4]. In the brief discussion that follows, a distinction will be made between: (a) industrial activities, such as uranium mining, that enhance exposures of workers or the general public to Rn and its daughters; and (b) exposures of the public ordinarily occurring in the general environment.

14.3.1. Industrial Activities

From the standpoint of radiological impact, the most widely recognized problem from Rn and its daughters is occupational exposure in the uranium mining industry (discussed in Chapter 12, Mining and Milling of Uranium Ores). In the United States a few thousand miners now require routine radiological monitoring, as do workers in uranium mills.

Tailings from uranium mills can constitute a source of environmental Rn, of most significance for those nearby. Radon from tailings may also contribute a numerically larger population dose to distant areas, although the added dose to individuals in such areas is far smaller than the dose from normally occurring Rn. In some of the Rocky Mountain states, considerable attention has been given in recent years to Rn from tailings. A 1969 study [15] indicated that background ^{222}Rn concentrations in four areas in Colorado and Utah averaged from 0.3 to 0.8 pCi/liter (0.011 to 0.030 Bq/liter). Directly over tailings, levels were higher by a factor of about 10; analysis of typical spatial distributions showed that radon concentrations in areas beyond about one-half mile were not greatly affected. The slight increase in radon concentrations at large distances may be estimated to cause a small population exposure that, if integrated over the 80,000-year half-life of ^{230}Th (which—over the long term—is the source of Rn in a tailing pile), can contribute numerically large population exposure commitments. The higher range of estimated resulting health effects average about 0.001 fatal cancers per nuclear electric GW_e-yr for each year in a ^{230}Th half-life [16], but tailings stabilization practices could reduce even this high estimate by a substantial factor. The EPA is now considering possible environmental standards for inactive uranium mill tailings [17].

Some houses have been built on foundations incorporating tailings. In late 1972, the AEC published remedial action criteria [18] for the area around Grand Junction, Colorado. These are aimed at eliminating some of the more important sources of exposure to the general public from tailings. Remedial action consists of removing tailings, ventilation, shielding, or use of sealants. This action, based on recommendations of the Surgeon General, is "suggested" when a radon daughter concentration exceeding 0.01 WL (or a level of 0.05 mR/hr external γ-radiation) is measured, and is "indicated" when 0.05 WL (or 0.10 mR/hr) is exceeded. In some cases, indoor levels around Grand Junction considerably exceeded these criteria, and the same is true in Butte, Montana, according to recent measurements by the EPA and state authorities.

A substantial incidence of indoor daughter concentrations greater than 0.01 WL has been found in residences built on or with materials from the phosphate mining industry. Florida produces most U.S. phosphates, and in 133 structures on reclaimed mining lands one-third were found to have average daughter concentrations greater than 0.01 WL, with a few greater than 0.05 WL [19]. Considering the risks associated with these levels and the control techniques available, the EPA—in response to a request from the Governor of Florida—recommended that remedial action be taken in residences with average daughter concentrations exceeding 0.02 WL; the EPA also recommended that such action be evaluated, on a structure-by-structure basis, for lower levels, but noted that they would be unlikely to be justified for concentrations less than 0.005 WL. The EPA is now participating in investigations of radon daughter levels in Tennessee and Alabama residences employing phosphate slag as a structural material.

14.3.2. General Outdoor and Indoor Environment

Radon occurs in air because of the presence of Ra in soils and rocks. Rates of emanation and radon concentrations have been measured by Pearson [20]. Near Chicago, Rn was present one meter above the ground at levels that varied diurnally from about 0.1 to 1.4 pCi/liter (3.7 to 51.8 mBq/liter). Emanation rates in regions where ^{238}U mining is commercially feasible were found to be larger by as much as two orders of magnitude than rates near Chicago [20]. However, there is not necessarily a direct correlation between emanation rates and radon concentrations near the ground.

Scattered observations indicate that indoor concentrations of Rn are generally higher than local outdoor concentrations. DOE's Environmental Measurements Laboratory measured radon concentrations in 21 homes in the New York–New Jersey area, finding the geometric mean of the annual average first-floor radon concentrations to be 0.83 pCi/liter, (0.031 Bq/liter) five times the comparable outdoor levels at these locations, 0.18 pCi/liter (6.7 mBq/liter) [21]. Indoor daughter concentrations were found to average 0.004 WL. Measurements elsewhere have found similar results. The fact that levels are higher indoors indicates that indoor levels arise to a significant extent from radon sources that are effectively internal to the building. These may include building structural materials, domestic water supplies, and soil or rock underlying or near the structure. With internal sources, the rate at which indoor air is exchanged for outdoor air has a direct impact on resulting indoor radon concentrations. For this reason, measures that would reduce building energy use by reducing infiltration or mechanical ventilation might increase average radon daughter exposures [22]. Preliminary measurements taken by Lawrence Berkeley Laboratory in a number of energy-efficient houses (including some employing solar gain as a heat source) found radon concentrations in the range of 1 to 20 pCi/liter (0.037 to 0.740 Bq/liter) with windows closed [23]. The high end of this range corresponds to about 0.1 WL, for a typical indoor equilibrium ratio, and could result in annual exposures approaching the occupational limit. Indoor Rn and daughter levels, however, are far from fully characterized and are receiving intense study as part of increased attention to indoor air quality and ventilation requirements, as well as radon control techniques.

Measurement of Rn and its daughters has become a most important area of work in characterizing the natural radiation environment, and work on Rn and its daughters has dominated the most recent symposium on the natural radiation environment [24].

14.4. RADON MEASUREMENT TECHNIQUES

In this section, we discuss measurements of radon in air and measurements of source strength. Subsequent sections treat the measurement of radon daughters in air and personnel dosimetry.

Some of the discussion here of pre-1972 techniques relies on a summary of radon instrumentation by Breslin [25]. ICRP Publication 24 [11] contains a briefer summary of methods for measuring radon and daughters in mines. Methods used for measurements in the general environment are described in reports of a 1977 radon workshop [26] held at the Environmental Measurements Laboratory (EML) and of the third symposium on the natural radiation environment [24].

Many of the basic measurement techniques described here were developed in response to recognition of the significance of elevated radon daughter exposures among miners. More recent techniques typically either adapt older methods to more general environmental applications or incorporate newer semiconductor and microcomputer technology in more complex instruments. Standard measurement methods for mines are specified in an American National Standards Institute standard [27]. More recently, the Department of the Interior has issued a report on calibration of Rn and radon daughter measurements [28].

14.4.1. Radon in Air

The measuring of radon gas activity should be distinguished from the activity of its daughters. This type of measurement is made by some of the more sophisticated ventilation engineers in studying mine air quality, by researchers investigating indoor Rn and daughter concentrations, and by those concerned with natural levels of radon gas in the air. It is used to measure ^{226}Rn from the emanation of its daughter ^{222}Rn and when ^{222}Rn is used as a tracer in atmospheric studies.

There are two quite different approaches to the specific measurement of ^{222}Rn. In one, some degree of equilibrium may be assumed to have become established between Rn and its daughters. Alternatively, the daughters may be removed from the sample gas, after which either the decay of Rn itself is detected or the daughters are allowed to ingrow again for counting.

Measurements of ambient radon activity can be made with simple instruments for routine monitoring, or with more complex instruments, for research.

The most common instrument for measuring radon gas is a scintillation or *Lucas chamber counter* [29]. The chamber itself is a small metal, glass, or plastic cell with a flat transparent bottom. Its shape is usually hemispherical, but can be cylindrical or conical as well. The inside is lined with silver-activated zinc sulfide [ZnS(Ag)] scintillator, and the scintillations are detected through the flat window by a photomultiplier. Filtered air samples can be drawn into the chamber with a pump, or alternatively by evacuating the chamber and then admitting a filtered sample through a valve. The filter removes long-lived α-particle activity (which could contaminate the chamber) as well as the short-lived radon daughters, while radon gas is admitted to the chamber. The radon gas subsequently reaches equilibrium with its daughters. The ZnS(Ag) photomultiplier combination measures the activity of ^{222}Rn, ^{218}Po, and ^{214}Po. Samples may be counted days after collection. Samples taken in the field in plastic bags, glass flasks, or metal cans may be transferred into scintillation chambers for laboratory analysis. In some cases, the Rn in a sample is concentrated by collection on charcoal (discussed later) or onto cooled glass beads. The collected radon may then be transferred, on de-emanation, to a Lucas chamber. Typical chamber volumes are in the 100 to 200-ml range, and are thus small enough to be easily portable. Figure 14.4 shows the original chamber of Lucas.

A difficulty in the original design, since overcome, was optimizing and stabilizing the detection efficiency for the daughters, which are usually electrically charged. When a conducting layer on the inside of the window is used, the charged daughters distribute themselves uniformly on the window and the ZnS(Ag) walls. According to Kristan and Kobal [30], this conducting layer is unnecessary if the photomultiplier is operated with a grounded photocathode, but this is disputed by Lucas [31]. Interferences from ^{220}Rn, or even ^{219}Rn, are possible because both are also α-emitters; these are usually small because of the short half-lives (55 sec for ^{220}Rn and 4 sec for ^{219}Rn). There are no other significant interferences. The sensitivity of the method depends on the sample volume and other details. As originally used in mines, the sensitivity was about 10 pCi/liter (0.37 Bq/liter) [25], although sensitivities one to two orders of magnitude better are achievable with larger flasks and longer counting times. With use, a chamber will suffer a gradual increase in background due to deposition of long-lived radionuclides.

Scintillation chambers are now commercially available, although they are not difficult to construct. An easily fabricated plastic scintillation flask is described by

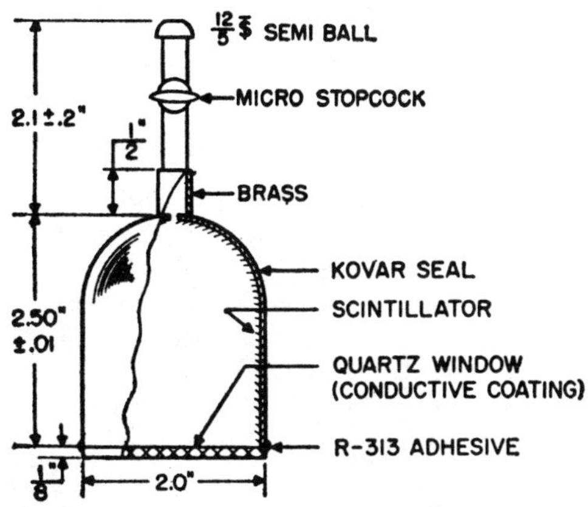

Figure 14.4. Lucas scintillation chamber. The photomultiplier tube, which views the ZnS(Ag) scintillations through the quartz window, is not shown (reproduced from Lucas, Ref. 29, with permission by the American Institute of Physics © 1957).

George [32]. Detailed calibration procedures for scintillation chambers are contained in [33] in the discussion of determining ^{226}Ra in water by a ^{226}Rn emanation technique.

An alternative to the Lucas chamber is the *two-filter method* [34]. A metal cylinder with a filter on each end is the basic sampling unit. Sample air is pumped through the cylinder at a known rate (typically for 5 min at about 10 liters/min). The upstream filter removes all particles, most importantly all radon daughters. Radon gas passes through, and inside the cylinder a small fraction decays to ^{218}Po. The ^{218}Po that is deposited on the downstream filter is immediately counted for α-particle activity. (Polonium-218 half-life is only 3 min.) Thomas and LeClare [34] used a ZnS(Ag)/photomultiplier detector system. Alpha-particle counts may be converted to radon concentration considering geometrical sizes, sampling rate, and counting time. For a 52-in. long, 3.6-in. diameter tube, this technique is sensitive at the level of a few pCi/liter. Calculations indicate that sensitivities as low as 0.1 pCi/liter (3.7 mBq/liter) could be achieved with a larger chamber, higher flow rate, and other small changes [34]. A variation of the method (discussed later) has been used for continuous measurements of environmental radon concentrations in the vicinity of 0.1 pCi/liter (3.7 mBq/liter), and sensitivities as low as 0.01 pCi/liter (0.37 mBq/liter) are claimed [35].

One disadvantage of the two-filter technique is the degree of care needed in sampling because of the small amount of activity present on the downstream filter. Another is that at relative humidities below about 25% the method has been found to yield results up to 20% too high [34]. The reproducibility of the method has been studied by Breslin [25], who finds 10 to 20% replication errors. Figure 14.5 [36] shows results from a study of the accuracy of the two-filter method, in which it is compared to measurements of flask samples analyzed in the laboratory using a pulse-type ionization chamber. The average precision of this method is about ±20%.

Both the Lucas-chamber and two-filter methods appear to be satisfactory for routine use in uranium mines. Furthermore, with long counting times or with large

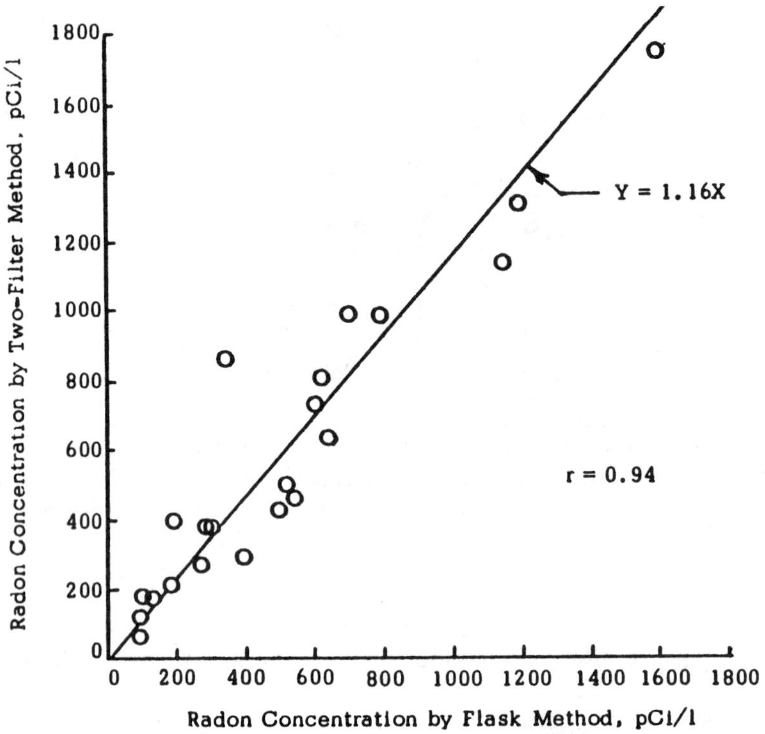

Figure 14.5. Comparison of field measurements by the two-filter method and the flask method for radon determination (reproduced from Breslin, Ref. 36).

volumes and low rates, the sensitivity required for environmental and research-type studies may be achieved.

Another technique, with sensitivity similar to that of these methods, is the use of a pulse-type ionization chamber. A sample of gas, collected through a filter to eliminate radon daughters, is admitted to the ion chamber and allowed to come to equilibrium before counting. This technique is described in detail in the EML *Procedures Manual* [37].

Field samples are often taken in a variety of containers, then transferred in the laboratory into detection chambers based on scintillation or pulse ionization. Alternatively, sampling may be performed by pumping through activated charcoal. In some cases the charcoal is cooled to low temperatures, such as that of dry ice (−78°C), where Rn is readily adsorbed. The sample in the charcoal trap may be counted directly by γ-spectroscopy using a sodium iodide detector or transferred for counting in a detection chamber. [See American Public Health Association (APHA) Intersociety Committee methods described later, as well as discussion of emanation measurements.] Field radon measurements integrated over a day or so may be accomplished simply by drawing air slowly (through a filter) into a sampling container such as a plastic bag [38]. The container may then be sent to a laboratory for analysis. Devices for filling containers at a fixed rate are commercially available.

In many cases, the information of interest is the radon concentration over periods of days or more, measured either continuously or by integration. Continuous techniques, often desired for research applications, can adapt the basic methods just described to automatic operation. Alternatively, integration techniques using thermoluminescent dosimeters or film may be simpler and less expensive to implement.

Continuous monitors based on ZnS scintillation chambers have been used by a number of workers and are commercially available. Air is pumped relatively slowly through a filter that removes daughters, then through the scintillation chamber. The phototube detects scintillations due to ^{222}Rn decay in the chamber, as well as decay of daughters formed therein. These tend to plate out on the walls of the chamber, so that the count total for any period arises partially from earlier radon concentrations. An instrument of this type has recently been used for continuous measurements in an energy-efficient house where the ventilation rate was being varied [39]; because of daughter plateout, the instrument responds accurately to changes in radon concentration in about 3 hr. Thomas and Countess [40] have employed data analysis techniques that correct for previous deposition. These instruments have capabilities in the 1-pCi/liter (0.037-Bq/liter) range.

Continuous monitors based on the two-filter method have also been developed. Several such instruments, using an automatic filter moving assembly for the second filter, have been used to monitor radon concentration as a function of altitude [35]. Both automatic and fixed filter instruments have been fabricated by EML [41].

Other continuous monitors utilize electrostatic collection of daughters by a detection surface that is placed at a negative potential. The ^{218}Po formed in the volume is usually positively charged and is attracted to the detector for subsequent measurement. Care is usually required, however, that high humidity does not promote charge recombination and loss of detection efficiency. An instrument designed by Wrenn et al. [42] employs a hemispherical ZnS(Ag)/photomultiplier detector, covered with aluminized Mylar to act as an electrode and placed in a volume defined primarily by a hemispherical polyurethane foam wall through which Rn can diffuse passively. The count rate arises primarily from daughters decaying on the electrode, and sensitivity better than 1 pCi/liter (0.037 Bq/liter) can be achieved with 40-min counting periods. Another instrument employs a semiconductor detector held at a high voltage and placed in a spherical volume through which filtered and dried air is pumped [43]. Such a device can also operate passively, as with the chamber just described. With α-spectrometry and a flow rate of $\simeq 100$ liter/hr, the detection limit is stated as 0.02 pCi/liter (0.74 mBq/liter) for a 1-hr counting period. Because this spectroscopic instrument need only measure ^{218}Po with its 3-min half-life, response to variations in radon concentrations can be much more rapid than in instruments that count total α-particles.

Crude, but highly sensitive, continuous monitors simply measure daughters collected from the air onto a filter. See the discussion later of APHA Intersociety Committee methods.

Integrating monitors employing α-particle sensitive thermoluminescent dosimeter (TLD) lithium fluoride chips have been developed by George and others at the EML [44,45]. They are inexpensive devices that require no attention during monitoring periods of a week or more. One is based on the two-filter principle: daughters collected on the second filter are detected by the TLD chip for readout in the laboratory after the sampling period [44]. Another device [45] is based on passive diffusion of radon through a filter and dessicant, then collection of daughters formed in the sensitive volume on a negative electrode, such as one of the continuous monitors just discussed. Polonium decay irradiates the TLD chip, which is subsequently read out in the laboratory. This passive monitor (see Fig. 14.6) can be

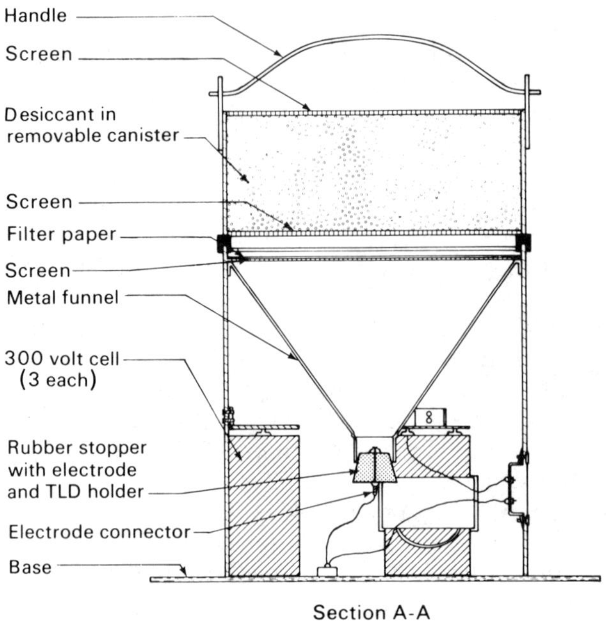

Figure 14.6. Passive environmental radon monitor (adapted from George, Ref. 45).

fabricated by commercial manufacturers for no more than a few hundred dollars apiece, and hundreds have been built. Both the two-filter and passive monitors have nominal sensitivities of a fraction of a pCi/liter for week-long samples, although practical sensitivity may be limited to the order of 1 pCi/liter (0.037 Bq/liter). Such passive monitors may also employ other thermoluminescent materials, particularly CaF_2(Dy).

Integral measurements may also be made using film and track-etch recording. These are typically small passive portable devices and are discussed later as dosimeters. One such technique, however, may be used for on-site monitoring, that is, the combination of a fast Polaroid film and a ZnS(Ag) scintillator developed by Bedrosian [46]. The device consists of a fast Polaroid film in a holder containing two ZnS(Ag) disks, one covered by filter paper and the other not covered. The uncovered ZnS(Ag) phosphor responds to α-particles from Rn and daughters in ambient air; the covered phosphor responds only to α-particles from Rn, which diffuses through the filter. The images on the film, exposed by the scintillations from the ZnS(Ag), are measured with a reflection densitometer. Bedrosian claims a lower limit of sensitivity for radon gas of 200 pCi/liter (7.40 Bq/liter) after 15 hr of exposure, and for radon daughters of less than 1 WL after 30 hr of exposure. Such a technique is useful where levels are high, because the results are available within a few minutes after the end of an exposure, and the system is very inexpensive. Of course, it yields no information about the extent to which the ambient air being sampled is in equilibrium.

The APHA Intersociety Committee's tentative method for Rn [47] includes a charcoal adsorption/emanation technique. Figure 14.7 shows the collection apparatus. Filtered air is dehumidified in a drying column (e.g., Drierite) and any

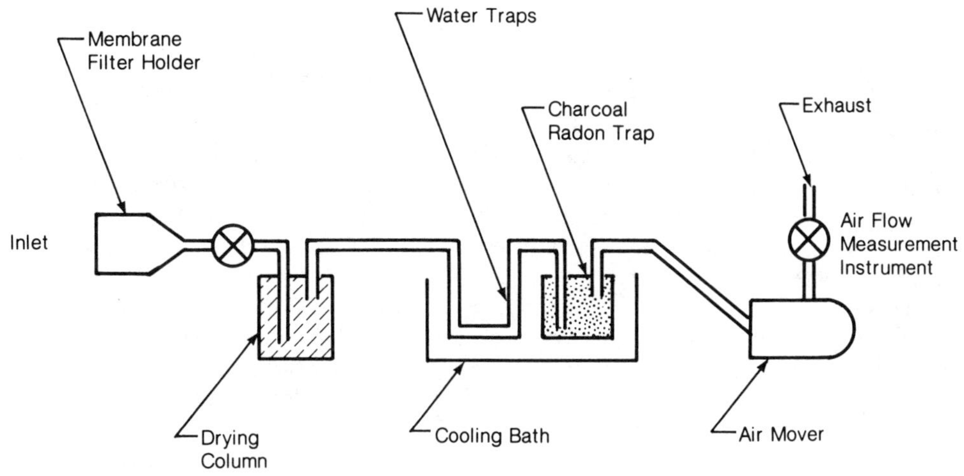

Figure 14.7. Sampling apparatus for collection of gaseous airborne radon. (reproduced from the American Public Health Association Intersociety Committee, Ref. 47, with permission of the American Public Health Association © 1977).

remaining water is trapped before the gas flows through the cooled activated charcoal. The cooling mixture is dry ice mixed with 1:1 chloroform and carbon tetrachloride [48].

Two counting techniques for this collection system are described in [47]. In the first, the collection trap is placed in a de-emanation and transfer apparatus; Rn is transferred to a Lucas chamber, allowed to reach equilibrium with its daughters during a 4-hr wait, and counted. In the second, the collection trap is counted directly with a NaI(Tl) crystal, a photomultiplier, and a multichannel analyzer. The 0.61-MeV and 1.76-MeV γ-rays from ^{214}Bi are counted after equilibrium has been established. Both methods are quite sensitive. The NaI(Tl) system can detect $\simeq$0.12 pCi/liter (4.44 mBq/liter) of Rn with a $\pm$10% error at a 95% confidence level [47], whereas the Lucas chamber method has been used to measure levels as low as about 0.010 pCi/liter (0.37 mBq/liter) with errors in the $\pm$0.005-pCi/liter (0.185-mBq/liter) range [48]. The main disadvantage is, of course, that the apparatus is relatively sophisticated, expensive, and not very portable. With the NaI(Tl) γ-spectroscopy technique essentially no interferences are caused by the sample being monitored (although terrestrial γ-radiation provides background). The interferences in the Lucas chamber are also small.

Two other highly-sensitive methods for qualitative evaluation recommended by the APHA Intersociety Committee use air filters to collect daughters [47]. Neither is applicable for quantitative measurements, because they rely upon an estimate of radioactive equilibrium between Rn and its daughters in the sampled air, and as such each is only useful for very approximate environmental measurements.

The first of these methods uses *α-particle counting.* Radon daughters are collected on a 0.8-μm filter, which is nearly 100% efficient and also has very little self-absorption. Because of the short daughter half-lives, the collection and decay rates reach equilibrium after a few hours. A 4-hr sampling time is recommended, at about 10 liters/min. Polonium on the filter is counted with a ZnS(Ag) photomultiplier system, or alternatively with an internal gas proportional counter.

Using a counting time of 10 min, the lower detection limit is quoted as being about 0.03 pCi/liter (1.11 mBq/liter), with a counting error of $\pm 10\%$ at 95% confidence. Interferences can occur from the long-lived daughters of ^{220}Rn, but re-counting the filter after a delay of 4 hr corrects for this problem. Other α-particle emitters could collect on the filter, but are not usually present with sufficient activity to constitute an important background. The most important uncertainty in this type of measurement is due to assumptions about the equilibrium state of the sampled air. In most environments the state of equilibrium would have to be established, unless only a rough estimate is required.

In the second Intersociety Committee air-filter method, *β-ray counting* is employed [47]. Because β-rays have a greater range than α-particles, one can use a thicker filter and greater sampling volumes for counting without problems of self-absorption or dust-loading. A positive-displacement blower forces air at 500 liters/min through a 2-in. glass fiber filter. The activities to be counted are the β-rays from ^{214}Pb and ^{214}Bi. After 20 min of sampling and 1 min for transfer, a 10-min β-ray count is taken (through 75 mg/cm^2 of absorber) using an internal gas proportional counter or a G-M counter. A concentration of 0.001 pCi/liter (0.037 mBq/liter) Rn of radon in equilibrium with its daughters will yield 332.2 disintegrations of ^{214}Pb and 416.9 of ^{214}Bi in this time interval. A second count after 5 hr permits subtraction of possible ^{220}Rn daughters. A more complex analysis procedure uses another 10-min count after 1 hr to determine the ^{214}Pb:^{214}Bi ratio, and hence the possible extent of disequilibrium in the original sample [49].

14.4.2. Source and Emanation Measurements

Scintillation chamber methods and γ-spectrometry of radon daughters are often used to measure the rate of production of Rn in a material, the actual Rn content, or the rate at which Rn leaves the material. Some of the methods used for the analysis of soil, building materials, and water are presented here.

The ^{226}Ra content of *soil, rock,* or *building materials* may be measured by enclosing the material in a suitable sealed container, permitting ingrowth of ^{222}Rn and its daughters to equilibrium, then using a NaI scintillation counter γ-spectrometer to measure the 1.76-MeV ^{214}Bi γ-rays. A second measurement with an unsealed container can be used to determine the fraction of ^{222}Rn able to leave the sample matrix [50]. Preliminary uranium surveys use aerial NaI spectrometry, counting the ^{214}Bi γ-ray; this survey measures ^{222}Rn directly rather than U. The Rn emanation rate may be measured directly by placing the sample in a sealed chamber for a known period, then transferring the Rn to a scintillation counter cell or other detector. This technique, as well as that with NaI measurements, is used at Lawrence Berkeley Laboratory for building materials, such as concrete [51].

Radon emanating from soil is often collected *in situ* by placing cannisters of activated charcoal on the ground for hours or even days. On return to the laboratory, the Rn collected may be measured by de-emanation and transfer or by γ-spectrometry using an NaI detector. For the latter technique, Countess [52] cites 20% counting uncertainty for a 10-min counting period and a total of 100 pCi (3.70 Bq) of ^{222}Rn collected. Radon emanating from a surface may also be collected in a container placed on the surface, then measured by drawing a sample [53]. These techniques are limited to measurements from small flat areas for periods of time that are short relative to radon's half-life.

Some efforts are now being devoted to developing *in situ* source monitoring techniques. For example, Lawrence Berkeley Laboratory is testing a "sniffer" that uses an open scintillation counter flask to identify localized sources of Rn in homes, such as cracks or penetrations in basement floors.

The ^{222}Rn content of *water* may be measured by bubbling carrier gas through a sample, thereby transferring Rn to a scintillation counter cell or other detector. A somewhat simpler measurement detects low-energy γ-rays, ^{214}Po (0.24, 0.30, and 0.35 MeV) and ^{214}Bi (0.60 MeV), using a shielded NaI detector. For the latter techniques, lower detection limits of 30 pCi/liter (1.11 Bq/liter) may be obtained using a 10-min counting period and sample volumes of 1 liter [54]. Finally, a 10-ml sample may be used with automatic commercially available liquid scintillation counter systems to obtain a lower detection limit of 10 pCi/liter (0.37 Bq/liter) for a 40-min counting period [55]. This method detects α-particles and β-rays from radon and its daughters.

14.5. DAUGHTER MEASUREMENT TECHNIQUES IN AIR

In this section, methods for measuring the potential alpha energy concentration (PAEC) in units of working level (WL) are treated first. For radiation protection purposes, PAEC is the quantity of interest. For research studies and surveys it is often desirable to measure concentrations of individual daughters, and such instruments are also discussed. Some of the working level monitors also measure individual daughter concentrations. Results from individual daughter measurements can be used to calculate PAEC in WL.

14.5.1. Potential Alpha Energy Concentration (Working Level)

Two types of potential α-particle energy concentration (PAEC) or working level (WL) measurement are used in mines and other situations of interest:

1. Measurements of PAEC or WL at a given time and place.
2. Measurements of integrated exposure either of a person or for an area; these are usually expressed as Working Level Hours (WLH) or Working Level Months (WLM).

Item 1 is emphasized here since item 2 is discussed in the chapter on dosimeters (Chapter 9).

One WL is defined as "any combination of radon daughters in one liter of air that will result in the ultimate emission of 1.3×10^5 MeV of potential alpha energy" [3]. Of the daughters, ^{214}Pb and ^{214}Bi are β-ray/γ-ray emitters. But, even though only the α-particles from ^{218}Po and ^{214}Po decay need be measured, the energy sum considered in the PAEC contains contributions from the ^{214}Po α-particles that arise from decay of the β-decaying daughters in the sample air. Figure 14.8 [2] shows the growth of PAEC in initially pure Rn.

The use of an air filter is common to many methods. The filter collects radon daughters, after which decays are counted. If only α-particles are counted, then what matters is the activity of ^{218}Po and ^{214}Po, the α-particle emitting daughters. Figure 14.9 [1] shows the buildup and decay of α-particle activity from individually

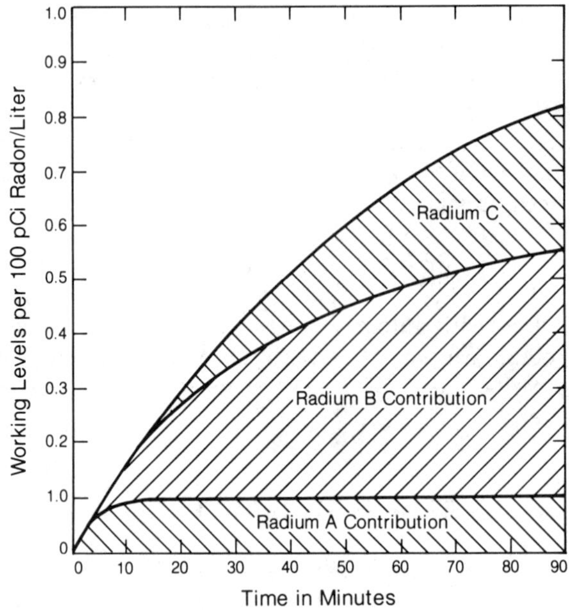

Figure 14.8. Growth of potential alpha energy concentration (in working level) in initially pure radon (e.g., freshly filtered air). Note that the earliest contribution to PAEC is from Po-218 (RaA), then Pb-214 (RaB), and still later Bi-214 (RaC). (reproduced from Evans, Ref. 2, with permission of the Health Physics Society © 1969).

isolated radon daughter isotopes, each with an initial decay rate of 10 disintegrations per minute [4.5 pCi (0.167 Bq)].

The most common method for measuring PAEC or WL is the *Kusnetz method.* Originally developed in 1956 [56], it has been the mainstay of radon daughter monitoring in the uranium mines ever since, and is now recommended by the American National Standards Institute (ANSI) as the "standard method" [27].

The Kusnetz method employs an air sampler (pump and filter) and an α-particle counter, usually an ionization chamber or a ZnS(Ag) scintillation counter. The basis of the method is that, about an hour after sampling, counts arise only from ^{214}Po formed on the filter by decay of the previous daughters, and the rate corresponds reasonably well to the potential α-particle energy concentration at the time of sampling. Air is sampled at 2 to 20 liters/min for 5 min. After a delay of from 40 to 90 min (most commonly 40), the count rate in counts per minute is measured, using a ratemeter. After determining dpm by correcting for the efficiency of filter collection and α-particle detection, a scale factor (tabulated versus delay before counting) is used to relate dpm to the working level in the original sampled air. This factor is relatively constant and independent of daughter equilibrium, assuming a constant delay of at least 40 min is always used [57].

The main feature which commends the Kusnetz method is its relative insensitivity to the concentration ratios of the three daughters ^{218}Po, ^{214}Po, and ^{214}Bi. For example, assume a 40-min delay before counting; if ^{218}Po:^{214}Pb:^{214}Bi concentrations are in the ratios of 100:100:100, 100:90:80, 100:45:35, and 100:15:6, the intrinsic error in determining WL from the data is only +7%, +8%, +2%, and −7%, respectively

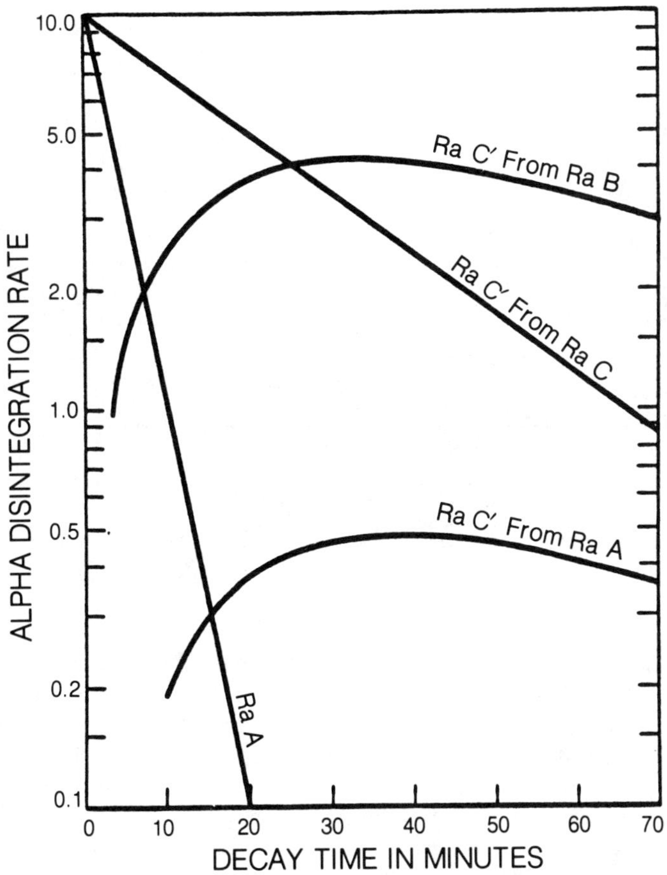

Figure 14.9. Buildup and decay of alpha activity from individual initially isolated radon daughter isotopes, RaA through RaC, with an initial disintegration rate for each isolated isotope of 10 disintegrations per minute (reproduced from U.S. Public Health Service, Ref. 1).

[58]. Groer [59] has shown that in very "young" air, such as air in which only ^{218}Po has had much chance to grow from the parent ^{222}Rn, the Kusnetz method underestimates the true WL by as much as 25%. Even so, because these errors are smaller than typical uncertainties in the way sampling represents true concentration, the Kusnetz method is accurate enough for most purposes.

The minimum sensitivity of the Kusnetz method as just described has been studied by Breslin, George, and Weinstein [36] and is shown in Fig. 14.10. The precision is about ±15% at 0.3 WL but degenerates rapidly at lower levels. This increased error is dominated by statistical fluctuations in the rate measurement [25] and is unavoidable except by using a scaler, discussed next. A full discussion is given by Loysen [60] of the various sources of error in the Kusnetz method.

The *modified Kusnetz method* substantially improves sensitivity by using a scaler to count for a fixed period, rather than using a ratemeter. In a laboratory environment, Breslin [25] counted for 4 min (from +38 to +42 min) using an α-particle scintillation counter and scaler. Reproducibility was found to be ±4.2%, ±14%,

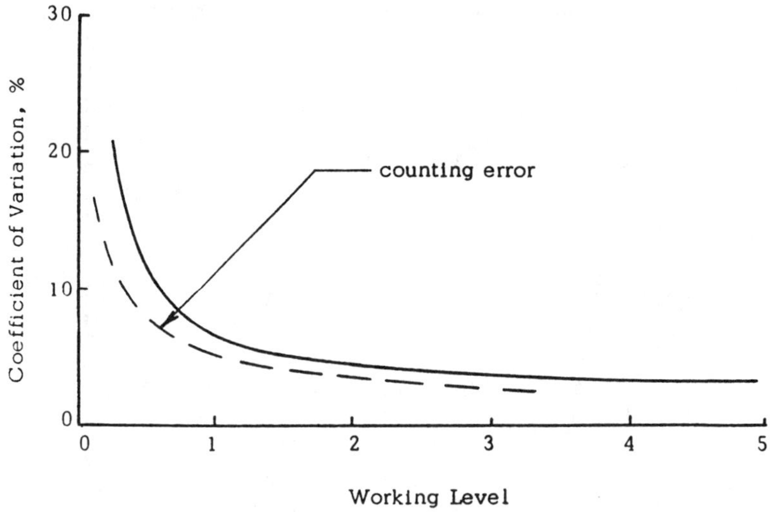

Figure 14.10. Precision for measurements of PAEC (working level) by Kusnetz method (reproduced from Breslin, George, and Weinstein, Ref. 36).

and ±35% at mean levels of 0.041, 0.0029, and 0.00046 WL, respectively, sensitive enough for almost any application.

Although the sensitivities just cited corresponded to a 5-min sample at 10 liters/min, most measurements are performed using lightweight pumps that sample at 2 liters/min [58], which degrades sensitivity. Membrane filters are widely used because of their 99+% retention of submicron particles and because most particles are deposited at the filter surface, minimizing the self-absorption of α-particles [27]. Glass fiber filters, equally efficient as collectors, suffer from more penetration and more self-absorption, but tests indicate that commercial glass filters have few such problems [58]. Moisture on the filter face can cause absorption problems for any of the filters as well as pressure problems in pumping. Air pump flowmeters may cause problems because their calibration is density-dependent and therefore sensitive to the different altitudes at which measurements are made. The details of this problem are discussed in the Bureau of Mines Handbook [58], which also discusses calibration procedures. An advantage of the Kusnetz method is that the α-particle detection system, which responds only to 7.69-MeV α-particles, need not have an energy-independent response. Perhaps the biggest drawback of the Kusnetz method is the minimum 45-min delay from start to finish. This inherent difficulty has stimulated the development of other WL techniques.

A method developed by Rolle in 1969 [61] and described further in a 1972 paper [62] makes possible much more refined measurements, using equipment identical to that of the Kusnetz method. Rolle describes how the choice of counting time affects systematic error, and indicates that intrinsic errors can be kept below about ±12% with counting for 10 min after about a 5-min wait. Rolle also discusses in detail the way volumetric and radiometric errors limit ultimate uncertainties to the ±20% range.

The methods just described all require a considerable time delay between the start of sampling and the end of counting: the Rolle method takes about 20-min, the Kusnetz upwards of 45. This has motivated the development of several instruments with more complex measurement schemes that take less time. These instruments typically use a surface-barrier detector for α-particle detection, often distinguishing between α-particles emitted from ^{218}Po and ^{214}Po (see discussion of α-spectrometry in the next section, Individual Radon Daughter Concentrations). The more complex devices have an additional detector for measuring the β-emitting daughters. An early instrument of the latter type [63] has been referred to as an instant working level meter (IWLM), but actually measures individual daughter concentrations, from which the potential α-particle energy concentration is calculated in WL. These instruments are described here. Background on the principles involved is presented more fully in the discussion of individual daughter measurements.

Schiager began development, at Colorado State University [64], of a WL meter employing a surface-barrier detector to measure ^{218}Po and ^{214}Po separately during a single very short counting period. A 1-liter sample is collected on a filter paper tape using a manual pump (the tape is advanced for each measurement). The potential alpha energy concentration (in WL) is calculated (electronically) from the ^{218}Po count plus about ten times the ^{214}Po count total. Figure 14.11 shows how such a procedure yields results within 10 to 15% of the actual PAEC over most of the range from equilibrium to complete disequilibrium. The largest uncertainty, for low daughter concentration, arises from poor counting statistics. For the completed prototype, the standard deviation of measurements taken with a 17-sec counting period (and a total measurement procedure of about 2 min) was 29% at 0.6 WL and 71% at 0.25 WL. Schiager [65] estimates that increasing the air sample size (to 3 liters), the counting period (to 100 sec), and the detector size would all improve the sensitivity by about a factor of 20, making it a useful survey instrument for measurements in mines and mills; the full measurement would take 3 to 4 min.

James and Strong [66] use a surface-barrier detector to count gross α-particles on a filter during the collection period and during an additional counting period, of

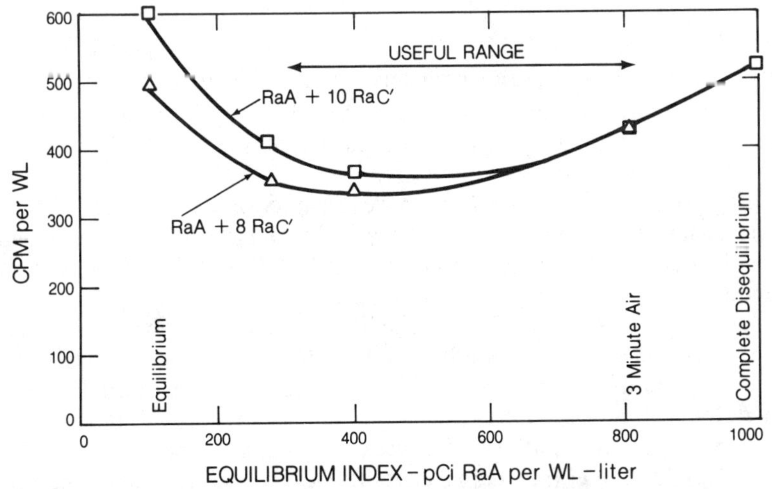

Figure 14.11. Theoretical survey meter response (cpm per WL) as a function of the degree of equilibrium of airborne radon progeny (pCi RaA per WL-liter) (reproduced from Schiager, Ref. 64).

equal length, that immediately follows the collection period. To a rough approximation, the first count gives the ^{218}Po concentration, and the second—for a range of equilibrium conditions—gives the daughter concentration in WL. For count periods of 5 min, a total measurement time of 11 min, and a 10 liters/min sampling rate, the statistical uncertainty is given as 1% at 1 WL, which compares favorably with more complex methods. Cliff, James, and Strong [67] have optimized this approach, ending the counting collection period at 2.5 min, and beginning the second count period (also 2.5 min) 2 min afterwards. Results of tests with a 0.85 liter/sec sampling rate indicate statistical uncertainties of about 20% at 0.005 WL. Counting during sampling has the advantage over the Kusnetz method that information on the short-lived ^{218}Po concentration is not lost before counting begins.

Instruments that measure total α-particles for three count periods or that measure ^{218}Po and ^{214}Po α-particles for two periods are treated in the discussion of individual daughter measurements. They can be used to obtain the potential α-particle energy concentration with no assumptions about the state of equilibrium between ^{218}Po, ^{214}Pb, and ^{214}Bi. Because of its very short half-life, ^{214}Po is always assumed to be in equilibrium with its precursor, ^{214}Bi.

The instrument class called the instant working level meter (IWLM) based on the work of Groer, Evans, and Gordon [63] uses three counting channels, namely two α-particle channels (from a surface-barrier detector) and one β-ray channel (from a plastic scintillator). Groer et al. [68] improved this device by using a 12 liter/min sampling rate and a very thin (0.003 in.) β-ray detector. After a 1-min background counting period and a 2-minute sampling period, the filter paper tape is advanced to a position such that the collection area is between the detectors; a 2-min counting period then begins. Radon daughter concentrations and potential α-particle energy concentration (in WL) are calculated electronically and are available on demand. The useful range of the instrument is given as 0.01 to 100 WL, with intercomparison tests indicating the 10 to 20% discrepancies in the range 0.3 to 1 WL [68]. This instrument has been developed further for the EPA to an "environmental working level monitor" with 3-min background, collection, and counting periods, and a sampling rate of 30 to 40 liters/min [69]. A commercial version is now available.

Shreve, Miller, and Cleveland [70] have developed a different type of IWLM, with (total) α-particle and β-ray detectors. For a range of realistic equilibrium conditions, the α-particle and β-ray total counts (for sampling, delay, and counting times of about 2 min each) give a good measure of the PAEC, both theoretically and according to comparison tests.

Integrating working level monitors have been developed, using TLD chips to accumulate counts from the α-particle decay of material collected on an air filter. A "Radon Progeny Integrating Sampling Unit" (RPISU) based on this principle and sampling 2 to 3 liters/min was developed by Schiager [71] for the State of Colorado and (ultimately) the EPA. Modified versions of the initial instrument have been employed for indoor studies in Florida's phosphate regions [19] and in Colorado's uranium milling regions [71]. An instrument that is similar in principle has been developed by the DOE EML [72], and is capable of measuring average concentrations as low as 0.0005 WL for week-long sampling periods. Small portable or fixed devices that integrate radon daughter levels are treated in the section on dosimetry.

The WL may also be determined by a measurement of the daughter β-ray activity. A β-ray channel is included in the instant working level monitors described previously. Simple β-ray counting has also been used in mining situations as a continuous working level monitor [73].

14.5.2. Individual Radon Daughter Concentrations

The ability to determine activities of individual radon daughters in an unknown atmosphere is often of great use for research, for cases of extreme disequilibrium, and for calculation of the potential α-particle energy concentration (in WL), as in the case of the IWLM [63] discussed previously. In an equilibrium atmosphere, of course, all of the short-lived daughters have activities equal to that of the parent; but complete equilibrium is never the case in practical situations, so that the relative activities differ from one. Even with its short (3-min) half-life, the first radon daughter, ^{218}Po, is not always in equilibrium with the parent Rn. The subsequent daughters are nearly always only partially ingrown, especially in well-circulated air. This is true in the general environment as well as in uranium-mine atmospheres. Another problem is the determination of the uncombined ^{218}Po fraction, that is, the fraction in the sample atmosphere that is free and not combined with particulate matter.

One technique [49] for measuring the ^{214}Pb:^{214}Bi ratio using the APHA air-filter/β-ray counting technique has already been discussed. This technique is complex and not very sensitive unless the ratio is very large or very small. For these reasons it is not often used. The technique of James and Strong [66] and Cliff, James, and Strong [67] measures disequilibrium, to some degree, by using two counting intervals.

Two more general approaches are the Tsivoglou method, involving several counting intervals, and the use of α-spectroscopy.

The *Tsivoglou method* [74] is one of the oldest techniques for radon daughter determinations. It employs an air filter and an α-particle counter/rate meter apparatus similar to that required for the Kusnetz method for potential α-particle energy concentration. Technical considerations required to obtain good data are also similar. One key difference not required for the Kusnetz method is that the α-particle counter response must be energy-independent. The Bureau of Mines *Handbook* [58] discusses calibration procedures in detail. See also [28]. After the air sample is taken (typically for 10 min at 5 liters/min), the α-particle count rate is measured at three later times: after delays of 5, 15, and 30 min. Daughter concentrations are then determined by solving three simultaneous equations. However, Breslin, George, and Weinstein [36], with 100-liter sample volumes at concentrations of 0.5 to 5 WL, found poor reproducibility for identically repeated measurements. Measurements for ^{214}Pb:^{218}Po ratios and ^{214}Bi:^{218}Po ratios had replication errors in the 15 to 25% and 25 to 35% ranges, respectively, largely due to use of a rate meter.

The *modified Tsivoglou method* uses of count totals over relatively long periods rather than rates from a meter. Thomas [75], using a 5-min sampling time at 10 liters/min, has taken count totals in three time intervals: 2 to 5, 6 to 20, and 21 to 30 min, resulting in notably improved precision. Breslin [25] calculates precisions for ^{218}Po, ^{214}Pb, and ^{214}Bi measurements are 4%, 4%, and 12%, respectively, at 0.3 WL and a radon daughter ratio of 100:30:10

This modified Tsivoglou technique can also be used for measurements of potential α-particle energy concentration. The precision is comparable to that obtained with the modified Kusnetz method. The added complication avoids small uncertainties arising from assumptions about the degree of disequilibrium.

A theoretical analysis of the Tsivoglou method, extending the treatment to any number of general counting times, has been given by Martz et al. [76]. A more complicated analysis technique has been developed by Raabe and Wrenn [77], who generalize the Tsivoglou method by performing a regression analysis to fit mathematically the observed count totals during various counting periods. Thus one is limited to three counting periods and three simultaneous equations and simultaneous determinations of ^{220}Ra daughters are possible. Measurements over 7 time intervals (the last 3 hr after taking a 1-min, 7.5-liter sample) are clearly more accurate than those of the Tsivoglou method, but the complicated analysis is probably only justified when research work requires high sophistication and accuracy or when ^{220}Ra daughter concentrations are to be determined.

Cliff [78] has reports that three-count methods can be improved by using the sampling period as one of the counting periods. Analyzing an instrument of the type described by James and Strong [66], he shows that for a total measurement period of 35 min—the same as for the Tsivoglou method—the sensitivity of the ^{218}Po determination may be improved by a factor of 3 to 5, depending on the state of equilibrium, by counting from 0 to 5, 6 to 25, and 26 to 35 min; sampling for 10 min and lengthening the total measurement period slightly results in substantially greater improvement. In each case, the determination of ^{214}Po and and ^{214}Pb is also improved, although not as much. For a 42-min measurement, sensitivities for each daughter are less than 1 pCi/liter (0.037 Bq/liter) for a wide range of equilibrium conditions.

Alpha spectroscopy is another useful approach for measuring individual radon daughter concentrations. Instead of using the air filter/α-particle counting technique, one can substitute an α-spectrometer for the α-particle detector system. Several types of α-spectrometers have been developed, all of which have been described in Chapter 7. Semiconductor detector systems now have the best resolution. Surface-barrier or diffused-junction detectors are often used for daughter measurements. The α-particle groups requiring resolution and measurement have energies of 6.00 MeV (^{218}Po) and 7.69 MeV (^{214}Po).

The method developed by Martz et al. [76] uses a semiconductor detector and multichannel analyzer, with a 4.3-liter/min collection rate onto a membrane filter. Separate determinations of the ^{218}Po and ^{214}Po α-particle activities are made twice, 5 and 30 min after sampling. Only two simultaneous equations are needed (compared to the three required in the Tsivoglou method) to determine the relative daughter concentrations. The main advantage of this method is its improved accuracy in determining the short-lived (3-min) ^{218}Po, since its α-particle is counted directly. In experimental comparisons with the Tsivoglou method, Martz et al. [76] found that the spectroscopic method was significantly more precise for both ^{218}Po (8% standard deviation compared to 29%) and ^{214}Po (12%).

Others [79–81] have suggested a *modified Martz* technique that integrates ^{218}Po and ^{214}Po counts over longer times. Because of progress in detectors and electronic data analysis, this approach, and others using α-spectrometry, can now be implemented in convenient, dependable, and compact instruments suitable for much more widespread application. Tremblay et al. [82] suggested that a straightforward extension of integral spectrometric techniques (comparable to Cliff's extension of the Tsivoglou method) would be to use the sampling period as one of the count periods. Other schemes exist, such as that of the instant working level monitor described previously [68,69], where—during a single counting period—^{218}Po α-particles, ^{214}Po α-particles, and $^{214}Pb + ^{214}Bi$ β-rays are measured. Keefe et al. [83] present analysis techniques that optimize sampling and counting times for the IWLM to yield maximum radon daughter sensitivities.

Gamma-ray spectroscopy is an alternative means of monitoring the β/γ radon daughters, ^{214}Pb and ^{214}Bi. As discussed previously, NaI detectors are often used for measurements of Rn in water after equilibrium is reached, and also for radon emanation measurements. Alone it is not sufficient to characterize all of the daughter concentrations in air, because ^{218}Po decay cannot be detected, unless more than one counting interval is used; a laboratory γ-ray technique for air filters is discussed by Irfan and Fagan [84]. Interferences from other naturally occurring γ-ray emitters makes general application difficult. However, the ^{214}Bi 1.76-MeV γ-ray is distinctive even in such situations, so it is frequently measured for uranium surveys.

Lead-210 is a long-lived radon daughter, effectively the endpoint of the short-lived daughter sequence. A number of specific detection techniques have been developed, including some proposed as indicators of exposure to short-lived daughters (e.g., Ref. 85). Lead-210 ($t_{1/2}$ = 22 yr) emits a low-energy β-ray. Chemical separation of ^{210}Pb, followed by β-ray counting, is described in Ref. 47. A radiochemical technique for ores and airborne particles is given by Sill and Willis [86]. Possible use of liquid scintillation counter spectrometry for determining integrated exposures by examining samples taken from air filters in mines is discussed by Rodgers [87].

Before leaving the subject of individual radon daughter measurements a brief discussion of determinations of *uncombined ^{218}Po fraction* and of radioactive *particle size distributions* is appropriate. The ICRP [10] has noted that the fraction of ^{218}Po that is not combined with particulate matter seems to play a major role in the radiobiological impact of the radon daughters in mines. This has motivated attempts to measure this uncombined fraction (f). These measurements are performed primarily in research rather than in routine monitoring.

Duggan and Howell [88] describe a system in which two filters sample the atmosphere side by side. One of them is "preceded by a diffusion battery designed to remove most of the unattached daughters but hardly any of the attached ones" [88]. Polonium-218 is distinguished from ^{214}Po by α-spectroscopy. Another method has been described by Fusamura and Kurosawa [89] in which gas is passed through a diffusion tube, and the differing diffusion coefficients are relied upon to bring about a partial separation. In this way, f values from 6 to 25% were measured to better than a factor of 2.

Another device based on diffusion has been described by Mercer and Stowe [90] and is shown in Fig. 14.12 [91]. "Air enters through an orifice in the center of the upper of two discs and flows radially outward between the discs and down past the edge of the lower disc." [91] As much as 80% of unattached ^{218}Po atoms (and typically 60 to 70%) can be collected on the discs [90]. Using this system, George and Hinchliffe [91] measured f values below 0.10 with precisions from of ± 0.007 to ± 0.014. George and Breslin have employed wire screen collectors [92] for measurement of uncombined ^{218}Po fractions in New York City air [93] and in New York–New Jersey homes [21]. A similar technique has been used for measurements in uranium mines [94]. Daughter activity has rarely been measured by particle size. George [93] used diffusion batteries, consisting of disks containing collimated holes (following Ref. 95), to separate daughter activity into several size fractions;

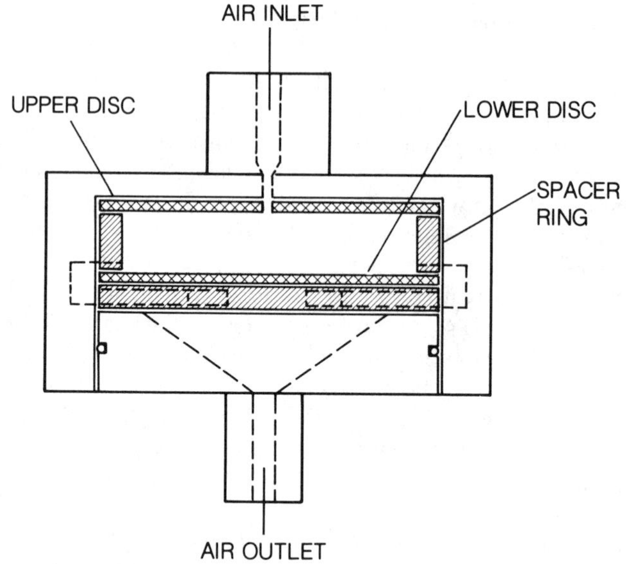

Figure 14.12. Diffusion sampler for uncombined radon daughters (reproduced from George and Hinchliffe, Ref. 91, with permission of the Health Physics Society © 1972).

high flow-rate diffusion batteries were used in Ref. 21. Prodi [96] recently described a method employing continuous inertial separation and track etch detection.

14.6. DOSIMETERS

A variety of instruments for measurement of integrated exposures have been devised, ranging from small dosimeters that may be carried on the person of a worker to instruments, large and small, for measuring the average concentration in an area over an extended period. The larger integrating instruments for measuring Rn or radon daughter concentrations were discussed previously. Here the treatment concentrates on small devices, primarily for use in mines, but also for the general indoor and outdoor environment.

14.6.1. Personnel Dosimeters

The usefulness of dosimetry for workers is obvious (and discussed in Chapter 9): a portable instrument that could accurately integrate WL exposure over time would help radiologically monitoring of occupationally exposed individuals (e.g., miners) by providing direct recording of individual exposures. Present dosimetry is done largely by measuring daughter PAEC and correlating this with the amount of time spent in each of the various working areas by a worker. This method has served a valuable purpose over a long period and has certain advantages over dosimeters, among which is that measurements are made by trained personnel with less inconvenience for the miner. Such daughter measurements will always be required to supplement any personnel dosimetry system that achieves wide acceptance.

The desired requirements for a personnel dosimeter for occupational use are: that it be sensitive down to an integrated exposure of about 1 WL-hr; that it be capable of weekly or biweekly readout; that it properly sample the air being breathed; and that it be light, rugged, and fail-safe.

A number of efforts over a several-year period around 1970 were devoted to development of personnel dosimeters employing film, track-etch, or TLD alpha detectors for recording Rn or radon daughter exposures in active or passive systems. Some properties of these dosimeters are given in Table 14.2, where Breslin [25] summarizes results of several evaluations of these devices performed at EML by While [97–99]. Eberline's alpha trace-count film dosimeter [100] and the NYU ZnS/film system [101], similar in principle to the Polaroid monitoring technique of Bedrosian [46], were designed for radon dosimetry. Laboratory standardization tests [97] of one have responses varying by up to a factor of 10 in repeated trials; response of the other varied by up to a factor of 3 and increased by a factor of 3 when the concentration increased by a factor of 10.

Of the six daughter dosimeters, five gave substantially less than satisfactory performance in the EML tests, and one proved somewhat better [97–99]. The daughter dosimeters performed with varying degrees of success in the laboratory, but the harsh conditions in mines (mechanical abuse, high humidity, mine dust and mud, corrosion, pump failures) uniformly caused difficulty. For the units from Oak Ridge National Laboratory [102], MIT [103], Colorado State University [104], and General Electric [105,106], reproducibility and mine comparison results were usually poor. The sixth dosimeter, the "MOD" unit from EML, profited from previous efforts, particularly the TLD units [103, 104], but was still not fully satisfactory [99], largely because of pump failures and TLD response to external γ-particle and β-ray radiation. Subsequent attention to these problems has produced a unit that appears to be dependable in mine conditions and with a range extending as low as 0.2 WL-hr [107]. The same sampling head (including a filter and two TLDs, one a control shielded from the filter) and pump have been incorporated in the area-integrating daughter monitor described previously [72]. The 1-kg weight of the pump, which is worn on the belt, may impede acceptance by miners.

The Bureau of Mines has developed a personnel dosimeter employing a surface-barrier semiconductor detector in the sampling head [108]. Alpha-particle induced pulses are counted in the unit's circuitry for subsequent readout. The unit shows good agreement in recent comparison tests for exposures of the order of 1 WL-hr.

Gammage, Kerr, and Huskey [109] have suggested use of ceramic beryllium oxide (BeO) dosimeters that are read for both thermoluminescence and thermally stimulated electron emission. Because of differing thermoluminescence and electron-emission response to α- and γ-radiation, a single detector could be used to measure α-particle emitters. The BeO detectors are, however, delicate.

The use of cellulose nitrate plastic for integrated track-etch radon daughter monitoring has been investigated further by Benton and his collaborators at the University of San Francisco. They [110] found that an active dosimeter, in which airborne daughters are collected on a filter facing a plastic detector, gives useful information—under optimum conditions—for an exposure range extending below 1 WL-hr. On the other hand, a passive dosimeter and a passive area monitor, each using two detectors (one protected from daughters by a filter), had low sensitivity and poor reproducibility, respectively.

Table 14.2. Radon and Radon Daughter Dosimeters (from Ref. 25)

Source	Ref.	Atmospheric Component Detected	Type of Detector	Location		Pump Air Flow (liters/min)	Effective Range of Measurement	Total Weight (oz.)	Period of Operation (hr)
				Pump	Detector				
Health & Safety Lab.	99	Rn dtrs	TLD (LiF)	belt	hat	0.100	3–1000 WL-hr	28	> 9
Oak Ridge Natl. Lab	99,102	Rn dtrs	α track etch	hat	hat	0.021	1–400 WL-hr	2	> 13
Mass. Inst. of Tech.	98	Rn dtrs	TLD (CaF_2:Dy)	belt	hat	0.100	$0.026–2 \times 10^5$ WL-hr	7	9
Colo. State Univ.	97,104	Rn dtrs	TLD (LiF)	belt	hat	0.170	$0.025–6 \times 10^6$ WL-hr	24	40
Eberline Inst. Co.	97	Rn dtrs	α track count	belt	belt	0.015	4–600 WL-hr	12	10
General Elect. Co.	97,105	Rn and Rn dtrs	α track etch	–	hat	passive	5–100 WL-hr	< 1	unlimited
Eberline Inst. Co.	97,100	Rn	α track count	–	hat	passive	4–1000 pCi-hr/cm^3	1	unlimited
New York Univ.	97,101	Rn	scint. + film	–	–	passive	10–200 pCi-hr/cm^3	4	unlimited

14.6.2. Integrating Area Monitors

Area monitoring in principle employs the integrating instruments discussed under Rn and daughter measurement techniques, the personnel dosimeters just discussed, or small passive monitors based on α-particle track-etch techniques. The track-etch technique was used in some of the prototype dosimeters discussed previously, but increasing attention is given to its use in small integrating devices, particularly for area radon monitoring. Workers at General Electric and elsewhere have examined use of track-etch radon detectors for uranium exploration. Fleischer et al. [111] set forth this idea; Ward, Fleischer, and Mogro-Campero [112] discussed the use of a membrane to exclude ^{220}Rn, as well as radon daughters; and the application of such techniques to indoor area monitoring is presently being evaluated in laboratories around the world.

14.7. SUMMARY AND CONCLUSIONS

This section outlined the various techniques for measuring ^{222}Rn and its daughters in uranium mines and in environmental media. The several quite different measurement problems in each area are summarized separately.

Radon in Air. Two simple methods exist for Rn measurements in the concentration range down to below 10 pCi/liter (0.370 Bq/liter). These are the Lucas ZnS scintillation chamber and the two-filter method. Both appear to be satisfactory for routine use in uranium mines, and each can be modified to achieve sensitivities below 1 pCi/liter (0.037 Bq/liter). For environmental measurements, where sensitivities below 0.1 pCi/liter (3.70 mBq/liter) are sometimes required, several more elaborate methods have been developed, often involving Lucas chambers as the detector. Such systems are often used for laboratory analysis of air samples taken in the field. Continuous radon monitors are now available, typically employing ZnS scintillation cells or two-filter systems. Inexpensive integrating monitors have been developed that employ TLD chips sensitive to α-particles, and work continues on passive film or track-etch integrating techniques. Many of the new developments are intended for monitoring in the general environment.

Source and Emanation Measurements. Radium-226 content of source materials may be measured by NaI γ-spectrometry. Radon emanation rates may be crudely measured by charcoal collection or accumulation in a container, followed by de-emanation and transfer to a Lucas cell or by NaI spectrometry. The radon content of water may be determined by these methods or by liquid scintillation counting systems.

Potential Alpha Energy Concentration (Working Level) Measurement. The Kusnetz method, used for many years as a standard technique in the uranium mines, is not sensitive enough to measure daughter concentration in the range below about 0.3 WL. The modified Kusnetz method and the method developed by Rolle are both sensitive enough ($\leq$ 0.01 WL) to meet almost any need. However, neither of the methods gives an immediate answer, the delays being about 45 min (modified Kusnetz) and 20 min (Rolle).

More rapid working level monitors have recently been developed, in some cases to the point of commercialization. Schiager's device is portable, simple, and relatively inexpensive; with commercial development, it could offer sensitivity and convenience suitable for mine applications. The two-count gross-α method of James, Strong, and Cliff is apparently being applied commercially. And the more complex "instant working level meter" developed by Groer and collaborators gives useful information down to the region of 0.01 WL and is available in a commercial version, making it suitable for general environmental application.

Individual Radon Daughters. A number of methods, all variations of the long established Tsivoglou technique or of α-spectrometry, rely upon measuring activity collected on an air filter. Determining the activity ratios of several daughters requires several counting periods, measurement of several parameters, or a combination of the two. The Tsivoglou method and variants rely on gross α-particle counting during several time periods. Martz and other workers have combined this approach with α-spectrometry to increase the sensitivity. And Groer and others have combined α-spectrometry with β-ray counting to calculate daughter concentrations based on a single counting period. The last has been developed into a fully automatic instrument. Some effort has also been devoted to measurement of daughter γ-rays, a technique commonly used for detecting ^{214}Bi in uranium exploration. Finally, some devices now exist for measurement of unattached fractions and daughter particle size distributions; these usually depend on daughter diffusion characteristics.

Personnel Dosimeters. The TLD dosimeter developed at EML for miners appears to have been successful, although this should not discourage development efforts aimed at a somewhat lighter, and perhaps simpler, device. Such efforts might be based on passive track-etch techniques. The decision has not yet been made as to whether personnel dosimeters or variants of the Kusnetz method combined with worker area time records will be used.

Integrating Area Monitors. Integrating working level monitors, using TLD chips, have been applied in indoor radon studies, as well as other applications. In addition, substantial work is now being devoted to adapting track-etch techniques developed for uranium exploration to more general application, including indoor studies.

In conclusion, much instrumentation in this area of measurements is much more satisfactory than it was a few years ago. Further development can still be expected in order to make use more convenient and to design instruments more appropriate for specific applications, particularly for indoor monitoring of sources and airborne concentrations. Nevertheless, the extent to which any particular instrument is used depends critically on regulatory monitoring requirements, whether for mines or for other situations.

REFERENCES

1. U.S. Public Health Service, "Control of Radon Daughters in Uranium Mines and Calculations on Biologic Effects," U.S. PHS Publication No. 494, 1957.

2. R. D. Evans, "Engineers Guide to the Elementary Behavior of Radon Daughters," *Health Phys.*, **17**, 229, 1969.
3. R. L. Rock, D. K. Walker, R. W. Dalzell, and E. J. Harris, *Controlling Employee Exposure to Alpha Radiation in Underground Uranium Mines,* U.S. Bureau of Mines Handbook Vols. I and II, 1970 and 1971.
4. United Nations Scientific Committee on the Effects of Atomic Radiation "Sources and Effects of Ionizing Radiation," United Nations, 1977.
5. *Federal Register,* **33**(252), 1968.
6. *Federal Register,* **34**(11), 1969.
7. *Federal Register,* **36**, 12,921, 1971.
8. B. Altshuler, N. Nelson and M. Kuschner, "Estimation of the Lung Tissue Dose from Inhalation of Radon and Daughters," *Health Phys.*, **10**, 1137, 1964.
9. Federal Radiation Council, *Guidance for the Control of Radiation Hazards in Uranium Mining,* FRC Report No. 8 (rev.), 1967.
10. International Commission on Radiological Protection, "Recommendations of the International Commission on Radiological Protection," ICRP Publication No. 6, Pergamon Press, New York, 1964.
11. International Commission on Radiological Protection, "Radiation Protection in Uranium and Other Mines," ICRP Publication No. 24, Pergamon Press, New York, 1976.
12. National Bureau of Standards, *Maximum Permissible Body Burdens and Maximum Permissible Concentrations of Radionuclides in Air and in Water for Occupational Exposure,* National Bureau of Standards Handbook 69, 1959.
13. International Commission on Radiological Protection, "Report of Committee II on Permissible Dose for Internal Radiation," ICRP Publication No. 2, Pergamon Press, New York, 1959.
14. National Council on Radiation Protection and Measurements, "Natural Background Radiation in the United States," NCRP Report No. 45, 1975.
15. S. D. Shearer and C. W. Sill, "Evaluation of Atmospheric Radon in the Vicinity of Uranium Mill Tailings," *Health Phys.*, **17**, 77, 1969.
16. R. O. Pohl, "Radioactive Pollution," *ASHRAE J.*, **18**(9), 47, 1976.
17. *Federal Register,* **44**(230), 68,290, Nov. 28, 1979.
18. U.S. Atomic Energy Commission, "Grand Junction Remedial Action: Proposed Criteria," *Federal Register*, **37**(203), 1972.
19. R. J. Guimond, W. H. Elett, J. E. Fitzgerald, Jr., S. T. Windham, and P. A. Cung, "Indoor Radiation Exposure Due to Radium-226 in Florida Phosphate Lands" U.S. Environmental Protection Agency Report EPA-520/4-78-031, 1979.
20. J. E. Pearson, "Natural Environmental Radioactivity from Radon-222," U.S. Public Health Service Publication No. 999-RH-26, 1967.
21. A. C. George and A. J. Breslin, "The Distribution of Ambient Radon and Radon Daughters in Residential Buildings in the New York–New Jersey Area," in Ref. 24.
22. R. J. Budnitz, J. V. Beck, C. D. Hollowell, W. W. Nazaroff, A. V. Nero, and A. H. Rosenfeld, "Human Disease from Radon Exposures: The Impact of Energy Conservation in Residential Buildings," *Energy and Buildings*, **2**, 209, 1979.
23. C. D. Hollowell, M. L. Boegel, J. G. Ingersoll, and W. W. Nazaroff, "Radon-222 in Energy Efficient Buildings," *Trans. Am. Nucl. Soc.*, **33**, 148, 1979.

24. T. F. Gesell and W. M. Lowder, Eds., *Proceedings of the Symposium on the Natural Radiation Environment III,* Technical Information Center, CONF-780422, Vol. 1, 1980.
25. A. J. Breslin, "Monitoring Instrumentation in the Uranium Mining Industry," DOE Environmental Measurements Laboratory, 1972, unpublished report.
26. A. J. Breslin, Ed., "Radon Workshop—February 1977," DOE Environmental Measurements Laboratory Report HASL-325, 1977.
27. American National Standards Institute, "Radiation Protection in Uranium Mines," ANSI N13.8, 1973.
28. R. T. Beckman, "Calibration Procedures for Radon and Radon Daughter Measurement Equipment," Mining Enforcement Safety Administration Report MESA-IR-1005, 1975.
29. H. F. Lucas, "Improved Low-Level Alpha Scintillation Counters for Radon," *Rev. Sci. Instrum.*, **28**, 680, 1957.
30. J. Kristan and I. Kobal, "Modified Scintillation Cell for the Determination of Radon in Uranium Mine Atmosphere," *Health Phys.*, **24**, 103, 1973.
31. H. F. Lucas, "Alpha Scintillation Radon Counting," in *Proceedings of the Workshop on Methods for Measuring Radiation In and Around Uranium Mills,* E. D. Harwood, Ed., Atomic Industrial Forum Program Report, Vol. 3, No. 9, 1977, p. 69.
32. A. C. George, "Scintillation Flasks for the Determination of Low Level Concentrations of Radon," in *Proceedings of the Ninth Midyear Health Physics Symposium, Denver, Colorado, 1976.*
33. *Standard Methods for the Examination of Water and Wastewater,* 15th ed., American Public Health Association, Washington, DC 1980.
34. J. W. Thomas and P. C. LeClare, "A Study of the Two-Filter Method for Radon-222," *Health Phys.*, **18**, 133, 1970.
35. H. Newstein, L. D. Cohen, and R. Kablin, "An Automated Atmospheric Radon Sampling System," *Atmos. Environ.*, **5**, 823, 1971.
36. A. J. Breslin, A. C. George, and M. S. Weinstein, "Investigation of the Radiological Characteristics of Uranium Mine Atmospheres," Health and Safety Laboratory Report HASL-220, 1969.
37. J. H. Harley, Ed., EML *Procedures Manual,* Environmental Measurements Laboratory Report HASL-300, 1972, revisions supplied annually. Earlier editions were published as Report NYO-4700.
38. C. W. Sill, "An Integrating Air Sampler for Determination of Radon-222," *Health Phys.*, **16**, 371, 1969.
39. W. W. Nazaroff, M. L. Boegel, C. D. Hollowell, and G. W. Roseme, "The Use of Mechanical Ventilation with Heat Recovery for Controlling Radon and Radon-Daughter Concentrations," Lawrence Berkeley Laboratory Report LBL-10222, 1980.
40. J. W. Thomas and R. J. Countess, "Continuous Radon Monitor," *Health Phys.*, **36**, 734, 1979.
41. R. T. Graveson, "A Survey of Radon Monitoring Techniques," in *Proceedings of the Radon Workshop, February 1977,* Health and Safety Laboratory Report HASL-325, 1977, p. 20; and J. W. Thomas, "An Environmental Radon and Thorium Monitor," p. 22.
42. M. W. Wrenn, H. Spitz, and N. Cohen, "Design of a Continuous Digital-Output Environmental Radon Monitor," *IEEE Trans. Nucl. Sci.*, **NS-22**, 645, 1975.

43. J. Porstendorfer, A. Wicke, and A. Schraub, "Methods for a Continuous Registration of Radon, Thoron, and Their Decay Products In- and Outdoors," p. 1293 of Ref. 24.
44. A. C. George, A. J. Breslin, and S. F. Guggenheim, "A Cumulative Environmental Radon Monitor," *Proceedings of Ninth Midyear Health Physics Symposium, Denver, Colorado, 1976.*
45. A. C. George, "A Passive Environmental Radon Monitor," Radon Workshop, February 1977, DOE Environmental Measurements Laboratory Report HASL-325, 1977.
46. P. H. Bedrosian, "Photographic Technique for Monitoring Radon-222 and Daughter Products," *Health Phys.*, **16**, 800, 1969.
47. *Methods of Air Sampling and Analysis,* 2nd ed., Intersociety Committee on Methods of Air Sampling and Analysis, American Public Health Association, Washington, DC, 1977.
48. G. E. Jones and L. M. Kleppe, "A Simple and Inexpensive System for Measuring Concentrations of Atmospheric Radon-222," Lawrence Berkeley Laboratory Report UCRL-16952, 1966.
49. L. B. Lockhart, R. L. Patterson, and C. R. Hosler, "Determination of Radon Concentrations in the Air through Measurement of Its Solid Decay Products," U.S. Naval Research Laboratory Report NRL-6229, 1965; "The Extent of Radioactive Equilibrium Between Radon and Its Short-Lived Daughter Products in the Atmosphere," U.S. Naval Research Laboratory Report NRL-6374, 1966.
50. A. R. Smith, Lawrence Berkeley Laboratory, private communication, 1980.
51. J. G. Ingersoll, Lawrence Berkeley Laboratory, private communication, 1980.
52. R. J. Countess, "^{222}Rn Flux Measurement with a Charcoal Counter," *Health Phys.*, **31**, 455, 1976.
53. M. Wilkening, "Measurement of Radon Flux by the Accumulation Method," in *Proceedings of the Workshop on Methods for Measuring Radiation in and around Uranium Mills,* E. D. Harwood, Ed., Vol. 3, No. 9, Atomic Industrial Forum, Washington, DC, 1977, p. 131.
54. R. J. Countess, "Measurement of ^{222}Rn in Water," *Health Phys.*, **34**, 390, 1978.
55. H. M. Prichard and T. J. Gesell, "Rapid Measurements of ^{222}Rn Concentrations in Water with a Commercial Liquid Scintillation Counter," *Health Phys.*, **33**, 577, 1977.
56. H. L. Kusnetz, "Radon Daughters in Mine Atmospheres: A Field Method for Determining Concentrations," *Am. Indust. Hyg. Assn. Q.* (now Journal), **17**, 85, 1956.
57. N. H. Harley and B. S. Pasternak, "The Rapid Estimation of Radon Daughter Working Levels when Daughter Equilibrium is Unknown," *Health Phys.*, **17**, 109, 1969.
58. R. L. Rock, D. K. Walker, R. W. Dalzell, and E. J. Harris, *Controlling Employee Exposure to Alpha Radiation in Underground Uranium Mines,* U.S. Bureau of Mines Handbook, Vol. I and II, 1970 and 1971.
59. P. G. Groer, "The Accuracy and Precision of the Kusnetz Method for the Determination of the Working Level in Uranium Mines," *Health Phys.*, **23**, 106, 1972.

60. P. Loysen, "Errors in Measurement of Working Level," *Health Phys.*, **16**, 629, 1969.
61. R. Rolle, "Improved Radon Daughter Monitoring Procedure," *Am. Ind. Hyg. Assn. J.*, **30**, 153, 1969.
62. R. Rolle, "Rapid Working Level Monitoring," *Health Phys.*, **22**, 1972.
63. P. G. Groer, R. D. Evans, and D. A. Gordon, "An Instant Working Level Meter for Uranium Mines," *Health Phys.*, **24**, 387, 1973.
64. K. J. Schiager, "Radon Progeny Inhalation Study," Colorado State University Report COO 1500-17, 1970.
65. K. J. Schiager, "The 3R-WL Air Working Level Survey Meter," *Health Phys.*, **33**, 595, 1977.
66. A. C. James and J. C. Strong, "A Radon Daughter Monitor for Use in Mines," in *Proceedings of the 3rd International Congress of the International Radiation Protection Association,* U.S. AEC Report CONF-730907, 1973, p. 932.
67. K. D. Cliff, A. C. James, and J. C. Strong, "Further Developments of a Radon Daughter Monitor for Use in Mines," National Radiation Protection Board Annual Report, England, 1979.
68 P. G. Groer, D. J. Keefe, W. P. McDowell, and R. F. Selman, "An Instant Working Level Meter with Automatic Individual Radon Daughter Readout for Uranium Mines," p. 950 of Ref. 66.
69. D. J. Keefe, W. P. McDowell, and P. G. Groer, "The Environmental Working Level Monitor," Argonne National Laboratory Report prepared for Office of Radiation Programs, U.S. EPA Report No. P7628C, 1978.
70. J. D. Schreve, R. W. Miller, and J. E. Cleveland, "A New Instrument for Quick Determination of Radon and Radon-Daughter Concentrations in Air," in *Proceedings of Fourth International Congress of the International Radiation Protection Association,* 1977, p. 111.
71. K. J. Schiager, "The Evaluation of Radon Progency Exposures in Buildings: A Report on Equipment and Techniques," Colorado State University, 1971; K. J. Schiager, "Integrating Radon Progeny Air Sampler," *Am. Indust. Hyg. Asso. J.*, **35**, 165, 1974; see also p. 83 of Ref. 26, and G. A. Franz, B. Peterson, H. O. Langer, and C. Thornberg, "Integrated Working Level Samplers Using Thermoluminescent Dosimeters: Operation and Interpretation," p. 175 of Ref. 53.
72. S. F. Guggenheim, A. C. George, R. T. Graveson, and A. J. Breslin, "A Time-Integrating Environmental Radon Daughter Monitor," *Health Phys.*, **36**, 452, 1979; see also Ref. 71.
73. R. F. Droullard and R. F. Holub, "A Continuous Working Level Monitor," p. 43 of Ref. 26.
74. E. C. Tsivoglou, H. E. Ayer, and D. A. Holaday, "Occurrence of Nonequilibrium Atmospheric Mixtures of Radon and Its Daughters," *Nucleonics*, **11**, 40, 1953.
75. J. W. Thomas, "Measurement of Radon Daughters in Air," *Health Phys.*, **23**, 783, 1972.
76. D. E. Martz, D. F. Holleman, D. E. McCurdy and K. J. Schiager, "Analysis of Atmospheric Concentrations of RaA, RaB, and RaC by Alpha Spectroscopy," *Health Phys.*, **17**, 131, 1969.
77. O. G. Raabe and M. E. Wrenn, "Analysis of the Activity of Radon Daughter Samples by Weighted Least Squares," *Health Phys.*, **17**, 593, 1969.

78. K. D. Cliff, "The Measurement of Low Concentrations of Radon-222 Daughters in Air, with Emphasis on RaA Assessment," *Phys. Med. Biol.*, **23**, 55, 1978.
79. G. D. Kerr, "Measurement of Radon Progeny Concentrations in Air by Alpha-Particle Spectrometry," Oak Ridge National Laboratory Report ORNL-TM-4924, 1975.
80. N. Jonassen and E. I. Hayes, "The Measurement of Low Concentrations of the Short-Lived Radon-222 Daughters in the Air by Alpha Spectroscopy," *Health Phys.*, **26**, 104, 1974.
81. W. W. Nazaroff, Lawrence Berkeley Laboratory, private communication, 1980.
82. R. J. Tremblay, A. Leclare, C. Mathieu, R. Pepin, and M. G. Townsend, "Measurement of Radon Progeny Concentration in Air by α-Particle Spectrometric Counting During and After Air Sampling," *Health Phys.*, **36**, 401, 1979.
83. D. J. Keefe, W. P. McDowell, P. G. Groer, and R. T. Witek, "Optimizing Measurement Sensitivity to Facilitate Monitoring Environmental Levels of Rn-Daughter Concentrations," *IEEE Trans. Nucl. Sci.*, **NS-25**, 787, 1978.
84. M. Irfan and A. J. Fagan, "Measurement of Radon Daughters in Air Using Gamma Spectrometry," *Nucl. Instrum. Meth.*, **166**, 567, 1979.
85. C. Pomroy and M. Measures, "External Gamma Counting and Bioassay Measurements for ^{210}Pb on Retired Uranium Miners," *Health Phys.*, **37**, 409, 1979.
86. C. W. Sill and C. P. Willis, "Radiochemical Determination of Lead-210 in Uranium Ores and Air Dusts," *Anal. Chem.*, **49**, 302, 1977.
87. J. C. Rodgers, "Radon Progeny Exposure Measurement Using Lead-210," U.S. Department of Health, Education and Welfare Publication No. NIOSH 75-116, 1974.
88. M. J. Duggan and D. M. Howell, "The Measurement of the Unattached Fraction of Airborne RaA," *Health Phys.*, **17**, 423, 1969.
89. N. Fusamura and R. Kurosawa, "Determination of f-Value in Uranium Mine Air," in *Assessment of Airborne Radioactivity,* International Atomic Energy Agency, Vienna, 1967, p. 213.
90. T. T. Mercer and W. A. Stowe, "Deposition of Unattached Radon Decay Products in an Impactor Stage," *Health Phys.*, **17**, 259, 1969.
91. A. C. George and L. Hinchliffe, "Measurements of Uncombined Radon Daughters in Uranium Mines," *Health Phys.*, **23**, 791, 1972.
92. A. C. George, "Measurement of the Uncombined Fraction of Radon Daughters with Wire Screens," *Health Phys.*, **23**, 390, 1972.
93. A. C. George, "Indoor and Outdoor Measurements of Natural Radon and Radon Daughter Decay Products in New York City Air," in *Proceedings of the Second International Symposium on Natural Radiation Environment,* Houston, Texas, U.S. AEC Report CONF-720805, 1972.
94. M. Raghavayya and J. H. Jones, "A Wire Screen–Filter Paper Combination for the Measurement of Fractions of Unattached Radon Daughters in Uranium Mines," *Health Phys.*, **26**, 417, 1974.
95. D. Sinclair and G. S. Hoopes, "A Novel Form of Diffusion Battery," *AIHA J.*, **36**, 39, 1975.
96. V. Prodi, "Size Characterization of Alpha-Emitting Aerosols by Means of an Inertial Spectrometer and Track Etching Detectors," presented at *Specialist Meeting on the Assessment of Radon and Daughter Exposure and Related Biological Effects,* Rome, Italy, March 3–7, 1980, in press.

97. O. White, Jr., "An Evaluation of Six Radon Dosimeters," Health and Safety Laboratory Report HASL TM 69-23 A, 1969.
98. O. White, Jr., "Evaluation of MIT and ORNL Radon Daughter Dosimeters," Health and Safety Laboratory Report HASL TM 70-3, 1970.
99. O. White, Jr., "Evaluation of MOD and ORNL Radon Daughter Dosimeters," Health and Safety Laboratory Report HASL TM 71-17, 1971.
100. E. L. Geiger, "Radon Film Badge," *Health Phys.*, **13**, 407, 1967.
101. C. Costa-Ribeiro, J. Thomas, R. J. Drew, M. E. Wrenn, and M. Eisenbud, "A Radon Detector Suitable for Personnel or Area Monitoring," *Health Phys.*, **17**, 193, 1969.
102. J. A. Auxier, K. Becker, E. M. Robinson, D. R. Johnson, R. H. Boyett, and C. H. Abner, "A New Radon Progeny Personnel Dosimeter," *Health Phys.*, **21**, 126, 1971.
103. R. D. Evans and R. J. Kolenkow, "Radium and Mesothorium Poisoning and Dosimetry and Instrumentation Techniques in Applied Technology," Massachusetts Institute of Technology Report MIT-952-5, Part I, 1968.
104. D. E. McCurdy, K. J. Schiager, and E. D. Flack, "Thermoluminescent Dosimetry for Personnel Monitoring of Uranium Miners," *Health Phys.*, **17**, 415, 1969.
105. D. B. Lovett, "Track Etch Detectors for Alpha Exposure Estimation," *Health Phys.*, **16**, 623, 1969.
106. R. L. Rock, D. B. Lovett, and S. C. Nelson, "Radon-Daughter Exposure Measurement with Track-Etch Films," *Health Phys.*, **16**, 617, 1969.
107. A. J. Breslin, S. F. Guggenheim, A. C. George, and R. T. Graveson, "A Working Level Dosimeter for Mines," Environmental Measurements Laboratory Report EML-333, 1977.
108. J. Durkin, "Electronic Radon Daughter Dosimetry," *Health Phys.*, **37**, 757, 1979.
109. R. B. Gammage, G. D. Kerr, and L. Huskey, "Exploratory Study of the Use of TSEE Dosimeters in Radon Monitoring," *Health Phys.*, **30**, 145, 1976.
110. A. L. Frank, E. V. Benton, and R. Casson, "Radon and Daughter Measurements with Active and Passive Devices Using Track-Etch Detectors," p. 6 of Ref. 26.
111. R. L. Fleischer, H. W. Altes, S. C. Furnan, P. B. Price, and R. M. Walker, "Particle Track Etching," *Science*, **178**, 255, 1972.
112. W. J. Ward, R. L. Fleischer, and A. Mogro-Campero, "Barrier Technique for Separate Measurement of Radon Isotopes," *Rev. Sci. Instrum.*, **48**, 1440, 1977.

15 URANIUM AND THORIUM

Uranium is familiar to most of the public. Its use in nuclear weapons brought it to public attention nearly four decades ago. Now it is the fuel for an important energy source, the nuclear power reactor. Thorium (Th), which has similar energy possibilities, is detected by similar methods, and occurs naturally in similar contexts, is also discussed in this chapter.

Methods for measuring concentrations of U and Th in the environment are emphasized, including methods for determining low levels in air, water, and certain other environmental media. Bioassay techniques are not included. The discussion reviews the physical characteristics of important U and Th isotopes, relevant radiation protection guides, ranges of concentrations in the natural environment, and various measurement techniques. Additional information on the measurement techniques treated in this chapter is found in Chapter 14, Radon, in which the ^{238}U decay chain and the instrumentation for ^{222}Rn measurements are discussed; other chapters that contain information related to U and Th detection are Chapter 6 (Gamma and X-Ray Spectrometry), Chapter 7 (Measurement of Alpha and Beta Particles), and Chapter 13 (Radium).

15.1. PHYSICAL CHARACTERISTICS

Uranium occurs naturally in three isotopic states, ^{234}U, ^{235}U, and ^{238}U. The two heavier isotopes originate in primordial material, and each is the parent of a decay chain. Uranium-234 occurs in the ^{238}U decay chain. The ^{235}U:^{238}U ratio is the same essentially everywhere on earth: 0.72% by mass. The daughter ^{234}U occurs at an abundance of 0.0057%, when it is in secular equilibrium with its precursor ^{238}U.

An additional uranium isotope of potential importance, ^{233}U, is produced in nuclear reactors by ^{232}Th neutron capture. Reactors also produce a small, but radiologically significant, amount of ^{232}Th, and other U and Th isotopes of less consequence.

Some physical properties of the more important uranium isotopes are shown in Table 15.1 [1]. With the exception of ^{237}U, these isotopes are α-particle emitters, and γ-rays accompany some of the α-particle decays. The half-lives of the three naturally occurring isotopes are all quite long. Fission of ^{235}U (^{233}U also fissions) is responsible for the unique role of U in nuclear reactors and weapons. However, the ability to fission has very little direct consequence as a radiological hazard, but it does find occasional use in detection of U.

The isotopes of Th with atomic weights from 223 to 234, are almost all α-particle emitters. Thorium-232 (with a half-life of 1.4×10^{10} yr) accounts for approximately 100% of naturally occurring Th and is the parent of a long decay chain ending at ^{208}Pb. Thorium-231 and ^{227}Th also occur naturally as a result of their production in the ^{235}U decay chain and ^{230}Th is produced in the ^{238}U decay chain. The historic name *thoron* is often applied to ^{220}Rn (a gas), which is *not* an isotope of Th, although it is an α-particle emitting daughter in the ^{232}Th chain.

15.2. RADIATION PROTECTION GUIDES

Potential harm from uranium ingestion depends on whether the U is soluble or insoluble. These differences have been taken into account in the establishment of radiation protection guides. The guides for natural U (in the isotopic mixture indicated in Table 15.1) and Th are discussed here.

Table 15.2, from the 10 CFR 20, presents the maximum permissible concentrations (MPC) of U and Th in air and water for workers and members of the general public. The occupational limits assume a 40-hr work week, and the general public limits are for areas to which all individuals have access. Limits for exposure of populations would normally be a factor of 3 less than the individual limits. The critical organs and maximum permissible body burdens indicated are from the International Commission on Radiological Protection (ICRP) Publication No. 2 [2].

The comparison of U and Th guides with those for ^{226}Ra in Table 15.2 is especially revealing. The MPC_{water} is much less restrictive for soluble U and Th than for Ra, while the values for MPC_{air} are comparable. Uranium, Th, and ^{226}Ra are not a problem in airborne suspension, except in rare circumstances involving dust storms. Uranium and Th are rarely a public health problem in drinking water; Ra occasionally is found in natural concentrations which exceed its much smaller MPC_{water}.

Table 15.1. Some Properties of Uranium and Thorium Isotopes (data presented in Lederer et al. Ref. 1)

Isotope	Half-Life	Natural Abundance	Member of Which Decay Chain	Immediate Daughter	Main Decay Products and Energies
U-238	4.5×10^{9} yr	99.27%	U-238 chain	Th-234 (24.1 days)	(77%) 4.20 MeV α (23%) 4.15 MeV α + 49 keV γ
U-237	6.75 days			Np-237 (2.14×10^{6} yr)	(95%) 0.245 MeV β
U-236	2.34×10^{7} yr			Th-232 (1.4×10^{10} yr)	(74%) 4.49 MeV α (26%) 4.44 MeV α + 49 keV γ
U-235	7.0×10^{8} yr	0.72%	U-235 chain	Th-231 (25.5 hr)	(57%) 4.40 MeV α + 205 and 163 keV γ (18%) 4.37 MeV α + 195 keV γ (25%) others
U-234	2.45×10^{5} yr	0.0054%	U-238 chain	Th-230 (8.0×10^{4} yr)	(72%) 4.77 MeV α (28%) 4.72 MeV α + 53 keV γ
U-233	1.59×10^{5} yr			Th-229 (7.3×10^{3} yr)	(84%) 4.82 MeV α (13%) 4.78 MeV α + 42 keV γ
U-232	72 yr			Th-228 (1.9 yr)	(69%) 5.32 MeV α (31%) 5.26 MeV α + 58 keV γ
Th-234	24 days		U-238 chain	Pa-234 (7 hr)	(72%) 0.198 MeV β (21%) 0.104 MeV β

Table 15.1. (Continued)

Isotope	Half-Life	Natural Abundance	Member of Which Decay Chain	Immediate Daughter	Main Decay Products and Energies
Th-232	1.41×10^{10} yr	100%	Th-232 chain	Ra-228 (5.76 yr)	(77%) 4.02 MeV α (23%) 3.96 MeV α + 59 keV γ
Th-231	26 hr		U-235 chain	Pa-231 (3.3×10^4 yr)	(52%) 0.302 MeV β (20%) 0.218 MeV β (22%) 0.138 MeV β
Th-230	8.0×10^4 yr		U-238 chain	Ra-226 (1.6×10^3 yr)	(76%) 4.69 MeV α (23%) 4.62 MeV α + 68 keV γ
Th-229	7.3×10^3 yr			Ra-225 (14.8 days)	(11%) 4.90 MeV α + misc. γ (56%) 4.85 MeV α + misc. γ
Th-228	1.91 yr		Th-232 chain	Ra-224 (3.66 days)	(73%) 5.42 MeV α (27%) 5.34 MeV α + 84 keV γ
Th-227	19 days		U-235 chain	Ra-223 (11.4 days)	(24%) 6.04 MeV α (23%) 5.98 MeV α + 32 and 62 keV γ (20%) 5.76 MeV α + misc. γ

Table 15.2. Maximum Permissible Concentrations in Air and Water for Uranium and Thorium

	MPC[a] for: Workers/General Public		MPBB[b]
Isotope	Air (10^{-12} Ci/liter)	Water (10^{-9} Ci/liter)	Occupational (10^{-6} Ci)
Natural uranium			
soluble	0.1/0.005	1000/30	5000
insoluble	0.1/0.005	1000/30	
Natural thorium			
soluble	0.06/0.002	60/2	
insoluble	0.06/0.002	600/20	
Radium-226			
soluble	0.03/0.003	0.4/0.03	100
insoluble	0.05/0.002	900/30	

[a]Code of Federal Regulations, 10 CFR 20, Appendix B. The MPC for workers assumes a 40-hr work week. The critical organ for soluble uranium in air and water is the kidney and GI tract, respectively; for insoluble uranium in air or water, it is the lung and GI tract, respectively [2].
[b]Reference 2.

15.3. SOURCES AND LEVELS OF ENVIRONMENTAL URANIUM

Uranium occurs widely as a trace constituent of soils and rock. The isotopic ^{235}U:^{238}U ratio of 0.72% is constant almost everywhere; an interesting exception is the natural Oklo reactor in Africa, where fissile uranium concentrations were great enough a billion years ago to sustain a chain reaction [3]. The presence of ^{238}U is manifest by the universal emanation of ^{222}Rn from rocks and soils (see Chapter 14).

The uranium content of rock is usually given in parts per million (ppm), by weight. One microgram of pure ^{238}U has an activity of 0.33 pCi (0.012 Bq). Hence one ppm of pure ^{238}U corresponds to 0.33 pCi/g (0.012 Bq/g); however, if ^{234}U is in equilibrium, then it adds another 0.33 pCi/g (0.012 Bq/g), as does every other of the ^{238}U chain.

In most rocks, uranium content ranges from 0.5 to 4 ppm, with ^{232}Th about 4 to 5 times more abundant [4]. True granite is the most uranium-rich of the common rocks, having typically 5 ppm of U. In uranium mining areas, concentrations of ~30 ppm are sometimes found [5]. Volcanic rocks have U in concentrations of about 0.1 to 0.4 ppm [6], and Atlantic beach sands about 0.1 to 1.5 ppm [7]. River waters in Japan [8] have uranium concentrations of a fraction of 1 μg/liter (~0.33 pCi/liter or 0.012 Bq/liter). The National Council on Radiation Protection and Measurements (NCRP) Report No. 45 [9] summarizes U and Th concentrations in major rocks and soils. Higher concentrations occur, even in nonuranium mining regions: for example, uranium concentrations from 8 to 36 pCi/g (0.296 to 1.332 Bq/g) have been found in Florida phosphate materials [10], corresponding to mass concentrations of 20 to 100 ppm. Costa-Ribeiro et al. [11] discuss thorium content of a monazite ore body.

In material that has remained totally undisturbed, ^{234}U will be in equilibrium (equal radioactivity) with its precursor, ^{238}U [12]. However, this is not always the case, perhaps because ^{234}U seems to exist preferentially in the hexavalent rather than the quadrivalent state [13], at least in altered materials. Nonequilibrium has been reported at about the -15% level in the upper several meters of a rocky granitic outcropping [14]. Several investigators have found enhanced ^{234}U in deep underground waters, in some cases with more than 7:1 for the ^{234}U:^{238}U activity ratio [15]. Soils, conversely, have been found to be slightly depleted of ^{234}U on the average. Enhancements have also been observed in normal fecal samples, probably because of enrichment of ground water being ingested [13].

Studies have been made of the uranium content of typical diets. In the United Kingdom [16], the concentrations in nearly all foods were found to range from 0.0002 to 0.001 ppm. The intake of U in food was found to be about 1.0 μg/day, which is similar to the value of 1.3 to 1.4 μg/day found by Welford and Baird for the United States [17]. These values correspond to daily intakes of about 1 pCi/liter (0.037 Bq/liter), very small compared with the amount permitted in a liter of drinking water (30.0 nCi or 1.1 kBq) [9]. Moreover, domestic drinking water in the United States contains U nearly always at levels well below 0.1 μg/liter [17], but some drinking supplies have concentrations as high as 5 to 8 μg/liter. These levels correspond to activity concentrations on the order of 1 pCi/liter (0.037 Bq/liter), 4 orders of magnitude below the MPC_{water}.

The NCRP Report No. 45 [9] summarizes airborne uranium concentrations at Argonne National Laboratory (Chicago) [18], New York City [19], and New York State [20], found to be on the order of 0.1 aCi/liter (3.7 nBq/liter). Values an order of magnitude higher have been found downwind of a coal-fired power plant (see Ref. 9, p. 78). Slightly smaller concentrations than these have been found for Th in air [9]. With the exception of power-plant effluents, the observed U and Th appear to be in airborne soil. In any case, all of the observed airborne concentrations are orders of magnitude lower than the general public concentration limit of 5 fCi/liter (0.18 mBq/liter) and 3 fCi/liter (0.11 mBq/liter), respectively, for U and Th.

Uranium levels are typically not much affected by man's activities. The most important exceptions are in and around uranium mining and milling facilities and potentially at nuclear fuel reprocessing plants (particularly if nuclear power plants were ever to depend significantly on ^{233}U). Even in these cases, the main radiological and environmental effects are seldom from uranium itself. Discussions of the radiological problems associated with such facilities are found in Chapters 10–12.

From the data just presented, it can be seen that U is *not* often a radiologically important nuclide. This does not imply that uranium measurements are unimportant: around uranium mills where U in air and/or water may be significant, environmental measurements are required to monitor any possible long-term buildup trends in the environment. Also, the radiological significance of Ra, Rn, and the radon daughters resulting from U decay is widespread.

15.4. MEASUREMENT TECHNIQUES

Among the environmental media requiring measurement of uranium activity are rocks, silt and soil, natural and waste waters, food, and air (or air filters). Because the uranium concentration in most samples is small, and because the decay α-

particles have such short ranges in matter ($\simeq$5 to 10 mg/cm^2), direct α-particle counting of environmental samples is seldom practical.

The techniques which have been developed fall into several categories, each discussed separately in the following sections. These categories are:

1. Gamma spectroscopy.
2. Neutron-induced methods.
3. Radiochemical analysis followed by α-particle counting or spectrometry.
4. Track-etch methods.
5. Fluorometric determinations.
6. Other methods.

The number and variety of uranium determination methods is so large that to survey them all would be impractical. Many of these methods arise largely because of uranium exploration efforts and in connection with fuel analyses in the nuclear industry. This discussion mentions briefly those techniques which are found in the procedures manuals of some of the larger analytical laboratories, or which are representative of a large class of techniques. Measurements of the naturally occurring isotopes are emphasized; many of these measurements may also be applied to other α-particle emitting isotopes. Because of the similarity of the ^{232}U half-life to that of ^{239}Pu, direct α-particle detection techniques for ^{233}U are similar to those described in Chapter 16, Plutonium.

15.4.1. Gamma Spectroscopy

Table 15.1 indicates that several of the natural uranium α-particle decays are accompanied by γ-ray emission. Also, when ^{238}U or ^{235}U is in equilibrium with its daughters in the long decay chains, several γ-ray emitting nuclides can be detected. The unique signature of a chain can increase sensitivity in the presence of higher backgrounds, compared to the measurement of only a single γ-ray.

A detailed discussion of γ-ray spectroscopic measurements is presented in Chapter 6. It suffices to mention here that the two main techniques are thallium-activated sodium iodide (NaI(Tl) scintillation systems and lithium-drifted germanium Ge(Li) semiconductor detector systems, the latter with significantly better energy resolution but poorer counting efficiency than the former. Better energy resolution reduces the effective background.

Of course, the farther down the chain a nuclide appears, the greater the chance for disequilibrium with the chain head. Measurements of ^{222}Rn daughters as indicators of ^{238}U are especially susceptible to difficulty, because radon is a noble gas that can relatively easily move from its site of formation, increasing the probability of disequilibrium.

The most prominent γ-ray in the ^{238}U chain is the 1.76-MeV γ-ray from ^{214}Bi, a daughter of radon. A study of the ^{238}U chain reveals that among the nuclides prior to ^{226}Ra, there are no γ-rays which combine a significant branching ratio ($> 1\%$) and an energy above 100 keV. Only ^{214}Pb and ^{214}Bi have γ-rays with energies above 300 keV. The ^{226}Ra γ-ray has an energy of 185.7 keV. The ^{235}U chain has several more prominent γ-rays, but their relative intensity is limited by the ^{235}U:^{238}U isotopic ratio of 0.72%. Indeed, ^{235}U itself has a 185-keV γ-ray (54%), but it is very difficult to resolve in the presence of the ^{226}Ra γ-ray at 185.7 keV.

When NaI(Tl) is used, the γ-ray emission most commonly analyzed is the 1.76-MeV γ-ray of ^{214}Bi because its high energy makes it stand out from nearly all of the other naturally occurring γ-ray lines. Similarly, the 2.62-MeV γ-ray from ^{208}Tl is commonly used as a signature for the ^{232}Th chain. Adams and Fryer [21] describe a portable NaI(Tl) spectrometer used for field surveys of the natural U, Th, and ^{40}K content in soils and rocks; accuracies in the few-percent range are obtained for uranium concentrations in the range of several ppm.

Wollenberg and Smith [22, 23] have also applied NaI(Tl) techniques to geological assays. They point out the possible difficulties with the loss of equilibrium in the ^{238}U series:

> Uranium can be leached from its host material, there are long half-lived decay members, and the gaseous member [radon] can be lost to the air—three conditions which may lead to lack of equilibrium. Most of the uranium-series γ-rays come from daughters of radium; therefore, some . . . assays may really reflect only radium content. This situation would seem most likely to occur with soil samples. [22]

Smith has indicated [24] that lack of equilibrium is often not a serious problem in solid rocky material because leaching and radon emanation may be unimportant. Sill [12] gives typical equilibrium ratios for ores.

Of course, Ge(Li) detectors, with their vastly superior resolution, are able to distinguish γ-ray lines which could never be resolved by a NaI(Tl) system. Figures 15.1 and 15.2 [25] compare NaI(Tl) and Ge(Li) spectra measured on the same

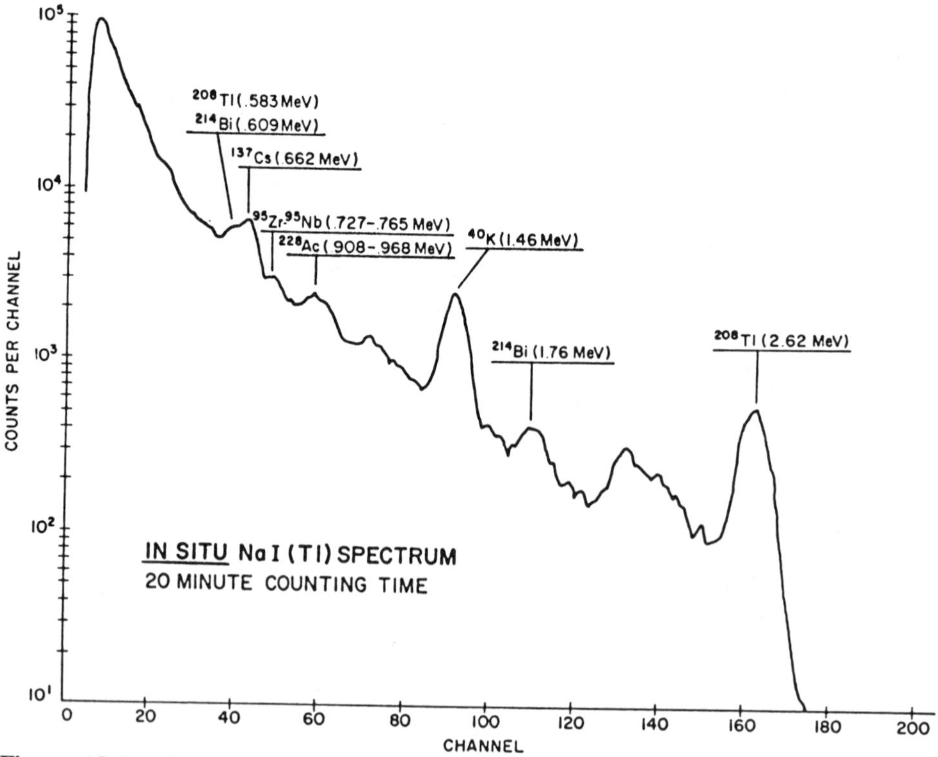

Figure 15.1. *In situ* spectrum, northeastern United States location, taken in 1971 with 10-cm × 10-cm NaI(Tl) crystal, 20-min counting time (reproduced from Beck, Lowder, and McLaughlin, Ref. 25).

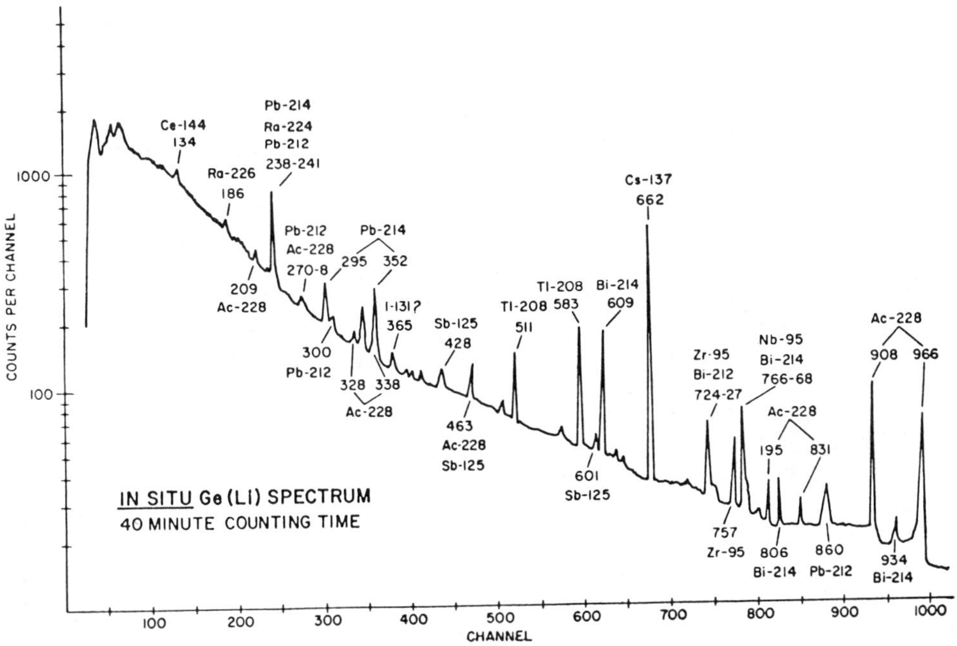

Figure 15.2. *In situ* spectrum, taken at the same location as in Figure 15.1 with 60-cm³ Ge(Li) detector, 40-min counting time. Photon energies in keV (reproduced from Beck, Lowder, and McLaughlin, Ref. 25).

sample. Note that the Ge(Li) spectrum, of which only the region below 1 MeV is shown, reveals fine structure not even suggested in the NaI(Tl) spectrum. Nevertheless, NaI(Tl) instruments have competed reasonably with Ge(Li) systems for approximate field surveys of U, especially at the lowest levels of sensitivity. There are several reasons for this: the ^{214}Bi line stands out prominently even with the poorer resolution of NaI(Tl); NaI(Tl) has greater detection efficiency; and the lines which might be studied with Ge(Li) are mainly at low energy, where backgrounds are higher, especially in the field. Smith and Wollenberg [26] discuss high-resolution spectrometry of the uranium and thorium decay series.

Both NaI(Tl) and Ge(Li) systems have sensitivities in the range of a fraction of one ppm, more than adequate for most uranium assay studies. An intrinsic advantage of γ-spectroscopic methods is the ability to measure large (many-gram) samples nondestructively. Adams and Gasparini [27] have written a good overview of the entire field of γ-spectral studies of rocks.

Gamma spectrometers may also be used for measurement of uranium lung burden by detection of the 185-keV ^{235}U γ-ray. See for example Ref. 28.

15.4.2. Neutron-Induced Methods

The fact that ^{235}U (and ^{233}U) will fission in the presence of thermal neutrons is exploited in nuclear reactor technology (uranium-238 requires fast neutrons). This is also the basis for several measurement techniques for extremely low concentrations of ^{235}U in various media. The fact that neutrons can be captured, particularly by ^{238}U and ^{232}Th, is also important for reactor operation. The photons emitted following capture may also be used as a basis for detection.

When a heavy nuclide ($Z \simeq 90$) undergoes fission, a small fraction of the fission products are neutron-rich and subsequently decay by neutron emission. Delayed neutron emission is unique to the fission process; there is hardly any other process that produces neutron emission. The only known exceptions are 4.2-sec ^{17}N and 0.17-sec ^{9}Li, neither of which causes a background in uranium measurements of this kind. Measurement of the delayed neutrons can be used as an indicator of the concentration of ^{235}U (or other fissionable radionuclides).

Amiel [29], irradiating samples in a slow-neutron flux of 4×10^{12} n/cm^2 · sec, reports sensitivities to less than 0.1 μg of U. With a (maximum) sample size of 200 g, this sensitivity approaches 1 ppb. For Th this method is about one order of magnitude less sensitive. The system, while expensive, has practically no interferences. Hamilton has reported an even lower detection limit, about 0.01 μg U with $\pm 10\%$ precision [19].

Hamilton has also reported a technique of much greater sensitivity [19]: uranium-laden dust (on an air filter or in solution) is placed on a polycarbonate surface which is irradiated for 1 hr in a neutron flux of 5×10^{11} n/cm^2 · sec. After storage for 24 hr to allow for the decay of short-lived activity, the polycarbonate is etched in 6N NaOH and examined under a conventional ×450 microscope. The total number of tracks is simply counted, each track being the record of a fission event in the detector. Sensitivities in the range of 1×10^{-14} g U are reported. Hamilton shows that this method is practically unique for ^{235}U, except in circumstances where very large amounts of ^{239}Pu are present. Carpenter and Cheek [30] have applied this technique to biological samples. A related method for U in water is described by McCorkell and Huang [31], who designed a discharge counter to replace visual methods. Various scanning techniques have been formulated for these track-etch techniques.

Gladney and Rook [32] describe a method for uranium based on combustion of the sample and gas-phase separation of the radioiodine resulting from fission.

Weaver [33] describes a neutron activation method, for determination of U in environmental samples, that uses a Ge(Li) detector to measure x-rays and low-energy γ-rays from ^{133}I, a uranium fission product. This method has a lower detection limit of about 25 ppb.

An alternative is to detect x-rays from neutron capture products. Mantel and Amiel [34] use a lithium-drifted silicon, Si(Li), spectrometer to measure L x-rays of neptunium and protactinium which are neutron capture products of U and Th, respectively. They find a limit of detection of 10^{-8} gm for U and 10^{-7} gm for Th in samples of 0.01 to 0.1 gm, for the neutron flux and counting geometry employed.

15.4.3. Radiochemical Analysis Followed by Alpha-Particle Counting or Spectrometry

Chemical analysis techniques are a mainstay for determining of extremely low levels of activity. The chemical separation of U or Th removes the background matrix in which the sample is found, and makes possible α-particle counting or α-spectroscopic determinations of great sensitivity.

The difficulty with α-particle counting and spectroscopy is that the short range of α-particles (5 to 10 mg/cm^2) requires the preparation of a nearly weightless sample.

Thus, the major disadvantages of these techniques are the difficulty and inconvenience of the analytical procedures, and the liability for error including variable yields. If properly exchanged, tracers (such as ^{232}U) may be used to monitor yields (see Chapter 7).

Electrodeposition is commonly used for preparing an α-particle emitting sample for counting. Care has to be taken when using electrodeposition, because yields can vary, particularly when used with field samples of variable composition. Only a few of the many electrodeposition techniques will be discussed here. In one U.S. Public Health Service (PHS) [35] procedure, uranium nitrate, dissolved in dilute ammonium oxalate, is electrodeposited on a nickel-plated copper disk. Iron, commonly present with U, does not interfere. Another PHS technique, by Levine and Lamanna [36], is based on solvent extraction of U with ethyl acetate, using aluminum nitrate as a salting agent. This procedure is simple, but less sensitive than the other. Talvitie [37] describes still another method for electrodeposition of Pu, Th, U, and Am (americium) as hydrous ions. The various activities are separated with an α-spectrometer with a claimed detection limit of about 0.02 pCi (0.74 mBq) of activity.

Another class of chemical techniques is coprecipitation. This technique, widely used for radium determination (see Chapter 13), can be used for simultaneous determinations of several α-particle activities when followed by α-particle spectroscopy. For example, Lieberman and Moghissi [38] describe a technique suitable for environmental samples including ashed food, vegetation, and soil. Uranium is coprecipitated with lanthanum fluoride and then measured using an α-spectrometer with a gas-flow proportional counter. Samples containing a few 10^{-15} Ci/liter of U and Th have been analyzed quantitatively [38]. Coprecipitation can be simpler and less expensive than electrodeposition, but more liable to error either from interferences or from incomplete dissolution of the ashed sample prior to the precipitation. An oxidation-reduction procedure [39] can help to avoid incomplete dissolution, but is tricky.

Sill [40] describes a coprecipitation method in which all of the elements from Ra through Cf (californium), except U, are precipitated by $BaSO_4$. Uranium must all be reduced to the quadrivalent state. Alpha spectroscopy can be used to distinguish among the several α-particle emitters. Alternatively, "because of their markedly different oxidation potentials, uranium and the first four transuranium elements can be separated efficiently from each other by proper choice of oxidant" [40]. The α-particles in the separated fraction can either be counted directly, or electrodeposited for precise α-spectroscopy. Preparation of actinide samples for the α-spectrometer without electrodeposition is described in Ref. 25.

Researchers at Idaho Falls have described methods for radium and thorium isotopes [41] and for long-lived ^{238}U series α-particle emitters [12] in ores, mill tailings, soils, and air dusts. Sill also describes a method for thorium and uranium isotopes in ores and mill tailings [42].

The Environmental Measurements Laboratory (EML) [43] recommends several radiochemical methods for uranium analyses. In one (for enriched uranium in urine), U is collected by alkaline phosphate precipitation and isolated by ion exchange, then electroplated, and its α-particles are counted. In another, useful for biological and environmental materials, U is acid leached, then separated by anion exchange chromatography and mercury cathode electrolysis. After electrodeposition, the samples are measured and α-spectrometry follows.

A radiochemical thorium determination in urine, water, and air filters is also described in the EML *Procedures Manual*. The sample is dissolved in sulphuric acid, and the Th extracted with a high-molecular-weight primary amine after iron, vanadium, and uranium are extracted. Thorium is back-extracted with hydrochloric acid, evaporated, ashed at 600°C, wet-ashed with nitric acid, and dissolved in diluted nitric acid. A sample evaporated on a planchet is counted (α- and β-ray) in a gas-flow proportional counter. A second sample may be used for spectrophotometric determination of total Th. Thorium recovery is high (96%).

Another thorium procedure is described by the Environmental Protection Agency (EPA) [44]. Thorium is precipitated with ferric hydroxide after being dissolved in nitric acid. It is then separated from iron, titanium, and zirconium by precipitation with a rare earth fluoride. It can then be counted for α-particles.

For some cases, liquid scintillator α-spectrometry may be employed for uranium determination. A method for uranium assay of phosphatic materials [45] utilizes solvent extraction followed by uranium stripping and reextraction into a liquid scintillator.

15.4.4. Track-Etch Methods

A general technique for the detection of heavily ionizing particles uses solid materials, such as polycarbonate film, which will record the passage of a particle. The material is thin-etched for visual or instrumental scanning. Such methods were noted previously for fission fragment track detection and are discussed in general by Fleischer, Price, and Walker [46]. A substantial effort has been expended by these authors and their collaborators in developing the basic idea of α-particle track detection into what is now a commercially practicable method for uranium exploration. This method utilizes small pieces of plastic, placed in a cup (often covered with a permeable membrane [47]), to detect α-particle decay of Rn and its daughters. Radon is able to migrate through the earth's crust and indicate the presence of subsurface U in situations where direct detection methods cannot be applied easily. The track-etch technique is not often used for monitoring of U at ordinary levels in environmental media. It is, however, used for environmental radon monitoring in air (see Chapter 14).

15.4.5. Fluorometric Determinations

The use of fluorometry is one of the oldest techniques for uranium determination. Only a few of the methods in common use are described here. It should be noted that a laser is now often used for excitation.

The EPA's Las Vegas laboratory [44] describes a fluorometry technique in which, after total dissolution, a uranium/uranium fluoride complex (which fluoresces under ultraviolet light) is formed. The principal interference is from iron, which unfortunately accompanies U in many sample media and which must be eliminated by careful chemical procedures.

The EML *Procedures Manual* [43] describes a method in which U is fused with sodium fluoride for ultraviolet fluorescence. This method, developed for urinalysis, is equally applicable to most other dissolved media. A sample-preparation method for bone, soil, food, and tissue is also described [43]. Uranium is leached from a dry-ashed residue and isolated by anion exchange chromatography plus mercury cathode electrolysis before fusion with sodium fluoride.

Rushing, Garcia, and Clark [48] present a fluorometric method similar to that discussed in the EML *Procedures Manual*. Their description goes into detail about some of the important problems with the technique. For example,

> ...in order to obtain the maximum fluorescence..., it is necessary that the uranium exist in the hexavalent condition in the final melt. While the fusion process will oxidize the uranium, it is best that reducing substances, such as carbon, be oxidized before the fusion step is undertaken. [48]

For thorium determination, a fluorometric procedure using morin [49] is available.

One potential drawback with fluorometric methods is that a standard radiochemical laboratory might not have the UV fluorometric equipment available. The equipment is, however, available commercially. Another problem, already mentioned, is interference from iron, especially in iron-rich media (not, however, a problem in the procedure described by King, Scott, and Disney [28]). Centanni, Ross, and DeSesa [50] have discussed some of the detailed considerations required for accurate fluorometric measurements, and Akaishi and Yabe [51] have suggested some precautions to take.

15.4.6. Other Methods

There are numerous other methods for uranium determination, which are not discussed here. Among these are polarography [52], spectrophotometry [53], coulometry [54,55], and x-ray fluorescence [56]. The book by Jeffrey [57] is a general reference on rock chemical analysis. Refer to the appropriate references for details.

15.5. SUMMARY AND CONCLUSIONS

This chapter has briefly discussed the typical levels of U in environmental media, the radiation protection guides, and some of the techniques that have been developed to measure uranium in the environment. Although the activity levels requiring measurement are small, the radiological hazard from U in the environment is also nearly always small.

The techniques for uranium measurement are varied, and among them they offer sufficient sensitivity to perform nearly all of the required measurements. Each class of methods has significant disadvantages, and the choice of method should undoubtedly be based as much on the judgment and skill of the analyst and the facilities available to him as on the intrinsic merit of the particular method. However, some general comments can be made in summary:

1. Gamma-spectroscopic methods are perhaps the easiest and cleanest of the techniques; they do however, rely on equilibrium in the long decay chains, which is frequently not the case.
2. Neutron-induced fission techniques offer high selectivity and sensitivity, but are unfortunately not accessible to many analytical laboratories.
3. Alpha spectrometry after radiochemical separation allows the simultaneous determination of several α-particle emitting radionuclides, which is advanta-

geous in many measurement situations. The analyst must take care to know chemical yields in the presence of possible unknown interferences.

4. Chemical analysis techniques followed by fluorometry, coulometry, or spectrophotometry require more analytical care than the other techniques, but otherwise seem to be acceptable. The existence of multivalent uranium compounds in a complex medium such as soil requires special consideration.

In conclusion, U in environmental samples can now be measured with any of several methods, and probably does not require extensive new research and development in the next few years. This is especially true given the small radiological importance of U and also the several other more pressing problems in the area of environmental radiation measurements.

REFERENCES

1. C. M. Lederer and V. S. Shirley, Eds., *Table of Isotopes,* 7th ed., Wiley, New York, 1978.
2. International Commission on Radiological Protection, "Permissible Dose for Internal Radiation," ICRP Publication No. 2, Pergamon Press, New York, 1959.
3. G. A. Cowan, "A Natural Fission Reactor," *Sci. Am.*, **235**, 36, 1976.
4. J. A. S. Adams and W. M. Lowder, Eds., *The Natural Radiation Environment*, University of Chicago Press, Chicago, 1964.
5. G. Phair and D. Gottfried, "The Colorado Front Range, Colorado, U.S.A., as a Uranium and Thorium Province," p. 7 of Ref. 4.
6. K. S. Heier and J. L. Carter, "Uranium, Thorium, and Potassium Contents in Basic Rocks and Their Bearing on the Nature of the Upper Mantle," p. 63 of Ref. 4.
7. A. Mahdavi, "The Thorium, Uranium, and Potassium Contents of Atlantic and Gulf Coast Beach Sands," p. 87 of Ref. 4.
8. Y. Miyake, Y. Sugimura, and H. Tsubota, "Content of Uranium, Radium, and Thorium in River Waters in Japan," p. 219 of Ref. 4.
9. National Council on Radiation Protection and Measurements, "Natural Background Radiation in the United States," NCRP Report No. 45, 1975.
10. C. E. Roessler, Z. A. Smith, U. E. Bloch, and R. J. Prince, "Uranium and Radium-226 in Florida Phosphate Materials," *Health Phys.*, **37**, 269, 1979.
11. C. Costa-Ribeiro, M. A. Barcinski, N. Figueiredo, E. Penna Franca, N. Lobão, and H. Krieger, "Radiobiological Aspects and Radiation Levels Associated with the Milling of Monazite Sand," *Health Phys.*, **28**, 225, 1975.
12. C. W. Sill, "Simultaneous Determination of ^{238}U, ^{234}U, ^{230}Th, ^{226}Ra, and ^{210}Pb in Uranium Ores, Dusts, and Mill Tailings," *Health Phys.*, **33**, 393, 1977.
13. C. W. Sill, EG&G, Idaho Falls, Idaho, private communication.
14. K. A. Richardson, "Thorium, Uranium, and Potassium in the Conway Granite, New Hampshire, U.S.A.," p. 39 of Ref. 3.
15. A. I. Spiridonov, A. N. Sultankhodzhaev, N. A. Surganova, and V. G. Tyminskii, "Some Results of Study of the Isotopes of Uranium ^{234}U in Ground Waters of the Tashkent Artesian Basin," *Uzb. Geol. Zh.*, **4**, 82, 1969.
16. E. I. Hamilton, "The Concentration of Uranium in Man and His Diet," *Health Phys.c,* **22**, 149, 1972.

17. G. A. Welford and R. Baird, "Uranium Levels in Humand Diet and Biological Materials," *Health Phys.*, **13**, 1321, 1967.
18. J. Sedlet, N. W. Golchert, and T. L. Duffy, "Environmental Monitoring at Argonne National Laboratory-1972," U.S. AEC Report ANL-8007, 1973.
19. E. I. Hamilton, "The Concentration of Uranium in Air From Contrasted Natural Environments," *Health Phys.*, **19**, 511, 1970.
20. P. McEachern, W. G. Myers, and F. A. White, "Uranium Concentrations in Surface Air at Rural and Urban Localities within New York State," *Envron. Sci. Tech.*, **5**, 700, 1971.
21. J. A. S. Adams and E. G. Fryer, "Portable γ-Ray Spectrometer for Field Determination of Thorium, Uranium, and Potassium," p. 577 of Ref. 4.
22. H. A. Wollenberg and A. R. Smith, "Studies in Terrestrial Radiation," p. 513 of Ref. 4.
23. H. A. Wollenberg and A. R. Smith, "Geological Factors Controlling Terrestrial Gamma-Ray Dose Rates," in *The Natural Radiation Environment II*, J. A. S. Adams, W. M. Lowder, and T. F. Gesell, Eds., U.S. ERDA Report CONF-720805, 1972 p. 457.
24. A. R. Smith, Lawrence Berkeley Laboratory, private communication.
25. H. L. Beck, W. M. Lowder, and J. E. McLaughlin, "In Situ External Environmental Gamma-Ray Measurements Utilizing Ge(Li) and NaI(Tl) Spectrometry and Pressurized Ionization Chambers," in *Rapid Methods for Measuring Radioactivity in the Environment*, International Atomic Energy Agency, Vienna, 1971, p. 499.
26. A. R. Smith and H. A. Wollenberg, "High-Resolution Gamma Ray Spectrometry of the Uranium and Thorium Decay Series," p. 181 of Ref. 23.
27. J. A. S. Adams and P. Gasparini, *Gamma-Ray Spectrometry of Rocks*, Elsevier Publishing Co., Amsterdam, 1970.
28. A. King, L. M. Scott, and J. L. Disney, "Comparison of Sodium-Iodide and Solid-State Detectors for the Measurement of Lung-Stored Uranium," *Health Phys.*, **34**, 112, 1977.
29. S. Amiel, "Analytical Applications of Delayed Neutron Emission in Fissionable Elements," *Anal. Chem.*, **34**, 1683, 1962.
30. B. S. Carpenter and C. H. Cheek, "Trace Determination of Uranium in Biological Material by Fission Track Counting," *Anal. Chem.*, **42**, 121, 1970.
31. R. H. McCorkell and Y. F. Huang, "Discharge Counter for the Determination of Uranium in Water by the Fission Track Method," *Rev. Sci. Instrum.*, **48**, 1005, 1977.
32. E. S. Gladney and H. L. Rook, "Simultaneous Determination of Tellurium and Uranium by Neutron Activation Analysis," *Anal. Chem.*, **47**, 1554, 1975.
33. J. N. Weaver, "Rapid, Instrumental Neutron Activation Analysis for the Determination of Uranium in Environmental Matrices," *Anal. Chem.*, **46**, 1292, 1974.
34. M. Mantel and S. Amiel, "Simultaneous Determination of Uranium and Thorium by Instrumental Neutron Activation and High Resolution X-Ray Spectrometry," *Anal. Chem.*, **45**, 2393, 1973.
35. U.S. Public Health Services, PHS Publication No. 999-RH-27, "Radioassay Procedures for Environmental Samples," U.S. Public Health Service, Bureau of Radiological Health, Rockville, MD 20852, 1967.

36. H. Levine and A. Lamanna, "Radiochemical Determination of Uranium in Environmental Media by Electrodeposition," U.S. Public Health Service, Publication No. 999-RH-11, 1965.
37. N. A. Talvitie, "Electrodeposition of Actinides for Alpha Spectrometric Determination," *Anal. Chem.*, **44**, 280, 1972.
38. R. Lieberman and A. A. Moghissi, "Coprecipitation Technique for Alpha Spectroscopic Determination of Uranium, Thorium, and Plutonium," *Health Phys.*, **15**, 359, 1968.
39. C. W. Sill and R. L. Williams, "Preparation of Actinides for Alpha Spectrometry Without Electrodisposition," *Analy. Chem.*, **53**, 412, 1981.
40. C. W. Sill, "Separation and Radiochemical Determination Using Barium Sulfate," *Health Phys.*, **17**, 89, 1969.
41. D. R. Percival and D. B. Martin, "Sequential Determination of Radium-226, and Thorium Isotopes in Environmental and Process Waste Samples," *Anal. Chem.*, **46**, 1742, 1974.
42. C. W. Sill, "Determination of Thorium and Uranium Isotopes in Ores and Mill Tailings by Alpha Spectrometry," *Anal. Chem.*, **49**, 618, 1972.
43. J. H. Harley, Ed., EML *Procedures Manual,* U.S. Health and Safety Laboratory Report HASL-300, 1972, with revisions supplied annually. Earlier editions of this manual were published as Report NYO-4700.
44. F. B. Johns, Ed., "Handbook of Radiochemical Analytical Methods," EPA National Environmental Research Center Las Vegas, EPA-680/4-75-001, 1975.
45. E. J. Bouwer, J. W. McKlveen, and W. J. McDowell, "Uranium Assay of Phosphate Fertilizers and Other Phosphatic Materials," *Health Phys.*, **34**, 345, 1978.
46. R. L. Fleischer, P. B. Price, and R. M. Walker, *Nuclear Tracks in Solids*, University of California Press, Berkeley, 1975.
47. W. J. Ward, R. L. Fleischer, and A. Mogro-Campers, "Barrier Technique for Separate Measurement of Radon Isotopes," *Rev. Sci. Instrum.*, **48**, 1440, 1977.
48. D. R. Rushing, W. J. Garcia, and D. A. Clark, "The Analysis of Effluents and Environmental Samples from Uranium Mills and of Biological Samples for Radium, Polonium, and Uranium," Vol. II in *Radiological Health and Safety in Mining and Milling of Nuclear Materials*, International Atomic Energy Agency, Vienna, 1964 p. 187.
49. C. W. Sill and C. P. Willis, "Fluorometric Determination of Submicrogram Quantities of Thorium," *Anal. Chem.*, **34**, 954, 1962.
50. F. C. Centanni, A. M. Ross, and M. A. DeSesa, "Fluorometric Determination of Uranium," *Anal. Chem.*, **28**, 1651, 1956.
51. J. Akaishi and A. Yabe, "Some Cautionary Problems of Fluorometric Determination of Uranium," Japan Atomic Energy Research Institute, Tokai, Report JAERI-13622, 1969.
52. H. E. Zittel and L. B. Dunlap, "Polarographic Determination of Uranium (IV) in Sodium Tripolyphosphate Supporting Electrolyte," *Anal. Chem.*, **34**, 1757, 1962.
53. K. Motojima, H. Yoshida, and K. Izawa, "Spectrophotometric Determination of Small Amounts of Uranium with 8-Quinolinol," *Anal. Chem.*, **32**, 1083, 1960.
54. C. M. Boyd and O. Menis, "Coulometric Determination of Uranium (IV) by Oxidation at Controlled Potentials," *Anal. Chem.*, **31**, 492, 1959.

55. W. D. Shults and P. F. Thomason, "Controlled-Potential Coulometric Determination of Copper and Uranium," *Anal. Chem.*, **31**, 492, 1959.
56. R. D. Giauque, F. S. Goulding, J. M. Jaklevic, and R. H. Pehl, "Trace Element Determination with Semiconductor Detector X-Ray Spectrometers," *Anal. Chem.*, **45**, 671, 1973.
57. P. G. Jeffrey, *Chemical Methods of Rock Analysis*, Pergamon Press, London, 1971.

16 PLUTONIUM

Plutonium (Pu) is one of the principal materials of both commercial and military nuclear power. It is produced primarily in fission reactors that contain uranium fuel, and its importance arises from the fact that a large portion of the Pu produced is fissile: like uranium (^{235}U), ^{239}Pu and ^{241}Pu can be caused to fission with neutrons, including those with low energy. Because such fission events also release neutrons, substantial amounts of energy can be extracted from Pu in a controlled or an explosive nuclear chain reaction.

Now that commercial nuclear reactors provide a noticeable fraction of U.S. and world electrical energy, these reactors account for most plutonium production. For the most part, this material now remains in the irradiated fuel after removal from reactors, but should this fuel be reprocessed, Pu could be recycled to provide part and even most of the fissile content of fresh fuel. For the current generation of water-cooled reactors, the amount of Pu recycled would be substantial. In fast breeder reactors, designed to produce more fissile material than they destroy, considerably larger quantities of Pu would be recycled. In other types of advanced reactors, particularly those that would depend heavily on thorium as the material from which fissile material (primarily ^{233}U) is produced, the amount of Pu to be handled would be considerably reduced.

Because Pu is a highly toxic substance, great care is taken to contain it at the sites and facilities where it is stored or handled. It is necessary that devices be available to monitor any releases from these facilities into environmental media and to measure concentrations of Pu in these media. The radiation protection standards are so strict for Pu that only small releases and low concentrations are tolerated. Monitoring instrumentation must therefore be extremely sensitive.

Although Pu is very toxic, the hazard from Pu arises largely from the α-particles emitted during radioactive decay. The most sensitive plutonium monitoring devices are based on the detection of this radiation. These devices are used for air monitoring and for measurement of Pu separated by radiochemical techniques from water, soil, and biological samples. Instruments for area survey or for monitoring of human subjects often utilize the electromagnetic radiation (x-rays or γ-rays) which follow radioactive decay. Activation and electrochemical methods are often used for Pu monitoring. Other techniques, such as mass measurement, can be used for highly sensitive analyses of a few samples, particularly in air sampling, but are not developed for routine use. For the most part, the techniques for monitoring Pu may also be used for monitoring other nuclear materials, including transuranics that may be associated with Pu.

Greater details of the measurement techniques treated in this chapter are discussed in Chapter 6, Gamma- and X-Ray Spectrometry; Chapter 7, Measurement of Alpha and Beta Particles; Chapter 13, Radium; Chapter 14, Radon; and Chapter 15, Uranium and Thorium. Budnitz reviewed techniques specifically for monitoring plutonium [1]. Much of this chapter is based on a lengthier treatment of uranium and plutonium monitoring instrumentation [2]. Detailed information on the properties, uses, and health effects of Pu may be found in Refs. 3–15.

16.1. PHYSICAL CHARACTERISTICS

The plutonium isotopes of greatest importance are those with mass number 238 to 242. Decay properties of these isotopes are given in Table 16.1. Plutonium-241 has the shortest half-life of these, 13 yr; the properties of its decay daughter, americium-241, are important because survey instruments often monitor ^{241}Am as a plutonium indicator. The other isotopes in Table 16.1 have half-lives from about 10^3 to 10^5 yr, so that after production they persist for long periods of time. The hazard presented by radionuclides is determined, not only by the toxicity of the nuclides being considered and by the amount produced in reactors, but also by the type of daughters arising from decay of reaction products and by the pathways which radionuclides can take to humans.

Plutonium used in commercial nuclear power operations would have several chemical forms. The most probable form for nuclear fuel is the dioxide. Like the uranium dioxide now used as fuel for light-water reactors, PuO_2 present in reactors would be a relatively insoluble ceramiclike material; its insolubility makes it relatively inaccessible to human beings, at least via ingestion. As a result, the primary hazard from the α-particles emitting isotopes, those of most concern, arises from the possibility of inhalation. However, low solubility also introduces difficulties for monitoring methods that require dissolution of a bulk sample. In some circumstances Pu may be present in a more soluble form than the dioxide; in this case, it can become accessible to humans via water-based pathways and, moreover, it can cause damage at lower concentrations because uptake is more efficient.

Toxicity depends on solubility which determines differences in uptake and retention. Airborne soluble Pu can reach the bloodstream by passage across the lung mucosa; in water it is absorbed from the gastrointestinal tract. In either case, it is largely deposited on bone surfaces, where its potential for causing cancer is considered greater than that of radium (Ra). If solubility of Pu decreases, so does the amount deposited in bone, so that the critical organ becomes the site of direct exposure: the lung or gastrointestinal tract. Substantial amounts of insoluble Pu deposited in lungs find their way to the pulmonary lymph nodes, which may therefore be the most sensitive site. See Refs. 3–8 for further discussion of health effects.

Table 16.1. Properties of the Plutonium Isotopes[a]

Isotope and Half-Life in Years	Intermediate Daughter and Half-Life in Years	Main Decay Radiation and Energies
^{236}Pu (2.85)	^{232}U (72)	69%, 5.77 MeV 31%, 5.72 MeV
^{237}Pu (0.12)	^{237}Np (2.1 × 10^6)	Electron Capture
^{238}Pu (86)	^{234}U (2.5 × 10^5)	72%, 5.50 MeV 28%, 5.46 MeV
^{239}Pu (2.44 × 10^4)	^{235}U (7.1 × 10^8)	73%, 5.16 MeV 15%, 5.15 MeV 12%, 5.11 MeV
^{240}Pu (6580)	^{236}U (2.4 × 10^7)	76%, 5.17 MeV 24%, 5.12 MeV
^{241}Pu (13.2)	^{241}Am (433)	β^- (0.031 MeV maximum)
^{242}Pu (3.8 × 10^5)	^{238}U (4.5 × 10^9)	76%, 4.90 MeV 24%, 4.86 MeV
^{241}Am (433)	^{237}Np (2.1 × 10^6)	86%, 5.49 MeV + 60 keV 13%, 5.44 MeV

[a]For more detailed information, see Ref. 16. For isotopic composition of spent reactor fuel, see Table 16.3.

16.2. RADIATION PROTECTION GUIDES

Radioactivity concentration guides given in Table 16.2 are based on consideration of the pathways for Pu and of the effect of decay on humans. These guides specify concentration limits for air and water which lead to the maximum permissible occupational exposures. They assume that workers are exposed to the stated concentrations for 40 hr/wk. The guides shown for members of the general public are for individuals; for population groups, the guides are lower by a factor of three. (For further discussion, see Ref. 17.) Plutonium particles may be difficult to monitor effectively at such low air concentrations simply because of the discrete nature of the activity: a substantial time might be required for a sampler to collect even a single particle, and the resulting delay may be inconsistent with a monitoring objective.

For the plutonium isotopes that emit α-particles, the concentration guides are derived from a comparison of soluble Pu with ^{226}Ra and its daughters. This leads to the soluble Pu limits given in the table and also to a maximum permissible occupational body burden of 0.04 μCi (1.48 kBq) for the (soluble) α-particle emitters shown [as compared with 0.10 μCi (3.7 kBq) for ^{226}Ra]. The limits for insoluble Pu are based on more general criteria for dose limitations, which limit whole-body

Table 16.2. Maximum Permissible Concentration[a] for Plutonium, Radiation, and Americium for Occupationally Exposed Workers and the General Public

	Exposed Workers/General Public	
Isotope	MPC in Air (10^{-15} Ci/liter)	MPC in Water (10^{-6} Ci/liter)
Plutonium-239[a]		
Soluble	2/0.06 (bone)	0.1/0.005 (bone)
Insoluble	40/1 (lung)	0.8/0.03 (G.I. tract)
Radium-226		
Soluble	30/3 (bone)	0.0004/0.00003 (bone)
Insoluble	50/2 (lung)	0.9/0.3 (G.I. tract)
Plutonium-241		
Soluble	90/3	7/0.2
Insoluble	$4 \times 10^4/10^3$	40/1
Americium-241		
Soluble	6/2	0.1/0.004
Insoluble	$10^2/4$	0.8/0.03

[a]The occupational maximum permissible concentrations (MPCs) would yield the dose equivalent limit if sustained 40 hr per week for a sufficient period. The MPC for individual members of the general public, however, is based on sustaining the stated level for 168 hr per week; the MPC for populations is lower by a factor of 3. (For ^{239}Pu and ^{226}Ra the critical organ is shown.) Plutonium-238, -240, and -242 have MPCs within 20% of the values for ^{239}Pu. See Code of Federal Regulations, Title 10, Part 20.

exposures to 5 rem/yr (50 mSv/yr) for radiation workers, 0.5 rem/yr (5.0 mSv/yr) for members of the public, and 0.17 rem/yr 1.7 mSv/yr) for population groups [16], or which have limits a factor of three higher for specific organs, including the lung. The maximum permissible insoluble plutonium lung burden corresponding to the occupational exposure of Table 16.2 is 0.016 μCi (0.59 kBq).

More specific limits apply to environmental emissions from commercial nuclear power operations. The Nuclear Regulatory Commission (NRC), in its review of licensing applications for light-water reactors, uses guidelines that typically limit exposure of members of the general public to approximately 0.005 rem/yr (0.05 mSv/yr); about 1% of the dose limitations just given. In addition the Environmental Protection Agency (EPA) has begun to develop standards which apply to the nuclear fuel cycle; for uranium-fueled light-water reactor power plants, 40 FCR 190 limits total fuel-cycle emissions of ^{239}Pu and other α-emitting transuranics to 0.5 mCi/GW$_e$-yr (18.5 MBq/GW$_e$-yr). Standards for facilities using recycled Pu have not yet been developed. More recent EPA regulations apply to exposures in the vicinity of plutonium-contaminated areas [18]. Concentration standards for soil may be formulated [19]. It has also been proposed [20] that waste streams containing more than 10 nCi/gm (370 Bq/gm) of Pu and other long-lived transuranics be treated as transuranic wastes.

The standards for other α-particle emitters are often as stringent as those for Pu. Many of these would have to be monitored at the same locations and, indeed, many Pu monitoring techniques apply equally well to these radionuclides. Of particular interest is ^{241}Am, the daughter of ^{241}Pu (^{241}Pu emits β-rays). Plutonium-241 has a short half-life (13 yr) and, within only a few decades, can contribute an amount of ^{241}Am activity that is comparable to the α-particle activities initially present in many plutonium mixtures.

16.3. SOURCES OF ENVIRONMENTAL PLUTONIUM

Plutonium does not occur naturally in significant quantities. Plutonium-239 is produced primarily by neutron capture on ^{238}U, followed by β-decay of ^{239}U and its daughter, neptunium-239; heavier isotopes are produced largely by subsequent neutron capture on Pu itself. Plutonium-238 is produced by a sequence of neutron captures beginning with ^{235}U. Production of Pu occurs almost entirely in nuclear reactors, whether commercial, research, or military. The isotopic mixture varies substantially, depending on the type of reactor and the purpose for producing the Pu. Table 16.3 gives the amount and composition of Pu that would be produced in commercial nuclear power plants of several types. A more recent discussion of production of Pu and other transuranics appears in Ref. 11. The efficacy of plutonium monitoring techniques can depend strongly on the isotopic mixture being examined. Many currently available techniques have, in fact, been developed for use with weapons-grade Pu, which is produced in such a way as to minimize the ^{240}Pu because this isotope releases many spontaneous neutrons, making efficient detonation of a weapon difficult. As a result, a typical sample consists of the following mixture: 93.6% ^{239}Pu; 5.9% ^{240}Pu; 0.4% ^{241}Pu; 0.01% ^{242}Pu; and 0.01% ^{238}Pu [19]. Mixtures from commercial reactors (Table 16.3) contain considerably less ^{239}Pu than do weapons-grade mixtures and more of the other isotopes.

The bulk of the world's plutonium inventory is contained in nuclear weapons, in nuclear reactors, and in materials associated with weapons and reactors. Environmental Pu originates from these inventories: from nuclear weapons already exploded, from leakage at facilities which handle or process Pu, or from special events such as the destruction (during take-off) of a plutonium-bearing satellite or the crash of military aircraft carrying nuclear weapons. Nuclear explosions, including atmospheric tests, have contributed about 440 kCi (16,000 TBq) of ^{238}Pu, ^{239}Pu, and ^{240}Pu to the worldwide plutonium inventory (in an activity ratio of about 3:58:39). The burnup during 1964 of a satellite with a power cell deriving its energy from ^{238}Pu decay contributed about 17 kCi (630 TBq) of this isotope to the global inventory. Plutonium-handling facilities are known to have released on the order of 10 Ci (0.37 TBq) to off-site environments [21]. Even more than 10 Ci has been released into local environments, either as the result of weapons tests or for purposes of studying plutonium transport and monitoring in the environment.

Most of this inventory was injected into the atmosphere in the mid-1960s, and this fact is reflected in the measured air concentrations and soil deposition rates. For example, during the 1970s, concentrations of ^{239}Pu in surface air at New York City were 0.1×10^{-18} Ci/liter (3.7 nBq/liter) or less, as compared with a peak value of 1.7×10^{-18} Ci/liter (62.9 nBq/liter) for 1963 [21]. The latter was one-tenth of the soluble plutonium concentration guide for large populations. Since 1963 measured concentrations have decreased substantially. At a given time con-

Table 16.3. Typical Isotope Composition of Plutonium Extracted from Nuclear Power Plants[a]

Isotope	Percent by Mass	Mass (kg)	Radioactivity (Ci)
Light-Water Reactor Fueled with Natural Plus Recycled Uranium			
238	2.4	6	1.0×10^5
239	58.4	142	8.7×10^3
240	24.0	59	1.3×10^4
241	11.2	27	2.8×10^6
424	3.9	10	3.7×10^1
Total	100	244	$1.2 \times 10^5(\alpha)$ $2.8 \times 10^6(\beta)$
Light-Water Reactor Fueled with Natural Uranium Plus Recycled Plutonium			
238	4.2	41	6.9×10^5
239	38.7	380	2.3×10^4
240	27.5	270	6.0×10^4
241	18.1	178	1.8×10^7
242	11.5	113	4.4×10^2
Total	100	982	$7.8 \times 10^5(\alpha)$ $1.8 \times 10^7(\beta)$
Liquid-Metal Fast Breeder Reactor Fueled with Depleted Uranium Plus Recycled Plutonium and Uranium			
238	0.1	1.4	2.4×10^4
239	71.7	1460.6	9.0×10^4
240	25.1	510.8	1.1×10^5
241	2.4	48.4	4.9×10^6
242	0.8	25.5	6.0×10
Total	100	2037	$2.3 \times 10^5(\alpha)$ $4.9 \times 10^6(\beta)$

[a]Yearly masses and radioactivity from fuel reprocessing for 1000 MW_e plants (80% plant factor), assuming 150-day decay time, except for LMFBR, which assumes 30 days. Data presented in Pigford and Ang [9].

centrations may vary by an order of magnitude from one place to another. Deposition rates have followed a similar pattern with the highest yearly value, 0.6 nCi/m^2 (22.2 Bq/m^2), occurring for 1963. Cumulative deposition in surface soils of the United States varies from 0.7 to 2.4 nCi/m^2 (25.9 to 88.8 Bq/m^2) [21], depending on the state, except Nevada. Cumulative deposition was higher in Nevada due to the presence of the Nevada Test Site. Most activity is contained in the top 5 cm of soil. Typical concentrations in surface soils are about 0.05 pCi/g (1.9 mBq/g) [21], about 1% of the natural actinide activity (due to U and Th), which in turn comprises about half of the natural α-particle activity in soil.

In the vicinity of nuclear facilities or of nuclear testing, concentrations can be much higher. For example, about half of the several curies released from the Rocky Flats plutonium facility is present on the surface in concentrations between

10 and 2000 nCi/m^2 (0.37 to 74.0 kBq/m^2) [22]. Many portions of the Nevada Test Site have ^{239}Pu soil concentrations of hundreds or thousands of pCi/g (3.7 to 37 Bq/g) [23]. In such areas, air concentrations can easily reach the general public concentration limits (see Ref. 24); resuspension from plutonium-bearing soil may be an important contributor to air concentrations in air [25,26]. The behavior of Pu released at Rocky Flats, Hanford, and Thule, Greenland, has been examined in detail [27,28,29].

Individuals who spend large amounts of time in the vicinity of such highly contaminated areas have the potential for accumulating the permissible body burden. General public body burdens of ^{239}Pu were found to be a few pCi (a few tenths of a Bq) in 1970–1977 [30], about a factor of 300 below the maximum permissible burden for populations. These burdens were accumulated primarily from inhalation of airborne Pu; due to low plant uptake [31], concentrations in food are about one-millionth the general public limits for water.

16.4. MEASUREMENT TECHNIQUES

The choice of monitoring techniques may depend strongly on applicable standards and on the conditions for monitoring. In the discussion that follows, this dependence will be apparent, but it is useful to note some specific considerations:

1. Because radiation protection standards depend on the environmental medium, the sensitivity required will vary. The monitoring technique also depends on the medium because the range of the principal radiations associated with Pu depends on the density of the medium.
2. Sensitivity requirements often determine the time required to measure Pu. Additional time may be needed if the technique requires chemical separation. Even more time may be needed, if it is necessary to discriminate between several radionuclides.
3. Both sensitivity requirements and time limitations depend on the conditions of monitoring—whether occupational or general public and whether routine or emergency.

A severe requirement on monitoring sensitivity is imposed by the fact that the radioactivity concentration guide for air in areas accessible to members of the general public is 6×10^{-17} Ci/liter (2.2 μBq/liter) for soluble Pu, the more restrictive form. A liter of air with this concentration will contribute only 2×10^{-6} decays/sec, so that many liters have to be sampled to achieve useful counting rates in a practical length of time.

The next sections discuss these classes of plutonium monitoring:

1. Air monitoring.
2. Radiochemical analysis of air, water, soil, and biological samples.
3. Field survey and personnel monitoring using x- or γ-ray detection.
4. Other methods.

16.4.1. Air Monitoring

The α-particles associated with the decay of plutonium isotopes are the signature used in monitoring for airborne Pu. The universal technique is to collect the radioactive particles by pumping air through a filter. This filter typically collects at least 99% of the Pu in the sampled air. Alpha particles are detected as they are emitted from the filter or the Pu can be chemically separated from the filter and deposited on a substrate suitable for more precise measurements. Techniques for α-particle monitoring are discussed in Chapter 7.

Analysts using these techniques for Pu must consider the α-spectra from plutonium decay and the spectra from interfering activities. Figure 16.1 shows α-particle energies and relative strengths for each of the plutonium mixtures listed in Table 16.3. It also gives the energies of important background α-particle groups

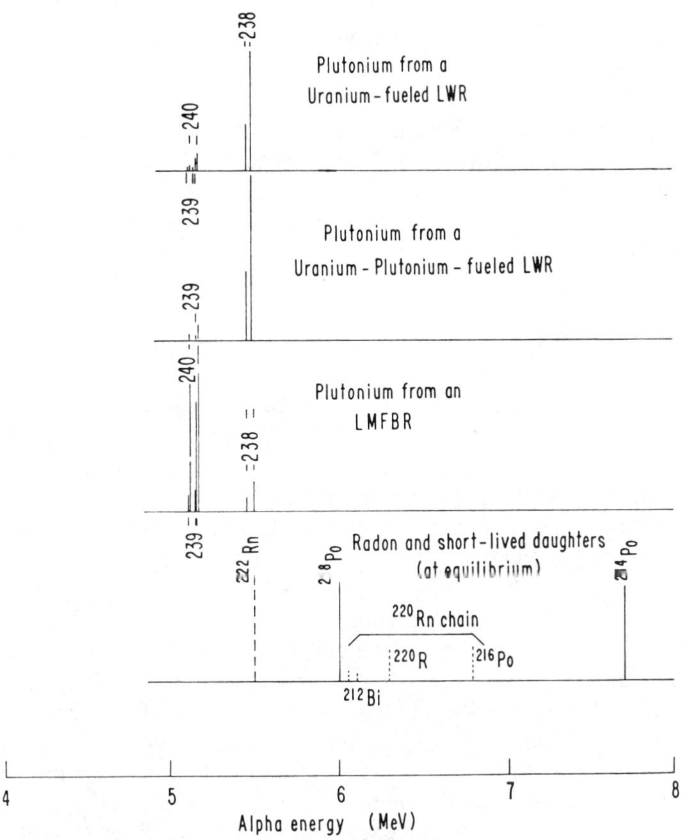

Figure 16.1. Relative strength of alpha groups from typical plutonium mixtures. Schematic spectra for the mixtures of Table 16.3 are given. Within each mixture, the height of each line is proportional to the strength of that group. (The mass numbers indicate which isotope accounts for each group.) Also shown are schematic spectra for the radon and thoron decay chains.

arising from daughters of ^{222}Rn, part of the ^{238}U decay series. The less important ^{220}Rn daughters, which are part of the ^{232}Th decay chain, are also shown. (See Chapter 14 for more detail on Rn and its daughters.) Isotopes of polonium (Po) constitute the most serious background because they attach themselves to particles, which are also collected on air filters. Radon typically occurs at concentrations of 0.01 to 1.0 pCi/liter (0.37 to 37.0 mBq/liter) in surface air [31], far above the concentration guides for Pu. The general public concentration guide for Rn is 1 pCi/liter (37 mBq/liter), as compared with the soluble plutonium guide of 6×10^{-5} pCi/liter (2.22 μBq/liter). Because of the comparatively high ambient concentration of radon daughters, measurement of total α-particle activity ordinarily can only put an upper limit on plutonium concentrations, unless the sample is held before counting, permitting decay of the Rn or its daughters.

The more useful air monitoring techniques discriminate against background α-particle activities, particularly the radon daughters. The remainder of this section discusses three techniques that leave the collecting filter intact, that is, direct α-spectroscopy, lifetime analysis (including holdup), and associated α-β discrimination. Radiochemical analysis is discussed later.

Direct Alpha Spectrometry. Because filters do not collect radon gas, it is clear from Fig. 16.1 that α-spectrometry (measurement of the energy of emitted α-particles) can discriminate against natural background, provided the monitoring system has sufficiently good energy resolution. In a typical continuous monitoring system, a semiconductor detector is used to measure the α-particle energy. Because of the detector arrangement for direct measurement from air filters, resolution is hundreds of keV rather than the tens of keV often available from such detectors. As a result the 6.0-MeV α-particle groups from ^{218}Po can interfere with the measurement of plutonium, they can even interfere with the ^{239}Pu 5.1-MeV alpha groups, as is shown in Fig. 16.2. The α-particle activity of commercial mixtures of Pu, and also of ^{241}Am, is even more difficult to measure because the dominant α-particle groups are at 5.5 MeV, even closer to the ^{218}Po group. The most useful plutonium monitoring systems have some method of discrimination, however crude. If counting occurs after collection, the ^{214}Po assumes more importance as background, and for long collection and/or holdup times, the daughters of ^{220}Rn dominate the background.

Monitoring systems from commercial manufacturers are available both as continuous monitors, where the detector system operates while the air sampling occurs, and as off-line monitors, where a filter from a separate sampling unit is inserted into a counting instrument. Sampling rates are typically 50 to 100 liters/min. The filter is most often 47 mm in diameter, and the detector is usually a large-area (up to 750 mm^2) diffused-junction or surface-barrier detector. Commercial units for continuous monitoring are usually small and self-contained, except for the vacuum pump. Rather than accumulate an energy spectrum, such as shown in Fig. 16.2, the units ordinarily count the α-particles within some energy "window" depending on the isotopes(s) of interest. Many units compensate for the ^{218}Po decays in this window (due to various energy-degrading processes), often by making a subtraction proportional to the number of counts in a ^{218}Po window. Such units can detect a 40-hr occupational concentration limit [2×10^{-15} Ci/liter (0.074 mBq/liter)] in a few hours [28]. In the event that daughters of ^{220}Rn present significant background, the same set of windows can be used for background compensation. The primary ^{220}Rn interference arises from the decay of ^{212}Bi, with α-particle groups at

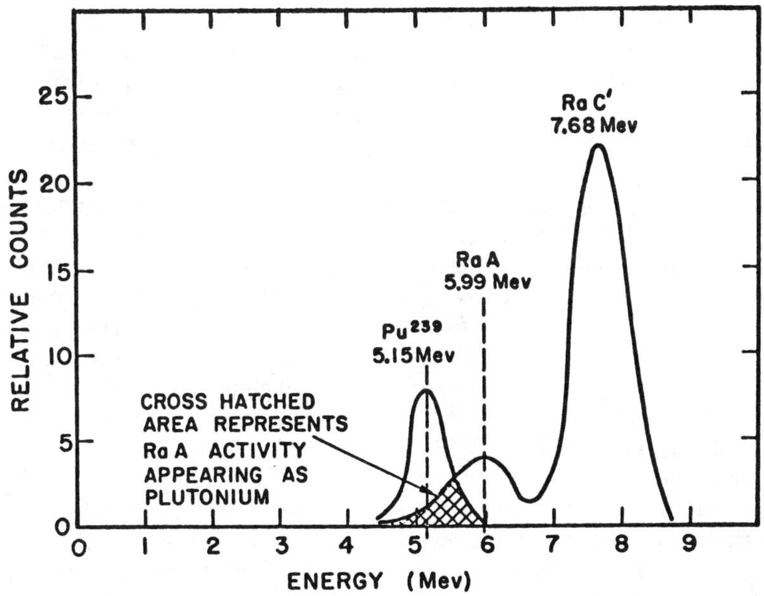

Figure 16.2. Radon daughter interference in continuous plutonium alpha detection. The illustration indicates how, in an ordinary alpha detection system, radon daughter activity can interfere with monitoring of plutonium. In this case, because of relatively poor resolution, ^{218}Po (Radium A) contributes counts in the region of the ^{239}Pu groups (reproduced from Phillips and Lindekin, Ref. 32, with permission of the Health Physics Society © 1963).

6.05 and 6.09 MeV, close to the ^{218}Po group. For counting after collection, ^{214}Po and the ^{220}Rn daughters can be more important than ^{218}Po.

Because of the relatively poor resolution of systems operating at atmospheric pressure, the semiconductor detector subsystem can be replaced by a gas proportional counter without substantially degrading the system resolution. This can be particularly useful for large filter areas. However, it is possible to improve the resolution and sensitivity of semiconductor systems by using evacuated counting chambers to reduce the energy spread of detected particles caused by energy loss in air (so that the only degradation remaining arises from energy deposition in the filter or collected particles) and by using more sophisticated data analysis techniques, including lifetime analysis, to discriminate against radon daughters [33]. Finally, the presence of other α-emitting species, such as curium, can complicate these low-resolution techniques by contributing α-particles to either the foreground or background windows.

Lifetime Analysis. The fact that the radon daughters detected in plutonium measurements are very short-lived can be used to discriminate against them. The ^{218}Po sequence is

$$^{182}Po \xrightarrow[3\ min]{\alpha} {}^{214}Pb \xrightarrow[27\ min]{\beta} {}^{214}Bi \xrightarrow[20\ min]{\alpha} {}^{214}Po \xrightarrow[164\ \mu sec]{\alpha} .$$

The most primitive time-discrimination technique simply holds up counting in a spectroscopic system until a large portion of the ^{218}Po has decayed. Each 30 min of holdup reduces the amount of ^{218}Po on a filter by a factor of about 1000, provided ^{222}Rn has not been adsorbed. Longer holdup reduces interference from ^{214}Po and even ^{212}Bi. A more sophisticated technique would use lifetime analysis to separate the short-lived activities (radon daughters) and long-lived activities (Po or other transuranics). In particular, the 3-min ^{218}Po activity can be distinguished by counting in consecutive time intervals and observing the decrease in the count rate in the plutonium window. However, if the ^{220}Rn daughters are present in significant amounts, a somewhat more complex discrimination scheme must be used because of the presence of ^{212}Bi. For this isotope, 6.1-MeV α-particles that appear in the plutonium window can be subtracted based on ^{212}Po concentration derived from the 8.78-MeV α-particle group from ^{212}Po, a daughter of ^{212}Bi.

In general, for any mixture of ^{220}Rn and radon daughters, the incorporation of lifetime analysis into a spectroscopic system can determine the daughter activities [33–37], substantially improving the sensitivity for detection of Pu and other long-lived transuranics. A system developed at Lawrence Livermore National Laboratory (LLNL) uses these techniques to measure a 40-hr concentration of plutonium in 30 min, an order of magnitude better than the simpler commercial monitoring systems [33]. The sensitivity of the LLNL system is another order of magnitude better for low specific activity isotopes, for example, ^{239}Pu. As is discussed later, the particulate nature of the activity can pose a basic limitation for high enough specific activity. The LLNL system has a continuous filter strip that collects radioactive particles in one area and is then stepped into an evacuated counting chamber; while counting occurs in the chamber, collection takes place on the next portion of the filter. The system samples air at 566 liters/min, and separating the chambers protects the array of semiconductor detectors from the sampled air, which in a reprocessing plant could be highly corrosive. Evacuating the counting chamber improves the energy resolution by a factor of approximately 5. The counting period is divided into two parts to discriminate against ^{218}Po counts in two windows: one window for Pu and Am (4.8 to 5.5-MeV α-particles) and the other for Cm (5.6 to 6.8-MeV α-particles). Bismuth-212 is subtracted on the basis of counts in a ^{212}Po window (8.0 to 8.8-MeV α-particles) if necessary.

Associated Alpha-Beta Measurements. In the associated β-ray method, α-particle counts due to Rn and Rn daughters are estimated by observation of the β-rays that effectively accompany the α-decay of ^{214}Po and ^{212}Po. These nuclides are formed by β-decay and almost immediately decay by α-particle emission because their half-lives are 164 and 0.3μsec, respectively. Observation of α-β coincidences can be used to estimate the amount of Rn and ^{220}Rn daughter activity that should be subtracted. Instruments employing this technique [33–36] typically count α-particles with a scintillator-photomultiplier counter and β-rays with a scintillation counter or Geiger-Müller tube. Subtracting a weighted α-β count rate from the α-particle count rate (without energy information) may suffer from several difficulties: (a) if there is a large radon background, it will be necessary to subtract two large numbers, yielding a very uncertain smaller number; (b) the relative detector efficiencies must be well known; and (c) the relative amounts of members of each decay chain must be known because in each case only one member is measured. Some of these difficulties would be alleviated by incorporating α-β subtraction in an

α-spectrometric system. However, because of the difficulties in making a reliable subtraction, this method is not widely used.

New Techniques for Plutonium Particulates. Most air monitoring techniques rely on the fact that plutonium particles can be collected on filter materials. More subtle use can be made of this particulate character, either as an integral part of the measurement technique or as a means to separate the Pu from other portions of the collected sample. Substantial attention has been given to the form in which Pu is present in nuclear fuel-cycle operations [38–43]. Much of the Pu, by mass, is present as particles with aerodynamic diameters from one to a few μm; however, a substantial portion may be smaller in size, particularly in chemical operations.

A technique studied at the Battelle Pacific Northwest Laboratory [44] uses a mass spectrometer to separate plutonium compounds from other substances obtained in direct air sampling. The plutonium is ionized by a filament onto which plutonium particulates impact after passing through a capillary tube. The ion source thus serves as the sampling unit (see also Ref. 45). The minimum detectable amount for such a system is much smaller than for other techniques because sensitivity is not limited by the radioactive rate or by inference from radon daughters. However, because the mass spectrometer ion source operates at high vacuum, only a small stream of air, on the order of 0.6 liter/min, can be sampled. At the 40-hr occupational concentration limit, 1-μm (aerodynamic diameter) ^{239}Pu particles occur only about once in several liters and much more rarely if the particles are larger. Under such conditions the unit may have to sample for long periods before encountering a plutonium particle.

The particulate character of plutonium may also be used as a discriminator. Impaction techniques, in which a stream of air rapidly changes direction and large particles impact onto a surface, have been used for size measurements [38]. Such techniques can also be used to separate Pu from the very small particles that carry most of the radon-daughter background [46]. The more advanced of such systems are *virtual impaction techniques*, in which, at each separation stage, the larger particles impact into a slowly moving airstream for filter collection and counting [47,48]. One such system, designed at Argonne National Laboratory [47], uses an α-particle detector to measure the collected particles, achieving a sensitivity about an order of magnitude better than the α-spectrometers described previously. Semiautomatic virtual impactors have been deployed for measuring conventional pollutants [49]. Virtual impaction concentrates the activity of interest into a small air-stream, a preconcentration that is useful for the mass spectrometric technique just described and for other techniques.

16.4.2. Radiochemical Analysis

This section discusses radiochemical analysis (chemical separations followed by counting) of air, water, soil, and biological samples. Chemical separation of the transuranic elements from the matrix in which they are found has two advantages: (a) these activities can be separated from one another and from background activities, particularly the uranium and thorium decay chains; and (b) for bulk samples, chemical separation removes the matrix that would interfere with detection of α-particles. Alpha particles of a few MeV have a range on the order of 10^{-3} cm in typical solid samples, and any sample with more than a small fraction of this thickness will seriously degrade energy resolution. Radiochemical analysis is often followed by electrodeposition of a thin layer of the radioisotopes, permitting high-resolution α-spectroscopy. However, it is also possible to prepare thin samples by filtration [50]. See Chapter 7 also for a general discussion of analysis techniques.

For off-line plutonium monitoring, chemical separation and α-spectrometry leads to the highest sensitivity, although it has the disadvantage that substantial time, equipment, and technical competence are required for the dissolution, separation, and counting procedures. Sample preparation depends on the type of sample, but the counting arrangement is relatively independent of sample type. Semiconductor detectors of the sort used in the direct α-particle monitoring are employed, but the system can be designed for much higher resolution, even tens of keV by use of evacuated counting chambers and very thin deposition layers.

A serious condition for the effectiveness of radiochemical analysis is that the Pu or other radioactive elements of interest actually be dissolved. Fired PuO_2, the principal form used for commercial nuclear power, is notoriously insoluble [51]. For this reason, vigorous dissolution techniques are often required. The technique must also be effective in extracting very small amounts of Pu: ^{239}Pu with an α activity of 1 pCi (0.037 Bq) has a mass of about 16 pg. The efficiency of plutonium extraction and deposition is often checked by the addition to the sample of a *tracer*, typically a known quantity of ^{236}Pu or ^{242}Pu. Neither ^{236}Pu or ^{242}Pu contributes substantially to the activity of either reactor-grade or weapons-grade Pu. The α-particles groups from these isotopes can be resolved from those of ^{238}Pu, ^{239}Pu, and ^{240}Pu in a high-resolution system, providing a measure of the extraction efficiency, assuming the tracer and the *unknown* plutonium isotopes are equally well dissolved. Sufficient tracer should be added so that counting statistics on the tracer groups does not lessen the precision of the final result [52].

A great deal of effort has been devoted to techniques for measurement of Pu in soils. The two classes of chemical separation used are: (a) total dissolution techniques, which bring the entire sample into solution; and (b) leaching, intended to selectively extract the substance(s) of interest from the sample matrix. Leaching is the less vigorous approach and hence is more liable to error; this technique should be checked by total dissolution. Samples are typically leached or dissolved with nitric acid plus some combination of hydrofluoric, hydrochloric, and perchloric acids. Specific elements are often separated from the resulting solution by ion-exchange techniques. Carrier solutions are then prepared for electrodeposition onto planchets, which may be analyzed in an α-spectrometer [53–55]. A total dissolution method for soil samples, after an initial treatment with nitric and hydrofluoric acids, uses a high-temperature fusion with anhydrous potassium fluoride and subsequent pyrosulfate fusion to dissolve the sample completely. Separation procedures yield four fractions that are ready for electrodeposition; (a) Th; (b) Pa, Np, and Pu; (c) U; and (d) Am, Cm, and Cf [56]. Hahn et al. evaluated the fusion method [57]. Sill, Hindman, and Anderson discussed its application for large samples [58].

An anion exchange method for water samples described by Wong [59] has recently been evaluated [60].

With careful high-resolution α-spectrometry, positive identification of particular radionuclides is possible with only a few counts in characteristic α-particle peak(s). For a 1-hr counting period, this implies a minimum sensitivity limit on the order of 0.2 pCi (7.4 mBq), the amount that an air filter collects in 1 hr at 1 liter/sec if Pu is present at the general public concentration limit of 6×10^{-17} Ci/liter (0.002 mBq/liter). In contrast, the α-particle air monitors described in the preceding section have sensitivities (under the same sampling conditions) at the occupational concentration of 2×10^{-15} Ci/liter (0.074 mBq/liter), corresponding to a collected activity of 8 pCi (0.296 Bq) or more. Longer counting periods, of course, improve the sensitivity of radiochemical analysis.

For media other than air, the available sensitivity is more than adequate: the general public concentration limit for plutonium in water is 50,000 pCi/liter (1.85 kBq/liter). The primary disadvantages of radiochemical analysis are the time, equipment, and personnel required. For many applications, waiting many hours or days for analytical results is not practical.

An overall review and evaluation of radiochemical techniques for environmental Pu is given in Ref. 61.

16.4.3. Field Survey and Personnel Monitoring

This section discusses field survey and personnel monitoring using x-ray or γ-ray detection. In many circumstances a relatively rapid determination of the radionuclide content of a sample or an area is required. This ordinarily precludes complex chemical analysis; however, direct measurement via α-particles is often impractical because of the short range of α-particles in matter. The electromagnetic radiation that often follows α-decay is a much more useful signature because it is less seriously degraded by passage through bulk material. This radiation includes both x- and γ-rays emitted from decay daughters left in an atomic or nuclear excited state.

Atomic excitations of U, the daughter of plutonium α-decay, yield L x-ray lines at 13.6, 17.3, and 20.2 keV [62], which occur in 5% of ^{239}Pu decays and in 12% of ^{240}Pu decays [62, 63]. Gamma rays following plutonium decay are rarely used for monitoring because of their much lower branching ratios; for ^{239}Pu, the most prominent γ-ray, 52 keV, is only a 0.02% branch.

Plutonium-241 decays to ^{241}Am; the decay of ^{241}Am has a 37% probability of producing L x-rays (13.9, 17.8, and 20.8 keV) and a 36% probability of leading to a 60-keV γ-ray [62]. Depending on the circumstances, this isotope can be used as a signature for Pu, or, if the ratio of Pu to Am is not known, it can constitute an interfering activity.

An important consideration in photon spectrometry is the geometry of the detector relative to the source. Detectors typically average over a large area unless collimators are used, and the interpretation of results depends on this averaging, on the isotopic composition of the contamination, and on its depth distribution in the survey area.

Photon Spectrometry with Scintillation Detectors and Gas Counters. Rapid area surveys are often done with portable instruments based on thin large-area scintillation detectors, most often thallium-activated sodium iodide [NaI(Tl)] crystals. The most straightforward instrument couples a NaI(Tl) crystal to a photomultiplier, allowing energy analysis of radiation interacting with the crystal. The archetype of this instrument, designed for monitoring of plutonium ground contamination by observation of americium photons, is the FIDLER (Field Instrument for Detection of Low Energy Radiation) [64] which uses a 12.7-cm-diam., 0.16-cm-thick crystal. The pulse windows are set for energies of 17 and 60 keV, the principal ^{241}Am photon groups, as indicated in Fig. 16.3. In practical applications the sensitivity of the FIDLER is found to be a few hundred nCi/m^2 (a few kBq/m^2) of ^{241}Am [65], although lower minimum detectable area distributions have been cited. Similar

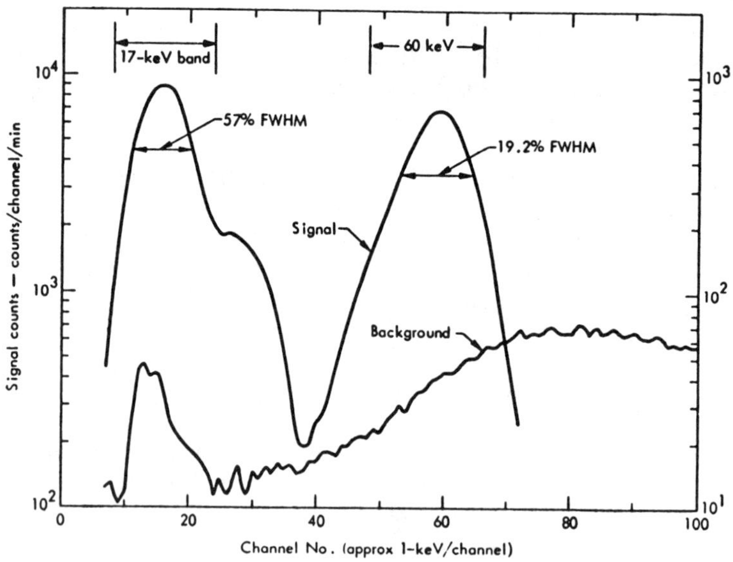

Figure 16.3. FIDLER calibration spectrum. The spectra are for ^{241}Am and background as measured with a 5-in.-diam by 10 1/16-in.-thick NaI(Tl) scintillation detector. Distance from ^{241}Am source to detector was 30.5 cm. Optimum discriminator settings for 17-keV band and 60 keV are shown (reproduced from Tinney, Koch, and Schmidt, Ref. 64).

instruments have been developed based on europium-activated calcium fluoride [CaF_2(Eu)] scintillators [62, 66], which—unlike NaI(Tl)—are not hygroscopic; one version of such an instrument can even detect α-particles [67].

Sophisticated systems have been developed to eliminate background due to high-energy photons that deposit only part of their energy in the thin scintillator. In the *phoswich* detector, illustrated in Fig. 16.4, a thin NaI(Tl) crystal is backed by a thick thallium-activated cesium iodide [CsI(Tl)] crystal, which stops high-energy photons. Since CsI(Tl) has a longer phosphorescent decay time than does NaI(Tl), its signal can be used to suppress the high-energy photon background in the thinner detector. This system has been used for plutonium lung assays, during which sophisticated detectors and shielding arrangements are feasible. Under these circumstances the phoswich detector can reduce background substantially and achieve sensitivities better than the minimum detectable ^{239}Pu amount for an unbacked NaI(Tl) detector [68]. Garg and Newton discuss a method to reduce low-energy background [69]. A portable survey instrument has recently been developed which uses a phoswich-type detector [70] resulting in a notable reduction in background as compared with the FIDLER.

Detectors relying on photon-induced ionization in a rare gas have somewhat better resolution than do thin NaI detectors. Large-area gas proportional counters have been used for plutonium lung assay, yielding minimum detectable ^{239}Pu activities of 6 nCi (0.22 kBq) and a resolution of 13% at 13.6 keV [71,72]. Gas proportional scintillation (as opposed to charge collection) counters have achieved considerably better resolution (about 7% in the L x-ray region, according to Refs. 73 and 74) and may be useful in the laboratory and for field surveys.

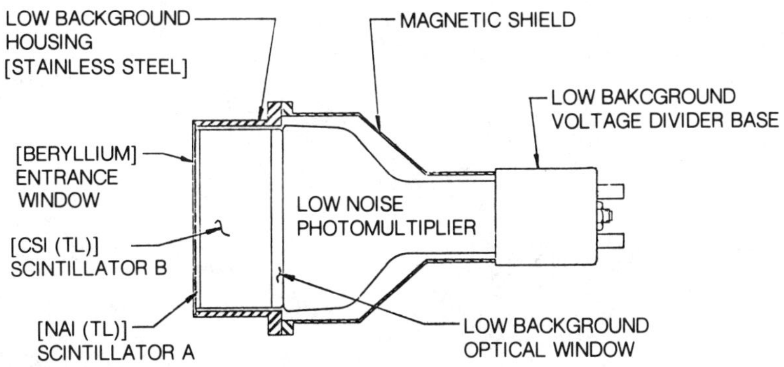

Figure 16.4. Phoswich detector. For low-energy electromagnetic radiation, whether x-rays or low-energy gamma rays, background may be reduced by using a thin CsI(Tl) scintillator as the primary detector and a thick NaI(Tl) scintillator in anticoincidence (to minimize the effect of high-energy radiation leaving small amounts of energy in the thinner detector).

Results of an interlaboratory comparison of the phoswich and gas proportional techniques suggests phoswich detectors to be more useful for lung assays [75].

Photon Spectrometry with Semiconductor Detectors. Semiconductor detectors made of silicon or germanium, often lithium drifted, offer 2 to 4% energy resolution for low-energy photons, which is superior to that of scintillators or gas proportional counters. On the other hand, semiconductor devices are typically very small, with active areas on the order of 10 cm^2, compared with the 100 cm^2 or more characteristic of the scintillators. An additional disadvantage of germanium detectors and Li-drifted detectors is that they must be cooled with liquid nitrogen during use, or, in the case of Ge(Li), at all times.

The sensitivity of semiconductor detectors is often superior to that of other systems, largely because of their better energy resolution (Fig. 16.5). A field comparison [46] between a FIDLER with an area of 130 cm^2, a thick Ge(Li) detector with a volume of 70cm^3 and a cross section of 15 cm^2, and a four-detector array of thin, high-purity (intrinsic) Ge, 0.25 cm thick with a total area of 33 cm^2, gave the following as minimum detectable concentrations of ^{241}Am under various conditions: 200 to 700 nCi/m^2 (7.4 to 25.9 kBq/m^2) for the FIDLER; 75 to 200 nCi/m^2 (2.8 to 7.4 kBq/m^2) for the Ge(Li) detector; and 40 to 100 nCi/m^2 (1.5 to 3.7 kBq/m^2) for the Ge array; all with 2000-sec counting times. Planar high-purity Ge arrays with total areas up to 100 cm^2 have been used for environmental survey or sample monitoring, but these arrays are not expected to have minimum detectable concentrations much less than the roughly 50 nCi/m^2 (1.85 kBq/m^2) or 5 pCi/cm^2 (0.185 Bq/cm^2) obtainable with a four-detector array.

The foregoing concentration limits, depending on the distribution in the sample, correspond to a few pCi of ^{241}Am per gram and somewhat more Pu. Consideration has been given to a more sensitive system [44] consisting of an array of Si detectors flanked by thick anticoincident NaI(Tl) detectors to suppress the background caused by high-energy photons. Such a system could be used for assay of soil samples without needing radiochemical separation and could have a minimum detectable concentration of about 1 pCi/g (0.037 Bq/g) for Pu alone and about 8 pCi/g

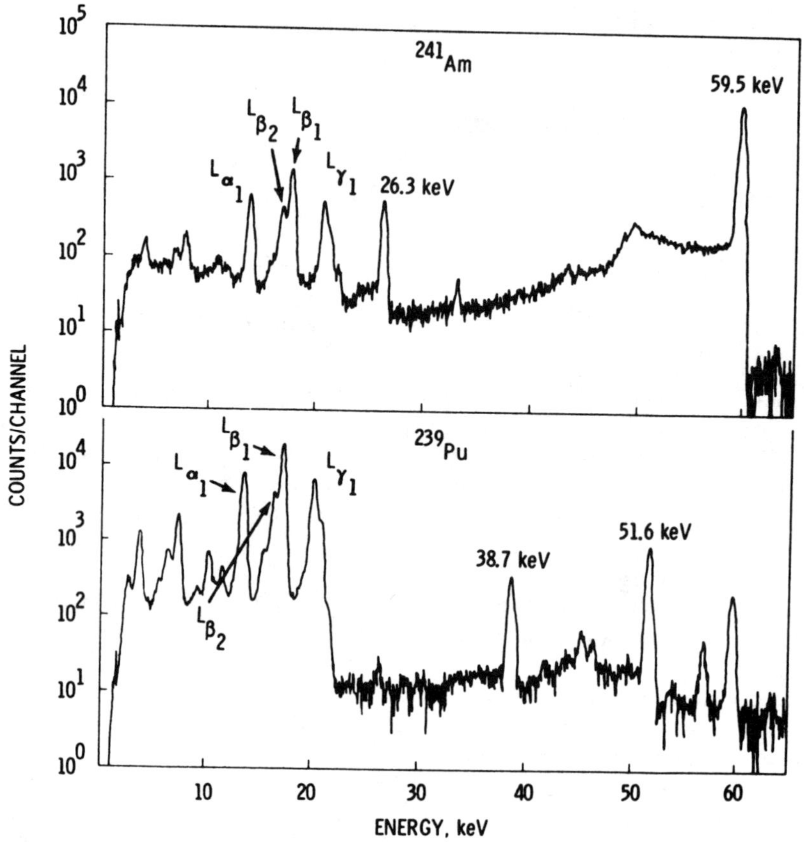

Figure 16.5. ^{239}Pu and ^{241}Am spectra taken with a Ge(Li) detector. Semiconductor detectors are increasingly being used for field measurement of plutonium/americium electromagnetic emissions. The illustration shows spectra accumulated with a 5-cm^2 Ge(Li) detector (reproduced from Brauer, Ref. 76, with permission of Institute of Electrical and Electronics Engineers © 1977).

(0.296 Bq/g) for Pu in the presence of Am; the sensitivity to Am would be considerably better than that for Pu alone. Other investigators have used simple Ge(Li) and Si(Li) detectors for soil assay, yielding results comparable to those from radiochemical analysis when concentrations are greater than 5 pCi/g (0.185 Bq/g) of Am and greater than 50 pCi/g (1.85 Bq/g) of Pu in the presence of americium [76].

Plutonium lung burdens have also been measured with semiconductor detectors, including Ge(Li) [77], Si(Li), and high-purity Ge [78], yielding sensitivity superior to scintillation detectors.

16.4.4. Other Methods

A number of methods for measuring Pu do not fit into any of the previous categories. These include several special methods for α-particle emitters, a number of electrochemical methods, and a few others. None of these will be described in detail.

For discussion of gross-α techniques consult Chapter 7. An α-particle technique that has been applied specifically to Pu is a zinc sulfide (ZnS) scintillation. Agapsowicz and Newton [79] used such a detector for monitoring airborne concentration of α-particle emitting aerosols. Use of ZnS for detecting soil Pu contamination is discussed in Ref. 80.

Liquid scintillation counting may be used for Pu. Hieber, Ide, and Boyd [81] applied this technique to environmental plutonium monitoring using a configuration suitable for multiple samples. Liquid scintillation counting for Pu in bone and tissue samples is discussed in Ref. 82. Other techniques are available but not often used, either because of their difficulty or their lack of sensitivity. These include: neutron activation techniques; photometric techniques, including x-ray fluorescence; and electrochemical techniques, including coulometry [83], polarography [84], and potentiometric and amperometric titration [85,86].

16.5. SUMMARY AND CONCLUSION

Plutonium is permitted in environmental media only at very low concentrations. The principal means for measuring plutonium isotopes detect α-particles emitted during decay, x-rays immediately following decay, or x-rays and γ-rays emitted following decay of ^{241}Am, the daughter of ^{241}Pu. In most cases substantial efforts are required to overcome the background due to uranium and thorium decay daughters, and (particularly in the case of air monitoring) the radon daughters.

Alpha particles are typically used as the signature for the presence of Pu in air; in monitoring α-particle energies directly from an air filter, instruments of moderate sensitivity are available to detect a 40-hr occupational concentration if present for a period of hours. However, the time required can be shortened by an order of magnitude by using specially designed and rather complex systems. Much greater sensitivity is possible if the α-particles are monitored after radiochemical analysis of the filter sample. Radiochemical analysis results in great sensitivity for measurement of Pu in bulk samples, whether water, soil, or biological samples: sub-pCi amounts can be measured in hour-long counting periods. X-rays and γ-rays, from Pu or Am, are typically used for survey efforts. Scintillation counters, gas counters, and semiconductor detectors are available for measuring radiation energies and can effectively measure soil distributions from several hundred nCi/m^2 to about 40 nCi/m^2 (1.48 kBq/m^2). Semiconductor arrays are the most sensitive. Fixed semiconductor detector systems may achieve great sensitivity, as low as 1 pCi/g (0.037 Bq/g). Portable semicontractor arrays can detect several pCi/g, which is not nearly as sensitive as radiochemical techniques. Radiochemical techniques are not generally portable and require skill and care in sample preparation.

Other types of monitoring techniques are available but not widely used. Perhaps the most significant improvement in monitoring techniques uses the fact that Pu is present in air as particles. At low concentrations, air sampling techniques may be plagued by the fact that Pu is concentrated into particles; on the other hand, impaction techniques may be used to concentrate these particles into a relatively small airstream relatively free of background activities. This could improve air monitoring capabilities.

Numerous plutonium monitoring techniques are available, and in many circumstances these techniques are adequate for measuring elevated levels in the environment. Substantial time is often required for measuring low concentrations, and this may constitute a serious problem. Although improvements may be made, it must be expected that this difficulty will persist, considering the low permissible concentrations for Pu and the correspondingly stringent requirements for monitoring capabilities.

REFERENCES

1. R. J. Budnitz, "Plutonium: A Review of Measurement Techniques for Environmental Monitoring," *IEEE Trans. Nucl. Sci.*, **NS-21**(1), 430, 1974.
2. A. V. Nero, "Measurement and Instrumentation Techniques for Monitoring Plutonium and Uranium Particulates Released from Nuclear Facilities," Lawrence Berkeley Laboratory Report LBL-5218, 1976.
3. H. C. Hodge, J. N. Stannard, and J. B. Husch, Eds., *Uranium, Plutonium, Transplutonium Elements*, Vol. 36 of Handbuch de Experimentellen Pharmakalogie, Springer-Verlag, New York, 1973.
4. B. J. Stover and W. S. S. Jee, *Radiobiology of Plutonium*, J. W. Press, Salt Lake City, 1972.
5. W. S. S. Jee, Ed., in *The Health Effects of Plutonium and Radium*, Proceedings of a Symposium Held in Sun Valley, Idaho, Oct. 6–9, 1975, J. W. Press, Salt Lake City, 1976.
6. International Commission on Radiological Protection, *The Metabolism of Compounds of Plutonium and Other Actinides*, ICRP Publication 19, Pergamon Press, New York, 1973.
7. National Academy of Sciences—National Research Council, *Health Effects of Alpha-Emitting Particles in the Respiratory Tract*, Environmental Protection Agency Report EPA 520/4-76-013, 1976.
8. B. W. Wachholz, Compiler, "Plutonium and Other Transuranium Elements: Sources, Environmental Distribution and Biomedical Effects," U.S. Atomic Energy Commission Report WASH-1359, 1974.
9. T. H. Pigford and K. P. Ang, "The Plutonium Fuel Cycles," *Health Phys.*, **29**(4), 451, 1975.
10. L. C. Hebel, Chairman of APS Study Group, "Report to the American Physical Society by the Study Group on Nuclear Fuel Cycles and Waste Management," *Rev. Mod. Phys.*, **50**(1), Part II, 1978.
11. *Transuranium Nuclides in the Environment*, Symposium Proceedings, San Francisco, 1975, STI/PUB/410, International Atomic Energy Agency, Vienna, 1976.
12. A. V. Nero, *A Guidebook to Nuclear Reactors*, University of California Press, Berkeley, 1979.
13. U.S. Nuclear Regulatory Commission, "Final Generic Environmental Statement on the Use of Recycle Plutonium in Mixed Oxide Fuel in Light Water Cooled Reactors," NRC Report NUREG-0002, 5 vols., 1976.
14. Energy Research and Development Administration, "Final Environmental Statement, Liquid Metal Fast Breeder Reactor Program," ERDA Report ERDA-1535, 3 vols., 1975, which also includes: Atomic Energy Commission, "Proposed Final Environmental Statement, Liquid Metal Fast Breeder Reactor Program," AEC Report WASH-1535, 7 vols., 1974 (NTIS).
15. *Proceedings of the Rocky Flats Symposium on Safety in Plutonium Handling Facilities*, USAEC Report CONF-710401, 1971 (NTIS).

16. C. M. Lederer, J. M. Hollander, and I. Perlman, *Table of Isotopes*, 6th ed., Wiley, New York, 1967.
17. National Council on Radiation Protection and Measurements, *Basic Radiation Protection Criteria*, NCRP Report No. 39, January 1971; International Commission on Radiological Protection, *Recommendations of the International Commission on Radiological Protection* (Adopted Sept. 17, 1965), ICRP Publication 9, Pergamon Press, New York, 1966.
18. Environmental Protection Agency, *Proposed Guidance on Dose Limits for Persons Exposed to Transuranium Elements in the General Environment*, Report EPA 520/4-77-016, 1977 (NTIS).
19. J. W. Healy, *A Proposed Interim Standard for Plutonium in Soils*, USAEC Report LA-5483-MS, Los Alamos Scientific Laboratory, 1974 (NTIS).
20. Transuranic Waste Disposal: Proposed Standards for Protection Against Radiation, and Exemptions and Continued Regulatory Authority, *Federal Register*, **39**(178), 32,921–32,923, Sept. 21, 1974.
21. M. E. Wrenn, "Environmental Levels of Plutonium and the Transplutonium Elements," p. 89 of Ref. 8.
22. P. W. Krey, "Remote Plutonium Contamination and Total Inventories from Rocky Flats," *Health Phys.*, **30**, 209, 1976.
23. L. L. Eberhardt and R. O. Gilbert, *Statistical Analysis of Soil Plutonium Studies*, Nevada Test Site, USAEC Report BNWL-B-217, Battelle Pacific Northwest Laboratory, 1971 (NTIS).
24. H. L. Volchok, R. H. Knuth, and M. T. Kleinman, "The Respirable Fraction of Plutonium at Rocky Flats," *Health Phys.*, **23**, 395, September, 1972.
25. P. L. Phelps and L. R. Anspaugh, "Resuspension Element Status Report," in *Dynamics of Plutonium in Desert Environments*, P. B. Dunaway and M. G. White, Eds., USAEC Nevada Operations Office, Report NVO-142, 1974, p. 221 (NTIS).
26. P. N. Lem, J. V. Behor, and F. N. Buck, "Resuspension of Plutonium from Contaminated Land Surfaces: Meteorological Factors," U.S. EPA Environmental Monitoring and Support Laboratory Report EPA-600/4-77-037, 1977 (NTIS).
27. C. A. Little and F. W. Whicker, "Plutonium in Rocky Flats Soil," *Health Phys.*, **34**, 451–457, 1978.
28. R. M. Emergy, D. C. Klopfer, and M. C. McShane, "The Ecological Export of Plutonium from a Reprocessing Waste Pond," *Health Phys.*, **34**, 255, 1978.
29. A. Aarkrog, "Environmental Behavior of Plutonium Accidentally Released at Thule, Greenland," *Health Phys.*, **32**, 271, 1977.
30. D. W. Moeller, "Environmental Surveillance for Nuclear Facilities," *Nucl. Safety*, **19**, 66, 1978.
31. National Council on Radiation Protection and Measurements, *Natural Background Radiation in the United States*, NCRP Report No. 45, 1975.
32. W. A. Phillips and C. L. Lindekin, "Plutonium Alpha Air Monitor Using a Solid State Detector," *Health Phys.*, **9**, 299, 1963.
33. J. F. Kordas and P. L. Phelps, "A Review of Monitoring Instruments for Transuranics in Fuel Fabrication and Reprocessing Plants," Lawrence Livermore Laboratory Report UCRL-52123 (Rev. 1), 1977; J. F. Kordas and P. L. Phelps, "An On-line High Sensitivity Measurement System for Transuranic Aerosols," *IEEE Trans. Nucl. Sci.*, **NS-24**(1), 670, 1977; J. F. Kordas and P.

L. Phelps, "A Review of Measurement Techniques for Stack Monitoring of Long-Lived Alpha Emitters," *IEEE Trans. Nucl. Sci.*, **NS-26**(1), 757, 1979.

34. E. C. Tsivoglou, H. E. Ayer, and D. A. Holaday, "Occurrence of Nonequilibrium Atmospheric Mixtures of Radon and Daughters," *Nucleonics*, **11**(9), 40, 1953.
35. D. E. Martz, "Analysis of Atmospheric Concentrations of RaA, RaB and RaC by Alpha Spectroscopy," *Health Phys.*, **17**(1), 131, 1969.
36. N. Jonassen and E. I. Hayes, "The Measurement of Low Concentrations of the Short-Lived Radon-222 Daughters in the Air by Alpha Spectroscopy," *Health Phys.*, **26**(1), 104, 1974.
37. R. J. Budnitz, "Radon-222 and Its Daughters: A Review of Instrumentation for Occupational and Environmental Monitoring," *Health Phys.*, **26**(2), 145, 1974.
38. J. C. Elder, M. Gonzales, and H. J. Ettinger, "Plutonium Aerosol Size Characteristics," *Health Phys.*, **27**(1), 45, 1974.
39. W. B. Seefeldt, W. J. Mechan, and M. J. Steindler, Characterization of Particulate Plutonium Released in Fuel Cycle Operations, ERDA Report ANL-75-78, Argonne National Laboratory, 1976 (NTIS).
40. E. W. Bretthauer, A. J. Cummings, and S. C. Black, "Characterization of Emissions from Plutonium-Uranium Oxide Fuel Fabrication," U.S. Environmental Protection Agency Report EPA-600/7-77-079, 1977 (NTIS).
41. L. M. McDowell and F. W. Whicker, "Size Characteristics of Plutonium Particles in Rocky Flats Soil," *Health Phys.*, **35**, 293, 1978.
42. O. G. Raabe, G. J. Newton, C. J. Wilkinson, S. V. Teague, and R. C. Smith, "Plutonium Aerosol Characterization Inside Safety Enclosures at a Demonstration Mixed Oxide Fuel Fabrication Facility," *Health Phys.*, **35**, 649, 1978.
43. S. M. Saunders, "Isolation and Characterization of Individual Plutonium-Bearing Particles in Atmospheric Effluents from a Nuclear Processing Plant," *Health Phys.*, **36**, 371, 1979.
44. J. M. Nielson, Pacific Northwest Laboratory Annual Report for 1975 to the ERDA Division of Biomedical and Environmental Research, Part 4, "Physical and Analytical Sciences and Analysis and Assessment," Battelle Pacific Northwest Laboratory, Report BNWL-2000, 1976. J. Stoffels, Radiological Sciences Department, Battelle Pacific Northwest Laboratory, private communication. See also Ref. 28.
45. W. D. Davis, "Continuous Mass Spectrometric Analysis of Particulates by Use of Surface Ionization," *Environ. Sci. Technol.*, **11**(6), 593 (1977). W. D. Davis, "Continuous Mass Spectrometric Determination of Concentration of Particulate Impurities in Air by Use of Surface Ionization," *Environ. Sci. Technol.*, **11**(6), 593, 1977.
46. G. K. Rusch, W. P. McDowell, and W. G. Knapp, "The ZPR-9 Airborne Plutonium Monitoring System," *IEEE Trans. Nucl. Sci.*, **NS-23**(1), 690, 1976.
47. T. J. Yule, "An On-Line Monitor for Alpha-Emitting Aerosols," *IEEE Trans. Nucl. Sci.*, **NS-25**(1), 762, 1978.
48. D. Talley, G. J. Newton, H. C. Yeh, and S. J. Waligora, "Development of a Five-Stage, Round Jet, Virtual Cascade Impactor," in *Inhalation Toxicology Research Institute Annual Report, 1975—1976,* ERDA Report LF-56, 1976, Lovelace Biomedical and Environmental Research Institute (NTIS).

49. B. W. Loo, J. M. Jaklevic, and F. S. Goulding, "Dichotomous Virtual Impactors for Large Scale Monitoring of Airborne Particulate Matter," Lawrence Berkeley Laboratory, Report LBL-3854, 1975.
50. C. W. Sill and R. L. Williams, "Preparation of Actinides for α Spectrometry Without Electrodeposition," *Anal. Chem.*, **53**, 412, 1981.
51. J. W. Healy, "Problems of Plutonium Measurement in the Environment," *IEEE Trans. Nucl. Sci.*, **NS-19**(1), 219 (1972).
52. International Atomic Energy Agency, "Rapid Methods for Measuring Radioactivity in the Environment," Symposium Proceedings, Neuherberg bei Munchen, 1971, STI/PUB/289, International Atomic Energy Agency, Vienna, 1971, comment by C. W. Sill, p. 199.
53. J. H. Harley, Ed., *HASL Procedures Manual*, USAEC Report HASL-300, U.S. Environmental Measurements Laboratory, U.S. Department of Energy, New York City, 1972 (updated to 1977).
54. M. G. White and P. B. Dunaway, Eds., *Nevada Applied Ecology Group Procedures Handbook for Environmental Transuranics*, ERDA Report NVO-166, 2 vols., Nevada Operations Office, Energy Research and Development Administration, 1976 (NTIS).
55. N. A. Talvitie, "Radiochemical Determination of Plutonium in Environmental and Biological Samples by Ion Exchange," *Anal. Chem.*, **43**(13), 1827, 1971.
56. C. W. Sill, K. W. Puphal, and F. D. Hindman, "Simultaneous Determination of Alpha-Emitting Nuclides of Radium Through Californium in Soil," *Anal. Chem.*, **45**(12), 1725, 1974.
57. P. B. Hahn, E. W. Bretthauer, P. B. Altringer, and N. F. Matthews, "Fusion Method for the Measurement of Plutonium in Soil: Single-Laboratory Evaluation and Interlaboratory Collaborative Test," U.S. Environmental Protection Agency Report, EPA-600/7-77-078, 1977.
58. C. W. Sill, F. D. Hindman, and J. I. Anderson, "Simultaneous Determination of Alpha-Emitting Nuclides of Radium Through Californium in Large Environmental and Biological Samples," *Anal. Chem.*, **51**, 1307, 1979.
59. K. M. Wong, "Radiochemical Determinations of Plutonium in Seawater, Sediments and Marine Organisms," *Anal. Chim. Acta*, **56**, 355, 1971.
60. C. T. Bishop, A. A. Glosby, R. Brown, and C. A. Phillips, "Anion Exchange Method for the Determination of Plutonium in Water: Single-Laboratory Evaluation and Interlaboratory Collaborative Study," U.S. EPA Report EPA-600/7-78-122, 1978 (NTIS).
61. D. E. Bernhardt, "Evaluation of Sample Collection and Analysis Techniques for Environmental Plutonium," U.S. EPA Office of Radiation Programs Technical Note DRP/LV-76-5, 1976.
62. H. L. Beck, J. E. McLaughlin, and K. M. Miller, "In Situ Determinations of Environmental Plutonium and Its Related Nuclides," *IEEE Trans. Nucl. Sci.*, **NS-23**(1), 676, 1976.
63. M. G. Strauss, "Considerations in Measuring Trace Radionuclides in Soil Samples by L X-Ray Detection," *IEEE Trans. Nucl. Sci.*, **NS-25**(1), 740, 1978.
64. J. F. Tinney, J. J. Koch, and C. T. Schmidt, "Plutonium Survey with An X-Ray Sensitive Detector," Lawrence Livermore National Laboratory, Report UCRL-71362, 1969.

65. J. A. Kirby, L. R. Anspaugh, P. L. Phelps, G. A. Armantrout, and D. Sawyer, "A Detector for In Situ Spectrometric Analysis of 241-Am and Plutonium in Soil," *IEEE Trans. Nucl. Sci.*, **NS-23**(1), 683, 1976).
66. B. S. Burton and M. N. Anastasi, *Development and Construction of an Instrument To Detect Fissionable Material*, Air Force Weapons Laboratory, Report AFWL-TR-71-145, 1972.
67. H. B. Piltingsrud and R. L. Farr, "A Beta-Alpha-Gamma X-Radiation (BAG-X) Detection Probe for Use with a Plutonium-Americium Survey Instrument," *Health Phys.*, **34**, 177, 1978.
68. T. Ishihar, "Plutonium Lung Monitor Using a Thin NaI(Tl) Crystal of Large Area," *Health Phys.*, **17**, 669, 1969.
69. S. P. Garg and D. Newton, "Reduction of the Low Energy Background of a Phoswich Detector," *Health Phys.*, **32**, 113, 1977.
70. M. A. Wolf, "Portable Survey Instrument for Low Levels of Transuranic Contaminants," Los Alamos Scientific Laboratory Report, LA-6860-MS, 1977; C. J. Umbarger and M. A. Wolf, "A Battery Operated Portable Phoswich Detector for Field Monitoring of Low Levels of Transuranic Contaminants," *Nucl. Instrum. Meth.*, **155**, 453, 1978.
71. T. Tomitani and E. Tanaka, "Large Area Proportional Counters for Assessment of Plutonium Lung Burden," *Health Phys.*, **18**, 195, 1970.
72. S. S. Yaniv, A. Brodsky, and D. Sashin, "Analysis of Plutonium-Americium Spectra in a Xenon Proportional Counter: Effect of Characteristic X-Rays," *Health Phys.*, **23**, 705, 1972.
73. A. J. P. L. Policarpo, "Improved Resolution for Low Energies with Gas Proportional Scintillation Counters," *Nucl. Instrum. Meth.*, **102**, 337, 1972.
74. H. E. Palmer, "Recent Advances in Gas Scintillation Proportional Counters," *IEEE Trans. Nucl. Sci.*, **NS-22**(1), 100, 1975.
75. D. Newton, F. A. Fry, B. T. Taylor, M. C. Eagle, and R. C. Sharma, "Interlaboratory Techniques for Measuring Lung Burdens of Low Energy Photon Emitters," *Health Phys.*, **35**, 751, 1978.
76. F. P. Brauer, "Measurement of Environmental ^{241}Am and the Plutonium:^{241}Am Ratio by Photon Spectrometry," *IEEE Trans. Nucl. Sci.*, **NS-24**(1), 591, 1977.
77. W. H. Tyree and R. W. Bistline, *Plutonium-239 Detection in Vivo with a Germanium-Lithium Array*, USAEC Report RFP-1488, Dow Chemical Company, Rocky Flats Division, 1970 (NTIS).
78. W. H. Tyree, *A High-Purity Germanium Detection System for in Vivo Measurement of Americium and Plutonium*, ERDA Report RFP-2525, Atomics International Division, Rocky Flats Plant, 1976 (NTIS).
79. D. E. Agapsowicz and G. J. Newton, "A Real-Time Alpha-in-Air Monitoring System," *Health Phys.*, **37**, 379, 1979.
80. A. J. Ahlquist, C. J. Umbarger, and A. K. Stoker, "Recent Developments for Field Monitoring of Alpha-Emitting Contaminants in the Environment," *Health Phys.*, **34**, 486, 1978.
81. R. Hiebert, H. Ide, and H. Boyd, "Low-Background Plutonium Assaying Using a Modified Commercial Liquid Scintillation Detector and Sample Changer," *Health Phys.*, **32**, 311, 1977.
82. W. J. McDowell and J. F. Weiss, "Liquid Scintillation Alpha Counting and Spectrometry and Its Application to Bone and Tissue Samples," *Health Phys.*, **32**, 73, 1977.

83. J. R. Stokely and W. D. Shults, "Controlled-Potential Coulometric Determination of Plutonium in the Presence of Iron," *Anal. Chem.*, **43**(4), 603, 1971.
84. K. Koyama, "Square Wave Polarography of Plutonium," *Anal. Chem.*, **32**(4), 523, 1960.
85. C. E. Pietri and A. W. Wenzel, "Improved Methods for the Constant-Current Potentiometric Determination of Plutonium," *Talanta*, **14**(2), 215, 1967.
86. C. E. Hedrick, "Improved Amperometric Procedure for Determining Plutonium," *Anal. Chem.*, **44**(2), 377, 1972.

17 STRONTIUM

There are two radioactive strontium (Sr) isotopes of significance in radiological measurements of the environment: strontium-89 and strontium-90. However ^{90}Sr is by far the more important of the two. Of all the fission products, ^{90}Sr is one of the most hazardous, ranking below only the iodine isotopes, some of the radioactive noble gases, and cesium-137 in importance.

The radiotoxicity of ^{90}Sr results from its decay characteristics and mode of metabolism in the body. The environmental significance of ^{90}Sr, is mostly due to its long half-life (28.1 yr). It is a pure β-particle emitter with only one decay mode, leading to yttrium-90 by emission of a negative β-particle with E_{max} = 546 keV. The most probable mode of decay of the ^{90}Y daughter nuclide is by β-particle emission (E_{max} = 2.27 MeV) with a 64-hr half-life, leading to stable zirconium-90. The β-particle energy spectra are shown in Fig. 17.1 and Fig. 17.2 [1]. Within tissue the maximum range of the ^{90}Y β-particle produced is about 10 mm, and in bone about 6 mm. The spectral energy distribution of β-particles means that the charged particle radiation from ^{90}Sr and ^{90}Y are completely absorbed within the body of even a small laboratory animal. Owing to the large difference between the half-life of ^{90}Sr and its daughter product ^{90}Y, the two radionuclides, when free of external influence, attain an equilibrium of activity (secular equilibrium) starting from pure ^{90}Sr within a few days. When an animal ingests a solution containing ^{90}Sr and ^{90}Y in equilibrium, the distribution of the two nuclides within the organism and its individual organs is not the same because the two are metabolized by different pathways. Consequently, the radioactive equilibrium within individual organs and tissues is disturbed. Because the mean residence time in bone is about 8 yr, Sr-Y equilibrium is assumed for dosimetric considerations once Sr reaches bone.

The other strontium isotope, ^{89}Sr, has a 52-day half-life, much shorter than that of ^{90}Sr. It decays by nearly pure β-particle emission to stable yttrium-89, the maximum beta energy being 1.463 MeV. Details of the radiochemical properties of Sr are reviewed in Ref. 2.

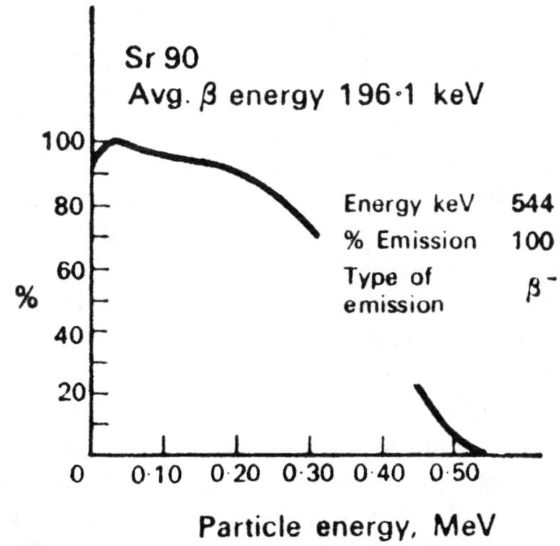

Figure 17.1. Spectrum of beta energies from decay of strontium-90 (reproduced from Mantel, Ref. 1, with permission of Pergamon Press © 1972).

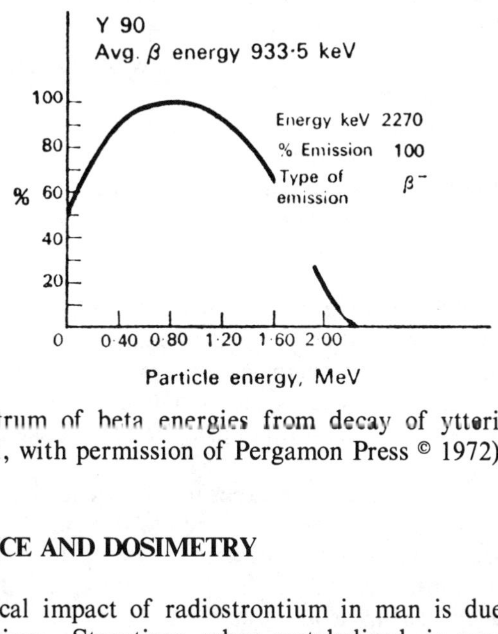

Figure 17.2. Spectrum of beta energies from decay of ytterium 90 (reproduced from Mantel, Ref. 1, with permission of Pergamon Press © 1972).

17.1. SIGNIFICANCE AND DOSIMETRY

The main radiological impact of radiostrontium in man is due to its biochemical resemblance to calcium. Strontium, when metabolized, is concentrated and localized in the skeleton [3], where it remains for several years. It subsequently decays to produce a radiation dose directly at the site. The 2.27-MeV maximum energy β-particle of the daughter ^{90}Y has a range of only about 1.0 g/cm^2.

The major hazard of ^{90}Sr is considered to be via the food chain. However, there appears to be a biological discrimination against radiostrontium since the strontium-to-calcium ratio in milk and cheese for example, is five to ten times lower than in the grass constituting the cattle food. In addition, the plant uptake of Sr from the soil is somewhat less than Ca uptake. However, this discrimination by plant species is partly a soil-mixing phenomenon [3].

In general, in each link of the food chain involving radiostrontium (soil, forage, crops, grazing cattle, milk, and man) there is a discrimination against Sr in favor of Ca. The mechanism of selective utilization of Ca in the presence of Sr is not completely understood. Three factors are noted: (a) preferential absorption of calcium from the gastrointestinal tract; (b) a larger secretion of calcium into the milk [3]; and (c) a larger excretion of Sr than of Ca in the urine. Phosphorus has been shown to decrease the absorption and retention of radiostrontium in animals [4]. The major site of strontium absorption in man is probably the upper small intestine. About 20% of an intravenously injected dose is absorbed if calcium intake is low; in a normal adult man the Sr uptake is about 15% [5].

In dosimetry studies the term *dose* has been used to express at least three different concepts. *Average skeletal dose* is a commonly used term derived from measurement of energy retention within radioactively labeled bones. The term refers to β-particle energy retained within individual bones, averaged over the skeleton. A large fraction of this energy is absorbed by the mineralized component of bone and by cells embedded in it (osteocytes). An assumption of uniform energy deposition implies that all marrow and bone cells receive the same dose from a given concentration of radionuclide in bone matrix, regardless of the spatial location with respect to bone volume. It has been shown that (a) after uniform labeling of ^{90}Sr in beagle bones cessation of active strontium administration causes nonuniform distribution, (b) the dose rates along the middle of bone can vary up to a factor of 3, and (c) dose rates do not correlate well with the strontium-90 content of the bone at the site of measurement [6,7].

Using another definition Goldman, Della Rosa, and Momeni [7] calculated doses from knowledge of the total body burden of ^{90}Sr-^{90}Y in the beagle and from dose rates measured by thermoluminescent dosimeters (TLD) exposed in the middle of long bones. Spiers et al. [8] defined dose in terms of absorbed dose to bone marrow or endosteal cells within actual bone structures and did not equate the relevant dose to the energy in the mineralized matrix.

Studies on the effects of radiostrontium administered in different ways to several types of experimental animals indicate that bone marrow is the most sensitive tissue at risk [5]. Mays and Lloyd [9] computed a value of one cancer case per million person-rads (see Chapter 2) exposed as a cumulative 50-yr risk to adults and suggested that juveniles may be 0.5 to 4 times as sensitive if bone-sarcoma induction is the endpoint.

In summarizing, for human radiation exposure experience based on acute external irradiation, a risk coefficient has been estimated of 10 to 40 cancer cases per million person-rads for leukemia and a risk coefficient possibly four times greater for all other malignant neoplasms. In addition a risk coefficient of 100 to 200 has been quoted for all malignant neoplasms in children [10].

Anyone studying the dosimetry of radiostrontium must take into account the decay and physical properties of Y. Both ^{90}Sr and ^{90}Y are bone-seeking radioisotopes. Yttrium-90 finds its way to the bones, after transient accumulation in the visceral organs. After incorporation both nuclides become so stably bound in the skeleton that only a minimal quantity is excreted [11].

Radionuclide absorption through the skin can also be of practical significance. It has been found that Sr is absorbed more than 200 times more readily through damaged than undamaged skin [12].

The factors used by the National Council on Radiation Protection (NCRP) [13] in converting strontium-90 activity to absorbed dose rate or committed absorbed dose are summarized in Table 17.1.

Table 17.1. Factors Used to Convert Concentrations to Absorbed Dose or Dose Rate in Tissue (Adapted from National Council on Radiation Protection and Measurements, Ref. 13)

Nuclide	Concentration Unit	Organ	Factor
		Absorbed Dose	
^{90}Sr	pCi/g calcium in body	Bone (narrow)	1.4 mrad/yr
		Bone (endosteal)	2 mrad/yr
		Committed Dose	
^{90}Sr	mCi/km^2 deposited	Bone (marrow)	0.8 mrad
		Bone (endosteal)	1.0 mrad

17.2. SOURCES OF ENVIRONMENTAL RADIOSTRONTIUM

The main sources of environmental radiostrontium are nuclear reactors with associated fuel reprocessing plants, radioactive waste disposal sites, and fallout after atmospheric nuclear bomb tests.

The operation of a nuclear reactor results in the direct and indirect production of ^{89}Sr and ^{90}Sr through neutron-induced fission of the fuel atoms. As shown by the following equations, these radiostrontium nuclides lie at or near the end of the cumulative fission chains representing mass numbers 89 and 90:

$$^{89}Se \xrightarrow[0.4\ sec]{\beta^-} {}^{89}Br \xrightarrow[45\ sec]{\beta^-} {}^{89}Kr \xrightarrow[3\ min]{\beta^-} {}^{89}Rb \xrightarrow[15\ min]{\beta^-} {}^{89}Sr \xrightarrow[50.4\ days]{\beta^-} {}^{89}Y\ \text{(stable)}$$

$$^{90}Se \xrightarrow[short]{\beta^-} {}^{90}Br \xrightarrow[16\ sec]{\beta^-} {}^{90}Kr \xrightarrow[32\ sec]{\beta^-} {}^{90}Rb \xrightarrow[4\ min]{\beta^-} {}^{90}Sr \xrightarrow[29yr]{\beta^-} {}^{90}Y \xrightarrow[64hr]{\beta^-} {}^{90}Zr\ \text{(stable)}.$$

Reactors can be designed and built so that most of the activities in the reactor fuel do not normally escape into the environment. At a reactor site, the average release rates of fission products are only very small fractions of the total fuel inventories. For example, at the Dresden Nuclear Power Station in Illinois the ratios of released to produced activities in 1968 [14] were 4×10^{-9} (for ^{90}Sr) and 6×10^{-8} (for ^{89}Sr). The average release rates in stack effluent were 700 pCi/sec (25.9 Bq/sec) (^{89}Sr) and 3 pCi/sec (0.111 Bq/sec) (^{90}Sr), while in liquid effluent the rates were 8000 and 900 pCi/sec (296 and 33 Bq/sec), respectively.

The 1977 report from the United Nations Scientific Committee on the Effects of Atomic Radiation (UNSCEAR) [15] indicates that ^{89}Sr and ^{90}Sr were released as particles in airborne effluents at the rate 0.60 Ci/MW_e-yr (22.2 GBq/MW_e-yr) and

0.14 Ci/MW$_e$-yr (5.1 GBq/MW$_e$-yr), respectively, from pressurized water reactors in 1974 while from boiling-water reactors the respective release rates were 330 Ci/MW$_e$-yr (12.2 TBq/MW$_e$-yr) and 5.0 Ci/MW$_e$-yr (185 GBq/MW$_e$-yr). The ^{89}Sr and ^{90}Sr discharged in liquid effluents was 0.014 Ci/MW$_e$-yr (0.52 GBq/MW$_e$-yr) and 0.0022 Ci/MW$_e$-yr (0.08 GBq/MW$_e$-yr) respectively, from pressurized-water reactors in 1974, while from boiling-water reactors the respective release rates were 0.80 Ci/MW$_e$-yr (29.6 GBq/MW$_e$-yr) and 0.098 Ci/MW$_e$-yr (3.62 Bq/MW$_e$-yr).

An Environmental Protection Agency (EPA) study on the discharge of radionuclides from the Oyster Creek Nuclear Generating Station into Barnegat Bay shows that one key radiation exposure pathway is fish consumption with the most important radionuclide being ^{90}Sr [16]. The observed concentrations of ^{90}Sr in the fish resulted in a dose equivalent of less than 1 mrem/yr (10 μSv/yr).

A primary potential source for environmental release is an operating nuclear fuel reprocessing plant where the fuel rods are dissolved and the radiostrontium is released from the cladding (see Chapter 11). Although no commercial plants are currently operating, precautions have been designed into the planned plants to assure practically complete containment in normal operation. For example, the 150-day storage of fuel elements before reprocessing allows 86% of the ^{89}Sr to decay (in contrast to 1% of the ^{90}Sr). At the now-abandoned Midwest Fuel Recovery Plant in Morris, Illinois, the design called for no environmental release of radioactive waste in either solid or liquid form. They had later suggested a limit on emissions of gaseous gross β-particle emitters (particles in air) of 7.0 μCi/sec (0.259 MBq/sec) based upon ^{90}Sr as the limiting radionuclide [17]. The total environmental emissions should be small. The size of possible releases without engineering barriers is shown in Table 17.2 which gives the amount of activity in spent fuel at various times after discharge.

The Nuclear Regulatory Commission stresses the difference in impacts depending upon whether or not plutonium is recycled [19]. For example, the integrated ^{90}Sr effluents from a light-water reactor industry, 1975–2000, are 0.18 Ci (6.66 GBq) for prompt uranium and plutonium recycle or uranium recycle only, and only 0.025 Ci (0.925 GBq) for material not recycled.

A study to determine environmental impact of an operating nuclear fuel reprocessing plant on the Idaho National Engineering Laboratory (INEL) site in southeastern Idaho was conducted recently [20]. It examined ^{90}Sr concentration in the bones of antelope foraging in the area. Metacarpal bones were collected from

Table 17.2. Summary of Activities Remaining in Spent Fuel for Strontium and Yttrium as a Function of Past Irradiation Time[a] (Adapted from Nuclear Regulatory Commission, Ref. 18)

			Curries/Metric Ton Uranium			
Z	Radionuclide	Half-Life	Discharge	160 days	1 year	10 years
38	Sr-89	50.4 days	7.18×10^5	8.51×10^4	5.53×10^3	5.15×10^{-16}
	Sr-90	28 yr	7.76×10^4	7.68×10^4	7.57×10^4	6.07×10^4
39	Y-90	64.2 hr	8.07×10^4	7.68×10^4	7.58×10^4	6.07×10^4

[a]Power = 30 MW; Burnup = 33,000 MWd per ton; Flux = 2.92×10^{13} n/cm^2·sec.

pronghorn antelope near the plant; control bones were collected from off-site animals at higher elevations. Average concentrations in metacarpals were 9.6 ± 2.8 pCi/g (0.355 Bq/g) (ash) within 10 km of the Idaho Chemical Processing Plant (ICPP), 4.0 ± 0.9 pCi/g (0.148Bq/g) for animals on the remainder of the INEL site, and 5.5 ± 1.0 pCi/g (0.203 Bq/g) for control animals. Antelope near the ICPP received approximately double the radiation dose [90 mrad/yr (0.90 mGy/yr)] to the bone compared to doses received by other INEL antelope [40 mrad/yr (0.04 mGy/yr)] as a result of ^{90}Sr in bone [20].

Although at the present time there are no operating spent fuel reprocessing plants in the United States, if any do start operating ^{90}Sr should be strictly monitored and controlled. The 1979 UNSCEAR report [15] gives the collective dose commitments from particles in airborne effluents and radionuclides in liquid effluents from fuel reprocessing plants. Table 17.3 summarizes the results. Of all the radionuclides released, Sr dominates the dose to bone [15].

The public is most aware of fallout as the source of radiostrontium. Widespread concern in the era of large-scale atmospheric nuclear bomb tests led to much research into the environmental behavior of radiostrontium. Fallout concentrations were measured extensively, and are still measured routinely [21]. The deposition was considerable. Quoting from the Federal Radiation Council [22],

> On January 1, 1963, the accumulated levels of Strontium-90 deposited over the United States varied from about 100 to 125 millicuries per square mile in the "wet" areas (areas of greatest annual precipitation) to 40 to 50 millicuries per square mile in the "dry" regions.

Table 17.3. Strontium Collective Dose Commitments from Fuel Reprocessing Plants (Adapted from UNSCEAR, Ref. 15)

Radionuclide	Normalized Activity Released (Ci/MW_e-yr)	Collective Dose (person-rad/MW_e-yr)			
		Total Body	Lung (whole)	Bone Marrow	Bone Lining Cells
In Airborne Effluents					
^{90}Sr	6×10^{-5}	—	0.02×10^{-4}	6.1×10^{-3}	8.4×10^{-3}
^{89}Sr	3×10^{-6}	—	0.03×10^{-6}	0.02×10^{-5}	—
		Total Body	Large (Lower) Intestine	Bone	Thyroid
In Liquid Effluents					
$^{90}Sr^a$	6.1	—	—	0.01	—
^{89}Sr	6.4	—	—	0.00001	—
$^{90}Sr^b$	0.013	—	—	0.4	—
$^{89}Sr^b$	0.0033	—	—	0.0003	—

[a]Release to salt water.
[b]Release to fresh water.

The recent Chinese nuclear test again deposited a small amount of ^{90}Sr as fallout [23].

In 1977, UNSCEAR reviewed Sr resulting from worldwide fallout [15]. Their report stated that most of the worldwide contamination of long-lived fission products has been from the stratospheric fallout resulting from large thermonuclear tests. It presents the annual deposition of ^{90}Sr in the northern and southern hemispheres for the period 1958–1975, cumulative deposition, and total atmospheric inventory in each hemisphere. "Deposition" refers to the activity of the radioactive material deposited on a specified area. The cumulative deposit is the activity present in a specified area at a given time; it is the result of past depositions and radioactive decay.

The deposition density of ^{90}Sr is measured monthly at a number of monitoring stations in both hemispheres [21]. (Deposition density is the activity deposited per unit area.) Studies on the latitudinal distribution of strontium-90 integrated deposition density show the deposition is largest in the midlatitude regions of each hemisphere.

The deposition of ^{90}Sr occurs over land and water, but that over land is the more important pathway for human exposure. The quantitative and qualitative aspects of the transfer of ^{90}Sr along terrestrial pathways is treated in detail by Francis [24], who found:

1. Soil texture has a greater influence on vertical distribution of ^{90}Sr than any other soil property.
2. There is little evidence that more than 10% of the soil radiostrontium is sorbed in nonexchangeable forms in most soils.
3. The exchangeable calcium status of the soil is the most important aspect of soil chemistry affecting plant uptake of Sr. Liming acid base-unsaturated soils is the only realistic soil treatment for reducing plant uptake of radiostrontium.

Measurements of the soil profile have shown that ^{90}Sr is present mostly in the upper layers, with 50% in the first 4 cm and practically all in the top 30 cm. Only small amounts of ^{90}Sr are leached out of the soil and conveyed into fresh-water rivers and lakes. Nevertheless, concentrations in drinking water are normally at least an order of magnitude greater than those found in foodstuffs.

In determining strontium content of foodstuffs, the ratio of strontium-90 to calcium content is usually determined. Studies of ^{90}Sr:Ca ratios in various foodstuffs over a significant period of time, showed that the ^{90}Sr:Ca ratio in milk products and meat reaches a peak during periods of high fallout rate and then decreases rapidly over the next 5 to 8 yr, while in fruit and vegetables it remains higher for longer periods [15]. It is expected that the long-term variation of the strontium-90 intake with time would depend on the composition of the diet.

In general, [15] ^{90}Sr in the diet comes mainly from milk products, grain products, fruits, and vegetables. Most of the ^{90}Sr from grain products is received during the years of heavier fallout; in the case of milk about 40% is received during the years of fallout and the rest over some 8 yr afterwards. For fruits and vegetables, most of the contribution seems to be delivered over a period of 12 to 14 yr after deposition. The concentration of ^{90}Sr in milk can provide a useful index of trends in a changing environment, not only for itself but also for other foods and for soils.

Studies of the fate of ^{90}Sr from fallout also reveal that Sr:Ca ratio in the total diet is an important measure of ^{90}Sr impact on man. "The (Sr/Ca) ratio in new bone being formed is one-fourth of that in the average diet because the body selectively discriminates against strontium." [25] Typical ratios of ^{90}Sr:Ca in total diet in 1963 [25] were 10 to 30 pCi/g Ca. One pCi of ^{90}Sr is the activity of 7×10^{-15} g. Activities in milk were 10 to 30 pCi/liter (0.37 to 1.11 Bq/liter); and in flour about 40 pCi/kg (1.48 Bq/kg). Strontium-89 in milk averaged 40 to 55 pCi/liter (1.48 to 2.04 Bq/liter) in 1963. These are peak values during the periods of greatest fallout.

Measurements of the ^{90}Sr:Ca ratios in bone samples indicated the extent to which humans have been contaminated. The review of data in the 1977 UNSCEAR reports [15] shows values of ^{90}Sr from 1962 to 1974 for adult vertebrae were 1 to 2 pCi/g Ca (0.037 to 0.074 Bq/g Ca). The stronium-90 concentration in the bones of children and adolescents is higher than in adults [26].

Some discussions of ^{90}Sr in food chains use the strontium unit (SU), defined as the ratio of 1 pCi (0.037 Bq) of ^{90}Sr per gram of Ca. At the 1965–1966 peak the bones reached a maximum of 7 SU in children under 4 yr old. A decade earlier, the value was 1 to 2 SU, while in adults maturing before 1945 it was only 0.1 SU. For comparison, 80 SU has been generally accepted as an amount which probably could be tolerated for a lifetime without medically detectable damage [27].

Strontium-89 is one of the main components of fallout activity in the first few months after a nuclear test. It has a half-life of 52 days and decays with the emission of a β-particle of maximum energy of 1.46 MeV. The ratio of activities ^{89}Sr:^{90}Sr at the time of fission is approximately 185. The total atmospheric impact of ^{89}Sr is estimated to have been 2.8×10^9 Ci (1.04×10^{20} Bq) [15].

In biological and food samples, measurements of ^{89}Sr must be sensitive to a few pCi/liter (a tenth of a Bq/liter) of milk, and perhaps fractions of a pCi/kg (hundredths of a Bq/kg) of other foods such as vegetables and baked foods. Samples of particles in air released from reactors or fuel reprocessing plants must be measured for concentration varying from a few pCi/sec (tenths of a Bq/sec) to perhaps a few Ci/sec (tens of GBq/sec).

Finally, standard gross β-particle measurements in both drinking water and respirable air are designed to place upper limits on ^{90}Sr in these media. For example, the standard method for gross β in drinking water [28] places a limit of 10 pCi/liter (0.37 Bq/liter) for ^{90}Sr. If ^{90}Sr is absent, then "the water supply would usually be regarded as radiologically acceptable provided that the gross beta concentration does not exceed 1,000 pCi/liter" (37 Bq/liter) [8].

17.3. RADIATION PROTECTION GUIDES

At the international level, the International Commission on Radiological Protection (ICRP) makes radiation protection recommendations that are used as an upper limit by most national bodies. See Chapter 2 for their recommendations that determine the limits discussed here. For occupational exposures [29,30], the annual limits on intake (ALI) for strontium compounds cleared from the lung over a period of a few days are 27 μCi ^{90}Sr (1.0 MBq) and 540 μCi (20.0 MBq) for ^{89}Sr. The concentrations for these same nuclides in air are 18.9 $\mu Ci/m^3$ (0.7 MBq/m^3) for ^{90}Sr and 810 $\mu Ci/m^3$ (30.0 MBq/m^3) for ^{89}Sr based on a 40-hr work week.

At the national level, the NCRP recommends quite low maximum permissible concentrations (MPC) in air and water. For occupational exposure, the NCRP [31] recommends the following MPCs on a 168-hr/wk basis for soluble Sr:

Isotope	168-hr/week Occupational MPC in Air (pCi/m^3)	168-hr/week Occupational MPC in Water (pCi/liter)
Strontium-90	100 (3.7 Bq/m^3)	1,000 (37.0 Bq/m^3)
Stronium-89	10,000 (370.0 Bq/m^3)	100,000 (3.7 kBq/m^3)

The maximum permissible body burdens (MPBB) are 2 μCi (0.07 MBq) for ^{90}Sr and 4 μCi (0.15 MBq) for ^{89}Sr, both for burden in the bone [31]. For the public, the MPC and MPBB levels are a factor of 10 smaller; and for a suitable average sample of a large general population, another factor of 3 smaller still.

The EPA has issued drinking water standards. The limit for ^{90}Sr is 8 pCi/liter (0.30 Bq/liter) [32]. For ^{89}Sr the limit is 80 pCi/liter (2.96 Bq/liter). In addition the EPA stipulates that the analytical technique for strontium determination shall have detection limits no greater than 2 pCi/liter (0.07 Bq/liter) for ^{90}Sr and 10 pCi/liter (0.37 Bq/liter) for ^{89}Sr.

The International Atomic Energy Agency (IAEA) has issued some reports that are pertinent to monitoring for radioactivity [33,34,35]. These should be consulted for details concerning general approach to such measurements. The NCRP has issued two reports that are useful references for radioactivity measurements [36,37]. A report of the American National Standards Institute, a national voluntary standards setting committee, is also pertinent [38].

17.4. MEASUREMENT CONSIDERATIONS

This section discusses strontium-90 and strontium-89 determination in air, water, milk, and other media. Either of two types of measurements is commonly required. In one, a quantitative measurement of strontium-90 specific activity [e.g., pCi/g (Bq/g)] is made, whereas in the second it is merely necessary to determine an upper limit of total β-particle activity to assure safety. In the second type, gross β is usually measured as a first screening step because it is easier than the more elaborate strontium-isolation chemistry. Only when the gross-β measurement exceeds guidelines, or when there is a strong suspicion of strontium contamination, is specific strontium analysis called for.

This section is concerned primarily with the quantitation of the specific activity of radiostrontium. Because Sr must be determined at activity levels well below the permissible gross-β levels in strontium's absence, radiochemical separation techniques are invariably necessary. There is an extremely wide variety of methods for radiostrontium, but all follow a similar outline: first the chemical concentration of ^{90}Sr, then purification and sample preparation, followed by counting.

Chemical techniques for radiostrontium separation have been tailored to the requirements of the various media to be measured. Because the chemical properties of yttrium are important, procedures generally depend upon multiple-step separations of Y from the bulk material. Media containing radiostrontium are frequently contaminated with radioactive cesium-137 (also a long-lived β-particle emitter), iodine-131, and perhaps other radionuclides as well, so that the chemistry can be complicated.

Separate treatments are necessary for the various media: air, water, milk, and other media (principally foodstuffs and soil). This is necessary because the chemical techniques differ. For example, in both milk and urine, Y forms an anionic complex which makes possible its separation by direct anion exchange. On the other hand, this procedure is inappropriate for extraction from air filters. Measurements of strontium-90 in biomedical samples such as urine, bone, or soft tissue are only briefly mentioned here.

17.4.1. Counting

Beta-particle counting is required in all methods. The final strontium sample has to be prepared for counting. For most instruments, it must be thin with reasonably uniform distribution over an area compatible with the instrument.

Counting is done by a number of instruments. These include internal or external thin-window gas proportional counters, G-M tubes, semiconductor detectors; and less frequently liquid or plastic scintillation counters. Most of these instruments require different sample-preparation methods. The problems and differences that characterize the various beta-particle counting techniques have been discussed in Chapter 7.

17.4.2. Strontium-90/Strontium-89 Separation

Chemical separation techniques cannot separate isotopes from each other. Consequently the end-product of chemical separation contains all strontium isotopes, including carrier Sr and in some cases ^{85}Sr added as a tracer for chemical yield.

There are two ways to measure ^{89}Sr in the presence of ^{90}Sr. The first is β-spectrometry, which relies upon the differences in the maximum energies of the β-spectra (E_{max} = 1.463 MeV for ^{89}Sr; E_{max} = 546 keV for ^{90}Sr; and E_{max} = 2.27 MeV for ^{90}Y). This measurement is not well suited when the ^{89}Sr:^{90}Sr ratio is either very large or very small.

The most commonly used spectrometric technique employs a succession of different absorbers to count selectively the highest-energy β-particles starting with ^{90}Y. If Y has been separated chemically, the strontium-89 β-particles are first counted. Detection efficiencies must be calibrated using known activities of the individual radionuclide species (or for ^{90}Y an equilibrium mixture of ^{90}Sr and ^{90}Y). Because geometrical effects can produce quite wide variations, it is not safe to rely on calculations of the fractional detection efficiency. Although this problem is most important in β-spectrometry, it is also important for determining β-spectral efficiency in any of the methods described here.

Liquid scintillation counter spectrometry, is a promising β-ray technique, because after the sample is dissolved or suspended in the liquid scintillator geometrical efficiency factors are well understood. Piltingsrud and Stencel [39] first separate Sr chemically. The $SrCO_3$ is then suspended in the liquid scintillator using a gelling agent, and counted, typically from 2 to 12 hr, using for example a Packard "Tri-Carb" system and a 512-channel analyzer. (See Chapter 20 for details.) For a 12-hr counting period, the minimum detectable activity is about 1 pCi (0.037 Bq). A typical spectrum is shown in Fig. 17.3; a computer unfolds the spectra at the three isotopes (^{90}Sr, ^{90}Y, ^{89}Sr). Kamada, Mita, and Saiki [40] describe in detail another technique for unfolding the β-ray spectrum.

The second, and today the more common technique for $^{89}Sr/^{90}Sr$ separation uses the ingrowth of the daughter ^{90}Y. Figure 17.4 shows growth of ^{90}Y and (^{90}Y + ^{90}Sr) activities with time from a pure strontium-90 sample. Counting rates (corrected for efficiency) are compared from two measurements, the first immediately after strontium chemical separation (within a few hours, for example) and the second 3 to 14 days later. Alternatively, a second chemical separation of the yttrium daughter can be performed after allowing for ingrowth. Both techniques are frequently used.

In comparing these last two methods, Bowman, Swindle, and Henderson [41] found that when the ^{89}Sr:^{90}Sr activity ratio varied from 0.3 to 1000, the two-separation technique was found to be preferable. The single-separation technique was favored when the ^{89}Sr:^{90}Sr ratio was between 0.1 and 0.3. They found that the single-separation method yields acceptable relative error levels only when the ^{89}Sr activity is close to twice the ^{90}Sr activity. The two-separation method gives a constant, acceptable relative error for ^{90}Sr, while the strontium-89 error varies from 50% for low strontium-89 activity to 10% at high strontium-89 activity. Because the ^{89}Sr:^{90}Sr ratio is generally not known prior to analysis, they recommend the two-separation method.

Measurements of *total radiostrontium* usually refer to the total activity of the strontium fraction immediately after Y is separated chemically. The time elapsed before counting is critical because ^{90}Y ingrowth starts immediately. After ^{90}Sr and ^{90}Y reach parent-daughter equilibrium the gross-β counting rate is almost twice the activity of the parent ^{90}Sr.

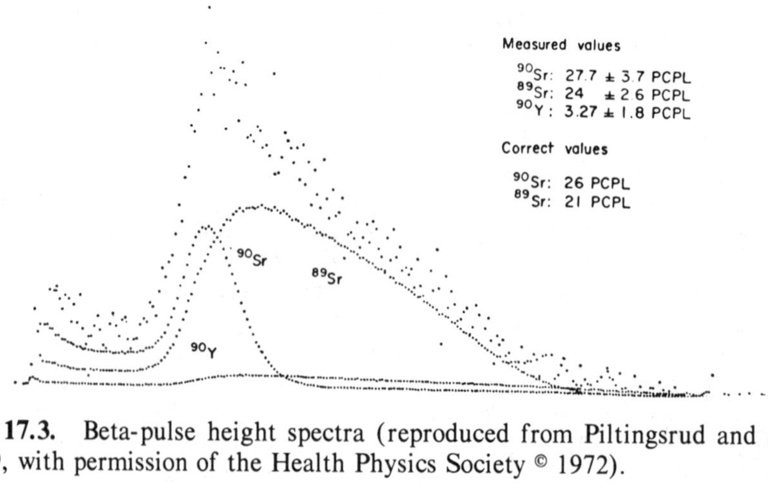

Figure 17.3. Beta-pulse height spectra (reproduced from Piltingsrud and Stencel, Ref. 39, with permission of the Health Physics Society © 1972).

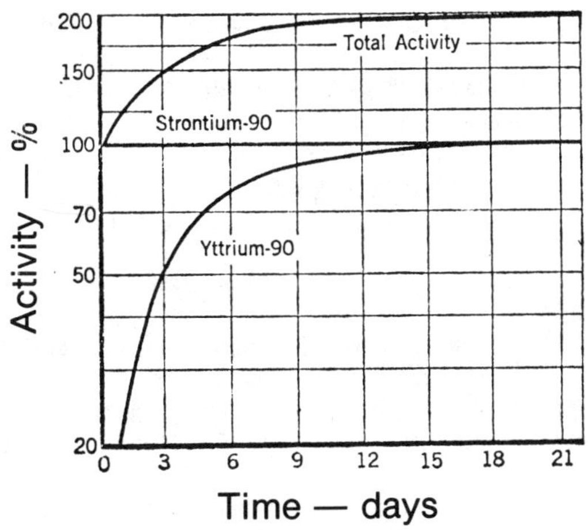

Figure 17.4. Yttrium-90 versus strontium-90 activity as a function of time (reproduced from Franson, Ref. 28, with permission of the American Public Health Association, American Water Works Association, and the Water Pollution Control Federation © 1981).

17.5. CHEMICAL TECHNIQUES

A good summary of the various types of chemical separation techniques is given by Bowen [42], quoted below. Although Bowen refers to measurements of seawater, his comments also apply to other liquids. He begins by discussing two attempts to extract ^{90}Y directly from environmental samples, using thenoyltrifluoroacetone (TTA) in benzene. He continues as follows:

> With the exception of these two ventures, all the procedures known to us for strontium-90 analysis in the oceans depend on the separation of most of the strontium, together with more or less of the calcium, followed by removal of all or most of the calcium and then by decontamination of the strontium fraction from other natural or artificial radionuclides by a series of ferric hydroxide and barium chromate scavenges; the cleaned strontium, usually after precipitation with ammonium carbonate for convenience, is then stored for enough time to allow yttrium-90 to come again to equilibrium with its parent, two weeks being a convenient time; yttrium-90 is separated with a minimum amount of iron, samarium or yttrium carrier, often with some additional scavenging steps, finally filtered, weighed and mounted for beta counting. Counting has been done using a variety of gas-filled Geiger-Müller or proportional counters, or with a solid plastic beta scintillator. The amount of data obtained to confirm the shape of the expected 64-hr half-life decay curve has varied widely, in response largely to economic factors; although "good radiochemical technique" dictates the use of repeated milkings of yttrium-90 and insistence on superimposable decay curves, this has been a nicety often omitted.
>
> In such a separation scheme, the number of variations available is not large:
>
> (a) Strontium may be initially precipitated as carbonate or as oxalate [43,44] and for the carbonate precipitation, sodium carbonate [45,46], sodium carbonate plus ammonium chloride [47], or ammonium carbonate [48] may be used.

(b) The oxalate precipitate may be decomposed by dry ignition [46] or by wet combustion with nitric acid.
(c) Strontium may be separated from such calcium and magnesium as accompany it in the initial precipitation by:
   (1) Selective complexation precipitation, followed by fuming nitric acid precipitations to obtain pure strontium nitrate [45,47].
   (2) Oxalate reprecipitation, followed by fuming nitric separations (Shvedov and Ivonova, referred to in Ref. 47).
   (3) Nitric acid precipitations without other treatment (cf. Ref. [44,46]; many others).
   (4) Selective elution serially, from cation exchange columns [49].
(d) The overall chemical yield of strontium in the separation procedure may be determined gravimetrically [43], by counting tracer stronium-85 (e.g., Ref. 44) by flame photometry or atomic absorption spectrophotometry, or volumetrically [49].
(e) The final separation of daughter yttrium-90 may be made using ferric hydroxide [45–47] or yttrium hydroxide [44], or samarium hydroxide. In each case the choice has been based on the availability—or ease of preparation—of radiochemically pure carrier and of the physical properties of its dried hydroxide as a counting medium.

With reference to item (a): although other reagents have been proposed (manganese dioxide by Shipman [50] or coprecipitation with rhodizonic acid), no data except those just cited have been reported using these or any other separation processes.

17.5.1. Air

In air, measurements of radiostrontium are nearly always made by examining particulate matter. The sampling technique is relatively straightforward: particulate matter is collected on a high-efficiency filter in a high-volume sampler. Air sampling rates of 1000 m^3/day are typical.

Chemical separation is required because of the normal presence of other β-ray emitting radionuclides in the particulate matter. For example, naturally occurring radon daughters can adhere to particles and produce sizeable β-ray activities. Methods of Air Sampling and Analysis [51] issued by the Intersociety Committee contains tentative methods for determining both ^{90}Sr and ^{89}Sr in air. The chemical separation is the same for both radiostrontium isotopes, and this method is likely to be adopted as a Standard Method in the future.

The filter, ashed in a muffle furnace, is subjected to the following chemical procedure [from 51]:

> Strontium is separated from calcium, other fission products, and natural radioactive elements. Nitric acid separations remove the calcium and most of the other interfering ions. Radium, lead, and barium are removed with barium chromate. Traces of other fission products are scavenged with iron hydroxide. After the strontium-90 + yttrium-90 equilibrium has been attained, the yttrium-90 is precipitated as the hydroxide and converted to the oxalate for counting. Strontium chemical yield is determined gravimetrically as strontium carbonate. Five tenths of a pCi (19 mBq) strontium-90 can be determined by this method with an error of less than $\pm 10\%$ at the 95% confidence level under usual conditions (counter background of less than 1 cpm and a 60-minute count).

Another technique for air filters was developed by Atomic Energy of Canada Limited (AECL) [52]. The first stages of the separation are described as follows:

> The ashed air filter is fumed with hydrofluoric and perchloric acids to remove silica. After extraction with nitric acid and fusion with sodium carbonate, an anion exchange column effluent is retained for the estimation of barium (-140), strontium (-89, -90) and cesium (-137).

An alternative separation method, also described in the AECL manual [52], is applicable when the ashed air filters contain little acid-insoluble material. In this method, "plutonium is separated from barium, strontium, and cesium by hexone extraction from a saturated ammonium nitrate solution." [52] (Hexone is methyl isobutyl ketone.) This method is shorter and easier than either of the others mentioned, making it attractive where the conditions make its use possible. Use of either AECL separation method must be followed by another stage to isolate radiostrontium from Ba and Cs.

Kreiger, Martin, and Frishkorn [53] have developed a radiochemical analysis technique for the determination of strontium, ruthenium, and cesium. This technique was developed in direct response to anticipated needs, since when full scale operation of fuel reprocessing plants becomes a reality, higher concentrations of ^{106}Ru will be introduced into the air. Minimum detection limits (3σ counting error) averaged 0.02 fCi/m^3 (0.7 $\mu Bq/m^3$) for ^{106}Ru and 0.01 fCi/m^3 (0.4 $\mu Bq/m^3$) for ^{137}Cs and $^{89/90}Sr$; chemical yields were 75, 80, and 75%, respectively.

17.5.2. Water

Measurements of Sr isotopes in drinking water and waste water only will be discussed. The limit for ^{90}Sr in drinking water is 8 pCi/liter (0.30 Bq/liter) when other sources of drinking water are not considered. Strontium-89 in drinking water is limited to 80 pCi/liter (3.0 Bq/liter).

Standard Methods [28] applies the same method for total radiostrontium and ^{90}Sr to either drinking water or filtered raw water. They state [28]:

> It is applicable to sewage and industrial wastes, provided that steps are taken to destroy organic matter and eliminate other interfering ions. In this analysis, a known amount of inactive strontium ions, in the form of strontium nitrate, is added as a "carrier." The carrier, alkaline earths and rare earths are precipitated as the carbonate to concentrate the radiostrontium. The carrier, along with the radionuclides of strontium, is separated from other radioactive elements and inactive sample solids by precipitation as strontium nitrate from fuming nitric acid solution. The strontium carrier, together with the radionuclides of strontium, is finally precipitated as strontium carbonate, which is dried, weighed to determine recovery of carrier, and then measured for radioactivity. The activity in the final precipitate is due to radioactive strontium only, because all other radioactive elements have been removed....
>
> Radioactive barium (^{140}Ba, ^{140}La) interferes in the determination of radioactive strontium inasmuch as it precipitates along with the radioactive strontium. This interference is eliminated by adding inactive barium nitrate carrier and separating this from the strontium by precipitating barium chromate in acetate buffer solution. Radium isotopes are also eliminated by this treatment....
>
> Two alternate procedures are given for the separation of ^{90}Y. In the first method, ^{90}Y is separated by extraction into tributyl phosphate from concentrated nitric acid solution. It is back-extracted into dilute nitric acid and evaporated to dryness for

beta counting. The second method consists of adding yttrium carrier, separating by precipitation as yttrium hydroxide, and finally precipitating yttrium oxalate for counting.

A similar technique, not identical in all respects but following the same general lines, is described in the AECL Manual [52].

The EPA has published an interim method for determining radiostrontium in drinking water [54]. The method consists of adding a strontium carrier to the drinking water sample, then collecting Sr as an insoluble carbonate. Afterwards it is separated from Ca as the nitrate. Impurities are removed by a hydroxide scavenge. After Ba is removed as the chromate, the Sr is purified as $SrCO_3$ for counting. The method gives total ^{89}Sr and ^{90}Sr activity. To determine ^{89}Sr, the planchet with the strontium sample should be stored at least 2 weeks so that the ^{90}Sr-^{90}Y activity will be in equilibrium. The Y is separated from the Sr, and ^{90}Sr activity is measured [54]. Strontium-90 activity can be subtracted from earlier total activity in order to determine ^{89}Sr. The tentative reference method for measurement of ^{89}Sr and ^{90}Sr in environmental waters is basically the foregoing procedure and is given in more detail in Ref. 55. The detection limit is 1 pCi/liter (0.037 Bq/liter) and the range of the method is 1 to 10^4 pCi/liter (0.037 to 370 Bq/liter) for a 1-liter sample.

17.5.3. Milk

Milk is one of the most important pathways for environmental strontium to man. The U.S. Public Health Service (PHS) Pasteurized Milk Network [56] developed a standard analytical procedure [55,57,58–61] which is typical of several other methods. The method removes Sr (and other alkaline and alkaline earth ions) by passing milk through a cation exchange resin, followed by removal of Y in an anion exchange resin.

This method from Ref. 56 follows:

> The procedure consists of storing the milk samples with formaldehyde preservative for the ingrowth of the yttrium-90 daughter of strontium-90, adding yttrium carrier, and then passing the milk consecutively through cation and anion exchange resin columns. The alkaline earth ions in milk are replaced by sodium ions in the cation exchange column, after which the yttrium is retained as an anionic complex—probably of citrate—in the anion exchange column. The effluent milk is discarded and the yttrium complex on the anion exchange resin is destroyed with hydrochloric acid. The yttrium, eluted with the acid, is precipitated as the oxalate, and the radio-yttrium is measured with a low-background beta counter....
>
> Initially, the maximum yttrium yield was 65% and average yields were 55%. Yttrium retention on the resin was improved from 80 to 90% by adding sodium citrate to the milk. The per cent eluted with 35 ml of hydrochloric acid was increased from 81 to 96% by decreasing the amount of anion exchange resin and by thoroughly stirring the resin during elution. Other losses were minimized by small procedural changes, so that the overall average yield was increased to 85%.

The U.S. PHS [60] described other methods for milk analysis. One uses ^{85}Sr as a tracer. Strontium is separated by successive carbonate precipitation and nitric acid washing, and the ^{90}Y ingrowth occurs in solution. A special feature is the separation of ingrown Y from lanthanum-140 contamination using tributyl phosphate. In another method from the *Procedures Manual* of the DOE Environment Measurements Laboratory (EML) [62], the milk sample is ashed to produce about a 10-g sample. Then the ash is dissolved in nitric acid, and a strontium carrier and ^{85}Sr are added. Phosphate is used to precipitate the alkaline earths. The phosphate precipitate is washed several times in dilute nitric acid solution which removes the Ca. Pb and Ba are removed by subsequent hydroxide and chromate scavenging steps. The pH during the chromate scavenging step is crucial as barium chromate will not precipitate in highly acidic solutions and Sr will partially precipitate in basic solutions [62].

Another method [60] removes milk proteins by precipitation with trichloroacetic acid (TCA); and precipitates the alkaline earths as oxalates. Subsequently, the oxalates are converted to nitrates and then the Sr is separated from Ca by differential solubility. When working in humid climates, care is required since the sample is counted as strontium nitrate on a stainless steel planchet, which can corrode the counting chamber if nitric acid is formed at high humidities. The *Handbook* of the EPA Environmental Monitoring and Support Laboratory (Las Vegas) [63] describes a method very similar to this last (TCA) technique.

17.5.4. Other Media

Chemical separation techniques have been developed for a wide variety of media besides air, water, and milk. The techniques are as varied as the problems, because media such as food or biological samples can have any of a large number of radioactive or stable impurities. Here only a few methods are mentioned.

With most biological samples, including bone, the first procedure is to ash the sample in a muffle furnace (500 to 800°C). This eliminates the large organic substrate, which might contain Sr or Y in complexed form. For samples volatile at dry-ashing temperatures, it is preferable to wet ash with nitric acid, followed by an equal-volume mixture of perchloric and nitric acids [64]. The EML *Procedures Manual* [62] outlines the procedures for both dry and wet ashing.

Those samples (e.g., soil, silt) which are not easily dissolved by acid digestion because of their high silica content are fused. The following fusion procedure is typical: the sample, with carrier plus 5 g of NaOH pellets per gram of sample, is fused over a Meeker burner for 20 to 30 min. The fused sample is dissolved, and then 0.5 g of anhydrous Na_2CO_3 is added and centrifuged. The precipitate is dissolved in 6 N HCl, ready for analysis [60].

After ashing or fusion, the radiochemical methods are quite similar to those already described, and the reader is referred to the appropriate references for details.

The PHS manual [60] contains radiostrontium techniques for urine, bone, bone ash, tobacco, vegetation, soil, and silt. The AECL manual [52] has techniques for animal and fish flesh, cereals, eggs, fats, and various vegetation as well as bone. The EML *Procedures Manual* [62] contains methods for vegetation, tissue, soil, bone tissue, and urine. The British Atomic Energy Research Establishment (AERE) has published radiostrontium methods in bone ash, milk ash, vegetable matter, human teeth, and soil [65]. Porter et al. [66] have developed a method using excess ethylenediaminetetraacetic acid (EDTA) which is applicable in samples with very large calcium:strontium ratios, and the strontium yield (as carbonate) is determined gravimetrically. It has been used for a wide variety of organic media. Finally, Baratta and Reavey [67] have described a method used by the U.S. Food and Drug Administration for food and other biological samples, which uses tributyl phosphate extraction and analysis of ^{90}Y. This is only a partial list of the many radiochemical analysis techniques in the literature; for more complete coverage, the reader is referred to the literature cited.

17.5.5. Yttrium Recovery after Ingrowth

The chemical separation of radiostrontium is often followed by a hold period allowing the ingrowth of ^{90}Y, yttrium separation, and counting of ^{90}Y. This method, as has been mentioned previously, is especially useful to determine ^{90}Sr in the presence of ^{89}Sr.

One critical aspect is the specificity of a scheme to separate ^{90}Y from ^{90}Sr. Goldin, Velten, and Frischkorn have developed such a scheme [64]. After purifying the strontium carbonate precipitate, and

> ingrowth of yttrium, the strontium carbonate precipitate is dissolved with hydrochloric acid, neutralized to methyl orange with ammonia, and buffered at pH 5. The yttrium is extracted into 2-thenoyltrifluoroacetone solution. After the organic phase is washed with water buffered at pH 5, the yttrium-90 is stripped by extraction with 1 *N* hydrochloric acid. The hydrochloric acid extract is evaporated on copper planchets, heated thoroughly to destroy residual organic matter, and counted, correcting for decay of the yttrium-90 between its extraction and the counting time.

Very little ^{89}Sr remains after this procedure: only $4 \times 10^{-4}\%$ of the ^{89}Sr is counted as ^{90}Y, while the yttrium-90 yields are typically 85% and the yttrium yield in the solvent extraction is greater than 95%.

An alternate chemical procedure is given by the EML *Procedures Manual* [62]:

> After ingrowth of yttrium, the strontium solution is heated to 90°, adjusted to pH 8 with NH_4OH, then 6 drops 30% H_2O_2 is added and heated for another hour to drive off excess H_2O_2. Then the precipitate and solution are cooled and centrifuged. The precipitate is washed by repeating the above procedure. It is then dissolved with HCl and yttrium oxalate is precipitated with additional oxalic acid solution, 2–3 drops NH_4OH adjusting pH to 2–3. The oxalate precipitate is dried and counted.

Calibrations must be made periodically to check the 64-hr half-life of ^{90}Y measured from the final yttrium sample. This half-life is in the range where such measurements are easy, yet they are omitted in some laboratories for reasons of economy or convenience. This check is one of the easiest ways to exercise internal quality control on a modest scale.

17.5.6. Interferences

When Sr is chemically separated, the major interferences in environmental samples are Ca, ^{140}Ba, and phosphates. Goldin, Velten, and Frishkorn [64] discuss the chemical considerations:

> *Calcium.* Strontium as an alkaline earth element is chemically similar to the other members of this family. Its resemblance to calcium, barium, and radium is especially marked. Calcium, which is widely distributed in nature, is one of the inert elements whose separation from radiostrontium has been most studied. Such separation is necessary if the strontium radiations are to be measured directly in samples of moderate or high calcium content, or if the final precipitate is weighed to determine

the chemical recovery of strontium. Occasionally calcium is present in such large quantities as to give unmanageable amounts of precipitate unless it is removed.

The precipitation of strontium (and barium) nitrate in concentrated nitric acid also separates strontium from calcium. The concentration of nitric acid is critical, as too dilute a nitric acid solution will solubilize too much strontium, while too strong a nitric acid will precipitate calcium with the strontium. The most favorable range is from 65 to 75% nitric acid. Concentrated nitric acid will suffice. If the sample is in aqueous solution, fuming nitric acid probably will be needed to take care of the dilution....

Barium and Radium. The fission product barium-140, also a member of the alkaline earth group, is carried along with strontium as carbonate. It must be removed whether the strontium or its daughter is to be counted, as there is a radioactive daughter, lanthanum-140, which will contaminate the yttrium. It is precipitated, with carrier barium, as chromate from a solution buffered at pH 5. Radium, which follows barium in this procedure, can be determined, although rather crudely, by alpha-counting the barium chromate precipitate....

Phosphates. Phosphates interfere by causing precipitation of strontium when the solution is made basic in the hydroxide scavenging step. Although the phosphate content of water is usually too low to be of consequence, this interference is very troublesome in the analysis of biological samples, whose ash is largely calcium phosphate. Phosphates are removed in the nitric acid treatment already described, being converted to phosphoric acid, which is removed with the supernate. Several treatments usually are necessary to ensure complete removal.

17.5.7. Calibration Techniques

There are a number of problems in making accurate measurements using the chemical techniques just discussed. Among them are losses of radiostrontium, losses of Y, and counting errors.

The use of standard radiostrontium preparations can reduce counting errors. The detection efficiency over a large range of ^{90}Sr and ^{90}Y containing samples must be understood. For this purpose secondary standards of low-level solid strontium carbonate sources have been prepared by the EML [30] by dilution of primary sources from the National Bureau of Standards (NBS). Standard sources for ^{90}Sr in milk have also been prepared [68].

Chemical yields can be determined using ^{85}Sr as a chemical tracer: it decays ($t_{1/2}$ = 64 days) by electron capture followed by emission of a 514-keV γ-ray. Its routine use is uncommon partly because of its own intrinsic radiological hazard, but it can serve a valuable function in periodic quality control. Other methods for determining chemical yield rely on classical chemical techniques, such as gravimetric measurements of carrier Sr or of possible interfering radionuclides.

Another quality control measurement, mentioned before, is the observation of yttrium-90's 64-hr half-life in counting of the yttrium fraction.

As in any analytical procedure, much of the success of radiostrontium assay depends upon the care of the analyst—an obvious point but one which cannot be overemphasized.

17.5.8. Quality Control

Quality control is one of the most important elements of any radionuclide-analysis program. Unfortunately, quality-control checks have usually revealed significant variations among laboratories analyzing for ^{90}Sr at low levels. These variations are a cause for concern because they demonstrate the fundamental difficulty in making routine, reliable, accurate ^{90}Sr measurements in environmental media.

One study is discussed here as an example—others show comparable results. In 1971, the EPA Office of Radiation Programs carried out a study of analysis capabilities in milk [64]. Thirty-five participating laboratories made triplicate determinations of unknown concentrations of ^{89}Sr, ^{90}Sr, ^{131}I, and ^{137}Cs at activity levels ($\pm 2\%$) of 31, 42, 69, and 52 pCi/liter (1.15, 1.55, 2.55, and 1.92 Bq/liter), respectively. Unfortunately, nine laboratories reported results whose mean exceeded the "control level" of about $\pm 10\%$ for ^{90}Sr. (For ^{89}Sr, ^{131}I, and ^{137}Cs the outlying laboratories numbered 4, 6, and 2 respectively.) For many of the outlying laboratories in the ^{90}Sr study, the range of reported results for the triplicate analyses exceeded 25%. This study shows that, while many laboratories accurately measure ^{90}Sr, a significant minority cannot.

17.6. SUMMARY AND CONCLUSIONS

This chapter examines the present status of radiostrontium measurements in environmental media. At the low levels which are of concern, the only methods for specific radiostrontium analysis involve analytical radiochemistry followed by β-particle counting.

A number of methods for chemical separation were compared. One difficulty in comparing different chemical methods is that so much depends upon the skill of the analyst himself. A choice among the various commonly used techniques may depend more on the particular equipment and skills available than on the differences among the techniques themselves. This point is well made by Bowen [42]:

> In the extensive series of studies of strontium-90 in sea-water, choice among these various possibilities has been made partly on the basis of prejudice, partly on economics, partly on the basis of which other radionuclides were to be sought in the same samples, partly on the basis of which alternative used fewest reagents known or suspected to be significant contributors to the radiochemical blank and partly on the basis of safety. In most of these variations the decisions have been legitimate ones, but we feel of even more importance have been questions which one can still only describe as those of art: in the hands even of very competent and experienced analysts, no one method, whether of trace element or of low-level radiochemical analysis, appears ever to have been uniformly superior in quality of results. For this reason, we strongly resist the suggestion that a single "standard" method be selected and recommended.

In support of his viewpoint, Bowen [42] cites interlaboratory calibration analyses in which nearly the full spectrum of possible radiochemical procedures was represented.

> ... it is clear, we believe, that in suitable hands each method was capable of delivering data of high quality; it is also clear from some of the entries ... that these well-established methods were capable in unsympathetic hands of yielding some pretty bizarre sets of numbers, indeed.

Apparently, none of the techniques suffers from critical intrinsic flaws; yet equally apparent, the results obtained with a particular method depend upon specific skill of the analyst. This situation is not satisfactory or necessary. Much can be done developing these or other methods to make them less dependent on individual talents. In summary: first, sensitivity of the methods commonly used is adequate; second, further research and development in this field is needed to simplify, standardize, and automate the methods. The market for a fully automated system may be insufficient to justify its development.

REFERENCES

1. J. Mantel, "The Beta Ray Spectrum and Average Energy of Several Isotopes of Interest in Medicine and Biology," *Int. J. Appl. Radiat. Isotopes*, **23**, 407, 1972.
2. D. N. Sunderman and C. W. Townley, *The Radiochemistry of Barium Calcium and Strontium*, NAS-NS 3010, National Academy of Sciences—National Research Council Subcommittee on Radiochemistry, 1960.
3. A. Engstrom, R. Bjormerstedt, C-J. Clemedson, and A. Nelson, *Bone and Radiostrontium*, Wiley, New York, 1957.
4. H. Spencer and L. Kramer, "Effect of Phosphorus on the ^{90}Sr Balance in Man," *Health Phys.*, **33**, 417, 1977.
5. M. Goldman and L. K. Bustad, Eds., *Biomedical Implications of Radiostrontium Exposure*, proceedings of a symposium at Davis, California, February 22-24, 1971, Technical Information Center, Oak Ridge, 1972; available as CONF-710201 from NTIS.
6. M. Goldman, R. J. Della Rosa, and M. H. Momeni, "Radiation Dose to Beagles from Continuum ^{90}Sr Exposure," p. 103 of Ref. 5.
7. M. H. Momeni, N. Jow, and E. Bradley, "Temporal and Spatial Marrow Dose for ^{90}Sr-^{90}Y in Beagles: A Measurement of Geometry–Distribution Factor," *Health Phys.*, **28**, 285, 1975.
8. F. W. Spiers, G. D. Zanelli, D. J. Darley, J. R. Whitwell, and M. Goldman, "Beta-Particle Dose Rates in Human and Animal Bone," p. 130 of Ref. 5.
9. C. W. Mays and R. D. Lloyd, "Bone Sarcoma Risk from ^{90}Sr," p. 352 of Ref. 42.
10. G. W. Dolphin, in General Discussion, p. 376 of Ref. 5.
11. A. F. G. Stevenson, "The Influence of Age and Sex on the Activity Ratio of Yttrium-90 to Strontium-90 in the Rat Skeleton After Incorporation of Strontium-90," *Health Phys.*, **29**, 285, 1975.
12. L. A. Ilyin, A. T. Inanmikov, Yu. D. Parfenov, and V. P. Stolyanov, "Strontium Absorption Through Damaged and Undamaged Skin," *Health Phys.*, **29**, 75, 1975.
13. National Council on Radiation Protection and Measurements, *Natural Background Radiation in the United States*, NCRP Report No. 45, 1975, 7910 Woodmont Ave. Washington, DC 20014.
14. B. Kahn, R. L. Blanchard, H. L. Krieger, H. E. Kolde, D. B. Smith, A. Martin, S. Gold, W. J. Averett, W. L. Brinck, and G. J. Karches, "Radiological Surveillance Studies at a Boiling Water Nuclear Power Reactor," Report BRH/DER 70-1, 1970, EPA Radiological Engineering Laboratory, 5555 Ridge Avenue, Cincinnati, OH 45213.

15. United Nations Scientific Committee on the Effects of Atomic Radiation, *Sources and Effects of Ionizing Radiation*, 1977 report to the General Assembly, with annexes, United Nations, New York produced by UNIDO, Vienna, 1977.
16. R. L. Blanchard and B. Kahn, "Abundance and Distribution of Radionuclides Discharged from a BWR Nuclear Power Station into a Marine Bay," *Nucl. Safety*, **20**(2), 190, 1979.
17. General Electric Co., "Applicant's Environmental Report, Midwest Fuel Recovery Plant, Morris, Illinois," Report NEDO-14504 and "Supplement 1," Report NEDO-14504-2, 1971, General Electric Company, Reactor Fuels and Reprocessing Department, San Jose, CA 95125.
18. U.S. Nuclear Regulatory Commission, "Environmental Survey of the Reprocessing and Waste Management Portions of the LWR Fuel Cycle," A Task Force Report, NUREG-0116 (supplement 1 to WASH-1248), Washington, DC 20555, April 1974.
19. Nuclear Regulatory Commission, "Final Generic Environmental Statement on the Use of Recycle Plutonium in Mixed Oxide Fuel in Light Water Cooled Reactors, Health, Safety, and Environment," Executive Summary, NUREG-0002, 1976, Office of Nuclear Materials Safety and Safeguards, Washington, DC 20555.
20. O. D. Markharn, D. K. Halford, R. E. Autennieth "Strontium-90 Concentrations in Pronghorn Antelope Bones Near a Nuclear Fuel Reprocessing Plant," *Health Phys.*, **38**, 811, 1980.
21. U.S. Department of Energy, "Fallout Program Quarterly Summary Reports," a series of quarterly reports available from U.S. DOE Environmental Measurements Laboratory, 376 Hudson Street, New York, NY 10014.
22. Federal Radiation Council, "Estimates and Evaluation of Fallout in the United States from Nuclear Weapons Testing Conducted Through 1962," FRC Report No. 4, 1963.
23. R. E. Simpson, E. J. Baratta, and J. T. Tanner, "Population Dose Commitment Associated with Various Radionuclides in Foods," abstracts of papers presented at Twenty-Third Annual Meeting of the Health Physics Society, *Health Phys.*, **35**, 919, 1978.
24. C. W. Francis, "Radiostrontium Movement in Soils and Uptake in Plants," U.S. Department of Energy Critical Review Series, TID-27564, 1978.
25. Federal Radiation Council, "Revised Fallout Estimates for 1964–1965 and Verification of the 1963 Predictions," FRC Report No. 6, 1964.
26. A. Bauman, N. Franic, M. Baumstark, and V. Popovic, "^{90}Sr in the Human Bone," *Health Phys.*, **32**, 321, 1977.
27. D. S. Grosch and L. E. Hopwood, *Biological Effects of Radiations*, 2nd ed., Academic, San Francisco, 1979.
28. *Standard Methods for the Examination of Water and Wastewater*, 15th ed., 1980, American Public Health Association, 1015 18th St., N.W., Washington, DC 20036.
29. International Commission on Radiological Protection, "Limits for Intakes of Radionuclides by Workers," ICRP Publication 30, Part I, *Ann. ICRP*, **2**, 1979.
30. International Commission on Radiological Protection, "Limits for Intakes of Radionuclides by Workers," ICRP Publication 30, Supplement to Part 1, *Ann. ICRP*, **3**, 1979.

31. National Council on Radiation Protection and Measurements, "Maximum Permissible Body Burdens and Maximum Permissible Concentrations of Radionuclides in Air and Water for Occupational Exposure," NCRP Report No. 22, 1959, 7910 Woodmont Ave., Washington DC 20014. Indicates Addendum 1 to NBS Handbook 69 issued in August 1963.
32. U.S. Environmental Protection Agency, "National Interim Primary Drinking Water Regulations," EPA-570/9-76-003, 1976, Office of Water Supply, Washington, DC.
33. International Atomic Energy Agency, "Monitoring of Airborne and Liquid Radioactive Released from Nuclear Facilities to the Environment," IAEA Safety Series No. 46, IAEA, Vienna, 1978.
34. International Atomic Energy Agency, "Radiological Surveillance of Airborne Contaminants in the Working Environment," IAEA Safety Series No. 49, IAEA, Vienna, 1979.
35. International Atomic Energy Agency, "Radiation Protection Procedures," IAEA Safety Series No. 38, IAEA, Vienna, 1973.
36. National Council on Radiation Protection and Measurements, "Instrumentation and Monitoring Methods for Radiation Protection," NCRP Report No. 57, 1978, 7910 Woodmont Ave., Washington, DC 20014.
37. National Council on Radiation Protection and Measurements, "A Handbook of Radioactivity Measurements Procedures," NCRP Report No. 58, 1978, 7910 Woodmont Ave., Washington, DC 20014.
38. American National Standards Institute, "American National Standard Specification and Performance of On-Site Instrumentation for Continuously Monitoring Radioactivity in Effluents," ANSI N13.10-1971, 1974, Institute of Electrical and Electronic Engineers, Inc. 345 East 47th Street, New York, NY 10017.
39. H. V. Piltingsrud and J. R. Stencel, "Determination of ^{90}Y, ^{90}Sr and ^{89}Sr in Samples by Use of Liquid Scintillation Beta Spectroscopy," *Health Phys.*, **23**, 121, 1972.
40. H. Kamada, M. Mita, and M. Saiki, "Measurements of ^{89}Sr and ^{90}Sr in Environmental Samples by a Low-Background Beta-Ray Spectrometer," in *Rapid Methods for Measuring Radioactivity in the Environment*, Proceedings of a Symposium in July 1971, International Atomic Energy Agency, Vienna, 1971, p. 427.
41. W. B. Bowman, II, D. L. Swindle, and D. L. Henderson "Radiochemical Determination of ^{89}Sr and ^{90}Sr: A Comparison of Methods Based on Error Analysis," *Health Phys.*, **31**, 495, 1976.
42. V. T. Bowen, "Analyses of Sea-Water for Strontium and Strontium-90," in *Reference Methods for Marine Radioactivity Studies*, International Atomic Energy Agency, Vienna, 1970, p. 93.
43. Y. Miyake, K. Saruhashi, Y. Katsuragi, "Strontium-90 in Western North Pacific Surface Waters," *Pap. Met. Geophys.* (Tokyo), **11**, 188, 1960.
44. G. G. Rocco and W. S. Broecker, "The Vertical Distribution of Cesium-137 and Strontium-90 in the Oceans," *J. Geophys. Res.*, **68**, 4501, 1963.
45. T. T. Sugihara, H. I. James, E. J. Troianello, V. T. Bowen, "Radiochemical Separation of Fission Products from Large Volumes of Sea Water Strontium, Cesium, Cerium and Promethium," *Anal. Chem.*, **31**, 44, 1959.

46. R. Higano and M. Shiozaki, "Radiochemical Analysis of Strontium-90 and Cesium-137 in Sea Water," *Hydrogr. Off.* (Japan), **1**, 137, 1960.
47. E. G. Azhazha, "Methods for the Determination of Strontium-90 in Sea Water," *Radioactive Contamination of the Sea* (V. I. Baranov and L. M. Khitrov, Eds., Akad. Nauk USSR–Oceanographic Commission, 1964; Transl. 1966 Israel Program Sci. Transl., U.S. Department of Commerce, p. 177.
48. D. C. Sutton and J. J. Kelly, "Strontium-90 and Cesium-137 Measurements of Large Volume Sea Water Samples," U.S. AEC Health and Safety Laboratory, Fallout Program Quarterly Summary Report HASL-196, 1968.
49. V. E. Noshkin and N. S. Mott, "Separation of Strontium from Large Amounts of Calcium, with Application to Radiostrontium Analysis," *Talanta*, **14**, 45, 1967.
50. W. H. Shipman, "Concentration of Strontium from Sea Water by Manganese Dioxide, Determination of Sr^{90} in Large Volumes of Sea Water," USNRDL-TR-995, 1966.
51. American Public Health Association, *Methods of Air Sampling and Analysis*, Intersociety Committee for a Manual of Methods for Ambient Air Sampling and Analysis, 1972, American Public Health Association, 1015 18th Street, N.W., Washington, DC 20036.
52. J. E. Guthrie and W. E. Grumitt, "Manual of Low Activity Environmental Analysis," Report AECL-1745, 1963, Atomic Energy of Canada Limited, Chalk River, Ontario, Canada.
53. H. L. Krieger, E. R. Martin, and G. W. Frishkorn, "Sequential Radiochemical Analysis for Ruthenium, Strontium, and Cesium in Environmental Air," *Health Phys.*, **30**, 465, 1976.
54. H. L. Krieger, "Interim Radiochemical Methodology for Drinking Water," EPA-600/4-75-008 (Revised), 1976, Environmental Protection Agency, Washington, DC 20460.
55. U.S. Environmental Protection Agency, "Measurement of Strontium-89 and Strontium-90 in Environmental Waters: A Tentative Reference Method," EPA-600/4-76-011, 1976, Environmental Monitoring and Support Laboratory, Las Vegas, NV 89114.
56. The U.S. Public Health Service's Pasteurized Milk Network reports its radionuclide measurements periodically in *Radiation Data and Reports.* Nuclides analyzed are strontium-89 and -90, iodine-131, cesium-137, and barium-140.
57. B. Kahn, G. K. Murthy, C. Porter, G. R. Hagee, G. J. Karches, and A. S. Goldin, "Rapid Methods for Estimating Fission Product Concentrations in Milk," Environmental Health Series, U.S. Department of HEW, Public Health Service, Publication PHS-999-R-2, 1963.
58. C. Porter, "Rapid Determination of Strontium-89 and Strontium-90 in Milk," p. 11 of Ref. 19.
59. C. Porter and B. Kahn, "Improved Determination of Strontium-90 in Milk by an Ion Exchange Method," *Anal. Chem.*, **36**, 676, 1964.
60. U.S. Public Health Services, "Radioassay Procedures for Environmental Samples," PHS Publication No. 999-RH-27, 1967, U. S. Public Health Service, Bureau of Radiological Health, Rockville, MD 20852.
61. C. Porter, D. Cahill, R. Schneider, P. Robbins, W. Perry, and B. Kahn, "Determination of Strontium-90 in Milk by an Ion Exchange Method," *Anal. Chem.*, **33**, 1306, 1961.

62. J. H. Harley, Ed., EML *Procedures Manual*, Report HASL-300, 1972, U.S. DOE Environmental Measurements Laboratory, 376 Hudson St., New York, NY 10014, updated annually, last update 1980. Earlier editions of this manual were published as Report NYO-4700.

63. F. B. Johns, "Southwestern Radiological Health Laboratory Handbook of Radiochemical Analytical Techniques," Report SWRHL-11, 1970. Available from SWRHL, now EPA Environmental Monitoring and Support Laboratory, Las Vegas, NV 89114.

64. A. S. Goldin, R. J. Velten, and G. W. Frishkorn, "Determination of Radioactive Strontium," *Anal. Chem.*, **31**, 1490, 1959.

65. A. Parker, E. H. Henderson, and G. S. Spicer, "Analytical Methods for the Determination of Radiostrontium in Biological Materials," Report AERE-AM 101, Atomic Energy Research Establishment, Harwell, U.K., 1965.

66. C. Porter, B. Kahn, M. Carter, G. Rehnberg, and E. Pepper, "Determination of Radiostrontium in Food and Other Environmental Samples," *Environ. Sci. Tech.*, **1**, 745, 1967.

67. E. J. Baratta and T. C. Reavey, "Rapid Determination of Strontium-90 in Tissue, Food, Biota, and Other Environmental Media by Tributyl Phosphate," *J. Agri. Food Chem.*, **17**, 1337, 1969.

68. M. Rosensteil, P. B. Hahn, and A. S. Goldin, "Interlaboratory Study of Strontium-90 Measurements in Milk—August 1963," *Health Phys.*, **11**, 966, 1965.

18 IODINE

Natural iodine has one stable isotope, ^{127}I; all other isotopes of iodine are radioactive. Man-made radioactive iodine is formed by neutron and gamma-ray reactions and the fission of heavy nuclei (U, Pu). Fission is an important source of radioiodine because this process occurs in nuclear reactors and in the explosion of nuclear bombs. Environmentally, fission is the most important source, because of all the sources of anthropoqenic radioiodine, fission products have the highest probability of reaching man.

Fission products, generally having short half-lives (measured in seconds or minutes), become the precursors of radioactive nuclides that decay, emitting β-particles (isomeric chains). The radioisotopes of iodine having mass numbers of 129 and 131 to 139 are formed in ten such isomeric chains. Iodine-128 ($t_{1/2}$ = 25 min) and ^{130}I ($t_{1/2}$ = 12.6 hr), are direct fission products and constitute less than 0.01% of the products on an atom basis [1]. Iodine radioisotopes with mass numbers 136 to 139 have half-lives of 2 to 86 sec, usually too short to be considered dangerous with respect to the time that passes from the instant of formation until uptake by the human body. Iodine-129 ($t_{1/2} = 1.7 \times 10^7$ yr) has a low specific activity and represents no immediate danger on release. However, because it has a long half-life, it will accumulate and be distributed globally. From the standpoint of short time scale radiation protection the iodine isotopes with mass numbers 131 to 135, which have half-lives of 1 hr to 8 days are of the greatest interest. The more important physical properties of these isotopes are summarized in Tables 18.1 and 18.2.

The two iodine isotopes of greatest significance in the environment are ^{131}I and ^{129}I. Their properties are sufficiently different that they are treated separately. The difference is due to the physical characteristics of the decays.

Table 18.1. Cumulative Yield of Iodine Radioisotopes on Fission of ^{233}U, ^{235}U, ^{238}U, and ^{239}Pu by Neutrons of Various Energies, Percent (Data Presented in Il'in et al., Ref. 1)

Iodine Isotope	Half-Life	Thermal Neutrons ^{235}U	Thermal Neutrons ^{233}U	Thermal Neutrons ^{239}Pu	Fission-Spectrum Neutrons ^{235}U	Fission-Spectrum Neutrons ^{238}U	Fission-Spectrum Neutrons ^{239}Pu	Fast Neutrons (E_n = 14 MeV) ^{238}U
129	1.6×10^7 yr							
131	8.06 days	3.1	2.9	3.77	3.11	3.33	4.85	4.26
132	2.3 hr	4.7	4.4	5.1	6.02	5.72	6.19	5.50
133	20.9 hr	6.9	3.37	5.2	6.02	5.72	6.19	5.50
134	54 min	7.8	8.1	7.1	7.83	6.90	6.27	6.00
135	6.7 hr	6.1	5.5	5.7	6.04	5.75	5.68	5.26

Iodine-131 has a half-life of 8.05 days. It decays by emission of one of several beta rays leading to excited states of ^{131}Xe; the excited states then decay to the ^{131}Xe (stable) ground state by gamma-ray emission. The most prominent beta ray (85% branching ratio) with E_{max} = 606 keV, is followed by a 364-keV gamma ray. This gamma ray is the principal emission used to detect ^{131}I. About 15% of the decays proceed by a weaker β-ray (E_{max} = 315 keV) followed by a more energetic gamma ray (638 keV).

Iodine-129 is a beta emitter with a 1.7×10^7-yr half-life. All the decays proceed by emission of a beta ray (E_{max} = 150 keV), followed by a 40-keV gamma ray (or K or L shell electron) leading to stable xenon-129. Neither the β-ray nor the γ-ray is very energetic and therefore both are difficult to detect. On the other hand, even small activities represent sizeable masses of ^{129}I in a material: one curie of pure ^{129}I would have a mass of about 6000 g. It is therefore possible to use a very sensitive technique such as neutron activation analysis on some ^{129}I samples.

The physical and chemical properties of iodine are important. At ordinary temperature elemental iodine is a solid (it melts at 114°C and boils at 184°C), but has a high vapor pressure [1.03 mm Hg (137 Pa) at 40°C] and readily sublimes. Vaporous iodine is readily adsorbed on condensation nuclei present in the air forming a fine mist. Iodine is highly reactive and attacks many substances, including those which are usually present in the air in the form of impurities. Iodine exhibits several oxidation states including: −1 (iodides), +5 (iodates), +7 (periodates). Listed in Table 18.3 are some ions and molecules commonly formed by iodine in nature [1].

Knowledge of the physical state and chemical forms of iodine radioisotopes in the air is critical to understanding the rate of precipitation of radioiodine to the surface of the earth and its behavior in the biosphere. It is also the key to developing and improving methods of trapping radioiodine to prevent its release into the environment from reactor waste. Finally, it is important in evaluating the extent of danger from accidental release of radioiodine.

Table 18.2. Ionizing Radiation of Iodine Isotopes (Data Presented in Il'in et al., Ref. 1)

Iodine Isotope	Beta Radiation MeV	Beta Radiation %	Photon Radiation MeV	Photon Radiation %
129	0.15	100	0.038	4
(K and L shell e^-s)		96		
131	0.812	0.7	0.722	3.0
	0.610	87	0.637	9.0
	0.340	9.3	0.364	78.4
	0.250	2.8	0.284	5.0
			0.080	0.74
132	2.156	18	0.950	23.8
	1.609	21	0.775	34.8
	1.490	12	0.720	18.5
	1.218	12	0.672	39.3
	1.037	15	0.667	24.0
	0.802	21	0.652	11.5
			0.647	23.2
			0.630	12.3
133	1.54	1	0.87	8.5
	1.23	93	0.53	94.5
	0.40	6		
134	2.41	25	1.07	14.3
	2.21	12	0.890	70.5
	1.68	7.5	0.848	95.2
	1.49	15	0.61	18.1
	1.25	23		
	1.05	1		
135	1.40	25	1.80	37.5
	1.00	40	1.28	37.5
	0.47	35	0.52	62.5

Table 18.3. Chemical Forms of Iodine (Data Presented in Il'in et al., Ref. 1.)

Valence	Ions and Molecules
−1	I^-, Hi, NaI, $HI \cdot n\,H_2O$
0	I_2
+1	ICl, IBr, HOI
+5	I_2O_5, IO_3^-, HIO_3, $NaIO_3$
+7	IO_4^-, HIO_4, $NaIO_4$
Organic Compounds	CH_3I, CH_2I_2, CHI_3, C_2H_5I, $C_2H_4I_2$

18.1. SOURCES OF ENVIRONMENTAL RADIOIODINE

Both ^{131}I and ^{129}I are important fission products in nuclear power reactors: the fission yields (iodine atoms per fission) from ^{135}U in power reactors are 2.9% for ^{131}I [2] and 1.0% for ^{129}I [3]. Because of its long half-life, ^{129}I will build up as the nuclear power industry grows. A good discussion of ^{129}I and its role in the nuclear power industry can be found in the article by Russell and Hahn [3].

With its 8-day half-life, there is no long-term buildup of ^{131}I. Of course, the inventory of ^{131}I will grow as the nuclear power industry grows. Routine releases of ^{131}I from operating nuclear power plants are small, but its release could be a serious problem near fuel reprocessing plants, especially if future breeder-reactor fuel is reprocessed without sufficient aging. Spent fuel must be retained for 150 days before reprocessing to eliminate ^{131}I by decay. At the reprocessing plant a very large fraction of the total ^{13}I inventory reaches the stack and it must be removed to prevent its release to the environment. Fortunately, this need was recognized early. Scrubbing techniques, filters, and other methods were introduced, so that even prior to the 1970s less than one-half the fuel iodine inventory was actually released [2].

In a normal operating nuclear power reactor, ^{131}I is released in quantities measurable in both the airborne emissions and liquid effluents [4]. The Electric Power Research Institute (EPRI) studied [5] airborne releases to the environment and calculated that the annual average ^{131}I releases to the atmosphere from a PWR or BWR are 0.21 curies and 0.36 curies, respectively. As part of this study they determined the major chemical forms of ^{131}I released, finding at BWRs that the average species distribution was [6]:

12%	particulate
28%	elemental
20%	HOI
40%	organic.

EPRI also found that at PWRs the average species distribution is [5]:

2%	particulate
27%	elemental
40%	HOI
31%	organic.

In the section on methods, we describe methods that differentiate among species; it should be noted that some of these might not detect a particular species or might underestimate it.

Fallout is another source of environmental hazard from radioiodine. During the large-scale atmospheric bomb tests of the early 1960s (and to a lesser extent during recent French and Chinese atmospheric tests), ^{131}I was in fallout throughout the northern hemisphere. Iodine-131 from atmospheric testing disappeared by May 1963 [7,8]. Its principal route to humans is through the grass-cow-milk food chain, which has been investigated considerably [9].

There are thus four potential main sources for fission-product iodine-129 and iodine-131 release to the environment:

1. Fallout from nuclear bomb tests in the atmosphere.
2. Release from the normal operations of nuclear power reactors.

3. Release from unlikely large accidents at nuclear power reactors.
4. Release from nuclear fuel reprocessing plants.

Other sources of lesser importance, but nevertheless requiring measurement capability, are from fuel transportation accidents, from leakage out of spent fuel storage facilities, and from medical and industrial uses.

A number of media require measurement: biological samples (urine, feces); whole-body or external (thyroid) counting; food (principally milk); soil samples; air filters; and water effluent streams. As mentioned earlier, to further complicate the measurement problem, radioiodines occur in any of several chemical forms: as elemental iodine, protein-bound iodine, ionic iodates and iodides, or various simple organic compounds such as methyl iodides. The measurement of radioiodines in bioassay samples (whole body, urine, feces) will not be discussed in this text.

Radioiodines are now in widespread use in nuclear medicine: for example, about 750,000 individuals received ^{131}I for thyroid analysis or therapy in 1966, with an average activity of 35 μCi per application [10]. This adds up to about 30 Ci of ^{131}I nationwide, and although some of it decays in the body a sizable fraction (50%) is expelled in human waste into municipal wastewater systems.

The development of diagnostic procedures using the isotope ^{123}I has enabled the reduction of individual doses per application considerably: the average ^{123}I thyroid dose is only 1 to 3% of the dose from a similar procedure using iodine-131 [11]. The probable increased use of ^{123}I will necessitate its measurement in various media (e.g., urine, wastewater). Iodine-123 has a 13.3-hr half-life, and its principal emission is a 159-keV gamma ray. Therefore it cannot normally be considered a serious environmental pollutant. Iodine-123 can be measured in environmental media with sufficient sensitivity by gamma spectroscopy (the reader is referred to Chapters 5 and 6 on measurement of gamma and x-rays).

Another isotope, ^{125}I ($t_{1/2}$ = 63 days) is increasingly used in medical research. It is used as a label in 5-iodo-2′-deoxyuridine (IUdR) that is incorporated into DNA in the place of thymidine. The distribution of the IUdR to the tissues and cells is practically identical to that of thymidine. It enters DNA less effectively. In mice, 5 to 10% of an intravenous dose of IUdR is incorporated into DNA and about 1 to 2% of an orally ingested quantity is incorporated. The unincorporated percentage, more than 90% of the iodine in IUdR, is thus made labile and enters the iodine pool. Unless the iodine uptake in the thyroid is blocked, approximately one-third of the radioiodine is finally concentrated in the thyroid gland [8].

Radioiodine presents a considerable range of specific activity levels in the environment requiring measurement. Levels, of course, depend on the medium in which the activity must be measured and its history. For example, in a study of the now-decommissioned Nuclear Fuel Services (NFS) reprocessing plant [12], the activity released during reprocessing of one metric ton of fuel (20,000 MW-days burnup, aged 150 days) was about 1.8 curies of ^{131}I and 0.022 curies of ^{129}I. The normal use of a silver-reacting scrubber reduces these activities by a factor of 200 before release to the stack, so that the released activities should be 9 and 0.11 mCi, respectively. Since levels such as this would be liberated daily at the (design) rate of 1 ton of fuel per day, activities in the millicuries range require measurement. This activity corresponds to specific activities of a few $\times 10^{-3}$ pCi/m^3 (a few $\times 10^{-4}$ Bq/m^3) of stack gas for both ^{131}I and ^{129}I.

In deer thyroids taken near the NFS reprocessing plant, ^{129}I concentrations in the range of 0 to 210 pCi/g (0 to 7.8 Bq/g) were observed [3]. Shortly after the detonation of nuclear weapons by the Chinese in 1965 [13], a study of bovine thyroids revealed ^{131}I concentrations in the range up to about 100 pCi/g (3.7 Bq/g). Concentrations at other times were at least an order of magnitude smaller.

Iodine-131 concentrations have typically been below 1 pCi/liter (0.04 Bq/liter) in milk samples routinely measured by, initially, the U.S. Public Health Service Pasteurized Milk Network and now by the Environmental Protection Agency (EPA) [14,15]. This is also the level at which it is necessary to measure ^{131}I in drinking water or other foodstuffs.

18.2. BIOLOGICAL CONSIDERATIONS FOR IODINE

In order to analyze and substantiate the principles for protecting man from exposure to the radioactive isotopes of iodine it is necessary to understand the role and metabolism of this element in the body. The thyroid gland is the critical organ for internal exposure to radioiodine. It is generally accepted that there is no fundamental difference in the biochemical behavior of stable iodine and its radioactive isotopes.

Because of the difference in the physical characteristics (half-life, decay pattern, and radiation energy), however, the quantitative indicators of the rate of transport of the radioactive isotopes of iodine and of the level of their accumulation in the body and the absorbed doses in organ and tissues (in units of intaken activity) differ for different isotopes. Because of rapid radioactive decay of the radioisotopes of iodine, some do not survive to the late phases of iodine metabolism, leading to apparent shifts in the maximum accumulation (in time and absolute magnitude) and equilibrium content of radioisotopes in various structural elements of the metabolic system. Therefore the metabolism of each radioisotope must be considered separately.

Iodine is an important bioelement without which the normal function of the body is impossible. It is present in the hormones of the thyroid gland. Iodine metabolism is thus different from that of the other halogens. From the points of intake [the lungs, the gastrointestinal tract [GIT], the integuments (skin)], iodine is absorbed by the blood as iodide and distributed to all organs and tissues, where it once again enters the blood and then is partially excreted through the kidneys, the GIT, the lungs, and the skin.

The concentration factor of iodide in all organs and tissues, with the exception of the thyroid gland, salivary gland, mammary glands, and gastric glands, is equal to unity. In the gastric glands, mammary glands, and salivary glands, iodide is eliminated by secretion. It is concentrated from the plasma with a coefficient of 6 [1].

Organic conversion rather than elimination of iodine takes place in the thyroid gland. Iodide forms a compound with an amino acid (tyrosine) and then is incorporated in the protein of the thyroid gland (thyroglobulin), which is a reservoir of the hormone. Organically combined iodine within thyroxine and other derivatives is released into the blood and then enters the tissues of the body. The hormone undergoes conversion in the cells of the tissue and is finally released again so that it reenters the blood stream, from which it is partially eliminated and once again partially accumulates in the thyroid gland, entering the hormone.

As already mentioned, iodine is distributed nonuniformly in the human body. According to the data used by the International Commission on Radiological Protection (ICRP) the thyroid gland contains about half the total iodine in the body

[16]. Most of the extrathyroidal iodine is contained in the muscles (40%), skin (5%), and bones (2%). The ICRP assumes [16] that of the iodine entering the body a fraction (0.3) is translocated to the thyroid while the remainder is excreted. The iodine in the thyroid apparently has a biological half-life of 120 days and is excreted from the gland in the form of organic iodine. This organic iodine is probably uniformly distributed among the organs and tissues of the body (other than the thyroid) and is retained there with a biological half-life of 12 days. One-tenth of the organic iodine is thought to be excreted in the feces and the rest retained in the body as inorganic iodine.

If a large amount of radioiodine were released to the atmosphere, some of the public might inhale or ingest amounts that could produce acute, continuing, or late thyroid effects. Acute effects include thyroiditis, induced about two weeks after exposure, and hypothyroidism (myxedema), arising within 3 to 6 months. Continuing and late effects include hypothyroidism arising after several years, adenomatous and fibrous nodules, and thyroid cancer. Of the acute and continuing effects, hypothyroidism following radioiodine exposure is by far the most common [17]. The NCRP has postulated a linear model with a threshold for hypothyroidism, assuming that a large number of cells would have to be altered to produce this condition due to the large reserve capacity of the thyroid gland. The NCRP suggests that in making a realistic estimate of the consequences of exposure to radioiodine one should assume that the threshold for the induction of hypothyroidism is 20 rad on a purely radiobiological basis; the more uniform distribution within the thyroid of the absorbed dose from external radiation might well increase the efficiency of inducing hypothyroidism, but further data are needed to establish or refute the point [17].

Iodine, and consequently radioiodine, passes readily across the placenta. The human fetal thyroid takes up iodine beginning about the 10th to 13th week of gestation. Between 14 and 22 weeks, iodine uptake ranges from 55 to 75%. In sheep during the last trimester it has been found that there is a state of relative fetal thyroidal hyperactivity when compared to the mother and that a similar state of fetal hyperactivity exists in the human fetus. Thus the absorbed dose to the fetal thyroid increases sharply as the fetus approaches term [17]. The only well-documented effect of radioiodine in the human fetus is that of hypothyroidism. The children in these reported instances developed well-recognized symptoms and signs of cretinism.

18.3. RADIATION PROTECTION GUIDES

The International Commission on Radiological Protection recommends [18] for nonstochastic effects a dose equivalent limit of 0.5 Sv (50 rem) to all tissues except the lens of the eye. For stochastic effects* the ICRP's recommended system for limiting exposure is based on the principle that the limit on risk should be equal whether the whole body is irradiated uniformly or whether there is nonuniform irradiation. This condition will be met if

$$\Sigma w_T H_T \leq H_{wb} \ ,$$

*Malignant and hereditary disease for which the probability of an effect occurring, rather than its severity, is regarded as a function of dose without threshold.

where w_T is a weighting factor representing the ratio of the stochastic risk arising from irradiating tissue (T) to the total risk when the whole body is irradiated uniformly ($w_T = 0.03$ for the thyroid).

H_T is the dose equivalent received by tissue (T), and H_{wb} is the stochastic dose equivalent limit for uniform irradiation of the whole body which is 50 mSv/yr (5 rem/yr).

The ICRP has also recommended [19] specific values for the annual limit on intake (ALI) and derived air concentration (DAC) for occupational exposure. The DAC for ^{131}I is 18,900 pCi/m^3 (7 $\times$ 10^2 Bq/m^3) for a 40-hr work week. The DAC for ^{129}I is 2,730 pCi/m^3 (1 $\times$ 10^2 Bq/m^3) and for ^{129}I the ALI is 5.4 μCi (2 $\times$ 10^5 Bq) with the thyroid as the important organ.

At the national level, the NCRP's recommended maximum permissible concentrations (MPC) of these nuclides for occupational exposure based 168-hr/wk basis [20] are:

Isotope	168-hour Occupational MPC in Air (pCi/m^3)	168-hour Occupational MPC in Water (pCi/liter)
Iodine-131	3,000	20,000
Iodine-129	600	4,000

With the thyroid as the critical organ, the NCRP maximum permissible body burdens (MPBB) are 0.7 μCi (26 kBq), for ^{131}I and 3 μCi (111 kBq) for ^{129}I [20].

These MPC and MPBB values are based upon direct ingestion of water or inhalation of air resulting in a dose to the thyroid (the principal site for iodine concentration in the body). The dose from ^{131}I decay is mostly (90%) from beta-ray emission, not from the gamma rays.

The U.S. Nuclear Regulatory Commission (NRC) regulations are set forth in 10 CFR 20 [21]. All NRC occupational limits are identical to those recommended by the NCRP. In 1978 the NRC issued design guidelines governing emissions from light-water power reactors under its jurisdiction [22]. These decreased the original MPC_a (MPC in air) for ^{131}I in the air around these reactors by between 4 and 5 orders of magnitude depending on location and form. The motivation for this large decrease was the desire to limit the ^{131}I dose to the thyroid to 15 mrem/yr (0.15 mSv/yr), assuming as the critical pathway, reactor-air-grass-cow-milk-infant. It is assumed that the cow used in calculating the pathway would eat grass exposed to ^{131}I from the reactor, and that the exposed individual (critical group assumed to be infants) would drink 1 liter of milk per day from that cow.

Although there is still not total agreement on the concentration factors in this pathway, the reasoning for the new factor follows. Two elements in the chain have been studied: air-to-milk and milk-to-thyroid.

Burnett [23] has deduced an ^{131}I concentration of 56,000 pCi/liter (2.1 kBq/liter) of milk, for an air concentration of 100 pCi/m^3 (3.7 Bq/m^3) (which is the MPC_a for the general public, one-thirtieth of the occupational MPC_a previously cited). Measurements by VanAs and Vleggaar [24] find a similar value for dry deposition and about twice as high a milk concentration when iodine is washed out of the air by rain.

Morley and Bryant [25] have studied the case of a 6-month-old infant consuming 0.7 liters of milk per day. An ^{131}I concentration in milk of 400 pCi/liter (14.8 Bq/liter) was found to produce a dose of 1500 mrem/yr (15 mSv/yr) to the child's thyroid. Using 1 liter/day consumption, this dose would be about 2100 mrem/yr (21 mSv/yr).

If an infant thyroid dose of only 5 mrem/yr (0.05 mSv/yr) (number assumed in this study) is desired, and assuming 1 liter/day consumption, the milk concentration must be only about 1 pCi/liter (0.04 Bq/liter) using Morley and Bryant's calculation. This necessitates a concentration limit in air of about 0.002 pCi/m^3 (2×10^{-5} Bq/m^3) using Burnett's calculation or about 0.001 pCi/m^3 (4×10^{-5} Bq/m^3) and using VanAs and Vleggaar's measurements for iodine washed out of the air by rain. This latter figure [0.001 pCi/m^3 (4×10^{-5} Bq/m^3)] is precisely a factor of 100,000 less than the original MPC_a of 100 pCi/m^3 (4 Bq/m^3). However since that study 15 mrem/yr (0.15 mSv/yr) has been settled upon as the limiting dose equivalent rate [24], resulting in a factor of about 30,000.

In addition to NRC standards, the EPA has issued general environmental standards covering the impact of nuclear power operations, except mining and milling operations [26]. These standards restrict the allowable emission for ^{129}I to 5 mCi/GW_e-yr (0.16 GBq/GW_e-yr).

The EPA has also issued drinking water standards for radioiodine. (National Interim Primary Drinking Water Regulations, Ref. 27.) These regulations state that "the average annual concentration of betaparticle and photon radioactivity from man-made radionuclides in drinking water shall not produce an annual dose equivalent to the total body or any internal organ greater than 4 millirem/yr" (0.04 mSv/yr). Table 18.4 summarizes the standards for selected isotopes of iodine. The two concentrations for ^{131}I and ^{129}I are 3 pCi/liter (0.11 Bq/liter) and 1 pCi/liter (0.037 Bq/liter) for water that will be used for drinking water.

These standards must be considered whenever making environmental measurements and judging the adequacy of monitoring instrumentation.

Table 18.4. Annual Average Drinking Water Concentrations Yielding 4 mrem/yr for a 2-Liter Daily Intake for Iodine Isotopes [40 CFR 141]

Radionuclides	Critical Organ	(pCi/liter)
$t_{1/2}$ = 24 hours		
^{129}I	Thyroid	3
^{129}I	Thyroid	1
^{131}I	Thyroid	3
^{132}I	Thyroid	90
^{133}I	Thyroid	10
^{134}I	Thyroid	100
^{135}I	Thyroid	30

18.4. MEASUREMENT CONSIDERATIONS: IODINE-131

The measurement of ^{131}I in environmental media is nearly always by detection of its 364-keV gamma ray (85% branching ratio); sometimes also of its 638-keV gamma rays (15%); and less often of its beta particle. Because of its presumed pathway to man, methods for assessing ^{131}I concentrations in milk have received considerable attention. The Environmental Measurements Laboratory has maintained the Pasteurized Milk Network since 1957; about 60 stations throughout the United States collect samples which are analyzed for ^{131}I, ^{137}Cs, ^{140}Ba, ^{89}Sr, and ^{90}Sr [15].

The use of low-background gamma-spectrometer systems is discussed elsewhere in this book (see Chapters 5 and 6) and will not be treated in detail here. The two dominant types of instruments are sodium-iodide scintillation counter spectrometers and germanium semiconductor spectrometers. The key to achieving great sensitivity for ^{131}I measurements is sample preparation prior to gamma-spectrometry measurements. This usually consists of chemical concentration procedures, which are discussed below for various media.

The minimum detectable concentration (MDC) of ^{131}I using gamma spectroscopy depends upon the method. For example, the Intersociety Committee [28] reports an MDC of about 10 pCi (0.37 Bq), while other analysts achieve an MDC of about 1 pCi (0.037 Bq) using very-low-background gamma systems and a 1000-min counting period.

With specially designed systems that detect beta-gamma coincidences, lower MDCs have been achieved. As an example, Brauer, Kaye, and Connally quote a remarkably small MDC of about 0.01 pCi using the following system [29]. They used a 5-in. × 5-in. NaI(Tl) well-type spectrometer, inside of which is placed a 0.75-in. × 0.020-in. plastic scintillator. Beta particles detected in the plastic phosphor are recorded in time coincidence with gamma rays detected in the NaI(Tl) well crystal; this arrangement lowers backgrounds considerably. Figure 18.1 shows a schematic diagram of the apparatus; a typical spectrum for ^{131}I is shown in Fig. 18.2.

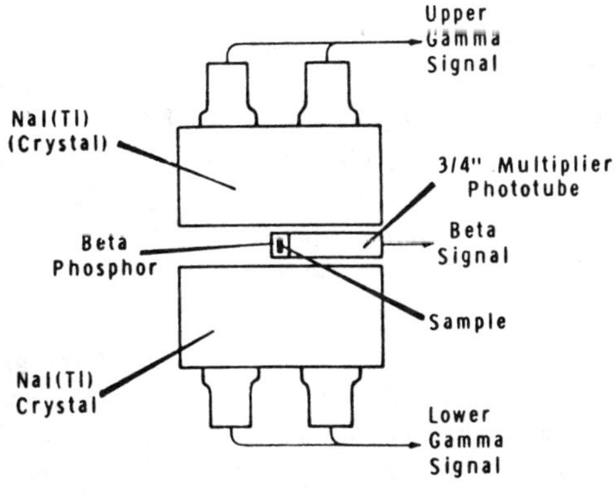

Figure 18.1. Low-level beta-gamma detector, showing beta phosphor inside well-type NaI(Tl) system (reproduced from Brauer, Kaye, and Connally, Ref. 29, with permission of the American Chemical Society © 1970.

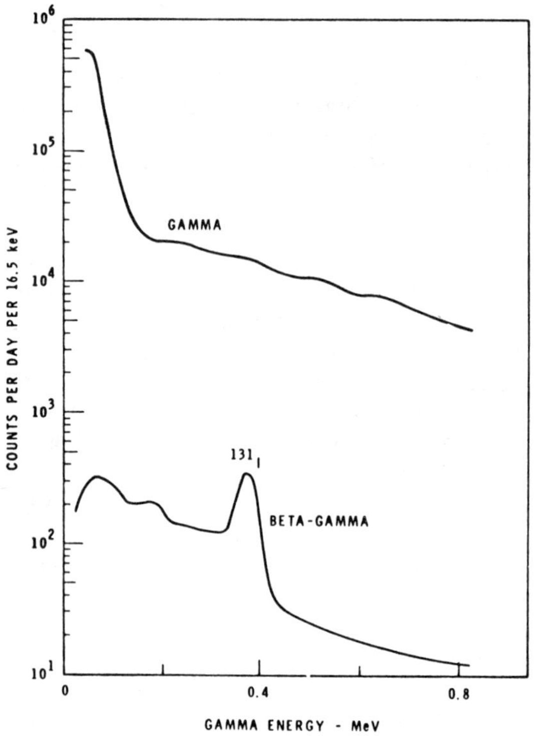

Figure 18.2. Gamma spectra of separated iodine sample taken in a 5-in. × 5-in. NaI(Tl) well system, with beta-ray detection with small plastic phosphor inside well (see Fig. 18-1). Gamma-ray and beta-gamma-ray coincidence spectra shown separately (reproduced from Brauer, Kaye, and Connally, Ref. 29, with permission of the American Chemical Society © 1970).

18.4.1. Measurement of ^{131}I in Air

The task of measuring ^{131}I activity in the atmosphere is complicated by the multiplicity of possible chemical forms (see introduction to this chapter). There are a number of similar air-sampling methods for radioiodine. The Intersociety Committee's "Tentative Method for Analysis for Iodine-131 Content of the Atmosphere" [28], is one example. Their procedure states [28]:

> The proposed method involves sampling iodine in its solid and gaseous states with an "absolute" particulate filter in series with an activated charcoal cartridge. The iodine-131 activity is then quantitatively determined by gamma spectroscopy. Iodine adsorbed on solid particles and possibly some vaporous iodine will be deposited on the "absolute" filter while the remaining vaporous iodine will be adsorbed on the activated charcoal...
>
> Minimum detectable quantity of iodine-131 by gamma spectroscopy is approximately 10 pCi (0.37 Bq). At a sampling rate of 0.01 m^3/min the practical minimum detectable quantity for a 7 day sample is 0.1 pCi/m^3 (0.0037 Bq/m^3)...

> The minimum detectable quantity [910 pCi (33.7 Bq) of iodine-131] can be determined with a precision of 10% at the 95% confidence level by employment of a well-shielded counting system and a sufficiently long counting time...
>
> Although collection of elemental iodine vapor on charcoal under the described conditions will approximate 100%, other vaporous iodine compounds such as organic iodides or hydrogen iodide may have much lower efficiencies.

The "Tentative Method" relies on an accurate flow measurement of sampled air through the system. The recommended charcoal cartridges are 5/8 in. in diameter and 1 1/2 in. deep, containing 3 g of 12 to 30-mesh activated charcoal. After collection, the sample is counted in a well-type NaI(Tl) scintillation counter coupled to a multichannel analyzer. In the presence of other gamma-ray emitters, decay measurements of the ^{131}I photopeak should be made over several days to observe the 8-day half-life. Calibration is by the use of a standard ^{131}I source in the same geometry as the unknown sample.

One of the difficulties with this method is that it employs a NaI(Tl) scintillation detector. The presence of other higher-energy gamma-ray emitters present substantial interference, making quantitative measurement difficult due to backgrounds under the ^{131}I peak. The use of a semiconductor detector would reduce this problem because of its superior resolution.

In the presence of an extremely large gamma-ray background, the Intersociety Committee [28] suggests that it may be necessary to employ chemical-separation techniques before counting.

A specially prepared filter for collection of iodine has been described by Baratta, Chabot, and Donlen [30]. The filter is impregnated with barium hydroxide-iodide-iodate, and after sample collection the iodine is concentrated chemically by dissolution and exchange between the radioiodine and the stable iodine.

Concentrations as low as 0.01 pCi/m^3 (3.7×10^{-4} Bq/m^3) might be detectable, using the advanced research procedures. For example, using the sampling rate of 0.01 m^3/min mentioned by the Intersociety Committee, a week of pumping will sample about 100 m^3 of air and collect 1 pCi (0.037 Bq) of ^{131}I, if the concentration is 0.01 pCi/m^3 (3.7×10^{-4} Bq/m^3). This is just about the minimum detectable activity which one can expect to measure in a very-low-background gamma-spectroscopy system even when counting for a 100-min period (see Chapters 5 and 6). Increasing the sampling rate by a factor of 3 would yield 3 pCi (0.11 Bq) of ^{131}I or alternatively reduce the minimum detectable air concentration by the same factor of 3. However, a high sampling rate tends to reduce the collection efficiency of the charcoal filters, especially for the nonelemental forms of iodine [31].

Charcoal can be specially activated by the addition of certain elements that increase its ability to hold iodine. Materials other than charcoal have also been used to collect iodine from the air. Silver-coated filter paper, charcoal-loaded filter paper, molecular sieves, silver-treated zeolites and scrubbers containing sodium thiosulfate or potassium hydroxide have all been used. Coated and loaded filter papers are inefficient for nonelemental iodine. Molecular sieves and silvered zeolite have proven quite effective not only for elemental iodine but also for many iodine compounds. It may be desirable to test several different methods for any given situation [32].

It is important to remember that when charcoal is saturated with all the radioactivity it can hold, the temperature rises quickly due to self-ignition. As any removal technique saturates, it no longer collects efficiently. Any analyst should regularly test for such saturation. The transactions of the Fourteenth ERDA Air Cleaning Conference [33] should be consulted for more information on radioiodine removal techniques. For a review of some of the important properties of radioactive iodine and charcoal the reader is referred to Lorenz, Martin, and Ngao [34].

Continuous monitors for ^{131}I, are also available. Wilhelm and Maknaus designed a system to be used as a continuous iodine monitor [35]. In this system, iodine accumulates on a special filter cartridge and is measured with a NaI(Tl) detector during the accumulation process. The results are displayed on a rate meter. An alarm sounds if total accumulated iodine activity (A) is too high in the filter cartridge (integral method) or a preset ratio of dA/dt is exceeded (differential method). They found that 2000 pCi/m^3 (74 Bq/m^3) can be detected with a measuring period of 1000 sec if the initial pulse rate is about 10 cps and if there is a negligible background rate. Winkelman and Jagersberger [36] report that it is possible to attain measurements for ^{131}I at a level of 10 pCi/m^3 (0.37 Bq/m^3) using a NaI scintillation detector surrounded by an iodine absorber and well designed shield. To achieve this sensitivity the background must not exceed 200 R/hr and a 40-keV window should be used.

Distenfeld and Klemish [37] designed a low-cost radioiodine air sampler. This is simply a specially designed canister and air pump. The canister has a prefilter to remove particulates before the absorber bed which is silver impregnated silica gel. Once samples are collected, they can be counted on either a NaI(Tl) or Ge detector.

In sampling for radioiodine it is important to consider the chemistry of iodine at very low concentrations, at high temperature, and with high radiation fields. I_2 is less stable at low concentrations and is both a better oxidizing and a better reducing agent when its concentration is low [38]. Both temperature and radiation field can perturb its chemistry. Different chemical forms require different kinds of collectors and absorbers. Some of these are listed below [39]:

Form of Iodine	Collector Used
Elemental	Copper or silver screens (freshly treated)
	Activated-charcoal-impregnated paper
	Scrubbers with a solution of $AgNO_3$, NaI (basic), $Na_2S_2O_3$
	Activated charcoal
	Molecular sieves
CH_3I	Scrubbers with a solution of $AgNO_3$
HOI	Aluminum or silicic acid impregnated with $AgNO_3$
	Activated charcoal
	Ag-X13 (silver zeolite) when humidity is low
	Alumina impregnated with 4-iodophenol (a special absorber for special investigations)
LiI or other basic iodide	These are particulates and can be collected on any high-efficiency filter paper.

By arranging in the proper sequence absorbers which are specific collectors for the different forms of iodine, estimates can be made of the quantities of the chemical species present. One of the most suitable absorbers for a mixture of iodine compounds is activated charcoal impregnated with 5% by weight of triethylenediamine or potassium iodide.

18.4.2. ^{131}I in Milk

The Public Health Service Pasteurized Milk Network developed a routine, standard analysis procedure for the determination of ^{131}I in milk [40,41]. After a composite is made from each sampling station in the network, the sample is counted with a gamma-ray scintillation counter spectrometer for 50-min. The three major gamma emitters commonly found in milk (^{131}I, ^{137}Cs, and ^{140}Ba) can all be determined with minimum detectable concentrations in the range of 10 pCi/liter (0.37 Bq/liter). A description of a gamma-spectroscopic technique with NaI(Tl) scintillator counter, as developed by the PHS, is described by an article by Hagee, Karches, and Goldin [42].

If γ-spectroscopy is not available, methods for determining ^{131}I in milk rely on chemical separation techniques to isolate individual radionuclides. Various separation techniques have been developed, generally using an anion exchange resin followed by gamma counting. They rely upon two facts: that "iodine-131 in market milk is mostly in the inorganic form as the iodide, very little being bound by proteins" [41]; and that "of the several fission products ..., iodine-131 is the only anionic radionuclide present in milk." [42]

The method described by Murthy [41] is typical: 50 ml of anionic resin (strongly basic quaternary type, 20 to 50 mesh size, in chloride form) with 1 gallon of milk flowing through the column at 40 to 50 ml/min will recover all but the protein-bound iodine. Single-channel counting with a 2-in. NaI(Tl) crystal counter is used, and concentrations in the range of 50 pCi/liter (1.85 Bq/liter) can be determined to ± 4 pCi/liter (± 0.15 Bq/liter) accuracy.

In a similar method described by Porter and Carter [43], the anion exchange resin retains $96 \pm 2\%$ of the radioiodine in the first liter of milk, with a loss of an additional 2% per liter up to 4 liters. Porter and Carter's method is designed for sample collection in the field, at the very low (1965) cost of about \$3 for the field sampling unit, plus \$0.50 per (nonreusable) vial with resin. Samples are analyzed and counted later in the laboratory.

Both of the above methods are "rapid," that is, determinations may be made in 2 to 3 hr.

Another method for fresh whole milk samples is described in the U.S. Department of Energy EML *Procedures Manual* [44]:

> Iodine-131 in whole milk samples is collected by the coagulation of protein-bound iodine with trichloroacetic acid and the simultaneous precipitation of ionic iodine as silver iodide. The filtrate from this separation and an aliquot of the untreated milk are then gamma counted on a NaI(Tl) scintillation crystal assembly.

This procedure has the advantage of also measuring any protein-bound ^{131}I. Subtraction of the corrected counting rate with and without treatment is necessary. Of course "the analysis may be made more specific by measuring with a gamma spectrometer." [44]

Concentrations as low as fractions of 1 pCi/liter (hundredths of a Bq/liter) of milk can be measured if the extraction procedure can remove enough iodine from enough milk. For example, using 10 liters of milk at 0.1 pCi/liter (3.7 mBq/liter), one can by brute force obtain 1 pCi (37 mBq/liter) of ^{131}I for later determination by gamma-ray spectroscopy. This activity is very close to the ultimate detection limits of the better low-background gamma-ray spectroscopic systems counting for 1000 min. Alternatively, one can use a commercially-available low-background beta counter of the type designed for tritium.

A second method is also recommended by DOE's EML [44]:

> Iodine-131 in milk is batch exchanged onto Dowex 1-X8 anion exchange resin. The resin is separated from the bulk milk sample, the iodine is stripped from the resin with 5 NaOCl, extracted with carbon tetrachloride, and precipitated as PdI_2. Recovery is determined gravimetrically. The precipitate is mounted and counted over a one week interval to check the decay of Iodine-131.

Rill et al. [45] have developed and tested a procedure for an instrumental method of measuring ^{131}I in milk. By using a beta-gamma coincidence method, they were able to keep background low and achieve specificity. All ^{131}I beta-ray decays are accompanied by gamma rays, and 81% are in coincidence with a 364-keV gamma ray. Extrapolation of the test results indicates that a detector with a 0.25-liter sample volume can measure ^{131}I in milk to 0.5 pCi/liter (18.5 mBq/liter) with no chemical processing of the sample.

Another β-γ coincidence system has been designed for ^{131}I in milk. Using a thin plastic scintillator and a NaI(Tl) well detector, Paperiello and Matuszek [46] reached a detection limit of 0.08 pCi/liter (3 mBq/liter) for a 100-min counting time following separation chemistry, with a background of 0.01 cpm and a system efficiency of 11%.

18.4.3. Measurement of ^{131}I in Water (Drinking and Waste)

Measurements of ^{131}I in drinking water are not made routinely as a part of public health practice, and no Standard Method is found in the American Public Health Association compilation [47]. Generally, ^{131}I does not occur in significant quantities in public water supplies frequently enough to call for routine monitoring. However the drinking regulations [27] do require the supplier of any community water system designated by the state as utilizing waters contaminated by effluents from nuclear facilities to monitor quarterly for ^{131}I radioactivity.

Contamination of waste waters is important, because of increasing use of the radioiodines in medicine and industry. The American Society for Testing and Materials (ASTM) [48] has published three different Standard Methods for the measurement of radioiodines in waste water. The methods are all designed for concentrations greater than 100 pCi/liter (3.7 Bq/liter).

The first ASTM method [49] employs "cation exchange followed by heterogeneous isotopic exchange of iodide with silver iodide." Only silver isotopes might interfere, and these are rarely encountered.

The second ASTM method [48]

> ...is based on the separation of iodine from other activities by distillation of elemental iodine into cold carbon tetrachloride. To assure chemical interchange with the

iodine carrier an oxidation-reduction cycle is made. The iodine is oxidized to iodate with permanganate, reduced to iodide with bisulfite, and distilled over as free iodine in the presence of nitrite.

The third ASTM method [50]

...is based on the separation of iodine isotopes from other radioisotopes by extraction into carbon tetrachloride... The iodine is oxidized to periodate in the presence of sodium hypochlorite and then reduced to iodide with sodium bisulfite in acid medium.

Recently interest has increased in on-line measurements of radioactivity to determine the correct conditions for operating nuclear power plants. Kuhlman, Conwell, and Goldberg [51] developed the use of a hyperpure germanium detector for process radiation monitoring in large nuclear power plants. They achieved a minimum detectable concentration of 313 pCi/liter (11.6 Bq/liter) with a 3-liter sample canister and a 38-cm^3 Ge detector (background was 20 mR/hr).

18.4.4. Measurement of ^{131}I in Other Media

Measurement of ^{131}I in media other than air, milk, and water is usually by routine gamma-spectroscopic analysis. Well-type germanium detectors offer better detection efficiencies and similarly low background counting rates compared with large coaxial detectors or opposed detector systems. In fact it has been found [52] that the minimum detectable activity for a 1440-min count in a well-type Ge(Li) detector is 0.16 pCi (5.9 mBq), 0.08 pCi (3.0 mBq) when an anticoincidence annulus is utilized for background reduction. The reader is referred to Chapters 5 and 6 for further discussions of these techniques.

18.4.5. Quality Control in Measurements

Quality control is one of the most important steps in the entire measurement process. The calibration of gamma-spectroscopic systems has been discussed in Chapter 6 and will not be treated here. The other key part of the iodine measurement system is chemical separation.

Quality-control checks have usually revealed significant variations among laboratories analyzing for ^{131}I at low levels. These variations are of concern since they demonstrate the fundamental difficulty in making routine, reliable, accurate ^{131}I measurements. The following study is an example. Others show comparable results.

In 1971, the EPA Office of Radiation Programs studied laboratory capability to determine ^{131}I in milk [53]. The 35 participating laboratories made triplicate determinations of unknown concentrations of ^{131}I, ^{89}Sr, ^{90}Sr, and ^{137}Cs at activity levels ($\pm$2%) of 69, 31, 42 and 52 pCi/liter, respectively. Unfortunately, six laboratories reported results whose mean exceeded the "control level" deviation of about $\pm$15% for ^{131}I. (For ^{89}Sr, ^{90}Sr, and ^{137}Cs the laboratories making too great an error numbered 4, 9, and 2, respectively.) For many of the laboratories exceeding the allowed error in the ^{131}I study, the range of reported results for triplicate analyses exceeded 30%. This study shows that, while many laboratories could determine ^{131}I well, a significant minority could not.

18.5. MEASUREMENT CONSIDERATIONS: IODINE-129

Iodine-129 decays with a 17-million-year half-life by emitting a low-energy beta ray (E_{max} = 150 keV) followed by a 40-keV gamma ray. Both the beta ray and gamma ray are of relatively low energy, and consequently difficult to detect. The MPC values for ^{129}I are all about a factor of 5 more restrictive than those for ^{131}I (because after ingestion or inhalation the ^{131}I decays relatively rapidly with its 8-day physical half-life whereas the ^{129}I lasts for the biological half-life of about 120 days). Combining these considerations, the measurement problem is seen to be fairly difficult.

Fortunately, however, the long half-life and low specific activity can be used to advantage. Since 1 curie (3.7×10^{10} Bq) of ^{129}I weighs about 6000 g, even minute activities represent sizeable numbers of atoms amenable to measurement techniques other than direct radioactive counting.

Three measurement techniques are frequently used: liquid scintillation counting, neutron activation analysis, and laser fluorescence. The first is useful when the activity to be measured is known to exceed a few pCi (a tenth of a Bq) of ^{129}I, whereas the latter is recommended for its very high sensitivity. In the first two techniques, iodine must be isolated chemically to free it from the sample matrix in which it is found. We shall not discuss here any of the several radiochemical techniques since they are identical to those used for ^{131}I and have already been discussed.

18.5.1. Liquid Scintillation Counting

A liquid scintillation counter system is used to detect the low-energy beta particles from ^{129}I. Iodine must be isolated chemically before counting by the scintillator system to provide a sample free of the many impurities that might cause quenching (self-absorption by the sample of the scintillation light).

Liquid scintillation counting is discussed in depth in Chapter 5 and Chapter 20, and that discussion will not be repeated here. The best systems can count minimum detectable activities (3σ) for tritium in the range of about 1 pCi/cm^3 (0.037 Bq/cm^3) in water, and the performance for ^{129}I should be comparable. The liquid scintillation counter method for ^{129}I described by Magno [54] and Gabay [55] and their co-workers, differs from that for tritium in that the sample is placed under a fluorescent light for about 2 hr to decolorize the iodine before counting. An error may be introduced unless the counting efficiency for ^{129}I is known as a function of energy, since the wide beta energy spectrum extends from the maximum (150 keV) down to zero energy. A standard ^{129}I solution is useful for this purpose. Standards are available from: the Office of Standard Reference Materials, U.S. National Bureau of Standards, Washington, DC 20234. Calibration should, if possible, be made experimentally rather than determined by calculation, since the ^{129}I beta is followed almost immediately by a 40-keV gamma ray which will also be detected in the scintillator some of the time, distorting the apparent beta energy spectrum.

Using a solvent extraction procedure on 4 liters of milk, followed by liquid scintillation counting, Gabay et al. [55] report a ^{129}I detection limit (95% confidence) of about 0.3 pCi/liter (11 mBq/liter), counting for 100 min at average background of 25 cpm.

18.5.2. Neutron Activation Analysis

Iodine-129 determination by neutron activation analysis relies upon the transformation of ^{129}I to ^{130}I by a neutron-capture reaction:

$$^{129}I(n,\gamma)^{130}I \xrightarrow[12.4\ hr]{\beta} {}^{130}Xe \quad .$$

Iodine-130 is counted in a gamma spectrometer by measuring its decay to ^{130}Xe; one of several beta particles is followed in each case by a series of gamma rays, including two that are easily measurable at 538 and 669 keV. Typically, ^{129}I is accompanied in any medium by much larger concentrations of stable, naturally occurring ^{127}I. This also is determined by a similar (n,γ) process:

$$^{127}I(n,\gamma)^{128}I \xrightarrow[25min]{\beta} {}^{128}Xe \quad .$$

When ^{128}I decays to ^{128}Xe the most easily detected emission is a 441-keV gamma ray (13%).

The intrinsic sensitivity limit of this method is caused by problems from the multiple activation of ^{127}I up to ^{130}I, where it masks ^{129}I:

$$^{127}I(n,\gamma)^{128}I(n,\gamma)^{129}I(n,\gamma)^{130}I \xrightarrow[12.4\ hr]{\beta^-} {}^{130}Xe \quad .$$

Typically, neutron activation analysis is capable of measuring $^{129}I/^{127}I$ ratios to much greater precision than the determination of the ^{129}I activity itself. This ratio may then be used together with other information on the ^{127}I content to determine ^{129}I very accurately.

Neutron activation analysis for ^{129}I was first suggested by Edwards [56], who suggest the use of ^{129}I as an environmental tracer. Investigations by Keisch [57], Cochran [12], and Magno [54] and their co-workers show that with chemical separation the technique suffers few interferences and may be used down to very low concentrations. As mentioned previously, iodine-isolation chemistry is required to obtain the best sensitivity; for example, if stable cesium-133 is present, it may interfere because the reaction $^{133}Cs(n,\alpha)^{130}I$ produces the same ^{130}I isotope as that from (n,γ) activation of ^{129}I.

Because of the multiple-capture production of ^{130}I from ^{127}I, there is an optimum neutron flux for the greatest sensitivity. Keisch, Koch, and Levine [57] have shown this to be about 10^{13} thermal neutrons $cm^{-2}sec^{-1}$. These investigators report that the theoretical limitation on the ^{129}I:^{127}I atom ratio is about 3×10^{-13} in a sample containing 1 g of iodine, corresponding to an incredibly small ^{129}I activity of about 5×10^{-17}Ci (1.9×10^{-6} Bq).

Ordinary gamma-ray spectrometry is done after activation using NaI(Tl) or Ge(Li) systems. Keisch, Koch, and Levine [57] also report use of a beta-gamma coincidence arrangement for ^{130}I to reduce backgrounds. The isotope ^{126}I may interfere with this technique since it is also a beta-gamma coincidence emitter. ^{126}I is produced from ^{127}I by the reaction

$$^{127}I(n,2n)^{126}I \xrightarrow[13\ days]{\beta} {}^{126}Xe$$

when the activating neutron beam contains neutrons of higher than thermal energy. Use of a well-thermalized neutron flux suppresses this background.

Brauer, Kaye, and Connally [29] report a variation on activation analysis, in which after irradiation the sample is measured with a mass-spectrometer for the xenon isotopes which are the endproducts of ^{130}I and ^{128}I decay. Both the ^{130}Xe:^{128}Xe and the ^{130}Xe:^{126}Xe ratios should be indicative of the ^{129}I:^{127}I ratio. Detection limits of the ratio in the range of $\sim 10^{10}$ have been reported.

18.5.3. Measurement Using Laser-Induced Fluorescence

In this technique a laser photon of appropriate frequency, impinging upon a sample, is selectively absorbed by the iodine molecule exciting it to a metastable electronic state. The excited iodine molecule can relax by collision processes to the ground state in which case it will not be detected, or it may relax by emitting a photon of longer wavelength. The fluorescence can be detected by a photoelectric detector. After appropriate corrections are made for competing collisional relaxation processes, collector efficiency, detector response characteristics, and other emissive interferences, the resulting fluorescence signal can be used to determine the I_2 concentrations in a gas sample. As this method is specific only for elemental iodine, all organic iodides must be transformed by a catalytic converter to elemental iodine prior to measurement.

Laboratory measurements with a prototype instrument have clearly demonstrated [58] that laser fluorescence will be sensitive and practical and could be used as a relatively inexpensive field monitoring instrument for $^{129}I_2$ down to or below the 100-pg/cm^3 level (0.0167 pCi/cm^3 or 0.62mBq/cm^3). Such an instrument will be capable of real-time instantaneous measurements or of operating in an integrating mode to accumulate loads over a period of time. The dynamic concentration range of this technique can be in excess of 10^5 and if required is capable of measurements over a range of 10^8, that is, 0.01 pCi/cm^3 level up to $\sim 10^{-2}$ g/cm^3 ^{129}I.

Measurements indicate that the signal from $^{127}I_2$ is detected with a sensitivity of less than 1% that of $^{129}I_2$ [58]. If a very large excess of $^{127}I_2$ over $^{129}I_2$ is present in the gas to be measured a significant portion of the ^{129}I present will be tied up in the mixed isotope molecule ^{127}I-^{129}I. The sensitivity for the mixed molecule was not measured by Baronavski and McDonald [58].

Fernandez et al. [59] report that a major disadvantage of this method is that the laser also responds to NO_2, which will be part of the effluents from a fuel reprocessing facility. Therefore some method separating $^{129}I_2$ from NO_2 present in the sample stream is necessary.

Hohimer and Hargis [60] report that severe quenching of iodine fluorescence by atmospheric molecules may limit the sensitivity of the fluorescence method for airborne radioiodine. They suggest intracavity absorption spectroscopy as an alternative method for the detection of airborne iodine-129. Intracavity absorption spectroscopy involves placing the weakly absorbing species within the cavity of a broadband dye laser and measuring its effect on the spectral output of the laser. The wavelength-dependent loss introduced by even a very weak absorber dramatically decreases the intensity of the laser output at the absorber wavelength. Hohimer and Hargis used external fluorescence detection with a continuous wave dye laser in this study [60] of intracavity absorption spectroscopy. They measured ^{129}I at concentrations in the range 5×10^{11} to 7×10^{15} molecules/cm^3, corresponding to 1.2×10^{-14} to 1.7×10^{-10} Ci/cm^3 (4×10^{-3} to 6 Bq/cm^3).

18.5.4. X-Ray Spectroscopy

Liquid scintillation counting techniques do not permit differentiation of both ^{129}I and ^{125}I. X-ray spectroscopy using a hyperpure germanium detector offers this capability. The K x-rays at 27.5 and 27.2 keV for ^{125}I and 29.7 and 29.4 keV for ^{129}I can easily be resolved with a hyperpure germanium diode with a resolution better than 200 eV FWHM. Of course the samples must be treated chemically in order to extract all iodine from a sample and convert it to an appropriate counting geometry. Preliminary calculations [61] indicated that the detection limit for ^{129}I is approximately 0.1 pCi/liter (3.7 mBq/liter) for a 4-liter sample. Honaker [62], using a hyperpure germanium well detector and a less rigorous chemical separation procedure, achieved about 1 pCi/liter (37 mBq/liter) sensitivity for ^{129}I.

18.6. SUMMARY AND CONCLUSIONS

The intent of this section has been to discuss the sources and principal measurement techniques for radioiodine (^{131}I and ^{129}I) in the environment.

Iodine-131 in Milk. Milk is the medium to which the greatest effort has been devoted. There are several laboratory methods for determining ^{131}I in milk with minimum detectable concentrations of about 10 pCi/liter (370 mBq/liter). With the very best equipment and long counting periods, fractions of 1 pCi/liter (mBq/liter) can be detected. Field sampling methods for milk are satisfactory, but laboratory analysis is required generally and 2 to 3-hour determinations are considered necessary unless a β-γ coincidence system is used (2 to 3-hour determinations are considered "rapid"). It appears that instrumentation is adequate to cover this measurement problem.

Iodine-131 in Air. The "Tentative Method" of the Intersociety Committee can detect levels at the level of about 0.1 pCi/m^3 (3.7 mBq/m^3). However, this is for a 7-day sample; for a 1-day sample the minimum detectable level would be seven times higher. The MPC level for the general public is 10 pCi/m^3 (370 mBq/m^3).

Detection limits perhaps one order-of-magnitude lower can be obtained using the very best low-background equipment. This method is not adequate when gaseous or liquid iodine compounds, such as organic iodides or hydrogen iodide are present, since they are not efficiently collected. Clearly, collection techniques for gases and liquids must be improved, since until all airborne radioiodine compounds can be collected, no measurement of them can be considered accurate.

Iodine-131 in Other Media. Gamma spectroscopy is generally adequate for activity determinations in most solid media, such as biological samples.

Iodine-129. The problem of sampling iodine-129 in air is similar to that for ^{131}I (see preceding). However, iodine-129 is not now often a problem of human radiological significance (even locally near nuclear fuel reprocessing plants), and measurements of it are principally to assure that it is not released to the environment in the first place. Restrictions on release are to prevent long-term environmental buildup. Instrumentation is adequate for measuring samples, but more effort must go into developing generally available techniques for real-time measurements.

General. Measurements of ^{131}I in air, milk, and solid biological samples can all be made with existing instrumentation to sufficient accuracy and low enough concentration for all practical purposes. The only exception is for those vaporous iodine compounds such as light organic iodides or HI which are not collected with adequate efficiency by established methods for determining ^{131}I in air. For ^{129}I, instrumental techniques are is also reasonably well developed.

REFERENCES

1. L. A. Il'in, G. V. Arkhangel'skaya, Yu. O. Konstantinov, and I. A. Likhtarev, *Radiative Iodine in the Problem of Radiation Safety*, AEC-tr-7536, 1974, translated from Radioaktivnyi iod v probleme radiatsionnoi bezopasnosti, Atomiz dat, 1972, 272 pp.
2. ORNL, "Siting of Fuel Reprocessing and Waste Management Facilities," Report No. ORNL-4451, 1971, Oak Ridge National Laboratory, Oak Ridge, TN 37830.
3. J. L. Russell and P. B. Hahn, "Public Health Aspects of Iodine-129 from the Nuclear Power Industry," *Radiol. Health Data Rep.*, **12**, 189, 1971.
4. U.S. Environmental Protection Agency, "Summary of Radioactivity Released in Effluents from Nuclear Power Plants from 1973 thru 1976," EPA-520/3-77-012, 1977, Office of Radiation Programs, Washington, DC 20460.
5. Electric Power Research Institute, "Sources of Radioiodine at Pressurized Water Reactors," EPRI NP-939, 1978, Project 294-1 Final Report, prepared by Science Applications, EPRI, 3412 Hillview Avenue, Palo Alto, CA 94304.
6. C. A. Pelletier, E. D. Barefoot, J. E. Cline, R. T. Hemphill, W. A.Emel, and P. G. Voilleque. "Sources of Radioiodine at Boiling Water Reactors," EPRI NP-495, 1978, Electric Power Research Institute, 3412 Hillview Avenue, Palo Alto, CA 94304.
7. Federal Radiation Council, "Estimates and Evaluation of Fallout in United States from Nuclear Weapons Testing Conducted through 1962," FRC Report No. 4, 1963, available through U.S. Government Printing Office, Washington, DC 20402; "Revised Fallout Estimate for 1964–1965 and Verification of the 1963 prediction," FRC Report No. 6, 1964, available from GPO.
8. National Council on Radiation Protection and Measurements, "Tritium and Other Radionuclide Labeled Organic Compounds Incorporated in Genetic Material," NCRP Report No. 63, 1979, 7910 Woodmont Avenue, Washington, DC 20024.
9. A. C. Ng, C. S. Colsher, D. T. Quinn, and S. E. Thompson, "Transfer Coefficients for the Radiation Dose to Man via the Forage-Cow-Milk Pathway from Radionuclide Released to the Biosphere," UCRL-53939, 1977, Lawrence Livermore Laboratory, Livermore, CA 94550.
10. A. W. Klement, C. R. Miller, R. P. Minx, and B. Shleien, "Estimates of Ionizing Radiation Doses in the United States, 1960 to 2000," Report ORP/CSD 72-1, 1972, U.S. Environmental Protection Agency, Office of Radiation Programs, Rockville, MD 20852.
11. B. M. Branson, H. N. Wellman, and J. G. Kereiakas, "An Ultra-Sensitive Iodine-123 Thyroid Uptake System Utilizing Peak: Scatter Organ Depth Correction," Int. J. Appl. Radiat. Isotopes, **22**, 502, 1971.

12. Northeastern Radiological Health Laboratory, "Radioactive Effluent from an Operating Nuclear Fuel Reprocessing Plant," Report No. BRH/NERHL 70-3, 1970, Winchester, MA 01890.
13. E. J. Baratta, "Uptake of Iodine-131 in Bovine Thyroids," in *Radioecological Concentration Processes*, Proceedings of a Symposium in Stockholm, April 25–29, 1966, Pergamon Press, Oxford, 1966 p. 609.
14. The U.S. Public Health Service Pasteurized Milk Network reported its radionuclide measurements periodically in *Radiation Data and Reports*. Nuclides analyzed were strontium-89 and 90, iodine-131, cesium-137, and barium-140. An extensive discussion of its procedures is in Vol. 9, pp. 731 ff., December, 1968.
15. H. N. Wellman and R. T. Anger, Jr., "Radioiodine Dosimetry and the Use of Radioiodines Other than Iodine-131 in Thyroid Diagnosis," *Seminars in Nuclear Medicine*, International Atomic Energy Agency, Vienna, 1971.
16. International Commission of Radiological Protection, "Report of the Task Group on Reference Man: Anatomical, Physiological and Metabolic Characteristics," ICRP Publication No. 23, 1975, Pergamon Press, Elmsford, NY 10523.
17. National Council on Radiation Protection and Measurements, "Protection of the Thyroid Gland in the Event of Releases of Radioiodine," NCRP Report No. 55, 1977, 7910 Woodmont Ave, Washington, DC 20014.
18. International Commission on Radiological Protection, "Recommendations of the International Commission on Radiological Protection," ICRP Publication 26, *Ann. ICRP*, **1**(3), 1977, Pergamon Press, Elmsford, NY 10523.
19. International Commission of Radiological Protection, "Limits for Intakes of Radioiodides by Workers," ICRP Publication 30 Part 1, *Ann. ICRP*, **2**(3/4), Pergamon Press, NY, 1978.
20. National Council on Radiation Protection and Measurements, "Maximum Permissible Body Burdens and Maximum Permissible Concentrations of Radionuclides in Air and in Water for Occupational Exposure," NCRP Report No. 22, 7910 Woodmont Ave., Washington, DC.
21. *Code of Federal Regulations*, Title 10, Energy, revised as of January 1, 1978, Chapter 1, Nuclear Regulatory Commission, Part 20, Standards for Protection Against Radiation, Appendix B, "Concentrations in Air and Water Above Natural Radiator Background," U.S. Government Printing Office, Washington, DC 20402.
22. *Code of Federal Regulations*, Title 10, Energy, revised as of January 1, 1978, Chapter 1, Nuclear Regulator Commission, Part 50, Licensing of Production and Utilization Facilities, Appendix I, "Numerical Guide for Design Objectives and Limiting Conditions for Operation to Meet the Criterion As Low as Practical for Radioactive Material in Light-Water-Cooled Nuclear Power Reactor Effluents," U.S. Government Printing Office, Washington, DC 20014.
23. T. J. Burnett, "A Derivation of the 'Factor of 700' for Iodine-131," *Health Phys.*, **18**, 73, 1970.
24. D. VanAs and C. M. Vleggaar, "Determination of an Acceptable Iodine-131 Concentration in Air when the Critical Intake is Through Milk," *Health Phys.*, **21**, 114, 1971.
25. F. Morley and P. M. Bryant, "Basic and Derived Radiological Protection Standards for the Evaluation of Environmental Contamination," in *Environmental*

Contamination by Radioactive Materials, International Atomic Energy Agency, Vienna, 1969.

26. U.S. Environmental Protection Agency, 40 CRF 190, "Environmental Radiation Protection Requirements for Normal Operations of Activities in the Uranium Fuel Cycle," Final Environmental Statement, 2 vols., EPA-520/4-76-016, 1976. Single copies are available from Director, Criteria and Standards Division (AW-460), 401 M Street, Washington, DC 20460.
27. U.S. Environmental Protection Agency, "National Interim Primary Drinking Water Regulations," EPA-570/9-76-003, Office of Water Supply, listed 40 CFR 141, 1976; available for GPD.
28. American Public Health Association, *Methods of Air Sampling and Analysis*, Intersociety Committee for a Manual of Methods for Ambient Air Sampling and Analysis, 1972. Available from American Public Health Association, 105 18th Street, N.W., Washington, DC 20036.
29. F. P. Brauer, J. H. Kaye, and R. E. Connally, "X-Ray and Beta-Gamma Coincidence Spectrometry Applied to Radiochemical Analysis of Environmental Samples," *Adv. Chem. Ser.*, **93**, 231, 1970.
30. E. J. Baratta, G. E. Chabot, and R. J. Donlen, "Collection and Determination of Iodine-131 in the Air," *Am. Indust. Hygiene Assn. J.*, **29**, 159, 1968.
31. R. R. Bellamy, "Elemental Iodine and Methyl Iodide Absorption on Activated Charcoal at Low Concentrations," in *Proceedings of the Thirteenth Air Cleaning Conference, San Francisco, 12–15 August 1974*, CONF-740807, edited by M. W. First, Vol. 2, 1975, p. 683.
32. Harry F. Schulte, "Current Technology in Sampling for Airborne Radionuclides," *IEEE Trans. Nucl. Sci.*, **N5-23**(3), 1197, 1976.
33. M. W. First, Ed., in *Proceedings of the Fourteenth ERDA Air Cleaning Conference Sun Valley, Idaho, 2–4 August 1976*. Sponsored by ERDA and Harvard Air Cleaning Laboratory, 2 vols., 1977, particularly session on IV-Radioiodine Removal and Retention.
34. R. A. Lorenz, W. J. Martin, and H. Ngao, "The Behavior of Highly Radioactive Iodine on Charcoal," p. 707, Ref. 31.
35. J. G. Wilhelm and H. Maknaus, "Continuous Monitoring of Radioactive Iodine Emission," p. 3 of Ref. 31.
36. I. Winkelmann and H. Jagersberger, "Monitoring of Airborne Release from LWR Power Plants in the Federal Republic of Germany," IAEA-SM-217/4; "Monitoring of Radioactive Effluents from Nuclear Facilities," Proceedings of a Symposium in Portoroz, Yugoslavia, 5–9 September 1977, International Atomic Energy Agency, Vienna, 1978, p. 198.
37. C. Distenfeld and J. Klemish, "A Low Cost Radioiodine Air Sampling System for Nuclear Incidents: Experience at Three Mile Island," *IEEE Trans. Nucl. Sci.*, **NS-27**(1), 675, 1980.
38. M. Kahn and J. Kleinberg, "Radiochemistry of Iodine," NAS-NS-3062, 1977, prepared for Subcommittee on Radiochemistry, National Academy of Sciences–National Research Council, available from NTIS.
39. International Atomic Energy Agency, "Radiological Surveillance of Airborne Contaminants in the Working Environment," Procedures and Data, Safety Series No. 49, IAEA, Vienna, 1979.
40. B. Kahn, G. K. Murthy, C. Porter, G. R. Hagee, G. J. Karches, and A. S. Goldin, "Rapid Methods for Estimating Fission Product Concentrations in

Milk," Environmental Health Series, U.S. Dept. of HEW, Public Health Service Publication No. PHS-999-R-2, 1963.

41. G. K. Murthy, "Rapid Determination of Iodine-131 in Milk," p. 7 of Ref. 40.
42. G. R. Hagee, G. J. Karches, and A. S. Goldin, "Determination of Iodine-131, Cesium-137, and Barium-140 in Milk by Gamma Spectroscopy," p. 17 of Ref. 40.
43. C. R. Porter and M. W. Carter, "Field Methods for Rapid Collection of Iodine-131 from Milk," *Public Health Rep.*, **80**, 453, 1965.
44. J. H. Harley, Ed., *HASL Procedures Manual*, Report HASL-300, 1972, U.S. DOE Environmental Measurements Laboratory, 376 Hudson Street, New York, NY 10014, updated through 1980. Earlier editions of this manual were published as Report NYO-4700.
45. G. K. Rill, D. F. Covell, H. I. Cordova, and D. L. Love, "Instrumental Method for Measurement of ^{131}I in Milk," paper No. 446 in *Proceedings of the Fourth International Congress of International Radiation Protection Association, Paris, 24–30 April 1977.*
46. C. J. Paperiello and J. M. Matuszek, "A Coincidence System for Environmental ^{131}I," *IEEE Trans. Nucl. Sci.*, **NS-22**(1), 642, 1975.
47. *Standard Methods for the Examination of Water and Wastewater*, 15th Ed., 1980, American Public Health Association, 1015 18th Street, N.W., Washington, DC 20036.
48. American Society for Testing and Materials, *Annual Book of ASTM Standards*, Part 23 (1971), 1916 Race Street, Philadelphia, PA 19103. The radioiodine methods are designated D2334-69.
49. The first ASTM method is based on W. J. Maeck and J. E. Rein, "Determination of Fission Product Iodine, Cation Exchange Purification and Heterogeneous Isotopic Exchange," *Anal. Chem.*, **32**, 1079, 1960.
50. The third ASTM method is based on L. E. Glendenin and R. P. Metcalf, *National Nuclear Energy Series*, Vol. 9, Book 3, p. 1625; G. J. Hunter and M. Perkins, "The Determination of Radioiodine," Report AERE-AM 64 1960, Atomic Energy Research Establishment, Harwell, U.K.
51. H. S. Kuhlman, R. L. Conwell, and A. J. Goldberg, "Development of an On-Line Hyperpure Germanium Process Radiation Monitoring System," *IEEE Trans. Nucl. Sci.*, **NS-25**(1), 296, 1978.
52. F. P. Brauer and W. A. Mitzlaff, "Evaluation of Well-Type Ge(Li) Detectors for Low-Level Radiochemical Analysis," *IEEE Trans. Nucl. Sci.*, **NS-25**(1), 398, 1978.
53. F. Knowles, "Interlaboratory Study of ^{131}I, ^{137}Cs, ^{89}Sr, and ^{90}Sr Measurements in Milk," Technical Experiment 71-MKAQ-1, 1971, available from EPA Office of Radiation Programs, 109 Holton Street, Winchester, MA 01890.
54. P. J. Magno, T. C. Reavey, and J. C. Apidianakis, "Iodine-129 in the Environment Around a Nuclear Fuel Reprocessing Plant," 1972, available from U.S. Environmental Protection Agency, Office of Radiation Programs, Washington, DC 20460.
55. J. J. Gabay, C. J. Paperiello, S. Goodyear, J. C. Daly, and J. M. Matuszek, "A Method for Determining Iodine-129 in Milk and Water," *Health Phys.*, **26**, 89, 1974.
56. R. R. Edwards, "Iodine-129: Its Occurrence in Nature and Its Utility as a Tracer," *Science*, **137**, 851, 1962.

57. N. Keisch, R. C. Koch, and A. S. Levine, "Determination of Biospheric Levels of Iodine-129 by Neutron Activation Analysis Research Laboratory, 1965, available from Texas A&M University, College Station, TX 77843.
58. A. P. Baronavski and J. R. McDonald, "A Radioiodine Detector Based on Laser Induced Fluorescence," NRL Memorandum Report 3514, 1977, Naval Research Laboratory, Washington, DC 20375.
59. S. J. Fernandez, G. D. Pierce, D. C. Hetzer, and B. G. Metes, "Methods Evaluation for the Continuous Monitoring of Carbon-14, Krypton-85, and Iodine-129 in Nuclear Fuel Reprocessing and Waste Solidification Facility Off-Gas," ICP-1187, Idaho National Engineering Laboratory, 1979, available from NTIS.
60. J. P. Hohimer and P. I. Hargis, "Intracavity Absorption with External Fluorescence Measurement for Detection of Radioiodine Isotopes," *Anal. Chem.*, **51**, 930, 1979.
61. J. C. Daly, C. J. Paperiello, S. Goodyear, and J. M. Matrezel, "The Determination of ^{125}I and ^{129}I Using an Intrinsic-Germanium Detector for X-ray Spectroscopy," *Health Phys.*, **29**, 753, 1975.
62. C. B. Honaker, "A Hyperpure Germanium Well Detector for the Radiochemical Analysis of Iodine-129," RHO-SA-65, 1978, Rockwell International Rockwell Hanford Operations, Richland, WA 99352.

19 KRYPTON

Growth in the nuclear power industry necessitates a careful assessment of any related potential for environmental pollution. Krypton-85 releases deserve special attention because of the inherent difficulty in their control [1] and the essentially nonreactive and mobile nature of the gas in the atmosphere. This chapter describes sources of ^{85}Kr, and environmental levels and discusses appropriate instrumentation for monitoring ^{85}Kr in the field and laboratory. Measurements of ambient and discharge levels may differ by two or more orders of magnitude and thus may require different sampling and measurement techniques.

19.1. PHYSICAL PROPERTIES

Krypton is a noble gas, colorless, tasteless and generally chemically inert. It is present (as stable Krypton-78, 80, 82, 83, 84, 86) in uncontaminated air at a concentration of 1.14 ppm = 1.14 ml/m^3 [2,3]. Of several radioactive isotopes, one (^{85}Kr) has a long enough half-life (10.76 yr) to be of some concern in the environment. Table 19.1 gives modes of ^{85}Kr decay. It shows that the principal decay mode is β-ray emission, with a maximum β-ray energy of 672 keV. The spectrum is shown in Fig. 19.1 [3]. The range for the maximum-energy beta ray is 0.3 g/cm^3 (3 mm in water, 250 cm in air). A small (0.43%) branching ratio also exists for emission of a 514-keV γ-ray.

Being chemically inert, Kr is not usually involved in biological processes. It is, however, absorbed by the body via inhalation and will dissolve in body fluids and tissues. Krypton is slightly soluble in blood, highly soluble in lipids, and diffuses rapidly in the body [7].

Table 19.1. Radiation of Krypton-85[a]

Radiation[b]	Mean Number per Decay	Endpoint Energy (MeV)	Mean Energy (MeV)	Energy (MeV)
	Primary Radiations from ^{85}*K Decay*			
1^-	0.0043	0.15	0.0437	
2^-	0.9957	0.672	0.025	
	Secondary Radiations from ^{85}*K Decay*			
Shell conversion electron	0.00003			0.49880
K x-ray	0.00001			0.01340
L_{mn} Auger electron	0.00004			0.00139
M_{xy} Auger electron	0.00008			0.00024

[a]Sources of data: Refs. 4, 5 and 6.
[b]1^- and 2^- refer to the spin state of nucleus during decay.

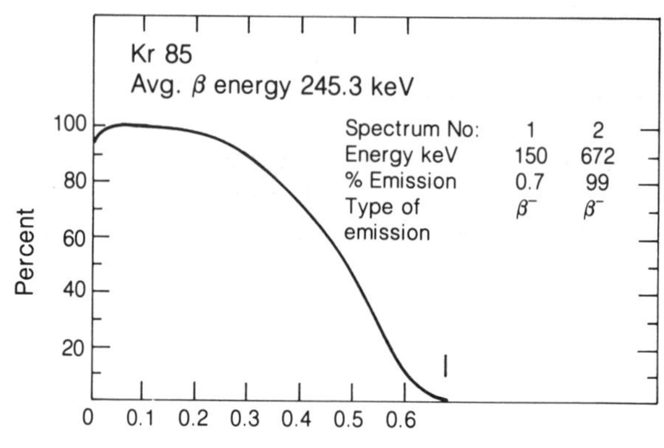

Figure 19.1. Beta spectrum of ^{85}Kr (reproduced from Mantel, Ref. 3, with permission of Pergamon Press © 1972).

19.1.1. Sources

Before the nuclear age, the natural level of ^{85}Kr was negligible. Krypton-85 is present to a small degree in the environment due to natural mechanisms such as the spontaneous and cosmic neutron-induced fissions of natural uranium and cosmic neutron capture reactions with atmospheric ^{84}Kr [8]. Diethorn and Stockho [9] discuss the expected steady-state environmental inventories from these sources. They calculated the equilibrium activity of atmospheric ^{85}Kr due to cosmic ray neutrons to be 10 Ci (0.37 TBq). The equilibrium activities due to spontaneous fission of natural U calculated were approximately 2 Ci (0.07 TBq) in the upper 3 m of the total land and water surface. Neutron-induced fissions were found to contribute equilibrium activities of about 10^{-5}Ci (0.37 MBq) in the oceans. These estimates of the natural inventory of ^{85}Kr are small compared to the world total ^{85}Kr inventory.

Krypton-85 is produced during fission, and has also been generated by nuclear weapons tests. It is estimated [1] that between 1945 and 1962 atmospheric tests generated an integrated amount of 5 MCi (185,000TBq), resulting in a current inventory of 2 MCi (74,000 TBq). Using a different method of calculation, the United Nations Scientific Committee on the Effects of Atomic Radiation (UNSCEAR) [10] estimated the production by nuclear explosions to be 3 MCi (111,000 TBq). Because Kr is a largely unreactive gas which remains to a great extent in the atmosphere, its concentration becomes fairly uniform throughout the earth's atmosphere within a few years.

The production of plutonium for use in weapons has also contributed to the worldwide inventory of ^{85}Kr. It has been estimated [11] that U.S. plutonium production reactors contributed about 15 MCi (555,000 TBq) as of 1966, and today that production has continued and the number is still large.

Although Krypton-85 is produced in a typical low-enrichment 1000-MW light-water reactor (LWR) at the annual rate of 300 Ci/MW (11.1 TBq/MW) of electric power [11], this is normally retained inside the fuel element cladding. Only a small fraction of all the noble gases released during normal reactor operations escape by leaking through the fuel cladding. The appearance of noble gas activity in effluent streams indicates fuel cladding failure. Krypton-85 provides only a few percent of the activity in pressurized-water reactor (PWR) and boiling-water reactor (BWR) releases [10]. This is also true during storage of spent fuel, except where fuel element cladding has failed. Cladding failure is of course more likely as the fuel elements get older; therefore, it is prudent to monitor for ^{85}Kr in the air surrounding stored spent fuel to detect fuel element degradation. The radiological impact from spent fuel storage is [12]:

1. Population dose due to the release of ^{85}Kr from leaking fuel elements.
2. Occupational exposure of plant personnel incurred while working in the vicinity of the spent fuel storage pool, for example, changing water purification filters and ion exchange filters. Although noble gases released from reactors may, under unusual circumstances, actually make the largest contribution to the dose for the local population, the contribution from ^{85}Kr to the collective dose in the local and regional areas is generally insignificant.

Since the releases from the fuel elements are small, ^{85}Kr releases from normally operating reactors are insignificant compared to the releases from fuel reprocessing plants. In fact, ^{85}Kr is one of the main radioactive fission products that might normally be released to the atmosphere in the reprocessing of nuclear reactor fuels. In fuel aged long enough for reprocessing (at least 150 days of storage for cool-down) it is present in the amount of about 11,200 Ci/Te (414 TBq/Te) of fuel [13]. In all reprocessing plants previously operated, the entire ^{85}Kr content of the fuel was released to the atmosphere. Since these plants operated with an output of from 1 to 6 tons per day, ^{85}Kr was a potential environmental pollutant that required measurement both in the plant stack during release and in the environment after dispersion. With an anticipated rise in nuclear power production, the worldwide inventory is expected to increase. One projection is shown in Fig. 19.2. The National Council on Radiation Protection and Measurement (NCRP) Report No. 44 estimates that

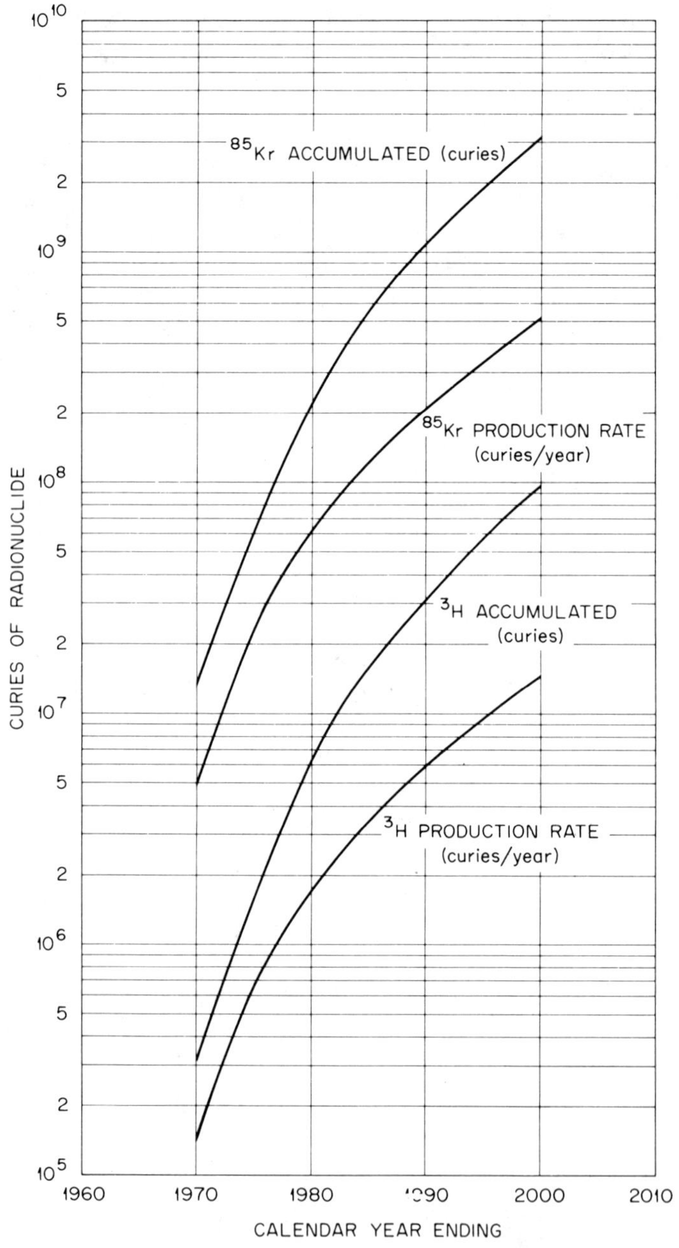

Figure 19.2. Estimated production of krypton-85 and ^{3}H (tritium) from the Nuclear Power Industry of the Free World (reproduced from Oak Ridge National Laboratory, Ref. 13).

^{85}Kr generation from the full fuel cycle during 1970 was 3.6 MCi (130,000 TBq) or about 70% of the quantity released by nuclear weapons tests from 1945 to 1962 [1].

Naval propulsion reactors also contribute to the inventory of ^{85}Kr [1]. Although not much information is available, it has been estimated that the inventory from this source was about 3 MCi (111,000 TBq) in 1970 [11]. This may continue to rise but will not be important in comparison with that from anticipated nuclear power reactors.

19.1.2. Dosimetry and Projected Doses

An American Physical Society study group cites ^{85}Kr as one of the dominant fuel-cycle gaseous effluents that affect long-term public exposures [14].

There are six radiation phenomena associated with ^{85}Kr:

1. Beta rays and photons emitted in air due to the radioactive decay of ^{85}Kr present in the atmosphere.
2. Bremsstrahlung emitted in air including both internal, emitted when the decay occurs, and external, emitted as the electrons move through the air.
3. Bremsstrahlung produced by the β-rays as they move through the skin and subcutaneous tissues.
4. Photons and β-rays emitted generally in the body due to ^{85}Kr absorbed in the tissues of the body.
5. Internal bremsstrahlung generated by β-rays as they pass through the tissues of the body.
6. Photons and β-rays generated by ^{85}Kr present in the air passages of the lungs.

Due to the low solubility of ^{85}K in water or reactivity with soils, vegetation, and so on, the primary concern is direct human exposure to a plume of ^{85}Kr. Doses to various parts of the body are summarized in Table 19.2 for such an immersion in a cloud of ^{85}Kr. Table 19.2 shows that the dose from β-rays to the skin is of primary dosimetric importance, the rest of the doses being 2 to 3 orders of magnitude lower.

The population dose commitments for each year of operation of the nuclear power industry can be calculated from the dose to the various organs, projected concentrations of ^{85}Kr, and the distribution of the world's population. NCRP did such a computation and their results are presented in Table 19.3. It shows that the most significant population dose commitment, that to the skin will grow by 2 orders of magnitude from 1970 to 2000.

The collective dose commitment from ^{85}Kr generated during nuclear power production is due almost entirely to release from reprocessing plants, the release from reactors being small in comparison. As the solubility of krypton in water is negligible, there is no sink for the krypton released into the atmosphere, and rather uniform concentrations are achieved in a period of about 2 yr.

Because ^{85}Kr released from reprocessing plants is so widely dispersed the collective dose commitment from ^{85}Kr released from reprocessing plants is significant worldwide [10]. In addition, the storage of spent fuel as a waste may produce concentrations that are not totally insignificant [15].

Diethorn and Stockho [9] reported that in the early 1970s the worldwide dose to an average man was about 10^{-3} mrad/yr (10^{-2} μGy/yr) to the bare skin, and about 10^{-4} mrad/year (10^{-3} μGy/yr) to the whole body. For comparison, the cosmic ray exposure per year is 50 to 100 mrads (500 to 1000 μGy).

Table 19.2. Equilibrium Absorbed Dose Rate to Body Organs Per Unit Air Concentration (rads/m^3)/(Ci/hr)[a] From Immersion in a Semi-Infinite Cloud of ^{85}Kr (Reproduced from National Council on Radiation Protection [1], with Permission of the NCRP © 1975)

	Organ							
Source	Skin[b]	Adipose Tissue	Lungs	Red Bone Marrow	Skeleton	Ovaries	Testes	Total Body
^{85}Kr in Air								
1 Photons in air	1.5×10^{-2}	1.2×10^{-1}	1.1×10^{-2}	1.4×10^{-2}	1.5×10^{-2}	4.7×10^{-3}	1.3×10^{-2}	1.2×10^{-2}
2 Betas in air	1.8	—	—	—	—	—	—	—
3 Bremsstrahlung in air	3.2×10^{-3}	2.1×10^{-3}	1.9×10^{-3}	3.6×10^{-3}	4.0×10^{-3}	1.1×10^{-3}	2.5×10^{-3}	2.3×10^{-3}
4 Bremsstrahlung in skin	5.8×10^{-6}	6.9×10^{-6}	4.0×10^{-6}	3.7×10^{-6}	1.1×10^{-5}	2.1×10^{-6}	9.9×10^{-6}	9.1×10^{-6}
^{85}Kr in the body								
5 Photons in the body[c]	2.2×10^{-6}	2.3×10^{-6}	2.2×10^{-6}	2.4×10^{-6}	2.3×10^{-6}	2.8×10^{-6}	3.1×10^{-6}	2.3×10^{-6}
6 Betas in the body[c]	3.7×10^{-4}	1.1×10^{-3}	3.7×10^{-4}	7.7×10^{-4}	3.7×10^{-4}	3.7×10^{-4}	3.7×10^{-4}	7.3×10^{-4}
7 Bremsstrahlung in the body[c]	4.0×10^{-7}	7.7×10^{-7}	6.9×10^{-7}	1.1×10^{-6}	9.1×10^{-7}	7.7×10^{-7}	9.1×10^{-7}	7.7×10^{-7}
8 Betas in airways of lung	—	—	1.8×10^{-2}	—	—	—	—	—
Total	1.8	1.5×10^{-2}	3.1×10^{-2}	1.8×10^{-2}	1.9×10^{-2}	6.2×10^{-3}	1.6×10^{-2}	1.5×10^{-2}

[a]For comparative purposes, since the Quality Factor (Q) equals one here, the absorbed dose in rads equals the dose equivalent in rems.
[b]The lenses of the eyes would be at a depth of 3 to 4 mm and hence would receive essentially no irradiation from the betas emitted in air. The dose rate would be expected to be much the same as the average dose rate to the skin due to photons in air (i.e., 1.8×10^{-2} from lines 1 and 3).
[c]These values may be subject to adjustment.

Table 19.3. Estimated World Population Dose Commitments From Annuals Worldwide ^{85}Kr Releases[a] (Reproduced from National Council on Radiation Protection [1], with Permission of the NCRP © 1975)

Year	Assumed World Population (Billions)	Annual ^{85}Kr Production (MCi)	Annual Population Dose Commitments (10^{-3} person-rem)					
			Skin		Skeleton		Red Bone Marrow	
			1 Year	68 Years	1 Year	68 Years	1 Year	68 Years
1970	3.7	6.5	18	240	0.4	4.8	0.18	2.4
1980	4.5	103	350	4,600	7.0	92	3.5	46
1990	5.6	461	1,900	25,000	39	510	19	250
2000	6.7	988	5,000	66,000	100	1,300	50	660

[a]Refer to Ref. 6 for estimates of the beam skin dose.

19.2. RADIATION PROTECTION GUIDES

The International Commission on Radiological Protection (ICRP) [15] has established a Derived Air Concentration (DAC) for 50 weeks exposure to ^{83}Kr at a 40-hr occupational exposure, an occupational limit of 1.35×10^8 pCi/m^3 (5M^{6}Bq/m^3). For individual nations applying these limits to "individuals of the general public," the applicable DAC values are a factor of 10 smaller 1.35 i$\times$ 10^7pCi/m^3 (0.5 MBq/m^3) and for exposure to a suitably large sample of the general public, another factor of 3 smaller still 4.5×10^6 pCi/m^3 (0.17 MBq/m^3). The National Council on Radiation Protection and Measurements [16] recommends the following maximum permissible concentrations in air (MPC_a):

1. 3×10^6 pCi/m^3 (0.11 MBq/m^3) for 168-hr work week occupational exposures.
2. 3×10^5 pCi/m^3 (0.011 MBq/m^3) for individuals in the general public.
3. 1×10^5 pCi/m^3 (0.037 MBq/m^3) for a suitably large sample of the general public.

The Nuclear Regulatory Commission's MPCs are the same as the NCRP MPC values [17]. This discussion will generally refer to the MPC for individuals (300,000 pCi/m^3 or 11 kBq/m^3), because exposure of a few individuals in the general public to 300,000 pCi/m^3 (11 kBq/m^3) is undoubtedly more likely than exposure of a larger population at the 100,000 pCi/m^3 (3.7 kBq/m^3) level. This MPC is equivalent to a fractional krypton-85 atomic concentration of 2.0×10^{-7} (^{85}Kr/total Kr) or 2.3×10^{-11} (^{85}Kr/total air). At the fence around a typical fuel reprocessing plant, concentrations of at most a few percent of MPC might occur periodically.

The U.S. Environmental Protection Agency has issued general environmental standards [17] which will limit, after 1983, the allowable emission of ^{85}Kr to 50,000 Ci/GW$_e$-yr (1850 TBq/GW$_e$-yr). The expected emissions from LWR reprocessing per GW$_e$-yr would be 372,000 Ci (13,800 TBq). If facilities limit the level in the discharge to the standard, then the ^{85}Kr dose rate will be reduced to about one-seventh of that without controls.

19.3. MEASUREMENT OF KRYPTON-85

In environmental samples, the measurement of ^{85}Kr generally relies on detection of the decay β-ray, with a detector sensitive to ionization energy loss. Recently, several systems have been designed which rely on detection of the 514-keV γ-ray (0.43% branching ratio), using solid or liquid scintillators and photomultipliers, gas-filled detectors, or solid-state detectors. Each of the methods has its advantages and disadvantages; some of them are discussed later. For the reader interested in greater detail, Budnitz [18] and Jaquish and Moghissi [19] reviewed the analytical methods and techniques for ^{85}Kr monitoring.

In this discussion measurements in the air around nuclear reactors and fuel reprocessing plants are emphasized. In nuclear reactors and fuel reprocessing plants, the principal beta-ray emitter besides ^{85}Kr is tritium (^{3}H); however, the tritium β-ray has such a low energy (E_{max} = 18.6 keV) that its interference in the

^{85}Kr measurement is not usually significant. (The reverse is *not* true, as discussed in Chapter 20, Tritium).

There are two quite different classes of ^{85}Kr measurement systems: those that operate in the field, designed to run unattended except for periodic maintenance; and those that require a laboratory. The latter are invariably the more sensitive and the more precise. Sample preconcentration may be necessary for either of these procedures.

When choosing a counting and/or preparation technique, it is essential to take into account the level to be measured. Recommendations for sampling and measurement techniques are contained in the International Atomic Energy Agency's Safety Series 46 report [20]. ANSI gives [21] recommendations for the selection of instruments for continuous monitoring and measurement radioactivity in effluents.

19.3.1. Sample Preparation

Krypton-85 must be separated and preconcentrated prior to analysis to permit measurement of ^{85}Kr at levels lower than the MPC_a, and to remove interfering radionuclides. Numerous techniques, based on the physical properties of krypton, have been developed to preconcentrate and separate ^{85}Kr. Among those are [18]:

1. Distillation of liquified gases (cryogenic distillation method).
2. Electrostatic diffusion.
3. Molecular diffusion (permselective membrane method).
4. Thermal diffusion.
5. Extraction by oils or solvents (solvent absorption method).
6. Chemical removal of other components of the sample.
7. Formation of chemical compounds with fluoride.
8. Formation of clathrates.
9. Pressurization of gas.
10. Adsorption on molecular sieves.
11. Adsorption on activated charcoal.
12. Gas chromatography (selective adsorption-desorption method).

Not all of these methods are suitable for monitoring though they may be applicable for large-scale separation systems, or special qualitative and quantitative analyses of noble gas mixtures. When measuring krypton-85 concentrations in air, the krypton must be separated from a large volume of oxygen and nitrogen and lesser amounts of argon, neon, methane, and xenon. The amount of krypton recovered must have sufficient radioactivity from krypton-85 to permit accurate counting. A volume of at least 0.5 to 2 ml is sought. In general, the methods used for monitoring employ some combination of the foregoing methods. In the following discussions the appropriate preconcentration/separation technique will be noted for each detection system.

19.3.2. Laboratory (and Stack) Techniques

A number of ^{85}Kr techniques that have been developed for use in the laboratory can be used as large monitoring systems for effluent stacks. They all have one major handicap, namely, that a large volume of gas cannot be sampled at some location and then transported for analysis. This limits sensitivity. However, the problem is counterbalanced because once in the laboratory the krypton can be separated/preconcentrated before counting, which is usually impossible to do in the field.

After sample preparation the two most common laboratory techniques use scintillation counters; one with plastic-scintillator shavings contained within a gas-tight vial, the other with a liquid scintillator as a solvent for the ^{85}Kr-containing gas. When the concentration is high enough, increasing numbers of researchers are measuring the ^{85}Kr 514-keV γ-ray.

Plastic Scintillator. The technique of Sax, Denny, and Reeves [2] (discussed here) is typical of plastic scintillator techniques.

Plastic scintillator shavins of 20 to 40 mesh (an inexpensive by-product from any of several scintillator manufacturers) are encased in a glass vial of volume about 4 ml. Pure krypton from the sample is obtained using vacuum-cryogenic techniques. After the vial is evacuated, it is filled to a known pressure [near 600 torr (80 kPa)] with the krypton sample. A scintillation counter then measures the pulse-height spectrum.

The counting efficiency (the fraction of the β-ray spectrum above the counting threshold) is determined using a known ^{85}Kr standard acquired from the National Bureau of Standards (NBS) [22]. The counting efficiency should be about 95%. Given about 0.5 ml of nearly pure krypton with the normal distribution of isotopes, except for ^{85}Kr, (equivalent to that present in about 0.5 m^3 of air), measurements can be made of normal ^{85}Kr background levels in the 10 pCi/m^3 (0.37 Bq/m^3) range to about $\pm$10% [2]. The limit of detection is about 1 pCi/m^3 (0.037 Bq/m^3) at the two-standard-deviation level with a 100-min counting time.

A monitoring system for continuous detection of ^{85}Kr has been designed [23] for monitoring stacks from nuclear facilities. It uses a plastic scintillator for beta-ray detection and most importantly a permselective membrane for selective ^{85}Kr enrichment; see Ref. 24. Howell, Cate, and Wong [23] reported a krypton enrichment factor of 30 and depletion factor for Xe of 0.0043, allowing them to achieve a 1×10^{-10} μCi/ml (3.7 μBq/ml) [= 100 pCi/m^3 (3.7 Bq/m^3)] sensitivity with a 100-min counting time.

Liquid Scintillator. A general description of liquid scintillation counting can be found in Chapter 4. However, there are some aspects of liquid scintillation counting unique to or particularly relevant to ^{85}Kr which will be discussed here.

The earliest descriptions of liquid scintillators used for ^{85}Kr counting are those of Horrocks and Studier [25], and Curtis, Ness, and Bentz [26]. Those early methods suffered from a limit on the amount of krypton which could be introduced into the solution.

Liquid scintillator techniques for ^{85}Kr rely upon the high solubility of krypton in certain liquid scintillators. Krypton is soluble in some aromatic solvents in the proportion of ~1 ml Kr/ml solvent [27,28].

Shuping, Phillips, and Moghissi [28] have reported a method in which about 10 ml gas is dissolved in about 25 ml of toluene-based liquid scintillator. The gas is introduced into an evacuated glass vial, filled with deaerated scintillation solution. "If the ^{85}Kr concentration in air is sufficiently high, the sample may be introduced directly into the solution and successfully counted." [28] More commonly, krypton is concentrated cryogenically. A rate of 1 cpm above background corresponds to

0.025 pCi (9.3 μBq) $^{85}Kr/m^3$ of air, so that levels of that order of magnitude are detectable. Accuracies in the ±4% range are achieved, when ^{85}Kr activities of ~10 pCi/m^3 (0.37 Bq/m^3) of air are present.

One of the more difficult problems in the procedure just described is the separation of krypton gas from the initial air sample. To solve this problem, Cummings, Shearin, and Porter [29] devised a complicated separation procedure for krypton involving "charcoal and molecular sieve cold traps, calcium sulfate, ascarite, and a titanium furnace (900°C) for the removal and separation of other air constituents from krypton." [29] It employs ^{83m}Kr as a radioactive tracer in the air sample to determine the ^{85}Kr yield directly. A description of a method for generating ^{83m}Kr (2-hr half-life) from ^{83}Rb in the laboratory has been given by Moghissi and Hupf [30].

Another ^{85}Kr system, described by Stevenson and Johns [31], uses a battery-operated air compressor to collect 1 m^3 of air in the field. (More recent applications use standard electrically powered air compressors). After a series of cryogenic absorptions and elutions in the lab, the krypton is dissolved and counted in a liquid scintillator. The krypton recovery ranges from 50 to 70% , measured volumetrically. These workers report a minimum detectable sensitivity (three standard deviations above background) of about 2 pCi (0.074 Bq) $^{85}Kr/m^3$ of air, with a 4-hr counting time.

Semiconductor Detectors. Semiconductor detectors are discussed in Chapter 4 and the reader should refer there for more detail. A Ge(Li) detector has been used to measure the 514-keV γ-rays from ^{85}Kr in the stack effluent of a BWR nuclear power plant [32]. A gas stream was sampled in the stack and passed through a series of pipes to a 10-liter Marinelli chamber where it was counted with a 12.4% relative efficiency Ge(Li) detector with a pulse-height analyzer for spectral measurements. Five other isotopes of krypton were detected besides ^{85}Kr; they were ^{85m}Kr, ^{87}Kr, ^{88}Kr, ^{89}Kr, ^{90}Kr as well as xenon isotopes. With an air flow of 57 liters/min, a pressure of 1 atmosphere, and a 40-hr counting time, the detection limit for ^{85}Kr was 3×10^{-7} Ci/m^3 [300,000 pCi/m^3 (11.1 kBq/m^3)]. Another group also using a Ge(Li) detector for online counting of effluent gases from nuclear facilities found a detection limit of 2×10^{-6} Ci/m^3 [2,000,000 pCi/m^3 (74 kBq/m^3)] for ^{85}Kr with a 50-min counting time and a 20 liter Marinelli counting chamber [33]. Of course, germanium detectors are not limited to use in on-line monitoring systems, but may be used in the laboratory. Preconcentration may be necessary because at present the systems are not sensitive enough for measuring ambient levels of ^{85}Kr.

19.3.3. Field Instruments

For measurements in the field, the ideal goal is an instrument that can record continuously, with reasonably short time-integration periods, at sensitivity levels well below the current average background of about 20 pCi/m^3 (0.74 Bq/m^3) of air. The ideal instrument must also be rugged enough to withstand temperature and humidity extremes, and shock, and should require little maintenance. Long-term (weeks to months) environmental monitoring demands sensitive detectors and systems that endure the environment with a minimum of attention.

A study [34] of several possible field instruments was carried out by the Northeastern Radiological Health Laboratory (NRHL), U.S. Public Health Service.

Four ionization chambers and four Geiger-Müller counters were examined, as well as a single β-ray scintillator. Instruments were calibrated in the laboratory, and the minimum detectable concentration was determined for each. Some of the instruments were then used in the field around a fuel reprocessing plant to determine performance under field conditions.

None of the field instruments came close to the sensitivity required for background measurements in the pCi/m^3 (hundredth of Bq/m^3) range so that NRHL concluded that laboratory techniques are necessary to detect ambient levels. However, all were sensitive enough to measure fractions of the MPC for individuals in the general public [300,000 pCi/m^3 (11.1 kBq/m^3)].

Recently, cryogenic methods have been used to liquefy the sample; the sample is held in a dewar until analysis. Where available, this method allows quite large samples to be collected in the field for laboratory analyses.

Flow-through Ionization Chambers. Four flow-through ionization chambers were tested in the laboratory using dry, radon-free air [34]. Their volume ranged from 0.5 to 4.3 liters. After calibration against an NBS standard [22], the minimum detectable concentrations (MDC, defined in Ref. 15 as twice the standard deviation in counting) varied from 130,000 to 40,000 pCi/m^3 (4.8 to 1.5 kBq/m^3), respectively. Calibrations showed errors (2σ) in the $\pm 7\%$ to $\pm 17\%$ range. The ionization chambers tested had several undesirable properties. First, they had to be used downstream of a radon holdup trap and filter (or the equivalent). Second, because of the use of the radon trap, the averaging time was about 30 min. Thirty minutes was much longer than the few minutes in which significant changes occur in ^{85}Kr concentration when a plume from a fuel reprocessing plant passes directly overhead. Third, to obtain reasonable accuracy, pressure, temperature, humidity, and external background variation had to be measured. Finally, maintenance requirements were high. To quote from ICRP Publication 30 [15]:

> It should be noted that although the ionization chamber systems can be made to operate in the field, the degree of care, number of precautions, and amount of operator training required to obtain usable data for environmental levels, and the questionable nature of the data obtained when the field levels passing the instrument are fluctuating rapidly, all weigh heavily in favor of using simpler systems for this purpose.

Geiger-Müller Counters. Several Geiger-Müller (G-M) tubes have also been tested [34]. They were calibrated against the flow-through ionization chambers just described, and their minimum detectable concentrations were determined. Calibration errors (2σ) were in the $\pm 13\%$ region for all of the G-M detectors. The most sensitive G-M detector was found to be the Eon 8008H, which is a double end-window pancake detector of 2-in. diameter with 3.5 mg/cm^2 mica windows. It had an MDC of 12,000 pCi/m^3 (0.44 kBq/m^3) with a counting time of 10 min (sample) and 10 min (background). Unfortunately, this detector was so fragile that its use in the field was precluded.

The other G-M detectors all had MDCs in the region of about 25,000 pCi/m^3 (0.93 kBq/m^3). These devices were an Amperex 18546 and an Eon 8001T (both one-window pancake detectors of 2-in. diameter); and an LND 719 and an Eon 5108E (both cylindrical probes, 5 1/2 in. long, 0.6 in. diameter, 30 mg/cm^2 wall thickness). The cylindrical instruments were durable against mild shock and rain in the field; the single-window detectors were less durable against shock. However, all performed in the field with sensitivities nearly identical to those measured in the laboratory.

Other researchers [35,36] reported that 10 mg/cm^2 stainless-steel window counters function reliably and that grouping several detectors into an array increased the overall sensitivity. The detector array consists of two pairs of β-ray-sensitive dual-window pancake types. One pair is shielded so that it provides a response to gamma photons allowing subtraction of the response to gamma-emitting radionuclides also possibly present. Using this array the authors of Refs. 35 and 36 found that the system can detect ^{85}Kr in the air at levels as low as 10,000 pCi/m^3 (0.37 kBq/m^3).

Proportional Counter. A multiwire multiplane proportional counter has been developed for beta-ray detection [37] using technology borrowed from experimental techniques of high-energy physics. For the ^{85}Kr in a semi-infinite cloud, the measured minimum detectable concentration for a 3-min count was 3×10^{-3} $\mu Ci/m^3$ [3,000 pCi/m^3 (0.111 kBq/m^3)]. It is anticipated that larger proportional counters operated for counting times of several hours would reduce this limit to below 1×10^{-4} pCi/cm^3 (3.7 $\mu Bq/m^3$) [100 pCi/m^3 (0.0037 kBq/m^3)] [37]. Although the detector was not taken into the field for monitoring concentrations of krypton-85 in air, the laboratory experiment indicated that it would be possible.

Scintillator Detector. Smith, Cochran, and Shleien [34] describe a comparison for relative sensitivity between a plastic scintillator (1.75 in. diameter $\times$ 1/16 in. thick) in conjunction with a single-channel analyzer system and the G-M detectors listed previously. An integral counting mode was employed. It was found that the best G-M detector was only a factor of 2.3 times more sensitive. This means that with reasonable improvements in photomultipliers, scintillators can be competitive with gas-filled detectors. In fact, Mayhugh, Lucas, and Utts [38] report that it is possible to use the phoswich configuration to produce a low-background beta-ray counting detector. Using $CaF_2(Eu)$ as a beta-ray detector (its chemical inertness allows direct contact with samples without any radiation entrance window) in anticoincidence with a NaI(Tl) detector reduces background about a factor of 2 because of Compton suppression and because the NaI(Tl) is an active shield. An additional advantage would be the simultaneous counting of α-, β-, and γ-rays.

19.4. SUMMARY AND CONCLUSIONS

Two important concentrations determine the sensitivity required of instruments for measuring ^{85}Kr in environmental air:

1. The maximum permissible concentration for individuals in the general public is 300,000 pCi/m^3 (11.1 kBq/m^3) [17].
2. The typical "background" level of ^{85}Kr in air today is in the range 14 to 18 pCi/m^3 (0.52 to 0.67 Bq/m^3) [8].

The ideal instrument is one capable of measuring the background level with great accuracy. Several of the laboratory techniques described have this sensitivity. The techniques that use liquid or plastic scintillation counters all suffer from a common problem: a complicated series of laboratory steps is required to concentrate the krypton from the original gas sample.

In contrast, the *field* monitoring instruments [34–36] are all several orders of magnitude less sensitive, with MDCs ranging from 12,000 to 28,000 pCi/m^3 (0.44 to 1.0 kBq/m^3) for the G-M detectors and from 39,000 to 190,000 pCi/m^3 (1.4 to 7.0 kBq/m^3) for the ion chambers. All of these instruments are thus capable of detecting ^{85}Kr at levels well below the 300,000 pCi/m^3 (11.1 kBq/m^3) MPC for individuals in the general public, but cannot measure ^{85}Kr in "background" samples.

At present, there does not exist a commercially available krypton-85 monitor which can be purchased off-the-shelf. The G-M detectors described in Refs. 34–36 can be put together from commercial sources to meet the need for monitoring in the field around fuel reprocessing plants and nuclear power reactors. Any commercial instrument should satisfy several requirements besides sensitivity:

1. Rugged construction, insensitivity to changes in temperature and humidity.
2. Some method of continuous recording of the data: at least a strip-recorder, or preferably a direct magnetic-tape record.
3. Some method of calibration: perhaps a frequent, rapid check with a high-energy source, and a less frequent check with a controlled gas of known ^{85}Kr concentration.

For the more sensitive measurements of activities below the natural background of about 15 pCi/m^3 (0.55 Bq/m^3), field instruments are presently out of the question. Some kind of cryogenic concentration method is required to increase the specific activity to the point where reasonable counting times ($\leq$ a few hours) are possible.

One key problem is the accuracy and reproducibility of the concentration procedure. Accurate determinations of the yield through a multistage procedure are notoriously difficult. This is true especially if a known volume of "ordinary" krypton is introduced as a carrier, since all krypton today unfortunately contains ^{85}Kr at levels in the ppm region. One solution is use of ^{83m}Kr as a tracer to determine the yield, if ^{83m}Kr is available.

A few technological developments will provide some increased sensitivity. The most important of these is the development of substantially better photomultiplier tubes [39]. These tubes have better pulse-height resolution capability enabling a more reliable discrimination between dark current and real signal outputs. This will improve the scintillation counting method.

The incorporation of integrated circuit electronics into G-M detector systems will increase the ruggedness and reliability of the instrument. Also, readout systems are being developed which will enable the data to be collected, recorded, and analyzed remotely, bypassing the strip-chart-recorder step entirely.

Finally, some of the instruments designed to measure tritium in environmental samples (see Chapter 20), with some modifications and changes in procedure, are also suitable (or adaptable) to measure ^{85}Kr.

REFERENCES

1. National Council on Radiation Protection and Measurements, "Krypton-85 in the Atmosphere—Accumulation, Biological Significance, and Control Technology," NCRP Report No. 44, 1975, 7910 Woodmont Avenue, Washington, DC 20014.
2. N. I. Sax, J. D. Denny, and B. R. Reeves, "Modified Scintillation Counting Technique for Determination of Low Level Krypton-85," *Anal. Chem.* **40**, 1915, 1968.
3. J. Mantel, "The Beta Ray Spectrum and Average Beta Energy of Several Isotopes of Interest in Medicine and Biology," *Int. J. Appl. Radiat. Isotopes*, **23**, 407, 1972.
4. C. M. Lederer and V. S. Shirley, Eds., *Table of the Isotopes*, 7th Ed., John Wiley and Son's, Inc., New York, 1978.
5. M. J. Martin, "Radioactive Atoms—Supplement 1," Oak Ridge National Laboratory Report ORNL-4923, 1973, Oak Ridge, Tennessee.
6. L. T. Dillman, "Radionuclide Decay Schemes and Nuclear Parameters for Use in Radiation Dose Estimation," nm/mird pamphlet No. 4, *J. Nucl. Med.*, Supplement No. 2, Vol. 10, 1969.
7. W. P. Kirk, "^{85}Kr: A Review of the Literature and An Analysis of Radiation Hazards," EPA Report NP-19251, 1972, Rockville, MD.
8. K. Rozanski, "Krypton-85 in the Atmosphere 1950–1977: A Data Review," *Environ. Int.*, **2**(3), 139–143, Pergamon, Great Britain, 1979.
9. W. S. Diethorn and W. L. Stockho, "The Dose to Man From Atmospheric Krypton-85," *Health Phys.*, **23**, 653, 1972.
10. United Nations Scientific Committee on the Effects of Atomic Radiation (UNSCEAR), *Sources and Effects of Ionizing Radiation*, 1977 report to the General Assembly, with annexes, United Nations, New York, UNIDO, Vienna, 1977; see particularly Annex D, "Radioactive Contamination Due to Nuclear Power Production," p. 165.
11. D. E. Bemhardt, A. A. Moghissi, and J. A. Cochran, "Atmospheric Concentrations of Fission Product Noble Gases," p. 4, *Noble Gases*, R. E. Stanley and A. A. Moghissi, Eds., CONF-730915, 1975, based on a symposium in Las Vegas, Nevada, September 24–28, 1973.
12. U.S. Nuclear Regulatory Commission, "Draft-Generic Environmental Impact Statement on Handling and Storage of Spent Light Water Power Reactor Fuel," Executive Summary and Text, NUREG-0404, Vol. 1, 1978, Office of Nuclear Material Safety and Safeguards, Washington, DC 20555.
13. Oak Ridge National Laboratory, "Siting of Fuel Reprocessing and Waste Management Facilities," ORNL-4451, 1971, Oak Ridge, TN 37830.
14. L. C. Hebel, Chairman, APS Study Group, "Report of the APS Study Group on the Nuclear Fuel Cycle and Waste Management," *Rev. Mod. Phys.*, **50**(1), part II, 1978.
15. International Commission on Radiological Protection, "Limits for Intakes of Radionuclides by Workers," ICRP publication 30, Part 1, *Ann. ICRP*, **2**(3/4), Pergamon Press, New York, 1979. Supersedes Publication 2.

16. National Council on Radiation Protection and Measurements, "Maximum Permissible Body Burdens and Maximum Permissible Concentrations of Radionuclides in Air and in Water for Occupational Exposures," NCRP Report No. 22, 1959, 7910 Woodmont Ave., Washington, DC 20014, including Addendum 1 issued in 1963.
17. *Code of Federal Regulations*, Title 40, revised as of January 1, 1978, subchapter F, Part 190, EPA Environmental Radiation Protection Standards for Nuclear Power Plants, U.S. Government Printing Office, Washington, DC 20402.
18. R. J. Budnitz, "Krypton-85: A Review of Instrumentation for Environmental Monitoring," *Noble Gases*, CONF-730915, R. E. Stanley and A. A. Moghissi, Eds., Environmental Protection Agency, 1975, p. 192.
19. R. E. Jaquish and A. A. Moghissi, "Survey of Analytical Methods for Environmental Monitoring of Krypton-85," p. 169 of Ref. 20.
20. International Atomic Energy Agency, "Monitoring of Airborne and Liquid Radioactive Release from Nuclear Facilities to the Environment," IAEA Safety Series No. 46, IAEA, Vienna, 1978.
21. American National Standards Institute, "ANSI Specification and Performance of On-Site Instrumentation for Continuously Monitoring Radioactivity in Effluents," ANSI N13.10, 1974, Institute of Electrical and Electronics Engineers, Inc., 345 East 47th Street, New York, NY 10017.
22. National Bureau of Standards, NBS Standard 204, 10/9/62, 60.8×10^6 dps/gram-mole.
23. R. H. Howell, J. C. Cate, and C. Wong, "Separation of HT, Noble Gases and HTO Vapor with Semipermeable Membranes, *Nucl. Instrum. Meth.*, **124**, 579, 1975.
24. S. Takamatsu, M. Ohno, T. Miyazawa, and O. Ozaki, "A Krypton-85 Monitoring System with Permselective Membranes," p. 367 of Ref. 32.
25. D. L. Horrocks and M. H. Studier, "Determination of Radioactive Noble Gases with a Liquid Scintillator," *Anal. Chem.*, **36**, 2077, 1964.
26. M. L. Curtis, S. L. Ness, and L. L. Bentz, "Simple Technique for Rapid Analysis of Radioactive Gases by Liquid Scintillation Counting," *Anal. Chem.*, **38**, 636, 1966.
27. W. F. Linke and A. Seidell, *Solubilities of Inorganic and Metal Organic Compounds*, Vol. II, 4th ed., American Chemical Society, Washington, DC, 1965, p. 342.
28. R. E. Shuping, C. R. Phillips, and A. A. Moghissi, "Low Level Counting of Environmental Krypton-85 by Liquid Scintillation," *Anal. Chem.*, **41**, 2082, 1969.
29. S. L. Cummings, R. L. Shearin, and C. R. Porter, "A Rapid Method for Determining ^{85}Kr in Environmental Air Samples," in *Proceedings of a Conference on Rapid Methods for Measuring Radioactivity in the Environment*, International Atomic Energy Agency, Vienna, 1971, p. 163.
30. A. A. Moghissi and H. B. Hupf, "A Krypton-83m Generator," *Int. J. Appl. Radiat. Isotopes*, **22**, 218, 1971.
31. D. L. Stevenson and F. B. Johns, "Separation Technique for the Determination of ^{85}Kr in the Environment," p. 157 of Ref. 29.
32. H. Iimura, A. Ihizuka, H. Koyama, Y. Ohtou, A. Kobayashi, H.Ohhashi, Y. Yuasa, and N. Nagao, "Development of a Monitoring Device for Noble Gas Nuclides in the Stack Effluent of a BWR Nuclear Power Plant," in *Proceed-*

ings of a Symposium on Monitoring of Radioactive Effluents from Nuclear Facilities, IAEA, Vienna, 1978, p. 279.

33. J. Morel, H. Goenvec, and J. Legrand, "Analyse et Controle par Spectrometrie Gamma des Rejets Gazeux en Prevenence D'installations Nucleaires," p. 347 of Ref. 32.
34. D. G. Smith, J. A. Cochran, and B. Shleien, "Calibration and Initial Field Testing of ^{85}Kr Detectors for Environmental Monitoring," Report BRH/NERHL 70-4, 1970, U.S. Public Health Service, Northeastern Radiological Health Laboratory, 109 Holton Street, Winchester, MA 01890.
35. R. T. Graveson, H. Beck, and D. C. Freeswick, "Continuously Recording Field Monitor for Krypton-85," *IEEE Trans. Nucl. Sci.*, **NS-22**, 654, 1975.
36. H. L. Beck and D. C. Freeswick, "A Continuous Monitoring System for ^{85}Kr in the Environment," *IEEE Trans. Am. Nucl. Soc.*, **21**, 92, 1975.
37. J. C. Gregory and T. A. Parnell, "Proportional Counter for Monitoring Low Levels of Krypton-85 and Other Beta and X-ray Emitting Nuclides," *IEEE Trans. Nucl. Sci.*, **NS-23**(1), 715, 1976.
38. M. R. Mayhugh, A. C. Lucas, and B. K. Utts, "Low Background Beta Counting with CaF_2(Eu) in a Phoswich Configuration," *IEEE Trans. Nucl. Sci.*, **NS-25**(1), 569, 1978.
39. C. C. Lo and B. Leskovar, "Studies of Prototype High-Gain Microchannel Plate Photomultipliers," *IEEE Trans. Nucl. Sci.*, **NS-26**(1), 388, 1979.

20 TRITIUM

Tritium, the only radioactive isotope of hydrogen, has been a ubiquitous contaminant resulting from nuclear programs. The nuclide is produced by the military nuclear program and, in the form of water, was distributed worldwide by the nuclear weapons tests that took place in the open atmosphere, particularly during the period 1954 to 1962. This latter source has diminished substantially as a result of the limited nuclear test ban agreement among the major powers, but tritium production by nuclear reactors is increasing rapidly and will in time become the dominant source [1]. Figure 20.1 illustrates the relative contributions from various sources to a potential exposure pathway to man. In the time since the radioactivity of tritium was discovered by Alvarez and Cornog in 1939 [2], this radionuclide has become important in many medical, scientific, and industrial fields. New uses of tritium in research and industry, along with expansion of the nuclear power industry, seem certain.

In the following discussion some indication is given of concentration levels that occur as natural background and those that can result from various operations, nuclear power related or medical. The primary focus is on measurements in air and water.

Several references are particularly useful for greater detail than is presented in this discussion. A monograph by Jacobs (Ref. 3, part of AEC Critical Review Series) is an excellent reference on the sources of tritium and its behavior in the environment. A more recent compilation is NCRP Report No. 61 [1]. The IAEA held symposiums in 1961 and 1978, the proceedings of which are basic references [4,5]. Bibliographies by Rudolph, Carroll, and Davidson [6] and Dixon et al. [7] are also useful. A tritium symposium was held in Las Vegas in September 1971 [8] in which the major research groups in the field summarized the status of instrumentation at that time.

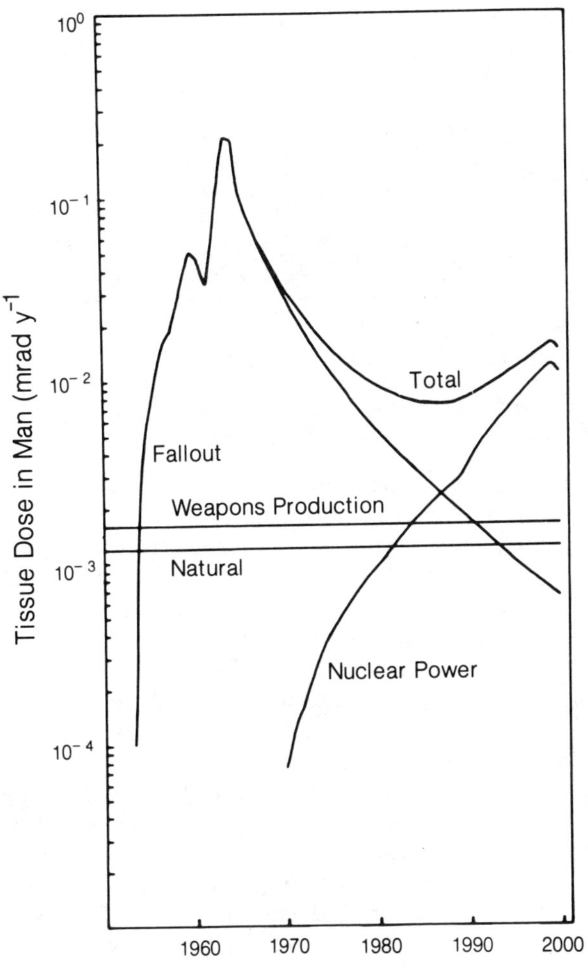

Figure 20.1. Projected tissue dose rate in man (reproduced from the National Council on Radiation Protection, Ref. 1, with permission of NCRP © 1979).

20.1. PHYSICAL CONSIDERATIONS

Tritium (the heaviest hydrogen isotope, at mass 3) decays only by β-particle emission with a half-life of 12.36 yr. The nuclide, often symbolized as ^3H or T, decays to form ^3He by emitting a beta particle with a maximum energy of 18.6 keV and an average energy of 5.7 keV. Figure 20.2 shows the beta-particle energy spectrum of tritium. The mean range of the beta particle in water is 0.69 μm and the maximum range approaches 6 μm in water or 0.5 cm in air. One gram of pure tritium would have an activity of 9620 Ci (356 TBq).

Gaseous T_2 at room temperature tends to form HT by reaction with gaseous hydrogen if present. The vapor, HTO, is formed readily and is in fact the most commonly encountered form of tritium in the environment [1]. Although mass effects do cause measurable differences in the physical and chemical properties of

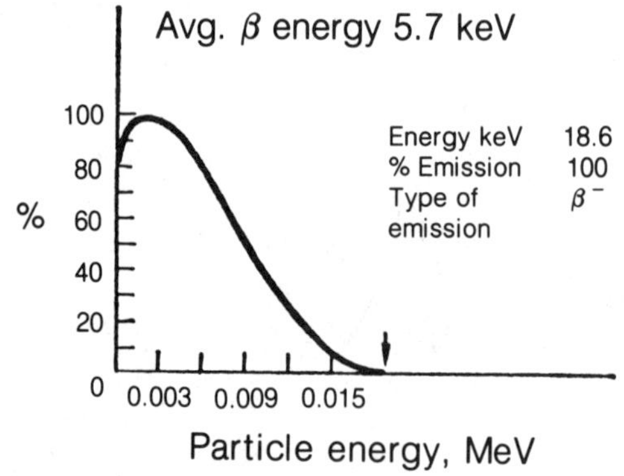

Figure 20.2. Beta energy spectrum for tritium (adapted from Fig. 7.1).

tritium, these effects do not in general invalidate the assumption that tritium follows closely the reactions of ordinary hydrogen.

The decay daughter nuclide, helium, is a noble gas. It will not form a stable chemical bond. When tritium is in a chemical C-T bond before decay, the daughter nuclide will be lost from the carbon, where tritium was bound, to produce a reactive carbonium ion which may cause molecular alteration. The maximum nuclear recoil energy of 3 eV is too low to break most chemical bonds. These effects do not measurably contribute to early cell death but may induce mutations [9].

In the literature on tritium, one frequently encounters the use of the Tritium Unit (TU) or Tritium Ratio (TR), defined as a tritium concentration of one tritium atom in 10^{18} normal hydrogen atoms. A concentration of 1 TU in water corresponds to 3.23×10^{-3} pCi/cm^3 (0.12 mBq/cm^3) of water, or alternatively, 0.12 dps/liter.

Another notation which often appears is the designation "protium" for normal hydrogen (^{1}H), to distinguish it from deuterium (^{2}H) and tritium (^{3}H) or triton. The three isotopes are also commonly denoted by H, D, and T.

20.1.1. Sources

To get an idea of the concentrations that can be encountered in the environment, we will discuss the various sources of tritium in the environment, man-made and natural. Tritium is produced both by ternary fission and by neutron reactions with light elements, such as boron, lithium, and deuterium.

Natural. The production of tritium by natural processes was discovered by Libby [10]. Tritium appears naturally in the environment at an estimated worldwide level of 70 MCi (2600 PBq) corresponding to a production rate of 4 MCi/yr (150 PBq/yr) [1]. All but a fraction of 1% is in the form of HTO in the air and water. The fractional atomic concentration of HT and HTO relative to ordinary hydrogen is about 4×10^{-15} (HT/H_2) in the atmosphere; about 8×10^{-17} (HTO/H_2O) as vapor in air; and about 1×10^{-18} in surface water [3]. Most of the naturally occurring tritium is produced by interactions of primary cosmic ray protons in the upper atmosphere [7]. Cosmic ray production has been calculated to occur at a global average rate of 0.16 to 0.20 tritium per square centimeter of the earth's surface per second [1]. A value of 0.19 tritium/cm^2·sec would result in a steady-state global inventory of 26 MCi (960 PBq) [1]. Based on satellite measurements, it is estimated that phenomena such as a major solar flare could account for an additional 0.4 tritium/cm^{-2}·sec, or more than twice the normal production rate [1]. In recent years the ambient surface water tritium concentration has been 150 to 200 pCi/liter (5.6 to 7.4 Bq/liter) [11].

Man-Made. During the most recent three decades of the twentieth century, the natural inventory of tritium has been overshadowed by man-made sources, primarily from the military weapons program and more recently the nuclear power program.

Military Weapons Programs. The production of tritium by nuclear and thermonuclear explosions depends on the spectrum of fission and fusion energy yields, the type of explosive, and the characteristics of the explosion site. Production of tritium by atmospheric nuclear tests has dwarfed natural levels: the tritium yield of a thermonuclear reaction (fusion) has been estimated at about 20 MCi/megaton (740 PBq/megaton) [1]. Essentially 100% of the tritium produced in this way is converted to HTO vapor [12]. The concentration in stratospheric water of tritium rose from about 0.03 pCi/cm^3 (1.1×10^{-3} Bq/cm^{-3}) (in the period prior to nuclear testing) to about 20,000 pCi/cm^3 (740 Bq/cm^3) in 1960, an increase of almost 6 orders of magnitude [3]. Although it is chemically identical to normal hydrogen for most purposes, stratospheric tritium may not mix quickly and thoroughly with much of the naturally occurring water.

The tritium inventory due to weapons testing reached a maximum of about 3100 MCi (115,000 PBq) in 1963, an amount that will decay to the natural level of 70 MCi (2590 PBq) approximately by the year 2020 [1].

Underground tests of thermonuclear devices have been conducted on a wide scale by the major nuclear powers since the Test Ban Treaty agreement was signed in 1963. This has not added significantly to the inventory of tritium in the atmosphere and surface waters, but an inventory of uncertain magnitude exists underground at the test sites. The extent to which this inventory can possibly contribute to human exposure will depend on the degree of its isolation.

Other aspects in the military weapons program also contribute to the tritium inventory. These are primarily associated with the fabrication of weapons, such as the plutonium production reactors and military fuel reprocessing plants. A small contribution results from the naval shipyards that service nuclear-powered submarines.

Nuclear Power Program. Tritium is formed in nuclear power reactors by two processes in two different places. It is formed in the fuel elements by ternary fission, whereas in the coolant system it is formed by neutron reactions with light elements. Tritium is one of the most important radioactive nuclides released to the environment during the normal operation of light-water nuclear reactors (LWR) and fuel reprocessing plants. Tritium appearing in the primary coolant of most light-water-cooled reactors has been released to the environment in gaseous and liquid waste streams. Estimates of the production rate by ternary fission range from below 12 to greater than 20 Ci ($<$0.44 to $>$0.74 TBq) per megawatt (thermal) per day (MW_t/d) [1]. In addition, light-element neutron reactions contribute about 600 to 800 Ci/yr to the reactor coolant [1]. Other reactor types also contribute, such as heavy-water reactors (CANDU in Canada) or, in the future, Liquid-Metal Fast Breeder Reactors. Essentially all the tritium produced is ultimately released to the environment, either at the reactor site or during fuel reprocessing. Tritium from reactors appears predominately as tritiated water (HTO), and to a lesser extent as hydrogen gas (HT). With the use of Zircaloy-clad fuel rods in light-water reactors, tritium diffusion from the spent fuel should be very small (1%), so that most release would occur at the fuel reprocessing plant [3].

Reprocessing plants currently in operation in the United States primarily handle spent fuels from military operations. Essentially all of the tritium accompanying the spent fuel is released in the form of tritiated water. Some of the tritium may also be released as gas or removed with solid wastes. The former Energy Research and Development Administration, now the Department of Energy, estimated [13] that the annual radiological effluents from an ideal LWR reprocessing plant would include 1.1 MCi (41 PBq) of tritium.

At some time in the future, a controlled thermonuclear reactor (CTR) may represent a potential source of tritium to the environment, but large-scale use for heat or power production seems unlikely in the next 25 years [1]. CTRs will contain substantial inventories of tritium which can be released in elemental form (HT) or as tritiated water vapor (HTO).

The results of a preliminary analysis [14] indicated that within 50 km of the site, atmospheric concentrations of tritium will be the limiting factor for routine releases, while on a regional scale, the surface water concentration tends to become the limiting factor.

Miscellaneous Sources. Tritium is also a significant problem outside of the reactor industry. Its use is increasing in medical and biological research as a tracer and in industrial applications. A study [15] of tritium's use in luminous watch dials is interesting, if only because of its historical connection with the problem of the radium dial painters of several decades ago. A study in the United Kingdom by Lambert and Vennart [16] reports that tritium is widely used in industry and that "the radiation doses received by some workers are not inconsiderable." Lambert and Vennart found that a few percent of the workers studied received doses less than, but comparable with, maximum permissible concentration (MPC) levels. Background information on the use of tritium in biology and medicine can be found in the book by Feinendegen [17].

Consumer products containing tritium are also an important source of man-made tritium in the environment. In the Federal Republic of Germany about the same tritium activity is yearly added to consumer products as is yearly released from all its nuclear installations [18]. A recent commercial application of tritium in the United States involves the backlighting of liquid crystal displays (LCD) in digital electronic watches. One million curies (MCi) (37 PBq) of tritium were distributed in 1977 in this product [19]. This is a significant quantity of tritium, when compared to the free-world power reactor-produced tritium [3 MCi/yr (111 PBq/yr)] or to naturally produced tritium [46 MCi/yr) (1700 PBq/yr)]. This is the single largest commercial application involving tritium to date. The final disposition of these consumer products will likely be disposal in solid refuse; this includes burial in land fills and incineration. Burial in land fills may contain the tritium for its effective lifetime, whereas incineration will release all the tritium gas (as HTO) to the atmosphere.

A small source of tritium is wastes in commercial radioactive-waste land burial sites. In a study [20] of the low-level radioactive-waste trenches at West Valley, New York, the amount of tritiated methane, TCH_3, released to the environment was estimated at approximately 1 to 8 Ci/yr (0.037 to 0.30 TBq/yr) from various trenches. Other small releases result from the leakage of tritium targets used in low-energy particle acceleration in neutron generators. If no measures are taken to prevent these tritium releases, about half the activity of tritium targets will be distributed to the environment [21].

20.1.2. Dosimetry and Projected Doses

The radiotoxicity of tritium reflects the complex interaction of a number of parameters characterizing the radionuclide. Some of the factors [22] that enhance the radiotoxicity of tritium are (a) its relatively long radioactive half-life, (b) its ease of entry into man as HTO through the lungs, gastrointestinal tract, or skin, and (c) as an isotope of hydrogen, constituting 10% of the total body by weight. Conversely, other characteristics identify tritium as a nuclide of low radiotoxicity. For example, almost all tritium entering the body as HTO is metabolized as ordinary water, and thus tritium is typically described as having a biological half-life in man equal to that of water, about is 8 to 12 days. There is some evidence for a longer half-life in man which will be discussed later. In addition tritium is a pure beta particle emitter of low energy, thus it is not an important external exposure hazard to man. The maximum permissible body burdens (MPBB) and the maximum permissible concentration in air and water (MPC_a and MPC_w) for HTO are among the highest recommended by standards-setting bodies.

Experimental determinations of the Relative Biological Effectiveness (RBE) for beta particles from tritium have yielded values varying from 1 to more than 2. A quality factor of 1 is assigned by the ICRP [23] and NCRP [24].

One of the major factors affecting the environmental health aspects of tritium is its resident time in a human body. Hydrogen atoms in living organisms can be separated into two major categories or pools. The first pool, tissue-water hydrogen, is defined as all hydrogen atoms in water molecules present within the organism. The second pool, tissue-bound hydrogen, is defined as all hydrogen present in organic molecules, such as proteins, fats, and carbohydrates. The tissue-bound hydrogen may be subdivided into two components: exchangeable hydrogen and nonexchangeable hydrogen. Approximately one-third of the body burden in man is tissue-bound hydrogen. The NCRP [9] found that following an acute exposure of HTO to man, only 1 or 2% of the tritium becomes organically bound; however, for chronic exposure to HTO, the tritium specific activity in the tissue-bound hydrogen pool reaches 25 to 40% of that of the HTO. Typical half-times for an excretion function associated with the three body hydrogen-pools are (a) 8 to 12 days, body-water pool; (b) 30 to 40 days, exchangeable hydrogen in organic material; and (c) 300 to 600 days, nonexchangeable hydrogen in organic material. Rohwer and Wilcox [22] estimate that organic binding results in a small (+20%) addition to the total dose potential calculated due only to body-water tritium.

The average dose rate to a tissue in which tritium is uniformly distributed is a function of the concentration of tritium and the mean disintegration energy. Thus, for 1 μCi/kg (0.037 MBq/kg) ^{3}H concentration, the absorbed dose rate, assuming total absorption of the disintegration energy within the substance, is 12.1 μrad/hr (0.121 μGy/hr) or 106 mrad/yr (1060 μGy/yr) [1]. The annual dose due to body water from an intake of 1 mCi (37 MBq) based on a 3-liter/day water balance, is 97 mrad/mCi (26 μGy/MBq). Thus a dose of 5 rem (50 mSv) would be delivered by intake of ~50 mCi (1850 MBq) of tritiated water. It should be noted that the water balance in individuals and even in the same individual at different times varies widely and thus alters the dose estimates.

The dose from tritium intake by various pathways depends on relative contributions to total water intake. Therefore to determine doses following releases of tritium to the environment, knowledge of integral concentrations of tritium in water from the various pathways is required. For estimates of these quantities, one must consider the regional and global dispersion of tritium and its entry into the hydrological cycle. Using a theoretical model for the hydrological cycle, Soldat and Baker [25] calculate that for a continuous release rate of 1 Ci (0.037 TBq) of tritium as HTO per year to the atmosphere, the total annual dose to the world population in the year 2000 is calculated to be 0.024 person-rem (0.24 person-Sv). They also calculated collective dose and dose commitment; for a release rate of one million Ci/yr (37,000 TBq/yr) of tritium in the form HTO to the atmosphere for 40 yr, a collective dose of 2×10^4 person-rem (200 person-Sv) and a total accumulated dose commitment of 1×10^6 person-rem (10,000 person-Sv) are estimated. Studies [26] modeling the conversion of HT to HTO in different biospheric reservoirs predict that an atmospheric release of HT may yield up to 1.7 times the dose commitment obtained after release of the same amount of tritium as HTO. However, Jacobs, Easterly, and Phillips [27] found that there is little impact on projected radiation dose when the conversion rates for HT to HTO are included. A rather complex model projected tissue dose rates (shown in Fig. 20.3) assuming certain input from future nuclear power installations and different mixes (i.e., LWR, HTGR, LMFBR, CANDU, etc.).

The reader interested in modeling should refer to Ref. 1 which discusses several environmental models from very simple to quite complex.

20.2. RADIATION PROTECTION GUIDES

The standards recommended by various agencies give an appreciation for the levels that must be measured to insure compliance. As discussed in Chapter 2, other guideline-setting agencies adhere to the basic upper limits set by the ICRP. For 50 weeks at 40-hr occupational exposure, the ICRP [28] has established a Derived Air Concentration (DAC) of 21.6×10^6 pCi/m^3 (8×10^5 Bq/m^3) for inhalation of tritiated water, and for submersion in elemental tritium the DAC is 5.4×10^{13} pCi/m^3 (2×10^{10} Bq/m^3).

For 168-hr occupational exposure, the NCRP [29] has established maximum permissible concentrations in air and water (MPC_a and MPC_w), as well as maximum permissible body burdens (MPBB). For individuals in the general public, the applicable MPC values are a factor of 10 smaller than for occupational exposure, and for exposure to a suitably large sample of the general public another factor of 3 smaller still. The U.S. Nuclear Regulatory Commission (NRC) [30] has established MPC levels identical to those of the NCRP.

The occupational MPBB values are 1 mCi (37 MBq) of HTO in body tissue, or 2 mCi (74 MBq) in the total body. The MPC values recommended by NCRP are as follows:

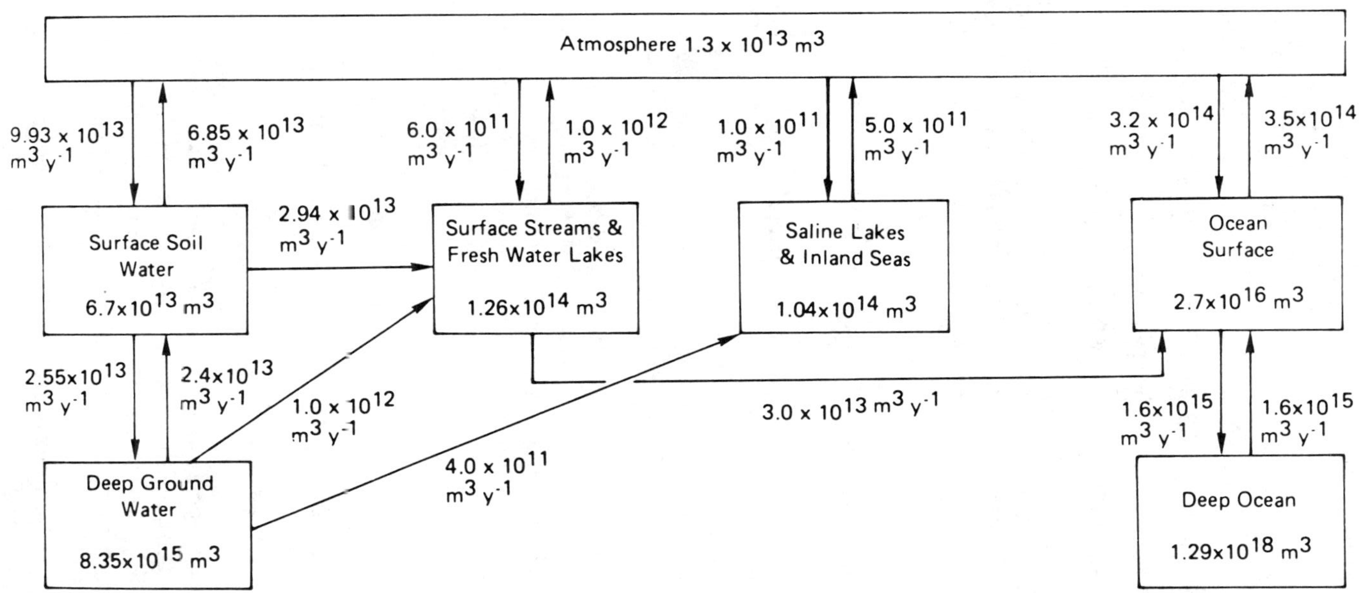

Figure 20.3. Seven-compartment model of world water cycle (reproduced from the National Council on Radiation Protection, Ref. 1, with permission NCRP © 1979).

	168-hour Occupational Limits	168-hour Limits, Individuals in General Public
As HTO in water	30,000 pCi/cm^3	300 pCi/cm^3
As HT gas in air	400 pCi/cm^3	40 pCi/cm^3
As HTO vapor in air	2 pCi/cm^3	0.2 pCi/cm^3

In addition, the U.S. Environmental Protection Agency (EPA) has established the MPC of tritium in drinking water to be 20,000 pCi/liter (740 Bq/liter) [31], corresponding to 20 pCi/cm^3 (0.74 Bq/cm^3) in the units of the foregoing tabulation.

20.3. MEASUREMENT CONSIDERATIONS

Measurements of tritium are difficult mainly because of the low energy of the tritium beta particles, which can be counted by only a few of the many detectors sensitive to ionization energy loss. This discussion is intended to cover the advantages and disadvantages of the various existing (or proposed) measuring systems for tritium. Measurements in the air and water as required around nuclear reactors and fuel reprocessing plants will be emphasized.

Four types of tritium measurements will be considered: determinations in water, in air as HTO, in air as HT, and in urine. Any of these measurements (except that in urine) may be applied to either environmental samples or process samples (water flow or gas stack). These various classes will be discussed separately where appropriate. Measurements in urine, will be considered only briefly.

Various interferences must be considered depending on where, when, and how the tritium concentration is being sampled or measured. In the gas phase around fuel reprocessing plants, the principal beta-particle emitter besides tritium is ^{85}Kr, whose higher-energy beta particle (E_{max} = 672 keV) can be a significant interference. In the water, ^{90}Sr (E_{max} = 546 keV) is present and may interfere.

20.3.1. Sampling and Sample Pretreatment

The collection of a representative sample for tritium analysis presents many of the same problems encountered with any other material. There are, however, peculiarities of tritium sampling that require special attention. The NCRP [32] lists three possible sources of error in the measurement of total activity or concentration:

1. Addition of atmospheric moisture to the collected sample through absorption, adsorption, or condensation.
2. Evaporation of water from the sample.
3. Exchange of organically bound hydrogen with hydrogen in atmospheric water.

As already discussed, tritiated water vapor and hydrogen gas are the most common forms of tritium in air and are the predominant forms of release from nuclear facilities. Tritiated methane is another form occurring naturally in the environment [33]. The collection of tritiated water vapor from air is of greatest importance, especially in the vicinity of nuclear reactors and nuclear fuel reprocessing plants. The more restrictive MPC recommended by the NCRP is for tritiated water vapor which is about 4 orders of magnitude lower than for tritiated hydrogen gas.

Collection of tritiated water vapor from air may be for measuring the specific activity of tritium in the moisture or the concentration per unit volume of air. It may be collected by:

1. Bubbling the air through nontritiated water or other appropriate solvents.
2. Passing the air through a desiccant.
3. Condensing or freezing.

The first sampling method, *bubbling*, is a widely applied and simple method for collecting HTO from air. Air is bubbled physically through a water-filled gas-washing bottle. When the isotopic ratio $HTO:H_2O$ in the water is low compared to that in the incoming air, the collection efficiency is high and predictable [34]. Efficiencies ranging from 90 to 98% have been reported [1,34] and may be determined by use of several bubblers in series. The collection efficiency for tritium gas by this method is less than 0.01% [1].

In the second method, *use of a desiccant*, air is drawn through a cell or column containing desiccant. The best adsorbent should satisfy several requirements: (a) a high capacity for moisture; (b) good moisture retention until intentionally desorbed; (c) low residual retention on desorption; and (d) mechanical stability and chemical inertness [1]. Typical desiccants include silica gel, molecular sieves (aluminosilicates), anhydrous calcium sulfate, and activated alumina. It should be noted that the quantity of water held by a given sorbent when equilibrium has been reached is related to the relative humidity of the air. Information on the adsorptive capacities of specific sorbents is usually provided by their manufacturers.

The amount of desiccant needed depends on the total moisture to be removed from the air and the adsorptive capacity of the desiccant. When it is difficult to estimate the amount of moisture to be adsorbed, sampling should be arranged so that less than half the collection capacity would normally be required. In some cases, it may be desirable to use a sorbent with an indicator which changes color as saturation is approached. Collection efficiency may be determined by tests with collectors in series. The efficiency of a well-designed cell is generally greater than 99% [35].

The third method, *condensation*, depends upon decreasing the air temperature as far as possible below the dew point. A cold trap using liquid nitrogen, liquid air, or various dry ice-solvent mixtures will collect moisture from an air sample. As mentioned earlier, air should be passed through the cold trap before the flow meter and pump to avoid the loss of tritiated vapor. Fine ice particles may form in the cold trap and their subsequent flaking could cause a decrease in collection efficiency. Glass wool, glass balls, or steel balls are sometimes placed in the trap to prevent this loss.

Of course, the reason for collecting the sample determines the choice of method, the precautions taken, and the auxiliary measurements needed. For measuring specific activity, only determination of the tritium concentration in the volume of moisture collected (by either of the last two methods listed) is needed. However, such a measurement on a sample collected using the first method, the absolute humidity of the air must be determined as well as the total volume of air from which the HTO was obtained. The absolute humidity can be determined from psychrometric tables, when the relative humidity (dew point) and air temperature are known. If the sampling period is short, then a single measurement of temperature and relative humidity may suffice. For long sampling periods, a recording hydrothermograph may be required. The air flow may be estimated from the average airflow rate and sampling time or measured directly with a wet-test meter. Meters should preferably (*always* in the case of a wet-test meter) be downstream of the collecting device to minimize contamination (or dilution) of the sampled moisture. The air pump should also be downstream of the collector.

To measure the concentration of tritium in air, the total tritium in the sample must be determined, together with the volume. The total tritium may be estimated from the total sample collected and the collection efficiency or from the specific activity of the collected moisture and the absolute humidity.

Once a sample has been collected, it should be carefully sealed in a container having low moisture absorptivity or permeability to reduce dilution, evaporation, or contamination.

For the collection and measurement of only HT or T_2 in air, any water vapor containing tritium (HTO) in the air must first be removed by one of the techniques discussed previously, then the gas stream may be sampled for gaseous tritium using the platinum-catalyzed reaction between oxygen and supplemental hydrogen to form collectible water vapor [36]. Hydrogen gas (including HT or T_2) in the air burns at ambient temperature, or slightly above, forming water which can be collected by the techniques already discussed. All existing HTO must be removed in order to prove that this induced reaction-made HTO is from the HT or T_2.

In collecting water samples, especially those with low tritium concentrations compared to the tritium concentration in air moisture, care must be exercised to minimize diffusion of HTO into the sample. Water samples containing low concentrations of tritium should be stored in tightly sealed containers. Since tritiated water vapor in the air may diffuse through polyethylene containers [1], it is best to use glass or metal containers if samples must be stored for several weeks or longer.

To test urine for tritium, only a few milliliters are required so a single voiding is sufficient. If the urine sample is being collected to monitor occupational exposure, the worker should remove any protective clothing and wash his hands thoroughly before the sample collection. Since several hours are needed after assimilation for HTO to be uniformly distributed in body fluids, any sample collected immediately after exposure will *not* represent the peak concentration. The sample should be collected at least two hours after exposure [1].

For information on collection of biota and soil samples or sampling tritiated particles, the reader is referred to Ref. 1.

Tritium in liquid or solid samples cannot be easily determined quantitatively by any kind of nondestructive analysis because its beta-particle radiation is of such low energy that most is absorbed in the sample. Most solid samples must be treated chemically or physically to be suitable for counting. The analysis of a solid environmental sample (soil, plant, or animal) for tritium usually requires extraction of water. Four methods have been found to be practical [1]:

1. Lyophilization (basically freeze drying).
2. Distillation at high temperature.

3. Dilution with solvent.
4. Azeotropic distillation.

Solid organically bound tritium can be measured if converted to a form suitable for counting to make measurement possible. The most widely used methods are wet or dry oxidation and solubilization in basic or acid solutions.

In cases where the specific activity of tritiated samples is so low that sufficient activity cannot be introduced into the counting equipment to give statistically adequate counts, enriching the tritium content of the counting sample may be feasible. For water samples, electrolysis is the most generally used method for enrichment. Thermal diffusion, gas chromatography, and sequential distillation may also be used [1].

20.3.2. Tritium in Water

In water, tritium occurs primarily as HTO (tritiated water). It is also occasionally found in tritiated compounds dissolved or suspended in water, but their measurement separate from water will not be discussed. However, frequently they are measured with HTO.

As mentioned previously, the maximum permissible concentration of tritium as HTO in liquid water is 3000 pCi/cm^3 (111 Bq/cm^3) for an individual in the general population. This MPC corresponds to 110 disintegrations per second per cm^3 (dps/cm^3). An even more restrictive MPC is in effect for drinking water, 20 pCi/cm^3 (0.74 Bq/cm^3). The main measurement problem is the low energy and range of the beta particles from tritium (the maximum range in water being 6 μm).

Liquid Scintillation Counting. Efforts to make sensitive, reliable measurements have essentially relied on liquid scintillation counting. There are many liquid scintillation systems described in the literature and several are available commercially. The properties of the various liquid scintillation systems will be summarized. Two excellent references on this subject are Horrocks and Peng [37] and Bransome [38]. A recent international conference on liquid scintillation counting should also be referred to [39].

Liquid scintillation counting of tritium relies on the fluorescence of dissolved scintillator molecules excited by the deposition of ionization due to beta particles. Energy transferred to solute molecules by the β-rays in turn excite fluorescent molecules. A photon is emitted when the excited fluorescent molecule returns to its ground state. For most practical scintillator solutes, the fraction of excited molecules which de-excite by photon emission is about 90%. The photons emitted are detected by a photomultiplier tube (PMT) or, in the case of tritium, generally by a pair of PMTs in time coincidence.

Horrocks describes this technique [40]:

> The main advantage obtained from the use of liquid scintillation counting as opposed to other conventional counting methods has to do with the fact that the sample containing the radioactivity is in intimate contact with, and in most cases dissolved in, the detector. This produces the ideal 4π geometry. A limiting factor in the liquid scintillation counting system is the photomultiplier tube [PMT]. [The best PMTs have photocathode efficiencies for conversion of photons into electrons that may exceed 50%.]

> The liquid scintillator solution consists of three main parts. The bulk of the solution is the solvent, usually an alkyl benzene (i.e., toluene) or dioxane. To the solvent an organic compound (or compounds) is added which is an efficient photon emitter and which emits photons in the wavelength region which is easily measured by the PMT. And finally the liquid scintillator solution contains the sample, although there are some uses where the excitations are generated by an external sample. The most troublesome problem encountered with liquid scintillation counting involves those problems associated with the introduction of the sample into the scintillator solution in a homogeneous (or near homogeneous) form without causing a drastic reduction in photon emission due to so called quenching processes. . . .
>
> Quenching is defined as any of those mechanisms which lead to a decrease in the number of photons emitted by a scintillator solution. There are two basic types of quenching: chemical and color. Chemical quenching involves the action of foreign materials upon the excited molecules, either solvent or solute molecules. Energy is removed from the excited molecules by a process which does not involve the emission of photons. Color quenching involves the reabsorption of emitted photons by foreign material which may be present in the liquid scintillator solution or sample. The reabsorbed photons will produce excited molecules which de-activate by non-radiative processes.
>
> The overall effect of these processes is the decrease in the average number of photons emitted per unit energy of exciting particle stopped in the solution. It is important to keep in mind that a 50% quenching of photon yield does not mean a 50% decrease in counting efficiency. Indeed the same amount of quenching will produce a greater change in counting efficiency for tritium than for ^{14}C.

Moghissi et al. have studied the significant factors which affect low-level liquid scintillation counting of tritium, both in homogeneous liquid systems such as solutions [41], and in emulsions [42]. From the paper on homogeneous solutions [41], several conclusions are important:

1. Plastic vials are preferable to glass vials; "glass seems to decrease the counting efficiency for tritium counting and increase the background as compared to plastic."
2. The solution which optimizes counting efficiency is 6 to 7 g of PPO, 1.2 to 1.5 g of *bis*-MSB, and 120 g of naphthalene per liter of dioxane. [PPO is 2,5-diphenyl-oxazole; *bis*-MSB is *p-bis-(o*-methyl-styryl)-benzene.] Various studies of concentrations were conducted to determine this optimum.
3. In solution, the optimum water content is about 1 ml water/4 ml solution. There are two competing processes here, namely that additional water increases quenching at the same time that it increases the proportion of water which contains the activity.
4. Phosphorescence is severe, and "in considering the long-lived component of phosphorescence, it is desirable to avoid any light which might excite the scintillation solution."
5. A constant check of possible instrument drift is required for precise measurements. "The recommended order of counting is: background, sample, standard calibration sample, duplicate of sample, background."
6. "The preparation of water samples is simple and consists of a single distillation." (In fact, some samples require no distillation at all.)
7. The *Y*-value (defined as minimum detectable activity at a confidence level of one sigma of statistical error for one-minute counting time) is about 1 pCi/cm^3 (0.037Bq/cm^3).

Even more sensitive systems using emulsions are available. Lieberman and Moghissi [42] report a *Y*-value of 0.7 pCi/cm^3 (0.026 Bq/cm^3) using the detergent Triton N101 (Rohm & Haas Co.) in scintillation solution at a ratio of 1:2.75. With this mixture a scintillation solution of *p*-xylene containing 7 g PPO and 1.5 g *bis*-MSB per liter can hold 40 to 50% water. This cocktail has a counting efficiency of 23% for the tritium beta-particles. By using three photomultipliers viewing a large-volume (250 ml) vial, Moghissi [43] has achieved a *Y*-value of 0.2 pCi/cm^3 (0.007 Bq/cm^3).

The calibration and standardization of systems as sensitive as these is a difficult problem. The National Bureau of Standards (NBS) has tritiated-water and tritiated-toluene standards [44]. The most frequently used calibration method is that with an "internal standard," in which a standard activity (e.g., from NBS) is used interchangeably with the unknown samples. The problem here is that the addition of a standard can change counting efficiency, because of differences in quenching. This problem can be overcome by the addition of a very small (microliter) amount of a high-activity standard to a previously counted sample, but the accurate measurement of a small size standard, usually made gravimetrically, is difficult and tedious.

In another attempt at calibration and standardization, Moghissi and Carter [45] used two samples which differed only in that one employed normal and the other tritiated naphthalene. This minimized most of the errors due to chemical differences.

Another method is to use one or more radioactive gamma-ray sources placed outside of the counting vial (the so-called external standard method). These can be used to adjust the gain on the photomultiplier tubes, or to partially compensate for differences in quenching characteristics among several samples in a train. Systems with automatic feedback control, using radioactive sources, are available from some of the commercial liquid-scintillation manufacturers.

Both *Standard Methods* [46] and the ASTM compilation [47] contain methods for calibration and standardization of using liquid scintillation counting for tritium analysis. The EPA has issued a tentative reference method [48] for tritium measurement in environmental waters.

Tritium in Flowing Water. Measurements in flowing water can, of course, be made by taking grab samples for batch counting. A continuous-flow system with continuous measurements frequently would be more desirable. The requirements of such a system are that it be sufficiently sensitive and precise; that it respond in a time comparable with the characteristic time for a change in tritium concentration in the flowing water; that it sample properly; and that it not be too expensive nor require excessive maintenance.

Ting and others [49, 50] have reported on a continuous-flow system, developed at Beckman Instruments for use in the secondary coolant of a pressurized-water reactor. When the device was field-tested, the tritium levels in the coolant at the San Onofre Nuclear Generating Station [50] were in the region of 1500 pCi/cm^3 (56 Bq/cm^3), measured with a few-percent precision.

The system works by mixing a toluene-based liquid scintillator (or a dioxane-based scintillator if very rapid mixing is required) with the liquid water sample in a mixing chamber; the mixed solution flows through a light trap, a liquid scintillation counter, and then out as waste. The critical part of the system is the mixing chamber. The liquid scintillator flows continuously into a vial; the water sample enters the vial by a dripping gravity-flow system, and is stirred for thorough mixing twice before exiting. Its time response is remarkably good: Figure 20.4 shows the response to 5 tritium samples of 10, 20, 30, 40, and 50 μl, injected instantaneously with a scintillator flow rate of 7.0 ml/min. Residence time was less than 2 min in the test, and the total response peak was over in about 4 min, as seen in Fig. 20.4. The efficiency for detection of the tritium beta particles was 32%. The sample flow rate was 0.5 ml/min, and the minimum detectable tritium concentration was about 25 pCi/ml (0.93 Bq/ml). This system shows promise for the continuous monitoring of tritium in flowing water. Its main drawback is high operation cost due to consumption of liquid scintillator. At 7 ml/min (about 10 liters/day), the annual cost of liquid scintillator would be in the range of several thousand dollars per year.

Osborne and Tepley [51] have largely overcome this last problem by designing a system that uses liquid scintillator at a rate of 2 μl/sec (0.12 ml/min) which has a minimum detectable concentration for HTO of 0.2 μCi/kg [= 200 pCi/cm^3 (7.4 Bq/cm^3)]. However due to the system complexity it may have only limited applicability, such as at the CANDU nuclear power installation for which it was designed.

Although this discussion stressed liquid scintillation counting, solid scintillators may also be used (discussed later).

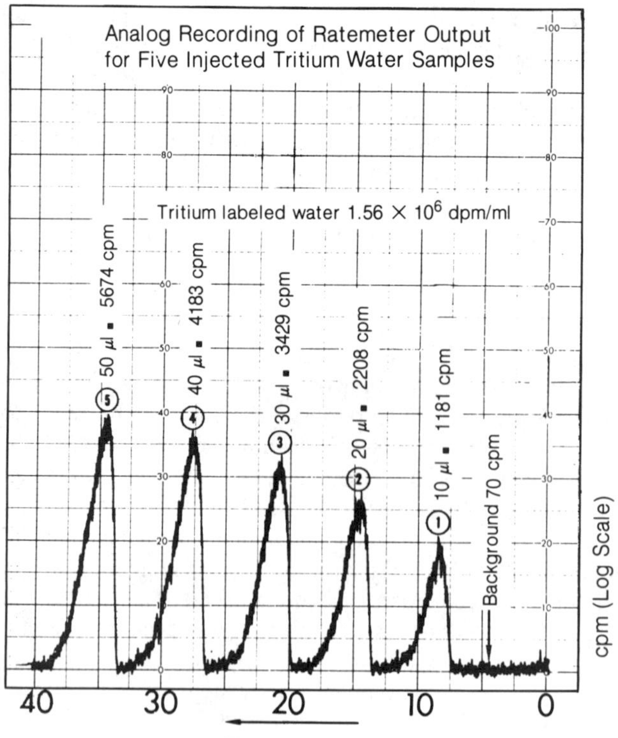

Figure 20.4. Response of the flowmeter described in Ref. 49. Analog recording of ratemeter output for five injected tritium water samples (reproduced from Ting and Litle, Ref. 49).

Urinalysis. The most common procedure for the determination of tritium body burden is radioassay of tritium (as HTO) in urine. Chronic exposure of workers to air containing HTO vapor at the NRC recommended occupational MPC level will produce a tritium activity of about 30,000 pCi/cm^3 (1100 Bq/cm^3) in urine [52,53].

Levels down to 10 pCi/cm^3 (0.37 Bq/cm^3) can be measured routinely with liquid-scintillation counting [54], using an emulsion technique similar to that described earlier [42]. Thus sensitivity is not a problem for urinalysis for tritium in the laboratory. For rapid measurements, which are sometimes required after possible large accidental exposures, Osborne has developed an automatic liquid scintillation urine analyzer [55]. This system makes a measurement two minutes after the sample is introduced, with a maximum sensitivity of about 1000 pCi/cm^3 (37 Bq/cm^3) and it will measure up to 10^7 pCi/cm^3 (4×10^5 Bq/cm^3) with a precision of about ±10%. This device is particularly useful to assess a possible HTO exposure immediately after an accident.

One difficulty with urinalysis is the possible quenching due to foreign substances in the sample. For this reason, the *p*-xylene-based emulsion technique is preferred over the homogeneous-solution method [54]. Alternatively, one can distill the urine sample before determining the tritium activity level with any of the conventional dioxane-based systems.

Working with an entirely different measurement system, Block, Hodgekins, and Barlow [56] have developed a proportional counter technique that can detect 1000 pCi/cm^3 (37 Bq/cm^3) and is relatively simple. The system is based on the reaction of water and calcium carbide to form acetylene as follows:

$$CaC_2 + 2H_2O \rightarrow C_2H_2 + Ca(OH)_2.$$

Tritiated water yields tritiated acetylene (C_2HT), which can be used as a counting gas in a proportional counter. This system is attractive particularly because of its adaptability for use in the field, despite the fact that its sensitivity is much less than that of the liquid scintillation systems.

Sandia Laboratories [57] have developed urinalysis capability using metallic calcium to separate the tritium and hydrogen from the water in urine. A sealed ion chamber is filled with the resulting evolved hydrogen. The main advantage of the technique is immediate analysis capability for use in emergency accident situations. The calcium cartridge is disposable, and the instrument makes determinations in 5 min, with ±15% accuracy (2σ level) down to levels of 10,000 pCi/cm^3 (370 Bq/cm^3). This instrument is somewhat portable (~18 lb) and is useful for field use in case large exposures occur in a remote location.

Solid Scintillation Counting. Although liquid scintillation counting has been the most commonly used technique for tritiated water measurements, plastic scintillation systems are also used. These systems have several drawbacks compared to the liquid-scintillation systems, but two advantages: first, plastic scintillation systems provide faster response than systems using liquid methods; and second, they are cheaper because liquid systems can consume large amounts of organic solvents and solutes, which must be disposed of after use.

In order to achieve high sensitivity with solid scintillation, a high surface-to-volume ratio is needed, and the early systems used small crystals of anthracene. More recently, various kinds of detectors using thin sheets or layers of anthracene or a plastic scintillator spread over lucite (for light piping) have been described.

Figure 20.5 shows one detector, in which 100 rods of lucite are coated with anthracene powder. The counter can detect 1000 pCi/ml (37 Bq/ml) of water (HTO) at a count rate of 10 cpm. To measure urine, the sample must be decolored by passage over a charcoal column. The response to tritium in air is 20 cpm for 1 pCi/ml (0.04 Bq/ml) of air. These could be detected in an unshielded laboratory where the normal external background γ-ray levels were at about 30 cpm.

Osborne [59] has developed the plastic detector shown in Fig. 20.6 with the sensitivity of 24 cpm for 1000 pCi/cm^3 (37 Bq/cm^3) of water. At this counting rate, small fractions of an MPC_w can be detected easily. For tritium in air, about 0.01 pCi/cm^3 (0.04 mBq/cm^3) can be detected. The background is about 40 cpm.

The main problem, to date, with plastic detectors has been that over a period of time the plastic surfaces tend to degrade because of sedimentation, bacterial growth, and discoloration. Any of these can produce large decreases in counting efficiency because of the difficulty in light collection through the thin light paths. The detector can sometimes be cleaned, but not always. (At present, this technique does not compare favorably with liquid scintillation counting for effluent water monitoring, although when specific activity is high and the water sample is pure it can be used satisfactorily.)

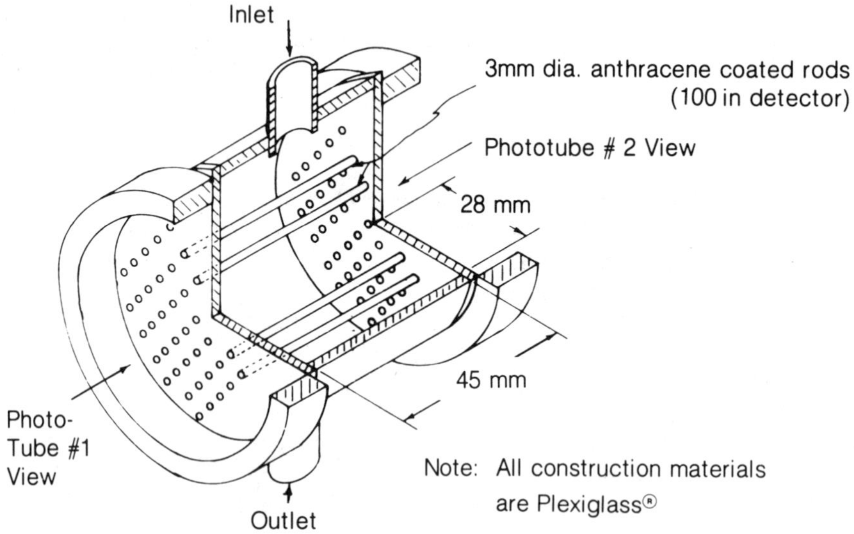

Figure 20.5. Schematic diagram of the scintillation detector (reproduced from Moghissi et al., Ref. 58, with permission of North-Holland Publishing Co. © 1969).

Figure 20.6. Detector for tritium in water. The scintillator sheets fit into a cubic lucite cell. The lucite plate shown here, on which the sheets are mounted, forms one vertical side of the cell (reproduced from Osborne, Ref. 59, with permission of North-Holland Publishing Co. © 1970).

20.3.3. Tritium in Air

As already discussed, tritium in air occurs primarily in two forms: as water vapor (HTO) and as hydrogen gas (HT). Tritiated organic compounds in the vapor phase and on particulate matter also occur occasionally.

There are two common approaches to measurements of HTO and T_2O in air. First, one can use an ionization chamber or gas proportional counter with filtered air introduced for internal counting; the tritium beta particles ionize the gas within the sensitive chamber volume. This method is sensitive to tritium in all of its gaseous chemical forms. Secondly, HTO vapor can be removed from the air with a bubbler or with some dessicant such as silica gel or a molecular sieve. This water must then be freed from the dessicant. The released tritiated water can be counted using one of the liquid or plastic scintillation techniques.

To measure tritium as HT or in tritiated organics internal counting may be used. Alternatively, it is possible to burn the gas, converting the tritium to HTO before dessication and counting. A number of detection devices have been adapted for measuring tritium. They are discussed in the following sections.

Ionization Chambers. The earliest tritium detector for stack effluent measurements was the Kanne chamber [60,61,62], developed in the early 1950s. The gas to be measured is drawn continuously through an electrostatic precipitator or ion trap, then into a large ionization chamber. The resulting current is measured with a picoammeter. Large chambers of volumes (18.5 and 51.5 liters) are in common use at Savannah River Laboratory [63]; they have minimum sensitivity limits (2σ) of 7 and 2 pCi/cm^3 (0.26 and 0.074 Bq/cm^3) in air, respectively. Marter and Patterson say [63]:

> Although these instruments are sensitive to unusual particulate air loading, external gamma radiation, and other radioactive gases, these problems are reduced by filtering the air intake, by locating the chambers in areas of low gamma background, and by using moisture traps between two chambers in series, respectively.

One important problem with ionization chambers is the background due to the presence of external γ-ray fields. There is no way for a single ion chamber to differentiate between external gamma rays and internal beta particles (or internal γ-ray or α-particles). The background is such that 1 mR/hr of gamma radiation is about equivalent to 100 pCi/cm^3 (3.7 Bq/cm^3) of tritium in air [64,65]. To reduce this problem, one common procedure is to place two identical chambers side by side, one sealed and the other open to the ambient air [66]. This reduces the interference by a factor of 50 to 100 [64]. The principal limitation lies in maintaining identical air masses in the two chambers in the presence of fluctuations in ambient pressure, temperature, and humidity.

If the gamma-ray flux is directional, chambers adjacent to one another may not sample the same flux. This problem is reduced in a design by Osborne and Cowper [67] in which the compensating chamber is placed inside the active chamber to provide equal response in all directions. A 40-liter chamber of this design has a minimum detectable limit (2σ) for tritium at levels as low as about 1 pCi/cm^3 (0.04 Bq/cm^3) of air. Osborne and Coveart [65], using the concentric ion chamber design, developed a portable tritium monitor capable of measuring 5 to 10×10^7 μCi/m^3 (0.19 to 3.7×10^5 MBq/m^3) with a 1.33-liter ion chamber and a solid state electrometer. The response time constant is 70 sec with an air flow rate 30 cm^3/sec. They achieve a compensation for gamma background irradiation from any direction of greater than 95°. They admit however, that the more elegant rotating chamber method described next can achieve better compensation in nonuniform fields but at the expense of greater weight and power consumption. Talbert and Hillbert [68] suggest an alternative configuration to achieve more effective gamma-ray subtraction than with stationary chambers in areas where external gamma-ray fields are nonuniform or may change spectrally or directionally. The technique [68] is dynamic with both chambers (a flow-through ion chamber and a sealed ion chamber) orbiting about a common axis parallel to, and equidistant from, the centerlines of the two chambers. Therefore, on the average each chamber occupies the same region in space for the same amount of time, allowing direct subtraction of the result from the gamma-ray compensating chamber from that of the flow-through ion chamber sampling radioactive gases. It was shown by Talbert and Hilbert [68] that in the presence of an external gamma-ray field of 50 mR/hr, the minimum detectable tritium concentration in the atmosphere is 25 μCi/m^3 (0.93 MBq/m^3) with very nearly 100% confidence in a 50-sec sampling time for 1-liter ion chambers. The corresponding numbers are 20 μCi/m^3 (0.74 MBq/m^3) in a 35 mR/hr field and 12 μCi/m^3 (0.44 MBq/m^3) in a 10 mR/hr field. In their discussion, Talbert and Hilbert [68] indicate that for concentric chambers as discussed earlier by Osborne and Cowper [67] in a field of 10 mR/hr the instrument reading would vary between +50 and −50 μCi/m^3 (+1.9 and −1.9 MBq/m^3).

Interference by other gaseous radionuclides is a major problem with ion chambers. Particles may be eliminated with filters, but radioactive noble gases (e.g., ^{41}Ar, ^{85}Kr) are unreactive and difficult to eliminate. Since each disintegration of ^{85}Kr produces many times the ionization of a tritium disintegration, this ionization chamber is inappropriate unless tritium is known or made to be the only significant radionuclide present. To overcome this problem, Osborne has reported the development of two identical chambers, except that one has "a self-renewing drying system [that traps HTO] so that continuous subtraction (tritium plus noble gases minus noble gases) can be accomplished." [69] Talbert [70] suggests another method to measure tritium in the presence of other beta-particle-emitting nuclides using a special instrument design and no special sampling system. The instrument employs a separate concentric chamber with a thin common wall opaque only to tritium betas. Currents produced in the two chambers are subtracted leaving only the current due to tritium.

Waters [71], in a review of general problems with ion chambers, reports that factors such as cigarette smoke, ions, aerosols, humidity, and memory effects can also cause erroneous readings in ion chambers. He suggests the use of a micrometer pore size filter followed by an electrostatic precipitator to prevent intake of particulate contamination. Materials such as rubber and some plastics [72] absorb significant amounts of tritiated water vapor, causing low readings. Furthermore, absorbed HTO may be re-emitted or exchanged with ordinary water vapor introducing errors in subsequent measurements. One satisfactory material is Teflon®, but Waters [72] reports that it is "difficult to use because of its tendency to cold-flow under pressure."

For monitoring tritium in air, the ionization chamber is the most commonly used instrument. Portable flow-through instruments of various sizes (typically with about 1-liter chamber volumes) have been a mainstay in health physics despite their problems.

Commercially available instruments for tritium monitoring are generally based on ion chambers using one of the designs discussed earlier.

Proportional Counters. The gas proportional counter differs from the ionization chamber in two important respects: it is a pulse-counting instrument, and it requires a particular filler gas to which the sample of (tritiated) air is added. It offers an advantage over the ion chamber in that energy discrimination is possible by electronic discriminating of the output pulse heights, since the heights are proportional to the energies deposited by the emitters in the chamber. The essential disadvantage of the gas proportional counter is the continual consumption of counting gas, which is discarded after being mixed with the tritiated sample and counted.

Driver [73] built a tritium counter of this type as early as 1956. Ehret [74] achieved a minimum detectable concentration (2σ) of 0.1 pCi/cm^3 (0.004 Bq/cm^3) of air, using a chamber of about 1-liter volume with methane as the counting gas, to which is added 20 to 30% air.

> Air is pumped through the counter by means of a small membrane pump. At the counter inlet, methane is added to the air within a small mixing chamber [74].

In this design, there must be a compromise between response time and inordinate consumption of methane. Using a methane flow rate of 25 liters/hr, a response time of a few minutes can be achieved. Ehret's design also included an anticoincidence proportional counter surrounding the tritium-sensitive counter, to discriminate against the long-range beta particles of ^{41}Ar and ^{85}Kr. This counter suffered from "memory effects" due to absorption of tritiated water vapor (HTO) in plastic tubes. Use of Teflon should eliminate this problem.

Block, Hodgekins, and Barlow [56] built a large-area, thin-window gas-flow proportional counter system that included two counters which were identical except that one had an aluminized Mylar window opaque to tritium beta rays while the other had a Formvar window thin enough to allow detection of tritium. This system can thus compensate not only for a gamma-ray background but also for the beta-particle activities of ^{41}Ar and ^{85}Kr. The window is porous to the counting gas (propane), and the flow occurs because of a positive pressure difference outside and inside the counter. The window is permeable to T_2O and HTO (but not HT or T_2). In a gamma-ray flux of 3 mR/hr, the counter had a minimum sensitivity in the region of about 1 pCi/cm^3 (0.04 Bq/cm^3).

In a more recent application of proportional counters, Howell, Cate, and Wong [75], utilizing a permselective membrane for sample enrichment, achieved a detection limit of 2.5 pCi/cm^3 (0.093Bq/cm^3) with a 250-cm^3-volume detector, a 250 cm^3/min gas flow, and system response time of about 5 min. Permselective membranes will be discussed in more detail in the section on special sampling techniques.

Solid Scintillators. For tritium counting, the use of solid scintillators, particularly for gases containing tritiated water vapor, has (as previously mentioned) a major contamination problem. Contamination can result from either tritium exchange with organic scintillators or absorption of tritiated water by hygroscopic crystals such as NaI(Tl). Colmenares et al. [76] used a Europium-doped calcium fluoride scintillator that they hoped, because of its low solubility in water, might overcome the objectionable properties of other solid scintillators. Their system was based on a CaF_2(Eu) scintillator-photomultiplier combination built into an all-metal housing, and an amplifying pulse shape analyzer. The analyzer used a combination of linear and digital integrated circuits for signal amplification, pulse shape discrimination, and single-channel analysis. For more detailed information on the design and construction features of both the detector and electronic components the reader is referred to Ref. 76 where the performance of the system was examined using tritiated standards. Two gas standards were used: (a) tritiated methane containing 0.01129 μCi/cm^3 (4.2 kBq/cm^3) (STP) of tritium, and (b) tritium in argon containing 3.27 μCi/cm^3 (120 kBq/cm^3) (STP). The second standard contained approximately 2 ppm of tritiated water which severely contaminated the entire system for several days after use. The minimum detectable concentration of tritium at an operating voltage of 917 V and a system calibration using tritiated methane gas standards over a tritium range from 0.35 nCi/cm^3 (0.013 kBq/cm^3) (STP) to 0.68 μCi/cm^3 (25 kBq/cm^3) (STP), was calculated by the statistical method of Altshular and Pasternak [77] for 95% confidence level, to be 48 pCi/cm^3 (1.8 Bq/cm^3) (STP) using an unshielded detector and a counting time of 1 min. Also by using specialized very thin film coatings on both the crystal and the walls of the sample cavity, Colmenares et al. [76] hoped to reduce contamination even further at moderately high tritium concentrations [3 μCi/cm^3 (110 kBq/cm^3) (STP)]; preliminary results were encouraging.

Singh and Kadwani [78] developed a flow-cell-type of scintillation detector system suitable for continuous monitoring of tritium in the ambient air. The detector recorded 2500 cps with a single photomultiplier tube type EMI 9514S for a 1 MPC_a (5 pCi/cm^3 or 0.19 Bq/cm^3 at STP) level of tritium activity release at 50 counts per second background and 2.5% counting efficiency. The cell, filled with plastic scintillator powder, did not show any background buildup over long periods of operation and had a low gamma-ray response. It should be noted that the system was not tested with a moderately high concentration of 3 μCi/cm^3 (110 kBq/cm^3) as was the previous system by Colmenares et al. In fact, it was designed so that the condensed sample would only have a few tenths μCi/cm^3 (a few kBq/cm^3) at 1 MPC release rates and all tests were carried out at this level. The system samples the air stream by condensing water and flowing this water through the detector.

Osborne [79] also designed a monitor for tritiated water vapor in air utilizing plastic scintillators. The tritiated water vapor in the sample air stream is collected in a water stream and measured in a flow scintillation counter. The detector, made from sheets of NE 102 scintillator 0.125 mm apart, provides 300 cm^2 of surface and has a count rate of 1 cps at the 10 pCi/cm^3 (0.37 Bq/cm^3) level corresponding to detection of 0.05% of the tritium disintegrations in the detector. The background counting rate in the absence of radioactive noble gases and in a gamma-ray field of less than 0.1 R/hr at the counter surface is 0.2 cps. In the monitor, the filtered sample air (flow rate 4000 cm^3/min) is mixed with a continuous flow of nontritiated water (4 cm^3/min). The two phases are then separated and the water phase which contains all but a few percent of the tritium is purged with nontritiated air to remove dissolved noble gases before passing to the detector. (This technique of using a "bubbler" is described further under Special Techniques). With this system 1 μCi/m^3 [1 pCi/cm^3 (0.037 Bq/cm^3)] tritium can be detected with a response time of several minutes.

Liquid Scintillation Counters. The use of silica gel as a dessicant to remove moisture (H_2O, T_2O, HTO) from air before counting for tritium with a liquid scintillator is a common technique for determining tritium in air. In the Intersociety Committee procedure [80], a 12 × 1.25 in. diameter aluminum cylinder is filled with about 180 g of silica gel, which can absorb moisture up to 40% of its own weight. Air is pumped at about 100 to 150 cm^3/min through the silica gel column, which collects essentially all of the moisture. With this much gel, a continuous sample can be taken over as long as a two-week period; of course, a smaller setup can be used for shorter sample-collection times.

Following sample collection, the silica gel is heated in a distilling flask to remove the moisture. The distillate is then counted using the liquid scintillation technique of Moghissi et al. [41]. A sample collected from air at 30°C (86°F) and 100% relative humidity will yield 6600 pCi/cm^3 (244 Bq/cm^3) when tritium (as HTO) is present at the MPC_a concentration of 0.2 pCi/cm^3 (0.007 Bq/cm^3) [80]. The system has negligible interference from other radioactive nuclides, because they do not absorb on silica gel and any which might be present would be eliminated in the distillation process. Also, tritium as HT gas is completely excluded by this method. The largest source of systematic error is the uncertainty in the volume of air sampled and the prevailing temperature, pressure, and relative humidity of the air.

Osloond et al. [35] have reported a detection limit (2σ) of 2×10^{-5} pCi/cm^3 (7×10^{-7} Bq/cm^3) of air by counting the silica gel directly (without distillation) in a liquid scintillation system.

The main drawback of gel sampling coupled with liquid scintillation counting is the fact that concentrations are integrated over a long time period (hours to days), and that each batch must be analyzed separately in the laboratory. Another drawback is that the dessicant must be renewed periodically, before it becomes saturated. The advantages are its excellent sensitivity and precision and lack of interferences. In fact, it has been approved as a "Tentative Method for Analysis for Tritium Content of the Atmosphere" by the Intersociety Committee for a Manual of Methods for Ambient Air Sampling and Analysis [80].

Other sampling methods coupled to liquid scintillation counting have been used for monitoring tritium in air. By using a bubbler exchanger (discussed under Special Techniques) to sample the air stream, Osborne and Tepley [51] achieved a minimum detection limit in air of 0.003 pCi/cm^3 (10^{-4} Bq/cm^3) with a scintillator flow of only 2 mm^3/sec.

Recently, Kato [81], using a liquid absorbent, ethylene glycol, and liquid scintillation counting, attained a minimum detectable concentration of tritium in air of 0.1 pCi/cm^3 (0.004 Bq/cm^3).

Special Techniques. Some special techniques have been devised sampling HTO vapor and HT in air. Some of the more important methods are outlined here.

Bubbler Exchanger. As already mentioned, Osborne [34,64] has removed tritiated water vapor (HTO) from air using a water-filled gas washing bottle or "bubbler." (See Ref. 34 for further details of the theory and practical realization of this method.) Moist air is bubbled through nontritiated water, and exchange occurs between the vapor and liquid phases. It can easily sample a known volume of air with overall collection efficiency close to 100%. For example, Osborne cites sampling for 10 hr at 10 cm^3/sec (or for 1 hr at 100 cm^3/sec) through 200 cm^3 of water with an overall collection efficiency of 96 to 97%. One advantage of this method is that tritiated hydrogen gas (HT or T_2) is eliminated. In a bubbler system studied at Los Alamos, the collection efficiency for HT was ~0.1% under conditions in which HTO collection efficiency was 90 to 95% [82]. Also eliminated are noble gases; Osborne [65] states "the ratio of noble gas activity to tritium activity in the water would be ~0.013% of the average of the ratio in the sampled air."

A typical bubbler system is somewhat less sensitive than are dessicant systems, with the minimum detectable activity in the region of 0.1 pCi/cm^3 (0.004 Bq/cm^3). This sensitivity however, may be acceptable for many applications, and so the method may be chosen over others because of its convenience. The bubbler is typically coupled to a liquid scintillation detector for continuous monitoring.

Combustion for HT Determinations. Two important techniques have been developed for the separate determination of tritium in air as HTO (water vapor) and HT gas.

In the first technique, Griffin, Cochran, and Bertuccio [36] have determined HTO and HT separately, using a technique whose flow train is shown in Fig. 20.7. Because HT (or T_2) in the gas to be measured is present at very low levels, tritium-free H_2 gas is added as a carrier. Any moisture (including HTO) is then removed in Air Dryer #1, consisting of Drierite-brand anhydrous calcium sulfate in Plexiglas cylinders (W. A. Hammond Drierite Company, Xenia, OH 45385). The

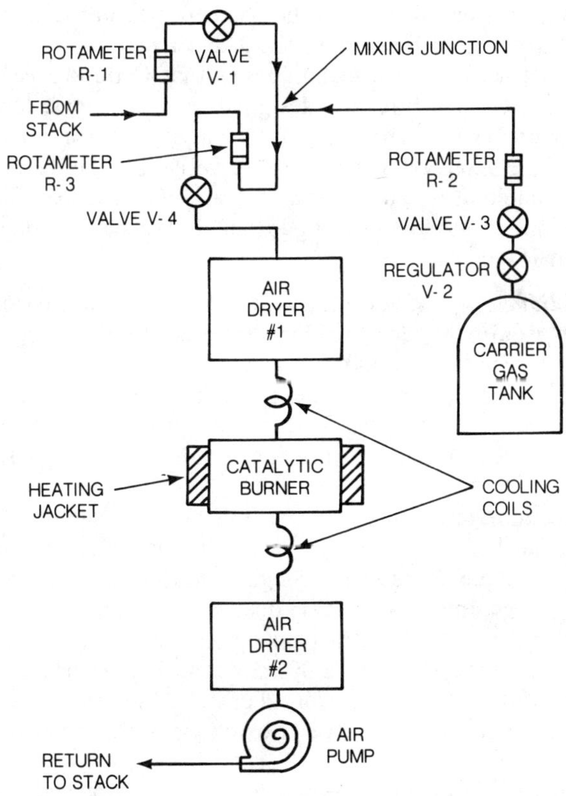

Figure 20.7. Sampling train for total airborne tritium (reproduced from Griffin, Cochran, and Bertuccio, Ref. 36).

dry gas is combusted in a catalytic burner, all combustible compounds are oxidized with the help of a 0.5% platinum on 1/8-in. diameter aluminum oxide pellets, at a temperature of 400°C maintained by a heating jacket. Not only HT, but also tritiated organics (such as CH_3T, etc.) are converted to HTO in this step. The HTO is then removed from the flow by Air Dryer 2. Both Drierite samples are dehydrated and the water vapor distilled under a palladium catalyst, operating at lower temperature than the platinum catalyst. The samples are then analyzed with a liquid scintillation counter. Levels of tritium (in stack gas) in the 0.2 to 2.0 pCi/cm^3 (0.007 to 0.07 Bq/cm^3) region were measured with accuracies of ±8.5%.

More recently Straight [83] adapted this basic method to field collection of gaseous tritium and tritiated water vapor with a unit suitable for operation in remote desert areas. Its primary feature is its capability of operating without supplemental hydrogen. The detection limit achieved with a liquid scintillation counter on a sample collected at a flow rate 0.5 liter/min and a one-week sampling period was 5×10^{-8} μCi/m^3 [5×10^{-8} pCi/cm^3 (1.9×10^{-9} Bq/cm^3)]. Thus for long sampling periods very high sensitivities can be attained.

Ostlund [84] has developed another technique. His procedure also converts HT and other tritiated compounds into HTO (water vapor) by a catalytic combustion process; palladium metal carried on a molecular sieve is used to combust hydrogen in air to better than 98%. The resulting water is absorbed in the same catalyst. Sampling times as short as 10 min can be used to measure tritium levels of about 1 pCi/cm^3 (0.04 Bq/cm^3), with errors in the 3 to 5% region.

One difference between the two techniques is that Ostlund's is better adapted to single measurements, while Griffin's is designed for continuous, long-term sampling followed by batch analysis of the integrated activity. Also, because in Ostlund's technique the catalyst based on a molecular sieve is also the dessicant, there is not a problem with contamination and "memory" which might exist in the other technique. Either technique has the advantage that it can make independent measurements of HTO and HT in air.

Enrichment Techniques. Most measurements of tritium in environmental samples ordinarily do not require an enrichment step prior to counting. However, techniques for tritium enrichment have been developed for those special cases in which tritium levels are extremely small. One example is in geological research, where studies have been made of underground natural waters whose ages are so great (many thousands of years) that tritium levels are minute [$\sim 10^{-3}$ pCi/cm^3 (4×10^{-5} Bq/cm^3)].

Among the procedures used to enrich tritium in water samples are chromatography, electrolysis, and thermal diffusion. Hoy [85] has developed a chromatographic system in which a sample is passed through a heated palladium column; tritium diffuses through the column faster than does normal hydrogen, and the first gas evolved is tritium-enriched. Tritium recovery of about 60% was obtained in the first 500 ml of gas evolved, and with a 30-min counting period a T:H ratio of 10^{-17} could be measured with ±41% (1σ) error. Perschke [86] was able to enrich tritium up to 40 times with 100% tritium recovery, using gas-solid chromatography on metal sieves.

Another approach is electrolysis. When water is electrolyzed, hydrogen is given off more readily than tritium; the relative loss rate can be as high as 35:1, varying with electrode material, surface conditions, electrolyte, and temperature. Hartley

[87] has made a detailed analysis of the parameters involved in electrolytic procedures. Ostlund and Werner [88] have achieved enrichments of about 100× by a series of successive electrolytic stages.

Coupling electrolysis with an ionization chamber, Yamaguchi [89] achieved a detection limit for the entire system of about 3×10^{-7} pCi/cm^3 (1×10^{-8} Bq/cm^3). However, he notes a couple of drawbacks: (a) a well-trained expert is required to handle the system and (b) the preparation of the enriched sample takes 7 to 10 days.

Thermal diffusion is another method used for tritium-hydrogen separation [90] but it is complicated and time consuming.

Recently it has been found possible [34] to selectively enhance the concentration of HTO over HT and other gases such as Ar, Kr, and Xe by using semipermeable membranes. The sensitivity of measurement and the degree of enhancement can be manipulated to provide the characteristics desired in such instruments as area monitors, stack monitors, or portable survey meters. In a two-stage dimethyl-silicone membrane system, using the enhanced output of the first stage as an input to the second stage, Howell, Cate, and Wong [75] obtained final enhancements of 1880 HTO to HT and 2440 HTO to Kr.

For further information on enrichment techniques, the interested reader is referred to the references already cited and Ref. 32.

20.4. SUMMARY

Four important regulatory guidelines determine the sensitivity required of instruments for measuring tritium in environmental media for radiation protection purposes. Three of these are the Maximum Permissible Concentrations (MPC) for an individual in the general population [29]:

3000 pCi/cm^3 (110 Bq/cm^3) (as HTO in water)

40 pCi/cm^3 (1.5 Bq/cm^3) (as HT gas in air)

0.2 pCi/cm^3 (0.007 Bq/cm^3) (as HTO vapor in air).

The fourth is the amount of tritium allowed in drinking water [31]: 20 pCi/cm^3 (0.74 Bq/cm^3).

The ideal instrument is one capable of measuring down to a fraction of the MPC levels in the appropriate medium. Two main types of measurements are needed for each medium, namely laboratory and field measurements. The requirements for a measurement technique are quite different in the two cases. Field instruments generally must be more rugged, require less maintenance, be less sensitive to temperature and humidity effects, and have some method of automatic readout (whether the measurements are continuous or batched). In the absence of these requirements, laboratory instruments are frequently more sensitive and more flexible.

20.4.1. Laboratory Measurements of HTO in Water

Liquid scintillation counting techniques are the established methods for this task. They are extremely sensitive [sensitivities below 1 pCi/cm^3 (0.037 Bq/cm^3) of water], have excellent reproducibility, suffer from few interferences (especially if distillation precedes counting), and are now commercially available in a variety of sophisticated, automatic systems. The disadvantages of liquid scintillation counting include both the high capital cost of equipment and high cost of operation. Plastic scintillator systems are less expensive and are faster, but at the expense of some sensitivity. An important disadvantage of techniques using plastic is that the scintillator surface tends to degrade in time.

20.4.2. Urinalysis

A well-developed technique for urinalysis uses liquid scintillation counting methods similar to those for HTO measurements in water. Automatic analysis equipment can reach sensitivities in the range of 10 pCi/cm^3 (0.4 Bq/cm^3) of urine. A proportional counter technique sensitive in the 1000 pCi/cm^3 (40 Bq/cm^3) range has also been developed for portable field use.

20.4.3. Field Measurements of HTO in Flowing Water

Systems for continuously measuring HTO in flowing water have been developed using a flow/mixing system and liquid scintillation counting. Refinements of this technique provide a reliable, sensitive instream monitoring capability for routine reactor effluent monitoring. Present sensitivities in the 200 pCi/cm^3 (7.4 Bq/cm^3) range seem adequate for effluent measurements. They can be improved at present by using more scintillator but only at a much higher cost.

20.4.4. Continuous Field Measurements of Total Tritium in Air

Flow-through ionization chambers and proportional counters have been a mainstay for tritium-in-air monitoring for nearly 25 years. Several commercially-available instruments now exist [91]. The ion chamber suffers from various types of interferences, most notably from external gamma-ray fields; however, well-developed compensation systems now seem capable of measuring down to a few pCi/cm^3 (a few tenths Bq/cm^3) in the presence of gamma-ray fluxes in the few-mrad/hour range. Single ion chambers cannot distinguish tritium from other gaseous emitters (e.g., ^{41}Ar, ^{85}Kr), and are sensitive to cigarette smoke, ions, and aerosols unless suitably protected (this protection is now available in some commercial instruments). Gas proportional counters can partially discriminate against other radionuclides and are less sensitive to aerosols. Their sensitivity [in the few pCi/cm^3 (a few tenths Bq/cm^3) range] is comparable to that of the ion chambers. From results obtained by Gregory and Parnell [92] using a multiwire multiplane proportional chamber, it seems possible to achieve much better sensitivities. To detect levels much below about a few pCi/cm^3 (a few tenths Bq/cm^3) in air, it is necessary, for the present, to remove tritiated water vapor from the air for liquid scintillation counting or use combustion techniques.

20.4.5. HTO in Air by Removal and Liquid Scintillation Counting

There are two well-developed methods for HTO (vapor in air) determinations at very low levels, namely techniques using silica gel, or bubblers, for removal of vapor. In each case, liquid scintillation counting measures HTO in the resulting water. With silica gel, sensitivities well below 0.01 pCi/cm^3 (0.4 mBq/cm^3) of air are achievable; with bubblers the sensitivities are typically in the region of about 0.1 pCi/cm^3 (4 mBq/cm^3). These are both *integrating* techniques, capable of field sampling, but the analysis is by-batch counting in the laboratory. Interferences are few and precision is high.

20.4.6. HT in Air (Combustion)

Combustion techniques followed by HTO counting can determine tritium as tritiated hydrogen gas (HT) at the level of a fraction of a pCi/cm^3 of air. If HTO vapor is removed by dessication first, separate HT and HTO measurements can be made.

20.5. CONCLUSIONS

At present, measurements at levels well below MPC values for individuals in the general public are possible in each of the important measurement situations. In particular, laboratory liquid-scintillation analyses of batch water samples (HTO in water, HTO vapor extracted from air) are extremely sensitive and suffer few interferences. Unfortunately, these measurements are time consuming and expensive. These same comments apply to the HTO-in-water system for in-stream effluent measurements.

In air samples, continuous measurements with ion chambers and proportional counters are only sensitive in the 2 to 10 pCi/cm^3 (0.07 to 0.4 Bq/cm^3) level, which is, however, well above the MPC value of 0.2 pCi/cm^3 (0.007 Bq/cm^3) for general public exposure. More sensitive measurements in air require more specialized techniques such as vapor removal and laboratory counting.

Clearly, further development is necessary before continuous, low-level monitoring of tritium in air can be performed. Finally, the ruggedness, reliability, and maintainability of many present field instruments must be improved.

REFERENCES

1. National Council on Radiation Protection and Measurements, "Tritium in the Environment," NCRP Report No. 61, 1979, 7910 Woodmont Ave., Washington, DC 20014.
2. L. W. Alvarez and R. Cornog, "Helium and Hydrogen of Mass 3," *Phys. Rev.*, **56**(2), 613, 1939.
3. D. G. Jacobs, "Sources of Tritium and Its Behavior Upon Release to the Environment," USAEC Critical Review Series, TID-24635, 1968, USAEC Division of Technical Information Extension, Oak Ridge, TN 37830.
4. International Atomic Energy Agency, *Tritium in the Physical and Biological Sciences, Proceedings of an* IAEA Symposium, 2 vols., Vienna, 1961.
5. International Atomic Energy Agency, *International Symposium on the Behavior of Tritium in the Environment* in San Francisco, CA, 16–20 October 1978, sponsored by IAEA and OECD Nuclear Energy Agency, IAEA Vienna, available from UNIPUB as STI/PUB/498, 1979.

6. A. W. Rudolph, T. E. Carroll, and R. S. Davidson, "Tritium and Its Effects in the Environment—A Selected Literature Survey," Report BMI-171-203, 1971, Battelle Memorial Institute, 505 King Ave., Columbus, OH 43201.
7. M. N. Dixon, C. F. Holoway, B. L. Houser, and D. G. Jacobs, "Indexed Bibliography on Tritium: Its Sources and Projections, Behavior Measurement and Monitoring Techniques, Health Physics Aspects and Waste Management," ORNL-5057, 1975, Oak Ridge National Laboratory, Oak Ridge, TN 37830.
8. A. A. Moghissi and M. W. Carter, Eds., *Proceedings of the Tritium Symposium*, 30 August to 2 September 1971, Las Vegas, Nevada, EPA National Environmental Research Center, Las Vegas, NV 89114, CONF-710809, 1973.
9. National Council on Radiation Protection and Measurements, "Tritium and Other Radionuclide Labeled Organic Compounds Incorporated in Genetic Material," NCRP Report No. 62, 1979, 7910 Woodmont Ave., Washington, DC 20014.
10. W. F. Libby, "Atmospheric Helium-3, and Radiocarbon from Cosmic Radiation," *Phys. Rev.*, **69**, 671, 1946.
11. D. C. Bogen, G. A. Welford, and C. G. White, "Tritium Distribution in Man and His Environment," paper IAEA-SM-232/75, in Ref. 5.
12. W. D. Bond, "Production of Tritium by Contained Nuclear Explosions in Salt: I, Laboratory Studies of Isotopic Exchange of Tritium in the Hydrogen-Water Systems," Report ORNL-3334, 1962, Oak Ridge National Laboratory, Oak Ridge, TN 37830.
13. Energy Research and Development Administration, "Report: Light Water Reactor Fuel Reprocessing and Recycling," ERDA-77-75, Preliminary Draft on Environmental Impact Statement, 1977. Available from NTIS.
14. T. J. Bander, D. S. Renne, and W. F. Sandusky, "An Analysis of Tritium Releases to the Atmosphere by a Controlled Thermonuclear Reactor," paper IAEA-SM-232/51, in Ref. 5.
15. A. A. Moghissi, E. D. Toerber, J. E. Regnier, M. W. Carter, and C. D. Posey, "Health Physics Aspects of Tritium Luminous Dial Painting," *Health Phys.*, **18**, 255, 1970. Also, S. L. Kaufman, M. C. Colvin, R. Lieberman, A. A. Moghissi, B. F. Rehnberg, and R. C. McMillan, "Investigation of Tritiated Luminous Compounds," Report EERL 71-2, 1971, EPA Eastern Environmental Radiation Laboratory, Montgomery, AL 36101.
16. B. E. Lambert and J. Vennart, "Radiation Doses Received by Workers Using Tritium in Industry," *Health Phys.*, **22**, 23, 1972.
17. L. E. Feinendegen, *Tritium Labelled Molecules in Biology and Medicine*, Academic, New York, 1967.
18. G. Wehner, "Discharges of Tritium to the Environment due to Unrestricted Use of Consumer Products Containing This Radionuclide," paper IAEA-SM-232/9, in Ref. 5.
19. F. Combs and R. J. Doda, "Large Scale Distribution of Tritium in a Commercial Product," paper IAEA-SM-232/57, in Ref. 5.
20. A. H. Lu and J. M. Matuszek, "Transport Through a French Cover of Gaseous Tritiated Compounds from Buried Radioactive Wastes," paper IAEA-SM-232/60, in Ref. 5.
21. P. F. Sauerman, W. Friedrick, J. Knieper, and H. Printz, "Fifteen Years of Experience in Handling Tritium Problems in Connection with Low-Energy Particle Accelerators," paper IAEA-SM-232/8, in Ref. 5.

22. P. S. Rohwer and W. H. Wilcox, "Radiological Aspects of Environmental Tritium," *Nucl. Safety*, **17**(2), 216, 1976.
23. International Commission on Radiological Protection, "Recommendations of the ICRP," ICRP Publication No. 26, *Ann. ICRP*, **8**(3), 1977.
24. National Council on Radiation Protection and Measurements, "Basic Radiation Protection Criteria," NCRP Report No. 39, 1970, 7910 Woodmont Ave., Washington, DC 20014.
25. J. K. Soldat and D. A. Baker, "Worldwide Population Doses from Tritium Released from Nuclear Facilities," paper IAEC-SM-232/69, in Ref. 5.
26. R. Bergman, U. Bergstrom, and S. Evans, "Environmental Transport and Long-Term Exposure for Tritium Released in the Biosphere," paper IAEA-SM-232/47, in Ref. 5.
27. D. G. Jacobs, C. E. Easterly, and J. E. Phillips, "The Influence of the Rate of Conversion of HT to HTO on Projected Radiation Doses from Release of Molecular Tritium," paper IAEA-SM-232/81, in Ref. 5.
28. International Commission on Radiological Protection, "Limits for Intakes of Radionuclides by Workers," ICRP Publication 30, Supplement to Part 1, *Ann. ICRP*, **3**(1–4), 1979.
29. National Council on Radiation Protection and Measurements, "Maximum Permissible Body Burdens and Maximum Permissible Concentrations of Radionuclides in Air and in Water for Occupational Exposures," NCRP Report No. 22, 1959, 7910 Woodmont Ave., Washington, DC 20014, including addendum 1 issued in August 1963.
30. U. S. Nuclear Regulatory Commission, 10 CFR 20 (Code of Federal Regulations, Title 10, Part 20), "Standards for Protection Against Radiation," Appendix B, Table II, Concentrations in Air and Water Above Natural Background.
31. Environmental Protection Agency, "National Interim Primary Drinking Water Regulations," EPA-570/9-76-003, 1976, Office of Water Supply. Available from GPO.
32. National Council on Radiation Protection and Measurements, "Tritium Measurement Techniques," NCRP Report No. 47, 1976, 7910 Woodmont Ave., Washington, DC 20014.
33. L. L. Burger, "Distribution and Reactions of Tritiated Hydrogen and Methane," paper IAEA-SM-232/54, in Ref. 3.
34. R. V. Osborne, "Sampling for Tritiated Water Vapor," in *Proceedings of the Third International Congress of the International Radiation Protection Association*, W. S. Snyder, Ed., CONF-730907-P2, 1974 p. 1428.
35. J. H. Osloond, J. B. Echo, W. L. Polzer, and B. D. Johnson, "A Tritium Air Sampling Method for Environmental and Nuclear Plant Monitoring," p. 531 of Ref. 8.
36. W. R. Griffin, J. A. Cochran, and A. A. Bertuccio, "A Sampler for Non-Aqueous Tritium Gases," p. 533 of Ref. 8.
37. D. L. Horrocks and C. T. Peng, Eds., *Organic Scintillators and Liquid Scintillation Counting*, Academic, New York, 1971.
38. E. D. Bransome, Ed., *The Current Status of Liquid Scintillators*, Grune & Stratton, New York, 1970.
39. P. Chin-Tzu, D. D. Horrocks, and E. Alpen, Eds., "Abstracts for International Conference on Liquid Scintillation Counting: Recent Applications and

Developments, University of California, San Francisco, California, 21–24 August 1979," *Int. J. Appl. Radiat. Isotopes*, **31**, 1–36, 1980.
40. D. L. Horrocks, "Measuring Tritium With Liquid Scintillator Systems," p. 143 of Ref. 8.
41. A. A. Moghissi, H. L. Kelley, J. E. Regnier, and M. W. Carter, "Low Level Counting by Liquid Scintillation—I. Tritium Measurement in Homogeneous Systems," *Int. J. Appl. Radiat. Isotopes*, **20**, 145, 1969.
42. R. Lieberman and A. A. Moghissi, "Low Level Counting by Liquid Scintillation—II. Applications of Emulsions in Tritium Counting," *Int. J. Appl. Radiat. Isotopes*, **21**, 319, 1970.
43. A. A. Moghissi, "Analysis and Public Health Aspects of Environmental Tritium," in *Advances in Chemistry Series*, No. 93, *Radionuclides in the Environment*, American Chemical Society, 1970, p. 419.
44. S. B. Garfinkel, W. B. Mann, R. W. Medlock, and O. Yura, "The Calibration of the National Bureau of Standards' Tritiated-Toluene Standard of Radioactivity," *Int. J. Appl. Radiat. Isotopes*, **16**, 27, 1965.
45. A. A. Moghissi and M. W. Carter, "Internal Standard with Identical System Properties for Determination of Liquid Scintillation Counting Efficiency," *Anal. Chem.*, **40**, 812, 1968.
46. American Public Health Association, *Standard Methods for the Examination of Water and Wastewater*, 15th ed., 1980, 1015 18th Street, Washington, DC 20036.
47. American Society for Testing and Materials, *Annual Book of ASTM Standards*, Part 23, 1971, 1916 Race Street, Philadelphia, PA 19103. The tritium methods are designated D2476-70.
48. Circled "Tentative Reference Method for Measurement of Tritium in Environmental Waters," EPA-600/4-75-013, 1975, Quality Assurance Branch, Environmental Monitoring and Support Laboratory, Las Vegas, NV 89114.
49. P. Ting and R. Litle, "Continuous Monitoring of Aqueous Tritium Activity," p. 170 of Ref. 8.
50. P. Ting and M. K. Sullivan, "Reactor Coolant Monitoring with a Discrete Sampling Flow Cell-Liquid Scintillation System," Report No. 558, 1971, Beckman Instruments, Inc., Fullerton, CA 92634.
51. R. V. Osborne and N. W. Tepley, "Monitoring Tritiated Water in Air and Water Effluents," in *Radiation Instrumentation*, W. W. Wadman, Ed., in *Proceedings of the Health Physics Society Eleventh Midyear Topical Symposium on Radiation Instrumentation*, San Diego, CA, January 16–19, 1979, p.203.
52. G. Cowper, "Health Physics Instrumentation," Report AECL-3864, 1971, Atomic Energy of Canada Limited, Chalk River, Ontario, Canada.
53. W. J. Brady and V. M. Mulligan, "Tritium Exposure Experience at the Nevada Test Site," presented at the Symposium on Instrumentation, Experience, and Problems in Health Physics Tritium Control, 16 February 1967, Sandia Corporation, Albuquerque, NM 87115.
54. A. A. Moghissi, R. Lieberman, M. W. Carter, and J. E. Regnier, "Improved Radiobioassay of Urine for Tritium," *Health Phys.*, **17**, 727, 1969.
55. R. V. Osborne, "Automatic Urinalyzer for Tritium AEP-5216," Report AECL-27-2 1968, Atomic Energy of Canada Limited, Chalk River, Ontario, Canada. Also, "Performance of an Automatic Analyzer for Tritium in Urine," *Health Phys.*, **18**, 87, 1970.

56. S. Block, D. Hodgekins, and O. Barlow, "Recent Techniques in Tritium Monitoring by Proportional Counters," Report UCRL-51131, 1971, Lawrence Livermore Laboratory, Livermore, CA 94550.
57. R. P. Baker and R. D. Richards. "Reliable and Versatile Instrumentation for Detection of Tritium Hazard Designed for Military Applications," unpublished report, Sandia Laboratories, Box 5800, Albuquerque, NM 87115; Sandia Report SC-M-68-235, 1968, "A New Family of Tritium Mounting Equipment," Sandia Laboratories, Box 5800, Albuquerque, NM 87115.
58. A. A. Moghissi, H. L. Kelley, C. R. Phillips, and J. E. Regnier, "A Tritium Monitor Based on Scintillation," *Nucl. Instrum. Meth.*, **68**, 159, 1969.
59. R. V. Osborne, "Detector for Tritium in Water," *Nucl. Instrum. Meth.*, **77**, 170, 1970.
60. W. R. Kanne, "Monitoring of Gas for Radioactivity," U.S. Patent 2,599,922, June 10, 1952, filed October 12, 1944.
61. J. E. Hoy, "Operational Experience with Kanne Ionization Chambers," *Health Phys.*, **6**, 203, 1961.
62. E. C. Morris, in *Assessment of Airborne Radioactivity*, International Atomic Energy Agency, Vienna, 1967, p. 491.
63. W. L. Marter and C. M. Patterson, "Monitoring of Tritium in Gases, Liquids, and in the Environment," *Trans. Am. Nucl. Soc.*, **14**, 162, 1971.
64. R. V. Osborne, "Monitoring Reactor Effluents for Tritium: Problems and Possibilities," Report AECL-4054, Atomic Energy of Canada Limited, Chalk River, Ontario, Canada, 1971, p. 496 of Ref. 8.
65. R. V. Osborne and A. S. Coveart, "A Portable Monitor for Tritium in Air," *Nucl. Instrum. Meth.*, **106**, 181, 1973.
66. J. D. Anthony, "Portable Tritium Monitor has Gamma Compensation," *Nucleonics*, **17**(4), 110, 1959.
67. R. V. Osborne and G. Cowper, "The Detection of Tritium in Air with Ionization Chambers," Report AECL-2604, 1966, Atomic Energy of Canada Limited, Chalk River, Ontario, Canada.
68. R. A. Talbert and R. D. Hilbert, "Gamma Insensitive Air Monitor for Radioactive Gases," *Nucl. Instrum. Meth.*, **96**, 61, 1971.
69. R. V. Osborne, Chalk River Nuclear Laboratories, private communication.
70. R. A. Talbert, "A Monitor for Tritium in Air Containing Other Beta Emitters," in *Proceedings of 23rd Conference on Remote Systems Technology*, American Nuclear Society, Washington DC, 1975, p. 89.
71. J. R. Waters, "Pitfalls and Common Errors in the Measurement of Tritium in Air," Report JLI-506, Johnston Laboratories, Inc., Cockeysville, MD 21030, presented at Health Physics Society Annual Meeting, Chicago, 2 July 1970.
72. D. O. Nellis, E. L. Hudspeth, I. L. Morgan, P. S. Buchanan, and R. F. Boggs, "Tritium Contamination in Particle Accelerator Operation," USDHEW, Public Health Service Publication 999-RH-29, Washington, DC 1967, p. 35.
73. G. E. Driver, "Tritium Survey Instruments," *Rev. Sci. Instrum.*, **27**, 300, 1956.
74. R. Ehret, "Proportional Flow Counters for Measurement of Tritium in Air," *Assessment of Airborne Radioactivity*, International Atomic Energy Agency, Vienna, 1957, p. 531.
75. R. H. Howell, J. C. Cate, and C. Wong, "Separation of HT, Noble Gases and HTO with Semipermeable Membranes," *Nucl. Instrum. Meth.*, **124**, 579, 1975.

76. C. Colmenares, E. G. Shapiro, P. E. Barry, and C. T. Prevo, "A Europium-Doped, Calcium-Fluoride Scintillator System for Low-Level Tritium Detection," *Nucl. Instrum. Meth.*, **114**, 277, 1974.
77. B. Altshuler and B. Pasternak, "Statistical Measures of the Lower Limit of Detection of a Radioactivity Counter," *Health Phys.*, **9**, 293, 1963.
78. A. N. Singh and M. G. Kadwani, "A Sensitive Detector System for the Continuous Monitoring of Tritium in the Air," *IEEE Trans. Nucl. Sci.*, **NS-21**(1), 188, 1974.
79. R. V. Osborne, "Central Tritium Monitor for CANDU Nuclear Power Stations," *IEEE Trans. Nucl. Sci.*, **NS-22**(1), 676, 1975.
80. *Methods of Air Sampling and Analysis*, Intersociety Committee for a Manual of Methods for Ambient Air Sampling and Analysis, American Public Health Association, 1015 18th Street, Washington, DC 20036, 1972.
81. T. Kato, "Measurement of Tritium in Air by Absorbent," *Nucl. Instrum. Meth.*, **163**, 463, 1979.
82. A. M. Valentine, "An Investigation of a Bubbler Tritium Sampler," Report LA-3916, 1968, Los Alamos Scientific Laboratory, Los Alamos, NM 87544.
83. R. J. Straight, "HT-HTO Sampling at the Nevada Test Site," paper IAEA-SM-232/87, IAEA/NEA International Symposium on the Behaviour of Tritium in the Environment, October 16–20, 1978, San Francisco, California, to be published.
84. H. G. Ostlund, "A Rapid Field Sampling for Tritium in Atmospheric Hydrogen," 1970, Report ML 70075, Rosenstiel School of Marine and Atmospheric Sciences, University of Miami, Miami, FL 33149. Also, H. G. Ostlund, M. O. Rinkel, and C. Rooth, "Tritium in the Equatorial Atlantic Current System," *J. Geophys. Res.*, **74**, 4535, 1969.
85. J. E. Hoy, "Tritium Enrichment by Gas-Solid Chromatography: Technique for Low-Level Analysis," *Science*, **161**, 464, 1968. Also, D. W. Hayes and H. E. Hoy, "A Chromatographic System for the Enrichment and Analysis of Low-Level Tritium Samples," p. 127 of Ref. 8.
86. H. Perschke, M. Crespi, and G. B. Cook, "A System for the Enrichment of Low-Level Tritium in Large Volume Hydrogen Samples by Gas Chromatography," *Int. J. Appl. Radiat. Isotopes*, **20**, 813, 1969.
87. P. E. Hartley, "Design and Performance of Tritium Measurement Systems Using Electrolytical Enrichment," *Nucl. Instrum. Meth.*, **100**, 229, 1972.
88. H. G. Ostlund and E. Werner, "The Electrolytic Enrichment of Tritium and Deuterium for Natural Tritium Measurements," p. 95, Vol. I of Ref. 4.
89. C. Yamaguchi, "Measurement of Low Level Tritium Concentration in Water by the Method of Electrolysis Enrichment and Ionization Chamber," KEK-76-13, 1976, National Laboratory for High Energy Physics, OHO-Machi, Tsukuba-gum, Ibaraki, Japan.
90. B. Verhagen, "Rapid Isotope Enrichment of Gases by Thermal Diffusion for Nuclear Dating," in *Radioactive Dating and Methods of Low-Level Counting*, International Atomic Energy Agency, Vienna, 1967, p 657.
91. D. L. Murphy, "Analysis of Low Energy Beta Emitters," *Radioelement Analysis: Progress and Problems*, W. S. Lyon, Ed., in *Proceedings of the Twenty-Third Conference on Analytical Chemistry in Energy Technology*, Ann Arbor Science Publishers, Ann Arbor, 1980.

92. J. C. Gregory and T. A. Parnell, "Proportional Counter for Monitoring Low Levels of Krypton-85 and Other Beta and X-Ray Emitting Nuclides," *IEEE Trans. Nucl. Sci.*, **NS-23**(1), 715, 1976.

APPENDICES

APPENDIX A

A.1. EXPLANATION OF INSTRUMENT NOTES

This appendix describes instruments that are currently available. Each note refers to one or a group of similar instruments. All the instruments included are listed in the Summary Sheets which appear in Appendix B.

The Summary Sheets are useful for locating an instrument as well as for making comparisons between instruments.

Table A.1 gives the format used for the Instrument Notes. The *generic name* of the instrument is that given by the manufacturer, such as survey meter, air monitor, liquid monitor. The heading *class* indicates whether the instrument is portable or only for use in a laboratory. The *principle of operation* states the kind of detector used, whether gas-filled (Geiger-Müller, ionization, or proportional), scintillator, semiconductor or other; also the associated electronics that are supplied or needed are indicated. *Sensitivity and range* gives information on the range of applicability, whether the instrument measures integrated dose and/or dose rate, the energy range of detection for the radiation measured, and the degree to which a reading represents the true value, as well as any energy dependence it may have. *Sampling* indicates whether the instrument functions continuously or intermittently, as well as whether sample preparation of special sampling procedures may be required. *Performance* gives specifications on temperature, humidity, timing and/or background radiation response; other information such as its behavior in a high radiation field, the signal size outputs, whether analog or digital output, and so forth, may also be found. *Requirements* provides the size, weight, and power requirements of the instrument. In *features*, available options are given, as well as any special information that does not fit into the previous categories. *References* gives the source of information for compiling the note. The *cost* is the latest price available. The *address* gives the current mailing address for the manufacturer's main office, the phone number, and sometimes the name of a person to contact for more information. The first page of each Instrument Note also gives a date (in the upper right-hand corner) indicating the last alteration based on information obtained from the indicated references.

Table A.1. Instrument Note Format

Generic Name of Instrument
Model Name and Number
Glossy Photograph (when available)
Class
Principle of Operation
Sensitivity and Range
Sampling
Performance
Requirements (Power, Size, Weight)
Features
References
Cost
Address

A.2. LOCATION OF INSTRUMENT NOTES

The instruments have been placed in seven sections:

Combination instruments
Instruments for measuring alpha radiation
Instruments for measuring beta and beta/gamma radiation
Instruments for measuring gamma and x-radiation
Instruments for measuring neutrons
Dosimeters
Instruments for measuring specific radionuclides.

In order to make a particular note easier to find, each note has a mnemonic expression in its upper right-hand corner. The first line of the mnemonic identifies the section. The identifiers for each section are:

Combination instruments—RAD-ALP,BET,GAM,XRA
Instruments for measuring alpha radiation—RAD-ALP
Instruments for measuring beta and beta/gamma radiation—RAD-BET,GAM
Instruments for measuring gamma and x-radiation—RAD-GAM,XRA
Instruments for measuring neutron—RAD-NEU
Dosimeters—RAD-DOS
Instruments for measuring specific radionuclides—RAD-NUC

Instruments appearing in the combination section may monitor radiations other than alpha, beta, gamma or x-radiation.

The second line of the mnemonic names the manufacturer. The notes in each section are arranged alphabetically according to manufacturer and a numeral identifies whether the note is the first, second (etc.) note for that particular manufacturer. In the NUC section the second line identifies the specific radionuclide measured and the third line identifies the manufacturer. The last line of the mnemonic gives the date of the last information received from the manufacturer.

The reader is also referred to the summary sheets (Appendix B) for a listing of the instruments in each section.

From these two sources of information the reader should be able to find instruments for a given application. Although the attempt was made to make this compilation complete, some manufacturers are not included because of incomplete response or lack of response. In some cases, the reader may wish to use this volume in conjunction with compilations by professional societies; although giving less detail on specific instruments, these sources typically include up-to-date lists of instruments and their manufacturers. For example, *Analytical Chemistry* publishes annually a Lab Guide listing instruments, equipment, and supplies available commercially in the United States. Another source is the annual Buyer's Guide published by *Nuclear News*.

A.3. COMBINATION INSTRUMENTS

This short section is intended only to point out where the information exists in this volume for understanding the operating characteristics, as well as the components (detectors, electronics, etc.), of combination instruments. Combination instruments are simply those instruments capable of measuring a variety of radiation types. Unfortunately, due to the widely varying properties of the different ionizing radiations, a single instrument cannot simultaneously determine and/or differentiate α, β, γ, x, and neutron radiation at the same time. For a discussion of the physical properties of radiation, as well as the basic detection methods, the reader is referred to Chapter 4, "Radiation Detection Methods," and the references mentioned therein.

Normally, combination instruments are composed of several components. Basically they tend to consist of a single electronics unit, with a variable high voltage supply and a count rate meter or other readout device, and an assortment of interchangeable probes or detectors. Some basic electronic components are briefly discussed in Chapter 4. To gain insight into the operating characteristics and possible deficiencies with specific detectors for the different types of radiation, the reader is referred to the following discussions and the references contained therein:

- Chapter 5, Measurement of Gamma and X-Rays
- Chapter 6, Gamma- and X-Ray Spectrometry
- Chapter 7, Measurement of Alpha and Beta Particles
- Chapter 8, Measurement of Neutrons
- Chapter 9, Dosimetry

The simplest combination instrument is an ionization or G–M detector having a thin window and a sliding thicker window cover to distinguish between beta and photon (x and gamma) radiation. Unfortunately, other such simple compact instruments do not exist for differentiating such combinations as neutrons and photons, alpha and neutrons, or making further differentiation in a more complex radiation environment. For this reason, it is wise for the user of radiation measuring instrumentation to be aware of the types and energy spectrum of radiation that he will normally encounter.

Information regarding the type of radiation normally encountered at several sources is discussed in several chapters. These sections attempt to give an insight into where radiation will appear and its type, as well as an idea of the energy spectrum one can normally expect to find. Such discussions appear in the following chapters:

- Chapter 3, Radiation Sources
- Chapter 10, Nuclear Power Plants
- Chapter 11, Nuclear Spent Fuel Processing
- Chapter 12, Mining and Milling of Uranium Ores

Radiation Counter Tubes
Amperex Electronic Corporation

Class	Detector
Principle of Operation	Geiger-Müller detectors used to count alpha, beta, gamma, and x-ray
Features	See Tables A.3.1 and A.3.2
References	Manufacturer's specifications
Address	Amperex Electronic Corporation 230 Duffy Avenue Hicksville, NY 11802 (516) 931-6200 Attn: Richard Gumbrecht

Table A.3.1. Cylindrical Counters

Model	Outside Diameter		Overall Length		Operating Voltage (V)	General Application	Price ($)
	(mm)	(in.)	(mm)	(in.)			
ZP1300/18529	7	0.276	27	1.063	500–650	γg count or current; $\leq$ 1000 R/hr, $\beta > 0.5$ MeV	75.00
ZP1310/18509	7	0.276	38	1.496	500–650	γg count or current; $\leq$ 300 R/hr, $\beta > 0.5$ MeV	68.00
ZP1320/18550	10	0.394	52	2.047	500–650	$\beta > 0.25$ MeV; γ-count or current	74.00
ZP1200/18503	17	0.667	55	2.165	375–600	γ	110.00
ZP1210/18520	22.2	0.874	170	6.693	375–475	γ	149.00
ZP1220/18545	22.2	0.874	270	10.63	380–480	γ	198.00
ZP1220/560N	8.7	11/32	88.9	3 1/2	850–1050	γ, High temp.	500.00
ZP1220/171G	8.7	11/32	482.6	19	700–850	γ	425.00
ZP1220/181G2	7.1	9/32	112.7	4 7/16	650–900	γ	325.00
ZP1220/90NB3	15.9	5/8	141.3	5 9/16	875–1075	β,γ	70.00
ZP1220/76NB3	15.9	5/8	192.5	7 9/16	875–1075	γ	40.00
ZP1220/912NB3	15.9	5/8	285.4	11 1/4	875–1075	β,γ	134.00
ZP1220/78L	15.9	5/8	279.4	11	450–700	γ	120.00
ZP1220/79L	15.9	5/8	362	14 1/4	450–700	γ	132.00

Table A.3.2. End-Window Counters

Model	Outside Diameter (mm)	Outside Diameter (in.)	Overall Length (mm)	Overall Length (in.)	Operating Voltage (V)	General Application	Price ($)
ZP1400/18504	17	0.669	55	2.17	375–600	β,γ	101.00
ZP1410/18505	25.9	1.02	57	2.24	450–700	α,β,γ	310.00
ZP1430/18526	34	1.039	57	2.24	450–750	α,β,γ	360.00
ZP1220/200NB	34.9	1 3/8	110.3	4 11/32	875–1075	α,β,γ	90.00

**Automatic Planchet Counting System
Baird Model 912-000**

Class	Laboratory
Principle of Operation	Gas flow proportional detector, automatic planchet sample changer, spectrometer, and printer
Sensitivity	Detector efficiency: Alpha, 35 to 40% Beta, 25% Typical background: Alpha, 0.2 cpm Beta, 2 cpm
Sampling	Batch-up to 100 samples automatically
Performance	Detector: 2-in. diam. with thin window (0.9 mg/cm^2) or windowless operation; plateau, 200 to 300 V; slope, 2% per 100 V Planchet changer: sampling capacity, 100 samples (planchet sizes, 1 in., 1 1/4 in., and 2 in., mixable); total sample changing time, ~12 sec Spectrometer: resolving time, 1 μsec pulse pairs; maximum continuous count rate, 1 MHz; count capacity, 999,999; preset count, all decades presettable 0 to 9 Time ranges: 0.1 to 999.9 sec or min Time capacity: 4 decades presettable 0 to 9 Current input mode: maximum gain, 10 V/μa; input sensitivity, 1μa full scale (FS); input impedance, 50 to 100 Ω Voltage input mode: maximum gain, 1000; input sensitivity, 10 mV (FS); input impedance, 10,000 Ω; voltage range 0 to 10 V High voltage supply: outputs: 200 to 1600 V dc at 5 mA maximum; 1600 to 3000 V dc at 5 mA maximum; polarity, positive
Requirements	Power: 110 V ac, 60 Hz Size: Changer Model 912-600, 43.3 cmW × 48.3 cmH × 53.3 cm D (17 in. × 19 in. × 21 in); Spectrometer Model 987-514, 43.7 cmW × 17.8 cmH × 38.1 cmD (17.5 in. × 7 in. × 15 in.); Printer Model 987-516; 21.6 cmW × 17.8 cmH × 38.1 cmD (8.5 in. × 7 in. × 15 in.) Weight: Changer, 61.7 kg (137 lb); Spectrometer, 20.4 kg (45 lb); Printer, 16.2 kg (36 lb)

Features	Lead shielding—can be used with G–M, scintillation or gas flow proportional detectors; counting gas required, but not included
References	Manufacturer's specifications
Cost	\$16,750 with matrix printer \$18,000 with teletype ASR 33 \$18,750 with calcium fluoride detector
Address	Baird Corporation 125 Middelsex Turnpike Bedford, MA 01730 (617) 276-6204 Attn: A. Doherty

Automatic Planchet Counting System
Baird Model 912-005

Class	Laboratory
Principle of Operation	Non-Gas, phoswich detector, automatic planchet sample changer, electronics and printer
Sensitivity	Detector efficiency: Alpha – 40% (^{210}Po) Beta – 45% (^{90}Sr), 12% (^{14}C) Gamma – 25% (^{137}Cs) Typical background: Alpha – $>$ 0.2 cpm Beta – $>$ 2.0 cpm
Sampling	Batch-up to 100 samples automatically 2 in., 1 in., 1 1/4 in. diameter planchets intermixable
Performance	Detector: phoswich, 2 in. diameter consisting of CaF^2 (Eu) sample detector and NaI(Tl) guard detector Planchet Changer: sampling capacity, 100 samples (planchet sizes 1 in., 1 1/4 in. and 2 in. mixable), total sampling time—8 seconds, shielding 4 in. virgin lead surrounding detector Spectrometer: contains high voltage power supply variable, 200–3000 Volts amplifier analyzer; coincidence detector module; register multiplexer; printer or TTY Interface; scaler timer; resolving time, 1 μsec pulse pairs; maximum continuous count rate, 1 megahertz count capacity: 999,999 preset counts: all decades presettable 0 to 9 Time Ranges: 0.1–999.9 seconds or minutes Time Capacity: 4 decades presettable 0 to 9 High Voltage Supply: output: 200–1600 V dc at 1 mA maximum: 1600–3000 V dc at 1 mA maximum: polarity, negative
Requirements	Power: 110 V ac, 60 Hz Changer Model 912-600: 61 cm W × 61 cm H × 61 cm D (24 in. × 24 in. × 24 in.) Spectrometer: 44.5 cm W × 18 cm H × 38.1 cm D (17 1/2 in. × 7 in. × 15 in.) Printer: 20.3 cm W × 18 cm H × 38.1 cm D (8 1/2 in. × 7 in. × 15 in.) Weight: Changer: 681 kg (1500 lb) Spectrometer: 20.4 kg (45 lb) Printer: 6.8 kg (15 lb)

Features	Non-Gas, scintillation detector (phoswich) modular
References	Manufacturer's specifications
Cost	$25,000 with printer
Address	Baird Corporation 125 Middlesex Turnpike Bedford, MA 01730 (617) 276-6204 Attn: A. Doherty

Laboratory Monitor
Baird Model 450

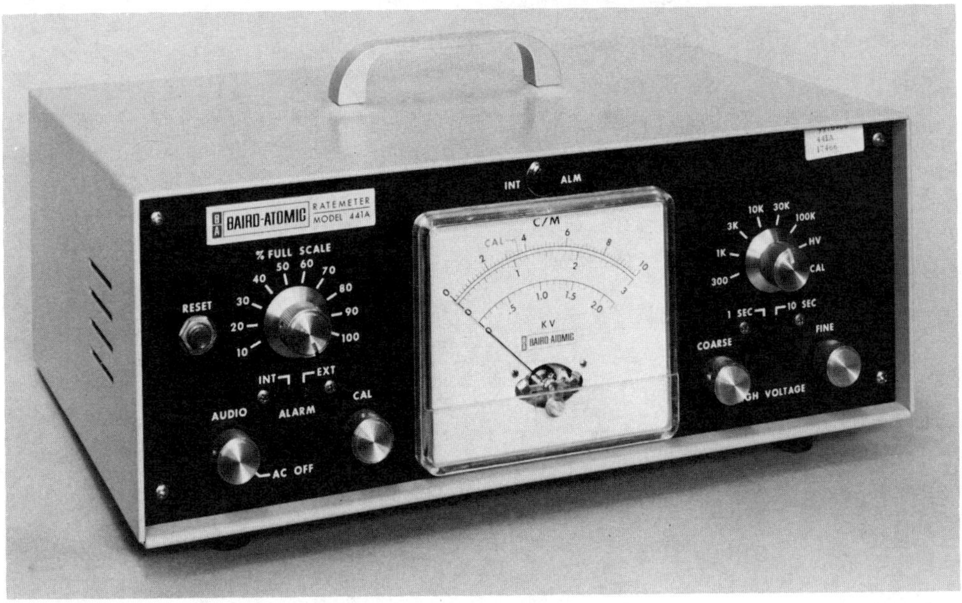

Class	Stationary
Principle of Operation	Uses G-M detector, gas flow counters, or scintillation detectors, amplifier, high voltage supply, meter readout, red alarm light, speaker
Sensitivity and Range	Range: 0–100, 300, 1,000, 3,000, 10K, 30K, 100K, 300K, 1M cpm Input sensitivity: minus 0.5 to 10 V, adjustable
Sampling	Continuous
Performance	High voltage supply: 0 to 1500 V, regulated 9.1% per 1% change in line voltage Recorder output: 100 mV Accuracy: better than 3% (FS), all ranges 3% at recorder outputs Time constants: 0.1, 1 sec, 10 sec
Requirements	Power: 110 V, 60 cycle, 10 W (220/50 optional) Size: 13.1 cm H × 28.2 cm W × 25.4 cm D (5.2 in. × 11.1 in. × 10 in.) Weight: 5.9 kg (13 lb)

Features	An alarm trip set for 10 to 100% full scale, all ranges; a red light flashes. Battery standby in case of power failure or portable use
References	Manufacturer's specifications
Cost	$1700.00
Address	Baird Corporation 125 Middlesex Turnpike Bedford, MA 01730 (617) 276-26000 Attn: A. Doherty

Linear Log Ratemeter
Baird Model 470

Class	Stationary
Principle of Operation	Uses G-M detector, gas flow detectors or scintillation detectors, high voltage supply, meter readout, speaker
Sensitivity and Range	Range: 0 to 10^6 cpm linear (9 ranges) 10 to 10^6 cpm log (5 decades) Input sensitivity: current input (G-M tube) 100 mA. Full scale preamplifier input (7 pin connector) 1 V full scale—impedance 5K
Sampling	Continuous
Performance	High Voltage Supply: 200 to 1500 V Maximum Load: 1.0 mA Load Regulation: 1% Ripple less than 2 volts Time Constants: 0.1, 1 and 10 seconds
Requirements	Power: 115 V ac ± 15%, 50/60 Hz Size: 21.5 cm W × 23 cm D × 16 cm H (8 1/2 in. × 9 in. × 6 1/4 in.) Weight: 3.6 kg (8 lb)
Features	Recorder output Background Suppression: 0 to 100% Discriminator: 0 to 100% Internal Speaker Multiple detector types
References	Manufacturer's specifications
Address	Baird Corporation 125 Middlesex Turnpike Bedford, MA 01730 (617) 276-6204 Attn: A. Doherty

Low Level $\alpha/\beta/\gamma$ Counting System
Canberra Models 2400 Automatic (α,β), 2401 (α,β), and 2408 (γ option)

Class — Laboratory

Principle of Operation — Sealed proportional sample counter with a sealed anti-coincidence guard detector to provide extremely low background sample counting. Provides automatic background crosstalk, concentration calculations and used defined experiment sequences.

It employs two main counting channels for alpha and beta events. Energy discrimination is accomplished through a programmable single channel analyzer. Cosmic and other background events are excluded from either sample channel through anti-coincidence circuitry connected to the guard detector.

When alpha, beta and gamma monitoring is required for the same sample, a Na(Tl) detector must be incorporated in the shield assembly.

Detectors and Performance

	Diameter	Efficiency	Background
Sealed	5.7 cm (2.25 in.)	Alpha ≥ 18% for ^{241}Am	≤ 0.1 cpm
		Beta ≥ 42% for ^{90}Sr	≤ 1.7 cpm
Flow	5.7 cm (2.25 in.)	Alpha ≥ 30% for ^{241}Am	≤ 0.1 cpm
		Beta ≥ 40% for ^{90}Sr	≤ 1.3 cpm

Guard Detector. Size 14.6 cm×16.2 cm (5.75 in.×6.375 in.)
Efficiency >99% cosmic detection.

Requirements

Power: 105–125 Volts or 210–250 Volts, 50–60 Hz, less than 90 watts.

Dimensions
2400: 61 cm × 76 cm × 70 cm (24 in. × 30 in. × 28 in.)
2401: 46 cm × 50 cm (18 in. × 20 in.)

Weight including shielding:
2400: 680 kg (1500 lb)
2401: 545 kg (1200 lb)

Flow Gas: Research grade proportional gas (90% Argon–10% Methane) (P-10)

Features	Gamma detector option (Model 2408). CRT data display and status functions. Automatic plateau and discriminator functions. Model 2400 also offers: 50 sample capacity (optional 100 sample capacity) Optical character reader planchet identification. Battery back-up of system parameters.
Cost	Model 2400 Automatic (α,β): $27500 Model 2401 ($\alpha$,$\beta$): $14500 Model 2408 ($\gamma$ option): $2450

Rascal
Ratemeter Scaler
Eberline Models PRS-1, PRS-1P, PRS-2, And PRS-2P

Class	Portable. May also be used as laboratory scaler.
Principle of Operation	Operates with most Eberline detector probes to measure alpha, beta, gamma and neutron radiation. Readout units internally selectable.
Sensitivity and Range	Functions as ratemeter or scaler. In ratemeter mode, integrates to preselected total counts and display rate. Digital readout with six decades. Decimal placement and calibration factor internally adjustable. For use with G–M or scintillation detectors; PRS-1P and PRS-2P also may be used with proportional detectors.
Sampling	Continuous
Performance	High voltage to probe: 500 to 1500 V for PRS-1 and PRS-2; 800 to 2400 V for PRS-1P and PRS-2P

	Temperature range: −18 to 60°C (0 to 140°F) No humidity dependence
Requirements	Power: Rechargeable Gel Cell, 75 hrs continuous operation/charge Size: 10 cm W × 24 cm L × 20 cm high (approx. 4 in. × 9 in. × 8 in.) Weight: 2.3 kg (5 lb.)
Features	"Smart" electronics integrates to preset level of precision desired then displays rate. Calibration factor for each probe and decimal point placement can be quickly set during probe change. Displays digital reading and applicable units.
References	See Eberline General Catalog
Cost	$1850 with Gel Cell
Address	Eberline Instrument Corporation P.O. Box 2108 Santa Fe, NM 87501 (505) 471-3232

Alpha, Beta, Gamma Survey Meters
Eberline Portable Models

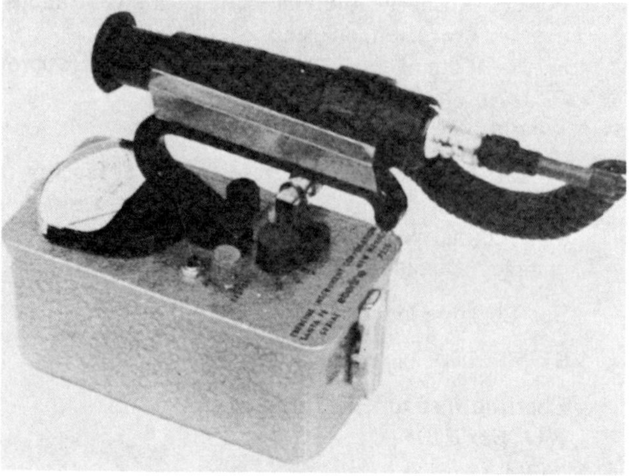

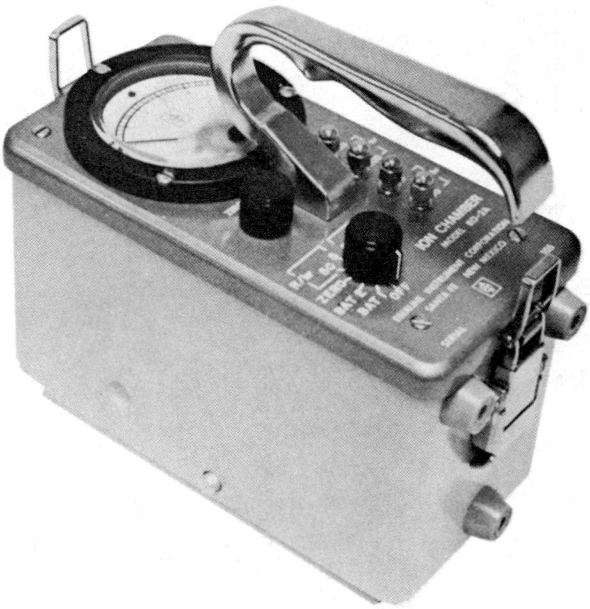

Description See Table A.3.3

Address Eberline Instrument Corporation
P.O. Box 2108
Santa Fe, NM 87501
(505) 471-3232

Table A.3.3. Eberline, Beta, and Gamma Survey Meters

	Model				
Specification	RO-2 (2A)	RO-3	RO-4	E-120/HP-270	E-520/HP-270
Radiation Measured	Beta, gamma x-ray	Beta, gamma x-ray	Beta, gamma x-ray	Beta, gamma x-ray	Beta, gamma x-ray
Detector	Juno type Ion Chamber 208-cm^3 volume	Cutie-pie type Ion Chamber 440-cm^3 volume	Juno type Ion Chamber 208-cm^3 volume	G-M, Energy Compensated	G-M, Energy Compensated
Range	0–50 mR/hr, 0–50 R/hr 4 ranges	0–5 to 0–5000 mR/hr, 4 ranges	0.1 mR/hr to 200 R/hr, 0.1 mR to 2000 mR (integrated range)	0–0.5 to 0–50 mR/hr, 3 ranges	0–0.2 to 0–2000 mR/hr, 5 ranges
Scale	Linear	Linear	Digital	Linear	Linear
Meter Accuracy	±5% FS	±5% FS	NA	±5% FS	±15% FS
Temperature	−40 to 60°C (−40 to 140°F)	−40 to 60°C (−40 to 140°F)	0 to 50°C (32 to 122°F)	−40 to 60°C (−40 to 140°F)	−40 to 60°C (−40 to 140°F)
Power	3 NEDA 1604 (9 V) batteries	3 NEDA (9 V) batteries	3 NEDA 1604 (9 V) and 2 NEDA 210 (30 V) batteries	Two "D" batteries	Two "D" batteries

Weight	1.75 kg (3.8 lb)	1.81 kg (4 lb)	1.59 kg (3.5 lb)	1.36 kg (3 lb)	2 kg (4.4 lb)
Size (W×L×H) Including Handle	10 × 21 × 20 cm (4 × 8 × 8 in.)	10 × 29 × 22 cm (4 × 11 × 9 in.)	10 × 21 × 19 cm (4 × 8 × 7 in.)	9 × 17 × 16 cm (4 × 7 × 6 in.)	10 × 20 × 17 cm (4 × 8 × 7 in.)
Features	Window 7 mg/cm^2 Mylar, wall 200 mg/cm^2 phenolic	Window 3.5 mg/cm^2 Mylar, wall 200 mg/cm^2 phenolic	Digital, measure integrated exposure and exposure rate	Reads out in mR/hr and cpm	Internal G-M tube for highest range
Energy responses ±20% or better	12 keV to 6 MeV	12 keV to 6 MeV	12 keV to 6 MeV	40 keV to 1.3 MeV	40 keV to 1.3 MeV
Cost	$950.00	$995.00	$900.00	$360.00	$550.00
Sampling	Continuous				
Features	Flat response (±20% or better) versus photon energy for measuring exposure or exposure rate. All also measure beta. RO instruments can be made sensitive to alpha.				

Survey Meter
Elscint Model GSM-1

Class	Portable, handheld
Principle of Operation	G-M detector, amplifier, meter readout, speaker
Sensitivity	Ranges: 0 to 0.5 mR/hr, 0 to 5 mR/hr, 0 to 50 mR/hr, 0 to infinity (logarithmic) Energy dependence: ±20% from 60 keV to 1.3 MeV
Sampling	Continuous
Performance	Accuracy: ±3% FS Temperature: −35 to 60°C (−31 to 140°F)
Requirements	Power: Four 1.5 V batteries, "C" type, 100 hr continuous service Size: 16.5 cm H × 10.2 cm W × 19.1 cm D (6 1/2 in. × 4 in. × 7 1/2 in.) Weight: 1.1 kg (2.4 lb)
Features	1. In very high radiation fields meter indicates full scale, not zero 2. Geiger tube may be operated with a cable at a distance up to 50 m from the reading instrument.
References	Manufacturer's specifications

Cost	$800.00, plus shipping
Address	Elscint, Inc. 138-160 Johnson Ave. Hackensack, NJ 07602 (201) 487-5885 Attn: Hanan Rotman

RAD-ALP,BET,GAM,XRA
Gamma 1
January 1983

Radiation Monitor
Multiple Alarm Radiation System
Gamma Industries Catalog No. 821-0091-002

Class	Stationary
Principle of Operation	G-M detector, amplifier, meter readout, audible alarm
Sensitivity and Range	Range: 0–2, 0–20, 0–200 mR/hr
Sampling	Continuous
Performance	Dead time: 75 μsec. Operating voltage: 500–600 V Meter scale: 0–20 mR/hr
Requirements	Power: 115 V ac, 60 Hz Size: 36.0 cm H × 33.0 cm W × 22.0 cm D (14 in. × 13 in. × 9 in.) Weight: 6.1 kg (13 1/2 lb)
Features	7-in. high flashing, rotating beacon warns of radiation in excess of the level selected. Pulsating 2900 Hz tone.
References	Manufacturer's specifications

Cost	\$710.00 with visual alarm \$790.00 with visual and audible alarm
Address	GAMMA Industries, A Division of Nuclear Systems, Inc. P.O. Box 2543 Baton Rouge, LA 70821 (504) 388-0800 Telex: 586473 Attn: Sales Manager

Automatic Alpha/Beta Counting System
Harshaw Model 12-A6

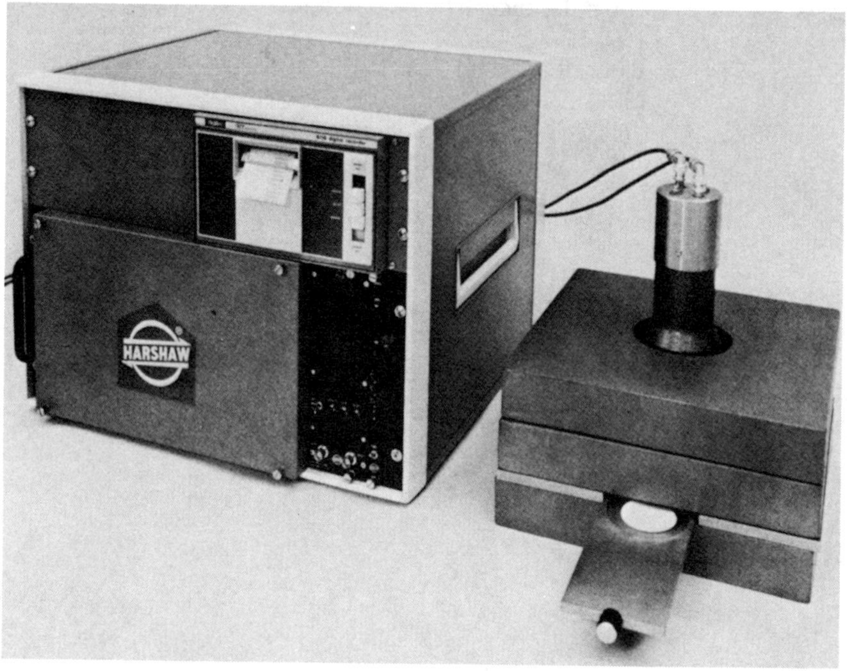

Class	Designed for laboratory use.
Principle of Operation	Utilizes a CaF_2(Eu)-NaI(Tl) Phoswich scintillator and pulse-shaping electronics to simultaneously and separately measure alpha, beta, and gamma radiation. Readout is LCD and printer.
Sensitivity Range and Performance	The following specifications are for separate and simultaneous alpha and beta counting: Detector diameter: 1 in.—alpha background, $>$ 0.05 cpm; beta background, $>$ 0.7 cpm 2 in.—alpha background, $>$ 0.1 cpm; beta background, $>$ 2 cpm Detector efficiency: alpha, $<$ 35% for ^{241}Am; $<$ 35% for ^{210}Po beta, $<$ 40% for ^{90}Sr/^{90}Y; $<$ 40% for ^{137}Cs; $<$ 40% for ^{14}C
Sampling	Automatically measures up to 100 planchets with samples.

Requirements	Power: 110 V, 60 Hz or 220 V, 50 Hz; 50 watts. Size: sample changer: 61 cm × 61 cm × 61 cm (24 in. × 24 in. × 24 in.) Weight: 675 kg (1500 lb) Electronics size: 50 cm × 50 cm × 38 cm (19.75 in. × 19.75 in. × 15 in.) Weight: 42 kg (93 lb)
Features	Model 12-A6 is housed in two cabinets, one is for electronics and the other for the sample changer and lead shielding. Reproducible background, pushbutton controls, high counting efficiency, detector uniformity, large Phoswich scintillator available.
References	Harshaw technical literature
Cost	12-A6: $26,6500.00 each
Address	Harshaw Chemical Company 6901 Cochran Road Solon, OH 44139 (216) 248-7400

Nuclear Radiation Detectors
Harshaw Phoswich Detectors

Class	Nuclear radiation detector (alpha, alpha/beta, alpha/gamma)
Principle of Operation	A Phoswich detector is a scintillator utilizing a thin primary crystal optically coupled to a thicker crystal of a different scintillation material. For alpha counting applications, CaF_2(Eu) is employed as the primary scintillator, with a thicker NaI(Tl) crystal operating as an anticoincidence shield. Signals originating in the crystals are electronically separated by the difference in scintillation decay times of each material. The calcium fluoride crystal is very efficient for alpha particles, yet transmits higher-energy gamma ray events to the NaI(Tl) crystal for electronic rejection.
Resolution	Typically 1 MeV for ^{241}Am
Diameters Available	1.9 to 12.7 cm (3/4 to 5 in.)
Features	Because calcium fluoride is relatively inert and nonhygroscopic, no entrance window is required between the crystal and sample in many applications. Passive shielding requirements are greatly reduced by the anticoincidence operation of the NaI(Tl) component.
Accessories Available	Low-background tube bases, low-background shields for table-top use, complete selection of nuclear instrument modules and pulse-shape analysis electronics.

References	Manufacturer's specifications
Cost	Refer to factory
Address	The Harshaw Chemical Company Crystal and Electronic Products Dept. 6801 Cochran Road Solon, OH 44139 (216) 248-7400

Remote Area Monitoring System
Health Physics Model 2010

Class: Stationary

Principle of Operation: Sealed tissue equivalent ionization chamber. With a tissue equivalent plastic detector-wall (A-150 or A-174 Shonka Plastic 0.07 in. thick). The gas used is a tissue equivalent mixture at a pressure of one atmosphere.

Sensitivity and Range:

Detector Volume (cm^3)	1031	
Range	0.1–9999.9 mrad/hr or 1–99999 mrad/hr	
Collection Voltage (V)	1000	
Detector Volume (cm^3)	25.8	5.2
Range	0.1–9999.9 rad/hr	0.01–999.99 krad/hr
Collection Voltage (V)	1000	300

Sampling: Continuous

Requirements: Power: 110 to 120 V ac, 47 to 440 Hz, 1 amp fuze
Dimensions:

Head Unit	
without bracket	with bracket
15 cm × 15 cm × 33 cm (6 in. × 6 in. × 17 in.)	15 cm × 20 cm × 43 cm (6 in. × 8 in. × 17 in.)

RAMS unit	
without bracket	with bracket
33 cm × 25 cm × 11.5 cm (13 in. × 10 in. × 4.5 in.)	33 cm × 25 cm × 28 cm (13 in. × 10 in. × 11 in.)

Weight:
Head unit: 4 kg (9 lb)
RAM unit: 5.9 kg (13 lb)

Features: Display: five digit LED 1 in. characters, behind red filter
Display update period: 3.95 seconds
Alarm: visual-two lamp red, one dim long life, Audio-Piezo electric, 85 dbA at 10 ft @ 2.8 kHz steady or intermittent

Options: Remote readout, remote alarms, remote detectors

Cost	$6000 per channel
Address	Health Physics Instruments, Inc. 124 Santa Felicia Drive Goleta, CA 93117 (805) 685-2612 Attn: J. Handloser

Radiation Survey Meter
Johnson and Associates Model GSM-10

Class	Portable, handheld
Principle of Operation	Alphas, betas, gammas, and neutrons are detected by various probes. Solid state. (See Tables A.3.4, A.3.5 and A.3.6).
Sensitivity and Range	Full scale ranges: 0.2, 2, and 20 mR/hr for beta-gamma; 500, 5000, and 50,000 cpm, linear ±4% Minimum measurable energy: Dependent upon the probe used. Approximately 15 keV with GP-200 thin window G–M tube Maximum measurable flux: Dependent upon the probe used. Approximately 1 R/hr with GP-90 X-50
Sampling	Continuous
Performance	Meter accuracy: ±2% FS Response time: Based on scale range being used. For 90% final reading: 2 sec on ×-1 scale, 1 sec on ×-10 scale and 0.9 sec on ×-100 scale. Input accessories: Calibration pulser, trans amp, register, and timer Output accessories: Recorder, speaker, alarm
Requirements	Power: Two "D" cells, 200 hr Size: 9 cm W × 9 cm H × 18 cm D (3.5 in. × 3.5 in. × 7 in.) Weight: 1.5 kg (3 lb)
Features	Battery check circuit, wide selection of detectors.
References	Manufacturer's specifications
Cost	$285.00
Address	William B. Johnson and Associates, Inc. Research Park, Boonton Avenue Montville, NJ 07045 (201) 334-9222

Table A.3.4. Scintillation Probes for Johnson Model GSM-10 Survey Meter[a]

	Model	
Specifications	ASP-2A	GSP-2A
Radiation Measured	Alpha	Gamma
Detector	ZnS	2×2 in. NaI(Tl)
Efficiency	25 to 40%	25%
Range	1 to 50,000 cpm	50,000 cpm
Accuracy	±5%	±5%
Sensitivity	100 pCi	100 pCi
Temperature	−40 to 65°C (−40 to 150°F)	−40 to 65°C (−40 to 150°F)
Power	Four 6-V penlight batteries (AA)	Four 6-V penlight batteries (AA)
Housing	Aluminum	Aluminum
Overall Size	22.9 cm long 6.0 cm o.d.	30.5 cm long 6.0 cm o.d.
	(9 in. long, 2 3/8 in. o.d.)	(12 in. long, 2 3/8 in. o.d.)
Weight	0.68 kg (1.5 lb)	0.91 kg (2 lb)
Price	$275.00	$595.00

[a]GSM-10 sells for $285.00. Add probe cost for complete instrument.

Table A.3.5. Geiger-Müller Probes for Johnson Model GSM-10 Survey Meter[a]

	Model				
Specifications	GP-90	GP-200	PPA-2	GP-90 X-50	GP-200 X-50
Radiation Measured	Beta Gamma	Alpha Beta Gamma	Alpha Beta Gamma	Beta Gamma	Alpha Beta Gamma
Detector	G-M Tube	G-M Tube	G-M Tube	G-M Tube	G-M Tube
Wall Thickness	30 mg/cm^2	1.4 mg/cm^2	1.5 mg/cm^2	30 mg/cm^2	1.5 mg/cm^2
Range Full Scale	0.2, 2, 20 mR/hr	0.2, 2, 20 mR/hr	0.2, 2, 20 mR/hr	10, 100, 1000 mR/hr	10, 100, 1000 mR/hr
Accuracy	±5% FS	±5% FS	±5% FS	±5% FS	±5% FS
Temperature	−48 to 60°C (−55 to 140°F)	−48 to 60°C (−55 to 140°F)	−34 to 60°C (−30 to 140°F)	−48 to 60°C (−55 to 140°F)	−48 to 60°C (−55 to 140°F)

Housing	Stainless Steel	Aluminum Mica	Lead Tungsten Aluminum	Aluminum	Aluminum Mica
Weight	0.20 kg (0.44 lb)	0.17 kg (0.38 lb)	0.57 kg (1.15 lb) 1.93 kg (4.25 lb) with shield	0.059 kg (0.13 lb)	0.059 kg (0.13 lb)
Overall Size	20.3 cm long 2.54 cm o.d. (8 in. long 1 o.d.)	15.2 cm long 3.81 cm o.d. (6 in. long 1 1/2 in. o.d.)	16.5 cm long × 8.9 cm W × 9.8 cm H (6 1/2 in. L × 3 1/2 in. W × 3 7/8 in. H)	8.9 cm long 0.95 cm o.d. (3 1/2 in. long (3 1/2 in. long 3/8 in. o.d.)	8.9 cm long 1.91 cm o.d. (3 1/2 long (3 1/2 in. long (3/4 in. o.d.)
Price	$65.00	$85.00	$225.00, Alum. $450.00, Tungsten	$110.00	$110.00

[a]GSM-10 sells for $285.00. Add probe for complete instrument.

Table A.3.6. Scintillation Probes for Johnson Model GSM-10 Survey Meter[a]

	Model		
Specifications	GSP-2	DGSP	FNSP-2A
Radiation Measured	Gamma	Low-energy gamma	Fast neutron
Detector	NaI(Tl)	NaI(Tl) cleaved crystal	Hornyak button
Range	0 to 50,000 cpm	0 to 50,000 cpm	0.2 MeV 20 MeV
Accuracy	±5%	±10%	+15% −30%
Temperature	−40 to 65°C (−40 to 150°F)	−40 to 65°C (−40 to 150°F)	−40 to 65°C (−40 to 150°F)
Power	Four 1.5-V penlight batteries (AA)	Four 1.5-V penlight batteries (AA)	Four 1.5-V penlight batteries (AA)
Housing	Aluminum	Aluminum	Aluminum
Weight	0.68 kg (1.5 lb)	0.91 kg (2 lb)	0.40 kg (0.875 lb)
Overall	25.4 cm long	25.4 cm long	27.9 cm long
Size	0.95 cm o.d. (10 in. long 3/8 in. o.d.)	6.35 cm o.d. (10 in. long 2.5 in. o.d.)	6.1 cm o.d. (11 in. long 2.4 in. o.d.)
Price	$595.00	$495.00	$395.00

[a]GSM-10 sells for $285.00. Add probe cost for complete instrument.

Radiation Monitor
Johnson Model RML-1A Count Rate Meter

Class	Laboratory
Principle of Operation	Accepts about 30 probes, amplifier, integrator, speaker, recorder, and meter readout
Sensitivity and Range	Energy range depends on probe, see previous note Full scale ranges: 500, 5,000, 50,000 cpm Minimum input sensitivity: 10 mV
Sampling	Continuous
Performance	Meter accuracy: ±2% FS Response time: 1.5 or 4.5 sec
Requirements	Size: 28 cm × 18 cm × 18 cm (11 in. × 7 in. × 7 in.) Weight: 3.4 kg (7.5 lb)
Features	Volume and discriminator control
Cost	$395.00
Address	William B. Johnson and Associates, Inc. Research Park, Boonton Avenue Montville, NJ 07045 (201) 334-9222

Portable Scaler
Johnson LS-4

Class	Laboratory and field
Principle of Operation	Uses Geiger, scintillation, or proportional counters, low voltage supply, variable high voltage, amplifier, LED readout. Solid state circuitry.
Sensitivity and Range	6 LED ranges: 0 to 999,999 counts, Resolution: 1.3 microseconds
Sampling	Preset time-elapsed count mode
Performance	Variable high voltage: 0 to 2500 V
Requirements	Power: 105–125 V ac, 1/2 amp, 60 cycle for line operation, 4 Ni-Cad batteries (rechargeable) for portable operation Size: 30.5 cm × 27.9 cm × 12.7 cm (12 in. × 11 in. × 5 in.) Weight: 4.1 kg (9 lb)
Features	Regulated and adjustable high voltage
References	Manufacturer's specifications
Cost	$1075.00
Address	William B. Johnson and Associates, Inc. Research Park, Boonton Avenue Montville, NJ 07045 (201) 334-9222

Mini-Ratemeter and Monitor
Johnson and Associates Model RML-3

Class	Laboratory and field
Principle of Operation	Uses Geiger, scintillation, or proportional counters
Sensitivity and Range	0 to 2,000 cps or 0 to 100,000 cpm
Performance	High-voltage power supply adjustable from 400 to 1400 V.
Features	Built-in speaker, On-Off switch. External power source may be used.
Requirements	Power: Two 9-V batteries Size: 14.5 cm × 16.5 cm × 18 cm (5 3/4 in. × 6 3/8 in. × 7 in.) Weight: 1.2 kg (2 lb. 10 oz.)
References	Manufacturer's specifications
Cost	$345.00
Address	William B. Johnson and Assocates, Inc. Research Park Montville, NJ 07045 (201) 334-9222 Attn: A. Purcaro

Alpha, Beta, Gamma Counters
LND Models

References Manufacturer's specifications

Address LND
3230 Lawson Blvd.
Oceanside, NY 11572
(516) 678-6141
Telex 14-4563

Table A.3.7. LND Alpha, Beta, and Gamma Counters

	Maximum Diameter		Overall Length				
Model	(cm)	(in.)	(cm)	(in.)	Operating Voltage	Application	Price
			End-Window Gas-Filled Counting Tubes				
704	0.87	0.344	3.81	1.5	450–650	α,β,γ	$ 96.00
705	0.87	0.344	3.81	1.5	450–650	α,β,γ	500.00
710	0.87	0.344	6.99	2.75	450–650	α,β,γ	82.00
7616	0.91	0.359	6.43	2.531	750–950	α,β,γ	92.00
712	1.52	0.600	5.40	2.125	450–650	α,β,γ	39.00
718	5.87	2.25	14.61	5.75	650–850	α,β,γ	194.00
722/5979	2.54	1	15.24	6	700–850	β,γ	60.00
7227	2.54	1	15.24	6	850–1000	x-ray	190.00
7228	2.54	1	15.24	6	1250–1450	x-ray	210.00
7229/8204M	2.54	1	15.24	6	660–800	α,β,γ	130.00
723	3.49	1.375	11.43	4.5	850–1000	α,β,γ	64.00
7231	3.33	1.312	3.49	2.5	450–700	α,β,γ	91.00
7232	3.49	1.375	12.38	4.875	900–1100	α,β,γ	95.00
7233	3.49	1.375	11.43	4.5	450–700	α,β,γ	95.00
724/7618	2.54	1.0	4.32	1.7	900–1100	β,γ	85.00
7242	2.54	1	2.54	1	450–650	β,γ	83.00
727/7840	3.49	1.375	11.43	4.5	1300–1600	α,β,γ	62.00
			Pancake Counting Tubes				
730	5.40	2.125	8.09	3.187	900–1100	α,β,γ	$ 210.00
731	5.40	2.125	8.09	3.187	900–1100	α,β,γ	120.00
7311/8767	5.40	2.125	7.70	3.032	850–1100	α,β,γ	107.00
732	5.40	2.125	8.09	3.187	900–1100	α,β,γ	350.00
7610	0.87	0.344	1.91	0.75	450–650	α,β,γ	230.00

Table A.3.7. (Continued)

Model	Maximum Diameter (cm)	Maximum Diameter (in.)	Overall Length (cm)	Overall Length (in.)	Operating Voltage	Application	Price
			Cylindrical Counting Tubes				
713	0.80	0.314	5.08	2.0	450–600	β,γ	$ 37.00
714	0.51	0.200	3.18	1.25	450–600	γ	37.00
716	0.51	0.200	1.91	0.750	450–600	γ	37.00
717	0.32	0.125	2.54	1.0	450–600	β,γ	280.00
719	1.59	0.625	27.94	11	900–1100	β,γ	75.00
720	1.59	0.625	9.84	3.875	900–1100	β,γ	42.00
7201	1.59	0.625	9.84	3.875	900–1100	β,γ	61.00
721	1.59	0.625	14.61	5.75	900–1100	β,γ	52.00
725	1.91	0.750	11.11	4.375	900–1100	β,γ	45.00
726/5980	0.87	0.344	5.08	2	660–800	γ	40.00
740	0.74	0.290	19.69	7.75	675–800	γ	90.00
			Medical Counting Tabs				
760	0.87	0.344	7.62	3.0	450–600	α,β,γ	$ 220.00
761	0.87	0.344	12.78	7	450–600	α,β,γ	490.00
762	0.87	0.344	21.84	8.6	450–600	γ	425.00
			Industrial and Cosmic Ray Counting Tubes				
78014	2.22	0.875	24.9	9.8	450–600	γ	$ 90.00
78024	2.5	1.0	68.58	27.0	1050–1400	γ	300.00
782	3.49	1.375	84.14	33.125	900–1100	γ	417.00
			Gas Sampling Proportional Counters				
495	1.91	0.75	22.86	9.0		low-level β	$ 705.00
4951	2.22	0.875	25.40	10.0		low-level β	705.00
4952	2.54	1.0	35.56	14.0		low-level β	705.00
4953	8.89	3.5	59.69	23.5		low-level β	1890.00
4932	2.24	0.88	26.67	10.5		low-level β	830.00
			Cosmic Ray Anti-Coincidence Guard Proportional Counters				
492	16.21	6.38	62.8	24.75			$6000.00
493	14.6	5.75	16.2	6.38	1450		1820.00

Survey Meter
Ludlum Models 2 and 3

Model 43-1
Large Area Alpha Scintillator

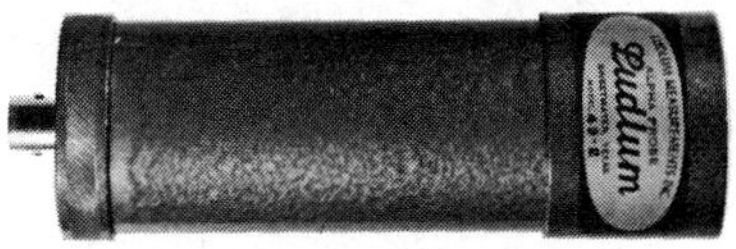

Model 43-2
Small Area Alpha Scintillator

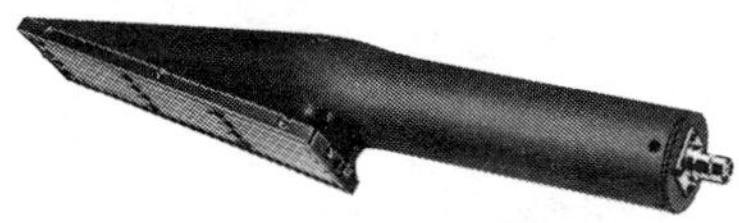

Model 43-5
Alpha Scintillator

Model 44-2
High Energy Gamma Scintillator

Model 44-6
Thin Wall Geiger-Müller Probe

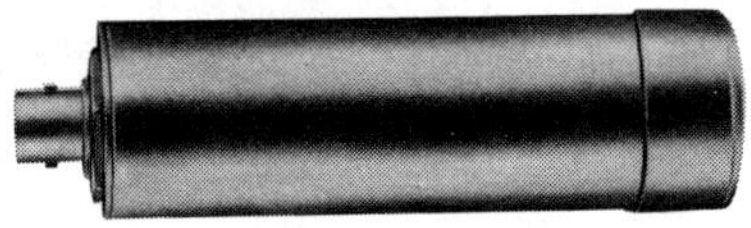

Model 44-7
End Window G-M

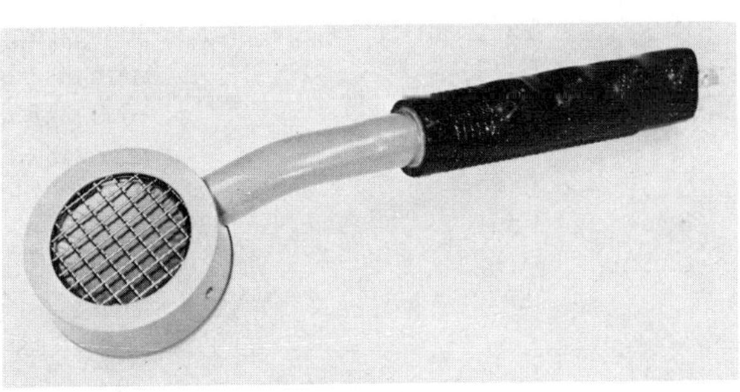

Model 44-9
Pancake G-M

Table A.3.8. Ludlum Probes

Model	Particle	Type	Tube	Window	Counting Area
43-1	Alpha	ZnS(Ag) Scint.	RCA 6199	2 mg/cm^2 Al Mylar 1 mg/cm^2	75 cm^2 4 in. diam.
43-2	Alpha	ZnS(Ag)	RCA 6199	1 mg/cm^2 Al Mylar	11.6 cm^2 1.5 in. diam.
43-5	Alpha	ZnS(Ag) Scint.	RCA 6199	1 mg/cm^2 Al Mylar	50 cm^2
44-2	Gamma	NaI(Tl) 1 in. × 1 in.			
44-3	Gamma	NaI(Tl) 1 mm × 1 in.		7 mg/cm^2 Al Mylar, 25 mg/cm^2 removable	
44-4	Beta Gamma	End window G-M	Amperex 150N	1.4 to 2 mg/cm^2 mica, 25 mg/cm^2 Mylar removable	25/32 in. diam.
44-6	Beta Gamma	Thin wall G-M	LND 725	30 mg/cm^2 stainless steel, 1000 mg/cm^2 removable	
44-7	Beta Gamma	End G-M	LND 723	1.5 to 2 mg/cm^2 mica	
44-9	Beta Gamma	Pancake G-M	LND 725		1.75 in. diam.

	Length		Diameter		Weight		
Efficiency	(cm)	(in.)	(cm)	(in.)	(kg)	(lb)	Cost
30% of 2π 50% of 2π for source 1/16 in. window	24.13	9.5	11.43	4.5	0.9	2	$350.00
50% of 2π	16.51	6.5	5.08	2	0.4	0.9	205.00
	19.81	7.8	5.84	2.3	0.9	2	325.00
31,000 cpm @ 0.1 mR/hr ^{226}Ra	19.05	7.5	5.08	2	0.4	0.9	235.00
	16.51	6.5	5.08	2	0.4	0.9	235.00
4000 cpm when exposed to ^{226}Ra flux of 1 mR/hr	19.05	7.5	3.05	1.2	0.45	1	175.00
1700 cpm = 1 mR/hr flux flux of ^{226}Ra	16.51	6.5	3.05	1.2	0.34	0.75	80.00
	13.46	5.3	3.56	1.4	0.23	0.5	85.00
	27.94	11	7.11	2.8	0.36	0.8	162.50

Class	Portable, handheld
Principle of Operation	Count ratemeter for a variety of G-M and scintillator probes
Sensitivity and Range	Sensitivity: See Table A.3.8 Range: Model 2: 0 to 5 mR/hr, ×0.1, ×1, ×10, linear Model 3: 0 to 2 mR/hr, ×0.1, ×1, ×10, ×100, linear 0 to 5,000 cpm, optional meter readout
Sampling	Continuous
Performance	Accuracy: ±5% FS Calibration Stability: Less than 5% variance to battery end point
Requirements	Power: Two "D" cells, 300 hr Size: 9 cm H × 9 cm W × 18 cm L (3.4 in. × 3.5 in. × 7.0 in.) exclusive of handle Weight: 1.6 kg (3.5 lb) less detectors
Features	Pushbutton reset, adjustable high-voltage output, speaker output
References	Manufacturer's specifications
Cost	Model 2: $280.00 Model 3: $295.00
Address	Ludlum Measurements, Inc. 501 Oak Street Sweetwater, TX 79556 (915) 235-5494

Survey Meter
Ludlum Models 12, 16

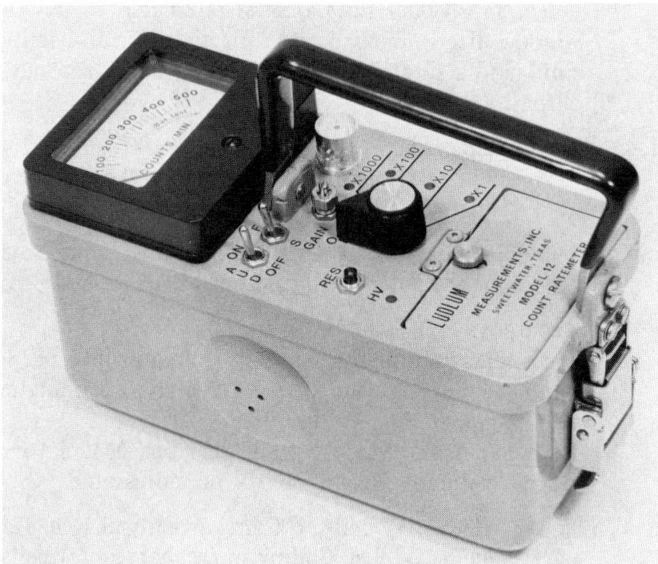

Model 12

Model 16

Class	Portable
Principle of Operation	Power supply for scintillation, proportional, or Geiger–Muller detectors. Model 12 is a count ratemeter, Model 16 includes a window discriminator switch for an upper discrimination level similar to a single-channel analyzer; meter readout, audio output.
Sensitivity and Range	Count range: 500 cpm, ×1, ×10, ×100, ×1000 Depends on the detector chosen, see other Ludlum notes.
Detectors	The detector models that may be used with the Model 12 and Model 16 are listed with Ludlum Models 2 and 3
Performance	Accuracy: ±5% of FS Discrimination range: Low discriminator level—2 to 60 mV Model 16, high discriminator level—2 to 60 mV Response time: Model 12—3 or 11 sec; Model 16—11 sec Output voltage: 700 to 2400 V dc, adjustable
Requirements	Power: Two "D" cells, 100 hr, or optional RM-42R mercury cells, Nickel-Cadmium (rechargeable) cells Size: 9 cm H × 9 cm W × 18 cm L exclusive of handle (3.4 in. × 3.5 in. × 7.0 in.) Weight: 1.6 kg (3.5 lb)
Features	Battery test, fluorescent mark to indicate "off" position
Reference	Manufacturer's specifications
Cost	Model 12: $395.00 Model 16: $445.00
Address	Ludlum Measurements, Inc. 501 Oak Street Sweetwater, TX 79556 (915) 235-5494

Neutron, Beta, Gamma, and Alpha Count Ratemeter
Ludlum Model 15

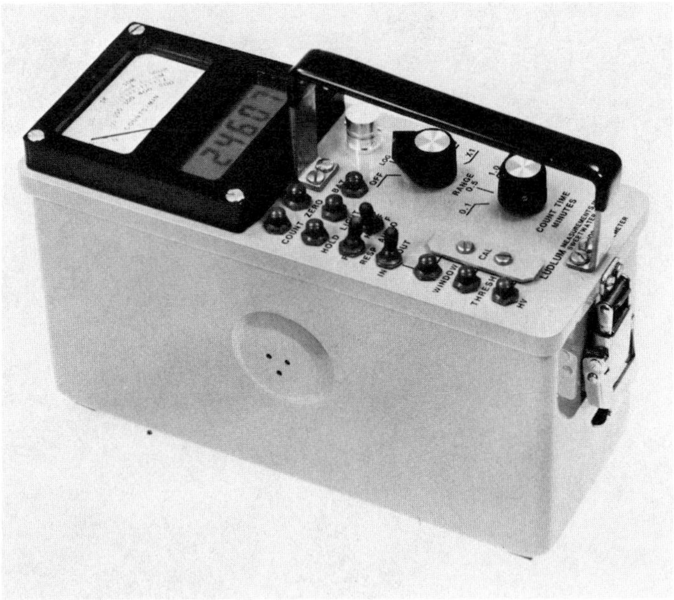

Class	Portable, handheld
Principle of Operation	Uses BF_3 detector and associated electronics for count ratemeter
Sensitivity and Range	Energy sensitivity: thermal Ranges: 500, 5000, 50,000, and 500,000 cpm
Sampling	Continuous
Performance	Meter accuracy: ±5% FS Response time: Fast, 8 sec Slow, 15 sec
Requirements	Power: Two "D" cells Size: 25 cm × 9 cm × 18 cm (9.8 in. H × 3.5 in. W × 7 in. L) Weight: 3.6 kg (8 lb)
Features	High-voltage adjustment, calibration adjustment for each scale
References	Manufacturer's specifications
Cost	$885.00

Address	Ludlum Measurements, Inc. 501 Oak Street Sweetwater, TX 79556 (915) 235-5494

Scalar Ratemeter ("Hand-Held")
Ludlum Model 2220

Class	Portable
Principle of Operation	Flexible controls allow operation of Geiger-Muller, scintillation or proportional detectors (to be selected from catalogue). Singe channel analyzer. Threshold and window positions adjustable with readouts (energy related) on the digital display. Built-in unimorph speaker with automatic pitch change in relation to meter reading. Four decade linear ratemeter and a log readout.
Sensitivity and Range	Depends on the detector chosen.
Performance	Accuracy: ±5% of full scale.
Features	Timer: Preset 0.1, 0.5, 1 minute, continuous. Count: Push button to reset scaler, timer, and start new count. H.V.: Press test button, read high voltage on scaler. Recessed multiturn screwdriver control. Threshold: Press test button, read setting on scaler. Recessed multiturn screwdriver control. Window: Press test button, read setting on scaler. Recessed multiturn screwdriver control. In-Out: Selects gross count or S.C.A. Hold: Stops counter/timer. Zero: Zeros meter. Response: 3 seconds slow and 11 seconds fast for meter response. Audio: On-Off toggle switch. Bat: Push button switch to test batteries.
Requirements	Power: Four "D" cell batteries Size: 24 cm × 9.5 cm × 12.5 cm (9 1/2 in. × 3 3/4 in. × 5 in.) exclusive of handle Weight: 2.3 kg (5 lb) including batteries
Price	$1,595.00
Address	Ludlum Measurements, Inc. 501 Oak Street Sweetwater, TX 79556 (915) 235-5494 Attn: Dwane Stevens

Waste Curie Monitor
National Nuclear Models WCM-10 and -20, and -30

Class	Laboratory
Principle of Operation	Six (Model WCM-10 and WCM-30) and Four (Model WCM-20) large area solid plastic scintillant detectors to scan the entire bag or barrel completely for deminimus segregation of waste as per 10 CFR 61.
Sensitivity and Range	Model WCM-10: detect 5 nCi with greater than 95% confidence in a 15 μR/hr background. In high sensitivity mode, less than 2 nCi. Model WCM-20: detect 50 nCi with greater than 95% confidence in a 15 μR/hr background. In high sensitivity mode, less than 20 nCi. Model WCM-30: detect 2 μCi with greater than 95% confidence in a 15 μR/hr background. In high sensitivity mode, less than 1 μCi.
Performance	Visual alarm. Counting time: 10 sec (100 sec at high sensitivity) (Models WCM-10 and WCM-30) 30 sec (300 sec at high sensitivity) (Model WCM-20)
Requirements	Power: 110 V/1.4 A 200 V/0.7 A Size: Model WCM-10: 127 cm L × 89 cm W × 91 cm H (50 in. L × 35 in. W × 36 in. H) Model WCM-20: 76 cm L × 76 cm W × 102 cm H (30 in. L × 30 in. W × 40 in. H) Model WCM-30: 51 cm L × 46 cm W × 51 cm H (20 in. L × 18 in. W × 20 in. H) Weight: Model WCM-10: 2860 kg (6300 lb) Model WCM-20: 908 kg (2000 lb) Model WCM-30: 682 kg (1500 lb)
References	Manufacturer's specifications
Cost	WCM-10 and WCM-20: $40,000 WCM-30: $30,000
Address	National Nuclear Corporation 1904 Colony Street Mountain View, CA 94043 (415) 962-9220 Attn: D. Kahn

Hand and Clothing Monitor
Nuclear Enterprises HFM6

Class	Stationary, personnel monitor
Principle of Operation	Scintillator detectors, amplifier, meter readout, system of lights giving status (safe, wash hands, etc.)
Specification	Model HFM6—both hands Range: 0 to 5 derived working limits Response time: 5 sec, can be varied 2 to 20 sec. Efficiency: α, 11% β, 26% Operating temperature: 5 to 40°C (−41 to 104°F) Accuracy: α, ±12% for one standard deviation at MPL level of 10^{-5} $\mu Ci/cm^2$ in 5-sec count time β, ±3% for one standard deviation at MPL of 10^{-4} $\mu Ci/cm^2$
Requirements	Power: (190 to 260 V, or 95 to 130 V, 50 to 60 Hz, at 25 V amp) Size: 75 cm W × 46 cm H × 42 cm D (29.5 in. × 18 in. × 16.5 in.) Weight: Refer to factory
Features	Independent α and β indication, with separate alarm presets; automatically corrects in high γ background (up to 1.5 mR/hr)
Cost	Refer to factory
Address	Spectrum Sciences 3050 Oakmead Village Dr. Santa Clara, CA 95051 (408) 727-1567 Telex: 176449 Semi Group Nuclear Enterprises Limited Bath Road, Beenham Reading, RG7 5PR, England (073) 521-2121

Portable Ratemeters
Nuclear Enterprises RMS, RM6

Principle of Operation	Accepts α scintillation and β,γ G-M probe, amplifiers, meter readout		
		Model RM5	Model RM6
Sensitivity and Range	Range:	0.5 to 5,000 cps, 3 decade, log	1 to 5,000 cps, 3 1/2 decade, quasi-log
	Output voltage:	300 to 1200 V	300 to 1400 V
	Minimum load impedance:	50 MΩ	--
	Input sensi-tivity:	50 mV	$< 10^{-11}$ coulombs (charge sensitive)
Performance	Sampling	Continuous	Continuous
	Operating temperature range:	−10 to 50°C (14 to 122°F)	0 to 40°C (32 to 104°F)
	Accuracy:	±20 FS	±20% FS
Requirements	Power:	Two type R-20 cells	Two type PP6 (or Varta 28) or, one type PP3 (or Varta 2022)
	Size:	24 cm L × 12 cm W × 14 cm H (9 in. × 4.7 in. × 5.5 in.)	24 cm L × 12 cm W × 16 cm H (9.5 in. × 4.7 in. × 4.9 in.)
	Weight:	1.6 kg (3.5 lb)	1.6 kg (3.5 lb)

Cost — Refer to factory

Address — Nuclear Enterprises Limited
Bath Road, Beenham,
Reading, RG7 5PR, England
(073) 521-2121
or
Spectrum Sciences
3050 Oakmead Village Dr.
Santa Clara, CA 95051
(408) 727-1567
Telex: 176449 Semi Group

Count Ratemeter
Nuclear Enterprises Model PCM 5

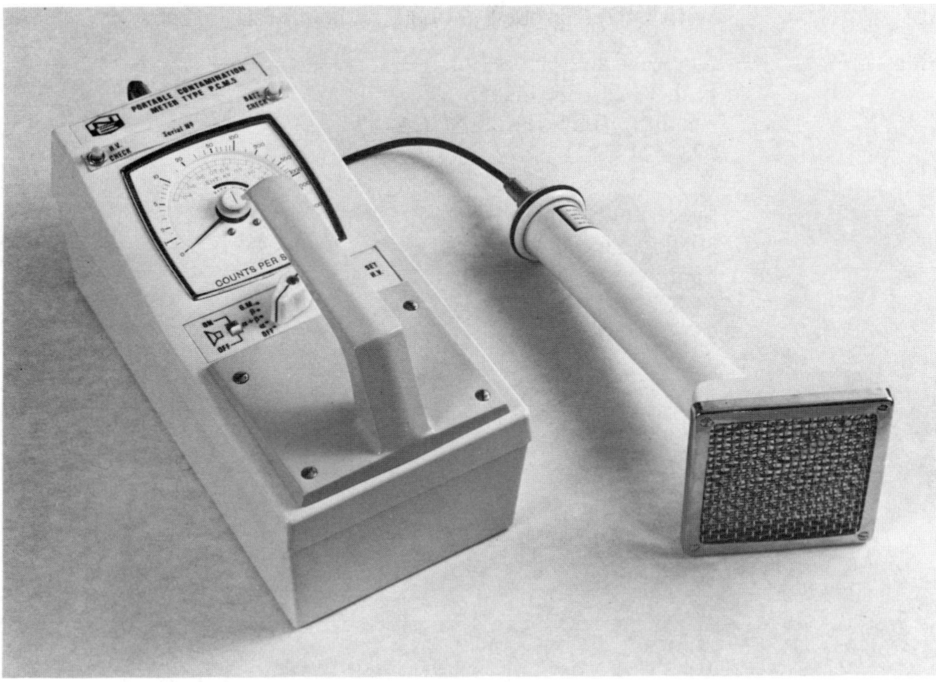

Class	Portable, handheld
Principle of Operation	Accepts α scintillation and β,γ G-M probe, amplifier, meter readout
Sensitivity	Input sensitivity: 50 mV Output voltage: 300-1200 V dc present; minimum load impedance 50 MΩ Ratemeter: 1 to 10, 10 to 100, 100 to 1000, 500 to 5000 cps Efficiency: depends on probes (see Probes)
Performance	Accuracy: ±25% for temperature −10 to 30°C; ±35% for −10 to 50°C Temperature: −10°C to +50°C (−14 to 122°F) Probes listed on Table A.3.9
Requirements	Power: Type "D" cells 1.5 V; 80 hr at 4 hr/day Size: 24 cm L × 12 cm W × 10 cm H (9.5 in. × 4.7 in. × 3.9 in.) Weight: 1.6 kg (3.5 lb)

Features	Loudspeaker, optional power regulator for 50 cps, 220 V ac
References	Manufacturer's specifications
Cost	With DP2/R probe $1949.00; without probe $1200.00
Address	Nuclear Enterprises Ltd. Bath Road, Beenham Reading, RG7 5PR, ENGLAND 073-521-2121 Spectrum Sciences 3050 Oakmead Village Dr. Santa Clara, CA 95051 (408 727-1567 Telex: 176449 Semi Group

Table A.3.9. Probes for Nuclear Enterprises PCM 5 Count Ratemeter

Type No.		Radiation Detected	Efficiency	Area (cm^2)
BP1	Geiger MX 16B	beta/gamma/x-ray	10 cps from 1 mR/hr 8 keV x-ray 40 cps from 1 mR/hr 20 keV x-ray 40 cps from 1 mR/hr ^{60}Co field	1.36
BP3/1	Geiger B12H	beta/gamma	5 cps from 10^4 $\mu Ci/cm^2$ $^{90}Sr/^{90}Y$ 135 cps from 1 mR/hr ^{60}Co field	
BP4	Scintillation, Anthracene	low-energy beta/gamma range, 120 keV upwards	15% for ^{14}C 30% for $^{90}Sr/^{90}Y$ betas	18
BP5	Scintillation Plastic NE 102A	beta/gamma	27.5% for $^{90}Sr/^{90}Y$ 120 cps for 1 mR/hr ^{60}Co field	49
BP6	Scintillation, Plastic NE 102A (thin)	low-energy beta/gamma	8% for ^{206}Tl source	100
BP7	Scintillation, Anthracene	low-energy beta/gamma	7% for ^{14}C betas	49
AP2	Zinc sulphide on perspex	alpha	25% for ^{239}Pu alphas	49
AP3	Zinc sulphide on perspex	alpha	25% for ^{239}Pu alphas	100
AP4	Zinc sulphide on perspex	alpha	27.5% for ^{239}Pu alphas	18
DP2/R	Zinc sulphide on NE 102A	alpha, beta, gamma	22.5% for alpha from ^{239}Pu 27.5% for beta from $^{90}Sr/^{90}Y$ 100 cps from 1 mR/hr ^{60}Co field	49
GP6	3/4 in. × 3/4in. (19 mm) NaI(Tl) crystal	gamma	850 cps for 1 mR/hr ^{60}Co field 1150 cps from 1 mR/hr ^{226}Ra field	
GP7	2 mm × 3/4 in. NaI(Tl) crystal	low-energy gamma and x-ray	3600 cps for 1 mR/hr 400 cps for 1 mR/hr ^{226}Ra	

Airborne Particulates Monitor
Nuclear Research Model NRC Lab Map-60

Class	Laboratory
Principle of Operation	Shielded sampler, detector, flow indicator and vacuum gauge, combination ratemeter, power supply, and pump motor
Sensitivity and Range	Detectors (see Table A.3.10)
Performance	Flowmeter 0 to 10 cfm
Requirements	Power: 115 V ac, single phase Size: 61 cm W × 105 cm H × 55 cm D (24 in. × 41.5 in. × 22 in.) Weight: About 90 Kg (200 lbs)
Features	Ratemeter high-level trip alarm
Cost	Price on request
Address	Nuclear Research Corporation Warrington Industrial Park P.O. Box H Warrington, PA 18976 (215) 343-5900

Table A.3.10. Detectors

Detector Model No.	Isotope	Time to Reach 100 cpm Net at 10^{-10} μCi/cm^3	Detector Type	Radiation Sensitivity	Background
MD-1B	$^{90}Sr + ^{90}Y$	15 min	G-M	β,γ	20 cpm
	^{60}Co	26 min			
	^{24}Na	35 min			
	^{137}Cs	35 min			
MD-3C	^{210}Po	21 min	ZnS(Ag)	α	5 cpm
	^{239}Pu	21 min			
MD-4C	^{204}Ti	21 min	Plastic Phosphor	β	20 cpm
	^{14}C	70 hr			
MD-5C	^{131}I	40 min	NaI(Tl)	γ	200 cpm

Hand and Foot Monitor
Nuclear Research Model HFA-200A

Class	Stationary
Principle of Operation	G-M tube monitoring of high-energy alpha, beta, and gamma radiation. Background subtraction for operation in 1 mR/hr Co-60 field. Analog readout with individual alarm adjustments.
Sensitivity	Three ranges each section: 0-500, 0-5000, 0-50,000 cpm Alarm adjustable over entire range
Sampling	Selectable 2 sec, 20 sec
Performance	Count rate accuracy: ±10% Temperature: 4.4 to 48.8°C (40 to 120°F)
Requirements	Power: 120 V ac, 1 A Size: 61 cm W × 142 cm H × 122 cm D (24 in. × 56 in. × 48 in.)
Features	Background subtraction Manual and automatic reset Automatic sequencing
References	Manufacturer's specifications
Cost	Price on request
Address	Nuclear Research Corporation Warrington Industrial Park P.O. Box H Warrington, PA 18976 (215) 343-5900

Portable Survey Meters
Nuclear Research Models CP-2, -5, 10

Class	Portable
Principle of Operation	Detector, amplifier, counter, meter readout
Model and Range	CP-2: 0.25, 250, 2500 mR/hr CP-5: 0-50, 500, 5000 mR/hr CP-10: 0-100, 1000, 10,000 mR/hr
Energy Sensitivity	Alpha from 2.0 MeV Beta from 35 keV Gamma 40 keV to 2 MeV
Energy Dependence	±15% for gamma and x-ray from 40 keV to 2 MeV
Sampling	Continuous
Accuracy	±10% FS
Requirements	Power: Four 22.5-V batteries, one 1.3-V battery; ~200 hr life; Size: 23 cm H × 29.2 L × 9 cm W (9 in. × 11.5 in. × 3.5 in.) Weight: 16 kg (3.5 lb)
Remarks	Designed and engineered to Oak Ridge National Laboratory specifications
Cost	CP-2: $595.00
Address	Nuclear Research Corporation Warrington Industrial Park P.O. Box H Warrington, PA 18976 (215) 343-5900

Radiation Laboratory Monitor
Nucleus Model L Ratemeter

Class	Laboratory
Principle of Operation	X-2 End window, G-M tube or scintillation detector, amplifier, meter readout, audible alarm.
Sensitivity and Range	Energy sensitivity: depends on probe Ranges: 500, 2000, 5000, 20,000 and 50,000 cpm full scale, linear. Voltage: 400–1600 Zener regulated, continuously variable Time constant: 2 sec
Sampling	Continuous
Requirements	Power: 110 V/60 Hz, 230 V/50 Hz, 15 watts Size: 12.7 cm H × 28 cm W × 16 cm D (5 in. × 12 in. × 6 1/4 in.) Weight: 5 kg (11 lb)
Performance	Front mounted speaker 6.4 cm (2 1/2 in.) with variable volume control
Options	Various G-M and scintillation detectors Recorder terminals
Cost	$375
Address	Nucleus, Inc. 461 Laboratory Road Oak Ridge, TN 37830 (615) 482-4041 Telex 557-482 Attn: R. Stevens

TRI-CARB^{R} Liquid Scintillation System
Models 4530, 4640, 4550, 4660

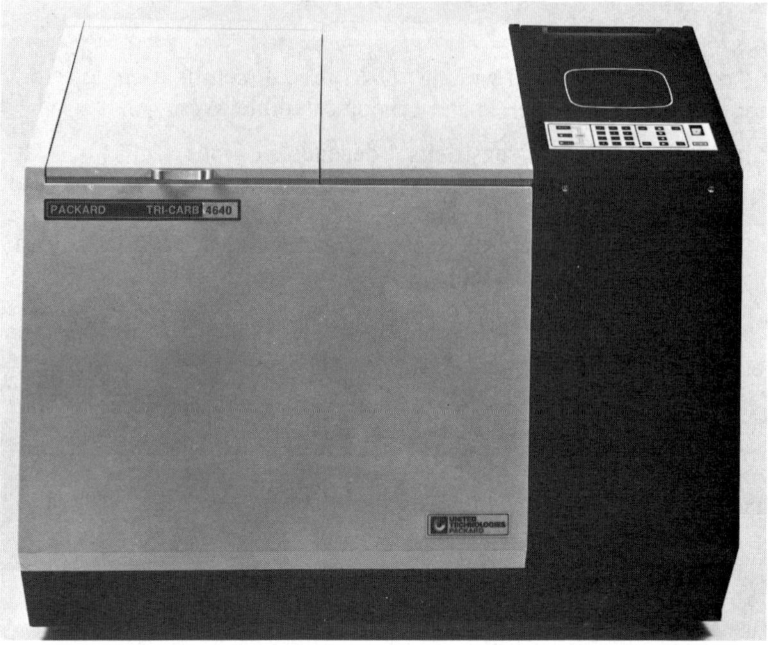

Class	Laboratory
Principle of Operation	Automatic liquid scintillation system with Spectralyzer^{TM}. Microprocessor controlled, cassette loaded, belt driven sample changer with forward and reverse controls. Video display. Count termination by time or 2 sigma %. Unique spectral analysis method independent of scintillation solution produdes SIE. ^{226}Ra external standard source to measure efficiency; analysis of spectrum produces SIE. Automatic Efficiency Control. Automatic normalization of system. Built-in Data Writer. PrioSTAT. Automatic data reduction.
Sensitivity and Range	18 to 2000 keV. Alpha, beta, gamma activity. Could also monitor x-rays.
Sampling	Model 4530: 300 (30 cassettes with 10 samples each) Model 4640: 460 (46 cassettes with 10 samples each) Model 4550: 504 (42 cassettes with 12 samples each) Small Vials Model 4660: 792 (66 cassettes with 12 samples each) Small Vials

Performance	Efficiency: Wide window 60% Min Tritium 0 to 2000 keV Wide window 95% Min Carbon 0 to 2000 keV E^2/B: 180 Min. Tritium 2 to 19 keV 380 Min. Carbon 4 to 156 keV
Requirements	Power: 117 V ac ± 10% or 234 V ac ± 10% 50 or 60 Hz 400 VA (max) Cooling system—1100 VA

Size:

Model 4530/4550	Model 4640/4660
114 cm (45 1/2 in.) Height	114 cm (45 1/2 in.) Height
131 cm (52 in.) Width	
79 cm (31 in.) Depth	

Weight:

Model 4530/4550	Model 4640/4660
358 kg (800 lbs)	385 kg (860 lbs)

Cost	Model 4530: $20,000 Model 4550: $20,000 Model 4640: $22,900 Model 4660: $22,900
Optional Features	DPM. Luminescence detection and correction. Spectraview. Integration and histogram. Temperature control. RS232 interface. Electrostatic controller.
Address	Packard Instrument Company 2200 Warrenville Road Downers Grove, IL 60515 (312) 969-6000

Laundry Monitor
Scintrex Model 139A

Class	Laboratory
Principle of Operation	Four banks of 510 mm diameter pancake, thin-window, G-M detectors (3 in each bank)
Sensitivity and Range	From 10 to 10^5 cpm. Logarithmic ratemeter
Performance	Accuracy: 10% of true value over the entire range Linearity: mean deviation $< \pm 4\%$ Stability: less than 2% above 1000 cpm less than 3% below 1000 cpm Recorder output: 0 to 6.2 V dc at 100 mA.
Requirements	Power: 120 V ac $\pm 10\%$, 60 Hz $\pm$ 5% Air supply: Oil free, dry instrument air at 515.5 to 1034 kPa. Temperature: 15 to 40°C (60 to 104°F) Humidity: up to 95% Ambient radiation field: up to 0.2 mR/hr Dimensions: 4.5 m H × 5.64 m L × 1.14 m D (14.8 ft × 18.5 ft × 3.7 ft) Weight: 500 kg (1100 lb)
Features	Individual logarithmic rate meter for each bank of G-M detectors. Clothing satisfying the maximum contamination level of < 0.15 μCi on 300 cm^2 of clothing is re-issued to the user. Automatic time constant selection. Rollers and conveyor speed: 6.1 m/min.
Cost	Refer to manufacturer
Address	Scintrex, Ltd. 222 Snidercroft Road Concord, Ontario L4K 1B5, Canada (416) 669-2280 Telex 06-964570 Attn: S. Nargolwalla

Survey Meter
Technical Associates Model TBM-3

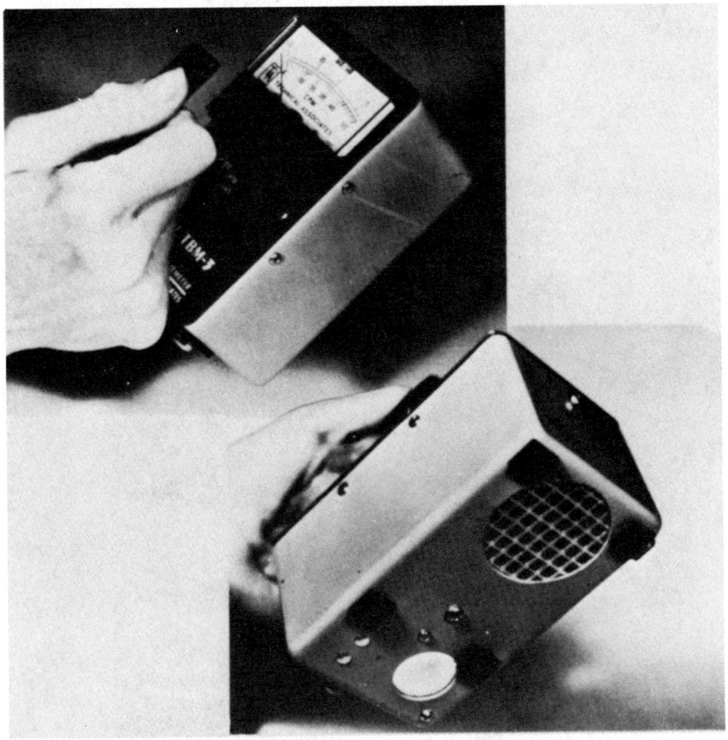

Class	Survey
Principle of Operation	Built-in pancake (2 in.) G-M thin window looks through hole on bottom of case. Very light one-hand operation.
Sensitivity and Range	0-500, 5000, 50,000 cpm; 0-0.15, 1.5, 15 mR/hr Sensitive to α, β, γ
Sampling	Continuous
Requirements	Power: 1 av transistor battery Size: 7.6 cm × 13.3 cm × 6 cm (3 in. × 5 1/4 in. × 2 1/4 in.) Weight: 22 oz
Cost	$345.00
Address	Technical Associates 7051 Eton Avenue Canoga Park, CA 91303 (213) 883-7043

Hand and Shoe Monitors
Technical Associates
HSM-10A, 10AM, 10BS, 10G

Class	Stationary, personnel monitor
Principle of Operation	Geiger tubes detect radiation, illuminated readout panel has projected messages that automatically warns user of contamination and contamination region, check incomplete and other messages (depending on model), counters or ratemeters also indicate count or count rate; alarms also.
Sensitivity and Range	Model HSM-10A: Beta-gamma (LED counters), background subtraction Model HSM-10AM: Beta-gamma (ratemeters) Model HSM-10BS: Alpha-beta-gamma (LED counters), background subtraction Model HSM-10G: Alpha and beta-gamma, 1. separately, but simultaneously; 2. dynamic digital background 2. subtraction
Sampling	Continuous
Performance	Accuracy, linearity ±5% FS
Requirements	Power: 105 to 125 V; 60 Hz; 300 watts (or 230 V, 50 to 60 Hz) Size: 172.1 cm H × 68.6 cm W × 70.0 cm D (67 3/4 in. × 27 in. × 24 in.) Weight: 272.2 kg (600 lb)
Features	1. Easy to read messages on illuminated panel in addition to count readout 2. Alpha-beta-gamma or beta-gamma (depend on model) 3. Adjustable alarm points 4. Background subtract on some models
References	Manufacturer's specifications
Cost	HSM-10A: $4879.00 HSM-10AM: $4450.00 HSM-10BS: $5375.00 HSM-10G: $7500.00
Address	Technical Associates 7051 Eaton Ave. Canoga Park, CA 91303 (213) 883-7043

Portable Survey Meters
Technical Associates CP-44, -5, -6, -7; ERJ-9

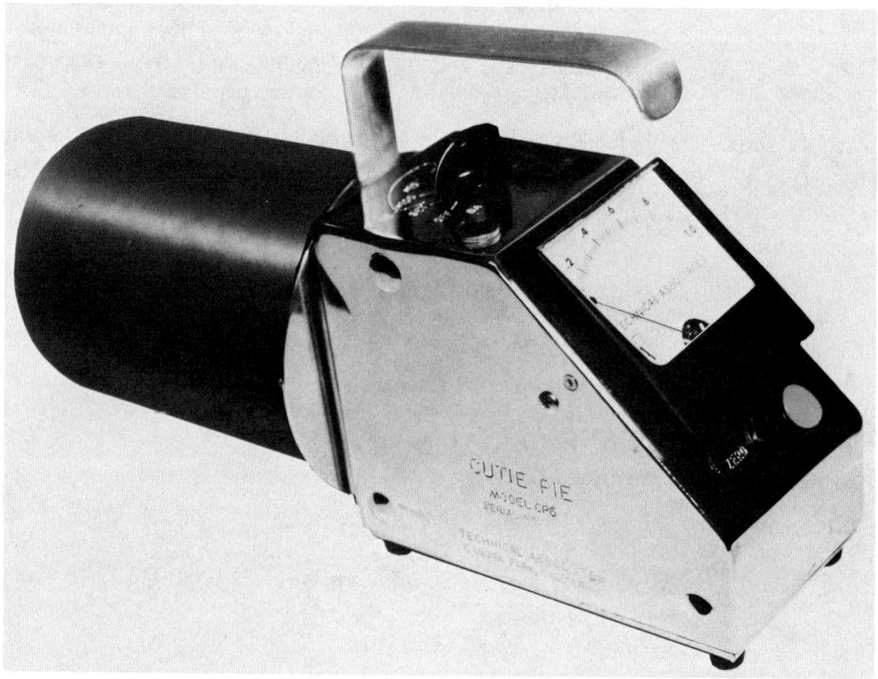

CP-6

Class	Portable
Principle of Operation	Ionization chamber, MOSFET amplifier, meter readout
Models	See Table A.3.11
References	Manufacturer's specifications
Address	Technical Associates 7051 Eaton Ave. Canoga Park, CA 91303 (213) 883-7043

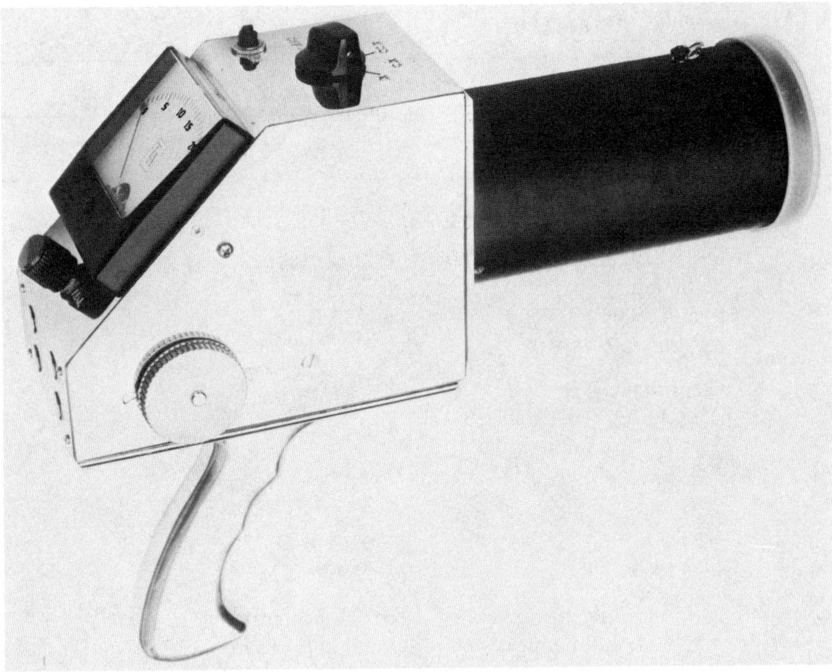

CP-44

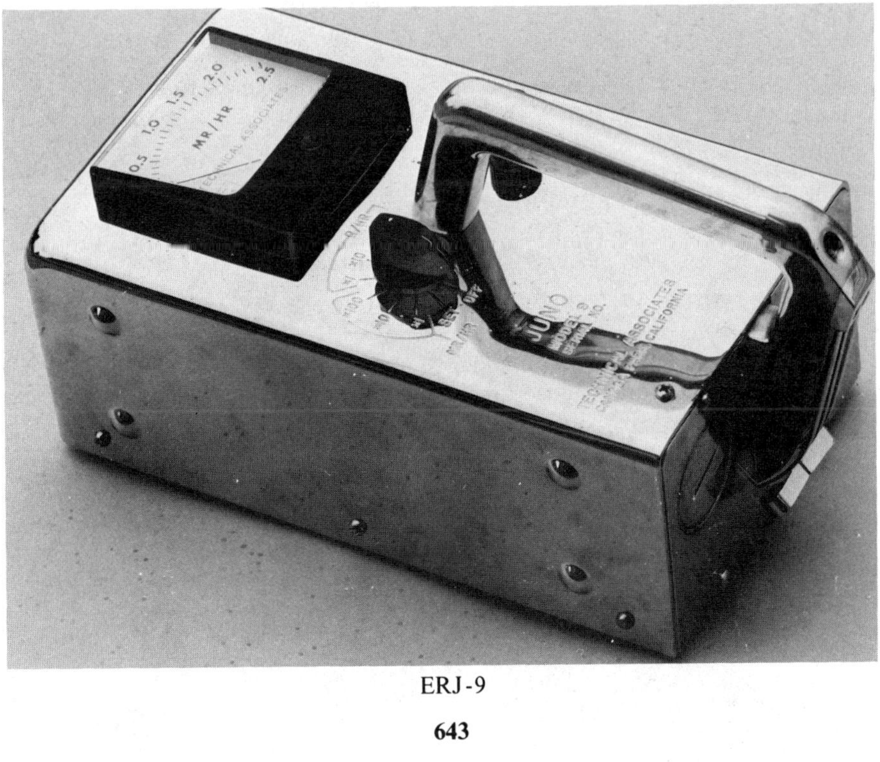

ERJ-9

Table A.3.11. Portable Survey Monitors

	Model	
Specifications	CP-44	CP-5
Type	Cutie-Pie	Cutie-Pie
Range	0–25, 250, 2500, 25,000 mR/hr; 0–500, 5000, 50,000 mR/hr (CP-44A)	6 linear ranges from 2.5 mR/hr to 250 R/hr
Sampling	Continuous	Continuous
Ionization Chamber Volume	589.9 cm^3 36 $in.^3$	589.9 cm^3 36 $in.^3$
Operating Temperature	−37.2 to 51.7°C −35 to +125°F	−37.2 to 51.7°C −35 to +125°F
Accuracy	±10% FS mR/hr ranges ±15% FS R/hr ranges	±10% FS (mR/hr) ±15% FS (R/hr)
Drift	< 1% per °F at room temperature	< 1% per °F at room temperature
Power	One "D" cell	One "D" cell
Size, Excluding Handle	33 cm × 16.5 cm × 10.2 cm (13 in. × 6 1/2 in. × 4 in.)	33 cm × 16.5 cm × 10.2 cm (13 in. × 6 1/2 in. × 4 in.)
Weight	2 kg (4.5 lb)	2 kg (4.5 lb)
Remarks	Flat response at all energies	Flat response at all energies
Cost	$570.00	$690.00

[a]ERJ-9-4R Juno same as ERJ-9 except 4 ranges: 0–50, 500, 5000, 50,000 R/min; $850.00.

CP-6	CP-7	ERJ-9[a]
Cutie-Pie	Cutie-Pie	Juno
6 linear ranges from 1 mR/hr to 100 R/hr	6 linear ranges from 0.5 mR/hr to 50 R/hr	6 linear ranges 0–2.5,25,250 mR/hr 0–2.5,25,250 R/hr
Continuous	Continuous	Continuous
1688.3 cm^3 103 $in.^3$	1688.3 cm^3 103 $in.^3$	589.6 cm^3 36 $in.^3$
−37.2 to 51.7°C −35 to +125°F	−37.2 to 51.7°C −35 to +125°F	−37.2 to 51.7°C −35 to +125°F
±10% FS (mR/hr) ±15% FS (R/hr)	±10% FS (mR/hr) ±15% FS (R/hr)	±10% FS (mR/hr) ±15% FS (R/hr)
< 1% per °F at room temperature	< 1% per °F at room temperature	< 1% per °F room temperature
One "D" cell	One "D" cell	One "D" cell
33 cm × 16.5 cm × 10.2 cm (13 in. × 6 1/2 in. × 4 in.)	33 cm × 16.5 cm × 10.2 cm (13 in. × 6 1/2 in. × 4 in.)	24.1 cm × 14.6 cm × 10.2 cm (9 1/2 in. × 5 3/4 in. × 4 in.)
2 kg (4.5 lb)	2 kg (4.4 lb)	2.7 kg (6 lb)
Flat response at all energies	Flat response at all energies	Flat response at all energies
$780.00	$880.00	$940.00

Portable Ratemeters
Technical Associates SML-2, FS-8, PUG-1, -1AB, -1E

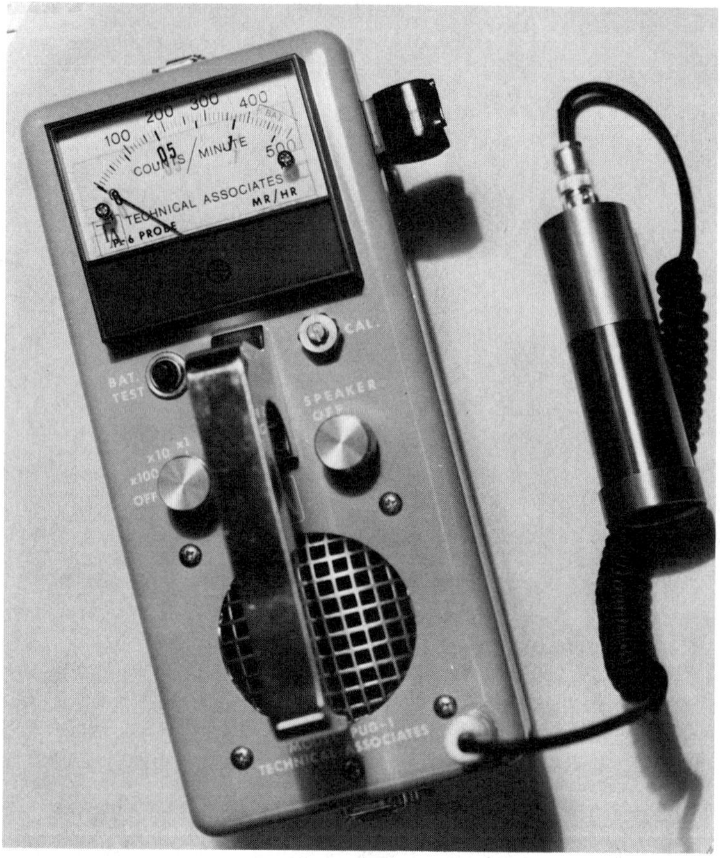

PUG-1

Class	Portable
Principle of Operation	Accepts G-M or scintillation detector, amplifier, counter, readout device
Models	See Table A.3.12
References	Manufacturer's specifications
Address	Technical Associates 7051 Eton Ave. Canoga Park, CA 91303 (213) 883-7043

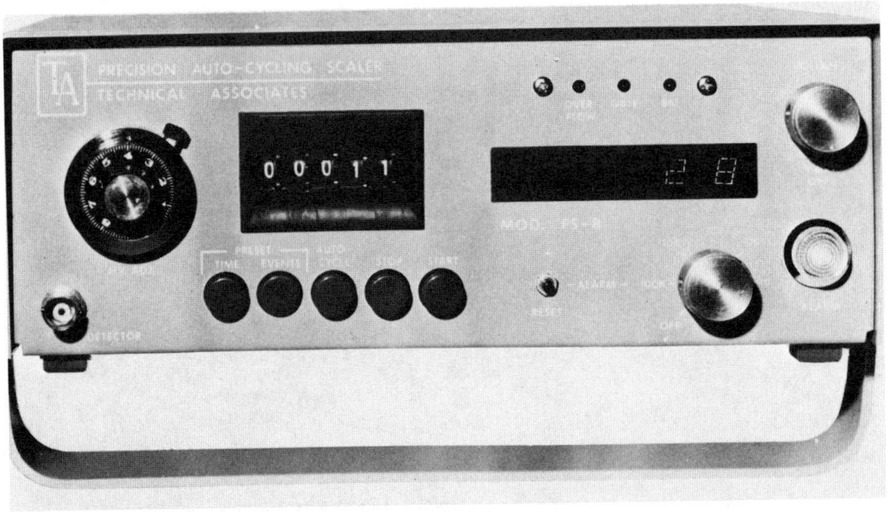
PRECISION AUTO-CYCLING SCALER
TECHNICAL ASSOCIATES

FS-8

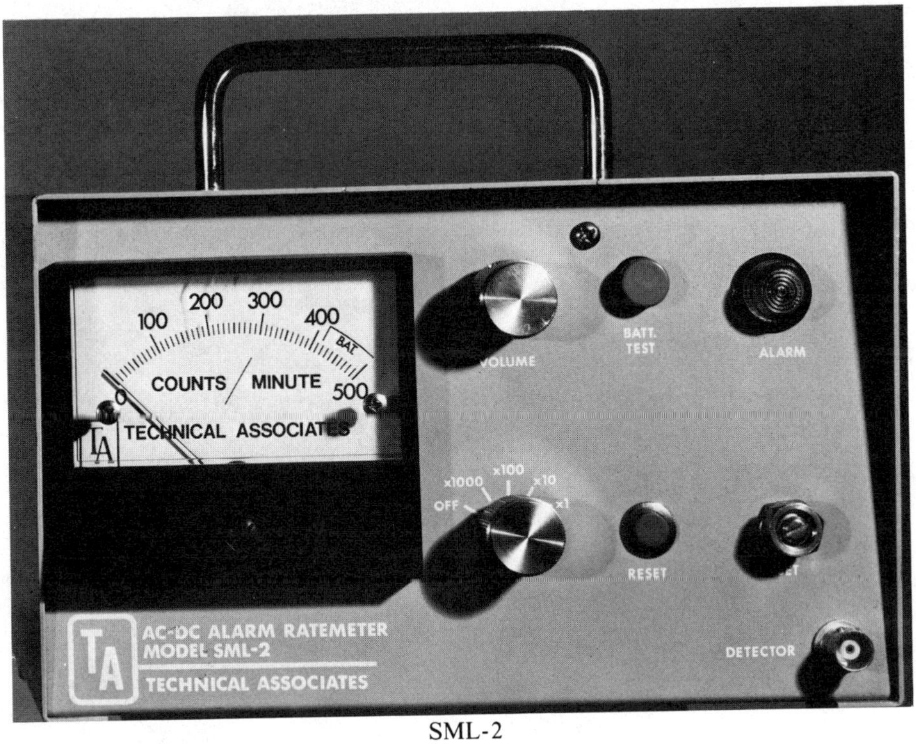
100
200
300
400
500
0
BAT.
COUNTS / MINUTE
TECHNICAL ASSOCIATES
VOLUME
BATT. TEST
ALARM
x1000
x100
x10
x1
OFF
RESET
AC-DC ALARM RATEMETER
MODEL SML-2
TECHNICAL ASSOCIATES
DETECTOR

SML-2

Table A.3.12. Portable Ratemeters

Specifications	Model: SML-2	Model: FS-8 Series	Model: PUG-1, PUG-1AB, PUG-1E
Type	Portable (rechargeable)	or laboratory (ac operation)	portable
Range	0–500, 5000, 50,000 500,000 cpm	Present count: 1 to 99,999 Preset time: 0.1 to 9999.9 sec Recycling time: 0.1 to 9999.9 sec	PUG-1: 0–500, 5000, 50,000, cpm PUG-1AB, PUG-1E: 0–500, 5000, 50,000, 500,000 cpm
Sampling	Continuous	Settable on preset count or time with or without recycling	Continuous
High Voltage		0 to 2000 V	
Input Sensitivity	100 mV	100 mV	
Accuracy	±2% FS (meter)	1% (time base)	±5% FS
Power	Four Ni-Cad "C" batteries (included) and ac 115 V 50–60 Hz (or 230 V)	Four Ni-Cad "D" batteries or 115 V, 50–60 Hz (or 230 V)	Three "D" cells
Size	20.3 cm W × 15.2 cm D × 12.7 cm H (8 in. × 6 in. × 5 in.)	24.1 cm × 26.7 cm × 8.9 cm 9 1/2 in. × 10 1/2 in. × 3 1/2 in.	10.2 cm H × 10.8 cm W × 25.4 cm L 4 in. × 4 1/4 in. × 10 in.
Weight	1.8 kg (4 lb)	3.0 kg (6 1/2 lb)	1.8 kg (4 lb)

RAD-ALP,BET,GAM,XRA
Technical Associates 4
January 1983

Remarks	Set table alarm (SML-2) Alarm indicators (SML-2) Meter readout Built-in speaker with volume control	Settable alarm Digital readout Built-in speaker with volume control FS-8W (FS-8 ac line operation only) $780.00	Zero suppress (PUG-1E) for use with scintillator probes
Cost	$550.00	$925.00 FS-8A (FS-8 with built-in amplifier, 1 mV sensitivity) $1050.00 FS-8C (FS-8 with built-in amplifier and single-channel analyses, 1 mV sensitivity) $1075.00. FS-8W (FS-8 ac line operation only) $780.00.	$240.00, $295.00, $520.00

Air Monitors
Technical Associates Model BAM-3A, 3B, 3H

Class	Laboratory
Principle of Operation	Motor-blower pulls in through filter and open gas chamber, viewed by an end-window G-M detector and sodium iodide detector, respectively. Log ratemeter, single-channel analyzer, and recorder.

Sensitivity and Range

BAM3A	BAM3B	BAM3H
10 to 100,000 cpm; (at 5 cfm) 6×10^{-10} μCi/ml·min exposure	10 to 100,000 cpm; 3×10^{-9} at 5 cfm (iodine) channel); 6×10^{-10} μCi/ml·min at 5 cfm (beta-gamma channel)	10 to 100,000 cpm; 6×10^{-10} μCi/ml·min at 5 cfm (beta particulates); 6×10^{-11} μCi/ml·min at 5 cfm (alpha particulates); 6×10^{-7} μCi/ml (gas)

Sampling	Continuous
Performance	Accuracy: ±10% except for resolution loss and statistical variation. Air flow: Up to 10 cfm
Requirements	Power: 115 V ac, 50 to 60 Hz Size: BAM3A: 91 cm L × 61 cm W × 107 cm H (36 in. × 24 in. × 42 in.) BAM3B: 91 cm L × 61 cm W × 107 cm H (36 in. × 24 in. × 42 in.) BAM3H: 107 cm L × 61 cm W × 107 cm H (42 in. × 24 in. × 42 in.) Weight: BAM3A: 288.3 kg (635 lb) BAM3B: 291 kg (640 lb) BAM3H: 436 kg (960 lb)
Features	Audible and visual alarms Can add features by adding modules
References	Manufacturer's specifications
Cost	BAM3A: \$5100; BAM3B: \$7660; BAM3H: \$10500
Address	Technical Associates 7051 Eton Avenue Canoga Park, CA 91303 (213) 883-7043

Alpha Air Monitor (Moving Filter)
Technical Associates Model SAAM-1, SAAM-2, SAAM-3

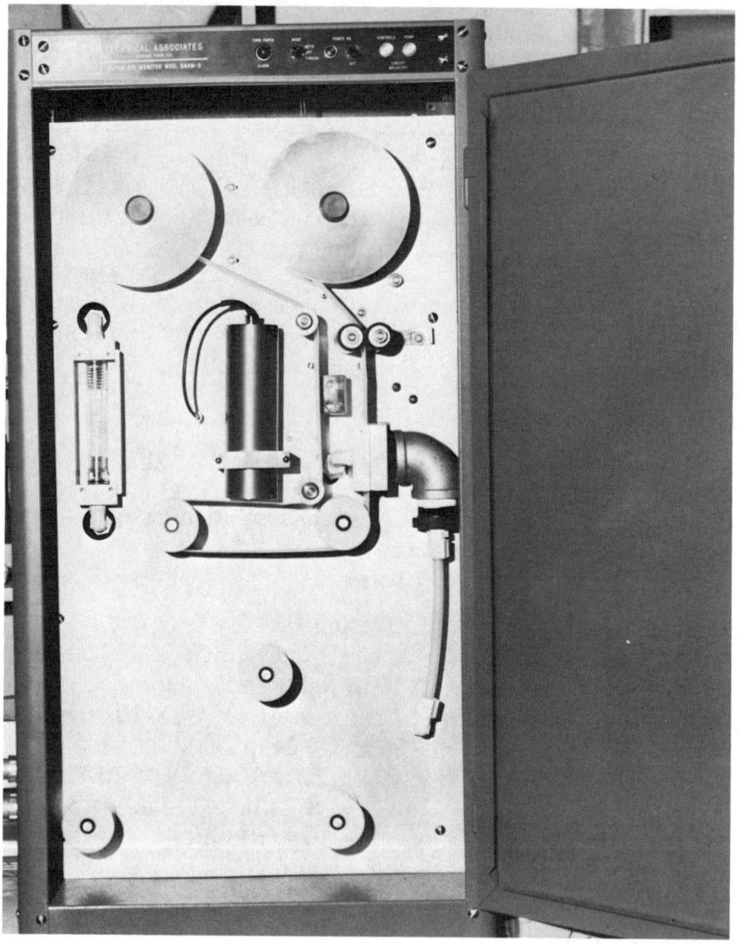

Class	Stationary
Principle of Operation	Air is drawn through section of filter tape. Tape with alpha deposits moves to alpha scintillation detector with predetermined time elapsing before count is taken (allows for decay of radon daughters). SAAM-1: Two channel moving filter air monitor, measures alpha and beta-gamma particulates SAAM-2: Three channel air monitor measures beta-gamma particulates iodine and the noble gases. SAAM-3: Integrated digital readout within preset time, or digital ratemeter readout and printed record
Sensitivity and Range	Sensitivity: 1% of most restrictive alpha contamination standards
Sampling	Air pumped through filter section for 30 min. Traversal of filter section to detector takes 4 to 5 hr.
Performance	Scalar accuracy for counting: ±1 count Timer accuracy: ±1%
Requirements	Power: 115 V ac, 50 to 60 Hz Size: 56 cm W × 91 cm L × 136 cm H (22 in. × 36 in. × 54 in.) (on wagon with 6 in. casters) Weight: 204 kg (450 lb)
Features	Automatic subtraction of radon daughter. May be left unattended for periods up to 30 days. Digital printer furnishes complete record
References	Manufacturer's specifications
Cost	SAAM 1: $8380 SAAM 2: $10,580 SAAM 3: $7260
Address	Technical Associates 7051 Eton Avenue Canoga Park, CA 91303 (213) 883-7043

Survey Meters
Victoreen Models: 425, 470A, 471,
471A, 471RF, 692 and Series 660

Class	Field
Principle of Operation	Depending on detector used, will measure alpha, beta, gamma or X-rays.
Sensitivity and Range	*Range* Model 425: 0–500, 5000, 50,000 and 500,000 cpm in four linear ranges Model 470A, 471A: 12 overlapping ranges: 0–3, 10, 30, 100, 300, 1000 mR/hr and R/hr for rate. 6 overlapping ranges: 0–3, 10, 30, 100, 300, 1000 mR in the "integrate" mode. Model 471, 471RF: 12 overlapping ranges: 0–1, 3, 10, 30, 100, 300 mR/hr and R/hr for rate. 6 overlaping ranges: 0–1, 3, 10, 30, 100, 300 mR in the "integrate" mode Model 692: 4 linear ranges: 0–10, 100, 1000 and 10000 mR/hr. *Series 660* Model 660-1 (Meter): 3 decade dependent on probe. Model 660-3 (Probe): Total exposure: 0.001 to 99.9 R Exposure rate: 0.01 to 999 R/min Model 660-4 (Probe): Total exposure: 0.1 mR to 9.99 R Exposure rate: 0.001 to 99.9 R/min Model 660-5 (Probe): Total exposure: 0.001 to 99.9 mR Exposure rate: 0.1 mR/hr to 9.99 R/hr

Performance

Model	Precision	Response time
425	5% full scale	~20 sec on 500 cpm 5 sec on 5000 cpm 2 sec on two highest ranges
470A 471A	10% full scale	8 sec on 3 mR/hr 3 sec on 10 mR/hr < 1 sec on other ranges
471 471RF	10% full scale	~8 sec on 1 mR/hr 3 sec on 3 mR/hr < 1 sec on other ranges
692	15% full scale	~5 sec on most sensitive ranges 2 sec on other ranges
660-3 (Probe) 660-4 (Probe) 660-5 (Probe)	Within 1% short term and 2%/year year at 20 ± 2°C	--

Requirements

Power

Model 425: 120 or 240 V ac, 50 or 60 Hz or 18 V at 1 Amp/hr with 3, 6 volt rechargeable "Lead Gel" cells.

Model 470A, 471A, 471, 471RF: Two 1.5 V "D" cells and four 22.5 V No. 505 batteries

Model 692: Two carbon-zinc D cells

Model 660-1: One 67.5 V collecting potential (shelf-life) and 4 rechargeable 1.25 V "D" size batteries

Size

Model 425: 9.5 cm H × 21.8 cm W × 23.5 cm L (3.75 in. × 8.5 in. × 9.25 in.)

Model 470A: 28 cm L × 12 cm W × 24 cm H (11 in. × 4.75 in. × 9.5 in.)

Model 471, 471A: 29 cm L × 13 cm W × 19.4 cm H (11.5 in. × 5.125 in. × 7.625 in.)

Model 471RF: 30.5 cm L × 13 cm W × 19.4 cm H (12 in. × 5.125 in. × 7.625 in.)

Model 692: 12.2 cm W × 17.1 cm H × 30.5 cm L (4.75 in. × 6.75 in. × 12 in.)

Model 660-1: 13.9 cm W × 21.5 cm L × 11 cm H (5.5 in. × 8.5 in. × 4.5 in.)

Model 660-3, 660-4 (Probes): 28.3 cm L × 6.36 cm cm W × 3.18 cm D (11 in. × 2.5 in. × 1.25 in.)

Model 660-5 (Probe): 38.3 cm L × 12.8 cm W × 4.83 cm D (15 in. × 5 in. × 1.9 in.)

Weight:

Model	Weight
425	2.7 kg (6 lb)
470A	1.8 kg (4 lb)
471A	2.2 kg (4.9 lb)
471	2.2 kg (4.9 lb)
471RF	2.4 kg (5.3 lb)
692	1.7 kg (3.7 lb)
660-1	3.2 kg (7 lb)
660-3,4,5	1 kg (2 lb)

References — Manufacturer's specifications

Cost

Model	Price ($)
425	595
470A	995
471A	1375
471RF	1395
692	795
660-1	1595
660-3,4	525
660-5	595

Address

Victoreen
10101 Woodland Ave.
Cleveland, OH 44104
(216) 795-8200
Attn: W. Snider

Radiation Survey Meters
Victoreen

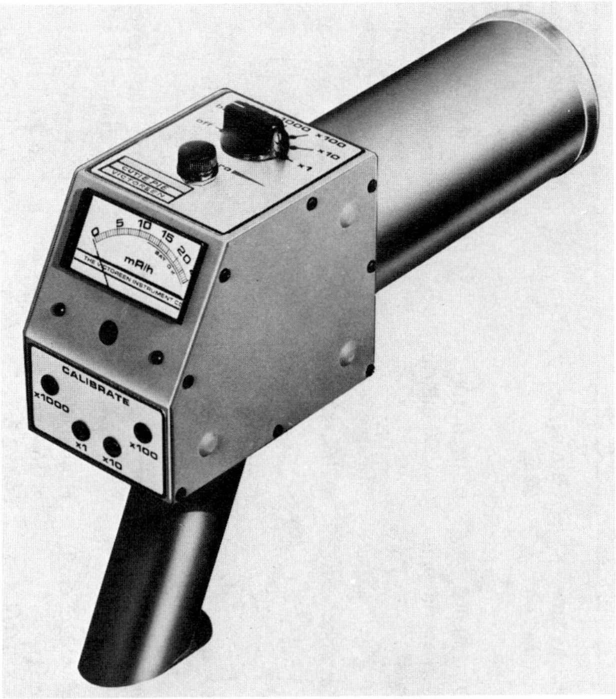

Class	Field
Specifications	See Table A.3.14
Address	Victoreen, Inc. 10101 Woodland Avenue Cleveland, OH 44104 (216) 795-8200 Attn: W. Snider

Table A.3.14. Victoreen Survey Meters

	Model			
Specifications	490 (Thyac III)	491	492	493
Radiation Detected	Alpha, beta, gamma depending on probes	Alpha, beta, gamma, x-ray, depending on probes	Gamma, x-ray above 60 keV	Alpha, beta, gamma, x-ray depending on probes
Detector Range	0–800, 8000 80,000 cpm & 0.2, 2, 20 mR/hr (489-4 probe)	0–150, 450, 150,000 cpm in seven overlapping linear ranges	0–10, 0–100, 0–1000 mR/hr	0 to 300, 3000, 30,000 cpm corresponding to 0–0.5, 5, 50 mR/hr with with 493-50 probe
Energy Dependence	Depends on probe	Depends on probe	±20%, 80 keV to 1.0 MeV	Depends on probe
Precision	±10% FS	±10% FS	±20% FS	±20% FS
Power Requirements	2 "D" cells	4 "D" cells	2 "D" cells	2 "D" cells
Features	Variable response time	Energy correction	Energy and temperature correction	Energy and temperature correction
Overall Dimensions	22.2 cm L × 11.4 cm W × 17.8 cm H (8 3/4 in. × 4 1/2 in. × 7 in.)	11.4 cm W × 22.2 cm L × 17.8 cm H (4 1/2 in. × 8 3/4 × 7 in.)	11.4 cm W × 22.2 cm L × 16.8 cm H (4 1/2 in. × 8 3/4 in. × 6 5/8 in.)	11.4 cm W × 22.2 cm L × 16.9 cm H (4 1/2 in. × 8 3/4 in. × 6 5/8 in.)
Net weight	1.82 kg (4 lb)	2.25 kg (5 lb)	1.6 kg (3.5 lb)	1.6 kg (3.5 lb)
Cost	$625.00 (less probe)	$710.00 (less probe)	$365.00	$350.00 (less probe)

Table A.3.14. (Continued)

	Model			
Specifications	498	740F	470A	885
Radiation Detected	Gamma, x-ray, beta	Alpha above 3.5 MeV, beta, above 70 keV, gamma above 7 keV; ion chamber	Alpha above 8 MeV, Beta above 120 keV, gamma and x-ray above 10 keV; ion chamber	X-ray, gamma
Detector Range	0–1,10,100 mR/hr 0–1 R/hr	0–25,250,2500, 25,000 mR/hr	Exposure rate mode: 0–3,10,30,100,300,1000 mR/hr or R/hr (rate); Integrate mode: 0–3, 10,30,100,300,1000 mR	0 to 999 mR
Energy Dependence	Depends on probe	10% 7 keV to 100 keV; ±15% 40 keV to 2 MeV	±15% 10 to 300 keV (bare chamber) ±10% 40 keV to 2 MeV (sleeve on)	50 keV to 1 MeV
Precision	±20% FS	±10% FS	±10% FS	±15%
Power Requirement	2 "D" cells	Four 22.5-V batteries	Two 1.5-V "D" cells and four 22.5-V No. 505 cell	One 9-V batteries
Features	High range	Rugged proven design	Measures complete energy spectrum over 2π solid angle	
Overall Dimensions	11.4 cm L × 22 cm W × 18 cm H (4 1/2 in. × 8 3/4 in. × 7 in.)	29 cm L × 9 cm W × 23 cm H (11 1/2 in. × 3 1/2 in. × 9 in.) (including probe)	27.9 cm L × 12.1 cm W × 24.1 cm H (11 in. × 4 3/4 in. × 9 1/2 in.)	6 cm × 11.4 cm × 2.5 cm H (2 1/2 in. × 4 1/2 in. × 1 in.)
Net weight	1.8 kg (4 lb)	1.7 kg (3.75 lb)	1.8 kg (4 lb)	0.17 kg (6 oz)
Cost	$475.00	$995.00	$995.00	$220.00

Radiation Survey Meters
Warrington Models: 2590, 2650, and 2650AR Series

2651

2652

RAD-ALP,BET,GAM,XRA
Warrington 1
January 1983

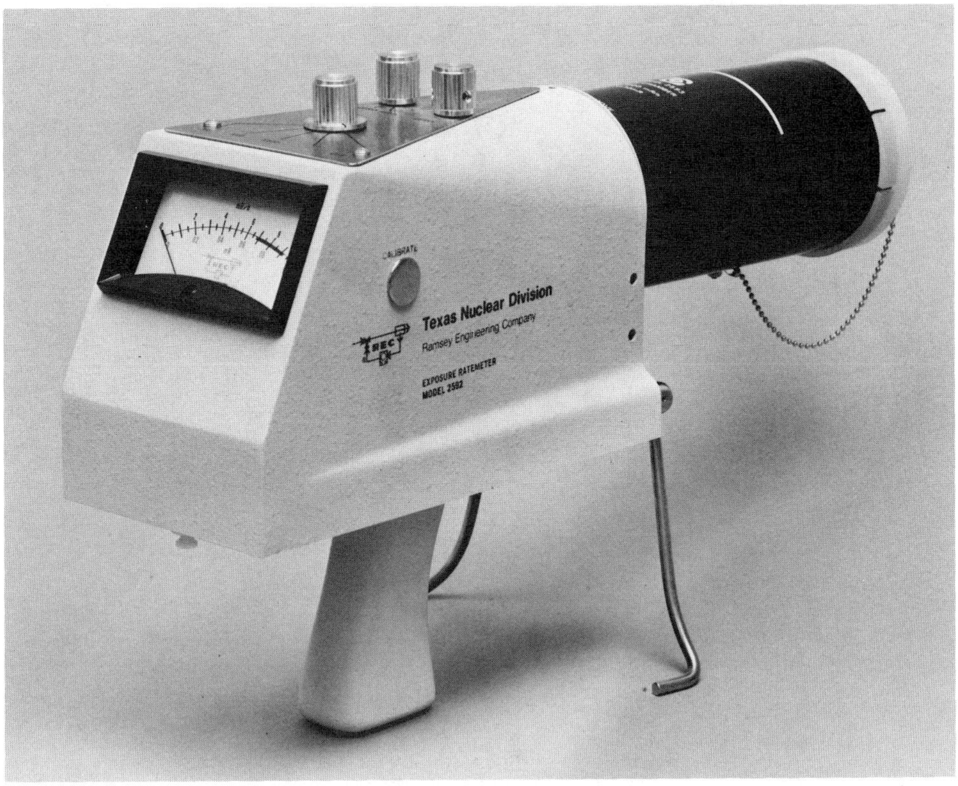

2595 and 2596

Principle of Operation	Ionization chamber, amplifier, oscillator meter readout
Specifications	See Table A.3.13
Address	Warrington Laboratories Executive Office Terrace 223 W. Anderson Lane, Suite 500 Austin, TX 78752 (512) 452-2838 Attn: J. Petermann

Table A.3.13. Radiation Survey Meters

	2590 Series		2550 Series*	
Specifications	2595 (low range)	2596 (high range)	2651	2652
Range				
Exposure rate	0–10,100,1000 mR/hr	0–10,100,1000 R/hr	0–0.1,0.3,1.0,3.0,10,30,100 mR/hr	
Accumulated Exposure	0–0.1,1,10 mR	0–0.1,1,10 R	Rate: 0–150,1500,15,000,150,000 cpm	
Sensitivity	Minimum calculated alpha energy: 1.7 MeV Minimum calculated beta energy: 0.025 MeV		With cap removed may measure alpha and beta radiations with energies as low as 40 keV	
Sensitive Volume	500 cm^3	5 cm^3		
Window Thickness	~1 mg/cm^2 aluminized Mylar		30 mg/cm^2	1.5 to 2.0 mg/cm^2
Accuracy	±10% FS calibrated with 1.17- and 1.33-MeV ray ^{60}Co		±10% FS	
Operating Temperature	0 to 55°C (32 to 131°F)		0 to 50°C (32 to 122°F)	
Time constants	3 sec	0.5 sec	Selectable 4 or 8sec on 0.1, 0.3, 1.0 mR/hr range; fixed 4 sec on 3 and 10 mR/hr range; fixed 2 sec on 30 and 100 mR/hr range	

Requirements			
Power	Three 1.5-V "D" cells, 480 hr continuous life		Four 1.5-V "D" cells, 300 hr @ 8 hr/day
Size	37.4 cm L × 10.2 cm W × 21.6 cm H (14.75 in. × 4 in. × 8.5 in.)		12.7 cm H × 10.2 cm W × 30.5 cm L (5 in. × 4 in. × 12 in.)
Weight	2 kg (4.5 lb)	2 kg (4.5 lb)	
Cost	Refer to manufacturer		Refer to manufacturer

*Model 2650 AR: this is a modification of the 2650 series which contains an audio-amplifier (speaker mounted internally and volume with separate on/off external switch) and rechargeable Ni-Cd batteries with an external detachable transformer.

Survey Meters
Models 306A and 308A

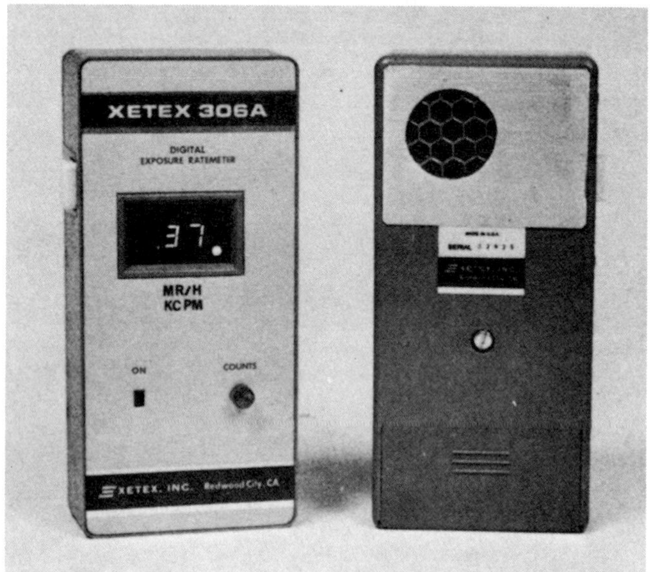

Class	Portable, handheld	
Principle of Operation	Halogen-quenched 1.2-in. diameter G-M detector with 1.5 to 2.0 mg/cm^2 window, digital logic and display (306A), and meter readout (308A)	
Sensitivity	α, β, γ and x-rays; 0.002 μCi lower detection limit	
	306A	**308A**
Range	0–9.99K cpm Overrange indicator	0–100, 0–1000, and 0–10,000 cpm range switch
Sampling	Push button	Continuous
Performance	Accuracy: ±10% FS Response time: 6 sec	±10% FS 12 sec (×1), 3 sec (×10 and ×100)
	Operating temperature: 0 to 50°C	0 to 50 °C
Requirements	Power: 4 "AA" alkaline penlite Power: batteries, life ~100 hr Size: 7.1 cm × 15.5 cm × 3.89 cm Size: (2.81 in. × 6.09 in. × 1.53 in.) Weight: 0.27 kg (9.5 oz) Weight: w/batteries	One 9 V transistor-type battery, life ~200 hr 7.1 cm × 15.5 cm × 4.7 cm (2.81 in. × 6.09 in. × 1.87 in.) 0.33 kg (11.5 oz) w/battery

Features

1. Autoranging, three-decade coverage (306A)
2. Zero stable (308A)
3. Battery test
4. LED "counts" lamp
5. Speaker (optional)*

Reference: Manufacturer's specifications

Cost:
306A: $345.00
308A: $345.00
306A-2: $380.00
308A-2: $380.00

Address:
Xetex, Inc.
660 National Ave.
Mountain View, Ca 94043
(415) 964-3261
Attn: Ken Sinclair

*Add -2 suffix for speaker option; e.g. 308A-2

Background Subtract Ratemeter
Xetex Model 315B

Class	Portable, handheld
Principle of Operation	Scintillation and proportional detectors, digital logic with background subtraction, meter display
Sensitivity and Range	Range: 0–1000 and 0–10,000 c/s *With Model 3150* Scintillation Probe (2 in. × 2 in. NaI-Tl)* X-ray and gamma ray Sensitivity: 1600 c/s at 0.1 mR/hr referred to ^{137}Cs Energy response: 30 keV up *With Model 3151* Proportional Probe (1 mg*·cm^{-2}) Alpha, Beta, and Gamma Efficiency: 30% of 2π emission, alpha source contacting grill
Sampling	Continuous
Performance	Amplifier gain: variable, max. gain = 2000 × Discriminator: differential type Background subtract: continuous Operating temperature: 0–50°C
Requirements	Power: four quick-charge "AA" ni-cd (150 hr on full charge) Size: 19 cm H × 7.5 cm W × 5 cm D (315B) (7 1/2 in. × 3 in. × 2 in.) 24.8 cm L × 7.3 cm Dia. (3150*) (9 3/4 in. × 2 7/8 in.) 19.7 cm H × 12.1 cm W × 2.5 cm D (3151*) (7 3/4 in. × 4 3/4 in. × 1 in.) Weight: 0.91 kgm (2 lb) (315B) 1.4 kgm (3 lb) (3150*) 0.31 kgm (0.7 lb) (3151*)
Features	1. Compares real-time count with stored background + preset count increment 2. Count increment of 1 to 99, switch selectable 3. Variable discriminator threshold and window 4. Compact, rugged polyester case
References	Manufacturer's specifications
Cost	$1075.00 (315B) $ 525.00 (3150*) $ 700.00 (3151*)

*Probes not included; order separately

Address

Xetex, Inc.
660 National Ave.
Mountain View, CA 94043
(415) 964-3261
Attn: Ken Sinclair

Digital Ratemeter/Survey Meter
Xetex Model 318A

Class	Portable, handheld
Principle of Operation	Ratemeter uses halogen-quenched filtered G-M tube detector, digital logic and 4 digit LED display; converts to survey meter with optional probe (Model 3180) using 1.2 in. diameter halogen-quenched G-M detector with 1.5–2.0 $mg \cdot cm^{-2}$ window
Sensitivity	*Ratemeter* (318A) Radiation detected: x-ray and gamma ray Range: 0 to 9999 mR/h, overrange indicator Response time: nominal 1 sec *Survey Meter (318A w/3180* probe)* Radiation detected: alpha, beta, gamma, and x-ray Range: 0 to 9999 c/s, overrange indicator Response time: nominal 1 sec
Sampling	Continuous
Performance	Accuracy: ±15% referred to ^{137}Cs (318A); ±10% FS (318A/3180*) Energy response: ±15% from 60 keV to 1.3 MeV (318A) Operating temperature: 0 to 50°C
Requirements	Power: four "AA" ni-cd (up to 8 hr per charge) Size: 15.5 cm H × 7.1 cm W × 3.9 cm Dia. (318A) (6.1 in. × 2.8 in. × 1.5 in.) 11.4 cm L × 4.1 cm D (3180*) (4 1/2 in. × 1 5/8 in.) Weight: 0.27 kg (0.6 lb) (318A) 0.22 kg (0.5 lb) (3180*)
Features	1. mR/hr and c/s flags 2. Large, bright LED display 3. Dual function for gamma/x-ray and contamination measurements 4. Unit charger (3181) or charging bin (3182) for up to 10 meters available
References	Manufacturer's specifications

*Not included, order separately

Cost	$410.00 (318A) $140.00 (3180*) $ 17.50 (3181*) $450.00 (3182*)
Address	XETEX, Inc. 660 National Ave. Mountain View, CA 94043 (415) 964-3261 Attn: Ken Sinclair

A.4. INSTRUMENTS FOR MEASURING ALPHA RADIATION

Included in the following pages are Instrument Notes of several different kinds. The reader should be aware of the following facts, in order to find all Instrument Notes appropriate for alpha detection and measurement:

1. Semiconductor detectors are useful for both alpha and beta counting and spectrometry. Instrument Notes for *silicon surface-barrier detectors* and *silicon lithium-drifted detectors* have been incorporated in this section. For more notes on semiconductors, the interested reader is advised to consult Gamma Spectrometry (Section A.6.1) in the Instrument Notes.
2. Many instruments are advertised as sensitive to alphas, betas, gammas, x-rays, and in some cases neutrons. Often these instruments consist of a variety of probes attachable to a central electronic counting unit. These instruments have all been placed in Section A.3, Combination Instruments, with the appropriate mnemonic (e.g., RAD-ALP,BET,GAM,XRA).
3. *Liquid-scintillation systems* can be used for alpha detection and measurements. However, all these instruments have been placed in the specific radionuclides section (A.9).

Alpha Air Monitor
Eberline Alpha-5A

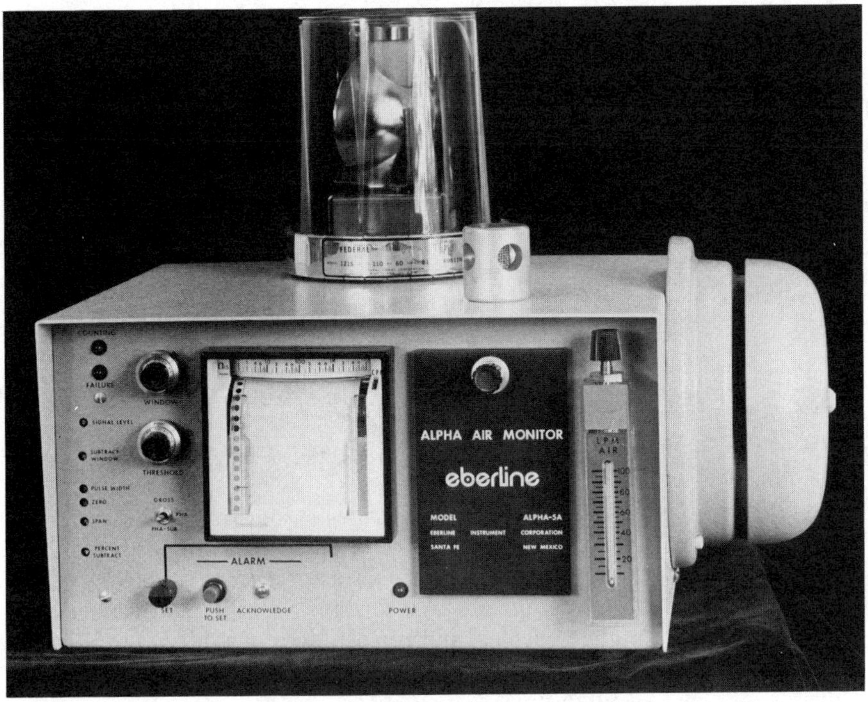

Class	Stationary
Principle of Operation	Silicon-diffused junction-type detector (490 mm^2 area), 2.54-cm diameter filter, air pump, single-channel analyzer, meter or strip readout.
Sensitivity and Range	Efficiency: About 50% of 2π (wide energy window) from 2.54 cm (1 in.) ^{239}Pu source Background: 400 cpm (wide energy window), with air flow of 30 liters/min (1.05 ft^3/min). This results in ~8 cpm in the window when the instrument is calibrated for ^{239}Pu. Range: 0–10000 cpm full log scale Response time: varies with count rate
Sampling	Continuous (air filter)

Performance	Detection limit: 4 MPC-hr of ^{239}Pu can be detected with an average radon-thoron background Temperature: −6.7 to 54.4°C (20 to 130°F) Threshold discriminator: adjustable from 0 to 10 MeV energy equivalent (0 to 1.0 V)
Requirements	Power: 105 to 125 V ac, 60 Hz at 0.1 A (220 V ac, 50 Hz optional) Size: 36 cm × 31 cm × 40 cm (14.25 in. × 12.25 in. × 15.75 in.) Weight: 8.9 kg (19.7 lb)
Features	1. Air pump must be provided separately. Model RAP-1 available as option. Air intake is screen protected. Air flow indicator goes from 10 to 100 liters/min (0.35 to 3.5 ft^3/min). 2. Single-channel analyzer has 10-turn adjustments for high and low settings of window. 3. Operation modes are gross, pulse-height-analyzing, or pulse-height-analyzing with background subtraction. 4. Subtraction circuit provides automatic subtraction of a percentage of pulses rejected by the window discriminator. Subtracted percentage adjustable from 10 to 60%. 5. Alarm adjustabale, 0 to 100% of full scale. Red alarm light, squealer, and remote alarm provided. 6. Calibration controls accessable through front panel. 7. 47 mm filter is standard. 8. Electronics are solid state. Charge-sensitive amplifier input. 9. Silicon detector is cleanable.
References	Manufacturer's specifications
Cost	$3150.00
Address	Eberline Instrument Corporation P.O. Box 2108 Santa Fe, NM 87501 (505) 471-3232

Scintillation Alpha Counter
Eberline Model SAC-4

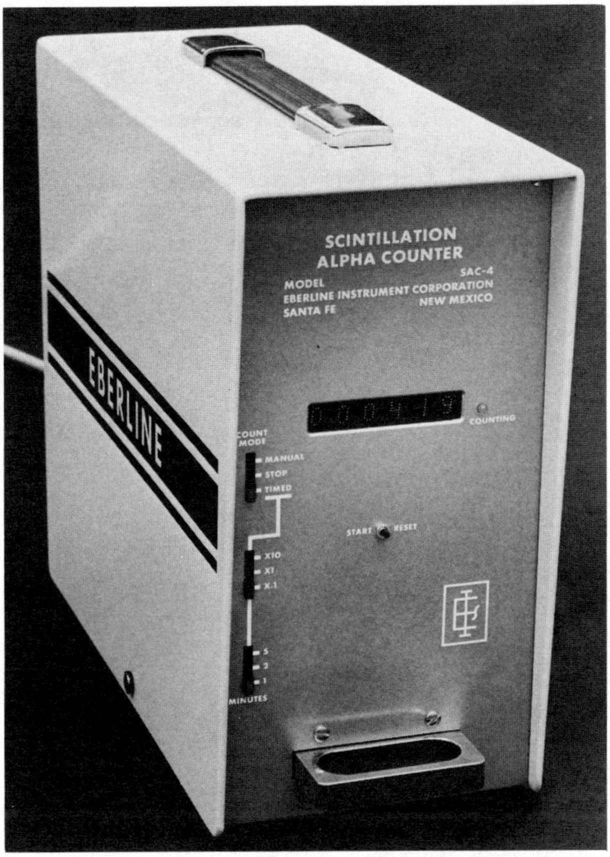

Class	Laboratory
Principle of Operation	Two-inch detector ZnS(Ag) powder on a plastic light pipe, high-voltage power supply, charge-sensitive input amplifier, timer, and 6-decade readout.
Sensitivity	Efficiency: 80% of 2π minimum from a 2.54-cm (1-in.) diameter ^{239}Pu source Background: < 0.3 cpm
Sampling	Maximum sample diameter 5.2 cm (2 1/32 in.); maximum thickness 0.95 cm (3/8 in.)

Performance	Plateau (detector): at least 200 V long with a slope less than 1% of 100 V. Preset timer from 0.1 to 50 min Temperature: 0 to 60°C (32 to 140°F) with less than ±50 V plateau shift
Requirements	Power: 115/230 ± 10% V ac, switch selectable, 50 to 60 Hz, 114 A, 3-wire Size: 29.2 cm × 15.2 cm × 35.6 cm (11 1/2 in. H × 6 in. W × 14 in. D) Weight: 7.6 kg (16 3/4 lb)
Features	Detector and drawer assembly easily decontaminated. Variable high-voltage supply.
References	Manufacturer's specifications
Cost	$2100.00
Address	Eberline Instrument Corporation P.O. Box 2108 Santa Fe, NM 87501 (505) 471-3232

Alpha Survey Meters
Eberline Portable Models

PAC-4G

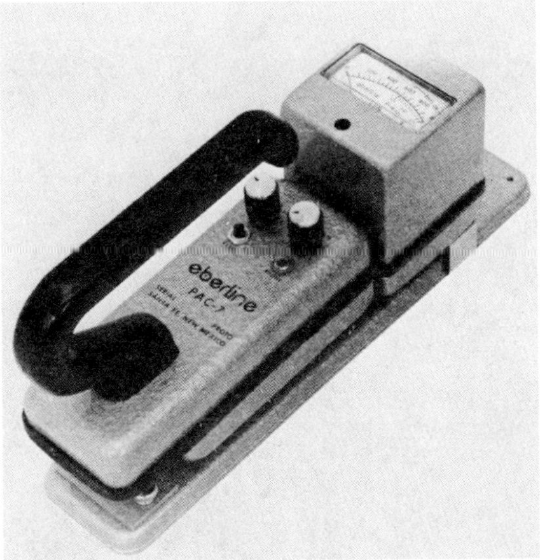

PAC-7

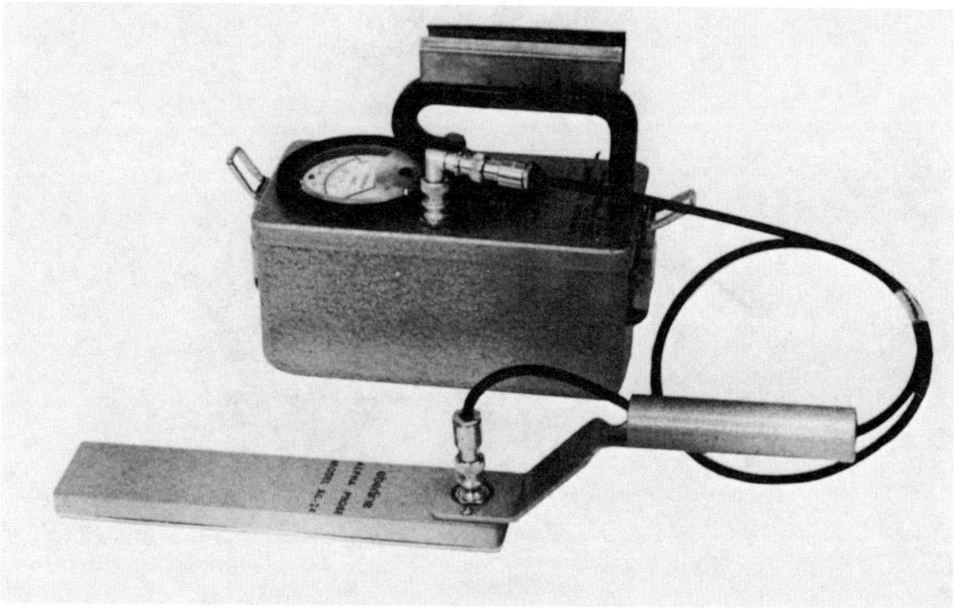

PAC-1SAGA

Class	Portable, handheld
Sampling	Continuous
Features	Some models have earphone and speaker outputs, regulated power supply, individual calibration controls for each range (see Table A.4.1).
Address	Eberline Instrument Corporation P.O. Box 2108 Santa Fe, NM 87501 (505) 471-3232

Table A.4.1. Eberline Portable Alpha Survey Meters

Specifications	PRS-1/AC-3	PRS-2/AC-3
Radiation Measured	Alpha	Alpha
Detector	Alpha scintillation probe	Alpha scintillation detector and G-M tube
Range	Six decade	Six decade
Scale	Digital	Digital
Accuracy	Limited only by counting statistics	
Temperature	−18 to 60°C (0 to 140°F)	−18 to 60°C (0 to 140°F)
Power	Rechargeable Gel-Cell battery	Rechargeable Gel-Cell battery
Weight	Inst. 2.2 kg (5 lb)	Inst. 2.3 kg (5 lb)
Size	20 cm H × 10 cm W × 24 cm L (8 in. × 4 in. × 9 in.)	20 cm H × 10 cm W × 24 cm L (8 in. × 4 in. × 9 in.)
Features	Ratemeter-Scalar Built-in speaker. User selects calibration constant and readout unit.	(See PRS-1/AC-3)
Remarks	PRS-1 may be used with many other detectors	PRS-2 may be used with many other detectors
Cost	$2350.00	$2350.00

Model		
PAC 4G-3	PAC-6	PAC-7
Alpha	Alpha	Alpha
Gas proportional alpha probe	Air proportional alpha counter	Air proportional alpha counter
0–5 × 10^5 cpm	0–500, 5000, 50,000 or 500,000 cpm	0–1000, 10,000 or 100,000 cpm
Linear	Linear	Linear
±10% FS	±5% FS	±5% FS
−40 to 60°C (−40 to 140°F)	−18 to 60°C (0 to 140°F)	−18 to 49°C (0 to 120°F)
5 "D" cells, 200 hr	2 "D" cells,	Single 9-V alkaline battery
4.25 kg (9.6 lb)	1.9 kg (4.25 lb)	0.55kg (1.2 lb)
22.9 cm H × 10.2 cm W × 25.4 cm L (9 in. × 4 in. × 10 in.)	10.2 cm W × 20.3 cm L × 25.4 cm L (4 in. × 8 in. × 7.75 in.)	5 cm W × 23 cm L × 19.7 cm H (2 in. × 9 in. × 5 in.)
Best gamma rejection	Good gamma rejection	Light weight, one hand operation
High-voltage selector option	Approximately 50% counting efficiency	Removable meter module
$1230.00	$870.00	$800.00

Portable Alpha Monitor
Harshaw Model 272

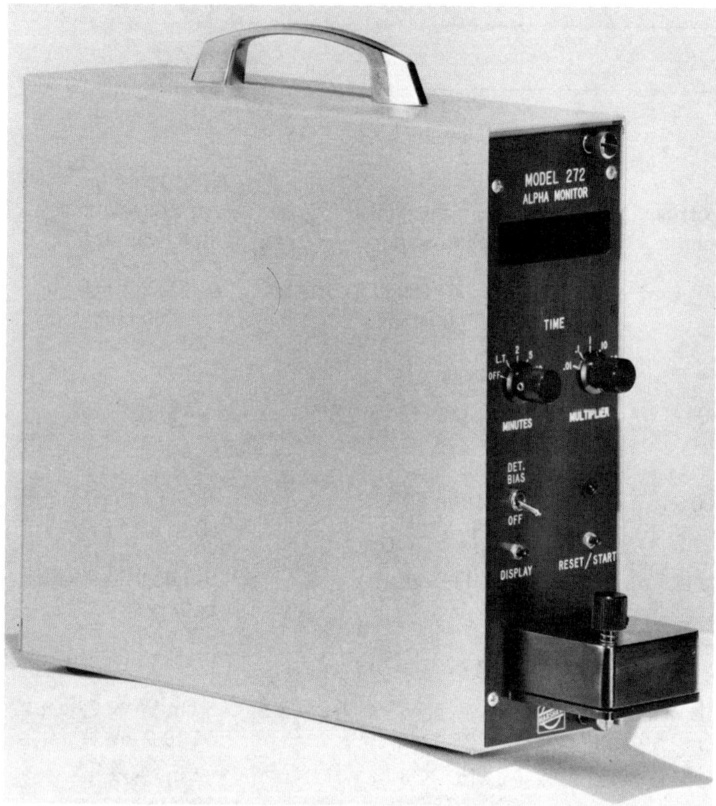

Class	Portable
Principle of Operation	Diffused-junction silicon detector (300 mm^2), pulse-height analyzer, four-digit LED display
Sensitivity and Range	Range: 0 to 9999 counts Time range: 0.02 to 1000 min Alpha sensitivity: 2 to 6 MeV
Sampling	Single sample counting
Performance	Efficiency: typically 25% for ^{241}Am alpha and standard sample holder. Background: < 1 count per hour Maximum sample size: 3.2 cm × 0.32 cm (1 1/4 in. o.d. × 1/8 in.)

Requirements	Power: Rechargeable Ni-Cd batteries + battery recharger (117 V ac) Size: 22.2 cm H × 8.3 cm W × 27.3 cm D (8 3/4 in. × 3 1/4 in. × 10 3/4 in.) Weight: 3.3 kg (7 1/4 lb)
Features	NIM Bin Version Optional Easily decontaminated housing and sample holder
References	Manufacturer's specifications
Cost	Refer to factory
Address	Harshaw Chemical Company Crystal and Electronic Products Dept. 6801 Cochran Road Solon, OH 44139 (216) 248-7400

Diffused-Junction Silicon Radiation Detectors
Harshaw Series DJ Silicon Radiation Detectors

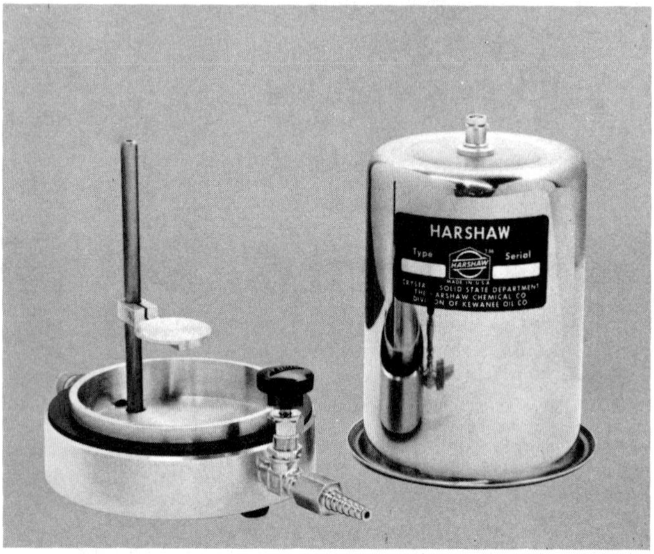

Class	Radiation detectors (alpha)
Principle of Operation	Diffused-junction (*p-i-n*) silicon detectors
Resolution	Values range from 25 to 60 keV for ^{241}Am (5.47 MeV) depending upon active area
Active Areas Available	100 mm^2, 400 mm^2, 450 mm^2, 490 mm^2, 700 mm^2
Depletion Depth	Typically 100 μ at 100 V positive bias (typically 200 μ for 700 mm^2 model)
Features	Detector surface may be cleaned using a cotton swab moistened with trichloroethylene, methanol, or denatured alcohol. Detectors are available with microdot or BNC connectors
Available Accessories	Model 101 Vacuum Chamber, NV-30 0–300 V dc NIM bias supply, NV-32 High Voltage NIM power supply
References	Manufacturer's specifications
Cost	Refer to Factory

Address Harshaw Chemical Company
Crystal and Electronic Products Dept.
6801 Cochran Road
Solon, OH 44139
(216) 248-7400

Phoswich Scintillation Detector
Harshaw Chemical Company

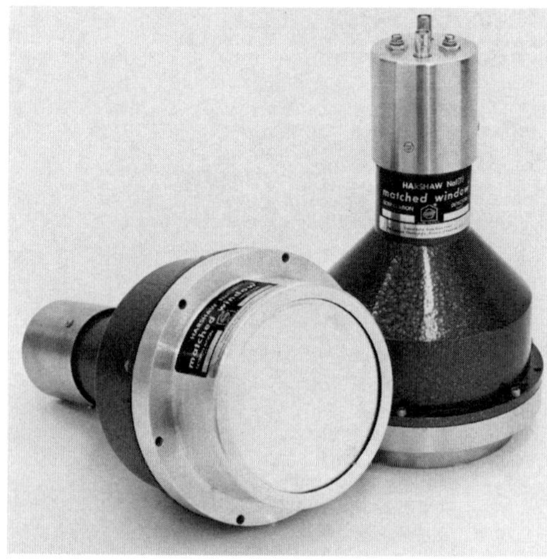

Class

Laboratory

Principle of Operation

The Phoswich detector is designed for high-sensitivity radiation counting where low-level, low-energy radiation is normally obscured by background radiation. The detectors consists of a thin primary scintillator and a thick secondary scintillator. Events that occur in the secondary crystal only, and those events that occur in both crystals simultaneously, are electronically rejected. Only the low-level radiation that is totally absorbed by the primary scintillator is counted. Events are sorted out using the Harshaw NC-25 A Pulse Shape Analyzer which recognizes events in the primary and secondary crystals by their different decay times.

By using a thin CaF_2(Eu) primary scintillator and a NaI(Tl) secondary scintillator, the Phoswich can detect alpha and beta particles in the presence of a gamma background.

Performance

Can suppress higher-energy backgrounds by factors of 3 to 10, compared to ordinary NaI(Tl) or CsI(Tl) detector.

Especially designed for low-energy x-rays from Americium-241, as used to determine plutonium.

Features

Pulse shape analysis system Model TASC-10 also available.

References	Manufacturer's specifications
Cost	Phoswich detector: ~$4500.00, depends on size
Address	Harshaw Chemical Company 6801 Cochran Road Solon, OH 44139 (216) 248-7400

Diffused-Junction Silicon Radiation Detector
Harwell Type SRD 25

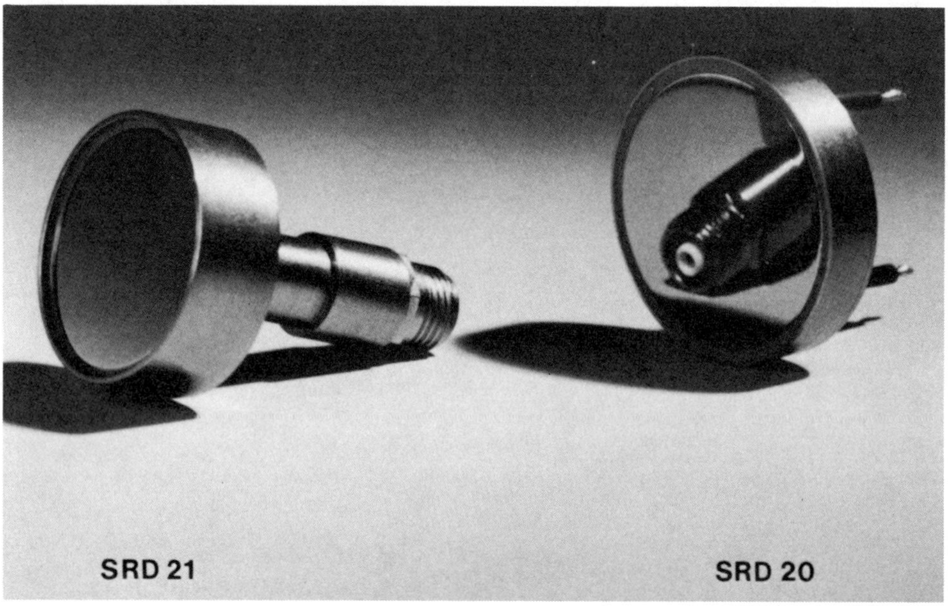

Class	Detector
Principle of Operation	Consists of silicon diode mounted within a stainless steel body and terminating at two wire pins, which are electrically isolated from the body. Diode mounted flush to form front surface
Sensitivity	Sensitive Area: 700 mm^2, thickness: 0.5 mm (nominal) Reverse current @ 15 V: < 2 μA @ 20°C Depletion depth @ 15 V: 60 μ (nominal)
Sampling	Not applicable without electronic accessories
Performance	Operating bias voltage: 15 V Total capacitance @ 15 V: 1300 pF (typical), 1700 pF (max.) Background: 2 counts/hour above 1.7 MeV, 20°C (typical)
Requirements	Power: Power supply separate Size: Detector dimension: diameter overall 3.3 cm; thickness: 11.9 mm
Features	The SRD 20 is the largest detector in a series of rugged α detectors from 10 to 700 mm^2 sensitive area
References	Manufacturer's specifications

RAD-ALP
Harwell (UKAEA) 1
January 1983

Cost	On application
Address	AERE, Harwell Instrumentation and Applied Physics Division Bldg. 347.3 Oxfordshire OX11 ORA England Tel: (0235) 24141 Telex: 83135 Attn: R. H. W. Sherwood (Ext. 4766)

Alpha Spectrometer
Harwell Model 36190

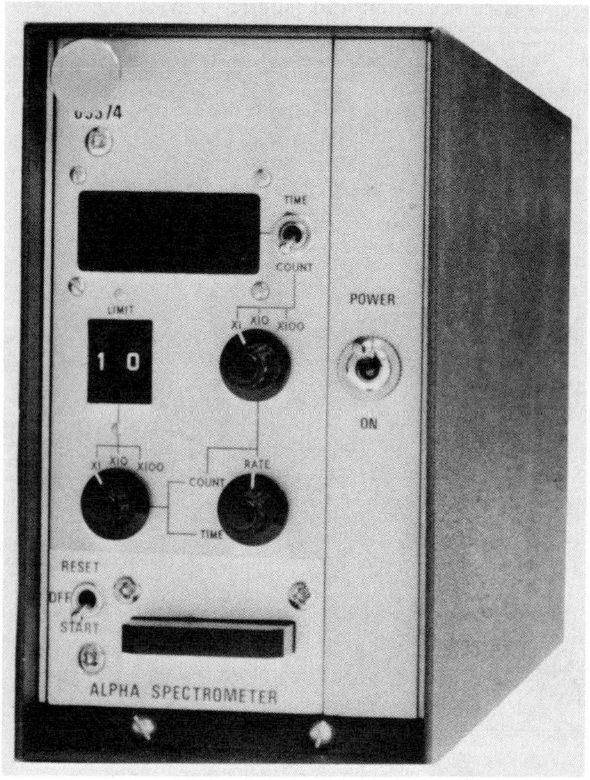

Class	Laboratory
Principle of Operation	Uses a silicon radiation detector to measure the α activity of samples such as filter paper used in the Harwell Personal Air Sampler Type 0434 (separate). Includes single-channel analyzer.
Sensitivity and Range	4-Decade LED display, with 1, 10, 100 prescale option. For α of 2.5 to 5 MeV, efficiency is 30% for ^{239}Pu. SCA threshold and channel width can be set in range 0.1 to 8 MeV.
Sampling	Discrete timing range: 1 to 9900 sec, 1 in. diam. tray for sample holder
Performance	Background: typical, 0.2 c/hr Detector: Harwell type SRD 22

Requirements	Power: 240 V ac main lines Size: 8.9 cm W × 17.8 cm H × 33.0 cm L (3.5 in. × 7 in. × 13 in.) Weight: 4 kg (8 1/2 lb)
Features	The SCA can be operated in the time limit, count limit or digital ratemeter mode. Display can be switched to indicate count time (seconds) or count rate (count per seconds)
References	Manufacturer's specifications
Cost	On application
Address	AERE, Harwell Instrumentation and Applied Physics Division Bldg. 347.3 Oxfordshire, OX11 ORA England Tel: (0235) 24131 Attn: R. H. W. Sherwood (Ext. 4776)

Alpha in Air Measuring System
Harwell Type 3280

Class	Laboratory, wall mounted
Principle of Operation	Air drawn through a fixed filter paper deposits alpha-active particulates which are viewed continuously by a silicon semiconductor alpha detector. 25 mm Whatman GF/A glass fiber filter paper is supplied in a cardboard box.
Sensitivity	An SRD 25 semiconductor detector gives a detection efficiency of about 30% (4π); the count sensitivity for ^{239}Pu is 0.4 cps at an exposure of 8 mpc-h (^{239}Pu mpc = 2 pCi/m^3) and an air flow rate of 37 liters/min. The instrument incorporates three-channel digital compensation against the interfering effect of radon/thoron decay products. In an average radon concentration of 100–200 pCi/m^3 reliable alarm operation can be achieved down to 5 mpc-hr.
Requirements	Power: 110 V/240 V Ac, 50/60 Hz, rated at 30 volt-amp Size: 27.8 cm × 56.0 cm × 29.0 cm Weight: 16 kg Air pump: normal flow rate 37 liters/min, rotameter indication.
Features	Compensated Pu digital pulse rate summed with a regular 10.67 pps pedestal pulse rate, with output interface to RS 422 for remote processing. Four pairs of isolated dry reed alarm contacts (open, de-energize on alarm) each for high activitity and equipment failure. The high activity alarm can be preset digitally over a wide range. Separate digital outputs from the four fixed counting channels. Low impedance MCA output for spectrum analysis. Local alarm is provided by a 12 W red beacon lamp and single or two-tone Yodalarm. Automatic self-test cycle is initiated every 160 minutes and whenever the Alarm Reset is operated.
Cost	Refer to manufacturer
Address	AERE, Harwell Instrumentation and Applied Physics Division Bldg. 347.3 Oxfordshire OX11 ORA England Tel: Abingdon (0235) 24141 Telex: 83135 Attn: R. H. W. Sherwood (Ext. 4766)

Liquid Monitor
National Nuclear Corporation Model LMS-1

Class	Portable
Principle of Operation	Alpha detecting unit, power supplies, amplifier, counter component, LED display
Sensitivity and Range	Maximum sensitivity: ~10 dpm/ml (depending on characteristic of solution to be monitored) Range: 0 to 99999 counts
Sampling	Single sample counting
Performance	Accuracy: typical ±20% with optically clear solution with pH 5 to 9 if containing concentrations over 500 ppm Operating temperature: 4.4 to 65.6°C (40 to 150°F)
Requirements	Power: 110 or 220 V Size: 12.7 cm D × 30.5 cm H (5 in. × 12 in.) Weight: 6.8 kg (15 lb), detector 6.8 kg (15 lb), electronics
References	Manufacturer's specifications

Cost	$10,500.00
Address	National Nuclear Corporation 1904 Colony Street Mt. View, CA 94043 (415) 962-9220

Air Particulate Monitors
Nuclear Measurements Corporation

Class	Laboratory
Principle of Operation	A gas flow proportional counter or scintillation detector, stationary or moving filter, air pump, counting ratemeter, meter or strip readout.
Specifications	See Table A.4.2
Features	1. Automatic air flow control to ±0.2 cfm within a flow rate range of 1 to 4 cfm. 2. Alarm adjustable 0 to 100% full scale, red alarm lamp, alarm bell. 3. Continuous duty, noncontacting Roots-type pump; belt-driven by split phase induction motor at 800 RPM. 4. Optional digital count ratemeter.
References	Manufacturer's specifications
Address	Nuclear Measurements Corporation 2460 North Arlington Avenue Indianapolis, IN 46218 (317) 546-2415 TWX 810-341-3137

Table A.4.2. Air Particulate Monitors

	Model					
Specifications	AM-2A	AM-3A	AM-2P	AM-3P	AM-22P	AM-33P
Radiation detected	α	α	α or β or α and β		α and β, separately but simultaneously	
Filter	2 1/4 in. diam. stationary	2 in. diam. moving	2 1/4 in. diam. stationary	2 in. diam. moving	2 1/4 in. diam. stationary	2 in. diam. moving
Range			3-cycle log 50–50,000 cpm			
Sensitivity, count buildup per hour[a]	374 cpm above 10-cpm background	278 cpm above 10-cpm background with filter speed of 1 in. per hour	692 cpm, α 702 cpm, β	518 cpm, α 525 cpm, β with filter speed of 1 in. per hour	692 cpm, α 695 cpm, β	518 cpm, α 525 cpm, β with filter speed of 1 in. per hour
Background	less than 1 cpm before collection		A 0.04 mR/hr background produces less than 0.1 cpm for α and 30 cpm for beta. A 10 mR/hr background produced 30 cpm plus 12 cpm/mR·hr for beta.			
Efficiency	20%	20% in SA-30 shield	α, 37% β, 37.5%	α, 37% β, 37% in SA-30 shield	α, 37% β, 37%	α, 37% β, 37% in SA-30 shield
Sampling	Continuous	Continuous	Continuous	Continuous	Continuous	Continuous

RAD-ALP
Nuclear Measurements 1
January 1983

Response time	Varies inversely with count rate from ~200 sec at 50 cpm to 1 sec at 50,000 cpm					
High	300 to 1300 V	300 to 1300 V	500 to 2000 V	500 to 2000 V	500 to 2000 V	500 to 2000 V
Power	90 to 130 V ac, 60 cycle, 5 to 7 A					
Size	53.3 cm D × 99.1 cm H × 99.1 cm W (21 in. × 39 in. × 39 in.)		30.5 cm D × 99.1 cm H × 76.2 cm W (12 in. × 39 in. × 30 in.)		53.3 cm D × 99.1 cm H × 99.1 cm W (21 in. × 39 in. × 39 in.)	
Weight	158.8 kg (350 lb)	158.8 kg (350 lb)	249.5 kg (550 lb)	249.5 kg (550 lb)	249.5 kg (550 lb)	249.5 kg (550 lb)
Cost	Special Request		\$7542.00	\$9672.00	\$8531.00	\$10,660.00
Requirements	A cylinder of P-10 gas					

[a]Assuming a 5-cfm regulated air flow and 10^{-10} $\mu Ci/cm^3$ concentration of activity.

RAD-ALP
Nuclear Measurements 1
January 1983

Continuous Alpha Air Monitor
Nuclear Research Model MAP-1000

Class	Mobile
Principle of Operation	Model MAP-1000 is a portable (readily transportable) continuous alpha air monitor which features automatic programmed sampling, delay for radon decay, counting, and printout.
Sensitivity and Range	Inlet air flow is such that background count rate does not register more than 36 cpm/m^3 with counting circuits set so that 200 alpha disintegrations per minute (normal uranium) per cubic meter of air will register 200 ± 20 on the printer. Temperature range: −18 to +55°C (−27.8 to 128°F)
Performance	The air monitor operates automatically in a time sequence as follows: 1. Sample for 30 min, change or transport paper, and repeat. Delay counting for at least 4.5 hr after sampling period 2. Count 30 min and print result 3. Average the 30-min counts for 8-hr interval 4. Print 8-hr average, time, and date.
Requirements	Power: 120 V ac, 60 Hz, single phase, 5 A max. Size: 66 cm max. W × 115 cm D × 122 cm H (26 in. × 45.3 in. × 48 in.) Weight: 159 kg (350 lb)
Features	Designed to Oak Ridge Gaseous Diffusion Plant requirements for monitoring uranium and other alpha emitters.
References	Manufacturer's specifications
Cost	Price upon request
Address	Nuclear Research Corporation Warrington Industrial Park P.O. Box H Warrington, PA 18976 (213) 343-5900

RAD-ALP
Nuclear Research Corporation 2
January 1983

Alpha Scintillation Survey Meter
Nuclear Research ASM-30

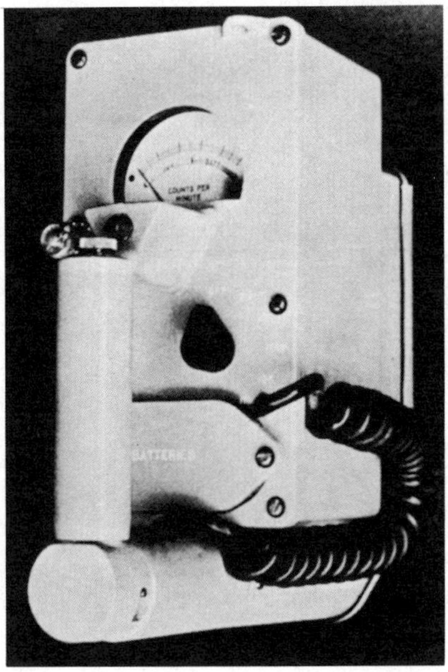

Class	Portable, handheld
Principle of Operation	Scintillation counter, [ZnS(Ag) Detector], counter, meter readout
Sensitivity and Range	Energy sensitivity: alpha, 3 MeV or greater Ranges: 10^3, 10^4, 10^5, 10^6 cpm Gamma Interference: above 5 R/hr
Sampling	Continuous
Performance	Meter accuracy: ±10% Temperature: 4.4 to 54.4°C (40 to 130°F)
Requirements	Power: two "D" cells Size: 21 cm L × 11 cm W × 15 cm H (8.5 in × 4.3 in. × 7.1 in.) Weight: 3.2 kg (7 lb)
Features	Earphone jack, battery check circuit. Designed and engineered to U.S. Navy specifications AIV/PDR-56B. Calibration alpha source

Cost	\$895.00 (w/carrying case and probe, add \$245.00)
Address	Nuclear Research Corporation Warrington Industrial Park P.O. Box H Warrington, PA 18976 (215) 343-5900

Silicon Surface-Barrier Detectors
Ortec Series A, B, C, D, E, R

Dimensions and Mounting Arrangements for Charged Particle Detectors
(Excluding G, P, and W Series)

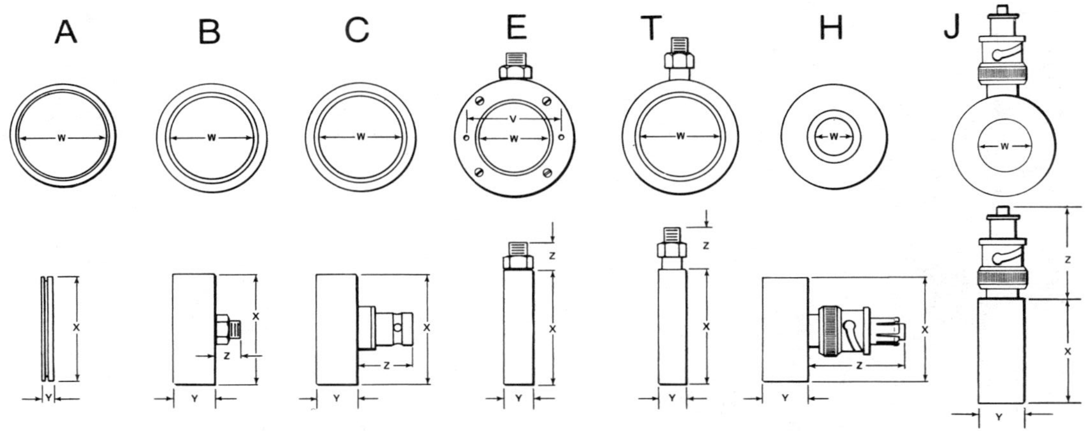

Detector Size (mm²)	W* (Nominal)	Type A Ring Mount		Type B Rear Microdot			Type C Rear BNC			Type E** Demountable Transmission				Type T Transmission Mount			Type H‡ Rear SHV			Type J*** Transmission SHV		
		X*	Y*	X	Y	Z*	X	Y	Z	X	Y	Z	V*	X	Y	Z	X	Y	Z	X	Y	Z
007†	3.0			8.1	6.5	7.1																
025	5.6	15.2	3.7	16.7	12.3	7.1	16.7	12.3	15.9	25.1	7.7	7.0	18.2	19.4	7.6	9.9	22.4	12.7	25.4	22.4	14.9	27.6
050	8.0	15.2	3.7	16.7	12.3	7.1	16.7	12.3	15.9	25.1	7.7	7.0	18.2	19.4	7.6	9.9	29.2	12.7	25.4	29.2	14.9	27.6
100	11.3	22.0	3.7	23.6	12.3	7.1	23.6	12.3	15.9	31.9	7.7	7.0	25.0	26.1	7.6	9.9	29.2	12.7	25.4	29.2	14.9	27.6
150	13.8	22.0	3.7	23.6	12.3	7.1	23.6	12.3	15.9	31.9	7.7	7.0	25.0	26.1	7.6	9.9				34.3	14.9	27.6
200	16.0	27.1	3.7	28.6	12.3	7.1	28.6	12.3	15.9	37.1	7.7	7.0	30.1	31.6	7.6	9.9				34.3	14.9	27.6
300	19.5	27.1	3.7	28.6	12.3	7.1	28.6	12.3	15.9	37.1	7.7	7.0	30.1	31.6	7.6	9.9						
400	22.6	30.5	3.7	32.0	12.3	7.1	32.0	12.3	15.9					34.8	7.6	9.9						
450	23.9	30.5	3.7	32.0	12.3	7.1	32.0	12.3	15.9	40.2	7.7	7.0	33.3	34.8	7.6	9.9						
600	27.6	34.1	3.7	36.1	12.3	7.1	36.1	12.3	15.9	43.8	7.7	7.0	36.9									
900	33.9	43.2	3.7	45.2	12.3	7.1	45.2	12.3	15.9	58.8	7.7	7.0	50.8	For A and B Series detectors 1500 or 2000 microns deep add 1.5 mm to Y dimension.								
2000	50.5	63.5	3.7	65.5	12.3	7.1	65.5	12.3	15.9													
Tolerances	±0.5	±0.3	±0.3	±0.3	±0.3	±0.3	±0.3	±0.3	±0.3	±0.5	±0.3	±0.3	±0.3	±0.3	±0.3	±0.3	±0.3	±0.3	±0.3	±0.3	±0.3	±1.0

*V, W, X, Y, and Z given in millimeters.

**E mounts are furnished on totally depleted detectors ⩾900 mm² in area, on X Series or D Series detectors which are ⩽10 microns in thickness, and on some special detectors. E mounts may be disassembled by removing screws. V dimension provides distance between mounting holes which are 1.7 mm in diameter and suitable for 080 screws.

†Built into a Microdot connector only. ‡Available only for A Series ⩾3000 μm. ***Available only for B Series ⩾3000 μm.

Class	Detectors
Principle of Operation	Silicon surface-barrier detectors
Features	1. Table A.4.3 shows range of prices from good resolution detectors to best resolution detectors. 2. Various mounts available: ring mount, rear microdot, rear BNC, transmission mount. 3. Front electrode is 40 $\mu g/cm^2$ gold, except for series R (ruggedized) which has 50 $\mu g/cm^2$ of aluminum. 4. Rear electrode is 40 $\mu g/cm^2$ of aluminum. 5. Resolutions are quoted on all detectors, measured at 22°C and 10^{-6} torr. 6. Many specialized detectors are also available.
References	Manufacturer's specifications
Address	ORTEC, Inc. 100 Midland Road Oak Ridge, TN 37830 (615) 482-4411

Table A.4.3. Ortec Silicon Surface-Barrier Detectors

Range of Guaranteed Max. Resolution (keV) α	β	Series	Range of Active Areas (mm)	Range of Depletion Depth (μ)	Description	Range of Prices
17–25	8–15	A	7	100 to 300		$ 210.00 – $440.00
14–24	6–19		25 to 200	100 to 2000	Partially depleted	145.00 – 3620.00
35–45	30–40		600 to 900	100	Partially depleted	770.00 – 1155.00
25–35	20–30		25 to 100	3000 to 5000	Partially depleted	2510.00 – 4890.00
21–30	16–35		450	100 to 1500		425.00 – 3560.00
16–29	7–25	B	25 to 200	150 to 2000	Totally depleted	230.00 – 4270.00
			300 to 450	150 to 2000	Totally depleted	540.00 – 4675.00
17–25	9–20	C	50 to 450	100 to 2000	Annular partially depleted	280.00 – 2965.00
		D	10	15	Planar totally depleted	1650.00
			25	15 to 100	Planar totally depleted	635.00 – 1870.00
			50	15 to 100	Planar totally depleted	815.00 – 2045.00
			100 to 300	20 to 100	Planar totally depleted	970.00 – 3140.00
12–20	4–14	E	7	100 to 300	Premium partially depleted	570.00 – 745.00
			25 to 450	100 to 500		440.00 – 1790.00
17–80	8–75	R	50 to 600	100 to 1000	Ruggedized partially depleted	305.00 – 2630.00
			900 to 2000	100 to 500	Ruggedized partially depleted	1560.00 – 3960.00

RAD-ALP
Ortec 1
January 1983

Alpha Counting and Spectroscopy System
ORTEC 576

Class	Laboratory
Principle of Operation	Two completely independent alpha spectrometry channels are contained in a double-wide NIM module. Each channel has a vacuum chamber with a sample holder and surface-barrier detector, a detector bias source, a preamplifier, a shaping and stretching amplifier, a biased amplifier, a test pulser, and a discriminator.
Sensitivity and Range	Energy ranges: 4 to 6 Mev, 5 to 7 MeV, and 6 to 8 MeV Energy output: 0.1 to 10 V
Sampling	Individual samples 2 at a time
Performance	System resolution*: (for 5.486 MeV ^{241}Am): $\leq$ 30 keV FWHM Efficiency*: $\geq$ 25% at 5.486 MeV, 1 1/2 in. diam. sample, 1-mm spacing Background*: < 20 counts/day above 3 MeV Sample size: Up to 1 1/2 in. diam. planchets or filter paper Sampling spacing: Adjustable from 1 to 15 mm from detector Nonlinearity: < 0.1% FS (0-10 V) Stability: System, < 0.1% within 10°C temperature change, pulser 50 ppm/°C
Requirements	Power: NIM-BIN Size: Double-wide NIM module
Features	1. Energy zero adjustments on front panel 2. Factory calibrated 3. Interval pulses for quick system checkout 4. NIM logic output from discriminators can be used for either counting or MCA routine 5. Choice of either vertical or horizontal sample holder
References	Manufacturer's specifications
Cost	576 Models with two 300-mm^2 detectors: \$3465.00 with two 450-mm^2 detectors: \$3805.00 with two 600-mm^2 detectors: \$4135.00 without detectors: \$2925.00
Address	ORTEC Inc. 100 Midland Road Oak Ridge, TN 37830 (615) 482-4411

*Based on 576-450 R, basic unit with 450 mm^2 Ruggedized™ low background detectors.

Silicon Surface Barrier Detectors
Paul Downey & Co.

Class	Detectors
Principle of Operation	Silicon surface barrier detectors for alpha,beta spectroscopy.
Specifications	See Table A.4.4.
Features	1. Energy resolution measured with ^{241}Am alpha at 5.477 MeV, and ^{137}Cs conversion electron at 624 keV. 2. Entrance window 40 μg/cm^2 of gold; back window is $<$ 40 μg/cm^2 Al. 3. Axial, transmission, ring mounts all available.
References	Manufacturer's specifications.
Address	Paul Downey & Co. 25 Lexington Avenue Trenton, NJ 08618 (609) 883-3287

RAD-ALP
Paul G. Downey 1
January 1983

Table A.4.4. Paul Downey & Co. Silicon Surface Barrier Detectors

Active Area (mm^2)	FWHM Resolution (keV) α	β	Depletion Depth (μ)	Price Range ($)
		Fully depleted		
25	18	9	300 to 1000	260 to 760
	24	19	300 to 1000	200 to 560
50	18	9	300 to 1000	340 to 1000
	24	19	300 to 1000	300 to 800
100	18	9	300 to 1000	420 to 1400
	26	20	300 to 1000	380 to 1200
150	20	13	300 to 1000	600 to 1600
	27	20	300 to 1000	500 to 1400
200	20	13	300 to 1000	700 to 1700
	30	25	300 to 1000	540 to 1100
300	21	17	300 to 500	760 to 1200
	30	25	300 to 500	580 to 940
400	28	21	300 to 500	880 to 1300
	35	30	300 to 500	700 to 1000
		Annular		
50	24	19	100 to 500	160 to 500
100	24	19	100 to 500	220 to 620
150	24	19	100 to 500	280 to 700
200	30	25	100 to 500	300 to 760
300	30	25	100 to 500	320 to 800
450	32	27	100 to 500	460 to 900
		Partially depleted		
25	15	7	100 to 500	140 to 440
	18	12	100 to 500	100 to 300
	24	19	300 to 500	100 to 160
50	15	7	100 to 500	180 to 680
	18	12	100 to 500	140 to 440
	24	19	100 to 500	120 to 260
100	17	11	100 to 500	220 to 640
	20	15	100 to 500	160 to 460
	24	19	100 to 500	140 to 300
150	17	11	100 to 500	260 to 700
	24	19	100 to 500	160 to 420
	30	25	100 to 500	140 to 340
200	20	15	100 to 500	260 to 640
	24	19	100 to 500	200 to 520
	30	25	100 to 500	160 to 400

Table A.4.4. (Continued)

Active Area (mm^2)	FWHM Resolution (keV) α	β	Depletion Depth (μ)	Price Range ($)
		Partially depleted		
300	20	15	100 to 500	300 to 780
	24	19	100 to 500	240 to 580
	30	25	100 to 500	200 to 460
450	22	17	100 to 500	500 to 1160
	26	21	100 to 500	380 to 820
	32	27	100 to 500	300 to 560
25	16	9	700 and 1000	600 and 720
	18	12	700 and 1000	420 and 500
	24	19	700 and 1000	300 and 380
50	16	9	700 and 1000	820 and 980
	18	12	700 and 1000	600 and 720
	24	19	700 and 1000	440 and 600
100	19	15	700 and 1000	900 and 1180
	20	15	700 and 1000	660 and 820
	24	19	700 and 1000	540 and 700
150	19	15	700 and 1000	1000 and 1240
	24	19	700 and 1000	660 and 860
	30	25	700 and 1000	560 and 760
200	20	15	700 and 1000	1040 and 1400
	24	19	700 and 1000	740 and 940
	30	25	700 and 1000	640 and 860
		Enviroguard		
50	19	15	100 to 500	220 to 700
	23	18	100 to 500	180 to 620
150	21	16	100 to 500	280 to 860
	24	19	100 to 500	240 to 800
300	24	19	100 to 500	460 to 1000
	30	25	100 to 500	400 to 900

Air Particulate
Selective Alpha Monitor
RADēCO/SAI Model 442A

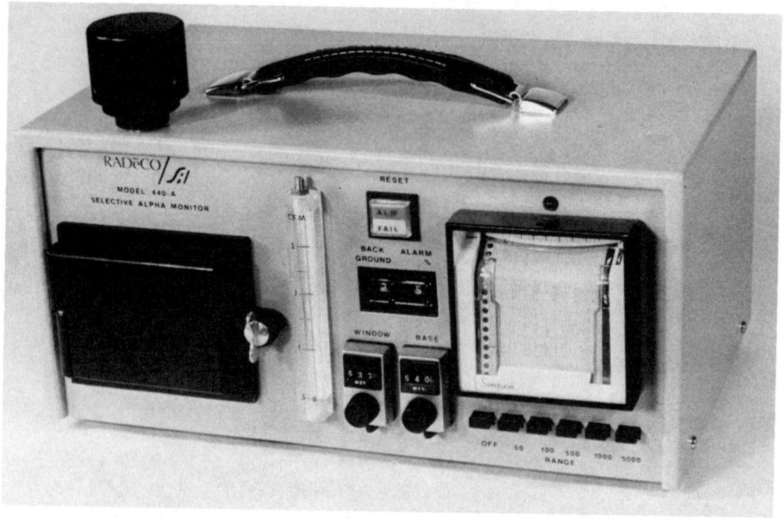

Class	Stationary
Principle of Operation	Filter, air pumps (extra), diffused-junction Si decontaminable detector, amplifier, single-channel analyzer, meter or recorder output. Capable of monitoring for ^{238}U, ^{235}U, ^{230}Th, and others.
Sensitivity and Range	Range: 0–100 with 0.5×, 1×, 5×, 10× and 50× cpm on meter Energy detected: baseline—0 to 1.0 MeV, adjustable window—0 to 10.0 MeV, adjustable Area of detector: 500 or 700 mm^2 Alpha efficiency: 23% of 2π with 0.6-MeV, window
Performance	Background: Window baseline—0 to 10 MeV, adjustable Window—0 to 3.0 MeV, adjustable Air flow: 1 to 6 cfm Operating temperature: 0 to 50°C (32 to 120°F) Filter: 47-mm diam. membrane type
Requirements	Power: 115 V, 60 Hz; 230 V, 50 Hz (special order); 100 V, 50 Hz (special order) Size: 40.6 cm W × 15.9 cm H × 27.9 cm D (16 in. × 6 1/4 in. × 11 in.) Weight: 7.2 kg (16 lb)

Cost	Model 442A: w/500 mm^2, \$2900.00 w/700 mm^2, refer to manufacturer
Features	1. Threshold adjustable audible alarm, adjustable high level visual alarm 2. Requires external air mover accessory 3. Fail-safe alarm, adjustable alarm delay 4. Short/long time constant (20 or 180 sec) 5. TTL pulse output for driving remote readout
References	Manufacturer's specifications
Address	RADēCO /SAI 4060 Sorrento Valley Boulevard San Diego, CA 92121 (619) 452-9150 TWX (910) 337-1774

Alpha Spectrometer
TENNELEC Model TC 256

Class	Stationary
Principle of Operation	Designed for silicon detector operation. Includes a vacuum chamber, charge sensitive preamplifier, linear amplifier, discriminator, biased amplifier, variable high voltage supply and variable test pulser. Samples can be planchets, filter paper and so on, up to 1.75 in. diameter. The spectrometer can be provided with either standard or cleanable front electrode or surface barrier silicon detectors.
Sensitivity and Range	*Energy Ranges* 3–8 MeV, 4–7 MeV, 3–5 MeV, 4–6 MeV, 5–7 MeV and 6–8 MeV. *Noise* Dependent on detector capacitance and leakage current. *Nonlinearity* $<$ 0.1% of full scale energy range selected *Detector Voltage* Variable from 0 to $\pm$150 V. Polarity selected by internal PCB mounted switch.
Requirements	*Power* +24 V, 125 mA; +12 V, 125 mA −24 V, 125 mA; −12 V, 35 mA *Weight* 2.9 kg (6.4 lb) *Dimensions* Standard double-width NIM Module: 7 cm × 22 cm (2.7 in. × 8.714 in.) per TID 20893 (Rev.)

Detectors

Silicon Surface Barriers	Cleanable Front Electrode
Area (mm^2)	
25	50
50	100
100	150
150	300
200	400
300	500
400	1000
500	
1000	

References	Manufacturer's specifications
Cost	$1695.00
Address	TENNELEC, Inc. P.O. Box D Oak Ridge, TN 37830 (615) 483-8405 Attn: Wayne Graves

Low Background α/β Counting Systems
TENNELEC LB 1000 Manual and LB 5100 Automatic Series

Class: Laboratory

Principle of Operation: The primary ionization generated by an alpha is much greater than for a beta and therefore less field strength (voltage) is required to meet the minimum pulse height requirement. However, the alpha and beta plateaus of a proportional counter overlap when using P-10 (90% Argon, 10% Methane) Gas. The detector is operated in the voltage range where the two plateaus overlap in order to simultaneously analyze the sample for alpha and beta radiation. Alphas and betas can be separated and analyzed concurrently by using a lower level discriminator set above noise and an upper level discriminator set between the beta continuum and the alpha peak.

Detector Characteristics: Background: 1.25 in. detector: $\leq$ 0.5 cpm alpha plus beta
2.25 in. detector: $\leq$ 1.3 cpm alpha plus beta

Sensitivity and Range: Efficiency: with 80 $\mu g/cm^2$ window: $\geq$ 30% for ^{14}C and ^{210}Po
with 80 $\mu g/cm^2$ window: $\geq$ 40% for ^{90}Sr–^{90}Y
with 500 $\mu g/cm^2$ window: $\geq$ 20% for ^{14}C and ^{210}Po
with 500 $\mu g/cm^2$ window: $\geq$ 40% for ^{90}Sr–^{90}Y

Alpha-Beta Crosstalk: For a typical 2.25 in. detector counting ^{210}Po and ^{90}Sr, with ^{90}Sr adjusted for maximum efficiency and 1% crosstalk into the alpha channel, there will be a $\leq$ 3% crosstalk for ^{210}Po into the beta channel.

Plateaus: ^{14}C: $\leq$ 1.5% slope/100–150 V plateau
^{90}Sr: $\leq$ 2.5% slope/100 V plateau

Uniformity: ±5% over 90% of window

Guard Detector Efficiency: $\geq$ 99% cosmic detection

Sample Count Rate

500,000 cpm with < 1.5% dead-time loss

Timebase Accuracy

0.002%, crystal timebase

Count Capacities

All three counters have capacity of up to 9,999,999 counts

Counting Time

Counting time per sample can be set at: For LB 1000 Series from 0.01 to 90,000 min or from 0.1 to 900,000 sec; For LB 5100 Series at any number between 00.01 min and 99,999.99 min.

Preset
In addition to being able to preset time per sample, independent preset counts may be entered for both alphas and betas.
Sample Changers

(Only in the LB 5100 Series) Capacity of 50 or 100 samples with a typical changeover time of 5 seconds.

Processor/Controller (LB 5100)

The focal point of the LB 5100 Automatic System is the LB 5100 Processor/Controller. In addition to the accummulation and sorting of data, controlling the sample changer, and so on, it provides for alpha and beta manipulation through operator input K factors. This allows the user to perform geometric efficiency correcctions, to transform the data from counts per minute to μcuries/unit volume or weight and to obtain 2 sigma percent results printed out for alpha and beta data.

Requirements

Power
LB 1000: 117 V ± 10%, 60 Hz ± 3%; ~1.0 amps or
230 V ± 10%, 50 Hz ± 3%; ~0.5 amps
LB 5100: 117 V ± 10%, 60 Hz ± 3%; ~4.5 amps or
230 V ± 10%, 50 Hz ± 3%; 2.25 amps
Weight
LB 1000 Total Shipping Weight: 664.5 kg (1465 lb)
LB 5100 Total Shipping Weight: 833.7 kg (1838 lb)
Size

LB 1000	LB 5100
Electronic Cabinet	
55.9 cm × 50.8 cm × 25.4 cm	52.7 cm × 52.7 cm × 49.5 cm
(22 in. × 20 in. × 10 in.)	(20.75 in. × 20.75 in. × 19.5 in.)
Sample Changer	
55.9 cm × 50.8 cm × 58.4 cm	88.9 cm × 66.0 cm × 76.2 cm
(22 in. × 20 in. × 23 in.)	(35.0 in. × 26.0 in. × 30.0 in.)
Terminal Printer	
—	53.3 cm × 53.3 cm × 15.2 cm
—	(21.0 in. × 21.0 in. × 6.0 in.)

Environment
Operating temperature: 16–27°C (65–80°F)
Relative humidity: 40–70%
Dust and radiation free

References

Manufacturer's specifications

Cost

LB 1000: $12,700–15,000
LB 5100: $25,900–30,000 (depending on accessories)

Warranty	One year
Address	TENNELEC, Inc. P.O. Box 2560 Oak Ridge, TN 37830-2560 (615) 483-8405 Attn: Phillip Cole

A.5. INSTRUMENTS FOR MEASURING BETA AND BETA/GAMMA RADIATION

Included in the following pages are Instrument Notes of several different kinds. The reader should be aware of the three following facts, in order to find all Instrument Notes appropriate for beta detection and measurement:

1. *Silicon surface-barrier detectors* and *silicon lithium-drifted detectors* are useful for both alpha and beta counting and spectrometry. Instrument Notes for all of these detectors have been incorporated in the alpha section (Section A.4, mnemonic RAD-ALP), and the interested reader is referred to that section. *These Notes are not included here.*
2. Many instruments (ionization chambers, G-M counters, gas proportional counters, etc.) are advertised by their manufacturers as useful for detecting and measuring *both betas and gammas*. However, many similar instruments, useful for betas, are advertised as if they were only gamma instruments. We have made a somewhat arbitrary decision to place the Instrument Notes according to the manufacturers' claims. Hence both this section RAD-BET and RAD-BET,GAM) and the section on gamma/x-ray detectors (Section A.6, (RAD-GAM,XRA) should be consulted together for a fuller overview of instruments of this type.
3. Many instruments are advertised as sensitive to alphas, betas, gammas, x-rays, and in some cases neutrons as well. Often these instruments consist of a variety of probes attachable to a central electronic-counting unit. These instruments have all been placed in Section A.3, Combination Instruments, with the appropriate mnemonic (RAD-ALP,BETA,GAM,XRA).
4. *Liquid-scintillation systems* are usually used for beta-detection and measurement. However, they have all been placed in the section Instrument Notes for Radionuclides (Section A.9).

Beta, Gamma Survey Meters
Eberline Portable Models

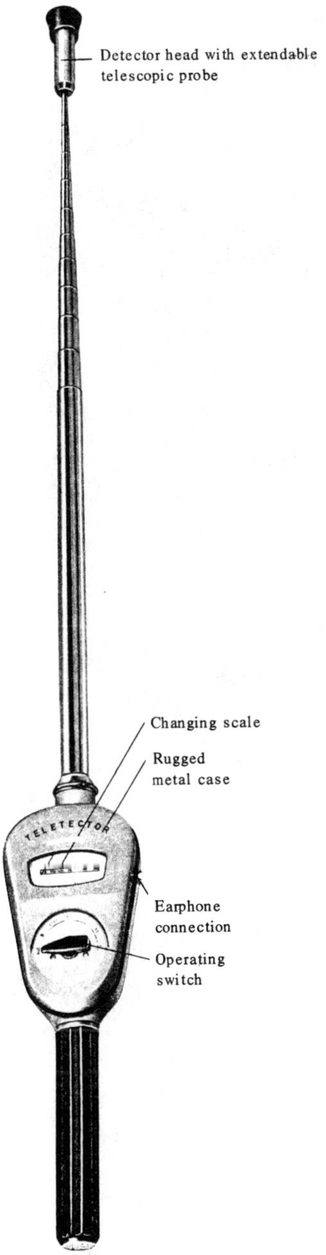

6112

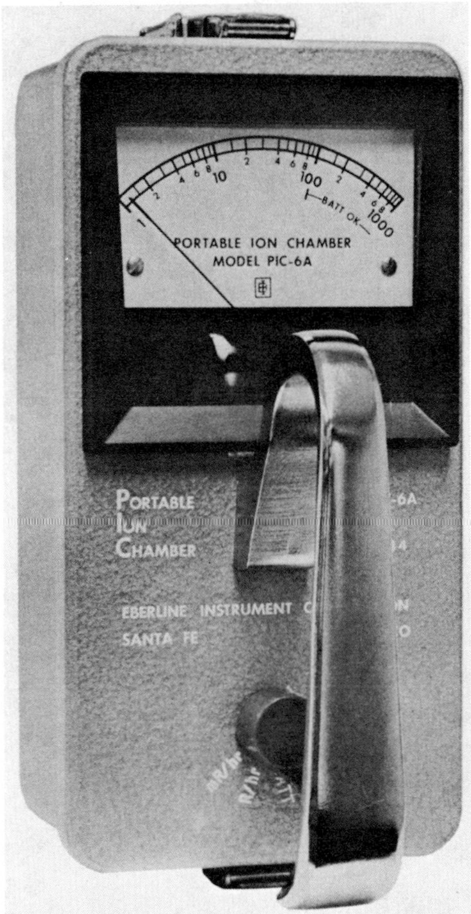

PIC-6A

E-130A

Class	Portable, handheld
Sampling	Continuous
Specifications	See Table A.5.1
Features	Some models have earphone and speaker outputs, regulated power supply, individual calibration controls for each range
Address	Eberline Instrument Corporation P.O. Box 2108 Santa Fe, NM 87501 (505) 471-3232

Table A.5.1. Eberline Portable Survey Meters

	Model			
Specifications	RT-1A	E-130 A	PIC-6A	E-520/HP-270
Radiation Measured	Gamma	Gamma	Gamma	Beta-gamma
Detector	G-M tube	G-M tube	Ionization chamber	G-M tube
Range	Approx. 25 chirps/min in a 10 mR/hr field	10, 100, 1000 mR/hr	1 mR/hr to 1000 R/hr (6 decades)	0 to 2000 mR/hr (5 ranges)
Scale	None	Linear	Linear	Linear
Accuracy		±5% FS	±20% of reading	±8% (0.2 mR/hr) ±15% FS (0 to 200 mR/hr) ±10% FS (0 to 2000 mR/hr)
Temperature	23 to 54°C (−10 to 130°F)	−40 to 60°C (−40 to 140°F)	−23 to 60°C (−10 to 140°F)	−40 to 60 °C (−40 to 140°F)
Power	9 V NEDA #1604 Batt. (nominal) 300 hr	2 "D" cells, 300 hr for C-Zn cell	2 #1604 NEDA-type batteries	2 "D" cells, 300 hr for C-Zn cells
Weight	0.17 kg (6 oz)	1.4 kg (3 lb)	1.48 kg (3 1/4 lb)	2 kg 4 3/8 lb)

Size	3 cm L × 6 cm W × 7.9 cm H (1 3/16 in. × 2 3/8 in. × 3 1/8 in.)	17.1 cm L × 8.6 cm W × 6.0 cm H (6 3/4 in. × 3 3/8 in. × 2.4 in.) handle included	21.3 cm L × 10.2 cm W × 14.3 cm H (8 3/8 in. × 4 in. × 5 5/8 in.)	10.2 cm W × 20.3 cm L × 17.1 cm H (4 in. × 8 in. × 6 3/4 in.) handle included
Features	Audible warning	Variable meter response time	Beta window in bottom	Both mR/hr and cpm scales
Remarks	Pocket monitor	Battery voltage regulated	Weatherproof housing	Utilizes 2 detectors
Cost	$165.00	$360.00	$725.00	$550.00 with HP-270 probe

Specifications	6112	E-120/HP-270	RO-7
Radiation Measured	Gamma	Beta, gamma	Gamma; mid- and high range probes also measure beta
Detector	G-M tubes	G-M tubes	Ion chamber
Range	1 mR/hr to 1KR/hr (5 ranges)	0.5, 5, 50 mR/hr 600, 6K, 60K	20,000 R/hr 200 R/hr 2 R/hr
Scale	Linear	Linear	Digital
Accuracy	±10% vs ^{137}Cs ±20% of reading	±5% FS	N/A
Temperature	−17.7 to 48.9°C (0 to 120°F)	−40 to 60°C (−40 to 140°F)	−10 to 55°C (15 to 130°F)
Power	Four 1.5-V "C" cells, 20 to 60 hr	2 "D" cells, 300 hr for C-Zn cells	3 9-V NEDA 1604 2 30-V NEDA 210

Table A.5.1. (Continued)

	Model		
Specifications	6112	E-120/HP-270	RO-7
Weight	3.18 kg (7 lb)	1.36 kg (3 lb)	1.1 kg (2.5 lb)
Size	91.4 cm L × 13.3 cm W × 8.3 cm H (36 in. × 5 1/4 in. × 3 1/4 in.) (probe retracted)	17.1 cm L × 8.6 cm W × (6.03 cm + 7.94 cm) H [6 3/4 in. × 3 3/8 in. × (2 3/8 in. + 3 1/8 in.)]	20.3 cm L × 10.8 cm W × 24 cm H (8 in. × 4 1/4 in. × 9 1/2 in.)
Features	Telescoping probe to 15 ft 1 in.	Calibration source included	Available with low-, medium- and high- range probes; underwater probe housing and with probe extender for use in air. Medium- and high-range probes also measure beta absorbed dose rate.
Remarks		E-120E has cpm meter face only	Has exchangeable probes

Cost	$2975.00	Less probe $280.00 with HP-270 $360.00	Instrument w/o detector, $730.00; Low-range detector (RO7 LD), $510.00; Medium-range detector (RO7 BM), $460.00; High-range detector (RO7 BH), $490.00; Underwater housing (RO7 UWH), $455.00

Stationary Air Monitor
Eberline Particulate Air Monitor AMS−3

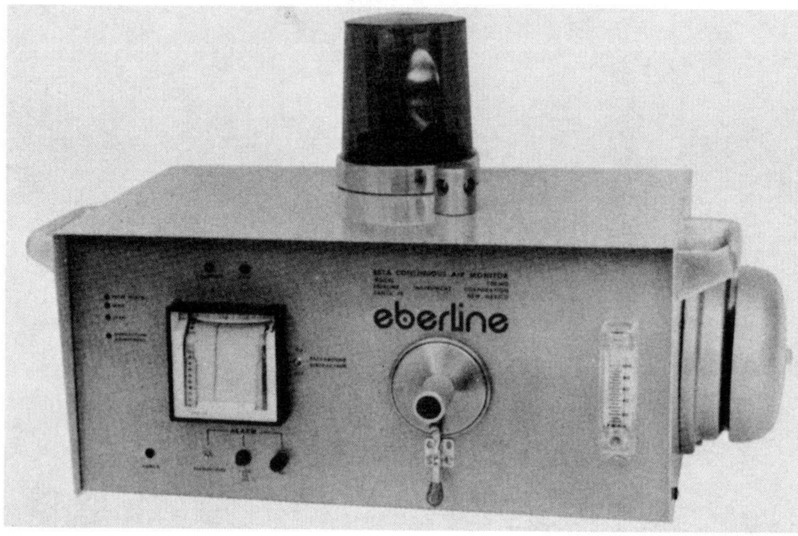

Class	Stationary
Principle of Operation	Air filter collects particulates. Collected beta and gamma radioactivity is detected and measured by a pancake-type geiger tube. Meter readout, and strip chart recorder. Air pump RAP-1 included.
Sensitivity and Range	Counting efficiency for 47-mm diam. source: ^{90}Tc (15%); ^{90}Sr, ^{90}Y (75%); ~2π efficiency Background: < 500 cpm when air flow is 1 cfm, measured in Santa Fe, NM. Gamma background subtracted automatically by second detector. Range: logarithmic from 10 to 100,000 cpm
Sampling	Continuous
Performance	Accuracy: Linearity 10% of reading Temperature: −29 to 60°C (−20 to 140°F)
Requirements	Power: 105 to 125 V ac, 50 to 60 Hz at approximately 0.3 A Size: 64 × 39 × 39 cm (25 in. L × 15 in. W × 15 in. D) Weight: 73 kg (160 lb)
Features	1. Model RAP-1 regulated air pump is furnished with AMS-3. 2. Can be calibrated for many different isotopes 3. Sensitivity 11 cpm for 10^{-12} μCi/cm^3 ^{90}SrY after four hours operation at 60 liters/min air flow rate.

4. Geiger-Müller detector is pancake-type, 1.75 in. diam., 1.4 to 2.0 mg/cm^2 mica window. Sccond detector provided for automatic gamma background subtraction.
5. Shield is equivalent to 2 in. of lead.
6. Takes 2 in. filter paper. A box of 50 filters (Type HV-70) is provided.
7. Air intake normally about 1 ft^3/min (28.3 liters/min). Air flow meter goes from 0 to 100 liters/min.
8. Alarm indicator flashes red light, bell, and relay contacts for remote alarm. Alarm point adjustable above 10 cpm.
9. Strip-chart recorder.

References — Eberline Catalog

Cost — $3300.00

Address — Eberline Instrument Corporation
P.O. Box 1208
Santa Fe, NM 87501
(505) 471-3232

Hand, Hand and Foot, Portal Monitors
Eberline PM-6

Class

Stationary

Principle of Operation

Eleven (11) gas flow proportional detectors, an ultrasonic motion sensor, a photoelectric occupant sensor, a microcomputer, support hardware, and an automatic change-over P-10 gas supply system.

Three modes of operation are available:

Mode 1: Preset All. Requires input of both alarm level and count time.

Mode 2: Maximum Sensitivity. Unit computes its own alarm level on a channel by channel basis.

Mode 3: Minimum Count Time. Unit computes its own counting time to be as short as possible based on prescribed alarm levels.

User may also select either walk-through or stop-count-go type of operation. Unit will automatically subtract ambient background.

Two types of alarm:
Personal Alarm. The display identifies the location of the alarm. An audible alarm (95 db @ 2900 Hz) sounds and a red light comes on.

Trouble Alarm. Activated by, counting gas low pressure, low count rate failure on any detector or high background alarm on any detector. A yellow light comes on.

Sensitivity

Depends on background level, false alarm rate, count time and/or walking speed and location of contaminating isotope.

A 60 nanocuries source of $^{90}Sr - ^{90}Y$ traversing the portal centerline at a speed of 2 miles per hour (3.2 Km/hr) has a 50% alarm probability with an approximate 3% false alarm rate if the unit is in a 0.02 mR/hr background with a 0.4 second count time.

If stop-count-go mode is selected with a count time of 5 seconds, the false alarm rate is approximately 0.3% for a source of 17 nanocuries.

Typical background response: 52000 cpm/mR·hr for ^{137}Cs
Typical background response: 6000 cpm/mR·hr for ^{60}Co

Requirements

Power: 105–125 V ac, 60 Hz
Size: 2.20 m × 1.22 m × 0.30 m (86.5 in. × 48 in. × 12 in.)
Weight: 159 kg (350 lb)

Detector Specifications

Probe: Aluminum housing
Protective Screen: Stainless steel mesh 72% open
Dimensions: 43.2 cm × 17.8 cm (17 in. × 7 in.) Sensitive area 503 cm^2 (78 $in.^2$)

Operating Voltage-Beta: 1000 3000 V
Gas Connections: Quick connect, Hansen B1-T10, 2 per detector.

Cost

$19000.00

Address

Eberline Instruments Corporation
P.O. Box 2108
Santa Fe, NM 87501
(505) 471-3232
Attn: R. D. Terry

Ion Chamber
Fisher Control Model NE 014

Class	Laboratory
Principle of Operation	Combination of reentrant cylinder, gamma chamber, and a parallel-plate beta chamber containing air at atmospheric pressure
Sensitivity and Range	Energy sensitivity: gamma, 100 keV to 3 MeV
Sampling	Continuous
Performance	Accuracy: Gamma, ±1%; beta, ±3% Response time: Zero
Requirements	Power: 100 pW per pA Size: 25.4 cm H × 25.4 cm D (widest part) (10 in. × 10 in.) Weight: 4.15 kg (9.1 lb)
Features	A.E.R.E. Type 1383A
Cost	Refer to manufacturer
Address	Fisher Control Ltd. Measurement and Analytics Division Century Works, Lewisham London, SE13 7LN, England Attn: A.E. Pegg Tel: 01-692 1271 Ext. 260 Telex 22469

Beta in Air Measuring System
Harwell Type 3230

Class	Laboratory, wall mounted.
Principle of Operation	Air drawn through a fixed filter paper deposits beta/gamma active dust which is viewed continuously by a gamma-compensated end-window G-M counter detector. 25 mm GF/A glass fiber filter paper is supplied in cardboard mount.
Sensitivity	Beta: 50 cps for 1 mpc-h (7000 dpm/m^3) at 37 liters/min Gamma: 18 cps at 1 mrad/hr
Requirements	Power: 110 V/240 V Ac, 50/60 Hz, rated at 30 volt-amp. Size: 27.8 cm × 56.0 cm × 29.0 cm Weight: 16 kg Air pump: normal flow rate 37 liters/min, ratemeter indication
Features	An automatic self-test is initiated every 110 minutes, and whenever the Alarm Reset is operated. Net beta digital pulse rate to RS 422 for remote processing, together with 4 pairs of dry alarm contacts for high sensitivity and equipment failure. Local alarm indication provided by 12 W beacon lamp and yodalarm
Cost	Refer to manufacturer
Address	AERE, Harwell Oxfordshire OX11 ORA England Tel: Abingdon (0235) 24141 Telex: 83135 Attn: R. H. W. Sherwood (Ext. 4766)

High Level Beta-Gamma Doserate Meter
Harwell Types 3247 and 3248

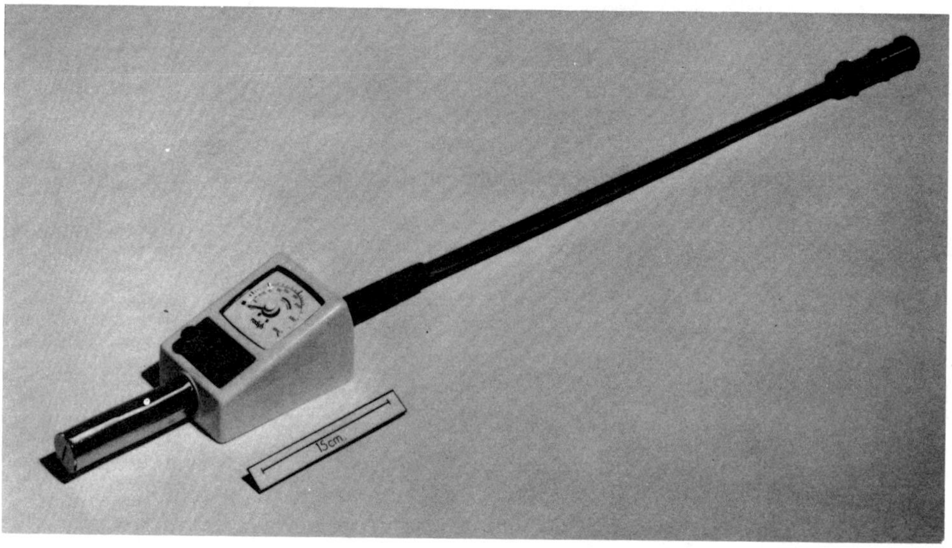

Class	Portable, handheld
Principle of Operation	Uses a G-M detector mounted on extendable tube (4 in.) for remote detection and survey of localized high-activity spills and log ratemeter with visual meter display.
Sensitivity	Energy range: (γ) 50 keV to 3 MeV Dose rate range: (γ) 0.1 to 100 rads/hr (type 3247); logarithmic; 5 to 5000 rads/hr (type 3248)
Sampling	Continuous
Performance	Dose rate accuracy: ±33.5% @ 20°C (type 3247) Operating temperature range: −5 to 45°F (23 to 113°F) Controls: function switch, 4-position switch: off, batt. check, set zero, on. Set zero, coin operated control for adjustment of electrical zero
Requirements	Power: Batteries, 2 U2 type (or equivalent), life 45 hr Size: Meter unit, 11.5 cm W × 10.2 cm H × 16.5 cm L (4 1/2 in. × 4 in. × 6 1/2 in.) plus 3.8 cm D × 14 cm L (1 1/2 × 5 1/2 in.) rear handles, extension rod 76.2 cm (2 ft 6 in.) long Weight: ~2 kg (4 1/2 lb) including probe cell

Features	Additional 2-m-long extension rods available. Failure indication: a small priming source (^{90}Sr) is mounted on inside wall of energy radiation compensating shield to enable the meter to be set to an electrical zero that is offset from the mechanical zero. Thus if the probe should be disconnected or inoperative, the meter pointer will fall back to mechanical zero so that it lies in FAULT sector of the scale.
References	Manufacturer's literature
Cost	On application
Address	AERE, Harwell Oxfordshire OX11 ORA England Tele: (0235) 24141 Attn: R. H. W. Sherwood (Ext. 4766)

Beta-Gamma Survey Meter
Health Physics Instruments Model 1075

Class	Portable, handheld
Principle of Operation	Uses two air equivalent ionization chambers, one β-γ and other γ plus ratemeters with amplifiers and two digital readouts
Sensitivity and Range	Energy range: 20 keV to 1.3 MeV (β-γ, if use energy response curve) 30 keV to 1.3 MeV (γ, if use energy response curve) Dose rate range: 0.001 to 200 rad/hr in three digital ranges: 200 rad/hr: 0.1 to 199.9 rad/hr 20 rad/hr: 0.01 to 19.99 rad/hr 2 rad/hr: 0.001 to 1.999 rad/hr
Sampling	Continuous
Performance	Accuracy: overall, $\pm$15% or two digits, which ever is greater (elevation, barometric corrections must be made) Temperature dependence: zero shift, $\pm$.005, 0 to 50°C; reading, $\pm$10% from 10 to 50°C; operating range, 10 to 50°C; storage limits, -20 to 65°C Pressure dependence: reading dependent on elevation and barometric pressure, detectors are unsealed ionization chambers Exposure rate response: 90% collection efficiency at 200 rad/hr Off scale indication: 1- - -; a numeral in the left-hand digit with all other digits blanked. Switching transient: range switching transient; 5 sec between adjacent ranges Detector volume: 300 cm^3 each Detector wall thickness: β-γ, 7 mg/cm^2; γ, 1 g/cm^2 Warm-up time: 30 sec Response time: 3.5 sec on 2 rad/hr range; < 1 sec on other ranges

Energy response:

Reading/True Dose rate

keV[a]	β-γ[b]	γ
22	0.72	0.05
32	0.82	0.31
46	0.87	0.56
72	0.94	0.76
120	1	0.88
^{60}Co	1	1

[a]Filtered x-rays (HF series filter).
[b]With 1/8 in equilibrium wall added.

Requirements
Power: batteries, 4 "D" cells; 1 "AA" (light)
Size: 17.9 cm H × 13.3 cm W × 30.5 cm L (7 in. × 5.25 in. × 12 in.)
Weight: 2.5 kg (5.5 lb)

Features
Carrying case in black anodized aluminum environmentally sealed

References
Manufacturer's specifications

Cost
$2400.00

Address
Health Physics Instrument, Inc.
124 Santa Felicia Drive
Goleta, CA 93117
(805) 685-2612

Hand Probes
Johnson and Associates Models HP-265 and PPA-2

Class	Laboratory
Principle of Operation	G-M tube which will work with any Geiger counter, Rate meter, Scaler, etc. where a 900 V supply is available and input sensitivity is not greater than 1 V.
Sensitivity and Range	Gamma sensitivity (^{137}Cs into window): ~3600 cpm/mR/hr (HP-265); ~6000 cpm/mR/hr (PPA-2) Beta Efficiency (with screen in place): $^{90}Sr-^{90}Y$ (E_{max} 0.54 to 2.2 MeV) ~45%; ^{99}Tc (E_{max} 0.29 MeV) ~30%; ^{14}C (E_{max} 0.15 MeV) ~10%
Performance	Operating voltage: 900 ± 50 V Plateau length: 100 V minimum Plateau slope: 0.1%/V maximum Dead time: 50 μsec max. (HP-265) 100 ec max. (PPA-2) Window thickness: 1.4 to 2.0 mg/cm^2 Window area: 15 cm^2 (1 3/4 in. diameter)
Requirements	Connector: BNC to accept coaxial cable Size: HP-265: 25.4 cm × 7 cm × 6.4 cm (10 in. × 2 3/4 in. × 2 1/2 in.) PPA-2: 16.5 cm × 9 cm × 10 cm (6 1/2 in. × 3 1/2 in. × 3 7/8 in.) Weight: 0.56 kg (1 1/4 lb)
References	Manufacturer's specifications
Cost	HP-265: $175.00 PPA-25: Refer to manufacturer
Address	William B. Johnson and Associates, Inc. Research Park Montville, NJ 07045 (201) 334-9222 Attn: A. Purcaro

Hand and Shoe Monitor
Johnson and Associates Model UM-4

Class	Laboratory
Principle of Operation	Uses G-M tubes for both hand and foot, amplifier, LED display, audible-visual alarm
Sensitivity and Range	Range: 0 to 100,000 cps
Sampling	Continuous
Performance	Accuracy: better than ±5% Display response: instantaneous
Requirements	Power: 115 V ac at 0.24 A Size: Electronic section, 43.2 cm × 38.1 cm × 21.3 cm; (17 in. × 15 in. × 8 3/4 in.) Base pedestal, 45.7 cm × 71.1 cm × 7.6 cm (18 in. × 28 in. × 3 in.) Weight: 56.7 kg (125 lb)

Features	Paper roll protects shoe detectors
Cost	Refer to manufacturer
Address	William B. Johnson and Associates, Inc. Research Park, Boonton Ave. Montville, NJ 07045 (201) 344-9222

Portal Monitors
National Nuclear Beta-10 and Gamma-10

Class	Laboratory
Principle of Operation	Beta-10: beta counting gas (P-10) Gamma-10: high performance solid scintillant
Sensitivity and Range	For a 0.4 second passing in 15 μR background:

Isotope	Beta-10	Gamma-10
^{137}Cs ($\beta+\gamma$)	100 nCi	200 nCi
^{60}Co ($\beta+\gamma$)	100 nCi	200 nCi
^{90}Sr (β only)	30 nCi	200 nCi

Performance	Passage time: 0.4; 0.8; 1.2 sec (digital) 0.2, 0.6, 0.8 sec (analog) Background update interval: 10, 20, 30, 40, 50, 60, 70, 80 sec (digital) 5, 20, 60 (analog)
Requirements	Power: 110 V/1.5 A; 220 V/0.7 A Size: 223.52 cm H × 33.56 cm D × 101.6 cm W (88 in. × 13.2 in. × 40 in.) Weight: Beta-10: 91 kg (200 lb) Gamma-10: 454 kg (1000 lb)
References	Manufacturer's specifications
Cost	Beta-10: $16000 (Foot monitor $2000) Gamma-10: $17000
Address	National Nuclear Corporation 1904 Colony St. Mountain View, CA 94043 (415) 962-9220 Attn: D. Kahn

Survey Meter
Nuclear Associates 05-700

Class	Portable, handheld
Principle of Operation	G-M tube, amplifier, meter readout.
Sensitivity and Range	Energy sensitivity: Beta, 175 keV minimum; Gamma, 20 keV through cobalt-60 gammas Ranges: 0.5, 5, 50 mR/hr (300, 3000, 30,000 cpm)
Sampling	Continuous
Performance	Meter accuracy: ±2% FS Temperature range: 15 to 75°C (59 to 167°F) Response time: 1 to 3 sec dependent on scale
Requirements	Power: Four 1.5-V "D" cells Size: 22 cm L × 11 cm W × 18 cm H (8.5 in. × 4.5 in. × 7 in.) Weight: 1.82 kg (4 lb)
References	Manufacturer's specifications
Cost	\$350.00 for 05-700 \$ 16.00 for 05-701 (optional headset)
Address	Nuclear Associates 100 Voice Road Carle Place, NY 11514 (516) 741-6360

Hand and Foot Monitor
Nuclear Research Model HFM-2

Class	Stationary
Principle of Operation	Detects and measures beta-gamma contamination on hands, shoes, hair, face, and clothing. Seven halogen-quenched G-M tubes (four for hands, two for shoes, one in external probe), hand and shoe detectors shielded with 0.5 in. lead, ratemeter amplifier, alarms.
Sensitivity and Range	Scaler range: 0 to 99,999 counts (3 four-digit registers, one for each hand, and one for shoes)
Performance	Preset time: 1 to 30 sec, variable
Requirements	Power: 105 to 125 V ac, 50 to 60 Hz Size: 173 cm H × 53 cm W × 56 cm D, (68 3/8 in. × 21 in. × 22 in.) shoe deck: 10 cm H × 53 cm W × 45 cm D (4 in. × 21 in. × 18 in.) Weight: 300 kg (650 lb) shipping.
Features	Allows self-examination for contamination, audible response in proportion to degree of contamination. External survey probe.
References	Manufacturer's specifications
Cost	Price on request
Address	Nuclear Research Corporation Warrington Industrial Park Box H Warrington, PA 18976 (215) 343-5900

Fixed and Portable Monitors
NRC Rem Rad Series DMW-200

Class	Portable (Model 200P) and fixed area (Model 200F) monitor
Principle of Operation	G-M tubes (which will not saturate even above 10,000 R)
Sensitivity and Range	DMW-200 P-1: 0.01 mR/hr to 999 R/hr DMW-200 P-2: 0.01 mR/hr to 9999 R/hr Temperature range: −40 to 50°C (−40 to 120°F)
Performance	Accuracy: ±10% of reading over entire range Energy dependence: ±15%
Features	*Model 200P* Controls: On/off; rate/dose; battery test; calibration check; alarm on/off; alarm set Display: Digital LCD with floating decimal point *Model 200F* 3 alarms: Failure alarm; warning alarm (adjustable); high alarm (adjustable) Outputs: Digital panel readout; remote digital readout; analog 4–20 mA
Cost	Price on request
Address	Nuclear Research Corporation Warrington Industrial Park Box H Warrington, PA 18976 (215) 343-5900

Beta-Gamma Air Monitor
RADēCO Model GM-222

Class	Laboratory continuous monitor for airborne beta-gamma emitters
Principle of Operation	Filter, G-M tube, amplifier, meter/recorder readout.
Sensitivity and Range	Range: 1K, 10K, 100K cpm Energy range: 150 keV up Air flow: 0.5 to 3 cfm, front panel indicator Probe: Detectors: 15-2, 3-4 or 5-6 mg/cm^2 Quench: halogen Operating voltage: 900 V dc Shield: 2 in. lead equivalent, nominal
Performance	Accuracy: ±10% of calibration Beta efficiency: $^{90}Sr-^{90}Y$, 35%; ^{14}C, 5% Time constant: continuously variable 3 to 15 sec Temperature: 4 to 49°C (40 to 120°F) Filter: 47 mm Air flow: ±2% of set level
Requirements	Power: 110 V, 60 Hz Size: 41.9 cm L × 22.2 cm W × 22.9 cm D (16 1/2 in. × 8 3/4 in. × 9 in.) Weight: 31.8 kg (70 lb)
Features	Threshold adjustable audible alarm; external air mover accessory; stainless steel air path; fail-safe indication.
Reference	Manufacturer's specifications

Cost	GM222A with 2 in. lead shield: $4700.00 with meter GM222: $5000.00 with recorder
Address	RADēCO 4060 Sorrento Valley Boulevard San Diego, CA 92121 (619) 452-9983

Beta Monitor
Scintrex Model 189D

Class	Laboratory and field
Principle of Operation	Monitor gamma compensated utilizing dual Geiger tube design
Sensitivity and Range	0.1 to 1200 rem/hr on a 4 decade logarithmic scale on two ranges
Performance	Linearity: maximum error is ±15% measured with ^{90}Sr ($+^{90}Y$) (2.274 MeV) through full range of instrument Energy dependence: maximum error less than ±30% as measured with ^{147}Pm (0.225 MeV—E_{max}), ^{204}Tl and ^{90}Sr Angular dependence: less than ±30% for ^{90}Y betas having an angle of incidence from 0 to ±15%°. Total detection angle is ±45°. Temperature dependence: maximum error is ±5% with respect to reading at +22°C (+72°F) through the range of −35 to +60°C (−31 to +140°F) Humidity dependence: less than ±1% up to 98% relative humidity Gamma compensation: maximum error ±20% for up to 20 R/hr of ^{137}Cs (0.662 MeV) gamma field at the minimum beta/gamma ratio of 2:1
Requirements	Size: 34.6 cm × 26.5 cm × 10.8 cm (13.8 in. × 10.6 in. × 4.3 in.) Weight: 1.95 kg (4.3 lb) with batteries Power: 4 "C" cells (1.5 V each)
References	Manufacturer's specifications
Cost	Refer to manufacturer
Address	Scintrex, Ltd. 222 Snidercraft Road Concord, Ontario L4K 1B5, Canada CANADA (416) 669-2280 Telex: 06-964570 Attn: S. Nargolwalla

High Range Survey Meters
Model CP-TP-54A
Model CP-TP-10K

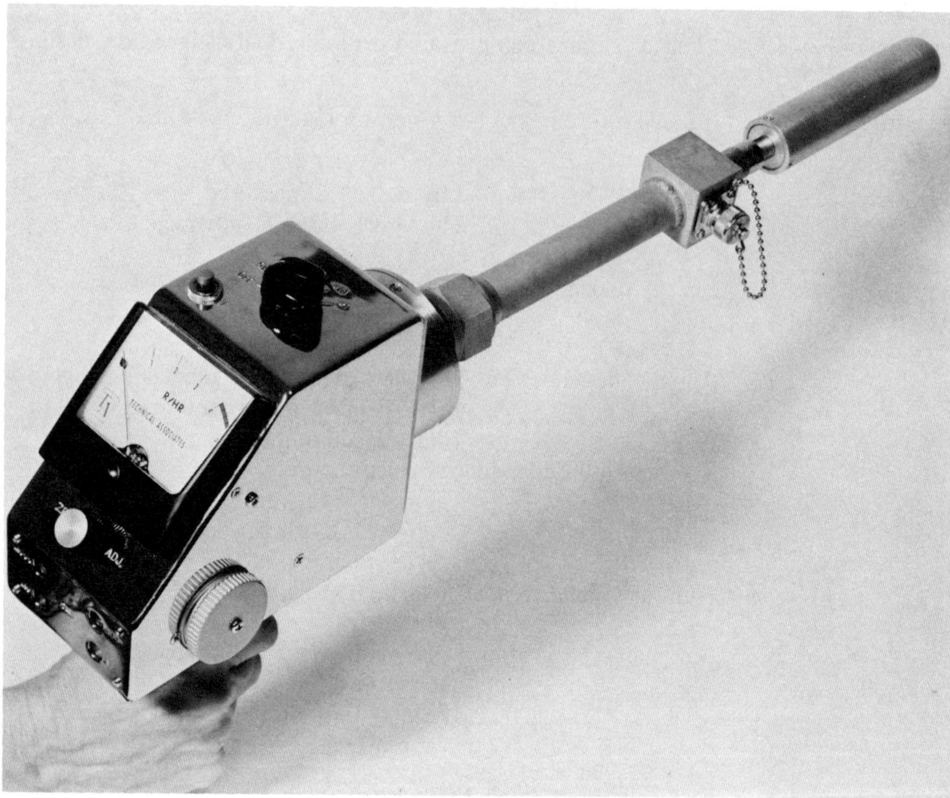

Class	Portable, handheld
Principle of Operation	Extendable air ion chamber, MOSFET amplifier, meter readout
Sensitivity and Range	CP-TP-54A: 0-0.5, 5, 50, 500 R/hr CP-TP-10K: 0-10, 100, 1000, 10,000 R/hr (available with beta-gamma or gamma only chamber)
Sampling	Continuous
Performance	Accuracy: ±5% FS Time constant less than 2 sec on all ranges
Requirements	Power: Single "D" cell Size: 33 cm × 16.5 cm × 10 cm (13 in. × 6 1/2 in. × 4 in.) excluding handle and with chamber close-coupled Weight: 1.8 kg (4 lb)

Features	High dose rate capability (to 10,000 R/hr) Interchangeable ion chambers Probe may be close-coupled, extended up to 200 in. on rod or extended on 25-ft cable Can monitor around a corner
References	Manufacturer's specifications
Cost	CP-TP-54A: $1325.00 CP-TP-10K: $1370.00 Includes 40-in. extension rod Additional 40-in. extension rods: Upon request
Address	Technical Associates 7051 Eaton Avenue Canoga Park, CA 91303 (213) 883-7043

Beta-Gamma Particulate Air Monitor
Technical Associates Model FM-5ABN

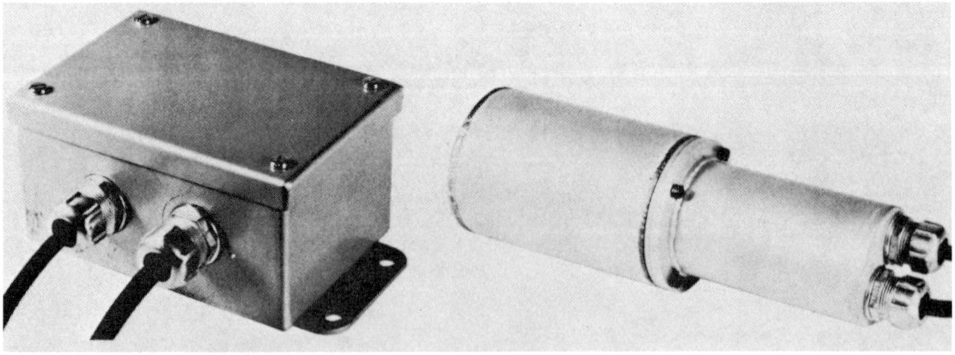

Class	Laboratory
Principle of Operation	Motor blower, filter paper, 2-in. pancake G-M detector, count rate meter, strip chart recorder, speaker, alarms
Sensitivity and Range	Range: log scale with response 10-100,000 cpm Limit of sensitivity: 10^{-4} μCi (betas > 0.1 MeV) or 10^{-8} μCi/ml · minute exposure
Sampling	Continuous
Performance	Background: typically 10 to 15 cpm Air flow adjustable to 100 liters/min
Requirements	Power: 115 V ac, 50 to 60 Hz Size: 43.2 cm W × 33.0 cm H × 34.3 cm D (17 in. × 13 in. × 15 1/2 in.) Weight: 18.2 kg (40 lb)
Features	Quick-change filter Audio and visual alarms
Reference	Manufacturer's specifications
Cost	FM-5ABN: $2280 FM-5ABN-LB: (with background subtraction) $2980
Address	Technical Associates 7051 Eaton Avenue Canoga Park, CA 91303 (213) 883-7043

Portal Monitors
Technical Associates Model PPM-21 and PPM-21A

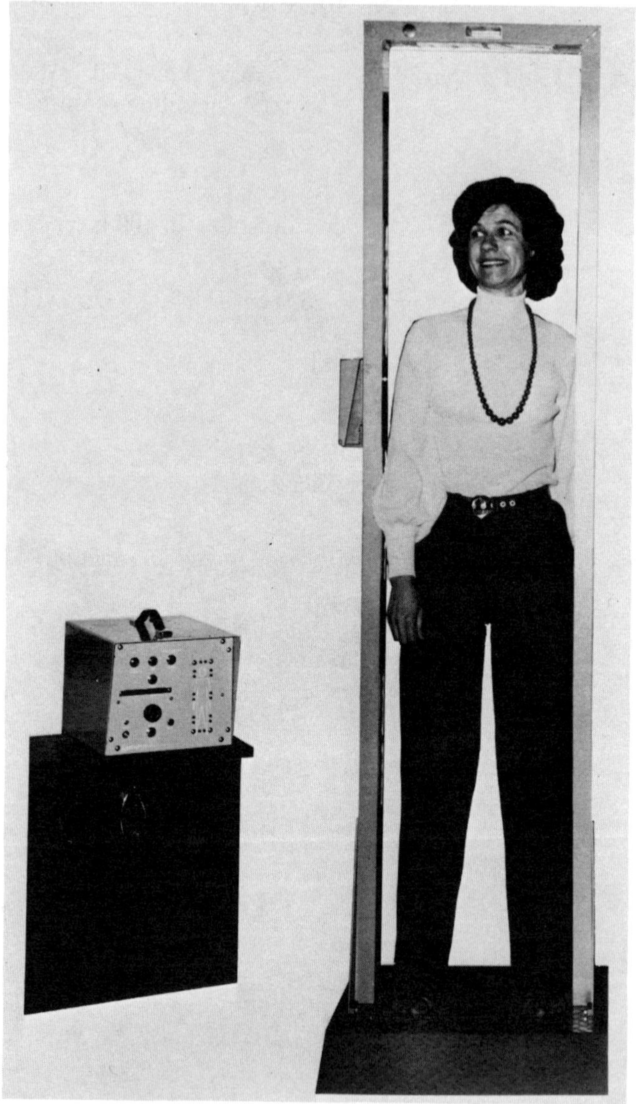

Class	Stationary
Principle of Operation	Geiger-Müller tubes detect beta-gamma. Scaler circuitry sums pulses. Indicator lights for contamination, system failure alarm, location of contamination, safe operation, etc. Timer and light beam switching included in PPM-21A.
Sensitivity and Range	8 channels (11 G-M detectors) Detector position light up—console lights indicate on human silhouette location of contamination. Presettable alarm level: 10 to 990 counts occurring within *recycling* time (presettable) for PPM-21 or within preset time (1 to 99 sec) for PPM-21A.
Sampling	PPM-21: Recycling count (digital ratemeter operation) PPM-21A: Light beam switching—scaler in preset time
Performance	Scaler accuracy for counting ± 1 count Timer accuracy: $\pm 1\%$
Requirements	Power: 115 V ac, 50 to 60 Hz (or 230 V) Size: 223.5 cm H × 63.5 cm W × 71.1 cm D (88 in. × 25 in. × 28 in.) console 22.9 cm H × 25.4 cm W × 33.0 cm D (9 in. × 10 in. × 13 in.) Weight: 81.6 kg (180 lb) total weight
Features	Scaler counting (high accuracy); low-level alarm Detector position light up Audio and visual alarms Light beam switch and timer warning system (PPM-21A) Dynamic background subtract standard
References	Manufacturer's specifications
Cost	PPM-21: \$2120.00 PPM-21A: \$2990.00
Address	Technical Associates 7051 Eaton Avenue Canoga Park, CA 91303 (213) 883-7043

Beta-Gamma Background Suppress Portal Monitor
Technical Associates Model PPM-25P, PPM-25GF

Class	Stationary
Principle of Operation	PPM-25P: G-M tubes detect radiation above a varying background level. Can activate presettable alarms. PPM-25GF: Gas flow detectors are used
Sensitivity and Range	PPM-25P: 8 channels (18 pancake—2 in. o.d.—G-M tubes) Detector position light up—console lights indicate on human silhouette location of contamination PPM-25GF: Nine gas flow detectors (screen in the arch and 2 in the foot deck, sensitivity 2× that of the PPM-25P. Pass-through time 0.2 sec less than for the PPM-25P.
Sampling	PPM-25P: Background is constantly monitored and updated except when a person walks within sensor range, then detectors switch to monitoring person. Increased radiation level is monitored for 2 sec or less
Requirements	Power: 115 V ac, 50 to 60 Hz Size: 223.5 cm H × 63.5 cm W × 71.1 cm D (88 in. × 25 in. × 28 in.) Weight: 81.6 kg (180 lb) total weight
Features	Very fast contamination checks possible with continuously updated background Audio and visual alarms Detector position light up
Cost	PPM-25P: $5940 PPM-25GF: $6860.00
Address	Technical Associates 7051 Eaton Avenue Canoga Park, CA 91303 (213) 883-7043

Liquid Scintillation System
Tracor Analytic Model 6895

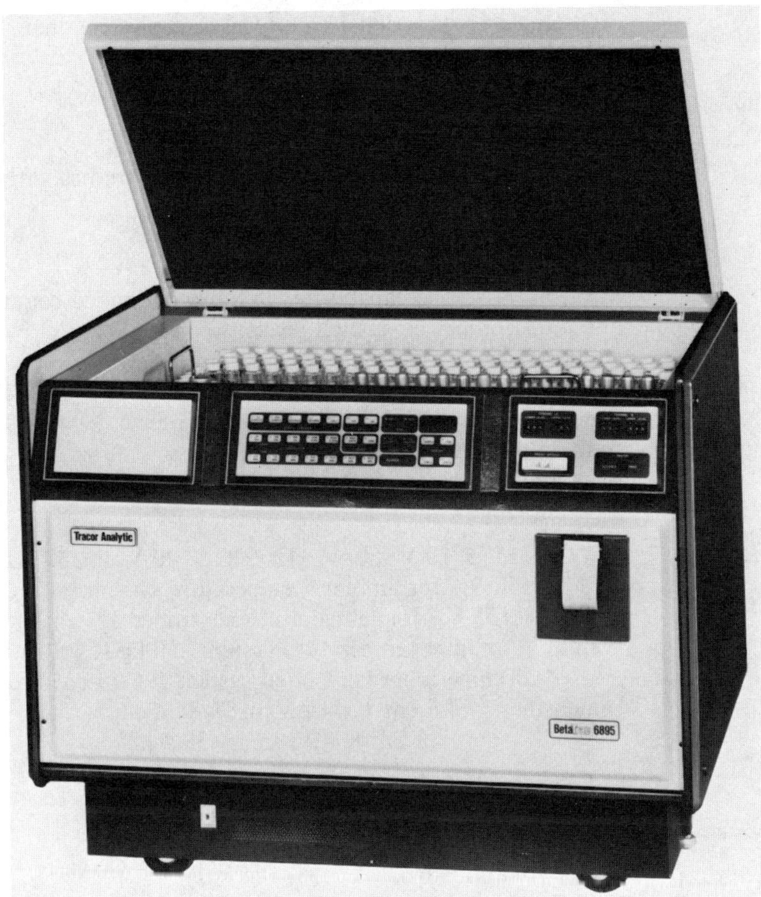

Liquid Scintillation System
Tracor Analytic Model 6895

Class	Laboratory
Principle of Operation	Detection of beta radiation. All systems include: microprocessor control; field upgradability; 300 sample capacity; calculation of cpm and quench correction ratios by SCR and/or ESR; two independent, simultaneous, sample analysis channels; two standardization channels; selectable time and count terminators with synchronous line timing; ^{133}Ba G.O. Geometry optimized external standard with positive mechanical positioning; battery back-up for memory protection in event of power interruption.

Sensitivity and Range	Detects and counts all alpha, beta and gamma emitting isotopes with an energy range up to 1700 keV
Sampling	300 sample bi-directional conveyor. Samples counted individually.
Performance	^{3}H efficiency greater than 60%. ^{14}C efficiency greater than 95%.
Options	DPM calculations for single and dual label studies with automatic cross contributor subtraction. Video display. Quench corrected printout. Multi-user/isotope capabilities to allow optimized counting of different isotopes or quench levels. Digital display. Temperature controlled from 0°C to ambient within 0.4°C. Paper tape punch and reader. Built-in printer. Background subtraction by sample measurement or programming.
Requirements	Power: 115 ± 10 V, 50/60 Hz; 230 ± 20 V, 50/60 Hz; 150 VA for ambient temperature systems; 1,125 VA for temperature controlled systems. Weight: Ambient temperature system 250 kg (550 lb). Temperature controlled system 296 kg (650 lb). Dimensions: 94.6 cm × 116.2 cm × 84.8 cm (37.3 in. × 45.8 in. × 33.4 in.)
Price	$14,250 to $25,395 depending on options
References	Manufacturer's specifications
Address	Tracor Analytic, Inc. 1842 Brummel Drive Elk Grove Village, IL 60007 (312) 364-9100

Condenser Roentgen-Meter
Victoreen Model 570

Class	Portable, handheld
Principle of Operation	External ionization chamber, meter readout
Sensitivity and Range	See Table A.5.3
Requirements	Power: 105 to 130 V ac, 50 to 60 Hz (220 V ac available on request) Size: 17.8 cm L × 10.2 cm W × 17.8 cm H (7 in. × 4 in. × 7 in.) Weight: 2.7 kg (6 lb)
Cost	$1500.00 plus chambers
Address	Victoreen, Inc. 10101 Woodland Avenue Cleveland, OH 44104 (216) 795-8200

Table A.5.3. Victoreen Condenser Roentgen Meter

Model Number	Range (Roentgen)	Intensity Limits[a] Model 570 (R/sec)	Energy Response		Material	Volume (cm^3) Approx.	Nominal Wall Thickness (mg/cm^2)	Approx. Thimble Dimensions Dia. × Length		Beta Cutoff (MeV)
			Tolerance	(keV eff)				(in.)	(cm)	
				Low Energy Chamber						
651	250	90	±5%	6-35	Mylar	0.20	6.6	5/8 × 5/8	1.59 × 1.59	0.066
				Medium Energy Chambers						
188	0.025	24×10^{-5}	±5%	30-660	Phenolic	1450	212	4 1/2 × 5 5/8	11.4 × 14.3	0.62
130	0.25	9×10^{-3}	±10%	42-660	Phenolic	177	212	2 × 5	5.08 × 12.7	0.62
227	1	0.195	±10%	30-660	Phenolic	41	212	1 1/4 × 3 1/4	3.18 × 8.25	0.62
633	2.5	0.3	±20%	30-660	Phenolic	17	212	1 1/4 × 1 5/8	3.18 × 4.13	0.62
228	5	5.1	±20%	30-660	Phenolic	8	212	3/4 × 2	1.91 × 5.08	0.62
326	10	5.1	±10%	30-660	Polystyrene	6	170	3/4 × 1 1/4	1.91 × 3.18	0.55
70–5	25	45	±5%	30-660	Nylon	1.83	67	1/2 × 1	1.27 × 2.54	0.27
131	100	75	±5%	30-660	Nylon	0.46	89	3/8 × 7/8	0.95 × 2.22	0.335
154	250	120	±5%	30-660	Nylon	0.183	89	3/8 × 7/8	0.95 × 2.22	0.335
				High Energy Chambers						
552	2.5	0.3	±10%	250-1300	Delrin	9.7	569	1 × 2 3/8	2.54 × 6.03	1.31
553	25	45	±5%	250-1300	Delrin	1.83	583	3/4 × 1 1/8	1.91 × 2.86	1.31
621	100	75	±5%	250-1300	Delrin	0.46	576	5/8 × 7/8	1.59 × 2.22	1.32

[a]These intensity limits are based on full scale reading and may be increased if readings less than one-half full scale are taken. A government report entitled "Measurement of High-Intensity Gamma Radiation" indicates that Victoreen Chambers have given valid readings in pulsed gamma fields of as high as 5×10^4 R/sec. Copies of this report are available from Victoreen or office of Technical Services, U.S. Dept. of Commerce.

Digital Radiation Monitor
Xetex Models 304A and 305B

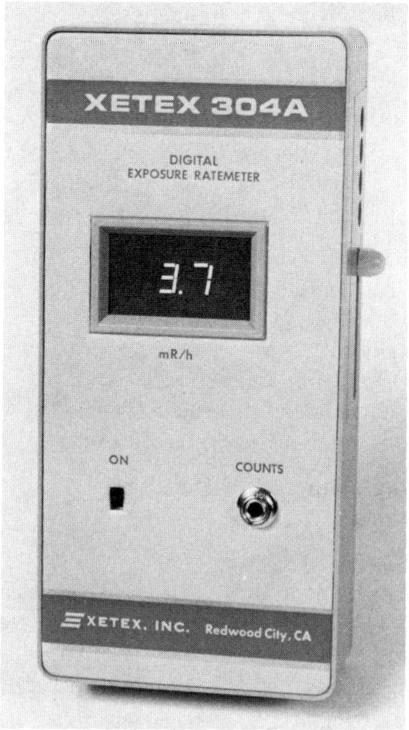

Digital Radiation Monitor
Xetex Models 304A and 305B

Class	Portable, handheld
Principle of Operation	Uses halogen-quenched, filtered G-M tubes as detectors, digital logic, LED display
Sensitivity	Beta-gamma
Range	304A: 0.1 to 99.9 mR/hr, overrange indicator; 305B: 0.1 mR/hr to 99.9 R/hr, overrange indicator
Sampling	304A-1 and 305B-1: push button 304A-2 and 305B-2: continuous 305B-3: continuous and push button

Performance	Accuracy: ±15% referred to ^{137}Cs Energy dependence: ±15% from 60 keV to 1.3 MeV gammas; betas greater than 0.25 MeV detected. Response time: nominal 1 sec
Requirements	Power: 305A-1 and 305B-2, 4 "AA" alkaline cells, 100 hr life; 304A-2, 305B-2 and 305B-3, 4 "AA" Ni-Cad cells, 8 hr/charge life Size: 7.1 cm × 15.5 cm × 3.9 cm (2.81 in. × 6.09 in. × 1.53 in.) Weight: 0.33kg (11.5 oz) w/batteries
Features	1. Autoranging 2. Six-decade coverage (305B) 3. Digital display 4. Overrange indication 5. Earphone and speaker (305B-3) 6. Charger included for 304A-2, 305B-2, and 305B-3
References	Manufacturer's specifications
Cost	304A-1: $312.00 304A-2: $390.00 305B-1: $430.00 305B-2: $520.00 305B-3: $550.00
Address	Xetex, Inc. 660 National Ave. Mountain View, CA 94043 (415) 964-3261

A.6. INSTRUMENTS FOR MEASURING GAMMA AND X-RADIATION

Included in the following pages are Instrument Notes for those instruments useful for *gamma and x-ray monitoring*. The reader should be aware of the following facts in order to find all Instrument Notes appropriate for gamma detection and measurement:

1. Instruments suitable for *gamma spectrometry* are also included here.
2. The filing system lists Instrument Notes *alphabetically by manufacturer*.
3. In the filing system, no distinction has been made between gamma, gamma/x-ray, and x-ray instruments. The choice of mnemonic (GAM,GAM-XRA, or XRA) has usually been made based upon the manufacturer's advertised specifications.
4. Many gamma instruments are also useful for beta radiation, and some for alpha radiation as well. If a manufacturer claims beta/gamma sensitivity, the instrument has been included in the beta section (Section A.5) with mnemonic RAD-BETA,GAM. If alpha/beta/gamma sensitivity is claimed, the instrument has been included in Section A.3, Combination Instruments, with mnemonic RAD-ALP,BETA,GAM,XRA. The reader should therefore consult those other sections together with this section for a fuller overview of instruments of this type.

Survey Monitor
Baird Model 904-460

Class	Portable, handheld
Principle of Operation	G-M detector, amplifier, meter readout
Sensisitivity and Range	Range: 1 to 100 mR/hr; 500 to 50,000 cpm
Performance	Accuracy: ±10% FS
Requirements	Power: Five batteries Size: 19.1 cm L × 10.2 cm W × 17.8 cm D (7.5 in. × 4 in. × 7 in.) Weight: 2.7 kg (6 lb)
Features	Includes radium source for calibration Model 904-460 detects alpha also
References	Manufacturer's specifications
Cost	$600.00
Address	Baird Corporation 125 Middlesex Turnpike Bedford, MA 01730 (617) 276-6000

Shielded Well Scintillation Counter
Baird Model 988-10C

Class	Laboratory
Principle of Operation	Well-type NaI(Tl) scintillation detector system 1 3/4 in. D × 2 in. thick NaI(Tl) crystal. Well size: 21/32 in. ID × 1 35/64 in. D
Sensitivity and Range	Efficiency: 35% for Iodine-131 Background: 250 cpm; 20 cpm when used with spectrometer
Sampling	Batch
Performance	Operating temperature: meets minimum requirements
Requirements	Power: 110 V ac, 60 Hz (220 V ac, 50 Hz optional) Size: 45.7 cm H × 32.4 cm D (18 in. × 12 3/4 in.) Weight: 131.5 kg (290 lb)
Features	Lead shield, tube mount; Model 810LX does not include detector

References	Manufacturer's specifications
Cost	$3550.00
Address	Baird Corporation 125 Middlesex Turnpike Bedford, MA 01730 (617) 276-6000

Gamma Counter
Beckman System 5500

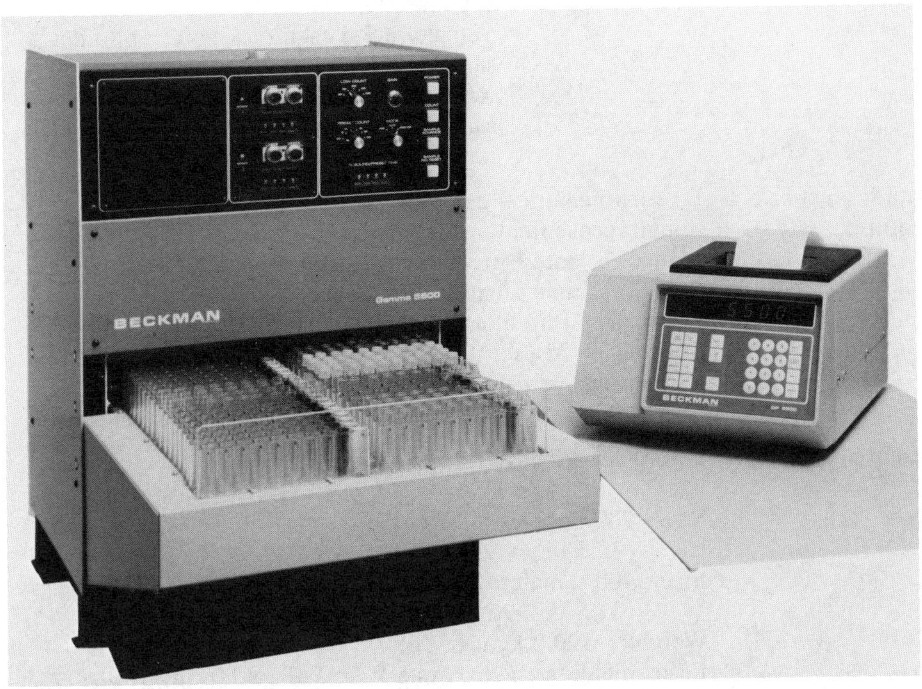

Class Laboratory

Principle of Operation The 5 cm (2 in.) crystal detector is positioned above the sample changer. A 7.6 cm (3 in.) crystal detector (option) will yield higher counting efficiencies for higher energy isotopes such as ^{131}I, ^{60}Co, ^{137}Cs, ^{51}Cr and ^{22}Na.

Sensitivity and Range

	5 cm (2 in.) Crystal	7.6 cm (3 in.) Crystal
Efficiency:		
^{125}I	≥ 72%	= 4%
^{137}Cs	≥ 30%	= 4%
Background:		
0 to 2 MeV	< 400 cpm	< 250 cpm
^{125}I window	< 50 cpm	< 0 cpm
Resolution:		
^{137}Cs	< 14%	< 12%

Performance	Sample capacity: 200 or 400 samples Available in one- or two-channel configurations Sample racks; each rack holds up to ten samples. Racks accommodate virtually any size sample tube up to 1.6 cm × 8.7 cm (0.63 in. × 3.4 in.)
Data Reduction Features	Five mathematical curve fit. Simultaneous final answers from two isotopes. Manual or automatic curving editing. Positive sample identification. Program call up by numbered rack card. Storage for 25 user defined programs. QC parameters calculated for standard curve. Up to 100 answers corrected for extraction efficiencies.
Requirements	Power: less than 2A at 120 V, 60 Hz less than 1.5A at 240 V, 50 Hz Size: 200 samples; 82.6 cm H × 55.9 cm W × 63.5 cm D (32.5 in. × 22 in. × 25 in.) Size: 400 samples; 82.6 cm H × 55.9 cm W × 87.6 cm D (32.5 in. × 22 in. × 25 in.) Weight: 120.2 kg (265 lb) Lead shielding (5 cm crystal): 52.6 kg (116 lb) Lead shielding (7.6 cm crystal): 107.5 kg (237 lb)
References	Manufacturer's specifications
Cost	Refer to manufacturer
Address	Beckman Instruments, Inc. Campus Drive at Jamboree Blvd. P.O. Box C-19600 Irvine, CA 92713 (714) 833-0751 Attn: T. McKillip

Germanium Gamma Spectrometry Detectors
Canberra Models 7229P/7229N/7229

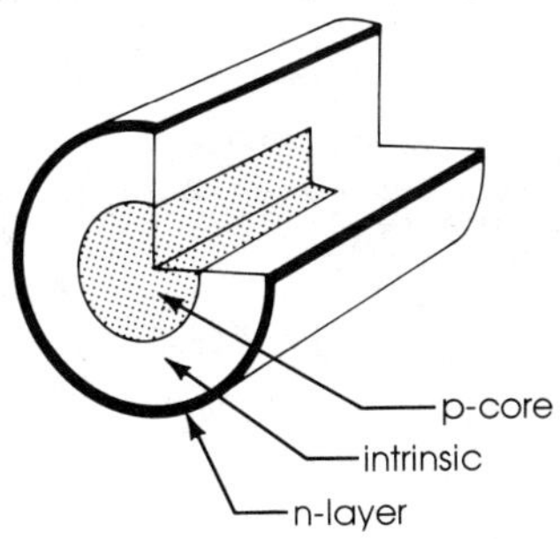

Class	Laboratory
Principle of Operation	Pure Ge and Ge(Li) semiconductor detectors used for gamma spectrometry.
Specifications	See Table A.6.1
Features	1. Various cryostat options and associated electronics also available. 2. Liquid nitrogen dewar, monitor and transfer device available.
References	Manufacturer's specifications
Address	Canberra Industries 45 Gracey Avenue Meriden, CT 06450 (203) 238-2351

Table A.6.1. Canberra Gamma Spectrometry Detectors

Specifications	Model 7229P and 7229N	Model 7229
Type	Pure Ge	Ge(Li)
Shape	Closed-end coaxial	Closed-end coaxial
Active area (cm^2)	Up to 28	Up to 28
Volume (cm^3)	10–150	10-150
Drifted depth (mm) (Depletion depth)	Up to 25	Up to 25
Resolution, FWHM		
(keV) for 1.33 MeV	1.8–2.2	1.8 to 2.3
(eV) for 122 keV	800–1300	
Efficiency[a]	5 to 30%	5 to 30%
Peak-to-Compton for 1.33 MeV	Up to 50:1	Up to 50:1
Price Range	\$8,500–\$30,000	\$5,100-\$24,780

[a]Efficiency measured relative to 3 in. × 3 in. NaI(Tl), at 1.33 MeV and 25 in. distance.

NaI(Tl) Gamma Spectrometry Systems
Canberra Model 802

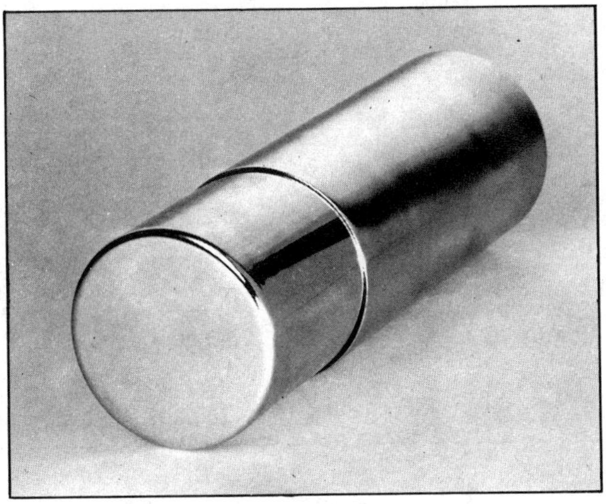

Class	Laboratory
Principle of Operation	NaI(Tl) crystal, photomultiplier tube, and tube base voltage divider assembled into a unit for gamma spectrometry.
Specifications	See Table A.6.2
Features	1. Entrance window 0.019 in. aluminum 2. Aluminum housing 3. Auxiliary electronics also available
Address	Canberra Industries 45 Gracey Avenue Meriden, CT 06450 (203) 238-2351 TWX: 710-461-0192 Attn: H. Palasek

Table A.6.2. Canberra NaI(Tl) Gamma Spectrometer Systems

	Crystal Diameter		Crystal Thickness		Tube Size		
Designation	(cm)	(in.)	(cm)	(in.)	(cm)	(in.)	Price for Assembly
802-1	3.8	1.5	2.5	1	5.1	2	$ 460.00
802-3	5.1	2	5.1	2	5.1	2	495.00
802-3W (well detector)	5.1 (1.59)	2 (0.625)	5.1 (3.65)	2 (1.438)	5.1	2	550.00
802-4	7.6	3	7.6	3	5.1	2	1,325.00
802-4W (well detector)	7.6 (1.67)[a]	3 (0.655)	7.6 (5.24)	3 (2.063)	5.1	2	1,470.00

[a]Size of well in the crystal

X-Ray Detectors
Canberra Industries

Class	Laboratory
Principle of Operation	Si(Li) detectors used for x-ray spectroscopy
Specifications	See Table A.6.3
Features	Liquid nitrogen cryostat, preamplifier, spectrum enhancer, system alarm, LN monitor, and bias supply also available
References	Manufacturer's specifications
Address	Canberra Industries 45 Gracey Avenue Meriden, CT 06450 (203) 238-2451

Table A.6.3. Canberra X-Ray Detectors

	Model		
Specifications	7313	7333	7383
Area (mm^2)	10	30	80
Depth (mm)	2	3	5
Resolution[a] at 5.9 keV FWHM	160 eV	170 eV	185 eV
Price range[b]	\$8000–8300	\$8000–8300	\$9100–9400

[a]Guaranteed worst case resolution @ 5.9 keV @ 1000 cps. Typical performance exceeds specifications.

[b]Includes Si(Li) detector, cryostat with choice of 0.5 mil Be window, with 7313, 7333 only.

X-Ray Scintillation Detectors
Canberra Models 1701 and 1702

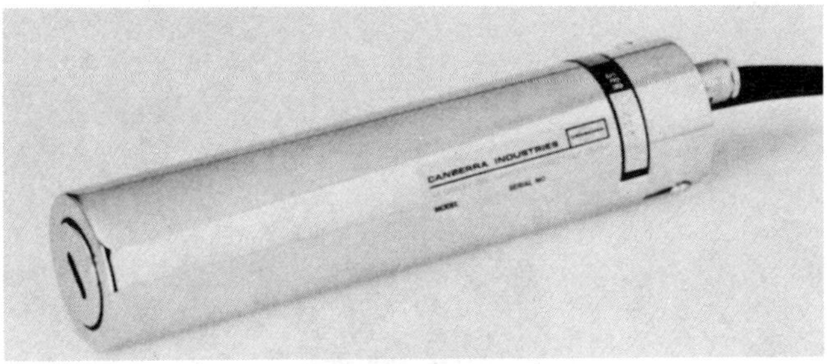

Principle of Operation	X-ray scintillation detector comprising a NaI(Tl) crystal and photomultiplier tube and a voltage divider network. A low noise preamplifier is built into Model 1702.
Specifications	*Models 1701 and 1702* Window: 0.005 in. Beryllium Transmission: 50% at 3.5 keV 88% at 6.0 keV Resolution: 55% or better at 6 keV Collimator: Removable brass with 0.125 in. × 0.5 in. slit Crystal: NaI(Tl) 1 in. diam., 1 mm thick Photomultiplier: Amperex XP 1010, selected *Model 1702 preamplifier* Type: Charge sensitive, FET input Power: 24 V Output: 0 to 20 V, 40-nsec risetime 60-μsec fall time, 50 Ω impedance
References	Manufacturer's specifications
Cost	1701, $ 910.00 1702, $1400.00
Address	Canberra Industries 45 Gracey Avenue Meriden, CT 06450 (203) 238-2351 TWX: 710-461-0192 Attn: H. Palasek

Pure Germanium Planar Detectors

Class	Laboratory
Principle of Operation	Pure Germanium Planar Detectors used for low-energy photon spectrometry.
Specifications	Active area: 25 to 1000 mm^2 Diameter: Up to 36 mm Depth: 5 to 13 mm Resolution, FWHM: 122 keV, 500–770 eV; 5.9 keV, 145–590 eV
Features	Cryostats and associated electronics also available.
References	Manufacturer's specifications
Cost	Price range: $4800–12000
Address	Canberra Industries 45 Gracey Avenue Meriden, CT 06450 (203) 238-2351

Germanium Detector
Canberra MAC Portable

Class	Field
Principle of Operation	Germanium detector. A stainless-steel dewar having two fill and vent ports arranged such that can be operated regardless of orientation. LN_2 holding time of 24 hours or more when used with average size detectors. The MAC is supplied with a sensor that signals when the LN_2 is depleted.
Sensitivity	Available as an option with most of the High Purity Germanium detectors offered by Canberra, including Planar, Coaxial or Reverse-Electrode Coaxial. Efficiency: 5%–30% Resolution: FWHM (at 1.3 MeV): 1.8–2.2 keV
Features	Standard preamplifier packaged behind the detector chamber within the confines of the 80 mm diameter snout. The preamplifier may employ either a room temperature on cooled FET depending on the detector type and performance level. The unit can also be installed in other apparatus such as whole-body counters, scattering chambers or low-level counting systems. Both manual and automatic refill systems are available.
Specifications	Weight: 6.5 kg (14.0 lb) empty LN_2 capacity: 1.4 liters Cool down time: 2 hours Electrical feedthrough: ceramic
Cost	$7500–$27,000 depending on efficiency and resolution.
Address	Canberra Industries 45 Gracey Avenue Meriden, CT 06450 (203) 238-2351 TWX: 710-461-0192 Attn: H. Palasek

Radioisotope Calibrator

Class	Laboratory calibration for gamma-emitting radioisotopes
Principle of Operation	Pressurized argon well-type ionization chamber and digital electrometer system.
Sensitivity and Range	Function: CRC-5/10, calibrator CRC-30, calibrator/computer Energy range: 25 keV to 3 MeV Sensitivity range: CRC-5/10, 0.1 μCi to 6 Ci max. Resolution: 0.1 μCi
Performance	Each calibrates over 200 radionuclides. For CRC-30, computer memory can accommodate up to 19 samples and concentration recall displays decay adjusted concentrations for push-button isotopes with a ticket printed patient dose hard copy record Accuracy: ±2% Operating temperature: 10 to 29°C (50 to 85°F)

Requirements	Power: 117 V ac ± 10%, 50/60 Hz, 220 V ac ± 10%, 50/60 Hz on request Size: CRC-5/10, 38.7 cm W × 16.5 cm H × 23.5 cm D (15 1/4 in. × 6 1/2 in. × 9 1/4 in.) CRC-30, 38.7 cm W × 17.2 cm H × 43.2 cm D (15 1/4 in. × 6 3/4 in. × 17 in.) Weight: CRC-5/10, 18.1 kg (401 lb); CRC-30, 23.6 kg (52 lb)
Features	1. Push button isotope selecion 2. Geometry independence 3. ^{99}Mo (equilibrium with ^{99}Tc) assay capability 4. Accommodates small animals for total body assay 5. CRC-5/10 field upgradable to computer/printer Model (CRC-30)
References	1. A. Suzuki, M. N. Suzuki, and A. M. Weis, "Analysis of a Radioisotope Calibrator," *J. Nucl. Med. Tech.*, **4**, 193–198, December 1976. 2. Manufacturer's specifications
Cost	CRC-5/10: $4525.00–$5300.00 CRC-30: $9225.00
Address	Capintec, Inc. 6 Arrow Road Ramsey, NJ 07446 (201) 825-9500 (800) 631-3826

Exposure/Exposure-Rate Meter
Capintec Model Diarad III

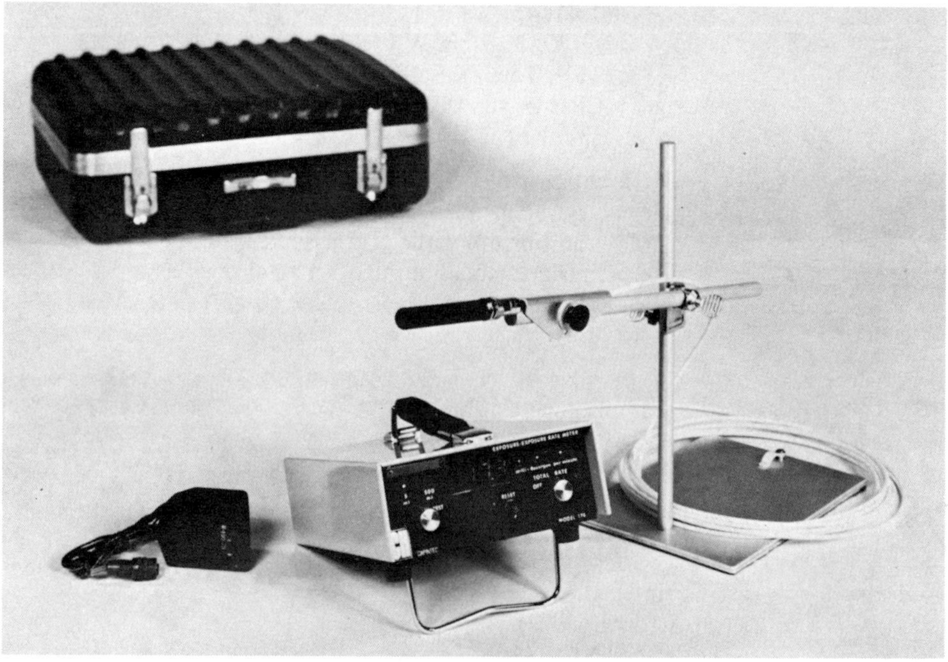

Class	Portable (battery operated) and line operated
Principle of Operation	Air ionization chamber detector and low impedance current/charge meter. The output digitally displayed in R, R/min, mR, and mR/min.
Sensitivity and Range	Radiation detected: x-ray Energy range: 50 to 250 KvP Resolution: 0.1 mR (mR/min) Interference: From gamma background only Accuracy: ±3% Precision: ±0.1% FS Linearity: 0.5% Stability: ±1%/yr
Sampling	Sample discrete field in diagnostic radiology
Performance	Temperature range: 15 to 29.4°C (59 to 85°F) Humidity range: Up to 60% RH without condensation unit is also usable 50 to 100°F and up to 95% RH

Requirements	Power: Line voltage, 115 V ac 50 to 60 Hz Size: 18 cm × 23 cm × 7 cm (7.1 in. × 9.1 in. × 2.8 in.) plus chamber and cable Weight: 4.5 kg (10 lb)
Features	Can be used to make measurements with other Capintec Ionization Chambers in other areas (e.g. mammography, electron beam therapy, intercavitary, etc.)
References	Product Bulletin and Instrument Manual
Cost	$2565
Address	Capintec, Inc. 6 Arrow Road Ramsey, NJ 07446 (201) 825-9500 (800) 631-3826

Exposure/Exposure-Rate Meter
Capintec Model Therarad I

Class	Laboratory and line operated portable
Principle of Operation	Air ionization chamber detector and rate/integrate electrometer digital display R, R/min, mR, and mR/min
Sensitivity and Range	Radiation detected: X-ray Energy range: Orthovoltage (0.05 keV), 0.6 to 35 MeV with suitable build-up cap Resolution: 0.001 R(R/min) Range: 2000 R(R/min) Interference: From gamma background only Accuracy: ±0.5% Linearity: ±0.1% Precision: ±1% FS Response time: 0.01 sec in integrate mode Sampling: Sample discrete field in radiation therapy
Performance	Temperature range: 15 to 29.4°C (59 to 85°F) Humidity range: Up to 60% RH; without condensation unit is also usable 50 to 100°F and up to 95% RH
Requirements	Power: 115 V ± 10 V; 60 Hz, 220 V on request. Size: 36 cm × 28 cm × 20 cm (14 in. × 11 in. × 8 in.) plus chamber and cable Weight: 7 kg (15 lb)
Features	Can be used to make measurements with other Capintec Ionization Chambers in other areas (e.g. mammography, electron beam therapy, interconnecting intracavitary measurements, etc.)
References	Product Bulletin and Instrument Manual
Cost	$2995.00
Address	Capintec, Inc. 6 Arrow Road Ramsey, NJ 07446 (201) 825-9500 (800) 631-3826

Exposure/Exposure-Rate Meter
Capintec Model Tomorad I

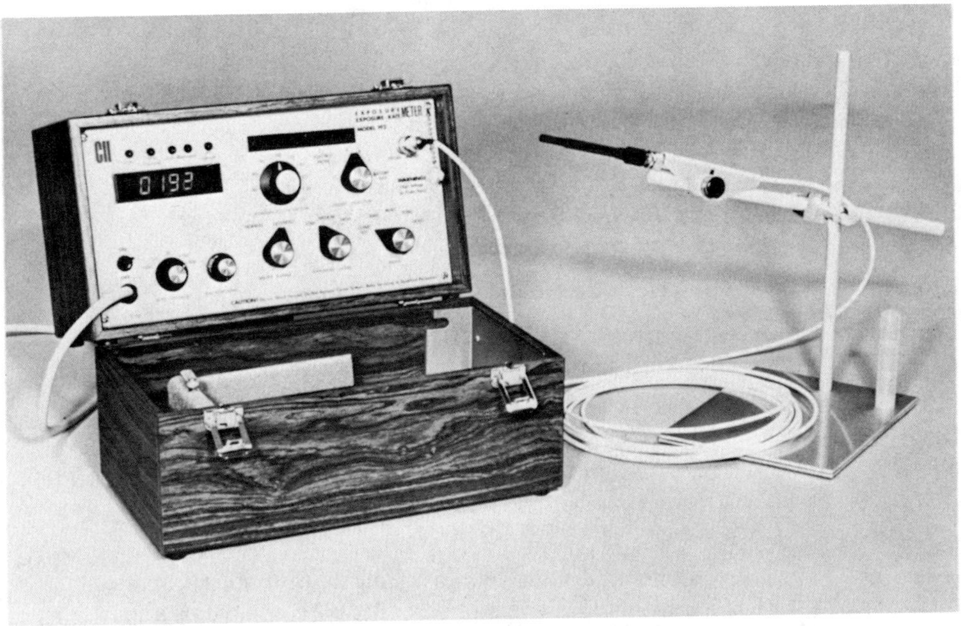

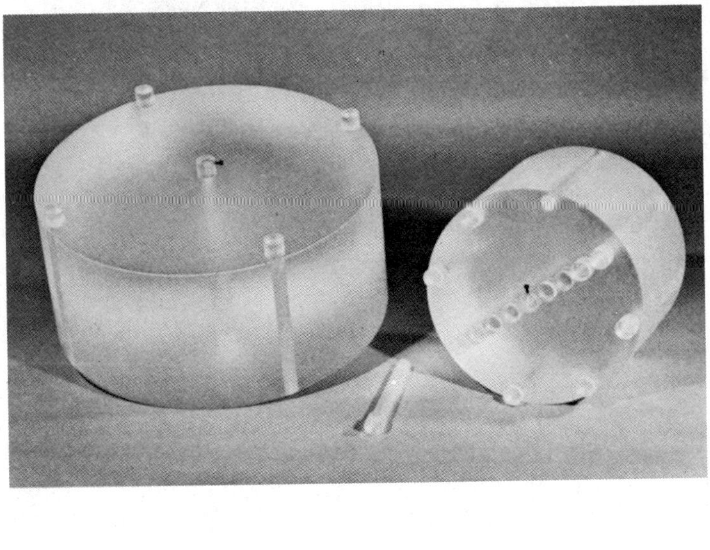

Class	Laboratory and line operated portable
Principle of Operation	Air ionization chamber detector and rate/integrate electrometer digital display R, R/min, mR, and mR/min
Sensitivity and Range	Radiation detected: x-ray Energy range: 40 to 200 keV effective for CT Scanner Measurement Resolution: 0.001 R cm, R cm/min Range: 2000 R cm, R cm/min Interference: From gamma background only Energy response: ±3%
Sampling	Samples discrete field from CT scanner
Performance	Temperature range: (50 to 85°F) Humidity range: Up to 60% RH; without condensation, unit is also usable 50 to 100°F and up to 95% RH
Requirements	Power: 115 V ±10 V, 60 Hz Size: 36 cm × 28 cm × 20 cm (14 in. × 11 in. × 8 in.) plus chamber and cable Weight: 7 kg (15 lb)
Features	Can be used to make measurements with other Capintec Ionization Chambers in other areas (e.g. mammography, electron beam therapy, intercavitary, etc.)
References	Product Bulletin and Instrument Manual
Cost	$3650.00
Address	Capintec, Inc. 6 Arrow Road Ramsey, NJ 07446 (201) 825-9500 (800) 631-3826

Digital Exposure and Exposure-Rate Meter

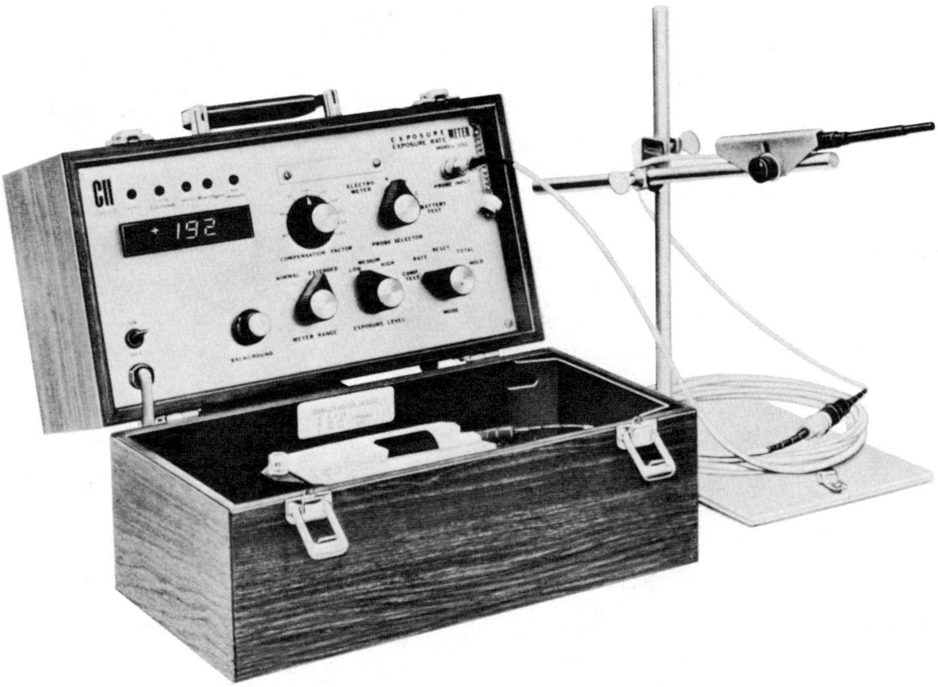

Class	Portable
Principle of Operation	System consists of ionization chambers (see Tables A.6.4 and A.6.5) and low input impedance current/charge meter. Direct reading digital output in mR, R, mR/minute and/or nanocoulombs and nanocoulombs/minute.

Sensitivity and Range

		174	192
	Range:	Rate = 20 mR/min to 200.0 R/min	2.000 mR/min to 2000 R/min
	(Full scale):	Integrate = 20 mR to 200.0 R	2.000 mR to 2000 R
	Resolution:	0.01 mR/min or 0.01 mR	0.001 mR/min or 0.001 mR
	Sampling:	Continuous	Continuous
Performance	Accuracy:	±0.5% of Electrometer	±0.5% of Electrometer
	Response time:	0.61 sec in integrate mode	0.01 sec in integrate mode
	Temperature stability:	±0.02% per °C	±0.02% per °C

Requirements	Rechargeable Ni-Cad batteries		
	Power:	117 V ac, 50/60 Hz (230 V ac optional)	117 V ac, 50/60 Hz (230 V ac optional)
	Size:	21.6 cm W × 10.2 cm H × 27.3 cm D (8 1/2 in. × 4 in. × 10 3/4 in.)	5.6 cm W × 26.0 cm H × 19.1 cm D (14 in. × 10 1/4 in. × 7 1/2 in.)
	Weight:	4.5 kg (10 lb) Including wood carrying case	
Features	1. Digital numerics display, 3 1/2 digits 2. Automatic range selection and display of units 3. Selection of probes for wide range of x-rays, gamma rays and electron energies 4. 174, 3-decade dynamic range for probe; 192, 4-decade dynamic range for probe 5. 174, Input selector for 2 probes; 192, Input selection for 3 probes and more on electrometer mode		
Cost	174, $1750.00* 192, $2190.00*		
Address	Capintec, Inc. 6 Arrow Road Ramsey, NJ 07446 (201) 825-9500 (800) 631-3826		

*Price does not include detector

Table A.6.4. Capintec Ionization Chambers

Model	Range (digital full scale)		Energy		Response	Collection Efficiency Maximum Exposure-Rate; Continuous Wave
	Rate (per min)	Integrate	KV	HVL		
PR-06C[a]	2–2000 R	2–2000 R	50–250	2 mm Al to ^{60}Co[b]	±3%	12 kR/min (99.8%)
PM-5[a]	2–2000 R	2–2000 R	50-250	2 mm Al to ^{60}Co[b]	±3%	11.4 kR/min (99.8%)
PM-30[a]	20 mR–20 R	20 mR–20 R	50-250	2 mm Al to 6 mm Cu	±2%	84 R/min (99%)
PR-05[c]	2–2000 R	2–200 R		^{137}Cs, ^{226}Ra, ^{60}Co	±2%	
PM-500	Suitable for 0.01 mR (mR/min) to 2 (R/min) applications					
PS-033	Suitable for soft x-ray, high-energy photon, and electron applications					

[a]Air-equivalent plastic (Shonka)
[b]For intracavitary application
[c]With equilibrium cap.

Table A.6.5. Ionization Chamber Specifications

	PR-06C	PR-05	PR-05P	PS-033
Sensitivity (nC/R) (nominal)	0.200	0.042	0.021	0.160
Resolution (max) (R, R/min)	0.001	0.001	0.01	0.001
Maximum display (R, R/min)[a]	2000	2000	2000	2000
Stability (% per year)	±0.5	±0.5	±0.5	±1
Stem effect (%)	< ±0.5	< ±1.0	< ±1.0	< ±0.5
Leakage current (A)	10^{-14}	10^{-14}	10^{-14}	10^{-14}
Recommended polarization voltage (V)	300	300	300	300
Chamber volume (mliter)	0.65	0.14	0.07	0.5
Wall thickness (mm)	0.25	1.2	1.2	0.0036[b]
Material	Air-eq. plastic	Air-eq. plastic	Air-eq. plastic	PET-Al[b]
Vented	Yes	Yes	Yes	Yes
Chamber connector	BNC-size triaxial	BNC-size triaxial	BNC-size triaxial	BNC-size triaxial
Averaging area (cm^2)	N/A	N/A	N/A	N/A

[a]With Model 192 Digital Dosimeter.
[b]Also available 0.13 mm Air-eq. plastic and 25 μ PET-Al.
[c]Air-equivalent plastic, polypropylene, graphite, and aluminum alloy.

Model				
PR-16C	PM-30	PM-500	PM-05	PC-4D
0.48	9.2	160	1.6	0.1 nC/R·cm
0.001	0.01 mr,mR/min	0.001 mR,mR/min	0.1 mR,mR/min	0.001 R·cm or R·cm/min
2000	20	2	200	2000 R·cm or R·cm/min
±1	±1	±2	±1	±1
< ±1.0	< 0.5	< 0.5	< 1.0	< 0.5 (for 0.3 m × 0.3 m field)
10^{-14}	10^{-14}	10^{-14}	10^{-14}	10^{-4}
300	300	300	300	300
1.5	28	530	5.3	3.0
0.25	0.50	1.2	0.3	0.33
Air-eq. plastic	Airreq. plastic	*c*	Air-eq. plastic	Air-eq. plastic
Yes	Yes	Yes	Yes	Yes
DNC-size triaxial	DNC-size triaxial	DNC-size triaxial	DNC-size triaxial	DNC-size triaxial
N/A	N/A	100	N/A	N/A

Source Position Indicator
Eberline Model SPI-2

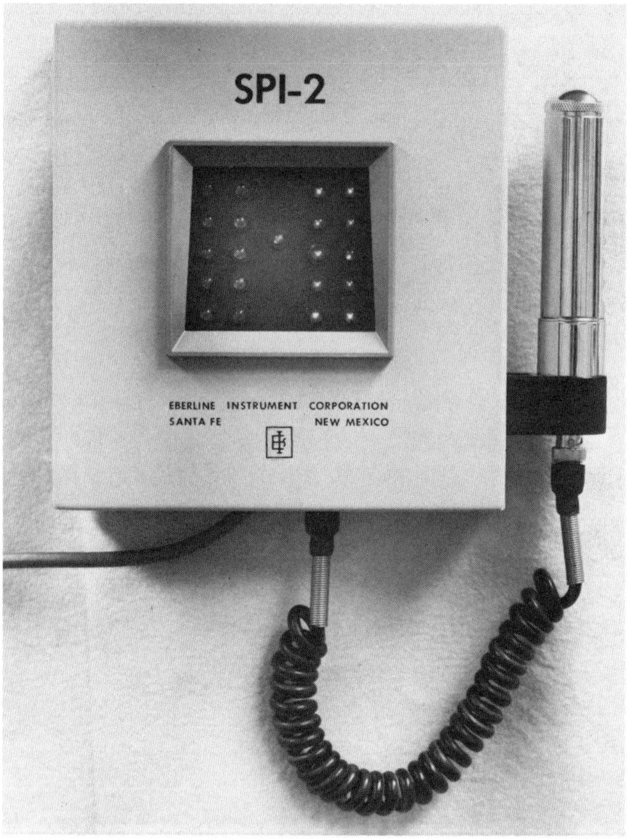

Class	Laboratory
Principle of Operation	Uses a halogen-quenched G-M tube mounted in an Eberline HP-270 hand probe to detect gamma radiation fields; designed primarily for use in areas to indicate when a radiation source is exposed (flashing red display) or when a radiation source is not exposed (flashing green display).
Sensitivity and Range	Alarm level adjustable for fields from 3 to 30 mR/hr, threshold is factory-set at approximately 5 mR/hr, unless special request is made. Minimum energy detected: approximately 70 keV Maximum energy detected: greater than 3 MeV Maximum allowable flux: 1000 R/hr

Sampling	Continuous
Performance	Accuracy: Alarms when exposure rate exceeds threshold Temperature: 4.5 to 43°C (40 to 110°F)
Requirements	Power: 110 to 125 V ac, 60 Hz at 0.2 A Size: 22.9 W × 21.6 H × 6.4 cm D (9 in. × 8.5 in. × 2 1/2 in.) Weight: 2.5 kg (5.5 lb)
Features	Go, no-go display; battery backup; measures scattered radiation from teletherapy machines; will not saturate in radiation fields as high as 1000 R/hr
References	Manufacturer's specifications
Cost	$715.00
Address	Eberline Instrument Corporation P.O. Box 2108 Santa Fe, NM 87501 (505) 471-3232

Survey Meter
Elscint Model GSM-1

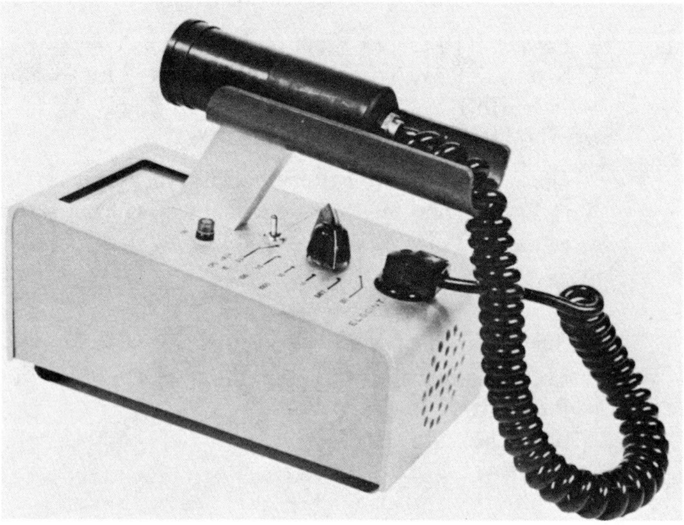

Class	Portable, handheld
Principle of Operation	G-M detector, amplifier, meter readout, speaker
Sensitivity and Range	Range: 0–0.5, 0–5.0, 0–50 mR/hr Energy dependence: ±20% from 60 keV to 1.3 MeV
Sampling	Continuous
Performance	Accuracy: ±2% FS Time constant: 2 to 4 sec Operating temperature: −20 to 60°C (−13 to 140°F)
Requirements	Power: Four 1.5-V "C" batteries. 90-hr continuous operation Size: 16.5 cm H × 10.3 cm W × 19 cm D (6 1/2 in. × 4 1/16 in. × 7 1/2 in.) Weight: 1.1 kg (2.4 lb)
Features	In very high radiation fields meter indicates fullscale, not zero; Geiger tube may be operated with a cable at a distance up to 50 m from the reading instrument.
References	Manufacturer's specifications
Cost	$800.00 plus shipping

Address

Elsinct, Inc.
138-160 Johnson Avenue
Hackensack, NJ 07602
(201) 487-5885
Attn: Hanan Rotman

Gamma Monitor
Fisher Controls Model GM-700

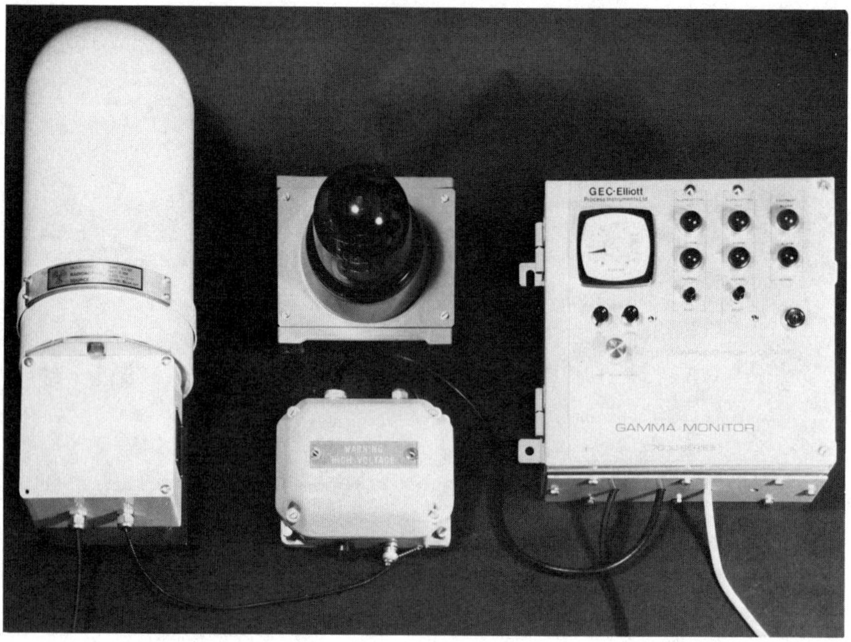

Class	Laboratory
Principle of Operation	Ion chamber, amplifier, meter readout, alarm options
Sensitivity and Range	Energy sensitivity: 90 keV to 1 MeV Range: 0.5 to 10^4 mR/hr, log
Sampling	Continuous
Performance	Accuracy: ±10% Temperature range: −10 to 55°C (14 to 131°F) Relative humidity: 5 to 100% Drift: Less than 10% of indicated value after 400 hr Less than 20% of indicated value after 500 hr
Requirements	Power: 25 W Size: Monitor, 30.5 cm × 30.5 cm (12 in. × 12 in.); Head amp., 17.8 cm × 12.7 cm (7 in. × 5 in.); Ion chamber, 55.9 cm × 19.1 cm (22 in. × 7 1/2 in.) Weight: Monitor, 12.7 kg (28 lb); Head amp., 4.1 kg (9 lb); Ion chamber, 12.7 kg (28 lb)

Features	Alarm level adjustments, recorder output
References	Manufacturer's specifications
Cost	Refer to manufacturer
Address	Fisher Controls Ltd. Nuclear Controls Division Century Works, Lewisham London SE13 7LN, England Attn: A. E. Pegg Tel. 01-692-1271 Ext. 260 International # +44 (1) 692-1271 Telex 22469

Four-Channel Gamma-Ray Spectrometer
Fisher Controls Type 0928

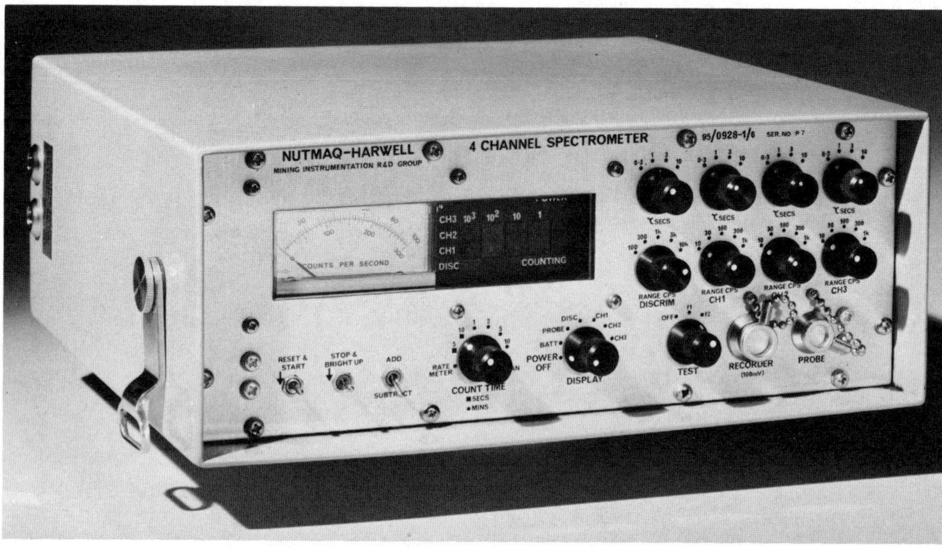

Class
: Portable, designed for field use

Principle of Operation
: Used with a variety of probes; it has a discriminator and three energy channels all of which may be present within the range 50 keV to 3 MeV. Counts derived from γ-ray energies exceeding the discriminator threshold or lying within the channel boundaries, may be displayed one at a time on the front panel meter and, simultaneously, on an external 4-channel recorder; alternatively they may be displayed one at a time as an accumulated count on a digital display.

Sensitivity
: Range (for channels 1, 2, and 3): 0–10, 0–30, 0–100, 0–300, 0–1000 cps
Range (discriminator): 0–100, 0–300, 0–1000, 0–10,000 cps
Sensitivity: depends on probe
Energy limits: Discriminator, 50 keV to 1 MeV
Channel 1; 1 MeV to 2 MeV
Channel 2; 1.25 MeV to 2.25 MeV
Channel 3; 2 MeV to 3 MeV

Features
1. Spectrum stabilized using 60-keV line from ^{241}Am reference source contained in probe
2. Front panel switch allows test pulser of 1 Hz or 136 Hz to be injected into digital counting circuits so that the correct operation of each counting chain may be monitored.

References	Manufacturer's specifications
Cost	Refer to manufacturer
Address	Fisher Controls Nuclear Controls Division, Century Works Lewisham, SE13 7LN, England 01-692-1271 Telex: 22469

Gamma Survey Meter
Fisher Controls Type 0593

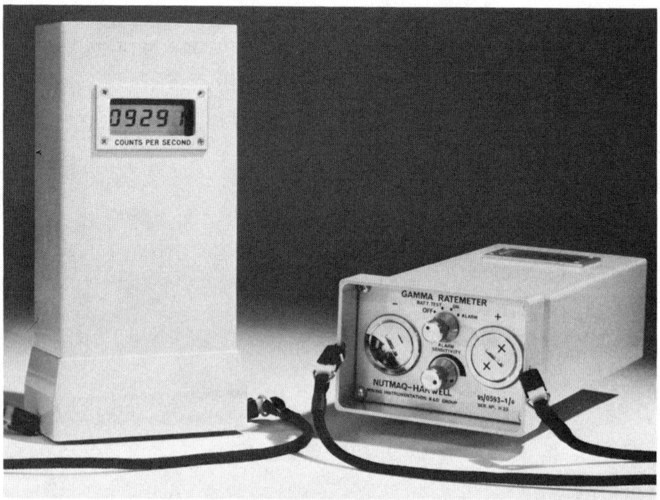

Class	Portable
Principle of Operation	γ-ray detector, solid-state circuitry, scaler, liquid crystal display, and audible alarm
Sensitivity and Range	Range: 0 to 5 digit readout
Sampling	Continuous
Performance	Count cycling time: 2 sec Operating temperature range: −5 to 50°C (23 to 122°F)
Requirements	Power: Six "C" cells Size: 7.4 cm × 12.4 cm × 24.0 cm (3 in. × 5 in. × 9 1/2 in.) Weight: 1.6 kg (3.5 lb)
Features	1. Audible alarm has an adjustable threshold and pitch related to count rate 2. Hermetically sealed 3. Fully compensated to accommodate voltage changes between 9 and 6.6 V 4. Battery test switch
References	Manufacturer's specifications
Cost	Refer to manufacturer

Address

Fisher Controls
Century Works
Lewisham, London SE13 7LN, England
Tele: 01-692-1271

Gamma Survey Meter
Fisher Controls Model NE-148A

Class	Portable, handheld
Principle of Operation	Sodium iodide scintillator, photomultiplier tube, ratemeter, meter readout
Sensitivity and Range	Energy sensitivity: 0.15 to 1.25 MeV Range: 0–30, 300, 3000 μR/hr (130, 1200, 10,000 cps) (radium equilibrium) Response time: 2, 0.2, 0.02 sec
Sampling	Continuous

Performance	Meter accuracy: ±10% FS Temperature range: −25 to 44°C (−13 to 111°F)
Requirements	Power: Two 1.5-V "D" cells, 400 hr Size: 21.5 cm × 6 cm × 14.6 cm (8.5 in. × 2.4 in. × 5.7 in.) Weight: 1.8 kg (4 lb)
Features	Audio output jack, waterproof to a depth of 5 ft., meter light A.E.R.E. Type 1597A
References	Manufacturer's specifications
Cost	Refer to manufacturer
Address	Fisher Controls Ltd. Measurements and Analytics Division Century Works, Lewisham London SE13 7LN, England Attn: A. E. Pegg Tel. 01-692-1271 Ext. 260 Telex: 22469

Survey Meter
Gamma Industries Models 150B, 155B, 250B, 252B

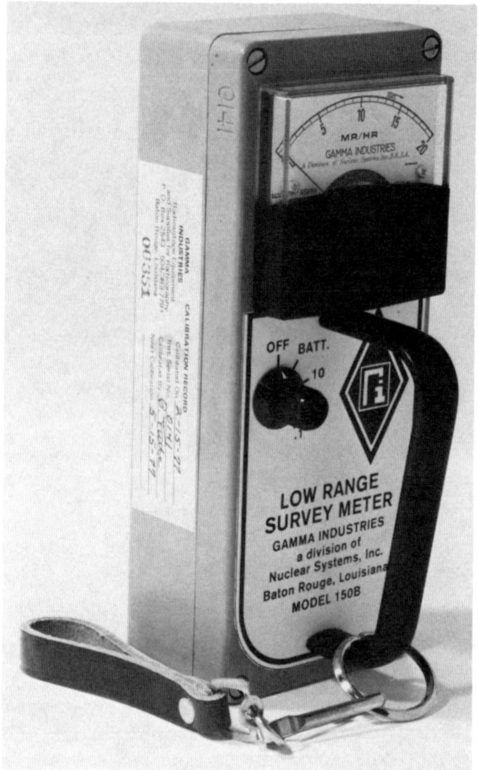

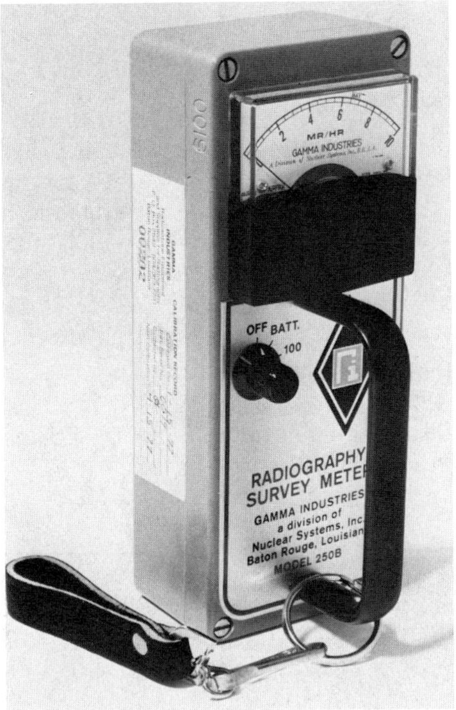
MR/HR
GAMMA INDUSTRIES
OFF
BATT.
100
RADIOGRAPHY
GAMMA INDUSTRIES
a division of
Nuclear Systems, Inc.
MODEL 250B

MR/HR
GAMMA INDUSTRIES
OFF
BATT.
10
LOW RANGE
SURVEY METER
GAMMA INDUSTRIES
a division of
Nuclear Systems, Inc.
MODEL 155B

Class	Portable
Principle of Operation	Geiger-Müller tube as detector, amplifier, electrometer, meter readout
Sensitivity and Range	see below

	150B, 155B	250B, 252B
Range:	0–2 mR/hr	0–10 mR/hr
	0–20 mR/hr	0–100 mR/hr
	0–200 mR/hr	0–1000 mR/hr

Sampling	Continuous
Performance	150B, internal probe 155B, external probe Accuracy: ±20% FS Operating temperature range: −17.8 to 75°C (0 to 167°F) Dead time: 50 μsec Energy range: 250 keV to 1.5 MeV

Requirements

	150B, 155B	250B
Power:	Two "D" cells	Two "D" cells
Size:	8.25 cm W × 20.32 cm L × 10.16 cm H (3 1/2 in. × 8 in. × 4 in.)	8.26 cm W × 20.32 cm L × 10.16 cm H (3 1/4 in. × 8 in. × 4 in.)
Weight:	1.135 kg (2.5 lb)	1.25 kg (2.75 lb)

	252B
Power:	Two "D" cells
Size:	20 cm L × 10 cm H × 8 cm W (8 in. L × 4 in. H × 3 in. W)
Weight:	1.1 kg (2.5 lb)

Features	1. Will not reverse in high-intensity gamma radiation fields 2. Low power consumption: > 400 hr from ordinary batteries, up to 1000 hr on alkalines.
References	Manufacturer's specifications
Cost	Model 150B: \$395.00 Model 155B: \$435.00 Model 250B: \$295.00 Model 252B: \$375.00
Address	Gamma Industries P.O. Box 2543 Baton Rouge, LA 70821 (504) 388-0800 Telex: 586473 Attn: Sales Manager

Portable, Low-Level Gamma and X-Ray Survey Meter
Harshaw Model 301

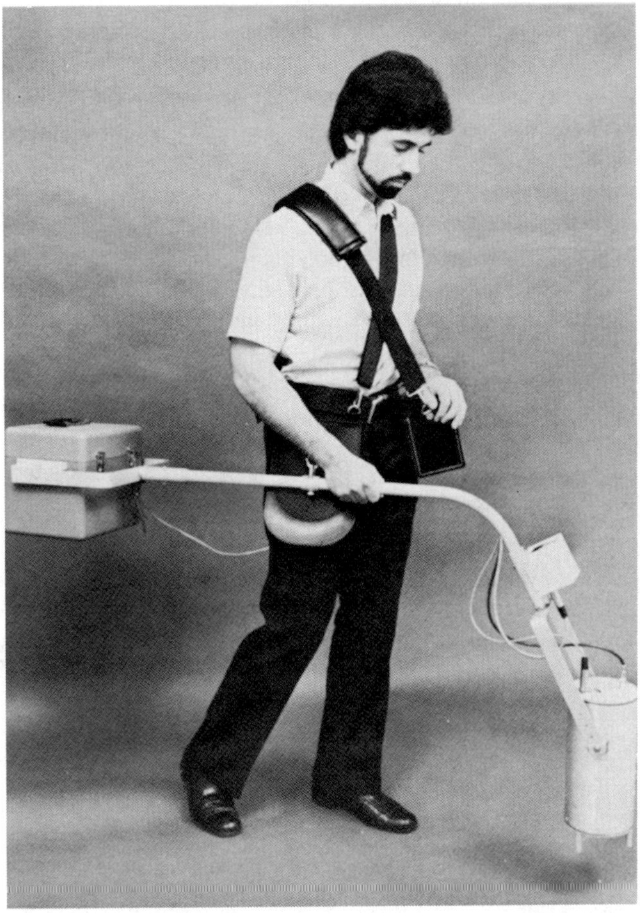

Class — Designed for rough handling in the field. Fully portable (Fig. A.6.2).

Principle of Operation — A Phoswich, a two-phosphor scintillation detector employs materials with different decay times and gamma sensitivities to distinguish between pulses from low-energy photon emitters and background events. A pulse shape analyzer and time-to-amplitude converter separates low-energy photon interactions from background events as they occur in the phosphors and processes the resulting signals to actuate output display and alarm devices.

Sensitivity and Range	The detector is optimized for high efficiency detection of 15 to 100 keV photons. Phoswich detector enables a background reduction of three to six times as compared to single NaI crystal types such as the FIDLER detector, by reduction of Compton events, photomultiplier tube noise rejection, and active guarding.
Sampling	Continuous. Microprocessor-based counting circuits provide continuous monitors via a logarithmic ratemeter with internally selectable time constants. A preset alarm threshold controls an audible "beeper" which responds to an average count rate. Static count measurements are made using operator-selected counting intervals. An eight-digit LCD provides numeric indication of counts, time, and background.
Requirements	Power: provided by battery pack Size: 20 cm × 28 cm × 23 cm (8 in. × 11 in. × 9 in.) Weight: 16.3 kg (36 lb)
Features	Light-weight. No line power required. Designed for remote area surveying. Battery charger included.
References	Harshaw Technical Literature *A Battery Operated Portable Phoswich Detector for Field Monitoring of Low Levels of Transuranic Contaminants* by C. J. Umbarger and M. A. Wolf of Los Alamos Scientific Laboratory. Nuclear Instruments and Methods, 155 (1978), pp. 453–457.
Cost	Model 301: $8000.00
Address	Harshaw Chemical Company 6801 Cochran Road Solon, OH 44139 (216) 248-7400

**Environmental Radiation Monitoring System–Air/Water
Harshaw Model 401 and 401W**

Class	Model 401W: Continuous, real-time gamma radiation monitor for water; Model 401: Air, or gaseous effluent pathways; non-portable.
Principle of Operation	NaI(Tl) scintillation detector drives stabilized electronic counting chain. Output meters calibrated in μR/hr or pCi/liter. Recorder outputs and alarms available.
Sensitivity and Range	Air radiation monitor range is 1 to 10 μR/hr. Water effluent radiation monitor sensitivity is 10 pCi/liter. Energy dependence: substantially independent of gamma ray energy in the range of 0.1 to 2.0 MeV.
Sampling	Continuous, real-time monitoring for both area and effluent systems.
Performance	Rugged POLYSCIN^{R} NaI(Tl) Scintillation Crystals are used. They are packed in foam plastic and housed in a weatherproof/waterproof container. An ^{241}Am pulser is incorporated in the scintillator for stabilizing the system against temperature changes and photomultiplier tube gain shift.
Requirements	Power: 115/230 V ac

Features	"Rugged and Reliable" environmentally protected scintillation detection. No gas counters. Counting instruments can be located up to 1 mile away from detector. Long-term accuracy assured by ^{241}Am pulser. Continuous performance verification by reference pulser alarm.
References	Harshaw Product literature available.
Cost	Model 401: $7950.00 Model 401W: $14980.00
Address	Harshaw Chemical Co. Crystal and Electronic Products Dept. 6801 Cochran Road Solon, OH 44139 (216) 248-7400

X- and Gamma Ray Monitor
Harshaw Model 2000P TLD System

Class	Laboratory
Principle of Operation	Thermoluminescence detector (Model 2000P) with an automatic integrating pico-ammeter (Model 2000B) and a strip-chart recorder (Model M-12).
Sensitivity and Range	1 mR to > 5000 R, using model 4040 bulb dosimeter 0.1 mR to > 500 R, using model 2040 bulb dosimeter
Performance	*Thermoluminescence detector* 1. Photomultiplier tube: Luminous sensitivity: 33 amperes per lumen (typical) 13 amperes per lumen (minimum) Dark current: 8×10^{-11} amperes (typical) 1.7×10^{-10} amperes (maximum) at 10°C (50°F) and sensitivity of 7 amperes per lumen 2. Phototube thermoelectric cooler: Nominal tube temperature: 10°C (50°F) Temperature stabilized to 0.56°C (1°F) Room temperature 16 to 30°C (60 to 85°F) 3. Heater current: Constant current in the range 0 to 10 amperes *Picoammeter* 1. Current measurement: Range: 10^{-13} to 10^{-4} amperes Accuracy: ±1% from 10^{-4} to 10^{-8} ampere; ±2% from 10^{-8} to 10^{-12} ampere at the linear output. Add ±3% to the above for log output and ±5% for meter Input current offset: $< 1 \times 10^{-14}$ ampere 2. Charge measurement: Range: 10^{-12} to 10^{-2} coulomb Accuracy: ±3% ± 1 digit Current suppression: 0 to 1.6×10^{-10} ampere for negative input current. Current enhancement 0 to 4×10^{-12} amperes Integrating period: manually selectable integration periods of 1 to 99 and infinity seconds

Requirements	Power: 105–125 V; regulated ±15 V dc at 200 milliamperes Size: *Thermoluminescence detector* 29 cm H × 21 cm W × 45.5 cm D (11.5 in. H × 8.25 in. W × 18 in. D) *Picoammeter* 28.75 cm H × 21 cm W × 45.75 cm D (11.5 in. H × 8.25 in. W × 18 in. D) *Recorder* 14 cm H × 9.2 cm W × 16.5 cm L (5.5 in. H × 3.625 in. W × 6.5 in. L) *Weight* Thermoluminescence detector, 20.5 kg (45 lb) Picoammeter, 11.35 kg (25 lb) Recorder, 2.04 kg (4.5 lb)
References	Manufacturer's specifications
Cost	$11000
Address	Harshaw 6801 Cochran Road Solon, OH 44139 (216) 248-7400 Attn: M. Cox/D. Elliot

RAD-GAM,XRA
Harwell (UKAEA) 1
January 1983

Area Gamma Alarm Monitor
Harwell Type 0740

Class	Laboratory, wall-mounted
Principle of Operation	G-M tube mounted internally, but which can be operated remotely up to 30 meters away from the monitor
Sensitivity and Range	Range: 0.01–300 mR/hr, logarithmic. Between 0.01 and 0.1 mR/hr it will give an "on scale" indication of background radiation without initiating the low count alarm. The meter will continue to indicate greater than full-scale at radiation levels above 300 mR/hr.
Sampling	Continuous
Performance	Operating temperature: −10 to 45°C (14 to 113°F) High radiation alarm: when the pre-set alarm level is exceeded the unit: 1. Sound a loud acoustic warning, generated by the Y03 Yodalarm 2. A 12 watt red beacon lamp fitted to the top of the monitor flashes continuously at 80 ppm 3. Four pairs of uncommitted light duty relay contacts (50 V, 250 mA), open and de-energize. Setting alarm level: can be set at any level between 0.1 and 300 mR/hr Sound mute button is provided to silence the acoustic alarm. Response time: not greater than 100 sec at 0.1 mR/hr and not more than 1 sec at 100 mR/hr Remote detector: a detachable probe containing a MX164 G-M tube can be operated at distances of up to 30 meters from the monitor.
Options	External recorder of 1 mA sensitivity and 100 ohm impedance. System may be computer operated.
Requirements	Power: 240 or 110 V ac, 50 Hz Size: 35 cm × 25 cm × 15 cm (14 in. × 10 in. × 6 in.) Weight: 7.5 kg (16.5 lb)
References	Manufacturer's specifications
Cost	Refer to manufacturer
Address	AERE, Harwell Oxfordshire OX11 ORA England Tel: Abingdon (0235) 24141 Telex: 83134 Attn: R. Sherwood (Ext. 4766)

X-Ray Proportional Counters
Harwell

Class	Detector
Principle of Operation	Tubular beryllium section of detectors gives reduced x-ray fluorescent emission.
Sensitivity	Resolution of a 2.5 cm (1 in.) diameter proportional Counter: ~0.23 keV at 1 keV ~2.3 keV at 100 keV Efficiency: greater than 50% at energies between 1.5 keV and 18 keV; less than 10% for energies greater than 70 keV. (Using a 25 μm Be window)

Specifications

Cylindrical		End-Window	
Diameter	Minimum wall thickness	Area	Minimum window thickness
~5 cm (2 in.)	~0.25 mm (0.010 in.)	6.1 cm^2	~0.12 mm (0.005 in.)
~2.5 (1 in.)	~0.12 mm (0.005 in.)	1.25 cm^2	~0.03 mm (0.001 in.)
~1.25 cm (1/2 in.)	~0.12 mm (0.005 in.)		

Features	Guaranteed for one year, with an expected life in excess of three years. Position-sensitive types available. Counters with gas fillings, body sizes and window thicknesses that differ from those stated here are also available.
Cost	Refer to manufacturer
References	Manufacturer's specifications
Address	AERE, Harwell Oxfordshire OX11 ORA England Tel: Abingdon (0235) 24141 Telex: 83135 Attn: R. Sherwood (Ext. 4766)

Tissue Equivalent Survey Meter
Health Physics Model 1010

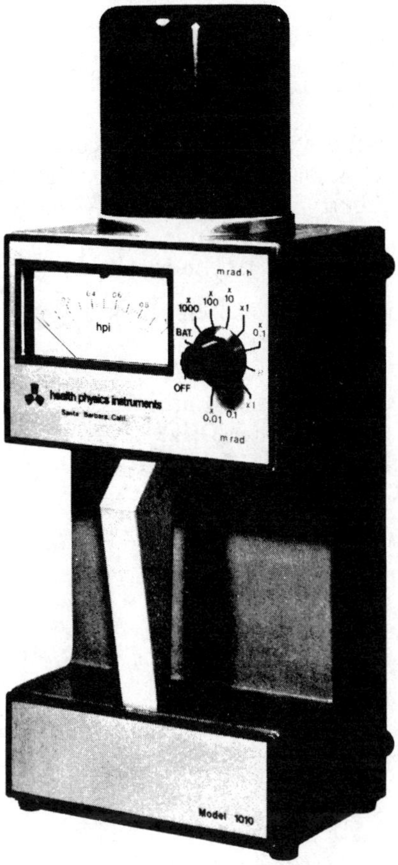

Class	Portable, handheld
Principle of Operation	Sealed tissue-equivalent proportional chamber detector, electrometer meter readout.
Sensitivity and Range	Range (rate): 0–0.1, 1, 10, 100, 1000 mrad/hr Range (dose): 0–0.01, 0.1, 1 mrad
Sampling	Continuous
Performance	Wall thickness: 0.1 in. Time constants: 0–0.1 mrad/hr, 10 sec 0–1 mrad/hr, 5 sec 0–10 mrad/hr, 2.5 sec 0–100 mrad/hr, < 1 sec 0–1000 mrad/hr, < 1 sec

	Energy response: ±5% from 100 keV to 1.25 MeV; with response to soft tissue−10% at 70 keV, 20% at 40 keV
Requirements	Power: Four "D" cells, 80 to 100 hours lifetime Size: 15 cm × 10 cm × 35 cm (5.75 in. × 4 in. × 13.75 in.) Weight: 2.4 kg (5.25 lb)
Features	1. Solid-state circuitry, IC-controlled high voltage supply 2. Chamber has center wire, filled with tissue-equivalent gas 3. Controls for high voltage, zero, and three calibration adjustment controls 4. Ion chamber cylindrical.
References	Manufacturer's specifications
Cost	$1950.00
Address	Health Physics Instruments, Inc. 124 Santa Felicia Drive Goleta, CA 93117 (805) 685-2612

Portable Remote Area Monitor
Health Physics Instrument Model 1072

Class	Portable, handheld
Principle of Operation	Uses an air-filled, unsealed ionization chamber as detector coupled with amplifier, ratemeter, and liquid crystal display readout with alarms
Sensitivity and Range	Dose rate range: 0.1 to 199.9 mR/hr (1072-1) 1 to 1999 mR/hr (1072-2)
Sampling	Continuous
Performance	Reproducibility: overall ±15% or two digits, whichever is greater Detector volume: 1 liter Temperature dependence: zero shift, ±5 in least significant digit 0 to 50°C Operating range: 10 to 50°C (50 to 122°F) Storage range: −20 to 65°C (−4 to 149°F) Pressure dependence: Reading dependent on elevation and barometric pressure Warm-up time: 30 sec Response time: 3 sec Exposure rate response: 90% collection at 2 R/hr off scale indication, 1- - - (a numeral in the left-hand digit with other digits blanked).
Requirements	Power: Batteries, eight "D" cells Size: 25 cm H × 16 cm W × 30 cm L (10 in. × 6 1/2 in. × 12 in.) Weight: 4 kg (8.8 lb)
Features	Audible and visual alarm LCD displays arrow for low battery condition
References	Manufacturer's specifications
Cost	$985.00
Address	Health Physics Instruments, Inc. 124 Santa Felicia Drive Goleta, CA 93117 (805) 685-2612

X-Ray Survey Meter
Johnson and Associates Model TVX-1

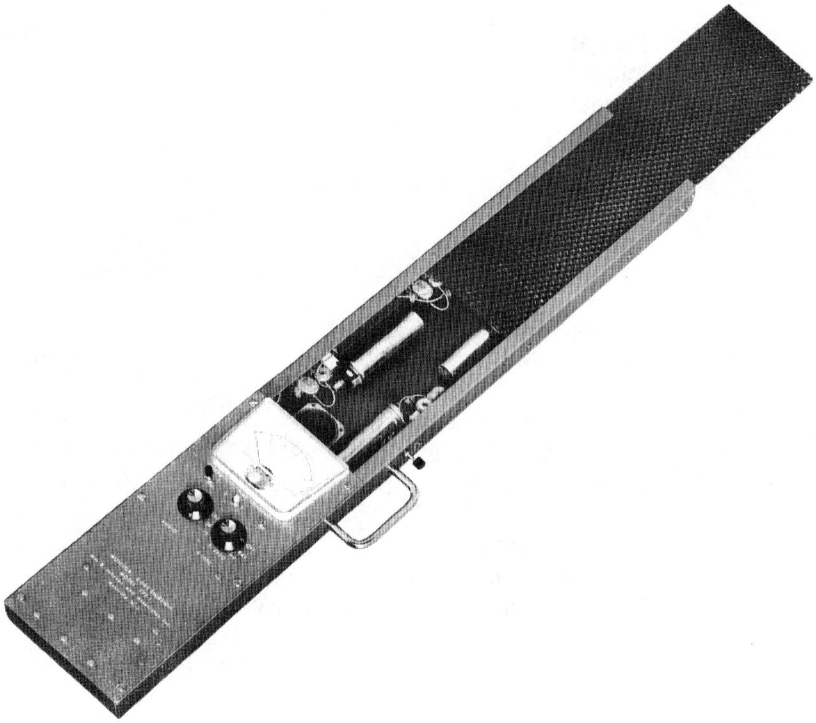

Class	Portable, handheld
Principle of Operation	Six G-M tubes spaced equidistantly to provide a large scanning area (20 in. × 5 in.) for the detection of x-rays from color TV receivers
Sensitivity and Range	Energy sensitivity: 15 keV to 10 MeV Range: 1.05 mR/hr to 1 R/hr; 3000, 10K, 30K, 100K cpm Minimum sensitivity: 0.05 mR/hr
Sampling	Continuous
Requirements	Power: Three 9-V batteries Size: 65 cm L × 11 cm W × 4 cm D (26 in. × 4.5 in. × 1.5 in.) Weight: 1.6 kg (3.5 lb)
Features	The TVX-1 graphite-coated lucite screen shields RF radiation while permitting low-energy x-rays to pass unattenuated.

References	Manufacturer's specifications
Cost	Refer to manufacturer, ~$1200.00
Address	William B. Johnson and Associates Research Park Boonton Ave. Montville, NJ 07045 (201) 334-9222

RAD-GAM,XRA
Johnson and Associates 2
January 1983

Survey Meter
Johnson Model GSM-12

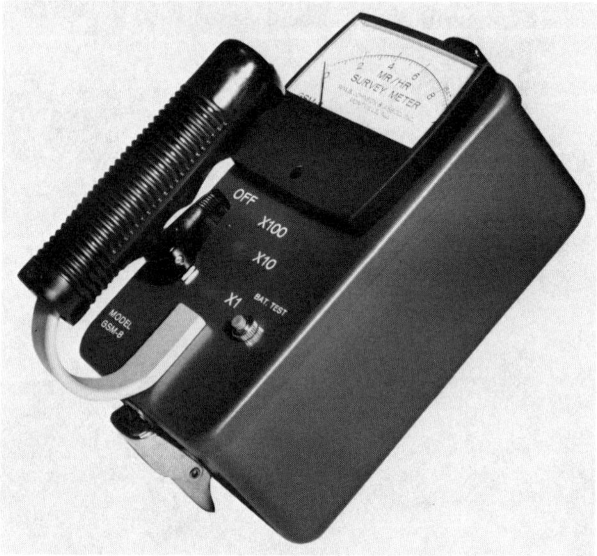

Class	Portable, handheld
Principle of Operation	Small halogen G–M detector, amplifier, meter readout
Sensitivity and Range	Gamma ranges, linear 0–10, 0–100, 0–1000 mR/hr
Sampling	Continuous
Performance	Linearity: ±5% FS Meter display—10-sec time constant Instrument saturation: over 10 R
Requirements	Power: 3.0 V dc (2 "D" cells) Size: 9 cm × 9 cm × 18 cm (3.5 in. × 3.5 in. × 7 in.) Weight: 1.5 kg (3 lb) (including batteries and detector) Temperature: −40 to +140°F
Features	Light weight portable, solid state, 3 ranges to 1 R/hr.
References	Manufacturer's specifications
Cost	$345.00

Address | Wm. B. Johnson and Associates, Inc.
Research Park
Boonton Ave.
Montville, NJ 07045
(201) 334-9222

Pulsed X-Ray Monitor
Lawrence Livermore National Laboratory

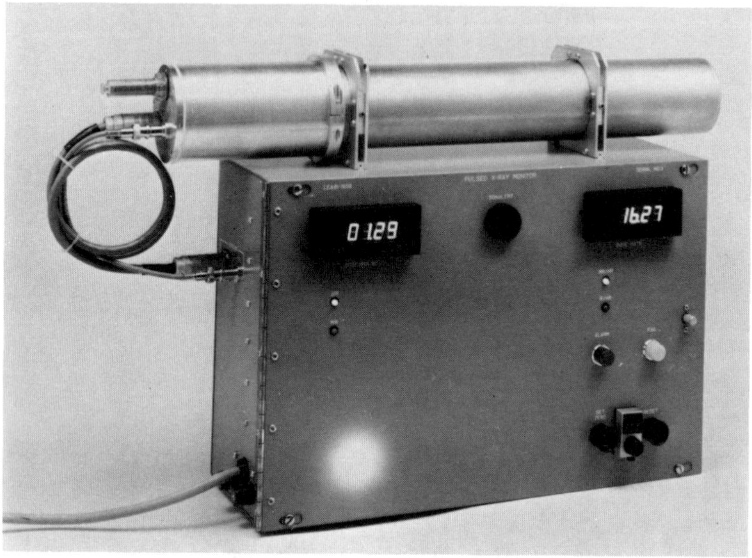

Class	Stationary
Stage of Development	Not commercially available
Principle of Operation	High pressure ionization chamber with MOSFET electrometer/charge-sensitivie amplifier. Measures dose rate and dose/pulse.
Sensitivity and Range	Low range: 0.1 to 100 mR/hr 0.08 to 100 μR/pulse High range: 0.1 to 1000 R/hr 0.1 to 1000 mR/pulse Linear pulse response up to 100 mR/pulse
Sampling	Continuous and dose/pulse
Requirements	Power: 117 V ac, 300 mA Size: 63.5 cm $\times$ 40.5 cm $\times$ 19 cm (25 in. $\times$ 16 in. $\times$ 7.5 in.) Weight: 12.8 kg (25 lb)
Features	1. Chamber and electrometer can respond linearly to short high intensity pulses, 30 ns wide at dose rate of 1.2×10^{10} R/h

2. Remote changing of sensitivity by $\times 10^4$ allows both background and beam spill measurements.
3. Chamber specified has 0.125 in. thick aluminum wall and HV electrode; other chambers available with thinner wall but reduced sensitivity.
4. High level and failure alarms and remote differential outputs.

References

1. R. C. Kaifer, T. E. Jenkins, and T. Straume, "Wide-Range Monitor for Pulsed X-Ray Sources," *IEEE Trans. Nuc. Sci. NS-29*, 808-811 (Feb. 1982).

Cost

Not commercially available, call for construction specifications

Address

Lawrence Livermore National Laboratory
Livermore, CA 94550
(415) 422-5245
Attn: R. C. Kaifer

X-Ray Proportional Detectors
LND Models

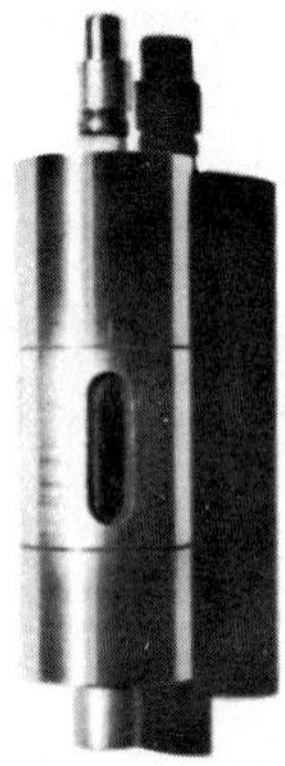
456

4561

4562

4565

4566

4567

4568

Class — Detectors

Specifications — See Table A.6.6

References — Manufacturer's specifications

Address — LND, Inc.
3230 Lawson Blvd.
Oceanside, NY 11572
(516) 678-6141

Table A.6.6. LND X-ray Detectors

Model	Maximum Diameter (cm)	Maximum Diameter (in.)	Overall Length (cm)	Overall Length (in.)	Operating Voltage	Price	Gas
456	2.54	1.00	10.16	4.00	1800	$765.00	Xenon
4561	3.00	1.181	9.50	3.740	1850	760.00	Xenon
4562	2.54	1.00	7.62	3.00	1300	760.00	Xenon
4565	3.33	1.31	10.16	4.00	1200	830.00	Xenon
4566	2.54	1.0	13.97	5.5	1800	830.00	Xenon
4567	6.35 × 6.35	1.25 × 1.25	6.67	2.625	1850	810.00	Xenon
4568	4.20 × 4.20	1.653 × 1.653	14.47	5.697	1900	875.00	Xenon

Gas-filled Counters
Ionization Chambers
LND Models

Class	Detectors
Specifications	See Tables A.6.7–A.6.9
References	Manufacturer's specifications
Remarks	Many more detectors with differing specifications for many uses.
Address	LND, Inc. 3230 Lawson Blvd. Oceanside, NY 11572 (516) 678-6141 Telex: 14-4563

Table A.6.7. LND X-Ray Proportional Counters

Model	Type	Remark	Price
400	End window	$\beta > 30$ keV γ or x-ray from 0.5 to 1.0 nm	$455.00
4011	End window	$\beta > 30$ keV γ or x-ray < 1.1 nm	550.00
4012	End window, removable		470.00
403	End window,	$\beta > 30$ keV γ or x-ray < 1.0 nm	600.00
4032	End window, removable		620.00
420	Quadralateral side window	β,γ or x-ray < 1.0 nm	660.00
421	Quadralateral side window	β,γ or x-ray < 1.0 nm	660.00
422	Quadralateral side window	β,γ or x-ray < 1.0 nm	825.00
423	Quadralateral side window	β,γ or x-ray < 1.0 nm	825.00

Table A.6.7. (Continued)

Model	Type	Remark	Price
424	Quadralateral side window	β,γ or x-ray < 1.0 nm	765.00
425	Quadralateral side window	β,γ or x-ray < 1.0 nm	840.00
426	Quadralateral side window	β,γ or x-ray < 1.0 nm	1440.00
428	Large Area	2-mil beryllium window	3280.00
450	Side window		620.00
451	Side window		850.00
4515	Windowless		730.00
452	Side window		620.00
4522	Windowless		620.00
453	Side window		800.00
4543	Side window		705.00
4548	Side window		830.00
4544	Side window		705.00
4545	Side window		830.00
45431	Side window		725.00
4549	Side window		900.00
4546	Side window		750.00
4545	Side window		830.00
45410	End window		830.00

Table A.6.8. LND Ionization Chambers

Model	Sensitivity (A/R·hr)	Range	Remark	Price
503	1×10^{-10}	10 mR/hr to 10^4 R/hr		\$850.00
512			High Pressure Argon	1335.00
52102	1×10^{-11}	10^{-3} to 10^3 R/hr		235.00
504	1.49×10^{-8}			2075.00
50413	1.2×10^{-9}			2075.00
50422	2.1×10^{-9}			1940.00
50437	6×10^{-9}			2000.00
50440	6.5×10^{-9}			2700.00
50442	1.2×10^{-9}			1800.00

Table A.6.9. LND Ionization Well Chambers

Model	Background	Price
507	1×10^{-14} A	\$1000.00
508	1×10^{-14} A	1000.00
509	1×10^{-14} A	1075.00

Survey Meter
Ludlum Model 5

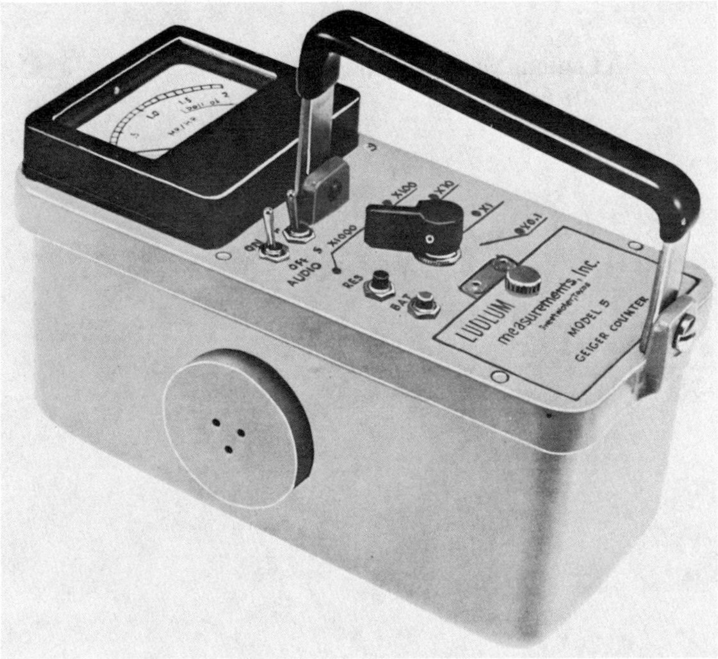

Class	Portable, handheld
Principle of Operation	Geiger-Müller detector, amplifier, electrometer, meter readout
Sensitivity	Energy range: lowest to highest dependent on thickness of entrance window and volume of the detector Activity range: 0–0.2 to 0–2000 mR/hr in 5 linear ranges Saturation: In excess of 200 R/hr
Detector	LND 712 with 0.032-in. cadmium shield for 0–20 mR/hr LND 714 for 0 to 200 and 0 to 2000 mR/hr scales
Requirements	Power: Two "D" cells, 300 hr Size: 9 cm H × 9 cm W × 18 cm L (3 1/2 in. × 3 1/2 in. × 7 in.), exclusive of handle Weight: 1.6 kg (3.5 lb)
Performance	Calibration – 5% Temperature: −23.3 to 48.9°C (−10 to 120°F) Calibration: Radium-226

Features	Built-in speaker; temperature compensated
References	Manufacturer's specifications
Cost	$375.00
Address	Ludlum Measurements, Inc. 501 Oak Street Sweetwater, TX 79556 (915) 235-5494

Survey Meter
Ludlum Model 14-C

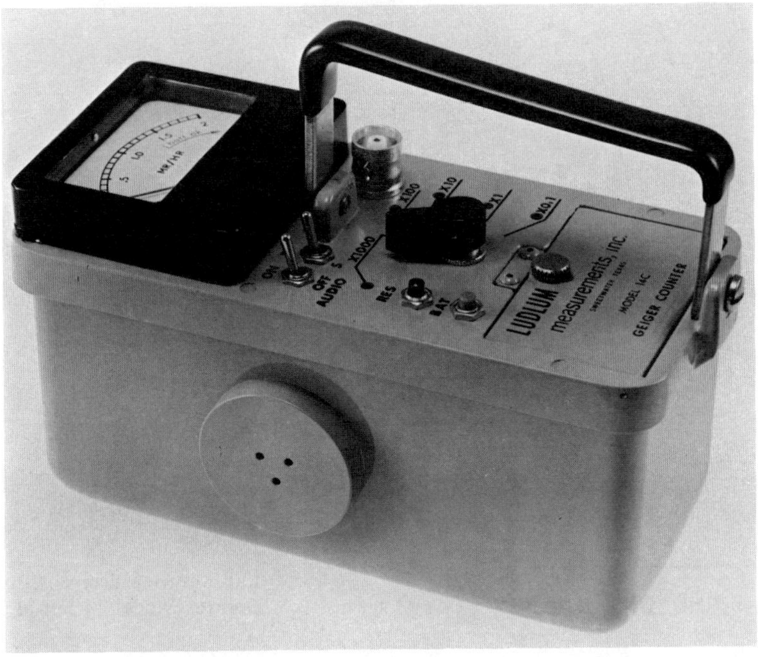

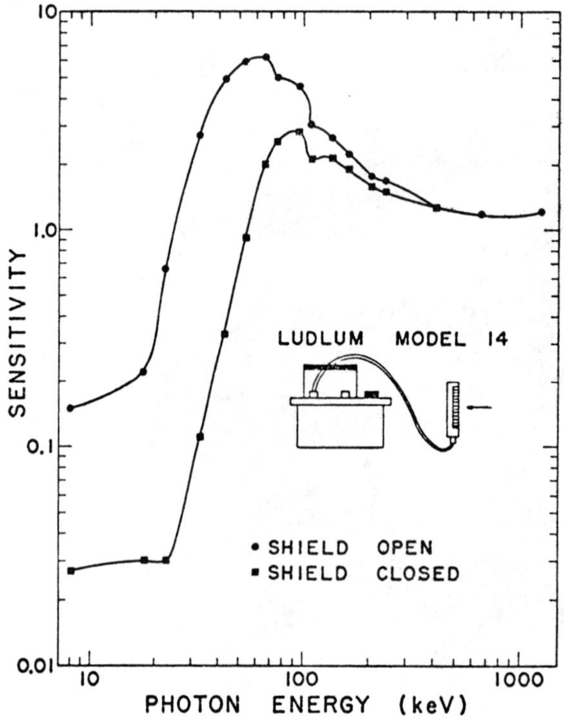

Ludlum Measurement, Inc. Geiger counter Model 14 Serial No. 85 *

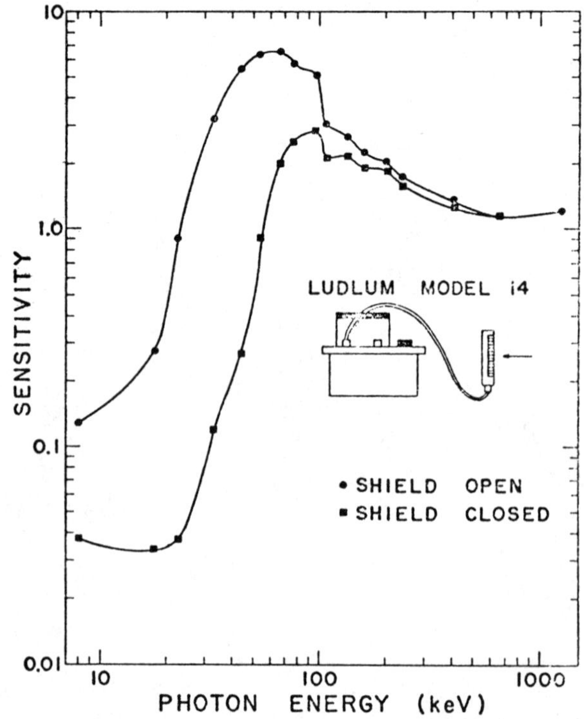

Ludlum Measurement, Inc. Geiger counter Model 14 Serial No. 117 *

RAD-GAM,XRA
Ludlum 2
January 1983

Class	Portable, handheld
Principle of Operation	Uses two G-M tubes, amplifier, electrometer, meter readout. Two G-M tubes: one internal, one external
Sensitivity and Range	Energy sensitivity: see Figs. A.6.2 and A.6.3 Full scale ranges: 0.2, 2, 20, 200, 2,000 mR/hr (internal G-M tube) Minimum detectable energy: 50 keV for internal G-M; external G-M depends on type selected
Sampling	Continuous
Performance	Meter accuracy: 2% Response time: Fast, 3 sec; Slow, 11 sec
Requirements	Power: Two "D" cells, 300 hr life Size: 9 cm × 9 cm × 18 cm (3.5 in. × 3.5 in. × 7 in.) Weight: 1.5 kg (3 1/2 lb)
Features	Speaker provided built-in
References	Manufacturer's specifications
Cost	$365.00 plus cost of external detector
Address	Ludlum Measurements Inc. 501 Oak Street Sweetwater, TX 79556 (915) 235-5494

Gamma Ray Assay Meter
National Nuclear Model HM-4 Samson

Class	Hand-held ^{235}U and Pu assay meter
Principle of Operation	Uses a collimated NaI detector, 1.9 cm × 1.9 cm (3/4 in. × 3/4 in.) with ^{241}Am stabilization of the gain. ^{241}Am and temperature stabilization of amplifier gain. Temperature compensation included.
Sensitivity and Range	10 to 1000 cps Integrated counts from 150 to ~700 keV
Performance	Counting times can be selected from 2 to 60 sec or set to infinity. Three decade logarithmic amplifier.
Requirements	Power: Six size "AA" rechargeable batteries. Nickel-cadmium batteries which provide 12 hours of continuous operation. Size: 8.25 cm W × 10.2 cm H × 25 cm L (3.25 in. W × 4 in. H × 9.75 in. L)
References	Manufacturer's specifications
Cost	$8000
Address	National Nuclear Corporation 1904 Colony Street Mountain View, CA 94043 (415) 962-9220 Attn: D. Kahn

High Sensitivity Door Monitor for Personnel Detection
National Nuclear Model DM-2

Class	Laboratory
Principle of Operation	Solid scintillation detectors. Lead shielding to accommodate low level background radiation. Additional shielding can be provided.
Sensitivity and Range	Sensitive to gammas and neutrons (from Pu). Detects 3 grams of ^{235}U enriched to 20% or more, in 3 mm brass at 50% confidence limit. False alarms, 0.1%. Detects 0.5 grams of Pu or 1 gram ^{233}U in 3 mm brass at 90% confidence limit. False alarms, 0.1%.
Performance	Detects 10 grams highly enriched U metal sphere or a 1 gram ^{239}Pu metal source in a 20 μR/hr background at 50% confidence level. Alarms on fissile material, loss of signal, and high level. Continuous, automatic background compensation. Adjustable time constants and sensitivity. Alarm activates local light and relay to remote station.
Requirements	Power: 110 V, 1.5 A or 220 V, 0.75 A Size: 223 cm H × 101 cm W × 33.6 cm D (88 in. H × 32 in. W × 16 in. D) Weight: 375 kg (800 lb)
References	Manufacturer's specifications
Cost	$14500
Address	National Nuclear Corporation 1904 Colony St. Mountain View, CA 94043 (415) 962-9220 Attn: D. Kahn

Ratemeter
Nuclear Associates Prima I, IIa

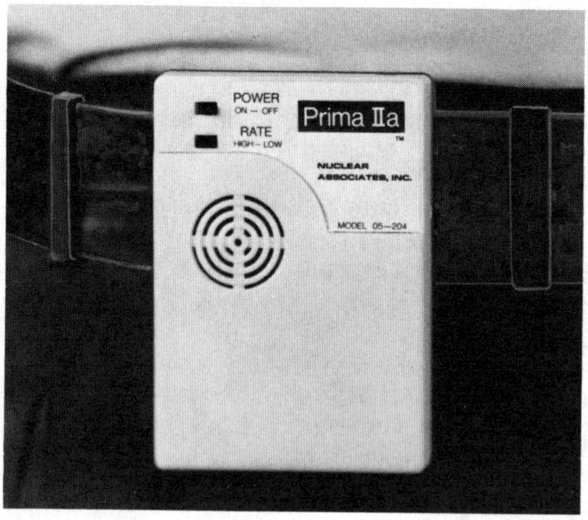

Specifications	See Table
References	Manufacturer's specifications
Address	Nuclear Associates 100 Voice Road Carle Place, NY 11514 (516) 741-6360

Specifications for Nuclear Associates Ratemeter

	Model	
Specifications	Prima I 05-101	Prima IIa 05-205
Radiation detected	Gamma, x-radiation	Gamma, x-radiation
Detector	G-M tube	G-M tube
Energy dependence	200 keV to 1 MeV	80 keV to 1 MeV
Ranges	Background to 2 R/hr	Background to 10 R/hr
Response time	Not applicable	Not applicable
Accuracy	±30%	±30%
Features	Audible output	Audible output
Power required	4.05-V mercury cell, 6-mo. life	9-V alkaline cell, 6-mo. life
Size	2 cm D × 13 cm L (0.75 in. × 5 in.)	6.5 cm × 9 cm × 2 cm (2.5 in. × 3.6 in. × 0.84 in.)
Weight	100 g (3.5 oz)	120 g (5 oz)
Price	$250.00	$165.00

Survey Meter
Nuclear Associates Model 05-571, 572

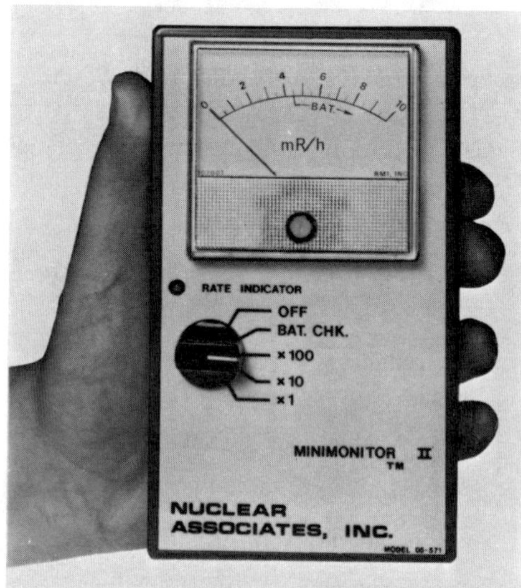

Class	Portable, handheld
Principle of Operation	G-M tube, amplifier, meter readout
Sensitivity and Range	Range: Model 571: 0–10, 100, 1000 mR/hr; Model 572: 0–500, 5000, 50,000 cpm
Sampling	Continuous
Performance	Accuracy: 571: ±20% of full scale; 572: ±10% of full scale Operating temperature: ±20 to 55°C (−4 to 131°F) Energy dependence: ±20%, 60 keV to 1.2 MeV Time constants: 571: 10, 2, and 0.3 sec; 572: 10 sec (×1), 2 sec (×10), 0.3 sec (×100)
Requirements	Power: Four "AA" alkaline batteries, ~500-hr life Size: 571: 15.2 cm × 8.9 cm × 3.8 cm 6 in. × 3 1/2 in. × 1 1/2 in.) 572: 15.2 cm × 8.9 cm × 5.1 cm (6 in. × 3 1/2 in. × 2 in.) Weight: 571: 0.43 kg (15 oz.) 572: 0.62 kg (22 oz.)
Features	571— A nonoverload circuit assures indicator operation in radiation field greater than 500 R/hr 572— High sensitivity (lower detection limit of 0.005 Ci) Large area, screened detector for surface contact measurements
References	Manufacturer's specifications
Cost	571: $295.00 572: $485.00
Address	Nuclear Associates 100 Voice Road Carle Place, NY 11514 (516) 741-6360

Hand/Shoe/Clothing Monitor
Nuclear Associates Model 05-696

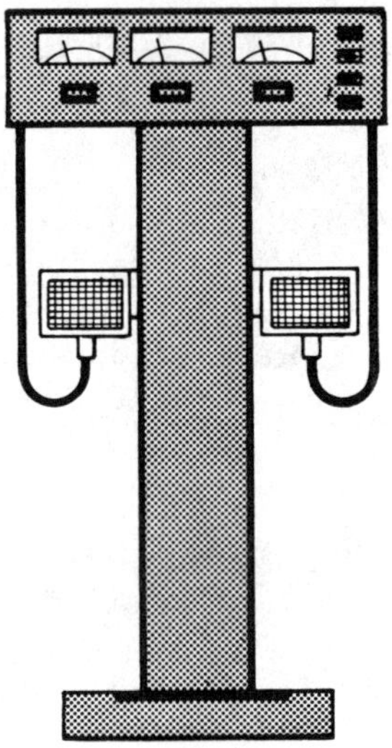

Class	Floor-mounted
Principle of Operation	Xenon-filled proportional counter with meter display
Sensitivity and Range	Three channels with 1 to 10^4 counts per cycle, logarithmic scales 1900 cpm per 10^{-4} μCi/cm^2 of I-125
Sampling	10-sec count cycle
Performance	Temperature range: -5 to $+50$°C (23 to 122°F)
Requirements	Power: 110 V ac, 60 Hz Size: 163 cm × 51 cm × 69 (64 in. × 20 in. × 27 in.)
Features	Four large-area 200 cm^2 window detectors Window thickness: 6 mg/cm^2—background subtract
References	Manufacturer's specifications
Cost	$8500.00
Address	Nuclear Associates 100 Voice Road Carle Place, NY 11514 (516) 741-6360

Primalert 10 Therapy Radiation Level Monitor
Nuclear Associates Model 05-433

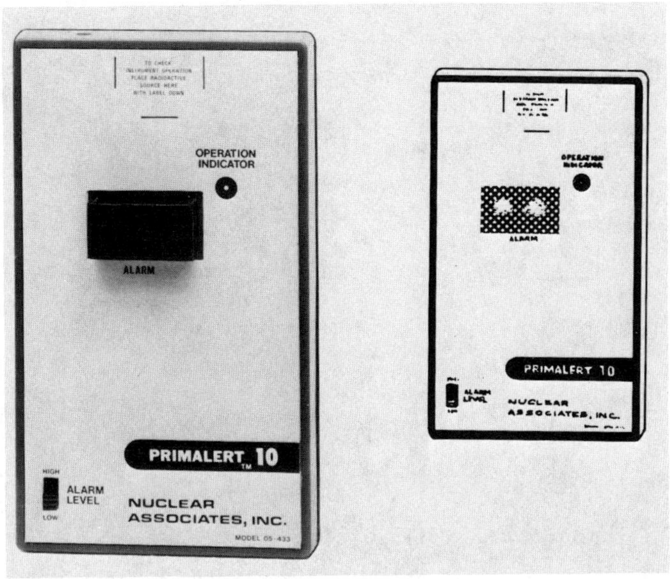

Class	Wall-mounted
Principle of Operation	Energy-compensated G-M detector and ratemeter with audible and visual alarm
Sensitivity and Range	Alarm level: 2.5 to 20 mR/hr
Sampling	Continuous
Performance	Energy response: −10%, +40% from 50 keV to 2 MeV referenced to Cs-137 Accuracy: ±20% (exclusive of energy dependence)
Requirements	Power: 110 V ac, 60 Hz Size: 15.2 cm × 8.9 cm × 3.8 cm (6 in. × 3.5 in. × 1.5 in.)
Features	Visual alarm and audible alarm Optional remote alarm, Model 05-434 (Cost: $110.00)
References	Manufacturer's specifications
Cost	$425.00
Address	Nuclear Associates 100 Voice Road Carle Place, NY 11514 (516) 741-6360

Primalert 35 Area Alarm Monitor
Model 05-437

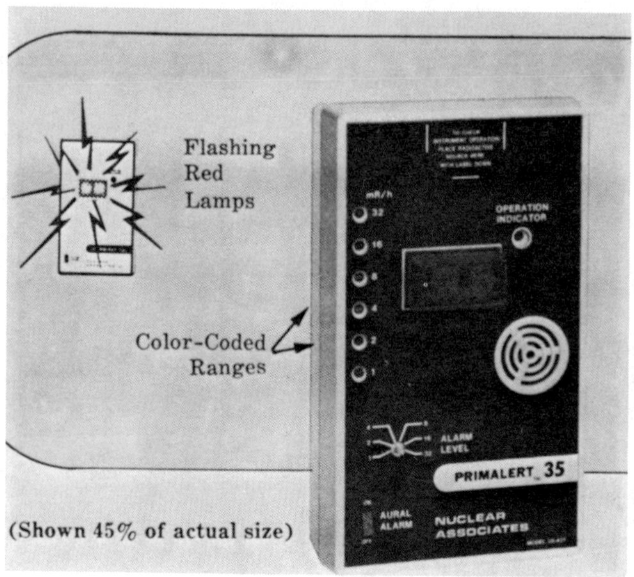

(Shown 45% of actual size)

Class	Wall-mounted
Principle of Operation	Energy-compensated G-M detector
Sensitivity	1, 2, 4, 8, 16, and 32 mR/hr Alarm can be set at any of the six levels
Sampling	Continuous
Performance	Energy response: ±20%, 80 keV to 1 MeV Accuracy: ±20% (exclusive of energy dependence)
Requirements	Power: 110 V ac, 60 Hz Size: 15 cm × 8.6 cm × 3.3 cm (5.9 in. × 3.4 in. × 1.3 in.)
Features	Audible and visual alarms Optional remote alarm, Model 05-434 (Cost: $110.00)
References	Manufacturer's specifications
Cost	$475.00
Address	Nuclear Associates 100 Voice Road Carle Place, NY 11514 (516) 741-6360

05-582 "Transport Index" Radiation Monitor (Prima 7) and Survey Meter For Measuring "Transport Index" of Packages Containing Radioactive Materials

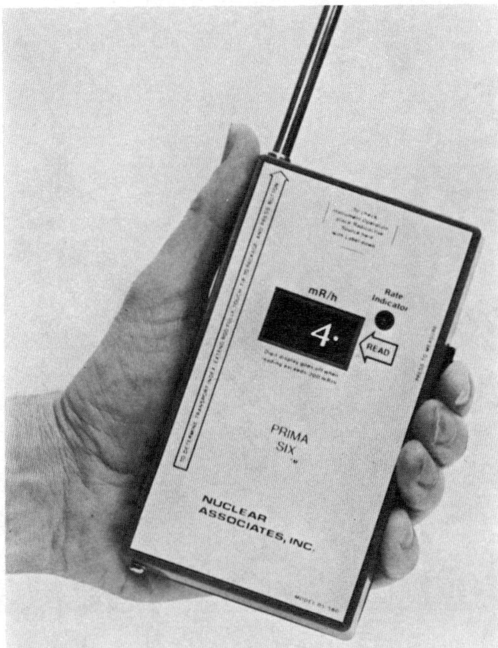

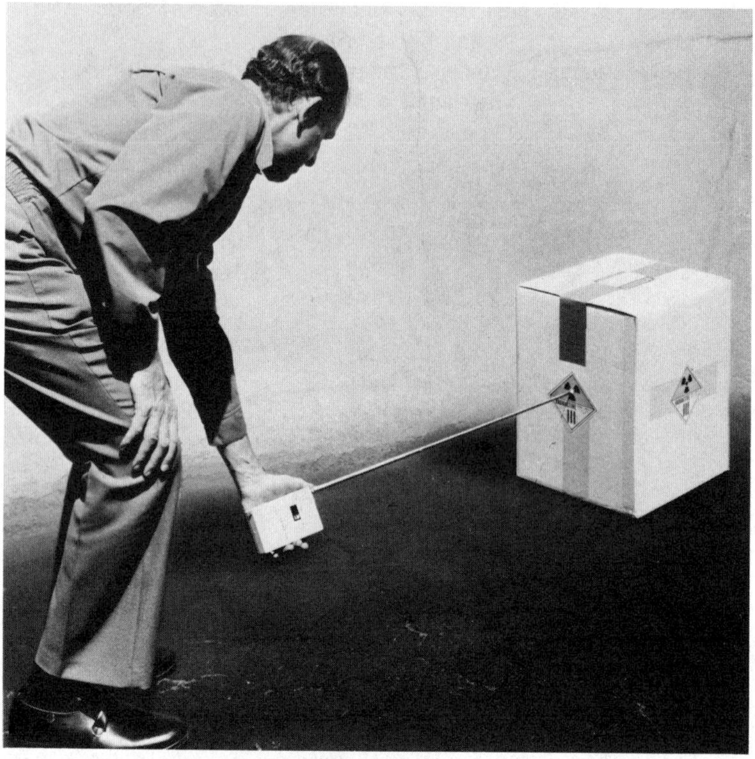

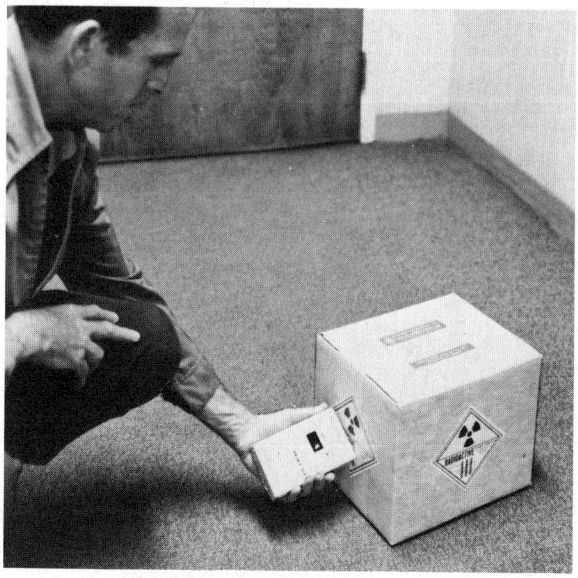

Class	Portable, handheld (with belt-clip)
Principle of Operation	Halogen-quenched G-M tube, LED readout (2-digit)
Sensitivity and Range	Range: 0 to 99 mR/hr, in 1-mR increments (with over-range indicator)
Sampling	Manual control
Performance	4-sec measurement cycle Accuracy: $\pm 10\%$ or 1 mR/hr, whichever is greater Control: Press-to-measure (4 sec); hold-to-read Energy response: $+40\%$, -12% from 50 keV to 1.2 MeV Response time: 4 sec Temperature: 0 to 50°C (32 to 120°F) with mercury cells -12 to 50°C (10 to 120°F) with alkaline cells
Requirements	Power: 4 "AA" mercury cells or 4 "AA" alkaline cells Size: 15.2 cm H $\times$ 8.9 cm W $\times$ 3.8 cm D (6 in. $\times$ 3 1/2 in. $\times$ 1 1/2 in.) Weight: 0.34 kg (12 oz)
Features	1. Accurate LED readout 2. Automatic counting cycle 3. "over-range" and "read" indicators 4. 3-ft length telescoping spacing rod 5. Calibration source ^{137}Cs, ~10 μCi, license-free
References	Manufacturer's specifications
Cost	$385.00
Address	Nuclear Associates 100 Voice Road Carle Place, NY 11514 (516) 741-6360

X/Gamma Survey Meter
Nuclear Associates 05-670

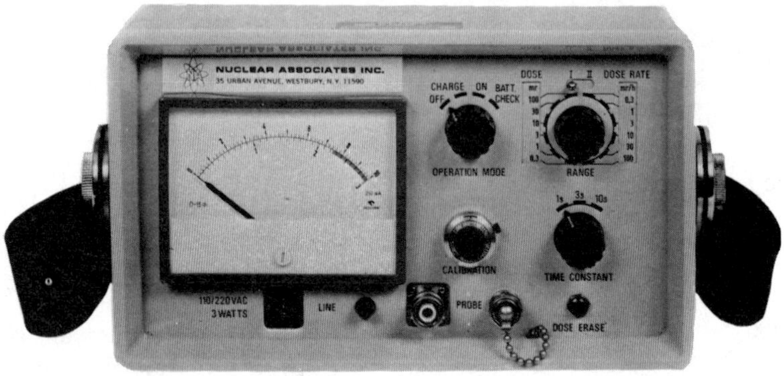

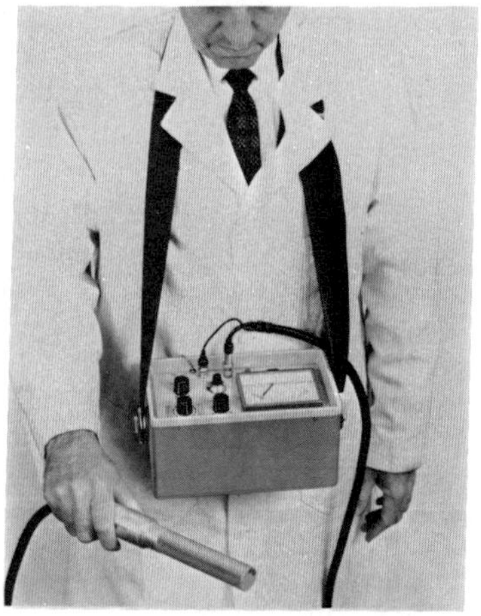

Class	Portable
Principle of Operation	Single plastic-wall ion chamber used both as a proportional gas amplification chamber and as an ionization chamber (by placing —200 V dc on the chamber), amplifier, electrometer, meter readout.
Sensitivity and	Energy dependence: See Fig.; $<\pm 10\%$ from 10 keV to 1.2 MeV

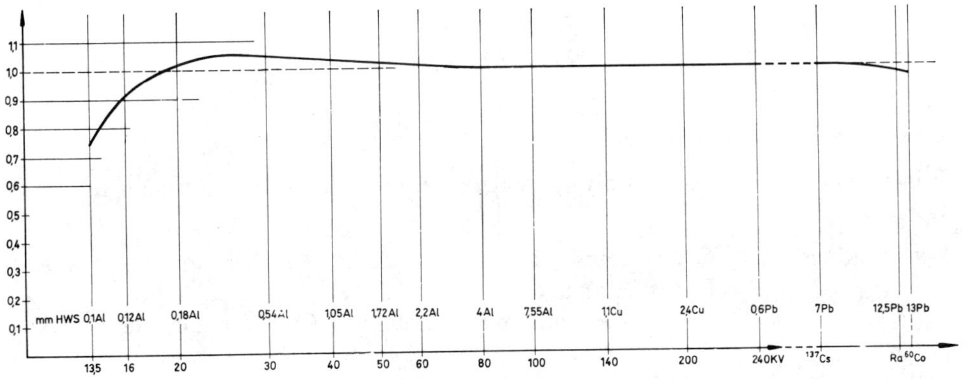

Energy Dependence X-Radiation Gamma Radiation

Range	Dose range: 15 linear scales, 0.3 mR to 3000 R FS Dose rate ranges: 15 linear scales, 0.3 mR/hr to 3000 R/hr FS
Sampling	Continuous
Performance	Instrument accuracy (less probe): 6% for dose rate measurement 8% for dose measurement Time constants: 1, 3, 10 sec
Requirements	Power: 110 V ac, 60 Hz Size: 14 cm H × 25 cm W × 14 cm D (5.5 in. × 10 in. × 5.5 in.) Probe cable 2 m L (6 ft) Weight: 3.6 kg (8 lb)
Features	Measures both dose and dose rate
References	Manufacturer's specifications
Cost	05-670 X/Gamma ray survey meter: $3895.00 05-671 Heavy-duty probe holder: $150.00 05-672 Probe extension cable, 20-ft: $195.00
Address	Nuclear Associates 100 Voice Road Carle Place, NY 11514 (516) 741-6360

NaI(Tl) Crystals
Nuclear Enterprises

Class	Detectors
Principle of Operation	NaI(Tl) crystal optically coupled to a photomultiplier tube
Specifications	Crystal sizes: Range from 1/2 in. × 1/2 in. to 5 in. × 6 in. Prices: Range from \$76 to \$3121 Demountable assemblies: Can accept crystal sizes from 1/2 in. × 1/2 in. to 6 in. × 4 in. Prices: Range from \$300 up Integral assemblies: Can except crystal sizes with diameter up to and including 3 in. Prices: Range from \$178 to \$918 NEI will consider manufacturing special or custom detectors (any quantity)
Other Features Also Available	Electronic accessories Low-background quartz windows and stainless steel mounts, Beryllium windows Anti-Compton setups Large well setups Cesium iodide (thallium) detectors also available
References	Manufacturer's specifications
Address	Nuclear Enterprises, Ltd. Bath Road, Beenham Reading, RG75PR ENGLAND Telephone: (073) 521-2121 Spectrum Sciences 3050 Oakmead Village Dr. Santa Clara, CA 95051 (408) 727-1567 Telex: 176449 Semi Group

Area Radiation Monitor System TA-900

Class	Rack or Nema cabinet-mount w/remote detector
Principle of Operation	Microprocessor based using three detector automatic ranging and digital readout in mR/hr, two significant digits and powers of ten.
Sensitivity and Range	0.01 mR/hr to 0.99 $\times 10^7$ mR/hr 80 keV to 1.5 MeV energy range Higher ranges available on request
Sampling	Every 4 sec
Performance	Accuracy: ±20% Energy dependence: ±20% Temperature: Detector, −40 to 75°C, 95% Rh; Control unit, −20 to +60°C, 95% Rh
Requirements	Power: 120 V ac, 30 W, plus external alarm power Size: Nema control units, 14 in. × 15 1/2 in. × 7 in. Detector: 9 1/2 in. × 11 1/2 in. × 4 3/4 in. Weight: 9 kg (20 lb)
Features	Outputs for remote alarms Microprocessor capability for background subtraction, and other special features.
Cost	Price on request
Address	Nuclear Research Corporation Warrington Industrial Park Box H Warrington, PA 18976 (215) 343-5900

Gamma Counters
N. Wood Models

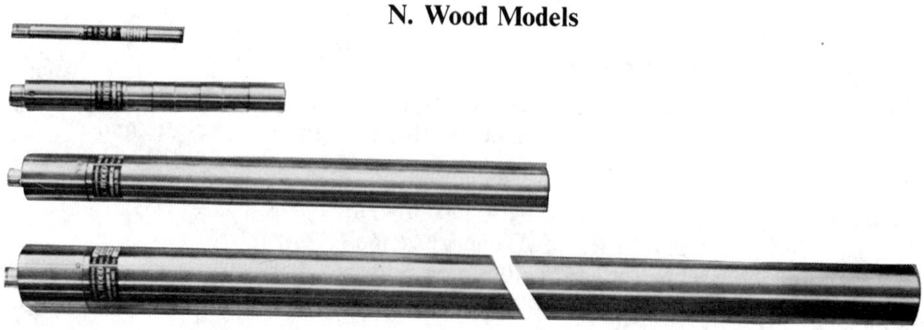

Class	Portable and/or laboratory
Principle of Operation	Proportional and G-M as specified (see Table A.6.10)
References	Manufacturer's specifications
Remarks	Counters can be made to specifications by special order
Address	N. Wood Counter Laboratory, Inc. 1525 E. 53rd Street Chicago, IL 60615 (312) 324-1114

Table A.6.10. N. Wood Gamma Counters

Model	Diameter (in.)	Length (in.)	Voltage (V)	Remarks	Cost[a]
				Proportional Gamma Counters	
D-5-3L	5/8 or 1/2	3	2000	Originally made for space research—adapted for use in	$ 95.00
D-10-5L	1	5	2050	portable instruments. Useful for all cosmic ray counting.	110.00
D-15-12L	1 1/2	12	2100	Filled to operate in the proportional region, these counters	125.00
D-15-34L	1 1/2	34	2100	have an almost unlimited life.	140.00
D-20-38L	2	38	2200	Larger models useful for cosmic ray telescopes, anticoincidence	200.00
D-2046L	2	46	2200	shields, carbon dating, etc.	240.00
				Geiger-Müller Counters	
D-5-3	5/8	3	1000	Used in portable instruments and probes—design adapted to associated	70.00
D-5-16	5/8	16	1000	counting equipment. Specify length up to 16 in.	80.00
D-10-5	1	5	1050	General purpose gamma counters. Can be made as flow counters.	90.00
D-10-12	1	12	1050	Specify 5 to 20 in. active length.	100.00
D-15-5	1 1/2	5	1100	General purpose gamma counters for research, geological survey,	100.00
D-15-12	1 1/2	12	1200	well logging, carbon dating, anticoincidence shields, and	110.00
D-20-12	2	12	1200	many industrial uses. 5 to 34 in. active length for counters	145.00
D-20-20	2	20	1200	1 1/2 in. in diameter. 12 to 46 in. active lenth for counters 2 in. in diameter.	165.00
D-20-34	2	34	1200	High-sensitivity G-M counters.	185.00

[a]Prices are from 1982

Si(Li) Photon Detectors
Ortec Series SLP

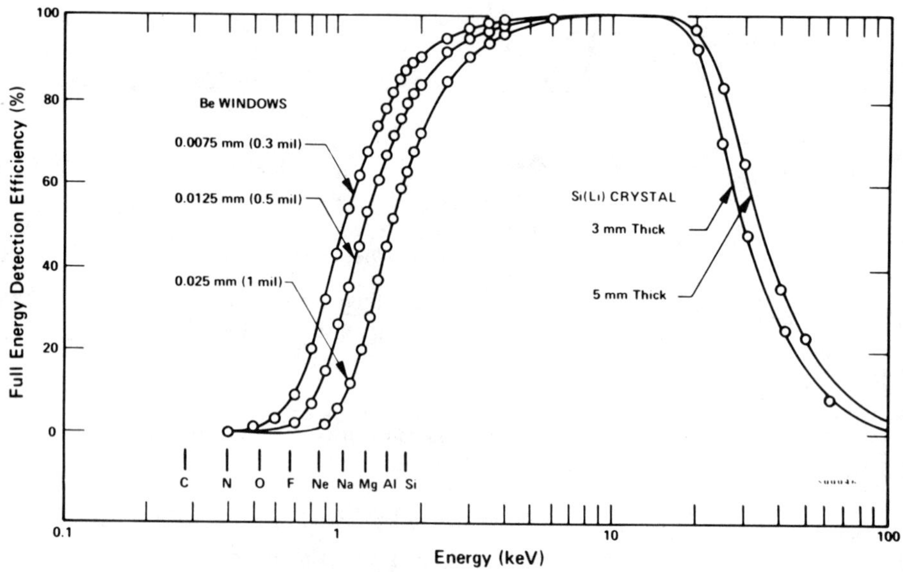

Class	Detectors
Principle of Operation	Silicon (lithium-drifted) detectors for low-energy photon spectrometry
Specifications	See Table A.6.11 Beryllium entrance windows: Standard: 1-mil Be window. Normally delivered with 4-, 6-, and 10-mm diam. detectors. 2-mil Be window, also. Optional: 0.5-mil Be window. Thin, unsupported window for 4- and 6-mm diam. detectors. ($275) 0.3-mil Be window. Ultra-thin, unsupported window for 4-mm diam. detectors. ($385)
Features	1. Price includes: a) Si(Li) crystal; Price includes: b) Cryogenic hybrid preamplifier; Price includes: c) Cryostat with entrance window; Price includes: d) Liquid nitrogen dewar 2. Figure shows effect of different beryllium windows and Si(Li) thicknesses.
References	Manufacturer's specifications
Address	EG&G ORTEC 100 Midland Road Oak Ridge, TN 37830 (615) 482-4411

Table A.6.11. Specifications for ORTEC Detectors

Cryostat Configuration	ORTEC Model No.	Dewar Size (liters)	Liquid Nitrogen Use Time (liters/day)	Guranteed Resolution[a] (eV) FWHM for ~Area/Diam. 12.5 mm^2/ 4 mm	30 mm^2/ 6 mm	80 mm^2/ 10 mm	200 mm^2/ 16 mm	Price Range
Vertical dipstick	CFG-SV	30	1.1	170 160	180 165	190 180	250 220	$6635–$9930 (Vertical dipstick, Horizontal dipstick, Light weight)
Horizontal dipstick	CFG-SH	30	1.1	170 160	180 165	190 180	250 220	
Light weight	7513	5	0.9	170 160	180 165	190 180	250 220	
Hand-held portable	CFG-GG-0.4	0.4	1.2	170 160	180 165	190 180	250 220	
Hand-held portable	CFG-GG-1.2	1.2	1.2	170 160	180 165	10 180	250 220	

[a]Detectors are available with a guaranteed resolution better than that shown in table. Total system resolution for ^{55}Fe 5.9 keV at 1000 counts/sec measured with an ORTEC 716A, a 459, and a dc-coupled 4096-channel analyzer. The FWTM is $\leq$ 2 FWHM. Guaranteed for one year. Upper and lower values are standard. Buyer must request the desired resolution.

Solid State Photon Detectors
Ortec Series: GEM, GMX, GLI, GWL, GLP and SLP

Class	Laboratory
Principle of Operation	Semiconductor photon detectors for spectroscopy in the energy span: 1 keV to 10 MeV. All the detectors are designed to operate at nearly liquid-nitrogen temperature. Detector system consists of a semiconductor detector element, a cryostat, a liquid-nitrogen dewar, a charge sensitive preamplifier, and a cable pack.
Performance, Sensitivity and Range	(see table)
Options	Different cryostat configurations. Different liquid-nitrogen transfer systems.
References	Manufacturer's specifications
Cost	

Series	Range ($)
GEM	9800 to 34950
GMX	11920 to 33130
GLI	8025 to 26965
GWL	20990 to 41450
GLP	6990 to 18390
SLP	6910 to 9330

Address

EG&G ORTEC
100 Midland Road
Oak Ridge, TN 37830
(615) 482-4411
Telex: 55-7450
Attn: S. Wagner

Performance, Sensitivity and Range:

Type of detector	Material	Geometry	Window Thickness (μm)	Standard Sizes	Standard Energy Resolution	Standard Peak/ Compton	Useful Energy Range
GEM Series[a]	p-type (High purity Ge)	Closed-end Coaxial	~600	5 to 40% Efficiency	1.75 to 2.20 keV at 1.33 MeV	24:1 to 60:1	40 keV–10 MeV
GMX Series[b]	n-type (High purity Ge)	Closed-end Coaxial	0.3 Ion Implanted	8 to 30% Efficiency	1.80 to 2.20 keV at 1.33 MeV	30:1 to 50:1	3 keV–10 MeV
GWL Series[c]	p-type (High purity Ge)	Well	0.3 Ion Implanted	70 to 120 cm^3 Active Volume	2.1 to 2.4 keV at 1.33 MeV	–	10 keV–10 MeV
GLP Series[d]	p-type (High purity Ge)	Planar	0.3 Ion Implanted	6 to 51 mm Diameter	165 to 760 eV at 5.9 keV 480 to 850 eV at 122 keV	–	3 keV–1 MeV
SLP Series[e]	Lithium-Drifted Silicon	Planar	0.1	4 to 16 mm Diameter	160 to 250 eV at 5.9 keV	–	$<$ 1 keV to 60 keV

[a]GEM Series: ion-implanted center contact—reliable and rugged; excellent peak symmetry; automatic high-voltage shutdown protects FET (optional).

[b]GMX Series: ultrathin ion-implanted entrance window (0.3 microns); spectroscopy from 5 keV to 10 MeV; over 25 times more resistant to neutron radiation damage than conventional coaxes; automatic high-voltage shutdown protects FET (optional); Be window.

[c]GWL Series: high absolute counting efficiency for small samples; ion-implanted blind well; spectroscopy from 10 keV to 10 MeV; near 4π geometry; well (1.45 $\times$ 4.0 cm) standard; automatic high-voltage shutdown protects FET (optional).

[d]GLP Series: ion-implanted front contact eliminates x-ray spectral interference; automatic high-voltage shutdown protects FET (optional); spectroscopy from 3 keV to 1 MeV.

[e]SLP Series: spectroscopy from ~1 keV to ~60 keV; superior resolution performance at low- and high-count rates; thin Be windows; automatic high-voltage shutdown protects FET (optional).

RAD-GAM,XRA
Ortec 2
January 1983

Auto-Gamma[R] System
Models 5650, 5780

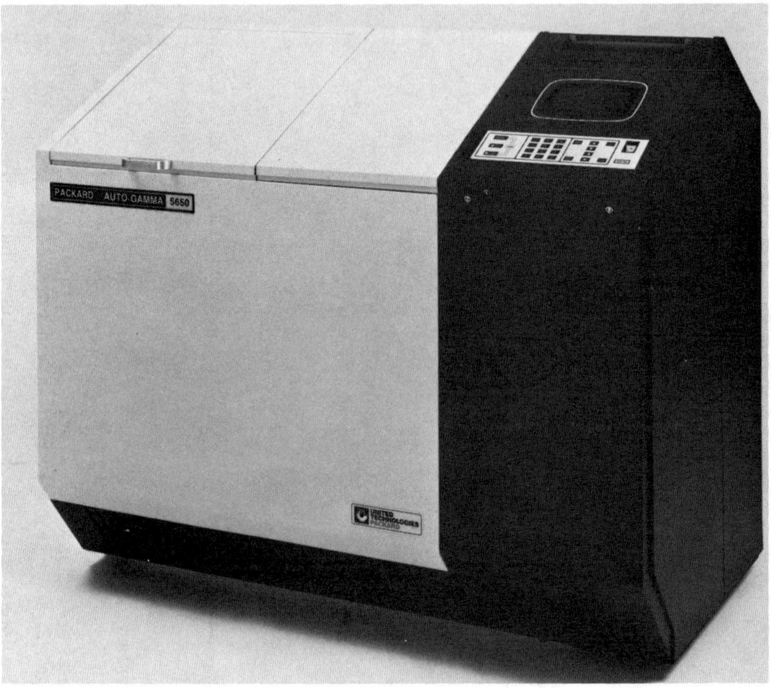

Class	Laboratory
Principle of Operation	Auto-Gamma radiation measurement system. 2 in. (5.08 cm) NaI(Tl) crystal detector with standard radiation, shielding up to 0.1 microcurie of I-131 (Clinical). 3 in. × 3 1/4 in. (7.6 cm × 8.25 cm) NaI(Tl) crystal detector with high energy radiation shielding up to 0.1 microcurie of Fe-59 (Research). 504 samples (Model 5650), 792 samples (Model 5780) on a belt driven sample changer with forward and reverse controls. Video display. Cassette type sample carriers with 12 positions. Spectralyzer™ storage of spectra for analysis in two energy regions. Multi-microprocessor system with 72 hours power fail protected memory capacity. Built-in Data Writer. PrioSTAT manual sample loader. Dual Assay.
Sensitivity and Range	Gamma and x-ray investigation. 15 to 200 keV
Sampling	504 samples (Model 5650) 1 to 42 cassettes 792 samples (Model 5780) 1 to 66 cassettes

Performance	*Clinical* Temperature: 15 to 35°C (59 to 95°F) Ambient Efficiency: I-125, 70% Typical Background: I-125, 20–25 cpm typical range* *Research* Temperature: 15 to 35°C (59 to 95°F) Efficiency: I-125, 70% typical; Cs-137, 20% typical Background: I-125, 33-37 cpm typical range* 15 to 1000 keV: less than 200 cpm* 15 to 2000 keV: less than 350 cpm*
Requirements	Power: 100 to 130 V ac at 50/60 Hz 200 to 260 V ac at 50/60 Hz 400 V ac maximum, 1100 VA with refrigeration Automatic restart after power failure

	Model 5650	Model 5780
Size:	114 cm (45 in.) Height	114 cm (45 in.) Height
	132 cm (52 in.) Width	157 cm (62 in.) Width
	79 cm (31 in.) Depth	79 cm (31 in.) Depth

	Model 5650	Model 5780
Weight:	385 kg (860 lbs)	430 kg (960 lbs)

Cost

Model 5650	Model 5780
$18,500 Clinical	$19,500 Clinical
$20,500 Research	

Optional Features	Multi-Assay RIA—15 protocols. Spectraview. Clock. Refrigeration. Third Region. RS232, TTY interface
Address	Packard Instrument Company 2200 Warrenville Road Downers Grove, IL 60515 (312) 969-6000

*measured in Downers Grove, USA

Multi-Prias™ Automatic Gamma Counter System
Models 5301, 5302, 5304

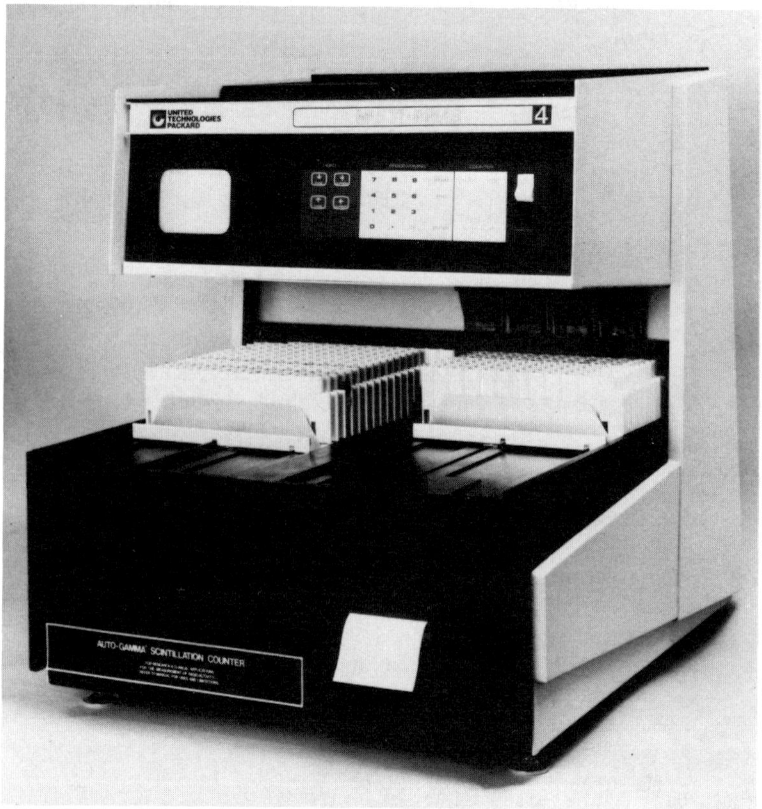

Class	Laboratory
Principle of Operation	Benchtop automatic gamma counter, expandable to 1, 2, and 4 detectors. 3 1/2 in. × 1 1/2 in. × 2 in. (8.9 cm × 3.8 cm × 5.08 cm) NaI crystal. Capacity for tubes 10 to 12 mm in diameter. Front panel pushbuttons and digital display for user entry. Video screen. Quality control program. Storage of up to 15 different data reduction or counting programs. Detector correlation. Count reproducibility. Contamination check. Self-calibration. Automatic background monitoring. RS232 compatible output. 25 cassettes.
Sensitivity and Range	Gamma and x-ray investigation. 15 to 500 keV. Capacity—unlimited

Sampling	Model 5301: 50 samples Model 5302: 100 samples Model 5304: 200 samples
Performance	Typical 78% I-125 efficiency Typical 70 cpm I-125 background
Requirements	Power: 100 to 130 V ac at 50 or 60 Hz 200 to 260 V ac at 50 or 60 Hz 300 VA maximum Auto restart after power failure Battery for 48 hour memory support Size: 70 cm (28 in.) Height 62 cm (24 in.) Width 70 cm (28 in.) Depth Weight: 684 kg (310 lb)
Cost	Model 5301: $13,500 Model 5302: $15,500 Model 5304: $19,500
Optional Features	Multi-Assay RIA—15 protocols. Automatic or keyboard data entry. Dual RIA. Extra custom window for sample contamination monitoring. Current loop output interface to teletypewriters. 80 column printout. Model 652 quiet impact teletypewriter RS232 operation.
Address	Packard Instrument Company 2200 Warrenville Road Downers Grove, IL 60515 (312) 969-6000

Gamma Counter
Packard Model Prias™PGD

Class	Laboratory
Principle of Operation	Bench-top automatic gamma counter. Capacity for tubes 10 to 16-mm in diameter. Front panel pushbuttons and digital display for user entry, review, and modification of program data. Storage of up to seven different data reduction or counting programs.
Sensitivity and Range	Range: Gamma and x-ray investigation
Sampling	240-sample capacity (408- and 504-sample capacity also available) Multiuser capability Samples loaded in cassettes Sample changer tray is removable
Performance	Temperature: Ambient system Efficiency and background: ^{125}I efficiency; 65% minimum Maximum background: 50 cpm
Requirements	Power: 117 or 234 ± 10% V ac; 50 or 60 Hz Size: 72 cm H × 63 cm W × 66 cm D (28 in. × 25 in. × 26 in.) Weight: 181 kg (400 lb)
Features	1. Pushbutton entry of programming data, up to seven programs can be stored. 2. Data output device integral to the system with results printed adjacent to the sample. 3. Screening function. 4. Preset counting conditions for ^{125}I, ^{57}Co or ^{76}Se, ^{51}Cr or ^{131}I, ^{22}Na or ^{59}Fe or ^{58}Co. 5. RS232 compatible output. 6. Power-fail protected memory.
Cost	$14,390
Address	Packard Instrument Company, Inc. 2200 Warrenville Road Downers Grove, IL 60515 (312) 969-6000

Germanium and Li Drifted Silicon Detectors
Princeton Gamma Tech

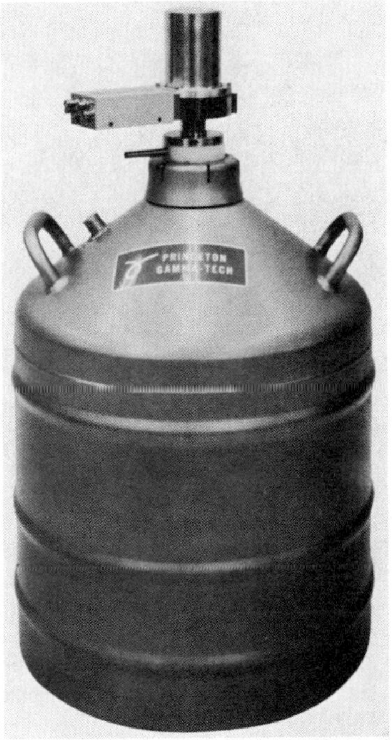

Class	Laboratory spectrometer
Principle of Operation	Ge(Li), Intrinsic (high purity) germanium and Si(Li) detectors, amplifiers and other electronics, cryostats, and accessories
Specifications	See Tables A.6.15 through A.6.17
Features	Many variations in full systems available. Special detector specifications available on request.
References	Manufacturer's specifications
Address	Princeton Gamma-Tech 1200 State Road Princeton, NJ 08540 (609) 924-8980

RAD-GAM,XRA
Princeton Gamma-Tech 1
January 1983

Table A.6.15. Specifications for Selected Coaxial Detectors[a]

Geometry	Range of Efficiency Relative to 3 in. × 3 in. NaI(Tl) at 1.33 MeV	Energy Resolution (keV)		Peak/Compton Ratio at 1.33 MeV	Price
		1.33 MeV	7.6 MeV		
Ge(Li) or IGC	5 to 50	1.7 to 2.3	4.5 to 6.0	20 to 75	$5,000 to $48,000
N1GC Intrinsic N	5 to 30	1.8 to 2.3	5.5 to 6.0	16 to 50	$10,000 to $45,000
Well detector	5 to 30	2.0 to 2.5	5.5 to 7.0	25 to 50	$12,000 to $50,000

[a]Also available as borehole probes, well detectors or in detector arrays (up to 18 detectors/array).

Table A.6.16. Standard Grade Planar Intrinsic (High-Purity) Germanium Detectors,[a] Straight Dipstick Cryostat,[b] Preamplifier, 30-Liter Dewar

Area (mm^2)	Resolution at 5.9 keV (eV)	Resolution at 122 keV (eV)	Be Window Thickness (in.)	Price
25	155	485	0.0013	$ 6500
25	145	480	0.0013	7000
50	165	490	0.0013	6800
100	175	495	0.0023	7600
200	220	540	0.0023	8000
300	250	550	0.006	8500
500	300	570	0.006	9300
1000	420	630	0.016	12400
1500	560	730	0.022	16200
2000	620	770	0.022	19400

[a]Also available in detector arrays (up to 18 detectors/array).
[b]A wide variety of other cryostats are available. Check with the manufacturer.

Table A.6.17. Si(Li) Detectors for Physics Applications[a]

Model	Active Area (mm^2)	Energy Resolution (eV FWHM at 5.9 keV)	Be Window Thickness (mm)	Price
LS13150	13	150	0.5	$8000
LS13155	13	155	0.5	7000
LS13160	13	160	0.5	6500
LS13170	13	170	0.5	5800
LS33155	30	155	0.5	9500
LS33160	30	160	0.5	8500
LS33165	30	165	0.5	8000
LS33170	30	170	0.5	7500
LS33180	30	180	0.5	6500
LS83175	80	175		8000
LS83180	80	180		7500
LS83190	80	190		6800

[a]A wide variety of cryostat configurations are available. Check with the manufacturer.

Germanium Well Detector
PGT Ge(Li) Well or High Purity Ge Well Detector

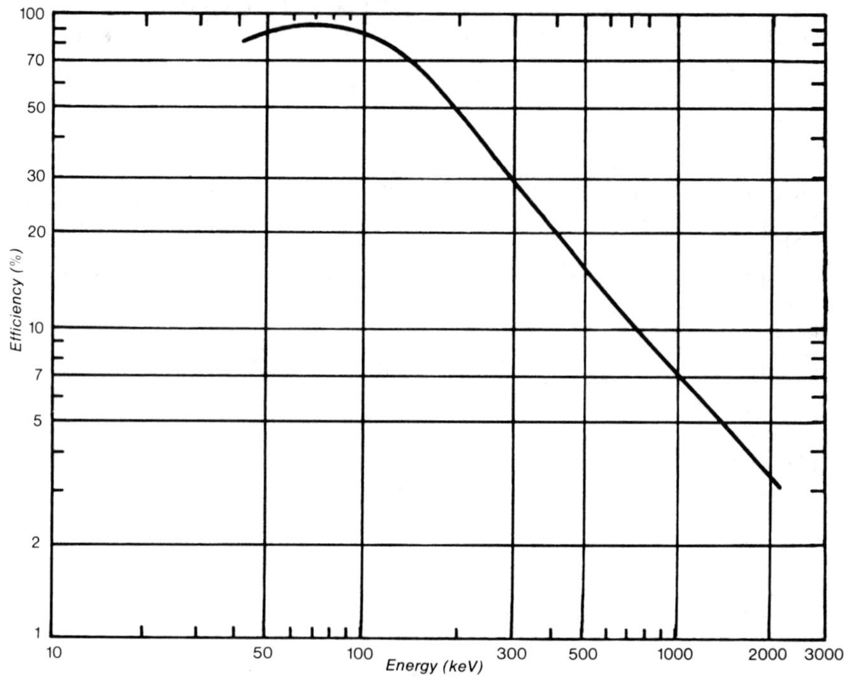

Class	Laboratory
Principle of Operation	A solid state semiconductor operated in reverse bias in order to be gamma spectrum detector. With associated electronics delivers a gamma spectrum.
Sensitivity and Range	γ energy range detected: 20 keV to 10 MeV Typical efficiency (see Fig. A.6.6 relative to 3 in. × 3 in. NaI @ 1.33 MeV): ~10 to 20% Typical Peak-To-Compton Ratio (@ 1.33 MeV): 40
Sampling	Continuous when operated
Performance	Usable well dimensions: 1–1.8 cm D × ≥ 4 cm L Detector volumes: 50–120 cm^3 Operating temperature: at spectrometer, liquid nitrogen temperature; high purity germanium well can be stored or cycled to room temperature Typical resolutions: 2.0 to 2.5 keV FWHM @ 1.33 MeV

Requirements	Power: Requires high voltage supply Size: Depends on dewar, check with manufacturer Weight: Depends on dewar, check with manufacturer
Features	Well detectors are normally supplied in vertical dipstick cryostats with 8-in. end caps to facilitate shielding. Other cryostats are available including low background types. Check with the manufacturer.
References	Manufacturer's specifications
Cost	Refer to manufacturer
Address	Princeton Gamma-Tech 1200 State Road Princeton, NJ 08540 (609) 924-8980

Scintibolic and Scintiflex Detectors
Quartz Products

Class	Laboratory
Principle of Operation	NaI(Tl) scintillators and photomultipliers
Sensitivity and Range	Energy range: 10 keV to 3 MeV, gammas
Performance	Accuracy: About 9% Temperature: −50 to 50°C (−58 to 122°F) Calibration: ^{137}Cs resolution, ^{60}Co peak-to-valley ratio, ^{55}Fe for x-rays
Requirements	*Scintibloc Crystals (Solids and Well Types)*

Diameter (mm):	32	38	44	51	76	76	127
Height (mm):	25	25	51	51	51	76	51
Energy resolution:	7 to 9% for Cs-137 (661 keV)						
Cost:	\$164.00–\$1433.00						

Scintibloc X Crystals

Diameter (mm):	25	32	44	76 to 250
Thickness (mm):	1 to 5	1 to 5	1 to 5	1 to 5
Energy resolution:	45 to 60% for Fe-55 (220 keV)			
Cost:	\$318.00–\$1433.00			

Scintibloc and Scintiflex Crystals (Solids and well-types)

Diameter (mm):	102	102	127	127	127	152	203	292	305
Height (mm):	76	102	76	102	127	102	102	102	305
Energy resolution:	7.5 to 9.5% for Cs-137 (661 keV)								
Cost:	\$1390.00–\$10200.00								

Note: Scintibloc refers to direct coupling
Scintiflex refers to demountable phototube

Features

1. Anti-Compton annuli (pair spectrometer annulus option) from 8 in. × 8 in. to 12 in. × 12 in.
2. Phoswich detector (NaI[Tl]/CsI[Tl]) for plutonium studies
3. Crystals up to 16 in. diameter

Other scintillating CsI(Tl), CsI(Na), anthracene, stilbene, ZnS(Ag) materials

References Manufacturer's specifications

Address Quartz Products Corporation
688 Somerset Street
P.O. Box 1347
Plainfield, NJ 07061
(201) 757-4545

X-Ray Monitor
Radcal Corporation Model 1015C

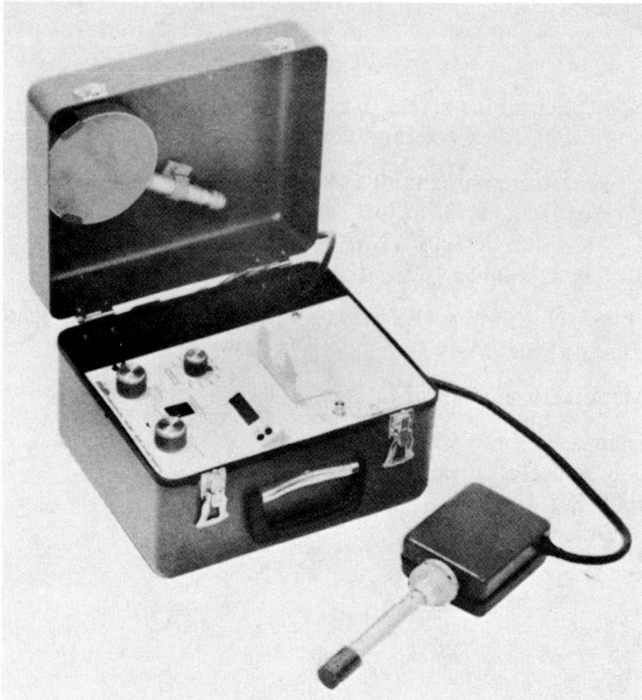

Class	Portable, designed for verifying proper operation of diagnostic x-ray machines.
Principle of Operation	Uses an air-equivalent ion chamber in conjunction with a current-to-frequency converter and digital-processing techniques to measure exposure, exposure rate, pulse exposure, and pulse width.
Sensitivity	Ranges: exposure rate: 1 mR/hr to 59,940 R/hr exposure (integrated): 2 μR to 999 R pulse exposure: 1 mR to 13 R pulse duration: 1 msec to 99.9 sec
Sampling	Continuous
Performance	Operating temperature: 5 to 45°C (41 to 113°F) Operating humidity: up to 95% RH, without condensation Barometric pressure: 600 to 1050 millibars Energy dependence: depends on chamber used

	Exposure rate dependence: depends on chambers Noise level: 1 digit, rms Resolution: 1% or 1 digit Uncertainty in pulse-amplitude discrimination level: 5%, rms
Requirements	Power: battery with lifetime ~750 operating hours Size: refer to manufacturer Weight: refer to manufacturer
Features	A number of options are available The 20-ft multiconductor cable between electrometer and readout units carries only digital signals, bias voltage, and power
References	Manufacturer's specifications
Cost	$2750.00 options and chambers available at additional cost
Address	Radcal Corporation, subsidiary of MDH Industries, Inc. 426 West Duarte Road Monrovia, CA 91016 (213) 357-7921 (800) 423-7169, outside of California Telex: 182910 Attn: D. Blackstone

X-Ray Monitor
Radcal Corporation Model 2025 AC

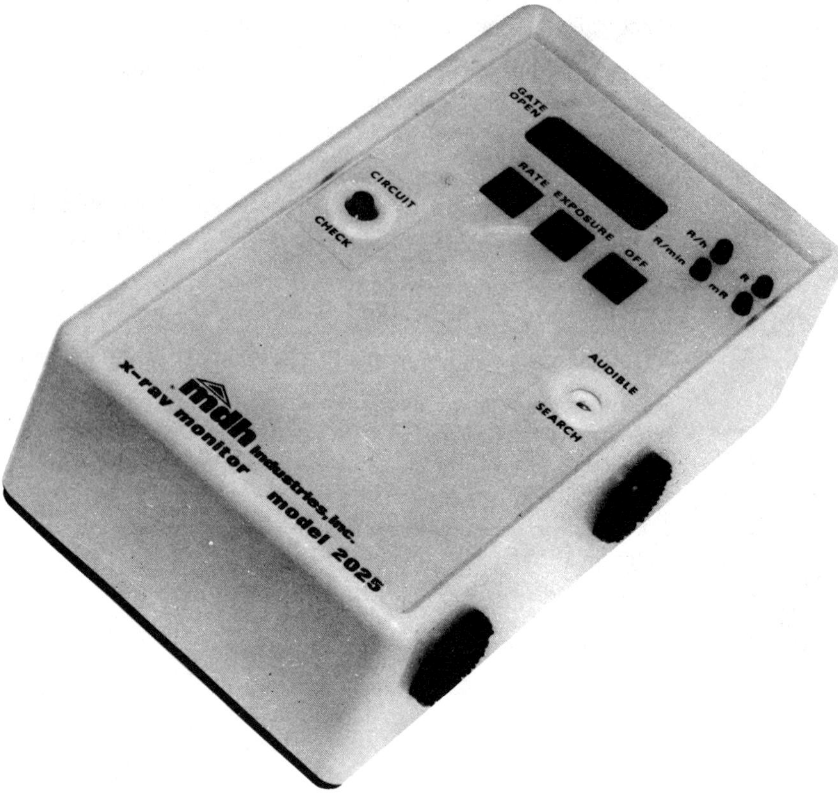

Class	Portable
Principle of Operation	Uses air-equivalent ion chambers with associated electronics for autoranging over 6 decades. Display is 3 1/2 digit liquid crystal.
Sensitivity	Exposure rate range: 0.1 mR/h to 1 MR/h Integrated exposure range: 1 μR to 20,000 R
Sampling	Continuous
Performance	Energy response: typically $\pm 5\%$ (refer to chamber specs) Accuracy: $\pm 4\%$ of reading, ± 1 digit Repeatibility: $\pm 1\%$ of reading, ± 1 digit Bias voltage: 300-V battery

Requirements	Power: four "C" cells, life over 1000 hr; Bias Battery, 1 year Size: Carry case, 36 cm × 43 cm × 13 cm (14 in. × 17 in. × 5 in. nominal); carries total system Weight: system weight, 5.7 kg (12 1/2 lb)
Features	Uses several different chambers for different purposes and energy ranges. Included is a circuit check to be sure electronics are performing as intended.
References	Manufacturer's specifications
Cost	$1050.00 (exclusive of electrometer/ion chambers)
Address	Radcal Corporation, subsidiary of MDH Industries, Inc. 426 West Duarte Road Monrovia, CA 91016 (213) 357-7921 (800) 423-7169, outside of California Telex: 182910 Attn: D. Blackstone

Ion Chambers
Radcal Corporation Models: 10X5 Series

Specifications (see table)

References Manufacturer's specifications (to be used with Radcal Corporation Model 1015C X-ray Monitor)

Cost $425 to $825

		Specifications	
		Measurement Modes	
Model	Performance	Exposure Rate	Exposure
10X5-0.6	Readout at therapy rates and integrated exposure levels	0.01 to 999 R/min	0.2 mR to 999 R
10X5-6	In-beam measurements	0.001 to 650 R/min	0.02 mR to 99.9 R
10X5-60	In-beam measurements, additional sensitivity, size restrictions	0.0001 to 65 R/min	0.002 mR to 9.99 R
10X5-6M	In-beam measurements, low-energy sources	0.001 to 650 R/min	0.02 mR to 99.9 R
10X5-0.18	In-beam measurements, high-intensity γ-rays	1 R/hr to 650 kR/hr	0.002 R to 9.99 kR
10X5-180	Leakage measurements	1 mR/hr to 650 R/hr	0.002 mR to 9.99 R
10X5-1800	Low level radiation	0.1 mR/hr to 65 R/hr	0.0002 mR to 0.999 R
10X5-10.3 CT	Rapid measurements of radiation output of CT equipment	2 mR/min to 1300 R/min	0.04 mR to 200 R

Address

Radcal Corporation, subsidiary of MDH Industries, Inc.
426 West Duarte Road
Monrovia, CA 91016
(213) 357-7921
(800) 423-7169, outside of California
Telex: 182910
Attn: D. Blackstone

Specifications (continued)

Measurement Accuracy

Energy Dependence	Exposure-Rate Dependence	Noise	Resolution	Calibration
±5%, 40 to 1333 keV	±2%, 10 mR/sec to 2000 R/sec	0.4 mR/sec (rms)	1% or: 0.2mR, 10 mR/min	±2% or better
±5%, 20 to 1333 keV	±5%, 0.4 mR/sec to 80 R/sec	0.02 mR/sec (rms)	1%	±5% or better
±5%, 20 to 1333 keV (with build-up)	±5%, 2 mR/min to 200 R/min	—	1% or: 0.002 mR, 0.1 mR/min	±5% or better
±5%, 10 to 40 keV	±5%, 0.4 mR/sec to 80 R/sec	0.02 mR/sec (rms)	1%	±5% or better
±5%, 45 to 1333 keV	±2%, 10 R/hr to 650 kR/hr	—	1% or: 0.002 R, 1 R/hr	±5% or better
±5%, 33 to 1333 keV	±5%, 20 mR/hr to 2000 R/hr	0.002 mR or 1 mR/hr (rms)	1%	±5% or better
±5%, 33 to 1333 keV	+0% or −5%, 0.1 mR/hr to 20 R/hr, −10% to 65 R/hr	0.0002 mR or 0.1 mR/hr (rms)	1% or: 0.002 mR, 0.1 mR/hr	±5% or better
±5%, 3 to 20 mm Al	±2%, 2 mR/sec to 40 R/sec	—	—	±5% or better

Electrometer/Ion Chambers
Radcal Corporation Models: 20X5 Series

		Specifications	
		Measurement Modes	
Model	Performance	Exposure Rate	Exposure
20X5-0.6	Therapy rates and integrated exposure measurements	0.010 R/min to 10,000 R/min	1 mR to 2000 R
20X5-3	In-beam measurements	0.001 R/min to 1000 R/min	0.1 mR to 200 R
20X5-60	In-beam measurements, additional sensitivity, size restrictions	0.1 mR/min to 100 R/min	0.01 mR to 20 R
20X5-6M	In-beam measurements, low-energy sources	0.001 R/min to 1000 R/min	0.1 mR to 200 R
20X5-0.18	In-beam measurements, high-intensity γ-rays	1 R/hr to 1000 kR/hr	0.01 R to 20 kR
20X5-180	Leakage measurements	0.001 R/hr to 1000 R/hr	0.01 mR to 20 R
20X5-1800	Low-level radiation	0.1 mR/hr to 65 R/hr	0.001 mR to 1.999 R
20X5-10.3CT	Rapid measurements of radiation output of CT equipment	1 mR/min to 1000 R/min	0.1 mR to 200 R

Specifications (continued)

Measurement Accuracy

Energy Dependence	Exposure-Rate Dependence	Noise	Resolution	Calibration
±5%, 40 to 1333 keV	–	–	1 mR or 10 mR/min or .5% of reading	–
±5%, 20 to 1333 keV	–	–	0.1 mR or 1 mR/min or .5% of reading	–
±5%, 20 to 1333 keV (with build-up)	±5%, 2 mR/min to 200 R/min	–	1% or 0.1 mR/min	±5% or better
±5%, 10 to 40 keV	–	–	1 mR/min, 0.1 mR or 0.5% of reading	–
±5%, 45 to 1333 keV	±2%, 10 R/hr to 1000 kR/hr	–	1% or 0.01 R, 1 R/hr	±5% or better
±5%, 30 to 1333 keV	–	–	1 mR/min, 0.01 mR or 0.5% of reading	–
±5%, 33 to 1333 keV	+0% or −5% 0.1 mR/hr to 20 R/hr, −10% to 65 R/hr	0.001 mR or 0.1 mR/hr (rms)	1% or 0.001 mR or 0.01 mR/hr	±5% or better
±5%, 3 to 20 mm Al	±5%, 0.4 mR/sec to 16 R/sec	–	–	±5% or better

Specifications	See previous page
References	Manufacturer's specifications (to be used with Radcal Corporation Model 2025AC X-ray Monitor)
Cost	\$655 to \$1060
Address	Radcal Corporation, a subsidiary of MDH Industries, Inc. 426 West Duarte Road Monrovia, CA 91016 (213) 357-7921 (800) 423-7169, outside of California Telex: 182910 Attn: D. Blackstone

X-Ray Proportional Counters
Reuter-Stokes

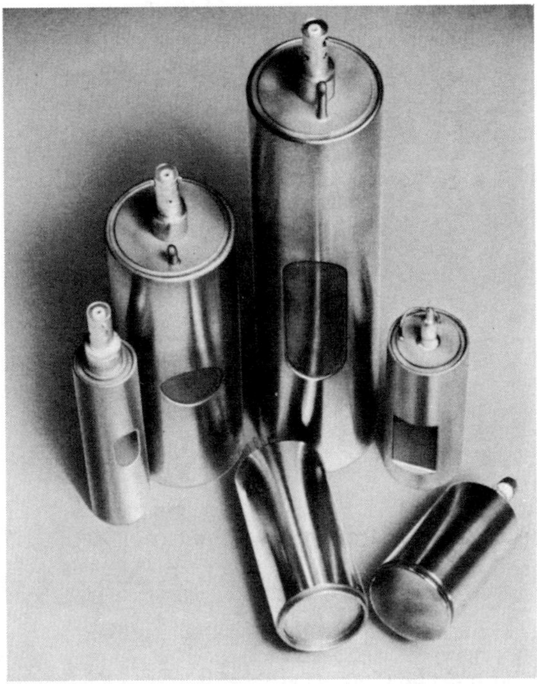

	RS-P3-1605-261	RS-P3-1605-285
Energy range	3 to 35 keV	1.5 to 8 keV
%FWHM	10% at 14.4 keV	17% at 5.98 keV 12% at 14.4 keV
Window	Be, 1 ft D 0.01 in. thick, 46 mg/cm^2	Be, 2 in. L × 1.7 in. W 0.005 in. thick, 23 mg/cm^2
Temperature range	−25 to 80°C (−13 to 176°F)	−25 to 55°C (−13 to 131°F)
Fill gas	Xe, CO_2 1 atm.	97% Xe, 3% CO_2 76 cm Hg
Power required	1500 to 2200 V dc	1500 to 2200 V dc
Size	19.05 cm L × 5.16 cm D (7.5 in. × 2.1 in.)	19.3 cm L × 5.1 cm W (7.6 in. × 2 in.)
Weight	0.28 kg (0.62 lb)	0.5 kg (1.25 lb)

Applications	Mossbauer effect studies	Industrial XRF analysis
Cost	$600.00 base	$1302.00
Features	Optional gas, windows, and gas pressures	

	RS-P3-0803-287	RS-P3-1608-271
Energy range	3 to 35 keV	6 keV – 15 keV
%FWHM	11% at 14.4 keV	17% at 5.98 keV
Window	Be, 1 in. D, 0.005 in. T	Be, 0.010 in.
Temperature range	−25 to 80°C (−13 to 176°F)	−25 to 55°C (−13 to 131°F)
Fill gas	Kr, CO_2, 152 cm Hg	Xe, CO_2, 76 cm Hg
Power required	1700 to 2200 V dc	2000 to 2500 V dc
Size	14.3 cm L × 2.54 cm D (5.5 in. × 1 in.)	26.7 cm L × 2.5 cm W × 8.3 cm D (10.5 in. × 1 in. × 3 1/2 in.)
Weight	0.1 kg (0.25 lb)	0.5 kg (1 lb)
Applications	Spectroscopy diffraction	Industrial XRF
Cost	$600.00	$1000.00

Features: Optional gas, windows, and gas pressures

Remarks: Other detectors meeting different specifications are available. Please contact the manufacturer to learn about these.

Address: Reuter-Stokes
18530 S. Miles Parkway
Cleveland, OH 44128
(216) 581-9400

Gamma Ionization Chambers
Reuter-Stokes RS-C4-1606 Series

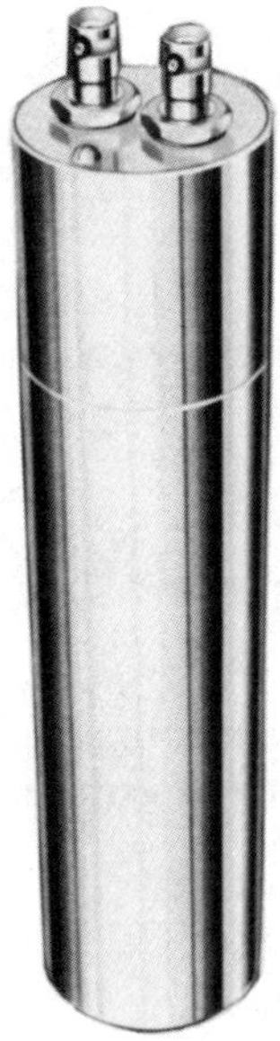

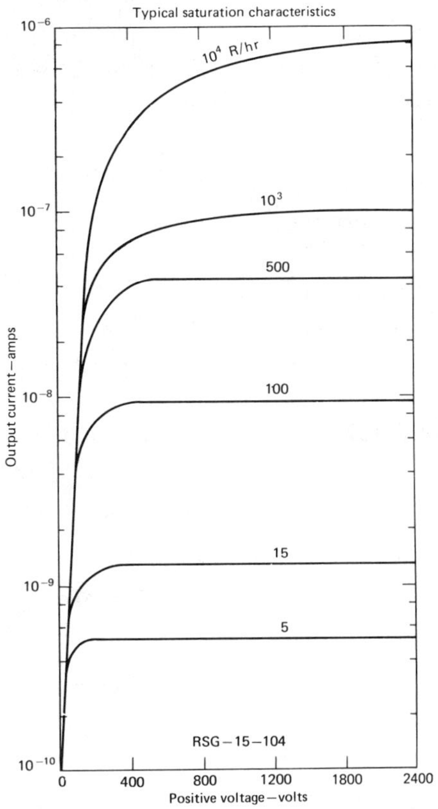

Class	Portable
Principle of Operation	Ionization chamber
Sensitivity and Range	Energy sensitivity: 0.4 MeV to 7 MeV Operating characteristics: See Table A.6.18 Flux range: 5 mR/hr to 10^6 R/hr
Sampling	Continuous
Performance	Accuracy: ±3% Tracerable to NBS—sensitive volume: 160 cm^3 Temperature: 200°C (392°F) max
Requirements	Power: 2000 V dc max Size: 23 cm L × 5 cm D (9.1 in. × 2 in.) Weight: 0.34 kg (0.75 lb)
Features	Stainless steel walls and all seals are ceramic to metal. Many other detectors available for x-ray and gamma.
References	Manufacturer's specifications
Cost	$1090.00 to $1575.00
Address	Reuter-Stokes 18530 S. Miles Parkway Cleveland, OH 44128 (216) 581-9400

Table A.6.18. Typical Operating Characteristics

Model No.	Sensitivity (amp/R·hr)	Flux Range (R/hr)
RS-C4-1606-209 (456 cm Hg-Xe)	8×10^{-10}	5×10^{-3} to 5×10^{2}
RS-C4-1606-203 (760 cm Hg-N_2)	1.2×10^{-10}	10^{-2} to 10^{4}
RS-C4-1606-205 (76 cm Hg-N_2)	1.2×10^{-11}	10^{-1} to 10^{5}
RS-C4-1606-207 (6.8 cm Hg-N_2)	1.2×10^{-12}	10^{0} to 10^{6}
RS-C4-2510-101 (xenon)	3.5×10^{-9}	500×10^{-6} to 25

Gamma Ionization Chamber
Reuter Stokes Model RS-C3-1006-202

Class	Monitoring plants and environs.
Principle of Operation	Gamma ionization chamber
Sensitivity and Range	Typical Gamma sensitivity: 1×10^{-11} amp/R·hr ±20% Gamma flux range: 10^0 to 10^8 Energy response: 60 keV to 3 MeV
Performance	Resistance at 25°C: High voltage electrode to shell: 10^{12} ohms (min) Signal electrode to shell: 10^{13} ohms (min) Capacitance: High voltage electrode: 240 pf Signal electrode: 150 pf Maximum ratings: Inter electrode voltage: 900 V Temperature: 300°C Humidity: 100%
Requirements	Maximum chamber diameter: 3.25 cm (1 1/4 in.) Maximum chamber length: 22.3 cm (8 3/4 in.) Net weight: 0.3 kg (11 oz.) Fill gas: nitrogen Connectors: type HN
References	Manufacturer's specifications
Cost	~$3,500.00
Address	Reuter-Stokes, Inc. 18530 South Miles Parkway Cleveland, OH 44128 (216) 581-9400 Telex: 98-5253 Attn: T. Fleming

Environmental Gamma Monitor
Reuter-Stokes Model RSS-111

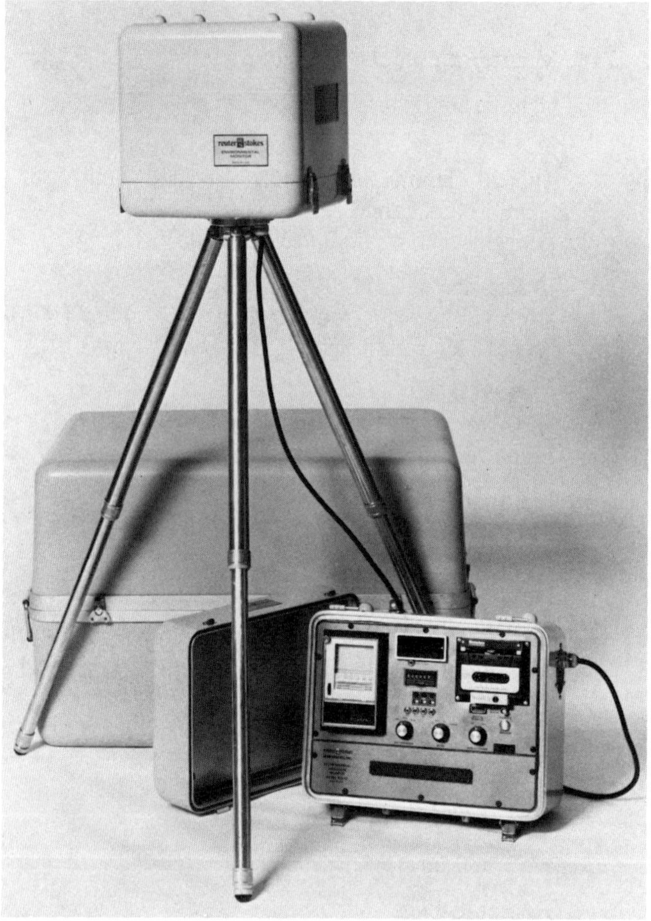

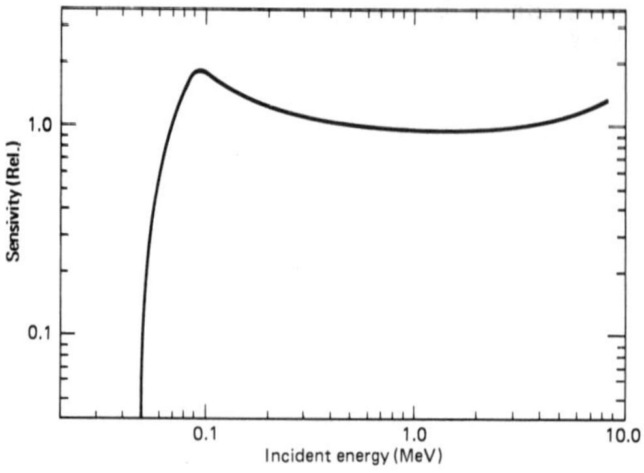

Class	Stationary or portable field instrument
Principle of Operation	High-pressure, spherical, argon-filled ionization chamber, solid-state electrometer. Separate housing for control circuitry, recorder, panel meter, batteries.
Sensitivity and Range	Sensitivity: 20 mV/μR·hr Energy range: 50 keV to 5 MeV (gammas) Flux range: 1 to 500 μR/hr (standard, to 5000 μR/hr optional)
Sampling	Continuous
Performance	Accuracy: $\pm$5% at 10 μR/hr Temperature: -24 to 54°C(-11 to 129°F) Recording: Strip-chart recorder, optional magnetic-tape or telemetry readout Response time: 5 sec Humidity: 0 to 100% (waterproof)
Requirements	Power: Ni-Cad rechargeable batteries for electrometer; 300-V power cell for ion chamber; ac operation also available Size: Two weatherproof housings (1-ft cubes = 30-cm cubes) Weight: Sensor housing 10.5 kg (23 lb), tripod 2.7 kg (6 lb), remote housing 17.2 kg (38 lb)
Features	1. Spherical ion chamber yields flat response. 2. Solid-state electrometer with MOSFET circuitry. 3. Weatherproof construction. 4. Optional ac power supply hookup. 5. Optional dose integrator, magnetic tape output, and tape reader. 6. Automatic range changing.
References	Manufacturer's specifications
Cost	$8727.00 with strip-chart recorder; digital displays and parallel digital output; optional readouts additional
Address	Reuter-Stokes 18530 South Miles Parkway Cleveland, OH 44128 (216) 581-9400

Environmental Radiation Monitoring System
Sentri 1011

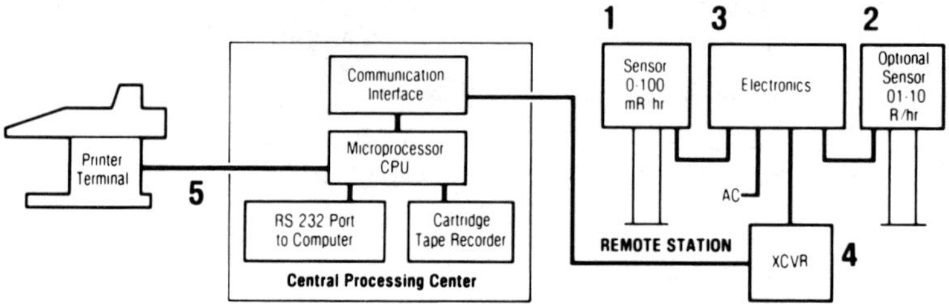

Class	Gamma detectors
Principle of Operation	High-pressure spherical argon-filled ionization chamber, solid-state electrometer, remote electronics station, central processing center.
Sensitivity and Range	Background to 10 R/hr
Sampling	Continuous
Performance	Accuracy: ±10% Temperature: −24 to 54°C (−11 to 129°F) Response time: 5 sec Humidity: 0 to 100% (waterproof)
Requirements	Power: 300 V dc Weight: Sensor housing 10.4 kg (23 lb) Remote housing 23 kg (50 lb)
Features	Low/intermediate sensors and electrometer; high-level electrometer; remote station: transmitter/receivers; 16 remote stations; central processing center.
References	Manufacturer's specifications
Cost	On request from manufacturer.
Address	Reuter-Stokes 18530 S. Miles Parkway Cleveland, OH 44128 (216) 581-9400 Telex: 98-5253 Attn: T. Fleming

Gamma Survey Meters
Scintrex Models 189-A, B and C

Class	Field
Principle of Operation	Hand portable G-M tube, energy-compensated
Sensitivity and Range	Model 189-A: 1 mrem/hr to 50 rem/hr Model 189-B: 10 rem/hr to 10,000 rem/hr Model 189-C: 0.2 rem/min to 200 rem/min
Performance	Linearity: maximum error ±15% measured on ^{137}Cs (0.662 MeV) through full range of instrument Energy Dependence: Model 189A; maximum error ±20% (0° and 90° radiation incidence) with respect to ^{137}Cs (0.662 MeV) Model 189B and C; maximum error at ±90° incidence: +46% at 183 keV, −47% at 107 keV; at 0° incidence: +42% at 183 keV, −30% at 107 keV. Maximum error with respect to response at ^{137}Cs (0.662 MeV). Angular Dependence: refer to manufacturer's specifications Exposure Rate Limitation: Model 189A: will withstand 32 kR/hr; Model 189B,C: will withstand 100 kR/hr Response time to geometrical 90% step: Model 189A; low scale: 5 sec for 500 mR/hr high scale: 6 sec for 50 R/hr Model 189B; 2 sec for 6 kR/hr Model 189C; 2 sec for 0.1 rem/hr Temperature Dependence: Model 189A; maximum error +8% with respect to reading at +22°C (+72°F) through the range −35 to +60°C (−31 to +140°F) Model 189B and C; maximum error +8.2%, −3.4% with respect to the reading at +22°C (+172°F) through the range −35 to +60°C (−31 to +140°F) Humidity dependence: Less than 1% up to 98% relative humidity Start-up drift: no observable drift after 2 seconds of start-up

Requirements	Power: 4 C-cells (1.5 V each) Size: 36.4 cm × 26.5 cm × 10.8 cm (14.6 in. × 10.6 in. × 4.3 in.) Weight: with batteries, 1.95 kg (4.3 lb)
Options	Extension tube: carrying strap and case
Cost	Refer to manufacturer
Address	Scintrex, Ltd. 222 Snidercroft Road Concord, Ontario L4K 1B5, Canada (416) 669-2280 Telex: 06-964570 Attn: S. Nargolwalla

Area Monitors
Model FML-5A
Model FML-5B

Class	Rack mount or case
Principle of Operation	Geiger detectors, log count ratemeter, high-level and system failure alarms
Sensitivity and Range	FML-5A: 6 decades, 0.01 mR/hr to 10 R/hr standard, other ranges to 1000 R/hr FML-5B: 4 decades, 1 mR/hr to 1 R/hr standard, or other 4 consecutive ranges
Sampling	Continuous
Performance	Energy dependence: ±20% from 100 keV to 1.2 MeV
Requirements	Power: 110 V ac, 60 Hz Size: 4 cm W × 3.4 cm H × 5.3 cm D (10 1/4 in. × 8 3/4 in. × 13 1/2 in.) Weight: console plus detector 9.5 kg (21 lb)

Features	All plug-in modular (can add channels) Antisaturation High- and low-level alarms Rack mount or case Rugged, heavy-duty probes with energy compensation Local or remote monitoring up to 500-ft cable Activation of movable check source in probe (option)
References	Manufacturer's specifications
Cost	FML-5A: $995.00 includes probe, first channel $715.00 per additional channel FML-5B: $975.00 includes probe, first channel $675.00 per additional channel
Address	Technical Associates 7051 Eaton Avenue Canoga Park, CA 91303 (213) 883-7043

High Range Underwater Monitor
Technical Associates Model CP-MU, CP-MU-1000

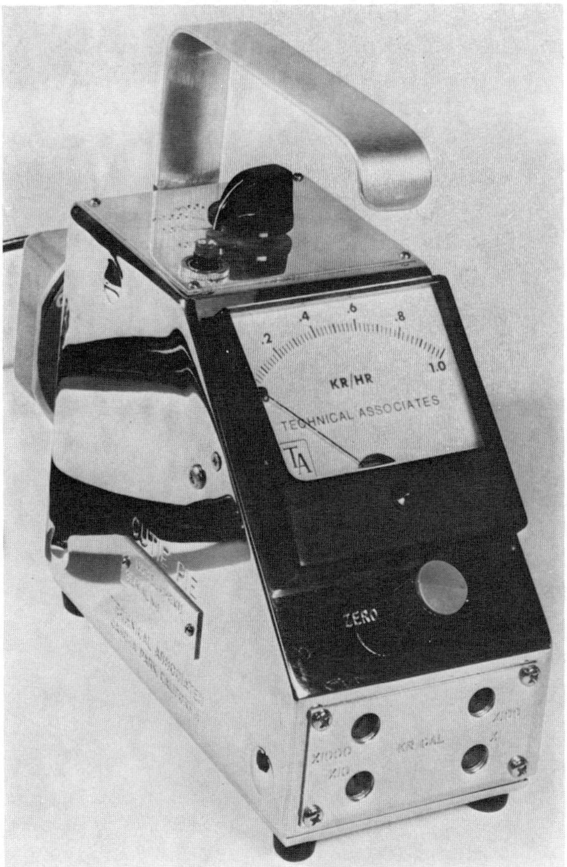

Class	Portable, handheld
Principle of Operation	Remote air ion chamber with MOFSET amplifier, meter readout
Sensitivity and Range	CP-MU: 0 to 1 kR/hr 0 to 10 kR/hr optional other ranges (lower) 0 to 10^5 R/hr 0 to 10^6 R/hr CP-MU-1000: 0–1 R/hr, 0–10 R/hr, 0–100 R/hr, 0–1000 R/hr
Sampling	Continuous

Performance	Accuracy: ±15% FS Time constant: 2 sec
Requirements	Power: Single "D" cell Size: 22.9 cm L × 14.0 cm H × 10.2 cm W (9 in. × 5 1/2 in. × 4 in.) excluding handle; cable (probe) length: 60 ft (longer optional) Weight: 1.8 kg (4 lb)
Features	Underwater monitoring of reactor spent fuel and so on, or hot cell monitoring to very high levels. Stability and accuracy (adjustment pot for each range) Portability
References	Manufacturer's specifications
Cost	CP-MU: \$1390.00 CP-MU-1000: \$1505.00
Address	Technical Associates 7051 Eton Avenue Canoga Park, CA 91303 (213) 883-7043

NaI(Tl) Scintillation Crystals
Teledyne Isotopes

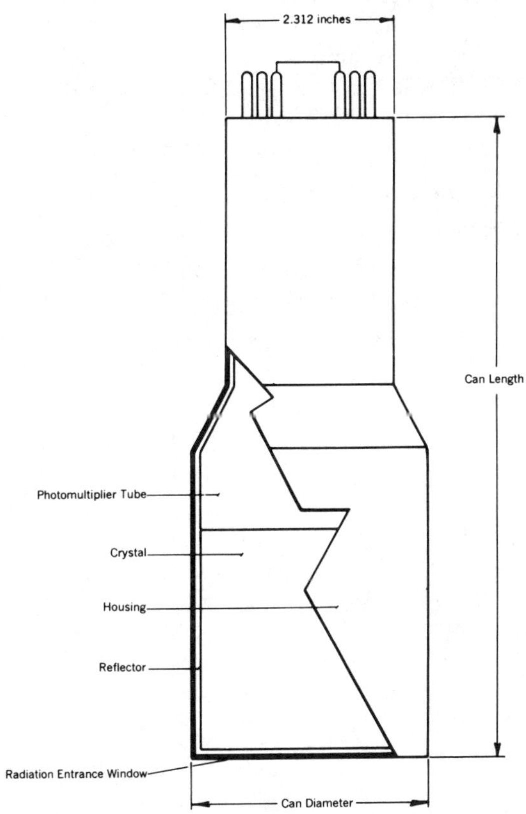

Class	Laboratory
Principle of Operation	NaI(Tl) crystals, photomultipliers, mounts, and assembled systems for photon spectrometry.
Specifications	NaI(Tl) crystals: 1. Wide variety of standard cylindrical crystals, from 3/4 in. D × 3/4 in. L to 9 in. × 4 in. Custom sizes available 2. Well shapes, hole through, and thin crystals also available 3. *Response time* can be pre-determined to satisfy customer needs through the use of specific photomultiplier tubes 4. Table A.6.19 shows a few representative crystals without photomultiplier tubes Assemblies: 1. Many types of assemblies available 2. All "Integral Series" assemblies have photomultiplier attached to NaI(Tl) and all hermetically sealed 3. Magnetic shield included 4. Tube base available as extra option 5. Aluminum housing (0.020 in.) standard: copper or stainless steel also available 6. Phototubes used are RCA 6342AVI, 2060; S83006EM1; Amperex 2012, SRC52B01, B75B01 7. Well-type and other geometries also available
References	Manufacturer's specifications
Address	Teledyne Isotopes 50 Van Buren Avenue Westwood, NJ 07675 (201) 664-7070

Table A.6.19. Representation Crystals

Designation	Crystal Diameter (cm)	Crystal Diameter (in.)	Crystal Thickness (cm)	Crystal Thickness (in.)	Crystal Price
S-42	2.54	1	1.27	1/2	$ 90.00
S-44	2.54	1	2.54	1	90.00
S-64	3.81	1.1/2	2.54	1	110.00
S-88	5.1	2	51	2	300.00
S-1212	7.6	3	7.6	3	700.00
ST-41-A	2.52	1	1 mm		120.00
ST-46-A	2.54	1	6 mm		120.00
ST-81-A	5.1	2	1 mm		190.00
ST-86-A	5.1	2	6 mm		190.00

Table A.6.20.

Designation with photomultiplier	Crystal Diameter (in.)	Crystal Thickness (in.)	Price
S-64-I	1.1/2	1	$280.00
S-78-I	1.3/4	2	360.00
S-84-I	2	1	310.00
S-88-I	2	2	390.00
S-1208-I	3	2	810.00
S-1212-I	3	3	950.00

NaI(Tl) Spectrometer Systems
Tennelec, Inc.

Class	Laboratory spectrometers
Principle of Operation	NaI(Tl) crystal, photomultiplier tube for gamma spectroscopy. Table A.6.21
Features	Electronic accessories also available. Individual NaI(Tl) crystals also available. All assemblies have magnetic shielding, hermetic seals.
References	Manufacturer's specifications
Address	Tennelec, Inc. P.O. Box 2560 Oak Ridge, TN 37830-2560 (615) 483-8405

Table A.6.21. Specifications for Tennelec NaI(Tl) Spectrometer Systems

Crystal Size Diameter × Thickness [mm (in.)]	Model Type	Performance Data	Photomultiplier Tube	Price
		Thin-Window NaI(Tl) X-Ray Detectors (^{55}Fe Peak/Valley Ratio, 5.9 keV)		
25 × 1.5 (1 × 0.2)	TD 25 SX 1/5	15:1	Philips XP1010	$ 430.00
32 × 1.5 (1.25 × 0.2)	TD 32 SX 1/5	15:1	Philips XP1010	500.00
44 × 1.5 (1.75 × 0.2)	TD 44 SX 1/5	15:1	Philips 153 AVP 02	600.00
		Standard NaI(Tl) Assemblies (^{137}Cs Energy Resolution, 662 keV)		
25 × 25 (1 × 1)	TD 25 S 25	8.5%	RCA 6199	$ 140.00
32 × 25 (1.25 × 1)	TD 32 S 25	8.5%	RCA 6199	150.00
38 × 25 (1.5 × 1)	TD 38 S 25	8.5%	RCA 6342A	165.00
44 × 51 (1.75 × 2)	TD 44 S 51	8.5%	RCA 6342A	270.00
51 × 51 (2 × 2)	TD 51 S 51	8.5%	RCA 6342A	320.00
76 × 51 (3 × 2)	TD 76 S 51	8.5%	RCA 8054	700.00
76 × 76 (3 × 3)	TD 76 S 76	8.5%	RCA 8054	850.00
102 × 102 (4 × 4)	TD 102 SE 102	9.0%	RCA 8055	1490.00
127 × 127 (5 × 5)	TD 127 SE 127	10.0%	RCA 8055	2290.00
		Well-Type Assemblies (^{137}Cs Energy Resolution, 662 keV)		
44 × 51 (1.75 × 2)	TD 44-SP-51	9.0%	RCA 6342A	$ 330.00
51 × 51 (2 × 2)	TD 51-SP-51	9.0%	RCA 6342A	390.00
76 × 76 (3 × 3)	TD 76-SP-76	9.0%	RCA 8054	990.00
102 × 102 (4 × 4)	TD 102-SPE-102	10.0%	RCA 8055	1645.00
127 × 127 (5 × 5)	TD 127-SPE-127	11.0%	RCA 8055	2645.00

Geiger-Müller Detectors
Texas Nuclear Model 9361

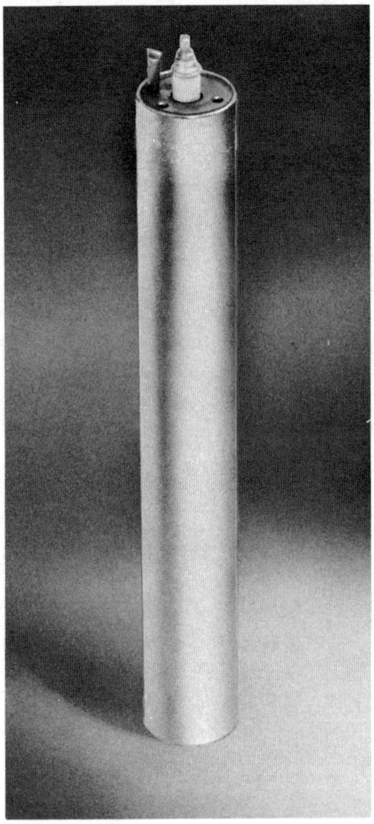

VOLTAGE PLATEAU OF THE 9361

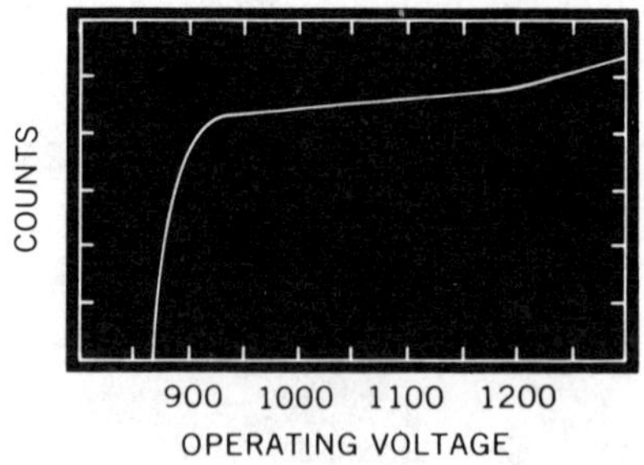

Class	Detectors
Principle of Operation	Geiger-Müller detector
Sensitivity and Range	Plateau length: ~200 V Plateau slope maximum: ~6%/100 V Pulse output: 2 to 20 V
Sampling	Continuous
Performance	Anode material: 0.0051 cm (0.002 in.) tungsten Insulators: Alumina ceramic Cathode: Bismuth screen Envelope material: Stainless steel Operating temperature: Up to 200°C (392°F) Filling gas: Specially purified organic-quenched gas
Requirements	Size: 2.54 cm (1.0 in.) outside diameter; 19.53 cm (7 11/16 in.) length; 15.2 cm (6 in.) active length Operating voltage: ~1100 V Weight: 1.36 kg (3 lb)
Remarks	Basic design available in other active lengths and large diameters.
Cost	Refer to factory
Address	Texas Nuclear Division Ramsey Engineering Company P.O. Box 9267 Austin, TX 78766 (512) 836-0801

G-M Tubes
Partial Listing
Manufactured by TGM

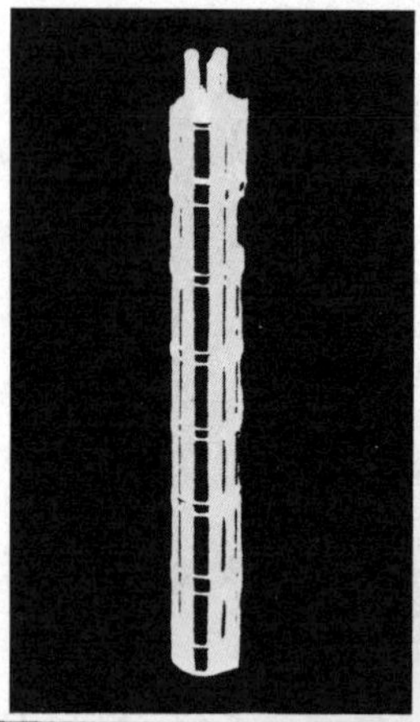
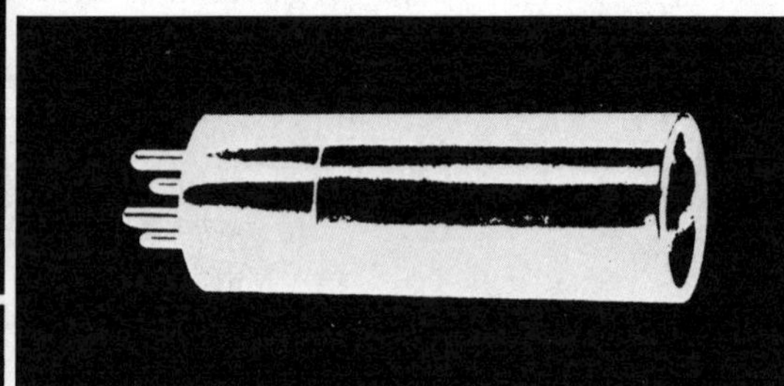
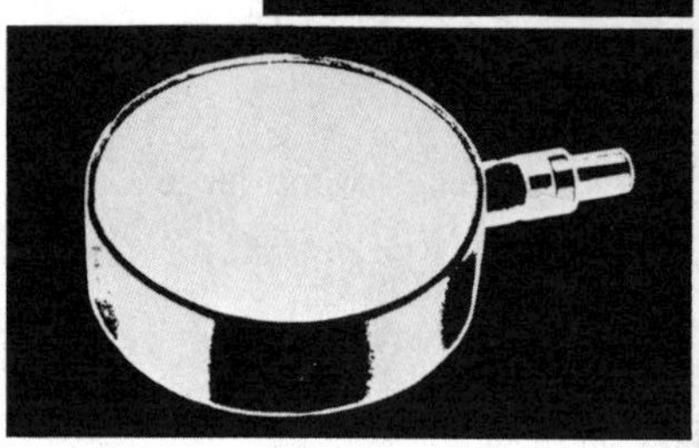
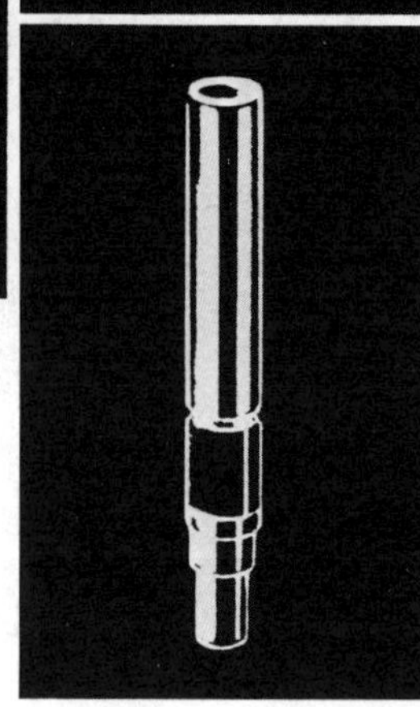

Class	Portable
Principle of Operation	Geiger-Müller tubes for replacement purposes and new equipment design.
Specifications	See Table A.6.22
References	Manufacturer's specifications
Address	TGM Detectors Inc. 34 Bear Hill Road Waltham, MA 02154 (617) 890-2090

Table A.6.22. Geiger-Müller Detectors

	Cathode Wall [mm (in.)]						
Model	Inside Diameter	Active Length	Window/Wall Thickness	Operating Voltage	Recovery Time (μsec)	Remarks	Price
			α,β,γ End Window, Cylindrical G-M Tubes				
			Window				
A201	20 (0.781)	102 (4.0)	1.4–2.0 mg/cm^2	900	100	detects x-ray	$120.00
K201	20 (0.781)	102 (4.0)	1.4–2.0	1300	100	detects x-ray	140.00
N201	20 (0.781)	102 (4.0)	1.4–2.0	900	100		60.00
N202	3.18 (0.125)	7.94 (0.313)	1.0–1.2	750	50		70.00
N204/7618	20 (0.781)	12.7 (0.5)	1.4–2	900	150		75.00
N210	29.4 (1.157)	48 (1.875)	1.4–2	900	200		65.00
N210BNC	29.4 (1.157)	48 (1.875)	1.4–2	900	200		73.00
N222	6.35 (0.25)	23.5 (0.926)	1.4–2	900	100		105.00
H220	29.4 (1.157)	25.4 (1.0)	1.4–2	1300	100	organic-quenched	80.00
H220/7840	29.4 (1.157)	12 (0.50)	1.8–2.2	1400	80	organic-quenched	75.00

Table A.6.22. (Continued)

Model	Cathode Wall [mm (in.)]						
	Inside Diameter	Active Length	Window/Wall Thickness	Operating Voltage	Recovery Time (μsec)	Remarks	Price
			γ *Cylindrical G-M Tubes*				
			Wall				
N302	6.4 (0.250)	7.9 (0.310)	0.51 mm (0.020 in.)	900	20		40.00
N308	3.18 (0.125)	12.7 (0.50)	0.94 (0.037)	600	50		27.00
N309	15.2 (0.60)	17.5 (2.687)	0.23 (0.009)	900	80		32.00
N310	15.2 (0.60)	148 (5.81)	0.23 (0.009)	900	80		35.00
N314	18.4 (0.726)	148 (5.81)	0.31 (0.012)	900	110		35.00
NP315	6.4 (0.250)	96 (3.8)	0.51 (0.020)	700	70	Pt-plated wall, high temperature	140.00
N320	15.2 (0.60)	38 (1.5)	0.23 (0.009)	500	100		27.00
NP358	15.2 (0.600)	108 (4.25)	0.23 (0.009)	900	100	Pt-plated wall, high temperature	200.00
NP334	18.4 (0.726)	108 (4.25)	0.31 (0.012)	900	130	Pt-plated wall, high temperature	230.00
TGM-16	25 (0.980)	305 (12)	0.51 (0.020)	1200	60	organic-quenched	115.00

RAD-GAM, XRA
TGM 1
January 1983

			α,β,γ End-Window, G-M Tubes				
N1002/8767	44.5 (1.75)	10.7 (0.42)	1.6–2 mg/cm²	900	30		$100.00
N1003	29.4 (1.157)	9.5 (0.375)	1.6–2	900	40		90.00
N1005	44.5 (1.75)	10.7 (0.42)	1.6–2	900	40		135.00
N1006	44.5 (1.75)	10.7 (0.42)	1.6–2	900	30		125.00
			β,γ Thin-Wall Cylindrical G-M Tubes				
			Wall				
N106	15.2 (0.60)	89 (3.5)	30–40 mg/cm²	900	75		$35.00
N106HT	15.2 (0.60)	89 (3.5)	30–40	900	75	200°C	52.00
N107	15.2 (0.60)	227 (8.94)	30–40	900	75		53.00
N107HT	15.2 (0.60)	227 (5.68)	30–40	900	75	200°C	65.00
N108	15.2 (0.60)	144 (8.94)	30–40	900	74		50.00
N112/6833	18.4 (0.726)	54.8 (2.156)	30–40	900	150		40.00
N112/HS	18.4 (0.726)	54.8 (2.156)	30–40	900	170		48.00
N119HS	18.4 (0.726)	142.9 (5.675)	30–40	900	150		75.00
N114/OCD-D-103	15.2 (0.60)	19 (0.75)	30–40	900	60		32.00
N115	4.8 (0.190)	6 (0.25)	80–100	500	20		32.00
N116	4.8 (0.190)	16 (0.625)	80–100	500	28		32.00
N117	7.9 (0.310)	28 (1.102)	30–50	500	45		38.00

Automatic Gamma Counter
Tracor Analytic Model 1185

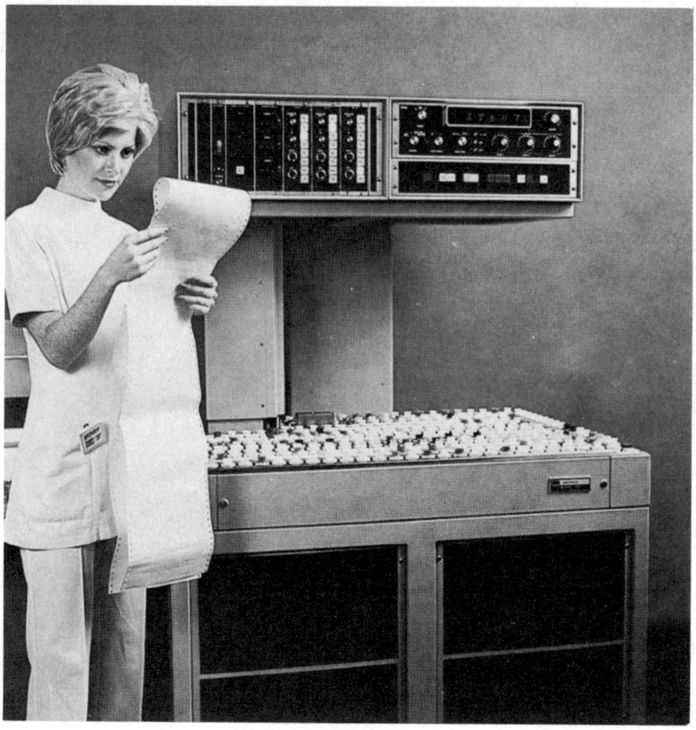

Class	Laboratory
Principle of Operation	A well-type scintillation detector and lead shield and a NaI(Tl) crystal with nearly 4π detection geometry.
Sensitivity and Range	Sensitivity expressed in terms of E^2/B ratio is at least 200 with a 2-in. crystal for ^{125}I Dynamic range: 100:1
Sampling	300-sample capacity, counted individually
Performance	Typical ^{125}I counting efficiency is 80% with 25 cpm background when a 2-in. crystal is used
Requirements	Power: 115 V ± 10 V, 60 Hz; or 230 V, 50 Hz Size: 162.6 cm H × 111.8 cm W × 85.1 cm D (64 in. × 44 in. × 33.5 in.) Weight: 907.2 kg (2000 lb)

Features	Automatic background subtraction, two or three analysis channels with both preset and adjustable windows, automatic calibration, automatic correction for spillover in dual labeled experiments, ratemeter. RIA data reduction available.
References	Manufacturer's specifications
Cost	$17,680–$32,200 depending on options
Address	Tracor Analytic, Inc. 1842 Brummel Drive Elk Grove Village, IL 60007 (312) 364-9100

Automatic Gamma Counter
Tracor Analytic Models 1190, 1191 and 1290

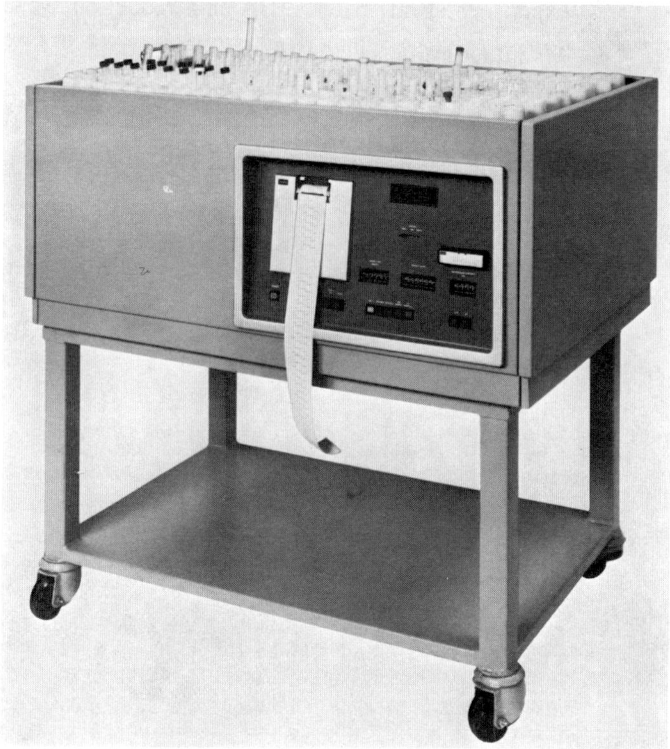

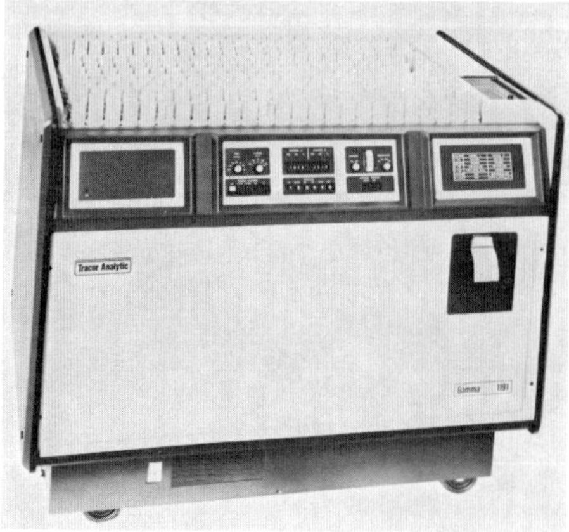

Class	Laboratory
Principle of Operation	NaI detectors. Both models 1190 and 1191 use a 300 sample bi-directional conveyor while model 1290 is a tray based bench top system where up to 648 samples in triplicate trays can be loaded at one time. Teletype or Texas Instruments printers are optional.
Sensitivity and Range	Detects and counts most gamma emitting isotopes such as ^{125}I, ^{57}Co, ^{75}Se and ^{131}I. Model 1290 cannot detect energies above 200 keV.

Performance

Model	Typical efficiency for ^{125}I
1190	70%
1191	75%
1290	80%

Features

Model 1190: Bench top design with capability to accept commonly used test tubes up to 15 mm × 125 mm; plug-in application modules for optimum isotope selection. ^{125}I module included; ^{57}Co, ^{75}Se and ^{131}I available as options; digital preset count terminator up to 999,999 counts and digital preset time terminator up to 99,999 minutes; digital background subtract of up to 9999 cpm; printout by a built-in printer is standard.

Model 1191: Capability to accept tube sizes of up to 16 mm × 150 mm; simultaneous time channel pulse height analysis of gross counts; net cpm with thumbwheel selected automatic ratio correction for dual label, low count reject, K-factor normalization, and percentage of standard calculation; instant print function without count termination.

Model 1290: "Hands-off" pipetting, mixing, incubation, centrifugation, decanting and counting of trays, 36 samples each. Simultaneous counting of up to 6 samples for fastest throughput. Up to 23 separate radioassay batches, having intermixed bound or free protocols, can be loaded. Each tray programmable for tray I.D. number, standard or sample tray. Can count ^{125}I and ^{57}Co, single or dual label. Preset time terminator for 0.01 minutes and up. Background subtraction.

Requirements

Power: 115 ± 10 V, 50/60 Hz or 230 ± 10 V, 50/60 Hz
Dimensions and Weight:

Model	Dimensions	Weight
1190	77 cm × 106 cm × 55 cm (30 in. × 42 in. × 22 in.)	227 kg (500 lb)
1191	116.8 cm × 93.9 cm × 85.0 cm (46 in. × 37 in. × 33.5 in.)	157.5 kg (350 lb)
1290	44 cm × 110 cm × 62 cm (17.2 in. × 43.4 in. × 24.6 in.)	115–131 kg (255–290 lb)

Price

Model 1190: $10,900 to $21,275
Model 1191: $15,000 to $25,595
Model 1290: $14,000 to $32,050
depending on options

References

Manufacturer's literature and specifications

Address

Tracor Analytic, Inc.
1842 Brummel Drive
Elk Grove Village, IL 60007
(312) 364-9100

Precision Electrometer and Detectors
Victoreen Model 500 and 500-3,4,5,6,12,13,14

Class	Laboratory
Principle of Operation	MOS-FET, dual polarity electrometer for use with a large number of commercially available coaxially and triaxially terminated ionization chambers
Sensitivity and Range	Linearity: within 0.01% full scale Accuracy: within 0.5% full scale for charge (nC) or current (pA) measurements Dependent on probe used

Detectors

	Ion Chambers		
Models:	550-6,5	5500-4,3	550-12,13,14
Applications	Teletherapy, Supervoltage Orthovoltage Diagnostic	Scatter Diagnostic Orthovoltage Supervoltage	Mammography Superficial Low-Energy X-Ray
Probe Volume (cm^3)	0.33; 3.3	33; 330	33; 3.3; 0.33
Sensitivity (nA/R·sec)	0.1; 1	10; 100	10; 1; 0.1
Continuous Rate Limit (R/sec)	2300; 10	1.5; 0.03	4.4; 100; 100
Exposure Pulse Limit (mR)	80; 6	2; 0.3	3.6; 17; 17
Wall Thickness (mg/cm^2)	133; 166	166	25
Leakage Current (A)	$\pm 4 \times 10^{-15}$	$\pm 4 \times 10^{-15}$	$\pm 4 \times 10^{-15}$

Performance	Display: 4 1/2 digit LED readout Display units: R, mR, R/min, mR/min, nC and pA Decay time constant of integration capacitor: greater than 2×10^6 sec Rate response time: 2.3 sec for readings 0 to 90% Temperature stability: within 0.03% per °C

Requirements	Power: 100 to 130 V, 200 to 260 V; 50 to 60 Hz Size: 15.4 cm H × 30 cm W × 27.5 cm D (6 in. H × 12 in. W × 11 in. D) Weight: 5.5 kg (12.2 lb)
References	Manufacturer's specifications
Cost	Electrometer: $2,995 Probes: $695
Address	Victoreen 10101 Woodland Ave. Cleveland, OH 44104 (216) 795-8200 Attn: W. Snider

Radiation Survey Meters
Victoreen

Specifications	See Table A.6.23
References	Manufacturer's specifications
Address	Victoreen, Inc. 10101 Woodland Ave. Cleveland, OH 44104 (216) 795-8200

Table A.6.23. Victoreen Radiation Survey Meters

	Model	
Specifications	692	808D
Radiation Detected	X-ray and gamma above 30 keV	Gamma
Range	0–10, 0–100, 0–1,000 0–10,000 mR/hr	0.1 to 100 mR/hr (3-decade log scale)
Detector	Ion chamber	G-M tube
Energy Dependence	±10% from 45 keV to 1.3 MeV	±20% from 80 keV to 2 MeV
Precision	±15% FS	±15% FS
Power Requirement	2 "D" cells	90 to 130 V ac, 60 Hz or 12 V dc; 220 V ac available
Features	Specifically designed for nondestructive testing	Fail-safe alarms Visual and audible alarms
Overall Dimensions	30.5 cm L × 12.1 cm W × 17.1 cm H (12 in. × 4 3/4 in. × 6 3/4 in.)	10.8 cm D × 35 cm W × 23.5 cm H (4 1/4 in. × 13 3/4 in. × 9 1/4 in.)
Net weight	2.25 kg (5 lb)	5.1 kg (11 1/4 lb)
Cost	$795.00	$1850.00

Area Monitor
Xetex Model 501A-1,2,3

Class	Laboratory and field
Principle of Operation	Halogen-quenced G-M detector, digital logic, digital alarm level set, LED display
Sensitivity and Range	Range: 0.1 to 99.9 mR/hr (Model 501A-1) 1 to 999 mR/hr (Model 501A-2) 0.01 to 9.99 R/hr (Model 501A-3) Alarm range: 1 to 99 mR/hr in 1 mR steps (501A-1) 10 to 990 mR/hr in 10 mR steps (501A-2) 0.1 to 9.9 mR/hr in 100 mR/hr steps (501A-3) Gamma and x-ray
Sampling	Continuous (8-sec update period: 501A-1) Continuous (6-sec update period: 501A-2) Continuous (2.5-sec update period: 501-3)
Performance	Accuracy: ±15% referred to Cs-137 Energy dependence: ±15% of true dose rate from 70 keV to 1.3 MeV Operating temperature: −20 to 50°C (−4 to 122°F)
Requirements	Power: 115/230 V, 50/60 Hz, 15 W Size: 10.8 cm H × 16.5 cm W × 20.3 cm D (4 1/4 in. × 6 1/2 in. × 8 in.) Weight: 2.54 kg (5.6 lb) w/battery

Features	Autoranging, digital display, standby battery power, alarm tone and flashing display; all of the above models available with a remote probe*, and up to 250 ft. of cable. Also range to 99.9 R/hr available.
References	Manufacturer's specifications
Cost	$770.00 (local detector) $830.00 (remote detector)
Address	Xetex, Inc. 660 National Ave. Mountain View, CA 94043 (415) 964-3261

*Add suffix R for remote probe; e.g., 501A-1R.

Accelerator Monitor
Xetex Model 303A

303A ENERGY DEPENDENCE

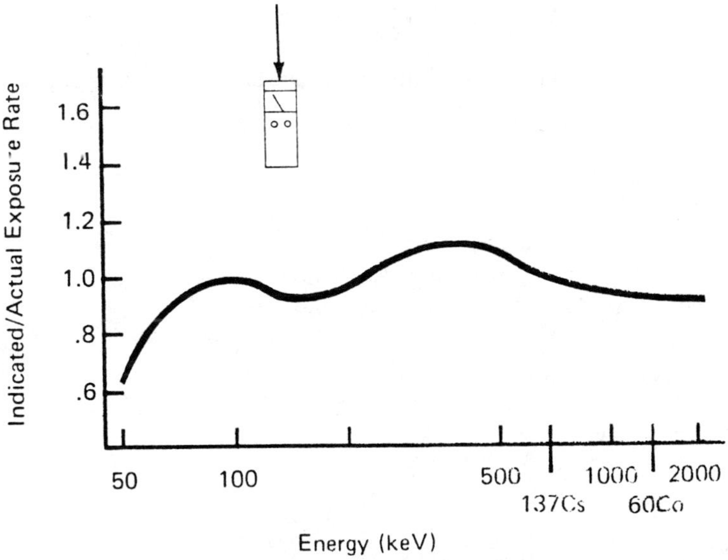

Class	Portable, subminiature
Principle of Operation	Solid state detector and associated electronics with meter readout
Sensitivity and Range	Radiation detected: x- and γ-rays Range: 0–10, 0–100, and 0–1000 mR/hr Zero drift: Less than 5% per hour on most sensitive scale (0–10 mR/hr) Response time: 4 sec or less
Sampling	Continuous
Performance	Operating temperature: 0 to 50°C (32 to 122°F) Energy dependence: ±15% Pulse capability: responds linearly to pulse peaks in excess of 1000 R/hr Accuracy: ±10% of full scale referred to ^{137}Cs
Requirements	Power: Battery 9-V transistor radio type (Mallory MN 1604); Life >200 hr Size: 7.14 cm × 15.5 cm × 3.89 cm (2.81 in. × 6.09 in. × 1.53 in.) Weight: 440 g (15.5 oz) including battery
Features	1. Conducting film on inside of box provides electrostatic shielding to prevent RF pickup 2. Calibration is facilitated by a single externally accessible multiturn potentiometer for each range 3. Power off and battery check positions provided on main switch 4. Low-power integrated circuits are used throughout
References	Manufacturer's specifications
Cost	$380.00
Address	Xetex, Inc. 660 National Ave. Mountain View, CA 94043 (415) 964-3261

High Level Probe
Xetex Model 302B

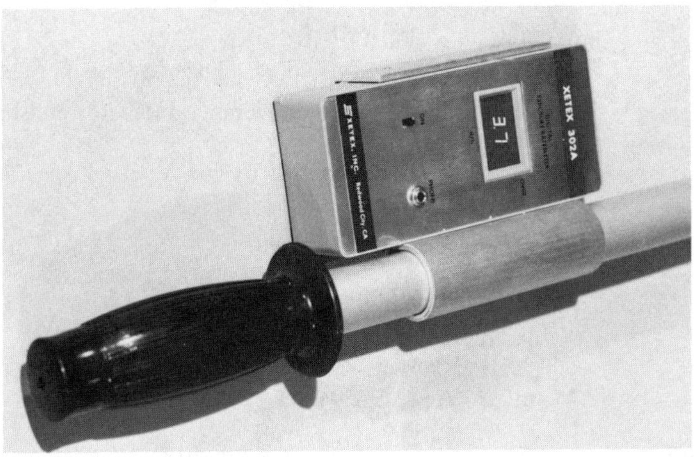

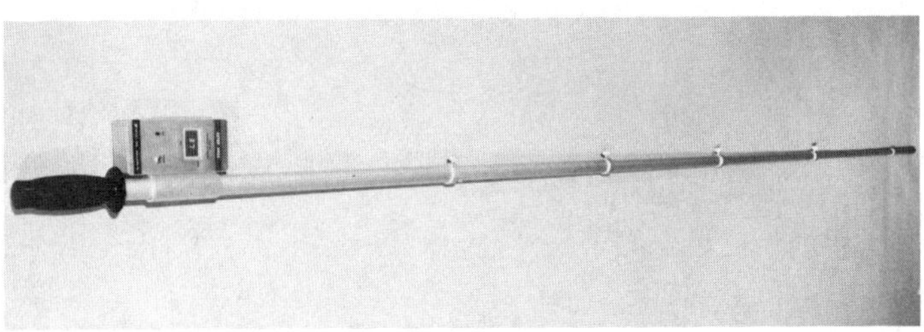

Class	Portable, handheld
Principle of Operation	Uses halogen-quenched G-M tube in waterproof housing, digital logic, and LED readout
Sensitivity and Range	x-, γ-rays Range: 0.01 to 1000 R/hr
Sampling	Continuous
Performance	Accuracy: $\pm 15\%$ referred to ^{137}Cs Energy dependence: $\pm 15\%$ from 70 keV to 1.3 MeV Response time: 3 sec or less Operating temperature: -20 to 50°C (-4 to 122°F)

Requirements	Power: Four "AA" Ni-Cad batteries, 8-hr life Size: Extended Probe—3.66 m (12 ft), collapsed probe—1.14 m (45 in.) Weight: 2.27 kg (5 lb)
Features	Autoranging Large digital display with overange indicator and battery check Telescoping probe with positive locks Rugged, lightweight
References	Manufacturer's specifications
Cost	$1210.00, including carrying case
Address	Xetex, Inc. 660 National Ave. Mountain View, CA 94043 (415) 964-3261

Emergency Digital Dosimeter
Xetex Model 418A

Class	Portable, personnel
Principle of Operation	Energy compensated halogen-quenched G-M tube detector and all solid-state electronics with 3 digit LED display.
Sensitivity and Range	Radiation detected: x-ray and gamma ray Range: 0–512 R in 1 R increments Accuracy: ±20% of indicated dose referred to 137-Cs Alarm range: 1 to 512 R presettable in 1 R steps
Sampling	Continuous
Performance	Energy dependence: ±20% from 100 keV to 3 MeV Rate response: specified accuracy to 1000 R/h Operating temperature: −20 to 50°C Case: waterproof, cast aluminum
Requirements	Power: 9 V battery (100 h operation with MN1604 alkaline) Size: 14.5 cm H × 8.0 cm W × 2.9 cm D (5.7 in. × 3.15 in. × 1.13 in.) Weight: 0.44 kgm (15.5 oz)
Features	1. High-level audible alarm with test button 2. Dose data and alarm set memory with protected resets 3. Continues to operate for rates > 1,000 R/h 4. Battery test button with go-nogo LED lamp 5. Heavy-duty belt clip 6. Optional leather carrying case with shoulder-length strap (Model 4180)
References	Manufacturer's specifications
Cost	$900.00 (418A) $ 40.00 (4180)
Address	Xetex, Inc. 660 National Ave. Mountain View, CA 94043 (415) 964-3261 Attn: Ken Sinclair

Area Monitor
Xetex Model 501A-5R

Class	Laboratory and field
Principle of Operation	Large (2 in. × 2 in.) NaI(Tl) detector, digital logic, digital alarm level set, LED display
Sensitivity and Range	Range: 1 to 999 kcpm Alarm range: 10 to 990 kcpm Gamma and x-ray
Sampling	Continuous (1 sec update period)
Performance	Sensitivity: 100 kcpm at 0.1 mR/h referred to 137-Cs Energy Response: from 30 keV Operating Temperature: −20 to 50°C
Requirements	Power: 115/230 V, 50/60 Hz and 12 V dc, 15 W Size: 10.8 cm H × 16.5 cm W × 20.3 cm D (4 1/4 in. × 6 1/2 in. × 8 in.) 24.8 cm L × 7.3 cm Dia. (probe) (9 3/4 in. × 2 7/8 in.) Weight: 3.9 kg (8.6 lb) with probe and battery
Features	Autoranging, digital display, standby battery power, alarm tone, flashing display, auto-reset; provided with 12 in. (coil cord) and 12 ft. probe cables
References	Manufacturer's specifications
Cost	$1275.00
Address	Xetex, Inc. 660 National Ave. Mountain View, CA 94043 (415) 964-3261 Attn: Ken Sinclair

RAD-GAM,XRA
Xetex 6
January 1983

Digital Dosimeter Telemetry System
Xetex Teledose (Model 503A)

Class	Portable, personnel
Principle of Operation	System consists of 5 Dosimeter/Transmitter (D/T) units and one Central Base Station (BS) unit. D/T Units: Energy-compensated halogen-quenched G-M Tube and all solid state digital electronics pulse keying FM transmitter on 1 mR dose increments BS Unit: Five channel FM receiver with 4 digit LED display for each channel
Sampling	Continuous
Sensitivity and Range	*D/T Units* Radiation detected: x-ray and gamma ray Dose range: 0–9999 mR Accuracy: ±15% of indicated dose referred to 137-Cs RF output power: 100 mW Emission: F2, pulse keying of a FM tone Frequency: 173.20375 MHz Antenna: Loop (helical whip optional) Transmitting range: 50m and beyond (unobstructed) *B/S Unit* Receiver type: superheterodyne Operating frequency: 173.20375 MHz Sensitivity: 1 μV Antenna: telescoping whip Dose displays: 0–9999 mR for each channel
Performance	*System* Energy dependence: ±15% from 60 keV to 1.3 MeV Rate response: specified accuracy to 10 R/h (all channels operating) Event signals: for each 1 mR dose increment, beep at D/T units and switch selectable chime for one or all channels at receiver Operating temperature: 0 to 50°C
Requirements	*D/T units* Power: 9 V battery (100 h operation with MN1604 alkaline in normal use) Size: 10.5 cm H × 7.6 cm W × 3.2 cm D (Aluminum case) (4 1/8 in. × 3 in. × 1 1/4 in.) Weight: 0.28 kgm (10 oz.)

B/S Unit
Power: 115/230 V, 50/60 Hz, 15 W, and 12 V dc, 12 W
Size: 27.3 cm H × 20.0 cm W × 27.9 cm D (Aluminum case)
(10 3/4 in. × 7 7/8 in. × 11 in.)
Weight: 4.1 kgm (9 lb)

Features

1. Real time dose transmission
2. Complete mobility
3. Wide variety of options:

Option	Description
5031	Remote antenna/preamp
5032	Dosimeter/transmitter alarm
5033	Doserate computer
5034	Receiver dose alarm
5035	Receiver doserate alarm

References — Manufacturer's specifications

Cost

$7875.00 (503A)
550.00 (5031)
253.00 (5032)
1800.00 (5033)
360.00 (5034)
450.00 (5035)

Address

Xetex, Inc.
660 National Ave.
Mountain View, CA 94043
(415) 964-3261
Attn: Ken Sinclair

RAD-GAM,XRA
Xetex 7
January 1983

Automatic Dosimeter System
Xetex Autodose (Models 440A/540A)

Class	Portable, personal
Principle of Operation	Dosimeter with energy compensated halogen quenched G-M tube detector, all solid state electronics, 4 digit LED display; provided with computer-controlled readout and record keeping system
Sensitivity and Range	*Dosimeter* (440A) Radiation detected: x-ray and gamma ray Range: 0–9999 mR in 1 mR increments Alarm range: 4–2048 mR in binary steps *Computer* (540A) Program: menu driven; read, issue, test, and store Users: up to 4000
Sampling	Continuous; readings push button (440A) and computer controlled (540A)
Performance	*Dosimeter* (440A) Energy dependence: ±15% from 60 keV to 1.3 MeV gamma Operating temperature: −20 to 50°C Maximum rate: 50 R/h (typical) for stated accuracy *Computer* (540A) Software: CP/U 2.2* Storage: dual floppys Display: 12 in. CRT and high-speed printer Interfaces: RS-232C Operating temperature: 10 to 40°C
Requirements	*Dosimeter* (440A) Power: 9 V battery (2 months normal use with MN1604 alkaline) Size: 13.0 cm H × 5.7 cm W × 2.5 cm D (aluminum case) (5 1/8 in. × 2 1/4 in. × 1 in.) Weight: 0.21 kg (7.5 oz) *Computer* (540A) Power: 115/230 V, 50/60 Hz, 250 W Size: 31.5 cm H × 51 cm W × 48 cm D (12 1/2 in. × 20 in. × 18 3/4 in.) Weight: 19.5 kg (43 lb)

*CP/M is registered trademark of Digital Research

Features	*Dosimeter* (440A) 1. Audible chirping with dual rate switch for 1 or 32 chirps per mR; continues above 1500 R/h 2. Go-nogo battery test lamp (LED) 3. Dose memory and protected reset *Computer* (540A) 1. Integrated desk-top 2. Selectric** style 87 key keyboard 3. Compact, rugged construction
References	Manufacturer's specifications
Cost	$205.00 (440A in 500 quantity) $14,980.00 (540A w/printer)
Address	Xetex, Inc. 660 National Ave. Mountain View, CA 94043 (415) 964-3261 Attn: Ken Sinclair

**Selectric is registered trademark of IBM

A.7. INSTRUMENTS FOR MEASURING NEUTRONS

In this section are found Instrument Notes for those commercial instruments which are capable of measuring neutrons. The filing system lists Instrument Notes *alphabetically by manufacturer.* These notes include instruments that are appropriate for neutron spectroscopy as well as dose-equivalent measurements, neutron flux measurements, and so on.

Neutron Dose-Equivalent Monitor
Eberline Radiation Monitor RM-16
Neutron REM Detector, NRD

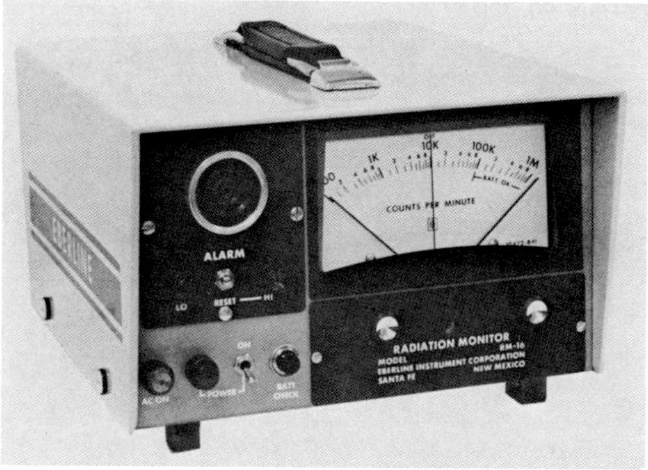

Class	Stationary
Principle of Operation	Polyethlene (cadmium-loaded) sphere of 9 in. dia., BF_3 tube inside, used with ratemeter.
Sensitivity	NRD: 1. Energy response close to rem-equivalent from thermal to about 10 MeV 2. Range 1 to 10,000 mrem/hr RM-16: 1. Input sensitivity is charge sensitive, adjustable from approximately 2×10^{-13} to 10^{-11} Coulomb. 2. Range of four logarithmic decades. 3. Window width is adjustable from 0 to at least double the input sensitivity.
Sampling	Continuous
Performance	NRD: 1. Gamma rejection up to 500 R/hr dependent on high voltage 2. About 50 cpm corresponds to 1 mrem/hr 3. Plateau approximately 300 V with a slope of 3% per 100 V 4. Directional response within ±10% 5. Operation voltage in typically 1600 to 2000 V

RM-16:
1. Response time varies continuously with count rates
2. Linearity within ±10% of reading typically, ±25% maximum
3. High voltage regulated and adjustable from less than +200 V to +2500 V
4. Temperature range is −1 to 60°C (30 to 140°F)

Requirements

	NRD	RM-16
Power:	105 to 125V, 60 Hz	105 to 125V, 50 to 60 Hz
Size:	22.9 cm D × 25.1 cm overall height (9 in. × 9 7/8 in.)	
Weight:	6.25 kg (13 3/4 lb)	

Features

1. Detector can be wall-mounted. Specify wall mounting bracket.
2. Cable up to 100 ft long from detector to electronics package
3. Alarm indication, high, red light and more penetrating tone; low blue lamp

References

Manufacturer's specifications

Cost

NRD Sphere w/cable: $990.00
RM-16 Rate Meter: $1500.00 w/meter face 10472-B42

Address

Eberline Instrument Corporation
P.O. Box 2108
Santa Fe, NM 87501
(505)471-3232

**Neutron Monitoring Instrumentation
PRS-2P/NRD**

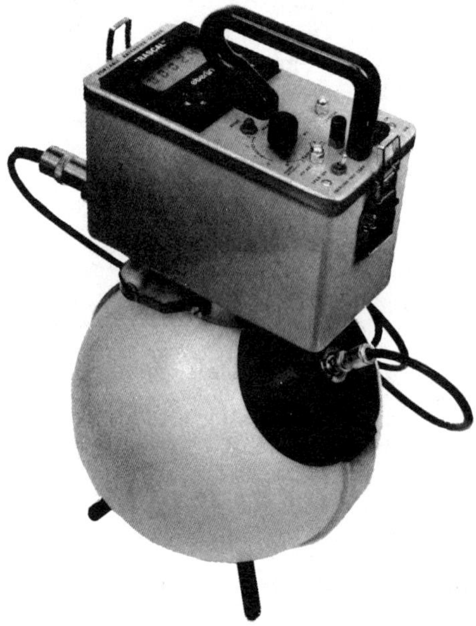

Class	Portable
Principle of Operation	BF_3 tube detector in spherical moderator measures neutron dose equivalent
Sensitivity and Range	0.2 mrem/hr to 10 rem/hr Typical rem response factors based on NBS Calibration:

Energy (keV)	2	24	144	235	515	754	1054
Cpm per mrem/hr	215	240	80	70	35	35	35

Sampling	Continuous
Performance	May be used to measure approximate dose equivalent. Rejects gamma up to 500 R/hr depending on high voltage setting. Directional response within ±10%. Operational temperature from −18 to 60°C (0 to 140°F).
Requirements	Power: Rechargeable Gel-Cell, 75 hr/charge Size: 10 cm W × 24 cm L × 20 cm H (4 in. × 9.5 in. × 7.75 in.) Weight: 2.3 kg (5 lb)

Features	Operates as count rate meter with preselected level of precision or as scaler. Multiplier factor in count rate mode converts reading to mrem/hr, digital display.
References	Eberline General Catalog
Cost	$2770.00
Address	Eberline Instrument Corporation P.O. Box 2108 Santa Fe, NM 87501 (505) 471-3232

Neutron Dose Equivalent Monitor
Harwell Type 0949-4

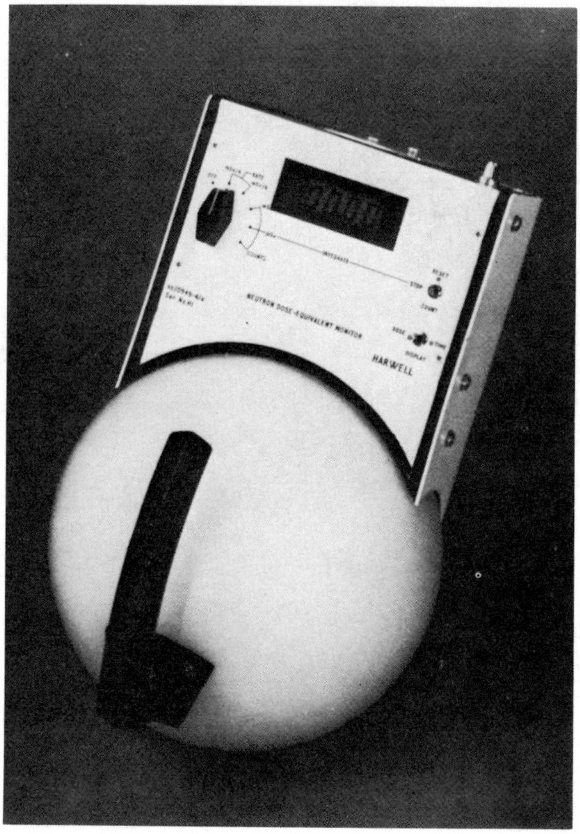

Class	Portable
Principle of Operation	The neutron detector is a polyethylene sphere, with an intermediate perforated cadmium layer, at the center of which is a spherical proportional detector with an intermediate electrostatic screen combined with a ratemeter that has a liquid crystal display
Sensitivity and Range	Energy range: thermal to 11 MeV Doserate range: 1: 1 mrem/hour to 10 rem/hr Doserate range: 2: 0.1 mrem/hr - 1 rem/hr Integrated dose range: 1: 0.1 mrem to 1 rem Integrated dose range: 2: 0.001 to 10 mrem Integrated dose range: 3: 1–10,000 counts (1 count ≡ 0.12 μrem)

Sampling	Continuous
Performance	20.8 cm diameter polyethylene sphere Accuracy: Within a factor of 2 (i.e., +100% to −50%) of the 1 ICRP recommendations (1963) for neutron energy Gamma sensitivity range: The gamma rejection ratio is > 3000:1 for gamma energies up to 7 MeV at an exposure of 1 R/hr. Response time: 0.001 rem/hr to 10 rem/hr: 1.7 sec Response time: 0.1 mrem/hr - 1 rem/hr: 17 sec Polar response: To broad beam radiation, within ±2% irrespective of instrument orientation.
Requirements	Power: Two PP6 9–V batteries, approximately 60-hr life (intermittent use) Size: 32 cm × 25 cm × 21 cm (13 in. × 10 in. × 8 in.) Weight: 6 kg (13 lb)
Features	Battery condition is continuously monitored and a low voltage is indicated by the appearance of two pulsating colon on the LCD.
References	Manufacturer's specifications
Cost	Available on request
Address	AERE, Harwell Instrumentation and Applied Physics Division Bldg. 347.3 Oxfordshire, OX11 ORA England Tele: (0235-24141, Ext. 4766) Attn: R. H. W. Sherwood

REM Meter for Pulsed Sources of Neutrons
Lawrence Livermore National Laboratory

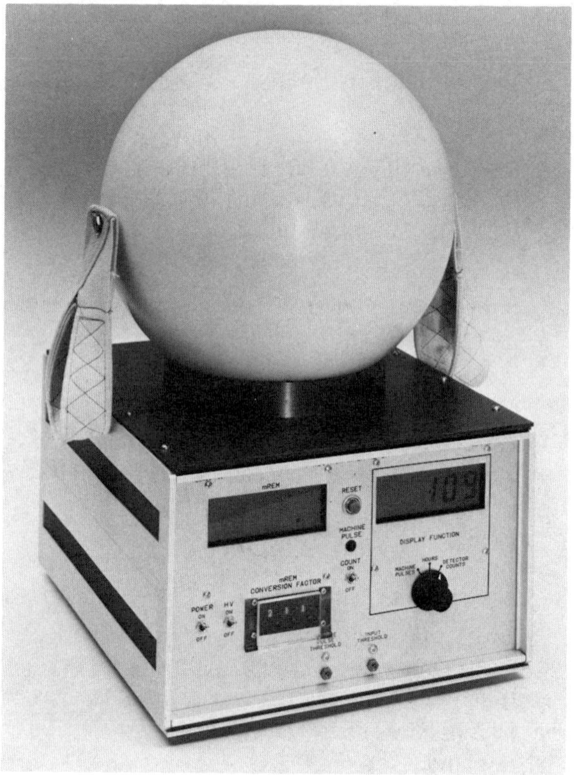

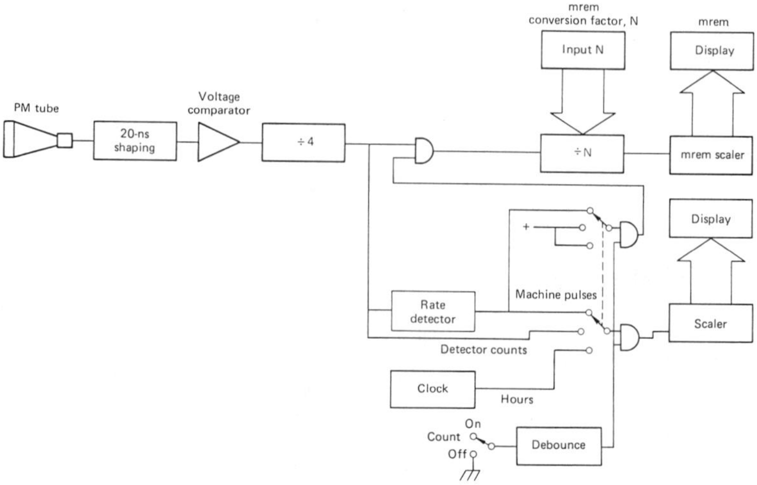

Class	Laboratory/Semi Portable
Stage of Development	Not commercially available
Principle of Operation	Fast glass scintillator thermal neutron detector enclosed in a polyethylene sphere. Neutrons thermalized by the polyethylene are detected by the scintillator. The short resolving time of the detector and electronics allows high count rates, such as those encountered from pulses of neutrons, to be counted with minimum counting losses.
Sensitivity and Range	Sensitivity: 9000 counts per millirem. Range: from background to 60 mrem per second (215 rem per hour) with less than 1% counting losses. Register will record up to 2 rem in 0.1 mrem increments.
Sampling	Continuous
Performance	Response as a function of neutron energy is similar to commercially available remmeters that have much lower count rate capabilities.
Requirements	Power: 110 V ac or rechargeable gel cell batteries. Size: 29 cm × 33 cm × 48.5 cm (11.5 in. w × 13 in. d × 19 in.) to top of sphere. Weight: 21 kg (45 lb)
Features	1. Continuously calculates mrem from detector counts and displays integrated value. 2. Can be operated manually or in a mode in which data acquisition is gated by the incident pulses of neutrons. 3. A second register can record detector counts, number of source pulses or elapsed time. 4. Calibration factors are easily entered via a 3 digit switch.
References	J. H. Thorngate, G. F. Hunt and D. W. Rueppel, "REM Meter for Pulsed Sources of Neutrons," Report UCID-18792, Lawrence Livermore Laboratory.
Cost	Not commercially available, call for construction specifications.
Address	Lawrence Livermore National Laboratory Livermore, CA 94550 Attn: J. H. Thorngate (415) 422-5216

Neutron Radiation Detectors, BF_3-He^3

Class	Portable, Laboratory
Principle of Operation	Neutron Detectors
Specifications	See Tables A.7.1 through A.7.6
References	Manufacturer's specifications
Address	LND 3230 Lawson Blvd. Oceanside, NY 11572 (516) 678-6141 Telex 14-4563

Table A.7.1. Specifications for LND Neutron Detectors

Model	Outside Diameter (cm)	Outside Diameter (in.)	Overall Length (cm)	Overall Length (in.)	Operating Voltage	General	Price
			Quadrilateral BF_3 Counting Tubes				
220	1.9 × 1.2	0.75 × 0.75	26.0	10.25	1100	Spectrometry	$850.00
221	2.9 × 2.5	1 × 1	26.0	10.25	1200	Spectrometry	850.00
223	3.8 × 3.8	1.5 × 1.5	36.2	14.25	1500	Spectrometry	1100.00
224	5.1 × 5.1	2 × 2	36.2	14.25	1500	Spectrometry	1100.00
			Cylindrical ^3He Counting Tubes				
250	2.54	1.0	132.72	52.25	1300	Flux mapping	$ 660.00
2501	0.76	0.300	13.02	5.125	1900	Time of flight	550.00
251	1.35	0.531	15.22	6.032	1000	Monitoring	380.00
2510	1.31	0.515	32.22	12.687	2200	Time of flight	690.00
2511	1.31	0.515	21.75	8.562	2100	Time of flight spectroscopy	690.00
252	1.31	0.515	36.75	14.562	2200	Time of flight spectroscopy	670.00
2520	2.62	1.031	26.04	10.25	1275	Monitoring	570.00
2521	2.62	1.031	24.45	9.625	1275	High temperature	980.00
2522	2.62	1.031	27.14	14.875	3400	Time of flight spectroscopy	730.00
2523	2.62	1.031	42.38	16.687	3400	Time of flight spectroscopy	1100.00
2524	2.62	1.031	57.62	22.687	3400	Time of flight spectroscopy	1465.00
2525	2.62	1.031	29.21	11.50	1300	Industrial gaging	1340.00
2526	2.62	1.031	34.29	13.50	1300	Industrial gaging	1590.00
2527	2.62	1.031	39.37	15.50	1300	Industrial gaging	1830.00
253	5.16	2.031	26.04	10.25	1500	Monitoring	1250.00
2531	5.16	2.031	29.53	11.625	4800	Spectrometry	1250.00
254	5.16	2.031	17.46	6.875	4800	Diffraction	1380.00

Table A.7.1. (Continued)

Model	Outside Diameter (cm)	Outside Diameter (in.)	Overall Length (cm)	Overall Length (in.)	Operating Voltage	General	Price
			Quadrilateral ^3He Counting Tubes				
262	2.54 × 2.54	1 × 1	26.04	10.15	1500	Spectrometry	$1040.00
263	5.1 × 5.1	2 × 2	36.8	14.50	2000	Spectrometry	2800.00
275	1.9 × 5.1	0.75 × 2	11.1	4.375	1700	Beam monitor	1000.00
27510	1.9 × 1.97			4.75	1900–2140	Beam monitor	2110.00
			Cylindrical BF_3 Counting Tubes				
200	2.24	0.875	132.40	52.125	1200	Flux mapping	$570.00
201	1.35	0.531	15.40	6.062	1600	Monitoring	265.00
2011	1.35	0.531	19.70	7.75	1200	Monitoring	265.00
20110	1.32	0.519	19.70	7.75	1050	Survey instruments	300.00
202	2.62	1.031	30.80	12.125	1100	General purpose	330.00
2027	2.62	1.031	30.80	12.125	1100	General purpose	330.00
2021	2.70	1.062	14.61	5.75	1600	AV/PDR-70	330.00
20210	2.62	1.031	35.48	13.97	1700	General purpose	360.00
2022	2.62	1.031	20.00	7.875	1700	General purpose	330.00
2023	2.62	1.031	40.32	15.875	1200	General purpose	360.00
2029	2.62	1.031	40.32	15.875	1200	General purpose	345.00
2024	2.62	1.031	60.64	23.875	1200	General purpose	390.00
2025	2.62	1.031	30.48	12	1200	Reactor source range	330.00
2026	2.62	1.031	76.52	30.125	1200	Reactor source range	415.00
2028	2.62	1.031	37.78	14.875	1600	Diffraction	950.00
2033	5.16	2.031	40.32	15.875	2200	General Purpose	300.00

203	5.16	2.031	40.32	15.875	2200	General Purpose	425.00
2031	5.16	2.031	27.94	11	3000	AN/PDR-58	415.00
20310	5.16	2.031	64.45	25.374	2700	General purpose	415.00
2034	5.16	2.031	30.56	12.041	1300	General purpose	425.00
212	5.16	2.031	33.63	12.031	2500	Diffraction	1100.00
2121	5.16	2.031	39.69	28.25	2500	Diffraction	1015.00

Table A.7.2. LND Neutron Detectors

Model	Max. Diameter (cm)	Max. Diameter (in.)	Overall Length (cm)	Overall Length (in.)	Operating Voltage	Sensitivity (cps/nv)	Price
			^{10}B Lined Proportional Counters				
230	1.35	0.531	As Specified		700	0.6	$1050.00
232	2.54	1.0	20.00	7.875	800	1.0	830.00
23210	2.54	1.0	38.74	15.25	950	4.0	830.00
23211	2.54	1.0	76.2	30.0	800	10	975.00
23011	1.27	0.50	29.61	11.656	600–900	0.6	830.00
23110	1.27	0.50	69.06	27.188	600–900	7.5	800.00
23113	1.27	0.50	160.82	63.313	600–950	7.5	1250.00
			Proton Recoil Fast-Neutron Detectors				
270	4.10	1.615	11.75	4.625	1500–3200		$970.00
2801	5.08	2.00	17.15	6.75	1400		625.00
2802	5.08	2.00	20.80	8.19	1400		850.00
2803	5.08	2.00	31.75	12.50	1400		800.00
2804	5.08	2.00	53.72	21.15	1400		920.00
281	3.81	1.50	25.40	10.0	2500		590.00
2821	2.54	1.0	17.15	6.75	1100		870.00
2822	2.54	1.0	20.80	8.19	1100		690.00
2823	2.54	1.0	31.75	12.50	1100		775.00
2824	2.54	1.0	53.75	21.15	1100		860.00
28114	5.1	2.0	70.00	27.75	2500		800.00
28115	5.1	2.0	85.34	33.6	2500		850.00

Table A.7.3. LND Fission Counters

Model	Max. Diameter (cm)	Max. Diameter (in.)	Overall Length (cm)	Overall Length (in.)	Operating Voltage	Remarks	Range	Price
3075	0.64	0.25	2.54	1.0	200–800	Stainless	10^3 to 10^8n	$1870.00
3076	1.27	0.5	13.97	5.5	200–800	Aluminum	10^3 to 10^8n	2000.00
3078	5.08	2.0	30.48	12	200–800	Aluminum	10^3 to 10^8n	2500.00

Table A.7.4. LND One-Dimension Position Sensitive Proportional Counters

Model	Max. Diameter (cm)	Max. Diameter (in.)	Overall Length (cm)	Overall Length (in.)	Operating Voltage	Anode	Price
25225	2.54	1.0	112.40	44.25	1000	quartz	$3500.00
25228	2.54	1.0	112.40	44.25	1000		3500.00
25229	2.54	1.0	43.18	17.00	1000	quartz	2500.00
25230	2.54	1.0	43.18	17.00	1000		2500.00

Table A.7.5. Gamma Compensated Neutron Ionization Chambers

Model	Gamma Sensitivity A/R/Hr	Size Diameter cm (in)	Size Length cm (in)	Neutron Sensitivity A/NV	Operating Voltage (V)	Price ($)
50444	1.0×10^{-12}	8.8 (3.5)	23 (9.25)	2.11×10^{-15}	200–800	1940.00
50445	3.0×10^{-12}	8.8 (3.5)	23 (9.25)	5.00×10^{-15}	200–800	1940.00
50446	6.0×10^{-12}	8.8 (3.5)	23 (9.25)	7.10×10^{-15}	200–800	1940.00
50447	7.2×10^{-12}	8.8 (3.5)	23 (9.25)	7.52×10^{-15}	200–800	1940.00
50448	9.0×10^{-12}	8.8 (3.5)	23 (9.25)	8.00×10^{-15}	200–800	1940.00
50449	1.0×10^{-12}	8.8 (3.5)	23 (9.25)	5.65×10^{-15}	200–800	1940.00
50450	3.0×10^{-12}	8.8 (3.5)	23 (9.25)	1.32×10^{-14}	200–800	1940.00
50451	6.0×10^{-12}	8.8 (3.5)	23 (9.25)	1.90×10^{-14}	200–800	1940.00
50452	7.2×10^{-12}	8.8 (3.5)	23 (9.25)	2.00×10^{-14}	200–800	1940.00
50453	9.0×10^{-12}	8.8 (3.5)	23 (9.25)	2.14×10^{-14}	200–800	1940.00

Table A.7.6. Boron Lined Neutron Ionization Chamber

Model	Gamma Sensitivity A/R/Hr	Size Diameter cm (in)	Size Length cm (in)	Neutron Sensitivity A/NV	Operating Voltage (V)	Price ($)
50454	7.50×10^{-13}	8.8 (3.5)	23 (9.25)	3.08×10^{-15}	200–800	1940.00
50455	2.25×10^{-12}	8.8 (3.5)	23 (9.25)	7.30×10^{-15}	200–800	1940.00
50456	4.50×10^{-12}	8.8 (3.5)	23 (9.25)	1.04×10^{-14}	200–800	1940.00
50457	5.40×10^{-12}	8.8 (3.5)	23 (9.25)	1.10×10^{-14}	200–800	1940.00
50458	6.75×10^{-12}	8.8 (3.5)	23 (9.25)	1.17×10^{-14}	200–800	1940.00
50459	7.50×10^{-13}	8.8 (3.5)	23 (9.25)	7.70×10^{-15}	200–800	1940.00
50460	2.25×10^{-12}	8.8 (3.5)	23 (9.25)	1.82×10^{-14}	200–800	1940.00
50461	4.50×10^{-12}	8.8 (3.5)	23 (9.25)	2.59×10^{-14}	200–800	1940.00
50462	5.40×10^{-12}	8.8 (3.5)	23 (9.25)	2.75×10^{-14}	200–800	1940.00
50463	6.75×10^{-12}	8.8 (3.5)	23 (9.25)	2.92×10^{-14}	200–800	1940.00

Survey Meter
Ludlum Models 2 and 3

Class	Portable
Principle of Operation	Count rate meter for a variety of G-M and scintillator probes.
Sensitivity and Range	Sensitivity: See Table A.7.7 Range: Model 2: 0 to 5 mR/hr, ×0.1, ×1, ×10, linear Range: Model 3: 0 to 2 mR/hr, ×0.1, ×1, ×10, ×100, linear 0 to 5,000 cpm, optional meter readout
Sampling	Continuous
Performance	Accuracy: ±5% FS Calibration stability: Less than 5% variance to battery end point
Requirements	Power: Two "D" cells, 300 hr Size: 9 cm H × 9 cm W × 18 cm L (3.4 in. × 3.5 in. × 7.0 in.) exclusive of handle
Features	Pushbutton reset, adjustable high voltage output, speaker output
References	Manufacturer's specification
Cost	Model 2: $280.00 Model 3: 295.00
Address	Ludlum Measurements, Inc. 501 Oak Street Sweetwater, TX 79556 (915) 235-5494

Ludlum Probes

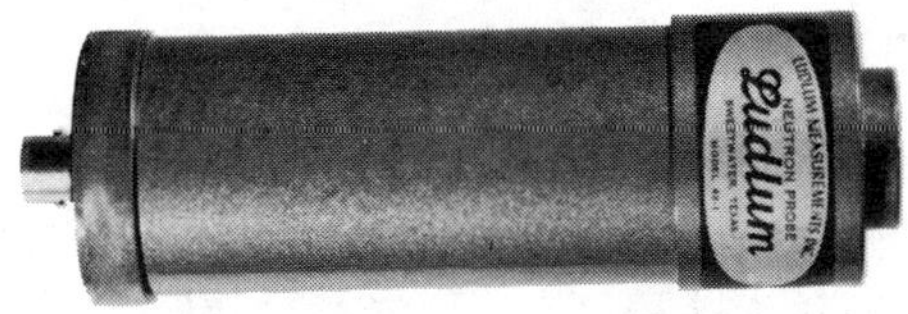

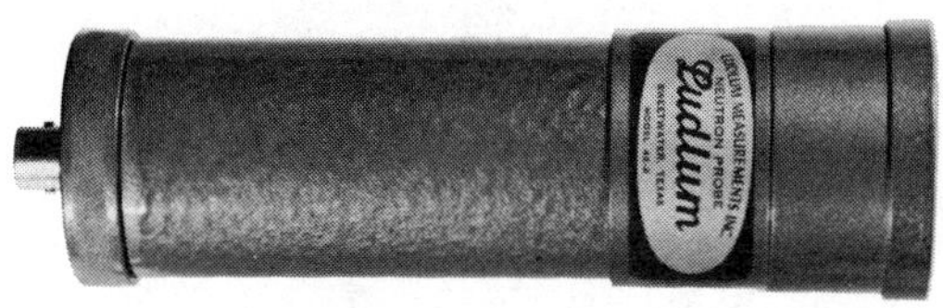

Class	Portable
Principle of Operation	ZnS(Ag) and ^{6}LiI(Eu) scintillators, and photomultiplier tubes, polyethylene sphere moderator
Specifications	See Table A.7.7
Reference	Manufacturer's specifications
Address	Ludlum Measurements, Inc. 501 Oak Street Sweetwater, TX 79556 (915) 235-5494

Table A.7.7. Ludlum Neutron Probes

Model	Type	Tube	Energy	Efficiency	Size	Weight	Cost
42-1	Scintillator, 1 in. diam., NE 400 Boron Glass Detector, with ZnS(Ag) activity, n/α reactor	RCA 6199	Thermal 0.025 eV	60 cpm = 1 n/cm^2·sec γ rejection > 10 R/hr	7 in. long 2 in. diam.	0.9 lb	$ 460.00
42-2	Scintillator, 2 in. diam., NE 404 Proton Recoil Detector, by ZnS(Ag)	RCA 6199	5 MeV	15 cpm = 1 n/cm^2·sec γ rejection > 1 R/hr	7.5 in. long 2.3 in. diam.	1 lb	380.00
42-4	Scintillator, 4mm, 4 mm Li^{6}I(Eu)	RCA 931A	Thermal to 7 MeV	Thermal 74 cpm = 1 mrem/hr 1 MeV 58 cpm = 1 mrem/hr γ rejection > 1.5 R/hr	10 in. sphere	18 lb	1070.00

Neutron Spectrometer
Ludlum Model 42-5

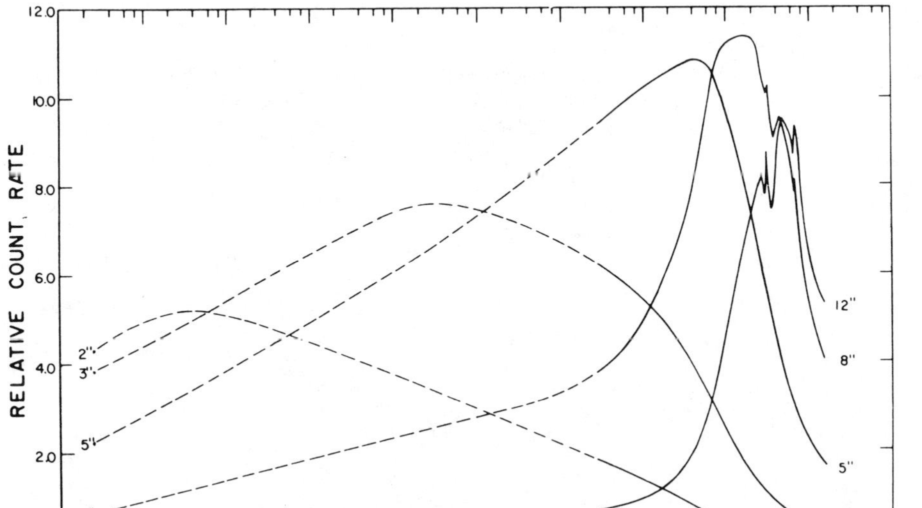

RAD-NEU
Ludlum 3
January 1983

Class	Mobile
Principle of Operation	Six different diameter spheres moderate various energy neutrons, LiI(Eu) crystal, light pipe, and RCA 6199 photomultiplier tube to be used with any counting system capable of accepting 2 mV minimum signals and providing 900 V dc to the detector.
Sensitivity and Range	Sphere diameter: 2, 3, 5, 8, 10, and 12 in. Energy ranges: See Fig. A.7.1
Sampling	Continuous
Performance	Accuracy, Efficiency, Response Time, and so on: see References
Requirements	Power: 900 V dc to be supplied by counting system Size: 112 cm H × 45 cm W × 52 cm D (45 in. × 18 in. × 21 in.) Weight: 55 kg (120 lb)
Features	X-ray inspection of all spheres insures void-free moderators. A crank-operated elevator assembly adjusts detector center from 44 to 73 in. working height.
References	1. Manufacturer's specifications 2. D.E. Hankins, "The Multisphere Neutron-Monitoring Technique," Los Alamos Report LA-3700, UC-41 Health and Safety, TID-4500, Los Alamos, NM, 1968.
Cost	$2550.00
Address	Ludlum Measurements, Inc. 501 Oak Street Sweetwater, TX 79556 (195) 235-5494

Radiation Survey Meter
Ludlum Model 15

Class	Portable, handheld
Principle of Operation	Uses moderated BF_3 detector for neutrons and end-window G-M detector for beta, gamma, and alpha, associated electronics of count rate meter, speaker, meter readout. The BF_3 detector may be used in or out of the moderator providing capability for both fast and slow neutrons.

Sensitivity and Range	Radiation detected: neutron, beta, gamma, and alpha Range: 1 to 500 cpm with multiples of ×1, ×10, ×100, ×1000
Sampling	Continuous: push-button switch for meter reset
Performance	Discrimination range: 2 to 60 mV Input impedance: 0.7 MΩ High voltage: variable from 700 to 2400 V Calibration stability: $<$ 15% variance to battery end point Meter Accuracy: ±5% FS Response time: Fast, 3 sec; Slow, 11 sec; Toggle switch selection Neutron detector Gamma rejection: Function of discrimination point. Normal 10 R/hr. Moderator: One-inch thick paraffin enclosed by a 0.030-in. thick cadmium thermal-neutron shield. Outer shell and fittings of aluminum End-window G-M detector Window: 1.5 to 2 mg/cm^2 mica Window diameter: 3.18 cm (1.25 in.) Wall: 0.318 cm (0.125 in.) stainless steel Dimension: 3.49 cm × 13.34 cm (1.375 in. × 5.25 in.)
Requirements	Power: Two "D" cells, life: 7100 hr Size: 25 cm × 9 cm × 18 cm (9.8 in. × 3.5 in. × 7 in.) Weight: 3.6 kg (8 lb)
Features	Calibration adjustment for each scale, temperature compensated. High-voltage selector switch selects fixed 900 V for G-M detector or variable high voltage for BF_3 and other detectors, calibration adjustment for each scale, temperature compensated.
Cost	$885.00
Address	Ludlum Measurement Inc. 501 Oak Street Sweetwater, TX 79556 (915) 235-5494

Neutron Counters
N. Wood Models

MN-1
MN-3
G-5-1
G-5-2
G-5-3
G-5-9
G-5-16
G-10-2
G-10-2A
G-10-5
G-10-12
G-10-12A
G-10-20
G-15-5
G-15-7H
G-15-34A
G-20-5
G-20-12

Class	Portable, Laboratory
Principle of Operation	BF_3 Neutron proportional counters
Specifications	See Table A.7.8
Address	N. Wood Counter Laboratory, Inc. 1525 E. 53rd Street Chicago, IL 60615 (312) 324-1114

Table A.7.8. N. Wood Neutron Counters

Model	Diameter (in.)	Active Length (in.)	Operating Voltage	General		Price[a-c]
MN-1	1/4	1/4 to 5	1100 to 2000 V	Pigtail lead.	Microneutron counters for	$160.00
MN-3	1/4	as	for 20 to 60 cm Hg	931/U connector	reactor instrumentation	280.00
		specified	pressure of BF_3.	(MN-3 has 5/16 in.		
				casing up to 56 in.		
				long)		
G-5-1	1		1400 to 2100 V	931/U connector	Small general	165.00
G-5-3	5/8	3	for 20 to 60 cm Hg	Caps same o.d.	purpose neutron	165.00
G-5-9	or	9		as counter	detectors	165.00
G-5-12	1/2 as	12				175.00
G-5-16	specified	16				185.00
G-10-2	1	2	1300 V for 20 cm Hg	Pigtail lead	Widely used in	165.00
G-10-2A	1	2	1800 V for 60 cm Hg	931/U connector	portable instruments.	180.00
G-10-5	1	4 1/2	1500 to 2200 V for 20	HN 82-805	Standard multiple	200.00
G-10-12	1	12 1/2	to 60 cm Hg. For	or	purpose neutron	220.00
G-10-20	1	20	lower oper. volt.	931/U connector	detectors	240.00
			specify small diam.			
			anode.			
G-15-5	1 1/2	5	1700 to 2250 V for 20	HN 82-805	Standard research	200.00
G-15-12	1 1/2	12	to 45 cm Hg. Higher	or	counters. High sensitivity	240.00
G-15-20	1 1/2	20	pressure may be	931/U connector	low background	260.00
G-15-34A	1 1/2	34	specified as well as	[G-15-34A was developed for cosmic radiation		300.00
			lower voltage.	studies (Simpson counter)]		

Table A.7.8. (Continued)

Model	Diameter (in.)	Active Length (in.)	Operating Voltage	General	Price[a-c]
G-20-5	2	5	1900 to 2300 V for 20 to 45 cm Hg. Higher pressures and lower voltages may be specified.	General-purpose high-sensitivity detectors. Useful in moisture meters and environmental monitoring. Connectors same as above models.	230.00
G-20-12	2	12			275.00
G-20-20	2	20			300.00
G-20-38	2	38			390.00

[a]Prices listed are for brass or copper cathode; prices for aluminum cathode, add $10.00; prices for stainless steel on request.

[b]Prices listed for Model G (Enriched BF_3, 97% Boron-10) are the same for Model F (Depleted BF_3, 11% Boron-10).

[c]Prices for 1983 will be approximately 10% higher.

Neutron Detectors (Portable and Slab)
National Nuclear Models SNAP-II, SD-I, and SNAP-III

Class	SNAP-II: Portable and shielded. SD-I: Large area detector. SNAP-III: Collimated neutron detector, 20:1. All are used to detect Pu.
Principle of Operation	SNAP-II: two ^{3}He detectors are used. SD-I: ^{3}He proportional counters embedded in a polyethylene moderator. SNAP-III: three ^{3}He detectors, 6 in. polyethylene moderator.
Sensitivity and Range	SNAP-II: measures Pu in the 10 to 2500 g range and uranium in the kilogram range. SD-I: Approximately 7% efficient for fission neutrons. SNAP-III: 20:1 directionality.
Requirements	Operating Voltage: SNAP-II: 1500 V SD-I: 1000 to 1200 V Preamplification: Required Cable/Connectors: Provided Size: SNAP-II: 24 cm diam. × 30.5 cm L (9.5 in. diam. × 12 in.) SD-I: 10.2 cm D × 64.7 cm H × 35.6 cm W (4 in. × 25.5 in. × 14 in.) SNAP-III: 40 cm × 48 cm × 68 cm (16 in. × 19 in. × 27 in.) Weight: SNAP-II: 6.8 kg (15 lb.) SD-I: 18.1 kg (40 lb.) SNAP-III: 150 kg (300 lb.)
References	Manufacturer's specifications
Cost	SNAP-II: $3500 SD-I: $6000 SNAP-III: $5000
Address	National Nuclear Corporation 1904 Colony Street Mountain View, CA 94043 (415) 962-9220 Attn: D. Kahn

Neutron Monitor
Nuclear Enterprise Model NM-2

Class	Portable
Principle of Operation	Combined moderator/attenuation assembly, BF_3 proportional counter, amplifier, meter readout visual.
Sensitivity and Range	Energy sensitivity: 0.025 eV to 10 MeV Range: 0.001 mrem/hr to 10 rem/hr, 5-decade log scale Gamma sensitivity: No significant response to fields up to 200 mR/hr Neutron Sensitivity: ~2.6 cps per mrem per hour
Sampling	Continuous
Performance	Detector output: normally 5 cps per 2.5 rem/hr Response: Correct rem-dose response is maintained within 20% over energy range Temperature range: -5 to 40°C (23 to 104°F) Digital Sampling time: 0–10 mrem/hr; 500 sec. 0–100 mrem/hr; 50 sec. 0–1000 mrem/hr; 5 sec. 0–10,000 mrem/hr; 0.5 sec. Pulse output: +2 V min. amplitude, 2 μsec wide pulse
Requirements	Power: 2 × 1.5 V cells U2, ~60 hour life Size: Diameter (excluding handle and legs) 22.0 cm (8.6 in.) Length (excluding handle) 40.2 cm (16 in.) Weight: 11.8 kg (25 lb)
Features	Remote scaler and recorder outputs provided
Cost	Available on request
Address	Nuclear Enterprises Bath Road, Beenham Reading RG7 5PR, England (073) 521-2121 or Spectrum Sciences 3050 Oakmead Village Drive Santa Clara, CA 95051 (408) 727-1567 Telex: 176449 Semi Group

Neutron Dose Equivalent Monitor
Nuclear Enterprises Model MK 7

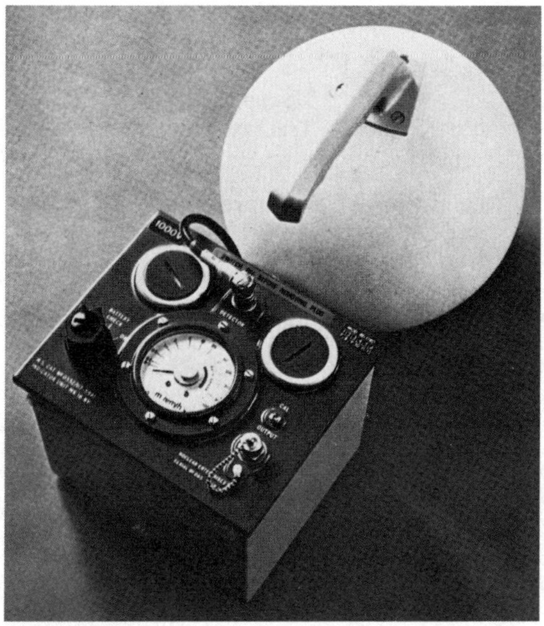

Class	Portable
Principle of Operation	The neutron detector is a polyethylene sphere with an intermediated perforated cadmium layer, at the center of which is a spherical ^{3}He proportional counter attached to a ratemeter with a visual log meter display
Sensitivity and Range	Energy range: thermal (0.025 eV) to 7 MeV Dose-equivalent rate range: 0.1 to 1000 mrem/hr
Sampling	Continuous
Performance	Correct rem response is maintained within a factor of 2%, in the thermal neutron energy range. At 14 MeV the response is down by a factor of 4; at higher energies the ratio falls but is typically 1000:1 at 10 R/hr. Gamma rejection ratio; i.e., counts per rem/hr: counts per Roentgen/hr: better than 3000:1(1R/hr, 100 keV). Output pulse: −5 to −15 V pulse 3.2 cps per mrem/hr

Requirements	Power: Batteries, 2 R20 (British type 02) Size: 38.0 cm L × 21 cm W × 25 cm H (15 in. × 8.3 in. × 10 in.) Weight: 7.3 kg (16.1 lb)
Features	Accessories: 1. 6-m extension cable between detector assembly and indicator unit Accessories: 2. Transport and storage case Output for remote scaler
References	Manufacturer's literature
Cost	Available on request
Address	Spectrum Sciences 3050 Oakmead Village Dr. Santa Clara, CA 95051 (408) 727-1567 Telex: 176449 Semi Group

Neutron Monitor
NRC RemRad Model NP-2 "Snoopy"

Class	Portable, handheld
Principle of Operation	A moderator/attenuator assembly, BF_3 detector, associated electronics used to convert detector pulse output to meter readout in mrem/hr. Scaler output for measuring total dose.
Sensitivity and Range	Sensitivity: 6000 counts/mrem Energy Range: Thermal (0.025 eV) to 15 MeV Ranges: 0–2, 20, 200, 2000 mrem/hr linear Gamma sensitivity: No response to gamma radiation in fields up to 500 R/hr of ^{137}Cs
Sampling	Continuous
Performance	Accuracy: ±10% of theoretical ICRP dose rate Temperature: −10 to 50°C (14 to 122°F) Detector resolving time: 1 μsec Directionality: +20% Detector operating voltage: 1750 to 195 V dc Background: Less than 5 cpm
Requirements	Power: Three "D" cells Size: 40 cm L × 24 cm Diam. (16 in. × 9.6 in.) Weight: 11.5 kg (25 lb)
References	1. J.O. Anderson and J. Braun, "A Neutron Rem Counter," AE-132 2. Evaluation of a Neutron Rem Dosimeter, Hazards Control Quarterly Report #18, UCLRL 12/67. 3. Manufacturer's Specifications
Cost	Price on request
Address	Nuclear Research Corporation Warrington Industrial Park Box H Warrington, PA 18976 (213) 343-5900

Fast Neutron Survey Meter
Model E-1 D

Class	Portable, handheld, fast neutrons
Principle of Operation	Hurst chamber
Sensitivity and Range	Energy range: 0.2 to 14 MeV Integrate dose range: 0.05 millirep (milliroentgen equivalent physical) Ranges: 5, 50, 500 millirep/hr NOTE: 1 cps $\simeq$ 1 millirep/hr
Sampling	Continuous
Performance	Accuracy: ±10% for Pu-Be neutrons
Requirements	Power: 6 "D" cells, 60 hr continuous Size: 25 cm × 12 cm × 15 cm (10 in. × 4 3/4 in. × 5 5/8 in.) Probe: 25 cm × 5 cm D (9 3/4 in. × 2 in.) Weight: 3.6 kg (8 lb)
Cost	Price on request
Address	Nuclear Research Corporation Warrington Industrial Park Box H Warrington, PA 18976 (215) 343-5900

Neutron Spectrometers
Ortec Spectrometers

Class

Laboratory

Principle of Operation

Two closely spaced detectors with a thin layer of neutron-sensitive material (^{6}LiF or ^{3}He) between, in a sandwich geometry. Neutrons incident on the layer give rise to charged particle pairs. The charged particles are detected in the adjacent surface barrier detectors, and the Q-value of that reaction provides a bias against background events.

Sensitivity

	525-782	525-780
Neutron-sensitive material:	150 $\mu g/cm^2$ of ^{6}LiF on foreground detector	^{3}He gas (5 atm pressure)
Energy Range:	3 MeV and above	1 to 6 MeV
Resolution: Thermal neutrons at 5 atm pressure	400 keV guaranteed 300 keV typical	100 keV guaranteed 84 keV typical
Efficiency: for thermal neutron	$\sim 10^{-3}$	$\sim 10^{-2}$
for 2.5 MeV neutron	$\sim 7 \times 10^{-7}$	$\sim 3 \times 10^{-6}$

Performance

Electronic noise: Typically $\geq$ 60 keV FWHM total system noise
Coincidence resolving time: 10 nsec/min, 30 to 40 nsec typical
Drift: about 5 keV/day at ambient temperatures

Requirements

Size: Two NIM bins, 48.4 cm W $\times$ 44.5 cm H $\times$ 43.2 cm D (19.1 in. $\times$ 17.5 in. $\times$ 17 in.)
780 Detector, 3.5 cm Diam. $\times$ 9.8 cm L (1.4 in. $\times$ 3.9 in.)
782 Detector, 3.5 cm Diam. $\times$ 2.4 cm L (1.4 in. $\times$ 0.9 in.)
Weight: 525-780, 50.5 kg (111 lb)
525-782, 44.8 kg (98.5 lb)

Features	Components: Detector head, detector bias supply. Preamplifier; linear amplifier; precision pulse generator; timing single channel analyzers; sum-invert amplifier; fast coincidence; bias amplifier, pulse stretcher, and linear gate; delay amplifier, bin and power supplies, cables.
References	Manufacturer's specifications
Cost	Model 525-780 with ^3He detector head: $22,980 Model 525-782 with ^6Li detector head: $22,670
Address	Oak Ridge Technical Enterprises Corporation 100 Midland Road Oak Ridge, TN 37830 (615) 482-4411

^{3}He Fast Neutron Spectrometer
Jordan Valley-Applied Radiation Model FNS-1

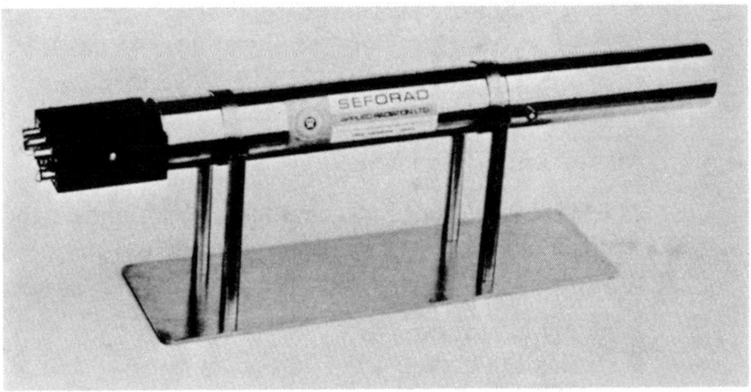

Class	Detector
Principle of Operation	The detector is an ion chamber filled with ^{3}He gas at a high pressure together with a dense gas mixture to stop the product particle from the reaction: $^3He + n \rightarrow {}^1H + {}^3H + 764$ keV. The electronics required is identical to that required by Ge(Li) spectrometry.
Sensitivity and Range	Energy range: 0.1 to 5.0 MeV Absolute efficiency at 1 MeV: 3×10^{-4} (point source at the side)
Sampling	Continuous
Performance	Energy resolution for thermal neutrons: better than 12 keV FWHM Energy resolution for 1 MeV neutron: better than 25 keV FWHM Output: Polarity, negative Amplitude, 50 mV/MeV Impedance, 50Ω Rise time, 0.2 to 5 μsec (10 to 90%) Shaping regimes, 4 μsec Gaussian Fall time, about 1 msec
Requirements	Power: preamplifier, ±24 V dc from NIM Standard Amplifier anode bias, +3000 V grid bias, +850 V Size: ion chamber, 5 cm D × 50 cm L (15 cm active) (2 in. × 20 in.) (6 in. active) preamplifier, 11 cm L × 8 cm W × 4 cm H (4.4 in. × 3.2 in. × 1.6 in.)

carrying case, 67 cm L × 15 cm W × 15 cm H (26 in. × 6 in. × 6 in.)
thermal neutron shield, 6 cm Diam. × 50 cm L (2.4 in. × 20 in.)
spectrometer stand (collapsible), 45 cm L × 15 cm W × 20 cm H (18 in. × 6 in. × 8 in.)
Weight: 7 kg (15 lb)

References	Manufacturer's specifications
Cost	\$11,000.00, includes ion chamber, preamplifier, thermal neutron shield, carrying case, pulse-height response spectrum to thermal neutrons.
Address	Peabody Scientific P.O. Box 2009 Peabody, MA 01960 (617) 535-0444 Marketers for Jordan Valley Applied Radiation

Criticality Dosimeter (CD)
Reactor Experiments Cat. No. 710

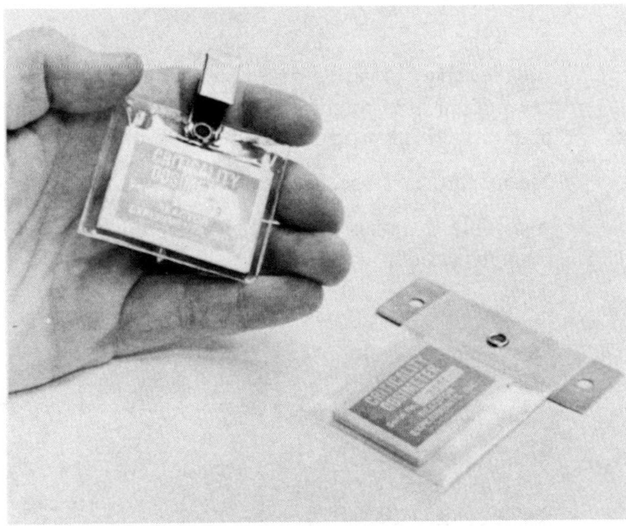

Class	Portable
Principle of Operation	Utilization of neutron activation foils; after criticality accident, incident spectra may be broken down.
Sensitivity and Range	Neutron energy intervals (foils are sensitive only to neutron) which may be obtained: Thermal to 2 eV, 0.3 to 2 eV, Thermal to 10 eV, 0.3 to 10 oV, 0.3 eV to 1 MeV, above 1 MeV, above 2.9 MeV (TLD for gamma doses greater than 0.5 R included in wall-mounted version)
Sampling	Continuous
Performance	Foils unaffected by temperature. Readout must be done on a G-M counter or multichannel analyzer. Foils include indium, gold, sulfur, cadmium-covered indium, cadmium-covered gold, and cadmium-covered copper, and a TLD using ^{7}LiF in the wall-mounted version

Requirements	Power: None required Size: 3.2 cm × 4.6 cm × 0.25 cm (1.26 in. × 1.8 in. × 0.1 in.) Weight: 14 g (1/2 oz)
Features	Inexpensive, sensitive dosimeter which is meant to be worn or used as an area monitor. The foils are sealed in a tough plastic package with a strong spring clip.
References	Manufacturer's specifications
Cost	Available in boxes of 10—Price per box Pocket version: 1–9 boxes: \$135.00 10–24 boxes: \$110.00 25–49 boxes: \$105.00 each Wall-mounted version: 1–9 boxes: \$170.00 10–24 boxes: \$140.00 25–49 boxes: \$135.00
Address	Reactor Experiments, Inc., 963 Terminal Way San Carlos, CA 94070 Don Eldred, Technical Sales (415) 592-3355

Reactor Experiments, Inc.

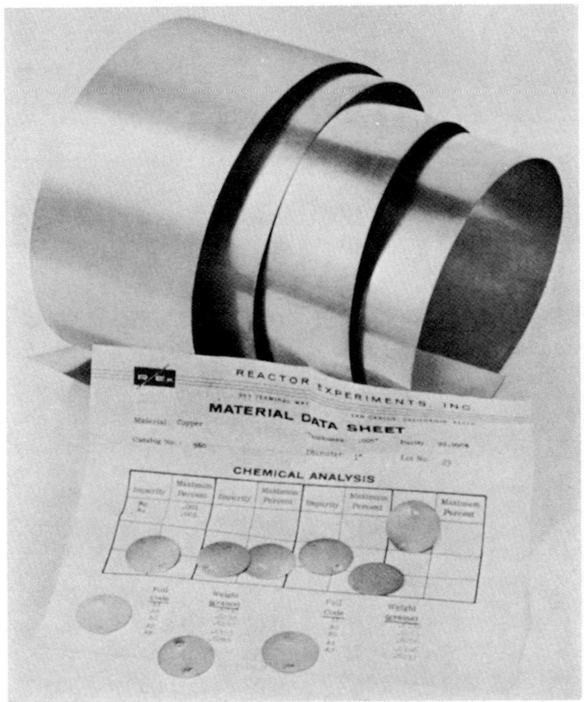

Class	Portable
Principle of Operation	Neutron dosimeter foils 12.7 mm (0.5 in.) diameter
Specifications	See Table A.7.9
Features	Each foil is thoroughly cleaned and accurately weighted to 0.0001 g with a weight code embossed directly on the foil and 28 page "Foil Instruction Manual." Flux wires, fission foils, and cadmium tubing is also available.
Address	Reactor Experiments, Inc. 963 Terminal Way San Carlos, CA 94070 (415) 592-3355

Table A.7.9. Reactor Experiments, Inc. Foils

Material	Cat. No.	Thickness (mm)	(in.)	Purity (percent)	Price per Box of 10
Aluminum	505	0.76	0.030	99.963	$105.00
Aluminum	505A	0.13	0.005	99.991	85.00
Aluminum	505B	0.05	0.002	99.999	100.00
Ammonium-Sulfate	538A	3.2	0.125	99.998	110.00
Cadmium Cover Set	531A	1.0	0.040	99.9	$145.00
Cadmium Cover Set	531	0.5	0.020	99.9	130.00
0.47% Cobalt-Aluminum	533A	0.13	0.005	99.861	260.00
Cobalt	506A	0.05	0.002	99.896	$145.00
Copper	509A	0.25	0.010	99.998	100.00
Copper	509	0.13	0.0005	99.998	105.00
Copper	509B	0.013	0.0005	99.997	$105.00
Dysprosium	581	0.025	0.001	99.995	185.00
7.9% Dysprosium-Aluminum	507B	0.013	0.004	99.9	280.00
0.155% Gold-Aluminum	546A	0.1	0.005		$265.00
Gold	521B	0.025	0.001	99.995	215.00
Hafnium	553	0.05	0.002	99.957	210.00
Indium	501	0.03	0.005	99.99	125.00
Iron	515	0.13	0.005	98.643	$155.00
5.2% Lutetium-Aluminum	508	0.1	0.004	99.963	325.00
Magnesium	514	0.13	0.005	99.781	130.00
81.3% Manganese-Copper	516	0.05	0.002	99.722	$110.00
Molybdenum	527	0.075	0.003	99.985	90.00
Nickel	513A	0.25	0.010	99.981	100.00
Nickel	513B	0.075	0.003	99.981	$100.00
Niobium	575	0.13	0.005	99.837	130.00
Rhodium	540A	0.025	0.001	99.97	250.00
Scandium	580	0.13	0.005	99.64	365.00
Silver	511	0.13	0.005	99.96	110.00
Sodium Chloride	543B	2.0	0.075	99.99	$110.00
Sulfur (1 in. diam.)	522A	4.75	0.187	99.99	105.00
Sulfur	522B	2.0	0.075	99.98	100.00
Tantalum	534	0.13	0.005	99.98	100.00
Thorium	512B	0.1	0.004	99.885	200.00

Table A.7.9. (Continued)

Material	Cat. No.	Thickness (mm)	Thickness (in.)	Purity (percent)	Price per Box of 10
Tin	551	0.1	0.004	99.57	$ 90.00
Titanium	535	0.25	0.010	99.876	180.00
Tungsten	526	0.15	0.006	99.96	210.00
Uranium (natural)	503	0.175	0.007	99.97	230.00
Uranium (depleted)	504L	0.025	0.001	99.958	335.00 for 5 foils
Vanadium	517	0.075	0.003	99.968	125.00
Zinc	552	0.13	0.005	99.99	120.00
Zirconium	541	0.13	0.005	99.82	120.00

Reuter-Stokes Neutron Detectors

Table A.7.10. Reuter-Stokes Neutron Detectors

Model	Maximum Diameter (cm)	Maximum Overall Length (cm)	Thermal Neutron Flux Range (nV)	Voltage Range (volts)
		$^{10}BF_3$ Proportional Counters		
RS-P1-0403-102	1.31	30.63	...to 3×10^4	1300 to 1600
RS-P1-0809-101	2.62	30.63	...to 3×10^4	1300 to 1600
RS-P1-0809-201	2.62	30.80	...to 3×10^4	1300 to 1600
RS-P1-0812-201				
RS-P1-1613-144	5.16	40.32	...to 4×10^3	2100 to 2800
RS-P1-1613-203	5.16	40.32	...to 4×10^3	2100 to 2800
RS-P1-1613-205				
		Helium-3 Proportional Counters		
RS-P4-0203-201 (40 atm.)	0.68	12.4	...to 5×10^4	1850 to 2100
RS-P4-0406-212 (10 atm.)	1.31	22.1	...to 10^4	1000 to 1200
RS-P4-0806-207 (4 atm.)	2.62	24.45	...to 10^4	1050 to 1250
RS-P4-0810-104 (10 atm.)	2.62	34.3	...to 3×10^3	1200 to 1500
RS-P4-0810-204 (10 atm.)	2.62	34.3	...to 3×10^3	1200 to 1500
RS-P4-0810-205 (4 atm.)	2.62	37.6	...to 4×10^3	3100 to 3900
RS-P4-1808-207 (4 atm.)	5.75	26.57	...to 2×10^3	1250 to 1450
		Helium-4 Proportional Counters		
RS-P5-0806-201 (10 atm.)				
RS-P5-0806-203 (15 atm.)				
		Fission Counters		
RS-P6-0805-134	2.62	20.00	...to 10^7	200 to 800
RS-P6-1608-110	5.16	30.80	...to 10^6	300 to 800
RS-P6-2403-114	6.35	11.91	...to 10^9	300 to 700
		B-10 Lined Proportional Counters		
RS-P7-0805-127	2.62	23.16	...to 10^5	600 to 850
RS-P7-0812-117	2.62	39.04	...to 2.5×10^4	600 to 850
RS-P7-0826-126	2.62	74.6	...to 10^3	600 to 850
		Compensated Ion Chambers		
RS-C2-1210-135	3.89	34.45	10^4 to 10^{11}	200 to 700
RS-C2-2514-115	8.02	60.32	2.5×10^2 to 2.5×10^{10}	300 to 800

Table A.7.10. (Continued)

Model	Maximum Diameter (cm)	Maximum Overall Length (cm)	Thermal Neutron Flux Range (nV)	Voltage Range (volts)
	Uncompensated Ion Chambers			
RS-C2-0808-129	2.62	23.02	10^4 to 10^{11}	100 to 700
RS-C2-2510-104	8.02	33.18	2×10^4 to 2×10^{10}	200 to 800
RS-C2-2511-137	8.02	36.83	10^3 to 10^9	500 to 1000
RS-C2-2599-125	8.02	317.82	5×10^2 to 5×10^{10}	200 to 1000
RS-C2-2599-132				
	Fission Chambers			
RS-C3-2510-114	8.02	33.18	...to 10^6	400 to 700
	In-Core Fission Chambers			
RS-C6-0201-231	0.478	5.56	...to 1×10^{14}	100
RS-C6-0201-288	0.648	5.24	...to 2×10^{14}	25 to 100

Reuter-Stokes
18530 S. Miles Parkway
Cleveland, OH 44128
(216) 581-9400
Telex: 98-5253
Attn: T. Fleming

Neutron Dose-Rate Monitor
REM-PUG

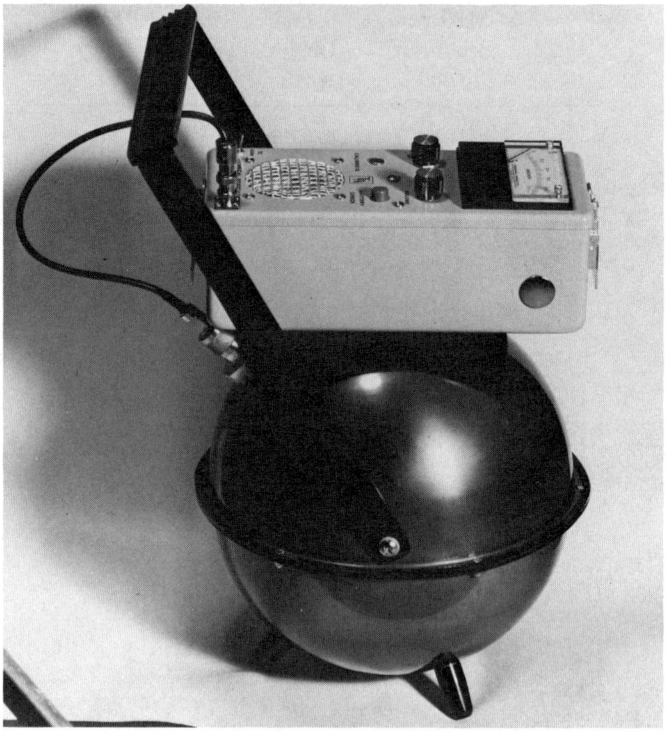

Class	Portable, survey
Principle of Operation	BF_3 proportional neutron detectors 10 in. polyethylene spherical moderator (LASL designed)
Sensitivity and Range	0.1 mrem/hr to 10 rem/hr log response
Sampling	Continuous
Performance	Follows rem curve closely
Requirements	Power: 3 "D" cells Size: 25 cm × 25 cm × 38 cm (10 in. × 10 in. × 15 in.) Weight: 10.7 kg (23.5 lb)
Features	Speaker; thermal to 7 MeV and higher; good rem response
Cost	$1700.00

Address	Technical Associates 7051 Eton Avenue Canoga Park, CA 91303 (213) 883-7043

Neutron Monitor and Neutron Probes

Class	Portable; probes used with survey meters, scalers, ratemeters
Principle of Operation	ZnS(Ag) scintillators and photomultiplier tubes, polyethylene moderator (PUG-1N)
Specifications	See Table A.7.11
References	Manufacturer's specifications
Address	Technical Associates 7057 Eton Avenue Canoga Park, CA 91303 (213) 883-7043

Table A.7.11. Technical Associates Monitors and Probes

Model	Type	Energy	Efficiency	Size	Weight	Cost
PNS-19	Scintillation; hornyak button; n,p reaction	Fast energy response ±30% from 0.25 to 14 MeV	5 cpm/neutron per cm^2/sec based on 5-MeV neutrons	2 in. diam. × 8 in. L	1 lb	$345.00
PNS-20	Scintillation; Li-6 in ZnS(Ag) matrix; neutron, alpha reaction	Thermal	60 cpm/neutron per cm^2/sec	2 in. diam. × 8 in. L	1 lb	525.00
PUG-1N Complete fast-slow neutron monitor, includes 4 ranges	Detection system includes PNS-20 plus moderator with polyethylene inner sleeve and boron-loaded plastic outer shield	Epithermal to fast		Moderator approx. 5 in. diam. × 6 in. L	8 3/4 lb total wt.	
PUG-1AB Survey Meter with attached detection system						

Neutron Detectors
Series 9300 Texlium Detectors

Class	Detector for neutron spectrometry
Principle of Operation	Texlium detectors utilize the reaction $^3He + ^1n \rightarrow ^1p + ^3H +$ 764 keV. Various gas pressures Sensitivity: thermal to fast neutrons
Sampling	Continuous
Performance	Use time: ~3 μsec maximum Operating voltage: see Figure Slope: 2%/100 V Operating temperature: −70 to 200°C (−94 to 392°F) Model numbers:

Filling Pressure	4-inch	6-inch	12-inch
1 atm	9301	9315	9329
2 atm	9303	9317	9331
4 atm	9305	9319	9333
6 atm	9307	9321	9335
10 atm	9311	9325	9339

Requirements	Power: Requires 900 to 1400 V dc between 2 and 10 atm Size: 2.54 cm (1 in.) outside diameter; (4.18 in.) + active length
Features	The 9300 series are supplied with HN type detector. Detectors with diameter ranging from 1/4 in. to 2 in. with special envelope material such as aluminum or copper, and other custom designs are available upon special request.
References	Manufacturer's specifications
Cost	Consult factory
Address	Texas Nuclear Division Ramsey Engineering Company Box 9267 Austin, TX 78766 (512) 836-0801 Tele: 77-6413

Neutron Detector
Texas Nuclear Model 9341

Class	Detector
Principle of Operation	A neutron detector using the reaction $^{3}He(n,p)T$.
Sensitivity and Range	Sensitivity: thermal and fast neutrons Energy range: 100 keV to 1 MeV, up to about 10 MeV using a pulse rise time discriminator. Active length: 15 cm (6 in.) Resolution: 7 to 10%
Performance	Gamma rejection: up to several R/hr Operating voltage: ~1900 V Temperature: −70 to 200°C (−94 to 392°F) Filling gas: 4 atm ^{3}He + 2 atm Kr Plateau length: ~500 V
Requirements	Power: Model 9341, ~1900 V dc Size: 25 cm (10.18 in.) long, including HN connector Model 9341, 1 in. diameter
References	Manufacturer's specifications
Cost	Consult factory
Address	Texas Nuclear Division Ramsey Engineering Company Box 9267 Austin, TX 78766 Tele: 77-6413

High Sensitivity Neutron Survey Meter
Victoreen Model 488A

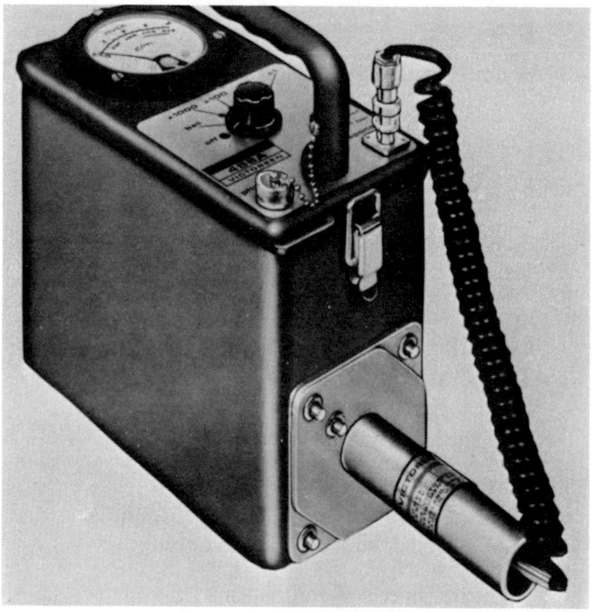

Class	Portable, handheld
Principle of Operation	Uses a 4-in. boron-lined neutron proportional counter, moderator, amplifier, ratemeter, meter readout
Sensitivity and Range	Range: Thermal to 14 MeV Thermal neutrons, four ranges: 0–800 cpm 0–8,000 cpm 0–80,000 cpm 0–800,000 cpm These ranges correspond to 0 to 12 thermal neutrons/cm^2·sec. ×1, ×20, ×100, ×1000 Gamma Sensitivity: Completely insensitive to gamma radiation fields to 500 R/hr
Sampling	Continuous
Performance	Precision: ±10% FS deflection Time response: Switch selected, 1.5 sec, 5 sec, 15 sec Moderator: high density polyethylene encased in 1/32 in. cadmium shield 2.54 cm (1 in.) wall thickness, overall moderator diameter 7.9 cm (3 1.8 in.)

	Drift: less than 5% over 24 hr Temperature operating range: −30 to 50°C (−20 to 120°F) Temperature dependence: less than 0.05%/°C Humidity range: 0 to 99%, noncondensing
Requirements	Power: Two "D" Cell Size: 30.5 cm L × 11.4 cm W × 25.4 cm H (12 in. × 4 1/2 in. × 10 in.) Weight: 3.85 kg (8.5 lb)
Features	The detector assembly may be utilized in three modes: either bare tube, tube with polyethylene moderator, or tube with moderator and cadmium shield. With these combinations the 488A detects thermal neutrons, moderated fast and thermal neutrons, and only fast neutrons respectively.
Cost	$2495.00
Address	Victoreen, Inc. 10101 Woodland Avenue Cleveland, Ohio 44104 (216) 795-8200

A.8. DOSIMETERS

The following pages contain descriptions of instruments used for monitoring doses received by personnel who are near sources of radiation. In most cases the instrument is to be worn on the person and integrates the exposure received over fairly long time periods. Descriptions are arranged alphabetically by manufacturer and include film badges, thermoluminescent detectors, ionization chambers, and chirping G-M tubes.

Stephen Pocket Dosimeter
Baird Model 907-800 Series

Class	Portable Personnel Monitor
Principle of Operation	An ionization chamber containing a moving quartz fiber electroscope, the indicating element, whose image is projected on the reticle by means of a self-contained microscope thereby measuring accumulated dose.
Sensitivity	55% efficient for 30 keV effective energy to 100% efficient for 80 keV and above for aluminum case. For brass case—55% efficient for 50 keV to 100% for 150 keV and above. Calibration accuracy: ±10% referred to radium or ^{60}Co Spectral dependency: ±20% from 80 keV to 2 MeV
Sampling	Continuous, cumulative
Performance	Detector—ionization chamber activated by the passage of ionizing radiation through the ionization chamber giving a reading on the reticle which is directly proportional to the X or gamma radiation exposure.
Requirements	Dosimeter charger ready to apply the necessary voltage charge to the positive electrode Dimensions: 11.6 cm × 1.38 cm (4 5/8 in. × 9/16 in.) Weight: aluminum 32 g (1.1 oz.) brass 40 g (1.375 oz.)
Features	Compact, light, rugged Plastic, protective end caps
Cost	Refer to manufacturer
Address	Baird Corporation 125 Middlesex Turnpike Bedford, MA 01730 (617) 276-6204 Attn: A. Doherty

Direct Reading Pen Dosimeters
Models 862, 883, 866, 608, 619, 686

Class — Personal dosimeter

Principle of Operation — Ionization chamber, quartz fiber electrometer, capacitor, and microscope

Sensitivity and Range

Model	Range	Number of Divisions	Smallest Scale Division
862*	0–200 mR	20	10 mR
883*	0–500 mR	25	20 mR
866	0–1 R	20	50 mR
865	0–1500 mR	30	50 mR
870	0–2500 mR	25	100 mR
611*	0–5 R	25	200 mR
608	0–10 R	20	500 mR
622	0–20 R	20	1000 mR
610	0–50 R	25	2000 mR
619	0–100 R	20	5 R
638	0–200 R	–	–
686	0–600 R	30	20 R

*equivalent models with mSv or mGy scales available

Sampling — Continuous

Performance — Calibration accuracy: ±10% @ midscale, ^{60}Co and ^{137}Cs
Rate dependence: ±5%, up to 10,000 R/hr
Energy dependence: ±15% of ^{60}Co response from 80 keV to ~1.3 MeV
Electrical leakage: 2% per day max. at room temperature
Operating temperature range: −10 to 50°C (+14 to 122°F)

Requirements — Power: Charger 909 on 1 battery
Size: 11.43 cm L (4 1/2 in.); 1.42 cm (9/16 in.) diameter
Weight: 35.44 g (1 1/4 oz)

Features — Hermetically sealed

References — Manufacturer's specifications

Cost — Consult factory

Address — Dosimeter Corporation
6106 Interstate Circle
Cincinatti, OH 45242
(513) 489-8100

Special Purpose Dosimeter
Models 884*, 609

Class
Personal direct reading dosimeters

Principle of Operation
Ionization chamber, quartz fiber electrometer, capacitor, microscope.

Sensitivity and Range

Model	Range	Scale Divisions	Calibration Accuracy
TE fast-neutron and gamma, Model 884	0–200 mrad	10 mrad	±10% (γ), ±20% (n)
Thermal neutron, Model 609	0–120 mrem	5 mrem	±20% (n) (thermal neutrons)

Sampling
Continuous

Performance
Rate dependence: ±5%, up to 10,000 R/hr
Energy dependence: ±20%, 80 keV to 2 MeV
Electrical leakage: 2% per day max.
Zero voltage: less than 200 V dc, max.
Operating temperature range: −10 to 50°C (+14 to 122°F)
Altitude: up to 7.620 m (25,000 ft)
Humidity: up to 98%

Requirements
Power: charged by external power source, charger Model 909
Size: 1.42 cm Diam. (9/16 in.), 11.43 cm max. length (4 1/2 in.)

Features
Hermetically sealed aluminum wall construction

References
Manufacturer's specifications

Cost
Consult factory

Address
Dosimeter Corporation
6106 Interstate Circle
Cincinatti, OH 45242
(513) 489-8100

*Recommended for those who know their fast neutron energy spectrum and can calibrate the dosimeters themselves.

Personnel Radiation Monitor
Model RT-1A

Class	Personal dosimeters
Principle of Operation	Halogen-quenched Geiger tube detects exposure rate, aural indications of field strength
Sensitivity and Range	Indications begin at ~0.1 mR/hr and increase proportionally in rate with increasing field. In high fields, individual chirps cease and a continuous tone is emitted. Operation has been tested up to 5000 R/hr. Sensitivity: 150 chirps per millroentgen (25 chirps per minute in 10 mR/hr field)
Sampling	Continuous
Performance	Operating temperature range: −23 to 54°C (−10 to 130°F)
Requirements	Power: 9-V NEDA #1604 battery, ~300-hr life Size: 3.02 cm thick × 6.03 cm W × 7.9 cm H (1 3/16 in. × 2 3/8 in. × 3 1/8 in.) Weight: 0.17 kg (6 oz)
Features	1. Improved resistance to humidity 2. Gasket-sealed aluminum case provides rugged, weather-resistant package
References	Manufacturer's specifications
Cost	$165.00
Address	Eberline Instrument Corporation P.O. Box 2108 Santa Fe, NM 87501 (505) 471-3232

TLD Dosimetry Service
Eberline

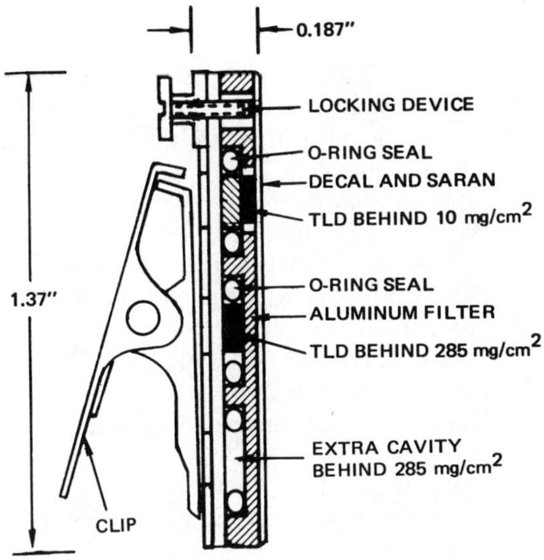

Class	Thermoluminescent Dosimetry Service (TLD)
Principle of Operation	TLD dosimeters with various parameters are supplied, returned for processing, readout, and results reported. Readout is by heating and observing thermoluminescence.
Sensitivity and Range	Exposure (mrem): 5 10 50 300 3000 30,000 Standard deviation: 20% 10% 4% 4% 4% 4%
Features	1. Badge has two TLD chips, behind 10 and 285 mg/cm^2, to enable calculation of beta or nonpenetrating x-ray dose. 2. Dosimeter contains natural lithium. 3. Wrist, ankle, ring badges also available. 4. Environmental badges (5 LiF chips) available in weatherproof bag. 5. Quarterly exchange is typical. Other cycles available.
References	Manufacturer's specifications
Cost	Consult manufacturer
Address	Eberline Instrument Corporation P.O. Box 2108 Santa Fe, NM 87501 (505) 471-3232

TLD Materials

Class	Detector
Principle of Operation	After the TLD has been exposed to radiation, it can be heated and it will luminesce. By proper calibration, the amount of light output is proportional to the radiation intensity. Materials Available: 1. *Lithium Fluoride (TLD) as—* Powder—TLD-100, 600, 700 Extrusions—TLD-100, 600, 700 in several geometries High Sensitivity Ribbons—TLD-100, 600, 700 in several geometries Cleaved Crystals—TLD-100, 600, 700 in several geometries 2. *Calcium Fluoride as—* CaF_2:Dy crystals in several geometries CaF_2:Dy powder CaF_2:Mn powder CaF_2:Mn chips in several geometries 3. *Lithium Borate as—* $Li_2B_4O_7$:Mn powder $Li_2B_4O_7$:Mn solid dosimeters in several geometries 4. *Calcium Sulfate as—* $CaSO_4$:Mn powder $CaSO_4$:Dy powder 5. *TLD Configurations including—* Bulb dosimeters Card dosimeters
Specifications	See Table A.8.1
References	Manufacturer's specifications
Cost	Consult factory
Address	The Harshaw Chemical Company 6801 Cochran Road Solon, OH 44139 (216) 248-7400

Table A.8.1. Specifications for Harshaw TLD Materials

Property/Type	LiF	$Li_2B_4O_7$:Mn	CaF_2(TLD-200)	$CaSO_4$:Mn[a,b]	$CaSO_4$:Mn[a,b]	$CaSO_4$:Dy
Density (gm/cm^3)	2.64 (powder ~1.3)	~2.4 (ribbons) ~1.2 (powder)	3.18	3.18	2.61	2.61
Effective atomic no. (Z) for photoelectric absorption	8.2	7.4	16.3	16.3	15.3	15.5
TL emission spectra	3500 to 6000 Å (4000 max.)	5300 to 6300 Å (6050 max.)	4400 to 6000 Å (5000 max.)	Peaks at 4835 Å at 5765 Å	4500 to 6000 Å	4800 Å 5700 Å
Temperature of main TL glow peak	195°C (383°F)	200°C (392°F)	260°C (500°F)	180°C (356°F)	110°C (230°F)	220°C (428°F)
Efficiency at ^{60}Co relative to LiF	1.0	0.15	10	30	70	20
Energy response 30 keV/^{60}Co	1.25	0.9	~13	~12.5	~10	~12.5
Useful range	mR–3×10^5 R	< 50 mR–10^6 + R	mR–3×10^5 R	10 μR–10^6 R	μR–10^4 R	μR–~10^5 R
Fading	Negligible[c] 5%/yr at 20°C	5% in 3 months	10% in first 24 hr 15% total in 2 weeks	10% in first 24 hr 16% total in 2 weeks	50% in first 24 hr	2% in 1 month 8% in 6 months

RAD-DOS
Harshaw 1
January 1983

Table A.8.1. (Continued)

Property/Type	LiF	$Li_2B_4O_7$:Mn	CaF_2(TLD-200)	$CaSO_4$:Mn[a,b]	$CaSO_4$:Mn[a,b]	$CaSO_4$:Dy
Physical forms	Powder extruded Teflon, glass encapsulated bulbs, cleaved crystals	Powder, solid chips, Teflon	Powder, Teflon, pressed, glass encapsulated bulbs	Powder, crystals, glass bulbs	Powder, Teflon	Powder, pressed ribbons

[a]Harshaw $CaSO_4$ is 80 to 200 mesh and dispensable; Con-Rad material very fine.
[b]The high-sensitivity TLD materials are all best handled, used, and stored in opaque containers to prevent fading from light exposure.
[c]Post-irradiation, pre-evaluation anneal for 10 minutes with LiF and 20 minutes with CaF_2(TLD-200) at 100°C normalizes these materials and eliminates fading. Other TLD materials include BeO:Mn; $CaSO_4$ (rare earth), Al_2O_3(Mn), SrF_2(Mn), $CaSO_4$(Tm), Mg_2SiO_4(Tb).

TLD Dosimetry System
Harshaw Model 2271

Class	Laboratory
Principle of Operation	Thermoluminescent dosimetry (TLD) system. TLD chips are read out by heating which generates luminescence, detected by a phototube. System is automated. TLD elements are lithium fluoride (LiF) or calcium fluoride (CaF).
Sensitivity and Range	*Exposure (Dose) Range:*

Gamma or x-rays	10 mR to 10^5 rad
Beta	40 mrad to 10^5 rad
Thermal neutrons	5 mrem to 10^5 rem
Environmental	> 1 mrad

Detectability (No background, 90% confidence level):

^{60}Cobalt gamma	10 mR ± 5 mR
Natural uranium beta	40 mrad ± 20 mrad
Thermal neutron	5 mrem ± 2.5 mrem
Environmental	1 mR ± 0.5 mR

Sampling	TLD card containing 2 elements can be read out in 40 sec
Performance	*Card configuration:*

Type G-1 gamma (no neutrons)	2 TLD-100, 1/8 in. × 1/8 in. × 0.035 in.
Type G-7 gamma (neutrons)	2 TLD-700, 1/8 in. × 1/8 in. 0.0035 in.
Type BG-1 beta/gamma (no neutrons)	2 TLD-100, 1/8 in. × 1/8 in. × 0.035 in.
Type BG-7 beta/gamma (neutrons)	2 TLD-700, 1/8 in. × 1/8 in. × 0.035 in.)
Type NG-67 neutron/gamma	One each TLD-600 and TLD-700, both are 1/8 in. × 1/8 in. × 0.035 in.
Typc G-2 gamma (environmental)	2 TLD-200 (CaF_2:Dy) 1/8 in. × 1/8 in. × 0.035 in.)
Type K-8 gamma (kilorad levels)	2 TLD-800 ($Li_2B_4O_7$:Mn) 1/8 in. × 1/8 in. × 0.035 in.

Uniformity of response:

^{60}Cobalt gamma	±15% at 100 mR
Natural uranium beta	±20% at 100 mrad
Thermal neutron	±25% at 100 mrad

Maximum error (90% confidence level):

^{60}Cobalt gamma	±7 mR or ±15% whichever is greater
Natural uranium beta	±20 mrad or ±20% whichever is greater
Thermal neutrons	±20 mrem or ±25% whichever is greater

Residual TL effect (following a 100 mR evaluation):

Equivalent gamma	5 mR
Equivalent beta	10 mrad

Reproducibility (repeatability):

Same card with 100 mR ^{60}Co	±5%
Same card with 100 mrad ^{U}nat	±10%
Reference light source	±3%

Evaluation time (two TLDs per card):

Both TLDs	< 40 sec per card
One TLD	< 30 sec per card

Features

1. System can accommodate 250 TLD cards (2,3 or 4 elements each) automatically
2. BCD identification of each card also automated
3. Manual mode also available
4. TLD reader, integrating picoammeter, flow regulator, flow meter, teletypewriter all included
5. Other systems are available: 3000, 2000, 2000C, 2000P. They can evaluate powders, solids, or bulk.

References — Manufacturer's specifications

Cost — Refer to manufacturer

Address — Harshaw Chemical Company
6801 Cochran Road
Solon, OH 44139
(216) 248-7400

Dosimeter for TLD Cards
Harshaw Model 8000 TLD System

Class	Laboratory
Principle of Operation	Badge loaded with thermoluminescence material. After exposure card heated and the light emmited is recorded by photomultiplier tubes, and stored in a memory. Card can be reused. Two minutes to need each card.
Sensitivity and Range	Exposure Range: x- or γ-rays 0.01 to 2000 rem; Low energy β-rays 0.05 to 2000 rem; High energy β-rays 0.01 to 2000 rem; Thermal neutrons 0.005 to 100 rem; Fast neutrons 0.01 to 100 rem; Environmental 0.001 rem and up Detectability: 1.25 MeV γ-rays 5 mrem; Uranium β-rays 20 mrem; 200 KeV β-rays 20 mrem; Thermal neutrons 2.5 mrem; Environmental γ-rays 0.5 mrem
Performance	Linear response with exposure in the range 0.001 to 100 rem to 5% or: ±5% or 5mrem, whichever is greater. Maximum time to process and evaluate each 4-element TLD card: 30 seconds. Maximum capacity without reloading: 480 TLD cards/holders.
Requirements	Power: 100–120 volts or 200–240 volts ±10%; 50 Hz or 60 Hz ±5% Temperature: 0 to 50°C (32 to 122°F) Size: 75 cm W × 105 cm H × 35 Cm D (30 in. × 41 in. × 14 in.) Weight: 80 kg (175 lb)
References	Manufacturer's specifications
Cost	~$100,000, depending on configurations.
Address	Harshaw 6801 Cochran Road Solon, Ohio 44139 (216) 248-7400 Attn: M. Cox/D. Elliot

Neutron Dosimeter
Harshaw DN105 PIN Diode

Class	Laboratory
Principle of Operation	The dosimeter operation is based on the fact that neutron damage in silicon tends to inhibit current flow. The diode is designed to be operated on the forward bias direction so that a current of 25 mA flows. A precision voltmeter is used to measure the voltage, V_f required to establish the current. After irradiation the value of the voltage is found to increase by ΔV_f. Using calibration data similar to Fig. 1, the neutron dose in tissue may be determined.
Sensitivity and Range	Mean value of sensitivity, V_f/D* 1.2 ± 0.1 mV/rad for AmBe neutrons
Sampling	Continuous
Performance	Mean value of V_f* 1.20 ± 0.10 V Randomness of V_f in group standard deviation 0.10 V
Requirements	Size: See figure.

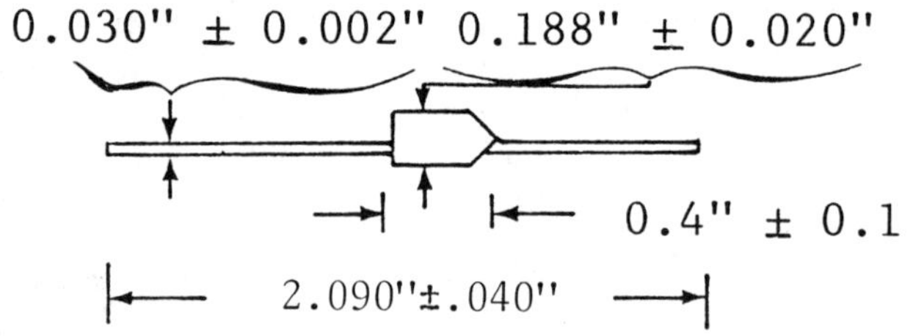

Although many instrument combinations may give good results, Harshaw recommends the following NIM module:

NC-28 Neutron Diode Analyzer
NS-12 Scaler
NQ-75 6-wide miniature NIM bin
737-216 Interconnect cable

*Specifications at 25 mA.

References	Manufacturer's specifications
Cost	Refer to factory
Address	The Harshaw Chemical Company 6801 Cochran Road Solon, OH 44139 (216) 248-7400

X-Ray Dosimeter
Nuclear Associates Model 06-002, -012

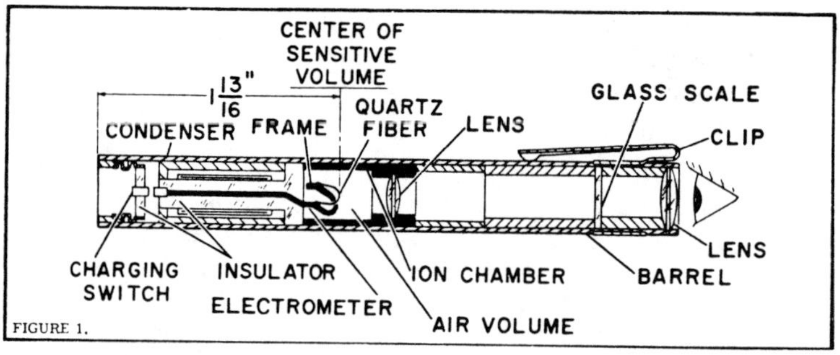

FIGURE 1.

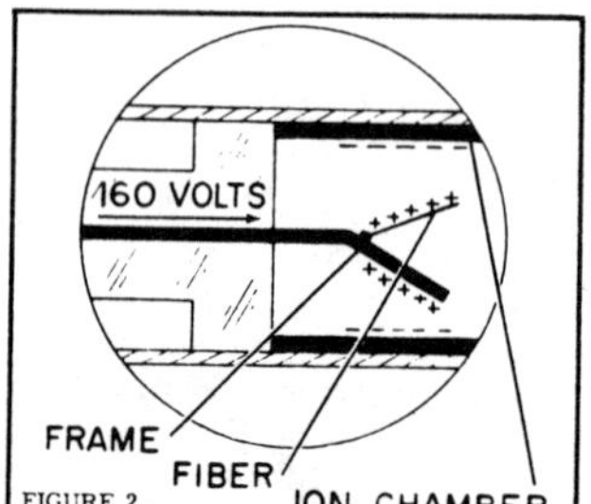

FIGURE 2.

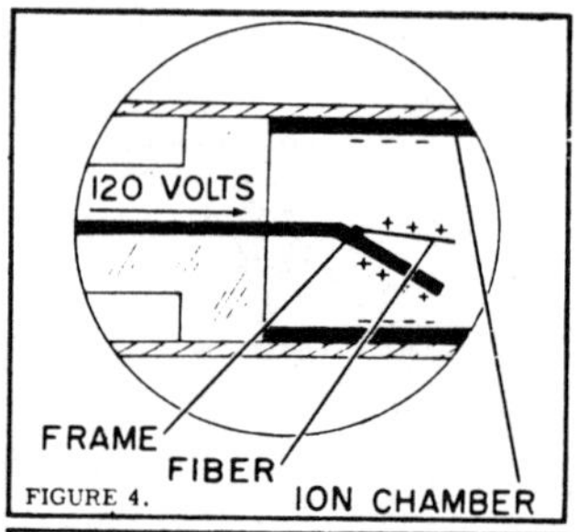

FIGURE 4.

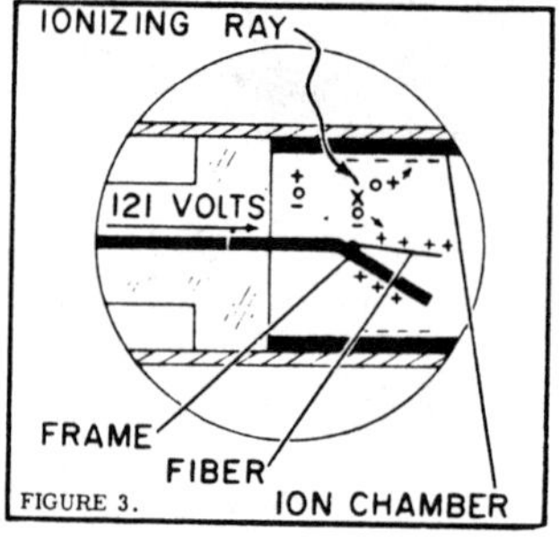

FIGURE 3.

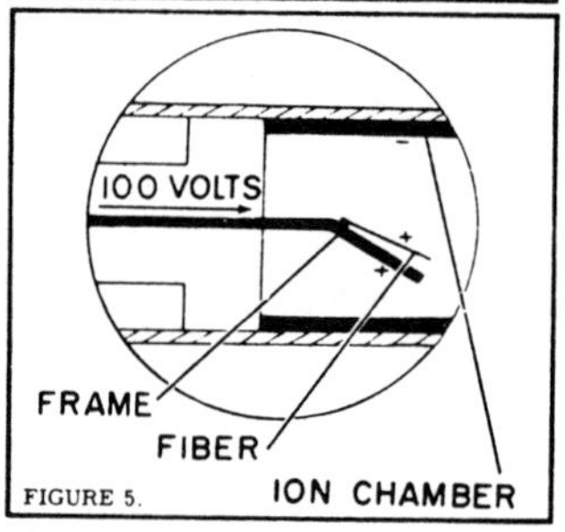

FIGURE 5.

Low Energy
Direct Reading
Dosimeter

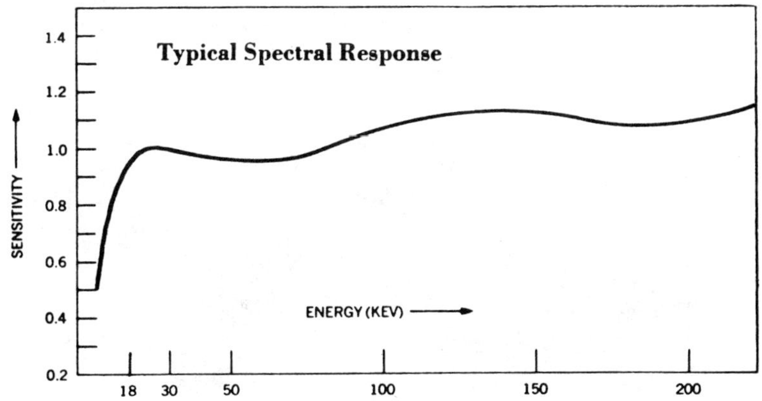

Class	Portable, pocket
Principle of Operation	0.03 in. plastic walls allow low-energy x-rays to penetrate the ion chamber.
Sensitivity and Range	Model / Range 06-002 0 to 200 mR 06-012 0 to 1.5 R
Sampling	Continuous
Performance	Accuracy: ±15% of true dose for 30-keV x-rays Energy dependence: ±10% of 30 keV reading for 15 to 100 keV ±20%/−10% of 30 keV reading for 30 to 225 keV Leakage: 2% of full scale/24 hr
Requirements	Power: Use charger Size: 10 cm L × 23.6 cm Diam. (4 in. × 9.16 in.) Weight: About 56 g (2 oz)
References	Manufacturer's specifications
Cost	06-002 $ 97.00 06-012 $195.00
Address	Nuclear Associates 100 Voice Road Carle Place, NY 11514 (516) 741-6360

Direct Reading Dosimeters
Nuclear Associates

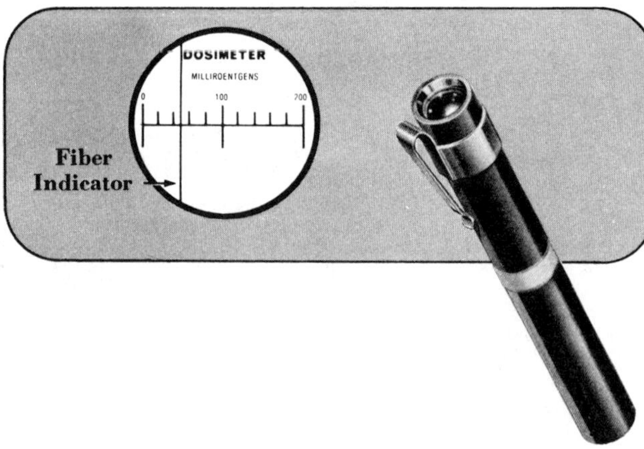

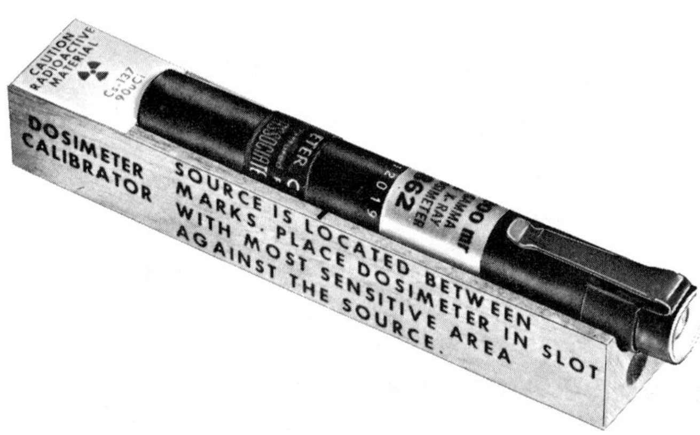

RAD-DOS
Nuclear Associates 2
January 1983

Class
Portable, pocket, fountain pen size

Principle of Operation
A quartz fiber electrometer, compound microscope calibrated readout.

Sensitivity and Range
Energy Sensitivity:

Model	Range	Cost 1–9 each
06-869	200 mR	$77.00
-883	500 mR	80.00
-866	1 R	110.00
-608	10 R	110.00
-638	200 R	110.00
-686	600 R	110.00

Sampling
Continuous

Performance
Accuracy: ±10% of dose
Leakage: 2% per 24 hr

Requirements
Power: Use charger
Size: 10 cm L × 1.5 cm D (4 in. × 9/15 in.)
Weight: About 56 g (2 oz)

Features
Optional Charger, Model 06-912: $91.00
Optional Calibrator, Model 06-200: $42.50

References
Manufacturer's specifications

Cost
See chart above

Address
Nuclear Associates
100 Voice Road
Carle Place, NY 11514
(516) 741-6360

Personal Digital Dosimeter
Nuclear Associates Model 06-502

Class	Portable, belt clip
Principle of Operation	Halogen-quenched G-M tube and digital display
Sensitivity and Range	Range: 0 to 999 mR—digital display 43 beeps per mR
Sampling	Continuous
Performance	Accuracy: ±10% with ^{60}Co at 1 R/hr
Requirements	Power: 9-V transistor battery Size: 6.35 cm × 11.43 cm × 2.54 cm (2 1/2 in. × 4 1/2 in. × 1 in.) Weight: 170 g (6 oz)
Features	Wide range: 0 to 999 mR
References	Manufacturer's specifications
Cost	$220.00
Address	Nuclear Associates 100 Voice Road Carle Place, NY 11514 (516) 741-6360

Direct Reading Dosimeter
Nucleus Model 06-884, 886

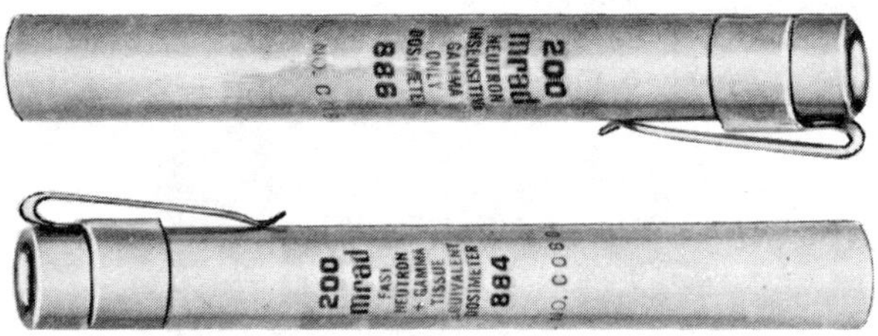

Class	Portable, pocket
Principle of Operation	Tissue-equivalent dosimeter,* measures both fast neutron and gamma dose in rads, based on collisions between fast neutrons and atoms of soft tissue with the empirical formula $C_5H_{40}O_{18}N$.
Sensitivity and Range	Neutron energies: 20 keV to 15 MeV Exposure: 20 mrads to 600 rads Gamma energies: 80 keV to 3 MeV
Sampling	Continuous
Performance	Accuracy: 20% of actual dose for neutrons 10% of actual dose for Cobalt-60 and Cesium-137 gammas 15% of actual dose for gammas from 80 keV to 3 MeV
Requirements	Power: Use charger Size: 10 cm (4 in.) in length Weight: 56 (2 oz)
Features	To measure the first-collision dose of fast neutrons, subtract the neutron-insensitive dosimeter reading (Model: 06-886, $115.00) from the tissue-equivalent dosimeter reading
Cost	$235.00 each
Address	Nucleus, Inc. P.O. Box 2561 Oak Ridge, TN 37830 (615) 482-4041

*Dosimeters that do not have the tissue equivalent feature see also available at lower cost as Models 06-002, 050, and 009.

Thermoluminescent Dosimetry Service (TLD)
Radiation Detection Company

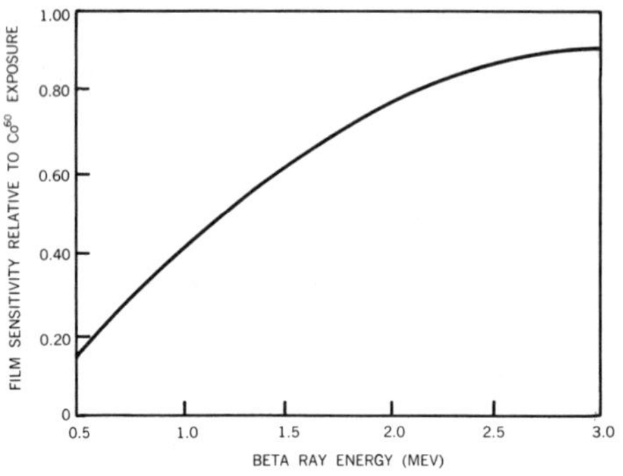

Beta Ray Sensitivity of Front Film in R-D Film Packet

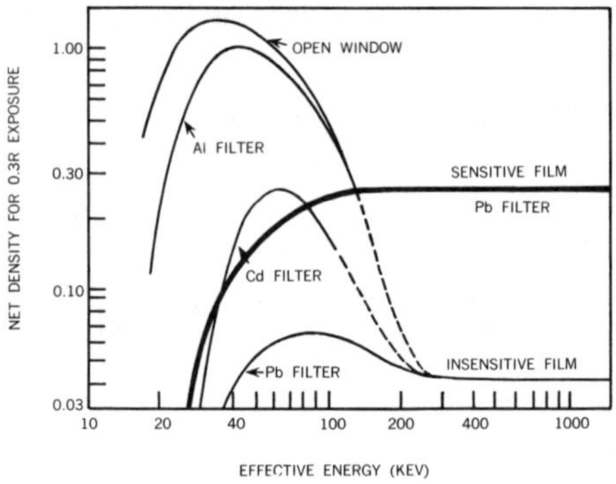

Film Densities under R-D Badge Filters for 0.3R Exposure

Class	TLD Badge Service
Principle of Operation	Badge is distributed to user, processed after exposure, exposure reading reported. Thermoluminescent dosimetry powder is sensitive to gamma, x-ray, and beta radiation.
Sensitivity and Range	(see table below)

Radiation	Energy	Dose Range
x-rays and Gamma-rays	10 keV and up	10* mrem to 10^5 rem
Beta	1MeV and up**	50 mrem to 10^5rem

*30 mrem for finger rings
**Special calibration can be provided for lower energies

Features	1. Badges cycle monthly 2. Reports mailed to user within five days of receipt of badges 3. Ring dosimeters also available 4. Albedo neutron TLD badge also available
References	Manufacturer's specifications
Cost	Upon request
Address	Radiation Detection Company 162 Wolfe Road Sunnyvale, CA 94086 (408) 735-8700 Attn: N. Goldstein

Class	Portable
Principle of Operation	Utilization of neutron activation foils; after criticality accident, incident spectra may be broken down.
Sensitivity and Range	Neutron energy intervals (foils are sensitive only to neutron) which may be obtained: Thermal to 2 eV, 0.3 to 2 eV, Thermal to 10 eV, 0.3 to 10 eV, 0.3 eV to 1 MeV, above 1 MeV, above 2.9 MeV (TLD for gamma doses greater than 0.5 R included in wall-mounted version)
Sampling	Continuous
Performance	Foils unaffected by temperature. Readout must be done on a G-M counter or multichannel analyzer. Foils include indium, gold, sulfur, cadmium-covered indium, cadmium-covered gold, and cadmium-covered copper, and a TLD using ^7LiF in the wall-mounted version
Requirements	Power: None required Size: 3.2 cm × 4.6 cm × 0.25 cm (1.26 in. × 1.8 in. × 0.1 in.) Weight: 14 g (1/2 oz)
Features	Inexpensive, sensitive dosimeter which is meant to be worn or used as an area monitor. The foils are sealed in a tough plastic package with a strong spring clip.
References	Manufacturer's specifications
Cost	Available in boxes of 10—Price per box Pocket version: 1–9 boxes: $135.00 10–24 boxes: $110.00 25–49 boxes: $105.00 each Wall-mounted version: 1–9 boxes: $170.00 10–24 boxes: $140.00 25–49 boxes: $135.00
Address	Reactor Experiments, Inc., 963 Terminal Way San Carlos, CA 94070 Don Eldred, Technical Sales (415) 592-3355

Alarming Digital Dosimeter
PDR-1B, PDR-1C, PDR-1BE

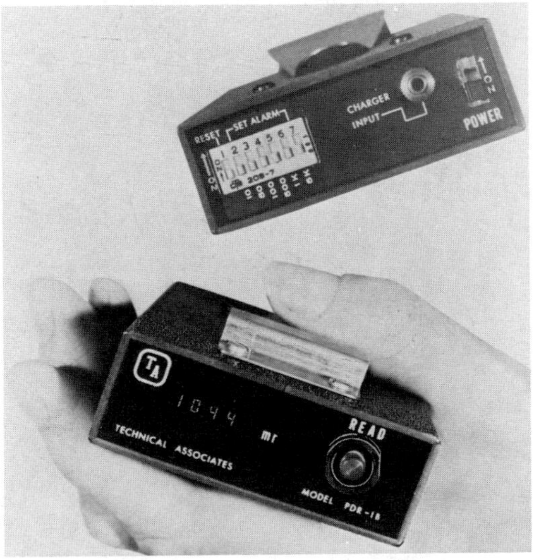

Class	Portable
Principle of Operation	PDR-1B, PDR-1C: G–M tube, audio output (chirp) and digital readout indicates dose accumulated—settable alarms PDR-1BE: detector is on cable to be worn on finger.
Sensitivity and Range	One chirp each mR—PDR-1B One chirp each tenth mR—PDR-1C Digital dose from 1 to 9999 mR—PDR-1B Digital dose from 0.1 to 999.9 mR—PDR-1C
Sampling	Continuous
Requirements	Power: 4 Ni-Cad miniature rechargeable batteries, with charger Size: 5.7 cm × 2.1 cm × 10.8 cm (2.25 in. × 0.813 in. × 4.25 in.) Weight: 196 g (7 oz)
Features	Ni-Cad rechargeable batteries Can be run off ac Memory protection—bias cells take over to protect memory in event of discharge or failure of main batteries Settable alarms

References	Manufacturer's specifications
Cost	PDR-1B: $275.00 PDR-1C: $295.00 PDR-1BE: $375.00
Address	Technical Associates 7051 Eton Avenue Canoga Park, CA 91303 (213) 883-7043

TLD Dosimetry System

Class	Thermoluminescent dosimetry (TLD) system
Principle of Operation	TL dosimeters with various parameters are read out by heating and observing thermoluminescence. Standard Badge (see Fig. A.8.4) has following filters: 1. 2 mm plastic + 1 mm Al 2. open 3. 1 mm plastic + 1 mm Cu + 1 mm Al 4. 1 mm plastic + 1 mm Pb + 1 mm Al
Sensitivity and Range	Personnel—3 mR sensitivity over 3 to 1000 R. Environmental—1 mR sensitivity over 1 to 1000 R.
Features	Model 8300 Reader 1. Integrator linear to ±1.5% over entire range 2. 6-decade display 3. Glow curve output (0 to 15 V) provided, with automatic range changing 4. Switch can enable reading of LiF, Ca:SO_4:Dy, and finger dosimeters TL Dosimeters RadiGuard card dosimeters (44.75 mm long, 31.75 wide, 0.4 mm thick) are available in $CaSO_4$:Dy-0.4H for environmental, $CaSO_4$:Dy-0.4 for personnel, beta/gamma, and in $^6LiF/CaSO_4$:Dy-0.4 for neutron monitoring.
References	Manufacturer's specifications
Cost	Model 8300 Reader $7990.00 (Other models, automatic and manual, are also available).
Address	Teledyne Isotopes 50 Van Buren Avenue Westwood, NJ 07675 (201) 664-7070

Direct Reading Dosimeters
Victoreen Model 541R/541L

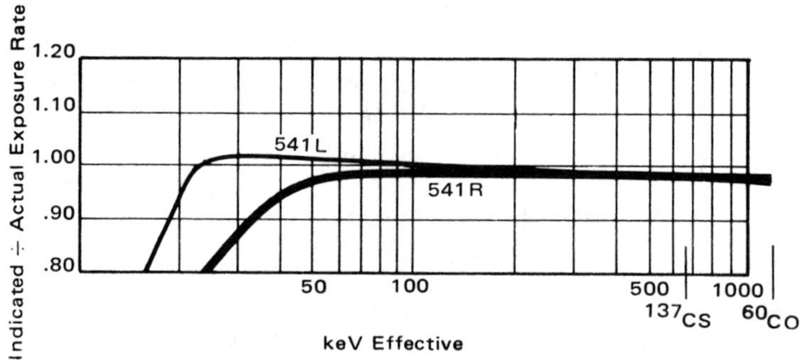

Class	Portable
Principle of Operation	Pocket ion chamber directly read on built in scale
Sensitivity and Range	Radiation detected: gamma, x-ray Energy response: 541R, 30 keV to 2 MeV 541L, 17 keV to 200 keV (see figure) Range: 0 to 200 mR Geotropism: less than 1 minor scale division
Performance	Leakage rate: 2% per 24 hours Precision: within 10% FS Environmental effects: temperature limits 10 to 50°C (14 to 122°F) humidity limit: 0 to 99%, noncondensing Typical energy response: see figure
Requirements	Power: Model 2000A charger Size: 10.2 cm L × 1.27 cm D (4 in. × 1/2 in.) Weight: 0.014 kg (0.5 oz)
Features	1. Uses Model 2000A charger 2. Meets ANSI N13.5—1972
Cost	541R: $70.00 541L: $97.00
Remarks	The mR scale is built into the dosimeter and is illuminated by an external light source

Address

Victoreen, Inc.
10101 Woodland Avenue
Cleveland, OH 44104
(216) 795-8200

Personal Dosimeter, Reader/Charger
Victoreen Model 687D Minometer II

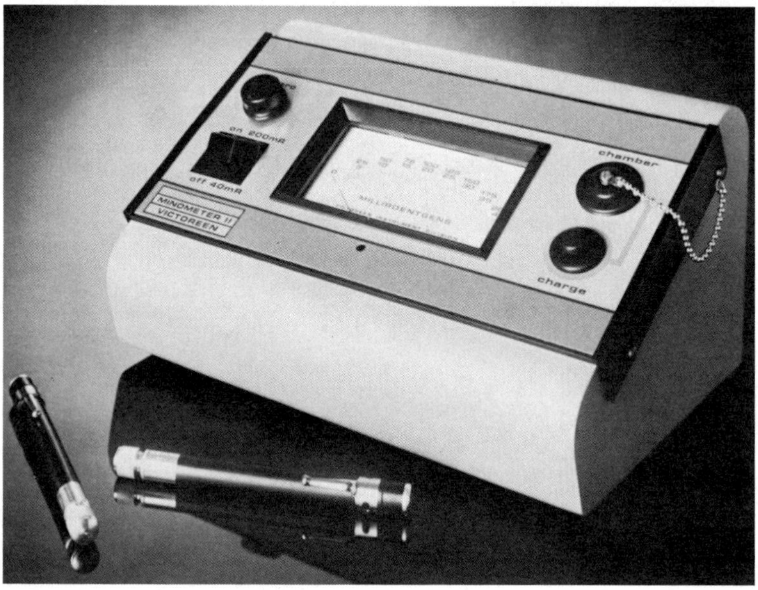

Class	Laboratory
Principle of Operation	Radiation penetrating a pocket chamber capacitor causes current to leak off in proportion to the amount of radiation. The resulting voltage drop is calibrated in milliroentgens and is read on the Minometer II.
Sensitivity and Range	Full scale ranges: 0–40, 0–200, 0–100, 0–500 mR depending upon dosimeter used
Performance	Instrument precision: Within 10% of full scale at standard temperature. Operating temperature: 10 to 35°C (50 to 95°F) Display: 10.2 cm (4 in.) meter.
Specifications	See Table A.8.3
Requirements	Power: 100 to 130 V ac, 50 or 60 Hz, 15 VA Size: 27.3 cm L × 19.1 cm W × 13.3 cm H (10.75 in. × 7.5 in. × 5.25 in.) Weight: 4.1 kg (9 lb)
Features	Reader and charger for pocket chambers

Cost $1350.00

Address Victoreen, Inc.
10101 Woodland Avenue
Cleveland, OH 44104
(216) 795-8200

Table A.8.3. Specifications for Victoreen Dosimeters

	Model	
Specifications	362	365
Radiation detected	Gamma & x-ray	Gamma & x-ray
Gamma energy dependence	within ±15% from 30 keV to 2 MeV	within ±15% from 40 keV to 2 MeV
^{137}Cs calibration	±15%	±15%
Leakage	less than 5%/day	less than 5%/day
Range	0 to 200 mR	0 to 500 mR
Overall dimensions	14 cm L × 1.27 cm Diam. (5 1/2 in. × 1/2 in.)	14 cm L × 1.27 cm Diam. (5 1/2 in. × 1/2 in.)
Temperature dependence:	within 0.16%/°C	Same
Humidity limits:	0 to 60% normal, up to 80% for limited time	Same
Weight	0.014 kg (0.5 oz)	0.014 kg (0.5 oz)
Price	$20.75	$20.75

Thermoluminescent Dosimeters
LiF and CaF_2:Mn

Typical LiF Supralinearity (^{137}Cs)

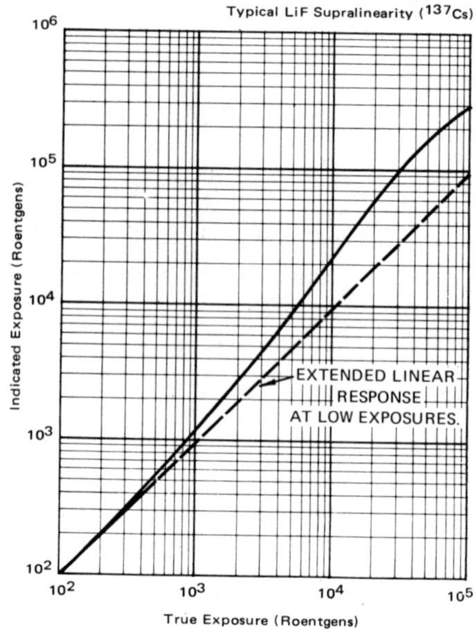

LiF Energy Response

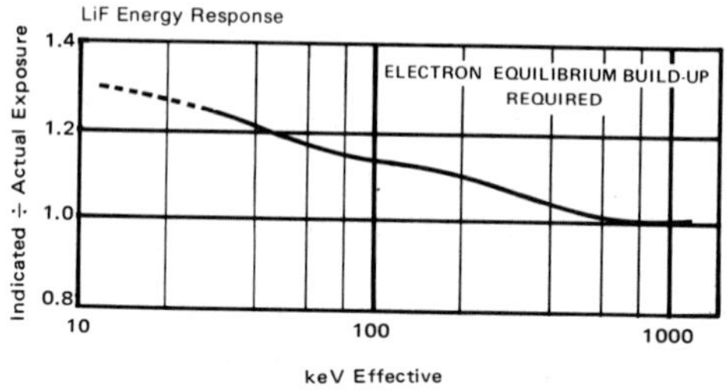

Energy Response, Bulb with Shield

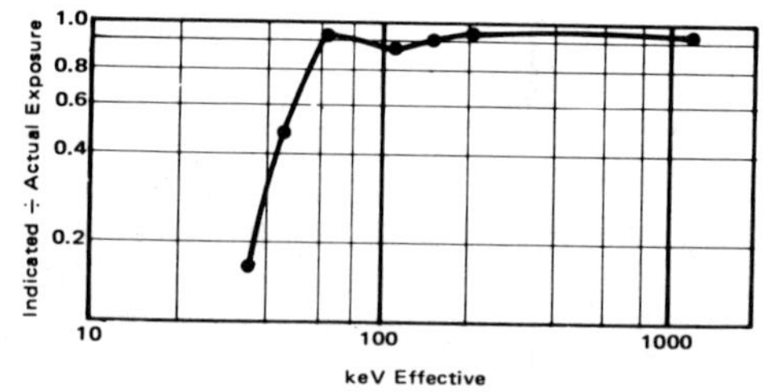

CaF_2:Mn Energy Response (Unshielded)

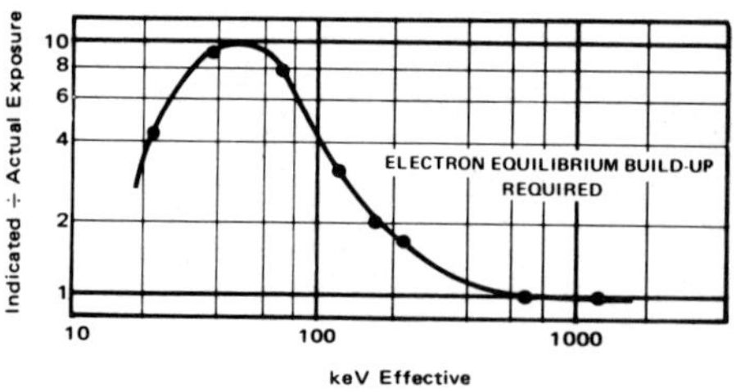

Energy Response, Chip with Shield

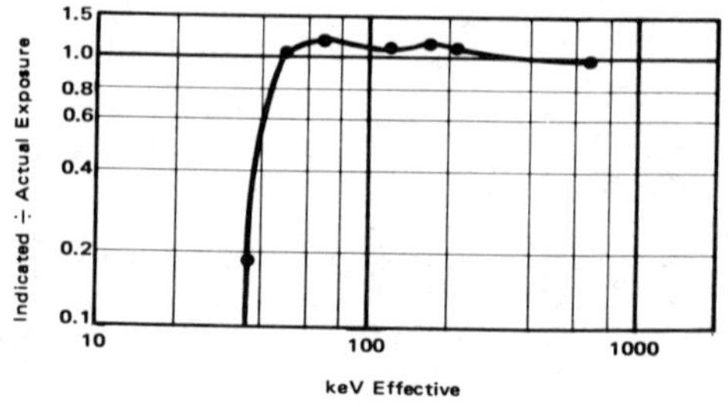

Class	Detectors
Principle of Operation	Based upon principle of thermoluminescence whereby radiation dose is related to light output as material (LiF or CaF_2:Mn) is heated.
Sampling	Integrated dose until moment of readout
Specifications	Tables A.8.4 and A.8.5
Requirements	TLD reader to readout dose accumulated Power: } related to TLD reader Size: } related to TLD reader Weight: } related to TLD reader
Features	TLD readers are also available from Victoreen
Remarks	Lithium Fluoride is supralinear above approx. 400 R See Figure. Gas Flow required below 400 mR. LiF sensitivity depends upon previous exposure history and the annealing technique used. For short-term measurement cycles (within 1 week), the recommended anneal cycle consists of: a high temperature dedose, a controlled cooling or quench plus a post irradiation anneal to empty unstable low temperature traps. Victoreen TLD Readers include automatic programming of this anneal cycle.
References	Manufacturer's specifications
Cost	Contact manufacturer
Address	Victoreen, Inc. 10101 Woodland Avenue Cleveland, OH 44104 (216) 795-8200

Table A.8.4. Specifications for Victoreen LiF Dosimeters

Model Numbers				Dose Range in		Dose Rate Dependence	
LiF	^{6}LiF	^{7}LiF	Configuration	Model 2800	Model 2810	(R/sec)	Size in mm
2600-6	2600-7	2600-8	Mini	1 mR to 15,000 R	1 mR to 10^5 R	Negligible to 10^9	12×1.4
	2600-54	26000-55	Chip	1 mR to 3000 R	< 1mR to 10^5 R	< 10% to 10^{11}	$3.2 \times 3.2 \times 0.9$
2600-70			Chip	1 mR to 3000 R	< 1 mR to 10^5 R	< 10% to 10^{11}	0.7×4.5
2600-58	26000-56	26000-57	Square Rod	1 mR to 3000 R	< 1mR to 10^5 R	< 10% to 10^{11}	$6 \times 1 \times 1$
4×10^{-8}	7.0×10^{-8}	9.0×10^{-11}	(Thermal neutron response) R/n·cm^2: ^{60}Co Rads				
2.0×10^{-10}	2.0×10^{-10}	2.0×10^{-10}	(Fast neutron response): PoBeR/n·cm^2				

Table A.8.5. Specifications for Victoreen CaF_2:Mn Dosimeters

		Dose Range				
Model	Configuration	Model 2800	Model 2810	Dose Rate Dependence (R/sec)	Linearity	Size in mm
2600-12	Mini-Round Glass tube	< 10 mR to 6,000 R	< 10 mR to 100,000 R	Negligible to 10^9	5% to 10^5 R	12 × 1.4
2600-14	Chip					3.2 × 3.2 × 0.9
2600-45	Bulb	< 5 mR to 6,000 R	< 0.1 mR to 100,000 R	< 10% to 10^{11}	3% to 10^4 R	4.3 × 11
2600-50	Square Rod	< 10 mR to 6,000 R	< 10 mR to 100,000 R			6 × 1 × 1
Neutron response R/n·cm^2:		^{60}Co Rads	Thermal = 1.2 × 10^{-10}			
Neutron response R/n·cm^2:		^{60}Co Rads	Fast = (PoBe) = 2.3 × 10^{-10}			

Mini-Gard Dosimeter and Reader
Xetex Model 403B/404A

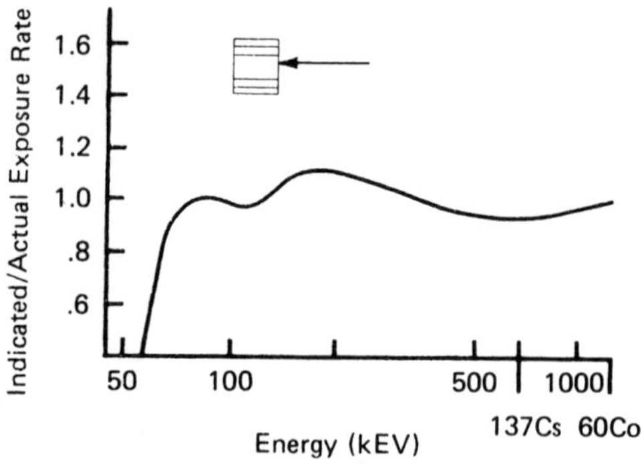

Class	Personal, portable
Principle of Operation	A small halogen-quenched, filtered G-M tube detector that reads out in chirps or continuously when preselected level reached. Also has a semiconductor memory that can be read out separately with a reader.
Sensitivity and Range	Radiation detected; x-, γ-rays Dose range: 1 to 4096 mR Energy dependence: $\pm 15\%$ from 70 keV to 1.3 MeV Maximum rate: 50 R/hr for rated accuracy
Sampling	Continuous intermittent readout
Performance	Alarm range: 4 to 1024 mR in binary steps Accuracy: $\pm 15\%$ for dose alarm or stored dose (referenced to ^{137}Cs) Memory: stores dose data until protected button depressed
Requirements	Power: Dosimeter, one 9-V battery (MN1604); Reader, 115-V, 60 Hz, 10 W. Size: Dosimeter, 57 mm × 110 mm × 25 mm (2 1/4 in. × 4 11/32 in. × 1 in.); Reader 21.8 cm × 15.5 cm × 10.2 cm (8.6 in. × 6.1 in. × 4 in.) Weight: Dosimeter, 0.17 kg (6 oz) w/battery Reader, 1.1 kg (2.5 lb)
Features	Accumulated dose can be erased, but will be retained even if turned off Dosimeter is housed in a rugged metal box with a spring clip for attachment to the belt or pocket.

References	Manufacturer's specifications
Cost	Dosimeter: $180.00 Reader: $360.00
Address	Xetex, Inc. 660 National Ave. Mountain View, CA 94043 (415) 964-3261

Digital Dosimeter
Xetex Model 409A-1,2 and Model 402A-1, 2

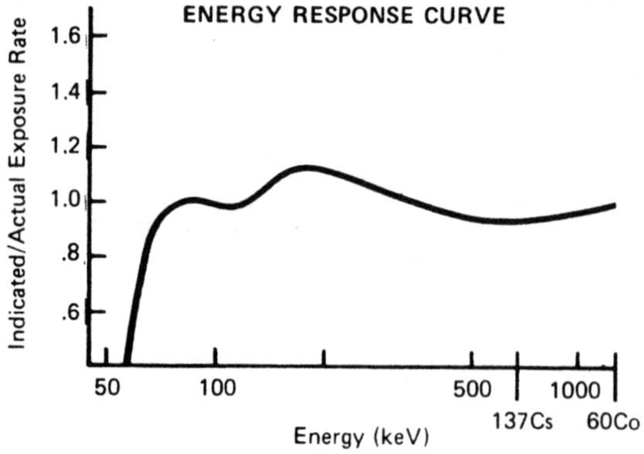

Class	Portable
Principle of Operation	A halogen-quenched, filtered G-M tube with associated electronics and four digit LED display (409A).
Sensitivity	Radiation detected: x- and γ-ray

Model	Dose Increment (mR)	Dose Range (mR)	Alarm Range (mR)	Rate* Capability (R/hr)
409A-1	1	1–9999	4–1024	50
409A-2	0.01	0.01–99.99	0.04–10.24	1
402A-1**	1	1–9999	4–1024	50
402A-2	0.01	0.01–99.99	0.04–10.24	1

Sampling	Continuous
Performance	Accuracy:† ±15% of indicated dose referred to ^{137}Cs Energy dependence:† 15% from 60 keV to 1.3 MeV Operating temperature:† 0 to 50°C (32 to 122°F)
Requirements	Power: MN1604 alkaline battery (life ~300 hr) Size: 7.6 cm × 10.5 cm × 3.2 cm (3 in. × 4 1/8 in. × 1 1/4 in.) Weight: 0.19 kg (7 oz) including battery

*For stated accuracy.
**Models 402A-1 and 402A-2 are slave display units only. They have no integral detectors and do not respond to radiation.
†409A only.

Features	Units continue to respond at high rates. Internal unit provides audible signal (beep) for each dose increment and continuous tone for alarm.
References	Manufacturer's specifications
Cost	409A-1: $280.00 409A-2: $290.00 402A-1: $160.00 402A-2: $160.00
Address	Xetex, Inc. 660 National Ave. Mountain View, CA 94043 (415) 964-3261

Digital Dosimeter
Xetex Model 415A-1, -4

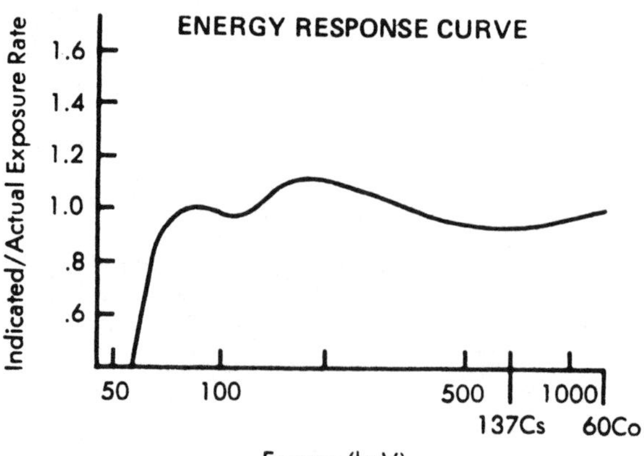

Class	Portable
Principle of Operation	Energy compensated halogen-quenched G-M tube detector and all solid state electronics with four-digit LED digital display.
Sensitivity and Range	Detector dimensions: 8 mm eff. length and ~5.5 mm diameter Radiation detected: x- and gamma-ray Range: 0 to 9999 mR in 1-mR increments Accuracy: ±15% of indicated dose referred to ^{137}Cs Alarm range: 4–2040 mR in binary steps (415A–4)
Sampling	Continuous, readings push button
Performance	Energy dependence: ±15% from 60 keV to 1.3 MeV gammas Operating temperature: −20 to 50°C (−4 to 122°F) Maximum rate: 50 R/hr (typical) for stated accuracy
Requirements	Power: Battery (6 months life w/MN1604 alkaline in normal use) Size: 57 mm × 110 mm × 25 mm (2 1/4 in. × 4 11/32 in. × 1 in.) Weight: 0.17 kg (6 oz)
Features	1. Audible signal chirping with dual rate switch to select 1 or ~40 chirps per mR; continues above 1500 R/hr 2. Go-no-go test lamp (LED) lights when battery is still good 3. Will not read zero in a high radiation field but will continue chirping above 1500 R/hr. 4. Controls: on-off switch, battery-test/display button, chirp rate switch, reset button. 5. Memory: dosimeter stores dose until protected button is depressed. 6. Low battery alarm (415A-4)
References	Manufacturer's specifications
Cost	415A-1: $195.00 415A-4: 285.00
Address	Xetex, Inc. 660 National Ave. Mountain View, CA 94043 (415) 964-3261

"Accu-Chirp" Radiation Monitor
Xetex Model 416A

Class	Portable, personal
Principle of Operation	Uses a halogen-quenched G-M tube as detector and gives an audible signal, chirping with dual rate switch to select approximately 1 or 40 chirps per mR (at ^{137}Cs energy)
Sensitivity and Range	Radiation detected: x-ray and gamma Energy range: 20 keV up Saturation rate: Continues to operate above 1500 R/hr
Sample	Continuous
Performance	Operating temperature: −20 to 50°C (−4 to 122°F) Controls: On-off switch, battery test button, chirp rate switch
Requirements	Power: One 9-V battery (MN1604) Size: 57 mm × 110 mm × 25 mm (2 1/4 in. × 4 11/32 in. × 1 in.) Weight: 0.17 kg (6 oz)
Features	Packaged in a rugged, anodized aluminum case with positive clip. Low cost
References	Manufacturer's specifications
Cost	$135.00
Address	Xetex, Inc. 660 National Ave. Mountain View, CA 94043 (415) 964-3261

A.9. INSTRUMENTS FOR MEASURING SPECIFIC RADIONUCLIDES

Included in the following pages are Instrument Notes for those instruments useful for *radionuclide measurements* of various kinds. The reader should be aware of the following facts, in order to find all Instrument Notes appropriate for radionuclide measurements.

1. The filing system lists Instrument Notes *alphabetically by manufacturer*.
2. Instruments for general *spectrometry* are found elsewhere: under RAD-ALP, RAD-BET, or RAD-GAM, as appropriate. These instruments are often similar to instruments found in this section (RAD-NUC), the distinction usually being based on the manufacturer's advertised specifications. Thus if a manufacturer claims that an instrument is useful for a specific radionuclide, or for a broad class of nuclides (e.g., "gases"), it is found here; general spectrometry instruments are elsewhere.
3. All *liquid-scintillation* systems are in this section, rather than under beta detection (RAD-BET).

HT/HTO Monitor
AECL/Chalk River

Class	Laboratory area monitor.
Stage of Development	Not commercially available.
Principle of Operation	HTO is separated from HT by permeation distillation. Each species is then measured separately in a flow through ionization chamber.
Sensitivity and Range	For each species, 0.1 to 10^6 MBq/m^3. HT contribution to HTO signal, less than 1 part in 5×10^3. HTO contribution to HT signal, less than 1 part in 10^2. Time response to HTO: ~10 min. Time response to HT: ~2 min.
Requirements	Power: 115 Vac, 60 Hz. Water: 1 liter/min at 60 psig. Compressed Air: 12 liter/min at 10 psig. Size: Fills one relay rack. Weight: Several hundred pounds.
Features	Real time separate determination of HT and HTO. Extensive fault detection.
Reference	R.G.C. McElroy, R.V. Osborne, and R.A. Surette, "A monitor for the separate determination of HT and HTO." IEEE Trans. Nucl. Sci. *NS29*, 816–818 (1982).
Address	R.G.C. McElroy or W.A. Seddon Atomic Energy of Canada Ltd. Chalk River Nuclear Laboratories Chalk River, Ontario KOJ 1JO Canada (613) 584-3311 (613) 687-5581 Telex: 053-34555

Liquid Scintillation Systems
Beckman LS5800/LS9800 Series

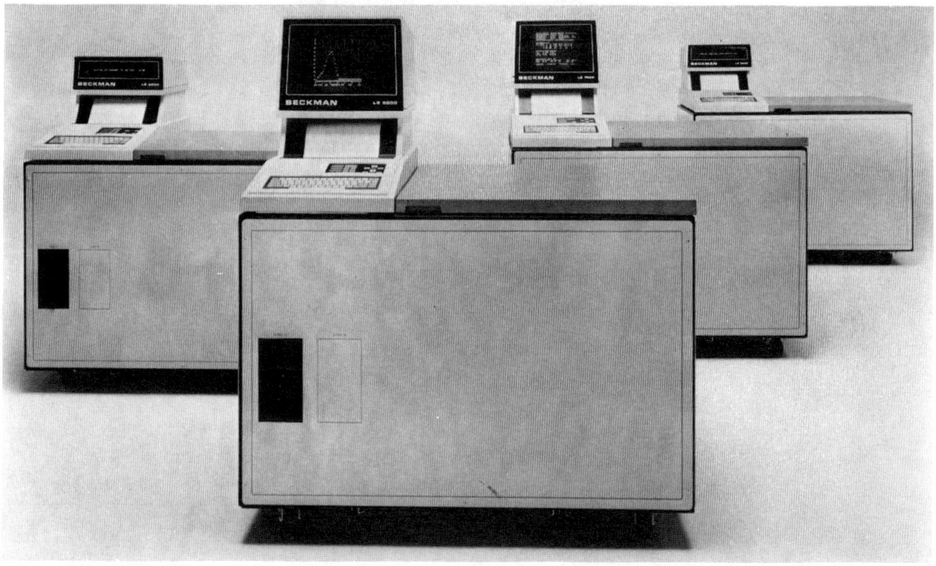

Class:	Laboratory
Principle of Operation	A radioactive decay event in a liquid scintillation cocktail produces a large number of photons. These are detected by multiplier photo tubes. A coincidence circuit using true multiplier phototubes will prevent any single event recorded by only 1 phototube to be recorded as a count. This reduces background noise.
Sensitivity and Range	Counting efficiency $^{3}H > 60\%$ $^{14}C > 95\%$ ^{14}C above $^{3}H > 70\%$ in window; exceeding 1% ^{3}H end point.
Performance	LS5800 Series: two sets of racks designed for standard and miniature vials. Sample Capacity: 336 to 648 samples. Command card programming. LS9800 Series: Sample Capacity: 300 (miniature or standard vials). Bidirectional serpentine belt. Command tower programming with second tower editing.

Features

Both LS5800 and LS9800 series have:
- Microprocessor control.
- Keypad editing.
- Display: digital, marquel or CRT
- Three counting channels.
- Auto-calibration with self-correction and verification.
- Sample channels ratio, selectable
- CPM/counts calculation program with normalization factor, replicate averaging, and half-life correction.
- Spectrum search.
- High count rate terminator.
- Time-of-day clock.
- Diagnostics

Options

Refer to manufacturer.

Requirements

Instrument	Weight kg (lb)	Length cm (in.)	Height cm (in.)	Depth cm (in.)	Power
LS6800	318	119.4	124.5	76.2	120V,4A, 60Hz
Series	(700)	(47)	(49)	(30)	240V,2A, 50Hz
LS 7800/LS9800	318	119.4	139.7	76.2	120 V,4A, 60Hz
	(700)	(47)	(55)	(30)	240 V,2A, 50Hz
With	355				120V,12A, 60Hz
Refrigeration	(780)				240V,6A, 50Hz
LS1800 Series	191	91.4	66	80	120V,3A, 60Hz
Basic Instrument	(420)	(36)	(26)	(31.5)	240V,1.5A, 50Hz
LS2800	21	48.3	45.7	66	120V,1A, 60Hz
Readout Device	(47)	(19)	(18)	(26)	240V,0.5A, 50Hz
LS3800	27	48.3	58.4	66	120V,1A, 60Hz
Readout Device	(60)	(19)	(23)	(26)	240V,0.5A, 50Hz

LS5800	43.7	48.3	71.1	66	120V,2A, 60Hz
Readout Device	(96)	(19)	(28)	(26)	240V,1A, 50Hz

Lead Shielding	LS5800 Series	LS9800 Series
Weight	133 kg (292 lb)	104 kg (230 lb)
Minimum Thickness	5.1 cm (2 in.)	5.1 cm (2 in.)

References Manufacturer's specifications.

Cost Refer to manufacturer.

Address Beckman Instruments, Inc.
Campus Drive at Jamboree Blvd.
P.O. Box C-19600
Irvine, CA 92713
(714) 833-0751
Attn: T. McKillip

Airborne Particulate, Iodine and Noble Gas Monitor
PING-3

SELF-CONTAINED FOR MONITORING
STACK-EFFLUENTS AND WORK AREAS.

FEATURES INCLUDE MICROCOMPUTER,
DIGITAL DISPLAYS, BACKGROUND SUBTRACTION,
HIGH ACCURACY AND SENSITIVITY,
ONE 3-INCH SHIELD (4 DETECTORS),
SIMPLE TO OPERATE AND MAINTAIN.

Class	Laboratory
Principle of Operation	Measures Airborne Radioactive Particulate, Iodine and Noble Gas (PING) by scintillator. Air is pulled through a filter to remove particulate, charcoal to remove iodine then a chamber for measuring noble gas.
Sensitivity and Range	Particulate: 5 to 10 cpm/hr for 10^{-11} μCi/cm^3 and air flow of 60 liters/min. Iodine: 3 cpm/hr for 10^{-11} μCi/cm^3 ^{131}I and air flow rate of 60 liters/min. Noble Gas: 25 cpm for 10^{-6} μCi/cm^3 ^{133}Xe at 14 psia (sensitive detector). Total range: 1×10^{-7} to 4×10^2 μCi/cm^3 for ^{133}Xe (low and medium range detectors).
Sampling	Continuous
Performance	Retains history of averages for last 23 ten-minute intervals, 24 one-hour intervals and 24 one-day intervals. Operating temperature: 10 to 49°C (50 to 120°F)
Requirements	Power: 115, 60 Hz, 6 amps Size: 36 in. W × 30 in. D × 41 in. H (91 cm × 76 cm × 104 cm) Weight: 682 kg (1500 lbs)
Features	Digital microcomputer system with automatic background subtraction for monitoring work areas and effluents; eight hour battery power back-up.
References	Eberline Catalog
Cost	$44,450
Address	Eberline Instrument Corporation P.O. Box 2108 Santa Fe, New Mexico 87501 (505) 471-3232

Radon Daughter Working Level Counter
Eberline PRS-2 and SPA-1

Class	Portable
Principle of Operation	A small portable air pump draws an air sample through a filter which is then measured for alpha activity. ZnS(Ag) crystal and photomultiplier detects alpha activity. Electronics unit contains high voltage supply and ratemeter scaler with speaker (audio).
Sensitivity and Range	Lower limit of sensitivity: 0.001 WL: upper 100. ZnS(Ag) crystal active area: 2.54 cm (1 in.) Ratemeter-scaler readout: six characters
Sampling	Counts individual air filter samples for determining working levels (WL) or radon daughters by the Modified Kusnetz or Rolle Methods
Performance	Response time: depends on pre-set number of counts selected (count rate mode) or integration time selected (scaler mode) Detector efficiency: 25% or better of 4π emission rate of an Eberline disc alpha standard Operating temperature: −18 to 60°C (0 to 140°F) High voltage supply: 500 to 1500V Largest filter size accommodated: 2.8 cm (1 1/8 in.) Detector connector: Eberline Type C J-1, waterproof connector mates with CP 1
Requirements	Power: Rechargeable Gell-Cell battery provides 75 hours continuous operation. Battery charger included. Recharges battery in 14 hours. Size: 19.7 cm H × 24.1 cm L × 10.2 cm W (7 1/2 in. × 9 1/2 in. × 4 in.) Weight: 2.3 kg (5 lb)
Features	1) Designed to comply with ANSI N13.8-1973. 2) Can be used as an "instant" WL meter using Rolle method. 3) Certified ^{230}Th Alpha Standard (Model DNS-4) available. 4) PRS-2 can be used with other detectors to measure radon gas, gamma radiation, beta contamination and alpha contamination.

References	Manufacturer's specifications.
Cost	$2470.00
Address	Eberline Instrument Corporation P.O. Box 2108 Santa Fe, NM 87501 (505) 471-3232

Radon Gas Monitor
Eberline Model RGM-2

Field	Designed for permanent installation but may be moved.
Principle of Operation	Instrument contains an Eberline Scintillation Detector for radon gas detection. A pump is included for continuous gas flow. A microcomputer records data for unattended operation. The RGM-2 supersedes RGM-1.
Sensitivity	Typically 4 cpm per pCi/liter with one cpm background. Lower Limit of Sensitivity: about 0.1 pCi/liter.
Sampling	Continuous
Performance	The instrument is designed to operate continuously under field conditions. Normal servicing consists of changing the dust filter on the sampling air line, changing the dust filter on the cooling air intake (as required during summer months) and replacing the printout paper in the printer.
Requirements	Operates on either 115 or 220 Vac, 50/60 Hz or 12 Vdc. Weight: approximately 50 kg (110 lb.) Dimensions: 61 cm × 76 cm × 28 cm (24 in. × 30 in. × 11 in.) deep with legs provided for permanent mounting out-of-doors.
Features	Improved alpha plateau for more stable operation. Weatherproof for direct mounting out-of-doors. Microcomputer for data storage and calculations. Automatic background subtraction. Electrostatic printout, easily read and reliable. Prints average pCi/liter hourly plus 24-hour average. Prints monitor identification, day, time, calibration factor, and background subtracted. Remote alarm contacts provided. Internal battery back-up (2 hrs) Technical manual supplied with each instrument.
References	Manufacturer's specification.
Cost	$8800.00
Address	Eberline Instrument Corporation P.O. Box 2108 Santa Fe, NM 87501 (505) 471-3232

Radon Gas Detector
Eberline Model SAC-R5[a]

Class	Detector
Principle of Operation	Five inch diameter scintillator detector in a light tight enclosure counts scintillation cells (SC-3, SC-5, SC-6) or is used with a separate sheet of ZnS scintillation material, to count alpha emission from filter papers and swipe sample.
Sampling	Maximum sample size: 12.4 cm D × 22.2 cm H (4 7/8 in. × 8 3/4 in.).
Performance	Photomultiplier tube: Nominal 5 in., 10-stage end window tube. Background: 1 cpm when voltage set properly on alpha plateau. Operating voltage: Positive 2500V maximum input resistance: 120 MΩ
Requirements	Power: Positive high voltage supply. Size: 16.5 cm Dia × 45 cm H (6.5 in. × 17.7 in.) Weight: 3.9 kg (8.5 lbs)
Features	1) Eberline can provide calibration of Radon Gas Counting Systems 2) Optional Accessories ZnS(Ag) Scintillation Sheet: 3 in. × 3 in., Eberline 10798-A12 (25 per package). Certified Thorium-230 Alpha Standard, DNS-4. 3) May be used with Eberline PRS-2, MS-2 or other scalers.

[a]Eberline also provides continuous Radon Daughter Working Level Monitor Model LM-1/WLR-1.

Scintillation Cells

	SC-3	SC-5	SC-6
Volume, liters	0.1	0.5	1.4
Response Factors*			
cpm per pCi	5.9	4.6	4.3
cpm per pCi/liter	0.59	2.3	6.0
Size Overall Height	3.4 in. (8.5 cm)	4.4 in. (11 cm)	8.7 in. (22 cm)
Size Overall Diameter	2.5 in. (6.4 cm)	4.5 in. (11.5 cm)	4.5 in. (11.5 cm)
Weight	0.3 lbs (140 gm)	0.85 lbs (385 gm)	1.5 lbs (680 gm)

Pressure Limits: From vacuum to approximately ambient. Do not pressurize.
Hose Fittings: For 5/16 I.D. hose.
*Nominal response factors after radon daughters reach equilibrium with ^{222}Rn.

Cost SAC-R5 $1330; SC-3 $80; SC-5 $85; SC-6 $105

Address Eberline Instrument Corp.
P.O. Box 2108
Santa Fe, NM 87501
(505) 471-3232

Uranium Enrichment Monitor
Eberline Model SAM-2

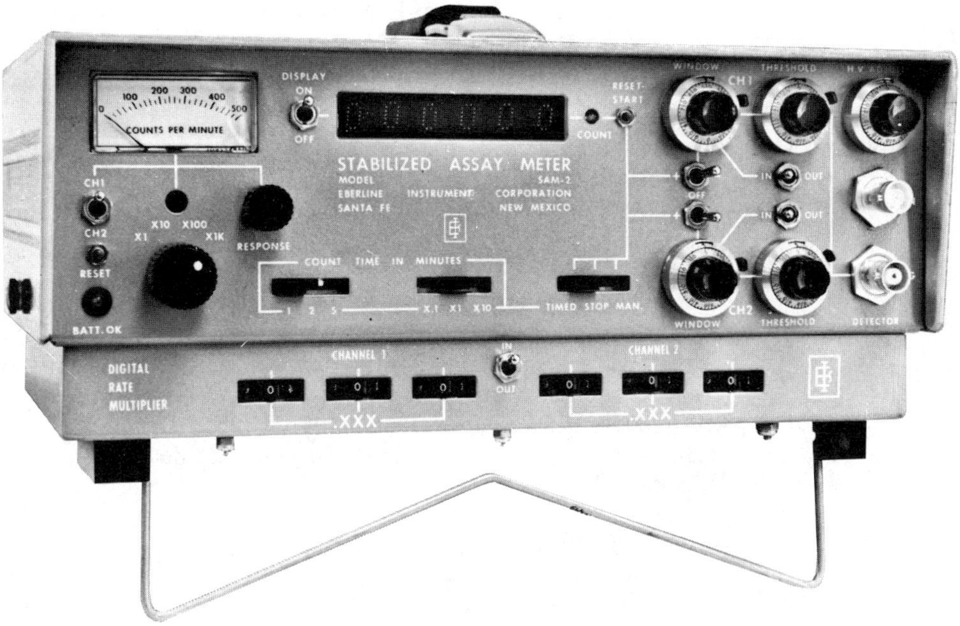

Class	Portable
Principle of Operation	Scintillation crystal NaI (Tl) doped with ^{241}Am, photomultiplier tube, two single channel analyzers, amplifier, high voltage gain control circuit, scaler, timer, ratemeter and digital readout.
Sensitivity and Range	Energy range: Optimized for the detection of the 185-keV gamma emitted by ^{235}U, may be used for other energies. Rate: 500, 5000, 50,000, and 500,000 cpm, linear Threshold: 0 to 1.0 V, calibrated Window: 0 to 1.0 V, calibrated
Performance	Pulse height resolution: (with RD-19) ~15% FWHM at 185 keV Temperature: 0 to 60°C (32 to 140°F) Timer: From 0.1 to 50 min Response time: 1 to 13 sec continuously variable, reset switch Linearity: ±5% FS Paired pulse resolution: ~4 μsec

Requirements	Power: 115 or 230 V ac switch selectable, 0.25 A Battery: 7.5 to 14 V at 1.0 A Size: Detector, 6.6 cm Diam. × 24.5 cm L (2.6 in. × 9.6 in.) Electronics, 11.5 cm H × 24.5 cm D × 26.8 cm W (4.5 in. × 9.6 in × 10.5 in.) Weight: Detector, 1.1 kg (2.5 lb) Electronics, 3.2 kg (6.8 lb)
Features	Mode switch: Timed-stop-manual Test circuit, voltage stabilization circuit, read either channel, add both channels, substract one channel, scaler and ratemeter displays, preset counting times. Accessories: Battery pack, lead shield, carrying case. Model RD-19 Stabilized Scintillation Detector: 2 in. Diam. × 1/2 in. thick NaI(Tl) crystal doped with ^{241}Am, with 2 in., 10-dynode, S11 response photomultiplier tube. Dynode resistor network and magnetic shield built into detector. Weighs 2 1/2 lb. Model RD-20 Stabilized Scintillation Detector: Same as Model RD-19 but with thin window crystal and lower gamma equivalent energy of the ^{241}Am for low-energy gamma measurements. Weighs 2 1/4 lb. Model RD-21 Stabilized Scintillation Detector: 5 in. diameter photomultiplier tube and crystal version of the RD-20, also for low-energy gamma measurements. Neutron Probe: Model RD-18 Directional Fast Neutron Detector (not stabilized). Model BP-4 Battery Pack: 12 V, 7.5 amp-hours, rechargeable battery in carrying case. Includes charger and connecting cables. Weighs 8 lb. Lead Shield for RD-19: 3/16 in. thick aluminum encased lead shield with collet-type retention to the detector. Has easily interchangeable end plates for varying collimation. Weighs 5 1/2 lb.
References	Manufacturer's specifications.

Cost	SAM-2: $2355.00 RD-19: $1215.00 RD-20: $1475.00 RD-22: $1210.00
Address	Eberline Instrument Corp. P.O. Box 2108 Airport Road Santa Fe, NM 87501 (505) 982-1881

RAD-NUC
EPA,EERF 1
January 1983

^{85}Kr Liquid Scintillation System
EPA Eastern Environmental Radiation Facility

Class	Laboratory
Stage of Development	Not commercially available.
Principle of Operation	Krypton is separated from gas samples with charcoal and molecular sieve cold traps, calcium sulfate, ascarite (NaOH preparation) and a titanium furnace (900°C). Sample is then dissolved in liquid scintillator and counted by a photomultiplier system.
Sensitivity	Minimum detectable concentration: $\sim$2.5 pCi/m^3 air at 2σ counting error.
Performance	Accuracy: Individual measurements of background ^{85}Kr (in the 15 pCi/m^3 range) are determined to about $\pm$10%.
Comments	^{83m}Kr is used as a tracer in the krypton separation process. Krypton recoveries are in the 80-90% range. About 1 m^3 of air can be processed in six hours.
References	S. L. Cummings, R. L. Shearin, C. R. Porter, "A Rapid Method for Determining ^{85}Kr in Environmental Air Samples," page 163 in *Rapid Methods for Measuring Radioactivity in the Environment*, International Atomic Energy Agency, Vienna (1971).
Address	C. R. Porter EPA, Eastern Environmental Radiation Facility P.O. Box 3009 Montgomery, AL 36109 (205) 272-3402

Krypton-85 Ion Chamber Systems

Class — Field instruments.

Stage of Development — Not commercially available; but can bc put together from commercial sources and the reference listed below.

Principle of Sensitivity — Flow-through ionization chambers for ^{85}Kr measurements.

Chamber Volume	Minimum Detectable Concentration ^{85}Kr (pCi/m^3)
0.5 liters	130,000
1.0 liters	190,000
2.8 liters	39,000
4.3 liters	39,000

Sampling — Continuous

Performance — Accuracy: $\pm 7\%$ to $\pm 17\%$ (2σ) error on calibration.

Comments

1. Dry, radon-free air required.
2. Radon holdup trap needed.
3. Needs frequent maintenance.
4. ^{85}Kr concentration is averaged over about 20–30 min interval due to radon trap.

References

D. G. Smith, J. A. Cochran, B. Shleien, "Calibration and Initial Field Testing of ^{85}Kr Detectors for Environmental Monitoring," Document BRH/NERHL 70-4, Northeastern Radiological Health Lab (1970).

Address

FDA, Winchester Engineering and Analytical Center
109 Holton Street
Winchester, MA 01890
(617) 729-5700

Krypton-85 G-M Detectors

Class — Field instruments.

Stage of Development — Not commercially available; but can be put together from commercial sources and the reference listed below.

Principle of Operation — Five Geiger-Müller tubes for ^{85}Kr measurements.

Instrument (G-M Dectector)	Geometry G-M Detector	Dimensions	Window Thickness mg/cm^2	Minimum Detectable Concentration (pCi/m^3) (10 min sample/ −10 min background)
Eon 8008H	2 window pancake	2 in. diameter	3.5	12,000
Eon 8001T	1 window pancake	2 in. diameter	1.4	25,000
Amperex 18546	1 window pancake	2 in. diameter	3.5	27,000
Eon 5108E	cylindrical	5 in. length, 1/2 in. diameter	30.	28,000
LND 719	cylindrical	5 in. length, 1/2 in. diameter	30.	24,000

Accuracy — All probes calibrated ±13% against standard ion chamber.

Comments —

1. Cylindrical probes are the most rugged.
2. Two-window pancake was too fragile for field use.
3. All probes are sensitive enough for use near a fuel-reprocessing plant.
4. Time-averaging is very short (a few minutes).

References —

D. G. Smith, J. A. Cochran, B. Shleien, "Calibration and Initial Field Testing of ^{85}Kr Detectors for Environmental Monitoring," Document BRH/NERHL 70–4, Northeastern Radiological Health Lab (1970).

Address —

FDA, Winchester Engineering and Analytical Center
109 Holton Street
Winchester, MA 01890
(617) 729–5700

Radon Daughter Analyzer
Harshaw Model 100 and 101

Class	Field
Principle of Operation	Alpha and beta radiation in sampled air measured by a diffused junction detector and a CaF_2:Eu scintillator respectively. Radium "A", "B" and "C" determined.
Sensitivity and Range	Detectability: $< 10^{-3}$ working level Dynamic Range: $< 10^{-3}$ to 10^2 working level
Performance	Sampling volumes: up to 66 liters/min. Line-time measurement: 4+ minutes Printout of counts, concentration and working level.
References	Manufacturer's specifications
Cost	$17,000
Address	Harshaw Chemical Company 6801 Cochran Road Solon, Ohio 44139 (216) 248-7400 Attn: M. Cox/D. Elliot

Tritium Monitor
Harwell Model

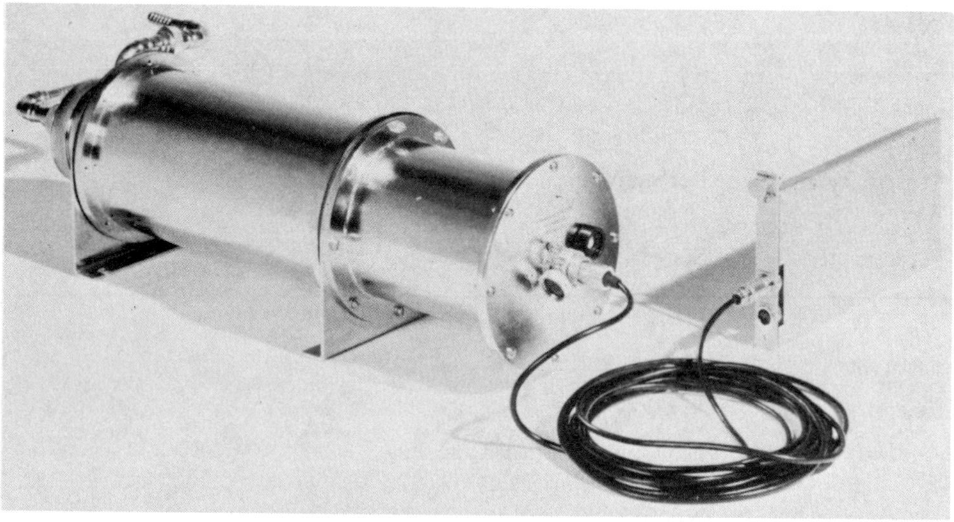

Class	Laboratory
Principle of Operation	An uncompensated ion chamber fitted with filter and ion trap to which is volted a digitizing heavy.
Sensitivity and Range	Detection range: 1 μCi/m^3 to 10^5 μCi/m^3
Sampling	Continuous
Performance	Sensitive detector volume: 2 liters Digitizing head, input current range: $< 10^{-14}$ to 2×10^{-10}A Output frequency range: 1 Hz to 10^5 Hz Temperature dependence: better than 1 μCi/m^3 per °C (5 to 35°C)
Requirements	Power: Line voltage Size: Ion chamber + digitizing head; 70 cm $\times$ 21 cm $\times$ 19 cm including outlet and inlet pipes
Features	The head is completely sealed and dessicated to eliminate ingress of moisture, but may easily be separated from ion chamber for servicing and when decontamination of the chamber is necessary.
Remarks	The chamber is uncompensated and when gamma backgrounds are significant should be enclosed in a lead shield.

References	Manufacturer's Specifications.
Cost	On application.
Address	AERE Harwell Electronics and Applied Physics Division Bldg. 347.3 Oxfordshire OX11 ORA ENGLAND Tele: (0235) 24141 Ext.: 4766 Attn: R. H. W. Sherwood

RAD-NUC
Harwell (UKAEA) 2
January 1983

Transportable Alpha-In-Air Monitor
Harwell Type 3276

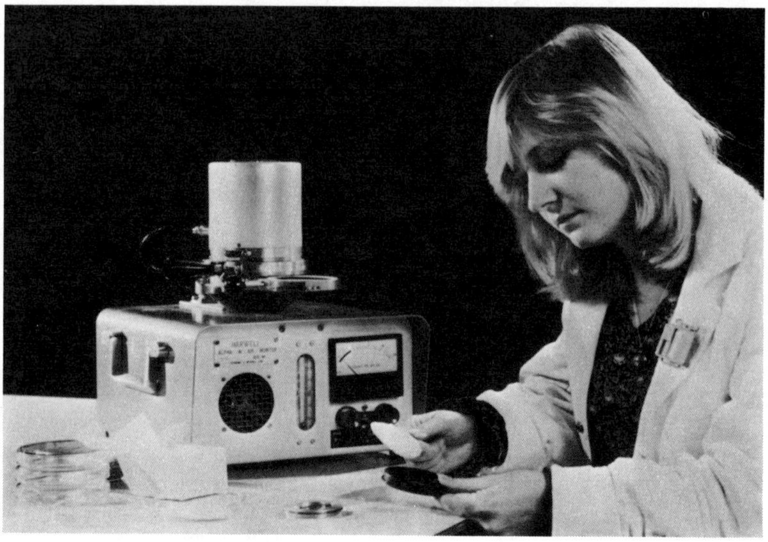

Class	Transportable
Principle of Operation	Diffused junction silicon detector views a filter paper through which air is continuously sampled, results in display of the activity of the collected particles on a visual meter of a ratemeter.
Sensitivity and Range	Count rate range: 0.1 to 100 counts per second Sensitivity: such that one standard deviation is about equivalent to +1 mpc-hr of Pu (occupational), under average conditions (i.e., ~225 pCi/m^3 of Rn and not more than 80% relative humidity).
Sampling	Continuous
Performance	Possible to set Pu alarm level. Air flow: 37 liters/minute (1.3 conf/minute) typical, adjustable Filter paper: 6 cm Dia glass fiber (Whitman GF/A 20-40 liter/min. Time Constant (meter indication/recorder output): 0.1 sec, 21 sec, 100 sec, or 200 sec.

Requirements	Power: Main ac input: 75 W, 240 V (or 110 V—50 Hz), as marked. Size: } Refer to manufacturer Weight: }
Features	Optional preset trip level alarm Also provision for connection to pen recorders, either logarithmic or linear mode or both. All events producing ionization in the detector can be monitored by a multichannel analyzer. Pulse amplitude selection and (Radon daughter) compensation. Selectable alarm level Remote alarm option
References	Manufacturer's literature.
Cost	On application.
Address	AERE, Harwell Oxfordshire OX11 ORA, ENGLAND Tele: (0235) 24141 Attn: R. H. W. Sherwood, Bldg. 347.3 Instrumentation and Applied Physics Division Ext. 4766

Air Monitor
Johnson and Associates Model TR-5

Class	Portable
Principle of Operation	Uses ion chamber combined with air pump causing airflow through filter and electrostatic precipitator, also a linear ratemeter with visual meter display
Sensitivity and Range	Ranges, tritium: 0–50, 0–500, 0–5000 Ci/m^3 γ: 0.25, 2.5, and 25 mR/hr Model TR-5M Ranges: tritium 0–25, 0–250, 0–2500 μCi/m^3 This is accomplished by doubling chamber volume and elimination of gamma compensation.
Sampling	Air flow 2 liters/min, continuous
Performance	Chamber volume: 2 liters Accuracy: ±15% of full scale (FS) Zero drift: 5% of FS—30 min. after 15 min. warmup;

Requirements	Power: 105–125V ac operation; batteries, NiCad (rechargeable) 4 hour life Size: 30.5 cm × 28.6 cm × 28.6 cm (12 in. × 11 1/4 in. × 11 1/4 in.) Weight: 6.8 kg (15 lbs.)
Features	Automatic change to battery in event of power failure Alarm: presettable, 20% to 100% FS
References	Manufacturer's literature
Cost	$2200.00
Address	Wm. B. Johnson and Associates, Inc. Research Park Montville, NJ 07045 (201) 334-9222

Portable Tritium Air Monitor
Johnston Laboratories Model 111

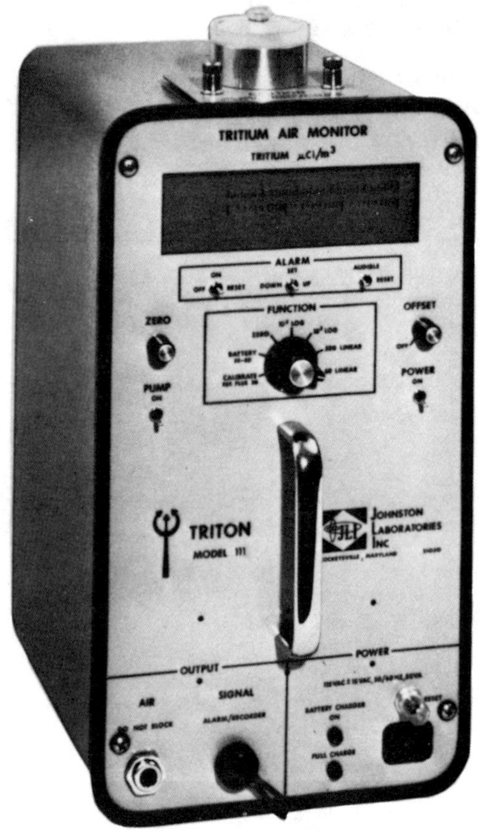

Class	Portable
Principle of Operation	An air pump draws air through ion chamber after filter which is read by electronics and output displayed on liquid crystal bar graph display.
Sensitivity	Tritium range: 0–50, 0–500 $\mu Ci/m^3$ (linear) 10^1–10^4, 10^3–10^6 $\mu Ci/m^3$ (log) Lower limit of sensitivity: 5 $\mu Ci/m^3$ (linear) 20 $\mu Ci/m^3$ (log)
Sampling	Continous air sampling

Performance	Gamma compensated to 5 mR/hr Accuracy: ±10% FS linear, ±20% of reading on log ranges Operating temperature: 0 to 50°C (32 to 122°F) Humidity: 0 to 90% R.H. Detectors, sample ion chambers: 2 each 0.4 liter—total 0.8 liter gamma compensation chambers: 2 each 0.4 liter—total 0.8 liter Filter system: 0.5 micron filters Air pump: positive displacement—continuous duty Zero Drift: 2%/24 hours after 5 min. warmup Time constant: 20 sec.on 0-50 μCi/m^3 range Air flow: 4 liters/minute External signals: Analog output (10V FS—10K)
Requirements	Power: 115 Vac, 50-60 Hz 25V ampere Size: 21.6 cm × 38.1 cm × 26.7 cm (8.5 in. × 15 in. × 10.5 in.) Weight: 9 kg (20 lbs.)
Features	Automatic change over to battery operation on power failure without false alarm; battery backing 6 hours 100% solid state: for ruggedness Rain-tight panel and case Air flow sensor built-in Battery
References	Manufacturer's specifications
Cost	$6500.00
Address	Johnston Laboratories 383 Hillen Road P.O. Box 20086 Towson, MD 21204 (301) 337-8700 Toll Free: (800) 638-8656 Attn: E. J. Pierzynski

Radionuclide Gas Monitor
Johnston Laboratories Triton 955B

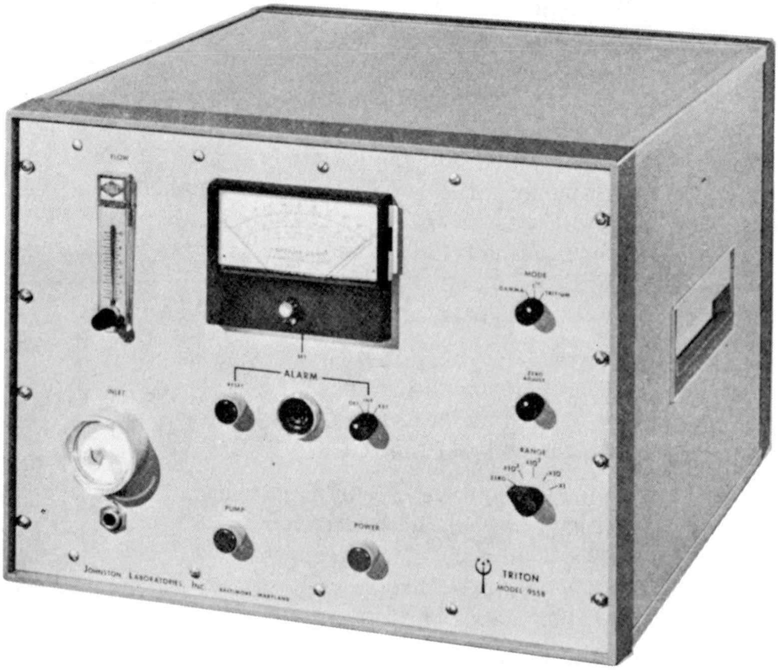

Class	Laboratory or stationary ionization chamber
Principle of Operation	Positive displacement pump draws air through submicron filter and electrostatic precipitator, then into flow-through ionization chambers. Signal amplified in electrometer circuit. Two identical sealed chambers measure external gamma flux, which is subtracted electronically.
Sensitivity	Tritium ranges: 4 ranges 0–10, 100, 1000, 10,000 $\mu Ci/m^3$ Carbon-14 ranges: 0–2, 20, 200, 2000 $\mu Ci/m^3$ Gamma ranges: 0.05, 0.5, 5, 50 mR/hour No differential measurement of different β-emitters.
Performance	Accuracy: $\pm$10% of full scale, reproductibility $\pm$2% Zero drift: After 15 minute warm-up, less than 2% of FS in 24 hours Time constants: 15 or 45 seconds Gamma compensation: Up to 5 mR/hour for uniform external field

	Air flow: 2 to 10 liters/minute, adjustable/0.5 micron filter Signal output: 0–10mV analog Humidity response: 0–95%, noncondensable
Requirements	Power: 115/230 Vac, 100W, 50/60 Hz Size: 50 cm × 37 cm × 55 cm (19.6 in. × 14.8 in. × 21.6 in.) Weight: 30kg (67 lb)
Features	Alarm for fail-safe operation, variable set point, Cleanable chambers, recorder output, remote operation mode Solid-sate circuity
References	1. Manufacturer's Specifications 2. J. R. Waters, Johnston Laboratories Internal Report #JLI-506, "Pitfalls and Errors in the Measurement of Tritium in Air".
Cost	Model 955B: $5600.00
Remarks	Very high sensitivity
Address	Johnston Laboratories 383 Hillen Road P.O. Box 20086 Towson, MD 21204 (301) 337-8700 Toll Free: (800) 638-8656 Attn: E. J. Pierzynski

Lucas Chamber
Johnston Laboratories Model LAC-2

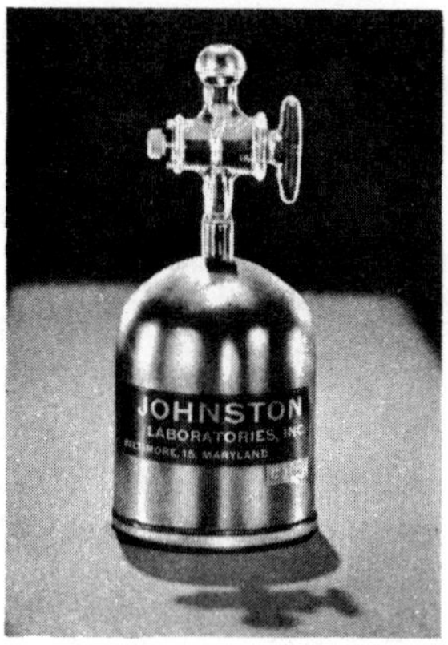

Class	Laboratory
Principle of Operation	Radon is concentrated and purified, and transferred to the Model LAC-2, Lucas chamber. Scintillations produced in the chamber are counted by a photomultiplier.
Sensitivity and Range	1 pCi of radon yields 5 cpm
Requirements	Size: 12.7 cm H × 5.1 cm Diam. (5 in. × 2 in.)
Features	
References	Manufacturer's specifications
Cost	LAC-2 Radon Counter (quartz window): $455.00
Address	Johnston Laboratories 383 Hillen Road P.O. Box 20086 Towson, MD 21204 (301) 337-8700 Toll Free: (800) 638–8656 Attn: E. J. Pierzynski

Xenon-133 Gas Monitor
Johnston Laboratories Model 133C

Class

Portable

Principle of Operation

Air is drawn through a front panel filter and then through an electrostatic precipitator and an ionization chamber. Beta decay electrons from ^{133}Xe produce a chamber current which is measured by an electrometer and displayed on a logarithmic scale meter. A red warning light and a brief tone—burst audible alarm are activated simultaneously when the concentration of ^{133}Xe exceeds the maximum permissible level (MPC) (1P μCi/m^3)

Range

Two logarithmic ranges
1. 0.1 to 10 MPC (1 to 100 μCi/m^3)
2. 1 to 100 MPC (10 to 1000 μCi/m^3)

Accuracy

15% of Reading

Response Time Constant: Approximately 25 seconds × 1 Range

Gamma-Ray Response: A level of about 1.4 mr/hr is required to produce a reading of 1 MPC. This means that a 20 mCi source of ^{99m}Tc placed six feet from the instrument will not cause a false alarm.

Features

Air Pump: Continous duty, positive displacement, diaphragm type. Flow is approximately 10 liter/min. Vinyl sniffer hose provided.

Particle Filter: 0.5 micron panel mounted see-through glass fiber filter.

Electrostatic Precipitator: Removes small particles and ions to prevent false alarms or spurious response to dust and cigarette smoke

Calibration Check: Instrument calibration can be easily checked with a small ^{99m}Tc source

Recorder Output: Designed for an inexpensive recorder available as an accessory.

Requirements

Power: 115 V, 50/60 Hz, 125 watts; also available for 100 V, 50/60 Hz or 230 V, 50/60 Hz.

Size: 28.5 cm × 26.5 cm × 43.5 cm (11 1/4 in. × 10 1/2 in. × 17 in.)

Weight: 9.6 kg (21 lb)

Price	$2,600.00 (includes recorder)
Address	Johnston Laboratories 383 Hillen Road P.O. Box 20086 Towson, MD 21204 (301) 337-8700 Toll Free: (800) 638-8656 Attn: E. J. Pierzynski

^{85}Kr Liquid Scintillation System
Mound Laboratory

Class	Laboratory
Stage of Development	Not commercially available
Principle of Operation	Sample gas is expended into evacuated vial, introduced at 25 torr pressure and dissolved into a 25-ml polyethylene vial filled with toluene-based liquid scintillator, and counted by a liquid scintillation counter.
Sensitivity	Minimum detectable concentration: $\simeq$15 pCi/cm^3 of Krypton or air Counting efficiency: 92.5% of ^{85}Kr betas are counted above threshold
Sampling	Limit of $\lesssim$ 25 mm of Hg can be introduced
Performance	Accuracy: Individual measurements to $\pm$5%
References	1. M. L. Curtis, S. L. Mess, and L. L. Bentz, "Simple Technique for Rapid Analysis of Radioactive Gases by Liquid Scintillation Counting," *Anal. Chem.*, **38,** 636, (1966); this paper describes the method. 2. R. E. Shuping, C. R. Phillips, and A. A. Moghissi, "Low Level Counting of Environmental ^{85}Kr by Liquid Scintillation," *Anal. Chem.*, **41,** 2082, (1969).
Address	W. E. Sheehan Monsanto Research Corporation Mound Laboratory P.O. Box 32 Miamisburg, OH 45342 (513) 865-3757

RAD-NUC
National Nuclear Corporation 1
January 1983

Effluent Monitor
NNC Model LMS-4

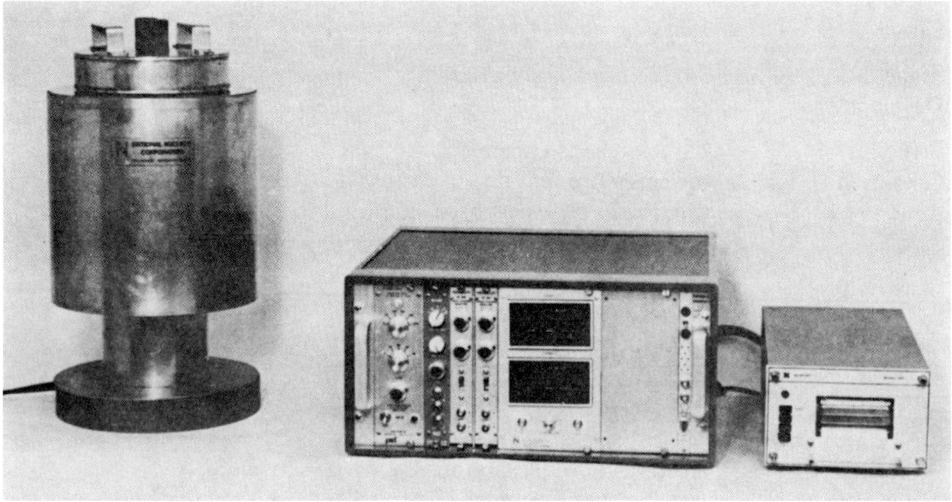

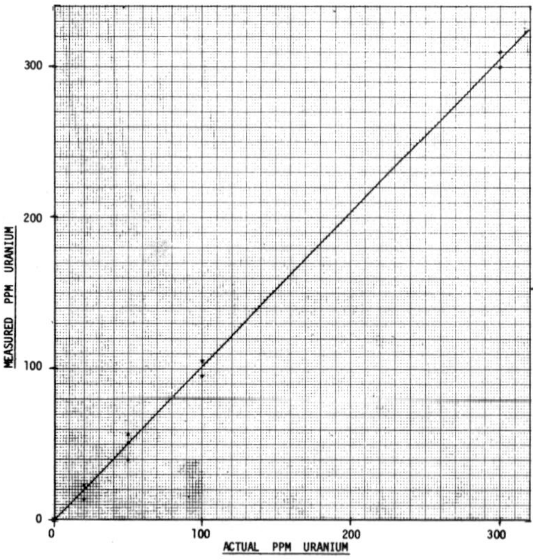

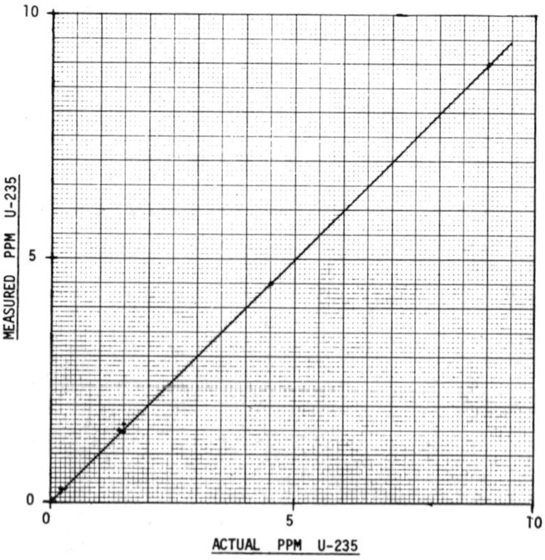

Class	Laboratory process monitor
Principle of Operation	Uses a sodium iodide detector with PHA to analyze certain broad well-identified γ-peaks in process solutions sampled with flow-through jars with data output on a scaler reader with printout of counts over a prescribed count period.

Sensitivity and Range

Nuclide Detected	γ-Peak Used	Sensitivity	Accuracy at 10 Times Detection Level
^{235}U	185 keV	0.03 ppm	0.05 ppm
^{238}U	50 keV (aged) 80 keV (< 30 days old)	10 ppm	15 ppm
Gamma	Various	40nCi/liter	60 nCi/liter

Sampling	Specific time period for disposable jar, 4 1/2 in. dia.
Performance	5 in. diameter NaI(Tl) detector
Requirements	Power: Ac line voltage + high voltage for detector Size: Refer to manufacturer; depends on options Weight: Refer to manufacturer; depends on options
Features	Shielding: 3 inch lead plus internal x-ray shield Easily replaceable, disposable jars for counting chamber
References	Manufacturer's specifications
Cost	\$15,000.00 → \$30,000.00 depending on options
Address	National Nuclear Corporation 1904 Colony St., Mountain View, CA 94003 (415) 962-9220

Personnel/Vehicle SNM Search and Survey Monitor
National Nuclear Model HM-3

Class	Portable
Principle of Operation	High gamma-ray sensitivity, NaI scintillating detector and photomultiplier combination.
Sensitivity and Range	Full scale readings up to 1000 or 10000 cps.
Performance	Alarm: Audible internal alarm. Two pin connector for external alarm. Four pin connector for remote alarm which mutes internal alarm Counting time: Internal jumper settings of 0.1 to 0.9 seconds.
Requirements	Power: 5, nickel-cadmium, 1.2 V, rechangeable batteries (about 100 hours of operation before recharging is necessary) Size: 20 cm × 12.7 cm × 15.2 cm (8 in × 5 in. × 6in.) Weight: 1.8 kg (4 lb)
References	Manufacturer's specification
Cost	$1700
Address	National Nuclear Corporation 1904 Colony Street Mountain View, CA 94043 (415) 962-9220 Attn: D. Kahn

β/γ-Coincidence Scintillation Spectrometry System

Class	Laboratory (ultra-low-level measurements).
Principle of Operation	The sample radioactivity to be measured is radiochemically separated and collected as a precipitate on filter paper. The separated source and filter are sealed between thin plastic films and placed on a plastic-wafer scintillator which is in turn mounted permanently on a plastic lightpipe and photomultiplier tube. The entire assembly is enclosed in a light-tight shield and inserted into the well of a 10 cm × 10 cm NaI(Tl) well crystal. The light pulses from the γ-rays incident on the crystal are accepted when in coincidence with the signal initiated by a β-particle impinging on the plastic-scintillator disc. The γ-ray spectrum is then analyzed using a 256-channel MCA and a least-squares computer program.
Sensitivity and Range	Lower limits of detection (estimated as 3 σ_{bkg}) for ^{131}I, ^{134}Cs and ^{228}Ra are 0.05 pCi, 0.08 pCi and 0.08 pCi, respectively.
Sampling	Up to 4 liters of milk or water are normally processed radiochemically for ^{131}I, ^{134}Cs or ^{228}Ra. Airborne ^{131}I is collected on TEDA-loaded charcoal cannisters. Vegetation, tissue, soil and sediment samples are limited to the quantity normally convenient for ashing or fusion as a preparative step prior to radiochemical separation.
Performance	Acquisition time: 100 − 400 min. Background: ^{131}I region−0.010 ± 0.005 cpm ^{134}Cs region−0.027 ± 0.007 cpm ^{228}Ra region−0.008 ± 0.005 cpm Efficiency: ^{131}I region−0.160 cpm/dpm ^{134}Cs region−0.145 cpm/dpm ^{228}Ra region−0.031 cpm/dpm
References	Paperiello, C. J. and J. M. Matuszek, "A β-γ Coincidence System for Environmental ^{131}I," *IEEE Trans. on Nucl. Sci. NS-22* (Feb. 1975).
Cost	$4,000 + multichannel analyzer and shielding
Address	Dr. J. M. Matuszek, Director Radiological Sciences Laboratory Division of Laboratories and Research New York State Department of Health Albany, N.Y. 12201 (518) 474-0004

Gas-Proportional β/γ-Coincidence Spectrometry System for Measurement of ^{133}Xe

Class	Laboratory (ultra-low-level meausurements)
Principle of Operation	The detector for low-level ^{133}Xe measurements consists of a β/γ-coincidence system using a 45-cm gas-proportional detector with thin aluminum walls and a 7.6 cm $\times$ 17.8 cm NaI(Tl) well detector. It includes a plastic scintillation anticoincidence cosmic-ray guard with all components enclosed in a 15.2-cm thick steel shield. the radiochemically separated and purified xenon sample is mixed with P-10 counting gas and loaded into the gas-proportional counting tube. The NaI(Tl) signal initiated by the 81-keV γ-ray of ^{133}Xe is recorded when recieved in coincidence with the proportional counter signal initiated by the β-particle emitted by ^{133}Xe. The coincident NaI(Tl) signals are analyzed using a 256-channel MCA. The spectral data are reduced by a least-squares computer program.
Sensitivity and Range	Lower limit of detection (estimated as 3 σ_{bkg}) for ^{133}Xe is 0.05 pCi.
Sampling	Xenon carrier gas is added to a grab sample of 1 to 3m^3 of air, and the xenon fraction is chromatographically separated for counting.
Performance	Acquisition time: 200–2000 mins. Background: (81-keV region): 0.018 $\pm$ 0.013 cpm Efficiency: (^{133}Xe): 0.26 cpm/dpm
References	Kunz, C. O. and C. J. Paperiello, "Xenon-133: Ambient Activity from Nuclear Power Stations," *Science 192*, 1235-37 (June 1976).
Cost	\$7500.00 + multichannel analyzer and shielding
Address	Dr. J. M. Matuszek, Director Radiological Sciences Laboratory Division of Laboratories and Research New York State Department of Health Albany, NY 12201 Tele: (518) 474-0004

Xenalert Room Air and Gas Trap Monitor
Model 36-751

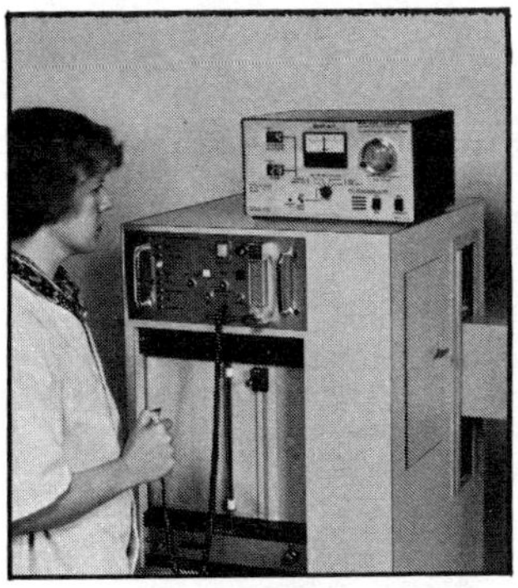

"XenAlert" monitors room air during lung function study.

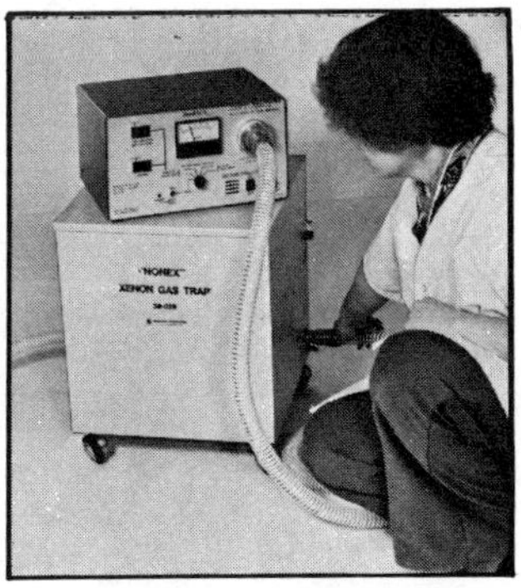

"XenAlert" displays gas trap output directly in MPC units.

Class	Bench-top air monitor
Principle of Operation	Pancake thin-window G-M tube
Sensitivity and Range	1, 10, 100, 1000 mpc for Xenon full scale Time constants: 40, 4, .4, and .04 secs. Integrates: 0-99 mpc-hr
Sample	Continuous
Performance	Accuracy: ±20% Reproducibility: ±5%
Requirements	Power: 110 Vac, 60 Hz Size: 17 cm × 31 cm × 27 cm (6.7 in. × 12.2 in. × 10.6 in.) Weight: 23 kg (10 lbs)
Features	Background subtraction 3/8 in. lead shielding, aural and visual alarm
References	Manufacturer's Specifications
Cost	$1975.00

Address

Nuclear Associates
100 Voice Road
Carle Place, NY 11514
(516) 741-6360

RAD-NUC
Nuclear Associates 2
January 1983

Iodine-125 Monitors
Models 05-676 and 05-695

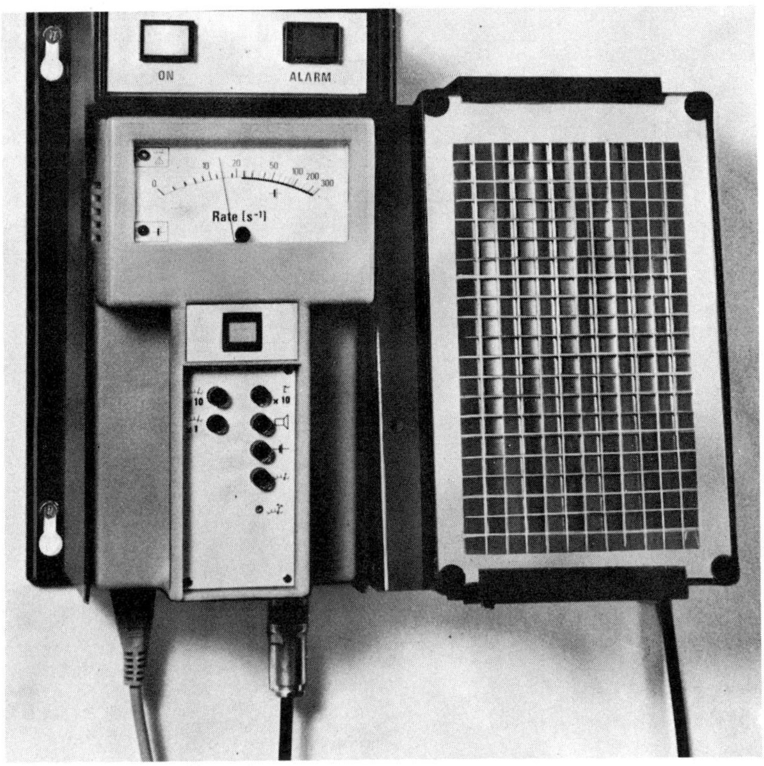

Class	Model 05-676	Model 05-695
	Portable Survey Instrument	Wall-Mounted Survey Instrument
Principle of Operation	Xenon-filled proportional counter	
Sensitivity and Range	0–300 and 0–3000 cps Time constant: 20 and 2 sec Model 05-695: 1900 cpm per 10^{-4} μCi/cm^2 of I-125 Model 05-676: 780 cpm per 10^{-4} μCi/cm^2 of I-125	
Sampling	Continuous	

Performance	Temperature range: −5 to +50°C (23 to 122°F)	
	Model 05-676	Model 05-695
	Power: Battery or 100 Vac converter optional	110 Vac
	Size: 22.9 cm × 22.2 cm × x 9.5 cm (9 in. × 8 3/4 in. × 3 3/4 in.)	30.5 cm × 28.6 cm × 15.2 cm (12 in. × 11 1/4 in. × 6 in.)
	Weight: 1.48 kg (3 1/4 lbs)	
Features	Sensitive and large area detector 100 cm^2	Same with 200 cm
References	Manufacturer's specifications.	
Cost	$2300.00	$2950.00
Address	Nuclear Associates 100 Voice Road Carle Place, NY 11514 (516) 741-6360	

RAD-NUC
Nuclear Measurement Corporation 1
January 1983

Air Radiation Monitor
Nuclear Measurements AM-21

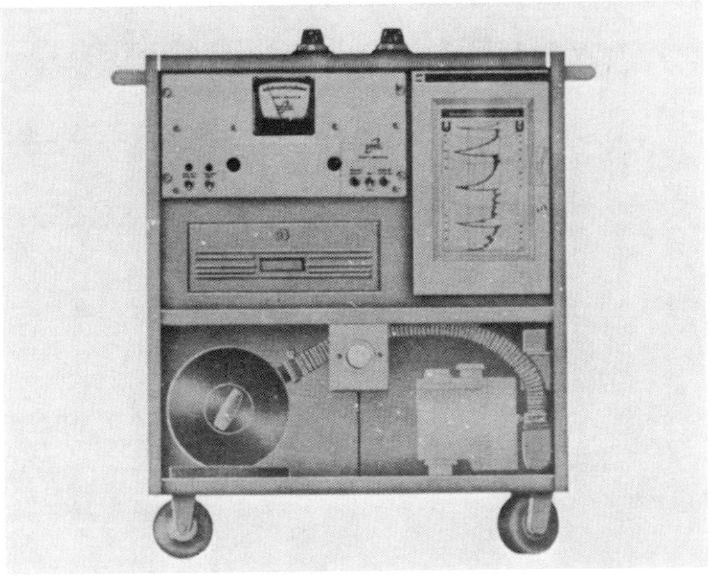

Class	Laboratory, used in operating plants and facilities.
Principle of Operation	The monitoring system consists of two parts, the pump system and the detecting system. Housed within a lead shield assembly is a gamma scintillation detector used for detecting gamma emitters below 1 MeV and a particulate collector; as the particulate matter builds up, the detector converts the ionizing events into pulses which are counted on ratemeter, also air is entering and leaving the lead housing at a flow rate from 3–10 cfm or 1–4 cfm through the filters for deposition of particulates.
Sensitivity and Range	Count rate will be 318 cpm for Iodine-131 at 10% yield, 5 cfm airflow rate, and 10^{-10} $\mu Ci/cm^3$ constant activity.
Sampling	Airflow through filters and charcoal cartridge
Performance	Ratemeter accuracy: ±10% of actual input cartridge Time Constant: 200 seconds at 50 cpm to 1 second at 50,000 cpm Stability: Drift not more than ±2% daily, ±20% annually except in narrow window spectrometry

Requirements	Power: 90–130 Vac, 60 Hz, 5-7 amp Size: 53 cm × 1 m × 1 m (21 in. × 39 in. × 39 in.) Weight: 250 kg (550 lbs)
Features	Graphic recorder provision, alarm controls
Cost	See table (with Esterline-Angus recorder)
Remarks	Other air monitors available similar to the above are given in the following table.
Address	Nuclear Measurement Corporation 2460 North Arlington Avenue Indianapolis, Indiana 46218 (317) 546-2415

RAD-NUC
Nuclear Measurement Corporation 1
January 1983

NMC Air Monitors

	Typical Background Counts for the given shield/detector system.		Efficiency		
Model	Moving Filter SA-30 Series	Moving Filter SA-30 Series	Fixed Filter SA-3L	Moving Filter SA-30 Series	Moving Filter SA-3L
AM-2A	-	-	20%	-	-
AM-3A	1 cpm	1 cpm	-	20%	11%
AM-2B	-	-	32%	-	-
AM-3B	21 cpm	40 cpm	-	32%	18%
AM-2D	-	-	10%	-	-
AM-3D	21 cpm	30 cpm	-	10%	5.7%
AM-2F	-	-	35%	-	-
AM-3F	30 cpm	30 cpm	-	32%	17%
AM-2G	-	-	19%	-	-
AM-3G	Gross 128 cpm Spec. 10 cpm	Gross 120 cpm Spec. 35 cpm	-	19%	8%
AM-21	-	-	17% Gross 3% Iodine	-	-
AM-31	Gross 97 cpm Spec. 10 cpm	Iodine 30 cpm Gross 115 cpm	-	10% Gross 15% Iodine	10% Gross 15% Iodine
AM-221	-	-	17% Gross 3% Iodine	-	-
AM-331	Iodine 10 cpm Gross 97 cpm	Iodine 30 cpm Gross 115 cpm	-	17% Gross 3% Iodine	17% Gross 3% Iodine
AM-2P	-	-	α37% β37.5%	-	-
AM-3P	$\alpha < 0.1$ cpm $\beta < 30$ cpm	$\alpha < 1$ cmp $\beta < 40$ cpm	- -	α37% β37%	α20% β22%
AM-22P	-	-	α37% β37%	-	-
AM-33P	$\alpha < 10$ cpm $\beta < 30$ cpm	$\alpha < 10$ cpm $\beta < 40$ cpm	-	α37% β37%	α35% β32%
PAPM-11	10 cpm	-	-	35%	-

RAD-NUC
Nuclear Measurement Corporation 1
January 1983

NMC Air Monitors

Model	Radiation Monitored	Price	Detectors	Recorder	Background, Fixed Filter SA−20 Series
AM-2A	Alpha	\$5200	SC-2A	1 Single Channel	1 cmp
AM-3A		6193	SC-2A	1 Single Channel	
AM-2B	Beta	\$6193	SC-2B	1 Single Channel	21 cpm
AM-3B	Beta	8293	SC-2B	1 Single Channel	
AM-2D	Gross Beta-Gamma	\$6193	DGM-2	1 Single Channel	21 cpm
AM-3D	Gross Beta-Gamma	8239	DGM-2	1 Single Channel	
AM-2F	Alpha, Beta, Gamma	\$7630	FGM-2	1 Single Channel	30 cpm
AM-3F	Alpha, Beta, Gamma	9790	FGM-2	1 Single Channel	
AM-2G	Gamma	\$7630	SC-2-2S	1 Single Channel	Gross 128 cpm
AM-3G	Gamma	9790	SC-2-2S	1 Single Channel	Spec. 10 cpm
AM-21	Iodine	\$8267	SC-2-1S	1 Single Channel	Gross 97 cpm Spec. 10 cpm
AM-221	Iodine + Alpha or Beta or Gamma	\$9957	SC-2-1S and DGM-2 or	1 Dual Channel	Iodine 10 cpm Gross 97 cpm
AM-331		12073	PC-21-FW		
AM-2P	Alpha or Beta or	\$7542	PC-21-FW	1 Single Channel	a < 0.1 cpm
AM-3P	Alpha + Beta	9672	PC-21-FW		b < 30 cpm
AM-22P	Alpha + Beta separatelly	\$8531	PC-21-FW	1 Dual Channel	a < 0.1 cpm
AM-33P	but simultaneously	10660	PC-21 FW		b < 30 cpm

Stack Monitor
Nuclear Research Model SM 101

Class	Mobile
Principle of Operation	Three channel radiation monitor: moving filter particluate, moving filter iodine and noble gas. Scintillation detector fully shielded with five decade analog readouts. Automatic air flow control 2-6 cfm.
Sensitivity and Range	Particulate: 3.5×10^2 cpm/μCi/cm^3: Cs-137 Iodine: 3×10^{12} cpm/μCi/cm^3: I-131 Noble Gas: 4×10^7 cpm/μCi/cm^3: Xe-133 4×10^7 cpm/μCi/cm^3: Kr-85
Sampling	Continuous
Performance	Accuracy: ±15% Temperature: 4.4 to 49°C (40 to 120°F)
Requirements	Power: 120 Vac, 20 amps (others available) Size: 39 cm L × 23.4 cm W × 35 cm H (60 in. × 36 in. × 54 in.) Weight: 681 kg (1500 lbs.)
Features	Remote control panel Remote readouts (optional) Check sources for each detector
Cost	Price on request
Address	Nuclear Research Corporation Warrington Industrial Park Box H Warrington, PA 18976 (215) 343-5900

Gas Proportional Counters
Packard Model 894

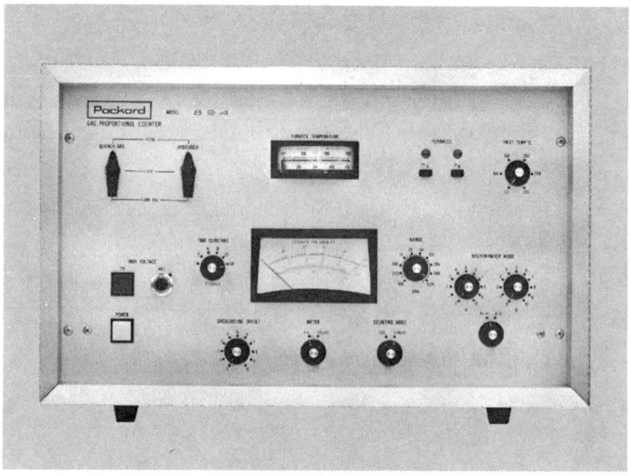

Class	Laboratory
Principle of Operation	Detects tritium or carbon-14 labelled organic compounds as they elute from a gas chromotographic column using a flow-through ion-multiplication detector. A copper oxide furnace oxidizes all organic compounds to carbon dioxide and water to eliminate condensation of high boiling point compounds in the proportional tube.
Sensitivity and Range	Sensitivity: Betas—200 dpm/peak carbon-14 labelled compounds Range: 10 linear ranges, 100 to 100,000 cpm full scale Five decade log scale Counting Efficiency: 58% Tritium 80% Carbon-14
Sampling	Continuous
Performance	Temperature: Operates at room temperature Inlet heating to 30°C Furnace heating to 800°C Time Constant: Six ranges, 1 to 50 sec
Requirements	Power: 120-220 V ac (switch), 50 or 60 Hz, 720 W Size: 35.6 cm H, 54.6 cm W, 61 cm D (14 in., 21.5 in., 24 in.) Weight: 90 kg (200 lb)

Features	Use with Model 7401 Gas Chromatograph or other gas chromatographs which have been fitted with stream splitters and heated exit ports. Background about 30 cpm (variable offset). Variable upper and lower discriminators. Strip chart-recorder and scaler output jacks. Right or left-hand capability to couple with any gas chromatograph
Cost	$8900
Address	Packard Instrument Company, Inc. 2200 Warrenville Road Downers Grove, IL 60515 (312) 696-6000

Trace™ Radioactivity Flow Monitor System
Packard Models 7130 and 7140

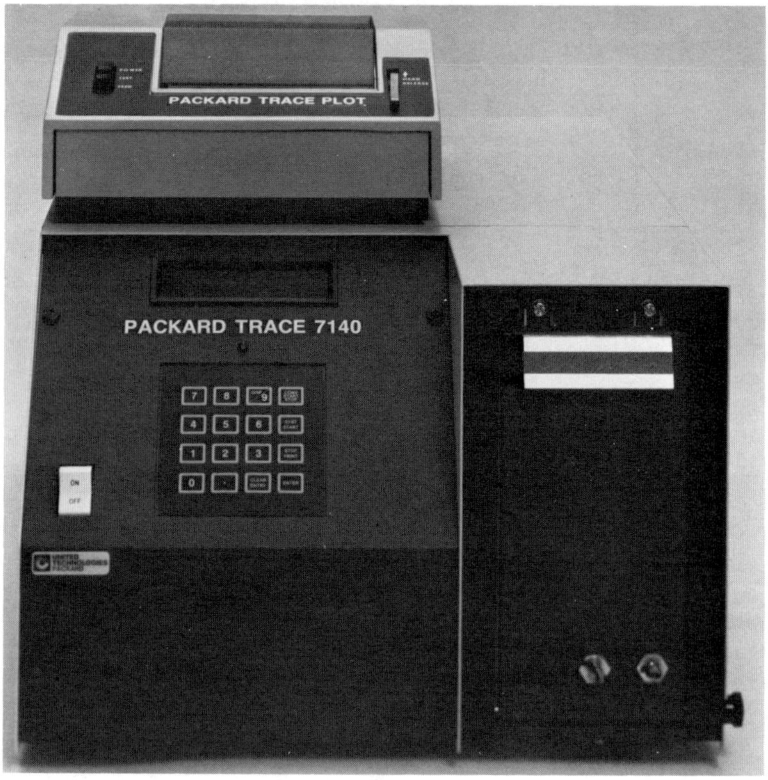

Class	Laboratory
Principle of Operation	Models 7130 and 7140: Monitor radioactivity in effluents from HPLC systems. Borosilicate glass flow cell packed with scintillation beads. Hinged panel mounted cell assembly. High sensitivity detector system. Sealed detector compartment. Fast release couplings from HPLC output and exit to waste or fraction collector. Alpha-numeric display. Measuring parameters entered through sensor keypad in simple question and answer routine. Microcomputer control and analysis. Detected activity processed through 2 channels for separation of radionuclides in dual label analysis. Analog output to potentiometric recorders and RS232 output.

Model 7140:
DPM calculation and spillover correction. Complete analysis of detected activity. Peak sensing and integration. Printer/plotter provides an interpolated histogram of detected activity during analysis. Program summary printed at completion of run. Manual or automatic run start from HPLC auto injector. Signal outputs to ancillary equipment for fraction collector or waste value. Dynamic efficiency depends on flow rate, cell size.

Sensitivity and Range

25 to 10^6 counts per minute

Performance

Sampling: 1 minute to 2000 minutes

Efficiency:	^3H static—7.5% Dynamic	8-11%*
	^{14}C static—75.0% Dynamic	72-78%*
Background:	$^3H < 25$ cpm	$^{14}C < 30$ cpm

Requirements

Power:	Model 7130: 100 Watts	
	Model 7140: 200 Watts	Freq. 50 or 60 Hz
	(approx.) 115/234 V	
Size:	Monitor	26.6 cm (10 1/2 in.) Height
		42 cm (16 1/2 in.) Width
		63 cm (24 3/4 in.) Depth
	Printer/Plotter	11.4 cm (4 1/2 in.) Height
		26.6 cm (10 1/2 in.) Width
		29 cm (11 1/2 in.) Depth
Weight:	Monitor	31.0 kg (68 lbs)
	Printer/Plotter	5.5 kg (12 lbs)

Cost

Model 7130: $9,000
Model 7140: $11,000

Optional Features

Model 589 Recorder, Single pen, 5 in. recorder.

Address

Packard Instrument Company
2200 Warrenville Road
Downers Grove, IL 60515
(312) 969-6000

Portable Tritium in Air Monitor
Scintrex Model AEP 5215A

Class	Laboratory and field area monitor.
Principle of Operation	Gamma-compensated ionization chamber to eliminate effects of high ambient gamma radiation fields. Meter scaled in MPC_a
Sensitivity and Range	0–10 MPC_a linear 0.8–10^2 MPC_a log 0.8 $\times 10^2$–10^4 MPC_a log 0.8 $\times 10^4$–10^6 MPC_a log Accuracy: deviation from calibrated sensitivity < 10% Sensitivity Drift: Linear range, +0.2% per °C Log range, −0.2% per °C Noise: 0.3 MPC_a rms on most sensitive range
Performance	Max. output current: 5 mA Sampling rate: 40 cm^3/sec normal with pump as supplied 250 cm^3/sec maximum with any pump. Max. relative humidity: 90% Pressure range: 20 kPa (~3 psi) above or below ambient pressure
Requirements	Power: Main batteries: two, 9 V carbon batteries (Eveready No. 246) or equivalent or line-operated supply, ±9V, 10 mA maximum Ion chamber: three 30 V carbon batteries (Everyready No. 413) or equivalent Size: Ionization chamber volume: 1.325 liter 37 cm H × 26 cm W × 33 cm D (14 1/2 in. H × 10 1/4 in. W × 13 in. D) Weight: 5.7 kg (~13 lb)
Feature	Calibration source: ^{63}Ni (1-10 μCi, 67 keV max. β energy)
Cost	Refer to manufacturer
Address	Scintrex, Ltd. 222 Snidercraft Road Concord, Ontario L4K 1B5, Canada (416) 669-2280 Telex 06-964570 Attn: S. Nargolwalla

Automatic Tritium in Urine Analyser
Scintrex Model 129A

Class	Laboratory
Principle of Operation	Urine is directly voided into the instrument. Minimum sample volume is 35 ml. Tritium concentrations in urine or other aqueous samples is measured by liquid scintillation counting.
Sensitivity and Range	Sensitivity: minimum of 5 μCi/liter = 0.1 MPBB (maximum permissible body burden) Range of display: 0.01 to 99.99 MPBB on a 4-digit LED display
Performance	Stability: About ±10% of actual sample activity Total processing and counting time: 66 sec (20 sec of counting time) Accuracy: Less than ±20% of actual sample activity for readings above 0.2 MPBB Alarm: contact shift supervisor if lamp illuminated above 20 MPBB
Requirements	Power: 120 Vac ±10%, 60 Hz ±5% Air Supply: Oil-free air, pressure range 552 to 1034 kPa at dew point less than 40°C (104°F) Water Supply: Pressure range 414 to 689.5 kPa Temperature range 5 to 45°C (40 to 113°F) Dimensions: Instrument: 76 cm D × 57 cm W × 1.95 m H) (2.5 ft D × 1.9 ft W × 6.4 ft H) Control Panel: 1.15 m D × 2.40 m W × 4.5 m H) (3.8 ft D × 7.9 ft W × 14.8 ft H) Weight: Instrument cabinet: 227 kg (500 lb) not including lead shield Lead shield: 159 kg (350 lb) Operating temperature: 15 to 40°C (60 to 104°F) Ambient Relative Humidity: up to 95% Ambient Radiation Field: up to 1 mR/hr Capacity of Standard tank: 14 litres
Features	Two sample processing programs 1. Assay sample followed immediately by assay of background 2. Assay any number of samples followed (10 minutes later) by assay of background A standby battery for use during a power failure Pre-settable counting time

Cost	Refer to manufacturer
Address	Scintrex, Ltd. 222 Snidercraft Road Concord, Ontario L4K 1B5, Canada (416) 669-2280 Telex 06-964570 Attn: S. Nargolwalla

Portable Tritium Monitor
Technical Associates Model PTG-5

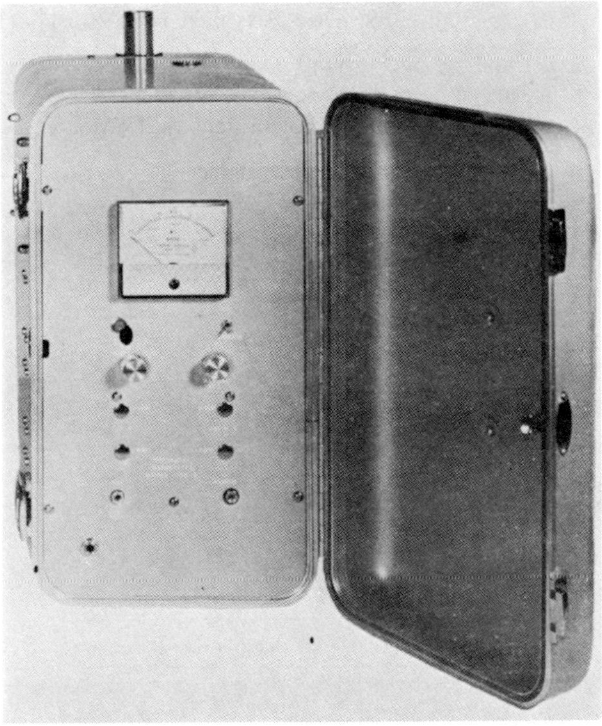

Class	Portable
Principle of Operation	Two ionization chambers, one a subtractive balanced chamber electrometer circuit to decrease background; deionized and filtered air intake. Will measure as free hydrogen or as volatile chemicals.
Sensitivity and Range	Ranges: log, 3×10^{-6} μCi/ml to 3×10^{-3} μCi/ml, 6 decades log, 3×10^{-3} μCi/ml to 3 μCi/ml, 6 decades
Sampling	Continuous
Performance	Background: Essentially eliminated by subtractive balanced chambers. Smoke, dust, and ion elimination: Filter and deionizer reduce effects to negligible level. Recorder: Plug is furnished for external 10mV recorder.

Requirements	Power: Mercury or NiCad Size: Front panel, 38.7 cm L × 21 cm W 15 1/4 in. × 8 1/4 in., excluding handle, 22.9 cm W × 40.6 cm L × 36.8 cm H (9 in. × 16 in. × 14 1/2 in.)
Features	Dynamic background compensation Can actuate alarm or recorder Calibration source, for in field calibration
References	Manufacturer's Specifications
Cost	$3625.00
Address	Technical Associates 7051 Eton Avenue Canoga Park, CA 91303 (213) 883-7043

**Iodine Localizer and Air Monitor
Technical Associates Model FMI-5**

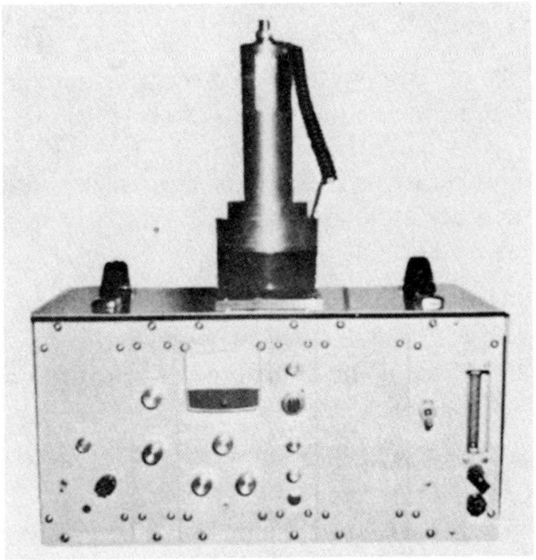

Class	Laboratory
Principle of Operation	As air monitor: Pump pulls air through charcoal filter, thin crystal gamma scintillation detection of I-125 or I-131 (or other isotopes), photomultiplier tube, single channel analyzer, count rate meter (scaler optional)—3 position switch on analyzer allows switching to I-131 or I-125 window settings or to analyzer mode where other settings can be dialed. As iodine localizer: Probe is removed from air monitor module and placed against neck for thyroid iodine localizing—3 position switch used to determine energy selection, count rate meter readout (scaler optional).
Sensitivity and Range	Sensitivity: 1 nanocurie I-125 with collimator and shielding Range: 0 to 500, 5000, 50,000, 500,000 cpm Energy: Better than 90% transmission through detector window for I-125
Sampling	Continuous
Performance	Accuracy: ±5% full scale count rate meter Optional scaler: ±1 count scaler; ±1% time base accuracy Pump: (optional) 0 to 80 cfh standard; other flow rates to 200 cfh available Variable high voltage: 0 to 2000 V

Requirements	Power: 115 Vac, 50-60 Hz Size: 17 in. W × 9 in. H × 13 1/2 in. D (43 cm × 22 cm × 34 cm) Weight: 22.7 kg (50) lbs exclusive of pump
Features	Dual purpose system: Air monitor and/or iodine localizer. Channel switching for I-125, I-131, and other isotopes. Plug-in modular. Can operate as scanning spectrometer with strip chart recorder (recorder optional). Has background suppress.
Cost	FMI-5: $2810.00 FMI-55: $3370.00 (scaler version) AP-2: Air pump complete with fitting, hoses, and orifice regulator to 3 cfm: $220.00
Address	Technical Associates 7051 Eton Avenue Canoga Park, CA 91206 (213) 883-7043

Liquid Scintillation Counting System
Model 6881

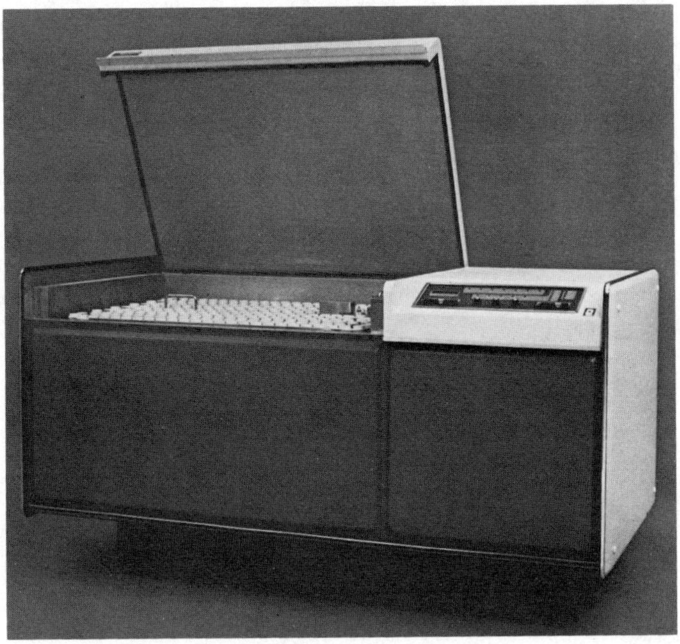

Class	Laboratory
Principle of Operation	Programmable liquid scintillation spectrometer system for counting single or dual labeled intermixed samples. Using the built-in multichannel analyzer, entire standards and sample spectra are measured and stored for determining the level of quench and optimum counting windows. System automatically optimizes counting windows, preset time, preset count, preferred method of quench correction, background subtract for 12 programs. The system uses two photomultiplier detectors, two simultaneous data channels and scalers/timer. Printout consists of: distintegrations per minute (optional) counts per minute in both channels, quench correction ratios (SCR and ESP) sample numbers and elapsed time.
Sensitivity and Range	Detects and counts all alphas, betas and gammas. Energy and Range: 0–1700 keV
Sampling	300 sample capacity, bi-directional conveyor with anti-jam mechanism on the conveyor and sample elevator.

Performance	^{3}H efficiency: greater than 60% E^2/B for ^{3}H greater than 200 ^{14}C efficiency: greater than 95% E^2/B for ^{14}C greater than 500
Requirements	Power: 115 ±10 Vac, or 230 ±10 Vac at 50–60 Hz (specify) Size: 34 in. (86.4 cm) D × 69.75 in. (117.2 cm) W × 36.5 in. (92.7 cm) H. Weight: 1000 lbs; 450 kg.
Features	Readout devices—specify Texas Instruments or Teletype Uses standard or minivials Mechanically positioned ^{133}Ba external standard under the vial. Refrigeration is optional. DPM calculations are optional.
References	Manufacturer's Specifications
Cost	$21,685 to $31,485 depending on options
Address	Tracor Analytic, Inc., 1842 Brummel Drive Elk Grove Village, IL 60007 (312) 364-9100

Liquid Scintillation Counting System
Tracor Analytic Model 6882

Class	Laboratory
Principle of Operation	Programmable liquid scintillation spectrometer system for counting single, dual or triple labeled intermixed samples. Using the built-in multichannel analyzer, entire standards and sample spectra are measured and stored for determining the level of quench and optimum counting windows. Twelve users can utilize any one of twelve different programs. The system automatically optimizes counting windows, changes the preset time, preset count, preferred method of quench correction, background subtract for 12 programs. The system uses two photomultiplier detectors, three simultaneous data channels and scalers/timer, two channel DPM, three channel CPM, printout consists of : DPM, CPM, quench correction ratios (SCR and ESP), sample numbers and elapsed counting time.
Sensitivity and Range	Detects and counts all alphas, betas, and gammas. Energy range: 0–1700 keV
Sampling	300 sample capacity, bi-directional conveyor with anti-jam mechanism on the conveyor and sample elevator
Performance	^{3}H efficiency; greater than 60% E^2/B for ^{3}H greater than 200. ^{14}C efficiency; greater than 95%, E^2/B for ^{14}C greater than 500.
Requirements	Power: 115 ± 10 Vac, or 230 ± 10 Vac at 50–60 Hz (specify) Size: 34 in. (86.4 cm) D × 69.75 in. (117.2 cm) W × 36.5 in. (92.7 cm.) H. Weight: 450 kg (1000 lbs)
Features	Readout devices—specify Texas Instruments or teletype Uses standard or minivials Mechanically positioned ^{133}Ba external standard under the vial Refrigeration is optional
References	Manufacturer's Specifications.
Cost	$28,150 to $34,150 depending on options.
Address	Tracor Analytic, Inc., 1842 Brummel Drive Elk Grove Village, IL 60007 (312) 364-9100

Liquid Scintillation Counting System
Tracor Analytic Delta/300 Model 6891

Class	Laboratory
Principle of Operation	Liquid scintillation spectrometer system for counting single or dual labeled samples. The system uses two photomultiplier detectors, two simultaneous data channels and scalers/timer. Printout consists of counts per minute in both channels, quench correction ratios (SCR and ESR), sample numbers and elapsed counting time.
Sensitivity and Range	Detects and counts all alphas, betas and gammas. Energy range: 0–1700 keV
Sampling	300 sample capacity, bi-directional conveyor with anti-jam mechanism on the conveyor and sample elevator
Performance	^{3}H efficiency: greater than 60%; E^2/B for ^{3}H greater than 160; ^{14}C efficiency: greater than 95%
Requirements	Power: 115 ±10 Vac; 230 ±10 Vac at 50–60 Hz (specify) Size: 30 in. (76 cm) D × 42 in. (107 cm) W × 22 in. (56 cm) H Weight: In USA; 225 kg (500 lbs)
Features	Built in printer plus optional Texas Instruments or Teletype. Uses standard or minivials. Mechanically positioned ^{133}Ba external standard under the vial. DPM calculations are optional.
References	Manufacturer's Specifications.
Cost	$13,000 to $19,800, depending on options.
Address	Tracor Analytic, Inc., 1842 Brummel Drive Elk Grove Village, IL 60007 (312) 364-9100

B.1. Combination Instruments

Instrument Note No.	Model	Use	Radiation Detected	Range	Page No.
Amperex Electronic Corporation					
Amperex 1	Numerous	Radiation counter tubes	α, β, γ	Various	578
Baird Corporation					
Baird 1	912-000	Automatic planchet counting system	α, β		581
Baird 2	912-005	Automatic planchet counting system	α, β, γ		583
Baird 3	450	Linear ratemeter (laboratory monitor)	α, β, γ	*Count*: 0 to 100 cpm through 0 to 10^6 cpm	585
Baird 4	470	Linear log ratemeter	α, β, γ	*Count*: 0 to 10^6 cpm (linear); 10 to 10^6 cpm (log)	587
Canberra Industries, Inc.					
Canberra 1	2400, 2401, 2408	Low level $\alpha/\beta/\gamma$ counting system	α, β, γ		588
Eberline Instrument Corporation					
Eberline 1	"Rascal" PRS-1, -1P, -2, -2P	Ratemeter scaler	α, β, γ, Neu		590
Eberline 2	RO-2/2A, RO-3, RO-4, E-120/HP-270, E-520/HP-270	Alpha, beta, gamma survey meters	α, β, γ	*Count*: 0 to 0.5, 5, 2000, 5000 mR/hr	592

B.1. (Continued)

Instrument Note No.	Model	Use	Radiation Detected	Range	Page No.
Elscint, Inc.					
Elscint 1	GSM-1	Survey meter	α, β, γ	*Energy*: 60 keV to 1.3 MeV; *Count*: 0 to 0.5, 5, 50 mR/hr	596
Gamma Industries					
Gamma 1	821-0091-002	Multiple alarm radiation system	α, β, γ	*Count*: 0 to 2, 20, 200 mR/hr	598
Harshaw Chemical Company					
Harshaw 1	12-A6	Automatic alpha/beta counting system	α, β		600
Harshaw 2		Phoswich detectors	α, β, γ		602
Health Physics Instruments, Inc.					
Health Physics 1	2010	Remote area monitoring system	α, β, γ		604
Wm. B. Johnson and Associates, Inc.					
Johnson and Associates 1	GSM-10	Radiation survey meter	α, β, γ, Neu	*Count*: 0 to 0.2, 2, 20 mR/hr for β, γ	606
Johnson and Associates 2	RML-1A Count Ratemeter	Radiation monitor	α, β, γ, Neu	*Count*: 0 to 500, 5000, 50000 cpm	611
Johnson and Associates 3	LS-4	Portable scaler	α, β, γ, Neu		612

Johnson and Associates 4	RML-3	Mini-ratemeter and monitor	α, β, γ		613
LND, Inc.					
LND 1	Various	Alpha, beta, gamma counter	α, β, γ		614
Ludlum Measurements, Inc.					
Ludlum 1	2, 3	Survey meters	α, β, γ	*Count*: Model 3: 0 to 2, ×0.1, ×10 mR/hr Model 2: 0 to 5, ×0.1, ×1, ×10 mR/hr	616
Ludlum 2	12, 16	Survey meters	α, β, γ	*Count*: 500 cpm, ×1, ×10, ×100, ×1000 Depends on detector	621
Ludlum 3	15	Neutron, beta, gamma and alpha count ratemeter	α, β, γ, Neu	*Count*: 0 to 500, 5000, 50000, 500000 cpm	623
Ludlum 4	2220	Hand-held scalar ratemeter	α, β, γ	*Count*: Depends on detector	625
National Nuclear Corporation					
National Nuclear 1	WCM-10, -20, -30	Waste curie monitor	α, β, γ		626
Nuclear Enterprises Limited					
Nuclear Enterprises 1	HFM-6	Hand and clothing monitor	α, β		627
Nuclear Enterprises 2	RM-5, RM-6	Portable ratemeters	α, β, γ	*Count*: RM-5: 0.5 to 5000 cps RM-6: 1 to 5000 cps	628

B.1. (Continued)

Instrument Note No.	Model	Use	Radiation Detected	Range	Page No.
Nuclear Enterprises 3	PCM-5	Count ratemeter	α, β, γ, x-ray	*Count*: 1 to 10, 10 to 100, 100 to 1000, 500 to 5000 cps	629
Nuclear Measurements Corporation					
Nuclear Measurements*	AM-21 Series	Air radiation monitor systems	See Appendix A.9 or Table B.7		1049
Nuclear Research Corporation					
Nuclear Research 1	NRC Lab Map-60	Airborne particulates monitor	α, β, γ, Nuc		632
Nuclear Research 2	HFA-200A	Hand and foot monitor	α, β, γ	*Count:* 0 to 500, 5000, 50000 cpm	633
Nuclear Research 3	CP-2, -5, -10	Portable survey meter	α, β, γ	Various	634
Nuclear Research*	SM-101	Stack monitor	See Section A.9 and Table B.7		1053
Nucleus, Inc.					
Nucleus 1	L, Ratemeter	Radiation laboratory monitor	Various	*Count:* 0 to 500, 2000, 5000, 20000, 50000 cpm	635

Packard Instrument Company					
Packard 1	TRI-CARB, Models 4530, 4640, 4550, 4660	Liquid scintillation system	α, β, γ, x-ray		636
Scintrex Ltd.					
Scintrex 1	139A	Laundry monitor	α, β, γ	*Count:* 10 to 10^5 cpm (log)	638
Technical Associates					
Technical Associates 1	TBM-3	Survey meter	α, β, γ	*Count:* 0 to 500, 5000, 50000 cpm; 0 to 0.15, 1.5, 15 mR/hr	639
Technical Associates 2	HSM-10A, -10AM, -10BS, -10G	Hand and shoe monitors	α, β, γ		640
Technical Associates 3	Cutie Pie: CP-44, -5, -6, -7; Juno: ERJ-9	Portable survey meters	α, β, γ	*Count:* Various	642
Technical Associates 4	SML-2, FS-8, PUG-1, PUG-1AB, PUG-1E	Portable ratemeter	α, β, γ	*Count:* Various	646
Technical Associates 5	BAM-3A, -3B, -3H	Air monitors	α, β, γ	*Count:* 10 to 100000 cpm	650
Technical Associates 6	SAAM-1, -2, -3	Alpha air monitor (moving filter)	α, β, Nuc		652

B.1. (Continued)

Instrument Note No.	Model	Use	Radiation Detected	Range	Page No.
Tracor Analytic, Incorporated					
Tracor Analytic 1*	6881	Liquid scintillation counting system	See Appendix A.9 or Table B.7		1065
Tracor Analytic 2*	6882	Liquid scintillation counting system	See Appendix A.9 or Table B.7		1067
Tracor Analytic 3*	Delta 300/ Model 6891	Liquid scintillation counting system	See Appendix A.9 or Table B.7		1068
Victoreen					
Victoreen 1	425, 470A, 471, 471A, 471RF, 692 and 660 series	Survey meters	α, β, γ	*Count:* Various	654
Victoreen 2	490, 491, 492, 493, 498, 740F, 470A, 885	Field survey meters	α, β, γ, x-ray	*Count:* Various	657
Warrington Laboratories					
Warrington 1	2590, 2650, 2650AR Series	Radiation survey meters	α, β	*Count:* Various	660

Xetex, Inc.					
Xetex 1	306A, 308A	Portable, hand-held survey meters	α, β, γ, x-ray	*Count:* 306A: 0 to 9.99 kcpm 308A: 0 to 1000, 0 to 10000 cpm	664
Xetex 2	315B	Portable, hand-held background subtract ratemeter	α, β, γ, x-ray	*Count:* 0 to 1000, 10000 cps	666
Xetex 3	318A	Portable, hand-held ratemeter/survey meter	α, β, γ, x-ray	*Count:* 0 to 9999 mR/hr	668

*The instrument note for models marked by an asterisk appears in a different section than that covered in the table.

B.2. Instruments for Measuring Alpha Radiation

Instrument Note No.	Model	Use	Range	Sensitivity	Page No.
Eberline Instrument Company					
Eberline 1	Alpha-5A	Alpha air monitor	0 to 10000 cpm; full log scale	*Mdl:* 4 MPC of ^{239}Pu in avg Radon-Thoron	671
Eberline 2	SAC-4	Alpha scintillation counter		α *Efficiency:* 80% 2π min. from 2.5 cm diam. ^{239}Pu source	673
Eberline 3	PRS-1/AC-3, PRS-2/AC-3, PAC-4G-3, PAC-6, PAC-7	Alpha survey meters	Various	Various	675
Harshaw Chemical Company					
Harshaw 1	272	Portable alpha monitor	*Energy:* 2 to 6 MeV *Count:* 0 to 9999 counts	*Efficiency:* 25% for Am	680
Harshaw 2	Series DJ	Diffused-junction silicon detectors	*Depletion depth:* typically 100 μm	*Resolution:* 25 to 60 keV for ^{241}Am	682
Harshaw 3		Phoswich scintillation detector			684
AERE Harwell (UKAEA)					
Harwell 1	SRD25	Diffused-junction silicon detector	*Depletion depth:* 60 μm (nominal)		686

Harwell 2	36190	Alpha spectrometer	4 decades with 1, 10, 100 prescale	α *Efficiency:* for 2.5 to 5 MeV α, 30% for ^{239}Pu	688
Harwell 3	3280	Alpha in air measuring system		*Efficiency:* 30% (4π) *Sensitivity:* 0.4 cps at 8 mpc-hr and flow of 37 liters/min.	690
National Nuclear Corporation					
National Nuclear 1	LMS-1	Portable liquid monitor	0 to 99999 counts	*Maxium sensitivity:* approx 10 dpm/ml	691
Nuclear Measurements Corporation					
Nuclear Measurements 1	AM-2A, -3A, -2P, -3P, -22P, -33P	Air particulate monitors	Various	*Efficiency:* α: 20 to 37% β: approx 37%	693
Nuclear Research Corporation					
Nuclear Research 1	MAP-1000	Continuous alpha air monitor			696
Nuclear Research 2	ASM-30	Alpha scintillation survey meter	*Energy:* > 3 MeV *Count:* 10^3, 10^4, 10^5, 10^6 cpm		697
EG&G ORTEC					
Ortec 1	Series A,B, C,D,E,R	Silicon surface-barrier detectors	*Depletion depth:* 15 to 5000 μm	*Resolution:* α: 14 to 80 keV β: 6 to 75 keV	699
Ortec 2	576	Alpha counting and spectroscopy system	*Energy:* 4 to 6, 5 to 7, 6 to 8 MeV	*Efficiency:* $\geq$ 25% at 5.486 MeV, 1-1/2 in. sample, 1 mm spacing	702

B.2. **(Continued)**

Instrument Note No.	Model	Use	Range	Sensitivity	Page No.
Paul G. Downey & Company					
Paul G. Downey 1		Silicon surface barrier detectors	*Depletion depth:* 100 to 1000 μm	*Resolution:* α: 15 to 35 keV β: 7 to 30 keV	703
RADēCO/SAI					
RADēCO 1	442A	Alpha air particulate monitor	*Energy:* 0 to 1.0 MeV adjustable *Counts:* 0 to 100 (with 0.5×, 1×, 5×, 10×, 50×) cpm	*Efficiency:* 23% of 2π with 0.6 MeV window	706
Technical Associates					
Technical Associates*	SAAM-1, -2, -3	Moving filter alpha air monitor	See Appendix A.3 and Table B.1		652
Tennelec, Incorporated					
Tennelec 1	TC 256	Alpha spectrometer	*Energy:* 3 to 8, 4 to 7, 3 to 5, 4 to 6, 5 to 7, 6 to 8 MeV	—	708
Tennelec 2	LB-1000 Manual LB-5100 Auto	Low background α/β counting systems	Various	Various	709
Warrington Laboratories					
Warrington*	2590, 2650, 2650AR Series	Radiation survey meter	See Section A.3 and Table B.1		660

*The instrument note for models marked by an asterisk appears in a different section than that covered in the table.

B.3. Instruments for Measuring Beta and Beta/Gamma Radiation

Instrument Note No.	Model	Use	Range	Sensitivity	Page No.
Eberline Instrument Corporation					
Eberline 1	RT-1A; E-130A; PIC-6A; E-520/HP-270; 6112; E-120/HP-270; RO-7	Portable, hand-held beta-gamma survey meters	Various	Various	713
Eberline 2	AMS-3	Particulate air monitor	*Count:* 10 to 10000 cpm (log)	*Efficiency:* with 47 mm diam. source, ^{90}Tc (15%), ^{90}Sr (75%), ^{90}Y (75%); approx. 2π	720
Eberline 3	PM-6	Hand, hand and foot portal monitors			722
Fisher Control Ltd.					
Fisher Control 1	NE 014	Ion chamber	*Energy:* 100 keV to 3 MeV; gamma		724
AERE, Harwell					
Harwell 1	3230	Beta in air measuring system	*Count:* β: 7000 dpm/m^3 at 37 liter/min. γ: 18 cps at 1 mrad/hr		725

B.3. **(Continued)**

Instrument Note No.	Model	Use	Range	Sensitivity	Page No.
Harwell 2	3247, 3248	Portable, hand-held, high level beta-gamma doserate meter	*Energy:* 50 keV to 3 MeV *Dose rate:* 3247: 0.1 to 100 rads/hr 3248: 5 to 5000 rads/hr		726
Health Physics Instrument, Incorporated					
Health Physics 1	1075	Portable, hand-held beta-gamma survey meter	*Energy:* 20 keV to 1.3 MeV *Dose rate:* 0.001 to 200 rad/hr		728
Wm. B. Johnson and Associates, Inc.					
Johnson and Associates 1	HP-265, PPA-2	Hand probe	*Gamma count:* HP-265: ~3600 cpm/mR/hr PPA-2: ~6000 cpm/mR/hr	*Beta efficiency:* ^{90}Sr, ^{90}Y, 45%	730
Johnson and Associates 2	UM-4	Hand and shoe monitor	*Count:* 0 to 100000 cps		731
National Nuclear Corporation					
National Nuclear 1	Beta-10, Gamma-10	Portal monitors	Various	Various	733

Nuclear Associates					
Nuclear Associates 1	05-700	Survey meter	*Energy:* β, 175 keV minimum; γ, 20 keV through ^{60}Co γ *Count:* 0.5, 5, 50 mR/hr (300, 3000, 30000 cpm)		734
Nuclear Research Corporation					
Nuclear Research 1	HFM-2	Hand and foot monitor	*Count:* 0 to 99999 counts		735
Nuclear Research 2	DMW-200 Series	Portable, fixed monitors	*Count:* DMW-200 P-1: 0.01 mR/hr to 999 R/hr DMW-200 P-2: 0.01 mR/hr to 9999 R/hr		736
RADēCO/SAI					
RADēCO 1	GM-222	Beta-gamma air monitor	*Energy:* > 150 keV *Count:* 1000, 10000, 100000 cpm	*Beta efficiency:* ^{90}Sr, ^{90}Y, 35%; ^{14}C, 5%	737
Scintrex, Ltd.					
Scintrex 1	189D	Portable, gamma compensated beta monitor	*Count:* 0.1 to 1200 rem/hr		739
Technical Associates					
Technical Associates 1	CP-TP-54A CP-TP-10K	Portable, hand-held high range survey meters	*Count:* 54A: 0 to 0.5, 5, 50, 500 R/hr 10K: 0 to 10, 100, 1000, 10000 R/hr		740

B.3. (Continued)

Instrument Note No.	Model	Use	Range	Sensitivity	Page No.
Technical Associates 2	F-5ABN	Beta-gamma air particulate monitor	*Energy:* βs > 0.1 MeV *Count:* 10 to 100000 cpm (log)		742
Technical Associates 3	PPM-21, PPM-21A	Portal monitors			744
Technical Associates 4	PPM-25P, PPM-25GF	Portal monitors			746
Tracor Analytic, Inc.					
Tracor Analytic 1	6895	Liquid scintillation system	*Energy:* 0 to 1700 keV βs	*Beta efficiency:* $^3H > 60\%$ $^{14}C > 95\%$	747
Victoreen, Incorporated					
Victoreen 1	570	Portable hand-held condenser, roentgen meter	Various	Various	749
Xetex, Incorporated					
Xetex 1	304A, 305B	Portable hand-held digital β-γ monitor	*Energy:* $\beta > 0.25$ MeV *Count:* 304A: 0.1 to 99.9 mR/hr 305B: 0.1 mR/hr to 99.9 R/hr	*Efficiency:* 60 keV to 1.3 MeV γ: 15%	751

B.4. Instruments for Measuring Gamma and X-Radiation

Instrument Note No.	Model	Use	Range	Resolution	Page No.
Baird Corporation					
Baird 1	904-460	Survey monitor	*Count:* 1 to 100 mR/hr; 500 to 50000 cpm		754
Baird 2	980-10C	Shielded well scintillation counter			755
Beckman Instruments, Incorporated					
Beckman 1	System 5500	Gamma counter		< 14% (^{137}Cs, 2 in. crystal)	757
Canberra Industries					
Canberra 1	7229, 7229N, 7229P	Pure Ge, Ge(Li) gamma detectors		*At 1.33 MeV:* 7229N, P: 1.8 to 2.2 keV 7229: 1.8 to 2.3 keV *At 122 keV:* 7229N: P 800 to 1300 eV	759
Canberra 2	802	NaI(Tl) spectrometry system			761
Canberra 3	7313, 7333, 7383	X-ray detectors		*At 5.9 keV (FWHM):* 7313: 160 eV 7333: 170 eV 7383: 185 eV	763
Canberra 4	1701, 1702	Scintillation detectors		*At 6 keV:* better than 55%	765

B.4. (Continued)

Instrument Note No.	Model	Use	Range	Resolution	Page No.
Canberra 5		Pure Ge planar detectors	Low Energy	*At 122 keV (FWHM):* 500-700 eV *At 5.9 keV (FWHM):* 145-590 eV	766
Canberra 6	MAC	MAC, portable		*At 1.3 MeV (FWHM):* 1.8 to 2.2 keV	768
Capintec, Inc.					
Capintec 1	CRC 5/10; CRC-30	Radioisotope calibrator	*Energy:* 25 keV to 3 MeV	0.1 μCi	769
Capintec 2	Diarad III	Exposure/exposure rate meter	*Energy:* x-ray	0.1 mR (mR/min)	772
Capintec 3	Therarad I	Exposure/exposure rate meter	*Energy:* x-ray *Count:* 2000 R (R/min)	0.001 R (R/min)	774
Capintec 4	Tomorad I	Exposure/exposure rate meter	*Energy:* x-ray (40 to 200 keV) *Count:* 2000 R cm (R cm/min)	0.001 R cm (R cm/min)	775
Capintec 5	174, 192	Digital exposure and exposure ratemeters	*Count Example:* 174: 20 mR/min to 200.0 R/min 192: 2.000 mR/min to 2000 R/min Depends on ionization chamber; see table in note	174: 0.01 mR (mR/min) 192: 0.001 mR (mR/min)	777

Eberline Instrument Corporation				
Eberline 1	SPI-2	Source position indicator	*Energy:* 70 keV to > 3 MeV	782
Elscint, Incorporated				
Elscint 1	GSM-1	Survey meter	*Energy:* ~60 keV to ~1.3 MeV *Count:* 0 to 0.5, 5, 50 mR/hr	784
Fisher Controls Ltd.				
Fisher 1	GM-700	Gamma monitor	*Energy:* 90 keV to 1 MeV *Count:* 0.5 to 10^4 mR/hr (depends on channel)	786
Fisher 2	0928	Four channel gamma spectrometer	*Energy:* 50 keV to 3 MeV (depends on channel) *Count* (Channels 1, 2, 3): 0 to 10, 30, 100, 300, 1000 cpm	788
Fisher 3	0593	Gamma survey meter	*Energy:* gamma	790
Fisher 4	NE-148A	Gamma survey meter	*Energy:* 0.15 to 1.25 MeV *Count:* 0 to 30, 300, 3000 mR/hr (130, 1200, 10000 cps)	792

B.4. (Continued)

Instrument Note No.	Model	Use	Range	Resolution	Page No.
Gamma Industries					
Gamma 1	150B, 155B, 250B, 252B	Portable survey meter	*Energy:* 250 keV to 1.5 MeV *Count:* 150B, 155B: 0 to 2, 20, 200 mR/hr 250B, 252B: 0 to 10, 100, 10000 mR/hr		794
Harshaw Chemical Company					
Harshaw 1	301	Portable, low level, gamma and x-ray survey meter	*Energy:* 15 to 100 keV		797
Harshaw 2	401, 401W	401: Air Monitoring System 401W: Water Monitoring System	*Energy:* 0.1 to 2 MeV *Count:* 401: 1 to 10 μR/hr 401W: 10 pCi/liter		799
Harshaw 3	2000P TLD System	Gamma and x-ray Monitor	*Count:* 0.1 to $>$ 5000 mR, depending on bulb dosimeter		801
AERE, Harwell					
Harwell 1	0740	Area gamma alarm monitor	*Count:* 0.01 to 300 mR/hr (log)		803

Harwell 2		X-ray proportional counters		*At 1 keV:* approx. 0.23 keV *At 100 keV:* approx. 2.3 keV	804
Health Physics Instruments, Inccrporated					
Health Physics 1	1010	Portable, hand-held, tissue equivalent, digital survey meter	*Energy:* 40 keV to 1.25 MeV *Rate:* 0 to 0.1, 1, 10, 100, 1000 mrad/hr *Dose:* 0 to 0.01, 0.1, 1 mrad		805
Health Physics 2	1072	Portable, hand-held, remote area monitor	*Count (Dose rate):* 1072-1: 0.1 to 199.9 mR/hr 1072-2: 1 to 1999 mR/hr		807
Wm. B. Johnson and Associates					
Johnson and Associates 1	TVX-1	Portable, hand-held, survey meter	*Energy:* 15 keV to 10 MeV *Count:* 0.05 mR/hr to 1 R/hr; 3000, 10000, 30000, 100000 cpm		808
Johnson and Associates 2	GSM-12	Portable, hand-held, survey meter	*Energy:* gamma *Count:* 0 to 10, 100, 1000 mR/hr		810
Johnson and Associates*	TR-5	Air monitor	See Section A.9 and Table B.7		1030

B.4. (Continued)

Instrument Note No.	Model	Use	Range	Resolution	Page No.
Johnston Laboratories					
Johnston*	Triton 955B	Radionuclide gas monitor	See Section A.9 or Table B.7		1034
Lawrence Livermore National Laboratory					
LLNL 1		Pulsed x-ray monitor	*Energy:* x-ray *Count:* Various (see note)		812
LND, Inc.					
LND 1	Numerous	X-ray proportional counters	*Energy:* x-ray		814
LND 2	Numerous	Gas-filled counters and ionization chambers	*Energy:* α, β, γ, x-rays		816
Ludlum Measurements, Incorporated					
Ludlum 1	5	Portable, hand-held, survey meter	*Energy:* depends on window *Count:* 0.0 to 2000 mR/hr in 5 linear ranges		819
Ludlum 2	14-C	Portable, hand-held, survey meter	*Energy:* see note for figure *Count:* 0 to 0.2, 2, 20, 200, 2000 mR/hr		821

National Nuclear Corporation				
National Nuclear 1	HM-4, Samscn	Gamma assay meter	*Energy:* 150 to ~700 keV *Count:* 10 to 1000 cps	824
National Nuclear 2	DM-2	Door monitor	*Energy:* gammas, neutrons	825
National Nuclear*	HM-3	Portable personnel/vehicle SNM search and survey monitor	See Section A.9 and Table B.7	1042
Nuclear Associates				
Nuclear Associates 1	Prima I, IIa	Ratemeter	*Energy:* gamma, x-rays *Count:* I: Background to 2 R/hr IIa: Background to 10 R/hr	826
Nuclear Associates 2	05-571, 05-572	Portable, hand-held, survey meters	*Energy:* 60 keV to 1.2 MeV *Count:* 571: 0 to 10, 100, 1000 mR/hr 572: 0 to 500, 5000, 50000 cps	828
Nuclear Associates 3	05-696	Hand/shoe/clothing monitor	*Count:* 3 channels, 1×10^4 counts per cycle (log)	830
Nuclear Associates 4	Primalert 10, 05-433	Radiation level monitor	*Count:* 2.5 to 20 mR/hr	831

B.4. (Continued)

Instrument Note No.	Model	Use	Range	Resolution	Page No.
Nuclear Associates 5	Primalert 35, 05-437	Area alarm monitor	*Count:* alarm at 1, 2, 4, 8, 16, and 32 mR/hr		832
Nuclear Associates 6	Prima 7, 05-582	"Transport Index" monitor	*Energy:* 50 keV to 1.2 MeV *Count:* 0 to 99 mR/hr		833
Nuclear Associates 7	05-670	X, gamma survey meter	*Energy:* 10 keV to 1.2 MeV *Count, dose (dose rate):* 0.3 mR to 3000 R (0.3 mR/hr to 3000 R/hr) in 15 linear scales		836
Nuclear Enterprises Ltd.					
Nuclear Enterprises 1		NaI(Tl) crystals			838
Nuclear Research Corporation					
Nuclear Research 1	TA-900	Area radiation monitor	*Energy:* 80 keV to 1.5 MeV *Count:* 0.01 to 0.99 $\times$ 10^7 mR/hr		839
N. Wood Counter Laboratory, Incorporated					
N. Wood 1	Numerous	Gamma counters, some portable	See table in note		840

EG&G ORTEC					
Ortec 1	Series SLP	Si(Li) photon detector		See table in note	842
Ortec 2	Series GEM, GMX, GLI, GWL, GLP, SLP	Solid state photon detectors		See table in note	844
Packard Instrument Company					
Packard 1	5650, 5780 Auto Gamma®	Radiation measurement system	*Energy:* 15 to 200 keV		846
Packard 2	5301, 5302, 5304 Multi-Prias®	Gamma counter system	*Energy:* 15 to 500 keV		848
Packard 3	Prias™ PGD	Gamma counter	*Energy:* gamma, x-ray		850
Princeton Gamma-Tech					
Princeton Gamma-Tech 1	Numerous	Germanium, Ge(Li) and Si(Li) detectors		See table in note	851
Princeton Gamma-Tech 2		Germanium well detector	*Energy:* 20 keV to 10 MeV		856
Quartz Products Corporation					
Quartz Products 1		Scintibloc and scintiflex detectors	*Energy:* 10 keV to 3 MeV, gamma	*Scintibloc:* 7 to 9% ^{137}Cs (661 keV) *Scintibloc X:* 45 to 60% ^{55}Fe (220 keV) *Scintibloc/Scintiflex:* 7.5 to 9.5% ^{137}Cs (661 keV)	858

B.4. (Continued)

Instrument Note No.	Model	Use	Range	Resolution	Page No.
Radcal Corporation					
Radcal 1	1015C	Portable x-ray monitor	*Count:* exposure rate: 1 mR/hr to 59940 R/hr pulse exposure: 1 mR to 13 R		860
Radcal 2	2025 AC	Portable x-ray monitor	*Count:* exposure rate: 0.001 to 600 kR/hr		862
Radcal 3	10X5 Series	Ion chambers	*Count:* exposure rate, exposure: numerous (see table in note)		864
Radcal 4	20X5 Series	Electrometer/ion chambers	*Count:* exposure rate, exposure: numerous (see table in note)		866
Reuter-Stokes					
Reuter-Stokes 1	RS-P3-1605-261, -271, -285, -287	X-ray proportional counters	*Energy:* 261: 3 to 35 keV 271: 6 to 15 keV 285: 1.5 to 8 keV 287: 3 to 35 keV		869

Reuter-Stokes 2	RS-C4-1606 Series	Gamma ionization chambers	*Energy:* 0.4 keV to 7 MeV	871
Reuter-Stokes 3	RS-C3-1006-202	Gamma ionization chambers	*Energy:* 60 keV to 3 MeV	873
Reuter-Stokes 4	RSS-111	Stationary or portable gamma monitor	*Energy:* 50 keV to 5 MeV *Count:* 1 to 500 μR/hr, standard (to 5 K μR/hr optional)	874
Reuter-Stokes 5	Sentri 1011	Environmental radiation monitoring system	*Sensitivity:* background to 10 R/hr	876
Scintrex				
Scintrex 1	189A, B, C	Gamma survey meters	*Count:* 189A: 1 mrem/hr to 50 rem/hr 189B: 10 rem/hr to 10000 rem/hr 189C: 0.2 rem/min to 200 rem/min	877
Technical Associates				
Technical Associates 1	FML-5A, FML-5B	Area monitors	*Energy:* 100 keV to 1 2 MeV *Count:* 5A: 0.01 mR/hr to 10 R/hr standard, 6 decades 5B: 1 mR/hr to 1 R/hr standard, 4 decades. Other ranges optional	879

B.4. (Continued)

Instrument Note No.	Model	Use	Range	Resolution	Page No.
Technical Associates 2	CP-MU, CP-MU-1000	Underwater monitor	*Count:* CP-MU: 0 to 1, 10, $10^5, 10^6$ kR/hr; CP-MU-1000: 0 to 1, 10, 100, 1000 R/hr; Other ranges optional		881
Teldyne Isotopes					
Teledyne 1	Numerous	NaI(Tl) scintillation crystals			883
Tennelec, Incorporated					
Tennelec 1	Numerous	NaI(Tl) spectrometer systems	*Energy:* x-ray, gamma		886
Texas Nuclear Division, Ramsey Engineering Company					
Texas Nuclear 1	9361	Geiger-Müller detectors			888
TGM Detectors, Incorporated					
TGM 1	Numerous	Geiger-Müller Detectors	*Energy:* α, β, γ		890
Tracor Analytic, Incorporated					
Tracor 1	1185	Automatic gamma counter			894

Tracor 2	1190, 1191, 1290	Automatic gamma counter	*Energy:* 1290: < 200 keV; most gamma	896
Victoreen				
Victoreen 1	Model 500 Series	Precision elecrometer and detectors	*Energy and Count:* see note (multiple)	899
Victoreen 2	692, 808D	Survey meters	*Energy:* 692: 45 keV to 1.3 MeV 808D: 80 keV to 2 MeV *Count:* 692: 0 to 10, 100, 1000, 10000 mR/hr 808D: 0.1 to 100 mR/hr (3 decade log scale)	901
Xetex, Incorporated				
Xetex 1	501A-1, 2, 3	Area monitor	*Energy:* 70 keV to 1.3 MeV, gamma, x-ray *Count:* 501A-1: 0.1 to 99.9 mR/hr 501A-2: 1 to 999 mR/hr 501A-3: 0.01 to 9.99 mR/hr	903
Xetex 2	303A	Portable miniature accelerator monitor	*Energy:* gamma, x-ray *Count:* 0 to 10, 100, 1000 mR/hr	905

B.4. (Continued)

Instrument Note No.	Model	Use	Range	Resolution	Page No.
Xetex 3	302B	Portable, hand-held, high level probe	*Energy:* gamma, x-ray *Count:* 0.01 to 1000 R/hr		907
Xetex 4	418A	Portable, emergency, digital dosimeter	*Energy:* 100 keV to 3 MeV *Count:* 0 to 512 R		909
Xetex 5	501A-5R	Area monitor, laboratory and field	*Energy:* gamma, x-ray, from 30 keV *Count:* 1 to 999 kcpm		910
Xetex 6	503A Xetex Teledose	Digital dosimeter, telemetry system	*Energy:* gamma, x-ray *Count:* 0 to 9999 mR		911
Xetex 7	Autodose 440A/540A	Portable, personal, automatic dosimetry system	*Energy:* 60 keV to 1.3 MeV, gamma *Count:* 0 to 9999 mR		913

*The instrument note for models marked by an asterisk appears in a different section than that covered in the table.

B.5. Instruments for Measuring Neutrons

Instrument Note No.	Model	Use	Energy Range	Counting Range	Page No.
Eberline Instrument Corporation					
Eberline 1	RM-16, NRD	Neutron ratemeter; neutron dose-equivalent monitor	NRD-1: Thermal to 10 MeV RM-16: 4 log decades	1 to 10000 mrem/hr	916
Eberline 2	PRS-2P/NRD	Neutron monitor		0.2 mrem/hr to 10 rem/hr	918
AERE, Harwell (UKAEA)					
Harwell 1	0949-4	Neutron dose-equivalent monitor	Thermal to 11 MeV	0.1 mrem/hr to 10 rem/hr (2 ranges)	920
Lawrence Livermore National Laboratory					
LLNL 1		REM meter for pulsed sources of neutrons	Thermal	Background to 60 mrem/sec	922
LND, Inc.					
LND 1	Numerous	Neutron counting tubes	Wide range available	Wide range available	924
Ludlum Measurements, Incorporated					
Ludlum 1	2, 3	Portable survey meters	Depends on probe; see Ludlum 2	Model 2: 0 to 5 mR/hr; (0.1×, 1×, 10×, linear) Model 3: 0 to 2 mR/hr; (0.1×, 1×, 10×, 100×, linear)	930
Ludlum 2	42-1, 42-2, 42-4	Neutron probes	Thermal to 7 MeV		931

B.5. (Continued)

Instrument Note No.	Model	Use	Energy Range	Counting Range	Page No.
Ludlum 3	42-5	Neutron spectrometer	See note	See note	933
Ludlum 4	15	Portable, hand-held survey meter	Fast and slow	1 to 500 cpm (1×, 10×, 100×, 1000×)	935
N. Wood Counter Laboratory, Inc.					
N. Wood 1	Numerous	Portable laboratory neutron counters			937
National Nuclear Corporation					
National Nuclear 1	SNAP-II, SD-1	Portable fixed neutron detector			941
Nuclear Enterprises					
Nuclear Enterprises 1	NM-2	Neutron monitor	0.025 eV (thermal) to 10 MeV	0.001 mrem/hr to 10 rem/hr	942
Nuclear Enterprises 2	MK-7	Neutron dose equivalent monitor	0.025 eV (thermal) to 7 MeV	0.1 to 1000 mrem/hr	943
Nuclear Research Corporation					
Nuclear Research 1	Rem Rad Model NP-2, "Snoopy"	Portable, hand-held neutron monitor	0.025 eV (thermal) to 15 MeV	0 to 2, 20, 200, 2000 mrem/hr (linear)	945
Nuclear Research 2	E-1D	Portable, hand-held fast neutron monitor	0.2 to 14 MeV	5, 50, 500 mrep/hr (1 mrep $\simeq$ approx. 1 cps)	946

EG&G ORTEC					
Ortec 1	525-780, 525-782	Neutron spectrometers	525-780: 1 to 6 MeV (resolution: 100 keV guaranteed) 525-782: > 3 MeV (resolution: 400 keV guaranteed)		947
Peabody Scientific (**Marketing for Jordan Valley Applied Radiation Ltd.**)					
Peabody 1	FNS-1	^{3}He fast neutron spectrometer	0.1 to 5.0 MeV (resolution: thermal neutrons, better than 12 keV FWHM; 1 MeV neutron, better than 25 keV FWHM)		949
Reactor Experiments, Inc.					
Reactor Experiments 1	710	Criticality dosimeter	Numerous		951
Reactor Experiments 2	Numerous	Neutron foils			953
Reuter-Stokes					
Reuter Stokes 1	Numerous	Neutron detectors	Various	Various	956

B.5. (Continued)

Instrument Note No.	Model	Use	Energy Range	Counting Range	Page No.
Technical Associates					
Technical Associates 1	REM-PUG	Portable neutron dose-rate monitor		0.1 mrem/hr to 10 rem/hr (log)	958
Technical Associates 2	PNS-19; PNS-20; PUG-1N; PUG-1AB	Neutron probes and monitors	Epithermal to 14 MeV	PUG-1AB: 0 to 500, 5000, 50000, 500000 cpm	960
Texas Nuclear Division, Ramsey Engineering Company					
Texas Nuclear 1	Series 9300	Texlium neutron detectors	Thermal to fast		962
Texas Nuclear 2	9341	Neutron detector	100 keV to 1 MeV; to 10 MeV optional w/pulse rise time discriminator		963
Victoreen, Inc.					
Victoreen 1	488A	High sensitivity neutron survey meter	Thermal to 14 MeV	0 to 800, 8000, 80000, 800000 cpm	964

B.6. Dosimeters

Instrument Note No.	Model	Use	Energy Dependence	Range	Page No.
Baird Corporation					
Baird 1	907-800 Series				967
Dosimeter Corporation of America					
Dosimeter 1	Numerous	Direct reading pen dosimeters	15% ^{60}Co from 80 keV to ~1.3 MeV	0 to 200 mR through 0 to 600 R	968
Dosimeter 2	884, 609	Special purpose dosimeters	886: γ only 884: fast neutron, γ 609: thermal neutron	886: 0 to 200 mrad 884: 0 to 200 mrad 609: 0 to 120 mrem	969
Eberline Instrument Company					
Eberline 1	RT-1A	Personal radiation monitor		~0.1 mR/hr to 5000 R/hr	970
Eberline 2		TLD dosimetry service	Various	Various	971
Harshaw Chemical Company					
Harshaw 1		TLD Materials	Various	Various	972
Harshaw 2	2271	TLD dosimetry system	Various	Various	975
Harshaw 3	8000 TLD System	Dosimeter for TLD cards	β, γ, X, neutron	Various	977
Harshaw 4	DN105 Pin Diode	Neutron dosimeter			978

B.6. (Continued)

Instrument Note No.	Model	Use	Energy Range	Counting Range	Page No.
Nuclear Associates					
Nuclear Associates 1	06-002, 06-012	Pocket x-ray dosimeter	±10% of 30 keV reading for 15 to 100 keV ±20% of 30 keV reading for 30 to 225 keV	002: 0 to 200 mR 012: 0 to 1.5 R	980
Nuclear Associates 2	Several	Pocket fountain pen-size dosimeters		200 mR to 600 R	982
Nuclear Associates 3	06-502	Portable belt-clip dosimeter		0 to 999 mR	984
Nucleus, Incorporated					
Nucleus 1	06-884, 06-886	Direct reading dosimeter	Neutrons: 20 keV to 15 MeV γ: 80 keV to 3 MeV	20 mrad to 600 rad	985
Radiation Detection Company					
Radiation Detection 1		TLD badge service (badges and finger rings)	β: > 1 MeV γ, x-rays: > 10 keV	10 mrem to 10^5 rem	986

Reactor Experiments, Incorporated					
Reactor Experiments 1	710	Criticality dosimeter	See Appendix A.7 and Table B.5		988
Reactor Experiments*		Neutron dosimeter foils	See Appendix A.7 and Table B.5		953
Technical Associates					
Technical Associates 1	PDR-1B; PDR-1C; PDR-1BE	Audio, digital dosimeters (1BE: finger ring)		0.1 to 9999 mR	989
Teledyne Isotopes					
Teledyne 1		TLD dosimetry system		Personnel: 3 to 1000 R Environmental: 1 to 10000 mR	991
Victoreen, Incorporated					
Victoreen 1	541R/541L	Direct reading dosimeters	541R: 30 keV to 2 MeV 541L: 17 to 200 keV	0 to 200 mR	992
Victoreen 2	687/362, 365	Personal dosimeter, reader charger	γ, x-ray	362: 0 to 200 mR 365: 0 to 500 mR	994
Victoreen 3	Numerous	TLD Dosimeters	Various	Various	996
Xetex, Incorporated					
Xetex 1	403B/404A "Mini-gard"	Personal, portable dosimeter	x-rays	1 to 4096 mR	1000

B.6. (Continued)

Instrument Note No.	Model	Use	Energy Range	Counting Range	Page No.
Xetex 2	409A-1, -2 402A-1, -2	Portable, digital dosimeter	Various	Various	1002
Xetex 3	415A-1, -4	Digital dosimeter	x- to γ-rays	0 to 9999 mR	1004
Xetex 4	416A "Accu-Chirp"	Portable, personal radiation monitor	$>$ 20 keV, x-, γ-rays		1006

*The instrument note for models marked by an asterisk appears in a different section than that covered in the table.

B.7. Instruments for Measuring Specific Radionuclides

Instrument Note No.	Model	Use	Nuclide Monitored	Range/Minimum Detection Limit (Mdl)	Page No.
Atomic Energy of Canada, Ltd., Chalk River Nuclear Laboratories					
AECL 1		HT/HTO monitor	^{3}H	*Count:* 0.1 to 10^6 MBy/m^3	1008
Beckman Instrument Corporation					
Beckman 1	LS 5800/LS 8900 Series	Liquid scintillation system	^{3}H, ^{14}C		1009
Eberline Instrument Co.					
Eberline 1	PING-3	Air particulates, iodine, noble gas monitor	Particulates, ^{131}I, noble gases	*Count:* Particulates: 5 to 10 cpm/hr for 10^{-11} μCi/cm^3 at 60 liter/min air flow; ^{131}I: 3 cpm/hr for 10^{-11} μCi/cm^3 at 60 liter/min air flow; Noble gases: 25 cpm for 10^{-6} liter/cm^3 at 14 psia	1012
Eberline 2	PRS-2, SPA-1	Portable radon daughter working level counter	Rn daughter	*Mdl:* 0.001 WL	1014
Eberline 3	RGM-2	Radon gas monitor	Rn	*Mdl:* ~0.1 pCi/liter	1016
Eberline 4	SAC-R5	Radon gas detector	Rn		1017
Eberline 5	SAM-2	Uranium enrichment monitor	Uranium	*Energy, optimum:* 185 keV (^{285}U γ) *Count:* 500, 5000, 50000, 500000 cpm (linear)	1018

B.7. (Continued)

Instrument Note No.	Model	Use	Nuclide Monitored	Range/Minimum Detection Limit (Mdl)	Page No.
Environmental Protection Agency, Eastern Environmental Radiation Facility					
EPA, EERF 1		^{85}Kr liquid scintillation system	^{85}Kr	*Mdl:* 2.5 pCi/m^3 air at 2σ counting error	1022
Food & Drug Administration, Winchester Engineering & Agricultural Center					
FDA, WEAC 1		^{85}Kr ion chamber system	^{85}Kr	*Mdl:* Depends on chamber 0.5 liter chamber: 130000 pCi/m^3 4.3 liter chamber: 34000 pCi/m^3 See note for more	1023
FDA, WEAC 2		Portable ^{85}Kr Geiger-Müller systems	^{85}Kr	*Mdl:* Depends on detectors EON-8008H: 12000 pCi/m^3 Amperex 18546: 27000 pCi/m^3 LND-719: 24000 pCi/m^3 See note for more	1024
Harshaw Chemical Company					
Harshaw 1	100, 101	Radon daughter analyzer	radon daughter	*Mdl:* $< 10^{-3}$ WL	1025
AERE Harwell					
Harwell 1		Tritium monitor	^{3}H	*Mdl:* 1 μCi/m^3 *Count:* 1 to 10^5 μCi/m^3	1026

Harwell 2	3276	Transportable alpha-in-air monitor	Pu	*Mdl:* approx 1 mpc/hr Pu (occupational) *Count:* 0.1 to 100 cps	1028
Wm. B. Johnson and Associates					
Johnson and Associates 1	TR-5	Air monitor	^{3}H, γ	*Count:* ^{3}H: 0 to 50, 500, 15000 Ci/m^3 γ: 0 to 0.25, 2.5, 25 mR/hr Model 5M has increased sensitivity; see note	1030
Johnston Laboratories					
Johnston 1	111	Portable tritium air monitor	^{3}H	*Mdl:* 5 μCi/m^3 (linear) *Count:* 0 to 50, 500 μCi/m^3 (linear); 10 to 10^4, 10^3 to 10^6 μCi/m^3 (log)	1032
Johnston 2	Triton 955B	Radionuclide gas monitor	^{3}H, ^{14}C, γ	*Count:* ^{3}H: 0 to 10, 100, 1000, 10000 μCi ^{14}C: 0 to 2, 20, 200, 2000 μCi/m^3 γ: 0 to 0.05, 0.5, 5, 50 mR/hr	1034
Johnston 3	LAC-2	Lucas chamber	Rn		1036
Johnston 4	133C		^{133}Xe	*Count:* 1 to 100 μCi/m^3 (log); 10 to 1000 μCi/m^3 (log)	1037

B.7. (Continued)

Instrument Note No.	Model	Use	Nuclide Monitored	Range/Minimum Detection Limit (Mdl)	Page No.
Monsanto Research Corporation, Mound Laboratory					
Monsanto 1		^{85}Kr liquid scintillation system	^{85}Kr	*Mdl:* approx. 15 pCi/m^3 air	1039
National Nuclear Corporation					
National Nuclear 1	LMS-4	Effluent monitor	^{235}U, ^{238}U, γ	*Energy:* ^{235}U: 185 keV ^{238}U: aged 50 keV, 80 keV *Mdl:* ^{235}U: 0.03 ppm ^{238}U: 10 ppm γ: 40 nCi/liter	1041
National Nuclear 2	HM-3	Portable personnel/vehicle SM	^{235}U, ^{239}Pu		1042
National Nuclear*	DM-2	High sensitivity door monitor	U, Pu, γ	See Appendix A.6 and Table B.4	825
New York State Department of Health, Radiological Sciences Laboratory					
New York Health 1		β/γ coincidence scintillation spectrometry system	^{131}I, ^{134}Cs, ^{228}Ra	*Mdl:* ^{131}I: 0.05 pCi ^{134}Cs, ^{228}Ra: 0.08 pCi	1043
New York Health 2		Gas proportional, β/γ coincidence spectrometry system	^{133}Xe	*Mdl:* 0.05 pCi	1044

Nuclear Associates					
Nuclear Associates 1	36-751 "Xenalert"	Air and gas trap monitor	^{133}Xe	*Count:* 0 to 1, 10, 100, 1000 mpc	1045
Nuclear Associates 2	05-676, 05-695	^{125}I monitors	^{125}I	*Count:* 0 to 300, 3000 cps	1047
Nuclear Measurement Corporation					
Nuclear Measurements 1	AM-21	Air radiation monitor systems	^{131}I, α, β, γ	Depends on detector selected	1049
Nuclear Research Corporation					
Nuclear Research 1	SM 101	Stack monitor	Particulate, ^{131}I, noble gas		1053
Nuclear Research*	NRC Lab MAP-60	Airborne particulates monitor	See Section A.3 and Table B.1		632
Packard Instrument Company					
Packard 1	894	Gas proportional counters	^{3}H, ^{14}C	*Counts:* 100 to 100000 cpm full scale, 10 linear ranges	1054
Packard 2	7130, 7140 "Trace"™	Radioactivity flow monitor system	^{3}H, ^{14}C		1056
Scintrex, Ltd.					
Scintrex 1	AEP 5215A	Portable tritium in air monitor	^{3}H	*Count:* 0 to 10 $(MPC)_a$ linear 0.8 to 10^2 $(MPC)_a$ log; 0.8×10^2 to 10^4 $(MPC)_a$ log; 0.8×10^4 to 10^6 $(MPC)_a$ log	1058

B.7. (Continued)

Instrument Note No.	Model	Use	Nuclide Monitored	Range/Minimum Detection Limit (Mdl)	Page No.
Scintrex 2	129A	Automatic tritium in urine analyzer	^{3}H	*Mdl:* 0.5 μCi/liter = 0.1 MPBB *Count:* 0.01 to 99.99 MPBB	1059
Technical Associates					
Technical Associates 1	PTG-5	Portable tritium monitor	^{3}H	*Count:* 3×10^{-6} to 3×10^{-3} μCi/ml (6 decades); 3×10^{-3} to 3 μCi/ml (6 decades)	1061
Technical Associates 2	FMI-5	Iodine localizer and air monitor	^{125}I, ^{131}I	*Mdl:* ^{125}I: 1 nCi *Count:* 0 to 500, 5000, 50000, 500000 cpm	1063
Tracor Analytic, Incorporated					
Tracor Analytic 1	6881	Liquid scintillation counting system	^{3}H, ^{14}C, α, β, γ	*Energy:* 0 to 1700 keV	1065
Tracor Analytic 2	6882	Liquid scintillation counting system	^{3}H, ^{14}C, α, β, γ	*Energy:* 0 to 1700 keV	1067
Tracor Analytic 3	Delta 300/ Model 6891	Liquid scintillation counting system	^{3}H, ^{14}C, α, β, γ	*Energy:* 0 to 1700 keV	1068

APPENDIX C

PREFIXES AND ACRONYMS

Multiplicative Prefixes

E	exa	10^{18}	d	deci	10^{-1}
P	peta	10^{15}	c	centi	10^{-2}
T	tera	10^{12}	m	milli	10^{-3}
G	giga	10^{9}	μ	micro	10^{-6}
M	mega	10^{6}	n	nano	10^{-9}
k	kilo	10^{3}	p	pico	10^{-12}
h	hecto	10^{2}	f	femto	10^{-15}
da	deka	10^{1}	a	atto	10^{-18}

ACRONYMS

ADC	analog to digital converter
BHR	U.S. Bureau of Radiological Health
BNFP	Barnwell Nuclear Fuel Plant
BRH	Bureau of Radiological Health
BWR	boiling-water reactor
CANDU	Canadian deuterium uranium reactor
CFR	Code of Federal Regulations
CRT	cathode ray tube
FET	field effect transistor
FWHM	full width at half-maximum
FWTM	full width at tenth maximum
HTGR	high temperature gas cooled reactor
ICRP	International Commission on Radiological Protection
LCD	liquid crystal display

LMBR	liquid-metal breeder reactor
LWR	light-water reactor
MADE	maximum dose equivalent
MDC	minimum detectable concentration
MFRP	Midwest Fuel Reprocessing Plant
MOS	metal oxide silicon
MPBB	maximum permissible body burden
MPC, MPC_a, MPC_w	maximum permissible concentration, air, water
MW_e	megawatts electrical
MW_t	megawatts thermal
NERHL	Northeastern Radiological Health Laboratory
NFS	Nuclear Fuel Service
NRC	Nuclear Regulatory Commission
PHA	pulse height analyzer
PHS	Public Health Service
PWR	pressurized-water reactor
RBE	relative biological effectiveness
STP	standard temperature and pressure (0°C and one atmosphere)
Te	metric tonne
TeHM	metric tonnes of heavy metal
TLD	thermoluminescence detector
UNSCAR	United Nations Scientific Committee on Atomic Radiation

SUBJECT INDEX

MANUFACTURER INDEX